MW00389294

INTRODUCTION TO ELECTRICITY

ROBERT T. PAYNTER
B.J. TOBY BOYDELL

Prentice Hall
Boston Columbus Indianapolis New York San Francisco Upper Saddle River Amsterdam
Cape Town Dubai London Madrid Milan Munich Paris Montreal Toronto Delhi
Mexico City Sao Paulo Sydney Hong Kong Seoul Singapore Taipei Tokyo

Editorial Director: Vernon Anthony
Acquisitions Editor: Wyatt Morris
Development Editor: Dan Trudden
Editorial Assistant: Yvette Schlarman
Director of Marketing: David Gesell
Marketing Manager: Harper Coles
Marketing Assistant: Crystal Gonzales
Senior Marketing Coordinator: Alicia Wozniak
Senior Managing Editor: JoEllen Gohr
Project Manager: Rex Davidson
Senior Operations Supervisor: Pat Tonneman
Operations Specialist: Laura Weaver
Art Director: Diane Ernsberger
Text Designer: Candace Rowley
Media Director: Allyson Graesser
Media Manager: Michelle Churma
Media Project Manager: Karen Bretz
Full-Service Project Management: Peggy Kellar
Composition: Aptara®, Inc.
Printer/Binder: R. R. Donnelley & Sons
Cover Designer: Brian Huber
Cover Image: Shutterstock
Cover Printer: Lehigh-Phoenix
Text Font: L Avenir Light

Credits and acknowledgments for materials borrowed from other sources and reproduced, with permission, in this textbook appear on the appropriate page within text.

All photos not specifically credited to other sources are from Robert Paynter.

Chapter opener photos are copyright © Fotolia (Safety section, chapters 1, 2, 15, 16, 17, 18, 22, 25, 26, and 27), copyright © iStock (chapters 8, 9, 10, 11, 12, 13, 19, 20, 21, 23, and 28), or copyright © Shutterstock (chapters 3, 5, 6, 7, 14, and 24).

Library of Congress Cataloging-in-Publication Data

Paynter, Robert T.
Introduction to electricity / Robert T. Paynter, B.J. Toby Boydell.
p. cm.
ISBN-13: 978-0-13-504087-4
ISBN-10: 0-13-504087-6
1. Electrical engineering. I. Boydell, B. J. Toby. II. Title.
TK146.P37 2011
621.3—dc22 2010034673

10 9 8 7 6 5 4 3 2 1

Prentice Hall
is an imprint of

www.pearsonhighered.com

ISBN 10: 0-13-504087-6
ISBN 13: 978-0-13-504087-4

PREFACE

Introduction to Electricity is written as a first text for students in electrical trade and electrical technology programs. The text assumes that the reader has no prior electrical training, and it is written at a level that makes it readily accessible to all postsecondary students.

The chapters follow a logical progression that begins with fundamental topics such as safety, electrical and electronic career paths, and test equipment, followed by electrical properties, units of measure, fundamental electrical laws, DC components and circuits, residential electrical wiring, magnetism, AC components and circuits, three-phase AC, transformers, generators and motors, alternators, power transmission and distribution, green power, and copper and fiber optic cabling. These topics prepare the reader for the courses that follow while providing a solid foundation for both traditional and emerging electrical career paths.

Chapter Overviews

Safety

This chapter introduces the reader to electrical safety. Among the topics are electrical shock hazards, shock-related injuries, arc flashes, grounding and bonding, lockout/tagout, fire hazards, and personal protective equipment (PPE).

Chapter 1 Getting Started

This chapter introduces the reader to electrical and electronic careers, basic electrical components (resistors, capacitors, inductors, transformers, switches, fuses, and circuit breakers), lab and test equipment, units of measure, and engineering notation.

Chapter 2 Basic Electrical Concepts

This chapter introduces basic electrical properties and their units of measure. Among the topics are atomic structure, electrical charges, coulombs and amperes, direct and alternating current, current and heat, voltage, resistance, conductors, and insulators.

Chapter 3 Basic Electric Components and Meters

This chapter introduces the reader to the physical characteristics and ratings of basic electrical components. Among the topics are wire sizes, ampacity, insulator voltage ratings, fixed and variable resistors, resistor standard values and color code, potentiometers and rheostats, battery types and ratings, DC voltage sources, switches, fuses and circuit breakers, analog and digital multimeters, and meter measurements.

Chapter 4 Ohm's Law and Power

This chapter introduces the reader to the most fundamental laws of circuit operation. Among the topics are Ohm's law relationships, predicting circuit behavior, Ohm's law calculations and troubleshooting, power, power calculations, power and heat, efficiency, power ratings, energy measurements, and basic circuit faults and symptoms.

Chapter 5 Series Circuits

This chapter introduces the reader to series DC circuit characteristics, calculations, and faults. Among the topics are circuit resistance; current, voltage, and power characteristics; circuit calculations; Kirchhoff's voltage law (KVL); voltage references; voltage dividers; common circuit elements; circuit measurements; fault symptoms and troubleshooting; and circuit grounds.

Chapter 6 Parallel Circuits

This chapter introduces the reader to parallel DC circuit characteristics, calculations, and faults. Among the topics are circuit voltage, current, and resistance; resistance and power calculations; Kirchhoff's current law (KCL); current sources; current dividers; practical parallel circuits; and circuit measurements, fault symptoms, and troubleshooting.

Chapter 7 Series-Parallel Circuits

This chapter introduces the reader to basic series-parallel DC circuits. The topics include connecting series circuits in parallel, connecting parallel circuits in series, series-parallel circuit calculations, equivalent circuits, loaded voltage dividers, voltmeter loading, the Wheatstone bridge, and series-parallel circuit fault symptoms and troubleshooting applications.

Chapter 8 Introduction to Residential Circuits: A Practical Series-Parallel Application

This chapter introduces the reader to residential electrical circuits. The topics include service drops and laterals, service meters, the service entrance, grounding, switches, receptacles, GFCIs, conductors, nonmetallic cables, and typical 120 V and 240 V circuits.

Chapter 9 Source and Load Analysis

This chapter introduces the reader to more advanced circuit analysis techniques. The topics include the effects of source resistance on voltage and current source operation, equivalent current and voltage sources, source conversions, Thevenin equivalent circuits, maximum power transfer, Norton equivalent circuits, and the superposition theorem.

Chapter 10 Magnetism

This chapter introduces the reader to the basic principles of magnetism. Among the topics are magnetic poles, flux and flux density, permeability, magnetic induction, magnetic classifications, electromagnetism, the left-hand rule, coils, ampere-turns, hysteresis, electromagnetic relays, magnetic shielding, and analog meter movements.

Chapter 11 Inductance and Inductors

This chapter introduces the reader to inductance, inductors, and the effects of inductance on DC circuit operation. The topics include the relationships between current and magnetism, induced voltage, inductor characteristics and operation, mutual inductance, series and parallel inductors, inductors in DC circuits, current rise and decay, RL time constants, and types of inductors.

Chapter 12 Capacitance and Capacitors

This chapter introduces the reader to capacitance, capacitors, and the effects of capacitance on DC circuit operation. The topics include capacitor construction, charge and discharge action, capacitor characteristics and ratings, series and parallel capacitors, capacitors in DC circuits, voltage rise and decay, RC time constants, and types of capacitors.

Chapter 13 Alternating Current (AC)

This chapter introduces the reader to alternating current (AC). Among the topics are sine waves, time and frequency calculations and measurements, amplitude calculations and measurements, effective (RMS) values and power, averaging and true-RMS meters, generating sine waves, phase and phase angles, rectangular and square waves, and duty cycle.

Chapter 14 Series Resistive-Inductive (RL) Circuits

This chapter introduces the reader to series AC resistive-inductive circuits. The topics include inductor voltage and current (phase), inductive reactance, series and parallel reactances, inductor power values, inductor quality (Q), series RL circuit voltages, impedance, phase angles, circuit calculations, power relationships and calculations, and power factor.

Chapter 15 Parallel Resistive-Inductive (RL) Circuits

This chapter introduces the reader to parallel AC resistive-inductive circuits. Among the topics are branch and total circuit currents; phase relationships; and current, voltage, impedance, and power calculations.

Chapter 16 Series Resistive-Capacitive (RC) Circuits

This chapter introduces the reader to series AC resistive-capacitive circuits. Among the topics are AC coupling and DC isolation, capacitor voltage and current (phase), capacitor resistance and reactance, series and parallel reactances, series RC circuit characteristics and calculations, impedance, phase angles, circuit calculations, power relationships and calculations, and power factor.

Chapter 17 Parallel Resistive-Capacitive (RC) Circuits

This chapter introduces the reader to parallel AC resistive-capacitive circuits. Among the topics are branch and total circuit currents; phase relationships; and current, voltage, impedance, and power calculations.

Chapter 18 RLC Circuits

This chapter introduces the reader to resistive-inductive-capacitive (RLC) circuits. Among the topics are series LC circuit phase and voltage characteristics, net series reactance, parallel LC circuit phase and current characteristics, net parallel reactance, resonance, resonant LC circuits, series RLC circuits, series RLC circuits, parallel RLC circuits, leading and lagging VAR, and power factor correction.

Chapter 19 Frequency Response and Filters

This chapter introduces the reader to circuit frequency response and passive filter characteristics and circuits. The topics include attenuation, frequency response curves, filter frequencies and bandwidth, filter response curves, filter quality (Q), low-pass filter circuits, high-pass filter circuits, bandpass and notch filter circuits, series and parallel bandpass filter calculations, and series and parallel notch filter calculations.

Chapter 20 Three-Phase Power

This chapter introduces the reader to three-phase alternating current characteristics and sources. The topics include generating three-phase power, phase sequences, inductive wye (Y) circuit characteristics and calculations, inductive delta (Δ) circuit characteristics and calculations, Y-Δ circuits, Δ-Y circuits, three-phase power and power factor, and power factor correction.

Chapter 21 Introduction to Transformers

This chapter introduces the reader to single-phase transformer operating characteristics and relationships. Among the topics are transformer voltage classifications; transformer action; transformer voltage, current, and power relationships; power losses; multiple-output and multiple-input transformers; autotransformers; harmonics; transformer hum; core structures; power derating; current transformers; and *NEC®* nameplate requirements.

Chapter 22 Three-Phase Transformers

This chapter introduces the reader to three-phase transformer operating characteristics and relationships. The topics include wye-wye (Y-Y) transformers, delta-delta (Δ-Δ) transformers, Y-Δ transformers, Δ-Y transformers, transformer banks, open-phase fault symptoms, open-delta connections, transformer power, dry-type and wet-type transformers, and three-phase transformer ratings.

Chapter 23 DC Machines: Generators and Motors

This chapter introduces the reader to the operating principles of DC generators and DC motors. The topics include generator elements, pole windings, series and parallel generator connections, output voltage control, compounding, DC motor action, torque and work, horsepower, series motors, motor current and torque, motor loading, shunt motors, compound motors, speed and direction control circuits, and kickback circuits.

Chapter 24 AC Machines: Alternators and Motors

This chapter introduces the reader to the operating principles of alternators and AC motors. The topics include rotating-armature and rotating-field alternators, single-phase and three-phase alternators, rotor and stator construction, armature winding, output voltage and frequency, voltage regulation, AC motor operation, motor speed, synchronous motors, induction motor speed and slip, split-phase induction motors, and three-phase synchronous motors.

Chapter 25 AC Power Transmission and Distribution

This chapter introduces the reader to commercial AC power generation, transmission, and distribution systems. The topics include an overview of transmission and distribution systems; hydroelectric plants; thermal power plants; wind power; overhead transmission lines; subtransmission lines; underground transmission lines; substations; and distribution grid voltages, transformers, and substations.

Chapter 26 Green Power

This chapter introduces the reader to green power generation and distribution systems. Among the topics are greenhouse gas sources, power requirements, gasification, syngas, biogas, combined cycle power, liquefied natural gas (LNG), power sources (nuclear, wind, solar, wave, and tidal), distributed generation, cogeneration, photovoltaic cells, fuel cells, "smart grid" technology, peaker plants, smart metering, and green lighting (CFLs, LED and OLED lighting).

Chapter 27 Introduction to Copper Cabling

This chapter introduces the reader to copper-based telecommunications cabling. Among the topics are commercial telecommunications systems, telephony, twisted pairs, electromagnetic interference (EMI), shielded and unshielded cables, cable categories, plenum and riser cables, wire color coding, modular plug and jack standards, 66 and 110 termination blocks, coaxial cable construction, RG ratings, and connectors.

Chapter 28 Introduction to Fiber Optic Cabling

This chapter introduces the reader to fiber optic telecommunications cabling. The topics include light frequencies and wavelength, reflection and refraction, types of optical fiber, optical fiber reflection, attenuation, connector and splice losses, simplex and duplex connectors, common connectors, splices, light sources (LEDs, lasers, and PIN diodes), and light detectors (photodiodes).

Introduction to Electricity also contains appendices that introduce the reader to working effectively with units of measure and with wire tables.

Features

From the start, our goal has been to produce a fundamentals book that students can really *use* in their studies. To this end, we have included the following learning aids and other features:

1. PERFORMANCE-BASED OBJECTIVES provide a handy overview of the chapter organization and a map to student learning.
2. KEY TERMS LISTS in the chapter openings identify new terms and the pages where they are introduced. This helps the student to quickly locate the discussion on each term listed.
3. MARGIN DEFINITIONS are provided in every chapter at the point where each new term is introduced.
4. HIGHLIGHT BOXES are included in each chapter.
   a. ***In the Field*** boxes help bridge the gap between theory and practice.
   b. ***Staying Safe*** boxes provide important safety information.
   c. ***Key Concept*** boxes reinforce important points in many of the chapters.
5. SUMMARY ILLUSTRATIONS provide a convenient description of circuit operating principles and applications. Many also provide comparisons between related circuits.
6. IN-CHAPTER PRACTICE PROBLEMS included in most examples provide students with an immediate opportunity to apply the principles and procedures being demonstrated. The answers to these problems are provided on the last page of each chapter.
7. IN-CHAPTER PROGRESS CHECKS end each section of the text, providing students with the opportunity to gauge their learning.
8. DETAILED CHAPTER SUMMARIES break the material in each chapter down into a bulleted list to provide an additional study aid.
9. CHAPTER REVIEW questions at the end of each chapter provide the students with an opportunity to test their understanding of the material.
10. PRACTICE PROBLEMS. An extensive set of practice problems is provided at the end of most chapters. The answers to many of the practice problems are provided in Appendix D.

Ancillaries

A number of ancillaries have been developed to help you and your students get the most out of *Introduction to Electricity*:

- ANNOTATED INSTRUCTOR'S GUIDE In addition to the material found in the standard version of the text, the Annotated Instructor's Guide (AIG) includes informative **tips** to share with students and suggested **discussion** topics. The AIG also identifies opportunities for the instructor to **demonstrate** skills and **hands-on** activities for the students. **Preview** margin notes connect material to material that is presented later in the text, while **review** margin notes do just the opposite. Annotations also identify and/or suggest **cross-curricular activities**, relevant **Internet searches**, hands-on **assessments** that measure student learning, and **career preparation** activities.

- **ONLINE POWERPOINTS®** A complete set of PowerPoints is provided to assist with classroom presentations and discussions.
- **ONLINE TESTGEN COMPUTERIZED TEST BANK** A computerized test bank is provided to assist in the development of quizzes and tests.
- **MYELECTRICITYLAB** MyElectricityLab is an online, interactive companion to *Introduction to Electricity* that is based on the learning objectives from the text. MyElectricityLab includes:
  - A full math tutorial that covers all the equations and formulas in the text. There are worked examples, sample problems, and e-book links for students who struggle with math.
  - Interactive animations that illustrate concepts that students struggle to visualize.
  - A customizable learning management system. Instructors can gauge student success through homework, testing, and quizzing that is all completed and automatically graded within the software.

More information can be found at www.myelectricitylab.com.

Download Instructor Resources from the Instructor Resource Center

To access supplementary materials online, instructors need to request an instructor access code. Go to www.pearsonhighered.com/irc to register for an instructor access code. Within 48 hours of registering, you will receive a confirming e-mail including an instructor access code. Once you have received your code, locate your text in the online catalog and click on the Instructor Resources button on the left side of the catalog product page. Select a supplement, and a login page will appear. Once you have logged in, you can access instructor material for all Prentice Hall textbooks. If you have any difficulties accessing the site or downloading a supplement, please contact Customer Service at http://247.prenhall.com.

In addition to these listed items, a **Lab Manual for Introduction to Electricity** (ISBN: 0135106222) is available for purchase and can be ordered through your Pearson representative. The lab manual contains over 45 exercises that were written to supplement the text. Among its features:

- The opening for each exercise ties the activity to the text material, identifies the relevant chapter objectives, and helps the student to connect the activity to working in the field.
- Early exercises include **detailed descriptions of the circuit connections** along with **step-by-step assembly instructions**, helping the student to build the circuits more quickly and efficiently. The circuit descriptions and assembly instructions become more general as students progress through the manual, moving them toward more independent lab activities.
- In the first half of the manual, **circuit diagrams showing how the circuit elements are connected** and how the circuit is tested are provided **along with the circuit schematics**, helping the students to make the connection between schematic diagrams and actual component layouts.

- The labs are intended for use with the **Lab-Volt® EMS** (electromechanical systems) line from Lab-Volt® Systems, Inc. with test equipment available from other providers. However, all labs can be adapted to use similar elements from other manufacturers. (See the lab manual appendix for suggested equipment and parts providers.)

Acknowledgments

A project of this size could not have been completed without help from a variety of capable and concerned professionals. A special thanks goes out to the following individuals for providing review and recommendations that helped shape the text:

Aurelio Jesus Aguilar, Houston Community College
Elmano Alves, Chaffey College
Greg A. Bailey, Des Moines Area Community College
Jeffery Cox, Central Community College
Wayne Gebhart, Pennsylvania College of Technology
Ricky C. Godbolt, Prince George's Community College
Jim Heidenreich, Cuyahoga Community College
Patrick Klette, Kankakee Community College
Jack Lane, Sheridan College
Deborah T. Lichniak, Thomas Nelson Community College
Niladri R. Mantena, Los Angeles Southwest College
Anthony P. Messuri, Youngstown State University
Roland E. Miller, NWCCD-Gillette College
JC Morrow, Hopkinsville Community College
Linda Pohlgeers, Cincinnati State Technical and Community College Gerald Rutherford
Murry Stocking, Ferris State University
Ken Warfield, Shawnee State University
Paul Westrom, New England Institute of Technology

We would also like to thank the staff at Pearson for their behind-the-scenes efforts on this text. The following people deserve special recognition:

Wyatt Morris, Acquisitions Editor
Dan Trudden, Development Editor
Rex Davidson, Project Manager

The following individuals also contributed to the development of the text:

Peggy Kellar, Project Manager, Aptara, Inc.
John DeDad, DeDad Consulting LLC

Finally, special thanks go out to our families and friends for their constant support and patience.

ROBERT (BOB) PAYNTER
B. J. TOBY BOYDELL

CONTENTS

INTRODUCTION TO ELECTRICITY

SAFETY

PURPOSE

Throughout this text, you will be introduced to the fundamental principles of electricity that electricians and technicians need to know. But before we even describe electricity, we need to talk about safety.

It is your individual responsibility to develop the skills and workplace habits that help ensure that you, and those who work with you, go home safely every day. During your training you will be working in the lab, where safe work habits are just as important as on the jobsite. Now is the time to start developing these habits.

There is, however, a problem with introducing safety this early in your education. You will not fully understand some of the terminology and concepts introduced here as we talk about safety. For this reason, we strongly recommend that you refer back to this introduction and revisit some of the websites we reference here as your knowledge base increases. As you will see, safe work practices are continually refined and redeveloped as your career unfolds.

KEY TERMS

The following terms are introduced in this chapter:

S.1 ELECTRICAL SHOCK HAZARD

According to the National Institute for Occupational Safety and Health (or NIOSH), electrocution is the third leading cause of work-related deaths among 16- and 17-year-olds, and the cause of 12% of all workplace deaths among young workers.

There is an old saying: "*It isn't the voltage that kills you . . . it's the current.*" This statement is true, but it can be misleading. Let's start by giving you a brief definition of some electrical terms.

- **VOLTAGE** is the force that generates electric current. Think of it as being like the pressure that causes water to flow in a pipe.
- **CURRENT** is the flow of electrical charge through a conductor. A conductor can be a copper wire, a metal pipe, a human body, or anything else that readily passes electrical current. A higher voltage generates more current in a given conductor, just as higher pressure causes more water to flow through a pipe.
- **RESISTANCE** opposes electrical current. Resistance limits current in the same way that the diameter of a pipe limits the flow of water. A material with *high* resistance is like a pipe with a *narrow* diameter. For a given pressure (voltage) less water (current) will pass through a narrow pipe (high resistance) than through a wider pipe (low resistance). A material with extremely high resistance that allows virtually no current at all is called an **INSULATOR**. An example of a good insulator is the plastic insulation on a copper wire. However, *any insulator can be forced to conduct electricity if the voltage is high enough.*

As you can see, there are relationships among voltage, current, and resistance. Based on these relationships we can establish an important point: *The higher the voltage that you are working with, the higher the possible shock current.* The human body has fairly high resistance. Nonetheless, it can still pass a dangerous amount of current, and certain conditions increase the body's susceptibility to shock.

One factor that can affect the body's susceptibility to shock is water. Water is a very good conductor (low resistance). If your body is wet (whether from rain, perspiration, or some other source) or if you are standing in water, then any shock that you receive will be much more severe than if you were working in perfectly dry conditions. Water and electricity are a dangerous combination.

The Effects of Electrical Shock

The effects of an electrical shock vary between individuals, but some general guidelines have been provided by the Occupational Health and Safety Administration (OSHA), a division of the U.S. Department of Labor. OSHA, first established in 1971, is tasked with ensuring safe working environments throughout the United States.[1]

[1]We strongly suggest that you visit their website (http://www.OSHA.gov) regularly as your education continues.

Table S.1 lists the effects of various current levels on the human body. Current is measured in units called **AMPERES** (or *amps*). A **MILLIAMPERE** (mA) is one one-thousandth (1/1000) of an ampere, so 1000 milliamperes equals 1 ampere. To put things into context, most of the electrical outlets in your home are rated to carry up to 15 amperes (15,000 mA) of current. The difference between no perceived effect and possible death is little more than one-tenth of an ampere.

TABLE S.1 • Effects of Electric Current in the Human Body

CURRENT	REACTION
Below 1 mA	Generally not perceptible
1 mA	Faint tingle
5 mA to 10 mA	Slight shock felt; not painful, but disturbing. May have trouble releasing the circuit.
10 mA to 20 mA	Painful shock, loss of muscular control. Individual cannot let go, but may be thrown away from the circuit if extensor muscles are stimulated.
20 mA to 100 mA	Extremely painful. Breathing is difficult.
100 mA to 800 mA	Cardiac and/or respiratory arrest. **Likely fatal!**

Shock-Related Injuries

Burns are the most common shock-related injury, and electrical burns are among the most serious types of burns. When electrical current passes through bone or tissue, it produces heat. This means that, along with burns that are visible, there may be internal burn damage as well. Don't be fooled if a person who receives a shock does not show obvious signs of injury. Get medical help immediately!

Current Pathways Through the Human Body

Together with burns, there may be other shock-related injuries. Severe muscle contractions can damage bones and cause a dangerous fall. A shock can also cause internal hemorrhaging, nerve damage, muscle damage (including the heart muscle), or even renal (kidney) failure.

Although it is not always possible, you should work with only one hand if you must test a live circuit. The reason for this has to do with the path that electricity will take through your body if you do get a shock. The pathway of the current through your body affects the type and severity of your injuries. Current pathways through the body are divided into three groups:

- *Touch potential* (hand to hand through the heart)
- *Step potential* (foot to foot through the lower body)
- *Touch step potential* (right hand to right foot, or left hand to left foot through the body).

Touch potential is the most dangerous. Because the current passes through the heart, there is a much higher potential for serious damage (or death). However, if the current does not pass through the heart (or another vital organ), there is a much better chance for survival.

Freezing

In some cases, an electrical shock causes muscles to contract and a person cannot let go of the circuit. It is critical to free someone who **FREEZES** to a circuit as quickly as possible, ***but do it safely***! If possible, power down the circuit. This will free the individual being shocked. If you cannot power down the circuit, **do not try to pull the person free using your hands**. If you touch someone who is being shocked, you will be electrocuted as well. Therefore, you must use something made of wood (like a two-by-four) or plastic (like PVC) that will not conduct electricity to pry or push the shock victim away from the circuit. Work quickly, but protect yourself as well; if you get hurt, there may be no one to get help.

How Can You Tell if a Shock Is Serious?

You can't. If you or someone else receives a shock, get medical help immediately.

Arc Flashes

An **ARC FLASH** is a sudden, explosive arc of electrical energy. It can result from a dropped tool, accidental contact with a high-voltage circuit, or improper use of test equipment.

Most residential wiring is 120 volts or 240 volts. This is not sufficient to cause an arc flash in most cases. Medium- and high-voltage circuits (600 volts and higher) are much more susceptible to arc flashes. The higher the voltage, the greater the danger of an arc flash.

Some electricians work around circuits that operate in the thousands of volts. At these levels, an arc flash can vaporize metal and project a *plasma* (ionized gas) arc that can start fires, destroy equipment, and kill people some distance away. The bottom line is that you have to be extremely careful when working with high-voltage circuits. Whenever possible, power down a circuit before you work on it. Do not work on live circuits unless it is absolutely necessary.

PROGRESS CHECK

1. Define the following terms:
   a. voltage
   b. current
   c. resistance
   d. insulator
2. What is the lowest current level that may cause death if you get a shock?
3. What is the most common shock-related injury?
4. List the three current pathways that a shock current can take through the body. Which is the most dangerous?
5. What is freezing? At what current range does freezing occur?
6. What is an arc flash? When is an arc flash more likely to occur?

S.2 GROUND FAULTS, GROUNDING, AND BONDING

Earlier, we mentioned that some of the concepts we are discussing may be beyond your understanding at this point. Here is a good example. The topics in this section are interrelated and very important, but they are also highly technical. If you do a web search on grounding and bonding, you will find Master Electricians arguing with each other over the relevant *NEC*® *(National Electrical Code*®*)*[2] directives. Compared to these debates, we barely scratch the surface of these topics in this section.

GROUND refers to the earth. In Great Britain, the terms *earth* and *earthing* are used instead of *ground* and *grounding*. **EARTH GROUND** is considered the electrical zero-volt reference in the same way that sea level is considered an altitude reference. Voltages are defined as being some value above (positive) or below (negative) ground. This means that any positive or negative voltage will generate a current to or from ground if a current path is provided.

It should be noted that the term "ground" is often used to describe something other than earth ground. *Chassis ground* and *signal ground* are two examples. In this discussion we are referring only to earth ground.

A **GROUND FAULT** is an unintended connection of an energized conductor to ground. For example, suppose that the "hot" (energized) wire in the power cord of an appliance has come loose and is making contact with the metal *chassis* (body) of the appliance. If the appliance is electrically isolated (no path to ground), nothing happens. But if you touch the appliance and *you* are grounded, your body provides a current path to ground and you get a shock! This is one reason why many appliances and other equipment have grounding prongs on their electrical plugs. The *ground prong* is connected internally to the metal chassis at one end and to earth ground (via the receptacle ground connection) at the other. **Never defeat or remove the ground prong on any plug!**

A **GROUND-FAULT CIRCUIT INTERRUPTER (GFCI)** is a device that senses any imbalance in circuit current that results from a ground fault. Normally, the current through the hot (energized) conductor equals the current through the neutral (return) conductor in any 120 V residential circuit. If a ground fault occurs, the current through the neutral conductor is lower than the current through the hot conductor. Any current imbalance between the two conductors greater than 5 milliamps (5/1000 of an amp) causes the GFCI to *open* (break the connection) very quickly and deenergize the circuit. GFCIs are covered in more detail in Chapter 8. A GFCI receptacle and a GFCI circuit tester are shown in Figure S.1. A GFCI circuit tester checks the polarity of the receptacle and can produce a ground fault to make sure that the GFCI is operating properly.

GROUNDING refers to the connection of equipment or an electrical system to earth ground. Earth ground can be established in a number of ways. Connection to a copper stake (*ground rod*) driven into the ground is one way. Connection to underground metal pipes (such as water pipes) or electrodes encased in the

[2]NFPA 70®, National Electrical Code and *NEC*® are registered trademarks of the National Fire Protection Association, Quincy, MA.

(a)

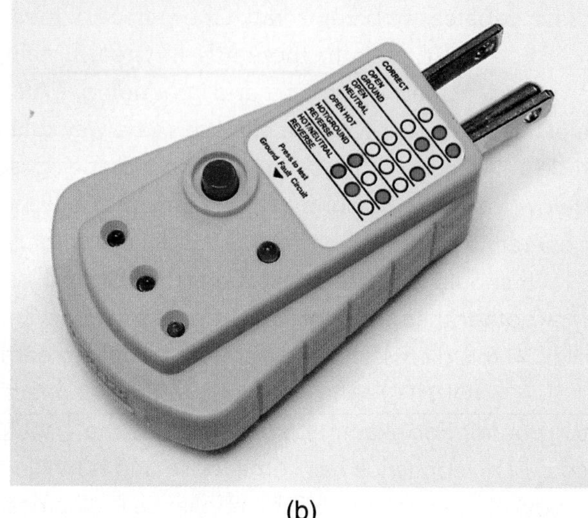

(b)

FIGURE S.1
Istock.com

concrete foundation of a structure are two others. The specific *NEC®* requirements on establishing a ground are far too detailed to cover here.

BONDING is the permanent joining of conductive non–current-carrying materials in a structure, such as water and/or gas pipes, metal conduits, or anything that could conduct current. Bonding is used to ensure a continuous, low-resistance path to ground for any short-circuit current. It also helps prevent arcing or sparking as a short circuit current goes to ground, and assures that any overcurrent protection devices (circuit breakers, fuses, etc.) will open quickly if a short circuit occurs.

There is one other important aspect to bonding. If all the metallic (conductive) materials in a structure are bonded together, they will all have the same voltage potential. Keeping everything at the same potential means that you cannot receive an electrical shock if you touch two different metal objects when a short-circuit current is present. To sum up, bonding is a critical component in the effective grounding of an electrical system.

One more point: In this section, we have touched on highly complex topics that are addressed by a multitude of regulations. Grounding and bonding are covered in *NEC® section 250* as well as IEEE® sections 142 and 1100. You may also want to reference *Soares Book on Grounding* published by the International Association of Electrical Inspectors (IAEI).[3]

[3]The IAEI website can be found at http://www.iaei.org.

PROGRESS CHECK

1. Define the following terms:
   a. ground
   b. earth ground
   c. ground fault
   d. insulator
2. What is the difference between grounding and bonding?
3. What items need to be bonded in any structure?
4. List three reasons why materials need to be bonded.
5. Which *NEC*® regulation covers grounding and bonding?

S.3 LOCKOUT/TAGOUT

Most electrical work is performed on *deenergized* circuits (i.e., circuits that have no power applied to them). It is important, however, to ensure that a circuit that is supposed to be powered down actually *is* powered down. **LOCKOUT/TAGOUT** is a common method of making sure that a circuit truly is deenergized and that it stays that way until work is completed. Some industries require only that a tag be attached to the circuit power switch or circuit breaker panel. Others require that the panel be locked and a tag attached. The person working on the circuit keeps the key to the lock, which guarantees that no one can restore power to the circuit while repair work is being performed. In cases where more than one person is working on a given circuit or machine, multiple locks may be used, with each worker keeping their own key.

Even when a circuit has been locked and tagged, you should still make sure that it is actually powered off, as follows:

1. Connect your meter to a circuit that you know is powered and take a reading. (This verifies that your meter is working properly.)
2. Use your meter to test the locked and tagged circuit to ensure the circuit truly is powered down.
3. Retest your meter on the powered circuit to confirm that the meter is still working properly.

If you follow this procedure, you can rest assured that you are working on a deenergized circuit. Figure S.2a shows a lockout/tagout kit. Figure S.2b shows an energy source that has been tagged and locked by several workers.

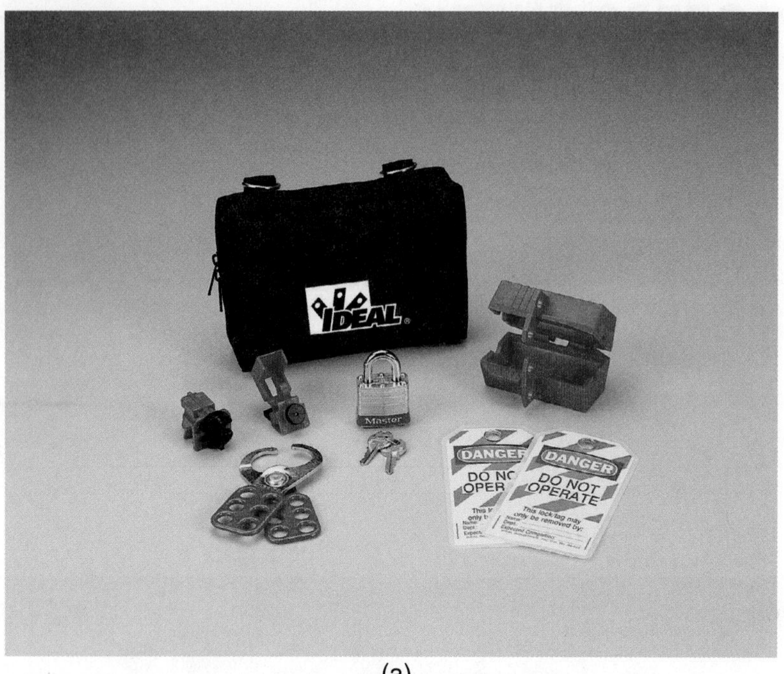

(a)

(b)

FIGURE S.2
Source: (a) Photograph courtesy of Ideal Industries, Inc.; (b) Istock.com

PROGRESS CHECK

1. What is the purpose of the lockout/tagout practice?
2. When a circuit is locked out, who keeps the key?
3. Explain the best procedure for making sure that a circuit is truly deenergized.

S.4 ELECTRICAL FIRE HAZARDS

Let's begin our discussion about electricity and fire hazards with the following facts about fires from the National Fire Protection Association (NFPA).[4]

- There were 52,500 home structure fires resulting from electrical failure or malfunction in the United Stated in 2006.
- These fires caused 340 deaths, 1400 injuries, and just under $1.5 billion in property damage.

[4]The NFPA website can be found at **http://www.nfpa.org.**

These figures do not include deaths and injuries to fire fighters and other emergency response personnel. According to the Federal Emergency Management Agency (FEMA), electrical distribution is the fifth-ranked cause of fires, the fourth-ranked cause of fire fatalities, and the second-ranked cause of property loss. There are many ways that electrical distribution can cause a residential fire. The sources of residential electrical fires are listed in Table S.2.

TABLE S.2 • Causes of Residential Fires Related to Electrical Distribution

CAUSE OF FIRE	PERCENT
Fixed wiring	34.7
Electrical cords and plugs	17.2
Lighting fixtures	12.4
Switches/receptacles	11.4
Lamps and bulbs	8.3
Circuit breakers/fuses	5.6
Meters/meter boxes	2.2
Transformers	1.0
Unknown or unclassified	7.3

Table S.2 identifies the sources of electrical fires, but does not explain *how* electricity can cause a fire. Electrical fires are ignited by arcing, short circuits, and/or excessive heating of conductors or electrical components. Research suggests that arcs and shorts seldom cause a residential fire. The NFPA®, however, has recognized that arcs caused by furniture resting on and crushing lamp and small appliance cords have, in fact, caused fires.

Industrial fires are another matter. Due to the much higher voltages that may be present, an industrial arc flash is quite capable of starting a fire. Any combustible material in the vicinity of an industrial arc flash may be ignited.

Excessive heating is the most common cause of electrical fires, and it is easy to cause a conductor to overheat. For example, looping a conductor upon itself several times can result in overheating when current passes through the conductor. Excessive heating can also result from excessive thermal insulation and overvoltage conditions (i.e., when the voltage on a line increases as the result of a fault or other condition). The most common cause of overheating, however, is poor electrical connections.

If an electrical connection is not secure, it can result in what is called a **HIGH-RESISTANCE CONNECTION**. A high-resistance connection heats up when current is present. (You will learn why in Chapter 4.) The heat causes oxidation and **CREEP** (the further loosening of connecting materials due to heating), which increases resistance, resulting in more heating, and so on. Eventually the connection reaches the point where it becomes a **GLOWING CONNECTION,** a result of very high temperatures. At this point it can easily ignite any nearby combustible material.

Putting It Out

The best and safest way to put out a fire is to use the proper fire extinguisher. Fires are classified into the five groups listed in Table S.3.

TABLE S.3 • Fire Classifications

FIRE CLASSIFICATION	FUEL	EXTINGUISHING AGENT
A	Common combustibles, like wood, paper, or cloth	Water, chemical foam, dry chemical
B	Flammable liquids, like oil, gasoline, paints, tar	Carbon dioxide (CO_2), dry chemical, aqueous forming foam (AFFF)
C	**Live electrical equipment**	**CO_2, dry chemical**
D	Combustible metals	Dry powder
K	Cooking oils and fats	Wet chemical

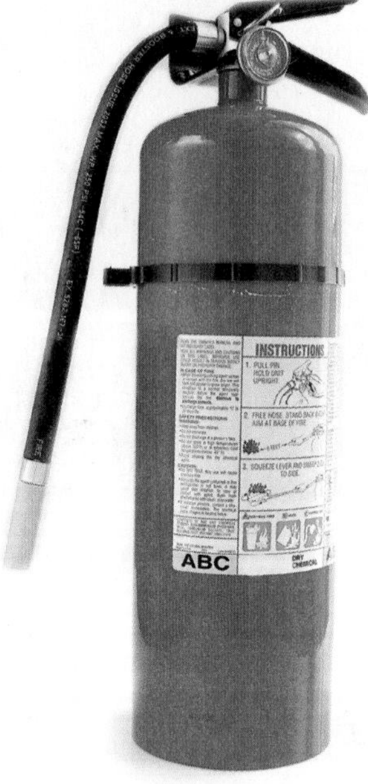

FIGURE S.3
Istock.com

As Table S.3 indicates, Class C fires involve live electrical equipment. You should always use an extinguisher with the proper classification. *Never use a Class-A type extinguisher on an electrical fire.* Using a water-based extinguisher could result in your death! Remember that water and electricity are a dangerous combination. Note that some extinguishers are rated for more than one type of fire. For example, a Class ABC extinguisher can be used on Class A, Class B, or Class C type fires. A multiuse fire extinguisher is shown in Figure S.3.

One more point: Fire extinguishers aren't very useful if you don't know where they are and how to use them. When you enter a new jobsite, one of the first things you should do is locate the fire extinguishers and make sure you know how to operate them. If a fire should start, you don't want to be scrambling to find a fire extinguisher.

PROGRESS CHECK

1. What is the most common electrical source of residential fires?
2. Why is an electrical arc more likely to start an industrial fire than a residential fire?
3. List three causes of conductors and components overheating.
4. What causes a high-resistance connection? What is a glowing connection?
5. List and describe the five classifications of fires.
6. Which type of fire extinguisher should never be used on an electrical fire?

S.5 PERSONAL PROTECTIVE EQUIPMENT

PERSONAL PROTECTIVE EQUIPMENT (PPE) refers to clothing and equipment that are used to protect a worker from injury on the jobsite. Employers are responsible for supplying the proper PPE and making sure it is used by their employees. Proper PPE requirements are outlined in *NFPA 70E®: Standard for Electrical Safety in the Workplace*, and in OSHA's Title 29 of the Code of Federal Regulations (CFR) Subpart 1, sections 1910.132 through 1910.138.

The type of PPE required is dictated by the type and location of the work being done. In this section we will cover some of the more common types of PPE, but this should not be considered a complete list.

Head Protection

Almost every jobsite requires some form of head protection. There are three electrical classifications for hard hats. These are:

- **Class C** (conductive): Class C offers low impact protection and is not designed for protection from contact with electrical conductors.
- **Class G** (general): As well as impact protection, Class G offers protection from contact with "low-voltage" conductors. One manufacturer proof-tests their Class G hard hats at 2200 volts, but does not certify protection at this voltage level.
- **Class E** (electrical): Class E hard hats are designed to reduce the danger of exposure to high-voltage conductors. One manufacturer proof-tests their Class E hard hats at 20,000 volts, but does not certify protection to this high a voltage.

Many types of hard hats include mounting slots for attaching a face shield. An electrician wearing a hard hat is shown in Figure S.4.

FIGURE S.4
Shutterstock.com

Eye Protection

The most common types of eye protection are safety glasses, goggles, and face shields. These are described as follows:

- **Safety glasses** are constructed using impact-resistant lenses with side-shields. Plastic frames are used to minimize shock hazards. They may be tinted to protect against low-voltage arcs and they may also have prescription lenses.
- **Goggles** fit snuggly to the face and provide better protection against spilled liquids than glasses. Goggles can also be worn over standard prescription glasses. Like safety glasses, they may be tinted for low-voltage arc protection.
- **Face shields** cover the entire face with an impact-resistant plastic shield. They offer the best protection against flying objects. Face shields may also be tinted for arc protection.

The type of eye protection used is determined by the type of work being performed.

Regardless of the type of eye protection used, it must be properly maintained and cleaned. It is very important not to impair your vision. Figure S.5 shows examples of safety glasses, goggles, and a tinted face shield mounted on a hard hat.

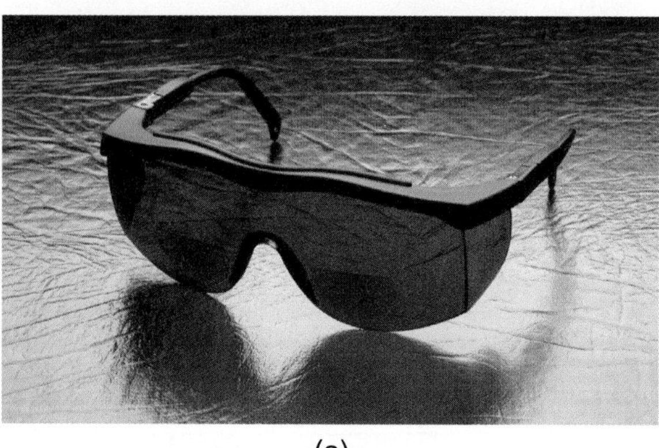

(a)

(b)

(c)

FIGURE S.5
Source: Photograph courtesy of Elvex Corporation-www.elvex.com.

Hand Protection

Rubber insulating gloves are worn under leather outer gloves to protect against electrical shock. Gloves are worn to protect the hands from cuts as well as electrical shock. The leather outer gloves help protect against punctures to the insulating gloves. Rubber insulating gloves are rated as indicated in Table S.4.

TABLE S.4 • Rubber Insulating Glove Classifications

CLASS	MAXIMUM AC VOLTAGE
00	500 V
0	1000 V
1	7500 V
2	17,000 V
3	26,500 V
4	36,000 V

Rubber gloves must be tested for holes and cuts on a regular basis. This is done by inflating the glove and checking for leaks. Some manufacturers employ two-layer construction where the inner layer is a different color than the outer. When the inner layer becomes visible the gloves should be discarded. The outer protective leather glove should also be inspected regularly. Figure S.6 shows an electrician wearing rubber gloves inside outer leather gloves.

FIGURE S.6
Istock.com

Hearing Protection

The National Institute for Occupational Safety and Health (NIOSH) estimates that 30 million workers in the United States are exposed to hazardous noise. Many electricians and technicians work in very noisy environments.

Earplugs and protective headphones are the most common types of hearing protection. Unfortunately, hearing protection can have a negative side effect. If you are wearing hearing protection you may not be able to hear warning signals. You must always stay alert when wearing any form of hearing protection. Figure S.7 shows examples of several types of hearing protection devices.

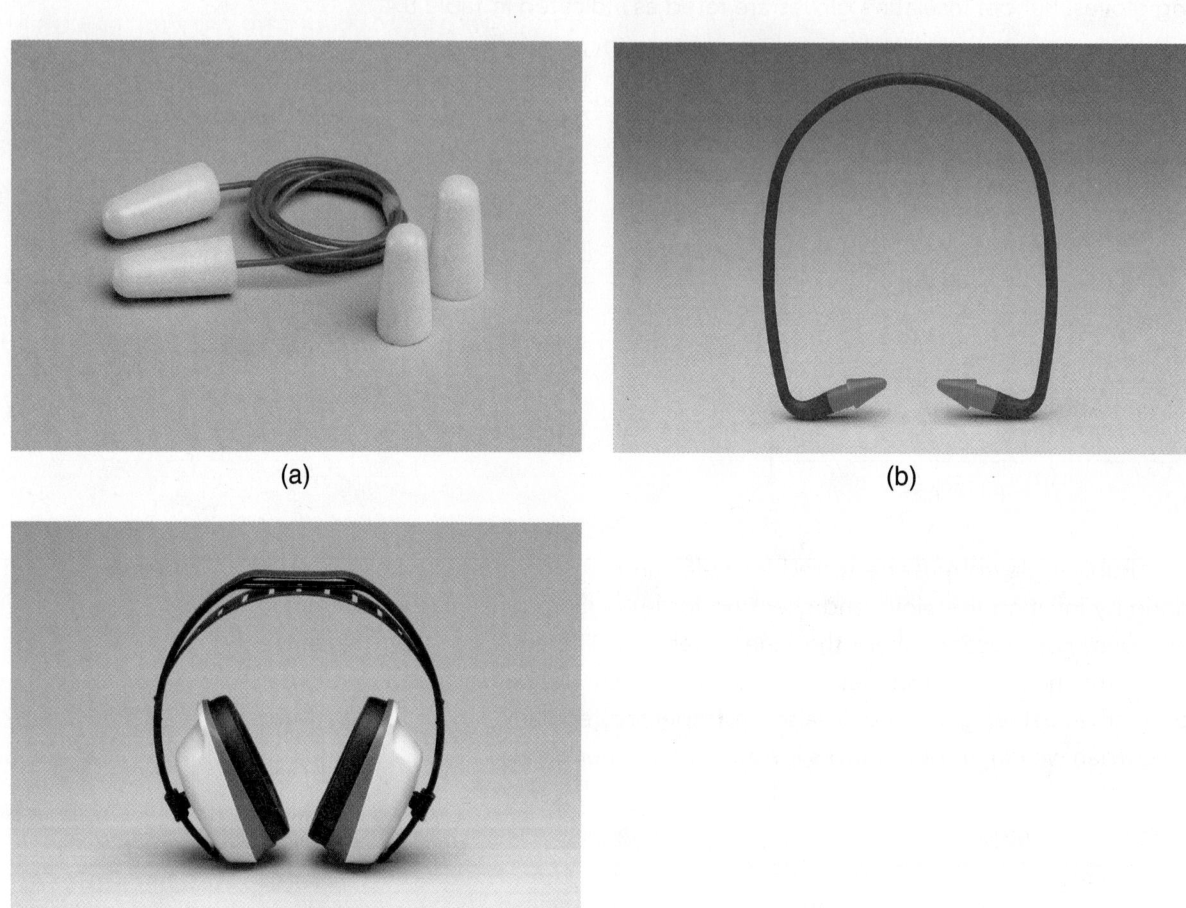

(a) (b) (c)

FIGURE S.7
Source: Photograph Courtesy of Elvex Corporation-www.elvex.com.

Protective Clothing

Certain types of clothing should never be worn when working with electricity. Synthetic materials like rayon, nylon, and polyester can easily catch fire from a spark or arc. Jewelry, such as watches, rings, or metal chains, must not be worn as it may make contact with the energized parts of electrical equipment. Long pants and long-sleeved shirts made of natural materials are a minimum requirement when working with electrical circuits, and nonconductive (electrically insulated) safety shoes or boots are mandated on many jobsites. Certain types of flame-resistant (FR) clothing may be required, especially when working with high-voltage equipment. All clothing should be kept clean of oil and grease and inspected regularly.

PROGRESS CHECK

1. What is PPE? Who has the responsibility to assure that proper PPE requirements are met?
2. List and describe the three classifications of hard hats.
3. List and describe the three main types of eye protection.
4. How many classifications of rubber insulating gloves are there? What does the classification of a given glove indicate?
5. How do you test a rubber glove for cuts or holes? Why do some gloves have two different colors of rubber?
6. What should an electrician never wear on the jobsite?

S.6 HAZARDOUS LOCATIONS

Electricians and electrical technicians work in just about every environment you can imagine. Certain types of jobsites offer a higher degree of risk than others. Hazardous locations are covered by *Article 500* of the *NEC® (NFPA 70)* and *NFPA 400 (Hazardous Materials Code)*. Some reasons a location may be hazardous include:

- **Dangerous Equipment** (both stationary and mobile): Electrical equipment can be dangerous, but there may be other dangerous equipment on a jobsite as well. You may be working near equipment that can crush or burn you. There may be moving equipment such as tow motors or robotics. You must familiarize yourself with the entire jobsite, not just the immediate area where you are working.
- **Toxic Materials or Atmospheres:** You must always be aware of your environment. The dangers of working in a chemical plant are very different from those in a wood-working facility. A lack of oxygen, or the presence of a toxic material or gas, may be a hazard in some locations. Many toxic gases present a fire or explosion hazard, as well as a breathing hazard. Some materials can pose a hazard that you wouldn't expect. For example, fine particles of grain, and some metals and plastics, can present an explosion hazard if exposed to a spark.
- **Confined Spaces:** Any space that is large enough for a worker to enter and perform their tasks but too small for normal occupancy, or has a restricted entry and/or exit, is considered a confined space. Storage tanks, boilers, ventilation duct-work, underground vaults, and trenches are just a few examples of confined spaces. Confined spaces are much more likely to develop dangerous atmospheres.

  Most jurisdictions require special permits to work in confined spaces. Atmospheric tests should be done prior to entering the confined space, and personnel must be stationed outside the confined space so they can respond to any emergency. If a person is in distress inside a confined space, *do not enter yourself*. You may be overcome as well. Seek help immediately.

- **Heights:** Electricians often work high off the ground. They may work on ladders, scaffolds, suspended platforms, utility poles, or in cherry pickers. When working more than six feet off the ground, a safety harness should be worn. The safety harness is connected to a short section of rope called a **LANYARD**. The lanyard is then attached to some secure structure to limit a fall. Figure S.8a shows a standard safety harness. Figure S.8b shows an electrician in a cherry picker wearing a safety harness.
- **Trenches:** Electricians are often called upon to bury electrical conduit. When working in a trench, there may be danger of the sides collapsing. Proper shoring techniques should always be followed to prevent injury.

These are only a few of the reasons that a jobsite may be considered hazardous. You should refer to the relevant OSHA and *NEC*® articles for more information on this topic.

(a)

(b)

FIGURE S.8
Shutterstock.com

PROGRESS CHECK

1. List and describe four reasons why a jobsite may be considered hazardous.
2. What is a confined space? What do many jurisdictions require for working in a confined space?
3. If someone is in distress in a confined space what should you do? What should you not do?
4. When should you wear a safety harness? What is a lanyard and what is it used for?

S.7 COMMON-SENSE RULES FOR WORKING SAFELY

There are many things that you can do to minimize workplace hazards. Following all local, state, and national codes is very important, but there are some common-sense steps that are important too. Here are a few suggestions that can help keep you safe:

- **Avoid Working Alone:** Whenever possible, avoid working by yourself. This is especially true if you are working with high-voltage circuits, or in an otherwise hazardous work environment. If something does happen, you will want someone there to help.
- **Learn First Aid:** If an accident occurs, you can't be of much help if you don't know what to do. As Table S.1 (page 3) indicates, an electrical shock can cause breathing and heart problems. Make sure that you can help when needed.
- **Avoid the Rush:** As you get more experience, you will learn certain short-cuts that may speed up your work. Never compromise safety to finish a job faster. As a sign on one work site reads, *"No job is so important that we cannot take the time to do it safely."*
- **Use the Right Tools:** It is always important to use the right tools for the job. Double-insulated tools provide the greatest protection from electrical shock. Figure S.9a shows a set of double-insulated tools. Figure S.9b highlights the double-insulated handles. *NFPA 70E*® mandates that insulated tools must be used when working on live circuits. It is also important to use tools correctly. For example, using a meter designed for low-voltage circuits on a high-voltage circuit presents a safety hazard. Finally, you

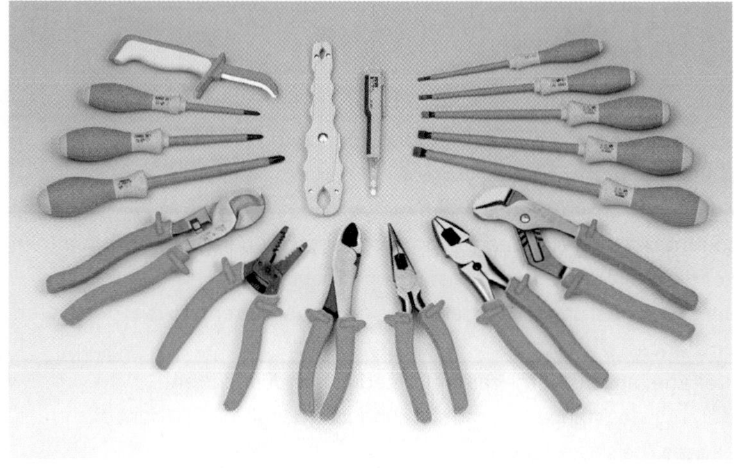

(a)

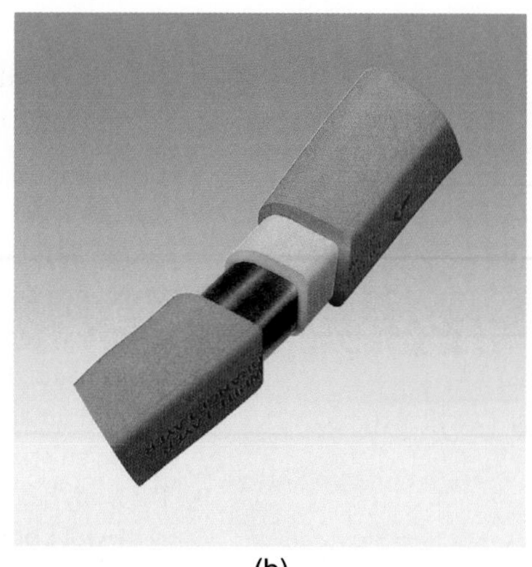

(b)

FIGURE S.9
Source: Photograph Courtesy of Ideal Industries, Inc..

should inspect your tools regularly to make sure they have not been damaged or compromised.

- **Know What You're Doing:** You may be exposed to situations and equipment that you have not encountered in your training. Don't take unnecessary risks. If you are unsure of how to proceed with a certain task, or how to use a specific piece of equipment, ask before taking action.
- **Stay Sober:** Alcohol and recreational drugs have no place on the jobsite. If your abilities are impaired, you are a danger to yourself and everyone you work with. No employer wants to see an employee injured or to face a lawsuit. The day you are found to be impaired on the job will likely be your *last* day on the job.
- **Stay Alert:** Chances are you will work alongside others. Don't take anything for granted. Look around and pay attention to what those around you are doing. Don't assume that they are working safely. Be proactive. Take responsibility for your own safety.

The National Institute for Occupational Safety and Health (NIOSH) is a department of the Centers for Disease Control and Prevention (CDC).[5]

PROGRESS CHECK

1. List five common-sense things that you can do to work more safely.
2. What is the safest type of tool for an electrician?
3. Why should you pay attention to what other workers are doing on your jobsite?
4. What is NIOSH?

S.8 SUMMARY

Electrical Shock Hazard

- Electrocution is the third leading cause of work-related deaths for 16- and 17-year-olds. It is the cause of 12% of all workplace deaths among young workers.
- Voltage is the force that generates electric current.
- Current is the flow of electrical charge through a conductor.
- Resistance limits current. At a given voltage, more current is generated through a low-resistance conductor than through a high-resistance conductor.
- An insulator has very high resistance and allows almost no current to be generated under normal circumstances.
  - Given enough voltage, any insulator can be forced to conduct current.
- The body has relatively high resistance, but can still conduct a dangerous amount of current.
  - A wet body increases the shock hazard.

[5]We strongly recommend that you visit the NIOSH website at http://www.cdc.gov/niosh for more information on workplace safety.

- Current is measured in amperes (amps).
  - A milliampere (mA) is one one-thousandth of an ampere.
- The amount of shock current determines the effect of a shock on the body. See Table S.1.
  - A shock current as low as 100 milliamperes may cause death under the right set of circumstances.
  - Most residential home outlets are rated for up to 15 amperes (15,000 mA).
- Electrical shock current creates heat when passing through bones and tissue.
  - Burns may be internal as well as external.
- Shocks can also cause dangerous falls, internal hemorrhage, and muscle and kidney damage.
- You should avoid working on live circuits wherever possible. When working with a live circuit, you should try to keep one hand out of the circuit.
- There are three classifications of current pathways through the body.
  - *Touch potential* (hand-to-hand, through the heart)
  - *Step potential* (foot-to-foot through the lower body)
  - *Touch-step potential* (right hand-to-right foot, or left hand to left foot, through the body)
  - Of the pathways listed, touch potential is the most dangerous.
- *Freezing* is when a shock current prevents you from letting go of a circuit.
  - If a person is frozen to a circuit you should free them as quickly as possible, but safely.
  - Try first to deenergize (disconnect) the source of the electrical shock.
  - If you cannot deenergize the circuit, use some nonconductive material (wood or plastic) to push or pull them from the circuit. Never use your hands.
- You cannot be sure how serious a shock is. You should get medical help immediately.
- An arc flash is an explosive release of energy resulting from an electrical short.
  - It can be from a dropped tool, accidental contact with a high-voltage circuit, or improper use of test equipment.
  - 120 V residential circuits normally do not produce an arc flash. Medium-voltage circuits (600 V) and higher are more susceptible to arc flashes.
  - A high-voltage arc flash can vaporize metal and project a very dangerous plasma (ionized gas) arc that may start fires, destroy equipment, and kill people some distance away.

Ground Faults, Grounding, and Bonding

- Ground refers to the earth.
  - Earth ground is considered the zero-volt reference.
- Voltage is defined as values above ground (positive) and below ground (negative).
  - Any positive or negative voltage will generate current to or from ground if there is a current path.
- A ground fault is the unintended connection of an energized conductor to ground.
  - If a ground fault occurs and you are in contact with the component, your body may provide the path to ground and you will receive a shock.
  - The ground prong on the plug of many devices connects the metal chassis of the device to earth ground through the wall receptacle.
  - You should never remove or defeat the ground prong on a plug.
- A ground-fault circuit interrupter (GFCI) protects against ground faults.
  - GFCIs sense any current imbalance greater than 5 mA (5/1000 of an amp) and break the connection.
  - A GFCI tester checks for polarity and produces a ground fault to test the GFCI.

- Grounding refers to the connection of equipment or an electrical system to earth ground.
- Earth ground is established in several ways.
    - By connecting to a copper stake driven into the ground.
    - By connecting to underground pipes like metal water pipes.
    - By connection to an electrode encased in the concrete foundation of a building.
- Bonding is the permanent joining of conductive non–current-carrying materials in a structure.
    - Bonding prevents arcs and sparks if there is a short circuit.
    - Bonding assures that overcurrent devices work quickly if a short circuit occurs.
    - Bonding makes sure that all conductive materials in a structure will be at the same voltage potential if there is a short.
- Grounding and bonding are complex issues that are governed by specific regulations. See *NEC® section 250* and IEEE 142 and 1100.

Lockout/Tagout

- The safest way to avoid shocks is to work on deenergized equipment.
- Lockout/tagout procedures are one common way to assure that a circuit is deenergized.
    - If a lockout is used, the person working on the circuit keeps the key so that no one can energize the circuit without their knowledge.
    - If more than one person is working on a circuit or machine, each person installs a lock and keeps the key.
- Even if a circuit has been tagged or locked out, you should still check to make sure that the circuit is actually deenergized. The following approach is the safest:
    - First test your meter on an energized circuit.
    - Test the deenergized circuit to make sure that it is powered down.
    - Retest your meter on an energized circuit to make sure that your meter is still working.

Electrical Fire Hazards

- The NFPA (National Fire Protection Association) is an important source for fire prevention information.
- Residential fires are caused by a number of electrical sources. See Table S.2. Fixed wiring is the most common source.
- There are three ways that electricity ignites a fire: shorts, arcing, and excessive heating of conductors or components.
    - Shorts and arcing are seldom the cause of residential fires as the voltage is too low, but the NFPA has shown that furniture crushing lamp and appliance cords has in fact caused some fires.
    - Industrial fires are often caused by arcs since the voltage is much higher.
- Excessive heat is the most common cause of electrical fires. There are several ways that a conductor or component can overheat.
    - Looping a conductor on itself several times will cause it to overheat.
    - Excessive thermal insulation is another cause of overheating.
    - Poor connections are the main source of overheating.

- A high-resistance connection causes a connection to heat up.
  - The resulting heat causes oxidation and creep. Creep is the further physical separation of connecting materials due to heat.
- Oxidation and creep cause more heating until the connection becomes a glowing connection.
  - A glowing connection is the result of extreme heat and is capable of igniting nearby combustible materials.
- The best way to put out a fire is to use the proper fire extinguisher.
- Fires and fire extinguishers fall into five classifications. See Table S.3.
- You should never use a water-based (Class A) extinguisher on an electrical (Class C) fire.
  - You should always locate the fire extinguishers on a new jobsite and make sure you know how to use them.

Personal Protective Equipment

- Personal protective equipment (PPE) is clothing or equipment that is used to protect a worker from injury on the jobsite.
  - The employer is responsible for determining the proper PPE requirements.
  - PPE is covered in *NFPA 70E*® and OSHA's Title 29 of the Code of Federal Regulations.
- Hard hats have three classifications.
  - Class C (conductive) offers impact protection but no electrical protection.
  - Class G (general) offers impact protection and low-voltage electrical protection.
  - Class E (electrical) offers impact protection and high-voltage electrical protection.
- There are three main types of eye protection.
  - Safety glasses are made using impact-resistant lenses (they may be prescription) and plastic frames. They can be tinted for low-voltage arc protection.
  - Goggles fit close to the face and offer better protection against spills. They may be tinted and can fit over standard prescription glasses.
  - Face shields cover the entire face with an impact-resistant plastic shield. They offer the best protection against flying objects. Face shields may also be tinted for arc protection.
- Rubber insulated gloves are worn inside outer leather gloves.
  - Rubber insulating gloves should be inspected for cuts and holes by inflating the glove and checking for leakage.
- Rubber insulating gloves fall into six classes. See Table S.4.
  - The glove ratings are based upon the level of voltage that the gloves can protect you from.
- The two main types of hearing protection are ear plugs and protective headphones.
  - You should be aware that hearing protection devices may prevent you from hearing warning signals. You must be alert when wearing hearing protection.
- Clothing made from synthetic materials and jewelry should never be worn by an electrician on the jobsite.
  - Synthetic material can catch fire, and metal jewelry is a shock hazard.
- The minimum requirement for an electrician is long-sleeved shirts and long pants made from natural materials.
  - Safety shoes or boots and flame-resistant (FR) clothing may be required on some jobsites.
  - All clothing should be kept clean of oil and grease and inspected regularly.

Hazardous Locations

- Hazardous locations are covered by *Article 500* of the *NEC®* *(NFPA 70)* and *NFPA 400 (Hazardous Materials Code)*. There are a number of reasons why a location may be considered hazardous.
- Dangerous equipment.
  - There may be machinery on the jobsite that may injure you.
- Toxic material or atmospheres.
  - Some jobsites may present both breathing and explosion hazards. You must be aware of the environment in which you are working.
- Confined spaces.
  - Any space too small for normal occupancy or that has restricted entry and/or exit is a confined space.
  - Storage tanks, boilers ductwork, and trenches are some examples.
  - Most jurisdictions require special permits to work in confined spaces.
  - Someone should be stationed outside the confined space to respond to any emergency. If one should occur, you should never enter the confined space yourself, but seek help immediately.
- Working at heights.
  - If you are working more than six feet off the ground, you should wear a safety harness.
  - The safety harness has a lanyard attached that in turn is connected to some secure structure to limit your fall.
  - Safety harnesses should be used when working in trenches as well.
- Working in trenches.
  - The walls of trenches may collapse. Proper shoring techniques must always be used when digging trenches.

Common-Sense Rules for Working Safely

- Although following all local, state, and national codes is very important, there are a number of common-sense things that can help you to work safe.
  - Don't work alone whenever possible.
  - Learn first aid.
  - Don't take dangerous shortcuts to speed up a job.
  - Always use the proper tools and keep them in good working order.
  - Don't proceed with a job unless you are confident that you know what you are doing.
  - Never use alcohol or drugs on the job.
  - Pay attention to other workers on the jobsite. Don't assume that they are working safely.
- Refer to the National Institute for Occupational Safety and Health (NIOSH) for more information on workplace safety.

CHAPTER REVIEW

1. The force that generates current in an electrical circuit is called ________.
   a) amperes
   b) voltage
   c) inductance
   d) power

2. Any material that allows no current to be generated when voltage is applied is a good ________.
   a) resistor
   b) conductor
   c) insulator
   d) capacitor

3. Water is a very good ________.
   a) insulator
   b) resistor
   c) inductor
   d) conductor

4. Current is measured in ________.
   a) watts
   b) amperes
   c) ohms
   d) volts

5. A shock current of as little as ________ may result in death under the right circumstance.
   a) 100 milliamperes
   b) 10 amperes
   c) 500 milliamperes
   d) 15 amperes

6. ________ is (are) the most common shock-related injury.
   a) Burns
   b) Falls
   c) Nerve damage
   d) Heart attacks

7. You should always try to work one-handed when testing a live circuit to avoid ________ potential since it is the most dangerous current pathway through the body.
   a) touch
   b) step
   c) touch-step
   d) all of the above

8. You should always use ________ to free someone who is frozen to a circuit.
   a) your feet
   b) a screwdriver
   c) your hands
   d) some nonconductive material

9. An explosive release of energy as a result of an electrical short is called a(n) ________.
   a) arc flash
   b) ground fault
   c) critical short
   d) terminal short

10. Earth ground is the ________.
   a) highest voltage in a circuit
   b) primary source of current
   c) zero-volt reference
   d) destination for all current

11. The unintended connection of an energized conductor to ground is referred to as a ________.
   a) ground fault
   b) chassis short
   c) shock fault
   d) critical short

12. A GFCI senses any current ________ between the hot and neutral lines and breaks the connection.
   a) level
   b) increase
   c) decrease
   d) imbalance

13. Ground is established in a circuit by connecting to ________.
   a) a copper stake driven into the ground
   b) an underground metal pipe
   c) an electrode encased in the foundation of the structure
   d) all of the above

14. The permanent joining of conductive non–current-carrying materials in a structure is referred to as ________.
   a) grounding
   b) bonding
   c) strapping
   d) all of the above

15. Lockout/tagout ________.
   a) is used to assure that a circuit remains deenergized
   b) can be used on both switches and breaker panels
   c) can be used by more than one worker at a time
   d) all of the above

16. If a circuit is locked, the key is kept by ________.
    a) the supervisor
    b) the worker who is working on the circuit
    c) the safety officer
    d) the employer

17. The most common source of residential electrical fires is ________.
    a) electrical appliances
    b) fixed wiring
    c) extension cords
    d) light fixtures

18. The most common cause of residential electrical fires is ________.
    a) excessive heating
    b) short circuits
    c) arcing
    d) crushed wiring

19. An electrical connection that is not tight may result in ________.
    a) a high-resistance connection
    b) a glowing connection
    c) a fire
    d) all of the above

20. A Class ________ fire involves live electrical equipment.
    a) A
    b) B
    c) C
    d) E

21. You must never use a Class ________ extinguisher on an electrical fire.
    a) A
    b) B
    c) C
    d) E

22. Determining the correct PPE on a given jobsite, and assuring that it is used, is the responsibility of ________.
    a) the *NEC*®
    b) the employee
    c) the employer
    d) the NFPA

23. You should use Class _______ head protection when working on high-voltage circuits.
    a) A
    b) E
    c) G
    d) HV

24. Eye protection devices are often tinted to protect the electrical workers from _______.
    a) arc flashes
    b) bright sunlight
    c) arc welding
    d) all of the above

25. The protective glove that offers the best high-voltage protection is Class _______.
    a) 00
    b) A
    c) H
    d) 4

26. Rubber insulating gloves are checked for cuts and holes by _______.
    a) inflating them
    b) filling them with water
    c) turning them inside out
    d) all of the above

27. One problem with using hearing protection is _______.
    a) it gets in the way
    b) it can cause short circuits
    c) you may not hear danger signals
    d) it prevents you from wearing head protection

28. An electrician should never wear _______ on the jobsite.
    a) a watch
    b) a rayon shirt
    c) a ring
    d) all of the above

29. A hazardous location is _______.
    a) a confined space
    b) a location with dangerous machinery
    c) a location with toxic materials
    d) all of the above

30. You should wear a safety harness if you are working more than _______ feet off the ground.
    a) 6
    b) 8

c) 10
d) 12

31. The short section of rope connected to the safety harness is referred to as a _______.
    a) lanyard
    b) sheet
    c) piton
    d) cleat

32. The safest type of electrical tools are _______ tools.
    a) low-impact
    b) ergonomic
    c) double-insulated
    d) right-handed

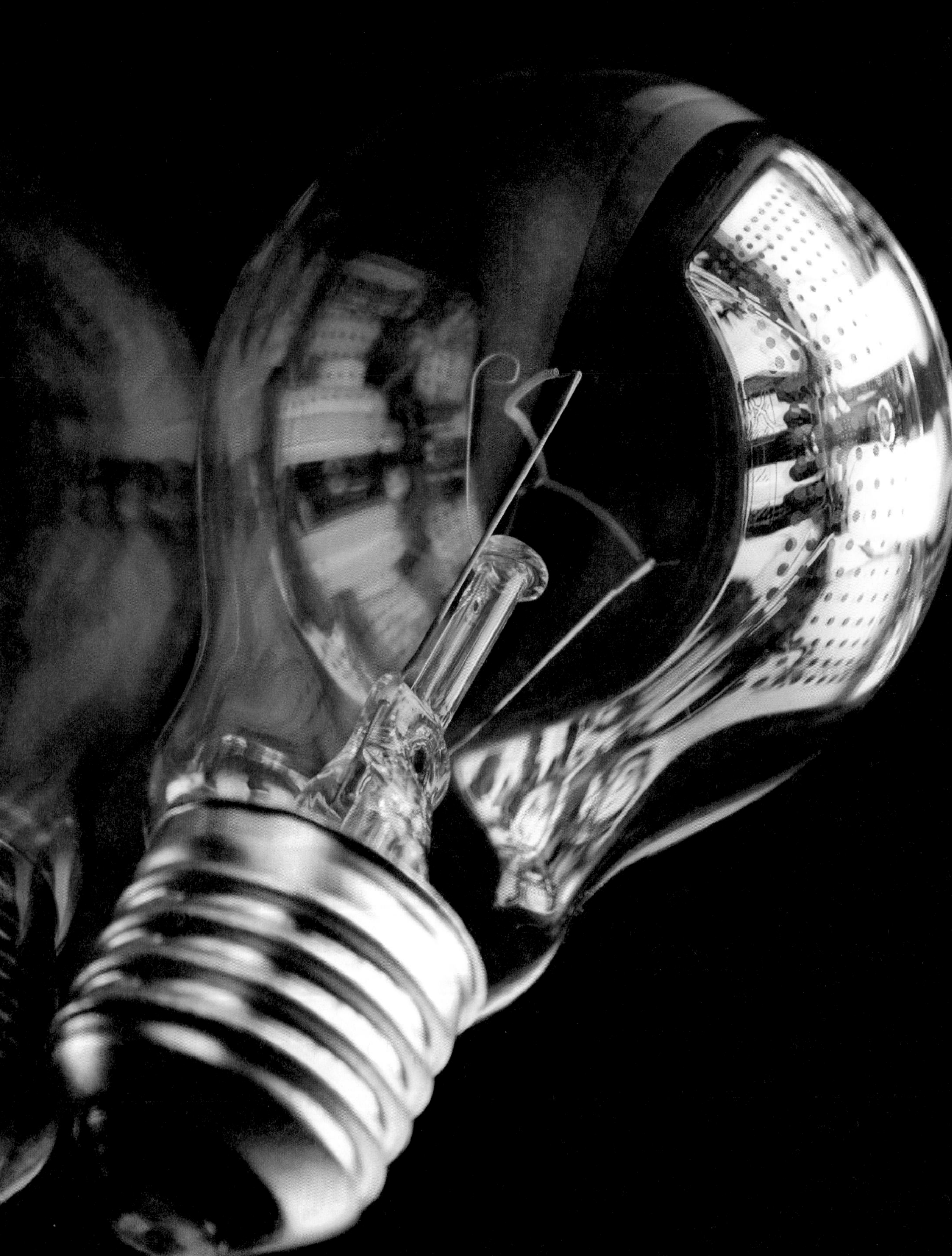

CHAPTER 1

GETTING STARTED

PURPOSE

The fact that you are reading this book says that you have chosen a career in electrical or electronics technology. Good choice! The Bureau of Labor Statistics (a division of the U.S. Department of Labor) predicts that the need for highly-trained professionals will continue to grow over the foreseeable future. As a result of industry growth and personnel changes, thousands of new positions continue to open every year. Taking this course is the first step toward filling one of those positions.

Over the next few chapters, we will begin our study of basic electrical circuits and principles. But first, let's take a look at some potential career paths and the tools that will help you get there.

KEY TERMS

The following terms are introduced and defined in this chapter on the pages indicated:

OBJECTIVES

After completing this chapter, you should be able to:

1. Compare and contrast the duties and working environments of residential and commercial electricians, telecommunications technicians, and outside linemen.
2. Compare and contrast the duties of electronics engineers, electronics technicians, and engineering technicians.
3. Describe the functions of communications, telecommunications, computer, industrial, biomedical, and aviation electronics systems.
4. Identify the most common electrical components.
5. Identify and describe the most commonly used pieces of electrical/electronic test equipment.
6. Describe the function, unit of measure, and schematic symbols for resistors.
7. Describe the function, unit of measure, and schematic symbols for capacitors.
8. Describe the function, unit of measure, and schematic symbols for inductors.
9. Identify the commonly used engineering notation prefixes, along with their symbols and value ranges.
10. Work with values expressed in engineering notation.

1.1 ELECTRICAL CAREERS

There are many career options for anyone who wants to work with electrical circuits and systems. The *Occupational Outlook Handbook* (OOH), a publication of the U.S. Department of Labor, provides detailed descriptions of 75 careers that require the ability to operate or maintain electrical systems. Of these, approximately 30 require a working knowledge of electrical and electronic components, circuits, and systems. The most common electrical careers are introduced briefly in this section.

Electricians

An **ELECTRICIAN** is someone who has been trained to install, operate, maintain, and repair electrical circuits and systems. Electricians work with various power circuits, control circuits, security circuits and systems, integrated voice, data, and video systems, and commercial and industrial motors and controls. All electricians must have a working knowledge of the basic principles of electricity and electronics, reading and interpreting blueprints, wiring techniques, safe working practices, first aid, hazardous materials, the National Electrical Code (*NEC*®), and all relevant state and local electrical codes and standards.

ELECTRICIAN Someone who has been trained to install, operate, maintain, and repair electrical circuits and systems.

Many electricians use the residential field as a launching point for their careers in the electrical construction industry, with the aim of increasing knowledge and expertise as well as wages.

IN THE FIELD

FIGURE 1.1 A residential electrician.
Istock.com

Residential Electricians

A **RESIDENTIAL ELECTRICIAN** installs, maintains, and repairs all the electrical circuits found in houses, apartments, and condominiums. These circuits include power receptacles, service panels, junction boxes, inside and outside lighting, and appliance power circuits. Electricians who work on construction sites set up

RESIDENTIAL ELECTRICIAN Someone who installs, maintains, and repairs all the electrical circuits found in houses, apartments, and condominiums.

temporary power circuits to be used during construction and install the permanent electrical circuits called for in the building plans and blueprints.

Residential electrician training in the United States includes two or more years of classroom instruction and on-the-job training in an apprenticeship program. This training includes:

- Installing, troubleshooting, and repairing common residential power, lighting, and communications circuits
- Using the *NEC*® to identify the proper conductors and wiring methods for use in any residential wiring application
- Installing and terminating copper and aluminum conductors
- Installing, terminating, and splicing twisted-pair cables
- Installing and terminating coaxial cables

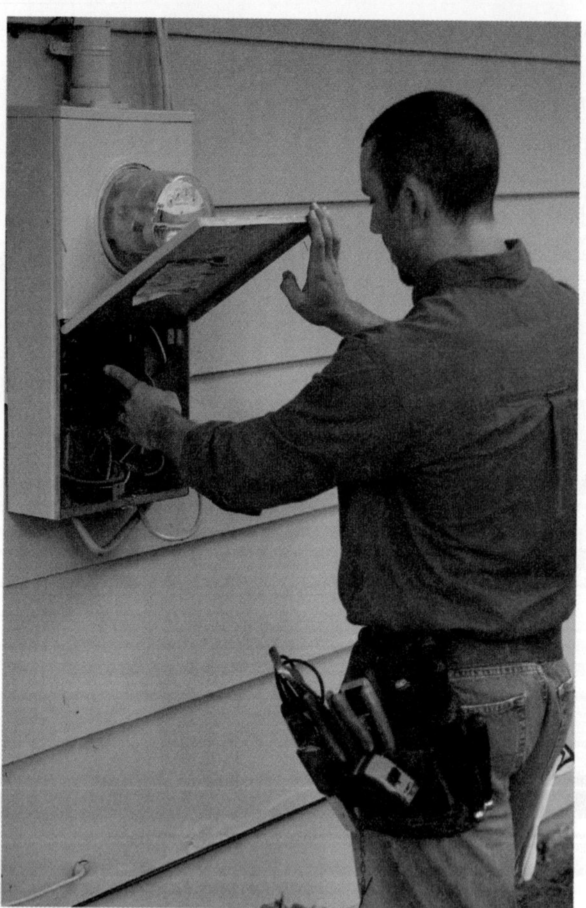

FIGURE 1.2 An electrician checks the electrical service entrance to a building.
Istock.com

INDUSTRIAL ELECTRICIANS
Electricians who install, maintain, and repair the circuits found in industrial and commercial buildings.

COMMERCIAL ELECTRICIANS
Electricians who install, maintain, and repair the circuits found in industrial and commercial buildings.

Industrial and Commercial Electricians

INDUSTRIAL ELECTRICIANS and **COMMERCIAL ELECTRICIANS** install, maintain, and repair the circuits found in industrial and commercial buildings, including power distribution circuits, motor and environmental power and control circuits, lighting circuits and fixtures, receptacles, service panels, junction boxes, power raceways, and fire alarm circuits.

Industrial and commercial electrician training in the United States may include five years of classroom instruction and on-the-job training in an apprenticeship program. This training includes:

- Using the *NEC*® to identify the proper conductors and wiring methods for use in any industrial or commercial application
- Installing and terminating copper and aluminum conductors
- Installing, terminating, and splicing high-voltage cables

Commercial and industrial electricians, whether working for an electrical contractor or in plant/facility maintenance, work with higher voltages and more complex systems. This working environment, along with higher responsibilities, results in a higher wage, increased benefits, and enhanced employment stability.

FIGURE 1.3 A commercial electrician.
Istock.com

Telecommunications Technicians

TELECOMMUNICATIONS TECHNICIANS install, maintain, troubleshoot, and repair residential and commercial data, audio, video, and security circuits. These circuits include low-voltage networks, voice and data transmission circuits, local area network (LAN) cables, audio and video distribution circuits, and security and access control circuits.

TELECOMMUNICATIONS TECHNICIAN Someone who installs, maintains, troubleshoots, and repairs residential and commercial data, audio, video, and security circuits.

Telecommunication technician training in the United States may include three years of classroom instruction and on-the-job training in an apprenticeship program. This training includes:

- Wire color coding, cable testing, and standards
- LAN installation and testing
- Telephony
- Working with twisted-pair, coaxial, and fiber-optic cables.

A residential, commercial, or telecommunications electrician may eventually consider pursuing a masters or contractors liscense and possibly starting his or her own business.

IN THE FIELD

FIGURE 1.4 A telecommunications electrician connecting wires to a termination block.
Fotolia.com

Outside Linemen

OUTSIDE LINEMAN Someone who installs and repairs power transmission and distribution lines and circuits.

OUTSIDE LINEMEN install and repair power transmission and distribution lines and circuits, including overhead high-voltage lines, transformers and connectors, power substations, and underground lines in residential neighborhoods. Outside linemen may work for independent contractors or for the local power utilities.

Outside lineman training in the United States may include 3 ½ years of classroom instruction and on-the-job training in an apprenticeship program. This training includes:

- Installing, splicing, and repairing high-voltage lines
- Evaluating and climbing wooden power poles
- Operating a bucket truck
- The proper use of rigging equipment

FIGURE 1.5 Outside linemen working on power lines near a utility pole.
Istock.com

PROGRESS CHECK

1. Which electricians might be found working in a residential neighborhood?
2. Which electricians would most likely be found working outdoors?
3. Which electricians would most likely work in a computer production facility?
4. Describe the training required for each of the following.
   a. Residential electricians
   b. Industrial and commercial electricians
   c. Telecommunications electricians
   d. Outside linemen

1.2 ELECTRONICS CAREERS

ELECTRONICS is the field that deals with the design, manufacturing, installation, maintenance, and repair of electronic circuits and systems. Audio and video systems, computers, radar systems, and industrial robotic machines are just a few examples of electronic systems.

ELECTRONICS The field that deals with the design, manufacturing, installation, maintenance, and repair of electronic circuits and systems.

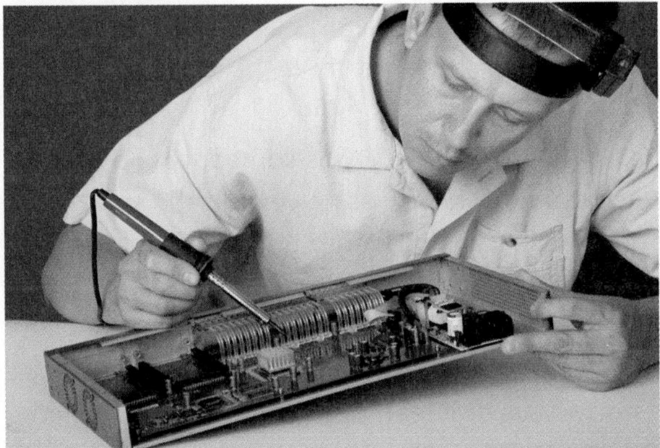

FIGURE 1.6 An electronics technician soldering components on a circuit board.
Istock.com

While a variety of trained professionals (such as technical writers and equipment assemblers) are employed in the electronics industry, the primary technical duties are performed by *electronics technicians* and *electronics engineers*.

Electronics Technicians

O*NET Online (a website provided by the U.S. Department of Labor) describes an **ELECTRONICS TECHNICIAN** as someone who locates and repairs faults in electronic systems and circuits using electronic test equipment, performs periodic maintenance, maintains records of maintenance and system failures, and installs

ELECTRONICS TECHNICIAN Someone who installs, maintains, and repairs electronic systems and circuits using electronic test equipment.

new systems. In contrast to the engineer, the electronics technician is typically responsible for circuit maintenance and repair.

Electronics technicians in the United States are typically graduates of a one- or two-year technical college or community college, or veterans who received their training in military training programs. While in school, technicians typically study DC circuits, AC circuits, solid-state components and circuits, digital electronics, and communications electronics.

FIGURE 1.7 A technician checking the connections in an electronic system.
Istock.com

Electronics Engineers

ELECTRONICS ENGINEER Someone who designs electronic components, circuits, and systems.

PROTOTYPE The first working model of a circuit or system.

O*NET Online describes an **ELECTRONICS ENGINEER** as someone who designs electronic components, circuits, and systems, develops new applications for existing electronics technologies, assists technicians in solving unusual problems, and supervises the construction and initial testing of system **PROTOTYPES** (initial working models).

In most cases, engineers in the United States are graduates of 4-year college or university degree programs. The courses that lead to an electronic engineering

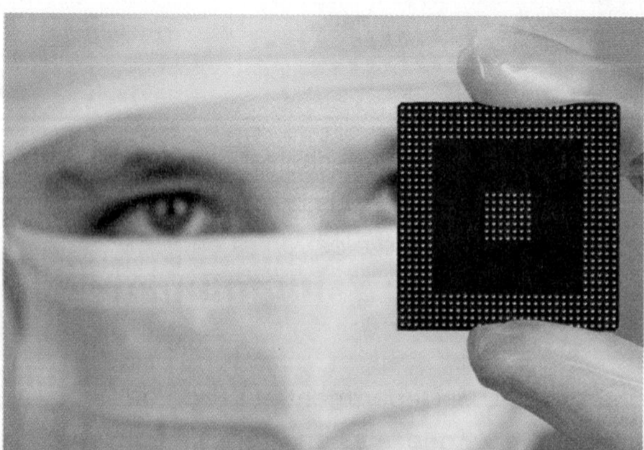

FIGURE 1.8 A technician inspecting a computer circuit prototype.
Istock.com

degree include calculus, physics and chemistry, statics and dynamics, and circuit analysis and design techniques.

Areas of Specialization

Most electronic engineers and technicians work with one (or more) of the many types of electronic systems. Some of the common areas of specialization are introduced here.

Communications Electronics

COMMUNICATIONS SYSTEMS are designed to transmit and/or receive information. Common communications systems include AM transmitters and receivers, FM transmitters and receivers, broadcast and cable television transmitters and receivers, and CB (citizens' band) transmitter/receivers, or *transceivers*.

COMMUNICATIONS SYSTEMS Electronic systems that transmit and/or receive information.

FIGURE 1.9 A telecommunications tower.
Istock.com

Telecommunications

TELECOMMUNICATIONS is the area of communications that deals with sending and receiving voice, data, and/or video information. The *Internet* is perhaps the

TELECOMMUNICATIONS The area of communications that deals with sending and receiving voice, data, and/or video information.

best-known example of a telecommunications system. Telecommunications terminals usually communicate with each other via telephone lines, satellite links, or cable (wire or optical).

FIGURE 1.10 A group of parabolic dishes used to communicate with satellites.
Pearson

Computer/Digital Electronics

COMPUTER SYSTEMS
Electronic systems that store and process information.

COMPUTER SYSTEMS are designed to store and process information. Personal computers, word processors, printers, and mainframes are a few examples of computer systems.

FIGURE 1.11 A desktop computer is only one type of computer system.
Pearson

Industrial Electronics

INDUSTRIAL ELECTRONICS
The area of electronics that deals with systems designed for use in industrial (manufacturing) environments.

INDUSTRIAL ELECTRONICS deals with systems that are designed for use in industrial (manufacturing) environments. Control circuits for industrial welders, robotics systems, and motor control systems are some examples of industrial circuits.

Biomedical Electronics

BIOMEDICAL ELECTRONICS
The area of electronics that deals with systems designed for use in diagnosing, monitoring, and treating medical problems.

BIOMEDICAL ELECTRONICS deals with systems that are designed for use in diagnosing, monitoring, and treating medical problems. Electrocardiographs, x-ray machines, diagnostic tread mills, and surgical laser systems are all examples of biomedical systems.

FIGURE 1.12 An industrial robotics system that is used in automobile production.
Istock.com

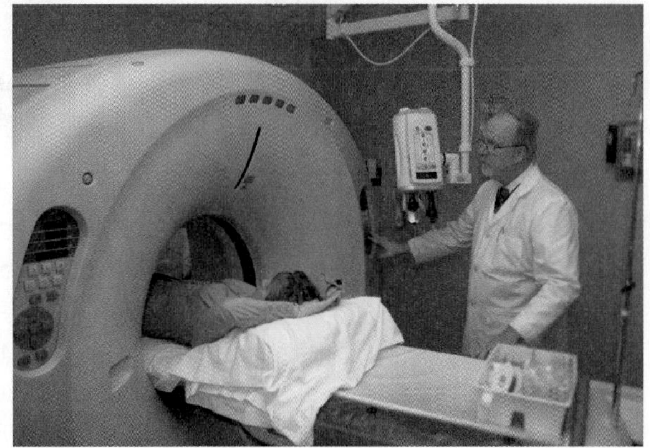

FIGURE 1.13 A medical scanner.
Istock.com

Aviation Electronics (Avionics)

AVIONICS is the area of specialization that deals with aircraft communications and navigation systems, aircraft and weather radar systems, and other on-board instruments and data systems. Because of the role that avionics plays in commercial air transportation, avionics programs are provided at schools accredited by the Federal Aviation Administration (FAA).

AVIONICS The area of electronics that deals with aircraft communications and navigation systems, aircraft and weather radar systems, and other on-board instruments and data systems.

FIGURE 1.14 The cockpit of a commercial jet.
Istock.com

PROGRESS CHECK

1. A company employs both technicians and engineers. Which would most likely be sent to a customer's site to resolve a technical problem?
2. Which of the electronics professionals described in this section must have a solid background in mathematics and physics?
3. Laser scanners like those found in grocery store checkout lines would most likely be classified as what type of electronic system?
4. Industrial robots would be classified as what type of electronic system?

1.3 COMPONENTS AND SYMBOLS

CIRCUIT A group of components that performs a specific function.

Electrical and electronic systems contain *circuits*. Loosely defined, a **CIRCUIT** is a group of components that performs a specific function. The function performed by a given circuit depends on the components used and how they are connected together.

Nearly all circuits, no matter how complex, are made up of the same basic types of components. While the number of components used and their arrangement varies from one circuit to another, the *types* of components used are almost universal. In this section, we will take a brief look at the basic types of components, the symbols used to represent them in electrical diagrams, and the units that are used to express their values.

Resistors

RESISTOR A component that is used to restrict (limit) current.

SCHEMATIC DIAGRAMS Drawings that show you how the components in a circuit are interconnected.

We have all heard the term *current* as it applies to a river or stream: It is the flow of water from one point to another. In electric circuits, the term *current* is used to describe the flow of electricity from one point in a circuit to another. A **RESISTOR** is a component that is used to restrict the current in a circuit, just as a valve can be used to restrict the flow of water through a pipe. Some resistors are shown in Figure 1.15a.

The resistor symbols shown in Figure 1.15b are used in **SCHEMATIC DIAGRAMS**, which show you how the components in a circuit are interconnected. The schematic symbol for a resistor indicates whether its value is fixed or variable. Detailed coverage of resistors is included in Chapter 3.

Capacitors

CAPACITOR A component that stores energy in an electric field; also called a **condenser.**

A **CAPACITOR** is a component that stores energy in an electric field. Capacitors are used in a wide variety of electrical and electronic circuits. Several capacitors and the common capacitor schematic symbols are shown in Figure 1.16.

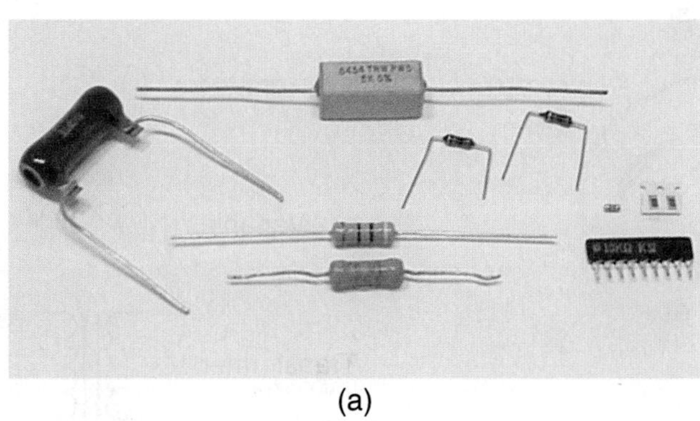

(a)

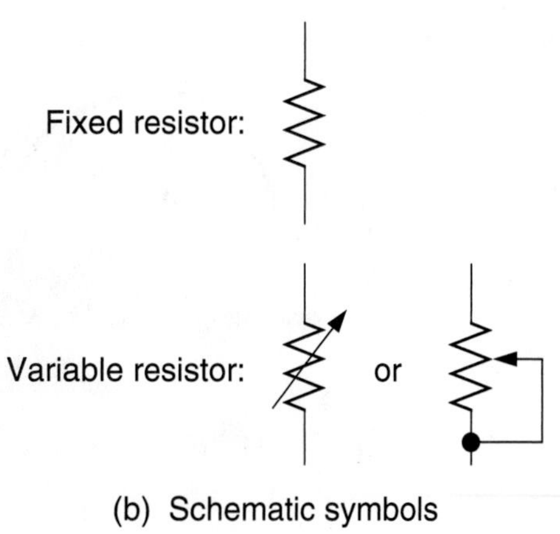

(b) Schematic symbols

FIGURE 1.15 Resistors.

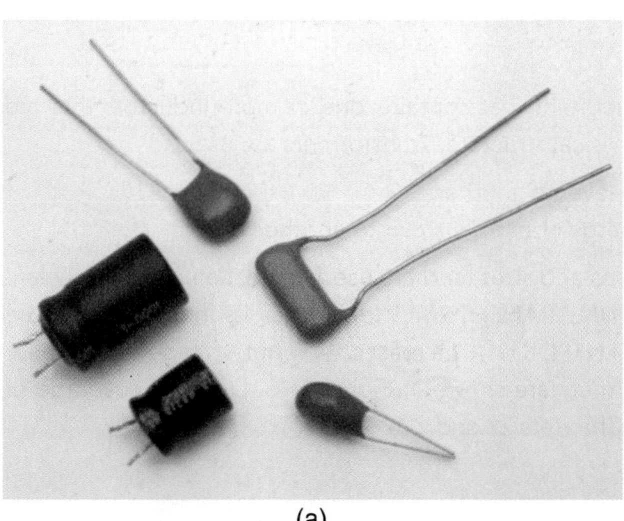

(a)

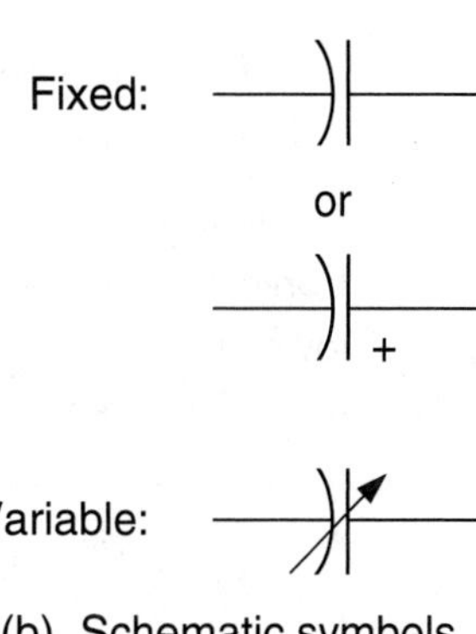

(b) Schematic symbols

FIGURE 1.16 Capacitors.

Capacitors are often referred to as **CONDENSERS** in older manuals. Detailed coverage of capacitors is provided in Chapter 12.

Inductors

An **INDUCTOR** is a component that stores energy in a magnetic field. Though both inductors and capacitors store energy, they are very different in terms of their construction and characteristics. Several types of inductors and the basic inductor schematic symbols are shown in Figure 1.17.

Inductors are often referred to as **COILS** or **CHOKES**. Detailed coverage of inductors can be found in Chapter 11.

INDUCTOR A component that stores energy in a magnetic field; also called a coil or a **choke**.

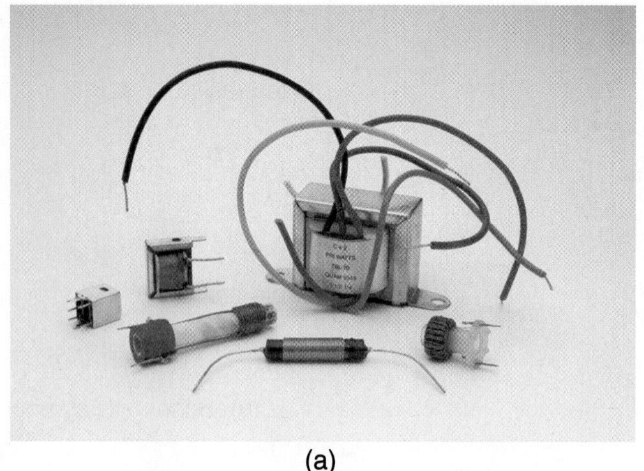

(a)

Fixed:

Variable:

Transformer:

(b) Schematic symbols

FIGURE 1.17 Inductors.

Transformers

TRANSFORMER A component that contains one or more inductors that are wrapped around a single physical structure.

A **TRANSFORMER** is a component that contains one or more inductors that are wrapped around a single physical structure. Transformers are used to

- Change one voltage level to another
- Transfer electrical energy from one circuit to another

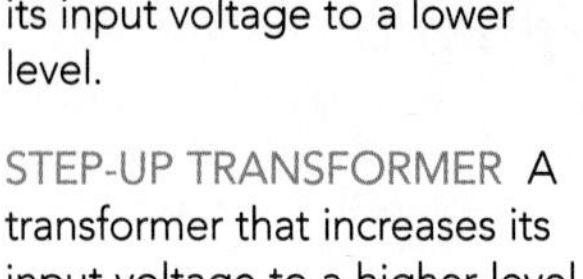

STEP-DOWN TRANSFORMER A transformer that decreases its input voltage to a lower level.

STEP-UP TRANSFORMER A transformer that increases its input voltage to a higher level.

Transformers come in all types and sizes, and are used in most electrical and electronic systems. A **STEP-DOWN TRANSFORMER** decreases its input voltage to a lower level. A **STEP-UP TRANSFORMER** increases its input voltage to a higher level. Some smaller transformers are shown in Figure 1.18. Detailed coverage of transformers is provided in Chapters 21 and 22.

Switches

SWITCH A component that is used to make or break an electrical connection.

A **SWITCH** is a component that is used to make or break an electrical connection. Several common types of switches are shown in Figure 1.19.

FIGURE 1.18 Transformers.

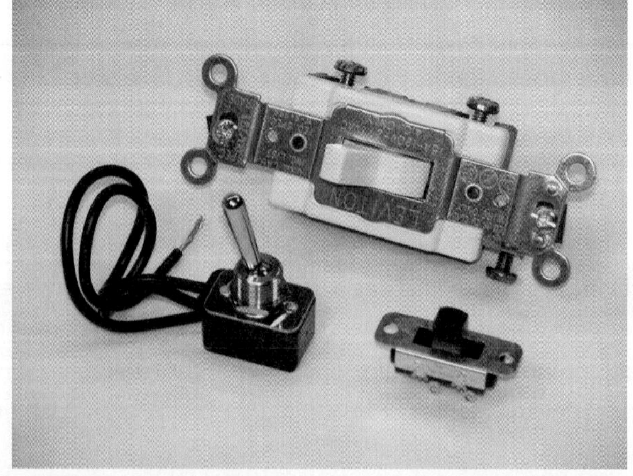

FIGURE 1.19 Switches.

Fuses and Circuit Breakers

A **FUSE** is a component that automatically breaks an electrical connection if the current increases beyond a certain value. If some condition, like an electrical short, causes the current in a circuit to suddenly increase, the circuit fuse "burns open" to break the current path and protect the circuit. Some typical fuses are shown in Figure 1.20.

FUSE A component that automatically breaks an electrical connection if the current increases beyond a certain value.

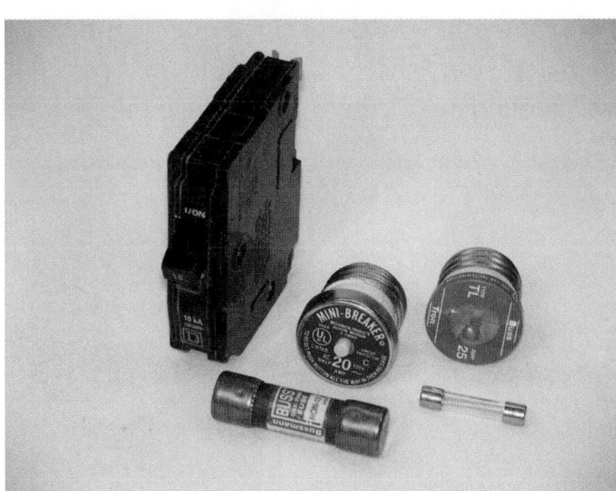

FIGURE 1.20 Fuses and circuit breakers.

A fuse can only protect a circuit once. If a fuse burns opens, it must be replaced. Unlike a fuse, a **CIRCUIT BREAKER** can be reset and used again after breaking a current path.

CIRCUIT BREAKER A component that can be reset and used again after breaking a current path.

If the current through a circuit breaker increases beyond its rated value, the breaker will "trip" (open) to break the current path. Once the source of the high current is repaired, the breaker can be reset (closed) and used again. Two circuit breakers are included in Figure 1.20. Note the mini-breaker (center-blue) that can be inserted in a fuse box.

PROGRESS CHECK

1. What do all circuits have in common?
2. Which of the components described in this section is/are used to limit current?
3. A capacitor stores energy in an _________ field. An inductor stores energy in a _________ field.
4. Which components are used to make or break an electrical connection?
5. What component must be replaced after it protects a circuit from a high-current condition?
6. Based on their schematic symbols, which two components in this section are most like each other?

1.4 TEST EQUIPMENT

Many circuit failures are not visible to the naked eye. For this reason, a wide variety of test equipment is used to measure circuit values and help diagnose circuit failures.

Digital Multimeter (DMM)

DIGITAL MULTIMETER (DMM) A meter that measures three of the most basic electrical properties: voltage, current, and resistance.

A **DIGITAL MULTIMETER (DMM)** is a meter that measures three of the most basic electrical properties: voltage, current, and resistance. Many DMMs are capable of making additional measurements, depending on the model. A typical DMM is shown in Figure 1.21.

FIGURE 1.21 A digital multimeter (DMM).
Courtesy of Altadox, Inc. (www.altadox.com)

Volt-Ohm-Milliammeter (VOM)

VOLT-OHM-MILLIAMMETER (VOM) A meter that uses an analog scale (rather than a digital scale) to provide a readout.

The DMM is a descendant of the **VOLT-OHM-MILLIAMMETER (VOM)**. The VOM is a meter that uses an analog scale (rather than a digital scale) to provide a readout. A typical VOM is shown in Figure 1.22. The DMM is preferred over the VOM because it has a numeric display and usually provides more accurate readings. However, there are some circuit tests that are easier to perform using a VOM.

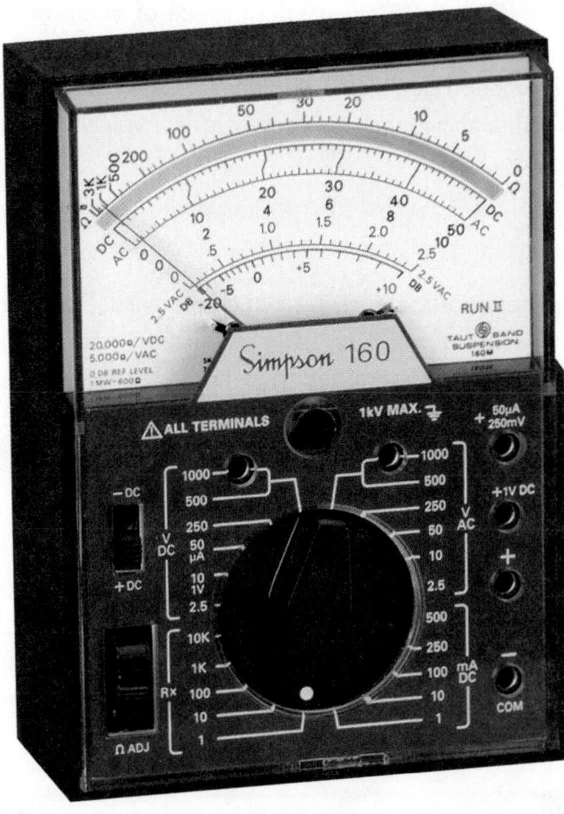

FIGURE 1.22 A volt-ohm-milliammeter (VOM). Photo courtesy of Simpson Electric Co.

Frequency Counters and Function Generators

FREQUENCY COUNTER A piece of equipment that indicates the number of times certain events occur every second.

The **FREQUENCY COUNTER** shown in Figure 1.23a is a piece of equipment that is used to indicate the number of times certain events occur every second. For

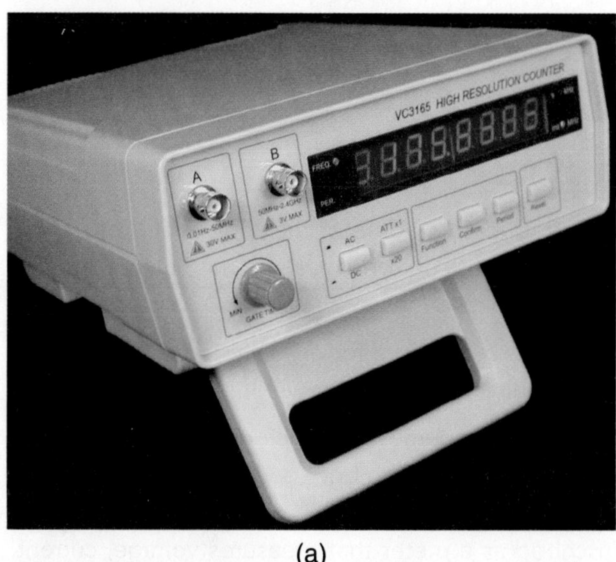

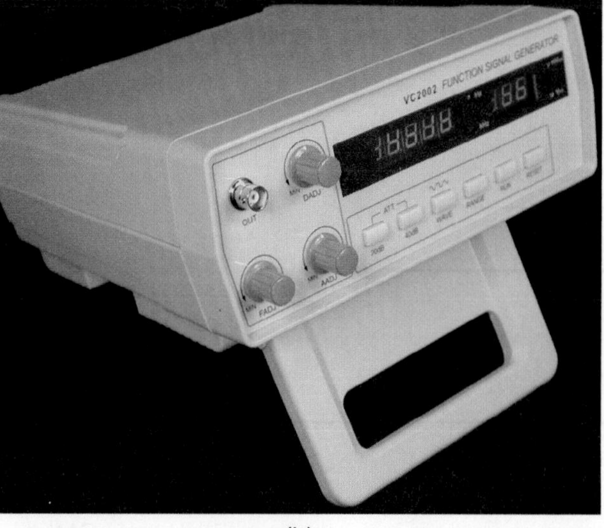

(a) (b)

FIGURE 1.23 A frequency counter and function generator. Courtesy of Altadox, Inc. (www.altadox.com)

FUNCTION GENERATOR A piece of equipment that generates a variety of waveforms.

example, Figure 1.23b shows a **FUNCTION GENERATOR**; a piece of equipment that is used to generate a variety of waveforms. A frequency counter can be used to determine how many times the function generator produces a complete waveform each second. Frequency counters and function generators are typically used to analyze circuits in school or industry labs.

Oscilloscopes

OSCILLOSCOPE A piece of test equipment that provides a visual display for voltage and time measurements.

The **OSCILLOSCOPE** is a piece of test equipment that provides a visual display for a variety of voltage and time measurements. A typical oscilloscope is shown in Figure 1.24. The oscilloscope may look a bit intimidating, but it is relatively easy to use with proper training. It is also one of the most versatile pieces of test equipment.

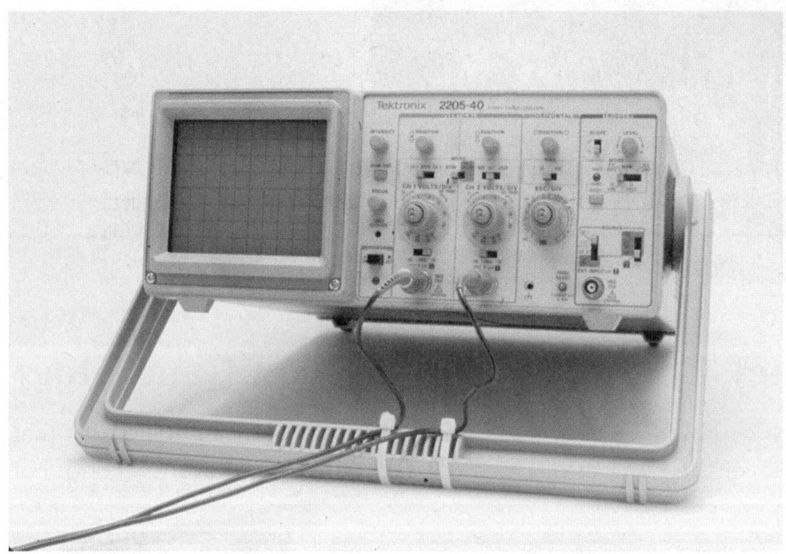

FIGURE 1.24 An oscilloscope.
(Copyright Tektronix. All Rights Reserved. Reprinted with permission.)

Units of Measure

In this section, you have been introduced to some of the tools that are used to measure electrical properties. Anything that can be measured has at least one unit of measure. For example,

- *Yards* and *meters* are two units of measure for *distance*.
- *Ounces* and *grams* are two units of measure for *weight*.

Like distance and weight, electrical properties have their own units of measure. Some of the properties that will be introduced in this book and their units of measure are listed in Table 1.1.

Earlier, the DMM was described as a meter that measures voltage, current, and resistance. As Table 1.1 shows, these measurements are expressed in *volts* (V), *amperes* (A), or *ohms* (Ω).

TABLE 1.1 • Electrical Properties and Their Units of Measure

PROPERTY	...IS MEASURED IN...	UNIT
Capacitance (C)		farads (F)
Current (I)		amperes (A)
Frequency (f)		hertz (Hz)
Impedance (Z)		ohms (Ω)
Inductance (L)		henries (H)
Power (P)		watts (W)
Reactance (X)		ohms (Ω)
Resistance (R)		ohms (Ω)
Time (t)		seconds (s)
Voltage (E or V)		volts (V)

An Important Point

When used correctly and maintained properly, your test equipment will provide you with accurate information for many years. However, if your test equipment is not properly calibrated, used, and maintained, it will provide faulty information and may even be damaged. For this reason, you must make sure that you understand *how* to use your test equipment *before* attempting to use it. When it comes to using any test equipment, the rule of thumb is simple: *If you aren't sure, ask before acting*! Begin by learning the proper measurement techniques, and then use your test equipment to take the measurements you need.

PROGRESS CHECK

1. What properties are measured using a DMM?
2. In terms of displaying values, what is the difference between a DMM and a VOM?
3. Which piece of test equipment is used for specific voltage and time measurements?
4. What three electrical properties are measured in ohms?
5. Based on the information provided in Table 1.1, answer the following question: The value displayed on a frequency counter is assumed to be in what unit of measure?

1.5 ENGINEERING NOTATION

ENGINEERING NOTATION
A shorthand method of representing large and small numbers.

You have probably heard people refer to 1000 meters as 1 *kilo*meter. And, you may have heard people use the term *mega*bucks to refer to millions of dollars. *Kilo-* and *mega-* are only two of the prefixes that are used in **ENGINEERING NOTATION**, a shorthand method of representing large and small numbers. The most commonly used prefixes and the ranges they represent are shown in Table 1.2.

TABLE 1.2 • Engineering Notation Prefixes, Symbols, and Ranges

PREFIX	SYMBOL	RANGE OF VALUES	
Tera	T	(1 to 999) trillion	(1 to 999) $\times 10^{12}$
Giga	G	(1 to 999) billion	(1 to 999) $\times 10^{9}$
Mega	M	(1 to 999) million	(1 to 999) $\times 10^{6}$
Kilo	k	(1 to 999) thousand	(1 to 999) $\times 10^{3}$
Milli	m	(1 to 999) thousandths	(1 to 999) $\times 10^{-3}$
Micro	μ	(1 to 999) millionths	(1 to 999) $\times 10^{-6}$
Nano	n	(1 to 999) billionths	(1 to 999) $\times 10^{-9}$
Pico	p	(1 to 999) trillionths	(1 to 999) $\times 10^{-12}$

The ranges of values listed in Table 1.2 are illustrated in Figure 1.25. As the figure shows, engineering notation prefixes identify the same number ranges as more familiar words like *millions*, *thousands*, and so on.

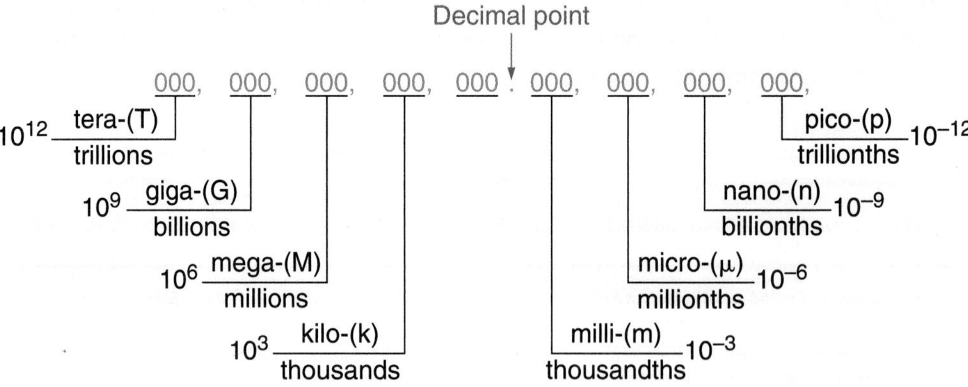

FIGURE 1.25 Metric prefixes.

Engineering notation prefixes are normally combined with units of measure (like those listed in Table 1.1). For example, here are some values and their engineering notation equivalents:

VALUE	ENGINEERING NOTATION EQUIVALENT
12,000 m (meters)	12 km (*kilo*meters)
284,000,000 Hz (hertz)	284 MHz (*mega*hertz)

0.095 s (seconds)	95 ms (*milli*seconds)
0.000047 F (farads)	47 μF (*micro*farads)

The advantage of using engineering notation is that it makes it much easier to convey large and small values. For example, it is much easier to say, "*I need a 47 μF capacitor*" than it is to say "*I need a 0.000047 farad capacitor.*" Recording values in engineering notation is also faster, simpler, and less prone to error.

Converting to and from Engineering Notation

Table 1.3 lists the *conversion factors* that will help you convert numbers to and from engineering notation form.

TABLE 1.3 • Engineering Notation Conversion Factors

PREFIX	SYMBOL	CONVERSION FACTOR	
Tera	T	1,000,000,000,000	10^{12}
Giga	G	1,000,000,000	10^{9}
Mega	M	1,000,000	10^{6}
Kilo	k	1,000	10^{3}
Milli	m	0.001	10^{-3}
Micro	μ	0.000001	10^{-6}
Nano	n	0.000000001	10^{-9}
Pico	p	0.000000000001	10^{-12}

To convert a number to engineering notation, simply *divide* the number by the conversion factor for the desired prefix. This technique is demonstrated in the next three examples.

EXAMPLE 1.1

We want to convert 5860 hertz (Hz) to kilohertz (kHz). The conversion factor for kilo (k) is 1000. Therefore,

$$5860 \text{ Hz} = \frac{5860}{1000} \text{ kHz} = 5.86 \text{ kHz}$$

PRACTICE PROBLEM 1.1

Convert 357,800 hertz (Hz) to kilohertz (kHz).

It is often easier to use the power-of-ten conversion factors listed in Table 1.3. Examples 1.2 and 1.3 use a power-of-ten conversion factor.

EXAMPLE 1.2

We want to convert 84,450,000 watts (W) to megawatts (MW). The conversion factor for mega (M) is 10^6. Therefore,

$$84{,}450{,}000\ \text{W} = \frac{84{,}450{,}000}{10^6}\ \text{MW} = 84.45\ \text{MW}$$

PRACTICE PROBLEM 1.2

Convert 2,900,000,000 watts (W) to gigawatts (GW).

EXAMPLE 1.3

We want to convert 0.00346 amperes (A) to milliamperes (mA). The conversion factor for milli (m) is 10^{-3}. Therefore,

$$0.00346\ \text{A} = \frac{0.00346}{10^{-3}}\ \text{mA} = 3.46\ \text{mA}$$

PRACTICE PROBLEM 1.3

Convert 0.0000492 amperes (A) to milliamperes (mA).

To convert numbers *from* engineering notation, simply *multiply* the value by the appropriate conversion factor. This technique is demonstrated in the next two examples.

EXAMPLE 1.4

We want to convert 6.8 megohms (MΩ) to ohms (Ω). The conversion factor for mega (M) is 10^6. Therefore,

$$6.8\ \text{M}\Omega = 6.8 \times 10^6\ \Omega = 6{,}800{,}000\ \Omega$$

PRACTICE PROBLEM 1.4

Convert 390 kilohms (kΩ) to ohms (Ω).

EXAMPLE 1.5

We want to convert 33 microfarads (μF) to farads (F). The conversion factor for micro (μ) is 10^{-6}. Therefore,

$$33\ \mu\text{F} = 33 \times 10^{-6}\ \text{F} = 0.000033\ \text{F}$$

PRACTICE PROBLEM 1.5

Convert 470 microfarads (μF) to farads (F).

Application: Meter Controls and Displays

Engineering notation is commonly seen on digital meter controls and displays. For example, look at the digital multimeter (DMM) faceplate in Figure 1.26. The **FUNCTION SELECT** sets the meter to measure volts (V), amperes (A), or ohms (Ω). When you look closely at the function select settings, you see that:

FUNCTION SELECT A control that sets a multimeter to measure volts (V), amperes (A), or ohms (Ω).

- Kilo- (k) is used in the ohms (Ω) section.
- Milli- (m) and micro- (μ) are used in the volts and amperes sections.

Engineering prefixes also appear in the displays of some meters. For example, a current reading in the mA range may be indicated by a small "m" appearing in the display.

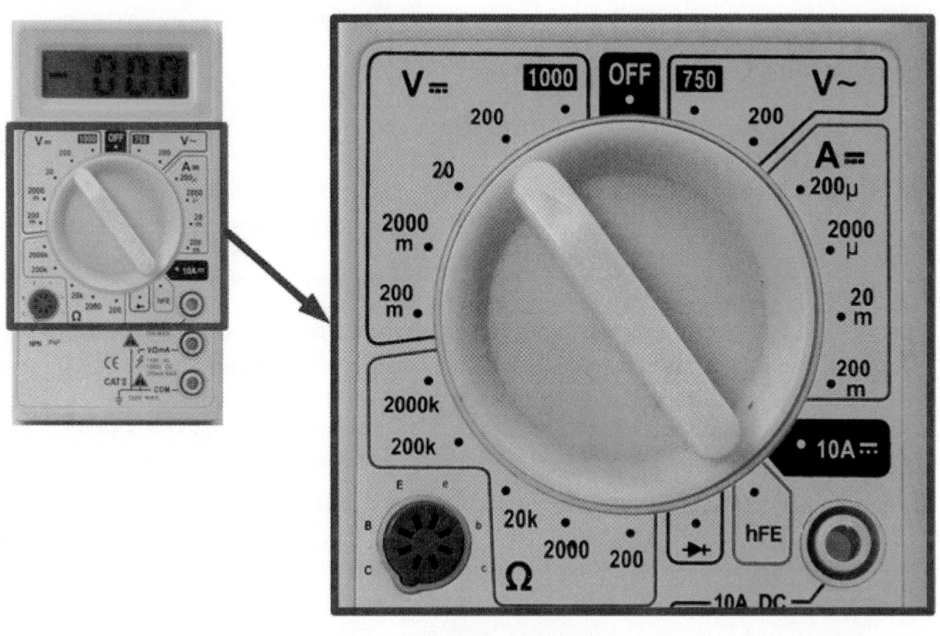

FIGURE 1.26 Meter function select.

PROGRESS CHECK

1. How do you convert a number to engineering notation?
2. How do you convert a number from engineering notation to standard form?
3. Take a look at the following relationships:

   1 A = 1000 mA   0.500 A = 500 mA
   0.050 A = 50 mA   0.005 A = 5 mA

   Based on these relationships, complete the following:
   a. 0.750 A = ________ mA
   b. 0.525 A = ________ mA
   c. 1.550 A = ________ mA
   d. 0.033 A = ________ mA
4. Take a look at the following relationships:

   1 kΩ = 1000 Ω   0.500 kΩ = 500 Ω
   0.050 kΩ = 50 Ω   0.005 kΩ = 5 Ω

   Based on these relationships, complete the following:
   a. 0.750 kΩ = ________ Ω
   b. 1.250 kΩ = ________ Ω
   c. 0.075 kΩ = ________ Ω
   d. 0.095 kΩ = ________ Ω

1.6 SUMMARY

Here is a summary of the major points that were covered in this chapter:

Electrical Careers

- Electricians install, operate, maintain, and repair electrical circuits and systems.
- Residential electricians install, maintain, and repair the electrical circuits found in houses, apartments, and condominiums.
- Industrial and commercial electricians install, maintain, and repair the circuits found in industrial and commercial buildings.
- Telecommunications technicians install, maintain, troubleshoot, and repair residential and commercial data, audio, video, and security circuits.
- Outside linemen install and repair power transmission and distribution lines and circuits.

Electronics Careers

- Electronics deals with the design, manufacturing, installation, maintenance, and repair of electronic circuits and systems.
- Electronics technicians locate and repair faults in electronic systems and circuits using electronic test equipment, perform periodic maintenance, maintain records of maintenance and system failures, and install new systems.
- Electronics engineers design components, circuits, and systems, develop new applications for existing electronics technologies, assist technicians in solving technical problems, and supervise the construction and initial testing of circuit prototypes.
- Communications systems are designed to transmit and receive information.
- Telecommunications is the branch of communications that deals with sending and receiving voice, data, and/or video information.
- Computer/digital systems are designed to store and process information.
- Industrial systems are designed for use in industrial (manufacturing) environments.
- Biomedical systems are designed for use in diagnosing, monitoring, and treating medical problems.
- Aviation electronics (avionics) deals with aircraft communications and navigation systems, aircraft and weather radar systems, and other on-board instruments and data systems.

Components and Symbols

- A resistor is a component that is used to limit the current in a circuit.
- A capacitor is a component that stores energy in an electric field.
- An inductor is a component that stores energy in a magnetic field.
- A transformer is used to change one voltage level to another, and to transfer electrical energy from one circuit to another.
- A switch is a component that is used to make or break an electrical connection.
- A fuse is a component that automatically breaks an electrical connection if the current exceeds a certain value.
- A circuit breaker performs the same function as a fuse, but can be reset and re-used after being activated.

Test Equipment

- A digital multimeter (DMM) measures voltage, current, and resistance.
- A volt-ohm-milliammeter (VOM) is a meter that uses an analog scale to provide a readout.
- A frequency counter is used to indicate the number of times certain events occur every second.
- A function generator is used to generate a variety of waveforms.
- An oscilloscope provides a visual display for a variety of voltage and time measurements.
- Anything that can be measured has at least one unit of measure (see Table 1.1).

Engineering Notation

- Engineering notation is a shorthand method of representing large and small numbers.
- Engineering notation prefixes represent ranges of numbers (see Table 1.2).
- To convert a number to engineering notation, divide the number by the prefix conversion factor (see Table 1.3).
- To convert a number from engineering notation, multiply the number by the prefix conversion factor (see Table 1.3).

CHAPTER REVIEW

1. A group of components designed to perform a specific function is referred to as a ________.
   a) system
   b) circuit
   c) program
   d) complex

2. The flow of charge from one point to another is called ________.
   a) electric force
   b) resistance
   c) inductance
   d) current

3. Resistors are used to ________ the flow of charge.
   a) enhance
   b) increase
   c) limit
   d) reverse

4. A visual representation of a group of components is referred to as a ________.
   a) schematic diagram
   b) system blueprint
   c) component diagram
   d) system diagram

5. Which of the following components store energy?
   a) inductors
   b) coils
   c) capacitors
   d) condensers
   e) both (a) and (c)

6. Transformers contain one or more ________.
   a) capacitors
   b) coils
   c) resistors
   d) voltage sources

7. The input voltage for a step-down transformer is ________ the output voltage.
   a) lower than
   b) higher than
   c) the same as
   d) the inverse of

8. A ________ can be reset after an overcurrent event occurs.
   a) circuit breaker
   b) slow-blow fuse
   c) circuit schematic
   d) fast-blow fuse

9. The ________ is usually preferred for most voltage measurements.
   a) DMM
   b) VOM
   c) oscilloscope
   d) frequency counter

10. The ________ is used to make voltage and time measurements.
    a) DMM
    b) frequency counter
    c) VOM
    d) oscilloscope

11. The letter used in engineering notation to represent one billion is ________.
    a) M
    b) T
    c) G
    d) k

12. The letter used in engineering notation to represent one billionth is ________.
    a) k
    b) μ
    c) n
    d) m

13. If you are installing a LAN system, you have likely been trained as a(n) ________.
    a) industrial electrician
    b) telecommunications technician
    c) electronics engineer
    d) outside lineman
    e) residential electrician

14. If you are maintaining a robotics and motor control system, you have likely been trained as a(n) ________.
    a) industrial electrician
    b) telecommunications technician
    c) electronics engineer
    d) outside lineman
    e) residential electrician

15. An inductor is also referred to as a ________.
    a) coil
    b) condenser
    c) choke
    d) transformer
    e) coil or choke

16. A ________ uses an analog scale readout to display a current reading.
    a) VOM
    b) function generator
    c) DMM
    d) frequency counter

17. In engineering notation, 300,000 Hz can be written as ________.
    a) 3 MHz
    b) 300 kHz
    c) 30 MHz
    d) 3 GHz

18. In engineering notation, 0.0002 F can be written as ________.
    a) 2 μF
    b) 0.2 μF
    c) 200 nF
    d) 200 μF

19. A component that stores energy in a magnetic field is the _______.
    a) inductor
    b) resistor
    c) capacitor
    d) condenser

20. A component that stores energy in an electric field is the _______.
    a) inductor
    b) resistor
    c) capacitor
    d) transformer

PRACTICE PROBLEMS

1. Convert each number to engineering notation.
   a) 38,400 m = ____________ km
   b) 234,000 W = ____________ kW
   c) 44,320,000 Hz = ____________ MHz
   d) 175,000 V = ____________ kV
   e) 4,870,000,000 Hz = ____________ MHz = ____________ GHz

2. Convert each number to engineering notation.
   a) 600 W = ____________ kW
   b) 1800 Ω = ____________ kΩ
   c) 3.2 A = ____________ mA
   d) 2,500,000 Ω = ____________ kΩ = ____________ MΩ
   e) 22,000 Ω = ____________ kΩ = ____________ MΩ

3. Convert each number to engineering notation.
   a) 0.22 H = ____________ mH
   b) 0.00047 F = ____________ μF
   c) 0.00566 m = ____________ mm
   d) 0.000045 A = ____________ μA
   e) 0.288 A = ____________ mA

4. Convert each number to engineering notation.
   a) 0.00000000244 s = ____________ ns = ____________ ps
   b) 0.000033 H = ____________ mH = ____________ μH
   c) 0.000000000051 F = ____________ nF = ____________ pF
   d) 0.935 s = ____________ ms = ____________ μs
   e) 0.00033 H = ____________ mH = ____________ μH

5. Convert each number to engineering notation.
   a) $2.2 \times 10^7\ \Omega$ = ____________ MΩ
   b) 3.3×10^3 W = ____________ kW
   c) 7×10^{11} Hz = ____________ GHz
   d) $5.1 \times 10^5\ \Omega$ = ____________ kΩ
   e) 8.5×10^4 W = ____________ kW

6. Convert each number to engineering notation.
   a) 4.7×10^{-5} F = ____________ μF
   b) 1×10^{-1} H = ____________ mH
   c) 6.6×10^{-4} s = ____________ ms
   d) 4×10^{-8} A = ____________ μA
   e) 2.2×10^{-10} F = ____________ pF

7. Convert each number from engineering notation to standard form.
   a) 36 kW = ____________ W
   b) 7.2 MHz = ____________ Hz
   c) 280 kΩ = ____________ Ω
   d) 452 mA = ____________ A
   e) 51 μF = ____________ F

8. Convert each number from engineering notation to standard form.
   a) 32.6 ms = ____________ s
   b) 175 kV = ____________ V
   c) 880 nH = ____________ H
   d) 3.3 MΩ = ____________ Ω
   e) 550 kHz = ____________ Hz

ANSWERS TO THE EXAMPLE PRACTICE PROBLEMS

1.1	357.8 kHz
1.2	2.9 GW
1.3	0.0492 mA
1.4	390,000 Ω
1.5	0.00047 F

CHAPTER 2

BASIC ELECTRICAL CONCEPTS

PURPOSE

When you mention the term *electricity,* most people form a mental image of lightning, a power transmission station, or some other evidence of the presence of electricity. Others form a mental image of an electric appliance, computer, or some other device that uses electricity. However, people rarely consider what electricity *is.*

In this chapter, we're going to discuss electricity, its sources and properties, and the relationships between them. As you will see, the study of electricity starts with a discussion of a basic building block of matter, the *atom.*

KEY TERMS

The following terms are introduced and defined in this chapter on the pages indicated:

OBJECTIVES

After completing this chapter, you should be able to:

1. Describe the relationship among matter, elements, and atoms.
2. Describe the structure of the atom.
3. Explain the concept of *charge* as it applies to the atom.
4. Define current and identify its unit of measure.
5. Describe the relationship among current, charge, and time.
6. Discuss the electron flow and conventional current approaches to describing current.
7. Contrast direct current (DC) and alternating current (AC).
8. Define *voltage* and describe its relationship to energy and charge.
9. Define *resistance.*
10. Compare and contrast conductors, insulators, and semiconductors.
11. Discuss the relationship between resistance and each of the following: *resistivity, material length, material cross-sectional area, temperature.*

2.1 MATTER, ELEMENTS, AND ATOMS

A battery has two terminals: a *negative* terminal and a *positive* terminal. While the terms *negative* and *positive* are commonly associated with electricity, most people don't know that they describe *charges* found in the atom. In this section, we look at matter, the structure of the atom, and the nature of charge.

Matter

MATTER is a word that is used to describe anything that has mass and occupies space. You, and everything around you, are classified as matter.

MATTER Anything that has mass and occupies space.

Matter exists as a solid, liquid, or gas in its natural form. For example, at room temperature:

- Copper is a solid.
- Water is a liquid.
- Carbon dioxide is a gas.

At the same time, matter can be made to change forms. For example, water (a liquid) becomes ice (a solid) when frozen and steam (a gas) when boiled.

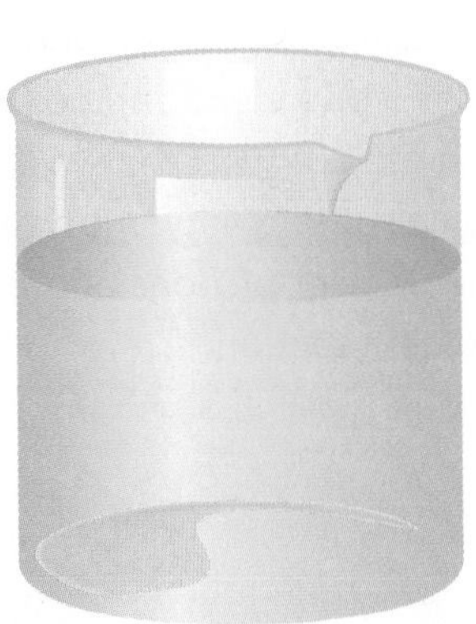

FIGURE 2.1 Water in its three states.

Elements

All matter is made up of various combinations of elements. An **ELEMENT** is a substance that cannot be broken down into two or more simpler substances. For example, *carbon* (an element) cannot be broken down into a combination of other elements. In contrast, *carbon dioxide* (which is *not* an element) can be broken down into carbon (C) and oxygen (O).

ELEMENT Any substance that cannot be broken down into two or more simpler substances.

As far as we know, all matter in the universe is made up of various combinations of the 109 known elements. While these elements each have unique physical and chemical properties, they all have one thing in common: They are all made up of atoms.

Atoms

ATOM The smallest particle of matter that retains the physical characteristics of an element.

BOHR MODEL The simplest model of the atom.

NUCLEUS The central core of an atom.

ELECTRONS Particles that orbit the nucleus of an atom.

PROTON One of two particles found in the nucleus of an atom; the other is the **neutron.**

The **ATOM** is the smallest particle of matter that retains the physical characteristics of an element. The simplest model of the atom is called the **BOHR MODEL**. The Bohr model of the *helium* atom is shown in Figure 2.2. It contains a central core (called the **NUCLEUS**) that is orbited by two particles, called **ELECTRONS**. The electrons revolve around the nucleus like planets orbit around the sun. The nucleus contains two other types of particles: **PROTONS** and **NEUTRONS**.

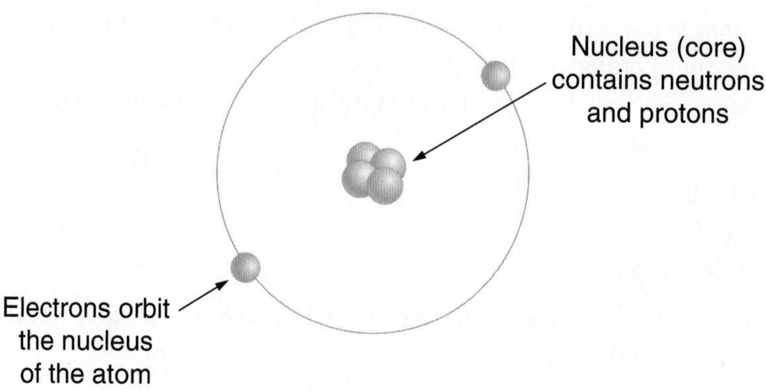
Nucleus (core)
contains neutrons
and protons
Electrons orbit
the nucleus
of the atom

FIGURE 2.2 The helium atom.

The number of electrons and protons contained in the atom varies from one element to another. This point is illustrated in Figure 2.3. As you can see, the hydrogen atom has one proton and one electron, the helium atom has two of each, and the lithium atom has three of each.

Figure 2.3 also serves to illustrate a very important point: In its natural state, an atom contains an equal number of protons and electrons. The importance of this point will be shown in the next section.

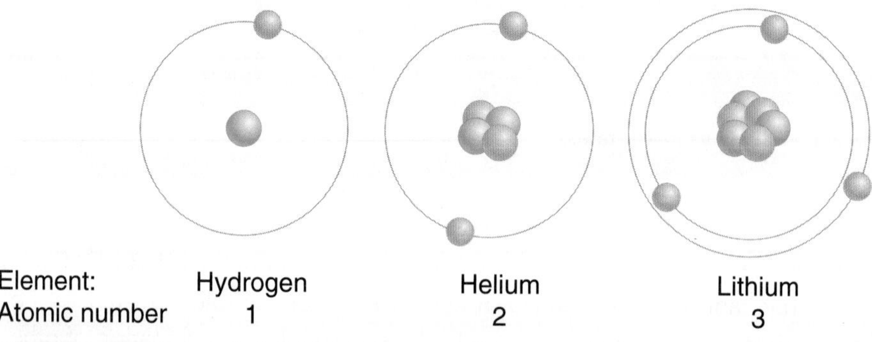
Element: Hydrogen Helium Lithium
Atomic number 1 2 3

FIGURE 2.3 Hydrogen, helium, and lithium.

ATOMIC NUMBER A unique number for every element on the periodic table, equal to the number of protons in the nucleus of the atom.

PERIODIC TABLE OF THE ELEMENTS A table that identifies all the known elements.

Atomic Number

Every element has a unique **ATOMIC NUMBER** that helps us locate that element on the **PERIODIC TABLE OF ELEMENTS** shown in Figure 2.4. The periodic table

atomic number → 14 | 28.09 ← atomic weight

Si ← symbol

Silicon ← name

1 1.01 **H** Hydrogen																	2 4.00 **He** Helium
3 6.94 **Li** Lithium	4 9.01 **Be** Beryllium											5 10.81 **B** Boron	6 12.01 **C** Carbon	7 14.00 **N** Nitrogen	8 15.99 **O** Oxygen	9 18.99 **F** Fluorine	10 20.17 **Ne** Neon
11 22.98 **Na** Sodium	12 24.30 **Mg** Magnesium											13 26.98 **Al** Aluminium	14 28.09 **Si** Silicon	15 30.97 **P** Phosphorus	16 32.06 **S** Sulfur	17 35.45 **Cl** Chlorine	18 39.94 **Ar** Argon
19 39.09 **K** Potassium	20 40.07 **Ca** Calcium	21 44.95 **Sc** Scandium	22 47.86 **Ti** Titanium	23 50.94 **V** Vanadium	24 51.99 **Cr** Chromium	25 54.93 **Mn** Manganese	26 55.84 **Fe** Iron	27 58.93 **Co** Cobalt	28 58.69 **Ni** Nickel	29 63.54 **Cu** Copper	30 65.39 **Zn** Zinc	31 69.72 **Ga** Gallium	32 72.61 **Ge** Germanium	33 74.92 **As** Arsenic	34 78.96 **Se** Selenium	35 79.90 **Br** Bromine	36 83.80 **Kr** Krypton
37 85.46 **Rb** Rubidium	38 87.62 **Sr** Strontium	39 88.90 **Y** Yttrium	40 91.22 **Zr** Zirconium	41 92.90 **Nb** Niabium	42 95.94 **Mo** Molybdenum	43 (98) **Tc** Technetium	44 101.07 **Ru** Ruthenium	45 102.90 **Rh** Rhodium	46 106.42 **Pd** Palladium	47 107.86 **Ag** Silver	48 112.41 **Cd** Cadmium	49 114.81 **In** Indium	50 118.71 **Sn** Tin	51 121.76 **Sb** Antimony	52 127.60 **Te** Tellurium	53 126.90 **I** Iodine	54 131.29 **Xe** Xenon
55 132.90 **Cs** Cesium	56 137.32 **Ba** Barium	57 138.90 **La**▸ Lanthanum	72 178.49 **Hf** Hafnium	73 180.94 **Ta** Tantalum	74 183.84 **W** Tungsten	75 186.20 **Re** Rhenium	76 190.23 **Os** Osmium	77 192.21 **Ir** Iridium	78 195.07 **Pt** Platinum	79 196.96 **Au** Gold	80 200.59 **Hg** Mercury	81 204.38 **Tl** Thallium	82 207.20 **Pb** Lead	83 208.98 **Bi** Bismuth	84 (209) **Po** Polonium	85 (210) **At** Astatine	86 (222) **Rn** Radon
87 (223) **Fr** Francium	88 (226) **Ra** Radium	89 (227) **Ac**▸ Actinium	104 (261) **Rf** Rutherfordium	105 (262) **Ha** Halnium	106 (263) **Sg** seaborgium	107 (262) **Bh** Bohrium	108 (265) **Hs** Hassium	109 (266) **Mt** Meitnerium									

58 140.11 **Ce** Cerium	59 140.90 **Pr** Praseodymium	60 144.24 **Nd** Neodymium	61 145 **Pm** Promethium	62 150.36 **Sm** Samarium	63 151.96 **Eu** Europium	64 157.25 **Gd** Gadolinium	65 158.92 **Tb** Terbium	66 162.50 **Dy** Dysprosium	67 164.93 **Ho** Holmium	68 167.26 **Er** Erbium	69 168.93 **Tm** Thulium	70 173.04 **Yb** Ytterbium	71 174.96 **Lu** Lutetium
90 232.03 **Th** Thorium	91 231.03 **Pa** Protactinium	92 238.02 **U** Uranium	93 237 **Np** Neptunium	94 244 **Pu** Plutonium	95 243 **Am** Americium	96 247 **Cm** Curium	97 247 **Bk** Berkelium	98 251 **Cf** Californium	99 252 **Es** Einsteinium	100 257 **Fm** Fermium	101 258 **Md** Mendelevium	102 259 **No** Nobelium	103 262 **Lr** Lawrencium

FIGURE 2.4 The periodic table of elements.

lists all of the known elements. Each block in the table represents one element. The atomic number for each element is found in the upper left-hand corner of its block. For example, the block for carbon (C) in the second row has the number 6 in its upper left-hand corner. That is the atomic number for carbon.

The atomic number for an element tells you how many protons it contains. For example, referring to Figure 2.3:

- The hydrogen (H) atom has 1 proton and its atomic number is 1.
- The helium (He) atom has 2 protons and its atomic number is 2.
- The lithium (Li) atom has 3 protons and its atomic number is 3.

Orbital Shells

SHELLS Electron orbital paths that circle the nucleus of an atom.

Electrons circle the nucleus of the atom in orbital paths that are referred to as **SHELLS**. For example, the copper atom in Figure 2.5 contains 29 electrons that travel around the nucleus in four orbital shells that are identified using the letters *K* through *N*. The maximum number of electrons that each shell can hold equals $2n^2$, where n is the number of the shell. For example, the third orbital shell (the M shell) can hold up to

$$2n^2 = 2(3^2) = 18$$

electrons.

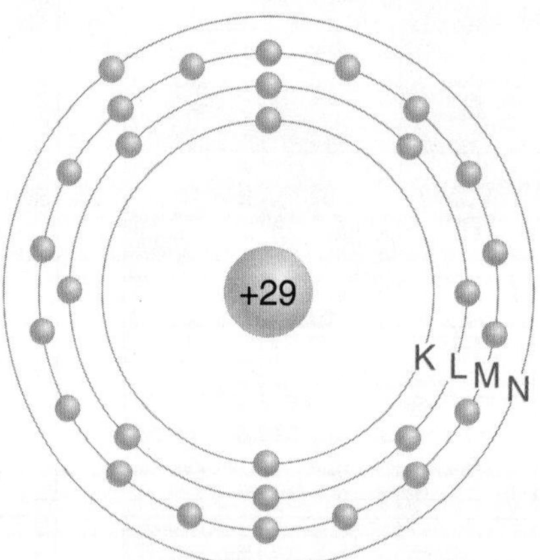

Note: The number written in the nucleus indicates the number of protons it contains.

FIGURE 2.5 A copper (Cu) atom.

The Valence Shell

VALENCE SHELL The outermost orbital path for a given atom.

The outermost shell of an atom is called the **VALENCE SHELL**. The electrons that occupy the valence shell of an atom are referred to as *valence electrons*.

The valence shell of a given atom cannot hold more than eight electrons. When the valence shell of an atom contains eight electrons, the shell is described as being *complete*. When it holds fewer than eight electrons, it is said to be *incomplete*. In the next section, you will see how the valence shell of an atom determines the electrical characteristics of the element.

PROGRESS CHECK

1. What is *matter*? What is an *element*?
2. What is an atom? Describe its physical make-up.
3. What is an *orbital shell*? What is the *valence shell*?
4. What is a *complete* valence shell?
5. Look at the sodium atom in Figure 2.6. How many electrons are in the valence shell of this atom?

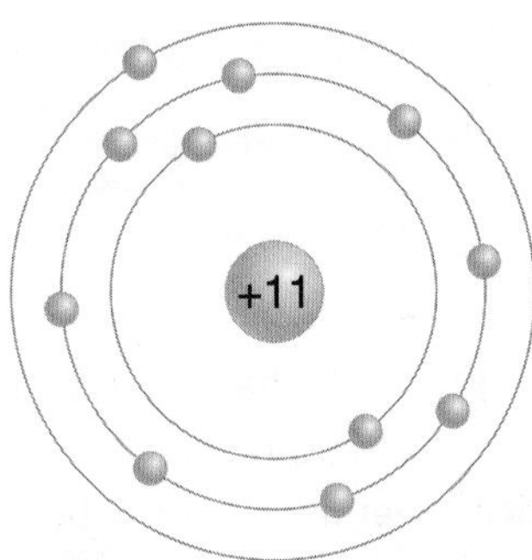

FIGURE 2.6 A sodium (Na) atom.

2.2 ELECTRICAL CHARGE

As stated earlier, a battery has a *negative* terminal and a *positive* terminal. These terms describe charges that are found in the atom. In this section, we look at the nature of charge.

Charge

CHARGE is a force that causes two particles to be attracted to, or repelled from, each other. There are two types of charges: **POSITIVE CHARGE** and **NEGATIVE CHARGE**. Note that the terms *positive* and *negative* are used to signify that the charges are opposites.

Because positive and negative charges are opposites, their effects cancel each other out. For example, an atom that contains two positive charges and two negative charges has a net charge of zero.

CHARGE An electrical force that causes two particles to be attracted to, or repelled from, each other.

POSITIVE CHARGE One of two electrical charges; the other is **negative charge**.

Protons and electrons are the sources of charge in the atom. As shown in Figure 2.7:

- Each proton has a positive charge.
- Each electron has a negative charge.

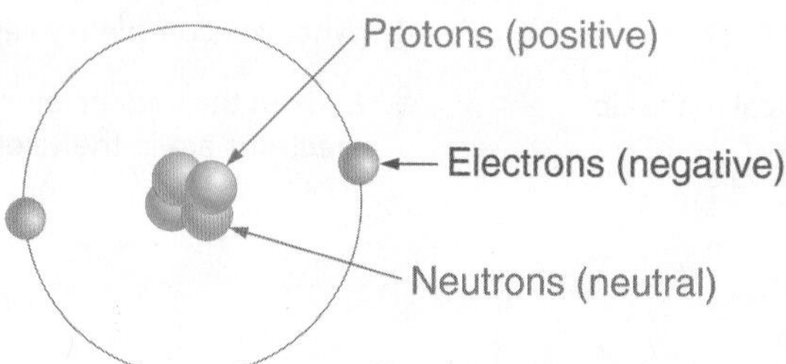

FIGURE 2.7 Atomic charges.

The atom in Figure 2.7 has an equal number of positive and negative charges that cancel each other out, resulting in a net charge of *zero*. The neutrons in the atom are electrically neutral (which is how they got their name).

Attraction and Repulsion

The most fundamental law of charges is that *like charges repel each other and opposite charges attract each other*. This law is illustrated in Figure 2.8. As you can see, two positive charges repel each other, as do two negative charges. However, a negative charge is attracted to a positive charge, and vice versa.

There are two forces that combine to keep an electron in its orbit. The force of attraction between a negatively charged electron and a positively charged proton is a centripetal (inward) force that offsets the centrifugal (outward) force of the orbiting electron. When these forces are balanced, the electron stays in its orbit.

Ions

An outside force can upset the balance between the forces acting on a given electron, forcing the electron to leave its orbit. This point is illustrated in Figure 2.9.

When the outside force causes the electron to leave its orbit, the atom is left with fewer electrons than protons, and thus has a net charge that is positive. For example, consider the copper atom shown in Figure 2.9. When the valence electron is forced to leave its orbit, the atom contains only 28 electrons (negative charges). At the same time, it still contains 29 protons (positive charges). Thus, the atom contains more positive charges than negative charges, and has a net *positive* charge. An atom with a net positive charge is referred to as a **POSITIVE ION**.

POSITIVE ION An atom with a net positive charge.

The forces associated with charges are represented as *outward* and *inward* forces

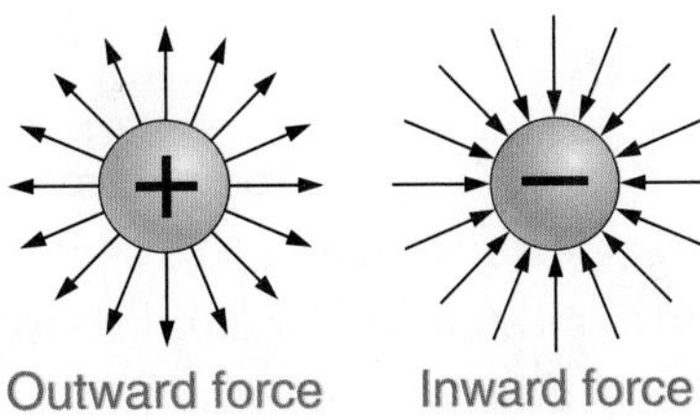

Two outward forces repel

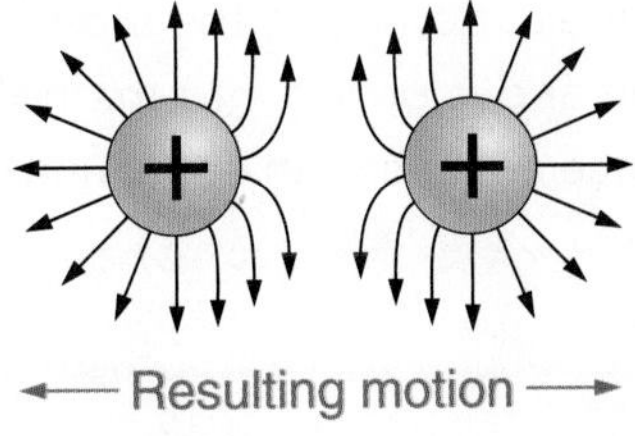

Two inward forces repel

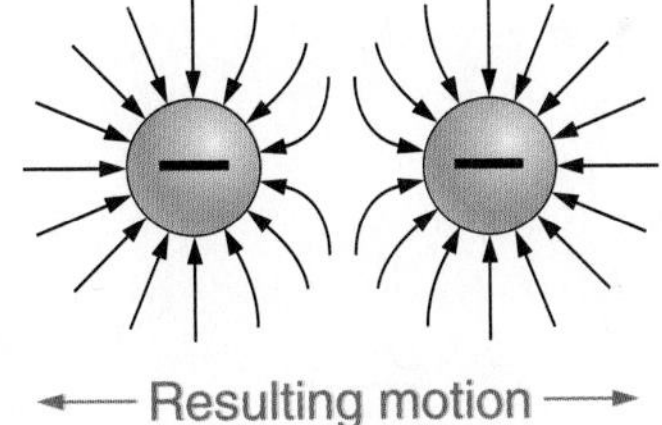

An outward force attracts an inward force

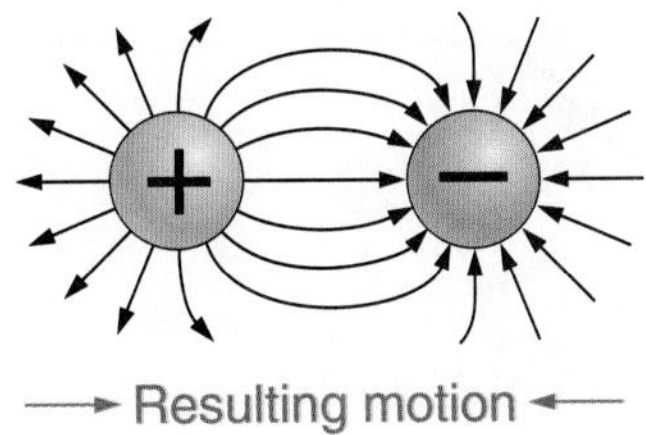

FIGURE 2.8 Attraction and repulsion of charges.

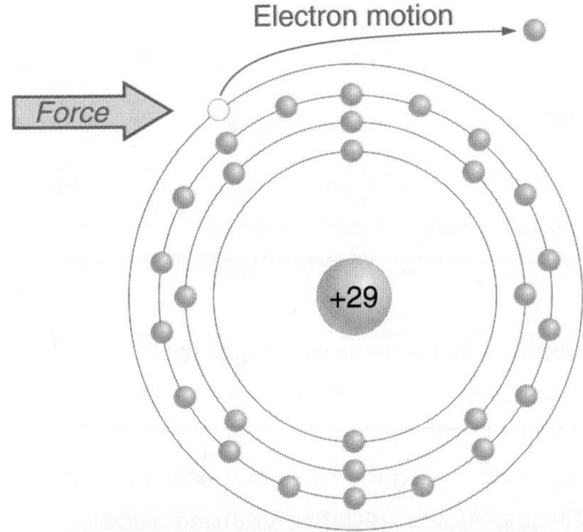

Note: The circle in the valence shell represents the gap left by the lost electron.

FIGURE 2.9 Generating a positive ion and a free electron.

Free Electrons

FREE ELECTRON One that is not bound to an atom and is free to drift from one atom to another.

When an outside force causes an electron to break free from its parent atom (as shown in Figure 2.9), that electron is referred to as a **FREE ELECTRON**. A free electron is one that is not bound to an atom, and thus is free to drift from one atom to another.

When a free electron drifts into the vicinity of a positive ion, the electron neutralizes that ion. For example, consider the free electron and the atom shown in Figure 2.10. When the electron nears the atom, it is attracted by the overall positive charge on the ion and is pulled into the hole in the valence shell. When this happens, the atom once again has an equal number of electrons and protons, and a net charge of *zero*.

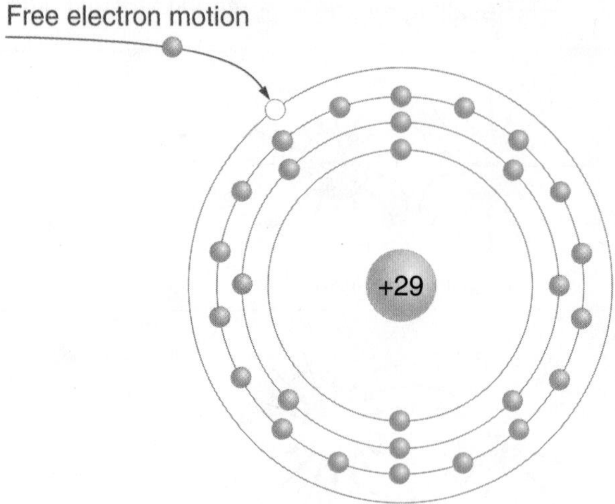

FIGURE 2.10 Neutralizing a positive ion.

Figure 2.10 and its related discussion serve to illustrate an important point: *Positive and negative charges always seek to neutralize each other.* As you will see in the next section, this is the basis of one of the primary electrical properties—*current*.

Putting It All Together

The atom consists of a central core, called the *nucleus*, and one or more electrons that orbit the nucleus. The nucleus contains *neutrons* and *protons*. Electrons and protons exhibit a property referred to as *charge*. Neutrons (the largest of the three particles) are electrically neutral.

Charge is a force that causes two bodies to be attracted to, or repelled from, each other. There are two types of charge, called *positive* charge and *negative* charge.

A positive charge is always attracted to a negative charge, and vice versa. Two like charges (i.e., two positive charges or two negative charges) repel each other.

Protons are positively charged particles and electrons are negatively charged particles. Since they have opposite charges, protons and electrons are attracted to each other. This attraction offsets the centrifugal (outward) force on the orbiting electrons, causing them to remain in their orbits.

An atom in its natural state contains an equal number of protons and electrons, and thus has a net charge of zero. If an outside force causes an atom to lose an electron, the atom is left with a positive net charge and is referred to as a *positive ion*.

Positive and negative charges always seek to neutralize each other. When a free electron (one that has broken free of its parent atom) drifts into the vicinity of a positive ion, the electron will fall into orbit around the atom. When this occurs, the net charge on the atom returns to zero.

PROGRESS CHECK

1. What are the two types of charges? How do they relate to each other?
2. What is the net charge on an atom in its natural state? Explain.
3. What is the fundamental law of charges?
4. What are *positive ions*? How are they created?
5. What is a *free electron*?
6. Describe what happens when a free electron drifts into the vicinity of a positive ion.

2.3 CURRENT

Current, voltage, and resistance are three of the fundamental electrical properties. Stated simply,

- **CURRENT** is the directed flow of charge through a conductor.
- **VOLTAGE** is the force that generates the current.
- **RESISTANCE** is an opposition to current that is provided by a material, component, or circuit.

CURRENT The directed flow of charge through a conductor.

VOLTAGE A force that generates the flow of electrons (current).

RESISTANCE The opposition to current provided by a material, component, or circuit.

Current, voltage, and resistance are the three primary properties of an electrical circuit. The relationships among them are defined by the fundamental law of circuit operation, called *Ohm's law*. Ohm's law is introduced in Chapter 4. In this section, we're going to take a close look at the first of these properties, *current*.

As you know, an outside force can break an electron free from its parent atom. In copper (and other metals), very little external force is required to generate free electrons. In fact, the *thermal energy* (heat) present at room temperature (22°C) can generate free electrons. The number of electrons generated varies directly with temperature. In other words, higher temperatures generate more free electrons.

The motion of the free electrons in copper is random when no directing force is applied. That is, the free electrons move in every direction, as shown in Figure 2.11. Since the free electrons are moving in every direction, the net flow of electrons in any one direction is zero.

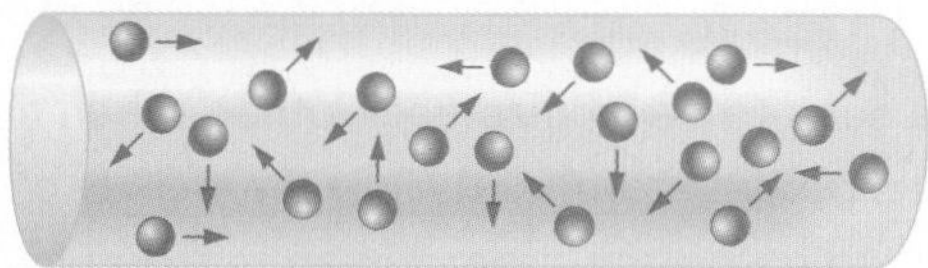

FIGURE 2.11 Random electron motion in copper.

Figure 2.12 illustrates what happens when an external force causes all of the electrons to move in the same direction. In this case, a negative potential is applied to one end of the copper and a positive potential is applied to the other. As a result, the free electrons all move from negative to positive, and we can say that we have *a directed flow of charge* (electrons). This directed flow of electrons is called *current.*

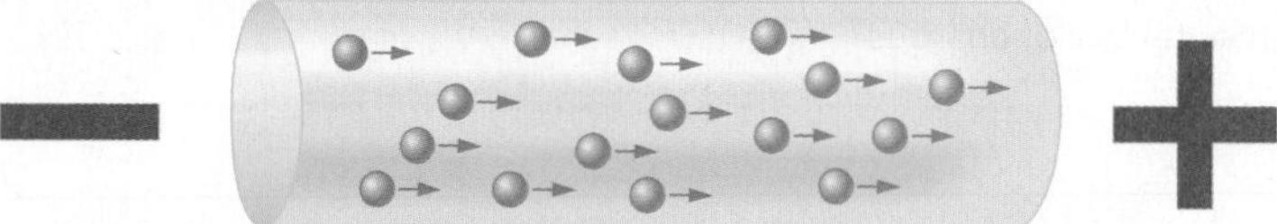

FIGURE 2.12 Directed electron motion in copper.

Let's look at what happens on a larger scale when electron motion is directed by an outside force. In Figure 2.13, the negative potential directs electron flow (current) toward the positive potential. The current passes through the lamp, causing it to produce light and heat. The more *intense* the current (meaning the greater its value), the greater the light and heat produced by the bulb.

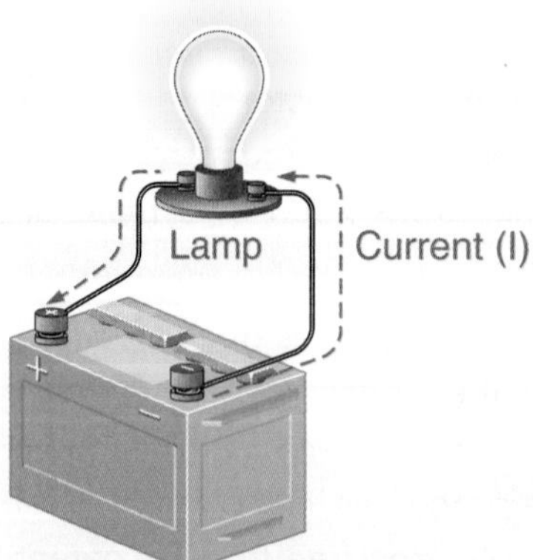

FIGURE 2.13 Current through a basic lamp circuit.

Current is represented in formulas by the letter *I* (for *intensity*). The intensity of current is determined by the amount of charge flowing per second: The greater the flow of charge per second, the more intense the current.

Coulombs and Amperes

The charge on a single electron is not sufficient to provide a practical unit of measure for charge. Therefore, the **COULOMB (C)** is used as the basic unit of charge. One coulomb equals the total charge on 6.25×10^{18} electrons.[1] When one coulomb of charge passes a point in one second, we have one **AMPERE (A)** of current. In other words,

COULOMB (C) The total charge on 6.25×10^{18} electrons.

AMPERE (A) The unit of measure for current; equal to one coulomb per second.

$$1 \text{ ampere} = 1 \text{ coulomb per second}$$

or

$$1\text{ A} = 1\text{ C/s}$$

The total current passing a point (in amperes) can be found by dividing the total charge (in coulombs) by the time (in seconds). By formula,

(2.1)

$$I = \frac{Q}{t}$$

where

I = the intensity of current, in amperes (A)
Q = the total charge, in coulombs (C)
t = the time it takes the charge to pass a point, in seconds (s)

This relationship is illustrated in Example 2.1.

EXAMPLE 2.1

Three coulombs of charge pass through a copper wire every second. What is the value of the current?

SOLUTION

Using equation (2.1), the current is found as

$$I = \frac{Q}{t} = \frac{3\text{ C}}{1\text{ s}} = 3\text{ C/s} = 3\text{ A}$$

PRACTICE PROBLEM 2.1

Ten coulombs of charge pass through a point in a copper wire every four seconds. What is the value of the current?

Example 2.1 is included here to help you understand the relationship between amperes, coulombs, and seconds. In practice, current is not calculated

[1] 6.25×10^{18} = 6,250,000,000,000,000,000 = 6.25 billion-billion.

using equation (2.1) because you cannot directly measure coulombs of charge. As you will learn, there are far more practical ways to calculate current.

Two Theories: Conventional Current and Electron Flow

CONVENTIONAL CURRENT A theory that defines current as *the flow of charge from positive to negative.*

ELECTRON FLOW A theory that defines current as *the flow of electrons from negative to positive.*

There are two theories that describe current, and you will come across both in practice. The **CONVENTIONAL CURRENT** theory defines current as *the flow of charge from positive to negative.* This theory is called "conventional current" because it is the older of the two approaches to current, and for many years was the only one taught outside of military and trade schools.

ELECTRON FLOW is the newer of the two current theories. Electron flow theory defines current as *the flow of electrons from negative to positive.* The two current theories are contrasted in Figure 2.14. Each circuit contains a battery and a lamp. Conventional current begins at the positive battery terminal, passes through the lamp, and returns to the battery through its negative terminal. Electron flow is in the opposite direction: It begins at the negative terminal of the battery, passes through the lamp, and returns to the battery through its positive terminal.

It is worth noting that the two circuits in Figure 2.14 are identical. The only difference between the two is how we describe the current. In practice, how you view current does not affect any circuit calculations, measurements, or test procedures. Even so, you should get comfortable with both viewpoints, since both are used by many engineers, technicians, and technical publications. In this text, we take the electron flow approach to current. That is, we will assume current is the flow of electrons from negative to positive.

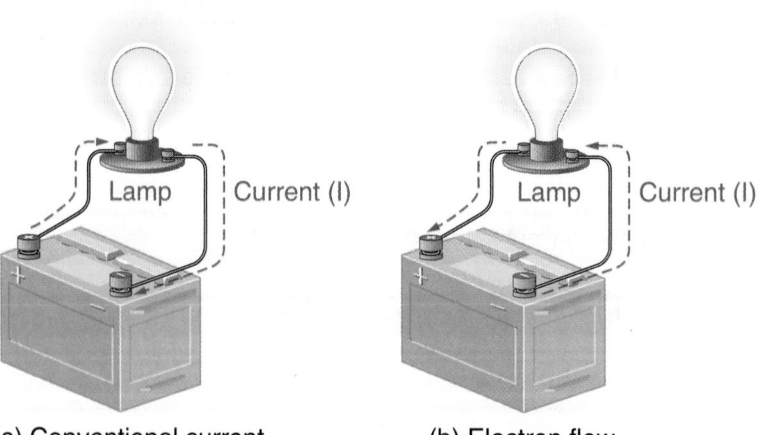

FIGURE 2.14 Conventional current and electron flow.

Direct Current (DC) Versus Alternating Current (AC)

DIRECT CURRENT (DC) Current that is in one direction only.

ALTERNATING CURRENT (AC) Current that continually changes direction.

Current is generally classified as being either **DIRECT CURRENT (DC)** or **ALTERNATING CURRENT (AC)**. The differences between direct current and alternating current are illustrated in Figure 2.15.

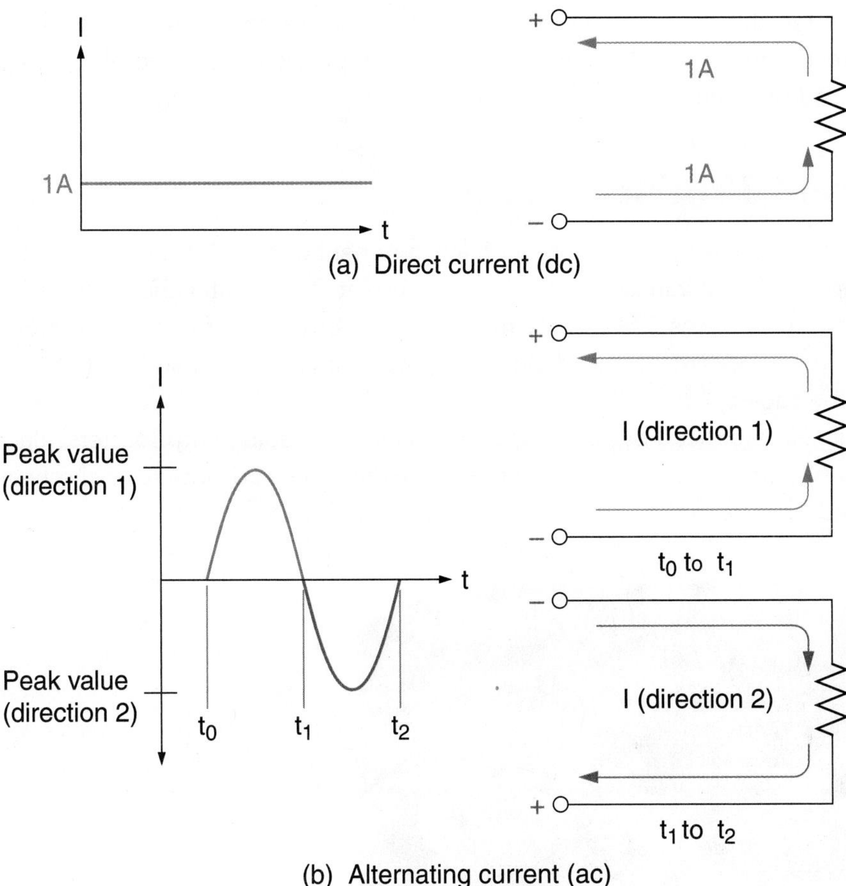

FIGURE 2.15 Direct current (DC) and alternating current (AC).

Because of their inherent advantages, alternating current systems are used in almost all utility power systems and end user distribution systems. There are two major reasons for the preference of AC systems: Lower generating costs and lower voltage losses when transmitted over long distances.

IN THE FIELD

Direct current is unidirectional. That is, the flow of charge is always in the same direction. The term *direct current* usually implies that the current has a fixed value. For example, the graph in Figure 2.15a shows that the current has a constant value of 1 A. While a fixed value is implied, direct current can change in value. However, the *direction* of the current does not change.

Alternating current is bidirectional. That is, the direction of current changes continually. For example, in Figure 2.15b, the graph shows that the current builds to a peak value in one direction and then builds to a peak value in the other direction. Note that the alternating current represented by the graph not only changes direction but is constantly changing in value.

Through the first twelve chapters of this text, we deal almost exclusively with direct current and DC circuits. Then our emphasis will shift to alternating current and AC circuits.

Current Produces Heat

Whenever current is generated through a component or circuit, heat is produced. The amount of heat varies with the level of current: The greater the current, the more heat it produces. This is why many high-current components, like motors, get hot when they are operated. Some high-current circuits get so hot that they have to be cooled.

The heat produced by current is sometimes a desirable thing. Toasters, electric stoves, and heat lamps are common items that take advantage of the heat produced by current.

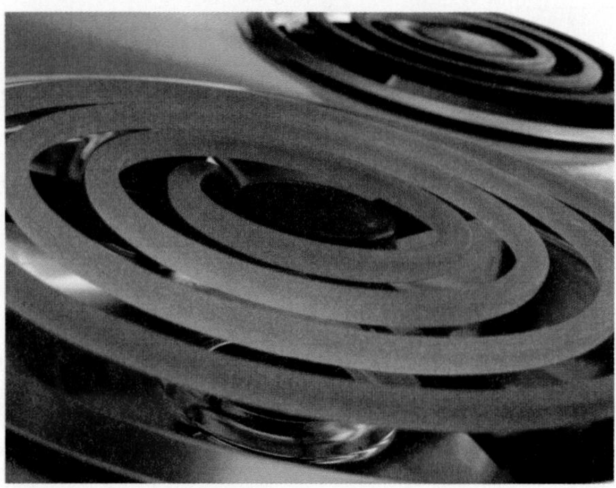

FIGURE 2.16 High current causes a stove heating element (burner) to glow red.
Istock.com

Putting It All Together

Free electrons are generated in copper at room temperature. When undirected, the motion of these free electrons is random, and the net flow of electrons in any one direction is zero.

When directed by an outside force, free electrons are forced to move in a uniform direction. This directed flow of charge is referred to as *current*.

Current is represented by the letter *I*, which stands for *intensity*. The intensity of current depends on the amount of charge moved and the time required to move it.

Current is measured in *amperes* (A). When one coulomb of current passes a point every second, you have one ampere of current.

There are two current theories. The *electron flow* theory describes current as the flow of charge (electrons) from negative to positive. The *conventional current* theory describes current as the flow of charge from positive to negative. Both approaches are widely followed. The way you view current does not affect the outcome of any circuit calculations, measurements, or test procedures.

Most electrical and electronic systems contain both *direct current* (DC) and *alternating current* (AC) circuits. In DC circuits, the current is always in the same direction. In AC circuits, current continually changes direction.

PROGRESS CHECK

1. How are free electrons generated in a conductor at room temperature?
2. What is *current*? What factors affect the *intensity* of current?
3. What is a *coulomb*?
4. What is the basic unit of current? How is it defined?
5. Contrast the *electron flow* and *conventional current* theories.

2.4 VOLTAGE

Voltage can be described as a force that generates the flow of electrons (current) through a circuit. In this section, we take a detailed look at voltage and how it generates current.

Generating Current with a Battery

The battery in Figure 2.17a has two terminals. The positive (+) terminal has an excess of positive ions and is described as having a *positive potential.* The negative (−) terminal has an excess of electrons and is described as having a *negative*

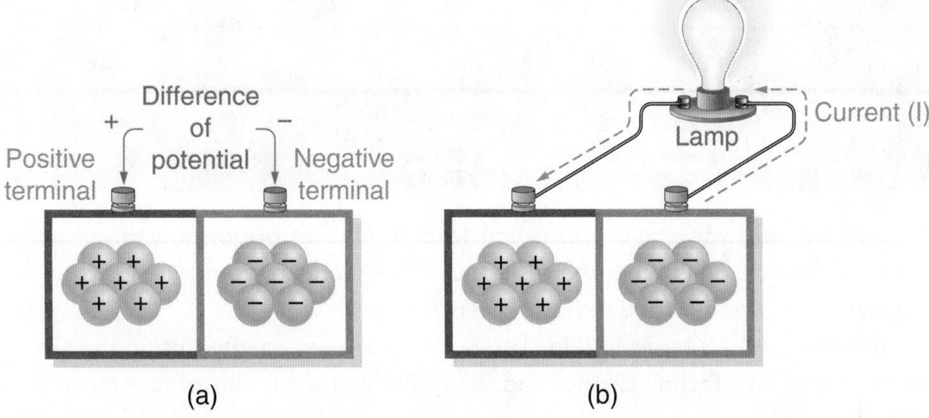

FIGURE 2.17 A difference of potential and a resulting current.

DIFFERENCE OF POTENTIAL The difference between the voltages at any two points in a circuit.

ELECTRICAL FORCE (E) The difference of potential across the terminals of a battery or other voltage source.

ELECTROMOTIVE FORCE (EMF) Another name for electrical force; the difference of potential across the terminals of a battery or other voltage source.

VOLT (V) The unit of measure for voltage; the amount of electrical force that uses one joule of energy to move one coulomb of charge.

potential. Thus, there is a **DIFFERENCE OF POTENTIAL**, or voltage (V), between the two terminals.

If we connect the two terminals of the battery with a copper wire and lamp (Figure 2.17b), a current is produced as the electrons are drawn to the positive terminal of the battery. In other words, there is a directed flow of electrons from the negative (−) terminal to the positive (+) terminal of the battery.

There are several important points that need to be made:

1. Voltage is a force that moves electrons. For this reason, it is often referred to as **ELECTRICAL FORCE (E)** or **ELECTROMOTIVE FORCE (EMF)**.
2. Current and voltage are not the same thing. Current is the directed flow of electrons from negative to positive. Voltage is the electrical force that generates current. In other words, current occurs as a result of an applied voltage (electrical force).

The **VOLT (V)** is the unit of measure for voltage. Technically defined, one volt is the amount of electrical force that uses one joule (J) of energy to move one coulomb (C) of charge. That is,

$$1 \text{ volt} = 1 \text{ joule per coulomb}$$

or

$$1\text{ V} = 1\text{ J/C}$$

PROGRESS CHECK

1. What is voltage?
2. How does voltage generate a current through a wire?
3. What is the unit of measure for voltage? How is it defined?

Critical Thinking

4. How would you define a *coulomb* in terms of voltage and energy?
5. How would you define a *joule* in terms of voltage and charge?

2.5 RESISTANCE

All elements provide some opposition to current. This opposition to current is called *resistance*. The higher the resistance of an element, component, or circuit, the lower the current produced by a given voltage.

OHM The unit of measure for resistance.

Resistance (R) is measured in **OHMS**. Ohms are represented using the Greek letter *omega* (Ω). Technically defined, one ohm is the amount of resistance that limits current to one ampere when one volt of electrical force is applied. This definition is illustrated in Figure 2.18.

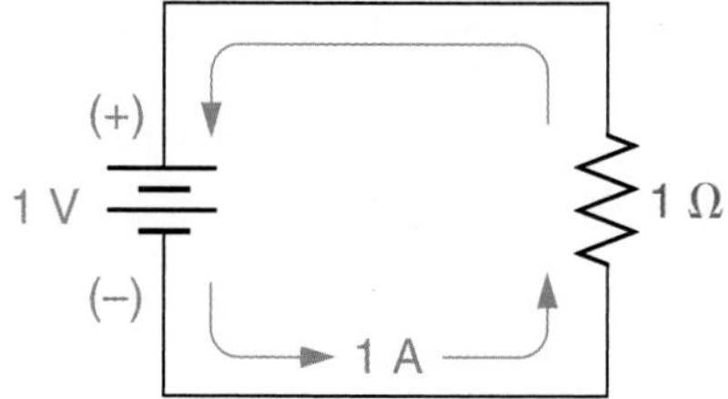

FIGURE 2.18 A basic circuit.

The *schematic diagram* in Figure 2.18 shows a battery that is connected to a resistor. As described in Chapter 1, a *resistor* is a component that provides a specific amount of resistance. As shown in the figure, a resistance of 1 Ω limits the current to 1 A when 1 V is applied. Note that the long end-bar on the battery schematic symbol represents the battery's positive terminal and the short end-bar represents its negative terminal.

Putting It All Together

We have now defined charge, current, voltage, and resistance. For convenience, these electrical properties are summarized in Table 2.1.

TABLE 2.1 • Basic Electrical Properties

PROPERTY	UNIT	DESCRIPTION
Charge (Q)	Coulomb (C)	The total charge of 6.25×10^{18} electrons.
Current (I)	Ampere (A)	The flow of charge. One ampere is equal to one coulomb per second.
Voltage (E or V)	Volt (V)	Electrical force. One volt is the amount of electrical force required to move a given amount of charge; equal to one joule per coulomb.
Resistance (R)	Ohm (Ω)	Opposition to current. One ohm of resistance limits current to one ampere when one volt is applied.

Many of the properties listed in Table 2.1 can be defined in terms of the others. For example, in our discussion on resistance, we said that *one ohm is the amount of resistance that limits current to one amp when one volt of electrical force is applied.* By the same token, we can redefine the ampere and the volt as follows:

1. One ampere is the amount of current that is generated when one volt of electrical force is applied to one ohm of resistance.
2. One volt is the amount of electrical force required to generate one amp of current through one ohm of resistance.

As you will see, these definitions relate closely to the most basic law of electricity, called *Ohm's law.* Ohm's law and its applications are covered in depth in Chapter 4.

PROGRESS CHECK

1. What is resistance?
2. What is the basic unit of resistance? How is it defined?
3. Define each of the following values in terms of the other two: Current, voltage, and resistance.

2.6 CONDUCTORS, INSULATORS, AND SEMICONDUCTORS

All materials are classified according to their ability (or inability) to conduct, as shown in Table 2.2.

TABLE 2.2 • Material Classifications

MATERIAL	CHARACTERISTICS
Conductor	Extremely low resistance; conducts with very little voltage applied
Insulator	Extremely high resistance; conducts only when an extremely high voltage is applied
Semiconductor	Resistance that falls about midway between a conductor and an insulator; limits the current at a given voltage

CONDUCTOR Any element or material that has extremely low resistance.

INSULATOR Any element or material that has extremely high resistance.

SEMICONDUCTOR Any element or material with resistance that falls about midway between that of a conductor and that of an insulator.

A good example of a **CONDUCTOR** is copper. Copper wire (which is the most commonly used conductor) passes current with little opposition. A good example of an **INSULATOR** is rubber. Rubber is used to coat the handles of many tools that are used in electrical work (such as pliers, screwdrivers, etc.) It takes an extremely high voltage to force rubber to conduct. A good example of a **SEMICONDUCTOR** is graphite (a form of carbon), which is used to make many common resistors. Graphite limits the amount of current that can be generated by a given amount of voltage. In this section, we take a look at some of the characteristics of conductors, insulators, and semiconductors.

Conductors

A conductor is a material that provides little opposition to the flow of electrons (current). Because the resistance of a conductor is low, very little energy is required to generate current through it. Thus, the materials with the lowest resistance make the best conductors.

In Section 2.1, you were told that the valence shell of an atom determines its electrical characteristics. The **CONDUCTIVITY** of an element is determined by:

CONDUCTIVITY The ease with which an element or compound conducts electricity.

1. **The number of valence shell electrons**. The fewer valence shell electrons that an atom contains, the easier it is to force the atom to give up free electrons. The best conductors contain one valence electron per atom.
2. **The number of atoms per unit volume**. With more atoms per unit volume, a given voltage can generate more free electrons. The best conductors contain a high number of atoms per unit volume.

Insulators

Insulators are materials that normally block current. For example, the insulation that covers a power cord prevents the current in the cord from reaching you when it is touched. Some elements (like neon) are natural insulators. Most of the insulators used to protect technicians in the field are compounds like rubber, Teflon, and mica (among others).

As you have probably guessed, conductors and insulators have opposite characteristics. In general, insulators are materials that have:

1. Complete valence shells (eight valence electrons).
2. Relatively few atoms per unit volume.

With this combination of characteristics, an extremely high voltage is required to force an insulator into conduction.

Semiconductors

Semiconductors are *materials that are neither good conductors nor good insulators*. For example, graphite (a form of carbon) does not conduct well enough to be considered a conductor. At the same time, it does not block current well enough

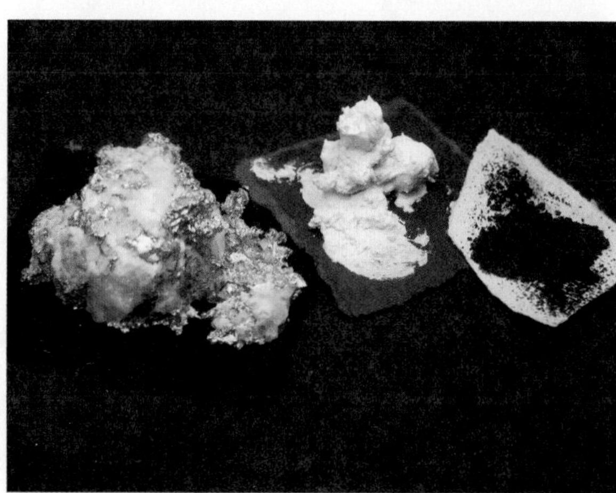

FIGURE 2.19 Copper, rubber, and graphite.

FIGURE 2.20 Some common insulated tools.
Istock.com

to be considered an insulator. Some other examples of semiconductors are *silicon, germanium,* and *gallium arsenide.*

In general, semiconductors have the following characteristics:

1. Valence shells that are half complete (i.e., contain four valence electrons).
2. A relatively high number of atoms per unit volume.

Conductor Resistance

The resistance of a conductor (at a specified temperature) depends on:

1. The resistivity of the conductor.
2. The length of the conductor.
3. The cross-sectional area of the conductor.

RESISTIVITY The resistance of a specific volume of an element or compound.

CIRCULAR-MIL-OHMS PER FOOT (CMIL-Ω/FT) The resistivity of a specific volume of a conductor. The specified volume has a cross sectional area of one circular mil and a length of one foot.

OHM-CENTIMETER (Ω-CM) The unit of measure for the resistivity of one cubic centimeter of a material.

RESISTIVITY is the resistance of a specific volume of an element or compound. The lower the resistivity of a conductor; the better it conducts. For example, the resistivity of copper is lower than the resistivity of aluminum. Therefore, copper is a better conductor than aluminum. In fact, a 1000 ft length of copper wire has just over half the resistance of a comparable length of aluminum wire.

Resistivity is commonly rated using one of two units of measure:

1. **CIRCULAR-MIL-OHMS PER FOOT (CMIL-Ω/FT).**
2. **OHM-CENTIMETERS (Ω-CM).**

The volumes used in these two ratings are illustrated in Figure 2.21. Note that the *NEC®* deals exclusively with circular mils.

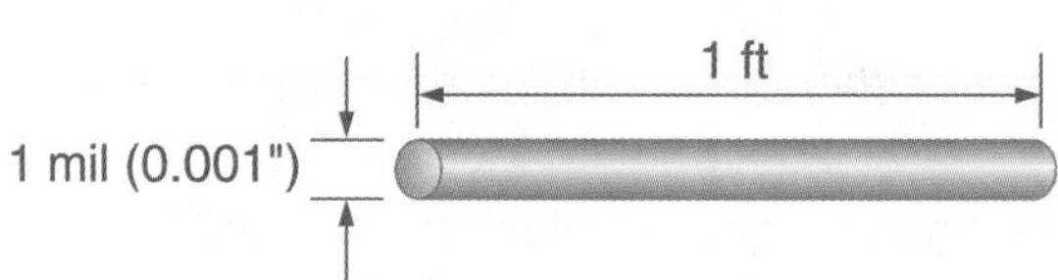

(a) One circular-mil-foot

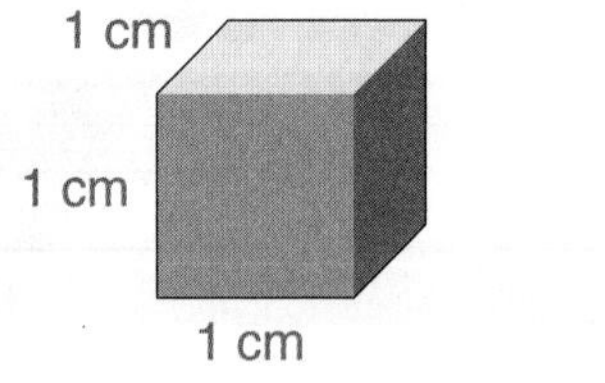

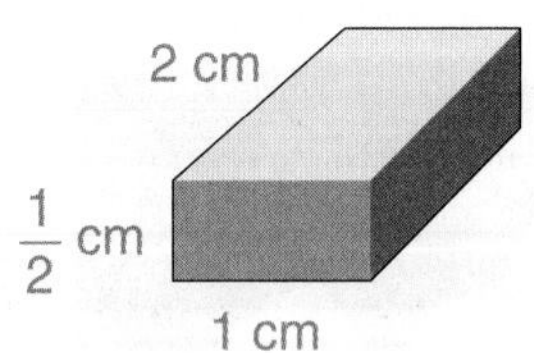

$V = 1\ \text{cm} \times 1\ \text{cm} \times 1\ \text{cm}$
$= 1\ \text{cm}^3$

$V = \frac{1}{2}\ \text{cm} \times 1\ \text{cm} \times 2\ \text{cm}$
$= 1\ \text{cm}^3$

(b) Two 1 cm^2 volumes

FIGURE 2.21 Volumes used to rate the resistivity of a conductor.

MIL A unit of length, equal to 0.001 inch.

CIRCULAR-MIL (CMIL) A measure of area found by squaring the diameter (in mils) of the conductor.

The **MIL** equals one-thousandth of an inch (0.001 in). The **CIRCULAR-MIL (CMIL)** is a measure of area found by squaring the diameter (in mils) of the conductor.

One **CIRCULAR-MIL-FOOT (CMIL-FT)** of material has *a diameter of one mil and a length of one foot* as shown in Figure 2.21a. When this volume of material is used to measure the resistivity of a given material, the rating is given in *circular-mil ohms per foot (cmil-Ω/ft).*

CIRCULAR-MIL-FOOT (CMIL-FT) A measure of volume with a diameter of one mil and a length of one foot.

The second unit of volume used for rating resistivity is the *cubic centimeter.* When this volume is used to measure the resistivity of a given material, the rating is given in *ohm-centimeters (Ω-cm).*

Table 2.3 shows the resistivity ratings of some elements that are commonly found in electrical and electronic circuits.

Table 2.3 • The Resistivity Ratings of Some Common Elements

ELEMENT	RESISTIVITY	
	cmil-Ω/ft	Ω-cm
Silver	9.9	1.645×10^{-6}
Copper	10.37	1.723×10^{-6}
Gold	14.7	2.443×10^{-6}
Aluminum	17.0	2.825×10^{-6}
Iron	74.0	12.299×10^{-6}
Carbon	2.1×10^{4}	3.50×10^{-3}

Calculating the Resistance of a Conductor

When the length and cross-sectional area of a conductor are known, the resistance of the conductor can be calculated using

(2.2)

$$R = \rho\frac{\ell}{A}$$

where

ρ = the resistivity of the material
ℓ = the length of the conductor
A = the cross-sectional area of the conductor

The Greek letter *rho* (ρ) in equation (2.2) is commonly used to represent resistivity. The following examples illustrate the calculation of the resistance of a conductor.

EXAMPLE 2.2

Calculate the resistance of a 1000 ft length of copper that has a diameter of 81 mils.

SOLUTION

According to Table 2.3, copper has a resistivity rating of 10.37 cmil-Ω/ft. With a diameter of 81 mils, the cross-sectional area is found as

$$\text{Area} = (81\text{ mils})^2 = 6561\text{ cmils}$$

Now, using equation (2.2), the resistance of the copper is found as

$$\begin{aligned} R &= \rho\frac{\ell}{A} \\ &= \left(10.37\frac{\text{cmil-}\Omega}{\text{ft}}\right)\left(\frac{1000\text{ ft}}{6561\text{ cmils}}\right) \\ &= 1.58\,\Omega \end{aligned}$$

With this kind of resistance, it is easy to see that copper provides very little opposition to current.

PRACTICE PROBLEM 2.2

Calculate the resistance of a 4000 ft length of copper that has a diameter of 64 mils.

What Equation (2.2) Shows Us

Every equation serves two purposes. First, it shows us how to calculate the value of interest. Second (and probably most important), it shows us the relationship between the value of interest and its variables. For example, equation (2.2) states that

$$R = \rho\frac{\ell}{A}$$

This relationship shows us that:

1. *The resistance of a conductor is directly proportional to its resistivity.* That is, increases (or decreases) in resistivity cause a similar change in resistance.
2. *The resistance of a conductor is directly proportional to its length.* That is, the longer the conductor, the higher its resistance.
3. *The resistance of a conductor is inversely proportional to its cross-sectional area.* Thus, the greater the diameter of a conductor, the lower its resistance.

Every conductor contains electrical resistance. As a result, a conductor can cause an unacceptable loss of voltage between its supply and load ends if the conductor is exceedingly long and/or not sized properly.

Temperature and Resistance

The calculation in Example 2.2 assumes an operating temperature of 20°C. The resistivity of a conductor is affected by operating temperature. The effect of temperature on the resistivity of a material depends on its temperature coefficient.

A **POSITIVE TEMPERATURE COEFFICIENT** is a rating that indicates that the resistance of a given material increases when temperature increases. A **NEGATIVE TEMPERATURE COEFFICIENT** is a rating which indicates that the resistance of a given material decreases when temperature increases, and vice versa.

Conductors have positive temperature coefficients. Thus, the resistance of a conductor increases when temperature increases.

POSITIVE TEMPERATURE COEFFICIENT A value that indicates how much the resistance of a material increases as temperature increases.

NEGATIVE TEMPERATURE COEFFICIENT A value that indicates how much the resistance of a material decreases as temperature increases.

PROGRESS CHECK

1. What is a *conductor*? An insulator? A *semiconductor*?
2. List the factors that determine the conductivity of an element.
3. What is *resistivity*?
4. Describe the units of measure that are commonly used to express resistivity.
5. What is a *positive temperature coefficient*?
6. What is a *negative temperature coefficient*?
7. What is the temperature coefficient for a conductor?

2.7 SUMMARY

Here is a summary of the major points that were made in this chapter:

Matter, Elements, and Atoms

- Matter is anything that has mass and occupies space.
  - Matter can be in the form of a solid, liquid, or gas.
- An element is a substance that cannot be broken down into two or more simpler substances.
- The atom is the smallest particle of matter that retains the physical characteristics of an element.
  - An atom is made up of neutrons, protons, and electrons.
  - The neutrons and protons make up the nucleus (core) of the atom.
  - The electrons orbit the nucleus.
- The atomic number of an element tells you how many protons it contains.
- Electrons travel in orbital paths that are called orbital shells.
- The valence (outermost) shell of an atom cannot contain more than eight electrons.
  - A "complete" valence shell contains 8 electrons.
  - An "incomplete" valence shell contains fewer than 8 electrons.

Electrical Charge

- Charge is a force that causes two particles to be attracted to, or repelled from, each other. The two types of charge are referred to as positive charge and negative charge.

- Protons have a positive charge and electrons have a negative charge.
- When an atom has an equal number of protons and electrons, the net charge of the atom is zero.
- Like charges repel each other. Unlike charges attract each other.
- When an atom loses an electron, it becomes a positive ion.
- A free electron is one that has broken away from its parent atom and is free to drift to another atom.
- Positive and negative charges always seek to neutralize each other.

Current

- Current is the directed flow of charge through a conductor.
- Electrons flow from negative to positive.
- The coulomb is the total charge of 6.25×10^{18} electrons.
- Current is measured in amperes.
  - One ampere equals one coulomb per second.
- Electron flow defines current as the flow of electrons from negative to positive.
- Conventional current defines current as the flow of charge from positive to negative.
- Direct current (DC) is unidirectional. Alternating current (AC) is bi-directional.
- Current produces heat when it passes through a component or circuit.

Voltage

- Voltage is a difference of potential that generates current. It is also referred to as electric force (E) or electromotive force (EMF).
- The unit of measure for voltage is the volt, equal to one joule per coulomb. The joule is the basic unit of energy.

Resistance

- Resistance is the opposition to current, measured in ohms (Ω).
- One ohm is the amount of resistance that limits current to one ampere when there is a one-volt difference of potential.

Conductors, Insulators, and Semiconductors

- A conductor has extremely low resistance and easily passes current.
- An insulator has extremely high resistance and allows current only when an extremely high voltage is applied.
- A semiconductor is neither a good conductor nor a good insulator.
- The conductivity of a material is determined by its number of valence shell electrons (per atom) and the number of atoms per unit volume.
- Conductors have few valence electrons per atom and a large number of atoms per unit volume.
- Insulators have complete valence shells and relatively few atoms per unit volume.
- Semiconductors have valence shells that are half complete.
- The resistance of a conductor, semiconductor, or insulator is determined by its resistivity, length, and cross-sectional area.

- Resistivity is the resistance of one circular-mil-foot or one cubic centimeter of a material.
- Conductors have low resistivity ratings while insulators have extremely high resistivity ratings.
- Conductors have positive temperature coefficients, which means that their resistance varies directly with temperature.

CHAPTER REVIEW

1. Matter that cannot be broken down into two or more simpler substances is referred to as a(n) ________ .
   a) solid
   b) model
   c) element
   d) property

2. The electron has a(n) ________ charge.
   a) positive
   b) negative
   c) neutral
   d) opposite

3. The atomic number of an atom tells you how many ________ it has.
   a) protons
   b) neutrons
   c) electrons
   d) shells

4. In its natural state, an atom has _________.
   a) a negative charge
   b) an equal number of electrons and protons
   c) an equal number of neutrons and protons
   d) a positive charge

5. A proton has a(n) _________ charge.
   a) positive
   b) negative
   c) neutral
   d) opposite

6. Electrons orbit the nucleus of an atom ________.
   a) at Bohr levels
   b) in shells
   c) in valences
   d) in negative paths

7. The maximum number of electrons that can be found in the second orbital path closest to the nucleus is ________.
   a) 9
   b) 8
   c) 6
   d) 12

8. If the outermost orbital path in an atom has six electrons it is ________.
   a) a positive ion
   b) a negative ion
   c) incomplete
   d) complete

9. The orbital path containing one or more electrons that is farthest from the nucleus is called the ________ shell.
   a) valence
   b) ionic
   c) negative
   d) positive

10. A positive charge and a negative charge always ________.
    a) repel each other
    b) add to each other
    c) attract each other
    d) combine to form a neutral charge

11. An outside force causes an atom to lose one electron. This produces ________.
    a) a positive ion
    b) a positive charge on the atom
    c) a free electron
    d) all of the above

12. If a free electron drifts in the vicinity of a positive ion, it ________.
    a) neutralizes the ion
    b) is repelled by the ion
    c) produces a negative ion
    d) causes all of the above

13. At room temperature the motion of free electrons in a material with no external force is ________.
    a) always from positive to negative
    b) always from negative to positive
    c) faster than at higher temperatures
    d) random

14. The symbol used for current in a formula is ________.
    a) C
    b) I
    c) A
    d) either I or A

15. There are 6.25×10^{18} electrons in one ________.
    a) ampere
    b) coulomb
    c) volt
    d) ohm

16. The flow of charge from positive to negative is called ________.
    a) negative current
    b) positive current
    c) conventional current
    d) intensity current

17. The major classifications of current are ________.
    a) positive and negative current
    b) direct and indirect current
    c) conventional and alternating current
    d) direct and alternating current

18. Whenever current is generated through a component or material ________ is produced.
    a) reactance
    b) heat
    c) static
    d) resistance

19. The force that causes the movement of charge in a material is called ________.
    a) voltage
    b) electrical force
    c) electromotive force
    d) all of the above

20. The volt is defined by the amount of ________ it uses to move charge.
    a) energy
    b) work
    c) current
    d) force

21. The ohm, the unit of resistance, is represented by the Greek letter ________.
   a) alpha, α
   b) phi, Φ
   c) omega, Ω
   d) beta, β

22. Insulators typically have ________ valence electrons.
   a) 2^n
   b) 4
   c) 8
   d) 6

23. Conductors tend to have ________.
   a) few valence electrons and many atoms per unit volume
   b) many valence electrons and many atoms per unit volume
   c) many valence electrons and few atoms per unit volume
   d) few valence electrons and few atoms per unit volume

24. The resistance of a material is inversely proportional to ________.
   a) its length
   b) its resistivity
   c) its cross-sectional area
   d) all of the above

25. The temperature coefficient of a conductor ________.
   a) increases with temperature
   b) decreases with temperature
   c) is negative
   d) is positive

PRACTICE PROBLEMS

1. One coulomb of charge passes a point every 20 seconds. Calculate the value of the current through the point.
2. A total charge of 2.5×10^{-3} C passes a point every 40 seconds. Calculate the value of the current through the point.
3. A total charge of 4.0×10^{-6} C passes a point every second. Calculate the value of the current through the point.
4. A total charge of 50×10^{-3} C passes a point every 2.5 seconds. Calculate the value of the current through the point.
5. Calculate the resistance of a 4 inch length of copper that has a cross-sectional area of 50 CM.
6. Calculate the resistance of a 12 inch length of aluminum that has a diameter of 0.003 inches.

7. Calculate the resistance of a 30 cm length of copper that has a cross-sectional area of 0.08 cm^2.
8. Calculate the resistance of a 10 cm length of iron that has a cross-sectional area of 0.05 cm^2.

ANSWERS TO THE EXAMPLE PRACTICE PROBLEMS

2.1	2.5 A
2.2	10.1 Ω

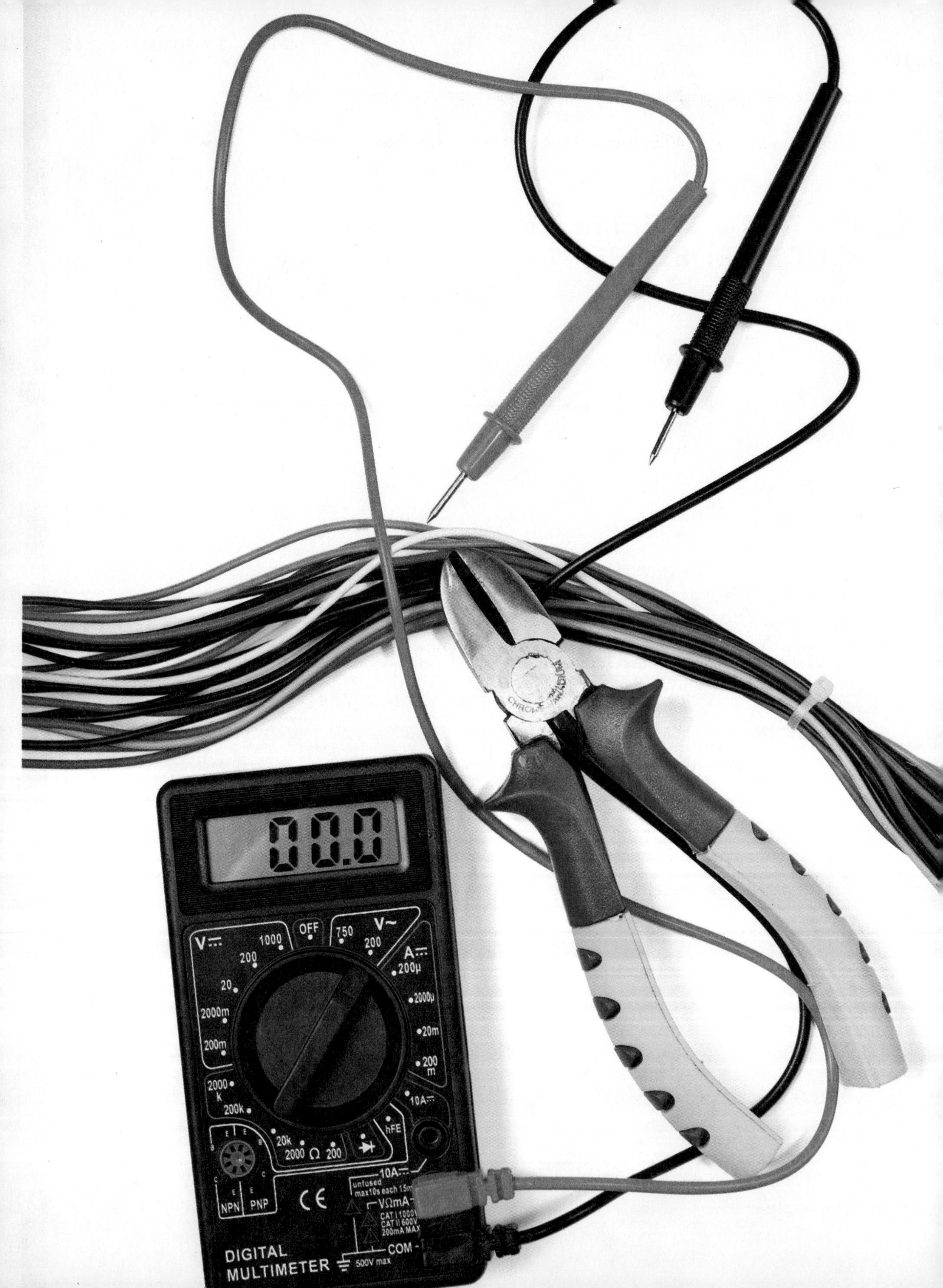
00.0
OFF
V⎓
1000
200
20
2000m
200m
2000 k
200k
750
V~
A⎓
200μ
2000μ
20m
200 m
10A⎓
hFE
20k
2000
Ω
200
NPN
PNP
CE
VΩmA
COM
500V max
DIGITAL MULTIMETER

CHAPTER 3

BASIC ELECTRIC COMPONENTS AND METERS

PURPOSE

There are two aspects to every science: *theory* and *practice*. Theory deals with the concepts of the science, while practice deals with using those concepts to accomplish something. In Chapter 2, you took the first step toward understanding the theoretical side of electricity. Now it is time to take the first step toward understanding its practical side.

In this chapter, you are introduced to the most basic electrical components and the ratings that describe their characteristics. You are also shown how to measure current, voltage, and resistance.

KEY TERMS

The following terms are introduced and defined in this chapter on the pages indicated:

AC adaptor, 97
American Wire Gauge (AWG), 68
Ammeter, 117
Ampacity, 68
Ampere-hour (Ah), 90
Analog multimeter, 108
Average breakdown voltage, 70
Battery, 89
Capacity, 90
Carbon-composition resistor, 72
Cartridge fuse, 101
Cell, 89
Common, 95
Digital multimeter (DMM), 108
Electrode, 89
Electrolyte, 89
Fixed resistor, 71
Fuse, 101
Gang-mounted potentiometers, 87
Impurity, 72
Integrated resistor, 73
Ionize, 102
Linear taper, 86
Momentary switch, 100
Multimeter, 107
Multiside board, 70
Multiturn potentiometer, 87
Negative voltage, 97
Normally-closed (NC) switch, 100
Normally-open (NO) switch, 100
Nonlinear taper, 86
Ohmmeter, 117
Overload, 104
Parallax error, 115
Polarity, 97
Pole, 99
Positive voltage, 97
Potentiometer, 85
Primary cell, 90
Printed circuit board, 69
Resolution, 87
Rheostat, 87
Rotary switch, 101
Secondary cell, 92
Solid wire, 67
Stranded wire, 67
Switch, 98
Taper, 86
Terminals, 99
Throw, 99
Tolerance, 74
Traces, 70
Type S fuse, 105
Type T fuse, 105
Variable resistor, 71
Voltmeter, 117
Wiper arm, 85
Wire-wound resistor, 73

OBJECTIVES

After completing this chapter, you should be able to:

1. Describe the commonly used types of wire and the American Wire Gauge (AWG) system of wire sizing.
2. Describe the ratings commonly used for conductors and insulators.
3. List and describe the various types of resistors.
4. Determine the range of values for a resistor.
5. Describe and use the standard resistor color code.
6. Describe the operation of commonly used potentiometers.
7. Describe the construction and characteristics of commonly used batteries.
8. Discuss the operation and use of a basic DC power supply.
9. List and describe the various types of switches.
10. Describe the construction, operation, ratings, and replacement procedures for fuses.
11. Interpret any voltage, current, or resistance reading on an analog meter scale.
12. Describe basic current, voltage, and resistance measurements.

3.1 CONDUCTORS AND INSULATORS

Any current-carrying conductor must be kept from coming into contact with anything (or anyone) around it. This is accomplished by coating the conductor with an insulator, as illustrated in Figure 3.1. As long as the insulator remains free of cuts or other breaks, the conductor remains isolated from the surrounding environment.

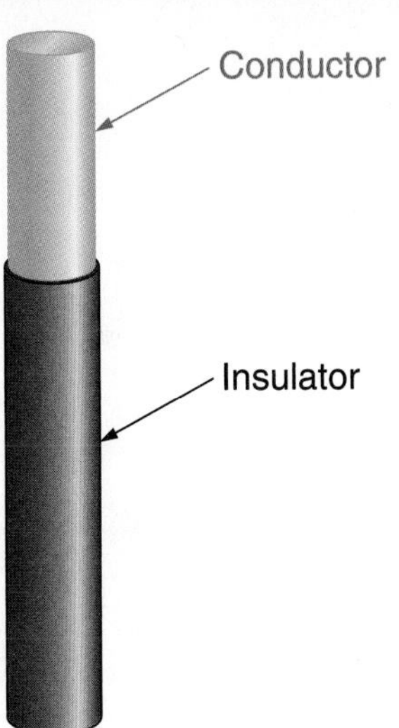

FIGURE 3.1 An insulated conductor.

Copper Wires

As you have probably figured out by now, copper is the most commonly used conductor. Copper wires are either *solid* or *stranded*, as shown in Figure 3.2. **SOLID WIRE** contains a solid core conductor. **STRANDED WIRE** contains a group of very thin wires that are wrapped together to form a larger diameter wire.

SOLID WIRE A wire that contains a single, solid conductor.

STRANDED WIRE A wire that contains any number of thin wires wrapped to form a larger wire.

Solid wires and stranded wires each have their strengths: Solid wires are easier to produce, while stranded wires are more flexible.

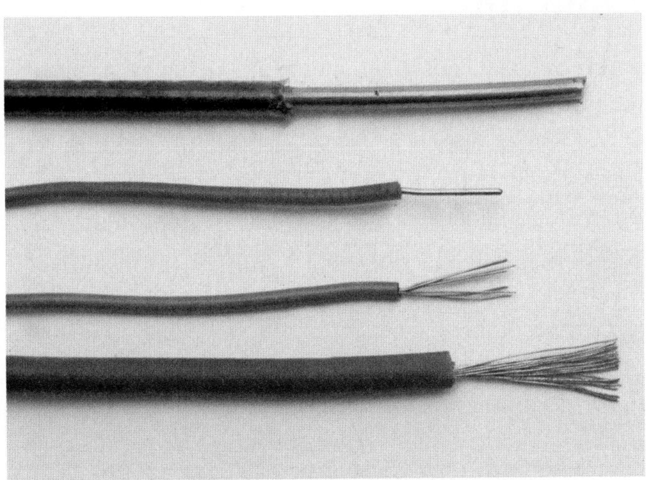

FIGURE 3.2 Solid and stranded wires.

Wires come in a wide range of diameters. There are approximately 42 sizes that fall between the diameters shown in Figure 3.3. Wire sizes are discussed in detail later in this section.

Why Copper Is Used

You may recall that the resistance of a conductor is determined (in part) by its resistivity. For example, the resistivity of copper is lower than the resistivity of aluminum, so the resistance of copper wire is lower than the resistance of a comparable aluminum wire.

Of all the readily available elements, only silver has a lower resistivity than copper. At the same time, silver is much more expensive than copper. Thus, copper is used because it provides the best balance between resistance and cost per unit length. (Cost is *always* a consideration in circuit design and construction.)

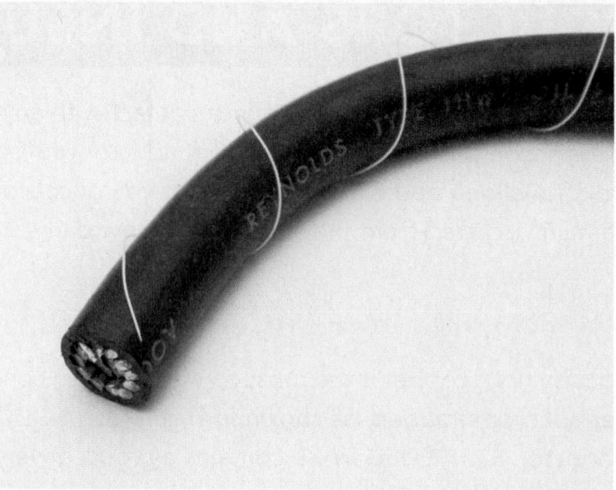

FIGURE 3.3 A 40-gauge wire wrapped around a 4/0 gauge wire.

Aluminum conductor alloys have been developed for a range of strengths, conductivities, and characteristics. Aluminum alloys used for building wire are formulated for connection performance and flexibility. The *NEC*® Sec. 110.14(A) requires terminals used for aluminum be so identified.

IN THE FIELD

Aluminum

Because of its low cost, aluminum would seem to be a better choice than copper for use in electrical circuits. However, aluminum has very low *elasticity*. When changes in temperature cause aluminum to expand and contract, it does not return to its original shape. This can cause problems when aluminum is connected to a copper connector (which has high elasticity). Even so, aluminum and copper-clad aluminum are used in electrical applications.

Wire Sizes

AMERICAN WIRE GAUGE (AWG) A system that uses numbers to identify industry-standard wire sizes.

Wires are produced in standard sizes. The size of a wire is determined by its cross-sectional area. The **AMERICAN WIRE GAUGE (AWG)** system uses numbers to identify these standard wire sizes. Table 3.1 lists the wires that are commonly used in residential and other electrical applications.

Ampere Capacity (Ampacity)

The lower the gauge of a wire, the greater its diameter. The greater the diameter of a wire, the higher its current capacity. In other words, lower gauge wires can handle more current than higher gauge wires.

AMPACITY The maximum allowable current that can be safely carried by a given wire gauge, measured in amperes.

Some wire gauges are rated according to their ampere capacity, or **AMPACITY**. Ampacity is the maximum allowable current that can be safely carried by a given wire gauge, measured in amperes. As current passes through a conductor, it produces heat. The greater the current, the more heat it produces. If the ampacity rating of a conductor is exceeded, the heat produced by the current may damage the conductor's insulation. As shown in Table 3.1, lower gauge wires have higher ampacity ratings.

Table 3.1 • The American Wire Gauge System

AWG NUMBER OR KCMIL	AREA (CMILS)	Ω/1000 FT	AMPACITY (A)*	APPLICATIONS/EXAMPLES
18	1624.3	6.385		Common extension cords
16	2582.9	4.016		
14	4106.8	2.525	15	House wiring/lighting
12	6529.0	1.588	30	
10	10,381	0.9989	40	
8	16,509	0.6282	60	High-current appliances
6	26,250	0.3951	60	
4	41,742	0.2458	100	Commercial wiring
3	52,634	0.1970	110	
2	66,373	0.1563	125	Car battery cable
1	83,694	0.1240	150	
1/0	105,530	0.0983	175	Power distribution
2/0	133,080	0.0780	200	
3/0	167,810	0.0618	225	
4/0	211,600	0.0490	250	
250	250,000	0.0390	300	
350	350,000	0.0309	350	
400	400,000	0.0245	400	

*For typical residential applications (3-wire, single phase, dwelling services and feeders.)

The ampacity ratings in Table 3.1 were obtained from the *2008 National Electrical Code* (*NEC*®). According to *Table 310.15(B)(6)* in the *2008 NEC*®, these ratings are valid for 3-wire copper, single-phase dwelling services and feeders.

Ampacity ratings are affected by the type of conductor, the type of insulation,[1] and the ambient temperature. As a result, the *NEC*® contains several ampacity tables. These and other wire tables provided in the *NEC*® are discussed in Appendix B.

STAYING SAFE Some single-phase dwelling unit electrical services may be derived from 3-phase power systems. Use caution here. *Table 310.15(B)(6)* does not apply to these services because the neutral conductor carries current even when the loads on each of the two phases are balanced.

PC Board Traces

Many modern electrical systems (such as smart home technology systems) include electronic circuits. Most of these circuits are built on **PRINTED CIRCUIT (PC) BOARDS**

PRINTED CIRCUIT (PC) BOARD A board that provides mechanical (physical) support for components and conductors that connect those components.

[1]See *Table 310.13* in the *2008 NEC*®.

MULTISIDE BOARD A PC board designed so that components are mounted on both sides.

like the one shown in Figure 3.4. Components are usually mounted on one side of the board, though **MULTISIDE BOARDS** have components mounted on both sides.

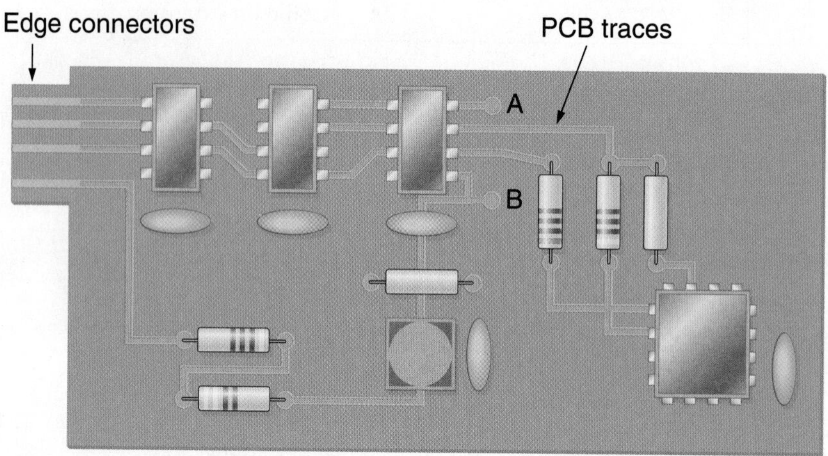

FIGURE 3.4 PC board traces.

TRACES The conductors that connect the components on a PC board.

The conductors that connect the components on a PC board are commonly called **TRACES**. Traces are usually made of copper. Like wires, traces have a limit on the amount of current they can carry. Under normal circumstances, a given trace is designed to carry a lot more current than is required by the circuit.

Insulator Average Breakdown Voltage

AVERAGE BREAKDOWN VOLTAGE The voltage that will cause an insulator to break down and conduct.

Any insulator can be forced to conduct if a high enough voltage is applied. The **AVERAGE BREAKDOWN VOLTAGE** rating of an insulator is the voltage that will cause the insulator to conduct, in kilovolts per centimeter (kV/cm). Some average breakdown voltage ratings are listed in Table 3.2.

Table 3.2 • Insulator Breakdown Voltage Ratings

MATERIAL	AVERAGE BREAKDOWN VOLTAGE (KV/CM)
Air	30
Rubber	270
Paper	500
Teflon	600
Glass	900
Mica	2000

The values listed in Table 3.2 are actually *maximum ratings*. That is, they are limits that cannot be exceeded if the insulator is to block current. For example, mica will block current as long as the applied voltage is less than 2000 kV/cm. However, if this value is exceeded, mica may no longer act as an insulator.

The Bottom Line

Conductors and insulators are an integral part of any electrical circuit. Conductors are used to pass current from one point to another, while insulators are used to prevent current from passing between points.

The American Wire Gauge (AWG) standard uses numbers to identify wires with specific cross-sectional areas. The lower the gauge of a wire, the greater its cross-sectional area, the lower its resistance, and the greater its ampere capacity (ampacity).

The average breakdown voltage of an insulator is the voltage at which it will break down and conduct, measured in kV/cm. Average breakdown voltage is an example of a *maximum rating.*

PROGRESS CHECK

1. Contrast solid and stranded wires.
2. Why is copper the preferred conductor in most applications?
3. What is the relationship between a wire and one that is three wire gauges higher?
4. What is the relationship between a wire and one that is ten wire gauges higher?
5. What is *ampacity*? Under what conditions is ampacity measured?
6. Define *average breakdown voltage.*

3.2 RESISTORS

As you know, a resistor is a component that is designed to provide a specific amount of resistance. Resistors are classified as being either fixed or variable. A **FIXED RESISTOR** is one that has a specific ohmic value that cannot be changed by the user. A **VARIABLE RESISTOR** is one that can be adjusted to any ohmic value within a specified range. Several fixed and variable resistors are shown in Figure 3.5. Note that variable resistors all have some mechanism for manually adjusting the component's resistance value.

In this section, we will focus on fixed resistors and their ratings. Variable resistors and their ratings are covered in Section 3.3.

FIXED RESISTOR One that has a specific ohmic value that cannot be changed by the user.

VARIABLE RESISTOR One that can be adjusted to any ohmic value within a specified range.

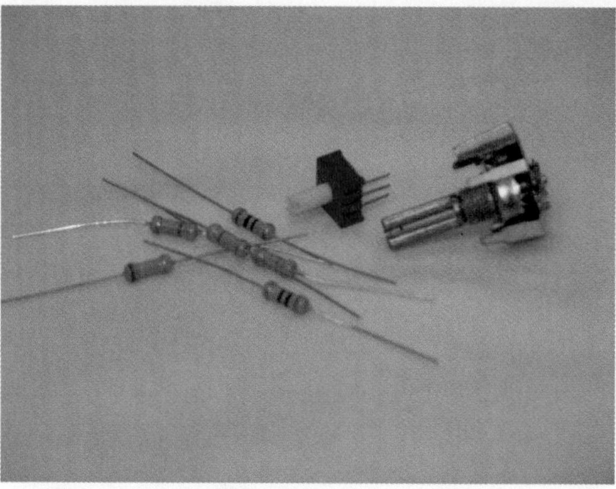

FIGURE 3.5 Fixed and variable resistors.

Carbon-Composition Resistors

CARBON COMPOSITION RESISTOR A component that uses carbon to provide a desired value of resistance.

The most commonly used resistor is the **CARBON-COMPOSITION RESISTOR**. The construction of the carbon-composition resistor is illustrated in Figure 3.6. As you can see, the resistor has two metal leads (conductors) that are separated by carbon. When current enters the resistor, it passes through the carbon that separates the leads, as shown in Figure 3.7. The relatively high resistivity of the carbon is the source of the resistor's opposition to current.

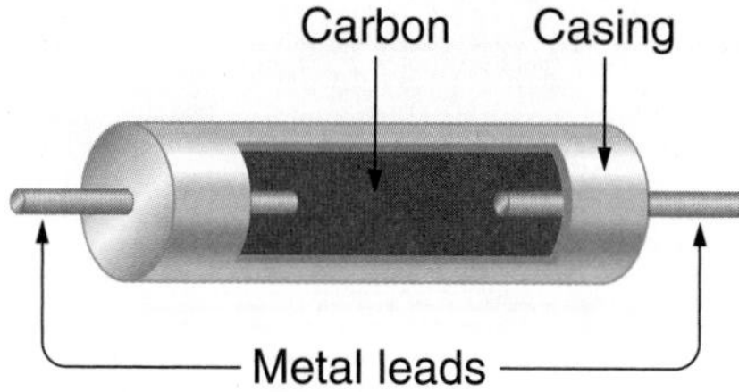

FIGURE 3.6 Carbon composition resistor.

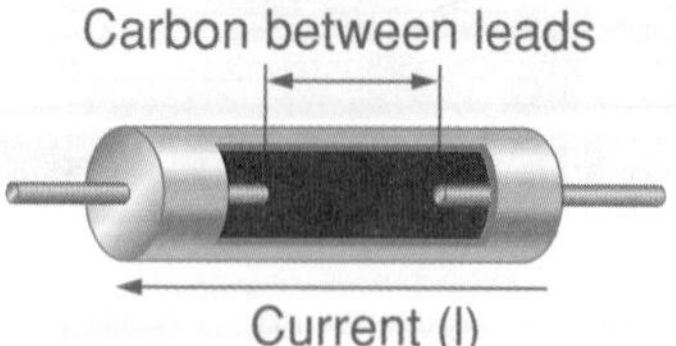

FIGURE 3.7 Current passing through a resistor.
Pearson

IMPURITY Any element that is added to a previously pure element to alter its electrical characteristics.

The value of a carbon-composition resistor is determined primarily by the purity of the carbon. By adding **IMPURITIES** (other elements) to carbon during the manufacturing process, the resistivity of the carbon can be increased or decreased, depending on the amount and type of impurity used. Controlling the

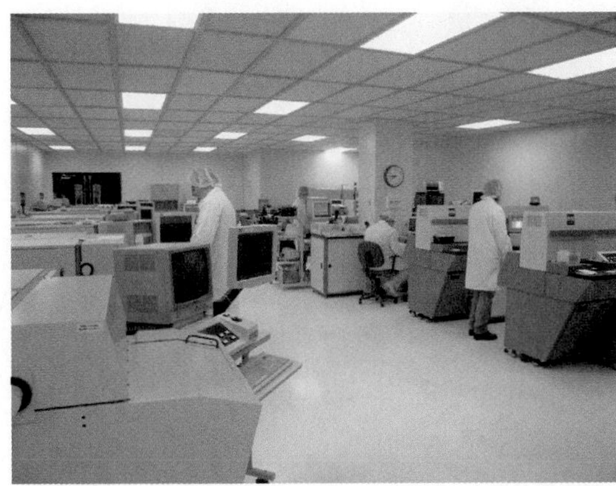

FIGURE 3.8 Resistor manufacturing environment.
Source: Photo courtesy of Vishay Intertechnology, Inc.

impurity levels makes it possible to produce carbon-composition resistors with a wide range of values.

Other Types of Resistors

A **WIRE-WOUND RESISTOR** uses the resistivity of a length of wire as the source of its resistance. A wire-wound resistor is illustrated in Figure 3.9. The value of the component is determined by the length of the wire between the leads. The longer the wire, the higher the resistance of the component.

WIRE-WOUND RESISTOR
A resistor that uses the resistivity of a length of wire as the source of its resistance.

FIGURE 3.9 Wire-wound resistor construction.

Wire-wound resistors are used primarily in high-power applications; that is, applications where the components must be able to dissipate (throw off) a relatively high amount of heat. A wire-wound resistor has more surface area than a comparable carbon-composition resistor, so it can dissipate more heat.

INTEGRATED RESISTORS are micro-miniature components that are made using semiconductors other than carbon. Integrated resistors have the advantage of being extremely small. These resistors are so small that several of them can be packaged in a single casing like the one shown in Figure 3.10a. The casing shown contains seven resistors that are connected between the pins as shown in Figure 3.10b. Integrated resistors are restricted to low-current applications.

INTEGRATED RESISTORS
Micro-miniature components that are made using semiconductors other than carbon.

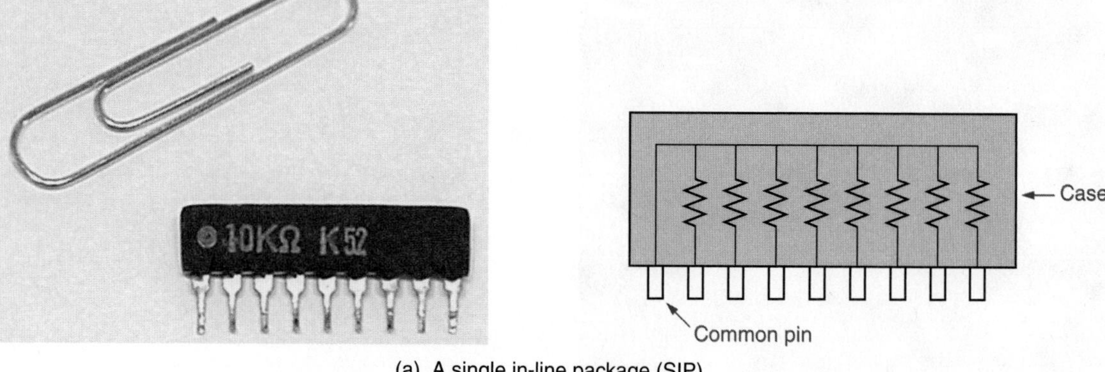

(a) A single in-line package (SIP)

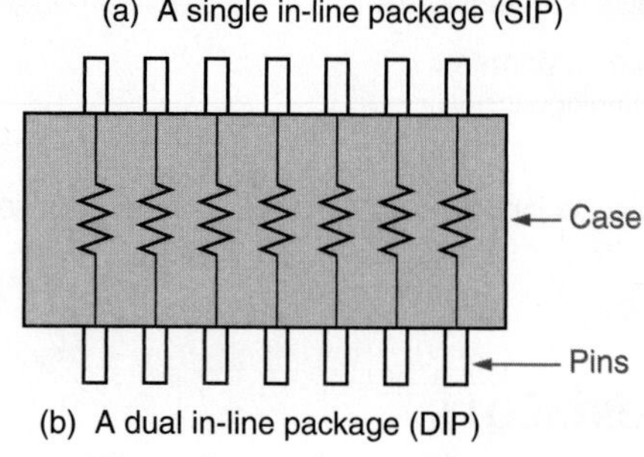

(b) A dual in-line package (DIP)

FIGURE 3.10 Integrated resistors.

Standard Resistor Values

Resistors are commercially produced in a variety of values. The standard resistor values are listed in Table 3.3. If you look at any row of numbers in the table, you'll see that it contains a series of values that all start with the same two digits. The only difference between the values is the power-of-ten multiplier. For example, the values shown in the first row (from left to right) could be written as follows:

$$10 \times 10^{-2}$$
$$10 \times 10^{-1}$$
$$10 \times 10^{0}$$
$$10 \times 10^{1}$$
$$10 \times 10^{2}$$
$$10 \times 10^{3}$$
$$10 \times 10^{4}$$
$$10 \times 10^{5}$$

As these values indicate, we can designate any resistor value using only the first two digits and a power-of-ten multiplier. As you will see, this is the basis for the resistor color code.

Resistor Tolerance

Even with advanced production techniques, there is almost always some difference between the actual and rated values of a resistor. For example, a resistor rated at

Table 3.3 • Standard Resistor Values (2% Tolerance and Higher)

Ω				KΩ			MΩ	
0.10	1.0	10	100	1.0	10	100	1.0	10
0.11	1.1	11	110	1.1	11	110	1.1	11
0.12	1.2	12	120	1.2	12	120	1.2	12
0.13	1.3	13	130	1.3	13	130	1.3	13
0.15	1.5	15	150	1.5	15	150	1.5	15
0.16	1.6	16	160	1.6	16	160	1.6	16
0.18	1.8	18	180	1.8	18	180	1.8	18
0.20	2.0	20	200	2.0	20	200	2.0	20
0.22	2.2	22	220	2.2	22	220	2.2	22
0.24	2.4	24	240	2.4	24	240	2.4	
0.27	2.7	27	270	2.7	27	270	2.7	
0.30	3.0	30	300	3.0	30	300	3.0	
0.33	3.3	33	330	3.3	33	330	3.3	
0.36	3.6	36	360	3.6	36	360	3.6	
0.39	3.9	39	390	3.9	39	390	3.9	
0.43	4.3	43	430	4.3	43	430	4.3	
0.47	4.7	47	470	4.7	47	470	4.7	
0.51	5.1	51	510	5.1	51	510	5.1	
0.56	5.6	56	560	5.6	56	560	5.6	
0.62	6.2	62	620	6.2	62	620	6.2	
0.68	6.8	68	680	6.8	68	680	6.8	
0.75	7.5	75	750	7.5	75	750	7.5	
0.82	8.2	82	820	8.2	82	820	8.2	
0.91	9.1	91	910	9.1	91	910	9.1	

100 Ω might have an actual value of 98 Ω or 104 Ω. Slight variations like these in component values are common occurrances.

Even though resistor values aren't always exact, they can be guaranteed to fall within a specified range of values. For example, a manufacturer may be able to guarantee that the actual value of every 100 Ω resistor will fall somewhere between 95 Ω and 105 Ω. This range of values is called the **TOLERANCE** of the component, and is given as a percentage of its rated value.

TOLERANCE The possible range of values for a component, given as a percentage of its nominal (rated) value.

Most common resistors have 2% or 5% tolerance ratings. This means that their actual values are guaranteed to fall within ±2% or ±5% of their rated values. Some older resistors have 10% and 20% tolerance ratings, but these components are not found in modern electrical and electronic systems.

Integrated resistors like those shown in Figure 3.10 often have very poor tolerance ratings—as high as 30% in some cases. As a result, their use is limited to circuits that can handle wide variations in resistance.

To determine the range of possible values for a resistor:

1. Multiply the rated value of the resistor by its tolerance to find its maximum variation in resistance.
2. Add the maximum variation to the rated value of the component to find its upper limit.
3. Subtract the maximum variation from the rated value of the component to find the lower limit.

This procedure is demonstrated in the following example.

EXAMPLE 3.1

Determine the range of possible values for a 470 Ω resistor that has a 5% tolerance.

SOLUTION

First, the maximum variation in resistance is found as

$$470\ \Omega \times 5\% = 23.5\ \Omega$$

The maximum variation is now added to the rated value of the component, as follows:

$$470\ \Omega + 23.5\ \Omega = 493.5\ \Omega \quad \text{(upper limit)}$$

The maximum variation is now subtracted from the rated value of the component, as follows:

$$470\ \Omega - 23.5\ \Omega = 446.5\ \Omega \quad \text{(lower limit)}$$

The range of possible values for this resistor is 446.5 Ω to 493.5 Ω.

PRACTICE PROBLEM 3.1

Determine the range of possible values for a 68 Ω resistor that has a 2% tolerance.

The higher the tolerance of a resistor, the wider its range of possible values. This point is illustrated in the following example.

EXAMPLE 3.2

Determine the ranges of possible values for a 33 kΩ resistor with a 2% tolerance and a 33 kΩ resistor with a 5% tolerance.

SOLUTION

For the 2% tolerance resistor:

$$(33\text{ k}\Omega) \times 2\% = 660\ \Omega$$
$$= 0.66\text{ k}\Omega$$
$$33\text{ k}\Omega + 0.66\text{ k}\Omega = 33.66\text{ k}\Omega \quad \text{(upper limit)}$$
$$33\text{ k}\Omega - 0.66\text{ k}\Omega = 32.34\text{ k}\Omega \quad \text{(lower limit)}$$

For the 5% tolerance resistor:

$$(33\text{ k}\Omega) \times 5\% = 1650\ \Omega$$
$$= 1.65\text{ k}\Omega$$
$$33\text{ k}\Omega + 1.65\text{ k}\Omega = 34.65\text{ k}\Omega \quad \text{(upper limit)}$$
$$33\text{ k}\Omega - 1.65\text{ k}\Omega = 31.35\text{ k}\Omega \quad \text{(lower limit)}$$

These results show that the 5% tolerance component has a wider range of possible values.

PRACTICE PROBLEM 3.2

Determine the range of possible values for a 910 Ω resistor with a 5% tolerance.

Lower tolerance components are considered to be the higher-quality components. In fact, the *ideal* (perfect) resistor, if it could be produced, would have a tolerance rating of 0%. This means its rated and measured values would always be equal.

Even though they are no longer produced, you may see resistors with tolerances of 10% in older circuits and systems. Later in this section, you'll be shown how the tolerance of a resistor is indicated.

The Resistor Color Code

In most cases, the value of a resistor is indicated by a series of color bands on the component. For example, look at the resistor shown in Figure 3.11. The resistor has four color bands. These bands are numbered as shown in the figure. Note that the fourth band is offset from the first three.

The first three bands on a resistor designate its rated value as follows:

Band 1: The color of this band designates the *first digit* in the resistor value.

Band 2: The color of this band designates the *second digit* in the resistor value.

Band 3: The color of this band designates the power-of-ten *multiplier* for the first two digits. (In most cases, this is simply the number of zeros that follow the first two digits.)

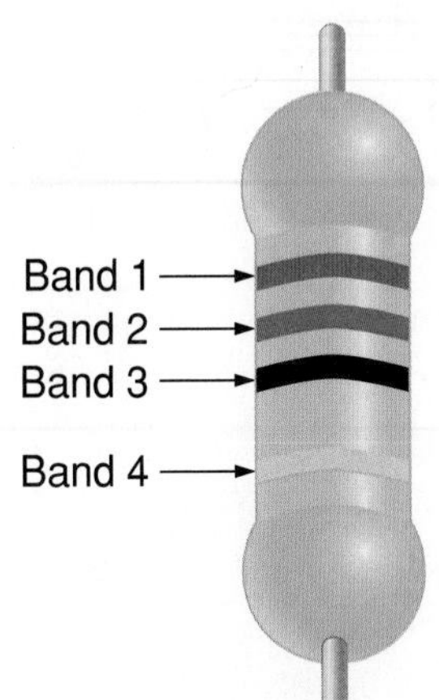

FIGURE 3.11 Resistor color bands.

The colors that most often appear in these three bands are coded as shown in Figure 3.12.

4-Band-Code

2%, 5%, 10%

560kΩ ± 5%

COLOR	1st DIGIT	2nd DIGIT	3rd DIGIT	MULTIPLIER	TOLERANCE
Black	0	0	0	1	
Brown	1	1	1	10	± 1% (F)
Red	2	2	2	100	± 2% (G)
Orange	3	3	3	1K	
Yellow	4	4	4	10K	
Green	5	5	5	100K	±0.5% (D)
Blue	6	6	6	1M	±0.25% (C)
Violet	7	7	7	10M	±0.10% (B)
Gray	8	8	8		±0.05%
White	9	9	9		
Gold				0.1	± 5% (J)
Silver				0.01	± 10% (K)

0.1%, 0.25%, 0.5%, 1%

237Ω ± 1%

5-Band-Code

FIGURE 3.12 The standard resistor color code.

The following series of examples shows how the first three color bands are used to indicate the value of a given resistor.

EXAMPLE 3.3

Determine the rated value of the resistor shown in Figure 3.13a.

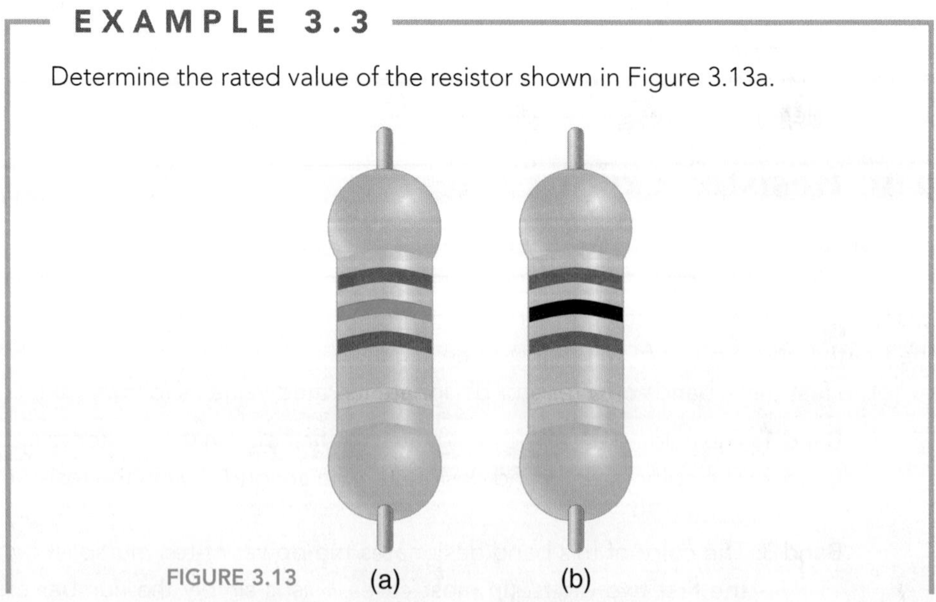

FIGURE 3.13

SOLUTION

Brown = 1, so the first digit in the resistor value is 1.
Green = 5, so the second digit in the resistor value is 5.
Red = 2, so the power-of-ten multiplier is $10^2 = 100$.
Combining these values we get

$$(15 \times 100)\Omega = 1500\ \Omega = 1.5\ \text{k}\Omega$$

Note that the multiplier band value (2) is equal to the number of zeros in the value of the resistor.

PRACTICE PROBLEM 3.3

Determine the rated value of the resistor in Figure 3.13b.

EXAMPLE 3.4

Determine the rated value of the resistor shown in Figure 3.14a.

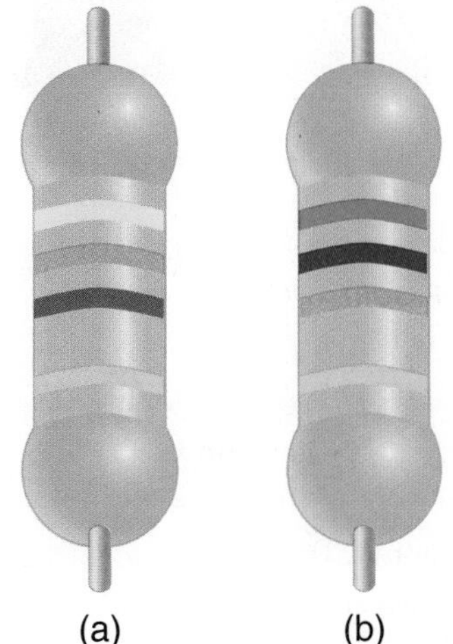

FIGURE 3.14 (a) (b)

SOLUTION

Yellow = 4, so the first digit in the resistor value is 4.
Orange = 3, so the second digit in the resistor value is 3.
Brown = 1, so the power-of-ten multiplier is $10^1 = 10$.
Combining these values we get

$$(43 \times 10)\ \Omega = 430\ \Omega$$

Again, the multiplier band told us the number of zeros following the first two digits of the component's value.

PRACTICE PROBLEM 3.4

Determine the rated value of the resistor in Figure 3.14b.

When the multiplier band on a resistor is black, you have to be careful not to make a common mistake. Entry-level technicians will often see a black multiplier band, see that black corresponds to *zero*, and assume that there is a zero after the first two digits. *A black multiplier band indicates that there are no zeros following the first two digits.* This point is illustrated in the following example.

EXAMPLE 3.5

Determine the rated value of the resistor shown in Figure 3.15a.

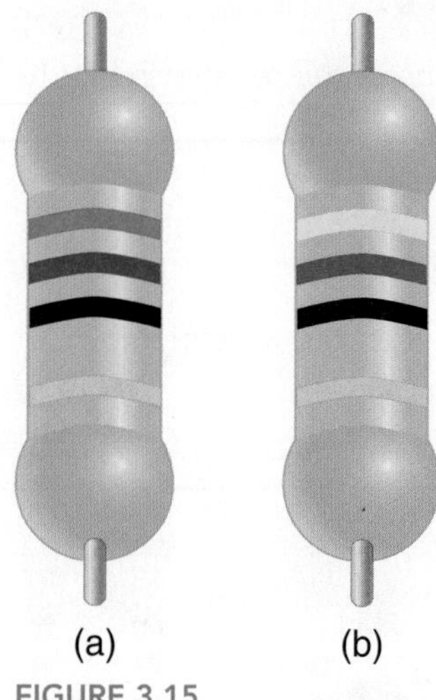

FIGURE 3.15

SOLUTION

Green = 5, so the first digit in the value is 5.
Blue = 6, so the second digit in the value is 6.
Black = 0, so the multiplier is $10^0 = 1$.
Combining these values we get

$$(56 \times 1)\Omega = 56\ \Omega$$

Note that no resistors are added after the first two digits. This is always the case when the multiplier band is black.

PRACTICE PROBLEM 3.5

Determine the rated value of the resistor in Figure 3.15b.

There are two other colors that may appear in the multiplier band. These colors and their multiplier values are as follows:

$$\text{Gold} = -1$$
$$\text{Silver} = -2$$

When the multiplier band is gold, a decimal point is added between the first two digits in the value. This is illustrated in the following example.

EXAMPLE 3.6

Determine the rated value of the resistor shown in Figure 3.16a.

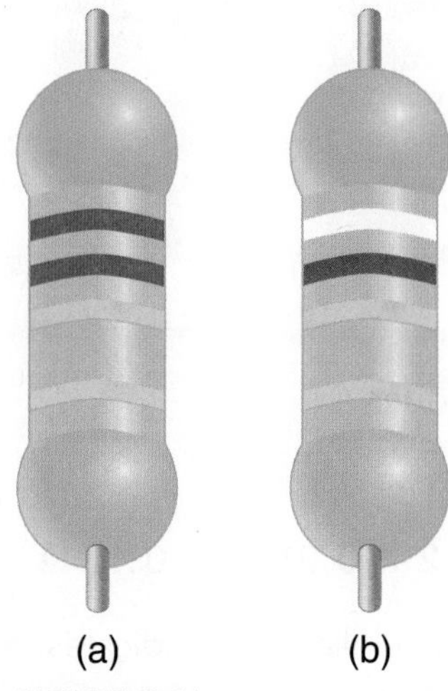

FIGURE 3.16

SOLUTION

Red = 2, so the first digit in the component value is 2.
Red = 2, so the second digit in the component value is (also) 2.
Gold = −1, so the multiplier is $10^{-1} = 0.1$.
Combining these values we get

$$(22 \times 0.1)\,\Omega = 2.2\,\Omega$$

As you can see, a gold multiplier band indicates that a decimal point is placed between the first two digits in the resistor's value.

PRACTICE PROBLEM 3.6

Determine the rated value of the resistor in Figure 3.16b.

When the multiplier band is silver, a decimal point is added in front of the two digits in the component value. This is illustrated in the following example.

EXAMPLE 3.7

Determine the rated value of the resistor shown in Figure 3.17a.

SOLUTION

Brown = 1, so the first digit in the component value is 1.
Red = 2, so the second digit in the component value is 2.
Silver = −2, so the multiplier is $10^{-2} = 0.01$.

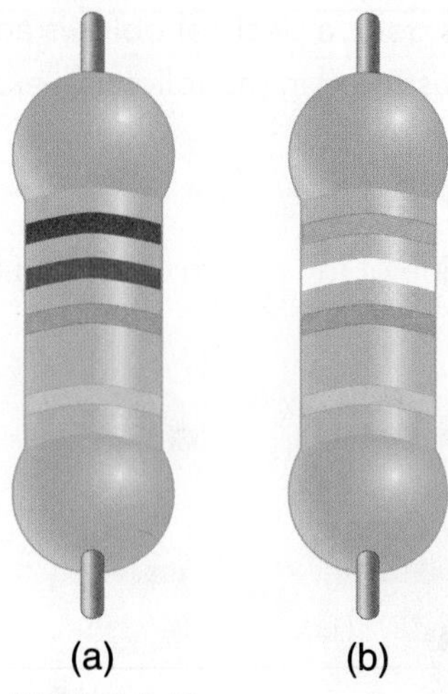

FIGURE 3.17

Combining these values we get

$$(12 \times 0.01)\,\Omega = 0.12\,\Omega$$

As you can see, a silver multiplier band indicates that a decimal point is placed in front of the first two digits in the resistor's value.

PRACTICE PROBLEM 3.7

Determine the rated value of the resistor in Figure 3.17b.

As often as not, you'll need to be able to determine the color code for a specific value of resistance so that you can locate the needed component. When this is the case, write the component value in standard form. The first two colors are determined by the first two digits. The color of the multiplier band is determined (in most cases) by the number of zeros that follow the first two digits. This is illustrated in the following example.

EXAMPLE 3.8

You need to locate a 360 Ω resistor. Determine the colors of the first three bands on the component.

SOLUTION

The first three bands are coded for 3, 6, and 1 (which is the number of zeros in the value). The colors that correspond to these numbers are as follows:

3 = orange
6 = blue
1 = brown

These are the colors (in order) of the first three bands, as shown in Figure 3.18.

FIGURE 3.18

PRACTICE PROBLEM 3.8

Determine the colors of the first three bands of a 24 kΩ resistor.

Resistor Tolerance

The fourth band on a resistor designates its tolerance. The colors used in the tolerance band of a four-band resistor are as follows:

$$\text{Red} = 2\% \qquad \text{Gold} = 5\% \qquad \text{Silver} = 10\%$$

The following example shows how the first four bands on a resistor are used to determine the range of possible values for the component.

EXAMPLE 3.9

Determine the range of possible values for the resistor in Figure 3.19a.

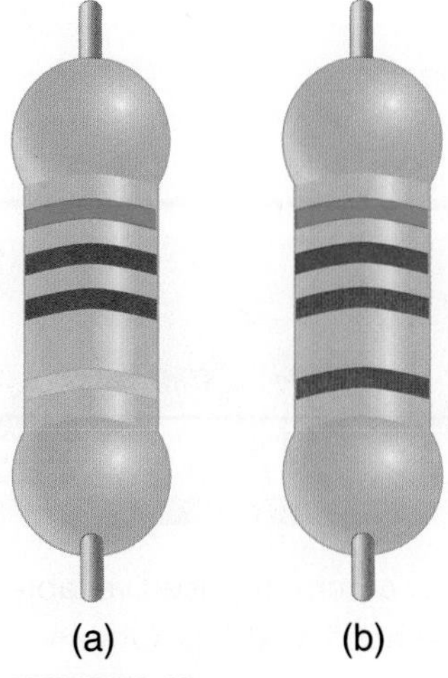

(a) (b)

FIGURE 3.19

SOLUTION

Green = 5, so the first digit is the component value is 5.
Brown = 1, so the second digit in the component value is 1.
Brown = 1, so the multiplier is $10^1 = 10$.
Gold = 5% (the tolerance of the component).
The first three bands indicate that the rated value of the component is 510 Ω, and

$$(510\ \Omega) \times 5\% = (510\ \Omega) \times 0.05 = 25.5\ \Omega$$

$$510\ \Omega + 25.5\ \Omega = 535.5\ \Omega \quad \text{(upper limit)}$$

$$510\ \Omega - 25.5\ \Omega = 484.5\ \Omega \quad \text{(lower limit)}$$

PRACTICE PROBLEM 3.9

Determine the range of possible values for the resistor in Figure 3.19b.

PROGRESS CHECK

1. Describe the physical construction of the carbon-composition resistor.
2. How is the value of a carbon-composition resistor indicated?
3. In what applications are wire-wound resistors primarily used?
4. What are integrated resistors?
5. What is the *tolerance* of a resistor?
6. How do you determine the range of possible values for a resistor?
7. What is indicated by the color of each of the first three bands on a resistor?
8. What is indicated by a black multiplier band on a resistor?
9. What is indicated by a gold multiplier band on a resistor?
10. How do you determine the color code for a specific standard resistor value?
11. What does the fourth color band on a resistor indicate?
12. List the colors that are commonly found in the fourth band and give the value that is indicated by each of them.

3.3 POTENTIOMETERS AND RHEOSTATS

The potentiometer is the most commonly used variable resistor. In this section, we will discuss the potentiometer and its ratings. Other variable resistors will be introduced (as needed) later in the text.

Potentiometers

The **POTENTIOMETER** is a three-terminal resistor whose value can be adjusted by the user. To help in our discussion, the terminals of a typical potentiometer are identified as shown in Figure 3.20. The potentiometer (or "pot") is designed so that the resistance between the middle terminal (B) and each of the outer terminals (A and C) changes when the control shaft is turned. Turning the control shaft in one direction reduces the resistance between terminals A and B (designated R_{AB}), and increases the resistance between terminals B and C (designated R_{BC}). Turning the control shaft in the opposite direction has the opposite effect. By setting the control shaft to a specific position, R_{AB} and/or R_{BC} can be set to any desired value within the limits of the component.

POTENTIOMETER A three-terminal resistor whose value can be adjusted by the user.

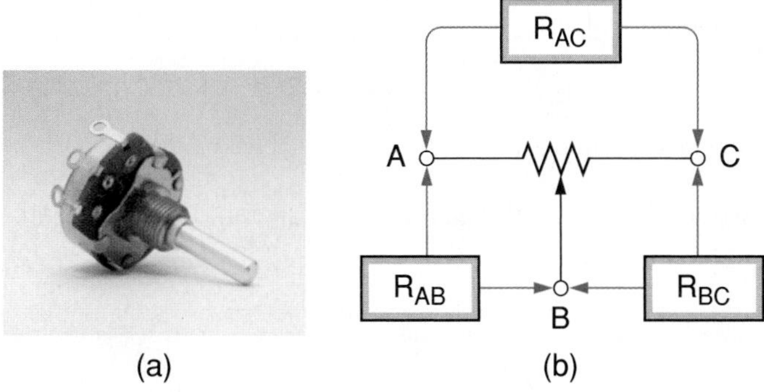

FIGURE 3.20 Potentiometer.

Potentiometer Construction and Operation

The construction of a typical pot is illustrated in Figure 3.21. The A and C terminals are shown to be connected to the ends of a length of carbon. The B terminal, which is referred to as the **WIPER ARM**, is connected to a sliding contact. When the control shaft is turned, the sliding contact moves along the surface of the carbon. Figure 3.22 shows how turning the control shaft affects the component resistance values.

WIPER ARM The potentiometer terminal that is connected to a sliding contact.

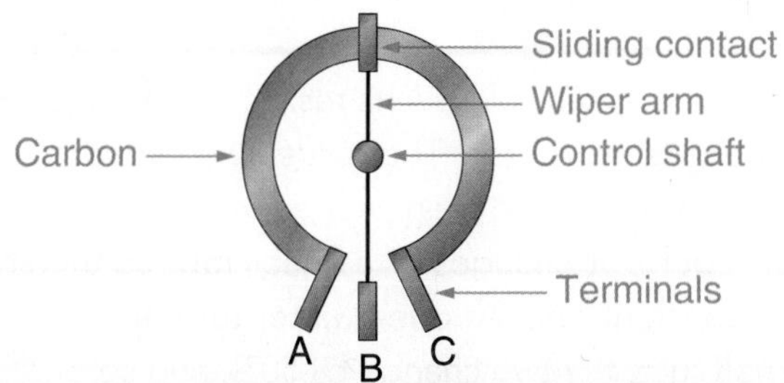

Note: Most potentiometers are constructed so that the moving contact can rotate approximately 350°.

FIGURE 3.21 Potentiometer construction.

When the sliding contact is in the center position (Figure 3.22a), there are equal amounts of carbon between the wiper arm and each of the outer terminals, and $R_{AB} = R_{BC}$. If we assume that the potentiometer is rated at $R_{AC} = 100\ \Omega$, then R_{AB} and R_{BC} must each be half this value, or $50\ \Omega$.

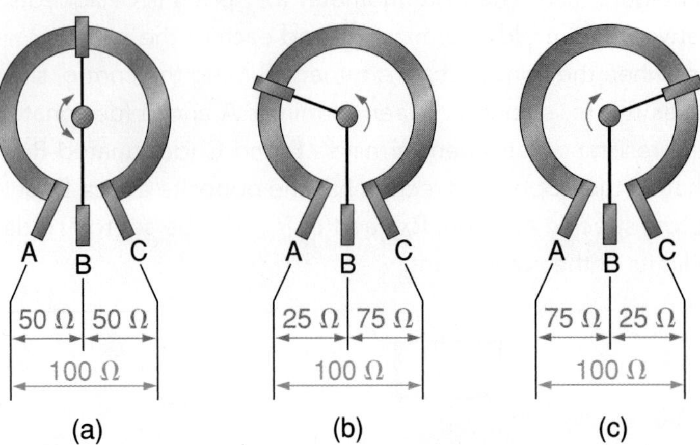

FIGURE 3.22 The effect of turning the control shaft on the component resistances.

The relationship between the values of R_{AB}, R_{BC}, and R_{AC} can be stated mathematically as

$$R_{AC} = R_{AB} + R_{BC} \tag{3.1}$$

This relationship holds true regardless of the position of the sliding contact.

If the control shaft is turned counterclockwise, the sliding contact moves as shown in Figure 3.22b. The amount of carbon between the wiper arm and terminal A decreases, so R_{AB} decreases. At the same time, the amount of carbon between the wiper arm and terminal C increases, so R_{BC} increases. Note that the sum of the two resistances is still equal to R_{AC} (as shown in the figure).

As Figure 3.22c shows, turning the control shaft clockwise causes the sliding contact to move toward terminal C. As a result, R_{BC} decreases and R_{AB} increases. As always, the sum of the two equals R_{AC}.

TAPER A measure of the rate at which potentiometer resistance changes as the control shaft is rotated between its extremes.

LINEAR TAPER Potentiometer resistance that changes at a linear (constant) rate as the shaft is turned.

NONLINEAR TAPER Potentiometer resistance that changes at a nonlinear (changing) rate as the shaft is turned.

Potentiometer Taper

The **TAPER** of a potentiometer is a measure of how its resistance changes as the control shaft is rotated between its extremes. There are two types of taper: **LINEAR TAPER** and **NONLINEAR TAPER**.

The resistance of a *linear-taper* pot changes at a linear rate as the shaft is turned. That is, it changes at a constant rate. At one-quarter turn, the two resistances change by 25%, at one-half turn, the two change by 50%, and so on.

The resistance of a *nonlinear-taper* pot changes at a nonlinear rate as the shaft is turned. That is, it changes at varying rates. For example, if you are turning the control shaft from its left extreme to its right extreme, the change in resistance is initially low and increases as you continue to turn the shaft.

Potentiometer Resistance Ratings

The resistance rating of a pot is the resistance between its outer terminals. Since the resistance between the wiper arm and either outer terminal can be adjusted to equal R_{AC}, the resistance rating of a pot is the maximum possible resistance setting for the component.

The resistance rating of a pot is normally printed on the component. The resistance may be written as a straightforward value (such as *5* kΩ, *10* kΩ, etc.) or in a code that is similar to the resistor color code. For example, a pot may be marked as shown in Figure 3.23. This simple code works as follows:

- In this case, the first two digits are the first two digits in the component value.
- The third digit is the power-of-ten multiplier.

Thus, the pot shown in Figure 3.23 has a value of $50 \times 10^2 = 5000\ \Omega = 5\ \text{k}\Omega$.

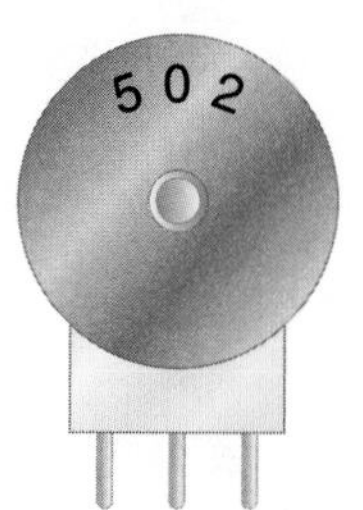

FIGURE 3.23 Potentiometer value code.

MULTITURN POTENTIOMETER A potentiometer whose control shaft can be rotated more than 360°.

RESOLUTION A measure of the rate at which potentiometer resistance changes, expressed in ohms per degree (Ω/°).

GANG-MOUNTED Two potentiometers that share a common shaft that controls both.

Multiturn Potentiometers

The control shaft on most potentiometers can be turned one full rotation (360 degrees) or less. However, some pots are geared to allow rotations that are greater than 360 degrees. The advantage of these **MULTITURN POTENTIOMETERS** is that they have better **RESOLUTION** than single-turn pots. That is, rotating the control shaft causes a much smaller change in resistance. This makes it easier to set the component to an exact value.

Gang-mounted Potentiometers

In some cases, potentiometers are **GANG-MOUNTED**. Two gang-mounted pots are shown in Figure 3.24. The pots on the left share a common control shaft. Therefore, when the value of one pot is varied, the value of the other is varied by the same amount. The gang-mounted pots on the right have separate control shafts. The outer shaft controls one of the pots while the inner control shaft controls the other. Normally, a knob is connected to the control shafts. When turned, the knob rotates the inner control shaft. When pushed in and turned, the knob rotates the outer control shaft.

FIGURE 3.24 Gang-mounted potentiometers.

Rheostats

A **RHEOSTAT** is a two-terminal variable resistor that is used to control current. The schematic symbol for a rheostat is shown in Figure 3.25a. In terms of its physical construction, a rheostat is typically a high-power resistor meaning the component can dissipate a significant amount of heat.

RHEOSTAT A two-terminal variable resistor that is used to control current.

It is not uncommon for a potentiometer to be wired as a rheostat. When wired as shown in Figure 3.25b, the resistance between the outer terminals of the pot (R_{AC}) is found as

$$R_{AC} = R_{AB}$$

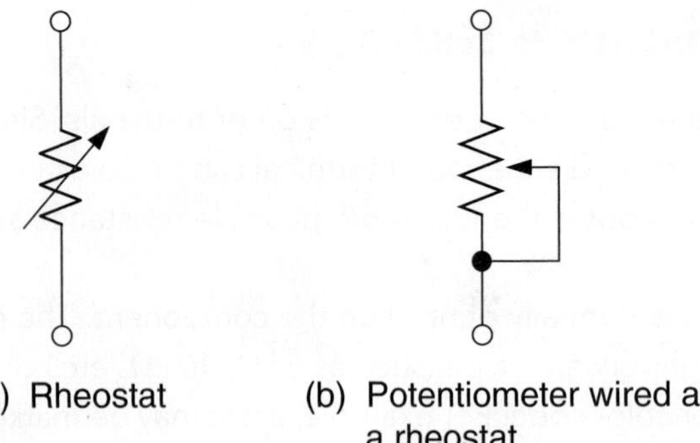

FIGURE 3.25 Rheostat symbol and potentiometer implementation.

By adjusting the wiper arm, the value of R_{AC} can be varied between approximately 0 Ω and the rated value of the potentiometer. (The potentiometer resistances described above are identified in Figure 3.20b.)

The Bottom Line

Potentiometers are three-terminal components whose resistance can be varied by the user. When the control shaft on a pot is turned, the resistance between the middle terminal (called the wiper arm) and each of the outer terminals is varied. The sum of the two resistances (R_{AB} and R_{BC}) always equals the resistance between the two outer terminals (R_{AC}).

The resistance of a linear-taper pot varies at a constant rate as the control shaft is rotated. The resistance of a nonlinear-taper pot varies at a changing rate as the control shaft is rotated. The resistance rating of a pot is usually printed on the component.

The control shaft of a multiturn potentiometer can be rotated more than 360°, making it easier to adjust its resistance to an exact value. Gang-mounted potentiometers share a common control shaft.

A rheostat is a two-terminal variable resistor that is used to control current. Rheostats are typically high-power components, which means they can dissipate a significant amount of heat. A potentiometer can be wired as a rheostat by connecting the wiper arm to one of the end terminals of the component.

PROGRESS CHECK

1. What is a potentiometer?
2. Briefly describe the operation of a potentiometer.
3. Describe the construction of a typical potentiometer.
4. Describe the relationship between the three resistances (R_{AB}, R_{BC}, and R_{AC}) of a potentiometer.
5. List and describe the two types of potentiometer taper.
6. What is a *multiturn potentiometer*? What advantage does a multiturn pot have over a standard potentiometer?
7. What is a rheostat? How is a potentiometer wired to function as one?

3.4 BATTERIES

In simple DC circuits, voltage is obtained from a **BATTERY**. A battery converts chemical, thermal, or light energy into electrical energy. Most common batteries produce a difference of potential (voltage) through a continuous chemical reaction. In this section, we discuss a variety of batteries and their characteristics.

BATTERY A device that converts chemical, thermal, or light energy into electrical energy.

Cells

A battery is made up of one or more cells. A **CELL** is a single unit that is designed to produce electrical energy through thermal (heat), chemical, or optical (light) means. In a typical battery, cells produce a voltage through chemical means. A basic cell is represented in Figure 3.26.

CELL A single unit in a battery that is designed to produce electrical energy through thermal (heat), chemical, or optical (light) means; most batteries contain multiple cells.

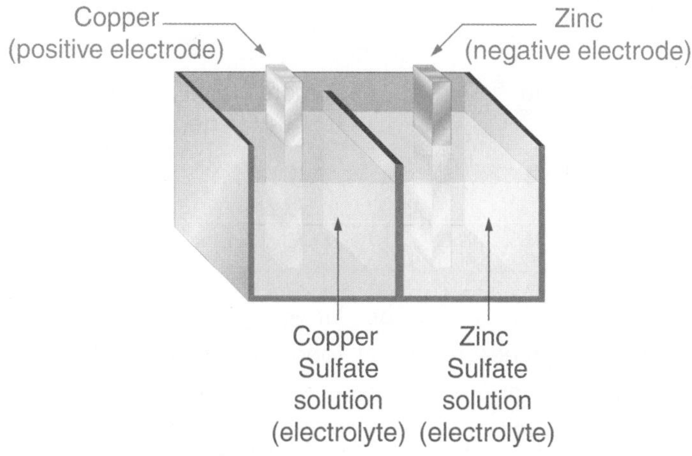

FIGURE 3.26 A chemical cell.

The cell shown in Figure 3.26 has two terminals that are called **ELECTRODES** and a chemical called the **ELECTROLYTE**. The electrodes are the battery terminals. Even though the electrodes in Figure 3.26 are identified as copper and zinc, other elements are commonly used as well. Each electrode is constantly in contact with the electrolyte, a chemical that interacts with the electrodes and serves as a conductor between them. A diaphragm (porous wall) prevents the copper sulfate and zinc sulfate from mixing, while allowing electrical charges to pass between them.

ELECTRODES The terminals of a chemical cell.

ELECTROLYTE A chemical that interacts with the electrodes and serves as a conductor between them.

As a result of the chemical interactions between the electrodes and the electrolyte, the negative electrode gains electrons and the positive electrode loses electrons. Thus, a difference of potential (voltage) is developed between the battery terminals.

The schematic symbol for a single cell is shown in Figure 3.27a. A battery may contain a single cell or a series of cells (as shown in Figure 3.27b). Note that the voltage across the terminals of the battery is equal to the sum of the cell voltages. Thus, the six 1.5 V cells in Figure 3.27b combine to form a 9 V battery. More commonly, a 9 V (or any other) battery is represented in a schematic diagram as shown in Figure 3.27c.

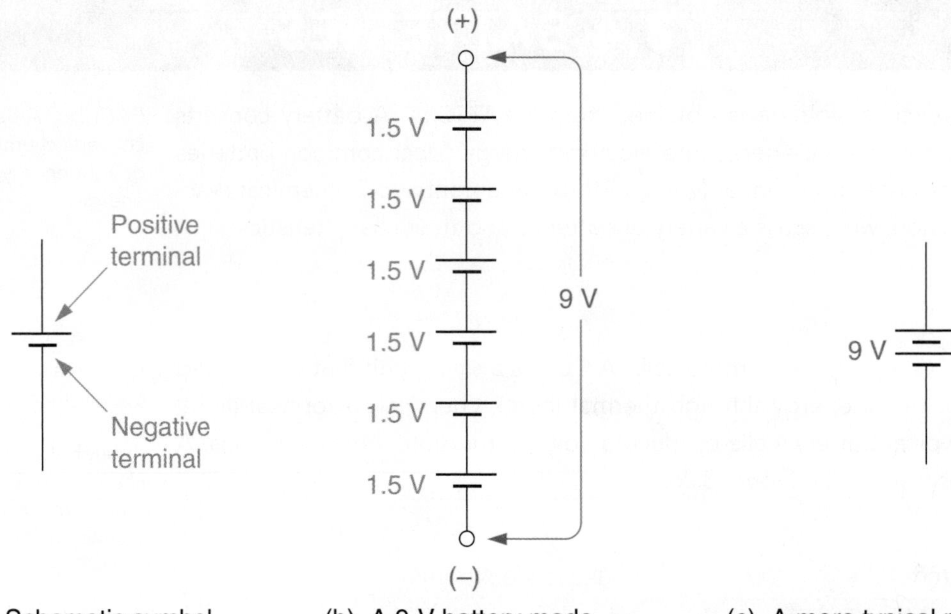

(a) Schematic symbol for a cell (b) A 9-V battery made of six 1.5-V cells. (c) A more typical representation of a 9-V battery.

FIGURE 3.27 Cells connected to form a 9-V battery.

Battery Capacity

CAPACITY A battery rating that indicates how long the battery will last at a given output current, measured in ampere-hours (Ah).

AMPERE-HOUR (Ah) The product of current and time.

Batteries are generally rated in terms of their output voltage and capacity. The **CAPACITY** rating of a battery is a measure of how long the battery will last at a given output current, measured in ampere-hours (Ah).

The **AMPERE-HOUR (AH)** is a unit of measure that equals the product of discharge current (the current supplied by the battery) and time. For example, a battery with a capacity of 1 Ah will last for:

1. 1 hour at a discharge rate of 1 A.
2. 2 hours at a discharge rate of 500 mA.
3. 10 hours at a discharge rate of 100 mA.
4. 100 hours at a discharge rate of 10 mA.

Note that the product of current (in amperes) and time (in hours) equals one (1) ampere-hour (Ah) in each of these examples. Also note that the battery lasts longer at lower output current values.

Typical capacity ratings vary with size and battery makeup. Some typical battery capacity ratings are approximately 2.4 Ah for AA-cells, 7.8 Ah for C-cells, and 12 Ah for D-cells.

Primary Cells

PRIMARY CELL A cell that cannot be recharged; also called a *voltaic cell* or *dry cell*.

In some cells, the chemical action between the electrolyte and the electrodes causes permanent changes in the chemical structure of the device. This type of cell is called a **PRIMARY CELL**. A primary cell is one that cannot be recharged. Primary cells are often referred to as *voltaic cells* or *dry cells* (because they contain dry electrolytes).

In general, batteries are named after one or more of their electrode or electrolyte elements (or compounds). Some commonly used batteries are described as follows:

- *Carbon-zinc* batteries contain one or more 1.5 V primary cells. They are extremely common and inexpensive, but have relatively low capacity (Ah) ratings. As a result, they drain relatively quickly in most applications. Carbon-zinc batteries are typically found in portable stereos, clocks, flashlights, and so on.
- *Alkaline* batteries are similar in many ways to carbon-zinc batteries. Alkaline batteries contain one or more 1.5 V primary cells. At the same time, they have relatively high capacity, long shelf life, and are relatively immune to the effect of mechanical stresses (such as vibrations).
- *Silver oxide* batteries have higher capacity (Ah) ratings than either carbon-zinc or alkaline batteries. This type of primary cell, which is rated at 1.55 V, is also more stable than carbon-zinc or alkaline batteries. As the battery drains, the voltage produced by the cell does not drop off as quickly as with other primary cells. These batteries are well suited for use in high-current devices, such as electronic shutter cameras.
- *Zinc-air* batteries have significantly higher capacity (Ah) ratings than other similar-sized batteries. A zinc-air battery actually interacts with oxygen in the air, using the oxygen as its positive electrolyte. As a result, the battery weighs much less than batteries of equal size, and must be kept sealed until used. The combination of high capacity and small size makes these batteries best suited for use in pagers and hearing aids.
- *Lithium* batteries are relatively small components that contain 3 V primary cells. Lithium batteries have high capacity (Ah) ratings and long shelf life. These batteries are typically used in high-current applications like those mentioned earlier.

Figure 3.28 shows several of the batteries described above.

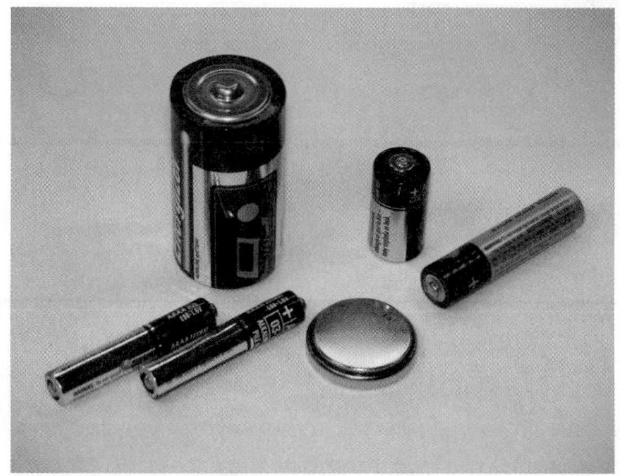

FIGURE 3.28 Primary (single-use) batteries.

SECONDARY CELL A cell that can be recharged; also called a *rechargeable cell* or *wet cell.*

Secondary Cells

A **SECONDARY CELL** is a type of cell that can be recharged. By generating a current through a secondary cell (in the opposite direction of the normal cell current), the chemical structure of the cell is restored, and the battery is recharged. Secondary cells are commonly referred to as *rechargeable cells* or *wet cells* (because they contain liquid electrolytes). The most commonly used rechargeable batteries are described as follows:

There are two types of secondary batteries: *Float* and *cycling*. Float batteries, such as car batteries, are designed to deliver high current for a short period of time while maintaining a near constant output voltage. A cycling battery, or *deep cycle* battery, is designed to deliver lower currents over much longer periods of time and experience a much greater decrease in terminal voltage (when discharging) than a floating battery. Deep cycle batteries are typically used in applications such as powering small electric vehicles (EV's).

IN THE FIELD

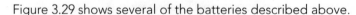

- *Lead-acid* batteries contain 2.1 V secondary cells. Lead-acid cells have relatively high capacity ratings (up to 20 Ah), and can last for hundreds of charge/discharge cycles. Typically, lead-acid batteries can be used for up to four years before having to be replaced.
- *Nickel-cadmium (Ni-Cd).* These batteries contain 1.15 V secondary cells. Ni-Cd cells have capacity ratings up to 5 Ah, and can last for hundreds of charge/discharge cycles. At equal capacity ratings, Ni-Cd cells are capable of higher output currents than lead-acid batteries and can be used for up to ten years before being replaced.
- *Nickel metal hydride and lithium ion* batteries are newer replacements for Ni-Cds. These batteries are smaller, lighter, and have higher capacity (Ah) ratings than comparable Ni-Cds.

Figure 3.29 shows several of the batteries described above.

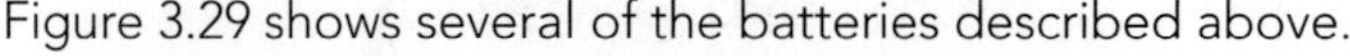

FIGURE 3.29 Secondary (rechargeable) batteries.

Connecting Batteries

In many applications, more than one battery is needed to supply the needed voltage or current. Figure 3.30 shows one of the two methods that are used to connect two or more batteries.

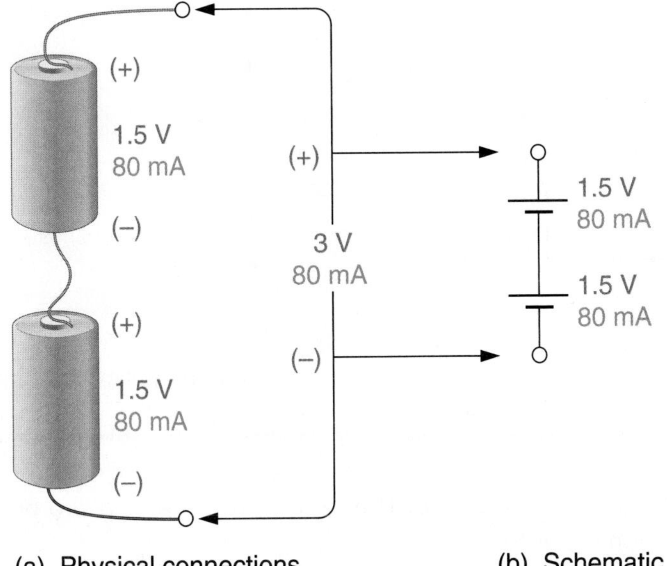

FIGURE 3.30 Series-connected batteries.

When two (or more) batteries are connected as shown in Figure 3.30, the batteries are said to be connected in *series*. Note that the negative terminal of one battery is connected to the positive terminal of the other. Here's what happens when two or more batteries are connected in this fashion:

1. The total voltage provided by the batteries is equal to the sum of their individual voltages.
2. The maximum current that can be supplied by the two does not increase. In other words, the maximum possible output current for the two is equal to the maximum possible output current for either battery alone.

Thus, the batteries in Figure 3.30 provide a 3 V output at a maximum current of 80 mA.

A practical application for series-connected batteries is shown in Figure 3.31. The flashlight holds two D-cells that are connected in series. These cells, rated at 1.5 V each, combine to generate 3 V to light the bulb.

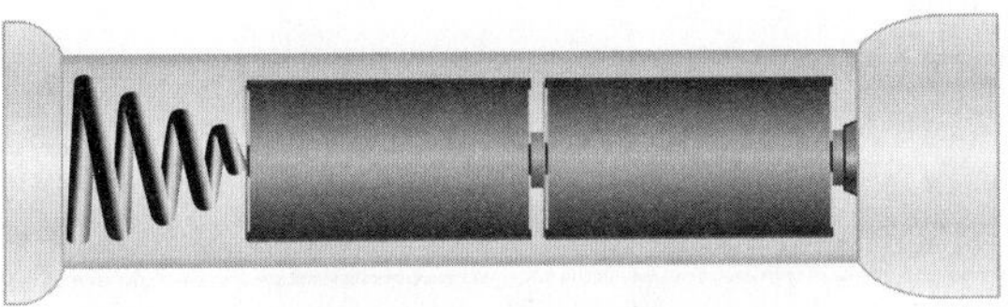

FIGURE 3.31 A flashlight containing series-connected batteries.

The other method of connecting two (or more) batteries is shown in Figure 3.32. When two or more batteries are connected in this fashion, the batteries are said to be connected in *parallel*. Note that the two batteries are connected

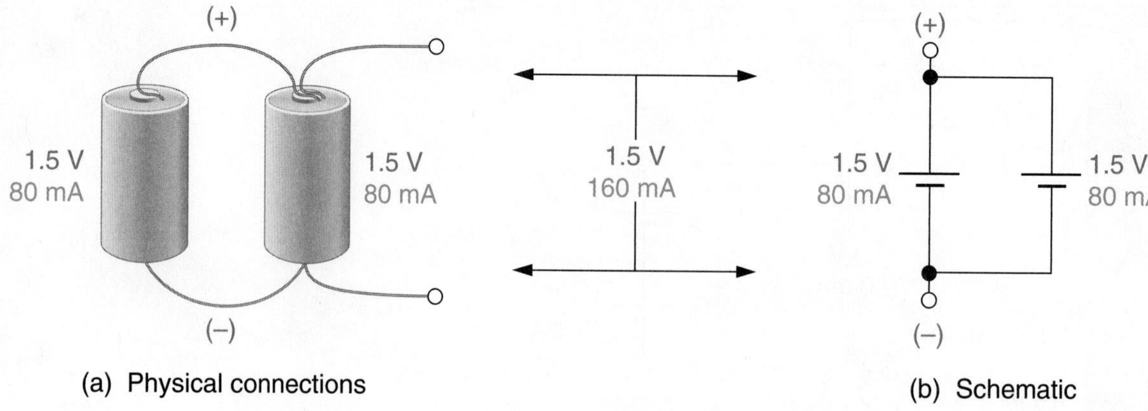

FIGURE 3.32 Parallel-connected batteries.

negative-to-negative and positive-to-positive. The results of connecting two or more batteries in this fashion are as follows:

1. The total voltage is equal to the individual battery voltages. In other words, the total battery voltage does not increase as a result of the connection.
2. The maximum possible output current is equal to the sum of the battery currents.

Thus, the connection shown in Figure 3.32 provides a 1.5 V output with a maximum output current of 160 mA.

The bases for the results of series and parallel battery connections are discussed in future chapters. However, the information presented here can help you understand why many devices (such as flashlights) contain more than one battery. If the batteries are connected in series, the purpose is to increase the total voltage. If they are connected in parallel, the purpose is to increase the maximum possible current.

PROGRESS CHECK

1. What is a battery?
2. What is a cell?
3. Briefly describe the operation of a cell.
4. How are cells connected to form a battery?
5. What are the differences between primary and secondary cells?
6. What is battery capacity?
7. What is the unit of measure for battery capacity?
8. What is the result of connecting batteries in series? In parallel?

3.5 DC POWER SOURCES

Even though batteries can provide DC voltages, their use in many circuits is impractical. For example, let's say we have a circuit that needs a 24 VDC input. In order to obtain this value with common "D" cells, we would have to connect 16 of them in series, which is not very practical.

In this section, we take a brief look at two devices that use an AC line input (120 VAC) to produce a DC output.

Variable DC Power Supplies

One method of obtaining a variety of DC voltages is to use a **DC POWER SUPPLY**. A DC power supply has variable DC outputs that can be adjusted to any value within its design limits. The front panel of a typical DC power supply is represented in Figure 3.33. DC power supplies are produced by a variety of manufacturers, and not all of them have the same features. However, the controls shown in Figure 3.33 are typical of those found on most DC power supplies.

DC POWER SUPPLY A device with variable DC outputs that can be adjusted to any value within its design limits.

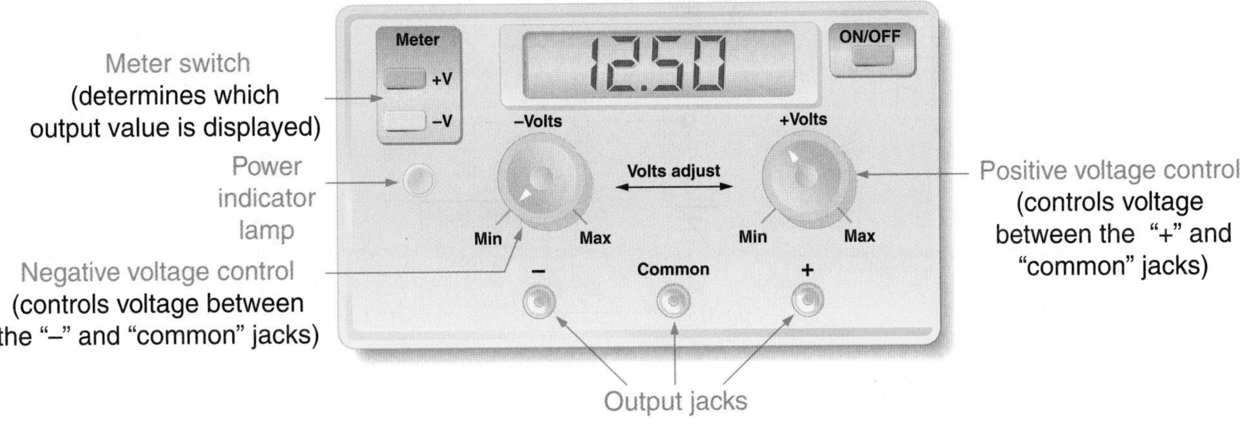

FIGURE 3.33 DC power supply control panel.

The DC Outputs

There are three DC outputs on the control panel represented in Figure 3.33. A simplified representation of these outputs is shown in Figure 3.34. The power supply contains two internal voltage sources with a **COMMON** connection point between the two. By turning the +Volts control, the output from voltage source A is varied. By turning the −Volts control, the output from voltage source B is varied.

COMMON A power supply terminal that is connected to both DC outputs.

Normally, the DC power supply is connected to a circuit in one of three fashions. The first is shown in Figure 3.35a. When connected as shown, the power supply output current takes the path indicated by the arrows. In this case, the voltage source (B) is not a part of the circuit, and the maximum power supply output voltage is 20 VDC.

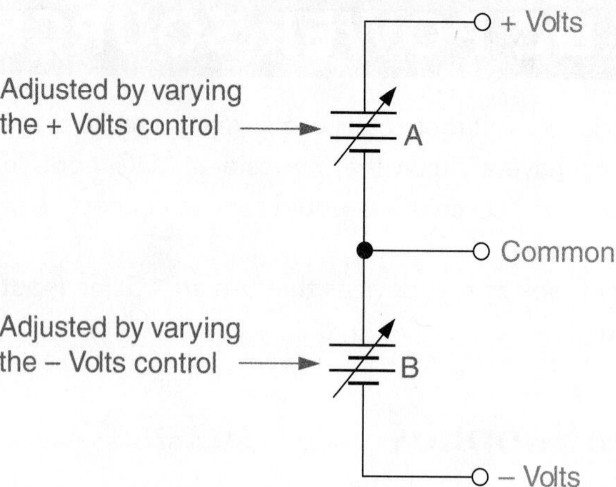

FIGURE 3.34 Simplified representation of a DC power supply.

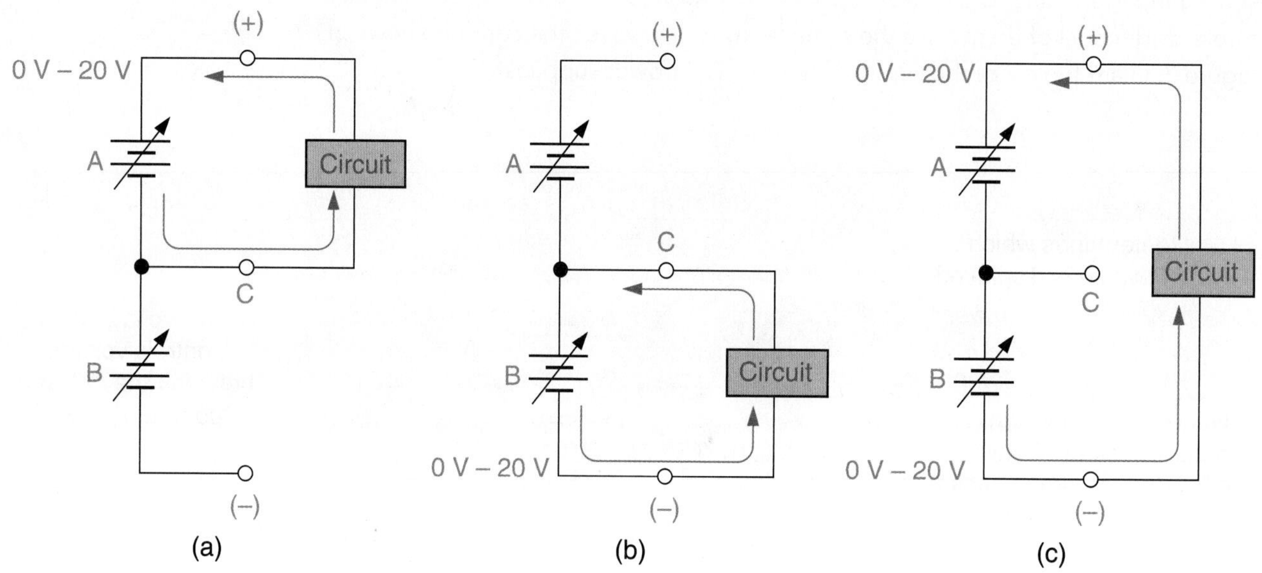

FIGURE 3.35 Power supply connections.

When connected as shown in Figure 3.35b, the power supply output current takes the path indicated by the arrows. In this case, the voltage source (A) is not a part of the circuit, and the maximum power supply output voltage is (once again) 20 VDC.

When connected as shown in Figure 3.35c, the power supply output current takes the path indicated. In this case, both voltage sources are a part of the circuit. Since the voltage sources are connected in series, the maximum power supply output voltage is equal to the sum of the maximum source voltages, 40 VDC.

In most cases, a DC power supply is connected as shown in Figure 3.35a or 3.35b. The connection shown in Figure 3.35c is only used when the required voltage exceeds the maximum possible output from either of the voltage sources.

Positive Voltage Versus Negative Voltage

Voltages are normally identified as having a specific **POLARITY**; that is, as being either *positive* or *negative.* For example, consider the positive and negative terminals of the voltage source shown in Figure 3.36. We could describe this source in either of two ways. We could say that:

1. *Side A is positive with respect to side B,* meaning that side A of the source is more positive than side B.
2. *Side B is negative with respect to side A,* meaning that side B of the source is more negative than side A.

POLARITY The electrical orientation of a voltage, identified as *negative* or *positive.*

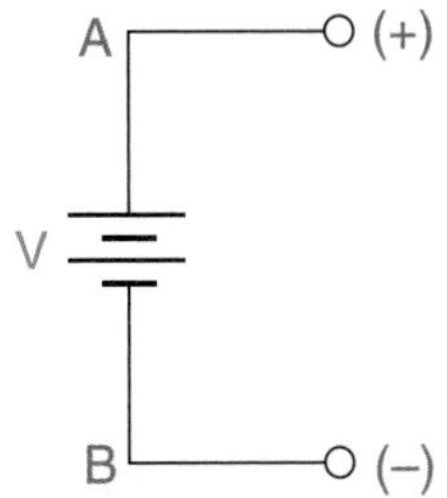

FIGURE 3.36

Both of these statements are correct, but they do not assign a specific polarity to the source. That is, we have not agreed on whether it is a positive voltage source or a negative voltage source.

In order to assign a polarity to any voltage, we must agree on a reference point. If we agree that side A of the voltage source in Figure 3.36 is the reference point, then we would say that the output is a **NEGATIVE VOLTAGE**, because side B is more negative than our reference point. If we agree that side B of the source is our reference point, then we would say that the output is a **POSITIVE VOLTAGE**, because side A of the source is more positive than our reference point.

NEGATIVE VOLTAGE A term indicating that a voltage is more negative than the circuit reference point.

POSITIVE VOLTAGE A term indicating that a voltage is more positive than the circuit reference point.

The *common terminal* of the DC power supply in Figure 3.33 is common to both voltage sources, and therefore, is used as the reference point. That is, the output polarity of either source is described in terms of its relationship to this point. For example, the negative terminal of source A is connected to the common point. Since the top side of A is positive with respect to the common (reference) point, the output from A is a positive DC voltage. By the same token, the positive side of source B is connected to the common, so the output from source B is a negative DC voltage. In each case, the output voltage polarity describes its relationship to the common terminal. Note that the common point is at 0 V.

Using a DC Power Supply

A DC power supply is used to provide power temporarily for various circuits. That is, the device is not used as a long-term DC voltage source. When connecting a circuit to a DC power supply:

1. Adjust the voltage controls on the DC power supply to their 0 VDC positions.
2. Connect the DC power supply to the circuit.
3. Turn on the DC power supply.
4. Slowly increase the output from the power supply to the desired value.

AC Adaptor

A fixed-output **AC ADAPTOR** is a circuit that converts an AC line voltage to the DC operating voltage required to power an electronic system like a laptop computer or printer. Unlike the DC power supply discussed earlier, an AC adaptor provides a

AC ADAPTOR A device that converts an AC line voltage to the DC operating voltage required to power an electronic system.

fixed output voltage that cannot be varied. For example, the AC adaptor shown in Figure 3.37 provides an output of 19 V at 3.42 A.

FIGURE 3.37 An AC adaptor.

AC adaptor circuits are sealed in a case that prevents the user from accessing the internal circuitry. Consequently, a faulty AC adaptor is replaced by a new one, rather than repaired.

PROGRESS CHECK

1. What is a DC power supply?
2. What determines the *polarity* of a voltage?
3. What is the *common* terminal of a power supply?
4. List the steps that you should take when connecting a DC power supply to a circuit.
5. What is the primary difference between an AC adaptor and a DC power supply?

3.6 SWITCHES AND CIRCUIT PROTECTORS

In any circuit, there must be a complete path between the terminals of the voltage source. A complete path is required to allow the flow of charge (current) between the voltage source terminals. Switches and circuit protectors (such as fuses and circuit breakers) are devices that make or break the current path in a circuit under certain circumstances.

SWITCH A component that allows you to manually make or break the connection between two or more points in a circuit.

Switches

A **SWITCH** allows you to manually make or break the connection between two or more points in a circuit. A switch is connected in a circuit as shown in Figure 3.38.

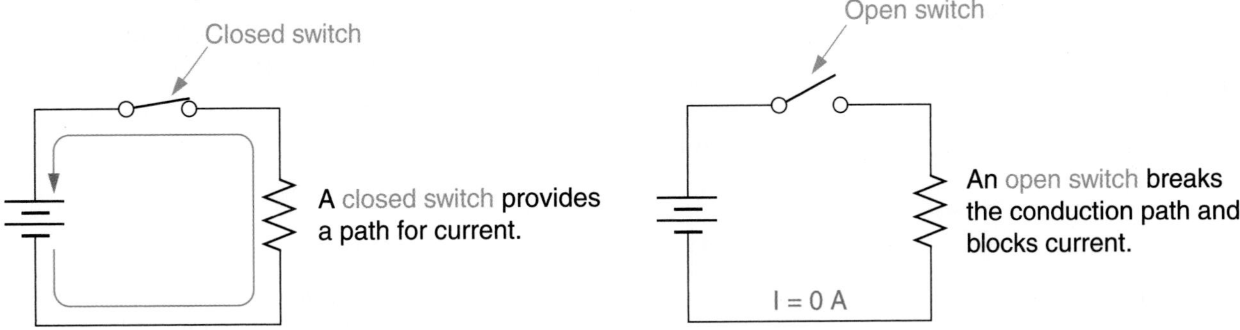

FIGURE 3.38 Closed and open switches.

A path for current is made when the switch is closed. The current path is broken when the switch is opened.

A switch is made up of one or more moving contacts, called **POLES**. A single-pole switch, like the one represented in Figure 3.38, has only one moving contact. A double-pole switch, like the one represented in Figure 3.39, has two moving contacts.

POLE The moving contact(s) in a switch.

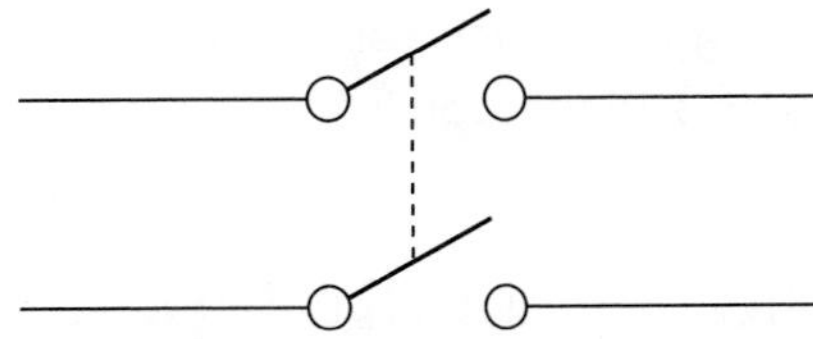

Note: The dashed line in the symbol indicates that the poles are not independent. Both are open or both are closed.

FIGURE 3.39 A double-pole switch.

The switches represented in Figures 3.38 and 3.39 can only be closed in one position. That is, they can only make or break a connection with a single non-moving contact. The non-moving contact is referred to as a **THROW**, and the switches shown are referred to as *single-throw* switches. In contrast, several *double-throw* switches are shown in Figure 3.40. Both of these switches have a moving contact that can be made or broken with either of two nonmoving contacts.

THROW The stationary contact(s) in a switch.

Switches are generally described in terms of the number of poles and throws they contain. For example, the switch represented in Figure 3.38 is described as a *single-pole, single-throw* (SPST) switch, meaning that it has a single moving contact that can be made (closed) in only one position. The switch in Figure 3.39 is described as a *double-pole, single-throw* (DPST) switch, meaning that it has two moving contacts that can be made (closed) in only one position. The switches shown in Figure 3.40 are the **single-pole, double-throw (SPDT)** switch and the **double-pole, double-throw (DPDT)** switch.

Any of the switches shown up to this point can easily be identified by the number of **TERMINALS** (wire connection points) on its case, as follows:

TERMINALS Wire connection points.

1. A SPST switch has *two* terminals.
2. A SPDT switch has *three* terminals.
3. A DPST switch has *four* terminals.
4. A DPDT switch has *six* terminals.

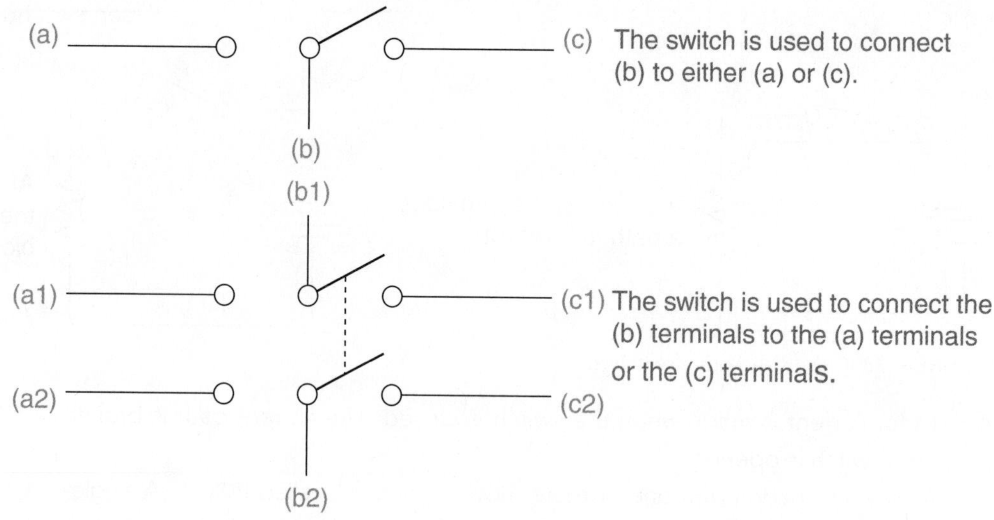

FIGURE 3.40 Two double-throw switches.

The number of terminals for a given switch can be seen by taking another look at the switches shown in Figures 3.38 through 3.40.

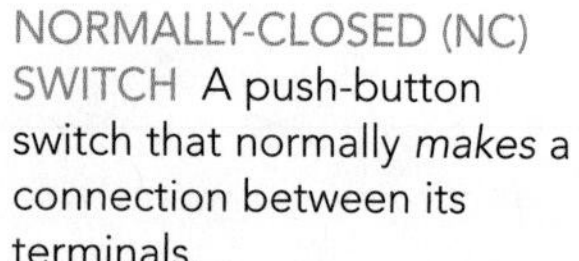
NORMALLY-CLOSED (NC) SWITCH A push-button switch that normally *makes* a connection between its terminals.

NORMALLY-OPEN (NO) SWITCH A push-button switch that normally *breaks* the connection between its terminals.

MOMENTARY SWITCH A switch that makes or breaks a connection only as long as the button is pushed.

FIGURE 3.42 The doorbell switch is a momentary switch.
Fotolia.com

Normally-Closed and Normally-Open Switches

Some switches are designed so that they normally make or normally break a connection. For example, the **NORMALLY-CLOSED (NC) SWITCH** is a push-button switch that normally *makes* a connection between its terminals. When the button is pushed, the switch breaks the connection. The schematic symbol for a normally-closed switch is shown in Figure 3.41a.

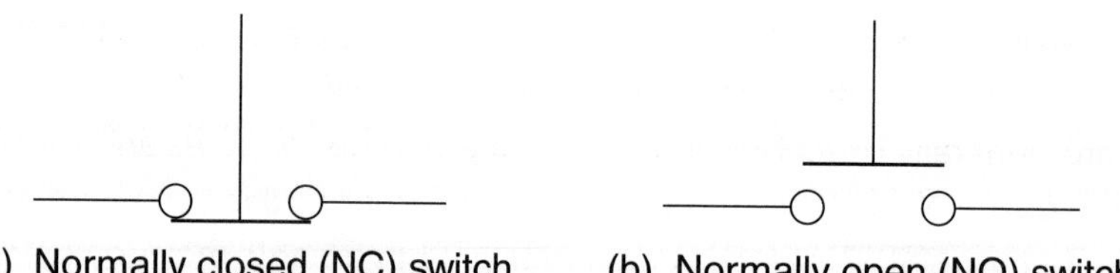

FIGURE 3.41 Normally-closed (NC) and normally-open (NO) switches.

The **NORMALLY-OPEN (NO) SWITCH** is the opposite of the NC switch. That is, the current path between its terminals is normally open. When the button is pushed, the switch makes the connection. The schematic symbol for this type of switch is shown in Figure 3.41b.

Some NC and NO switches are classified as **MOMENTARY SWITCHES**. That is, the connection is made or broken only as long as the button is pushed. Others maintain the made or broken connection until the button is pressed again. The doorbell switch shown in Figure 3.42 is an example of a momentary switch.

Rotary Switches

The simplest explanation of a **ROTARY SWITCH** is that it is a switch with one or more poles and any number of throws. The schematic symbol for a rotary switch is shown in Figure 3.43. By turning the control shaft, a connection is made between the pole and one of the throws.

ROTARY SWITCH A switch with one or more poles and multiple throws.

(a)

(a)
(b)
(c)
(d)

(b) Schematic symbol

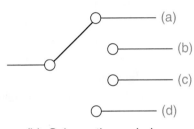

FIGURE 3.43 A rotary switch and its schematic symbol.
(a) Elma Corporation

Fuses

FUSE A component that allows current to pass under normal circumstances and opens automatically if the current exceeds a specified value.

CARTRIDGE FUSE A fuse that contains a thin conductor that melts at relatively low temperatures.

In some ways, a **FUSE** is like a normally-closed switch. A fuse is designed to allow current to pass under normal circumstances. However, unlike a normally-closed switch, a fuse is designed to open automatically if the current exceeds a specified value. The simplest type of fuse and its schematic symbols are shown in Figure 3.44.

As current passes through a conductor, heat is produced. The amount of heat varies directly with the value of current. That is, as current increases, so does the amount of heat it produces.

The fuse in Figure 3.44 is called a **CARTRIDGE FUSE**. It contains a thin conductor that is designed to melt at relatively low temperatures. If the current through the component reaches a specified level, the conductor heats to its melting point. When it melts, the connection between the ends of the fuse is broken, the fuse is destroyed, and it effectively acts as an open switch.

The purpose of a fuse is to protect a circuit or system from excessive current. For this reason, a fuse is normally placed between a power source and the circuit or system it is designed to protect, as shown in Figure 3.45. As you can see, the total circuit current provided by the power source in Figure 3.45 passes through the fuse. As long as the fuse is good, the current path is complete. However, if a problem develops that causes the circuit current to exceed the rating of the fuse, the component opens and the current path is broken. This protects the circuit from excessive current.

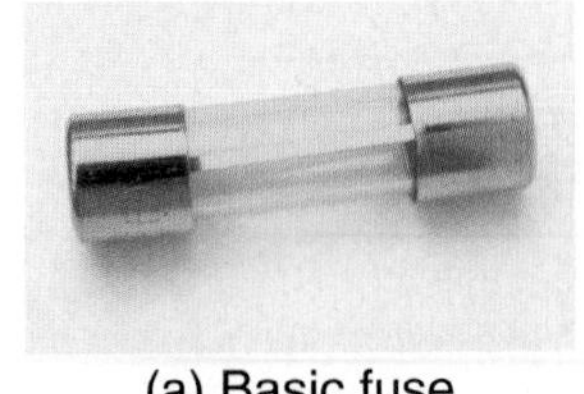

(a) Basic fuse

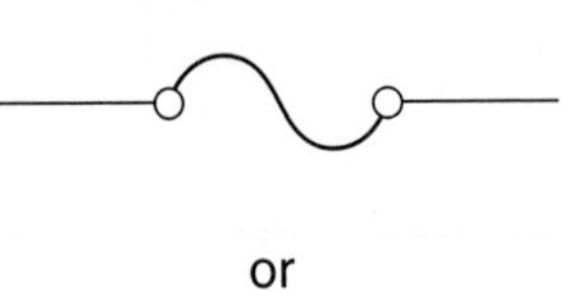

(b) Schematic symbols

FIGURE 3.44 A cartridge fuse.

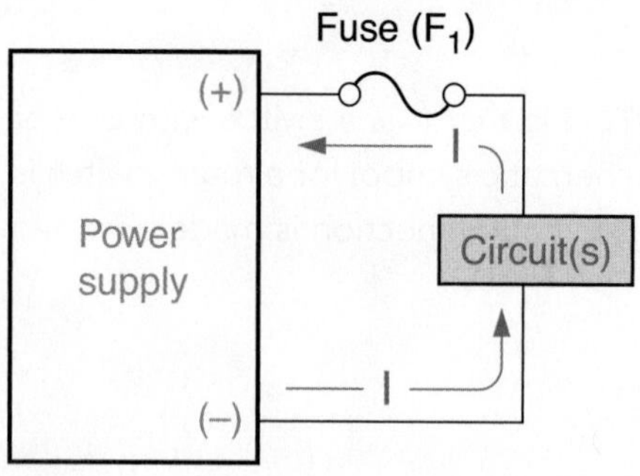

FIGURE 3.45 Fuse position in a basic circuit.

Fuse Ratings

Fuses have a current rating and a voltage rating. The *current rating* of a fuse is the maximum allowable fuse current. If the current rating of a fuse is exceeded, the fuse will "blow" (open).

In order for you to understand the voltage rating of a fuse, we must establish one of the characteristics of an *open circuit.* In an open circuit, the applied voltage is felt across the open. For example, consider the circuit shown in Figure 3.46a. The applied voltage has a value of 50 VDC. A difference of potential exists between points A and B. This difference of potential is equal to the applied voltage, 50 VDC. (The basis of this principle is covered in Chapter 4. For now, we are interested only in what happens when a circuit opens. Later, you will be shown why it happens.)

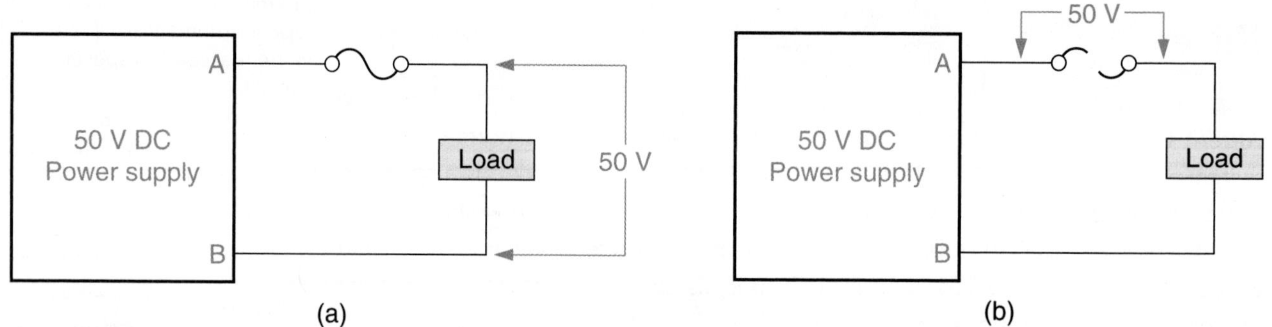

FIGURE 3.46 Circuit voltages with a good fuse and an open fuse.

The *voltage rating* of a fuse is the maximum amount of voltage that an open fuse can block. The explanation of this rating starts with the circuit in Figure 3.46b. As you can see, the fuse in the circuit has opened. Once it has opened, the applied voltage (50 VDC) is felt across the fuse.

You may recall that the *average breakdown voltage* of an insulator is the difference of potential that will cause the insulator to break down and conduct. The voltage rating of a fuse is basically the same thing. If the voltage across an open fuse exceeds its voltage rating, the air in the fuse may **IONIZE** (charge), causing the open fuse to conduct again. For example, consider the two circuits in Figure 3.47. In Figure 3.47a, the applied voltage is less than the voltage rating of the

IONIZE The process of becoming electrically charged.

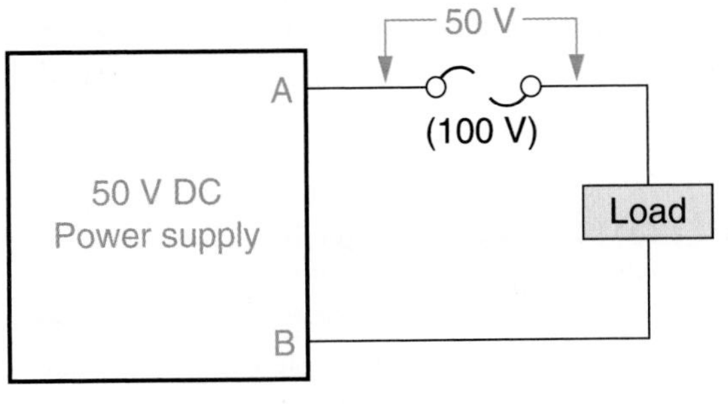

(a) The fuse blocks current

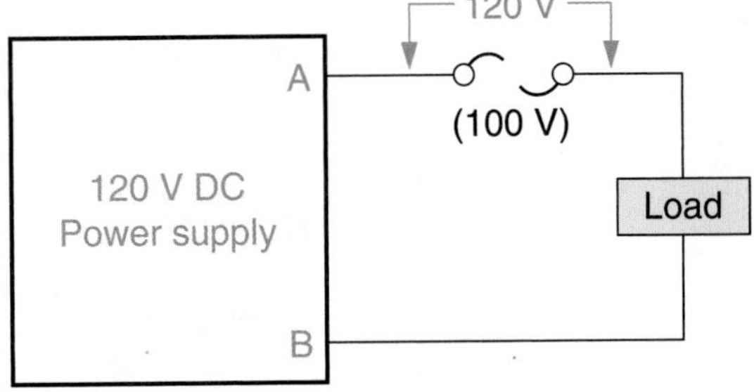

(b) The fuse no longer blocks current

FIGURE 3.47 Voltages across open fuses.

fuse. In this case, the open fuse does not allow conduction, and the circuit is protected. However, the applied voltage in Figure 3.47b is greater than the voltage rating of its fuse. Because of this, the open fuse ionizes and conducts. The result is that the circuit is no longer protected.

STAYING SAFE Before replacing a fuse both ends of the fuse should be tested to verify that the circuit is de-energized. Power is still present at the input side of a blown fuse until the circuit is de-energized. If a person comes in contact with both ends of a blown fuse in an energized circuit, they complete the circuit and can be severely injured.

Replacing a Fuse

When a fuse is blown, it must be replaced. When replacing a fuse, these guidelines must be followed:

1. Make sure that the circuit is de-energized. For example, if you're replacing the fuse for a central air conditioning unit, make sure that the unit is switched off at the thermostat.
2. Use an exact replacement. That is, use a fuse that is the same size and has the same current rating as the original fuse.
3. *Never replace a fuse with one that has a higher current rating!*

The first of these guidelines is designed to protect both you and the circuit. Replacing a fuse with the system energized can result in a surge of current. Not only could this surge hurt you, it could also severely damage the system. The second guideline needs no explaining. A replacement fuse should never allow a higher current than the original fuse.

NOTE: You can replace a fuse with one having a higher voltage rating than the original. For example, a 10 A/240 V fuse can be used in place of a 10 A/120 V fuse.

The third guideline is critical. *If you replace a fuse with one that has a higher current rating, you run the risk of damaging the circuit and potentially starting a fire.* As you know, current produces heat. If you replace a fuse with one that has a higher current rating, a problem could cause more current to be drawn through the circuit than it was designed to handle. The resulting heat could destroy the circuit and could even result in a fire.

Once a fuse has been replaced, one of two things will happen:

1. The circuit operation will return to normal, or
2. The new fuse will blow when power is restored.

If the circuit operation returns to normal, then the problem was likely nothing more than an old or defective fuse. If the new fuse blows, then there is a problem that is causing excessive current to be drawn from the power source. In this case, the problem must be diagnosed and repaired before attempting to replace the fuse again.

Types of Fuses: Electronic Applications

Three types of fuses are found in electronic systems: high-speed instantaneous, normal instantaneous, and time delay. As you have probably guessed, these categories rate fuses in terms of the time required for them to blow.

OVERLOAD An abnormal excessive-current condition.

High-speed instantaneous fuses react faster to an **OVERLOAD** (excessive current) condition than any other type of fuse. High-speed instantaneous fuses are also known as *fast-acting* or *quick-acting* fuses.

Normal instantaneous fuses are somewhat slower to respond to an overload condition. Normal instantaneous fuses are also known as *instantaneous* or *normal-blow* fuses.

Time delay fuses have the same time lag ratings as normal-blow fuses, but they can handle short-term overloads without blowing. Many applicances (such as microwave ovens) produce a current surge when they turn on. If a time delay fuse is in the circuit of such an appliance, the current surge will not blow the fuse. Note that time delay fuses are also known as *lag* or *slow-blow fuses*.

Fuses come in a variety of types and sizes, as described in Table 3.4. Note that the type, current, and voltage ratings are commonly identified on the body of the fuse.

Table 3.4 • Common Cartridge Fuses

TYPE	DESCRIPTION	TYPICAL CURRENT RATINGS AND RANGES
AGC	Standard blow	250 mA to 30 A
MDL	Slow blow	100 mA to 10 A
GJV	Standard blow WP[a]	250 mA to 10 A
SBP	Slow blow WP	500 mA to 7 A
ABC	Ceramic[b]	10 A, 12 A, 15 A, and 20 A
GMA	Standard blow	125 mA to 10 A
MDX	Slow blow	250 mA to 8 A
GMP	Standard blow WP	250 mA to 6 A

[a]WP stands for *with pigtails*. Pigtails are wires that are soldered to the ends of the components.
[b]Used in microwave applications.

Types of Fuses: Electrical Applications

Electrical fuses are contained in fuse boxes like the one shown in Figure 3.48. When a working fuse is installed, it connects the AC line input to the rest of the circuit.

FIGURE 3.48 A fuse panel.

There are several types of fuses: cartridge, type T, type S, and blade. The fuses described previously are all cartridge fuses. As shown in Figure 3.44, a cartridge fuse is one that has a tube-shaped body.

A **TYPE T FUSE** has an *Edison base*. As shown in Figure 3.49a, the Edison base resembles the base of a light bulb. This type of fuse screws directly into a standard fuse socket.

TYPE T FUSE A fuse with an *Edison base* (a base similar to the base of a light bulb).

(a) Type T (Edison base) fuses

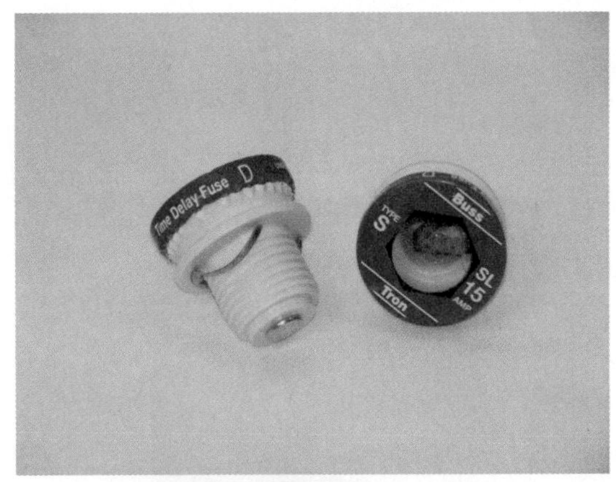

(b) Type S (rejection base) fuses

FIGURE 3.49 Common electrical fuses.

A **TYPE S FUSE** is shown in Figure 3.49b. The base on the type S fuse is called a *rejection base*. A rejection base requires the use of an adaptor to fit into a standard fuse socket. Once a rejection base adaptor is installed in a fuse socket, it cannot be removed.

TYPE S FUSE A fuse with a *rejection base* (one that requires the use of an adaptor to fit into a standard fuse socket).

Type S fuses with different current ratings have different threads. For example, a 30 A fuse will not fit in the socket adaptor for a 15 A fuse. This makes it impossible for a type S fuse to be replaced by one with a different current rating. Note that rejection bases are also referred to as *tamperproof bases*. The basic characteristics of type T and type S fuses are summarized in Table 3.5.

Table 3.5 • Type T and Type S Fuses

TYPE	DESCRIPTION	VOLTAGE / CURRENT RATINGS
T, S	Time delay/Heavy Duty	120 V/Up to 30 A
TL, SL	Time delay/Medium Duty	120 V/Up to 30 A

Blade fuses are commonly used in automotive applications. Several blade fuses and a fuse holder are shown in Figure 3.50.

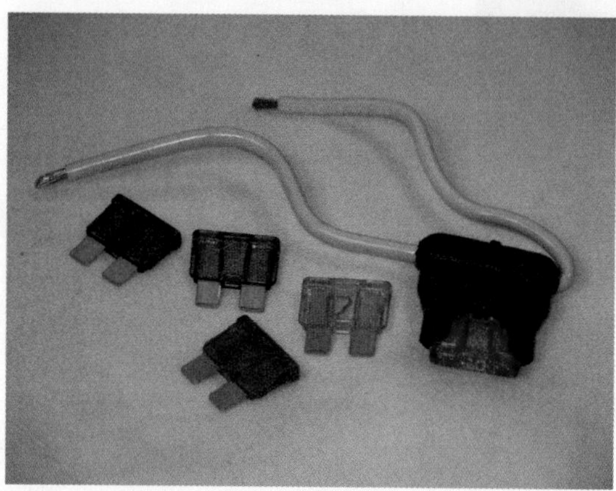

FIGURE 3.50 Blade fuses and fuse holder.

Circuit Breakers

Circuit breakers are designed to protect circuits from overload conditions, just like fuses. However, unlike fuses, circuit breakers are not destroyed when activated. A typical circuit breaker and its schematic symbol are shown in Figure 3.51.

(a) Circuit breaker

(b) Schematic symbol

FIGURE 3.51 A circuit breaker and its symbol.

In essence, a circuit breaker acts as a normally closed switch. When an overload condition develops, the circuit breaker opens, and the circuit is protected. Flipping the reset switch on the breaker returns it to its normally closed position, and the device can then be used again.

There are three basic types of circuit breakers. A *thermal-type* circuit breaker responds to the heat that is generated by an overload condition. The *magnetic-type* responds to a magnetic field that is produced by an overload condition. (The relationship between current and magnetism is covered in Chapter 10.) The *thermomagnetic-type* responds to both heat and a magnetic field, and is the circuit breaker equivalent of a slow-blow fuse.

The Bottom Line

Switches, fuses, and circuit breakers allow us to make or break the conduction path in a circuit under certain circumstances. Switches are manual devices that allow us to manually make or break a circuit. Fuses and circuit breakers normally allow conduction through a circuit. If an *overload* condition develops, a fuse or circuit breaker will open and interrupt the circuit path. In doing this, they protect a circuit or system from the excessive heat that can be produced in an overload condition.

PROGRESS CHECK

1. What is a switch?
2. What is a circuit protector?
3. What is a pole? What is a throw?
4. List and describe the four basic types of switches. State how each can easily be identified.
5. What is a *normally open* switch? A *normally closed* switch?
6. What is a *momentary* switch?
7. Describe rotary switches.
8. What is a fuse?
9. Describe the operation of a fuse.
10. List and describe the common fuse ratings.
11. List the guidelines for changing a fuse.
12. Compare and contrast type T and type S fuses.
13. What is a *circuit breaker*?
14. What is the primary difference between a circuit breaker and a fuse?

3.7 ANALOG AND DIGITAL MULTIMETERS

A **MULTIMETER** is an instrument used to measure current, voltage, and resistance. Some multimeters measure additional electrical properties (capacitance, for example), but they all measure amperes, volts, and ohms.

There are two types of meters: analog and digital. An **ANALOG MULTIMETER** uses a pointer that moves over a stationary scale to indicate the result of a measurement. A **DIGITAL MULTIMETER**, on the other hand, uses a digital readout to display the result of a measurement. An analog display and a digital display are shown in Figure 3.52.

STAYING SAFE Before connecting a meter or other circuit tester to any circuit, make sure the tester is designed to handle a voltage that is greater than any in the circuit being tested.

MULTIMETER An instrument used to measure current, voltage, and resistance.

ANALOG MULTIMETER A meter with a pointer that moves over a stationary scale to indicate the result of a measurement.

DIGITAL MULTIMETER A meter that uses a digital readout to display the result of a measurement.

NOTE: Analog meters cannot be used to measure alternating current (AC).

FIGURE 3.52 A digital meter (left) and an analog meter (right).
Source: (Left) Courtesy of Altadox, Inc. (www.altadox.com); (Right) Photo courtesy of Simpson Electric Co.

Reading an Analog Multimeter Display

The display on an analog multimeter contains several scales for displaying measured values. The meter display in Figure 3.53 is a typical analog multimeter display.

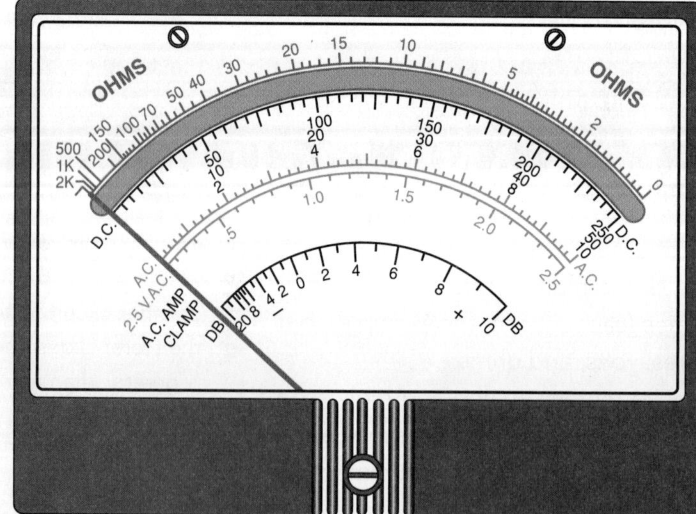

FIGURE 3.53 An analog meter display.

NOTE: The wide red space below the "OHMS" scale in Figure 3.53 normally contains a reflective (mirrored) surface.

The uppermost (red) scale is used to display resistance readings. The scale below it (black) is used to display DC voltage and current readings. The strip between the resistance and DC scales is a mirror that is used to prevent a specific type of error, called *parallax error*. (Parallax error is discussed further in a moment.) The bottom scale (green) is used to display AC voltage readings.

Function select switch

The function select switch on an analog multimeter is used to set the meter for the type of measurement being taken. A typical function select switch is shown on the front of the meter in Figure 3.54. This switch allows the user to select the measurement type and value range.

FIGURE 3.54 A meter function select switch
Fotolia.com.

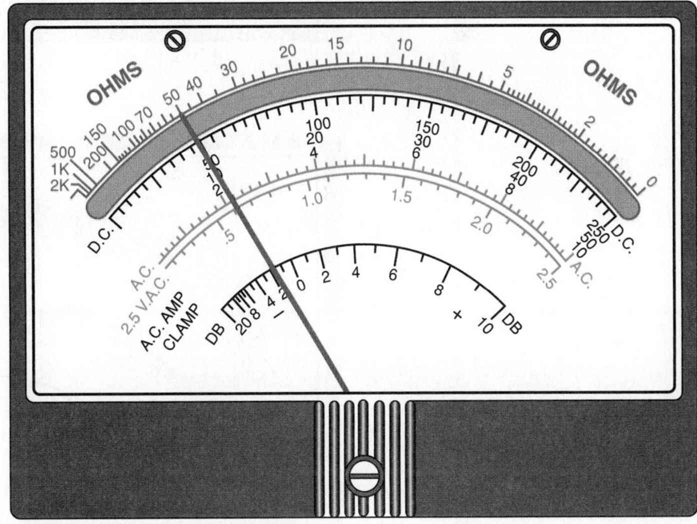

FIGURE 3.55 An example meter display.

Resistance readings

The value of a resistance reading is determined by the position of the pointer on the resistance scale and the setting of the function select switch. Figure 3.55 shows a meter display with the pointer positioned over the number 50. Assuming the function select switch is set to the ×10 ohms position, the combination of the meter reading and the resistance multiplier gives us

$$R = 50 \times 10\ \Omega = 500\ \Omega$$

Example 3.10 demonstrates how a resistance reading is interpreted.

NOTE: Analog resistance meters have a *zero adjust* control that is used to calibrate the meter. This control must be adjusted every time you switch from one resistance scale to another. The steps taken to zero-adjust a meter is outlined in its manual.

EXAMPLE 3.10

The meter display in Figure 3.56a was obtained with the function select switch set to the ×1 Ω position. What is the value of the resistance being measured?

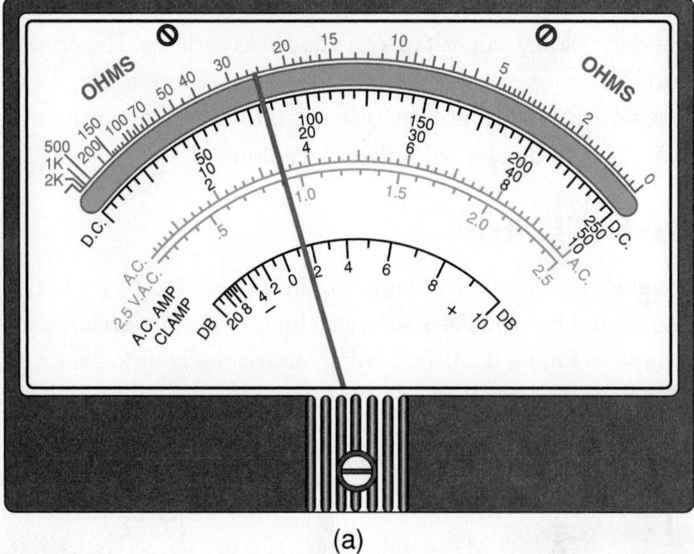

(a)

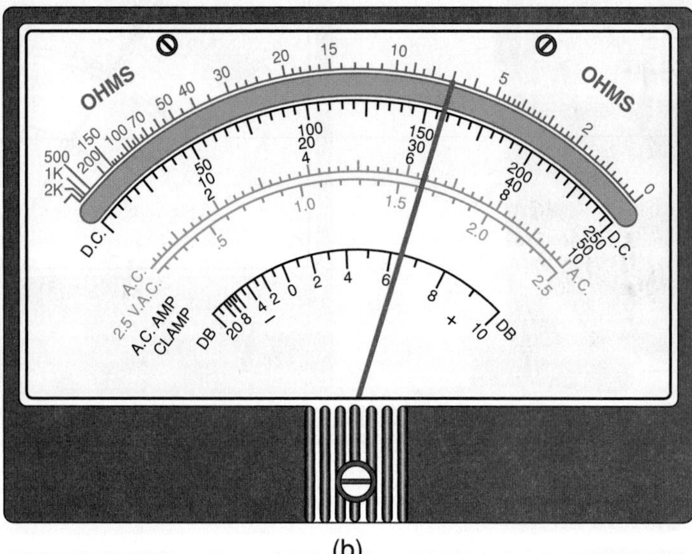

(b)

FIGURE 3.56

SOLUTION

The meter display in Figure 3.56a was obtained with the function select switch set to the $\times 1\ \Omega$ position. The pointer is positioned at 25 on the ohms scale. Therefore, the resistance being measured has a value of

$$R = 25 \times 1\ \Omega = 25\ \Omega$$

PRACTICE PROBLEM 3.10

The meter display in Figure 3.56b was obtained with the function select switch set to the $\times 1$ kΩ scale. What is the value of the resistance being measured?

DC Voltage and Current Readings

The value of a DC voltage or current reading is determined by the position of the pointer on the DC scale and the setting of the function select switch. Figure 3.57 shows a meter display. The value represented by the display depends on the position of the function select switch.

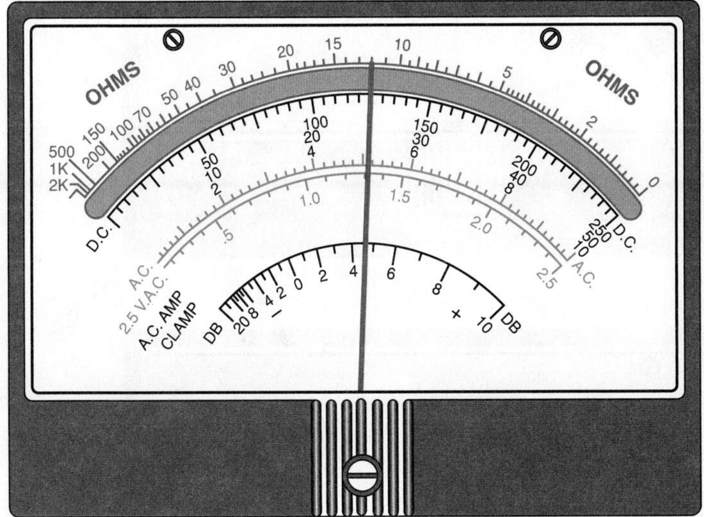

FIGURE 3.57 Analog meter display.

The DC scales indicate *full scale deflection* values; that is, the values represented when the pointer is positioned at the right end of the scale. For example, when the 50 V function is selected, the labeled divisions on the DC scale represent 10 V, 20 V, 30 V, 40 V, and 50 V (from left to right). However, when the 500 V function is selected, the same increments represent 100 V, 200 V, 300 V, 400 V, and 500 V. The following examples will help you better understand how the DC scales work.

EXAMPLE 3.11

The meter display in Figure 3.58a was obtained with the function select switch set to the 250 V position. What is the value of the voltage being measured?

SOLUTION

The meter display in Figure 3.58a was obtained with the function select switch set to the 250 V position. The labeled divisions represent 50 V increments when the 250 V scale is being used. The needle is positioned halfway between the 100 V and 150 V increments, so the voltage being measured is 125 V.

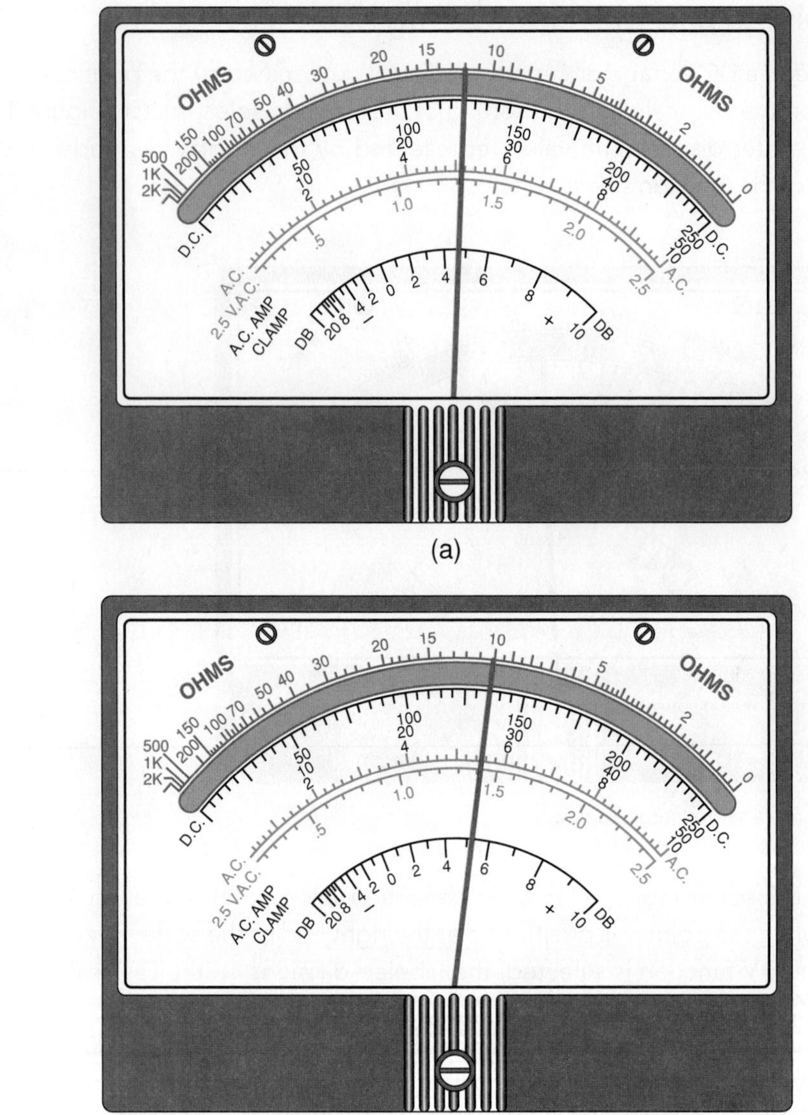

FIGURE 3.58

PRACTICE PROBLEM 3.11

The meter display in Figure 3.58b was obtained with the function select switch set to the 10 V position. What is the value of the voltage being measured?

In some cases, the scales represent voltage increments that are smaller than those shown. This is demonstrated in Example 3.12.

EXAMPLE 3.12

The meter display in Figure 3.59a was obtained with the function select switch set to the 1 V position. What is the value of the voltage being measured?

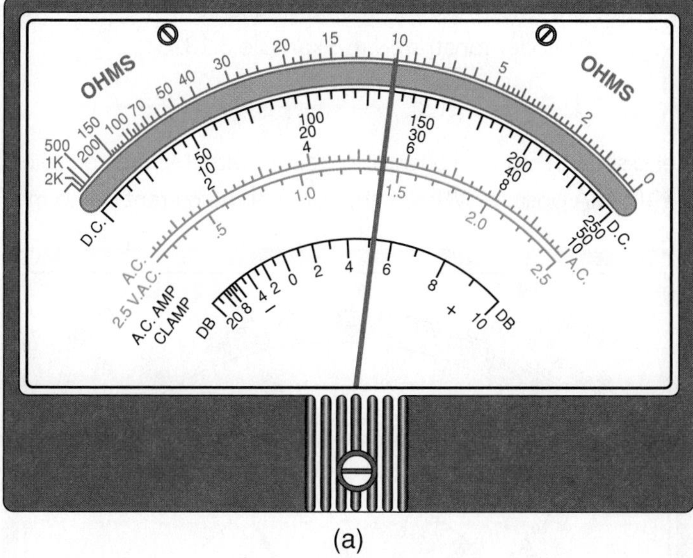

(a)

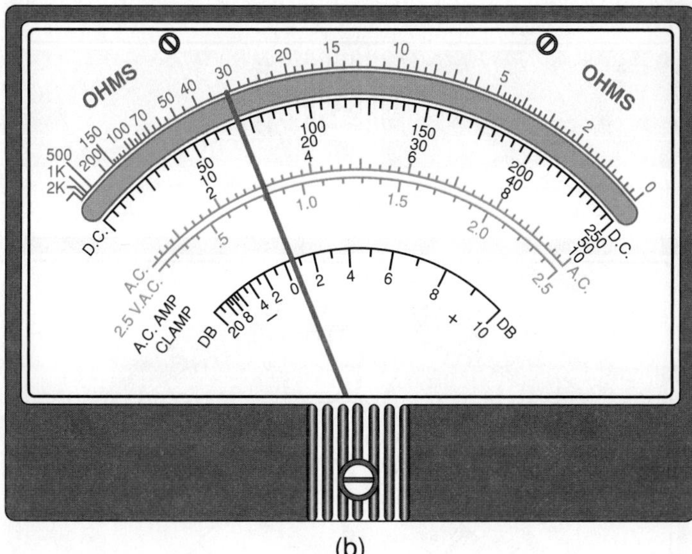

(b)

FIGURE 3.59

SOLUTION

When the 1 V scale is used, the lowest set of labels for the DC scale represent the following values (from left to right): 0.2 V, 0.4 V, 0.6 V, 0.8 V, and 1.0 V. With the needle pointed as shown, the voltage being measured is 0.54 V.

PRACTICE PROBLEM 3.12

The meter display in Figure 3.59b was obtained with the function select switch set to the 2.5 V position. What is the value of the voltage being measured?

The DC scales are interpreted in the same fashion when the meter is set for *current* readings. This is demonstrated in Example 3.13.

EXAMPLE 3.13

The meter display in Figure 3.60a was obtained with the function select switch set to the 100 mA position. What is the value of the current being measured?

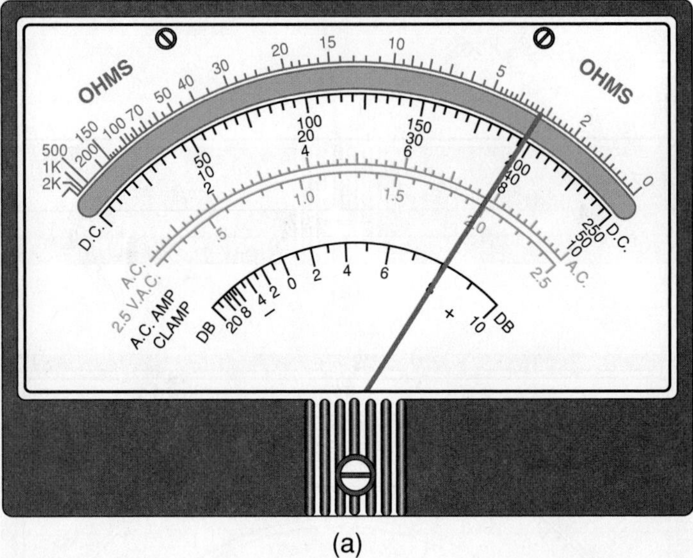

(a)

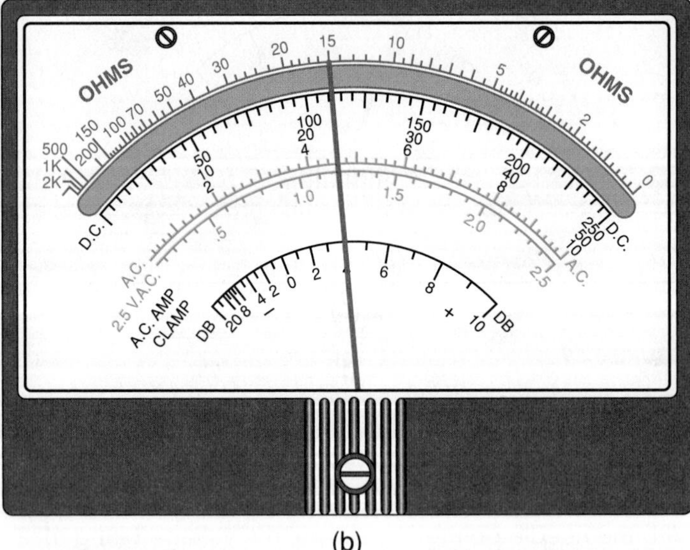

(b)

FIGURE 3.60

SOLUTION

When the function select is set to this position, the lowest set of labels for the DC scale represent the following values (from left to right): 20 mA, 40 mA, 60 mA, 80 mA, and 100 mA. With the needle pointed as shown, the current being measured is 78 mA.

PRACTICE PROBLEM 3.13

The meter display in Figure 3.60b was obtained with the function select switch set to the 1 mA position. What is the value of the current being measured?

AC Voltage Readings

The AC scales (with the exception of the 2.5 VAC scale) use the same numbers as the DC scales. However, they are measured using the AC scale positioned *below* the numbers. For example, take a look at the meter display shown in Figure 3.61. The reading displayed is being taken with the function select set to the 50 V position. If you mistakenly look at the DC scale, you get a reading of 4 V. However, the AC scale below the numbers indicates that the correct reading is 5 V. When an AC voltage is measured with the function select set to 2.5 V, the lower of the two AC scales is used.

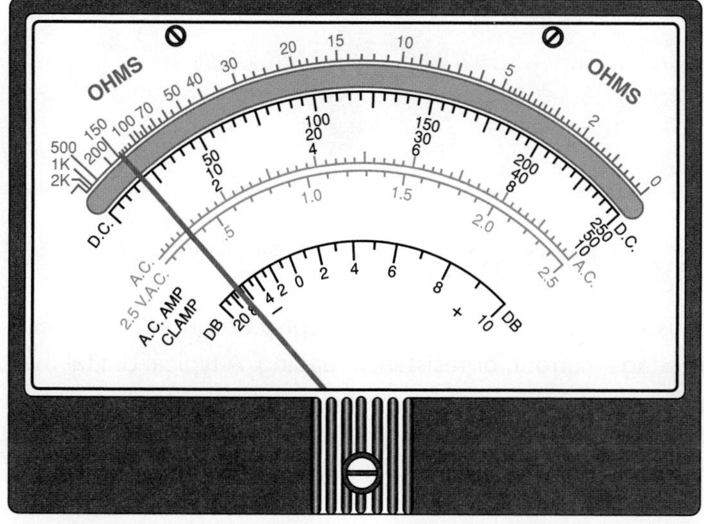

FIGURE 3.61 An example AC voltage reading.

Parallax Error

If two people stand on opposite sides of a clock (as shown in Figure 3.62) and look up at the clock at the same time, they'll see different times. To the person on the left, it might appear to be a couple of minutes *after* 6:00. To the person on the right, it might appear to be a couple of minutes *before* 6:00.

The change in readings that result from looking at the scale from an angle is called **PARALLAX ERROR**. Many analog meter scales have a small mirror located between the ohms and DC scales that prevents parallax error. When reading the display, look at the mirror. If you can see the needle's reflection in the mirror, change your line of sight so that the reflection disappears. When you cannot see the reflection, you are looking straight on the display, and no parallax error is introduced.

PARALLAX ERROR Any change in readings that results from looking at a scale from an angle.

FIGURE 3.62 Paralax error causes two people on opposite sides of a clock to see different times.

Reading a Digital Display

Parallax error is not an issue when reading a digital display because it provides a numeric readout of a voltage, current, or resistance reading. A typical digital multimeter (DMM) readout is shown in Figure 3.63. As you can see, the reading is clear and not subject to viewing errors.

FIGURE 3.63 A digital meter display.

PROGRESS CHECK

1. What does an analog multimeter use to display the results of circuit readings?
2. What does "full scale deflection" mean?
3. How do you determine the value of a resistance reading on an analog meter?
4. How do you determine the value of a voltage reading on an analog meter?
5. What is parallax error?

3.8 MEASURING VOLTAGE, RESISTANCE, AND CURRENT

In the lab manual that accompanies this text, there are exercises that deal with measuring current, voltage, and resistance. This section serves as an introduction to these exercises.

Multimeters

There are many types of meters that are used to measure various electrical properties. For example, a **VOLTMETER** is used to measure voltage, an **OHMMETER** is used to measure resistance, and an **AMMETER** is used to measure current. As you learned earlier, a multimeter is a single piece of test equipment that can act as any of these three meters. That is, it can be used to measure current, voltage, or resistance. A typical digital multimeter (DMM) is shown in Figure 3.64.

VOLTMETER A meter used to measure voltage.

OHMMETER A meter used to measure resistance.

AMMETER A meter used to measure current.

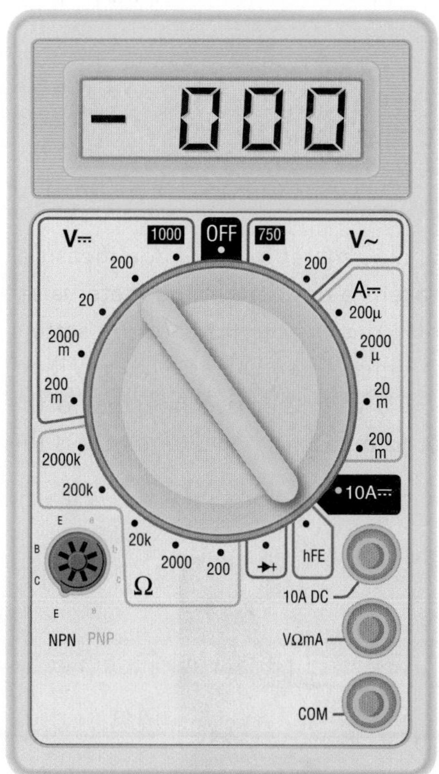

FIGURE 3.64 A digital multimeter (DMM).

FIGURE 3.65 A volt-ohm-milliameter (VOM).
Source: Photo courtesy of Simpson Electric Co.

An analog multimeter is commonly referred to as a *volt-ohm-milliameter (VOM)*. A typical VOM is shown in Figure 3.65. Comparing the DMM and VOM in Figures 3.64 and 3.65, you can see that they have the types of displays described earlier in the chapter.

In the last section, you learned how to read the display on an analog or digital multimeter. In this section, you are shown how to connect these meters to circuits and components to take the desired current, voltage, and/or resistance measurements.

Measuring Voltage

Voltage is measured by connecting a voltmeter "across" a component. This means that the two voltmeter leads are connected to the two sides of the component as shown in Figure 3.66.

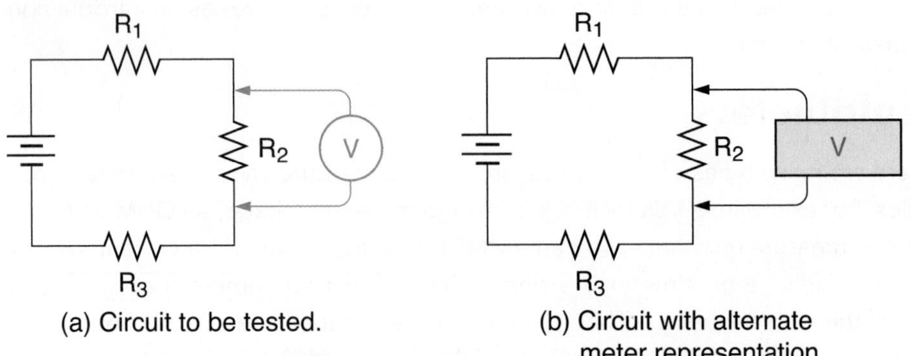

Using the meter in Figure 3.64, this connection is made as follows: The "VΩmA" input lead is connected to the side of the resistor that leads back to the voltage source, the "COM" jack is connected to the other side of the resistor, and the "V" function is selected.

FIGURE 3.66 Voltmeter connections.

STAYING SAFE Standard voltmeters (like those presented here) cannot be used in high-voltage applications. Attempting to do so can result in serious injury or death. Special high-voltage meters must be used in high-voltage applications.

When using an analog meter for a voltage measurement, care must be taken to ensure that the meter is set for the proper polarity. This is accomplished by setting the polarity switch to either the +DC or −DC position. Figure 3.67 shows how a voltmeter is connected to the two sides of a resistor to measure the voltage across the component.

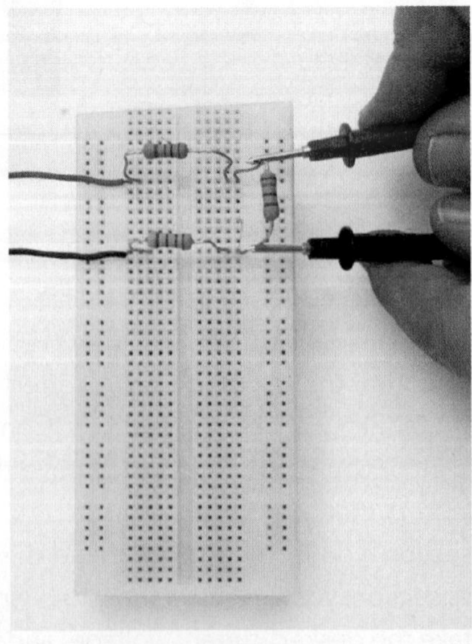

FIGURE 3.67 Measuring voltage.

Measuring Resistance

Measuring resistance can get a little tricky because resistance cannot be measured with power applied to the circuit. In most cases, the safest and most reliable method of measuring resistance is to simply remove the component from the circuit. Once it is removed from the circuit, the meter is connected *across* the component (just like a voltmeter) as shown in Figure 3.68.

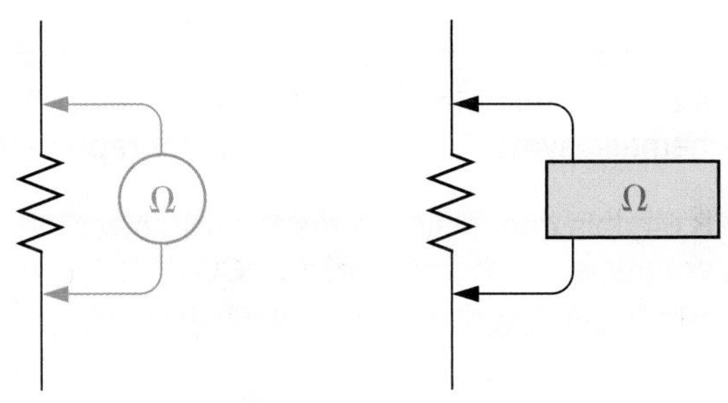

(a) Resistor with ohmmeter schematic symbol. (b) Circuit with alternate meter representation.

FIGURE 3.68 Ohmmeter connections.

Using the meter in Figure 3.64, this connection is made as follows: The "VΩmA" input lead is connected to either side of the resistor, the "COM" input lead is connected to the other side of the resistor, and the "Ω" function is selected.

An important point needs to be made at this time: *If you hold both leads of a component while measuring its resistance, you may get a faulty reading.* When you hold both ends of a component while measuring its resistance, the resistance of your skin can affect the reading. To avoid this problem, you should hold only one component lead, as shown in Figure 3.69.

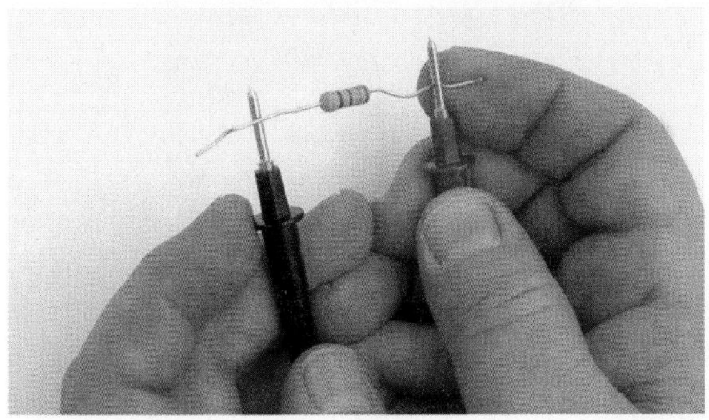

FIGURE 3.69 Measuring resistance.

Measuring Current

Current can be measured by inserting an ammeter *in the current path.* Let's say that we want to measure the current in the circuit shown in Figure 3.70a. You must insert the meter so that the current passes through the meter. This is accomplished by *breaking the current path* and inserting the meter across the break, as shown in Figure 3.70b. The photo in Figure 3.71 shows the actual connection of the DMM for measuring current.

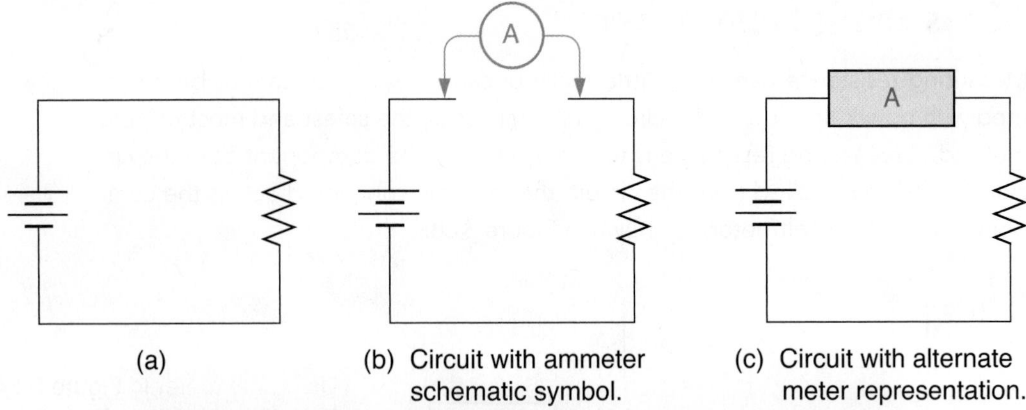

Using the meter in Figure 3.64, this connection is made as follows: The "VΩmA" input lead is connected to the source, the "COM" input lead is connected to the resistor, and the "mA" function is

FIGURE 3.70 Ammeter connections.

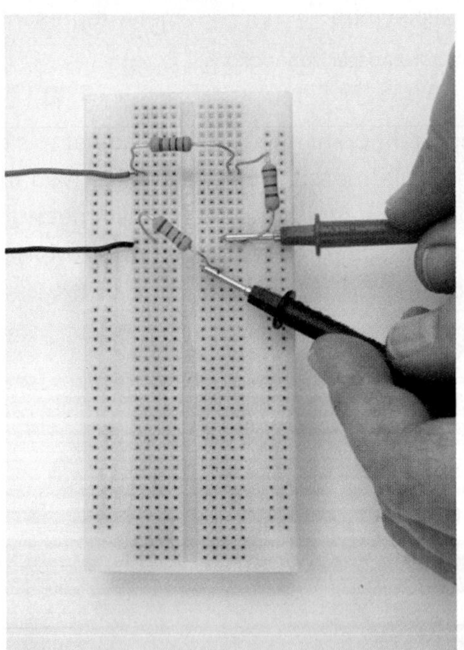

FIGURE 3.71 Measuring current.

When measuring direct current (DC) with a VOM, you must ensure that the proper meter polarity is observed (as it is for voltage measurements). This is accomplished by connecting the red lead on the positive side of the circuit break and the black lead on the negative side of the break.

Measuring Current with a Clamp-on Meter

A *clamp-on meter* can be used to measure an alternating current (AC) without breaking the conduction path. A typical clamp-on meter is shown in Figure 3.72a.

When AC passes through a conductor, a magnetic field forms around that conductor. When a clamp-on meter is connected as shown in Figure 3.72b, it measures the magnetic field strength around the conductor, and provides a readout indicating the value of the conductor current.

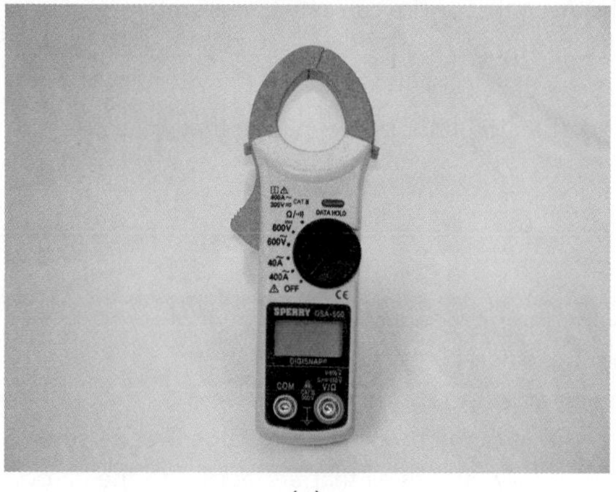

(a)

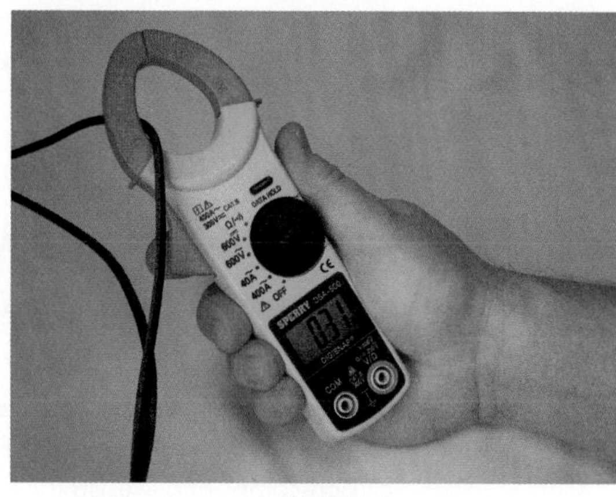

(b)

FIGURE 3.72 Clamp-on meter.

One Final Note

There is a lot to know about the proper use of test equipment. Make sure you take the time to get thoroughly familiar with any type of meter before attempting to use it. This will help you to avoid getting faulty readings and possibly damaging the meter.

PROGRESS CHECK

1. How is a voltmeter connected in a circuit?
2. How is an ohmmeter connected to a component?
3. What precautions must be taken when measuring resistance?
4. What does an ammeter measure? How is it connected in a circuit?
5. What precaution must be taken when using a VOM to measure current or voltage?
6. What type(s) of current can be measured with a clamp-on meter?

3.9 SUMMARY

Here is a summary of the major points that were covered in this chapter:

Conductors and Insulators

- Theory deals with the concepts of a science. Practice deals with using those concepts to accomplish something.
- Copper is the most commonly used conductor because it provides the best balance between resistance and cost.

- Solid wire contains a solid conductor. Stranded wire contains a group of thin wires wrapped together to form a larger diameter wire.
  - Solid wire is cheaper to produce.
  - Stranded wire is more flexible.
- The *American Wire Gauge* (AWG) system sizes wires by cross-sectional area.
- Ampacity is the maximum amount of current that a wire can carry, measured in amperes.
- Circuits are usually mounted on printed circuit boards (PC boards). The copper conductors that interconnect the components are called *traces*.
- The average breakdown voltage rating of an insulator is the voltage that will cause it to break down and conduct, given in kV/cm.

Resistors

- A resistor is a component that is designed to provide a specific amount of resistance.
  - The value of a fixed resistor cannot be varied by the user.
  - The value of a variable resistor can be adjusted by the user.
- The most common type of resistor is the carbon-composition resistor. The value of a carbon-composition resistor is determined by impurities that are added to the carbon between its leads.
- Wire-wound resistors use the resistivity of a length of wire to produce the desired resistance.
  - Wire-wound resistors are used primarily in high-power applications.
- Integrated resistors are extremely small components. Several of these resistors are usually housed in a single case.
- The value of any standard-value resistor can be given using the first two digits of its value and a power of ten.
- The tolerance of a resistor is the guaranteed range of measured values for the component, given as a percentage of the rated value.
- The resistor color code is used to indicate the rated value and tolerance of a resistor.
- Resistors have power ratings. That is, they are rated in terms of the amount of heat they can dissipate.
  - The size of a carbon-composition resistor indicates its power rating.

Potentiometers and Rheostats

- A potentiometer is a three-terminal resistor whose value can be adjusted (within set limits) by the user. A potentiometer is usually referred to as a *pot*.
- The center terminal of a pot is called the wiper arm. By adjusting the control shaft, the resistance between the wiper arm and each of the outer terminals is varied.
- The sum of the two wiper arm resistances (R_{AB} and R_{BC}) is equal to the resistance between the two outer terminals (R_{AC}).
- The taper of a potentiometer is a measure of how its resistance varies as the control shaft is turned.
  - The resistance of a linear-taper pot varies at a constant rate.
  - The resistance of a nonlinear-taper pot varies at a changing rate.
- Pots are rated in terms of their maximum outer terminal resistance and power dissipation capability.
- Standard pots are either carbon-composition or wire-wound.
- Multiturn pots are high-resolution pots.
- Gang-mounted potentiometers share a common control shaft.

- A rheostat is a two-terminal variable resistor that is used to control current.
  - A potentiometer can be wired to work as a rheostat by connecting its wiper arm to one of its outer terminals.

Batteries

- A battery is a component that produces a difference of potential from a continuous chemical reaction.
- Batteries are made up of one or more cells. A cell is a single unit that is designed to produce electrical energy through thermal, chemical, or optical means.
- Primary cells (or dry cells) are not rechargeable. Secondary cells (or wet cells) can be recharged.
- The capacity of a battery is a measure of how long it will last at a given output current, measured in ampere-hours (Ah).
- Batteries are connected in series to increase their total output voltage. They are connected in parallel to increase their total output current.

DC Power Sources

- A DC power supply is a piece of lab equipment that provides variable DC voltages.
- Voltages are generally assigned a polarity; that is, they are identified as being positive or negative.
- The *common* terminal on a DC power supply is the terminal to which polarity is referenced.
- In order to have current in a circuit, there must be a complete path between the terminals of the voltage source.
- A DC converter is a circuit (or device) that converts an AC input to a DC output.

Switches and Circuit Protectors

- A switch allows you to manually make or break the current path in a circuit.
- A switch contains:
  - One or more movable contacts, called *poles.*
  - One or more non-moving contacts, called *throws.*
- Some switches are classified as *normally open* (NO) or *normally closed* (NC).
  - A normally open switch closes when activated.
  - A normally closed switch opens when activated.
- A momentary switch is one that opens or closes only as long as the button is pressed.
- A fuse is a circuit protector that is designed to break a current path (open) when its current exceeds a specified value.
- The current rating of a fuse indicates the maximum allowable fuse current.
- The voltage rating of a fuse indicates the maximum voltage that an open fuse is designed to block.
- Never replace a fuse with one that has a higher current rating.
- Fuses are classified according to the amount of time required for them to *blow* (open).
- A *type T* fuse has an *Edison base,* which resembles the base of a light bulb.
- A *type S* fuse has a *rejection base*. A rejection base must be fitted with an adaptor to fit in a fuse box. Once inserted, the adaptor cannot be removed.
- Circuit breakers are normally closed switches that are designed to open under overload conditions. Switching an open circuit breaker "off" and then "on" restores it to normal operation.

Analog and Digital Multimeters

- A multimeter is an instrument that is used to measure current, voltage, and resistance. Some can measure additional values.
  - An analog multimeter uses a pointer that moves over a stationary scale to indicate the result of a measurement.
  - A digital multimeter uses a digital display to indicate the result of a measurement.
- Parallax error occurs when an analog scale is viewed at an angle, causing an inaccurate reading.
- An ammeter is used to measure current. The meter is placed in series with the circuit under test.
- A voltmeter is used to measure voltage. The meter is connected across the component or circuit under test.
- An ohmmeter is used to measure resistance. The meter is connected across the component or circuit under test.
  - Power must be removed before a resistance reading is taken.

CHAPTER REVIEW

1. The material used to isolate a conductor from its environment is referred to as ________.

   a) an isolator b) an insulator
   c) a semiconductor d) a conduit

2. Copper is the most commonly used conductor material because ________.

   a) it is the cheapest
   b) it has the lowest resistance
   c) it is the best compromise between price and low resistance
   d) it is the most flexible

3. Aluminum is not used as often as copper in wires because ________.

   a) it has poor elasticity
   b) it has high resistivity
   c) it is too expensive
   d) it is not flexible enough

4. 1000 feet of 12 gauge wire has ________ the resistance of 1000 feet of 6 gauge wire.

   a) six times b) four times
   c) one quarter d) ten times

5. Wire gauge is based upon ________.

   a) the area in cmils and length of the wire
   b) ampacity
   c) ampacity and wire length
   d) the area in cmils

6. The ampacity rating of a wire is determined by ________.
   a) the gauge of the wire
   b) the temperature of the wire
   c) the ambient temperature
   d) the wire material
   e) all of the above

7. Conductors on printed circuit boards are called ________.
   a) traces
   b) PCBs
   c) inductors
   d) commons

8. An insulator can begin to conduct if ________.
   a) a high enough voltage is applied
   b) a high enough current is applied
   c) it is cold enough
   d) if it is too thick

9. The resistance value of a carbon-composition resistor is determined during the manufacturing process by ________.
   a) the diameter of the leads
   b) the purity of the carbon
   c) the density of the carbon
   d) the diameter of the resistor

10. Wire-wound resistors have ________ than carbon-composition resistors.
    a) lower resistance
    b) higher resistance
    c) lower power handling
    d) higher power handling

11. The first three color bands of a ________ resistor are gray, blue, and red.
    a) 860 Ω
    b) 95 kΩ
    c) 8.5 kΩ
    d) 8.6 kΩ

12. A high-tolerance resistor is ________ than a low-tolerance resistor.
    a) better
    b) worse
    c) has a higher accuracy
    d) none of the above; it depends on the value.

13. A 22 kΩ resistor with a 5% tolerance may have a variation in resistance that is ________ a 47 kΩ resistor with a 2% tolerance.
    a) higher than
    b) lower than
    c) equal to
    d) cannot be determined with the information supplied

14. A 330 kΩ resistor is measured and found to actually be 322 kΩ. The tolerance band on this resistor should be ________.
    a) red
    b) gold
    c) silver
    d) orange

15. A ________ resistor has color bands of orange-blue-silver.
    a) 3.6 Ω
    b) 3.5 Ω
    c) 0.45 Ω
    d) 0.36 Ω

16. The middle terminal on a potentiometer is called ________.
    a) the wiper arm
    b) the common terminal
    c) the ground terminal
    d) the reference terminal

17. For potentiometers, $R_{AC} = R_{AB} + R_{BC}$ when ________.
    a) the shaft is fully turned clockwise
    b) the shaft is fully turned counterclockwise
    c) the shaft is in the center
    d) all of the above

18. The taper of a potentiometer refers to ________.
    a) its resistance value
    b) the rate of change of resistance as the shaft rotates
    c) its conductivity
    d) its power handling

19. Multiturn pots have ________ than single-turn pots.
    a) higher power handling
    b) a more linear taper
    c) better resolution
    d) all of the above

20. A potentiometer can be wired as a rheostat by connecting ________.
    a) terminal B to C
    b) terminal B to A
    c) terminal A to C
    d) either B to C or A to C

21. Rheostats are used to ________.
    a) control voltage
    b) control current
    c) divide current
    d) divide voltage

22. Between the terminals of a battery is ________.
    a) an electrolyte
    b) a potential difference
    c) a voltage
    d) all of the above

23. Secondary cells ________.
    a) have a dry electrolyte
    b) can only be used once
    c) are rechargeable
    d) none of the above

24. Connecting two batteries in series __________.
    a) doubles the maximum current
    b) doubles the voltage
    c) doubles the voltage but halves the current
    d) doubles the current but halves the voltage

25. A battery rated at 12 Ah can provide 400 mA for ________.
    a) 30 hours
    b) 48 hours
    c) 4.8 hours
    d) 36 hours

26. The common terminal in a DC power supply is ________.
    a) the voltage reference for the positive voltage
    b) the voltage reference for the negative voltage
    c) at zero volts potential
    d) all of the above

27. The moving component in a switch is referred to as the ________.
    a) pole
    b) throw
    c) wiper
    d) tbreak

28. A DPST switch ________.
    a) has two poles and one throw
    b) has four terminals
    c) can only be closed in one position
    d) all of the above

29. The voltage rating of a fuse tells you ________.
    a) the minimum voltage that will cause the fuse to blow
    b) the maximum voltage that will cause the fuse to blow

c) the minimum voltage that will cause the blown fuse to conduct
d) the maximum voltage that will cause a blown fuse to conduct

30. Slow-blow fuses are used in ________.
    a) low current conditions
    b) high current conditions
    c) circuits that experience brief current surges
    d) circuits that have brief current drop-outs

31. Type S fuses ________.
    a) handle more power than type T fuses
    b) handle less power than type T fuses
    c) have a faster reaction time than type T fuses
    d) prevent you from installing a different rated fuse

32. Circuit breakers are classified as ________.
    a) magnetic-type
    b) thermal-type
    c) thermo-magnetic type
    d) all of the above

PRACTICE PROBLEMS

1. Determine the guaranteed range of values for a 110 kΩ resistor with a 2% tolerance.
2. Determine the guaranteed range of values for a 1.5 MΩ resistor with a 2% tolerance.
3. Determine the guaranteed range of values for a 360 kΩ resistor with a 5% tolerance.
4. Determine the guaranteed range of values for a 6.8 Ω resistor with a 10% tolerance.
5. For each of the color codes listed, determine the rated value of the component.

	Band 1	*Band 2*	*Band 3*	*Value*
a)	Brown	Blue	Brown	_____
b)	Red	Violet	Green	_____
c)	Orange	White	Yellow	_____
d)	Green	Brown	Black	_____
e)	Yellow	Violet	Gold	_____
f)	Green	Blue	Orange	_____

6. For each of the color codes listed, determine the rated value of the component.

	Band 1	*Band 2*	*Band 3*	*Value*
a)	Red	Black	Brown	_____
b)	Orange	Black	Silver	_____
c)	Green	Brown	Black	_____
d)	White	Brown	Green	_____
e)	Brown	Blue	Blue	_____
f)	Blue	Gray	Silver	_____

7. Determine the colors of the first three bands for each of the following resistor values.

a) 33 kΩ	e) 120 kΩ
b) 15 Ω	f) 10 MΩ
c) 910 kΩ	g) 1.8 Ω
d) 2.2 kΩ	h) 0.24 Ω

8. Determine the colors of the first three bands for each of the following resistor values.

a) 11 Ω	e) 160 kΩ
b) 20 MΩ	f) 0.13 Ω
c) 680 Ω	g) 9.1 MΩ
d) 3.3 Ω	h) 36 Ω

9. For each of the color codes listed, determine the rated component value and the guaranteed range of measured values.
   a) Red–Yellow–Yellow–Gold
   b) Orange–Blue–Gold–Red
   c) Gray–Red–Black–Gold
   d) Blue–Gray–Silver–Red

10. For each of the color codes listed, determine the rated component value and the guaranteed range of measured values.
   a) Yellow–Orange–Silver–Gold
   b) Brown–Gray–Yellow–Red
   c) Yellow–Violet–Orange–Silver
   d) Brown–Green–Silver–Gold

ANSWERS TO THE EXAMPLE PRACTICE PROBLEMS

3.1	66.6 Ω to 69.4 Ω
3.2	865 Ω to 956 Ω
3.3	1 kΩ
3.4	56 kΩ
3.5	47 Ω
3.6	9.1 Ω
3.7	0.39 Ω
3.8	Red–yellow–orange
3.9	5.04 kΩ to 6.16 kΩ
3.10	7 kΩ
3.11	5.4 V
3.12	0.7 V
3.13	0.44 mA (440 μA)

E
I
R

CHAPTER 4

OHM'S LAW AND POWER

PURPOSE

In Chapter 2, you were shown that current is the flow of charge in a circuit, voltage is the difference of potential (or *electromotive force*) that causes the current, and resistance is the circuit's opposition to current. You were also shown that the units of current, voltage, and resistance can be defined in terms of each other, as follows:

1. One *ampere* is the amount of current that is generated when a 1 V difference of potential is applied across 1 Ω of resistance.
2. One *volt* is the difference of potential required to generate 1 A of current through 1 Ω of resistance.
3. One *ohm* is the amount of resistance that limits the current generated by a difference of potential of 1 V to 1 A.

These statements clearly indicate that there is a relationship between current, voltage, and resistance. This relationship is defined by *Ohm's law.*

Ohm's law is one of the most powerful tools for working with electrical circuits. As you will see, it can be used to:

1. Predict how a circuit will respond to any change in voltage, current, or resistance.
2. Calculate the value of any basic circuit property (current, voltage, or resistance) when the other two values are known.
3. Determine the likely source of a circuit fault.

In this chapter, we are going to discuss Ohm's law and its applications. We will also discuss another primary circuit property, called *power,* and its relationship to current, voltage, and resistance. Finally, we take a look at some miscellaneous topics that apply to basic circuits.

KEY TERMS

The following terms are introduced and defined in this chapter on the pages indicated:

OBJECTIVES

After completing this chapter, you should be able to:

1. Describe the relationship among *voltage*, *current*, and *resistance*.
2. Predict how a change in either voltage or resistance will affect circuit current.
3. Use *Ohm's law* to calculate any one of the following values, given the other two: current, voltage, and resistance.
4. Describe how Ohm's law can be used in circuit troubleshooting.
5. List the characteristics of an *open circuit*.
6. List the characteristics of a *short circuit*.
7. Define *power* and describe its relationship to current and voltage.
8. Calculate power, given any two of the following: current, voltage, and resistance.
9. Describe the relationship between power and heat.
10. Determine the minimum acceptable power rating for a resistor in a circuit.
11. Define and calculate *efficiency* and the *kilowatt-hour* (kWh) unit of energy.
12. Calculate any two of the following values, given the other two: current, voltage, resistance, and power.
13. Calculate the range of current values for a circuit.
14. Define *load* and *full load*.

4.1 OHM'S LAW

In the early nineteenth century, Georg Simon Ohm, a German physicist, identified several cause-and-effect relationships among circuit current, voltage, and resistance. Specifically, he found that *current is directly proportional to voltage and inversely proportional to resistance*. This statement, which has come to be known as **OHM'S LAW**, tells us that:

OHM'S LAW A law stating that *current is directly proportional to voltage and inversely proportional to resistance*.

1. The current through a *fixed resistance*:
   a. increases when the applied voltage increases.
   b. decreases when the applied voltage decreases.
2. The current generated by a *fixed voltage*:
   a. decreases when the resistance increases.
   b. increases when the resistance decreases.

We'll start our discussion of Ohm's law by taking a closer look at these two statements.

The Relationship Between Current and Voltage

As stated earlier, voltage is the difference of potential that generates the flow of charge (current) in a circuit. According to Ohm's law, *the greater the difference of potential, the greater the resulting current*. This relationship is illustrated in Figure 4.1.

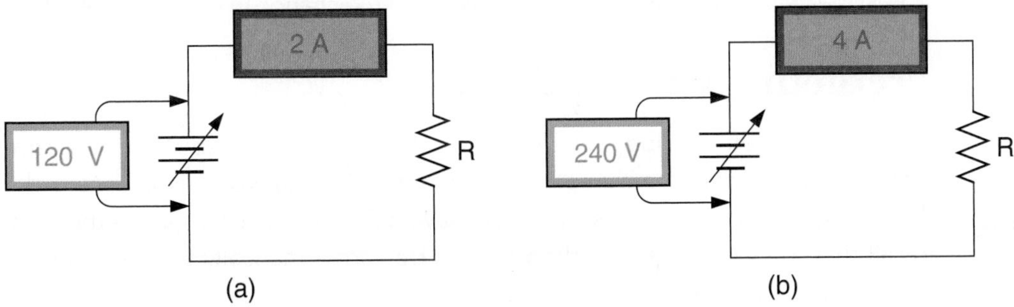

FIGURE 4.1 Voltage and current.

In Figure 4.1a, the meters show the value of the applied voltage (120 V) and the resulting value of circuit current (2 A). If the applied voltage doubles (as shown in Figure 4.1b), the circuit current also doubles. In other words, the current increases proportionally to any increase in voltage. By the same token, any decrease in the applied voltage causes a proportional decrease in current.

In Figure 4.1, *the current is changing as a result of a change in voltage*. In other words, the change in voltage is the *cause* and the change in current is the *effect*.

The Relationship Between Current and Resistance

As you know, resistance is the opposition to current in a circuit. According to Ohm's law, *the greater the opposition, the lower the current at a given voltage.* This relationship is illustrated in Figure 4.2.

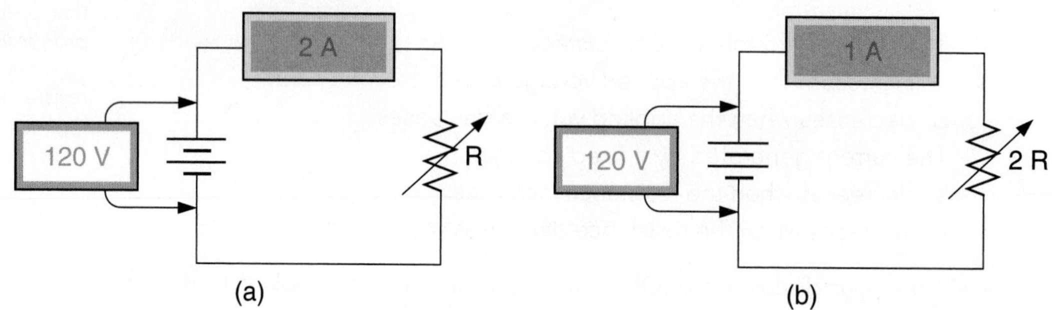

FIGURE 4.2 Resistance and current.

In Figure 4.2a, the meters indicate the values of the applied voltage (120 V) and the resulting current (2 A). If the circuit resistance is doubled as shown in Figure 4.2b, the circuit current drops to half its original value, even though the applied voltage has not changed. In other words, *the current has decreased proportionally to the increase in resistance.* By the same token, a decrease in the circuit resistance causes circuit current to increase proportionally.

In this case, *the current is changing as a result of a change in resistance.* In other words, the change in resistance is the cause, and the change in current is the effect.

Predicting Circuit Behavior

The cause-and-effect aspect of Ohm's law can be used to predict the response that current will have to a change in voltage or resistance. For example, consider the circuit in Figure 4.3. In this circuit, we have a DC power supply and a rheostat. Recall that the resistance of a rheostat can be varied manually to any value between 0 Ω and its maximum (rated) value.

Ohm's law allows us to predict the effect that adjusting the rheostat will have on the circuit current. If we increase the resistance of the rheostat, Ohm's law tells

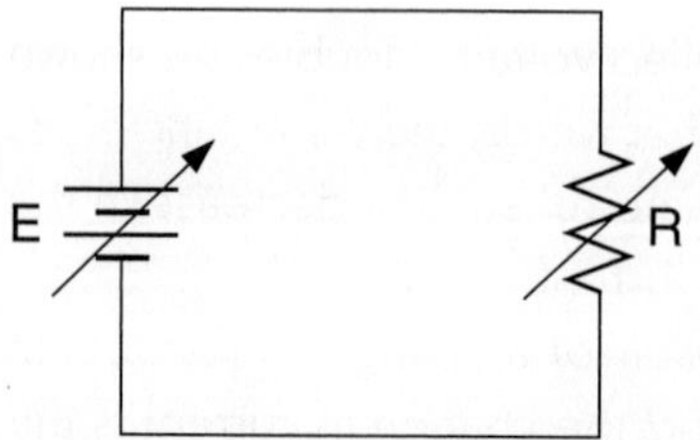

FIGURE 4.3

us that the circuit current will decrease. If we decrease the resistance of the rheostat, Ohm's law tells us that the circuit current will increase.

Ohm's law also allows us to predict the effect that adjusting the applied voltage will have on the circuit current. If we increase the applied voltage, Ohm's law tells us that the circuit current will increase. If we decrease the applied voltage, Ohm's law tells us that the circuit current will decrease.

As you can see, Ohm's law can be used to predict how a change in circuit voltage or resistance will affect circuit current. At the same time, there is a subtle message here: *When you adjust any circuit component, there will always be some type of response to the change.* Before you adjust any circuit component, you should always take a moment to consider the possible impact of doing so.

Using Ohm's Law to Calculate Circuit Values

Ohm's law also provides a specific mathematical relationship between the values of current, voltage, and resistance in a circuit. This relationship is given as

(4.1)

$$I = \frac{E}{R}$$

where

I = the circuit current
E = the applied voltage
R = the circuit resistance

Equation (4.1) allows us to calculate the value of current in a circuit when the values of the applied voltage and circuit resistance are known, as demonstrated in Example 4.1.

EXAMPLE 4.1

What is the value of current in Figure 4.4?

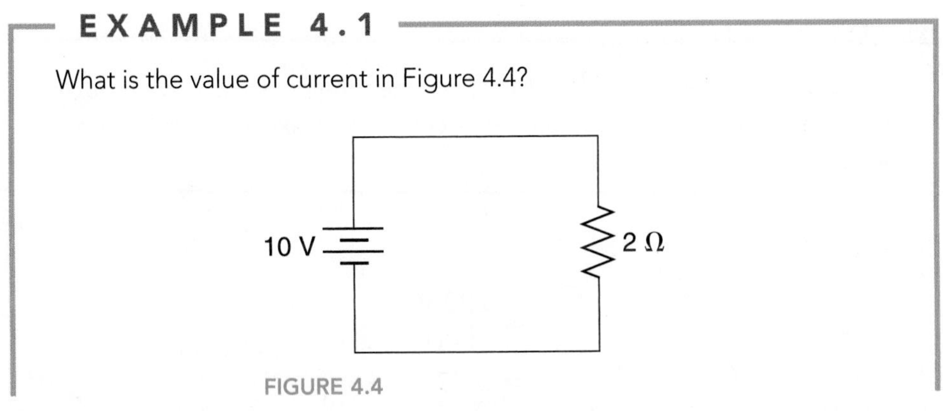

FIGURE 4.4

SOLUTION

Using Ohm's law, the value of the circuit current is found as

$$I = \frac{E}{R} = \frac{10\text{ V}}{2\ \Omega} = 5\text{ A}$$

PRACTICE PROBLEM 4.1

A circuit like the one in Figure 4.4 has values of E = 18 V and R = 100 Ω. Calculate the value of the current in the circuit.

Ohm's law can be applied to a variety of practical situations. The following example demonstrates one practical application for equation (4.1).

EXAMPLE 4.2

The power supply in Figure 4.5 has a maximum possible output of 220 V. Will the fuse open when the power supply output is increased to its maximum value?

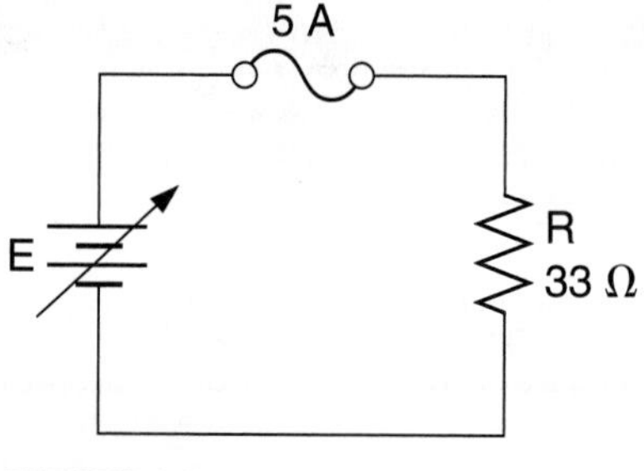

FIGURE 4.5

SOLUTION

To determine whether the fuse will open when the power supply output is increased to its maximum value, we calculate the maximum circuit current as follows:

$$I = \frac{E}{R} = \frac{220\text{ V}}{33\ \Omega} = 6.67\text{ A}$$

Since the fuse is rated at 5 A, it will open if the power supply is set to its maximum limit.

PRACTICE PROBLEM 4.2

A circuit like the one in Figure 4.5 has a 20 A fuse, a 33 Ω resistor, and a maximum power supply output of 600 V. Determine whether or not the fuse will open if the power supply output is adjusted to its maximum value.

Another form of Ohm's law allows us to calculate voltage when the values of current and resistance are known. If we multiply both sides of equation (4.1) by R, we get:

$$E = I \times R \tag{4.2}$$

The use of this relationship is demonstrated in Example 4.3.

EXAMPLE 4.3

What is the value of the applied voltage (E) in Figure 4.6?

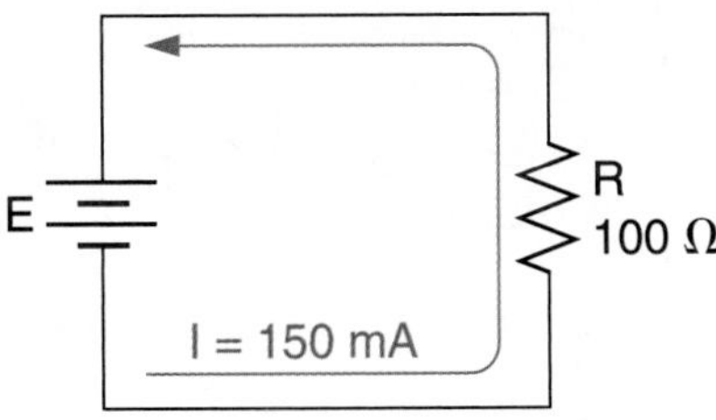

FIGURE 4.6

Using Ohm's law, the value of the applied voltage for the circuit is found as

$$\begin{aligned} E &= I \times R \\ &= 150 \text{ mA} \times 100\ \Omega \\ &= 15 \text{ V} \end{aligned}$$

PRACTICE PROBLEM 4.3

A circuit like the one in Figure 4.6 has values of I = 5 A and R = 33 Ω. Calculate the value of the applied voltage for the circuit.

The following example demonstrates a practical application of equation (4.2).

EXAMPLE 4.4

The boom box in Figure 4.7 has two 8 Ω speakers. Each speaker draws 1.25 A. What is the value of the voltage driving each of the speakers?

FIGURE 4.7
Istock

SOLUTION

Using Ohm's law, the voltage driving each speaker is found using the speaker resistance and current values as follows:

$$\begin{aligned} E &= I \times R \\ &= 1.25\text{ A} \times 8\ \Omega \\ &= 10\text{ V} \end{aligned}$$

Thus, the voltage at each boom box output is 10 V when the speakers are drawing 1.25 A.

PRACTICE PROBLEM 4.4

How much voltage is required to draw 16 A through a 7.5 Ω heating element?

A third form of Ohm's law allows us to calculate resistance when the values of voltage (E) and current (I) are known. If we divide both sides of equation (4.2) by I, we get:

(4.3)

$$R = \frac{E}{I}$$

This use of this equation is demonstrated in Example 4.5.

EXAMPLE 4.5

What is the value of the circuit resistance in Figure 4.8?

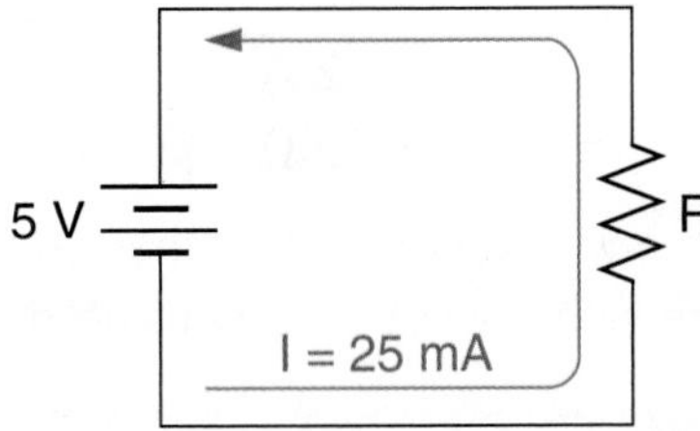

FIGURE 4.8

SOLUTION

Using Ohm's law, the value of R for the circuit is found as

$$R = \frac{E}{I} = \frac{5\text{ V}}{25\text{ mA}} = 200\ \Omega$$

PRACTICE PROBLEM 4.5

A circuit like the one shown in Figure 4.8 has values of E = 120 V and I = 400 mA. Calculate the value of the circuit resistance.

The following example demonstrates a practical application for equation (4.3).

EXAMPLE 4.6

What is the minimum allowable setting for the rheostat in Figure 4.9?

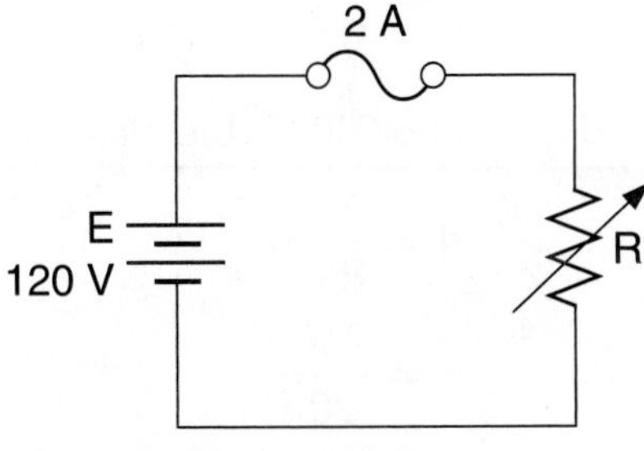

FIGURE 4.9

SOLUTION

The circuit has a 2 A fuse. Using I = 2 A and E = 120 V, the minimum allowable setting for the rheostat is found as

$$R = \frac{E}{I}$$
$$= \frac{120\text{ V}}{2\text{ A}}$$
$$= 60\ \Omega$$

If the rheostat is set to any value that is less than 60 Ω, the current will exceed 2 A, and the fuse will likely open. (Try using equation (4.1) to prove this to yourself.)

PRACTICE PROBLEM 4.6

A circuit like the one in Figure 4.9 has a 300 V source and a 15 A fuse. Determine the minimum allowable setting for the rheostat in the circuit.

Current Relationships

It was stated earlier that the current through a fixed resistance is directly proportional to the applied voltage. This relationship is reinforced by Example 4.7.

EXAMPLE 4.7

The circuit in Figure 4.10 has the values shown. What happens to the circuit current if the applied voltage doubles?

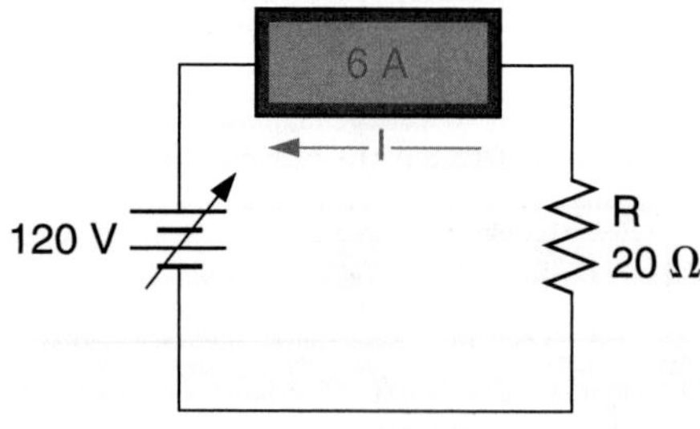

FIGURE 4.10

SOLUTION

If the applied voltage doubles (to 240 V), the circuit current increases to

$$I = \frac{E}{R}$$
$$= \frac{240\text{ V}}{20\ \Omega}$$
$$= 12\text{ A}$$

Thus, when the applied voltage doubles, the circuit current also doubles (as was stated on page 133).

PRACTICE PROBLEM 4.7

Verify that the circuit current in Figure 4.10 will drop to half its value if the applied voltage is cut in half.

You have been told that the current generated by a fixed voltage is inversely proportional to the circuit resistance. This relationship is reinforced by Example 4.8.

EXAMPLE 4.8

The circuit in Figure 4.11 has the values shown. What happens to the circuit current if the circuit resistance doubles?

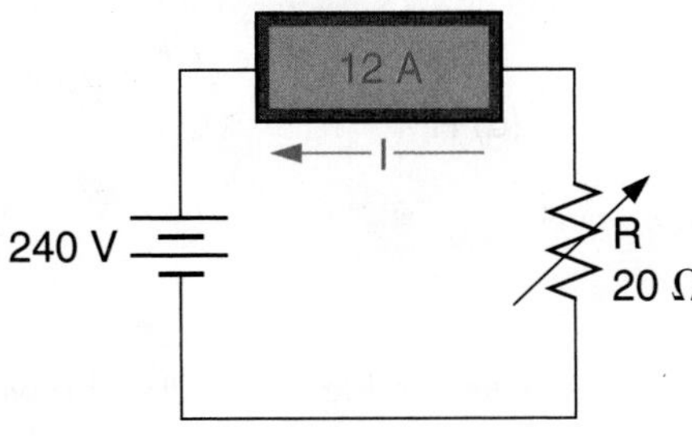

FIGURE 4.11

SOLUTION

If the circuit resistance doubles (to 40 Ω), the circuit current decreases to

$$I = \frac{E}{R}$$

$$= \frac{240\text{ V}}{40\ \Omega}$$

$$= 6\text{ A}$$

As you can see, doubling the circuit resistance causes the current to drop to half its original value (as was stated on page 134).

PRACTICE PROBLEM 4.8

Verify that the current in Figure 4.11 will double if the circuit resistance is cut in half.

Ohm's Law Wheel

The Ohm's law wheel can help you remember the various forms of Ohm's law, as illustrated in Figure 4.12. First, draw the wheel shown in Figure 4.12a. Then, cover the value of interest to see the correct equation. For example, when calculating

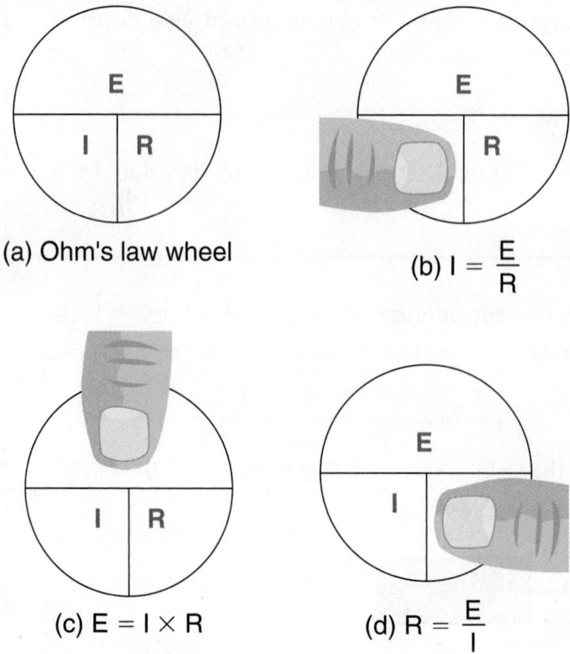

FIGURE 4.12 The Ohm's law wheel.

current (I), cover the letter I as shown in Figure 4.12b. The wheel then shows that I is found as E/R. The other two forms of Ohm's law can be identified as shown in the figure.

Troubleshooting with Ohm's Law

TROUBLESHOOTING The process of isolating and identifying a fault in a circuit.

TROUBLESHOOTING is the process of locating faults in a circuit or system. When a circuit or system fails to operate as expected, an electrical technician is called in to locate the source of the problem.

When a circuit doesn't operate as expected, Ohm's law is one of the best tools you have to help diagnose the problem. For example, Ohm's law tells you that:

1. Current *increases* as a result of:
   a. an increase in the applied voltage.
   b. a decrease in the circuit resistance.
2. Current *decreases* as a result of:
   a. a decrease in the applied voltage.
   b. an increase in the circuit resistance.

The first of these statements can be used to determine the possible causes of the problem shown in Figure 4.13a. According to Ohm's law, the measured current in the circuit should be approximately 1 A. However, the ammeter shows a reading of 4 A. This high current indicates (according to statement #1) that either the applied voltage is too high or the circuit resistance is too low.

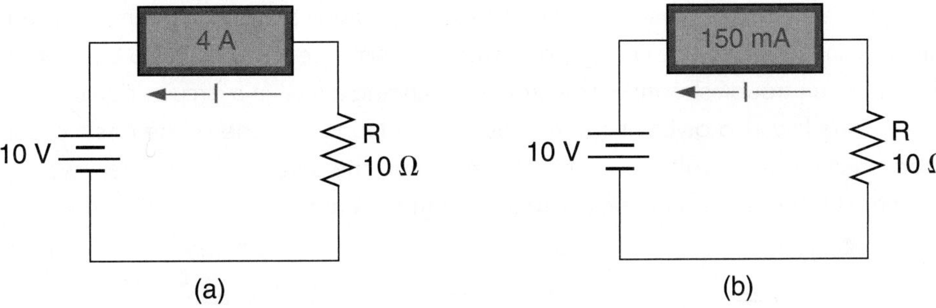

FIGURE 4.13

The second statement can be used to determine the possible causes of the problem shown in Figure 4.13b. Ohm's law tells us that the measured current in the circuit should be approximately 1 A. However, the ammeter shows a reading of 150 mA. This low current indicates (according to statement #2) that either the applied voltage is too low or the circuit resistance is too high.

In each of these cases, Ohm's law has served a very valuable purpose: *It has told you the types of problems you are looking for.* Once you know the types of problems you are looking for, a few simple measurements with a multimeter will tell you which of those problems is actually present.

One Final Note

In this section, you have been shown how Ohm's law can be used to:

1. Predict the response of a circuit to a change in voltage or resistance.
2. Calculate the value of current, voltage, or resistance in a circuit.
3. Determine the possible causes of a problem in a simple circuit.

As you continue with the rest of this chapter (and the chapters to follow), you will see that Ohm's law is the most frequently used electrical principle. For this reason, you should take the time to become thoroughly familiar with Ohm's law and all that it can do for you.

Summary

Ohm's law states that current is directly proportional to voltage and inversely proportional to resistance. These cause-and-effect relationships are summarized in Table 4.1.

TABLE 4.1 • Summary of Ohm's Law Relationships

CAUSE	...RESULTS IN...	EFFECT
Increased voltage		Increased current
Decreased voltage		Decreased current
Increased resistance		Decreased current
Decreased resistance		Increased current

The relationships given in Table 4.1 allow us to predict how the current in a given circuit will respond to any change in voltage or resistance. They can also be used to help in circuit troubleshooting, the process of finding a fault in a circuit or system.

Ohm's law also gives us the means of calculating any one of the basic circuit properties (current, voltage, or resistance) when the values of the other two are known. The three mathematical forms of Ohm's law are:

$$I = \frac{E}{R} \qquad E = I \times R \qquad R = \frac{E}{I}$$

PROGRESS CHECK

1. What is Ohm's law?
2. Describe the relationship between voltage and current.
3. Describe the relationship between resistance and current.
4. What happens to the current through a rheostat when you adjust its value. Explain your answer.
5. List the three Ohm's law equations.
6. What is *troubleshooting*?
7. List the possible causes of high circuit current.
8. List the possible causes of low circuit current.
9. When troubleshooting, what purpose does Ohm's law serve?
10. List the three most common applications of Ohm's law.

4.2 POWER AND EFFICIENCY

POWER The rate at which a component, circuit, or system uses energy.

Now that we have discussed current, voltage, and resistance, it is time to take a look at a fourth electrical property that is found in every circuit: **POWER**. Power is the rate at which a component, circuit, or system uses energy. Power consumption by electrical circuits always results in the production of some amount of heat. The more power a given circuit consumes, the more heat it produces. In this section, we will discuss power, how its value is calculated, and how it relates to heat. We will also take a look at a circuit rating that is based on power: *efficiency*.

Power

Technically defined, *power (P)* is the amount of energy used per unit time. When a voltage source uses one *joule* of energy per second to generate current, we say that it has used one **WATT (W)** of power. By formula,

WATT (W) The unit of measure for power. One watt equals one joule per second.

$$1 \text{ watt} = 1 \text{ joule per second}$$

This relationship is illustrated in Figure 4.14. As you can see, we have a 1 V source that is generating 1 A of current through 1 Ω of resistance. Now, consider the following relationships that were introduced in Chapter 2:

1 A = 1 coulomb per second
1 V = 1 joule per coulomb

The power supply in Figure 4.14 is using one joule of energy to move one coulomb of charge per second. Therefore, it is using one joule of energy per second. This rate of energy use is equal to one watt (W) of power. The relationship between power, voltage, and current can be seen by comparing the circuit in Figure 4.14 with those in Figure 4.15.

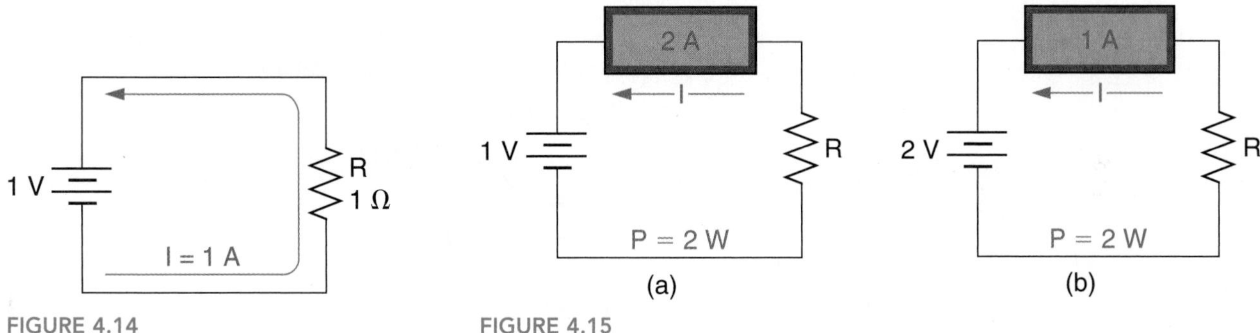

FIGURE 4.14

FIGURE 4.15

In Figure 4.15a, 1 V is being used to generate 2 A of current. In this case, the power supply is using one joule of energy per second to move *each* of the two coulombs of charge. Therefore, the power supply is using two joules per second, or 2 W.

In Figure 4.15b, 2 V is being used to generate 1 A of current. In this case, the power supply is using two joules of energy per second to move one coulomb of charge. Again, the power supply is using two joules per second, or 2 W.

Calculating Power (Watt's Law)

You have seen that power is related to both current and voltage. The mathematical relationship between the three values, which is known as Watt's law, is given as

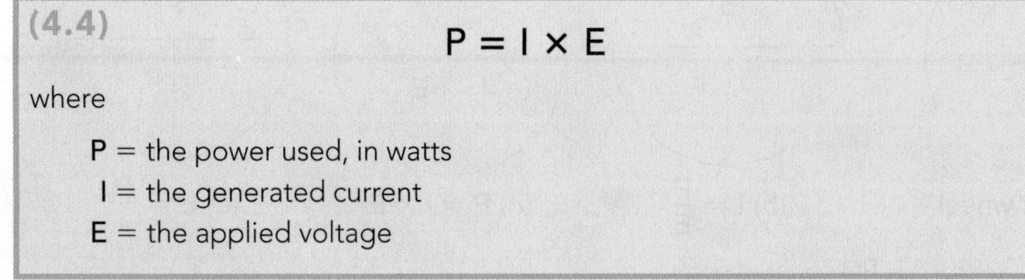

(4.4)

$$P = I \times E$$

where

P = the power used, in watts
I = the generated current
E = the applied voltage

Example 4.9 demonstrates the use of this equation.

EXAMPLE 4.9

How much power is being provided by the voltage source in Figure 4.16?

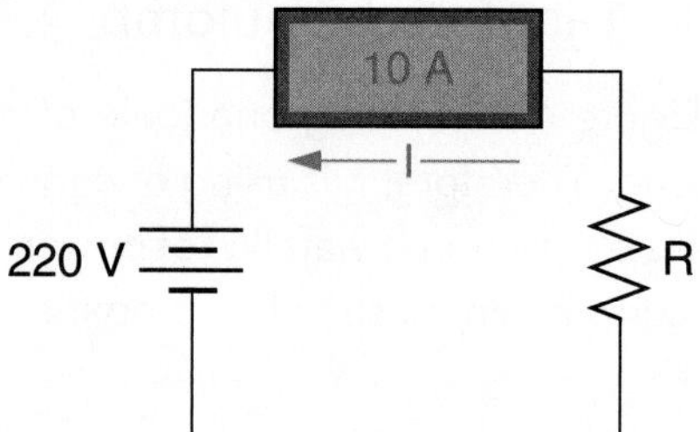

FIGURE 4.16

SOLUTION

The power being provided by the voltage source can be found as

$$\begin{aligned} P &= I \times E \\ &= 10\text{ A} \times 220\text{ V} \\ &= 2200\text{ W} \\ &= 2.2\text{ kW} \end{aligned}$$

PRACTICE PROBLEM 4.9

A 120 V source is generating 5.25 A of current. Calculate the power that is being provided by the source.

The Power Wheel

The power wheel can help you remember the various forms of the basic power equation, as illustrated in Figure 4.17. First, draw the wheel shown in Figure 4.17a. Then, cover the value of interest to see the correct equation. For example, when you want to calculate current (I), cover the letter *I* as shown in Figure 4.17b. The wheel shows that I is found as P/E. The other two forms of the basic power equation can be identified as shown in the figure.

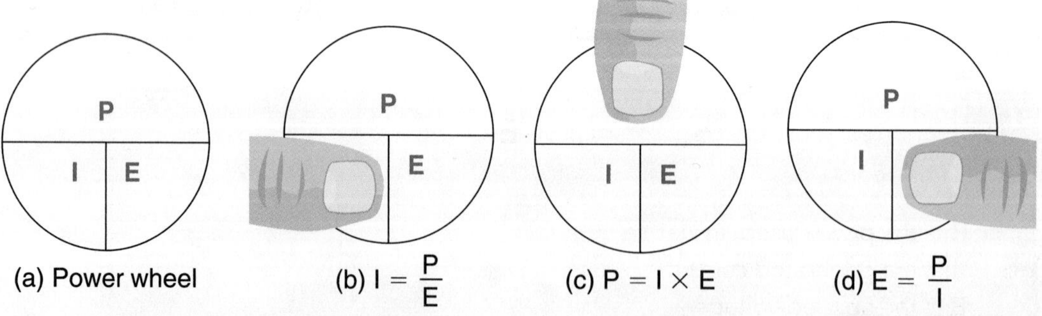

FIGURE 4.17 The power wheel.

Other Power Equations

We can use Ohm's law to derive two other useful power equations. By substituting (I × R) for the value of (E) in equation (4.4), we obtain an equation that defines power in terms of current and resistance, as follows:

$$P = I \times E$$
$$= I \times (I \times R)$$

or

(4.5)
$$P = I^2R$$

As Example 4.10 demonstrates, this equation can be used to calculate power when the values of current and resistance are known.

EXAMPLE 4.10

How much power is being provided by the voltage source in Figure 4.18?

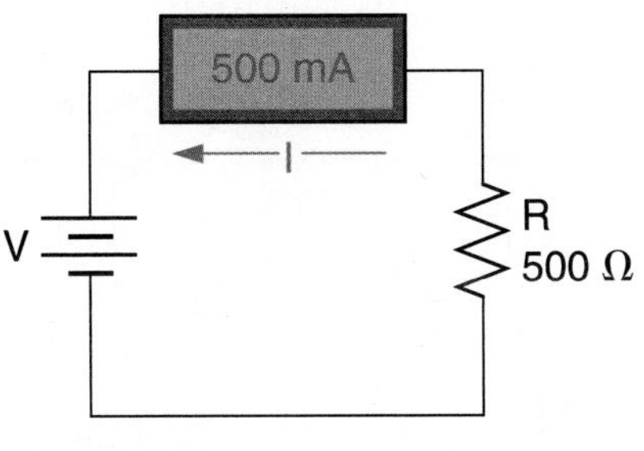

FIGURE 4.18

SOLUTION

The power being provided by the voltage source can be found as

$$P = I^2R$$
$$= (500\ \text{mA})^2 \times (500\ \Omega)$$
$$= 125\ \text{W}$$

PRACTICE PROBLEM 4.10

A circuit like the one in Figure 4.18 has values of $I = 0.75$ A and $R = 300\ \Omega$. Calculate the amount of power being provided by the source.

By substituting E/R for the value of I in equation (4.4), we obtain an equation that defines power in terms of voltage and resistance, as follows:

$$P = I \times E$$
$$= \frac{E}{R} \times E$$

or

(4.6)
$$P = \frac{E^2}{R}$$

As Example 4.11 demonstrates, this equation can be used to calculate power when the values of voltage and resistance are known.

EXAMPLE 4.11

How much power is being provided by the voltage source in Figure 4.19?

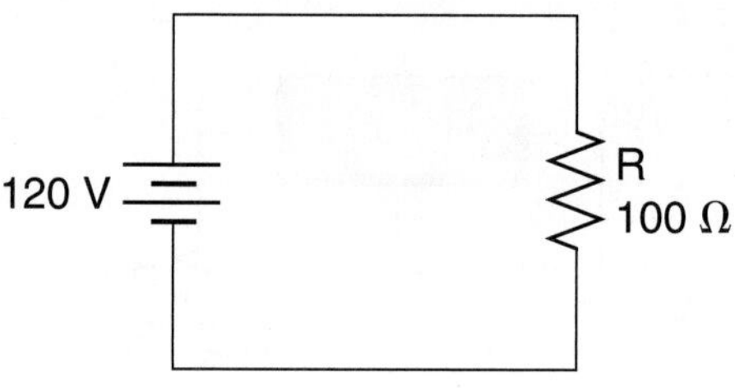

FIGURE 4.19

SOLUTION

The power being provided by the voltage source can be found as

$$P = \frac{E^2}{R}$$
$$= \frac{(120\ V)^2}{100\ \Omega}$$
$$= 144\ W$$

PRACTICE PROBLEM 4.11

A circuit like the one in Figure 4.19 has values of $E = 120\ V$ and $R = 470\ \Omega$. Calculate the power being provided by the source.

As you can see, power can be calculated using any two basic circuit properties and the appropriate power relationship. The basic power relationships are summarized as follows:

$$P = I \times E \qquad P = I^2R \qquad P = \frac{E^2}{R}$$

Power and Heat

It was mentioned earlier that power consumption by electrical components results in the production of heat. At this point, we're going to take a look at the connection between power and heat.

In basic science classes, we were all taught that matter cannot be created or destroyed. The same principle applies to energy. While energy may change forms, it cannot be created or destroyed.

Like energy, power is not created. It is merely transferred from one part of the circuit to another. This point is illustrated in Figure 4.20. As you can see, the DC power supply is drawing energy from the AC receptacle and transferring it to the resistor. Thus, the energy is merely being transferred from one place to another. It is not being created. At the same time, it cannot be destroyed, which leads us to the relationship between power and heat.

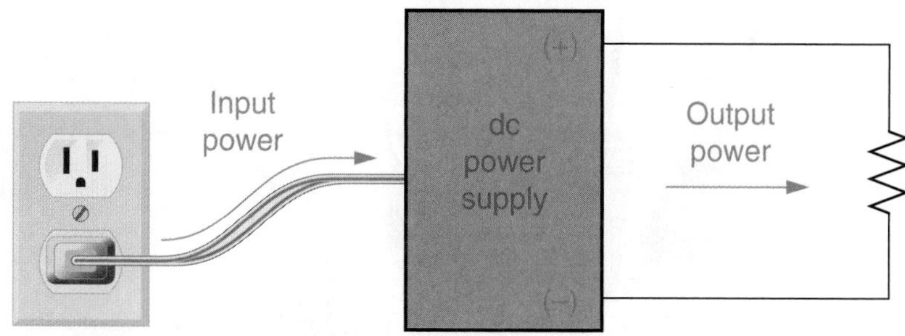

FIGURE 4.20

When energy is transferred to the resistor in Figure 4.20, the resistor absorbs that energy. Since this energy cannot be destroyed, the resistor converts it into another form: in this case, heat. Some other examples of energy conversion are as follows:

1. A toaster converts electrical energy to heat.
2. A light bulb converts electrical energy to light and heat.
3. A speaker converts electrical energy to mechanical energy (sound).

Note that any device designed to convert energy from one form to another is referred to as a **TRANSDUCER**.

TRANSDUCER A device that converts energy from one form to another.

As you know, power is defined as *energy used per unit time*. If a component such as a resistor is absorbing a given amount of energy per unit time, it must be capable of converting that energy into heat at the same rate. This is why the power rating of a component is critical. If you exceed the power rating of a component, that component will not be able to dissipate the heat it is generating quickly enough. As a result, the component will keep getting hotter, and will eventually be destroyed by the excessive heat.

Resistor Power Ratings

You have been told that the *power rating* of a component is a measure of its ability to dissipate heat, measured in watts (W). All resistors have power ratings. If the power rating of a resistor is exceeded, the resistor will likely be destroyed as a result of excessive heat.

The power rating of a carbon-composition resistor is indicated by its size. Figure 4.21 shows some pairs of common resistors. From left to right, the resistor pairs have ratings of 1 W, ½ W, ¼ W, and ⅛ W. The higher the power rating of a resistor, the more heat it can dissipate. Note that the other types of resistors (such as wire-wound resistors) usually have their power ratings printed on them.

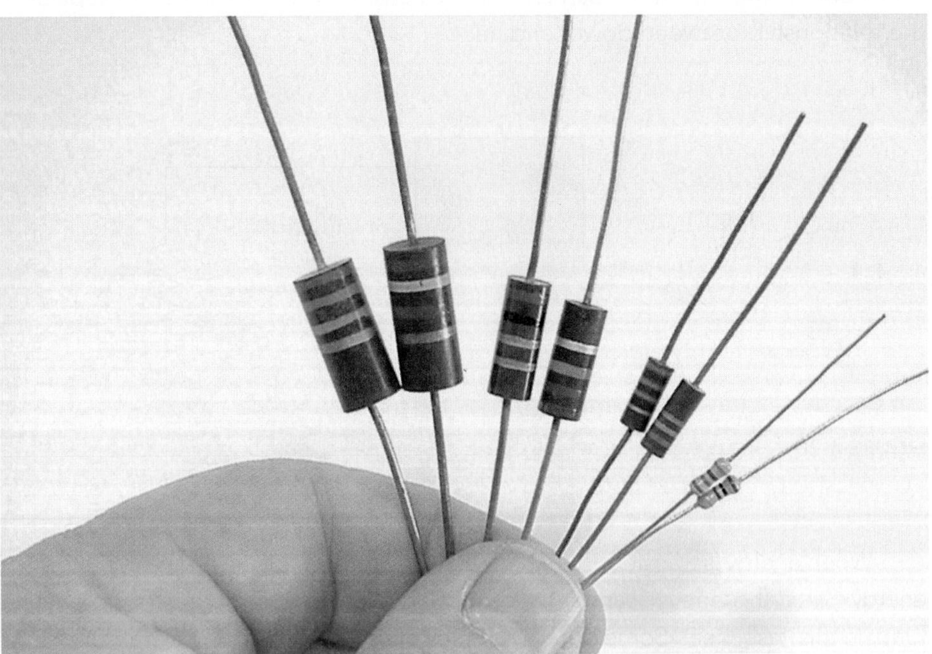

FIGURE 4.21 Pairs of common resistors.

When choosing a resistor for a particular application, the power rating of the component should be greater than the amount of power it is expected to dissipate. This ensures that the resistor will be able to dissipate the applied power without being destroyed. A practical application of this principle is provided in the following example.

EXAMPLE 4.12

What is the minimum acceptable power rating for the resistor in Figure 4.22?

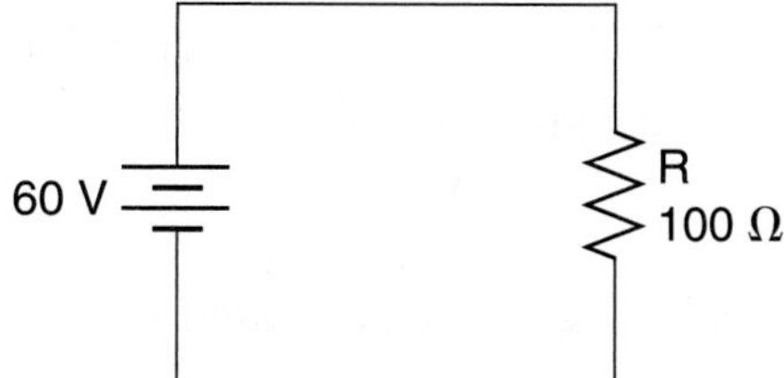

FIGURE 4.22

SOLUTION

The minimum acceptable power rating for the resistor can be found as

$$P = \frac{E^2}{R}$$
$$= \frac{(60\ \text{V})^2}{100\ \Omega}$$
$$= 36\ \text{W}$$

Since this power must be dissipated by the resistor, the power rating of the component must be greater than 36 W.

PRACTICE PROBLEM 4.12

A circuit like the one in Figure 4.22 has values of $E = 18$ V and $R = 10\ \Omega$. Determine the minimum acceptable power rating for the resistor.

Efficiency

Every circuit contains a variety of components that use some amount of power. Therefore, the output power from a circuit must be less than its input power.

The **EFFICIENCY** of a circuit is the ratio of its output power to its input power, given as a percent. By formula,

Power losses occur throughout a circuit. There are power losses in the load, its supply conductors, and its power supply. These losses can have many causes, including conductor resistance, friction, and mechanical loading.

IN THE FIELD

EFFICIENCY The ratio of circuit output power to input power, given as a percentage.

(4.7)

$$\eta = \frac{P_{OUT}}{P_{IN}} \times 100\%$$

where

η = the circuit efficiency, written as a percent
P_{OUT} = the circuit output power
P_{IN} = the circuit input power

Example 4.13 shows how the efficiency of a DC power supply is calculated.

EXAMPLE 4.13

A DC power supply has the following ratings: $P_{OUT} = 5\text{ W}$ and $P_{IN} = 12\text{ W}$. What is the efficiency of the power supply?

SOLUTION

The efficiency of the power supply can be found as

$$\eta = \frac{P_{OUT}}{P_{IN}} \times 100\%$$
$$= \frac{5\text{ W}}{12\text{ W}} \times 100\%$$
$$= 41.7\%$$

This means that 41.7% of the power supply's input power is converted to output power. The other 58.3% of the input power is used by the power supply itself.

PRACTICE PROBLEM 4.13

A DC power supply has the following maximum ratings: $P_{OUT} = 20\text{ W}$ and $P_{IN} = 140\text{ W}$. What is the efficiency of the power supply?

There are a few points that should be made about efficiency. First, it is not possible for circuit efficiency to be greater than 100%. An efficiency greater than 100% would imply that the output power is greater than the input power, which is impossible. (Remember: Energy cannot be created.) Also, since all electrical circuits absorb some amount of power, every practical efficiency rating is less than 100%.

Energy Measurement

Every appliance in your home uses energy. While this energy cannot be measured directly, it can be determined by measuring the amount of power used and the time over which it is used.

The practical unit of measure for the energy used by a utility customer is the **KILOWATT-HOUR (kWh)**. One kilowatt-hour is the amount of energy used by a 1000 W device that is run for one hour. The energy you use, in kilowatt-hours, is found as

KILOWATT-HOUR (kWh) The product of power (in kilowatts) and time (in hours), used as a unit of measure of energy.

(4.8)

$$E_{kWh} = \frac{P \times t}{1000}$$

where

P = the power used, in watts
t = the time, in hours

Note that the 1000 in the denominator is a conversion factor. It converts watts into kilowatts (so the answer is in kilowatt-hours). Example 4.14 demonstrates the use of equation (4.8).

EXAMPLE 4.14

How much energy is required to run twenty 60 W light bulbs for six hours?

SOLUTION

The amount of power used by twenty 60 W light bulbs is found as

$$20 \times 60\text{ W} = 1200\text{ W}$$

Using this combined power, the total energy used is found as

$$E_{kWh} = \frac{P \times t}{1000}$$
$$= \frac{1200\text{ W} \times 6\text{ hours}}{1000}$$
$$= 7.2\text{ kWh}$$

PRACTICE PROBLEM 4.14

How much energy is used by a 1500 W dishwasher, a 3600 W clothes dryer, and a 750 W air conditioner that are all being used for 2 hours?

The **SERVICE METER** at your home measures energy in kilowatt-hours. Analog service meters have a series of dials that are laid out as shown in Figure 4.23. Note that (from left to right), the first two dials indicate *thousands* of kWh, the third dial indicates *hundreds* of kWh, and so on. For example, the dial shown in Figure 4.23 indicates that 18,592 kWh of energy have been used. Note that the dials on an analog service meter alternate between clockwise and counterclockwise rotation.

SERVICE METER A meter that measures the energy (in kWh) provided by the power company.

FIGURE 4.23 A watt-hour meter.

The Bottom Line

Power is the amount of energy supplied or absorbed by a circuit or component. Technically defined, it is the amount of energy used per unit time.

The basic unit of power is the watt (W). One watt of power is equal to one joule per second. When a voltage source uses one joule of energy per second to generate current, we say that it is using one watt of power.

Power can be calculated when any two of the following values are known: current, voltage, and resistance. The basic power equations are as follows:

$$P = I \times E \qquad P = I^2R \qquad P = \frac{E^2}{R}$$

Energy cannot be created or destroyed. It is merely transferred from one device to another and/or converted from one form to another. When a voltage source provides a given amount of power to a resistive circuit, that power is absorbed by the circuit and then dissipated in the form of heat.

If the power rating of a resistor is lower than the amount of power it is absorbing, it cannot dissipate the total heat being generated. Thus, it continues to get hotter, and is eventually destroyed.

The efficiency of a circuit or component is the ratio of its output power to its input power, given as a percent. The higher the efficiency rating of a circuit or component, the less power it actually absorbs. In practice, no circuit or component is 100% efficient. They all absorb some amount of power, and therefore, have efficiency ratings that are less than 100%.

Utility companies measure residential energy consumption in kilowatt-hours (kWh). One kilowatt-hour is the amount of energy that a 1000 W device uses when run for one hour.

PROGRESS CHECK

1. What is *power*?
2. Explain the *watt* unit of power.
3. Describe the relationship between power and current.
4. Describe the relationship between power and voltage.
5. List the three basic power equations.
6. Discuss the relationship between power and heat.
7. What is a *transducer*?
8. What is *efficiency*? How is it calculated?
9. Why can't a circuit or device have an efficiency that is greater than 100%?
10. What unit is used by power companies to measure residential energy consumption? What does this unit equal?

4.3 BASIC CIRCUIT CALCULATIONS

In any circuit problem, you have two types of variables: those with values that are known (given in the problem) and those with values that are unknown. Any problem generally requires that you *define one or more of the unknown variables in terms of the known variables.* For example, consider the simple problem shown in Figure 4.24. In order to solve for the circuit current (the unknown value), we must define it in terms of voltage and resistance (the known values). In this case, the solution is simple: We just use Ohm's law in the form of

$$I = \frac{E}{R}$$

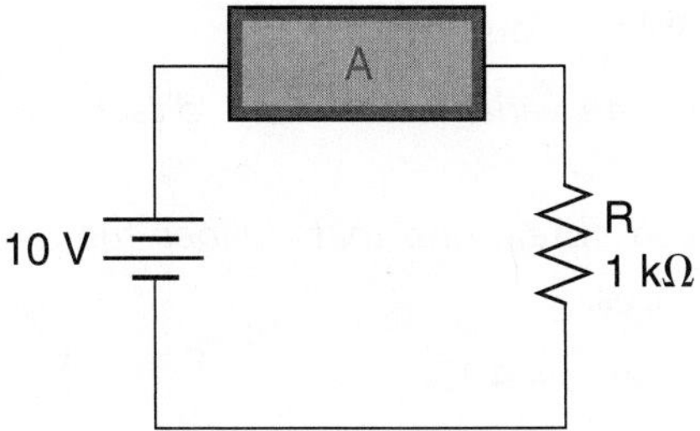

FIGURE 4.24 A circuit with an unknown value of current.

This equation defines our unknown value (current) in terms of our known values (voltage and resistance). Solving the equation solves the problem.

In many cases, the solution is not as obvious as it is in Figure 4.24. For example, consider the problem in Figure 4.25. In this case, we are being asked to solve for the value of the resistor. We are given the values of applied voltage and power. Now, consider the equations we have used so far:

$$I = \frac{E}{R} \qquad E = I \times R \qquad R = \frac{E}{I}$$

$$P = \frac{E^2}{R} \qquad P = I^2R \qquad P = I \times E$$

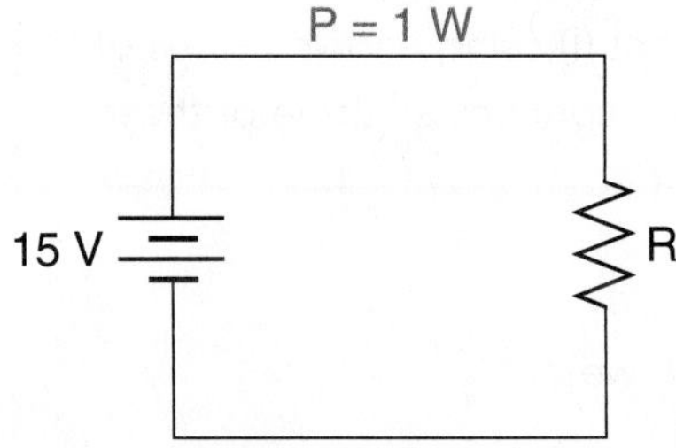

FIGURE 4.25 A circuit with an unknown value of resistance.

None of these equations directly defines resistance in terms of voltage and power. However, if you look closely at the equations, you will see that there is one equation that contains *all three* of our variables. That one is:

$$P = \frac{E^2}{R}$$

If we use algebra to rearrange the equation, we get

$$R = \frac{E^2}{P}$$

Now we have an equation that defines our unknown value (resistance) in terms of our known values (voltage and power), and we can solve the problem.

When you encounter any basic circuit problem, you should:

1. *Find an equation that contains all of the variables that are directly involved in the problem.*
2. *If necessary, rearrange the equation to obtain one that defines the unknown value in terms of the known values.*

The use of this procedure is demonstrated in Example 4.15.

EXAMPLE 4.15

How much current is passing through the circuit in Figure 4.26?

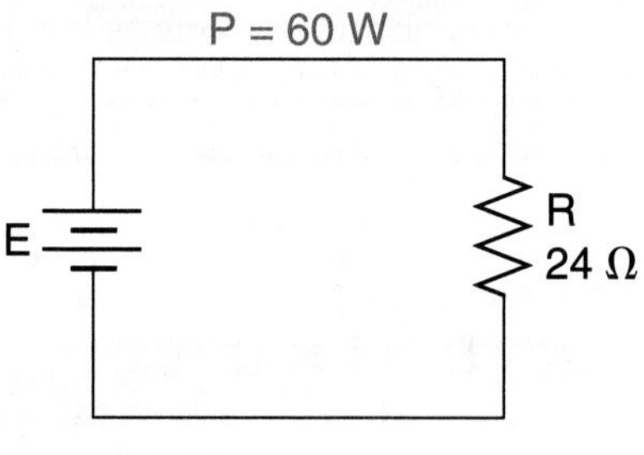

FIGURE 4.26

SOLUTION

We are trying to find the value of circuit current (I) using known values of power (P) and resistance (R). The equation that contains all three of these properties is:

$$P = I^2R$$

Rearranging this equation to solve for current, we get

$$I = \sqrt{\frac{P}{R}}$$

In this case,

$$I = \sqrt{\frac{P}{R}}$$
$$= \sqrt{\frac{60\ \text{W}}{24\ \Omega}}$$
$$= 1.58\ \text{A}$$

PRACTICE PROBLEM 4.15

A circuit like the one shown in Figure 4.26 has the following values: P = 75 W and R = 18 Ω. What is the value of the circuit current?

By following the procedure demonstrated in the example, you can solve for any one of the basic circuit properties (current, voltage, resistance, or power) when any *two* values are known.

Resistor Tolerance

Up to this point, we have ignored the effect that resistor tolerance can have on the measured current in a basic circuit. As the following example illustrates, the actual current in a circuit will fall within a range of values that is determined by the tolerance of the resistor.

EXAMPLE 4.16

What are the minimum and maximum values of current for the circuit in Figure 4.27?

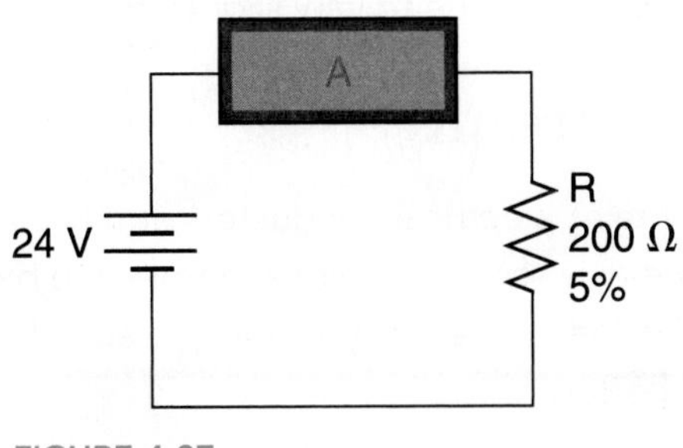

FIGURE 4.27

SOLUTION

The resistor in Figure 4.27 is a 200 Ω resistor with a 5% tolerance. The range of possible values for the component is found as

$$200\ \Omega \times 5\% = 10\ \Omega$$
$$200\ \Omega - 10\ \Omega = 190\ \Omega \quad \text{(minimum)}$$
$$200\ \Omega + 10\ \Omega = 210\ \Omega \quad \text{(maximum)}$$

The maximum circuit current is found using the minimum resistor value and Ohm's law, as follows:

$$I = \frac{E}{R}$$
$$= \frac{24\ \text{V}}{190\ \Omega}$$
$$= 126\ \text{mA}$$

The minimum circuit current is found using the maximum resistor value and Ohm's law, as follows:

$$I = \frac{E}{R}$$
$$= \frac{24\ \text{V}}{210\ \Omega}$$
$$= 114\ \text{mA}$$

These results indicate that the actual circuit current may fall anywhere between 114 mA and 126 mA.

PRACTICE PROBLEM 4.16

A circuit like the one in Figure 4.27 has a 48 V source and a resistor that is color coded as follows: Blue-Gray-Brown-Silver. Determine the range of current values for the circuit.

As you can see, the tolerance of a resistor can affect the actual value of current in a circuit. When working in lab, keep in mind that the tolerance of a resistor may cause a measured current value to vary slightly from its predicted value.

The Simplest Circuit

The simplest circuit contains electrical conductors and two components: one that supplies power and one that absorbs it. For example, the battery in Figure 4.28 is supplying power to the load. The load, in turn, is absorbing the power that is being supplied by the battery.

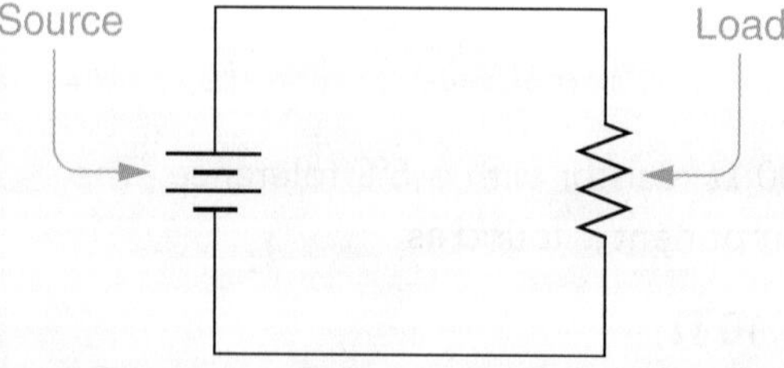

FIGURE 4.28 The components found in the most basic circuit.

In a circuit like the one in Figure 4.28, we generally refer to the components as the *source* and the *load*. The source supplies the power and the load absorbs (uses) it.

Every circuit or system, no matter how simple or complex, is designed to deliver power (in one form or another) to one or more loads. For example, consider the items shown in Figure 4.29. The boom box (a fairly complex electronic system) is designed to deliver power to its load: the speakers. The table lamp (a relatively simple electrical device) is designed to deliver power to its load: the light bulb. In both cases, delivering power to the load is the purpose of the system or device.

FIGURE 4.29 A boom box and lamp.
Fotolia.com

When the resistance of a load is variable (as shown in Figure 4.30), the minimum load resistance is referred to as a **FULL LOAD**. A full load is one that draws maximum current from the source. As Ohm's law tells us, maximum current is produced when the load resistance is at its minimum value.

FULL LOAD A load that draws maximum current from the source.

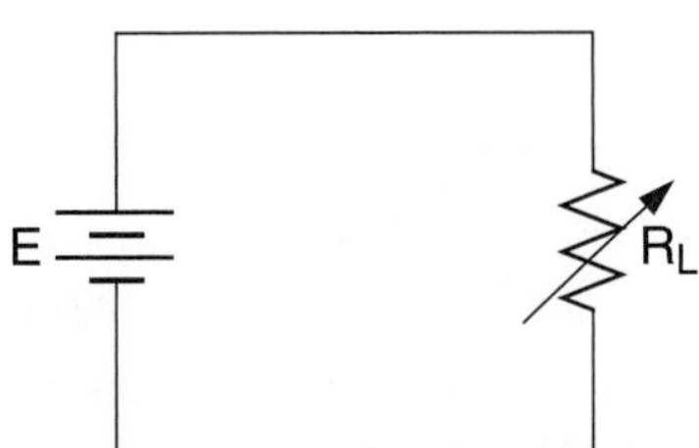

FIGURE 4.30 A source with a variable load.

As you will see, circuits always contain one or more loads. You will also find that our goal in analyzing these circuits is usually to determine the values of load current, voltage, and/or power.

PROGRESS CHECK

1. Describe the procedure for selecting the proper equation to solve a basic circuit problem.
2. What effect does resistor tolerance have on a measured value of circuit current?
3. What is a *source*? What is a *load*?
4. What is a *full load*?

4.4 BASIC CIRCUIT FAULTS AND SYMPTOMS

Two common types of circuit faults are referred to as the open circuit and the short circuit. In this section, we examine the symptoms associated with each of these faults.

Circuit Faults: The Open Circuit

OPEN CIRCUIT A complete break in the current path through all or part of a circuit.

An **OPEN CIRCUIT** is a physical break in a conduction path. Figure 4.31 illustrates the characteristics of an open circuit.

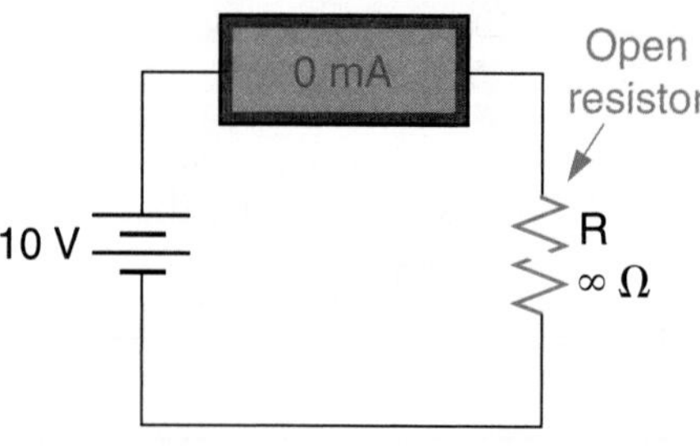

FIGURE 4.31

You already know that a fuse blocks current when it opens, because it breaks the required path for the flow of charge. We get the same result if any component physically opens: The path for conduction is broken and the current drops to zero.

Since an open component is a physical break in a circuit, the resistance of the open component is infinite (for all practical purposes). Thus, if you attempt to measure the resistance of an open component, its resistance will be too high for the ohmmeter to measure. Many DMMs display the letters "OL" (for Over Limit) when resistance is too high to be measured on the selected scale.

When a component opens in a basic circuit, the applied voltage is dropped across the open. This point is illustrated in Figure 4.32. When the voltmeter is placed across the open (as shown in Figure 4.32), it is effectively connected across the open terminals of the battery. Thus, we measure the full applied voltage across the open. As a summary, here are the characteristics of an open circuit:

1. The circuit current drops to zero.
2. The resistance of the component is too high to measure.
3. The applied voltage can be measured across the open.

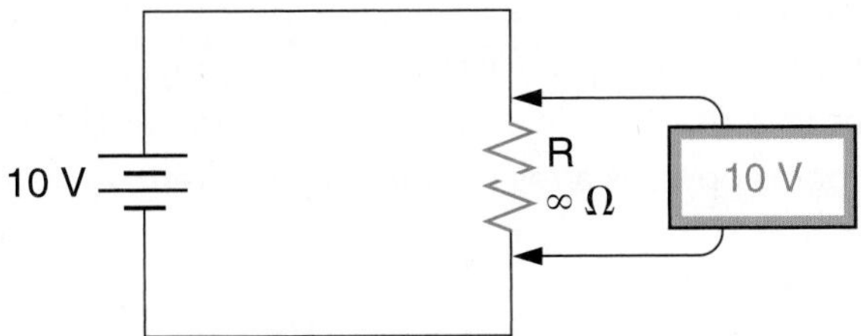

FIGURE 4.32 The voltage across an open component.

Any component can open. Normally, a component opens when the current becomes too high, causing excessive power to be absorbed. The excessive power causes the component to burn open. Note that an open in a circuit is repaired by replacing the open component.

Circuit Faults: The Short Circuit

A **SHORT CIRCUIT** is the opposite of an open. A short circuit is an unintentional low-resistance connection between two or more points in a circuit. For example, consider the circuit shown in Figure 4.33. Here we have created a short by connecting a wire across a lamp. Since current takes the path of least resistance (an old—but true—saying), we have effectively dropped the circuit resistance to 0.5 Ω as shown in the figure. Note that the circuit current bypasses the lamp, so the lamp does not light.

SHORT CIRCUIT An unintentional, extremely low resistance connection that allows an abnormally high current through all or part of a circuit.

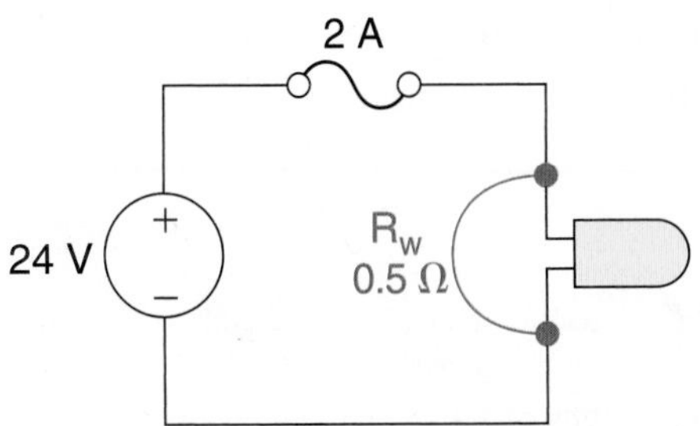

FIGURE 4.33 A shorted lamp.

Ohm's law tells us that a decrease in resistance causes an increase in circuit current. For example, with the total short-circuit resistance of 0.5 Ω, the current in Figure 4.33 is found as:

$$I = \frac{E}{R} = \frac{24\text{ V}}{0.5\ \Omega} = 48\text{ A}$$

In this case, the circuit current exceeds the rating of the fuse, and the fuse opens. At the same time, if there were no fuse in the circuit, the excessive current would eventually cause the short circuit current path (or the power supply) to burn open. In other words, the current in a shorted circuit is extremely high until the fuse (or some other component) burns open as a result of the high current.

Overcurrent protective devices are designed to interrupt any fault current level greater than the component rating. The *interruption rating* of a fuse or circuit breaker indicates the maximum current that the component can interrupt at a specified voltage.

IN THE FIELD

Ohm's law tells us that E = I × R. Since the resistance of a shorted component is extremely low, the voltage across the shorted component is extremely low. As a summary, here are the characteristics of a short circuit:

1. The resistance of the shorted component is extremely low.
2. The current through the short circuit current path is extremely high. This high current will likely cause a circuit fuse (or some other component) to burn open.
3. The voltage across the shorted component is extremely low.

One Final Note

We have now completed our coverage of circuits that contain only one resistive element. In Chapter 5, we will start to discuss circuits that contain more than one resistive element. As you will see, the circuits become more complex, but the principles and relationships that we have established so far still hold true.

PROGRESS CHECK

1. List the characteristics of an open circuit.
2. List the characteristics of a short circuit.

4.5 SUMMARY

Here is a summary of the major points that were covered in this chapter:

Ohm's Law

- Ohm's law states that *current is directly proportional to voltage and inversely proportional to resistance.*
- The current through a fixed resistance:
  - Increases when the applied voltage increases.
  - Decreases when the applied voltage decreases.
- The current generated by a fixed voltage:
  - Decreases when the circuit resistance increases.
  - Increases when the circuit resistance decreases.
- Ohm's law can be used to predict the response of circuit current to a change in voltage or resistance.
- When you adjust any circuit component, there is always some type of circuit response to the change.
- Troubleshooting is the process of locating faults in an electronic circuit or system.
- When troubleshooting, Ohm's law is used to determine the types of potential problems you are looking for.
- Ohm's law is commonly used to:
  - Predict the response of a circuit to a change in voltage or resistance.
  - Calculate the value of current, voltage, or resistance in a circuit.
  - Determine the possible causes of a problem in a circuit.

Power and Efficiency

- Power is the amount of energy used per unit time.
- Power is measured in watts (W). One watt of power is equal to one joule per second.
- Power is directly proportional to both current and voltage.
- In any circuit, power is not created. It is merely transferred from one part of the circuit to another.
- Any device that converts energy from one form to another is called a *transducer*.
- Resistors convert the energy they absorb to heat.
- When choosing a resistor for a given application, the power rating of the component must be greater than the anticipated power in the circuit.
- Efficiency is the ratio of output power to input power, given as a percent.
- It is not possible to have an efficiency rating that is greater than 100%.
- Utilities use the kilowatt-hour (kWh) as the unit of measure for energy. One kilowatt-hour is the amount of energy that a 1000 W appliance uses in one hour.

Basic Circuit Calculations

- To solve basic circuit problems:
  - Find an equation that contains all of the variables that are directly involved in the problem.
  - If necessary, rearrange the equation to obtain one that defines the unknown value in terms of known values.

- Since resistors have tolerances, the measured current in a circuit will actually fall within a range of values.
- A source delivers power. A load absorbs (uses) power.
- The purpose of a circuit or system is to deliver power (in one form or another) to one or more loads.
- A full load is one that draws maximum current from the source.
- An *open circuit* is a break in the conduction path of a circuit. An open circuit has the following characteristics:
    - Circuit current is zero.
    - The resistance of the open component is too high to measure.
    - The applied voltage can be measured across the open.
- A *short circuit* is a low-resistance condition that does not normally exist in a circuit. A short circuit has the following characteristics:
    - The resistance of the short circuit is extremely low.
    - The current through the short circuit is extremely high. This high current may cause a circuit fuse (or some other component) to burn open.
    - The voltage across the shorted component is very nearly zero.

CHAPTER REVIEW

1. Ohm's law is a powerful tool that is used to ________.
    a) predict how a circuit reacts to changes in current
    b) predict how a circuit reacts to changes in voltage
    c) predict how a circuit reacts to changes in resistance
    d) predict all of the above
2. Current is ________ proportional to voltage and ________ proportional to resistance.
    a) inversely, inversely
    b) inversely, directly
    c) directly, inversely
    d) directly, directly
3. The current in a circuit has increased, and the resistance has remained unchanged. This means that ________.
    a) the voltage has increased
    b) the voltage has decreased
    c) the power has decreased
    d) the voltage has increased and power has decreased
4. The voltage applied to a circuit has tripled, but the resistance has decreased by a factor of 3.5. In this case, the current has ________.
    a) increased by a factor of 6.5
    b) decreased by a factor of 6.5
    c) increased by a factor of 10.5
    d) decreased by a factor of 1.5
5. A circuit has the following values: $E = 1.6$ kV and $I = 8.8$ A. The circuit resistance is ________.
    a) 1.82 kΩ
    b) 182 Ω

c) 5.5 kΩ

d) 550 Ω

6. A circuit has 220 V applied to it which produces 11.8 A of current. If the value of the applied voltage doubles and the current does not change, the value of the circuit resistance changes to ________.

   a) 37.2 kΩ

   b) 9.3 Ω

   c) 93 Ω

   d) 18.6 Ω

7. A given circuit has values of E = 220 V and R = 45 Ω. If the fuse in this circuit should not blow under these conditions, the lowest value fuse you could use is ________.

   a) 1.5 A

   b) 2 A

   c) 5 A

   d) 7.5 A

   e) 10 A

8. A circuit dissipates 36 watts of power over a period of 3 minutes. This circuit has used ________ joules of energy.

   a) 108

   b) 12

   c) 129,600

   d) 6480

9. An air conditioner uses 2.358 megajoules of energy in 30 minutes. The power consumption for this device is ________.

   a) 1310 W

   b) 78.6 kW

   c) 1.179 MW

   d) 2620 W

10. The current through a component has increased by a factor of 3.5. The power dissipated by this component ________.

    a) increased by a factor of 7

    b) increased by a factor of 12.25

    c) decreased by a factor of 3.5

    d) decreased by a factor of 7

11. If the resistance of a circuit decreases and the applied voltage remains the same, then the circuit power ________.

    a) must increase

    b) must decrease

    c) remains the same

    d) cannot be determined with this information

12. A 550 V power supply delivers 8.5 A of current to a motor. The power supplied to the motor equals ________.

    a) 64.7 W

    b) 4675 W

c) 468 W

d) 647 W

13. A speaker has a resistance of 8 Ω. If 2.85 A of current is delivered to the speaker by the amplifier, the speaker is dissipating ______.

a) 23 W

b) 65 W

c) 1.02 W

d) 22.5 W

14. 118 V is measured across a heating element with a known resistance of 34.8 Ω. This means that this heating element produces ________.

a) 4.1 kW

b) 48 kW

c) 4 kW

d) 400 W

15. Any device that converts energy from one form to another is called a(n) ________.

a) transducer

b) inverter

c) converter

d) transposer

16. A component is rated at 1200 W maximum. It is connected to a power supply that delivers a maximum of 305 V. The maximum allowable current rating for the circuit breaker in this circuit is ________.

a) 3 A

b) 3.5 A

c) 8.0 A

d) 7.0 A

17. A circuit is rated at 4500 W maximum. The resistance of this circuit is known to be 110 Ω. The maximum allowable current rating for the circuit breaker in this circuit is ________.

a) 6.5 A

b) 7.0 A

c) 8.0 A

d) 6.0 A

18. A circuit that has low efficiency ________.

a) converts little of its input power to heat

b) converts most of its input power to output power

c) converts little of its input power to output power

d) converts most of its output power to heat

19. A 120 V power supply produces 6.5 A of current. The power delivered to the load is 260 W. This circuit has an efficiency rating of ________.

a) 33.3%

b) 66.6%

c) 7.1%

d) 11.7%

20. Two 60 W outdoor lights are left on all day for 30 weeks. The amount of energy used by these light bulbs is ________.
    a) 604.8 kWh
    b) 302.4 kWh
    c) 86.4 kWh
    d) 172.8 kWh

21. A 4.7 kΩ resistor dissipates 185 mW of power. The current through the resistor is ________.
    a) 39.4 μA
    b) 1.59 A
    c) 6.27 mA
    d) 29.5 mA

22. A component dissipating 1.86 W of power has 60 V across its terminals. The resistance of the component is ________.
    a) 32.3 kΩ
    b) 208 Ω
    c) 6.7 kΩ
    d) 1.94 kΩ

23. "Full load" occurs when ________.
    a) load resistance is at its maximum
    b) load resistance is at its minimum
    c) load power is at its minimum
    d) load voltage is at its maximum

PRACTICE PROBLEMS

1. For each combination of voltage and resistance, calculate the resulting current.
   a) E = 180 V R = 2.2 kΩ
   b) E = 80 V R = 18 Ω
   c) E = 120 V R = 470 Ω
   d) E = 16 V R = 33 Ω
   e) E = 150 V R = 6.8 Ω

2. For each combination of voltage and resistance, calculate the resulting current.
   a) E = 220 V R = 47 Ω
   b) E = 8.6 V R = 510 Ω
   c) E = 22.8 V R = 82 kΩ
   d) E = 10 V R = 1.5 MΩ
   e) E = 100 V R = 12 Ω

3. For each combination of current and resistance, calculate the voltage required to generate the current.
   a) I = 10 A R = 8.2 Ω
   b) I = 65 mA R = 100 Ω

c) I = 13 A   R = 2.7 Ω

d) I = 24 mA   R = 1.1 kΩ

e) I = 800 mA   R = 3.6 Ω

4. For each combination of current and resistance, calculate the voltage required to generate the current.

a) I = 20 A   R = 9.1 Ω

b) I = 3.3 A   R = 12 Ω

c) I = 14 mA   R = 180 Ω

d) I = 8 A   R = 12 Ω

e) I = 100 μA   R = 16 kΩ

5. For each combination of voltage and current, determine the resistance needed to limit the current to the given value.

a) E = 600 V   I = 4 A

b) E = 94 V   I = 200 mA

c) E = 110 V   I = 500 mA

d) E = 60 V   I = 2.2 A

e) E = 33 V   I = 330 μA

6. For each combination of voltage and current, calculate the resistance needed to limit the current to the given value.

a) E = 8.8 V   I = 24 mA

b) E = 12 V   I = 16 mA

c) E = 28 V   I = 8 mA

d) E = 180 mV   I = 260 μA

e) E = 1 V   I = 250 nA

7. For each combination of voltage and current, calculate the power that is being provided by the voltage source.

a) E = 600 V   I = 4 A

b) E = 94 V   I = 200 mA

c) E = 110 V   I = 500 mA

d) E = 60 V   I = 2.2 A

e) E = 33 V   I = 330 μA

8. For each combination of voltage and current given in problem 6, calculate the power that is being supplied by the voltage source.

9. For each combination of current and resistance, calculate the power that is being supplied by the voltage source.

a) I = 10 A   R = 8.2 Ω

b) I = 65 mA   R = 100 Ω

c) I = 13 A   R = 2.7 Ω

d) I = 24 mA   R = 1.1 kΩ

e) I = 800 mA   R = 3.6 Ω

10. For each combination of current and resistance given in problem 4, calculate the power that is being supplied by the voltage source.

11. For each combination of voltage and resistance, calculate the power that is being supplied by the voltage source.
    a) E = 180 V R = 2.2 kΩ
    b) E = 80 V R = 18 Ω
    c) E = 120 V R = 470 Ω
    d) E = 16 V R = 33 Ω
    e) E = 150 V R = 6.8 Ω
12. For each combination of voltage and resistance given in problem 2, calculate the power that is being supplied by the voltage source.
13. Calculate the minimum acceptable power rating for the resistor in Figure 4.34a.
14. Calculate the minimum acceptable power rating for the resistor in Figure 4.34b.
15. Calculate the minimum acceptable power rating for the resistor in Figure 4.34c.
16. Calculate the minimum acceptable power rating for the resistor in Figure 4.34d.

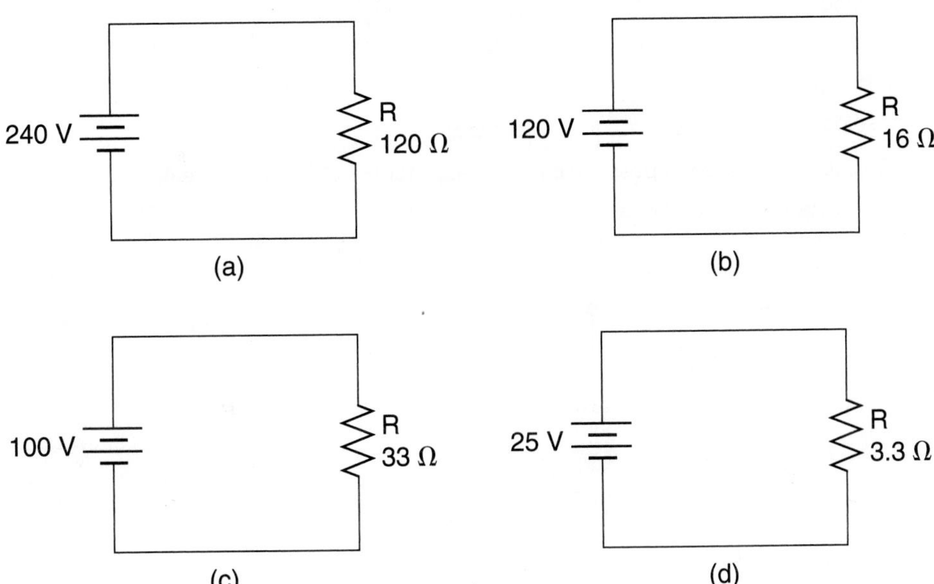

FIGURE 4.34

17. For each combination of input and output power, calculate the efficiency of the circuit.
    a) P_i = 150 W P_o = 36 W
    b) P_i = 600 W P_o = 44.5 W
    c) P_i = 85 W P_o = 6.7 W
    d) P_i = 1 kW P_o = 340 W
    e) P_i = 15 kW P_o = 600 W
18. A circuit working at 26% efficiency has an output power of 0.86 kW. Calculate the input power.
19. The input power to a circuit is found to be 1200 W. What is the load power if the circuit has an efficiency rating of 18.5%?
20. Calculate the energy used (in kWh) to run twelve 150 W light bulbs for eight hours.
21. Calculate the energy used (in kWh) to run a 500 W microwave oven and a 2400 W air conditioner for three hours.

22. Calculate the energy used to run the following combination of appliances for the times given: a 700 W microwave oven for 20 minutes, a 3800 W dishwasher for 30 minutes, a 2200 W air conditioner for 12 hours, and an 1800 W clothes dryer for 1 hour.

23. Complete the chart below.

	Current	*Voltage*	*Resistance*	*Power*
a)	10 mA	______	______	4 W
b)	______	32 V	______	16 mW
c)	______	______	3.3 kΩ	231 mW
d)	15 mA	45 V	______	______
e)	24 mA	______	1.2 kΩ	______

24. Complete the chart below.

	Current	*Voltage*	*Resistance*	*Power*
a)	______	______	1 kΩ	240 mW
b)	50 mA	2 V	______	______
c)	______	8 V	200 Ω	______
d)	______	16 V	______	800 mW
e)	35 mA	______	______	1.4 W

25. Calculate the range of possible current values for the circuit in Figure 4.35a.

26. Calculate the range of possible current values for the circuit in Figure 4.35b.

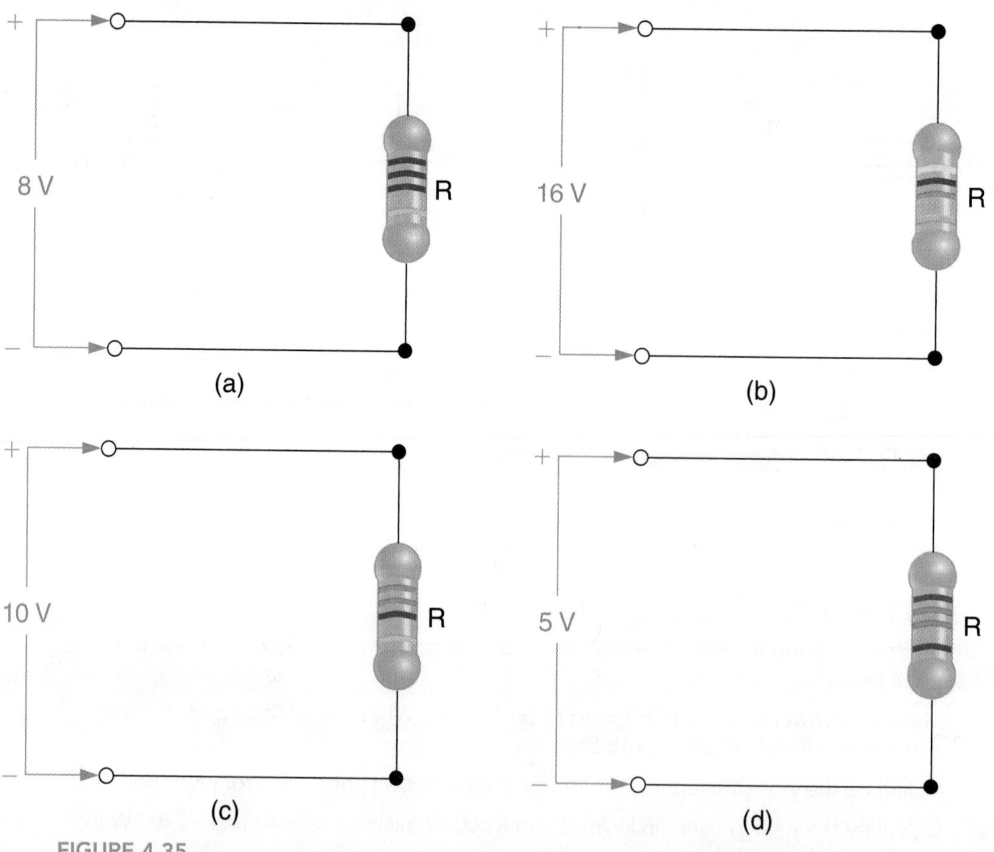

FIGURE 4.35

27. Calculate the range of possible current values for the circuit in Figure 4.35c.
28. Calculate the range of possible current values for the circuit in Figure 4.35d.

CHALLENGERS

29. Calculate the minimum allowable setting for the potentiometer in Figure 4.36.
30. Calculate the maximum allowable setting for the voltage source in Figure 4.37.

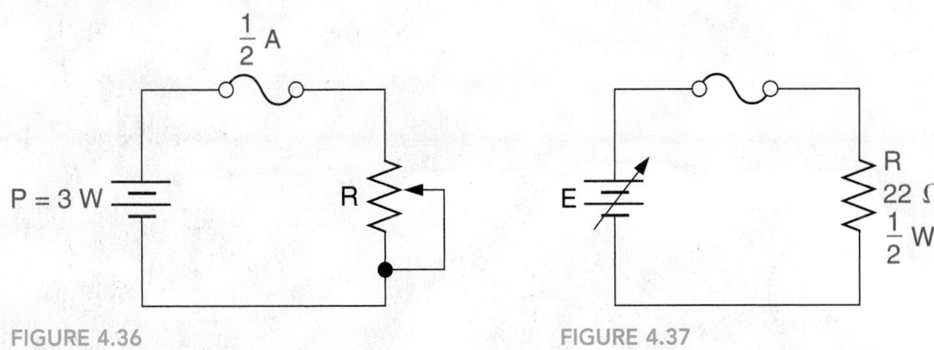

FIGURE 4.36

FIGURE 4.37

ANSWERS TO THE EXAMPLE PRACTICE PROBLEMS

4.1	0.18 A (180 mA)
4.2	I = 18.2 A (max), so the fuse won't open.
4.3	165 V
4.4	120 V
4.5	300 Ω
4.6	20 Ω
4.7	120 V/20 Ω = 6 A
4.8	240 V/20 Ω = 12 A
4.9	630 W
4.10	169 W
4.11	30.6 W
4.12	32.4 W
4.13	14.3%
4.14	11.7 kWh
4.15	2.04 A
4.16	64.2 mA to 78.4 mA

OFF

CHAPTER 5

SERIES CIRCUITS

PURPOSE

Now that we have covered the fundamental concepts of electricity, it is time to begin applying these concepts to some common types of circuits.

Two common circuit configurations are the *series circuit* and the *parallel circuit*. These two types of circuits are shown in Figure 5.1. As you can see, the two circuits look very different from each other. In fact, they not only look different, they are nearly opposites in terms of their operating characteristics. At the same time, electric circuits, no matter how complex, are designated as series circuits, parallel circuits, or a combination of the two.

In this chapter, we're going to discuss the operating characteristics of series circuits, along with some related topics. Parallel circuits are introduced in Chapter 6.

KEY TERMS

The following terms are introduced and defined in this chapter on the pages indicated:

Catastrophic failure, 206
Chassis ground, 208
Earth ground, 208
Ground (common), 189
Kirchhoff's Voltage Law (KVL), 188
Overload, 197
Series circuit, 175
Voltage divider, 191
Voltage-divider equation, 192
Voltage drop, 191

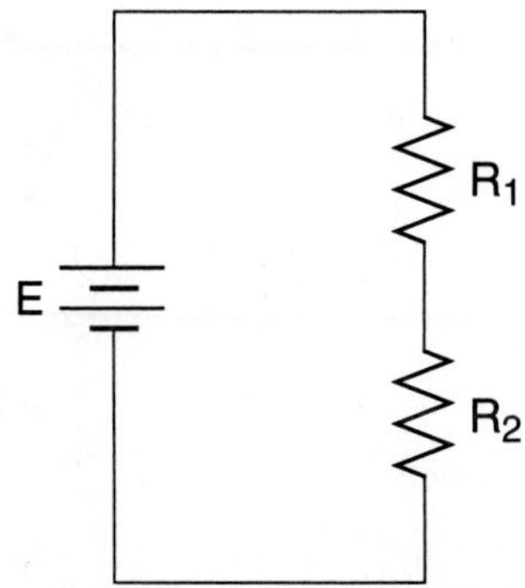

(a) Two-resistor series circuit

E
R_1
R_2

(b) Two-resistor parallel circuit

FIGURE 5.1 Simple series and parallel circuits.

OBJECTIVES

After completing this chapter, you should be able to:

1. State the current characteristic that distinguishes series circuits from parallel circuits.
2. Calculate any resistance value in a series circuit, given the other resistances.
3. Determine the total current in any series circuit.
4. Describe the voltage and power relationships in a series circuit.
5. State and explain Kirchhoff's voltage law (KVL).
6. Calculate the voltage from any point in a series circuit to ground.
7. Calculate the voltage across any component (or group of components) in a series circuit using the voltage-divider equation.
8. List the fault symptoms associated with an open component in a series circuit.
9. List the fault symptoms associated with one or more shorted components in a series circuit.
10. Discuss the effects of component aging and stress on circuit operation.
11. Compare and contrast *earth ground* with chassis ground.

5.1 SERIES CIRCUIT CHARACTERISTICS

Simply defined, a **SERIES CIRCUIT** is a circuit that contains only one current path. For example, consider the circuits shown in Figure 5.2. In each case, the current generated by the voltage source has only one path, and that path contains all of the components in the circuit. In contrast, the parallel circuit in Figure 5.1b contains two current paths between the terminals of the voltage source: one through R_1 and one through R_2.

Series circuit principles apply to

- Fuse and circuit breaker connections
- Electric service meters
- Switch (control) circuits
- Security circuits

among others.

IN THE FIELD

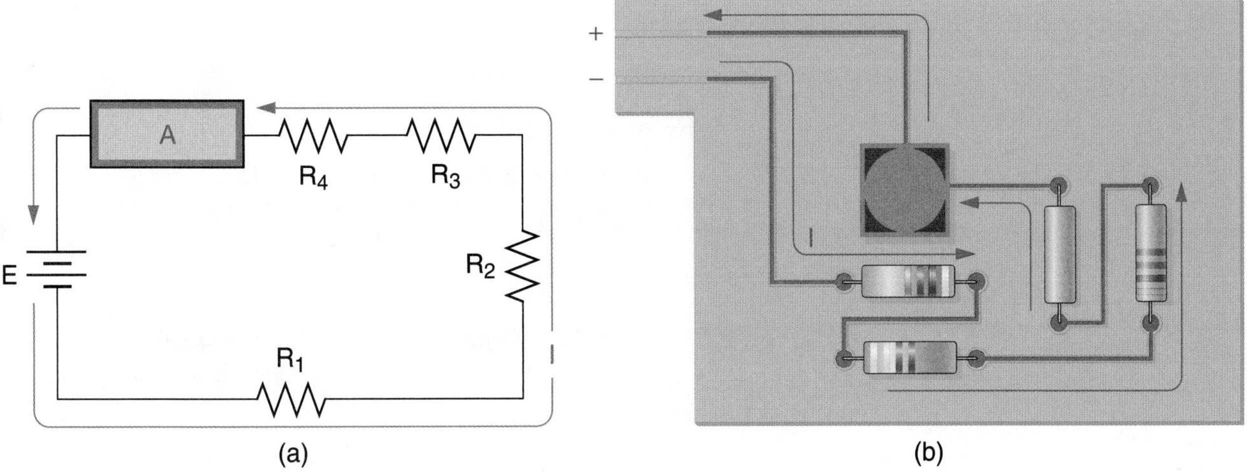

FIGURE 5.2 Example series circuit schematic and construction.

Resistance Characteristics

SERIES CIRCUIT A circuit that contains only one current path.

Figure 5.3 shows a series circuit that contains a battery and four resistors. Since the circuit current passes through all of the resistors, the total circuit resistance is equal to the sum of the individual resistor values. By formula,

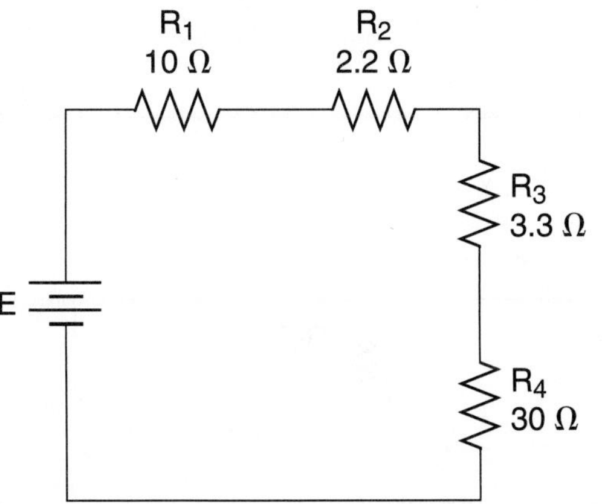

FIGURE 5.3 A four-resistor series circuit.

(5.1)

$$R_T = R_1 + R_2 + \cdots + R_n$$

where

R_T = the total circuit resistance

R_n = the highest numbered resistor in the circuit

The total resistance in a series circuit is calculated as demonstrated in Example 5.1.

EXAMPLE 5.1

What is the total resistance in the circuit shown in Figure 5.3?

SOLUTION

The total circuit resistance is found as

$$\begin{aligned} R_T &= R_1 + R_2 + R_3 + R_4 \\ &= 10\ \Omega + 2.2\ \Omega + 3.3\ \Omega + 30\ \Omega \\ &= 45.5\ \Omega \end{aligned}$$

PRACTICE PROBLEM 5.1

A circuit like the one shown in Figure 5.3 has the following values: R_1 = 680 Ω, R_2 = 180 Ω, R_3 = 220 Ω, and R_4 = 33 Ω. What is the value of the total circuit resistance?

When you need to find the value of an unknown resistance in a series circuit, you can calculate it by subtracting the sum of the known resistances from the total circuit resistance. This technique is demonstrated in Example 5.2.

EXAMPLE 5.2

What value of R_3 in Figure 5.4 will set the total circuit resistance to 120 Ω?

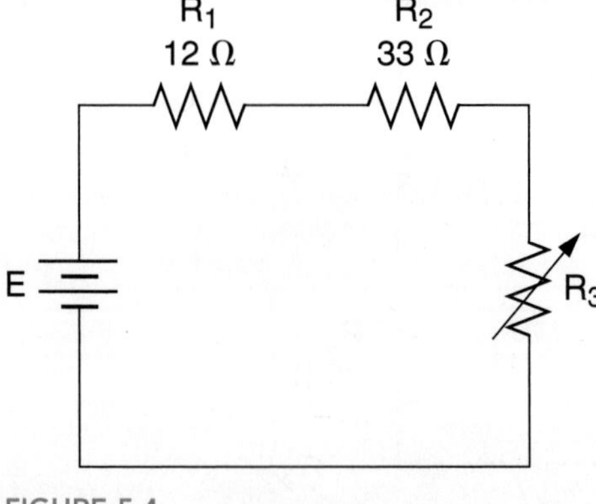

FIGURE 5.4

SOLUTION

The combined resistance of R_1 and R_2 is found as

$$R_1 + R_2 = 12\ \Omega + 33\ \Omega$$
$$= 45\ \Omega$$

R_3 must account for the difference between 45 Ω and the desired total of 120 Ω. Therefore, it must be adjusted to a value of

$$R_3 = R_T - (R_1 + R_2)$$
$$= 120\ \Omega - 45\ \Omega$$
$$= 75\ \Omega$$

PRACTICE PROBLEM 5.2

A circuit like the one shown in Figure 5.4 has values of $R_1 = 47\ \Omega$ and $R_2 = 11\ \Omega$. What rheostat setting will provide a total circuit resistance of 160 Ω?

As a summary, here are the series circuit resistance characteristics:

- The total resistance in a series circuit is equal to the sum of the individual resistor values.
- When the value of one resistor is unknown, it can be determined by subtracting the sum of the known resistor values from the total circuit resistance.

Current Characteristics

Because a series circuit contains only one current path, we can state that the current at any point in a series circuit must equal the current at every other point in the circuit. This principle is illustrated in Figure 5.5. As you can see, each of the four ammeters shows a reading of 1 ampere. The fact that the meters all show the same reading makes sense when you consider the fact that

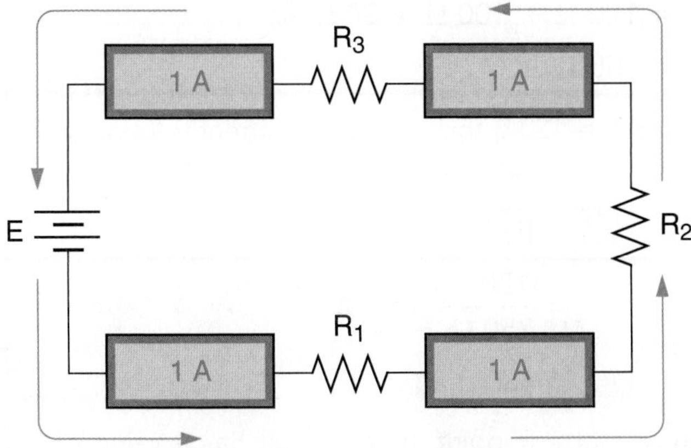

FIGURE 5.5 Current through a series circuit.

current is the flow of electrons. Because electrons are leaving (and entering) the source at a rate of one coulomb per second, they must be moving at the same rate at all points in the circuit.

Figure 5.5 also illustrates the fact that it makes no difference where you measure the current in a series circuit. Since the current is the same at all points, you will obtain the same reading regardless of where you place the meter in the circuit.

The actual value of current in a series circuit depends on the source voltage (E) and the total circuit resistance (R_T). When the source voltage and total circuit resistance are known, Ohm's law is used to calculate the total circuit current, as demonstrated in Example 5.3.

EXAMPLE 5.3

What is the value of the current through the circuit in Figure 5.6?

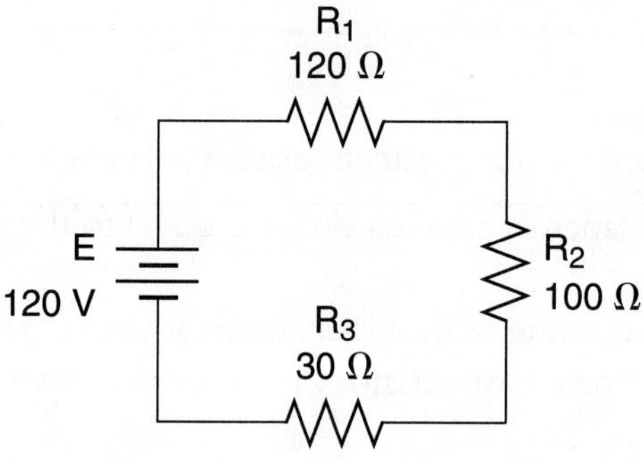

FIGURE 5.6

SOLUTION

To determine the total current (I_T) through the circuit, you begin by calculating the total circuit resistance, as follows:

$$\begin{aligned} R_T &= R_1 + R_2 + R_3 \\ &= 120\ \Omega + 100\ \Omega + 30\ \Omega \\ &= 250\ \Omega \end{aligned}$$

Now, using $R_T = 250\ \Omega$ and $E = 120$ V, the total circuit current is found as

$$\begin{aligned} I_T &= \frac{E}{R_T} \\ &= \frac{120\text{ V}}{250\ \Omega} \\ &= 480\text{ mA} \end{aligned}$$

Thus, the measured current at any point in the circuit has a value of 480 mA.

PRACTICE PROBLEM 5.3

A series circuit like the one in Figure 5.6 has the following values: $R_1 = 180\ \Omega$, $R_2 = 150\ \Omega$, $R_3 = 75\ \Omega$, and $E = 240$ V. What is the value of the circuit current?

In Example 5.2, you were shown how a rheostat can be adjusted to provide a specific value of total resistance. The following example shows how a potentiometer can be used to provide for a specific value of circuit current.

EXAMPLE 5.4

What adjusted value of R_3 in Figure 5.7 will set the value of the circuit current to 1.5 A?

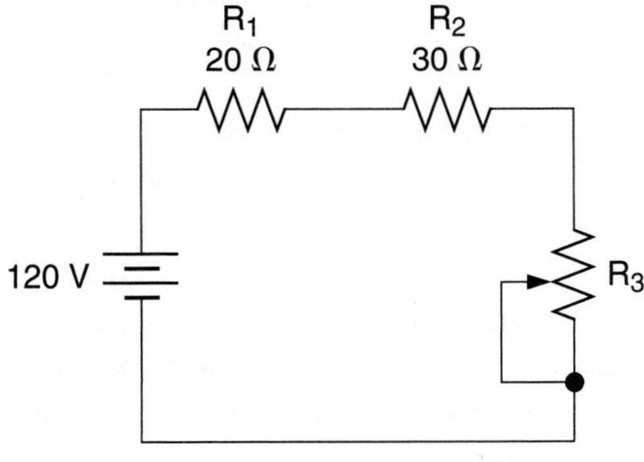

FIGURE 5.7

SOLUTION

To set the circuit current to 1.5 A, we need a total resistance of

$$R_T = \frac{E}{I_T} = \frac{120\text{ V}}{1.5\text{ A}} = 80\ \Omega$$

The sum of R_1 and R_2 is now subtracted from the value of R_T to obtain the required value of R_3, as follows:

$$R_3 = R_T - (R_1 + R_2) = 80\ \Omega - 50\ \Omega = 30\ \Omega$$

Adjusting the value of R_3 to 30 Ω will provide a total circuit resistance of 80 Ω. This resistance will set the circuit current to the desired value of 1.5 A.

PRACTICE PROBLEM 5.4

A circuit like the one in Figure 5.7 has the following values: $E = 240\text{ V}$, $R_1 = 22\ \Omega$, and $R_2 = 47\ \Omega$. What setting of R_3 is needed to limit the circuit current to 2.5 A?

As a summary, here are the series circuit current characteristics:

- The current through a series circuit is equal at all points in the circuit, and thus, can be measured at any point in the circuit.
- The actual value of current in a series circuit is determined by the source voltage and the total circuit resistance.

Voltage Characteristics

Whenever current passes through resistance, a difference of potential (voltage) is developed across that resistance, as given by the relationship

$$V = I \times R$$

Note that the letter V is used to represent the voltage across a circuit component. That helps us to distinguish component voltages from source voltages (which are represented by the letter E).

Because the current in a series circuit passes through all of the resistors, a voltage is developed across each resistor. For example, consider the circuit shown in Figure 5.8. The circuit current (200 mA) is shown to be passing through a 100 Ω resistor and a 300 Ω resistor. According to Ohm's law, the voltage across R_1 (which is designated as V_{R1}) is found as

$$\begin{aligned} V_{R1} &= I_T \times R_1 \\ &= 200\text{ mA} \times 100\ \Omega \\ &= 20\text{ V} \end{aligned}$$

The voltage across R_2 (which is designated as V_{R2}) is found as

$$\begin{aligned} V_{R2} &= I_T \times R_2 \\ &= 200\text{ mA} \times 300\ \Omega \\ &= 60\text{ V} \end{aligned}$$

If we were to connect several voltmeters to this circuit (as shown in Figure 5.9), we would obtain the readings shown. Note that the polarities of the component voltages are determined by the direction of the circuit current. Since electrons flow from negative to positive, the "entry" side of the component is more negative than the "exit" side. This is why the component voltages are given the polarity signs shown in the figure.

If you look closely at Figure 5.9, you'll see that the sum of the component voltages is equal to the source (or total) voltage. This relationship, which holds true for all series circuits, is given as

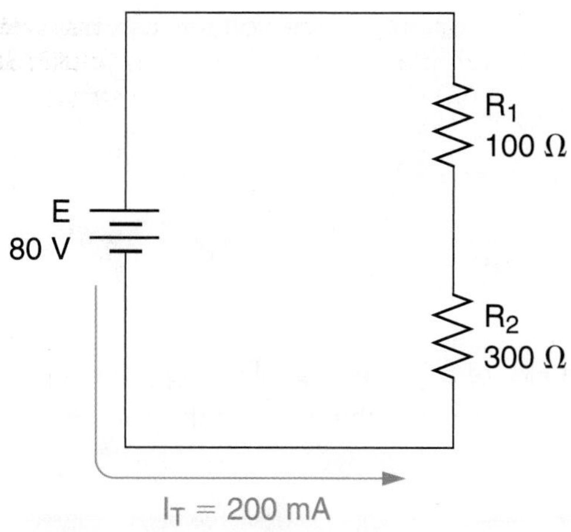

FIGURE 5.8 Series circuit current.

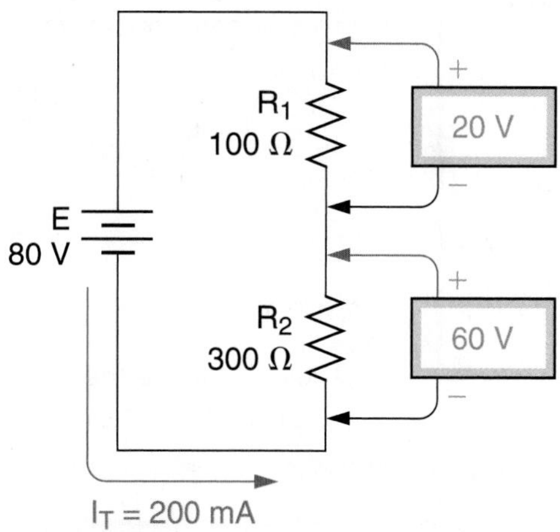

FIGURE 5.9 Series circuit current and voltages.

(5.2)

$$E = V_{R1} + V_{R2} + \cdots + V_{Rn}$$

where

E = the source (or total) voltage

V_{Rn} = the voltage across the highest numbered resistor in the circuit

An application of this relationship is demonstrated in Example 5.5.

EXAMPLE 5.5

What is the value of the source voltage in Figure 5.10?

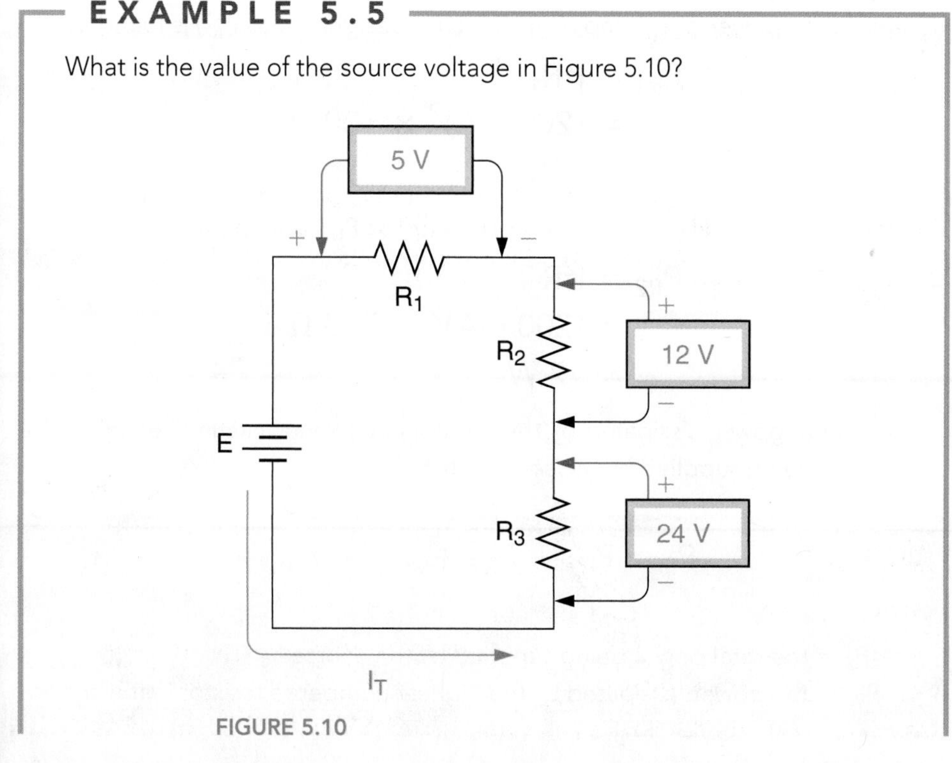

FIGURE 5.10

Using the component voltages, the value of the source voltage can be found as

$$\begin{aligned} E &= V_{R1} + V_{R2} + V_{R3} \\ &= 5\text{ V} + 12\text{ V} + 24\text{ V} \\ &= 41\text{ V} \end{aligned}$$

PRACTICE PROBLEM 5.5

A series circuit like the one shown in Figure 5.10 has the following values: $V_{R1} = 20$ V, $V_{R2} = 160$ V, and $V_{R3} = 88$ V. What is the value of the source voltage?

In the next section, we will take a closer look at the voltage relationships in series circuits. For now, remember that the sum of the component voltages in a series circuit is equal to the source (or total) voltage.

Power Characteristics

In Chapter 4, you learned that current passing through any resistance results in power being dissipated. When current passes through any resistance, some amount of power is dissipated by the component, as given in the relationship

$$P = I^2R$$

Since the current in a series circuit passes through all of the resistors, it would follow that they are all dissipating some amount of power. Referring back to Figure 5.8, the power dissipated by R_1 (which is designated as P_{R1}) is found as

$$\begin{aligned} P_{R1} &= I^2R_1 \\ &= (200\text{ mA})^2 \times 100\ \Omega \\ &= 4\text{ W} \end{aligned}$$

The power dissipated by R_2 (which is designated as P_{R2}) is found as

$$\begin{aligned} P_{R2} &= I^2R_2 \\ &= (200\text{ mA})^2 \times 300\ \Omega \\ &= 12\text{ W} \end{aligned}$$

The total power dissipated by the resistors in a series circuit is equal to the total power being supplied by the source. By formula,

(5.3)

$$P_S = P_{R1} + P_{R2} + \cdots + P_{Rn}$$

where

P_S = the total power being supplied by the source
P_{Rn} = the power dissipated by the highest numbered resistor in the circuit

For example, the total power being supplied by the source in Figure 5.8 is found as

$$\begin{aligned} P_S &= P_{R1} + P_{R2} \\ &= 4\text{ W} + 12\text{ W} \\ &= 16\text{ W} \end{aligned}$$

The following example verifies the relationship given in equation (5.3).

EXAMPLE 5.6

Demonstrate that the total power supplied by the source in Figure 5.11 equals the sum of the resistor power dissipation values.

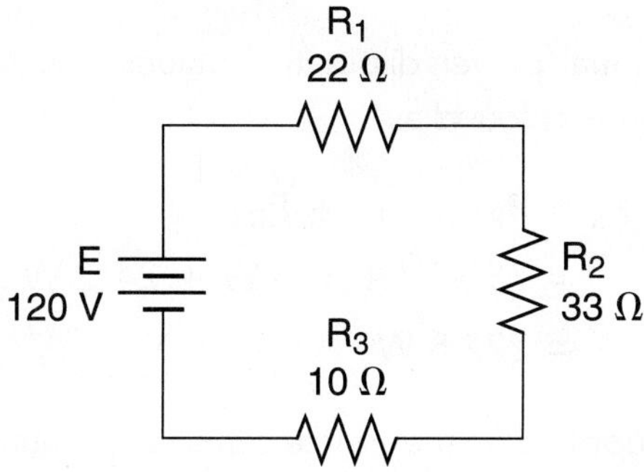

FIGURE 5.11

SOLUTION

Calculating the power values for the circuit in Figure 5.11 begins by calculating the total circuit resistance, as follows:

$$\begin{aligned} R_T &= R_1 + R_2 + R_3 \\ &= 22\ \Omega + 33\ \Omega + 10\ \Omega \\ &= 65\ \Omega \end{aligned}$$

Next, the circuit current is found as

$$\begin{aligned} I_T &= \frac{E}{R_T} \\ &= \frac{120\text{ V}}{65\ \Omega} \\ &= 1.85\text{ A} \end{aligned}$$

Once we know the value of the circuit current, we can find the power dissipated by each of the resistors in the circuit as follows:

$$\begin{aligned} P_{R1} &= I_T^2 \times R_1 \\ &= (1.85\text{ A})^2 \times 22\ \Omega \\ &= 75.3\text{ W} \end{aligned}$$

$$\begin{aligned} P_{R2} &= I_T^2 \times R_2 \\ &= (1.85\text{ A})^2 \times 33\ \Omega \\ &= 113\text{ W} \end{aligned}$$

and

$$\begin{aligned} P_{R3} &= I_T^2 \times R_3 \\ &= (1.85\text{ A})^2 \times 10\ \Omega \\ &= 34.2\text{ W} \end{aligned}$$

Adding the individual power dissipation values, the total power dissipated by the resistors is found as

$$\begin{aligned} P_S &= P_{R1} + P_{R2} + P_{R3} \\ &= 75.3\text{ W} + 113\text{ W} + 24.2\text{ W} \\ &= 222.5\text{ W} \end{aligned}$$

The total power supplied by the source can also be found as

$$\begin{aligned} P_S &= I_T^2 \times R_T \\ &= (1.85\text{ A})^2 \times (65\ \Omega) \\ &= 222.5\text{ W} \end{aligned}$$

These results verify the relationship given in equation (5.3).

PRACTICE PROBLEM 5.6

A circuit like the one in Figure 5.11 has the following values: $E = 180$ V, $R_1 = 180\ \Omega$, $R_2 = 220\ \Omega$, $R_3 = 140\ \Omega$. Demonstrate that the total power supplied by the source equals the sum of the resistor power dissipation values.

Series Circuit Analysis

The complete analysis of a series circuit involves determining the values of R_T, I_T, and P_T, along with the resistor voltage and power values. The following example demonstrates the complete analysis of a series circuit.

EXAMPLE 5.7

What are the current, voltage, and power values for the circuit in Figure 5.12?

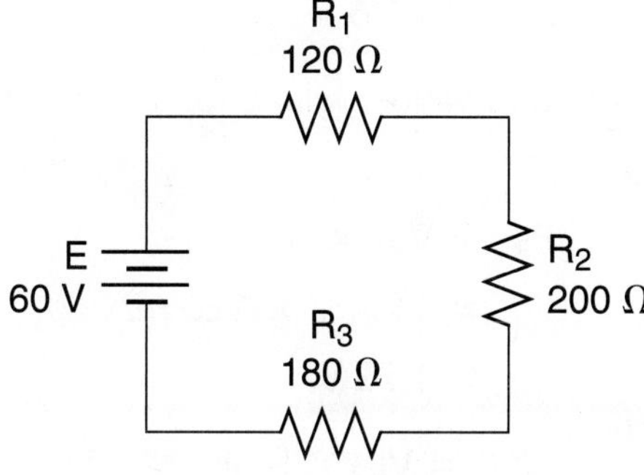

FIGURE 5.12

SOLUTION

The total resistance in the circuit in Figure 5.12 is found as

$$R_T = R_1 + R_2 + R_3$$
$$= 500\ \Omega$$

Next, the total circuit current is found as

$$I_T = \frac{E}{R_T}$$
$$= \frac{60\ \text{V}}{500\ \Omega}$$
$$= 120\ \text{mA}$$

The total power drawn from the source can be found as

$$P_S = E \times I_T$$
$$= 60\ \text{V} \times 120\ \text{mA}$$
$$= 7.2\ \text{W}$$

The voltage across each resistor can now be found using Ohm's law, as follows:

$$V_{R1} = I_T \times R_1$$
$$= 120\ \text{mA} \times 120\ \Omega$$
$$= 14.4\ \text{V}$$

$$V_{R2} = I_T \times R_2$$
$$= 120\ \text{mA} \times 200\ \Omega$$
$$= 24\ \text{V}$$

and

$$V_{R3} = I_T \times R_3$$
$$= 120\text{ mA} \times 180\ \Omega$$
$$= 21.6\text{ V}$$

The individual power dissipation values for the resistors can now be found as:

$$P_{R1} = V_{R1} \times I_T$$
$$= 14.4\text{ V} \times 120\text{ mA}$$
$$= 1.73\text{ W}$$

$$P_{R2} = V_{R2} \times I_T$$
$$= 24\text{ V} \times 120\text{ mA}$$
$$= 2.88\text{ W}$$

and

$$P_{R3} = V_{R3} \times I_T$$
$$= 21.6\text{ V} \times 120\text{ mA}$$
$$= 2.59\text{ W}$$

This completes the circuit calculations.

PRACTICE PROBLEM 5.7

A circuit like the one in Figure 5.12 has the following values: $E = 220$ V, $R_1 = 33\ \Omega$, $R_2 = 22\ \Omega$, and $R_3 = 10\ \Omega$. What are the circuit values of current, voltage, and power?

The component voltage and power dissipation values found in the example can be verified by comparing their sums to the source voltage and total power, as follows:

$$V_{R1} + V_{R2} + V_{R3} = 14.4\text{ V} + 24\text{ V} + 21.6\text{ V}$$
$$= 60\text{ V}$$

and

$$P_{R1} + P_{R2} + P_{R3} = 1.73\text{ W} + 2.88\text{ W} + 2.59\text{ W}$$
$$= 7.2\text{ W}$$

These values ($E = 60$ V and $P_T = 7.2$ W) verify that the component voltage and power dissipation values calculated in the example are correct.

Summary

A series circuit is one that contains a single path for current. The total resistance in a series circuit is equal to the sum of the individual resistor values. The total circuit current is the same at every point in the circuit, and is generally found by dividing the source voltage by the total circuit resistance.

When current passes through the resistors in a series circuit, a voltage is developed across each resistor. The sum of these voltages is equal to the circuit's source voltage.

The current through the resistors in a series circuit also causes each component to dissipate some amount of power. The sum of the component power dissipation values is equal to the total power supplied by the source. The resistance, current, voltage, and power characteristics of series circuits are summarized in Figure 5.13.

Series Circuit Characteristics

Sample schematic:

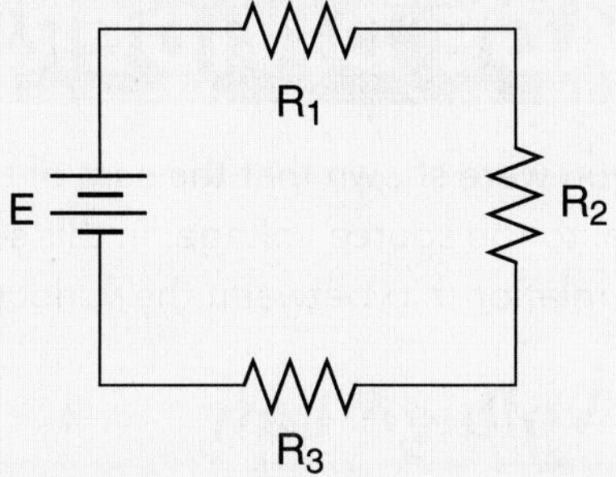

Resistance: The total resistance is equal to the sum of the individual resistances. By formula:

$$R_T = R_1 + R_2 + \ldots + R_n$$

Current: The current at any point in the circuit equals the current at all other points. The value of current depends on the source voltage and the total circuit resistance. By formula:

$$I_T = \frac{E}{R_T}$$

Voltage: The sum of the component voltages must equal the source voltage. By formula:

$$E = V_1 + V_2 + \ldots + V_n$$

Power: The total power used by the circuit must equal the power delivered by the source. By formula:

$$P_S = P_1 + P_2 + \ldots + P_n$$

FIGURE 5.13 Series circuit characteristics.

PROGRESS CHECK

1. What is a series circuit?
2. Describe the relationship between the total resistance in a series circuit and the individual resistor values.
3. Describe the relationship between current values at various points in a series circuit.
4. What determines the value of current in a series circuit?
5. Describe the procedure for setting the current in a series circuit to a specific value using a potentiometer.
6. How do you determine the polarity of the voltage across a given resistor?
7. Describe the relationship between the total voltage applied to a series circuit and the individual component voltages.
8. Every resistor in a series circuit dissipates some amount of power. Why?
9. Describe the relationship between the power being supplied by the source in a series circuit and the power dissipated by the individual components.

5.2 VOLTAGE DIVIDERS

In the last section, you were shown that the sum of the component voltages in a series circuit is equal to the source voltage. In this section, we are going to take a closer look at the relationship between the various voltages in a series circuit.

Kirchhoff's Voltage Law

As you know, the sum of the component voltages in a series circuit must equal the source voltage. This relationship, which was first described in the 1840s by a German physicist named Gustav Kirchhoff, has come to be known as **KIRCHHOFF'S VOLTAGE LAW (KVL)**. The following example illustrates a practical application of this law.

KIRCHHOFF'S VOLTAGE LAW (KVL). A law that effectively states that the sum of the component voltages in a series circuit must equal the source voltage.

EXAMPLE 5.8

What is the value of the load resistance in Figure 5.14?

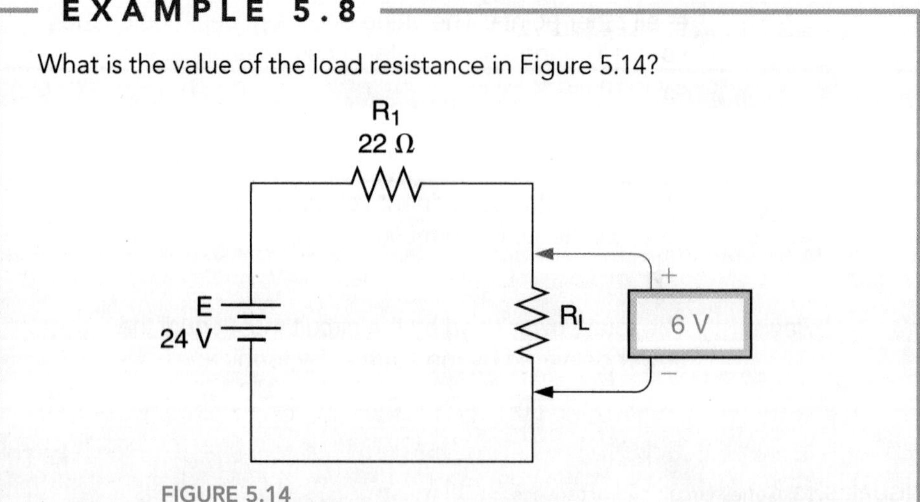

FIGURE 5.14

SOLUTION

KVL states that the sum of the component voltages must equal the source voltage. Therefore, the voltage across R_1 in Figure 5.14 must equal the difference between the source voltage (24 V) and the load voltage (6 V), as follows:

$$V_{R1} = E - V_{RL} = 24\text{ V} - 6\text{ V} = 18\text{ V}$$

Current is the same at all points in a series circuit, so the value of I_T can be found using Ohm's law and the values of V_{R1} and R_1, as follows:

$$I_T = \frac{V_{R1}}{R_1} = \frac{18\text{ V}}{22\ \Omega} = 409\text{ mA}$$

Now that we know the values of V_{RL} and I_T, the load resistance can be found as

$$R_L = \frac{V_{RL}}{I_T} = \frac{6\text{ V}}{409\text{ mA}} = 14.7\ \Omega$$

PRACTICE PROBLEM 5.8

A circuit like the one in Figure 5.14 has the following values: $E = 240$ V, $V_{RL} = 35$ V, and $R_1 = 82\ \Omega$. What is the value of the circuit current?

Voltage References

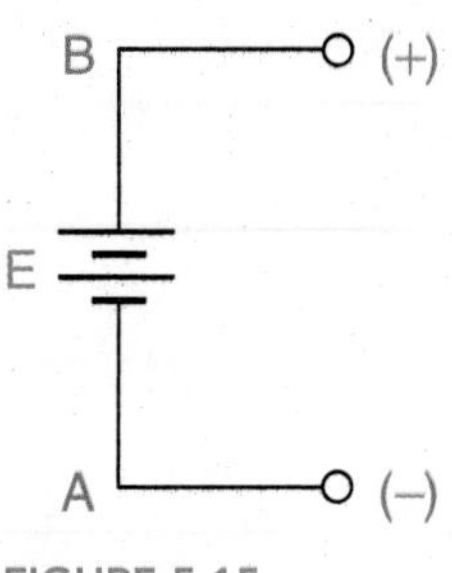

FIGURE 5.15

In Chapter 3, it was stated that we must establish a reference point if the polarity of a given voltage is to be agreed on. For example, consider the circuit in Figure 5.15. Is the source voltage positive or negative? If we reference the source voltage to point A, we say that the source voltage is *positive with respect to the reference.* If we reference the source voltage to point B, we say that the source voltage is *negative with respect to the reference.*

In any circuit, there is a point that serves as the reference for all voltages in the circuit. This point is referred to as **GROUND (or COMMON)**, and is designated by the symbol shown in Figure 5.16a. Note that ground is considered to be

GROUND (or COMMON) A point that serves as the reference for all voltages in the circuit.

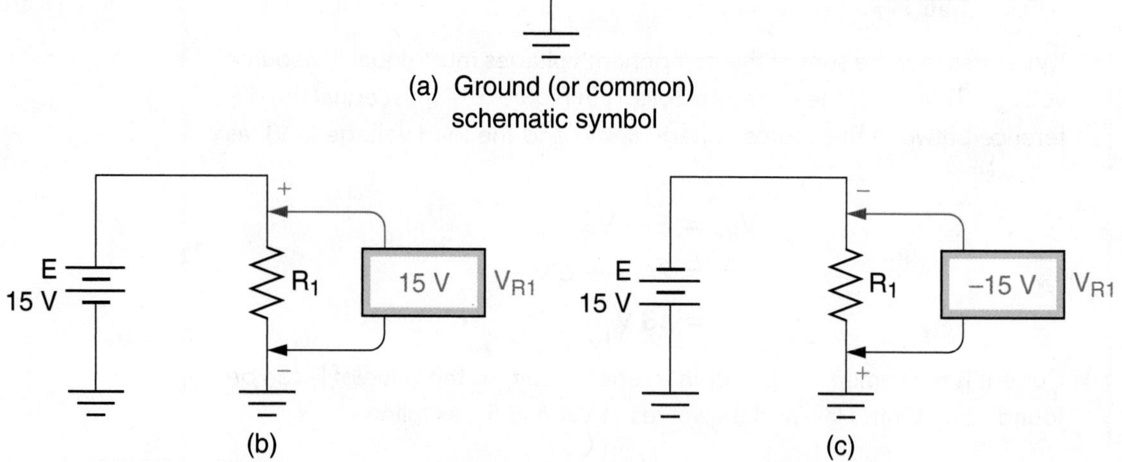

FIGURE 5.16 Negative ground and positive ground.

the 0 V point in a circuit, and that all voltages in the circuit are referenced to that point. For example, the negative sides of the source and the resistor in Figure 5.16b are both returned to ground. Thus, E and V_{R1} are both *positive* with respect to ground (the reference). The positive sides of the source and the resistor in Figure 5.16c are returned to ground, so E and V_{R1} are both *negative* with respect to ground.

When the ground symbol appears more than once in a circuit, it indicates that the various points are physically connected, as shown in Figure 5.17. Thus, the ground symbols in a given schematic indicate that two or more points are:

1. Considered to be at a 0 V potential.
2. Physically connected to each other.

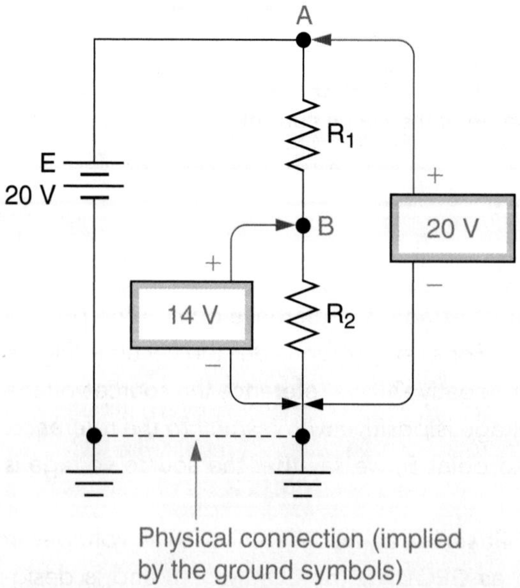

FIGURE 5.17 Voltages referenced to ground.

Note that the ground connection throughout a circuit is often referred to as a *ground-loop*.

In the circuit shown in Figure 5.17, the voltage from point A to ground is equal to the source voltage, 20 V. The voltage from point B to ground is equal to 14 V. Generally, we say that the 6 V difference between the voltages at points A and B is *dropped* across R_1. The remaining 14 V is dropped across R_2. (Don't forget: The term **VOLTAGE DROP** is commonly used to describe a change from one potential to a *lower* potential.)

VOLTAGE DROP A term that is commonly used to describe a change from one potential to a *lower* potential.

The Voltage-Divider Relationship

The term **VOLTAGE DIVIDER** is often used to describe a series circuit because the source voltage is divided among the components in the circuit. For example, consider the circuits shown in Figure 5.18. In each case, the source voltage (E) is divided among the resistors in the circuit. As is always the case, the sum of the resistor voltages in each series circuit is equal to the source voltage.

VOLTAGE DIVIDER A term that is often used to describe a series circuit because the source voltage is divided among the components in the circuit.

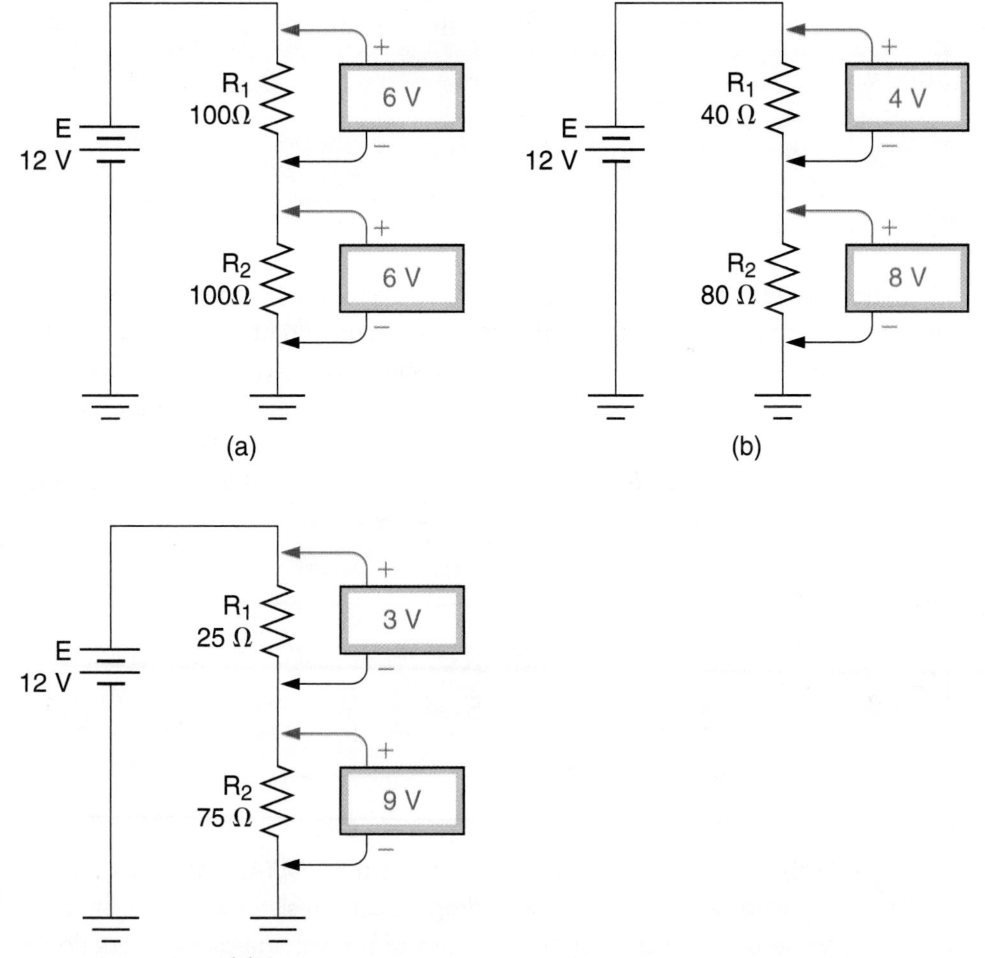

FIGURE 5.18 Component voltages add up to equal the source voltage.

If you look closely at the circuits in Figure 5.18, you'll see that the ratio of any resistor voltage to the source voltage is equal to the ratio of that resistor's value to the total circuit resistance. For example, in Figure 5.18a, R_1 has a value of 100 Ω. The total circuit resistance equals the sum of the two resistors, 200 Ω. Thus, R_1 counts for one-half of the total circuit resistance and it is dropping one-half of the source voltage. By the same token:

1. In Figure 5.18b, R_1 (40 Ω) accounts for one-third of the total circuit resistance (120 Ω), and it is dropping one-third of the source voltage.
2. In Figure 5.18c, R_1 (25 Ω) accounts for one-fourth of the total circuit resistance (100 Ω), and it is dropping one-fourth of the source voltage.

As you can see, the ratio of V_{R1} to E_S is equal to the ratio of R_1 to R_T in each of the circuits shown in Figure 5.18. This relationship, which applies to all series circuits, is stated mathematically as follows:

(5.4)

$$\frac{V_{Rn}}{E} = \frac{R_n}{R_T}$$

where

R_n = the resistor of interest
V_{Rn} = the voltage drop across R_n (with *n* being the component number)

The validity of equation (5.4) is demonstrated further in Figure 5.19. Here we have two circuits along with their calculated values of current, V_{R1} and V_{R2}. As you can see, the values of V_{R1} and V_{R2} for the two circuits are equal, even though the actual resistor values are different. This is because the voltages are determined by the resistor *ratios* rather than by the resistor values themselves. In both circuits, R_1 accounts for ¾ of the total resistance, and therefore drops ¾ of the source voltage.

If we multiply both sides of equation (5.4) by E, we get the following useful equation:

(5.5)

$$V_{Rn} = E \times \frac{R_n}{R_T}$$

VOLTAGE-DIVIDER EQUATION An equation that allows us to calculate the voltage drop across a resistor without first calculating the value of the circuit current.

Equation (5.5), which is commonly referred to as the **VOLTAGE-DIVIDER EQUATION**, allows us to calculate the voltage drop across a resistor without first calculating the value of the circuit current. The use of the voltage-divider equation is demonstrated in Example 5.9.

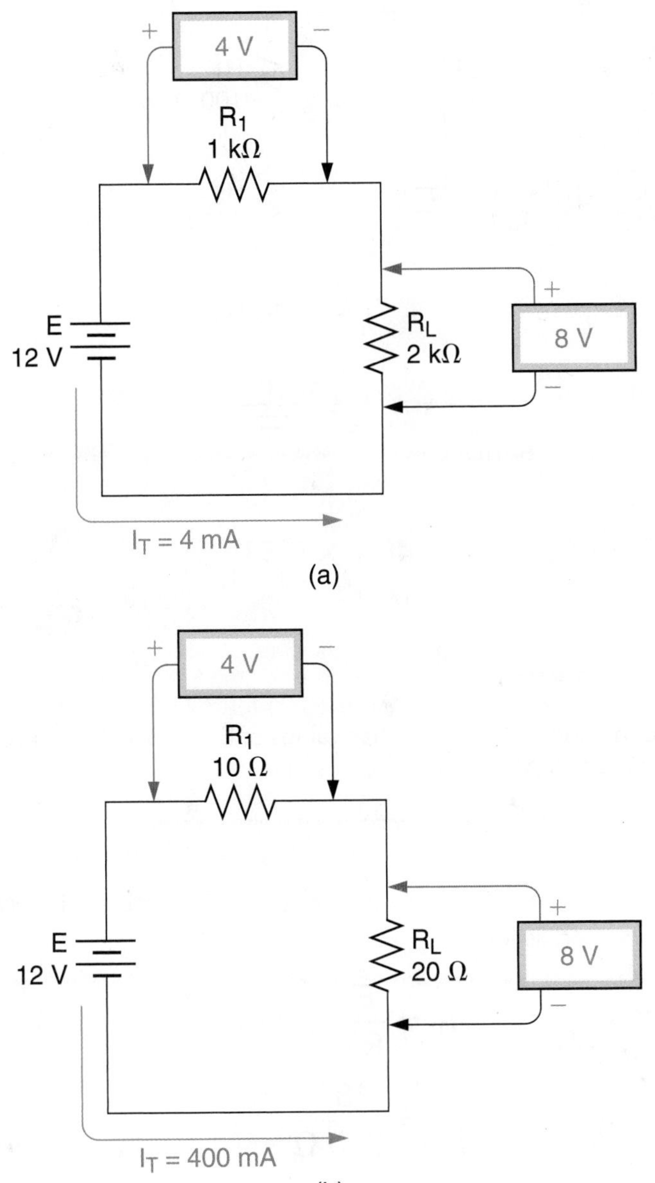

FIGURE 5.19 Equal voltage and resistance ratios.

EXAMPLE 5.9

What is the value of V_{R3} in Figure 5.20?

To calculate the voltage across R_3, use the values of V_{R3} and R_3 in place of V_{Rn} and R_n in the voltage-divider equation, as follows:

$$V_{R3} = E \times \frac{R_3}{R_T}$$

$$= 48\ \text{V} \times \frac{100\ \Omega}{300\ \Omega}$$

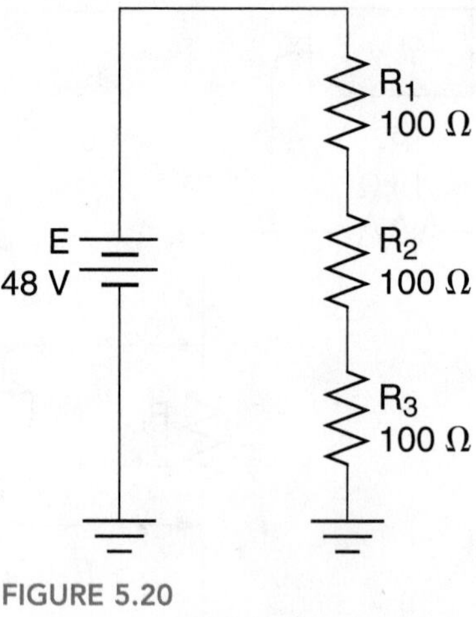

FIGURE 5.20

$$= 48\ \text{V} \times 0.333$$
$$= 16\ \text{V}$$

PRACTICE PROBLEM 5.9

A circuit like the one in Figure 5.20 has values of $E = 120$ V, $R_1 = 1$ kΩ, $R_2 = 220$ Ω, and $R_3 = 330$ Ω. What is the value of V_{R2}?

Had we not used the voltage-divider equation, the problem in Example 5.8 could have been solved as follows:

$$I_T = \frac{E}{R_T}$$
$$= \frac{48\ \text{V}}{300\ \Omega}$$
$$= 160\ \text{mA}$$

and

$$V_{R3} = I_T \times R_3$$
$$= 160\ \text{mA} \times 100\ \Omega$$
$$= 16\ \text{V}$$

As you can see, we obtained the same results with fewer steps using the voltage-divider equation.

The following example demonstrates how the voltage-divider equation can be used along with Kirchhoff's voltage law to calculate the values of the voltage drops in a circuit.

EXAMPLE 5.10

What are the values of V_{R1} and V_{R2} in Figure 5.21?

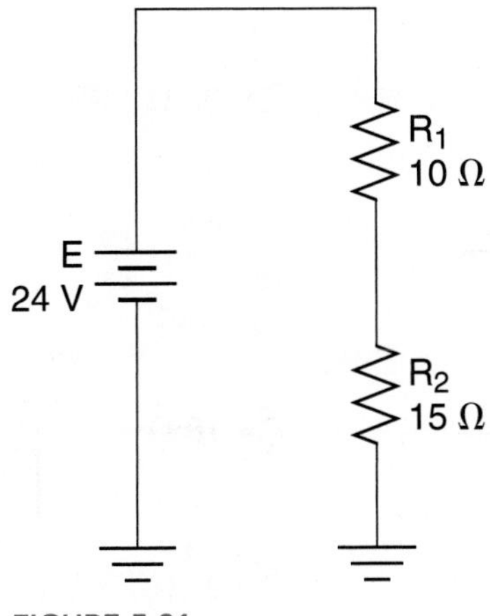

FIGURE 5.21

SOLUTION

The values of V_{R1} and V_{R2} can be calculated as follows:

$$\begin{aligned} V_{R1} &= E \times \frac{R_1}{R_T} \\ &= 24\text{ V} \times \frac{10\ \Omega}{25\ \Omega} \\ &= 24\text{ V} \times 0.4 \\ &= 6\text{ V} \end{aligned}$$

Since the sum of the voltages must equal the source voltage, V_{R2} can be found as the difference between E and V_{R1}, as follows:

$$\begin{aligned} V_{R2} &= E - V_{R1} \\ &= 24\text{ V} - 6\text{ V} \\ &= 18\text{ V} \end{aligned}$$

PRACTICE PROBLEM 5.10

A circuit like the one in Figure 5.21 has values of $E = 200$ V, $R_1 = 330\ \Omega$, and $R_2 = 470\ \Omega$. What are the values of V_{R1} and V_{R2} for the circuit?

The voltage-divider equation can also be used to find the voltage across a group of resistors, as demonstrated in Example 5.11.

EXAMPLE 5.11

What is the voltage from point A to ground (V_A) in Figure 5.22?

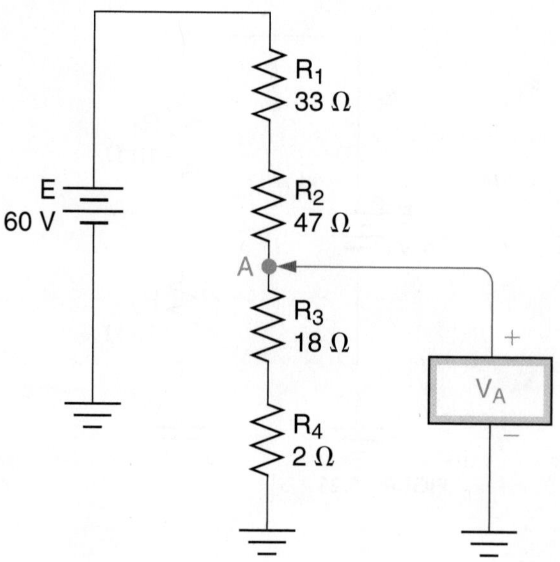

FIGURE 5.22

SOLUTION

The voltage from point A to ground (V_A) equals the sum of V_{R3} and V_{R4}. The total resistance between point A and ground has a value of

$$R_A = R_3 + R_4$$
$$= 20\ \Omega$$

Using this value in the voltage-divider equation, the value of V_A is found as follows:

$$V_A = E \times \frac{R_A}{R_T}$$
$$= 60\text{ V} \times \frac{20\ \Omega}{100\ \Omega}$$
$$= 12\text{ V}$$

PRACTICE PROBLEM 5.11

A circuit like the one in Figure 5.22 has the following values: $E = 180$ V, $R_1 = 22\ \Omega$, $R_2 = 47\ \Omega$, $R_3 = 33\ \Omega$, and $R_4 = 10\ \Omega$. What is the voltage from point A to ground?

Had we used the previously established approach to the problem, we would have calculated the values of R_T, I_T, V_{R3}, and V_{R4}. Then we would have added the values of V_{R3} and V_{R4} to obtain the result found in Example 5.11. In this case, the voltage-divider equation provides a more efficient approach to solving the problem.

One Final Note

In this section, we have established two relationships that apply to every series circuit: Kirchhoff's voltage law and the voltage-divider equation. As you will see, these relationships are used almost as commonly as Ohm's law.

PROGRESS CHECK

1. What does Kirchhoff's voltage law state?
2. What is *ground*? What is its symbol?
3. What is indicated by the appearance of the ground symbol at two or more points in a circuit?
4. What is meant by the term *voltage drop*?
5. What is the voltage-divider relationship?
6. How does the voltage-divider equation simplify circuit voltage calculations?

5.3 PRACTICAL SERIES CIRCUITS

The circuits shown in the first two sections of this chapter help to illustrate the basic operating principles of series circuits. However, they are not very useful beyond their role as educational tools. As you will see, practical series circuits contain a number of components that combine to deliver power to one or more specific loads.

Common Series Circuit Elements

OVERLOAD A high current condition.

Practical series circuits typically contain all or most of the four elements shown in Figure 5.23. The *source* provides power for the load. The *circuit protector* disconnects the source from the load in the event of an **OVERLOAD** (high current condition). The *switch* provides a means of manually switching the circuit on and off, and the *load* makes use of the power provided by the source. (Though not listed above, remember that circuits also contain *conductors* that connect the various elements.)

Though commonly found in electronics systems, series connections are found only in control circuits and other special-purpose electrical circuits (such as light switch circuits and overcurrent protection circuits).

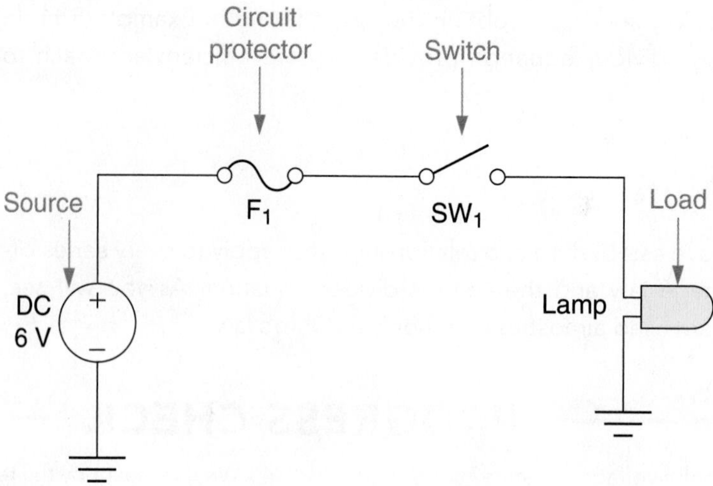

FIGURE 5.23 A practical series circuit.

Like the series circuits shown earlier, the circuit in Figure 5.23 has only one current path. When the switch is closed, electrons flow from the negative terminal of the battery and pass through the load, the switch, and the fuse (in that order). Assuming the switch and fuse are working properly, they have very little resistance. If we assume an ideal value of $R = 0\ \Omega$ for these two components, then the voltages across the fuse and the switch can be found as

$$\begin{aligned} V &= I \times R \\ &= I \times 0\ \Omega \\ &= 0\ V \end{aligned}$$

With 0 V across the fuse and 0 V across the switch, the voltage across the lamp is found as

$$\begin{aligned} V_{LAMP} &= E - V_{F1} - V_{SW1} \\ &= 6\ V - 0\ V - 0\ V \\ &= 6\ V \end{aligned}$$

As this result indicates, the load voltage equals the source voltage. In practice, the load voltage will be slightly less than the source voltage because the switch, the fuse, and the wires are not perfect conductors.

The relationship between the source voltage and the load voltage is not unique to the circuit in Figure 5.23. In any series circuit containing a fuse, a switch, and a load, the load voltage is approximately equal to the source voltage under normal circumstances. If there is a significant difference between the source and load voltages when the switch is closed, a problem likely exists somewhere in the circuit. This point is discussed in greater detail in Section 5.4.

Measuring the Source and Load Voltages

The component voltages in Figure 5.23 are measured as shown in Figure 5.24. In Figure 5.24a, a DMM is connected across the terminals of the source. Note that the black lead is connected to the negative terminal of the source, while the red lead is connected to its positive terminal, and that the DMM is set to measure voltage. In this case, the meter shows a reading of 6 V.

In Figure 5.24b, the voltmeter is connected across the load. As you can see:

- The red lead is connected to the side of the load that traces back to the positive terminal of the DC source.
- The black lead is connected to the side of the load that traces back to the negative terminal of the DC source.

When connected as shown, the voltmeter reads 5.9 V, which is slightly lower than the value of the source voltage.

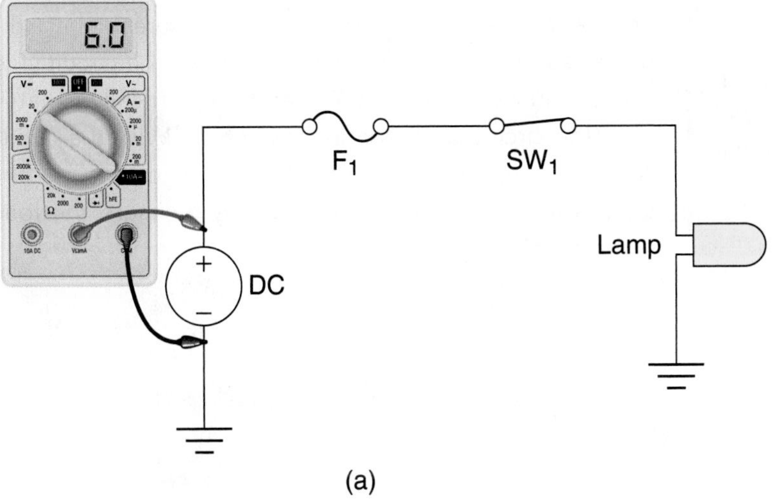

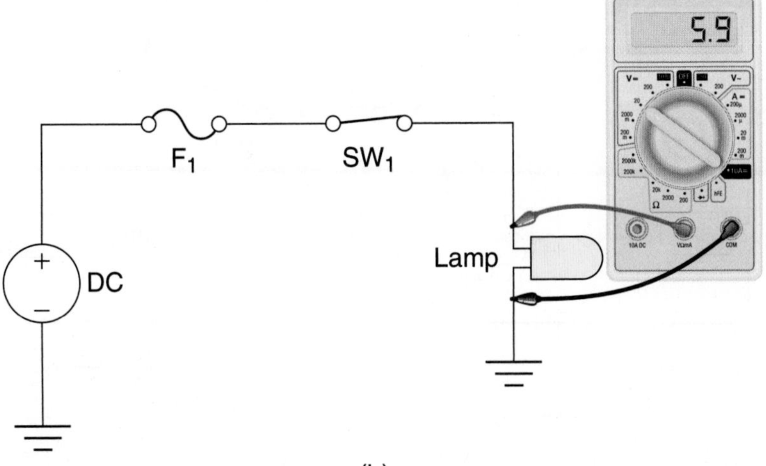

FIGURE 5.24 Measuring source and load voltages.

Measuring the Circuit Current

In Chapter 3, you learned that current is measured by breaking the current path and bridging the gap with the current meter. For the circuit in Figure 5.23, this can easily be accomplished in either of two ways:

1. Remove the fuse from the fuse holder and connect the ammeter to the ends of the fuse holder as shown in Figure 5.25a. When connected as shown, the ammeter completes the current path and provides the reading shown in the figure.
2. Connect the ammeter across the open switch as shown in Figure 5.25b. When connected as shown, the ammeter bypasses the open switch and completes the current path. The current passing through the meter provides the reading shown in the figure.

Measuring Circuit Resistance

STAYING SAFE A circuit must be de-energized before taking any resistance readings. Connecting an ohmmeter to an energized circuit will damage (and possibly destroy) the meter.

The resistance of a load can be measured when the power source is electrically isolated from the circuit. The power source in Figure 5.23 can be electrically isolated from the circuit as shown in Figure 5.26. With the circuit switch open, the ohmmeter is connected across the load. This allows the meter to safely measure the load resistance.

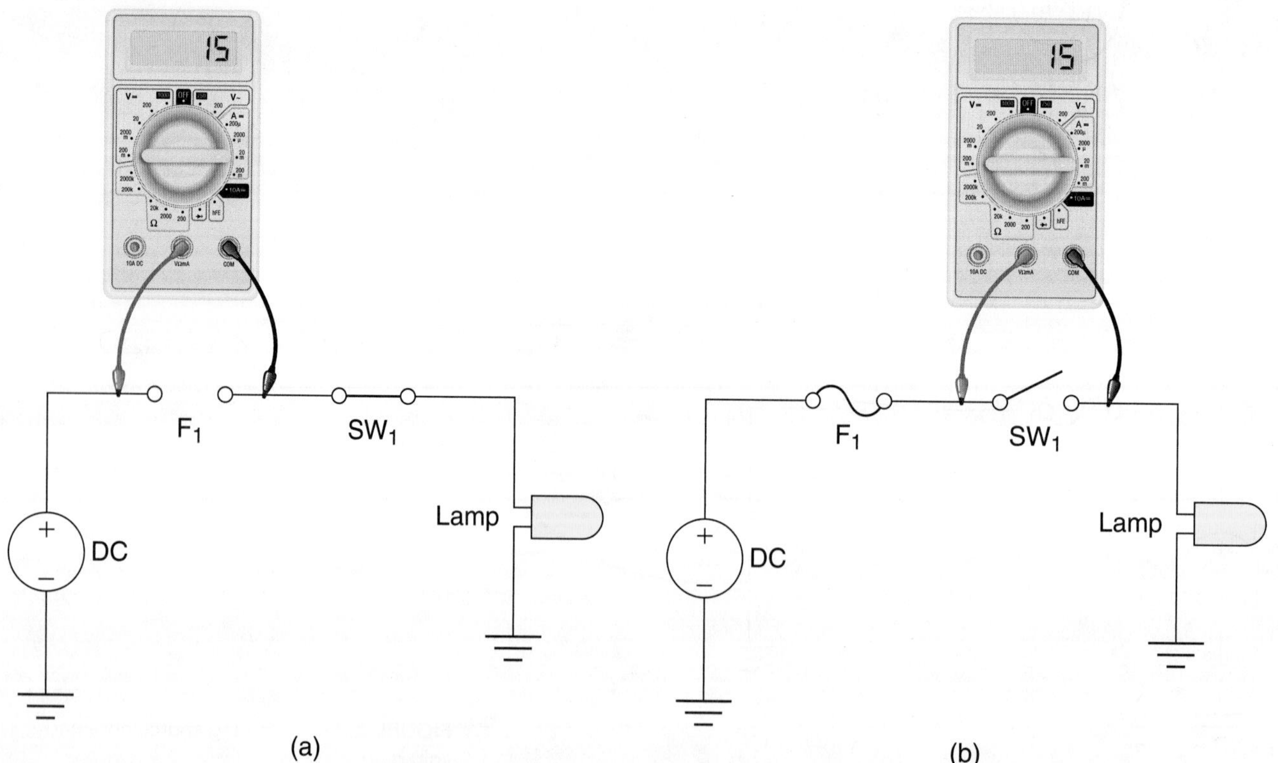

FIGURE 5.25 Measuring circuit current.

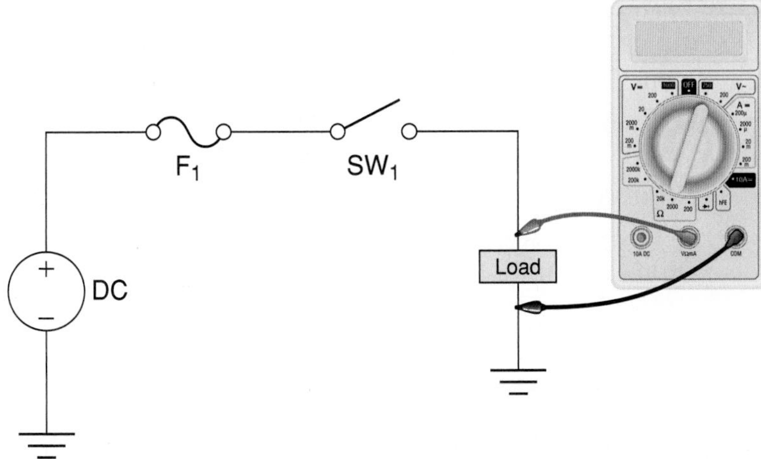

FIGURE 5.26 Measuring load resistance.

PROGRESS CHECK

1. What circuit elements are commonly found in practical series circuits?
2. A series AC circuit has a 120 V source, a fuse, a switch, and a space heater. What voltage would you expect to measure across the heater?
3. Describe the two procedures that can be used to measure the current through a practical series circuit.

5.4 SERIES CIRCUIT FAULT SYMPTOMS AND TROUBLESHOOTING

In this section, we are going to discuss the characteristic symptoms of several circuit faults and a basic approach to series circuit troubleshooting.

Open-Component Fault Symptoms

When one component in a series circuit opens, the results of the open are as follows:

1. The circuit current drops to zero.
2. The source voltage is dropped across the open component.

These two symptoms can be explained with the help of Figure 5.27.

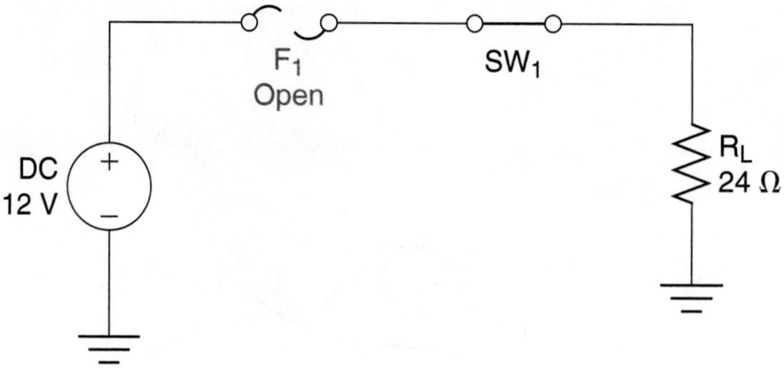

FIGURE 5.27 A series circuit with an open fuse.

In Figure 5.27, the fuse (F_1) is open. As you know, there must be a complete path between the terminals of the voltage source in order to have any current. Since the path in the circuit has been broken, $I_T = 0$ A.

Ohm's law tells us that

$$V = I \times R$$

Since $I_T = 0$ A in Figure 5.27, the voltage across R_L (V_{RL}) is found as

$$\begin{aligned} V_{RL} &= I_T \times R_L \\ &= 0\text{ A} \times 24\ \Omega \\ &= 0\text{ V} \end{aligned}$$

Since $V_{RL} = 0$ V, Kirchhoff's voltage law tells us that the source voltage must all drop across the open component, as follows:

$$\begin{aligned} V_{F1} &= E - V_{RL} \\ &= 12\text{ V} - 0\text{ V} \\ &= 12\text{ V} \end{aligned}$$

Thus, the voltage across the open component in a series circuit is equal to the source voltage. This holds true no matter how many components there are in the circuit.

Measuring the Voltage Across an Open Component: A Practical Consideration

When you measure the voltage across an open component in a series circuit with a voltmeter, the measured voltage may be slightly lower than the source voltage. The reason for this is illustrated in Figure 5.28.

When a voltmeter is connected across an open component (as shown in Figure 5.28), the meter completes the circuit. As a result, a low-value current is generated in the circuit, causing a voltage to be dropped across each of the other resistors. Since V_{R1} and V_{R2} (in this case) each have some measurable value, the voltage across the open is slightly less than the source voltage.

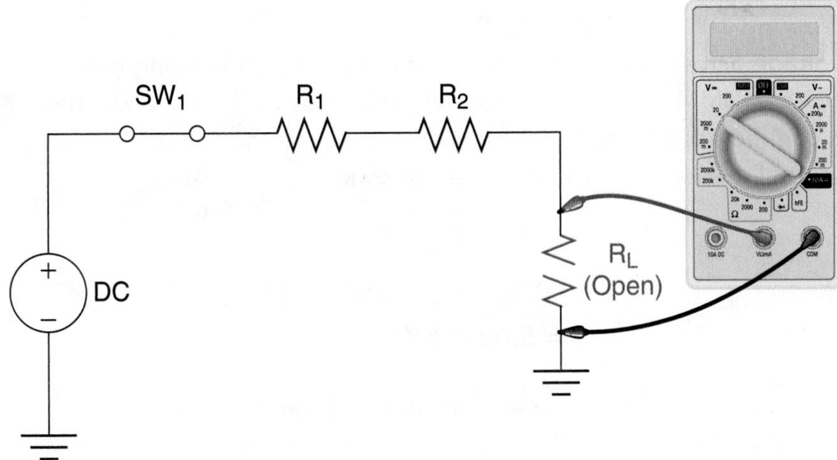

FIGURE 5.28 Measuring the voltage across an open component.

How much less? That depends on the values of the resistive components in the circuit and the internal resistance of the meter. When the internal resistance of the meter (R_{INT}) is much greater than the remaining circuit resistance, the presence of the meter has only a slight effect on the measured voltage across the open. This point is illustrated in Example 5.12.

EXAMPLE 5.12

Is the voltage reading shown on the meter in Figure 5.29 correct?

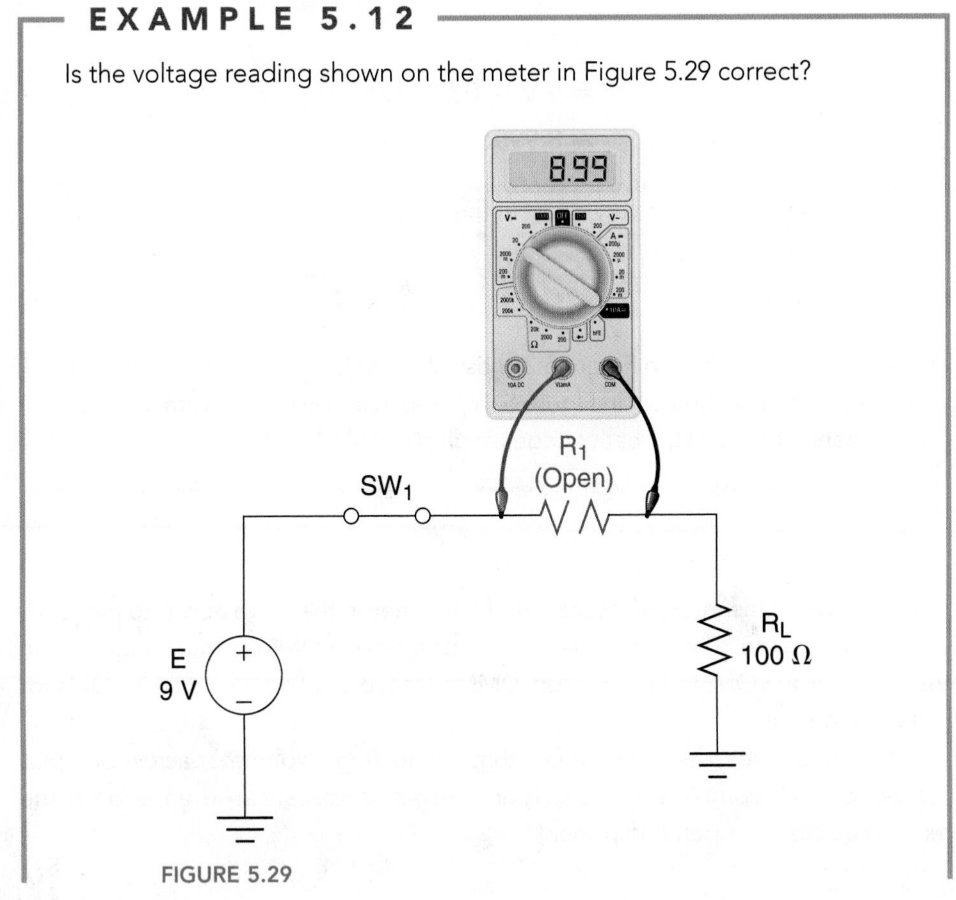

FIGURE 5.29

SOLUTION

The voltage across the open component in Figure 5.29 is being measured with a DMM that has an internal resistance of 5 MΩ. With the DMM connected, the total circuit resistance equals the sum of R_L and the internal resistance of the meter, as follows:

$$\begin{aligned} R_T &= R_L + R_{INT} \\ &= 100\ \Omega + 5\ \text{M}\Omega \\ &= 5.0001\ \text{M}\Omega \end{aligned}$$

Using this value of R_T, the voltage across the load can now be found as

$$\begin{aligned} V_{RL} &= E \times \frac{R_L}{R_T} \\ &= 9\ \text{V} \times \frac{100\ \Omega}{5.0001\ \text{M}\Omega} \\ &= 0.0001799\ \text{V} \qquad (179.9\ \mu\text{V}) \end{aligned}$$

Finally, the voltage measured across the open resistor is found as the difference between E and V_{RL}, as follows:

$$\begin{aligned} V_{R1} &= E - V_{RL} \\ &= 9\ \text{V} - 0.0001799\ \text{V} \\ &\cong 8.9998\ \text{V} \end{aligned}$$

The reduction in V_{R1} is very slight in this case.

PRACTICE PROBLEM 5.12

A circuit like the one in Figure 5.29 has values of E = 12 V and R_L = 910 Ω. Assuming the voltmeter in Figure 5.29 is replaced by a VOM with an input resistance of 200 kΩ, what voltage reading will the VOM provide?

The higher the internal resistance of a voltmeter, the more accurate the readings it will provide. The internal resistance of a typical DMM is much higher than that of a typical VOM. For this reason, DMMs are typically preferred over VOMs for voltage readings.

You should keep in mind the fact that connecting a voltmeter across an open component will complete the circuit and, in some cases, cause an error in the reading across the open component.

Shorted-Component Fault Symptoms

When a resistive component in a series circuit is shorted, the following occur:

1. The voltage across the shorted component drops to near zero.
2. The total circuit resistance decreases, causing the circuit current to increase.
3. The increase in current typically causes the fuse or circuit breaker to open.

The results of a resistive component shorting are illustrated in Figure 5.30.

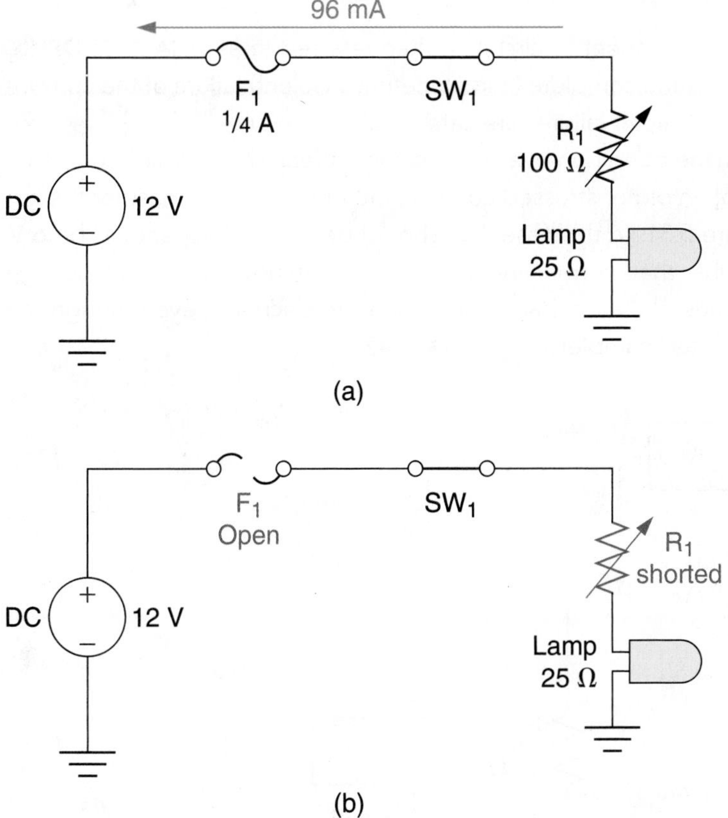

FIGURE 5.30 The effects of a shorted component.

Figure 5.30a shows the normal circuit conditions. In Figure 5.30b, R_1 is shown to be shorted. Assuming that $R_1 \cong 0\ \Omega$, the total circuit resistance drops from 125 Ω to 25 Ω (the resistance of the lamp). This drop in resistance causes the circuit current to increase to

$$
\begin{aligned}
I &= \frac{E}{R_L} \\
&= \frac{12\text{ V}}{25\ \Omega} \\
&= 0.48\text{ A} \qquad (480\text{ mA})
\end{aligned}
$$

This value exceeds the rating of the circuit fuse, causing it to open.

It should be noted that a short in any circuit can result in components being damaged from excessive current. For example, the sudden increase in circuit current in Figure 5.30 could have caused the lamp to burn out before the fuse opened. If a faulty circuit contains a burned component (like a lamp) and/or an open fuse, there may be a shorted component somewhere in the circuit.

Component Aging and Stress

CATASTROPHIC FAILURE The complete (and sometimes violent) failure of a component.

A shorted or open component or circuit is often referred to as a **CATASTROPHIC FAILURE**, because it is the complete (and sometimes violent) failure of the component. However, not all circuit failures are catastrophic in nature.

In many cases, the operation of a circuit can be affected to a lesser degree by the partial failure of an old or stressed component. For example, consider the circuit shown in Figure 5.31. In this case, V_{R1} should be 6 V and V_{RL} should be 6 V. However, V_{R1} is higher than normal and V_{RL} is lower than normal. In this case, age and/or stress has caused the resistance of the load to decrease, even though the component hasn't failed completely.

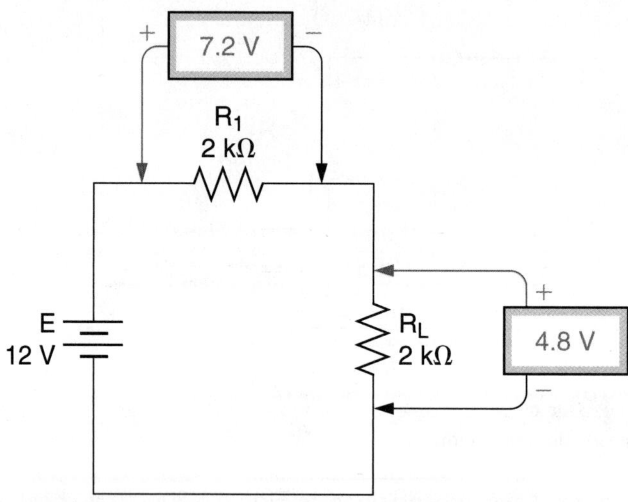

FIGURE 5.31 Voltage readings that result from component aging and/or stress.

Series Circuit Troubleshooting

As you know, troubleshooting is the process of locating and repairing one or more faults in a circuit. In many cases, circuit faults can be detected using your senses (sight, smell, hearing, and touch). In many other cases, a given circuit must be tested using standard test equipment in order to locate any faults. Let's start by identifying some basic resistor faults that can usually be detected with your senses.

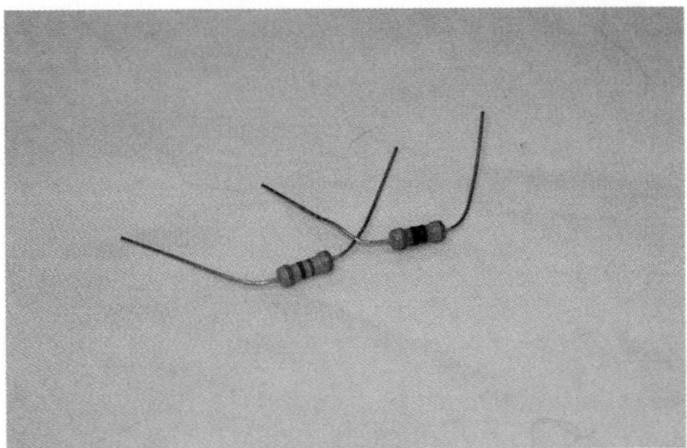

FIGURE 5.32 A good resistor (left) and a burned resistor (right).

When a component opens, it is normally the result of excessive current and the heat it produces. Therefore, an open component will often show signs of excessive heat, such as

1. Extreme discoloring.
2. A crack in the body of the component.

Component discoloring and body damage are both shown in Figure 5.32. When you see either of these signs, the component has burned open.

When the value of a resistive component slowly decreases (due to component aging or stress), the current through the component slowly increases. This causes an increase in the amount of power that the component is dissipating, which increases the stress and eventually leads to complete failure of the component. However, even before a resistive component fails completely, there may be signs that it is going bad:

1. The circuit containing the component may operate erratically or outside of its design specifications. That is, sometimes it may work and other times it may not.
2. The component may smell as if it is beginning to burn.

When a circuit begins to operate erratically or smell as if it is burning, odds are that the component is in the last stages of failing.

When there are no physical signs of damage to a circuit, any component failures must be diagnosed with the use of your test equipment. In most series DC circuits, any circuit faults can be diagnosed with the use of your DMM and your knowledge of how the circuit operated under normal conditions.

STAYING SAFE A failing component will usually be extremely hot and should not be touched. If a component shows signs of burning, it should be removed from the circuit and checked only after it has had time to cool.

One Final Note

Troubleshooting is a skill. Like any skill, it is not really learned by reading, but rather, through experience. The circuit failures described in this section have been provided to show you some of the basics of troubleshooting. However, your experiences will show you that very few circuit failures are as "cut-and-dried" as those presented here.

PROGRESS CHECK

1. When a single resistor in a series circuit opens, the source voltage is all dropped across the open component. Why?
2. Why is the measured voltage across an open component usually slightly less than the full source voltage?
3. Why are voltmeters with high internal resistance preferred over those with low internal resistance?
4. What are the fault symptoms associated with a shorted resistor in a series circuit?
5. What can happen to the other components in a series circuit as the result of one component shorting?
6. What are the symptoms normally associated with component aging and stress?
7. What are the physical signs of a component that has been destroyed by excessive heat?
8. What precautions should be taken when a component shows signs of burning?

5.5 EARTH GROUND VERSUS CHASSIS GROUND

Earlier in the chapter, you were introduced to ground as a 0 V reference and common connection point in a circuit. There are actually two types of ground connection, called earth ground and chassis ground. The symbols for each are shown in Figure 5.33.

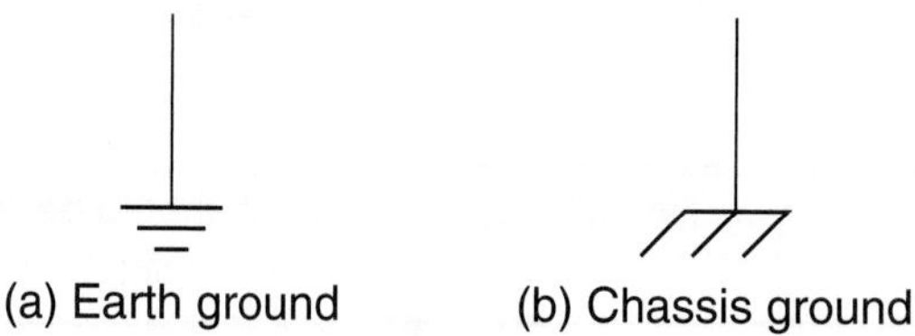

FIGURE 5.33 Ground symbols.

EARTH GROUND A physical (and electrical) connection to the earth.

CHASSIS GROUND A 0 V reference for the components in an electronic system that does not provide a connection between the system ground and the earth.

As its name implies, **EARTH GROUND** provides a physical connection to the earth. Earth ground is established via:

- A grounding electrode in an electrical wiring system
- A grounded wall outlet that provides AC power for an electronic system

An earth ground ensures that the circuit or system ground is at a 0 V potential with respect to the earth. Since the circuit is grounded to the earth, no difference of potential exists across you (or any piece of test equipment) that comes into contact with the ground connection.

A **CHASSIS GROUND** is not *returned* (connected) to earth. In other words, chassis ground provides a 0 V reference for the components in an electronic system, but it does not provide a connection between the system grounds and the earth. As a result, there may be a difference of potential between the circuit ground connections and the earth. This difference of potential could be felt by you (or any piece of test equipment) that happens to come into contact with the circuit ground.

STAYING SAFE You must always be careful when dealing with a chassis ground loop. It can hurt you if you happen to come into contact with it!

PROGRESS CHECK

1. What is the primary difference between earth ground and chassis ground?
2. Why must you be careful when working on an system that has a *chassis ground*?

5.6 SUMMARY

Here is a summary of the major points that were made in this chapter:

Series Circuit Characteristics

- The two most basic types of circuits are the series circuit and the parallel circuit.
- Most circuits can be classified as a series circuit, a parallel circuit, or some combination of the two.
- A series circuit is one that contains only one current path.
- The total resistance in a series circuit is equal to the sum of the individual resistance values.
- The current at any point in a series circuit equals the current at every other point in the circuit.
- Because current is constant throughout a series circuit, it makes no difference where in the circuit it is measured.
- The actual value of current in a series circuit is determined by the source voltage and the total circuit resistance.
- The current passing through the resistors in a series circuit causes a voltage to be developed across each resistor.
- The sum of the resistor voltages in a series circuit equals the source voltage.
- The current passing through a series circuit causes each resistor to dissipate some amount of power.
- The power dissipated by the resistors in a series circuit is equal to the total power supplied by the source.

Voltage Dividers

- According to Kirchhoff's voltage law (KVL), the sum of the component voltages in a series circuit must equal the total (or source) voltage.

- Ground is the accepted reference for all voltages in a circuit.
- The presence of the ground symbol in a schematic indicates that two or more points are
  - Considered to be at a zero-volt potential.
  - Physically connected to each other.
- The term *voltage drop* is often used to describe the voltage across a given component in a circuit.
- The term *voltage divider* is often used to describe a series circuit because the source voltage is divided among the resistors in the circuit.
- The ratio of a given resistor voltage to the source voltage is equal to the ratio of the resistor's value to the total circuit resistance.
- The voltage-divider equation allows you to calculate the voltage across a given resistor (or group of resistors) without calculating the value of the circuit current.

Practical Series Circuits

- Most practical series circuits contain a power source, a circuit protector, a switch, and one or more loads.
- In a series circuit containing a power source, a circuit protector, a switch, and a load:
  - Little or no voltage is dropped across the circuit protector or the switch.
  - Nearly all the source voltage (E) is dropped across the load.
- When measuring the voltage across a component in a series circuit, the voltmeter is connected so that
  - The red lead is connected to the side of the load that traces back to the positive terminal of the DC source.
  - The black lead is connected to the side of the load that traces back to the negative terminal of the DC source.
- The current through a series circuit can be measured by connecting an ammeter across the input and output terminals of
  - An empty fuse holder
  - An open switch
- The total resistance of a series circuit can be measured by
  - Opening the switch.
  - Connecting the ohmmeter to the output side of the switch and the ground side of the load.

Series Circuit Fault Symptoms and Troubleshooting

- When a component opens in a series circuit
  - The circuit current drops to zero.
  - The total source voltage is dropped across the open component.
- The measured voltage across an open component is usually lower than the source voltage, because the voltmeter completes the circuit.
  - When the circuit is completed, the low circuit current causes some of the source voltage to be dropped across the other resistive components in the circuit.
  - The higher the internal resistance of the meter, the closer a voltage reading comes to its ideal value.

- When a resistive component in a series circuit is shorted:
  - The voltage across the shorted component drops to zero.
  - The total circuit resistance decreases, causing the circuit current to increase.
  - The increase in current causes the fuse or circuit breaker to open.
- Opens and shorts are referred to as catastrophic failures, because they are complete (and sometimes violent) component failures.
- When a resistor burns open, it will generally show signs of excessive heat, such as:
  - Extreme discoloring
  - A crack in the body of the component
- Partial failure of a resistor usually has one or more of the following symptoms:
  - The circuit containing the resistor may operate erratically.
  - The component may smell as if it is beginning to burn.
  - The component may be unusually hot.
- Troubleshooting involves:
  - Taking the appropriate circuit measurements.
  - Thinking about what those measurements are telling you.

Earth Ground Versus Chassis Ground

- Earth ground provides a physical connection to earth.
- Chassis ground serves only as a 0 V reference within a circuit.
- There may be a difference of potential between a chassis ground and earth. Therefore, you must be careful not to assume that it is safe to touch a chassis ground connection.

CHAPTER REVIEW

1. The two basic circuit types are ________.
   a) series-parallel
   b) series and parallel
   c) series and reactive
   d) series and combination

2. Series circuits ________.
   a) have only one series component
   b) have only one current path
   c) have the same voltage across each component
   d) dissipate equal power across each component

3. The current passing through each component in a series circuit ________.
   a) is proportional to the resistance of each component
   b) is proportional to the voltage drop of each component
   c) is proportional to the power dissipated by each component
   d) is the same through each component

4. Three resistors are connected in series. One is 47 Ω, the second is 86 Ω, and the color code of the third is unreadable. The total circuit resistance is measured at 180 Ω. This means that the value of the unknown resistor is ________.
   a) 47 Ω
   b) 86 Ω
   c) 68 Ω
   d) cannot be determined

5. Three 1.1 kΩ resistors are in series with a rheostat. The rheostat must be adjusted to ________ in order to have a total circuit resistance of 5.28 kΩ.
   a) 18.9 kΩ
   b) 1980 Ω
   c) 19.8 kΩ
   d) 198 Ω

6. A 220 V source is connected to the circuit described in question 5. The current through the rheostat is ________ and the current through the 1.1 kΩ resistors is ________.
   a) 200 mA, 111 mA
   b) 417 mA, 417 mA
   c) 20 mA, 11.1 mA
   d) 41.7 mA, 41.7 mA

7. A 470 Ω and a 330 Ω resistor are connected in series with a 115 V power supply. The voltage across the 330 Ω resistor is ________.
   a) 47.5 V
   b) 67.6 V
   c) 47.4 V
   d) 67.5 V

8. Two resistors are connected in series. One of them is known to be 550 Ω and you measure 164.5 V across it. The other resistor is 470 Ω. This means that the applied voltage is ________.
   a) 140 V
   b) 348 V
   c) 298 V
   d) 305 V

9. A 680 Ω resistor is connected in series with a rheostat and an 85 V power supply. The resistor drops 76.8 V. The voltage drop across the rheostat is ________.
   a) 8.2 V
   b) 3.2 V
   c) 6.8 V
   d) you need to know the current to solve this problem

10. Three resistors are connected in series: 2.2 kΩ, 4.7 kΩ, and 5.6 kΩ. The 2.2 kΩ resistor will drop ________% of the applied voltage.
    a) 17.6
    b) 37.6
    c) 44.8
    d) 36.7

11. Two resistors are connected in series with a 40 V power supply ________. One of the resistors is known to be 8.8 kΩ. You measure 10 V across it. The value of the other resistor is ________.
    a) 24.6 kΩ
    b) 26.4 kΩ
    c) 35.2 kΩ
    d) 32.5 kΩ

12. A 115 V power supply is connected in series with a 220 Ω and a 560 Ω resistor. The 560 Ω resistor is dissipating 12.17 W. The total power supplied to the circuit is ________.
    a) 60 W
    b) 6.7 W
    c) 17 W
    d) cannot be determined with this information

13. The circuit described in question 12 has had the power supply adjusted to a new value. The 220 Ω resistor is now dissipating 4.46 W. The power supply is now applying ________ to the circuit.
    a) 142 V
    b) 111 V
    c) 126 V
    d) cannot be determined with this information

14. A 22 Ω and a 33 Ω resistor are connected in series. The ________ resistor dissipates more power than the ________ resistor.
    a) 22 Ω, 33 Ω
    b) 33 Ω, 22 Ω
    c) current is the same so both dissipate the same power
    d) cannot be determined with this information

15. Voltages are ________ with respect to ground.
    a) positive
    b) negative
    c) either positive or negative
    d) neither positive nor negative

16. KVL states that ________.
    a) the sum of all power dissipations in a series circuit equals the applied power
    b) the current is the same at any point in a series circuit
    c) the voltage drop is constant in any series circuit
    d) the sum of all voltage drops in a series circuit equals the applied voltage

17. When multiple components are connected to a common ground point, this is often referred to as a ________.
    a) ground-loop path
    b) common signal path
    c) system return path
    d) ground-loop short

18. A 22 kΩ resistor is in series with an unknown resistor and a 65 V power supply. The 22 kΩ resistor drops 46 V. The value of the unknown resistor is ________.
    a) 31.1 kΩ
    b) 15.6 kΩ
    c) 8.0 kΩ
    d) 9.1 kΩ

19. When measuring the voltage across a load, the black lead of the voltmeter is connected to ________.
    a) ground
    b) the positive terminal of the source
    c) the side of the load leading back to the negative terminal of the source
    d) the side of the load leading back to the positive terminal of the source

20. When there is an open in a series circuit ________.
    a) current drops to zero
    b) the supply voltage is dropped across the open
    c) power dissipation drops to near zero
    d) all of the above

21. A blown fuse in a series circuit is usually a sign that ________.
    a) there is a short
    b) there is an open
    c) either a or b
    d) neither a nor b

22. Shorts and opens are often referred to as ________.
    a) calamity failures
    b) catastrophic failures
    c) terminal failures
    d) system failures

23. You are measuring the voltage across a shorted component using a DMM. You do not measure the full applied voltage across the component, though it is close. The reason this happens is ________.
    a) the meter is faulty
    b) you forgot to calibrate the meter
    c) connecting the meter completes the circuit
    d) DMMs have low internal resistance

24. The difference between earth ground and chassis ground is ________.
    a) the number of ground connections
    b) one is a negative ground, the other is a positive ground
    c) earth ground may not be at 0 V potential
    d) chassis ground may not be at 0 V potential

PRACTICE PROBLEMS

1. Each resistor combination below is connected as shown in Figure 5.34. For each combination, calculate the total circuit resistance.

	R_1	R_2	R_3	R_4
a)	1 kΩ	220 Ω	330 Ω	1.1 kΩ
b)	10 Ω	18 Ω	47 Ω	200 Ω
c)	150 Ω	220 Ω	820 Ω	51 Ω
d)	10 kΩ	91 kΩ	5.1 kΩ	300 Ω

2. Each resistor combination below is connected as shown in Figure 5.34. For each combination, calculate the total circuit resistance.

	R_1	R_2	R_3	R_4
a)	1 MΩ	470 kΩ	270 kΩ	51 kΩ
b)	22 kΩ	39 kΩ	12 kΩ	75 kΩ
c)	360 Ω	1.1 kΩ	68 Ω	2.2 kΩ
d)	8.2 kΩ	3.3 kΩ	9.1 kΩ	5.1 kΩ

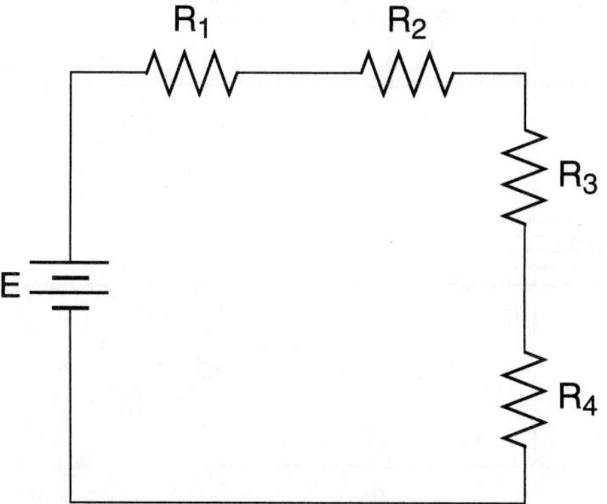

FIGURE 5.34

3. The resistor combinations below are for a circuit like the one in Figure 5.35. In each case, determine the unknown resistor value.

	R_1	R_2	R_3	R_T
a)	1.1 kΩ	330 Ω	_____	1.9 kΩ
b)	_____	47 kΩ	91 kΩ	165 kΩ
c)	33 kΩ	_____	6.2 kΩ	44.3 kΩ
d)	27 Ω	39 Ω	82 Ω	_____

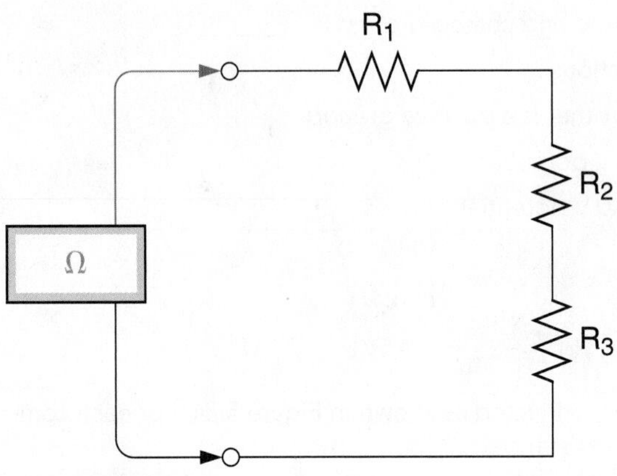

FIGURE 5.35

4. The resistor combinations below are for a circuit like the one in Figure 5.35. In each case, determine the unknown resistor value.

	R_1	R_2	R_3	R_T
a)	200 kΩ	33 kΩ	_____	308 kΩ
b)	_____	82 kΩ	75 kΩ	204 kΩ
c)	510 Ω	_____	68 Ω	611 Ω
d)	1.5 MΩ	2.2 MΩ	10 MΩ	_____

5. Determine the potentiometer setting required in Figure 5.36a to provide a total resistance of 52 kΩ.

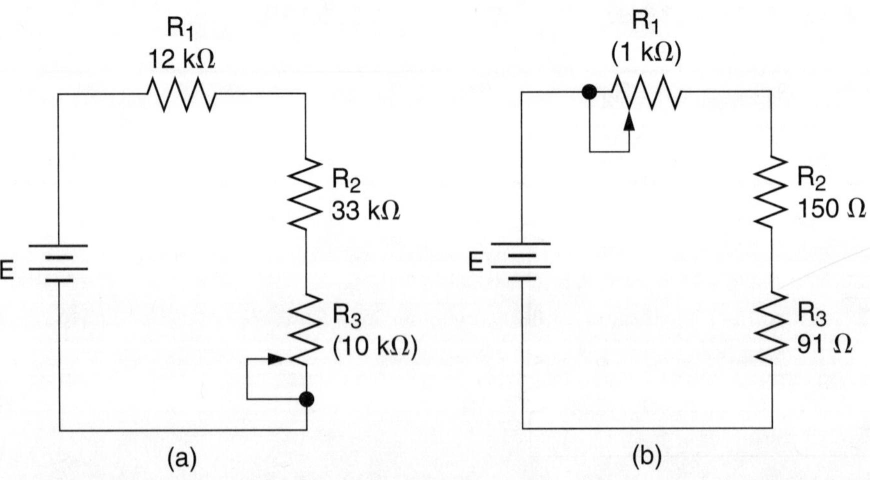

FIGURE 5.36

6. Determine the potentiometer setting required in Figure 5.36b to provide a total resistance of 872 Ω.
7. Calculate the total current for the circuit in Figure 5.37a.

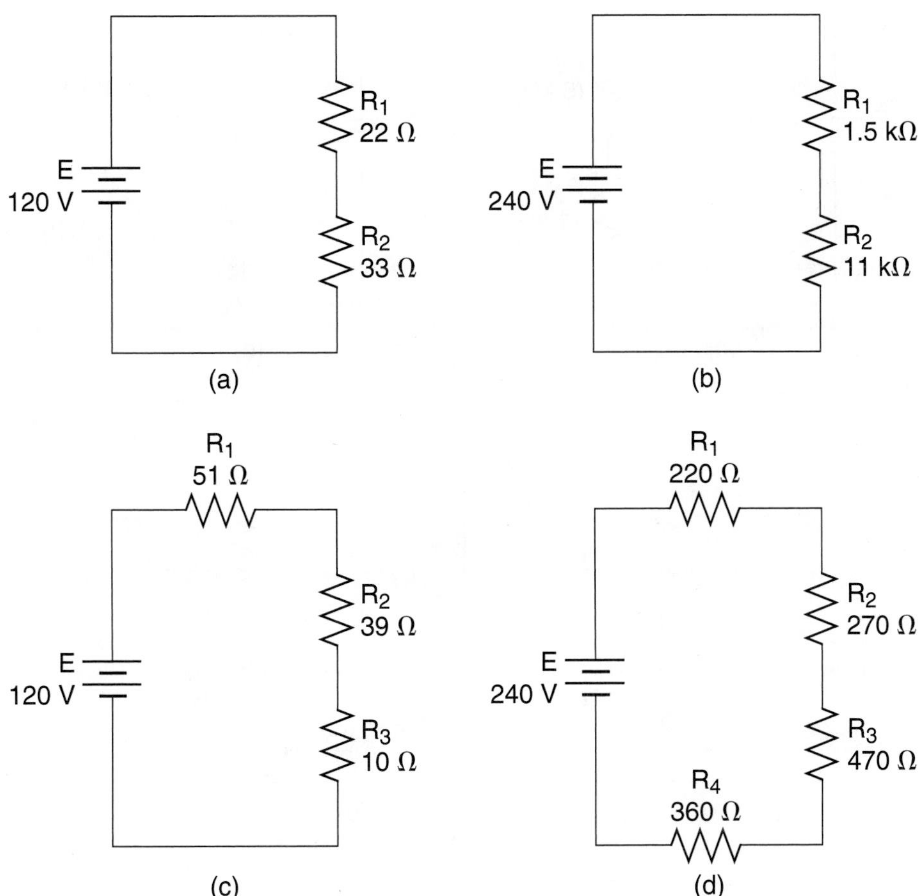

FIGURE 5.37

8. Calculate the total current for the circuit in Figure 5.37b.
9. Calculate the total current for the circuit in Figure 5.37c.
10. Calculate the total current for the circuit in Figure 5.37d.
11. Determine the rheostat setting needed to set the current in Figure 5.38a to 223 mA.

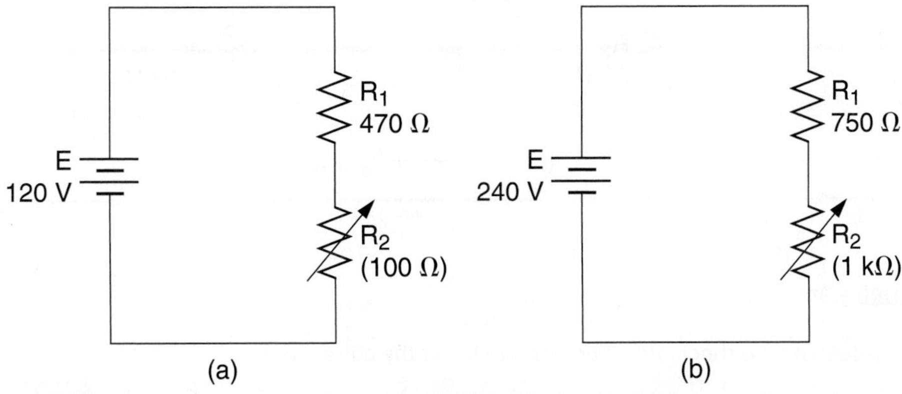

FIGURE 5.38

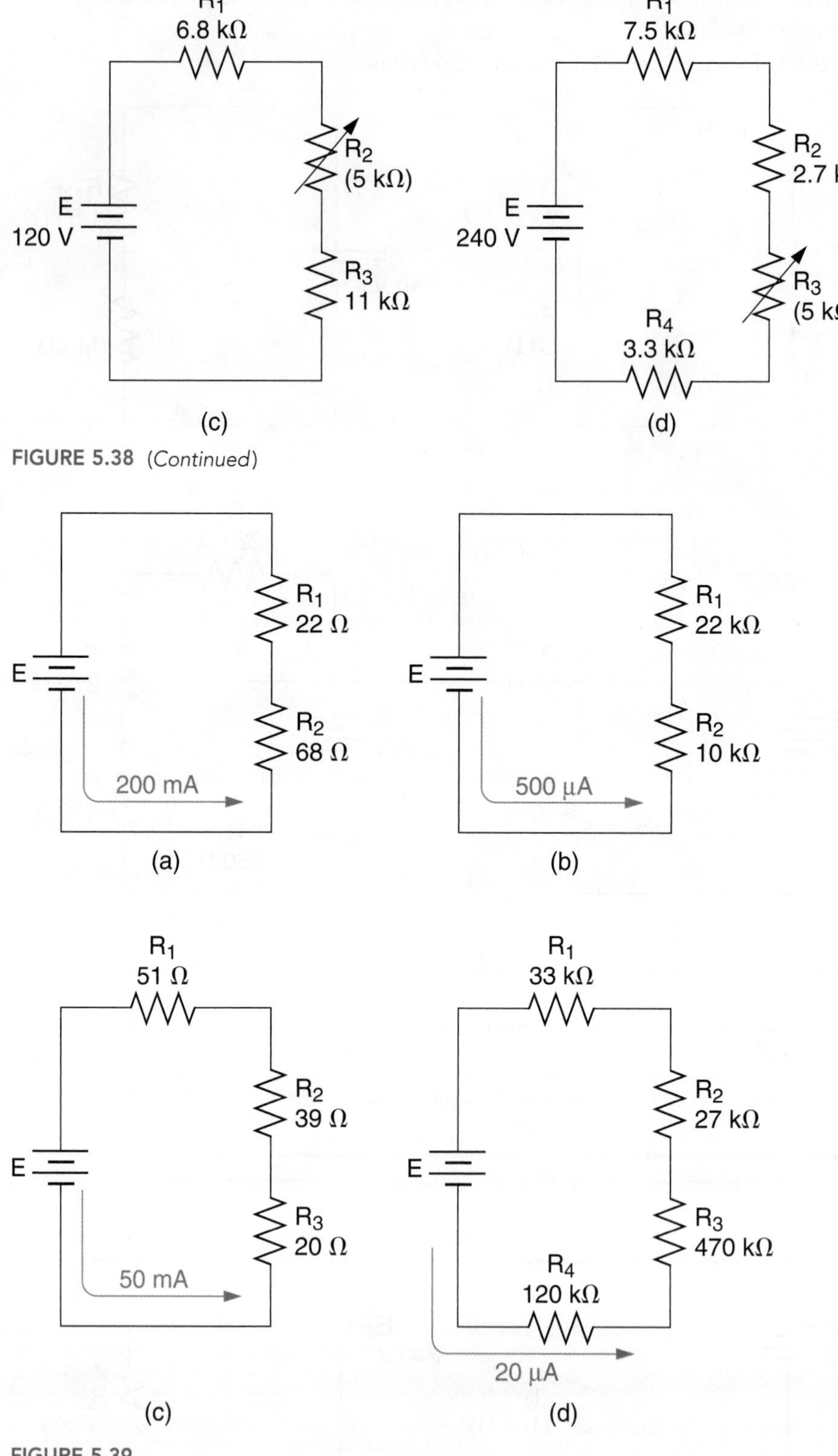

FIGURE 5.38 *(Continued)*

FIGURE 5.39

12. Determine the rheostat setting needed to set the current in Figure 5.38b to 2.2 mA.
13. Determine the rheostat setting needed to set the current in Figure 5.38c to 5.87 mA.

14. Determine the rheostat setting needed to set the current in Figure 5.38d to 13.8 mA.
15. Calculate the value of the source voltage in Figure 5.39a.
16. Calculate the value of the source voltage in Figure 5.39b.
17. Calculate the value of the source voltage in Figure 5.39c.
18. Calculate the value of the source voltage in Figure 5.39d.
19. For the circuit in Figure 5.40a, verify that the total power dissipated by the resistors is equal to the power being supplied by the source.

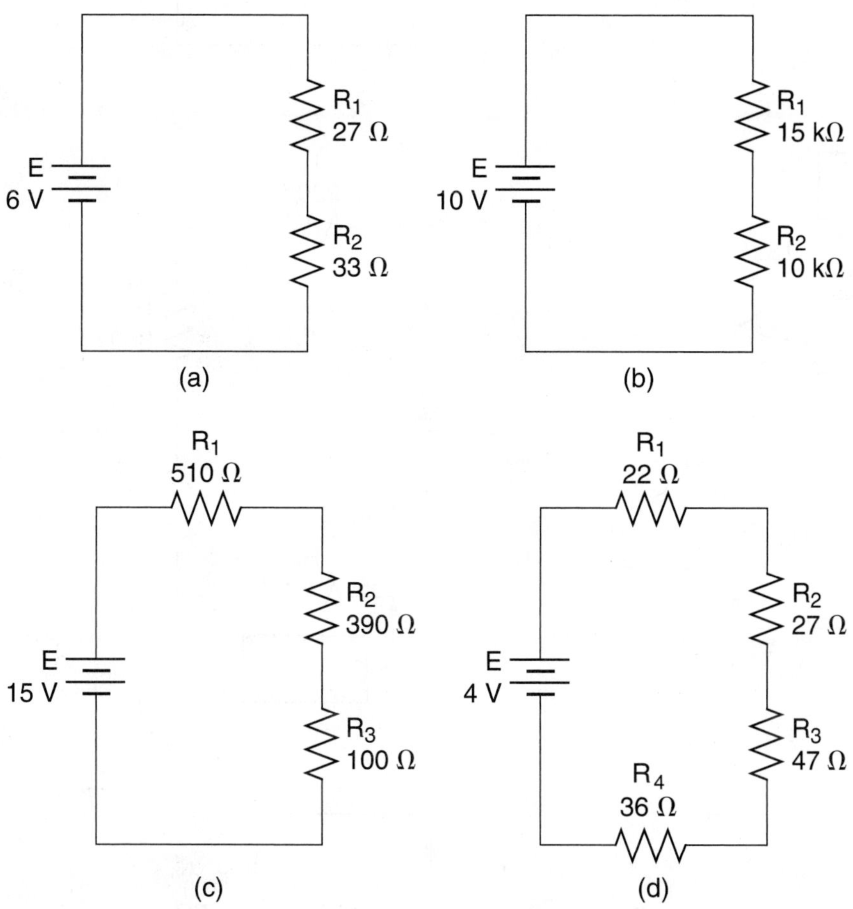

FIGURE 5.40

20. Repeat problem 19 for the circuit in Figure 5.40b.
21. Repeat problem 19 for the circuit in Figure 5.40c.
22. Repeat problem 19 for the circuit in Figure 5.40d.
23. Calculate the value of the current in Figure 5.41a.
24. Calculate the value of the current in Figure 5.41b.
25. Calculate the value of the current in Figure 5.41c.
26. Calculate the value of the current in Figure 5.41d.
27. For the circuit in Figure 5.42a, calculate the voltage from point A to ground.

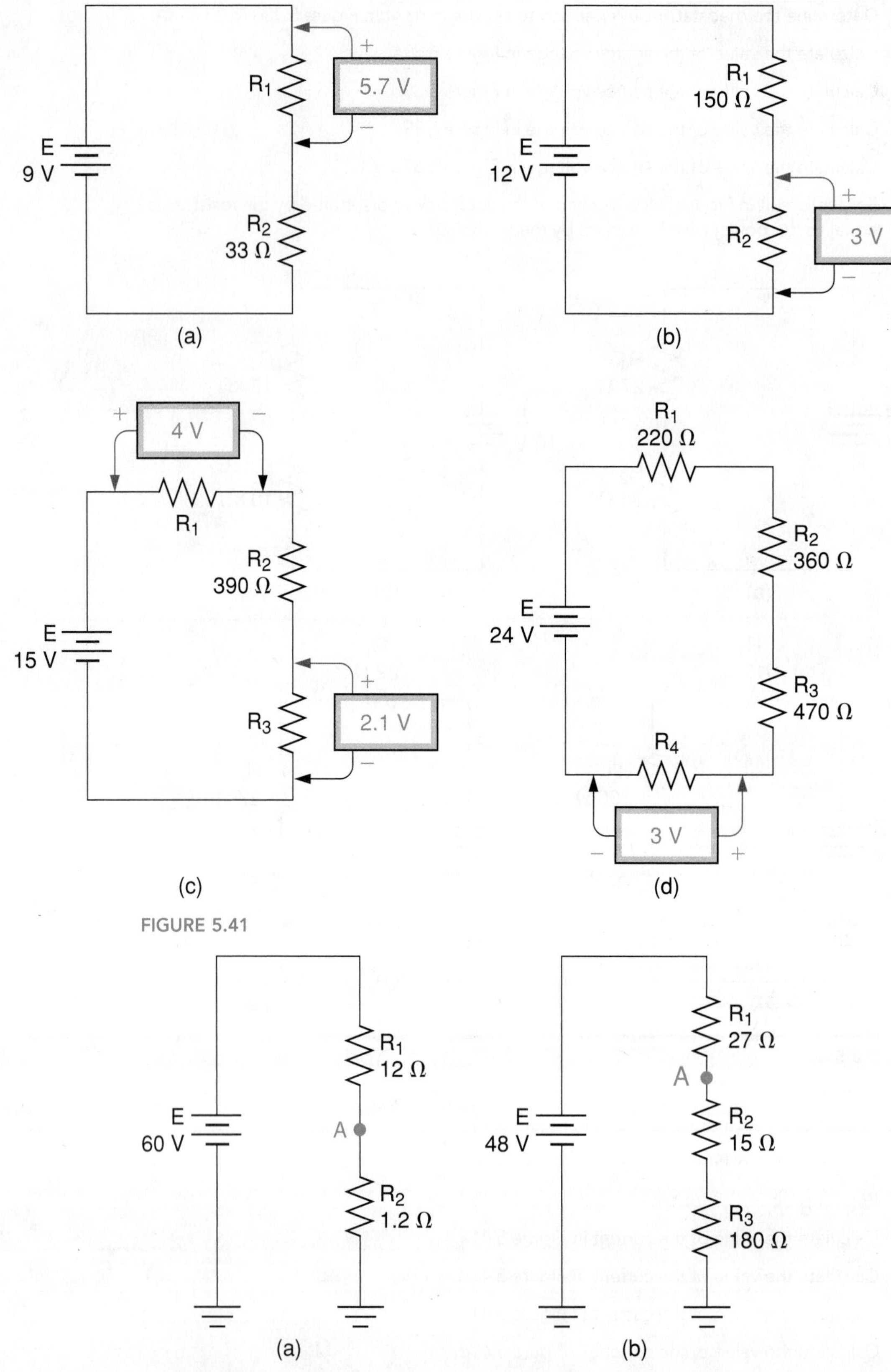

FIGURE 5.41

FIGURE 5.42

28. For the circuit in Figure 5.42b, calculate the voltage from point A to ground.
29. Using the voltage divider equation, calculate the voltage across each of the resistors in Figure 5.43a.

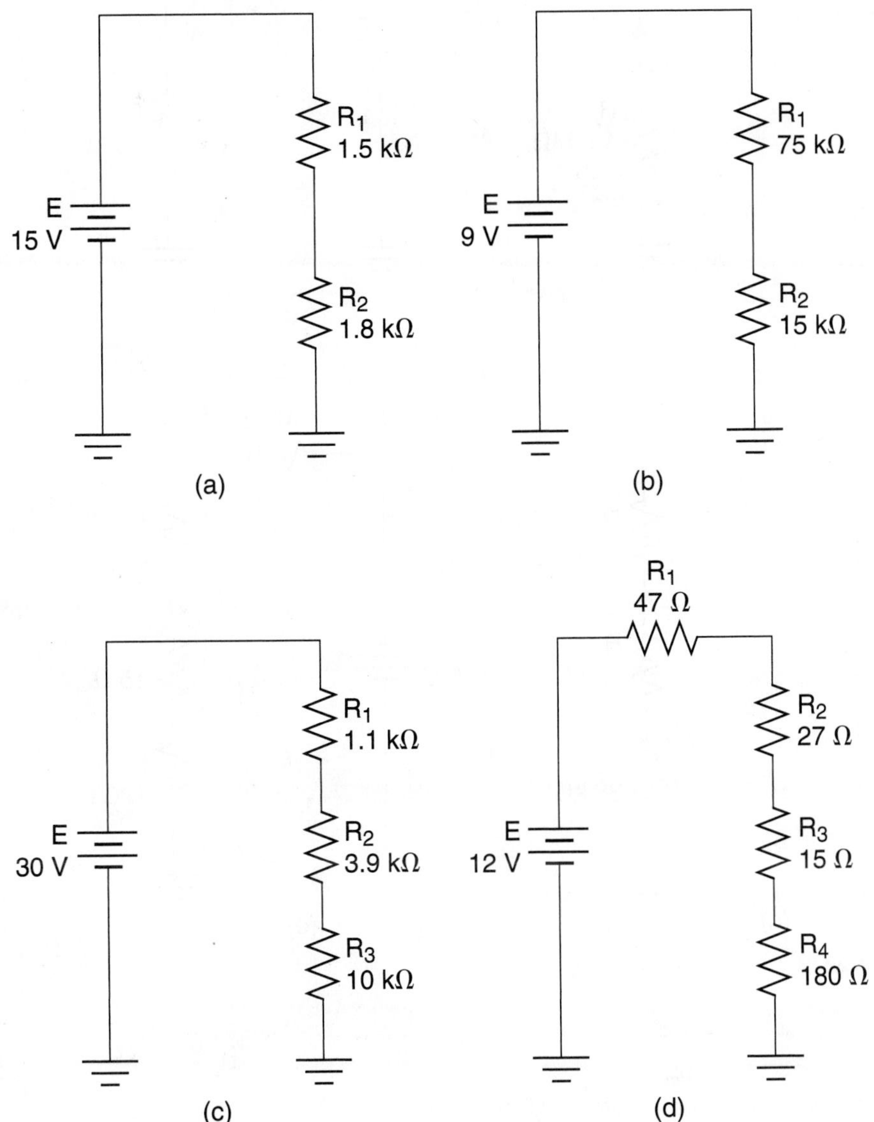

FIGURE 5.43

30. Repeat problem 29 for the circuit in Figure 5.43b.
31. Repeat problem 29 for the circuit in Figure 5.43c.
32. Repeat problem 29 for the circuit in Figure 5.43d.
33. Using the voltage-divider equation, determine the voltage from point A to ground for the circuit in Figure 5.44a.
34. Repeat problem 33 for the circuit in Figure 5.44b.

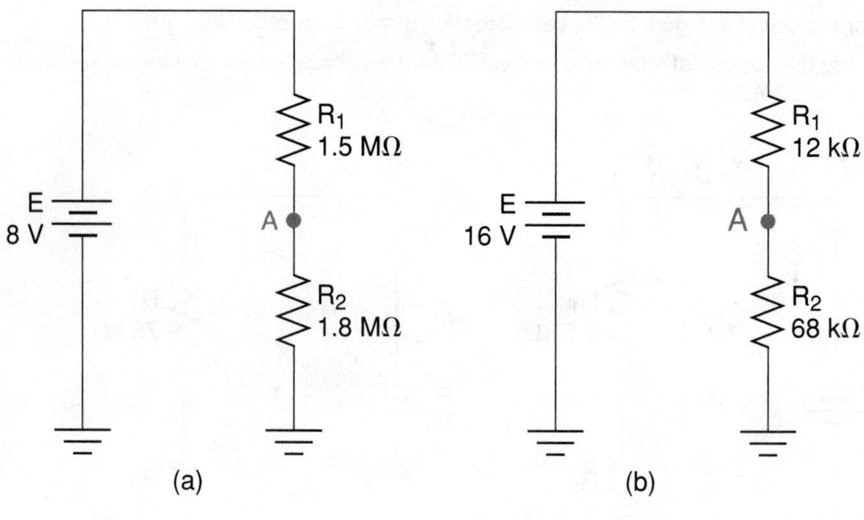

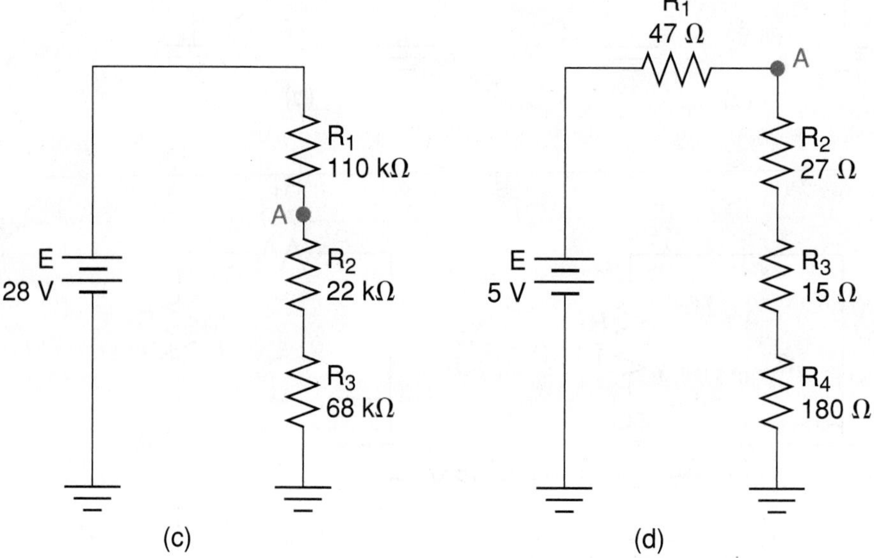

FIGURE 5.44

35. Repeat problem 33 for the circuit in Figure 5.44c.
36. Repeat problem 33 for the circuit in Figure 5.44d.
37. Calculate the voltage, current, resistance, and power values for the circuit in Figure 5.45a.
38. Calculate the voltage, current, resistance, and power values for the circuit in Figure 5.45b.
39. Calculate the voltage, current, resistance, and power values for the circuit in Figure 5.45c.
40. Calculate the voltage, current, resistance, and power values for the circuit in Figure 5.45d.

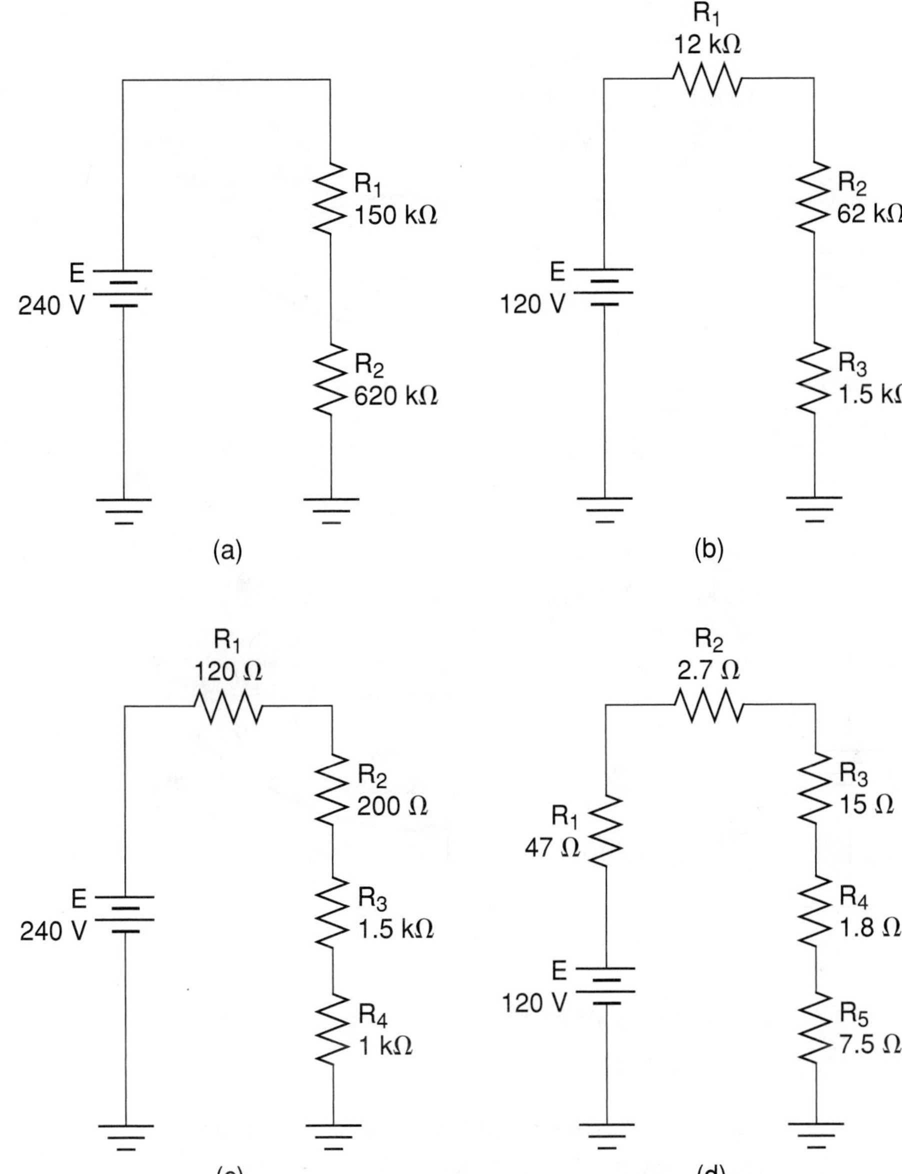

FIGURE 5.45

41. Assume that the voltage across the open resistor in Figure 5.46a is being measured with a DMM that has an internal resistance of 10 MΩ. Determine the meter reading.

42. The voltage across the open resistor in Figure 5.46a is being measured with a VOM that has an internal resistance of 200 kΩ. Determine the meter reading.

43. Repeat problem 41 for the circuit in Figure 5.46b.

44. Repeat problem 42 for the circuit in Figure 5.46b.

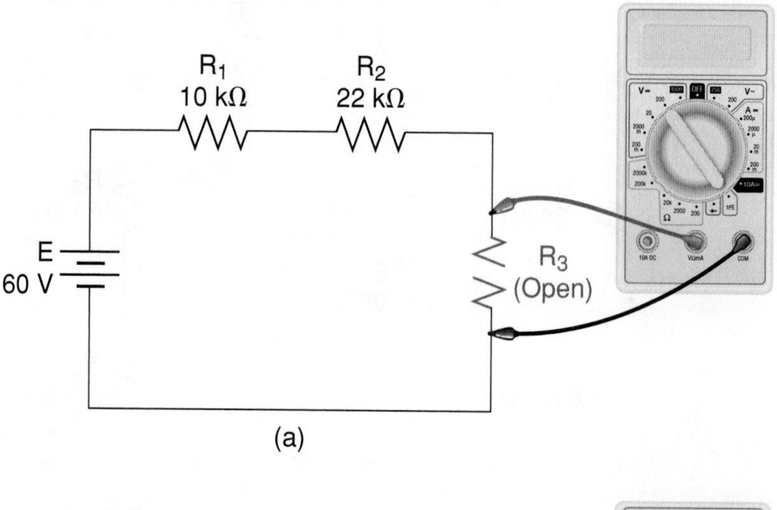

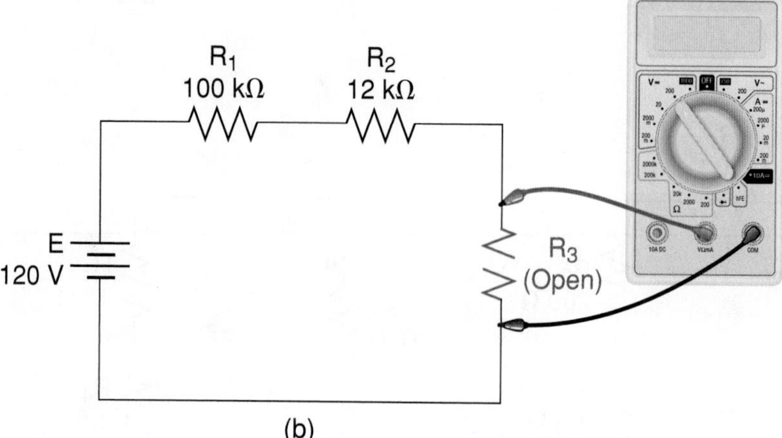

FIGURE 5.46

TROUBLESHOOTING PRACTICE PROBLEMS

45. Determine the fault (if any) that is indicated by the readings in Figure 5.47a.
46. Determine the fault (if any) that is indicated by the readings in Figure 5.47b.

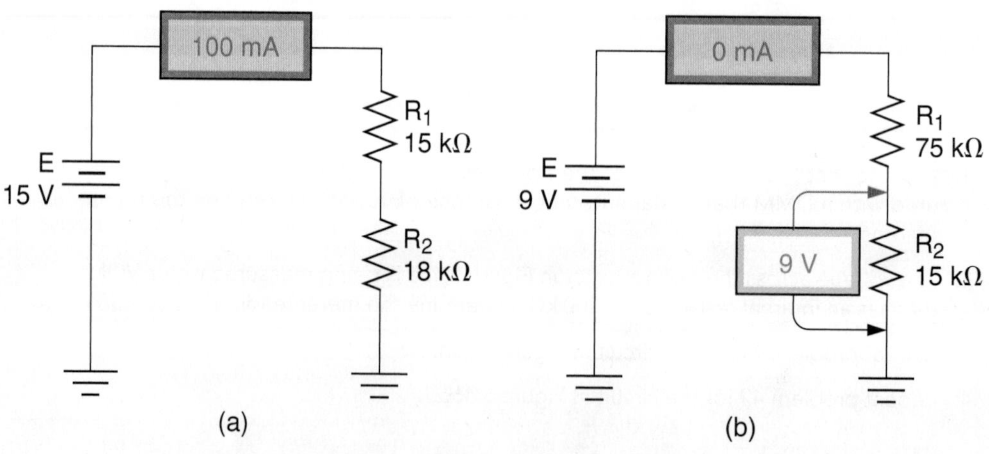

FIGURE 5.47

ANSWERS TO THE EXAMPLE PRACTICE PROBLEMS

5.1	1113 Ω (1.11 kΩ)
5.2	102 Ω
5.3	593 mA
5.4	27 Ω
5.5	268 V
5.6	P_{R1} = 20 W, P_{R2} = 24.4 W, P_{R3} = 15.6 W, and P_T = 60 W
5.7	R_T = 65 Ω, I_T = 3.38 A, V_{R1} = 111.5 V, V_{R2} = 74.4 V, V_{R3} = 33.8 V, P_{R1} = 377 W, P_{R2} = 251 W, P_{R3} = 114 W, P_T = 744 W
5.8	2.5 A
5.9	42.6 V
5.10	V_{R1} = 82.5 V, V_{R2} = 117.5 V
5.11	69.1 V
5.12	11.9 V

CHAPTER 6

PARALLEL CIRCUITS

PURPOSE

In this chapter, we will continue our coverage of basic circuits by discussing the operating characteristics of parallel circuits. As you will see, these circuits are nearly the opposite of series circuits in many ways.

KEY TERMS

The following terms are introduced and defined in this chapter on the pages indicated:

Branch, 228	Load, 254
Current divider, 232	Node, 241
Current source, 242	Overload, 254
Kirchhoff's current law (KCL), 240	Parallel circuit, 228
	Surge, 254

OBJECTIVES

After completing this chapter, you should be able to:

1. Identify a parallel circuit.
2. Describe the current and voltage characteristics of parallel circuits.
3. Calculate any current value in a parallel circuit, given the source voltage and the branch resistances.
4. Contrast the current and voltage characteristics of series and parallel circuits.
5. Calculate the total resistance in any parallel circuit, given the branch resistances.
6. Describe *Kirchhoff's current law* and solve the Kirchhoff's current equation for any parallel circuit.
7. Compare and contrast a *voltage divider* with a *current divider*.
8. Use the *current divider* equations to solve for either branch current in a two-branch circuit.
9. Calculate the current, voltage, power, and resistance of any parallel circuit.
10. Describe the procedure used to locate an *open* branch in a parallel circuit.
11. Describe the symptoms that are associated with a *shorted* branch in a parallel circuit.
12. Describe the sources and symptoms of a circuit *overload.*

6.1 PARALLEL CIRCUIT CHARACTERISTICS

PARALLEL CIRCUIT A circuit that provides more than one current path between any two points.

BRANCH Each current path in a parallel circuit.

A **PARALLEL CIRCUIT** is a circuit that provides more than one current path between any two points. Several parallel circuits are shown in Figure 6.1. As you can see, each circuit contains two or more paths for current. Note that each current path in a parallel circuit is referred to as a **BRANCH**. For example, the circuit in Figure 6.1a contains two branches, the circuit in Figure 6.1b contains three branches, and so on.

> The term *parallel*, as used here, does not necessarily mean they are mounted side-by-side. Instead, it means *alternate*. A break or opening in any branch of a parallel circuit does not stop the flow of electrons to the remaining branches.
>
> KEY CONCEPT

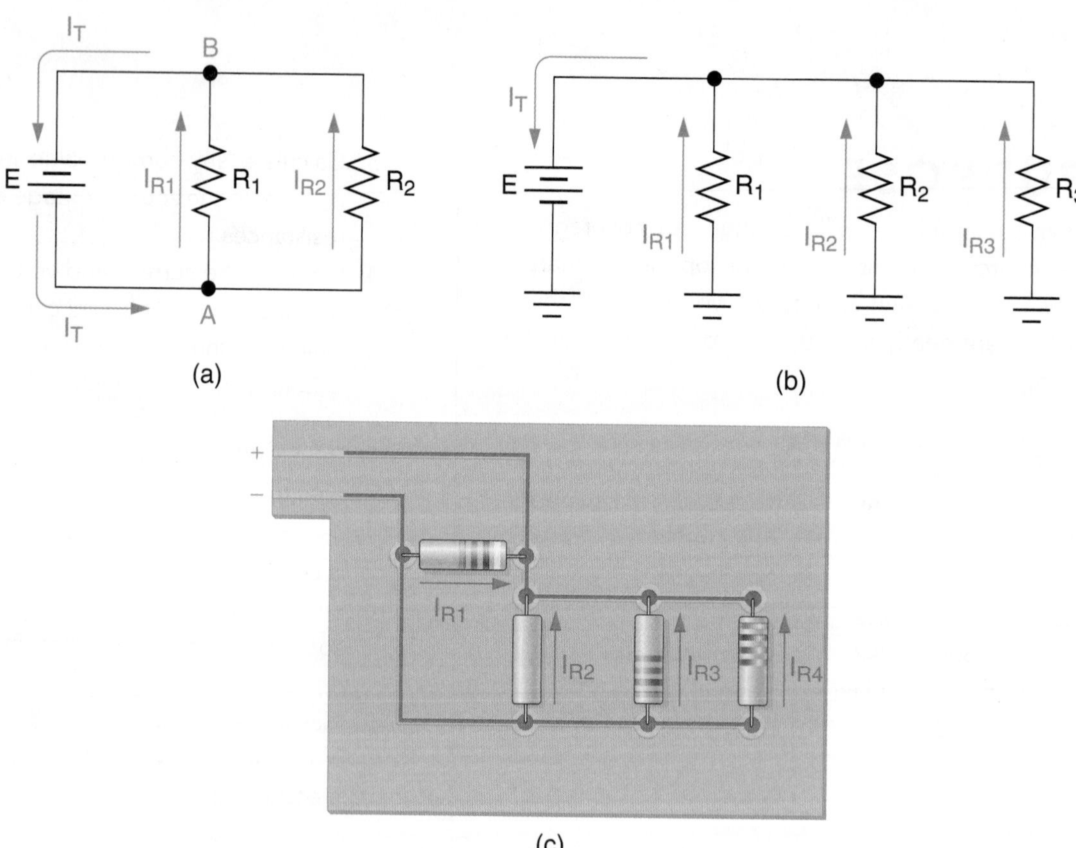

FIGURE 6.1 Parallel circuits.

Current Characteristics

Using the circuit in Figure 6.1a, let's take a look at the current action in a typical parallel circuit: The current leaves the negative terminal of the source. When it reaches the first connection point in the circuit (point A), the current splits. Part of the current passes through R_1 and the remainder passes through R_2. When these currents reach the second connection point (point B), they recombine into a single current and return to the source.

The total current in a parallel circuit (I_T) is equal to the sum of the individual branch current values. By formula,

(6.1)
$$I_T = I_{R1} + I_{R2} + \cdots + I_{Rn}$$

where

I_{Rn} = the current through the highest numbered resistor in the circuit

The total current in a parallel circuit is calculated as shown in Example 6.1.

EXAMPLE 6.1

What is the value of the circuit current in Figure 6.2?

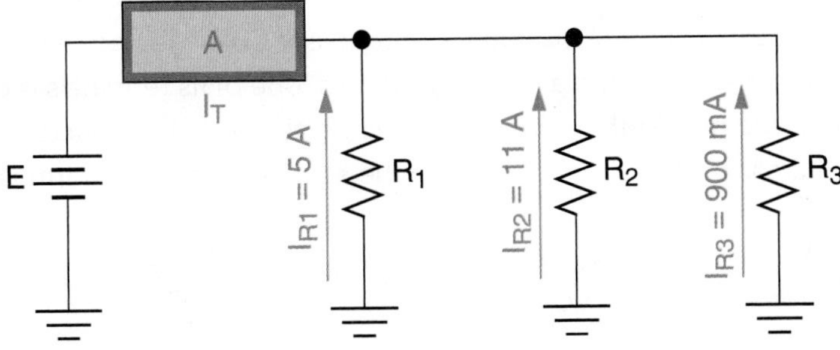

FIGURE 6.2

SOLUTION

Using equation (6.1), the circuit current is found as

$$\begin{aligned} I_T &= I_{R1} + I_{R2} + I_{R3} \\ &= 5\text{ A} + 11\text{ A} + 900\text{ mA} \\ &= 16.9\text{ A} \end{aligned}$$

PRACTICE PROBLEM 6.1

A circuit like the one in Figure 6.2 has the following values: I_{R1} = 6.4 A, I_{R2} = 3.3 A, and I_{R3} = 800 mA. What is the value of the circuit current?

When you think about it, the result in Example 6.1 makes sense. The branch currents must all come from the voltage source. Therefore, the total current being supplied by the source must equal the sum of all the branch currents.

When multiple power sources are connected in parallel, the voltage remains the same but the current and ampere-hour capacity increase.

KEY CONCEPT

Voltage and Current Values

When two or more components are connected in parallel, the voltage across each component is equal to the voltage across all of the others. This characteristic is illustrated in Figure 6.3.

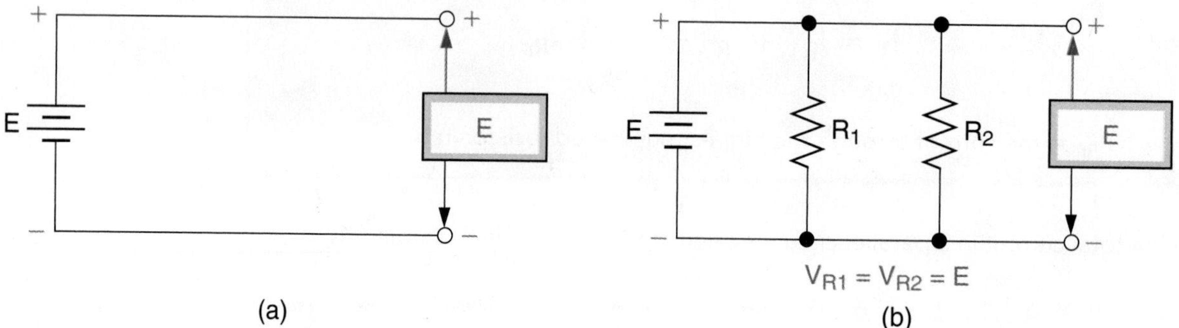

FIGURE 6.3 Branch voltages are equal.

As the first circuit shows, the source voltage can be measured between the (+) and (−) conductors. Each time a resistor is added, one of its terminals is connected to the (+) conductor and the other is connected to the (−) conductor. Therefore, the voltage across each resistor is equal to the source voltage. By formula,

(6.2)

$$E = V_{R1} = V_{R2} = \cdots = V_{Rn}$$

where

V_{Rn} = the voltage across the highest numbered resistor in the circuit

The current through each branch of a parallel circuit is determined by the source voltage and the resistance of the branch. Therefore, we can use Ohm's law to calculate the value of any branch current as follows:

$$I_{Rn} = \frac{E}{R_n}$$

Once we have calculated the value of each branch current in a parallel circuit, we simply add those values to determine the total circuit current. This approach to calculating circuit current is demonstrated in Example 6.2.

EXAMPLE 6.2

What is the value of the circuit current in Figure 6.4?

SOLUTION

The individual branch currents in Figure 6.4 are found as follows:

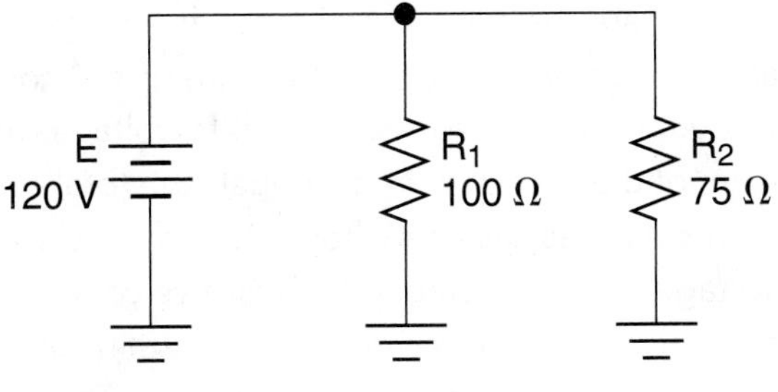

FIGURE 6.4

$$I_{R1} = \frac{E}{R_1} = \frac{120\text{ V}}{100\ \Omega} = 1.2\text{ A}$$

and

$$I_{R2} = \frac{E}{R_2} = \frac{120\text{ V}}{75\ \Omega} = 1.6\text{ A}$$

Once the branch current values are known, the total circuit current can be found as

$$I_T = I_{R1} + I_{R2} = 1.2\text{ A} + 1.6\text{ A} = 2.8\text{ A}$$

PRACTICE PROBLEM 6.2

A circuit like the one in Figure 6.4 has the following values: $E = 100$ V, $R_1 = 24\ \Omega$, and $R_2 = 36\ \Omega$. What is the value of the circuit current?

Before we move on, let's take a moment to compare the voltage and current characteristics of parallel circuits with those of series circuits. These characteristics are summarized in Table 6.1. As you can see, the voltage and current characteris-

TABLE 6.1 • A Comparison of Series and Parallel Circuit Characteristics

CIRCUIT TYPE	VOLTAGE CHARACTERISTICS	CURRENT CHARACTERISTICS
Series	$E = V_1 + V_2 + \cdots + V_n$	Equal at all points
Parallel	Equal across all branches	$I_T = I_1 + I_2 + \cdots + I_n$

tics of the parallel circuit are the opposite of those for the series circuit. In series circuits, currents are equal at all points and the sum of the component voltages equals the total (or source) voltage. In parallel circuits, voltages are equal across all branches and the sum of the branch currents equals the total current.

You may recall that a series circuit is often referred to as a voltage divider because the source voltage is divided among the resistive components. By the same token, a parallel circuit is often referred to as a **CURRENT DIVIDER** because the total circuit current is divided among the circuit branches. The concept of a current divider is discussed in greater detail later in this chapter.

CURRENT DIVIDER Another name for a parallel circuit, which divides the source current among its branches.

Resistance Characteristics

In Example 6.2, the total current being supplied by a 120 V source was found to be 2.8 A. If we apply Ohm's law to these values, the total circuit resistance is found as

$$R_T = \frac{E}{I_T} = \frac{120\ \text{V}}{2.8\ \text{A}} = 42.9\ \Omega$$

As you can see, this value is lower than the value of either resistor in the circuit.

In any parallel circuit, the total resistance is always lower than any of the branch resistance values. This relationship is not as strange as it first seems. Consider the circuits shown in Figure 6.5. In Figure 6.5a, a 10 V source is generating 1 A of current

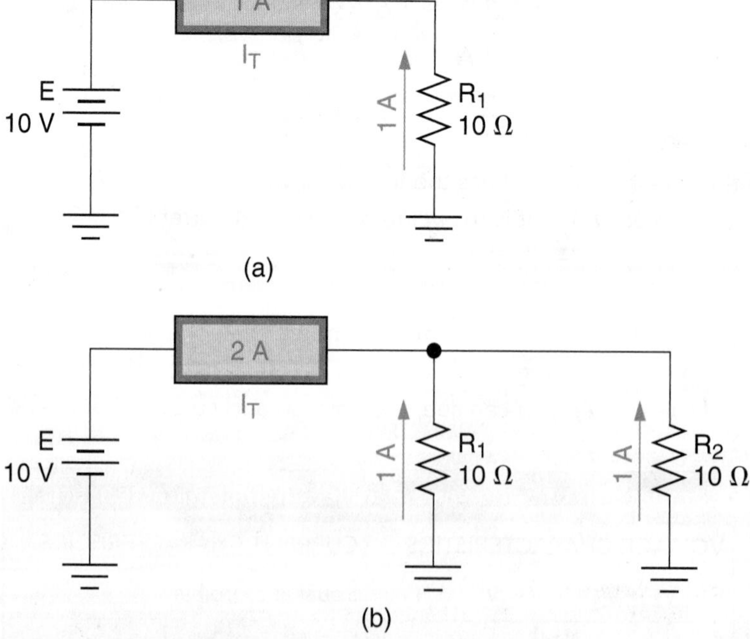

FIGURE 6.5 Adding a parallel resistor increases the total circuit current.

through 10 Ω of resistance. If a second branch is added to the circuit (as shown in Figure 6.5b), the same 10 V source is now generating another 1 A of current through the second branch. As you can see, the total current being generated has increased to 2 A even though the applied voltage has remained constant at 10 V. According to Ohm's law, this can only happen if the total resistance in the circuit decreases. Therefore, the addition of one or more parallel branches always results in a decrease in circuit resistance and an increase in circuit current. (Remember, resistance and current are inversely proportional whether in a series or a parallel circuit.)

Calculating the Total Parallel Resistance Using Ohm's Law

There are several approaches to calculating the total resistance in a parallel circuit. The first is to calculate the total circuit current and then use Ohm's law to calculate the total resistance. This approach is demonstrated in Example 6.3.

EXAMPLE 6.3

What is the value of the circuit resistance in Figure 6.6?

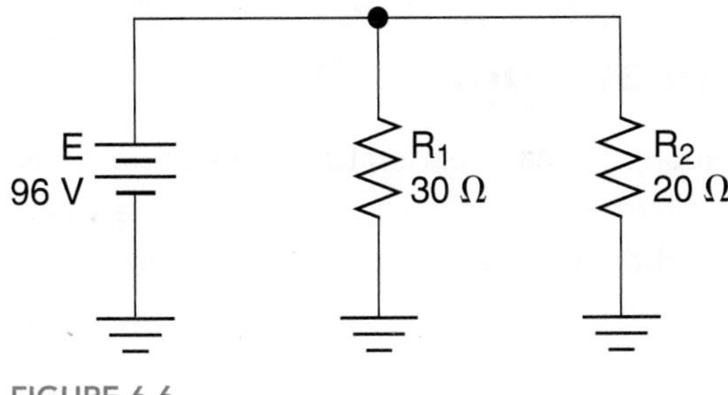

FIGURE 6.6

SOLUTION

To calculate the total resistance, calculate the branch currents as follows:

$$I_{R1} = \frac{E}{R_1}$$

$$= \frac{96 \text{ V}}{30 \ \Omega}$$

$$= 3.2 \text{ A}$$

and

$$I_{R2} = \frac{E}{R_2}$$

$$= \frac{96 \text{ V}}{20 \ \Omega}$$

$$= 4.8 \text{ A}$$

The circuit current is now found as

$$I_T = I_{R1} + I_{R2}$$
$$= 3.2\text{ A} + 4.8\text{ A}$$
$$= 8\text{ A}$$

and the total resistance is found as

$$R_T = \frac{E}{I_T}$$
$$= \frac{96\text{ V}}{8\text{ A}}$$
$$= 12\ \Omega$$

PRACTICE PROBLEM 6.3

A parallel circuit like the one in Figure 6.6 has the following values: $E = 240$ V, $R_1 = 180\ \Omega$, and $R_2 = 120\ \Omega$. What is the value of the circuit resistance?

Calculating Total Parallel Resistance: The Reciprocal Approach

There are three equations that can be used to calculate total parallel resistance using the values of the branch resistances. The first of these is the *reciprocal approach.*

You may recall that the total resistance in a series circuit equals the sum of the individual resistance values. For example, the total resistance of a two-resistor circuit is found using

$$R_T = R_1 + R_2$$

In a parallel circuit, the *reciprocal* of the total resistance equals the sum of the reciprocal branch resistances. For example, a two-resistor parallel circuit has a value of

(6.3)
$$\frac{1}{R_T} = \frac{1}{R_1} + \frac{1}{R_2}$$

If we take the reciprocal of each side of equation (6.3), we get

(6.4)
$$R_T = \frac{1}{\frac{1}{R_1} + \frac{1}{R_2}}$$

Equation (6.4) may look intimidating at first, but it's actually very easy to solve on any standard calculator. To solve the equation, you enter the values into your calculator in the same manner as you would any addition problem. However:

1. After entering each resistor value, you press the *reciprocal key*. Depending on the calculator, this key is labeled [1/x] or [x^{-1}].
2. After both resistance values have been entered, press the [=] key followed by the reciprocal key.

The use of this key sequence is demonstrated in Example 6.4.

EXAMPLE 6.4

What is the value of the circuit resistance in Figure 6.6?

SOLUTION

The total resistance in Figure 6.6 can be calculated using the following key sequence:

30 [1/x] [+] 20 [1/x] [=] [1/x]

Using this key sequence, the calculator gives you a result of 12 Ω, the same value calculated in Example 6.3.

PRACTICE PROBLEM 6.4

A circuit like the one in Figure 6.6 has the following values: $R_1 = 240\ \Omega$ and $R_2 = 360\ \Omega$. What is the value of the total circuit resistance?

The reciprocal approach to solving for total parallel resistance can be expanded to include any number of parallel branches, as demonstrated in Example 6.5.

EXAMPLE 6.5

What is the value of the circuit resistance in Figure 6.7?

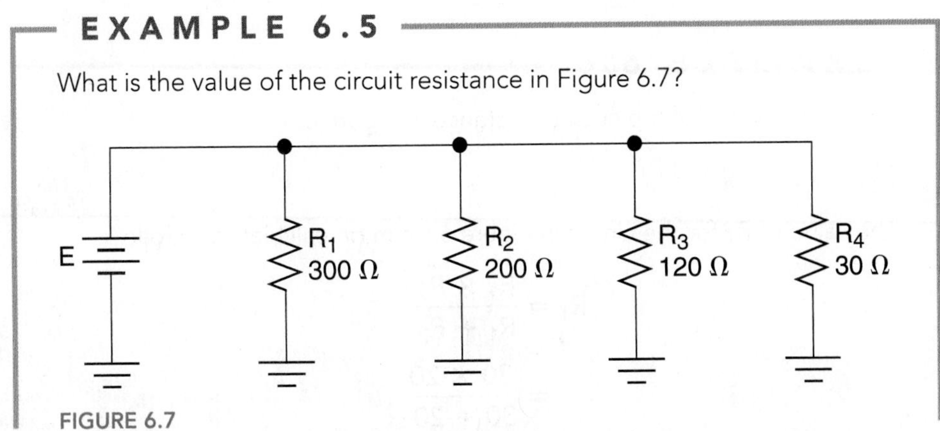

FIGURE 6.7

SOLUTION

The value of R_T for the circuit in Figure 6.7 can be calculated using this expanded version of equation (6.4):

$$R_T = \frac{1}{\frac{1}{R_1} + \frac{1}{R_2} + \frac{1}{R_3} + \frac{1}{R_4}}$$

The calculator sequence for solving this equation is

300 [1/x] [+] 200 [1/x] [+] 120 [1/x] [+] 30 [1/x] [=] [1/x]

Using this key sequence, the calculator gives you a result of 20 Ω.

PRACTICE PROBLEM 6.5

A circuit like the one in Figure 6.7 has the following values: $R_1 = 100\ \Omega$, $R_2 = 240\ \Omega$, $R_3 = 360\ \Omega$, and $R_4 = 180\ \Omega$. What is the value of the circuit resistance?

Calculating Total Parallel Resistance: The Product-Over-Sum Approach

Another approach that is commonly used to solve for total parallel resistance is the *product-over-sum* approach. This method allows you to solve for the total resistance of a two-branch circuit as follows:

(6.5)

$$R_T = \frac{R_1 \times R_2}{R_1 + R_2}$$

The use of this equation is demonstrated in Example 6.6.

EXAMPLE 6.6

What is the value of the circuit resistance in Figure 6.6?

SOLUTION

The value of R_T for the circuit in Figure 6.6 can be calculated as follows:

$$R_T = \frac{R_1 \times R_2}{R_1 + R_2}$$

$$= \frac{30 \times 20}{30 + 20}\ \Omega$$

$$= \frac{600}{50}\ \Omega$$

$$= 12\ \Omega$$

PRACTICE PROBLEM 6.6

Rework Practice Problem 6.4 using the product-over-sum approach.

There are a couple of points that should be made regarding equation (6.5):

1. The equation can only be used on parallel circuits containing two branches. As written, it will not work when there are more than two branches in the circuit.
2. Solving the product-over-sum equation actually requires more key strokes than solving the reciprocal equation. For this reason, you may find it easier to stick with the reciprocal approach to calculating the total resistance of any two-branch parallel circuit.

Calculating Total Parallel Resistance: Equal Branch Resistance Values

There is another "special-case" approach to calculating total parallel resistance. When any number of equal-value resistances are connected in parallel, the total resistance equals the resistance of any one branch divided by the number of branches. For example, consider the circuit shown in Figure 6.8. Here we have three 24 Ω lamps connected in parallel. The total resistance of this circuit can be found as

$$R_T = \frac{24\ \Omega}{3}$$

$$= 8\ \Omega$$

Note that the resistance of one branch (24 Ω) is being divided by the number of branches in the circuit (3). Again, this is a method that can only be used in a special case: when all of the branch resistances are equal.

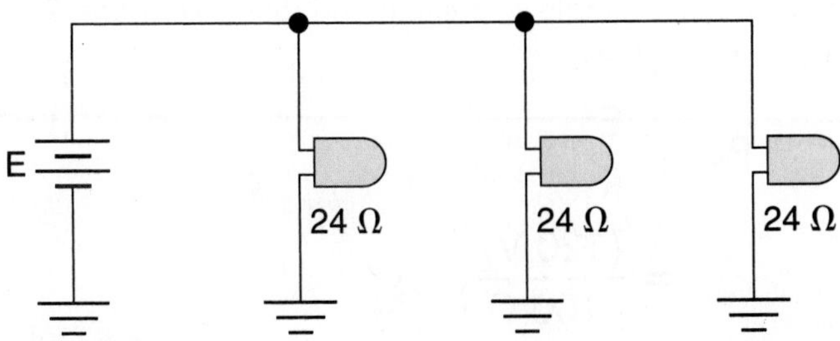

FIGURE 6.8 Equal-value resistances connected in parallel.

Putting It All Together

There are three approaches to calculating the total parallel resistance. These methods, along with their applications, are summarized in Table 6.2.

TABLE 6.2 • Solving for Total Parallel Resistance

METHOD	EQUATION	APPLICATIONS
Reciprocal	$R_T = \dfrac{1}{\dfrac{1}{R_1} + \dfrac{1}{R_2} + \cdots + \dfrac{1}{R_n}}$	All circuits
Product-over-sum	$R_T = \dfrac{R_1 \times R_2}{R_1 + R_2}$	Circuits containing only two branches
Equal resistances	$^*R_T = \dfrac{R}{n}$	Circuits containing equal branch resistance values

*R = the resistance of one branch; n = the number of branches

Power Characteristics

The power characteristics of parallel circuits are nearly identical to those of series circuits. For example, the total power in a parallel circuit is found as the sum of the power dissipation values for the individual components in the circuit. Also, component power dissipation values in any parallel circuit are calculated using the same standard power relationships we used in series circuits:

$$P = I \times E \qquad P = \frac{E^2}{R} \qquad P = I^2R$$

As you know, power is a measure of the rate at which a component absorbs (or uses) energy. That rate does not take into account how the component is connected in a circuit. For example, a 100 Ω resistor with 100 mA of current through it uses 1 W of power, regardless of whether that component is in a series circuit or a parallel circuit. The total power that is used by a circuit is found as the sum of the individual power values regardless of the circuit configuration.

In terms of power dissipation, there is one major difference between series and parallel circuits. In a series circuit, the higher the value of a resistor, the higher the percentage of the total power it dissipates. In a parallel circuit, the opposite is true. That is, the lower the value of a branch resistance, the higher the percentage of the total power it dissipates. For example, the circuit in Figure 6.4 (page 231) has the following power dissipation values:

$$P_{R1} = \frac{E^2}{R_1}$$

$$= \frac{(120\text{ V})^2}{100\ \Omega}$$

$$= 144\text{ W}$$

and

$$P_{R2} = \frac{E^2}{R_2}$$
$$= \frac{(120\ \text{V})^2}{75\ \Omega}$$
$$= 192\ \text{W}$$

As you can see, the lower value resistor is dissipating more power. This is due to the fact that power is inversely proportional to resistance when voltage is fixed.

One Final Note

You have been shown the basic voltage, current, resistance, and power characteristics of parallel circuits. Figure 6.9 provides a comparison of a basic parallel circuit and a basic series circuit.

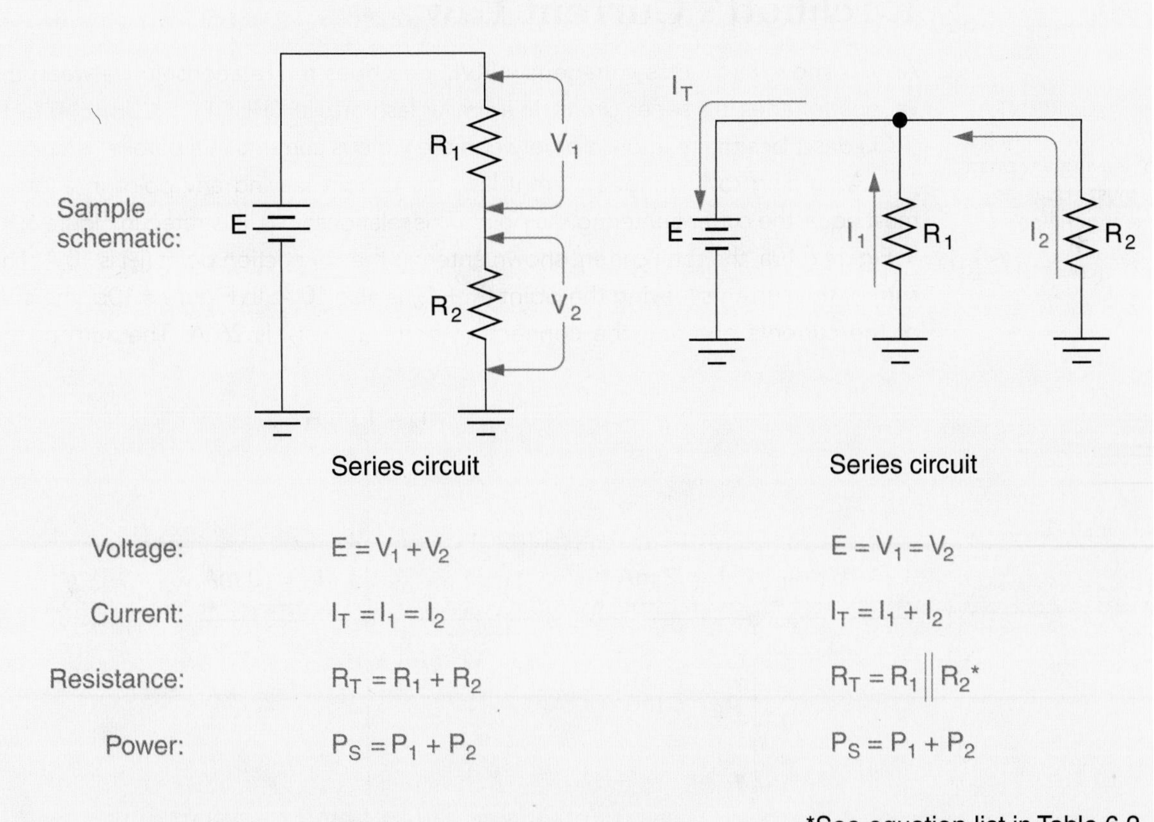

FIGURE 6.9 Series vs. parallel circuit characteristics.

PROGRESS CHECK

1. What is a *branch*?
2. Describe the current characteristics of a parallel circuit.
3. Describe the voltage characteristics of a parallel circuit.
4. Why are parallel circuits referred to as current dividers?
5. Why is the total resistance of a parallel circuit lower than any of its branch resistance values?
6. Compare and contrast series and parallel circuits in terms of their basic properties.

6.2 CURRENT DIVIDERS

In the last section, you were shown that the sum of the branch currents in a parallel circuit is equal to the total circuit current. In this section, we're going to take a closer look at the relationship between the various currents in a parallel circuit.

Kirchhoff's Current Law

KIRCHHOFF'S CURRENT LAW (KCL) A law stating that the current leaving any point in a circuit must equal the current entering the point.

As you know, Kirchhoff's voltage law (KVL) describes the relationship between the various voltages in a series circuit. In a similar fashion, **KIRCHHOFF'S CURRENT LAW (KCL)** describes the relationship between the various currents in a parallel circuit.

According to Kirchhoff's current law, the current leaving any point in a circuit must equal the current entering the point. This relationship is illustrated in Figure 6.10. In Figure 6.10a, the total current shown entering the connection point (I_T) is 10 A. The sum of the currents leaving the point ($I_1 + I_2$) is also 10 A. In Figure 6.10b, the sum of the currents entering the connection point ($I_1 + I_2$) is 26 A. The sum of the

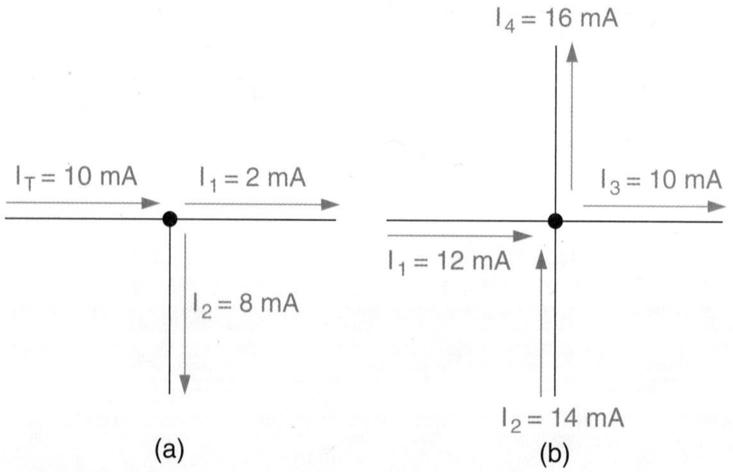

FIGURE 6.10 Examples of Kirchhoff's current law (KCL).

currents leaving the connection point ($I_3 + I_4$) is also 26 A. In each circuit, the current leaving the connection point equals the current entering that point. Note that points connecting two or more current paths (like those in Figure 6.10) are often referred to as **NODES**.

NODE A point connecting two or more current paths.

There are a variety of practical applications for Kirchhoff's current law. One such application is demonstrated in Example 6.7.

EXAMPLE 6.7

What rheostat setting is required to provide a value of $I_T = 2.8$ A in Figure 6.11?

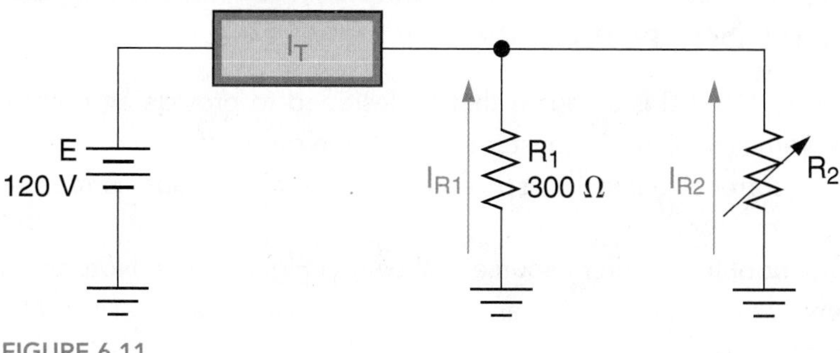

FIGURE 6.11

SOLUTION

The adjusted value of the rheostat in Figure 6.11 can be determined as follows: First, the value of I_{R1} is found as

$$I_{R1} = \frac{E}{R_1} = \frac{120\text{ V}}{300\ \Omega} = 400\text{ mA}$$

The desired value of I_{R2} can now be found as

$$I_{R2} = I_T - I_{R1} = 2.8\text{ A} - 400\text{ mA} = 2.4\text{ A}$$

Finally, Ohm's law is used to determine the rheostat setting, as follows:

$$R_2 = \frac{E}{I_{R2}} = \frac{120\text{ V}}{2.4\text{ A}} = 50\ \Omega$$

PRACTICE PROBLEM 6.7

A circuit like the one in Figure 6.11 has the following values: $E = 120$ V and $R_1 = 200\ \Omega$. What adjusted value of R_2 is required to set the total source current to 1.8 A?

In Example 6.7, a rheostat is used to set the value of the circuit current. There is a type of source, however, whose output current is fixed and relatively independent of the resistance values in the circuit. Such a source is referred to as a current source. At this point, we will take a brief look at this type of source.

Current Sources

CURRENT SOURCE A source that provides an output current value that remains relatively constant over a wide range of load resistance values.

A **CURRENT SOURCE** is a source that is designed to provide an output current value that remains relatively constant over a wide range of load resistance values. As such, a current source is also referred to as a *constant current source* or *regulated current source.*

The symbol for a current source is shown in Figure 6.12a. Note that the output current of the source is given and that the arrow in the symbol indicates the direction of the current.

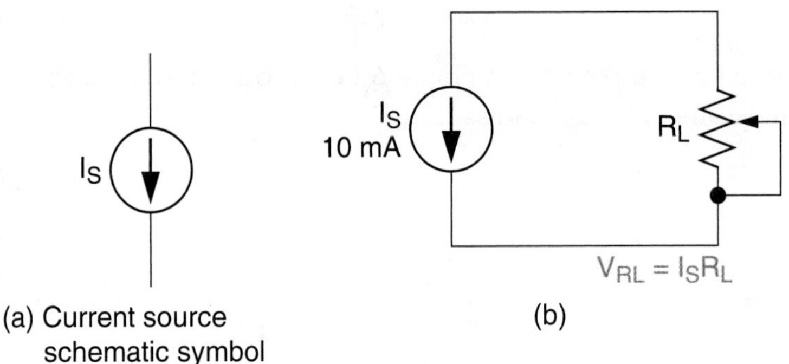

FIGURE 6.12 Current sources.

Figure 6.12b contains a 10 mA current source and a variable load resistance. When the load resistance changes, the circuit current remains at the rated value of the source. Thus, a change in the load resistance results in a change in load voltage (in keeping with Ohm's law) but no change in load current. This point is demonstrated in Example 6.8.

EXAMPLE 6.8

What happens to the load voltage in Figure 6.13 when the load resistance increases from 100 Ω to 400 Ω?

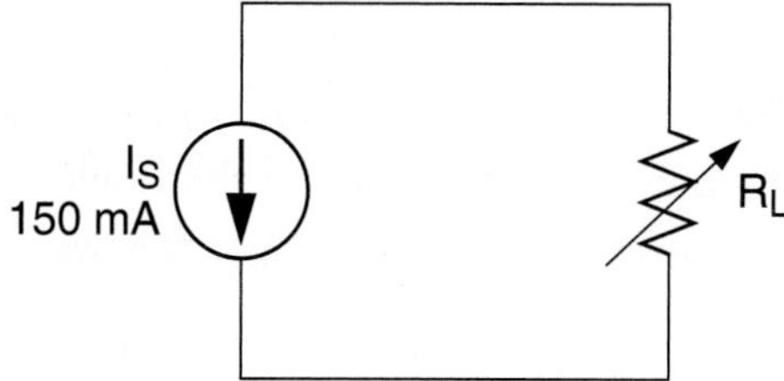

FIGURE 6.13

SOLUTION

When $R_L = 100\ \Omega$, the value of V_{RL} in Figure 6.13 is found as

$$V_{RL} = I_S \times R_L$$
$$= 150\text{ mA} \times 100\ \Omega$$
$$= 15\text{ V}$$

The value of the circuit current is fixed at 150 mA. Therefore, when R_L changes to 400 Ω, the load voltage changes to

$$V_{RL} = I_S \times R_L$$
$$= 150\text{ mA} \times 400\ \Omega$$
$$= 60\text{ V}$$

Note that the change in load resistance causes V_{RL} to change from 15 V to 60 V, but load current does not change. The value of the circuit current (in this case) is not affected by the load resistance.

PRACTICE PROBLEM 6.8

A circuit like the one in Figure 6.13 has a 10 A current source. What change in V_{RL} occurs if R_L is adjusted from 12 Ω to 32 Ω?

When a current source is used in a parallel circuit, the process for determining the values of the branch currents gets a bit more complicated. This point is illustrated in Example 6.9.

EXAMPLE 6.9

What are the values of I_{R1} and I_{R2} in Figure 6.14?

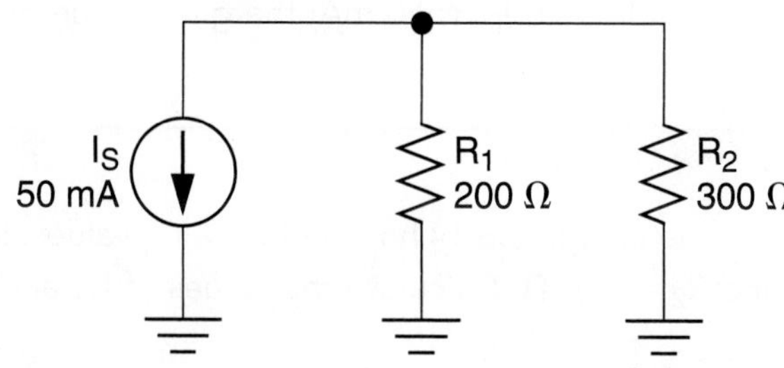

FIGURE 6.14

SOLUTION

To calculate the values of I_{R1} and I_{R2}, we need to know the value of the branch voltages. To determine the value of the branch voltages, we start by calculating the value of R_T for the circuit, as follows:

$$R_T = \frac{1}{\frac{1}{R_1} + \frac{1}{R_2}}$$

$$= \frac{1}{\frac{1}{200\ \Omega} + \frac{1}{300\ \Omega}}$$

$$= 120\ \Omega$$

Using Ohm's law, we can now find the value of the source voltage, as follows:

$$E = I_S R_T$$

$$= (50\ \text{mA})(120\ \Omega)$$

$$= 6\ \text{V}$$

Now, using $E = 6$ V, the values of the branch currents can be calculated using Ohm's law, as follows:

$$I_{R1} = \frac{E}{R_1}$$

$$= \frac{6\ \text{V}}{200\ \Omega}$$

$$= 30\ \text{mA}$$

and

$$I_{R2} = \frac{E}{R_2}$$

$$= \frac{6\ \text{V}}{300\ \Omega}$$

$$= 20\ \text{mA}$$

Note that the sum of I_{R1} and I_{R2} is 50 mA, the given value of the source current.

PRACTICE PROBLEM 6.9

A circuit like the one in Figure 6.14 has the following values: $I_S = 25$ mA, $R_1 = 100\ \Omega$, and $R_2 = 150\ \Omega$. Calculate the values of I_{R1} and I_{R2} for the circuit.

There is a simpler way to solve a problem like the one in Example 6.9: We can simply treat the circuit as a current divider.

Current Dividers

You may recall that a series circuit can be viewed as a voltage divider because the source voltage is divided among all the components in the circuit. In a similar fashion, a parallel circuit can be viewed as a *current divider* because the source current is divided among all the branches in the circuit. Just as we used the voltage divider equation to determine component voltage drops in series circuits, we can use a *current divider equation* to solve for branch currents in parallel circuits.

A comparison of voltage dividers and current dividers is provided in Figure 6.15. Note the difference in the resistance ratios used in the divider equations. For the voltage divider, the source voltage is multiplied by the ratio of the component resistance to the total circuit resistance. For the current divider, the source current is multiplied by the ratio of the total circuit resistance to the component (or branch) resistance, as follows:

(6.6)

$$I_n = I_S \frac{R_T}{R_n}$$

where

I_n = the current through R_n
I_S = the source current

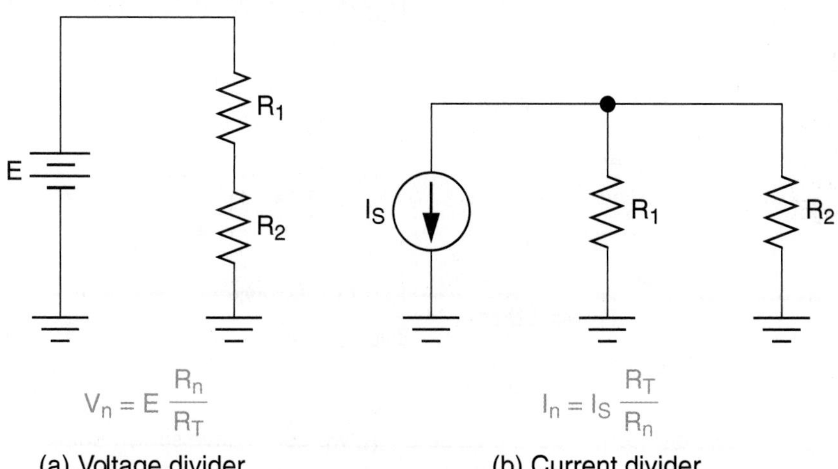

FIGURE 6.15 A comparison of voltage and current dividers.

Again, we have a situation where series and parallel circuits have opposite characteristics. Example 6.10 demonstrates the use of the current divider equation.

EXAMPLE 6.10

What are the values of I_{R1} and I_{R2} in Figure 6.16?

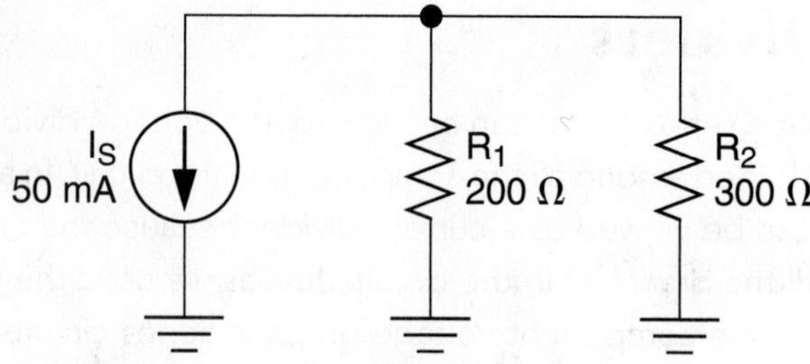

FIGURE 6.16

SOLUTION

Calculating the branch currents in Figure 6.16 begins with calculating the total circuit resistance, as follows:

$$R_T = \frac{1}{\dfrac{1}{R_1} + \dfrac{1}{R_2}}$$

$$= \frac{1}{\dfrac{1}{200\ \Omega} + \dfrac{1}{300\ \Omega}}$$

$$= 120\ \Omega$$

The value of I_{R1} can now be found as

$$I_{R1} = I_S \times \frac{R_T}{R_1}$$

$$= 50\text{ mA} \times \frac{120\ \Omega}{200\ \Omega}$$

$$= 30\text{ mA}$$

Finally, the value of I_{R2} can be found as

$$I_{R2} = I_S \times \frac{R_T}{R_2}$$

$$= 50\text{ mA} \times \frac{120\ \Omega}{300\ \Omega}$$

$$= 20\text{ mA}$$

Of course, once we determined the value of I_{R1}, we could have simply subtracted it from the source current to find the value of I_{R2} (according to KCL).

PRACTICE PROBLEM 6.10

Rework Practice Problem 6.9 using the technique shown in this example.

Solving Current Dividers: Another Approach

Equation (6.6) can be modified to provide us with another easy method of determining the values of the branch currents in a two-branch circuit. For example, the branch current values for the circuit in Figure 6.15b can be found as

(6.7)

$$I_{R1} = I_S \times \frac{R_2}{R_1 + R_2}$$

and

(6.8)

$$I_{R2} = I_S \times \frac{R_1}{R_1 + R_2}$$

Note that in each of the above equations the resistance of the other branch is used in the numerator of the resistance ratio. If we want to determine the value of I_{R1}, we use the value of R_2 in the fraction, and vice versa. Example 6.11 demonstrates the validity of these current-divider equations.

EXAMPLE 6.11

What are the values of I_{R1} and I_{R2} in Figure 6.16?

SOLUTION

The value of I_{R1} in Figure 6.16 can be found as

$$I_{R1} = I_S \times \frac{R_2}{R_1 + R_2}$$

$$= 50 \text{ mA} \times \frac{300\ \Omega}{500\ \Omega}$$

$$= 30 \text{ mA}$$

and the value of I_{R2} can be found as

$$I_{R2} = I_S \times \frac{R_1}{R_1 + R_2}$$

$$= 50 \text{ mA} \times \frac{200\ \Omega}{500\ \Omega}$$

$$= 20 \text{ mA}$$

If you look back at Examples 6.9 and 6.10, you'll see that these are the same results we obtained using the procedures established in those examples.

PRACTICE PROBLEM 6.11

Using the current divider equations, rework Practice Problem 6.9.

Once again, we could have solved for either branch current value and then subtracted that value from the source current to obtain the value of the other branch current. The results would have been the same.

The Bottom Line

The relationship between the various currents in a parallel circuit is described by Kirchhoff's current law (KCL). In effect, this law states that the branch currents in a parallel circuit must add up to equal the source current.

In most cases, the values of the branch currents in a parallel circuit can be found using Ohm's law. When the branch current values are known, they are simply added to find the value of the total circuit current.

A current source generates an output current that remains relatively constant for a wide range of load resistances. When a current source is used in a parallel circuit, the process of determining the values of the branch currents is greatly simplified by using the current divider equations.

PROGRESS CHECK

1. What is *Kirchhoff's current law?*
2. What is a *current source?*
3. How does a circuit with a current source respond to a change in load resistance?
4. Describe how you would solve for the various current values in a parallel circuit containing a voltage source.
5. Describe how you would solve for the various current values in a parallel circuit containing a current source.

6.3 PRACTICAL PARALLEL CIRCUITS

The circuits shown in the first two sections of this chapter help to illustrate the basic operating principles of parallel circuits. However, they are not very useful beyond their role as educational tools. As you will see, practical parallel circuits contain a number of branches that deliver power to two or more specific loads.

A Low-Voltage Lighting System

Practical parallel circuits are typically used to provide power to two or more independent loads. For example, the schematic for a low-voltage lighting circuit is shown in Figure 6.17. The source provides power for the lamps. Note that the source contains:

- An internal *circuit protector* that breaks the connection to the circuit in the event of an *overload* (high-current condition).
- A *light-activated* on/off switch that switches the power supply on at dusk and off at dawn.

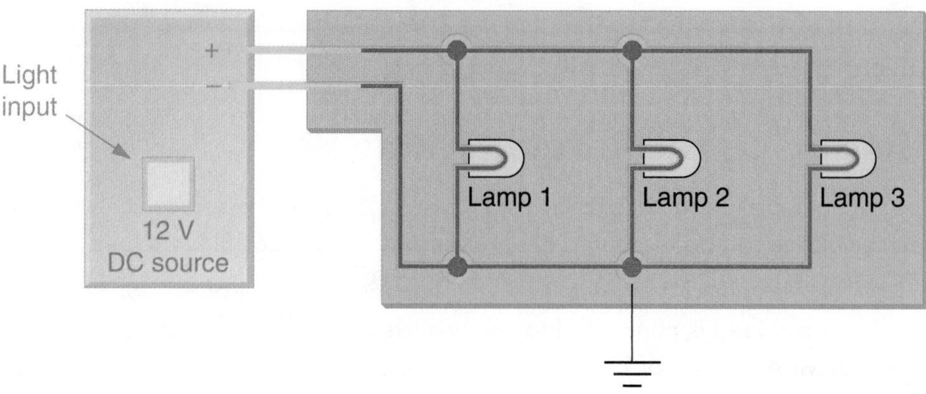

FIGURE 6.17 A low-voltage lighting circuit.

Circuit Calculations

Like the parallel circuits shown earlier, the circuit in Figure 6.17 has multiple current paths (branches). When the switch activates the power source, 12 V is applied to each of the lights in the circuit. Assuming that 7 W lamps are being used in the circuit, we can calculate the current through each light (I_L) as follows:

$$
\begin{aligned}
I_L &= \frac{P_L}{E} \\
&= \frac{7\text{ W}}{12\text{ V}} \\
&= 0.583\text{ A} \qquad (583\text{ mA})
\end{aligned}
$$

The total circuit current equals the sum of the individual branch (light) currents. With three lights in the circuit, the total current can be found as

$$
\begin{aligned}
I_T &= 3 \times I_L \\
&= 3 \times 0.583\text{ A} \\
&= 1.75\text{ A}
\end{aligned}
$$

We could also calculate the total circuit current by using total circuit power in the original current equation. The circuit contains three 7 W lights, so the total circuit power has a value of

$$P_T = 3 \times 7\text{ W}$$
$$= 21\text{ W}$$

and the total current has a value of

$$I_T = \frac{P_T}{E}$$
$$= \frac{21\text{ W}}{12\text{ V}}$$
$$= 1.75\text{ A}$$

Measuring the Circuit Voltage

The voltage applied to each branch in Figure 6.17 equals the source voltage. The source voltage can be measured easily as shown in Figure 6.18. Note that:

- The black lead is connected to the negative terminal of the source, while the red lead is connected to the positive terminal. In this case, the meter shows a reading of 12 V.
- Measuring the voltage across any branch in the circuit (in this case across any of the lights) gives you the same 12 V reading.

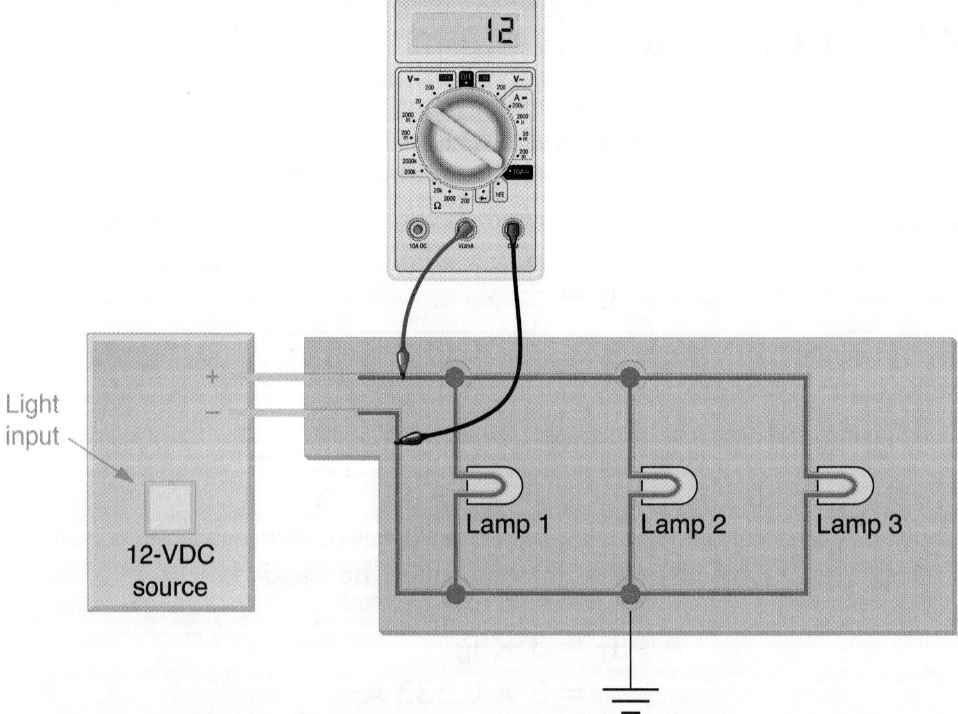

FIGURE 6.18 Measuring the lighting circuit source voltage.

Many parallel circuits contain switches and/or fuses in each branch. In Chapter 7, you will be shown how to measure branch current when this is the case.

PROGRESS CHECK

1. Three lights, a 100 W, a 150 W, and a 60 W, are connected in parallel to a 120 V source. What is the value of circuit current when all three lights are on?
2. Where would you measure the load voltage for the circuit described in Question 1?

6.4 PARALLEL CIRCUIT FAULT SYMPTOMS AND TROUBLESHOOTING

At this point, you have been exposed to the basic operating characteristics of parallel circuits. In this section, we discuss the characteristic symptoms of several circuit faults and the basic approach to parallel circuit troubleshooting.

STAYING SAFE When connecting batteries in parallel, make sure they are connected with the proper polarity to avoid overheating and damage.

Fault Symptoms: An Open Branch

You may recall that an open circuit fault in a series circuit is easy to diagnose because it results in a total loss of circuit current. The open component in a series circuit is identified by measuring the voltage across each of the components in the circuit. The voltage across the open component will equal the total voltage.

An open branch in a parallel circuit can be more difficult to diagnose than an open component in a series circuit. The reason for this is illustrated in Figure 6.19. As you can see, the voltage readings in the two circuits are identical. The fact that R_2 has opened has had no effect on the voltage measured across that branch. Therefore, voltage readings will not help to diagnose an open component in a parallel circuit (as they will in a series circuit).

One way of identifying an open branch is illustrated in Figure 6.20. An ammeter is connected as shown in the figure. Then, one of the branches is disconnected. If the disconnected branch is already open, there will be no change in the ammeter reading. If the branch is not open, the ammeter reading will decrease when the branch is disconnected. (This decrease is caused by the decrease in

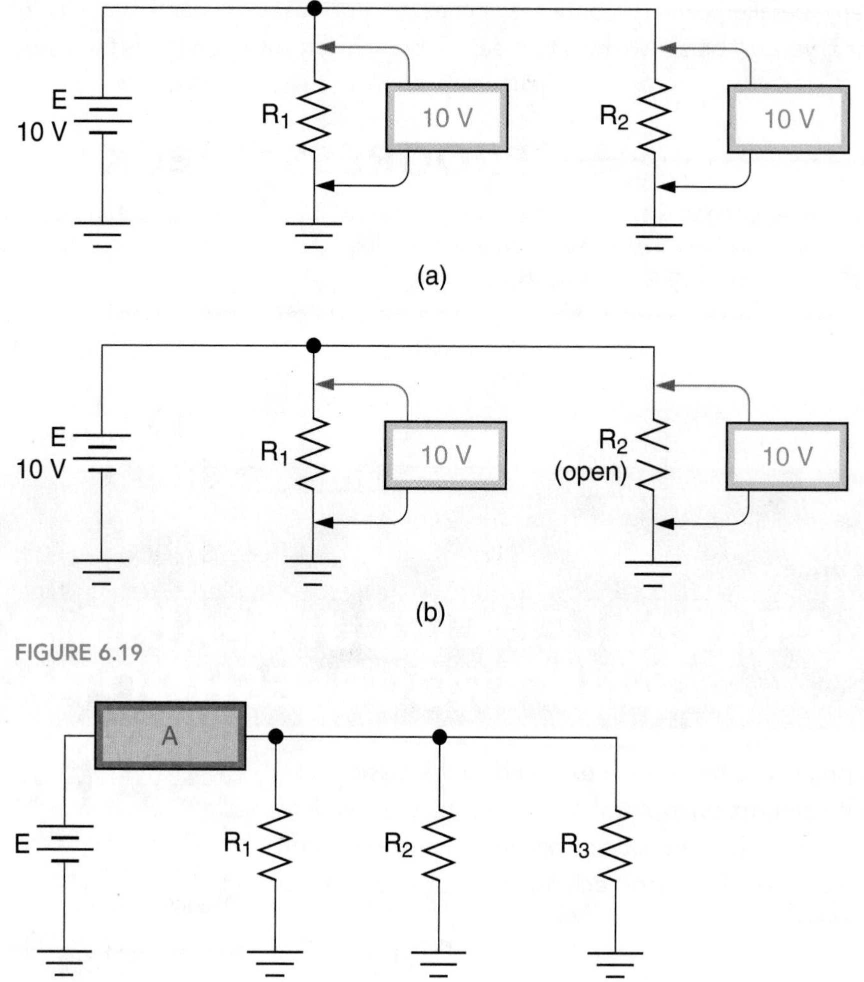

FIGURE 6.19

FIGURE 6.20 An ammeter measuring the total current in a parallel circuit.

current that results when a branch is removed from a parallel circuit. Remember, the more branches in a parallel circuit, the higher the total circuit current.

If disconnecting a branch causes a decrease in the ammeter reading, reconnect the branch and repeat the process with the next branch. Proceed until you hit the branch that does not cause a decrease in the reading when disconnected. That branch is open.

A Practical Consideration

In most cases, a parallel circuit is made up of components other than resistors. For example, consider the circuit shown in Figure 6.21. Here we have three DC motors that are connected in parallel to a voltage source. The source in this circuit provides the voltage required to drive each of the motors.

If a motor in Figure 6.21 opens, it simply stops working. At the same time, the other two motors continue to operate normally. Thus, there are no measurements, current or otherwise, required to find the open component. The motor that isn't running is the open component.

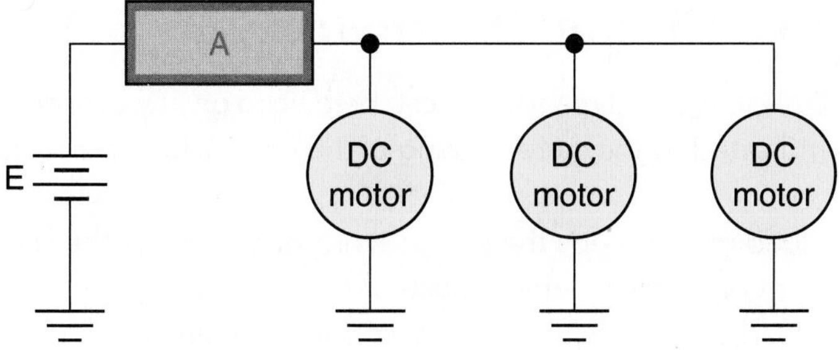

FIGURE 6.21 DC motors connected in parallel.

You may encounter situations where an open component is not so easy to identify. For example, suppose you have a storage tank (for some liquid) that contains three heating elements connected in parallel and you suspect that one of the elements has failed because the material is not heating to the correct temperature. Identifying the bad heating element could be accomplished by monitoring the circuit current as you disconnect each of the heating elements, one at a time. When disconnecting one of the elements does *not* result in a change in circuit current, you have found your open element.

Fault Symptoms: A Shorted Branch

When a branch in a parallel circuit shorts, the total circuit resistance drops to near 0 Ω. This point can be explained with the help of Figure 6.22. You know that the total resistance in a parallel circuit must be lower than any of the individual branch resistance values. When branch three in Figure 6.20 is shorted, the lowest branch resistance is near 0 Ω. Therefore, the total circuit resistance cannot be greater than this value.

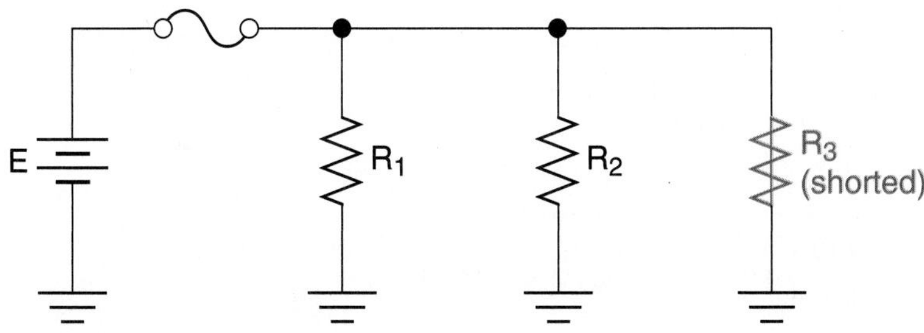

FIGURE 6.22

The low resistance of a shorted branch results in a drastic increase in the source current. In the case of the circuit in Figure 6.22, this increase in current will cause one of two things to happen. Either:

- The shorted component will burn open; or
- The circuit fuse will open.

If the fuse opens and is replaced before the short is corrected, the new fuse will also blow when the circuit is energized. Any circuit that continually blows fuses contains a shorted component or branch.

Fault Symptoms: Circuit Overload

LOAD The current demand on the output of a circuit.

OVERLOAD A load that exceeds the design limits of a circuit or the current rating of its overcurrent protective element.

SURGE A sudden and momentary current overload that is commonly caused by the startup of an electrical machine, such as a motor or compressor.

There is a limit on the **LOAD** (current demand) that can be placed on any circuit. In many cases, this limit is indicated by the current rating of the circuit's fuse or circuit breaker.

An **OVERLOAD** is a load that exceeds the design limits of a circuit or the current rating of its overcurrent protective element (fuse or circuit breaker). The simplest example of an overload occurs when you connect one too many appliances to a circuit. When this occurs, any (or all) of the following can occur:

1. The fuse blows (or circuit breaker opens).
2. The circuit conductors become abnormally hot.
3. Other items connected to the circuit do not operate properly. For example, a lamp connected to the circuit dims.

It should be noted that a short circuit or ground fault is not considered to be an overload.

Some overloads are temporary, self-correcting conditions. For example, an air conditioner causes a current **SURGE** when it first comes on. This surge, which is needed to bring the compressor motor up to speed, can be represented as shown in Figure 6.23. As you can see, the motor current initially surges to a high value and then settles down to its "normal" value. The overload caused by the initial surge can cause lights to dim momentarily when the system is activated.

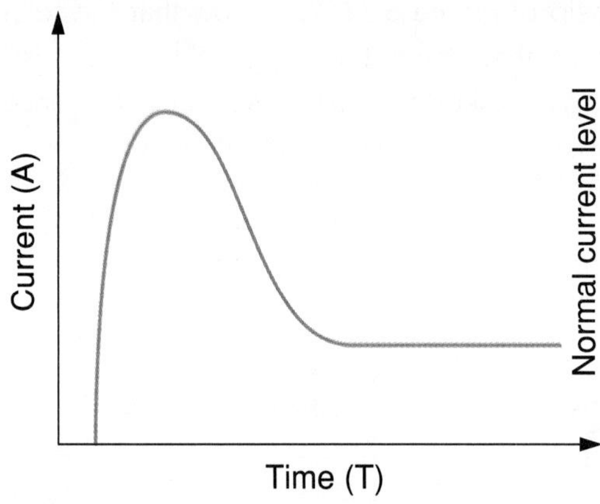

FIGURE 6.23 A current surge.

When not a temporary condition, an overload is typically a symptom of a circuit problem. For example, a motor with bad bearings will draw more current than one that is operating normally. The increased current can cause conductors to be hotter than normal and, after time, to draw enough current to trip the circuit breaker (or blow a fuse).

A Word of Caution

Some professionals you meet in the field may attempt to find a short like the one in Figure 6.22 by defeating the fuse. That is, they may connect a wire across the fuse holder to force the circuit to continue operating under conditions that would normally blow the fuse. The goal of this trick is to cause the shorted component to burn out, thus giving them an easy way to determine which branch is shorted. *This type of shortcut is extremely dangerous!* Not only can it cause a fire, it can also damage the circuit's power source and may cause personal injury to anyone working on the circuit. For your own safety, always use the established troubleshooting techniques rather than potentially dangerous shortcuts.

PROGRESS CHECK

1. An open branch in a parallel circuit is more difficult to diagnose than an open resistor in a series circuit. Why?
2. Write a list of the steps you would take to find an open branch in a three-branch parallel circuit.
3. If a branch in a parallel circuit shorts, the total resistance drops to near 0 Ω. Why?
4. In a parallel circuit that does not contain a fuse or circuit breaker, what are the possible outcomes of a shorted branch?
5. Why is it dangerous to bypass the fuse in any circuit?

6.5 SUMMARY

Here is a summary of the major points that were covered in this chapter:

Parallel Circuit Characteristics

- A parallel circuit is a circuit that provides more than one current path between any two points.
- Each current path in a parallel circuit is referred to as a branch.
- The total current in a parallel circuit is equal to the sum of the branch currents.
- The voltage across each branch in a parallel circuit is equal to the voltage across all the other branches.
- The current through each branch in a parallel circuit is determined by the source voltage and the resistance of the branch.
- In terms of their voltage and current characteristics, parallel and series circuits are nearly opposite:
  - Currents are equal throughout series circuits. Branch currents in parallel circuits may or may not be equal in value, and add up to equal the total circuit current (in keeping with KCL).
  - Component voltages add up to equal the applied voltage in series circuits, and are equal across all branches in parallel circuits.
- Parallel circuits are often referred to as *current dividers* because the circuit current is divided among the branches.
- The total resistance in a parallel circuit is always lower than any of the branch resistance values.
- There are several means by which the total resistance of a parallel circuit can be calculated. The resistance equations for parallel circuits are summarized in Table 6.2 (page 238).
- The power characteristics of parallel circuits are nearly identical to those of series circuits, with one exception.
  - The total power in a parallel circuit is found as the sum of the branch power values.

  - In parallel circuits, the lowest-resistance branch dissipates the greatest amount of power. In series circuits the highest-resistance component dissipates the greatest amount of power.

Current Dividers

- Kirchhoff's current law (KCL) states that the current leaving any point must equal the current entering that point.
- A current source is a source that is designed to provide an output current value that remains relatively constant over a wide range of load resistance values.
  - The schematic symbol for a DC current source indicates the value and direction of its output current.
- In any circuit containing a current source, a change in load resistance results in a change in load voltage. Load current remains relatively constant.
- The current-divider equation is a valuable tool for calculating the values of the branch currents in a parallel circuit that contains a current source.

Parallel Circuit Fault Symptoms and Troubleshooting

- An open branch in a parallel circuit is more difficult to diagnose than an open resistor in a series circuit.
  - An open branch can be identified using a technique that involves disconnecting the circuit branches one at a time while measuring the total circuit current.
- A shorted branch will cause the circuit fuse or circuit breaker to open.
- An overload is a condition in which the load on a circuit exceeds the circuit design limits or the rating of the circuit protective device (fuse or circuit breaker).
  - Some overload conditions are temporary and self-correcting, such as an overload that occurs as the result of an air conditioner switching on.
  - Some overload conditions are symptoms of problems, such as an overload caused by a motor with bad bearings.
- Bypassing the fuse in any circuit is extremely dangerous and should never be attempted.
- In a parallel circuit containing a fuse, the condition of the fuse is a good indicator of the type of problem you're looking for.
  - If the fuse has blown, you are likely looking for a short.
  - If the fuse is good, you are likely looking for an open.

CHAPTER REVIEW

1. A parallel circuit ________.
   a) has two or more current paths
   b) has a current that splits and then recombines
   c) has the same voltage across each current path
   d) all of the above

2. Each current path in a parallel circuit is often referred to as ________.
   a) a stick
   b) an ipath
   c) a branch
   d) a conduit

3. The current in each branch of a parallel circuit ________.
   a) is always the same
   b) is always different
   c) depends on the resistance in the branch
   d) always flows to ground

4. The voltage across each branch in a parallel circuit ________.
   a) is always the same
   b) is always different
   c) depends on the resistance in the branch
   d) is always positive with respect to ground

5. Three resistors are connected in parallel. The power supply delivers 26.5 A to the circuit. You measure 11.62 A in one branch and 5.77 A in the second. You cannot gain access to the third path to make a measurement. Nonetheless, you know that the third branch has a current of ________.
   a) 9.11 A
   b) 8.42 A
   c) 911 mA
   d) 842 mA

6. A 220 V source is connected to a circuit similar to the one described in question 5. The current in the first two branches is the same, but the resistance of the third branch has changed. The current in branch 3 is now 6.85 A. This means that the resistance of branch 3 is ________.
   a) 32.1 Ω
   b) 1.5 kΩ
   c) 321 Ω
   d) 150 Ω

7. A 470 Ω and a 330 Ω resistor are connected in parallel with a 115 V power supply. The voltage across the 330 Ω resistor is ________.
   a) 47.5 V
   b) 67.6 V
   c) 115 V
   d) 67.5 V

8. Two resistors are connected in parallel. One of them is known to be 220 Ω and you measure 1.39 A through it. The other resistor is 470 Ω. The current through the 470 Ω resistor is ________.
   a) 2.97 A
   b) 6.51 A
   c) 1.39 A
   d) 651 mA

9. 240 V is applied to a two-branch parallel circuit. Total circuit current is known to be 14.6 A. If one of the resistors is known to be 22 Ω, the other resistor must be ________.
   a) 65 Ω
   b) 650 Ω
   c) 22 Ω
   d) you need to know the current in that path to solve this problem

10. Parallel circuits are often called ________.
   a) parallelograms
   b) voltage dividers
   c) current combiners
   d) current dividers

11. Adding another branch to any parallel circuit ________.
   a) increases total circuit current
   b) increases current in each branch
   c) increases total circuit voltage
   d) increases total circuit resistance

12. The technique for solving parallel resistance that can be used in all cases is called the ________ method.
   a) inverting
   b) reciprocal
   c) product-over-sum
   d) general

13. A 115 V power supply is powering the lights in a warehouse. There are 46 lights all wired in parallel. If each light has a resistance of 135 Ω, the total circuit resistance is ________.
   a) 850 mΩ
   b) 8.5 Ω
   c) 2.93 Ω
   d) 29.3 Ω

14. The method that can be used to solve any two-branch parallel circuit is called the ________ method.
   a) reciprocal
   b) product-over-sum
   c) equal-branch
   d) both a) and b)

15. Three 330 Ω resistors connected in parallel are driven by a 12.5 V power supply. Each resistor dissipates ________ of power.
   a) 1.42 W
   b) 473 mW
   c) 14.2 W
   d) 4.73 W

16. Any point in a parallel circuit where current divides or recombines is often referred to as a ________.
    a) node
    b) nodule
    c) terminal
    d) terminus

17. The 115 V power supply driving the 46 light bulbs in question 13 must supply ________ of power to this circuit.
    a) 45 kW
    b) 98 W
    c) 288 W
    d) 4.5 kW

18. A 22 kΩ resistor is in parallel with an unknown resistor and a 65 V power supply. If the 65 V source is supplying 320 mW to the circuit, this means that the value of the unknown resistor is ________.
    a) 33 kΩ
    b) 3.3 kΩ
    c) 2.06 kΩ
    d 206 Ω

19. A 20 A current source is connected to a 1.0 kΩ rheostat. As the load resistance decreases, the load voltage ________ and the load current ________.
    a) increases, decreases
    b) decreases, increases
    c) decreases, remains the same
    d) increases, remains the same

20. A current source rated at 65 mA is connected to a rheostat set to 1860 Ω. The rheostat setting changes to 1240 Ω. The voltage across the rheostat changes from ________ to ________.
    a) 121 V, 80.6 V
    b) 80.6 V, 121 V
    c) 121 V, 86 V
    d) there is no change

21. A 600 mA current source is connected to a 33 Ω resistor and a 47 Ω resistor connected in parallel. The current through the 33 Ω resistor is ________.
    a) 353 mA
    b) 426 mA
    c) 248 mA
    d) they are both the same, as this is a 300 mA current source

22. Monitoring the total current entering a parallel network as you disconnect each branch, one at a time, is a technique for finding a(n) ________.
    a) ground fault
    b) open branch

c) shorted branch

d) shorted or an open branch

23. A parallel network of 6 heating elements all have 32 Ω of resistance. If two of the elements open, the voltage across the entire network ________.

a) decreases by 33%

b) increases by 17%

c) increases by 33%

d) does not change

24. If the breaker protecting a parallel network has tripped, it is a good bet that there is a(n) ________ in the network.

a) short

b) open

c) drastic increase in load resistance

d) short or an open

PRACTICE PROBLEMS

1. The table below lists the currents through a 3-branch parallel circuit. For each combination of branch currents, find the total circuit current.

	I_1	I_2	I_3	I_T
a)	1.2 A	2.8 A	600 mA	________
b)	5.5 A	360 mA	1.04 A	________
c)	12 A	250 mA	3.3 A	________
d)	500 mA	226 mA	1.08 A	________

2. The table below lists the currents through a 3-branch parallel circuit. For each combination of branch currents, find the total circuit current.

	I_1	I_2	I_3	I_T
a)	1.5 A	3.3 A	770 mA	________
b)	8 A	250 mA	5000 mA	________
c)	520 mA	2.8 A	0.85 A	________
d)	10 A	1.12 A	90 mA	________

3. Determine the values of the branch currents and the total current for the circuit in Figure 6.24a.

4. Determine the values of the branch currents and the total current for the circuit in Figure 6.24b.

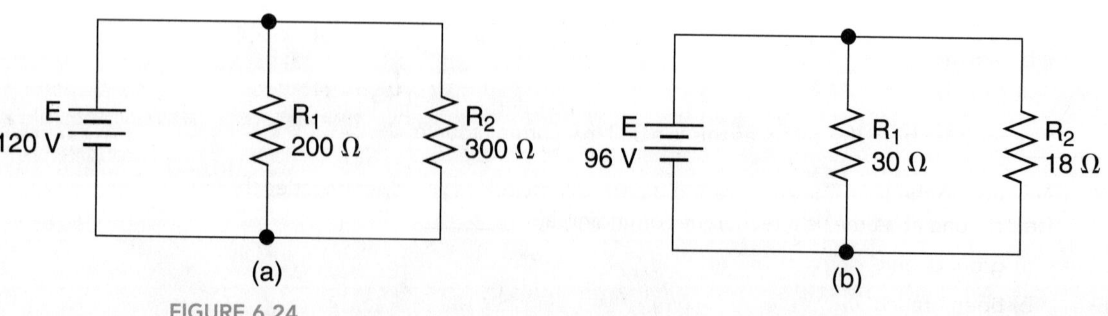

FIGURE 6.24

5. Determine the values of the branch currents and the total current for the circuit in Figure 6.25a.

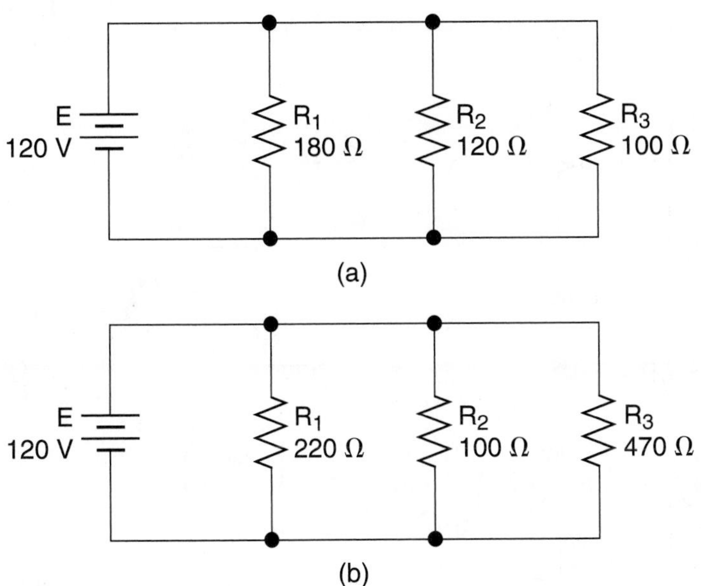

FIGURE 6.25

6. Determine the values of the branch currents and the total current for the circuit in Figure 6.25b.
7. Calculate the value of R_T for the circuit in Figure 6.24a.
8. Calculate the value of R_T for the circuit in Figure 6.24b.
9. Calculate the value of R_T for the circuit in Figure 6.25a.
10. Calculate the value of R_T for the circuit in Figure 6.25b.
11. Calculate the value of R_T for the circuit in Figure 6.26a.
12. Calculate the value of R_T for the circuit in Figure 6.26b.

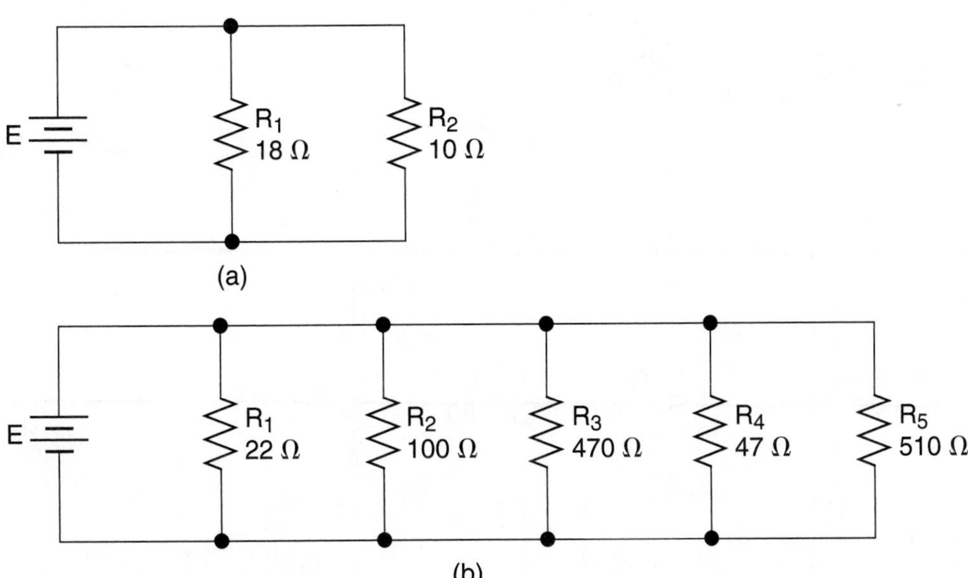

FIGURE 6.26

13. Calculate the value of R_T for the circuit in Figure 6.27a.
14. Calculate the value of R_T for the circuit in Figure 6.27b.

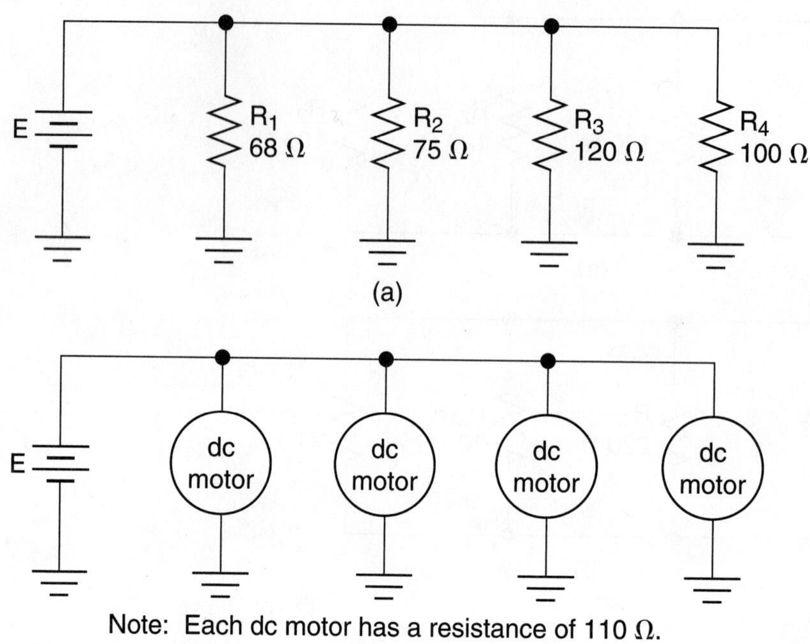

FIGURE 6.27

15. Determine the potentiometer setting required to set the total circuit current in Figure 6.28a to 500 mA.
16. Determine the potentiometer setting required to set the total circuit current in Figure 6.28b to 720 mA.

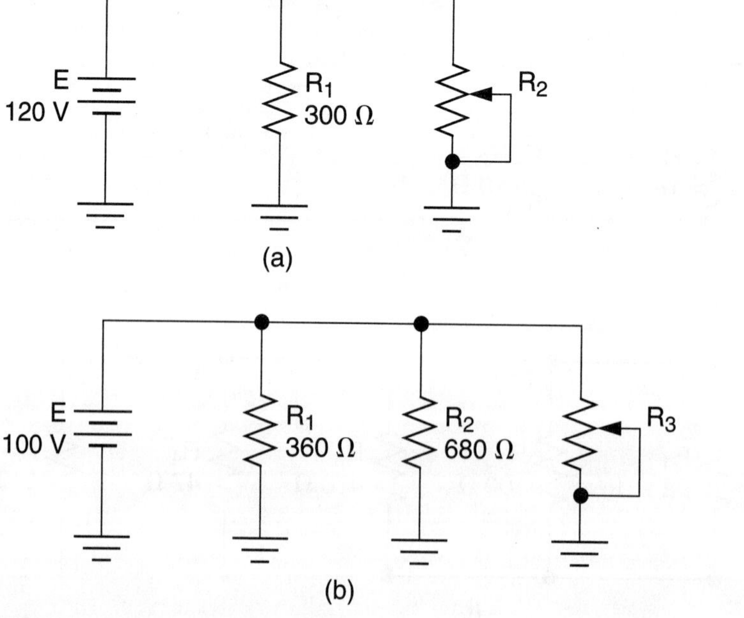

FIGURE 6.28

17. For the circuit in Figure 6.29a, determine the change in V_L that occurs when R_L is changed from 1.2 kΩ to 3.3 kΩ.
18. For the circuit in Figure 6.29b, determine the change in V_L that occurs when R_L is changed from 470 Ω to 220 Ω.

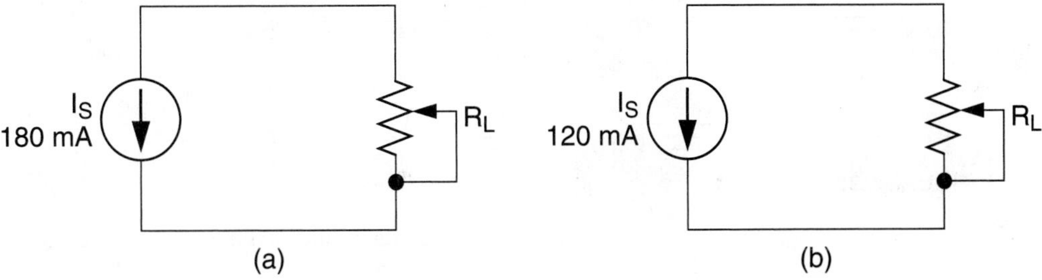

FIGURE 6.29

19. Determine the branch current values for the circuit in Figure 6.30a.
20. Determine the branch current values for the circuit in Figure 6.30b.
21. Determine the branch current values for the circuit in Figure 6.30c.
22. Determine the branch current values for the circuit in Figure 6.30d.

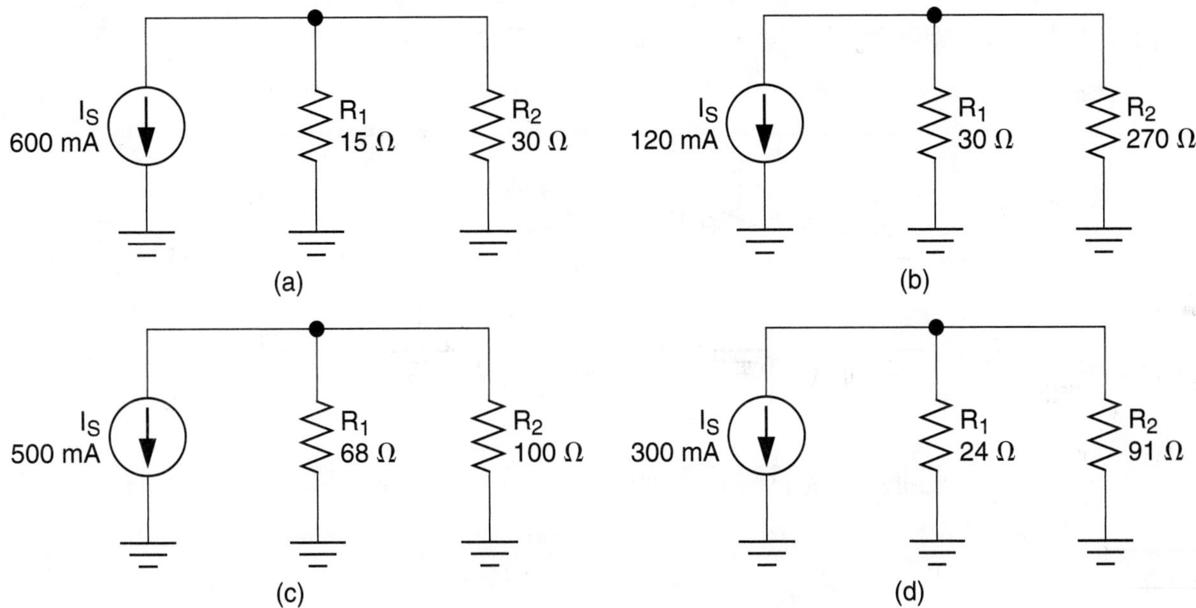

FIGURE 6.30

23. Calculate the current, resistance, and power values for the circuit in Figure 6.31.

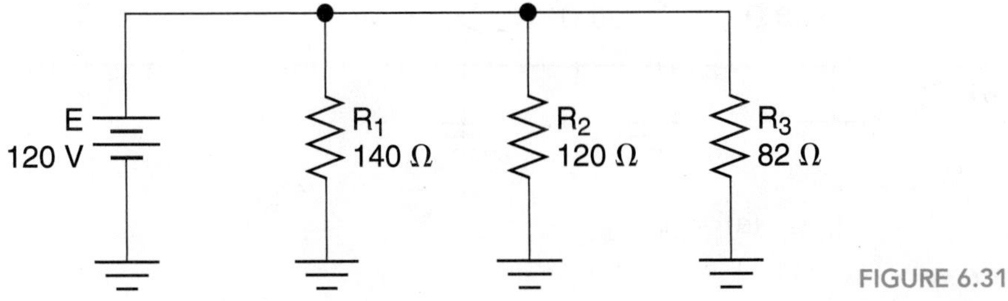

FIGURE 6.31

24. Calculate the current, resistance, and power values for the circuit in Figure 6.32.

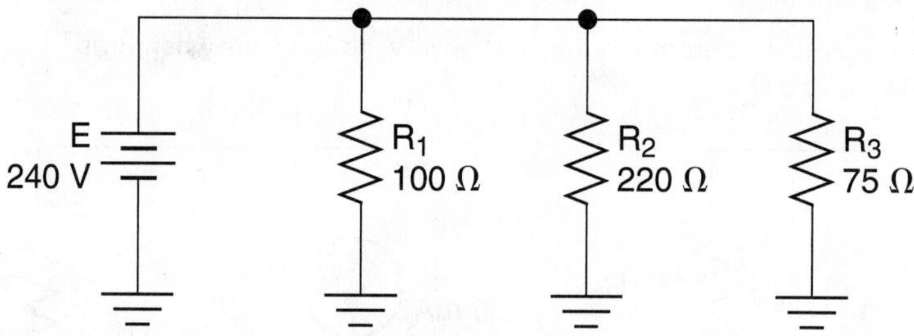

FIGURE 6.32

25. Calculate the current, resistance, and power values for the circuit in Figure 6.33.

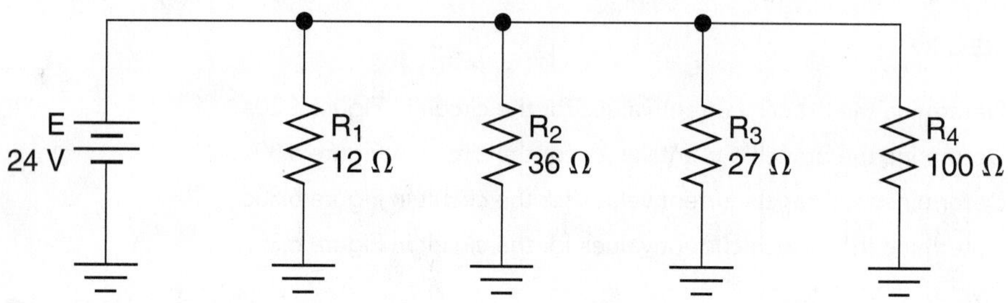

FIGURE 6.33

26. Calculate the current, resistance, and power values for the circuit in Figure 6.34.

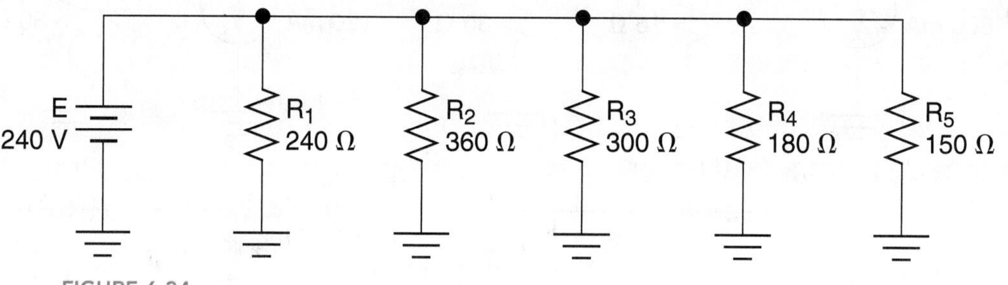

FIGURE 6.34

CHALLENGERS

27. For each of the circuits shown in Figure 35, find the missing value(s).

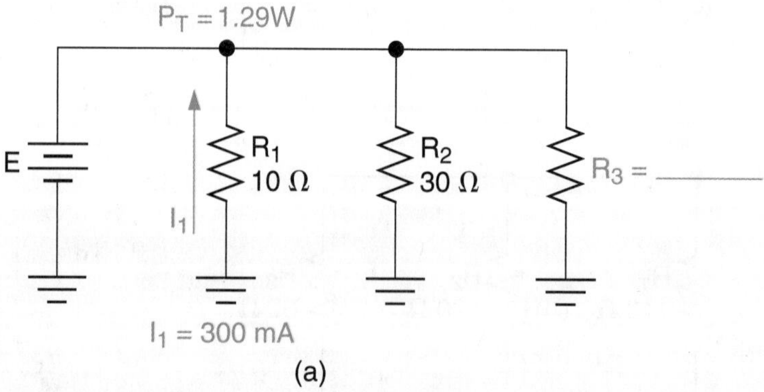

FIGURE 6.35

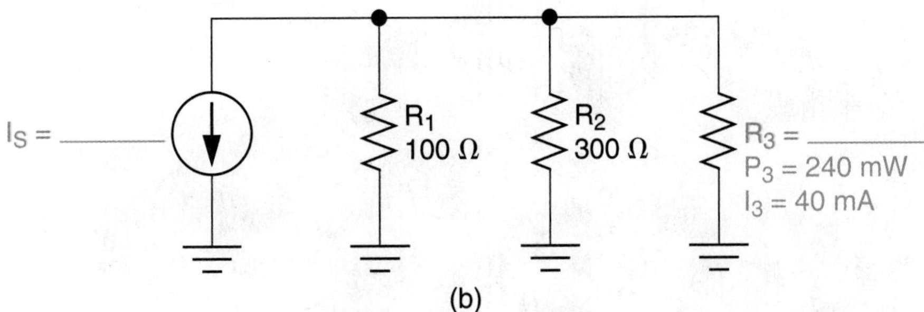

FIGURE 6.35

28. In terms of circuit fault diagnosis, explain why motors (like those shown in Figure 6.21) are always connected in parallel.

ANSWERS TO THE EXAMPLE PRACTICE PROBLEMS

6.1	10.5 A
6.2	6.94 A
6.3	72 Ω
6.4	144 Ω
6.5	44.4 Ω
6.6	144 Ω
6.7	100 Ω
6.8	V_{RL} changes from 120 V to 320 V
6.9	I_{R1} = 15 mA, I_{R2} = 10 mA
6.10	I_{R1} = 15 mA, I_{R2} = 10 mA
6.11	I_{R1} = 15 mA, I_{R2} = 10 mA

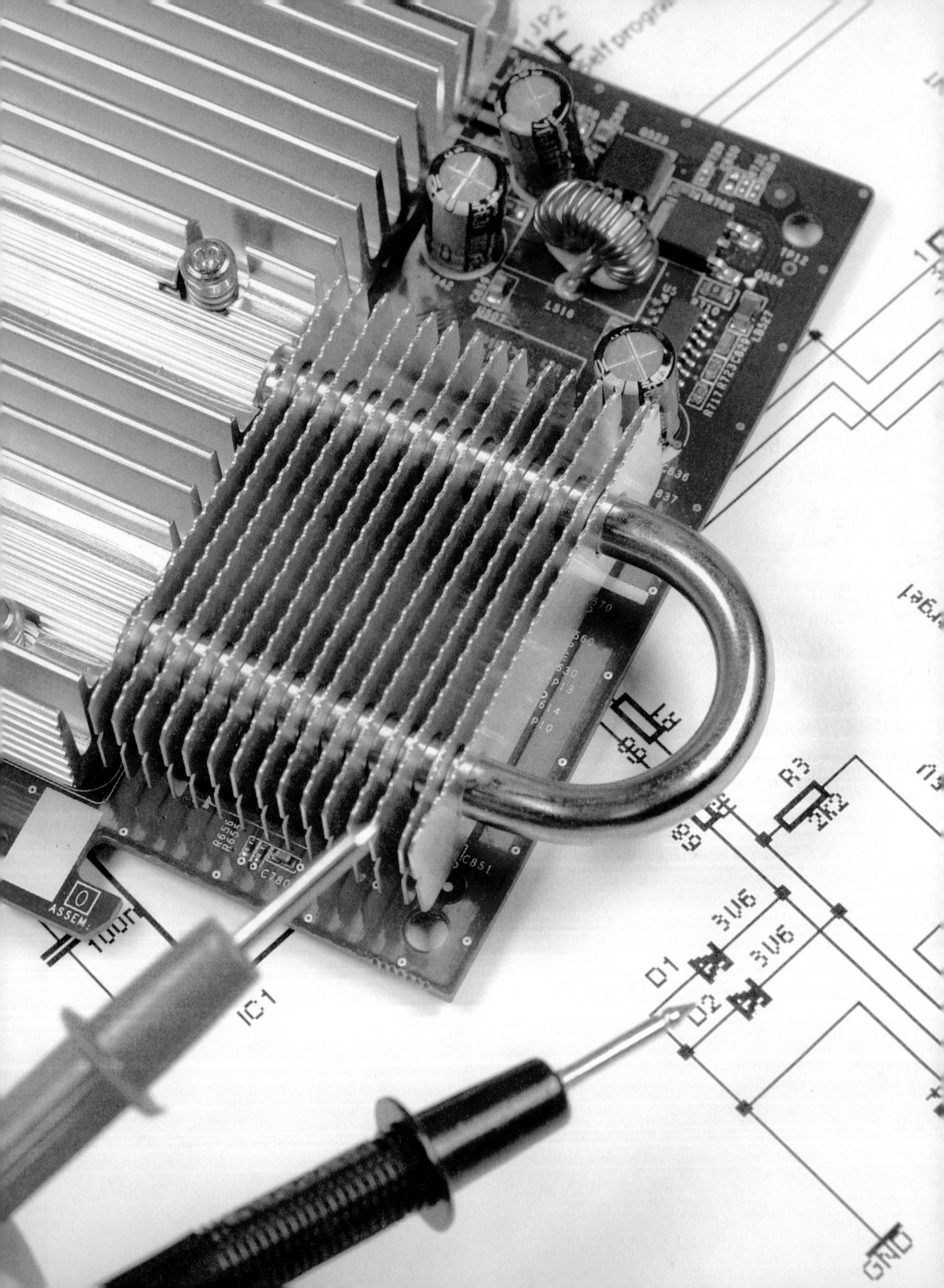

R3
2K2
3V6
3V6
D1
D2
IC1
GND
ASSEM
C780
C851
C836

CHAPTER 7

SERIES-PARALLEL CIRCUITS

PURPOSE

The circuits we've covered so far contain components that are connected either in series or in parallel. In practice, most circuits contain a combination of both series and parallel connections. Circuits that contain both series and parallel connections are referred to as **SERIES-PARALLEL CIRCUITS** or **COMBINATION CIRCUITS**. Several examples of series-parallel circuits are shown in Figure 7.1. As you can see, some of the components in each circuit are clearly in series or parallel, and some are not. This holds true for any series-parallel circuit.

In this chapter, you will be shown how to work with series-parallel circuits. You will also be introduced to some specific circuits that are commonly used in a variety of applications.

KEY TERMS

The following terms are introduced and defined in this chapter on the pages indicated:

Combination circuit, 268
Decade box, 301
Equivalent circuit, 276
Galvanometer, 298
Load-specific fault, 308
No-load output voltage (V_{NL}), 292
Output voltage, 292
Parallel-equivalent circuit, 284
Series-equivalent circuit, 284
Series-parallel circuit, 268
Thermistor, 303
Voltmeter loading, 295
Wheatstone bridge, 298

OBJECTIVES

After completing this chapter, you should be able to:

1. Describe and analyze the operation of series circuits that are connected in parallel.
2. Describe and analyze the operation of parallel circuits that are connected in series.
3. Derive the series-equivalent or parallel-equivalent for any given series-parallel circuit.
4. Solve a given series-parallel circuit for any current, voltage, or power value.
5. List the steps used to solve any value in a series-parallel circuit.
6. Discuss the effects of load resistance on the operation of a loaded voltage divider.
7. Discuss *voltmeter loading* and its effects on voltage measurements.
8. Describe the construction and operation of the Wheatstone bridge.

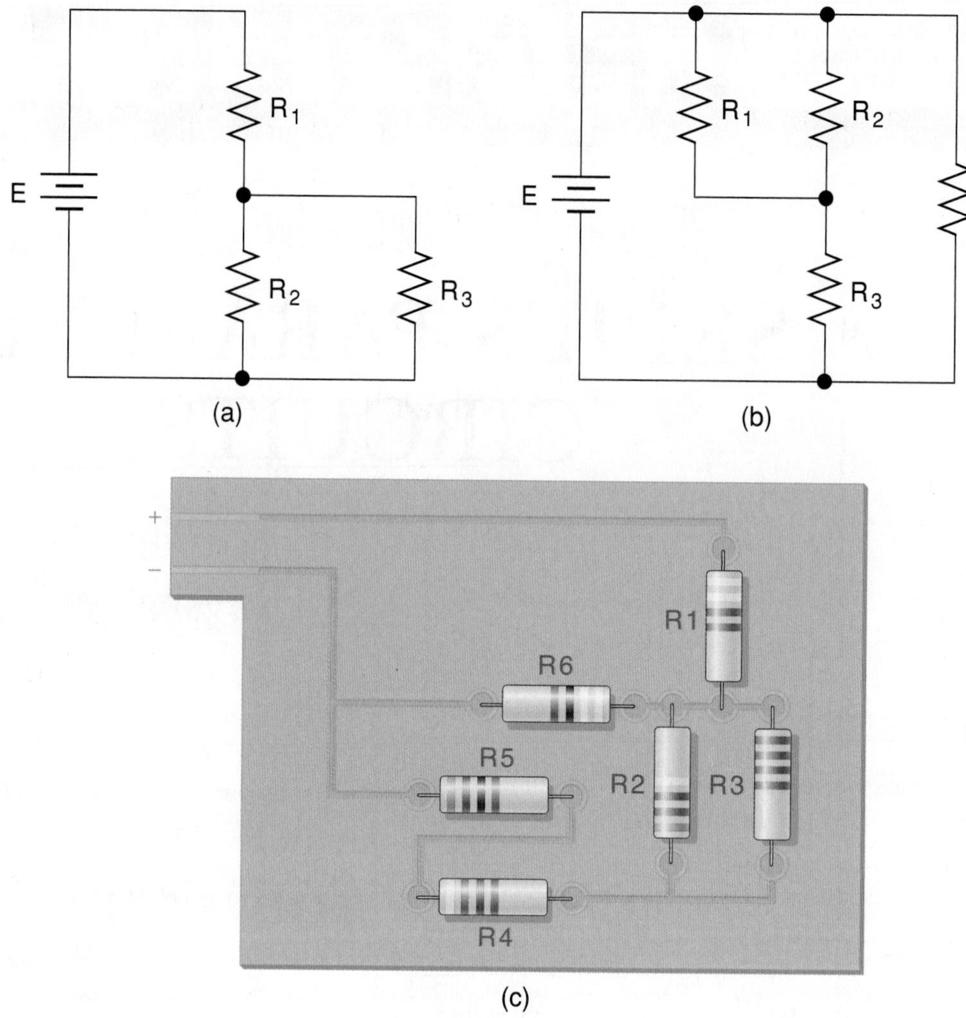

FIGURE 7.1 Examples of series-parallel circuits.

9. Discuss the use of the Wheatstone bridge as a resistance measuring circuit.
10. List the basic series-parallel fault symptoms.
11. Describe the general approach to troubleshooting series-parallel circuits.

SERIES-PARALLEL CIRCUIT
One that contains both series and parallel connections.

COMBINATION CIRCUIT
Another name for a series-parallel circuit.

7.1 INTRODUCTION TO SERIES-PARALLEL CIRCUITS

Series-parallel circuits vary widely in their functions and complexity. For example, an industrial robotic system can be viewed (loosely) as one very complex series-parallel circuit. At the same time, many series-parallel circuits have relatively simple configurations. For example, consider the circuits shown in Figure 7.2. The circuit in Figure 7.2a contains *two series circuits that are connected in parallel*. One series circuit consists of R_1 and R_2, while the other consists of R_3 and R_4. The circuit in Figure 7.2b contains *two parallel circuits that have been connected in series*. R_1 and R_2 make up one of the parallel circuits, while R_3 and R_4 make up the other.

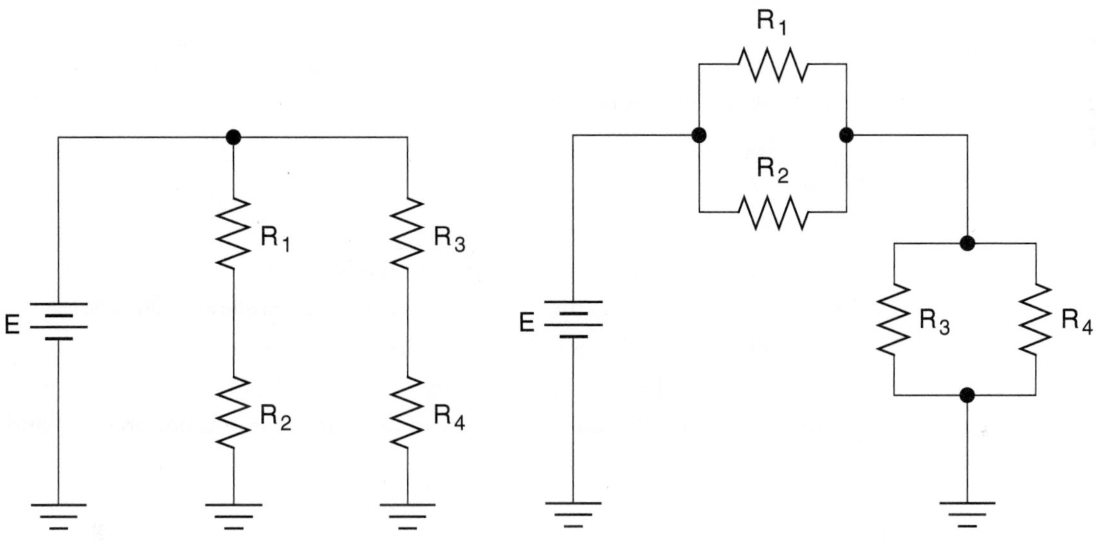

FIGURE 7.2 Relatively simple series-parallel connections.

Connecting Series Circuits in Parallel

Let's take a closer look at the circuit in Figure 7.2a. For the sake of discussion, the branches in the circuit are labeled A and B as shown in Figure 7.3. Since R_1 and R_2 form a series circuit, branch A retains all of the characteristics of any series circuit. Therefore:

1. The currents through the two components are identical ($I_{R1} = I_{R2} = I_A$).
2. The sum of the component voltages equals the source voltage ($V_{R1} + V_{R2} = E$).
3. The total resistance of the branch (R_A) equals the sum of the resistor values ($R_A = R_1 + R_2$).

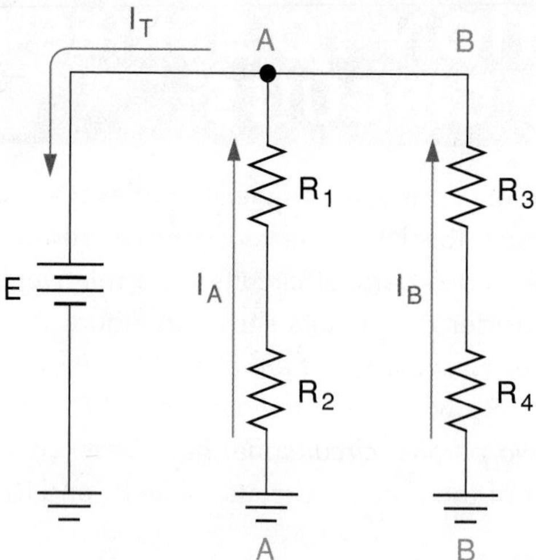

FIGURE 7.3 Series circuits connected in parallel.

Since R_3 and R_4 also form a series circuit, branch B retains all the characteristics of any series circuit. Therefore:

1. $I_{R3} = I_{R4} = I_B$
2. $V_{R3} + V_{R4} = E$
3. $R_B = R_3 + R_4$

Because these two branches form a parallel circuit, the combination of branch A and branch B retains the characteristics of any parallel circuit. Therefore:

1. The voltage across each branch is equal to the source voltage ($V_A = V_B = E$).
2. The sum of the branch currents is equal to the source current ($I_A + I_B = I_T$).
3. The total circuit resistance is equal to the parallel combination of R_A and R_B ($R_T = R_A \| R_B$).

Remember: The $\|$ symbol is used to indicate that the resistances are connected in parallel.

When two (or more) series circuits are connected in parallel, calculating the circuit values of voltage, current, and power begins with treating each branch as an individual series circuit. For each series circuit, the values of voltage, current, and power are calculated. The results can then be used to determine the overall circuit values as they are in any parallel circuit. This approach to solving circuit values is demonstrated in Example 7.1.

EXAMPLE 7.1

Calculate the voltage, current, and power values for the circuit in Figure 7.4.

SOLUTION

Calculating the voltage, current, and power values for the circuit in Figure 7.4 starts with the analysis of branch A. The total resistance of this branch is found as

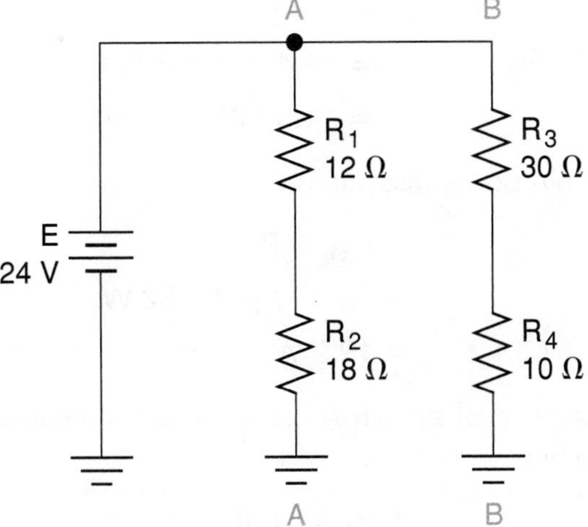

FIGURE 7.4

$$\begin{aligned} R_A &= R_1 + R_2 \\ &= 12\ \Omega + 18\ \Omega \\ &= 30\ \Omega \end{aligned}$$

The current through branch A can now be found using Ohm's law, as follows:

$$\begin{aligned} I_A &= \frac{E}{R_A} \\ &= \frac{24\text{ V}}{30\ \Omega} \\ &= 0.8\text{ A} \qquad (800\text{ mA}) \end{aligned}$$

Since I_A is the current through R_1 and R_2, the values of V_{R1} and V_{R2} can be found as

$$\begin{aligned} V_{R1} &= I_A \times R_1 \\ &= 0.8\text{ A} \times 12\ \Omega \\ &= 9.6\text{ V} \end{aligned}$$

and

$$\begin{aligned} V_{R2} &= I_A \times R_2 \\ &= 0.8\text{ A} \times 18\ \Omega \\ &= 14.4\text{ V} \end{aligned}$$

The power dissipated by R_1 and R_2 can now be found as

$$\begin{aligned} P_{R1} &= V_{R1} \times I_A \\ &= 9.6\text{ V} \times 0.8\text{ A} \\ &= 7.68\text{ W} \end{aligned}$$

$$\begin{aligned} P_{R2} &= V_{R2} \times I_A \\ &= 14.4\ \text{V} \times 0.8\ \text{A} \\ &= 11.52\ \text{W} \end{aligned}$$

and the branch A power dissipation (P_A) is found as

$$\begin{aligned} P_A &= P_{R1} + P_{R2} \\ &= 7.68\ \text{W} + 11.52\ \text{W} \\ &= 19.2\ \text{W} \end{aligned}$$

Now, the sequence of branch A calculations is repeated for branch B. First, the total branch resistance is found as

$$\begin{aligned} R_B &= R_3 + R_4 \\ &= 30\ \Omega + 10\ \Omega \\ &= 40\ \Omega \end{aligned}$$

Next, the branch current is found as

$$\begin{aligned} I_B &= \frac{E}{R_B} \\ &= \frac{24\ \text{V}}{40\ \Omega} \\ &= 0.6\ \text{A} \qquad (600\ \text{mA}) \end{aligned}$$

Using this value of I_B, we can calculate the resistor voltages in the branch as follows:

$$\begin{aligned} V_{R3} &= I_B \times R_3 \\ &= 0.6\ \text{A} \times 30\ \Omega \\ &= 18\ \text{V} \end{aligned}$$

and

$$\begin{aligned} V_{R4} &= I_B \times R_4 \\ &= 0.6\ \text{A} \times 10\ \Omega \\ &= 6\ \text{V} \end{aligned}$$

Finally, the power values for this branch are found as

$$\begin{aligned} P_{R3} &= V_{R3} \times I_B \\ &= 18\ \text{V} \times 0.6\ \text{A} \\ &= 10.8\ \text{W} \end{aligned}$$

$$\begin{aligned} P_{R4} &= V_{R4} \times I_B \\ &= 6\ \text{V} \times 0.6\ \text{A} \\ &= 3.6\ \text{W} \end{aligned}$$

and the branch B power dissipation (P_B) is found as

$$P_B = P_{R3} + P_{R4}$$
$$= 10.8 \text{ W} + 3.6 \text{ W}$$
$$= 14.4 \text{ W}$$

The branches are in parallel, so the total circuit current is found as the sum of the branch currents, as follows:

$$I_T = I_A + I_B$$
$$= 0.8 \text{ A} + 0.6 \text{ A}$$
$$= 1.4 \text{ A}$$

The total circuit resistance can be found using Ohm's law, as follows:

$$R_T = \frac{E}{I_T}$$
$$= \frac{24 \text{ V}}{1.4 \text{ A}}$$
$$= 17.1 \ \Omega$$

Finally, the total circuit power can be found as the sum of the branch power values, as follows:

$$P_T = P_A + P_B$$
$$= 19.2 \text{ W} + 14.4 \text{ W}$$
$$= 33.6 \text{ W}$$

PRACTICE PROBLEM 7.1

A circuit like the one in Figure 7.4 has the following values: $R_1 = 22 \ \Omega$, $R_2 = 33 \ \Omega$, $R_3 = 12 \ \Omega$, $R_4 = 10 \ \Omega$, and $E = 120$ V. Calculate the voltage, current, and power values for the circuit.

Some Observations About the Circuit in Example 7.1

Using the voltage and current values found in Example 7.1, we can make some important observations about series-parallel circuit characteristics. Figure 7.5a shows the voltages we calculated for the circuit in Figure 7.4. Note that the voltage across branch A equals the voltage across branch B. These voltages *must* be equal in value because the branches are in parallel. At the same time, voltage is divided among the resistors within each branch based upon their relative value. This is consistent with the voltage-divider characteristic of series circuits. Also note that

$$V_{R1} + V_{R2} = V_{R3} + V_{R4} = E$$

which satisfies Kirchhoff's voltage law.

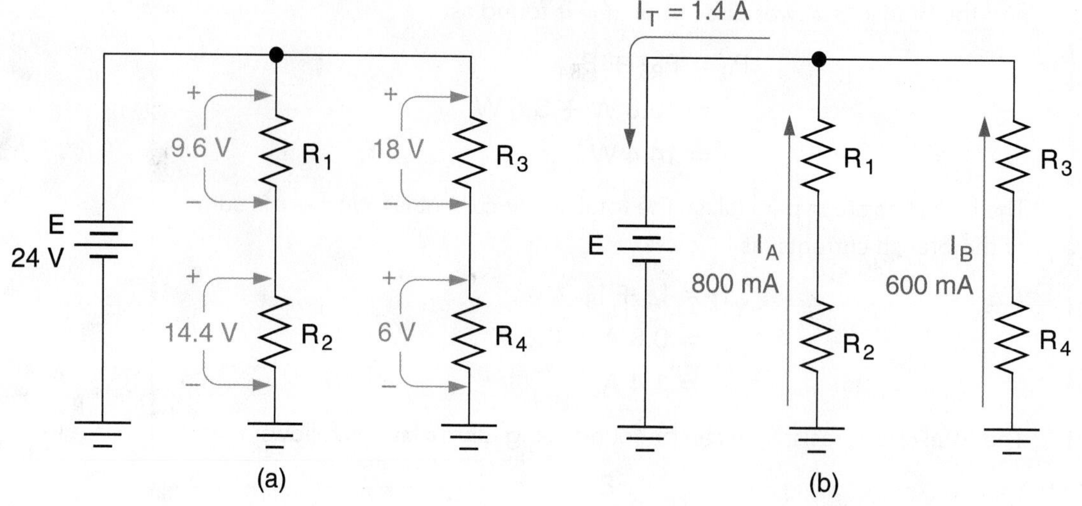

FIGURE 7.5 Calculated voltage and current values for the circuit in Figure 7.4.

The current characteristics of the circuit are illustrated in Figure 7.5b. As is always the case with series circuits, the values of I_{R1} and I_{R2} are equal, as are the values of I_{R3} and I_{R4}. The branch currents (I_A and I_B) are not equal in this case. At the same time, the sum of I_A and I_B is equal to the total source current (as is the case with any parallel circuit). This is consistent with Kirchhoff's current law.

The bottom line is this: Each branch is a series circuit that retains the voltage and current characteristics of any series circuit. The combination of the branches, however, is a parallel circuit that retains the voltage and current characteristics of any parallel circuit.

Connecting Parallel Circuits in Series

Figure 7.6 shows two parallel circuits (identified as *loop A* and *loop B*) connected in series with a single voltage source. Loop A (which consists of R_1 and R_2) retains the characteristics of any parallel circuit. Therefore:

1. The voltages across the two components are equal ($V_{R1} = V_{R2}$).
2. The sum of the component currents equals the total circuit current ($I_{R1} + I_{R2} = I_T$).
3. The total resistance of the loop is equal to the parallel combination of the two resistor values ($R_A = R_1 \parallel R_2$).

Note that R_A is used to denote the total resistance in loop A. By the same token, loop B (which consists of R_3 and R_4) retains the characteristics of any parallel circuit. Therefore:

1. $V_{R3} = V_{R4}$
2. $I_{R3} + I_{R4} = I_T$
3. $R_B = R_3 \parallel R_4$

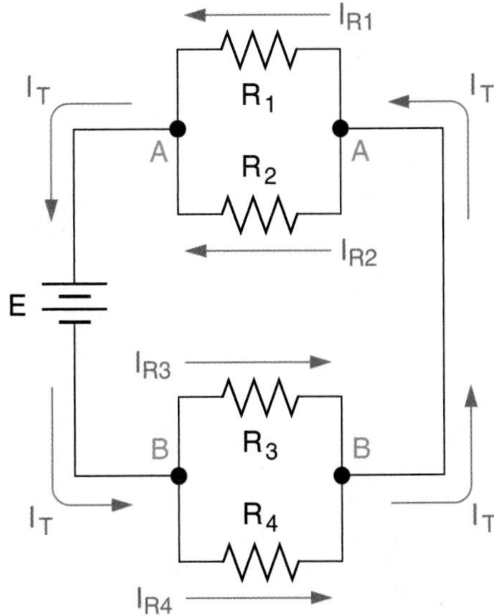

FIGURE 7.6 Parallel circuits connected in series.

Since these two loops form a series circuit, the combination of loop A and loop B retains the characteristics of any series circuit. Therefore:

1. The sum of V_A and V_B equals the source voltage ($V_A + V_B = E$).
2. The currents through the two loops must equal the source current ($I_{R1} + I_{R2} = I_{R3} + I_{R4} = I_T$).
3. The total circuit resistance equals the sum of R_A and R_B ($R_T = R_A + R_B$).

When two or more parallel circuits are connected in series (as shown in Figure 7.6), calculating the circuit values of voltage, current, and power starts with representing each of the parallel circuits as a single resistance. Next, the voltage across each equivalent resistance is calculated, followed by the current and power values for the various resistors. This approach to analyzing the circuit is demonstrated in Example 7.2.

EXAMPLE 7.2

Calculate the voltage, current, and power values for the circuit in Figure 7.7. The voltage, current, and power values for the circuit in Figure 7.7 are found as follows: First, the resistance of loop A (the parallel combination of R_1 and R_2) is found as

$$R_A = \frac{1}{\dfrac{1}{R_1} + \dfrac{1}{R_2}}$$

$$= \frac{1}{\dfrac{1}{200\ \Omega} + \dfrac{1}{300\ \Omega}}$$

$$= 120\ \Omega$$

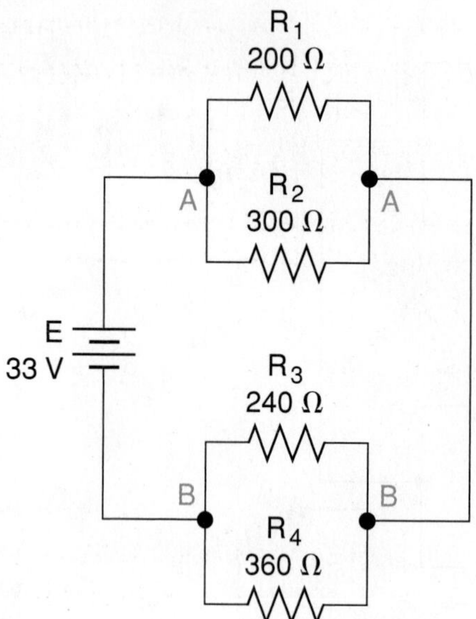

FIGURE 7.7

The resistance of loop B is now found as

$$R_B = \frac{1}{\frac{1}{R_3} + \frac{1}{R_4}}$$

$$= \frac{1}{\frac{1}{240\ \Omega} + \frac{1}{360\ \Omega}}$$

$$= 144\ \Omega$$

Using the calculated values of R_A and R_B, we can redraw the original circuit as shown in Figure 7.8a. Using this **EQUIVALENT CIRCUIT** and the voltage-divider relationship, we can calculate the voltage across each of the loops as follows:

EQUIVALENT CIRCUIT A circuit derived by combining groups of parallel and/or series components in one circuit to obtain an equivalent, but simpler, circuit.

$$V_A = E \times \frac{R_A}{R_A + R_B}$$

$$= 33\ \text{V} \times \frac{120\ \Omega}{120\ \Omega + 144\ \Omega}$$

$$= 33\ \text{V} \times 0.455$$

$$= 15\ \text{V}$$

and

$$V_B = E \times \frac{R_B}{R_A + R_B}$$

$$= 33\ \text{V} \times \frac{144\ \Omega}{120\ \Omega + 144\ \Omega}$$

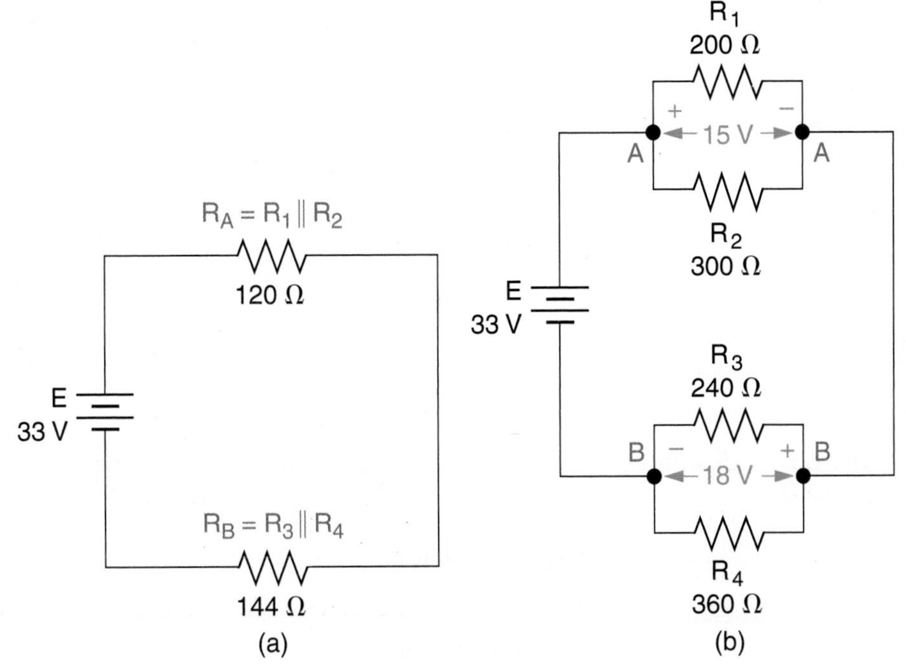

FIGURE 7.8

$$= 33\text{ V} \times 0.545$$
$$= 18\text{ V}$$

These calculations indicate that we have 15 V across loop A and 18 V across loop B, as shown in Figure 7.8b. We can now use Ohm's law to calculate the values of I_{R1} and I_{R2}, as follows:

$$I_{R1} = \frac{V_A}{R_1}$$
$$= \frac{15\text{ V}}{200\ \Omega}$$
$$= 75\text{ mA}$$

and

$$I_{R2} = \frac{V_A}{R_2}$$
$$= \frac{15\text{ V}}{300\ \Omega}$$
$$= 50\text{ mA}$$

Using Ohm's law, we can now calculate the values of I_{R3} and I_{R4}, as follows:

$$I_{R3} = \frac{V_B}{R_3}$$

$$= \frac{18\text{ V}}{240\ \Omega}$$

$$= 75\text{ mA}$$

and

$$I_{R4} = \frac{V_B}{R_4}$$

$$= \frac{18\text{ V}}{360\ \Omega}$$

$$= 50\text{ mA}$$

The power dissipated by each of the resistors in the circuit can now be found as

$$P_{R1} = V_A \times I_{R1}$$
$$= 15\text{ V} \times 75\text{ mA}$$
$$= 1.125\text{ W}$$

$$P_{R2} = V_A \times I_{R2}$$
$$= 15\text{ V} \times 50\text{ mA}$$
$$= 750\text{ mW}$$

$$P_{R3} = V_B \times I_{R3}$$
$$= 18\text{ V} \times 75\text{ mA}$$
$$= 1.35\text{ W}$$

and

$$P_{R4} = V_B \times I_{R4}$$
$$= 18\text{ V} \times 50\text{ mA}$$
$$= 900\text{ mW}$$

Finally, the total circuit current and total power can be calculated as follows:

$$I_T = I_{R1} + I_{R2}$$
$$= 75\text{ mA} + 50\text{ mA}$$
$$= 125\text{ mA}$$

and

$$P_T = P_{R1} + P_{R2} + P_{R3} + P_{R4}$$
$$= 1.125\text{ W} + 750\text{ mW} + 1.35\text{ W} = 900\text{ mW}$$
$$\cong 4.13\text{ W}$$

PRACTICE PROBLEM 7.2

A circuit like the one in Figure 7.7 has the following values: $R_1 = 10\ \Omega$, $R_2 = 15\ \Omega$, $R_3 = 18\ \Omega$, $R_4 = 36\ \Omega$, and $E = 24$ V. Calculate the voltage, current, and power values for the circuit.

Some Observations About the Circuit in Example 7.2

Using the voltage and current values found in Example 7.2, we can make some observations about parallel circuits that are connected in series. For example, Figure 7.9a shows the voltage values we calculated in Example 7.2. Note that the sum of the loop voltages ($V_A + V_B$) equals the source voltage (E), which satisfies KVL. Also note that each loop voltage is applied equally across two parallel resistors. That is, $V_A = V_{R1} = V_{R2}$ and $V_B = V_{R3} = V_{R4}$.

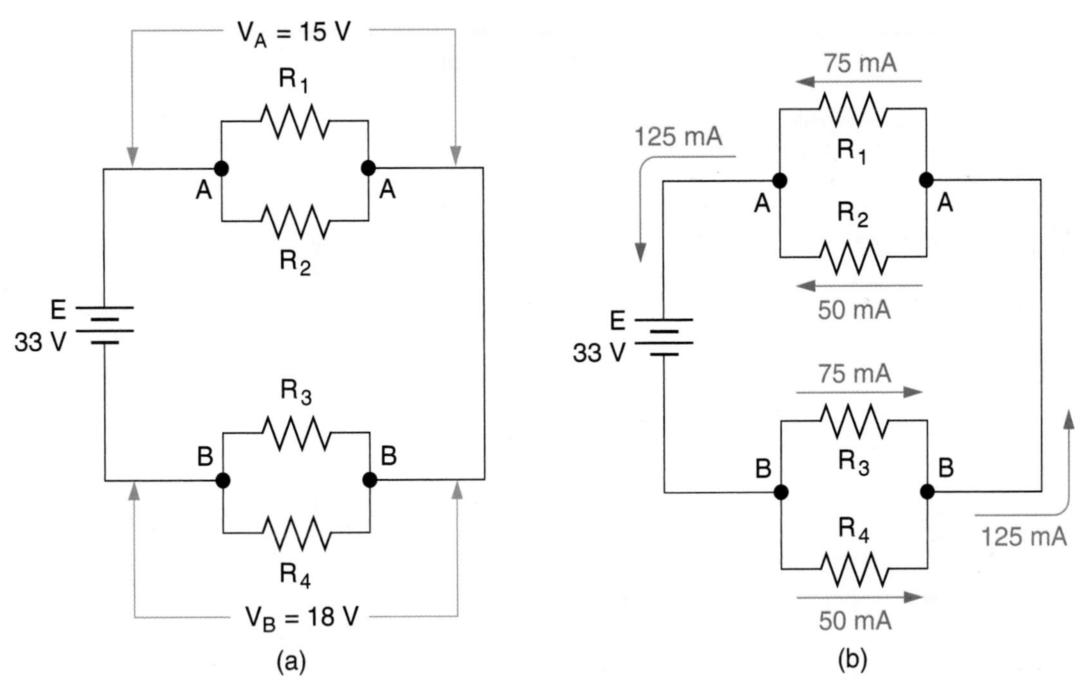

FIGURE 7.9 Calculated voltage and current values for the circuit in Figure 7.7.

The current characteristics of the circuit in Example 7.2 are illustrated in Figure 7.9b. As is always the case with parallel circuits, $I_T = I_{R1} + I_{R2}$ and $I_T = I_{R3} + I_{R4}$, which satisfies KCL. In fact, we could have calculated the total current in Example 7.2 using $I_T = I_{R3} + I_{R4}$. The result would have been the same as the one we calculated in the example.

So, here's the bottom line: Each loop is a parallel circuit that retains all of the voltage and current characteristics of any parallel circuit. At the same time, the overall circuit acts as a series circuit, and retains all of the voltage and current characteristics of any other series circuit.

Circuit Variations

Series-parallel circuits come in a near-infinite variety of configurations, from relatively simple to extremely complex. In this section, we have analyzed only two of the most basic configurations.

Throughout the remainder of this chapter (and the next), we will look at a variety of series-parallel circuits. We will focus on the approaches used to calculate the voltage, current, resistance, and power values for those circuits. As you will see, any series-parallel circuit can be dealt with effectively when you:

1. Use the proper approach.
2. Remember the basic principles of series and parallel circuits.

PROGRESS CHECK

1. List the current, voltage, and resistance characteristics of series circuits.

2. List the current, voltage, and resistance characteristics of parallel circuits.

*3. Refer to Figure 7.10a. Determine which (if any) of the following statements are *true* for the circuit shown.
   a. Under normal operating conditions, $V_{R3} = V_{R4}$.
   b. R_2 is in series with R_3.
   c. The value of E can be found by adding the values of V_{R1}, V_{R2}, and V_{R4}.
   d. The total power (P_T) for the circuit can be found by adding the values of P_{R1}, P_{R2}, and P_{R4}.

*4. Answer each of the following questions for the circuit shown in Figure 7.10b.
   a. Which two current values add up to equal the value of I_{R4}?
   b. Which two current values add up to equal the value of I_{R1}?
   c. Is R_1 in series with R_2? Explain your answer.
   d. Which three voltage values could be added to determine the value of E?

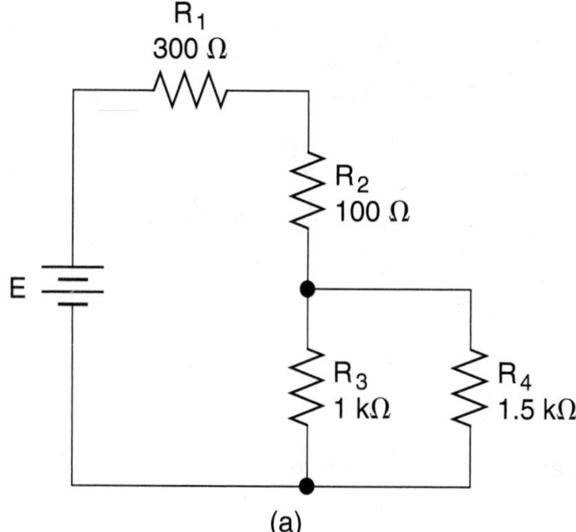

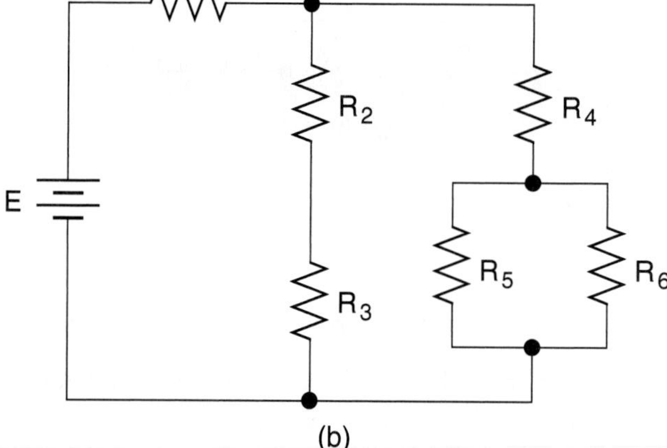

FIGURE 7.10

*The answers to these questions are included with the practice problem answers at the end of the chapter.

7.2 ANALYZING SERIES-PARALLEL CIRCUITS

The goal of any series-parallel circuit analysis is to determine the value of voltage, current, and/or power for one or more components. For example, consider the circuit shown in Figure 7.11. One practical circuit analysis goal would be to determine the value of load power for the circuit. Another might be to determine how load current responds to a change in source voltage.

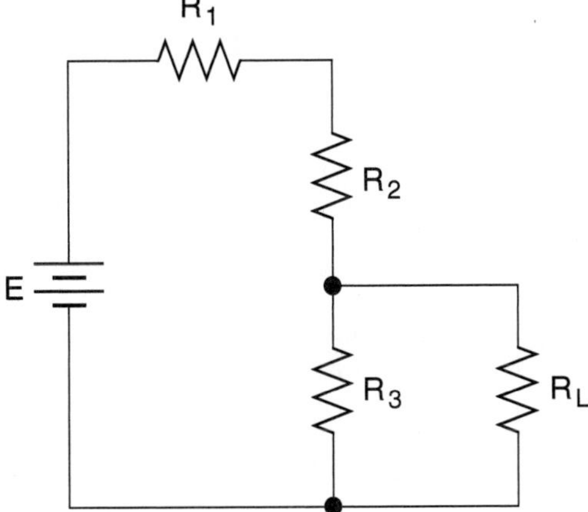

FIGURE 7.11 A loaded voltage divider.

Calculating a voltage, current, or power value in a series circuit or a parallel circuit is usually a simple, straightforward problem. However, this is not always the case with series-parallel circuits. When calculating series-parallel circuit values, we are sometimes required to perform both series-circuit and parallel-circuit calculations. For example, to calculate the value of load power (P_{RL}) for the circuit in Figure 7.11, we must first calculate the value of load voltage (V_{RL}). And to calculate the value of V_{RL}, we must first combine R_3 and R_L into a single equivalent resistance ($R_3 \| R_L$). This series of calculations is demonstrated in Example 7.3.

EXAMPLE 7.3

What is the value of load power for the circuit in Figure 7.12a?

SOLUTION

Calculating the value of load power for the circuit in Figure 7.12a begins with combining R_3 and R_L into a single equivalent resistance (R_{EQ}), as follows:

$$R_{EQ} = \frac{1}{\dfrac{1}{R_3} + \dfrac{1}{R_L}}$$

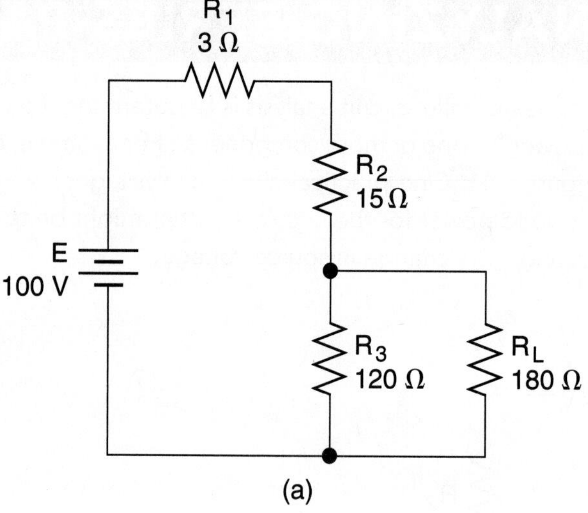

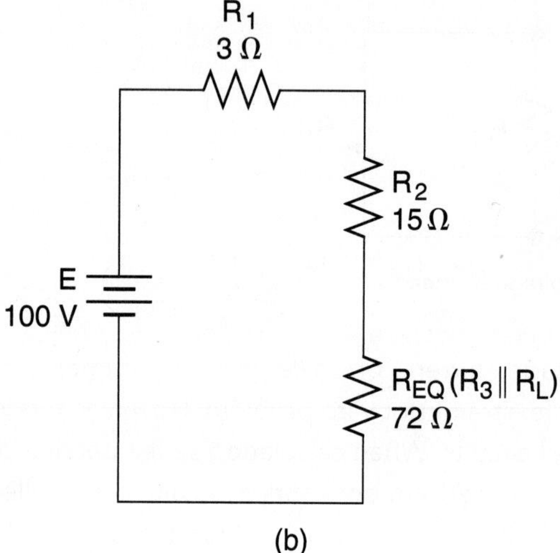

FIGURE 7.12

$$= \frac{1}{\dfrac{1}{120\ \Omega} + \dfrac{1}{180\ \Omega}}$$

$$= 72\ \Omega$$

When we replace the parallel circuit with its equivalent resistance, we get the circuit shown in Figure 7.12b. As you can see, we now have a simple series circuit. The voltage across the equivalent resistance (V_{EQ}) can now be found as

$$V_{EQ} = E \times \frac{R_{EQ}}{R_1 + R_2 + R_{EQ}}$$

$$= 100\ \text{V} \times \frac{72\ \Omega}{100\ \Omega}$$

$$= 72\ \text{V}$$

The voltage across the equivalent resistance (V_{EQ}) is equal to the voltage across the original parallel circuit. That is, $V_{EQ} = V_{R3} = V_{RL}$. Therefore, we can find the value of load power as follows:

$$P_L = \frac{V_{RL}^2}{R_L}$$

$$= \frac{(72\text{ V})^2}{180\ \Omega}$$

$$= 28.8\text{ W}$$

PRACTICE PROBLEM 7.3

A circuit like the one in Figure 7.12a has the following values: $R_1 = 120\ \Omega$, $R_2 = 150\ \Omega$, $R_3 = 300\ \Omega$, $R_L = 100\ \Omega$, and $E = 60$ V. Calculate the value of load power for the circuit.

Equivalent Circuits

The process of calculating a voltage, current, or power value in a series-parallel circuit usually begins with deriving a simplified equivalent of the original, just like the one in Figure 7.12a.

Equivalent circuits are derived by combining groups of parallel and/or series components to obtain an equivalent, but simpler, circuit. For example, consider the series-parallel circuit and its series equivalent shown in Figure 7.13.

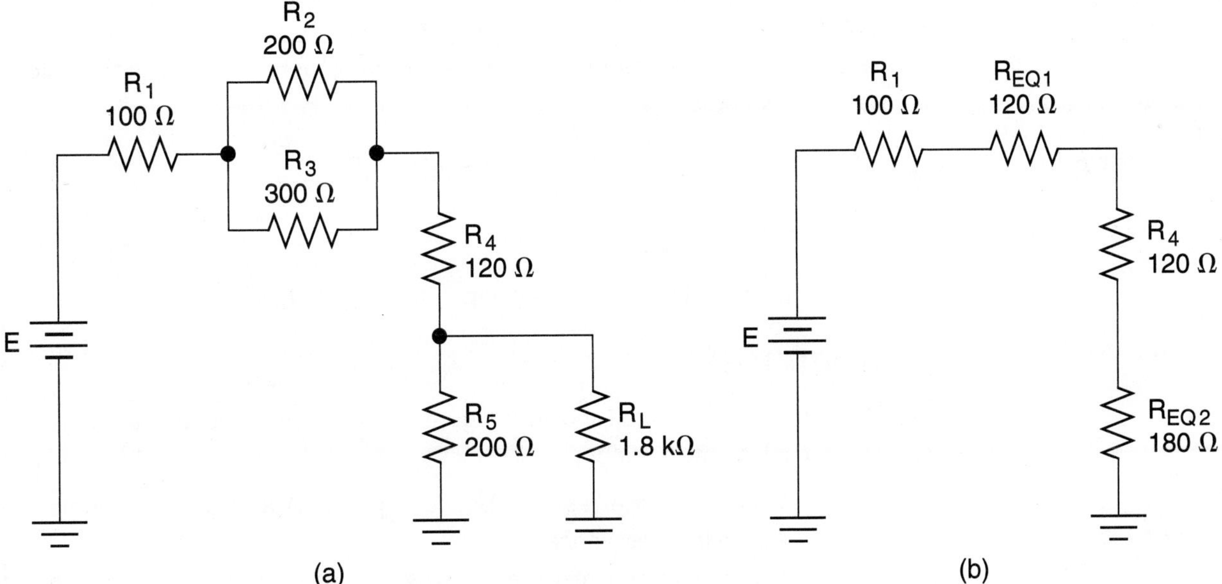

FIGURE 7.13 A series-parallel circuit and its series-equivalent circuit.

The equivalent circuit shown was derived as follows:

1. R_2 and R_3 were combined into a single equivalent resistance (R_{EQ1}), just like we did in Example 7.3. The value of R_{EQ1} was found as follows:

$$R_{EQ1} = \frac{1}{\frac{1}{R_2} + \frac{1}{R_3}} = \frac{1}{\frac{1}{200\ \Omega} + \frac{1}{300\ \Omega}} = 120\ \Omega$$

2. R_5 and R_L were combined into a second equivalent resistance (R_{EQ2}). The value of R_{EQ2} was found as

$$R_{EQ2} = \frac{1}{\frac{1}{R_5} + \frac{1}{R_L}} = \frac{1}{\frac{1}{200\ \Omega} + \frac{1}{1800\ \Omega}} = 180\ \Omega$$

SERIES-EQUIVALENT CIRCUIT An equivalent circuit that is made up entirely of components connected in series.

PARALLEL-EQUIVALENT CIRCUIT An equivalent circuit that is made up entirely of components connected in parallel.

Using these values, the equivalent circuit was drawn as shown in Figure 7.13b.

The process of deriving an equivalent circuit ends when the original series-parallel circuit has been simplified to a **SERIES-EQUIVALENT CIRCUIT** or **PARALLEL-EQUIVALENT CIRCUIT**; that is, a circuit that is made up entirely of components connected either in series or in parallel. This point is demonstrated in the following series of examples.

EXAMPLE 7.4

Derive an equivalent for the series-parallel circuit shown in Figure 7.14a.

SOLUTION

The series-parallel circuit shown in Figure 7.14a can be reduced to a simple series-equivalent circuit by:

1. Combining the resistors between points A and B into a single equivalent resistance (R_{AB}).
2. Combining the resistors between point C and ground into a single equivalent resistance (R_C).

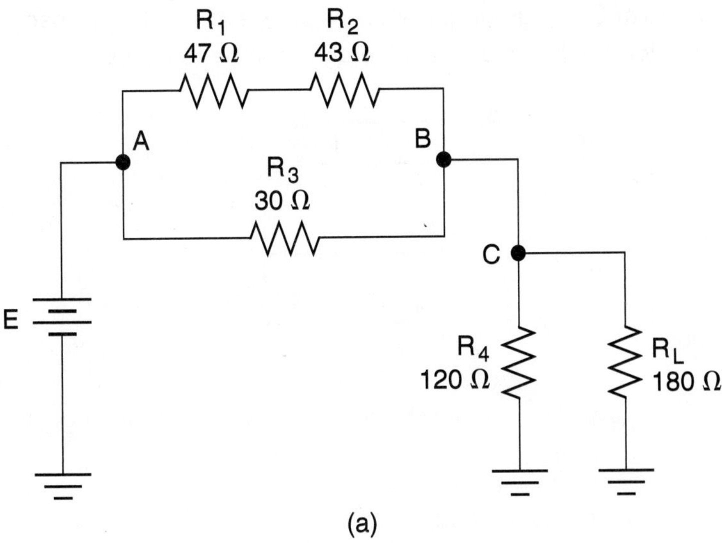

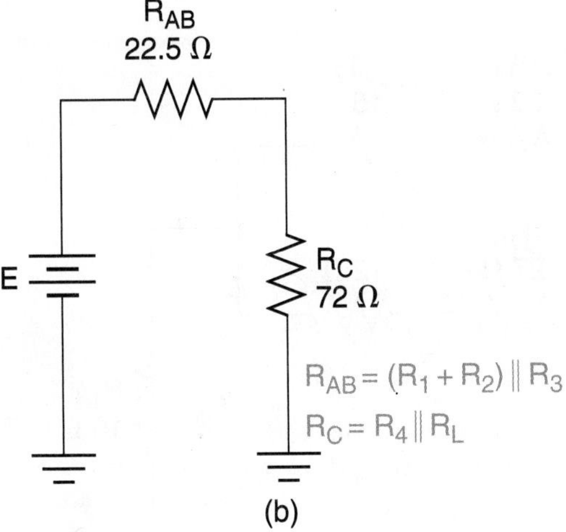

FIGURE 7.14

Between points A and B, R_1 and R_2 form a series circuit that is in parallel with R_3. The sum of R_1 and R_2 can be calculated as follows:

$$\begin{aligned} R_{EQ} &= R_1 + R_2 \\ &= 90\ \Omega \end{aligned}$$

Combining R_{EQ} with R_3 gives us

$$\begin{aligned} R_{AB} &= \frac{1}{\dfrac{1}{R_{EQ}} + \dfrac{1}{R_3}} \\ &= \frac{1}{\dfrac{1}{90\ \Omega} + \dfrac{1}{30\ \Omega}} \\ &= 22.5\ \Omega \end{aligned}$$

Between point C and ground, we have a parallel circuit that is made up of R_4 and the load. Therefore, the value of R_C can be found as

$$R_C = \frac{1}{\frac{1}{R_4} + \frac{1}{R_L}}$$

$$= \frac{1}{\frac{1}{120\ \Omega} + \frac{1}{180\ \Omega}}$$

$$= 72\ \Omega$$

Combining R_{AB} and R_C, we get the series-equivalent of the original circuit drawn in Figure 7.14b.

PRACTICE PROBLEM 7.4

Derive the series-equivalent for the circuit shown in Figure 7.15.

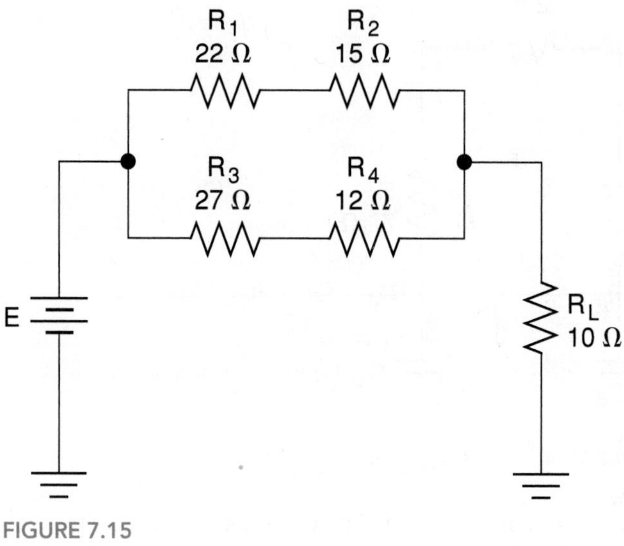

FIGURE 7.15

EXAMPLE 7.5

Derive the *parallel-equivalent* for the circuit shown in Figure 7.16a.

SOLUTION

The series-parallel circuit in Figure 7.16a can be reduced to a simple parallel-equivalent circuit by:

1. Combining the resistors between points A and B into a single equivalent resistance (R_{AB}).
2. Combining the resistors between points C and D into a single equivalent resistance (R_{CD}).

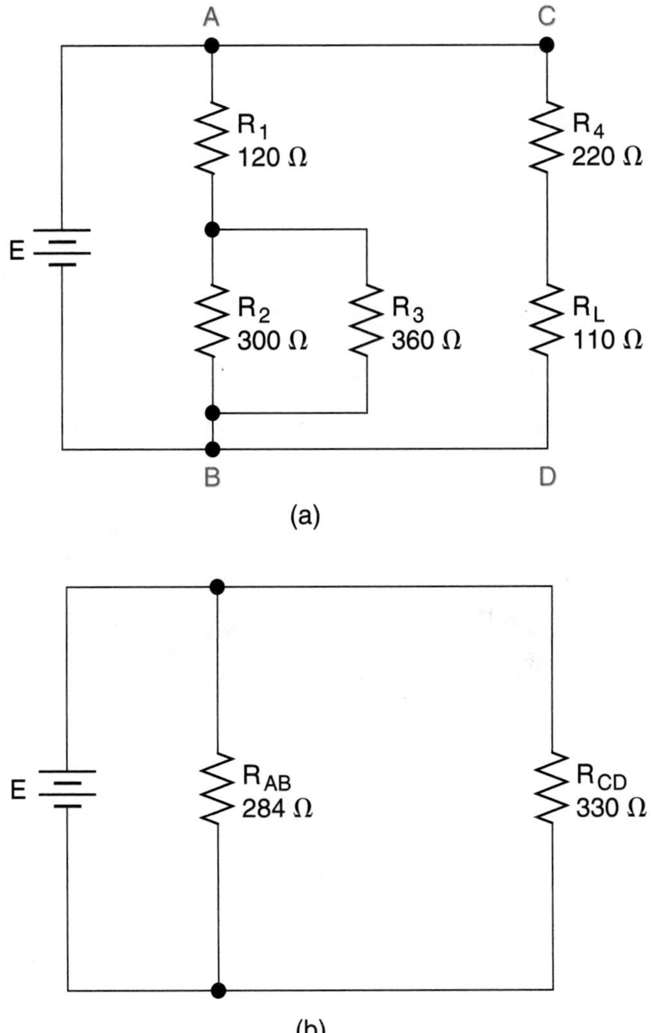

FIGURE 7.16

Between points A and B, R_1 is in series with the parallel combination of R_2 and R_3. Combining R_2 and R_3 into a parallel equivalent resistance (R_{EQ}) gives us:

$$R_{EQ} = \frac{1}{\frac{1}{R_2} + \frac{1}{R_3}}$$

$$= \frac{1}{\frac{1}{300\ \Omega} + \frac{1}{360\ \Omega}}$$

$$= 164\ \Omega$$

Adding R_{EQ} to R_1 gives us

$$R_{AB} = R_1 + R_{EQ}$$

$$= 120\ \Omega + 164\ \Omega$$

$$= 284\ \Omega$$

Between points C and D, R_4 is in series with the load resistance. Therefore, the value of R_{CD} can be found as

$$R_{CD} = R_4 + R_L$$
$$= 220\ \Omega + 110\ \Omega$$
$$= 330\ \Omega$$

Using the calculated values of R_{AB} and R_{CD}, the parallel-equivalent of the original circuit is drawn as shown in Figure 7.16b.

PRACTICE PROBLEM 7.5

Derive the parallel-equivalent for the circuit shown in Figure 7.17.

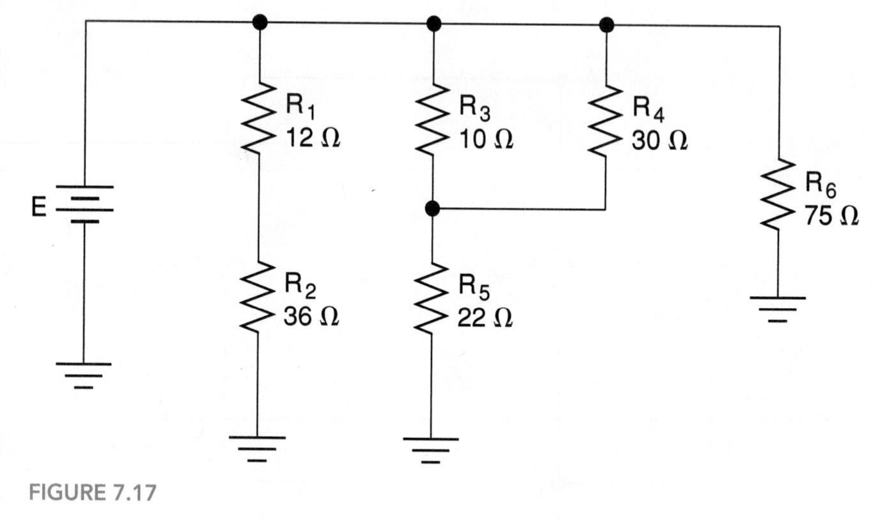

FIGURE 7.17

Series-Parallel Circuit Reduction

In some cases, you must reduce an entire circuit to its series-equivalent or parallel-equivalent in order to determine a given current, voltage, and/or power value. In others, a value can be determined by reducing only a portion of the original circuit. That is, we can solve for a given value without reducing the entire circuit to a series- or parallel-equivalent.

The key to series-parallel circuit calculations is to know how far the circuit must be reduced in order to solve the problem. For example, consider the circuit shown in Figure 7.18a. Let's say we want to determine the value of load power for this circuit. If we combine the load resistance and R_3 into a single equivalent resistance (as shown in Figure 7.18b), we have reduced the circuit enough to solve for the load power. This point is demonstrated in Example 7.6.

EXAMPLE 7.6

Calculate the value of load power (P_{RL}) for the circuit in Figure 7.18a.

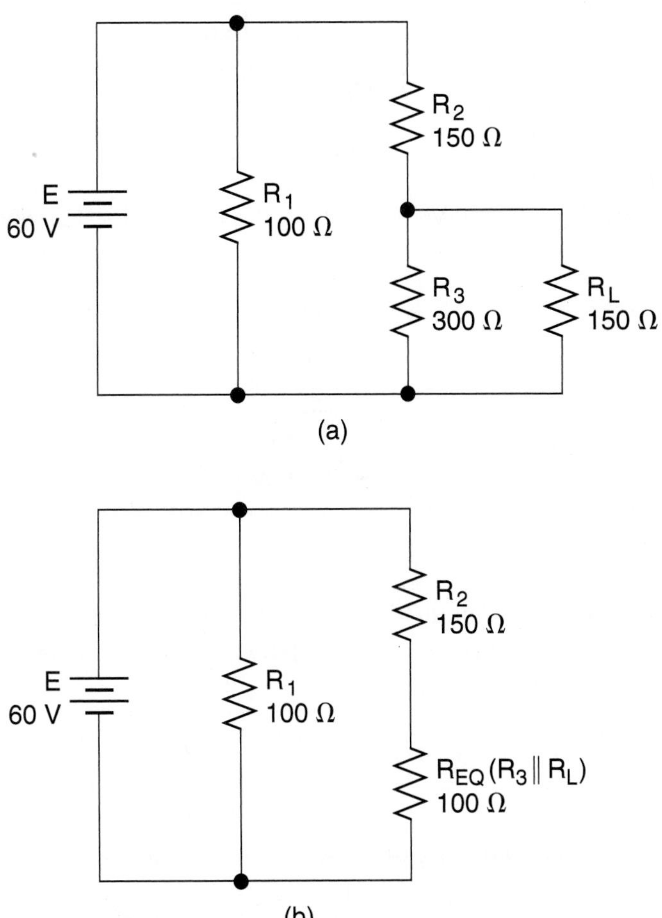

FIGURE 7.18

SOLUTION

Calculating the load power for the circuit in Figure 7.18a begins by combining R_3 and R_L into a parallel equivalent resistance (R_{EQ}) as follows:

$$R_{EQ} = \frac{1}{\frac{1}{R_3} + \frac{1}{R_L}}$$

$$= \frac{1}{\frac{1}{300\ \Omega} + \frac{1}{150\ \Omega}}$$

$$= 100\ \Omega$$

Since the source voltage is applied to the series combination of R_2 and R_{EQ}, the voltage across R_{EQ} can be found as follows:

$$V_{EQ} = E \times \frac{R_{EQ}}{R_2 + R_{EQ}}$$

$$= 60\text{ V} \times \frac{100\ \Omega}{250\ \Omega}$$

$$= 24\text{ V}$$

As you know, $V_{EQ} = V_{R3} = V_{RL}$. Therefore, the value of load power can be found as

$$P_{RL} = \frac{V_{RL}^2}{R_L}$$

$$= \frac{(24\text{ V})^2}{150\ \Omega}$$

$$= 3.84\text{ W}$$

PRACTICE PROBLEM 7.6

A circuit like the one in Figure 7.18a has the following values: $E = 48$ V, $R_1 = 20\ \Omega$, $R_2 = 30\ \Omega$, $R_3 = 120\ \Omega$, and $R_L = 180\ \Omega$. Calculate the value of load power for the circuit.

In Example 7.6, we didn't need to reduce the entire circuit in order to solve the problem. We only needed to simplify the branch containing R_2, R_3, and the load. At the same time, calculating such values as total resistance (R_T), current (I_T), or power (P_T) would have required a complete reduction of the circuit to its simplest parallel-equivalent.

Summary

As you can see, calculating component values of voltage, current, and/or power in most series-parallel circuits involves three steps:

1. Reduce the circuit to a series-equivalent circuit or a parallel-equivalent circuit.
2. Apply the basic series and parallel circuit principles and relationships to the equivalent circuit.
3. Use the values obtained from the equivalent circuit to calculate the desired value(s).

When you use this three-step approach, you will rarely have a problem solving for any single value in a series-parallel circuit.

PROGRESS CHECK

1. How are equivalent circuits derived?
2. What are the steps involved in solving most practical series-parallel circuit problems?

7.3 CIRCUIT LOADING

There are many circuits and concepts that you will see time and again throughout your career. One important concept is the relationship between a load and the operation of its source circuit. In this section, we will examine the effects that loads can have on circuit operation.

Loaded Voltage Dividers

Most circuits are designed for the purpose of supplying power (in one form or another) to one or more loads. For example, consider the circuit shown in Figure 7.19. Here we have a voltage divider that is being used to supply the voltages required for the operation of two loads (designated as Load A and Load B). The voltage and current ratings for each load are given in the illustration. Assuming that the circuit is operating properly, the voltage divider is providing the values of $V_A = 24$ V and $V_B = 48$ V required by the loads.

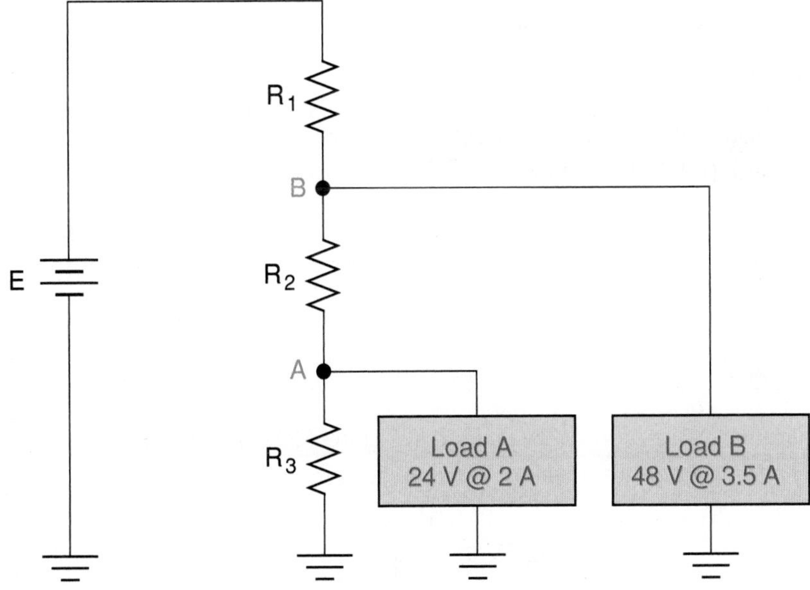

FIGURE 7.19 A loaded voltage divider.

In practice, voltage dividers are usually designed for the purpose of providing one or more specific load voltages (as shown in Figure 7.19). When used for this purpose, the voltage divider is referred to as a *loaded voltage divider.*

The operation and analysis of a loaded voltage divider is very similar to that of any other series-parallel circuit. However, there is one value of interest that is somewhat unique to this circuit. This value is called the **NO-LOAD OUTPUT VOLTAGE (V_{NL})**. The no-load output voltage is the voltage measured at the load terminals with the load removed. Note that the term **OUTPUT VOLTAGE** is commonly used to describe the voltage at the load terminals of a circuit.

NO-LOAD OUTPUT VOLTAGE (V_{NL}) The voltage measured at the load terminals of a circuit with the load removed.

OUTPUT VOLTAGE The voltage at the load terminals of a circuit.

The measurement of V_{NL} is illustrated in Figure 7.20b. As you can see, the load is removed from the circuit and the voltmeter is connected to the open load terminals.

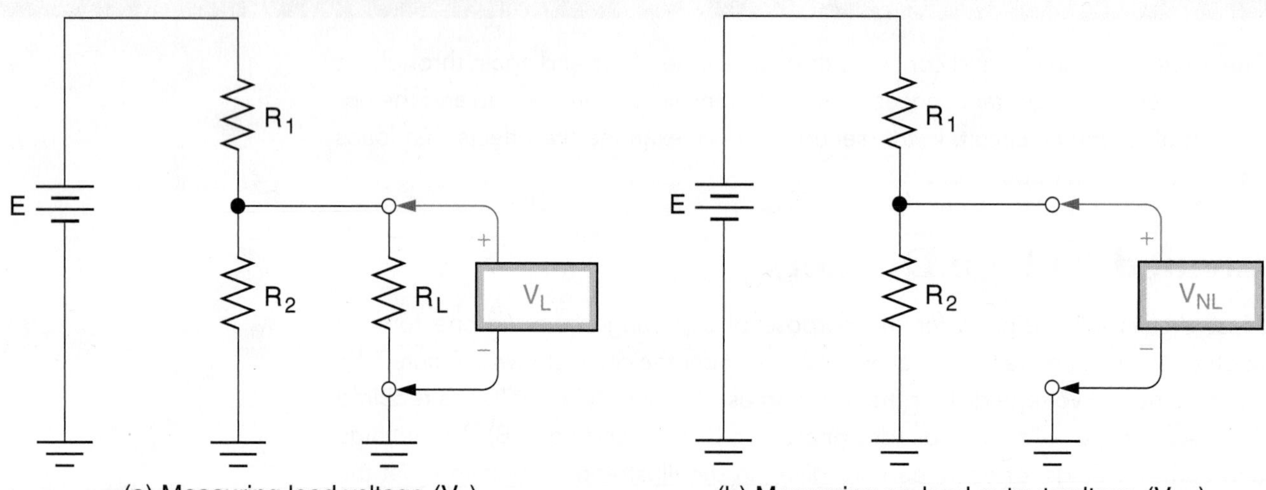

(a) Measuring load voltage (V_L)

(b) Measuring no-load output voltage (V_{NL})

FIGURE 7.20 Load voltage vs. no-load output voltage.

The Relationship Between V_{NL} and V_L

The no-load output voltage (V_{NL}) for a voltage divider is always greater than the circuit load voltage (V_L). This relationship is demonstrated in Example 7.7.

EXAMPLE 7.7

Determine the values of V_L and V_{NL} for the circuit in Figure 7.21a.

SOLUTION

To calculate the value of V_L for the circuit in Figure 7.21a, we start by combining R_2 and R_L into an equivalent resistance, as follows:

$$R_{EQ} = \frac{1}{\dfrac{1}{R_2} + \dfrac{1}{R_L}}$$

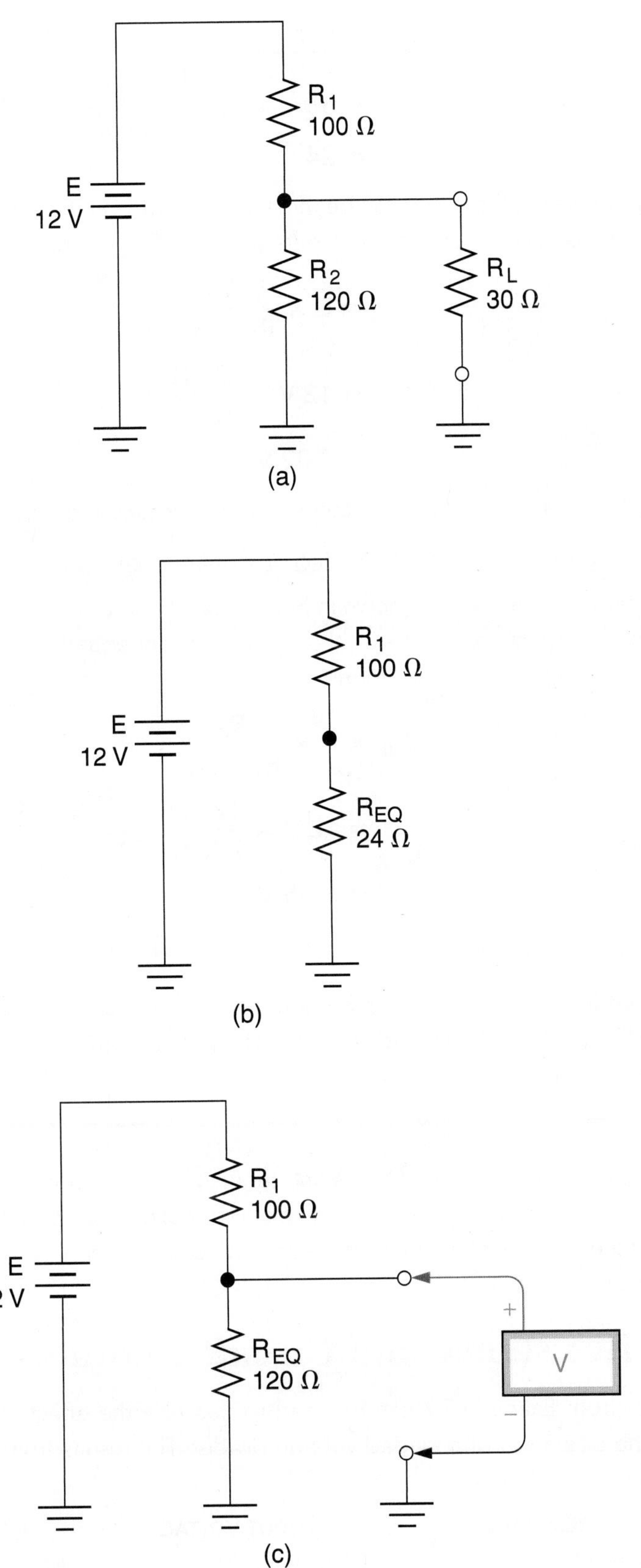
R1
100 Ω
E
12 V
R2
120 Ω
RL
30 Ω
(a)
R1
100 Ω
E
12 V
REQ
24 Ω
(b)
R1
100 Ω
E
12 V
REQ
120 Ω
+
V
−
(c)

FIGURE 7.21

$$= \frac{1}{\dfrac{1}{120\ \Omega} + \dfrac{1}{30\ \Omega}}$$
$$= 24\ \Omega$$

Using this value, the circuit can be redrawn as shown in Figure 7.21b. Using the voltage divider equation, the voltage across R_{EQ} can now be found as

$$V_{EQ} = E \times \frac{R_{EQ}}{R_1 + R_{EQ}}$$
$$= 12\ V \times \frac{24\ \Omega}{124\ \Omega}$$
$$= 2.32\ V$$

Since V_{EQ} is across the parallel combination of R_2 and the load,

$$V_L = V_{EQ} = 2.32\ V$$

If we remove the load (as shown in Figure 7.21c), the voltmeter is measuring the voltage across R_2. Using the voltage divider equation, the no-load output voltage (V_{NL}) can be found as

$$V_{NL} = E \times \frac{R_2}{R_1 + R_2}$$
$$= 12\ V \times \frac{120\ \Omega}{220\ \Omega}$$
$$= 6.55\ V$$

PRACTICE PROBLEM 7.7

A circuit like the one in Figure 7.21a has the following values: $E = 120$ V, $R_1 = 100\ \Omega$, $R_2 = 150\ \Omega$, and $R_L = 300\ \Omega$. Calculate the values of V_L and V_{NL} for the circuit.

As you can see, the value of V_{NL} is greater than the value of V_L. This relationship (which always holds true) can be used to explain several practical circuit relationships. The first of these is the relationship between load resistance and the operation of its source circuit.

Load Resistance and Circuit Operation

The results from Example 7.7 can be used to describe the effects of load resistance on the operation of a loaded voltage divider. The results from the example are summarized as follows:

LOAD RESISTANCE	OUTPUT VOLTAGE
$R_L = 30\ \Omega$	$V_L = 2.32$ V
$R_L = \infty\ \Omega$	$V_{NL} = 6.55$ V

As these results indicate, load voltage varies directly with load resistance. That is:

- An increase in load resistance causes an increase in output voltage.
- A decrease in load resistance causes a decrease in output voltage.

As you will see, this relationship has many practical applications.

Voltmeter Loading

VOLTMETER LOADING The reduction in component voltage that can result from connecting a voltmeter across a relatively high-resistance (or open) component.

When a voltmeter is connected to any resistive component in a series circuit, the voltmeter will always provide a reading that is lower than the normal component voltage. In some cases, the reading may be significantly lower. This is due to the effects of the meter's internal resistance, called **VOLTMETER LOADING**, as shown in Figure 7.22.

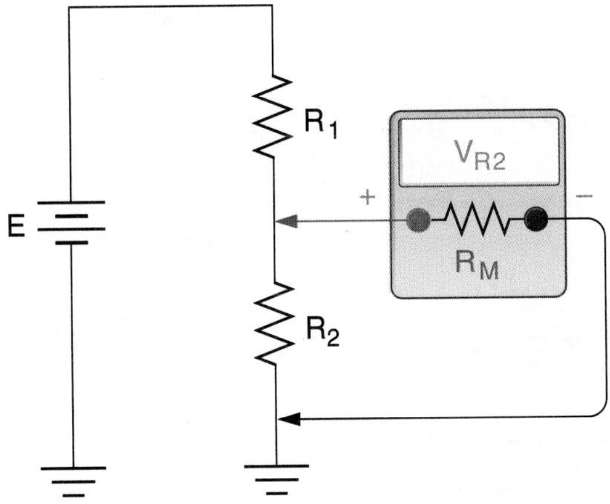

FIGURE 7.22 Voltmeter loading.

When the meter is connected as shown in the figure, it essentially converts the series circuit into a loaded voltage divider, with the meter's internal resistance acting as the load. If the internal resistance of the meter is low enough, the impact on the voltage being measured can be significant. This point is demonstrated in Example 7.8.

EXAMPLE 7.8

Determine the value of V_{R2} for the circuit in Figure 7.23a. Then, determine the value of V_{R2} when the voltmeter is connected as shown in Figure 7.23b.

SOLUTION

The value of V_{R2} for the series circuit is found as

$$V_{R2} = E \times \frac{R_2}{R_1 + R_2}$$

$$= 40\text{ V} \times \frac{100\text{ k}\Omega}{250\text{ k}\Omega}$$

$$= 16\text{ V}$$

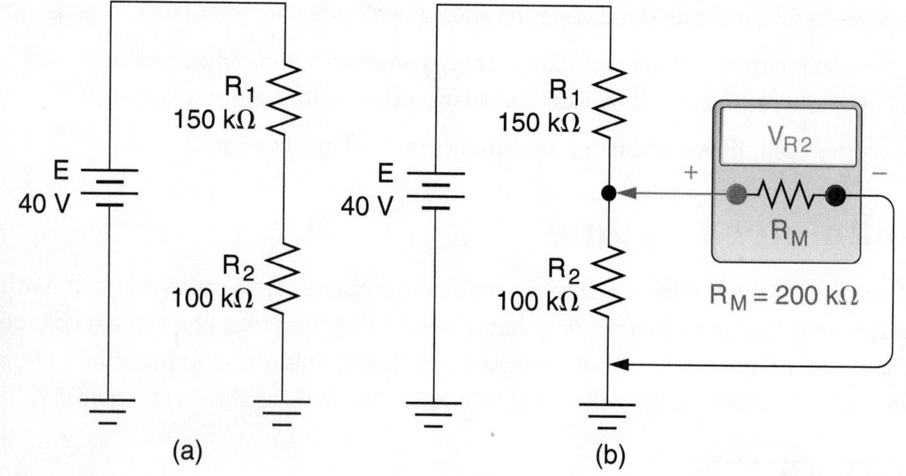

FIGURE 7.23

When the meter is connected as shown in Figure 7.23b, the meter's internal resistance (R_M) forms a parallel circuit with R_2. The resistance of this circuit is found as

$$R_{EQ} = \frac{1}{\frac{1}{R_2} + \frac{1}{R_M}}$$

$$= \frac{1}{\frac{1}{100\text{ k}\Omega} + \frac{1}{200\text{ k}\Omega}}$$

$$= 66.7\text{ k}\Omega$$

Taking this equivalent resistance into account, the value of V_{R2} can be found as follows:

$$V_{R2} = E \times \frac{R_{EQ}}{R_1 + R_{EQ}}$$

$$= 40\text{V} \times \frac{66.7\text{ k}\Omega}{216.7\text{ k}\Omega}$$

$$= 12.3\text{ V}$$

PRACTICE PROBLEM 7.8

The value of V_{R2} for the circuit in Figure 7.23a is measured using a DMM with an internal resistance of 10 MΩ. Calculate the value of V_{R2} with the meter connected. Compare your result to the values obtained in the example.

As you can see, connecting a voltmeter to a circuit can have a significant effect on the accuracy of the measurement. This impact is reduced by using a meter with high internal resistance. For example, consider the circuits shown in Figure 7.24. The internal resistance of the VOM in Figure 7.24a produces a parallel-equivalent resistance of 40.6 kΩ. At the same time, the DMM in Figure 7.24b produces a parallel-equivalent resistance of 50.7 kΩ. Since this value is approximately equal to the value of R_2 = 51 kΩ, the DMM presents virtually no load on the circuit. Therefore, the voltage reading in Figure 7.24b is extremely accurate.

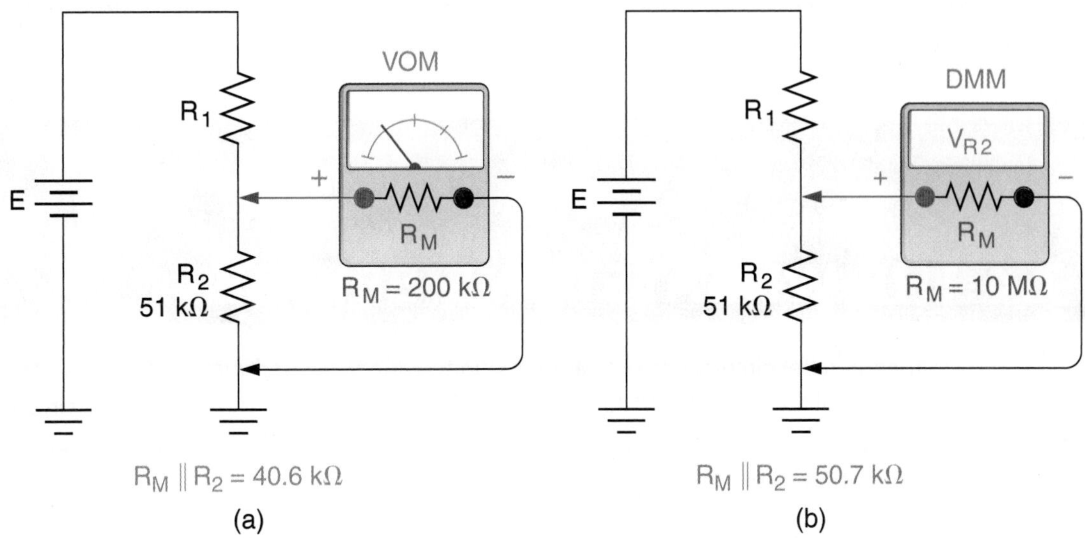

FIGURE 7.24 The loading effect of a VOM is greater than that of a DMM.

When you consider the impact of circuit loading, it is easy to understand why DMMs are preferred over VOMs for most voltage measurements. DMMs typically have much higher input resistance than VOMs. Therefore, voltage measurements taken with a DMM are much more accurate than comparable readings taken with a VOM.

Summary

In this section, you have seen how the presence of a load can affect the operation of a voltage divider. The output voltage produced by a voltage divider varies directly with the resistance of its load. This means that:

- An increase in load resistance causes an increase in output voltage.
- A decrease in load resistance causes a decrease in output voltage.

The no-load output voltage (V_{NL}) from a voltage divider is the maximum output voltage produced by the circuit.

The effects of circuit loading become apparent when a voltmeter is connected across any resistive component in a series circuit. When a voltmeter is connected across a resistive component, it forms a parallel circuit with that component, effectively lowering its resistance. This results in a reading that is lower than the true voltage across that component. The lower the internal resistance of the meter, the greater the error in the meter reading.

PROGRESS CHECK

1. What is a loaded voltage divider?
2. What is the no-load output voltage (V_{NL}) of a voltage divider?
3. For a loaded voltage divider, what is the relationship between V_L and V_{NL}?
4. For a loaded voltage divider, what is the relationship between output voltage and load resistance?
5. Discuss the effects of *voltmeter loading* on series circuit voltage measurements.

7.4 THE WHEATSTONE BRIDGE

WHEATSTONE BRIDGE A circuit containing four resistors and a meter "bridge" that provides extremely accurate resistive measurements.

Series-parallel circuits come in a near-infinite variety of configurations. In this section, we will focus on the **WHEATSTONE BRIDGE**. The Wheatstone bridge is a circuit containing four resistors and a meter "bridge" that provides extremely accurate resistive measurements.

Bridge Construction

The Wheatstone bridge consists of a DC voltage source, four resistors, and a meter, as shown in Figure 7.25. The resistors form two series circuits connected in parallel. The meter is connected as a "bridge" between the series circuits (which is how the circuit got its name).

GALVANOMETER A current meter that indicates both the magnitude and the direction of a low-value current.

The meter in Figure 7.25 is a **GALVANOMETER**. A galvanometer is a current meter that can indicate both the magnitude and the direction of a low-value current. As you will see, the galvanometer can be used to indicate the relationship between the voltages at points A and B in the circuit.

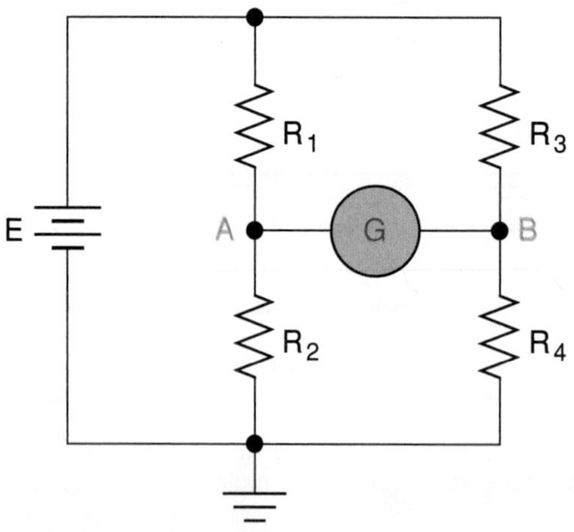

FIGURE 7.25 A Wheatstone bridge.

Bridge Operation

Each branch in the Wheatstone bridge is a voltage divider. If we remove the galvanometer (as shown in Figure 7.26), the voltage from point A to ground (V_A) and the voltage from point B to ground (V_B) can be found as follows:

(7.1)

$$V_A = E \times \frac{R_2}{R_1 + R_2}$$

and

(7.2)

$$V_B = E \times \frac{R_4}{R_3 + R_4}$$

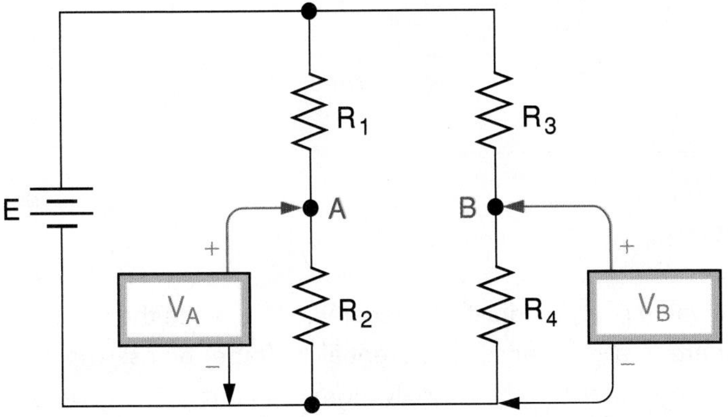

FIGURE 7.26

The operation of the Wheatstone bridge is based on the fact that it has three possible operating states; that is, three possible combinations of V_A and V_B. These operating states are as follows:

$$V_A = V_B$$
$$V_A > V_B$$
$$V_A < V_B$$

Each of these operating states is illustrated in Figure 7.27.

In Figure 7.27, R_4 is a variable resistor (rheostat). By adjusting the value of R_4, we can set V_B to any value (within limits). In Figure 7.27a, R_4 is adjusted so that $V_B = V_A$. When these voltages are equal, there is no difference of potential across the galvanometer, and the galvanometer current (I_G) equals 0 A. When in this operating state, the bridge is said to be *balanced*.

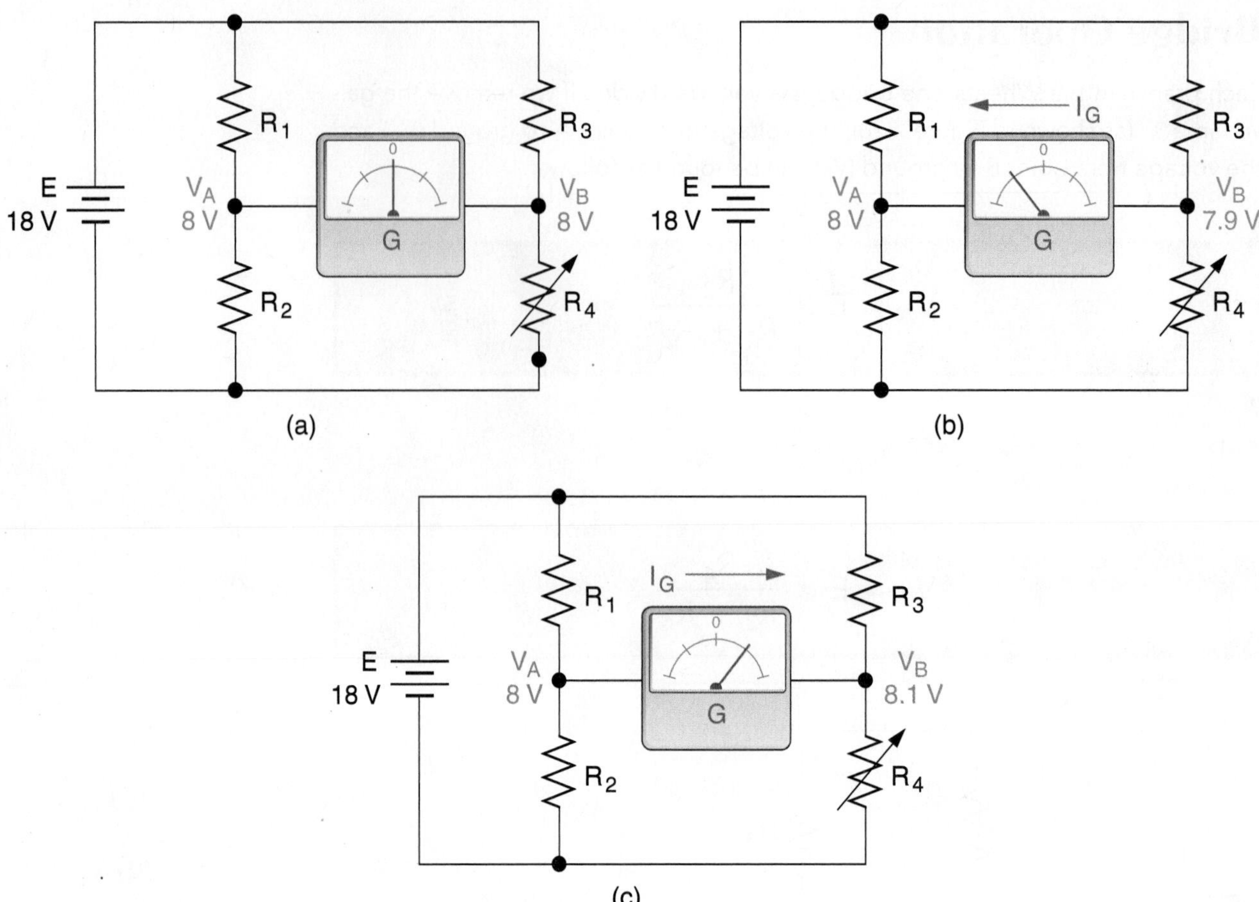

FIGURE 7.27 The operating states of a Wheatstone bridge.

In Figure 7.27b, the value of R_4 is adjusted so that $V_B < V_A$. Since the voltages are no longer equal, there is a difference of potential (voltage) across the galvanometer. As a result, a measurable value of galvanometer current (I_G) is generated. The galvanometer reading indicates the value and direction of I_G.

In Figure 7.27c, the value of R_4 has been adjusted so that $V_B > V_A$. Again, the voltages are unequal, and there is a difference of potential across the galvanometer. Again, a measurable value of I_G is generated with the galvanometer indicating its value and direction.

Resistance Ratios

By definition, a Wheatstone bridge is balanced when $V_A = V_B$. For these voltages to be equal, the resistance ratios in equations (7.2) and (7.3) must be equal. In other words, a bridge is balanced only when

(7.3)
$$\frac{R_2}{R_1} = \frac{R_4}{R_3}$$

Note that this relationship is derived by setting equations (7.1) and (7.2) equal to each other and simplifying the result. As Table 7.1 shows, each of the operating states is produced by a specific relationship between the resistance ratios.

TABLE 7.1 • Bridge Resistance Ratios and Operating States

RESISTANCE RATIO	RESULTING OPERATING STATE
$\frac{R_2}{R_1} = \frac{R_4}{R_3}$	$V_A = V_B$
$\frac{R_2}{R_1} > \frac{R_4}{R_3}$	$V_A > V_B$
$\frac{R_2}{R_1} < \frac{R_4}{R_3}$	$V_A < V_B$

Measuring Resistance with a Wheatstone Bridge

A Wheatstone bridge can be used as an effective resistance measuring circuit when modified as shown in Figure 7.28. In the bridge circuit shown:

- The unknown resistance (R_X) has been placed in the R_3 position.
- R_4 has been replaced by a decade box.

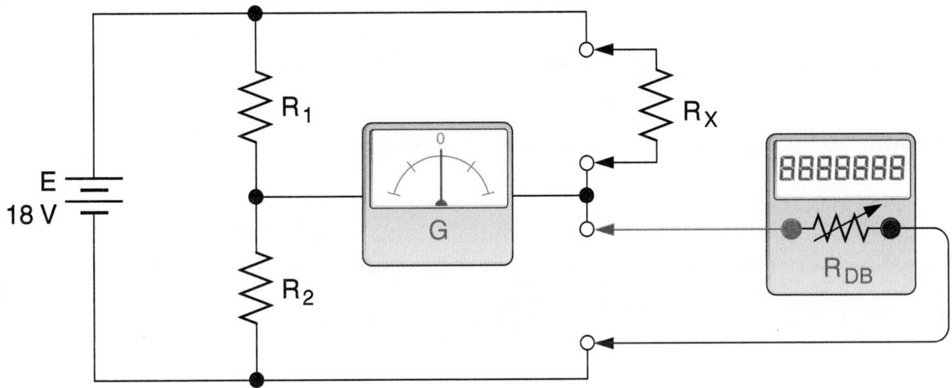

FIGURE 7.28 A resistance measuring circuit.

A **DECADE BOX** is a device that contains several series-connected potentiometers. By adjusting the individual potentiometers, the resistance of the device (R_{DB}) can be set to any value within its limits, usually between 1 Ω and 9.999999 MΩ. (The value of R_{DB} for any setting is indicated by the readouts on the decade box.)

DECADE BOX A device containing several series-connected potentiometers that allow the resistance of the device to be set to any value within its limits, usually between 1 Ω and 9.999999 MΩ.

When connected as shown in Figure 7.28, the value of the unknown resistance is found by balancing the bridge; that is, by adjusting the value of R_{DB} so that the galvanometer reads 0 A. When the bridge is balanced, the resistance ratios of the series circuits must be equal. Therefore, the value of R_X can be found using a modified form of equation (7.3), as follows:

(7.4)

$$R_X = R_{DB} \times \frac{R_1}{R_2}$$

Example 7.9 demonstrates the procedure for measuring an unknown resistance with a Wheatstone bridge.

EXAMPLE 7.9

What is the value of R_x in Figure 7.29?

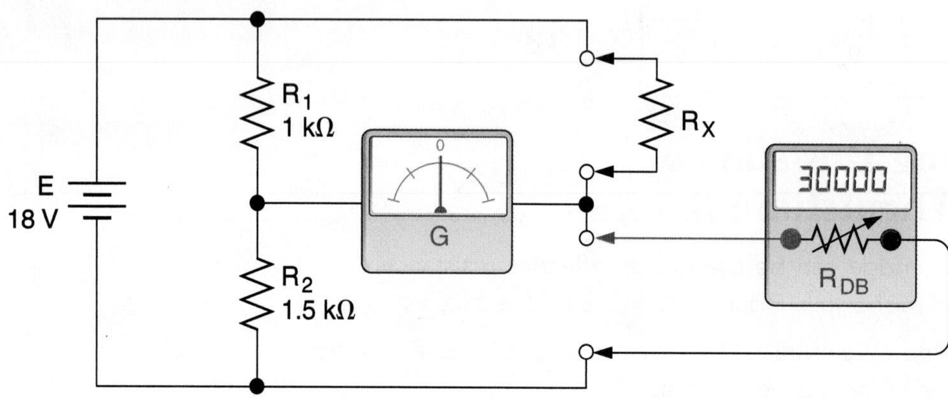

FIGURE 7.29

The bridge in Figure 7.29 is constructed and the decade box is adjusted until the bridge is balanced ($I_G = 0$ A). When the bridge is balanced, the readouts on the decade box indicate that the value of R_{DB} is 30 kΩ. Using the known resistance values, the value of R_X is found as

$$R_X = R_{DB} \times \frac{R_1}{R_2}$$

$$= 30\text{ k}\Omega \times \frac{1\text{ k}\Omega}{1.5\text{ k}\Omega}$$

$$= 20\text{ k}\Omega$$

PRACTICE PROBLEM 7.9

A circuit like the one in Figure 7.29 has values of $R_1 = 10$ kΩ and $R_2 = 20$ kΩ. The galvanometer indicates that the bridge is balanced when R_{DB} is set to 15 kΩ. What is the value of R_X?

A Circuit Modification

Calculating the value of R_X is easier when a Wheatstone bridge is constructed as shown in Figure 7.30. In this circuit, $R_1 = R_2 = 1$ kΩ. When R_1 and R_2 are

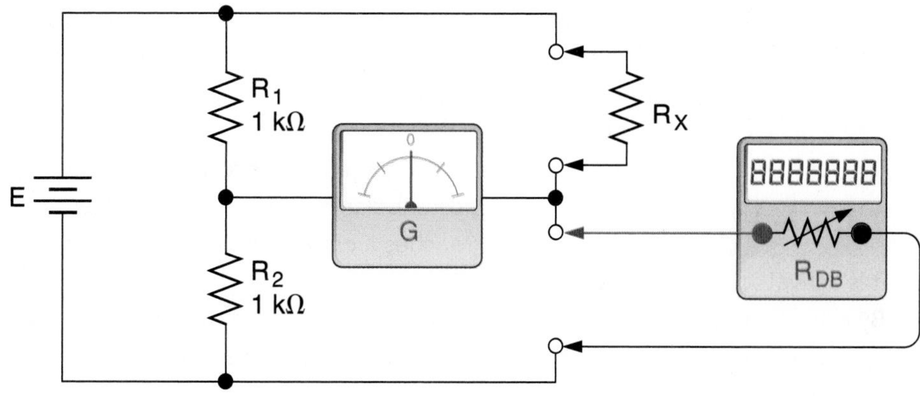

FIGURE 7.30

equal in value, the fraction in equation (7.4) equals 1, and the value of R_X can be found as

$$R_X = R_{DB} \quad \text{(when } R_1 = R_2\text{)} \tag{7.5}$$

This means that a bridge with equal values of R_1 and R_2 will balance when R_{DB} equals the value of R_X. This simplifies the overall procedure for measuring the value of an unknown resistance.

A Wheatstone Bridge Application

You have been shown how a Wheatstone bridge can be used to measure the value of an unknown resistance. A more practical application for the circuit is illustrated in Figure 7.31. Here we have a temperature alarm circuit that is made up of a bridge, a **THERMISTOR**, and an alarm trigger circuit.

THERMISTOR A component whose resistance varies with temperature.

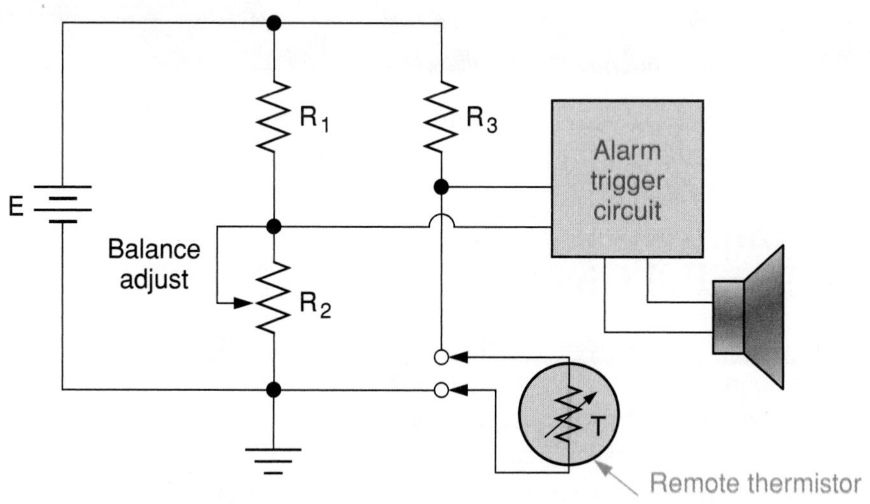

FIGURE 7.31 A temperature alarm circuit.

Thermistors are also commonly used in electrical circuits that are heat activated, such as the fire suppression systems (sprinklers) in commercial buildings.

A thermistor is a component whose resistance varies with temperature. The thermistor has a negative temperature coefficient, meaning that its resistance decreases as temperature increases. As the temperature applied to the thermistor in Figure 7.31 increases, its resistance drops and the bridge becomes unbalanced. If the voltage produced by the imbalance reaches a certain value, the alarm trigger circuit is activated, sounding the alarm.

By changing the thermistor in Figure 7.31 to another type of sensor, we could modify the circuit to respond to changes in pressure, light, or some other physical characteristic.

PROGRESS CHECK

1. What is a *galvanometer*?
2. What are the three operating states of a Wheatstone bridge?
3. Describe a *balanced* bridge.
4. Describe the operating states of a Wheatstone bridge in terms of resistance ratios.
5. Describe the procedure for measuring resistance with a Wheatstone bridge.

7.5 FAULT SYMPTOMS AND TROUBLESHOOTING

In this section, we will briefly discuss some common series-parallel circuit fault symptoms and the general approach to troubleshooting these circuits.

Fault Symptoms in Series-Parallel Circuits

When troubleshooting a series-parallel circuit, it is important to remember the symptoms of open and shorted components in both series circuits and parallel circuits. As a review, these symptoms are given in Table 7.2.

Even though series-parallel circuits are more complex, the symptoms associated with each of the faults listed in Table 7.2 remain the same. This is demonstrated in the following series of examples.

TABLE 7.2 • Symptoms of Open and Shorted Components in Series and Parallel Circuits

CIRCUIT TYPE	FAULT	SYMPTOM(S)
Series	**Open**	Circuit current drops to zero The applied voltage is dropped *across* the open component(s) The voltage across each remaining component drops to 0 V
	Short	Circuit current increases The voltage across the shorted component drops to 0 V The voltage across each remaining component increases
Parallel	**Open**	Current through the open branch drops to zero Total circuit current decreases Other branches continue to operate normally
	Short	The shorted branch causes the power supply fuse to blow The measured resistance across all branches is close to 0 Ω. (The faulty branch shorts out all parallel branches) The voltage measured across any branch is 0 V

EXAMPLE 7.10

R_1 in Figure 7.32 is open. Determine the effect of this open on the voltage and current values in the circuit.

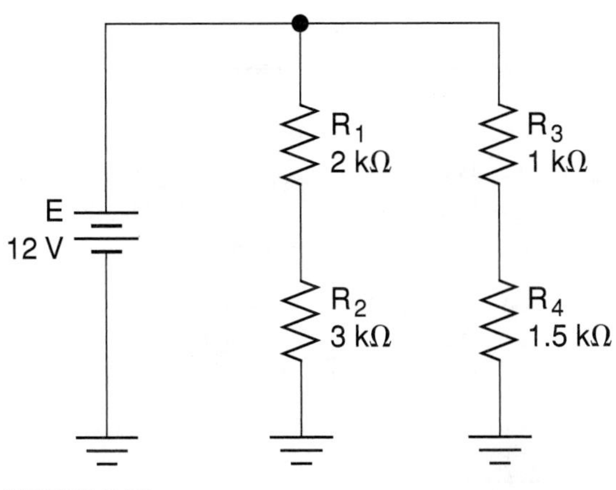

FIGURE 7.32

SOLUTION

Since R_1 is open, there is no current through that branch ($I_{R1} = I_{R2} = 0$ A). With no current through the branch, no voltage is developed across R_2. Therefore, $V_{R2} = 0$ V and $V_{R1} = E = 12$ V.

The problem in the circuit is isolated to the first branch; that is, it does not affect the operation of branch B. Therefore, the branch B values are all normal: $I_B = 4.8$ mA, $V_{R3} = 4.8$ V, and $V_{R4} = 7.2$ V.

Finally, the loss of I_A causes the total current in the circuit to decrease. The total current is now equal to the value of I_B, 4.8 mA.

PRACTICE PROBLEM 7.10

Assume that R_1 in Figure 7.32 is normal and that R_4 is open. Determine the effect that this fault has on the voltage and current values in the circuit.

The symptoms of the problem in Example 7.10 are illustrated in Figure 7.33. Compare the branch A readings to the series-open symptoms in Table 7.2. As you can see, the symptoms are exactly as described in the table:

1. Branch current has dropped to zero.
2. The applied voltage is dropped across the open component (R_1).
3. The voltage across the remaining component (R_2) is 0 V.

Compare the overall readings in the circuit to the parallel-open symptoms in Table 7.2. Again, the symptoms are exactly as described in the table:

1. Current through the open branch has dropped to zero.
2. The total circuit current has decreased.
3. The working branch has been unaffected by the open.

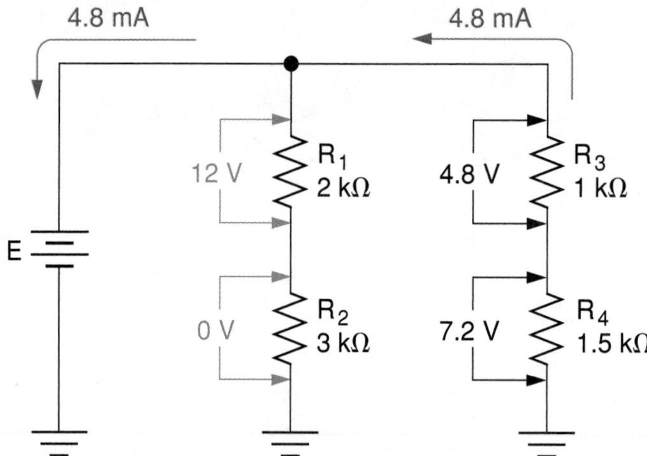

FIGURE 7.33 Calculated voltage and current values for the circuit in Example 7.10.

Even though the circuit is more complex than those covered in previous chapters, it still responds to the fault in a logical and predictable fashion. Let's go through another fault and its symptoms.

EXAMPLE 7.11

R_2 in Figure 7.34 is shorted. Determine the effect of this short on the voltage and current values in the circuit.

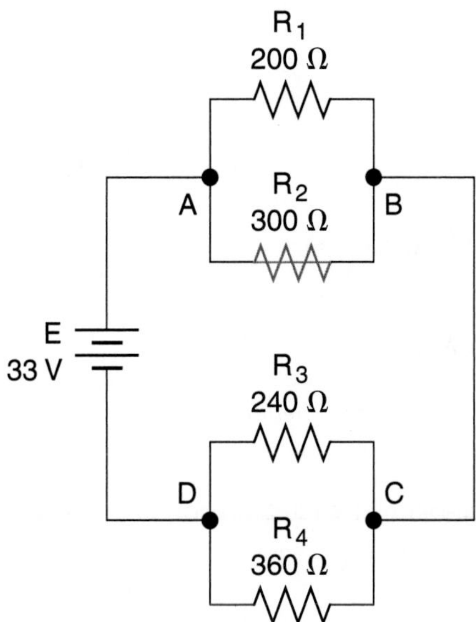

FIGURE 7.34

SOLUTION

Since R_2 is shorted, the combined resistance of R_1 and R_2 drops to approximately 0 Ω. Therefore, $V_{AB} = 0$ V and $V_{CD} = E$. The increase in V_{CD} causes the current through loop B to increase from 125 mA (normal) to 230 mA. Since the loops are in series, the total circuit current also increases to 230 mA.

PRACTICE PROBLEM 7.11

Assume that R_2 in Figure 7.34 is normal and that R_3 is shorted. Determine the effect that this fault has on the voltage and current values in the circuit.

The symptoms of the problem in Example 7.11 are illustrated in Figure 7.35. Compare the loop A readings to the parallel-short symptoms in Table 7.2. As you can see, the symptoms are identical to those described in the table:

1. The combined resistance of the loop is 0 Ω.
2. The voltage across both branches is 0 V.

Compare the overall readings in the circuit to the series-short symptoms listed in Table 7.2.

Again, the symptoms are identical to those listed in the table:

1. The total circuit current increased (from 125 mA to 230 mA).
2. The voltage across the shorted component (loop A) dropped to 0 V.
3. The voltage across the remaining component (loop B) increased.

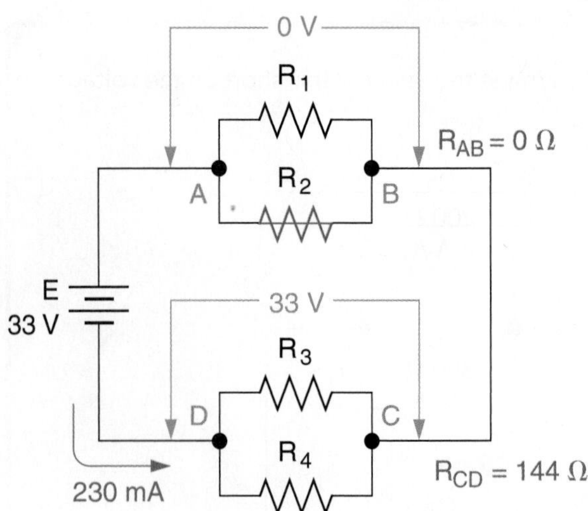

FIGURE 7.35 Calculated voltage and current values for the circuit in Example 7.11.

Once again, the series-parallel circuit responds to a fault in a logical and predictable fashion.

Series-Parallel Circuit Troubleshooting

LOAD-SPECIFIC FAULT In a circuit that has two or more loads, a fault that affects only one load.

In many practical situations, series-parallel circuit troubleshooting begins with determining whether or not a given fault is a **LOAD-SPECIFIC FAULT**; that is, whether or not the fault is affecting more than one load. Determining whether or not a fault is load-specific can lead you directly to the most likely cause(s) of the problem. For example, consider the circuit shown in Figure 7.36. Let's assume that Load 1 isn't working and the other two are. In this case, the problem must lie with the load itself and/or R_4. Why? Because these two components are the only ones that are unique to Load 1. In other words, they are the only ones that could cause a problem in the load 1 circuit without affecting the other two loads. Using the same logic:

1. A malfunction that is unique to Load 2 could only be caused by the load itself and/or R_5.
2. A malfunction that is unique to Load 3 could only be caused by the load itself and/or R_6.

As you can see, load-specific faults can be relatively easy to isolate.

Faults that are common to more than one load can also be isolated when you take the time to identify the components that are common to those loads. Usually one (or more) of these components is the source of the problem. For example, let's assume that Load 1 and Load 2 in Figure 7.36 are not working. The most likely cause of the problem would be R_1. Why? Because this component is part of the voltage divider that is common to both of these loads. Since the voltage source and the fuse also affect the operation of Load 3 (which is working), they are not likely to be the cause of the problem. Therefore, we would start by testing R_1. If all three loads were inoperative, we would assume that the voltage source or the fuse was the cause of the problem.

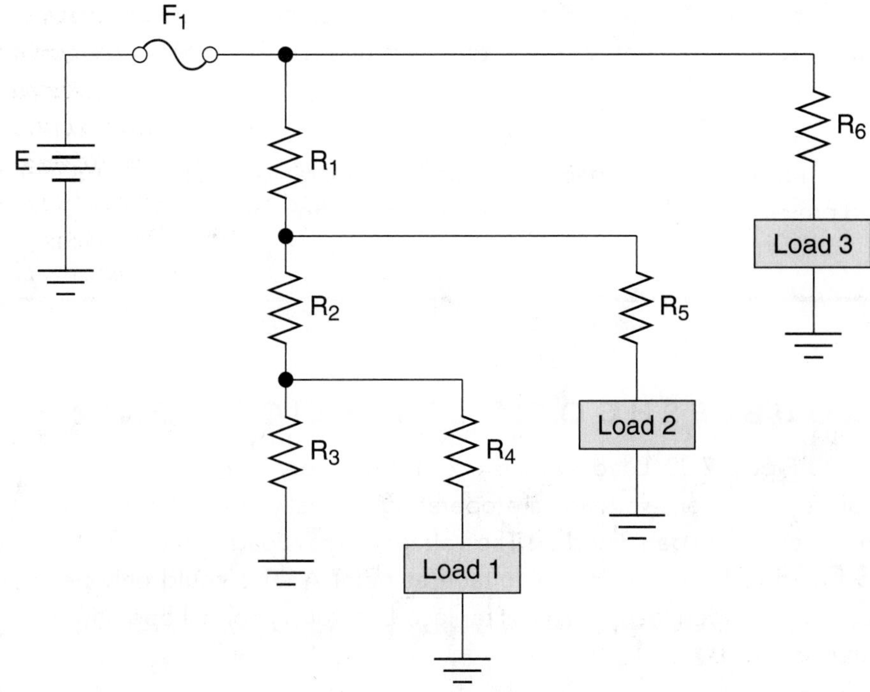

FIGURE 7.36 A loaded voltage divider.

As you can see, series-parallel circuit troubleshooting starts with observing symptoms and isolating the possible cause(s) of those symptoms. This process is demonstrated further in the following troubleshooting applications.

TROUBLESHOOTING APPLICATION 1

The circuit in Figure 7.37 is designed to provide the operating voltages required by three DC motors. However, motors 2 and 3 are not working. Since B is the common point for these two loads, the voltage at this point is measured and found to be 0 V. Since Motor 1 is working (indicating that there is a

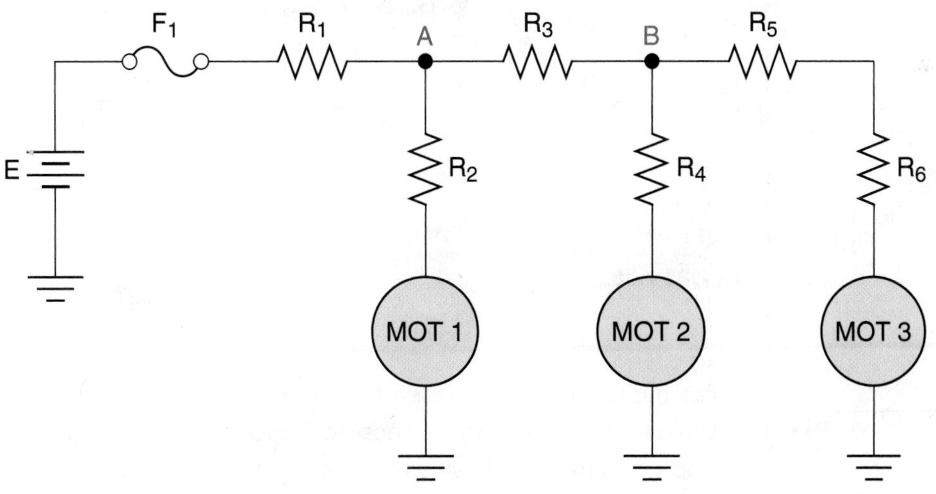

FIGURE 7.37

voltage at point A), R_3 is assumed to be open. Checking the resistance of this component verifies that it is open. Replacing the component restores normal operation to the circuit.

AN OBSERVATION:

For this circuit, motors 2 and 3 form the only *motor pair* that could be inoperative. In other words, there is no single fault that could prevent motors 1 and 3 (or motors 1 and 2) from operating without affecting the third motor.

TROUBLESHOOTING APPLICATION 2

Refer to Figure 7.38. Load 3 in this circuit is not operating and L_3 is not lit. Since the other two loads are operating normally, the only suspect components are Load 3 and L_3. The voltage across Load 3 is measured and found to be equal to the voltage at point A. This could only be caused by an open load. Load 3 is replaced, restoring normal operation to the circuit.

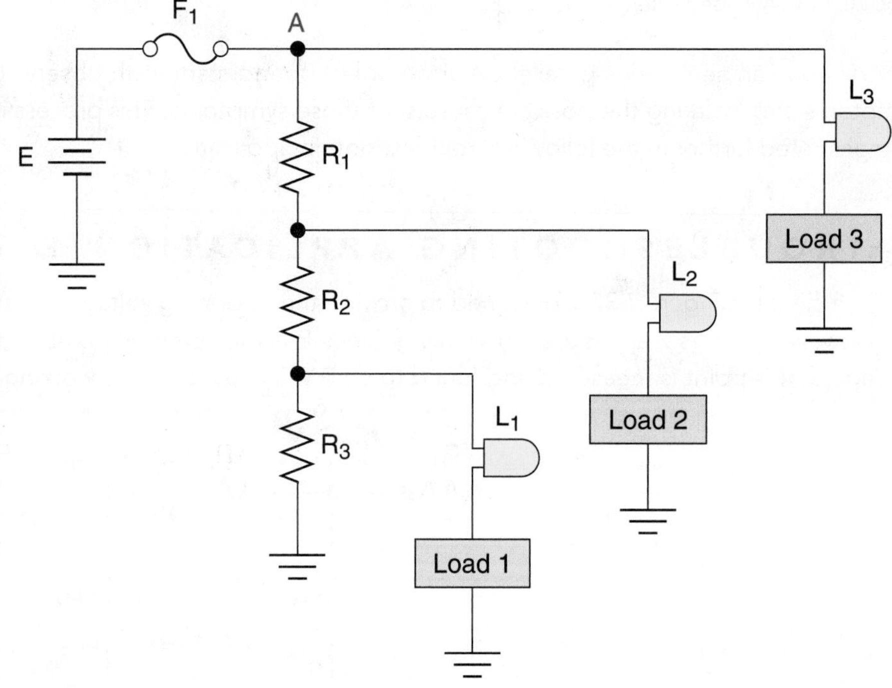

FIGURE 7.38

AN OBSERVATION:

Had the problem been caused by an open L_3, the voltage across the load would have been 0 V. This is because the voltage at point A would have been dropped across the open lamp, as is the case in any series connection.

TROUBLESHOOTING APPLICATION 3

The lamps and loads in Figure 7.38 are all inoperative. In this case, the voltage source and the fuse are the only components that are common to all of the inoperative loads. The fuse is assumed to be faulty and is replaced. Replacing the fuse restores normal operation to the circuit.

AN OBSERVATION:

In this case, it was not necessary to test the circuit voltages because only the source and the fuse could have caused all the loads to go inoperative at the same time. Had the loads remained inoperative after replacing the fuse, the next step would have been to test the circuit voltage source.

As you can see, practical troubleshooting begins before taking any measurements. While there are many exceptions to this, you can generally save yourself a great deal of time and trouble by observing the operation of a circuit and thinking your way through the problem before actually testing the circuit.

PROGRESS CHECK

1. What should be kept in mind when troubleshooting series-parallel circuits?
2. Refer back to Figure 7.7. What symptoms would you expect to see if R_1 opens?
3. Refer back to Figure 7.7. What symptoms would you expect to see if R_1 shorts?
4. What is the first step in troubleshooting a practical series-parallel circuit? Why is this step important?

7.6 SUMMARY

Here is a summary of the major points that were covered in this chapter:

Introduction to Series-Parallel Circuits

- Circuits that contain both series and parallel elements are referred to as series-parallel circuits or combination circuits.
- When series circuits are connected in parallel:
  - Each branch retains all the characteristics of any series circuit.
  - The combined branches retain all the characteristics of any parallel circuit.
- When parallel circuits are connected in series:
  - Each loop is a parallel circuit that retains all the characteristics of any parallel circuit.
  - The combined loops retain all the characteristics of any series circuit.

Analyzing Series-Parallel Circuits

- Any series-parallel circuit can be dealt with effectively when you:
  - Use the proper approach
  - Remember the basic principles of series circuits and parallel circuits

- In most cases, the goal of any series-parallel circuit analysis is to determine the value of voltage, current, and/or power for one or more components.
- The process of calculating series-parallel circuit values usually begins with deriving a simplified equivalent of the original circuit. The simplification process ends when the circuit has been reduced to either a series-equivalent or a parallel-equivalent circuit.

Circuit Loading

- Voltage dividers are usually designed for the purpose of providing one or more specific load voltages. When used for this purpose, the voltage divider is referred to as a loaded voltage divider.
- The no-load output voltage (V_{NL}) from a voltage divider is the voltage measured at the load terminals with the load removed.
  - The term *output voltage* is commonly used to describe the voltage at the load terminals of a circuit.
  - The value of V_{NL} is always greater than the value of load voltage (V_L) for the circuit.
- For a loaded voltage divider, load voltage varies directly with load resistance.
- A voltmeter may provide a reading that is significantly lower than the actual component voltage.
  - When a voltmeter is connected across a component in a series circuit, the resistor and meter form a parallel circuit lowering the equivalent resistance.
  - The effect of voltmeter loading depends on the meter's internal resistance. The lower the value of R_M, the greater the error in the voltage reading.

The Wheatstone Bridge

- The Wheatstone bridge is a circuit that can be used to provide extremely accurate resistance measurements.
  - The Wheatstone bridge is shown in Figure 7.25. The meter used is a galvanometer (a current meter that indicates both the magnitude and the direction of a low-value current).
- When the resistance ratios in a Wheatstone bridge are equal, the bridge is said to be balanced.
  - Balance in a Wheatstone bridge is indicated by a galvanometer reading of $I_G = 0$ A.
- A Wheatstone bridge can be used as a resistance measuring circuit when modified as shown in Figure 7.28. The value of an unknown resistance is measured as follows:
  - The unknown resistance is placed in the position labeled R_X.
  - The decade box is adjusted until the bridge is balanced.
  - The reading on the decade box is used (with the other known resistance values in the circuit) to calculate the unknown value.

Fault Symptoms and Troubleshooting

- When troubleshooting a series-parallel circuit, it is important to remember the symptoms of open and shorted components in both series and parallel circuits.
  - These symptoms are provided in Table 7.2.
  - Even though series-parallel circuits are more complex, the symptoms associated with each type of fault (listed in Table 7.2) remain the same.

- In many practical situations, series-parallel circuit troubleshooting begins with determining whether or not a given fault is *load-specific.*
  - Determining whether or not a fault is load-specific can lead you to the cause(s) of the problem.
- Circuit troubleshooting starts with observing symptoms and isolating the possible cause(s) of those symptoms. (Determining whether or not the problem is load-specific is part of this process.)

CHAPTER REVIEW

1. A series-parallel circuit is also referred to as a ________.
   a) complex circuit
   b) combination circuit
   c) bridge circuit
   d) all of the above

2. Two series circuits are connected in parallel. Branch A has a higher resistance than branch B. This means that the voltage across branch A is ________ branch B.
   a) higher than
   b) lower than
   c) the same as
   d) less than or equal to

3. When two series circuits are connected in parallel, the current through the branches ________.
   a) is always the same
   b) is always different
   c) depends on the resistance in the path
   d) always flows to ground

4. A series circuit having three resistors is in parallel with another series circuit with two resistors. All the resistors are 1 kΩ. The total resistance of the circuit is ________.
   a) 1.2 kΩ
   b) 200 Ω
   c) 120 Ω
   d) 2.0 kΩ

5. When two parallel circuits are connected in series ________.
   a) the voltage across both parallel circuits is equal
   b) the branch currents are all equal
   c) the sum of the currents through the branches of each parallel circuit equals the total circuit current
   d) the voltage across all branches is equal

6. Two three-branch parallel circuits are connected in series. Each branch contains a single 470 Ω resistor. The applied voltage is 240 V. The voltage across each parallel branch is ________.
   a) 240 V
   b) 80 V
   c) 40 V
   d) 120 V
7. The current through each branch in the circuit described in question 6 equals ________.
   a) 1.53 A
   b) 3.06 A
   c) 255 mA
   d) 766 mA
8. One of the resistors in the circuit described in question 6 has opened. The total resistance in this circuit is now________.
   a) 392 Ω
   b) infinite
   c) 157 Ω
   d) 235 Ω
9. The current through each branch of the parallel circuit that has the open, described in question 8, must be ________.
   a) 306 mA
   b) zero
   c) 128 mA
   d) cannot be solved with the information given
10. Four parallel circuits are connected in series. Each parallel circuit has five branches with two resistors in each branch. All the resistive components are identical. The applied voltage is 305 V and the total circuit current is 450 mA. The power dissipated by each parallel circuit equals ________.
    a) 137 W
    b) 27.5 W
    c) 34.3 W
    d) it depends on the value of the resistors
11. The power dissipated by each individual resistor in the circuit described in question 10 equals ________.
    a) 3.43 W
    b) 6.86 W
    c) 1.72 W
    d) it depends on the value of the resistors
12. A component in a series-parallel circuit opens. The total circuit resistance increases. This means that this circuit consists of ________.
    a) series circuits connected in parallel
    b) parallel circuits connected in series

c) either (a) or (b)

d) cannot be determined with the information given

13. A series-parallel circuit is comprised of either two two-branch parallel circuits connected in series (one resistor per branch), or two two-resistor series circuits connected in parallel. All resistive components have equal values. One of the components shorts out and the total circuit resistance drops by one-half. This means that the this circuit is comprised of ________.

    a) two two-branch parallel circuits connected in series
    b) two two-resistor series circuits connected in parallel
    c) either (a) or (b)
    d) cannot be determined with the information given

14. As a result of the short described in question 13, the power dissipated by the total circuit has ________.

    a) halved
    b) doubled
    c) increased by a factor of 4
    d) dropped to zero

15. During circuit analysis, combining groups of components to produce a simplified circuit produces a(n) ________ circuit.

    a) derived
    b) Kirchhoff
    c) derivative
    d) equivalent

16. Most circuits are designed to ________.

    a) provide power to a load
    b) dissipate power
    c) generate power
    d) increase power

17. A loaded voltage divider ________.

    a) is a low-current circuit
    b) is designed to produce a specific voltage for a load
    c) is a high-current circuit
    d) is a high-power circuit

18. V_{NL} is ________ than V_L.

    a) usually higher
    b) always higher
    c) usually lower
    d) always lower

19. Two 100 Ω resistors are connected to form a voltage divider. A 100 Ω load is connected. If the applied voltage is 120 V, the load voltage is ________.
    a) 60 V
    b) 40 V
    c) 50 V
    d) 30 V

20. If load resistance increases, then the loaded output voltage ________.
    a) increases
    b) decreases
    c) remains the same
    d) remains the same, but load current decreases

21. An ideal voltmeter would have ________.
    a) infinite current capacity
    b) zero resistance
    c) infinite resistance
    d) both (a) and (b)

22. Voltmeter loading always results in the ________.
    a) voltage measurement being higher than it should be
    b) voltage measurement being lower than it should be
    c) total circuit current being slightly higher than it should be
    d) total circuit current being slightly lower than it should be
    e) both (b) and (c)

23. The meter that is used as the "bridge" of the Wheatstone bridge is called a ________.
    a) wattmeter
    b) gaussmeter
    c) bridge meter
    d) galvanometer

24. The bridge in the Wheatstone bridge is ________ with the rest of the circuit.
    a) in series
    b) in parallel
    c) in series-parallel
    d) neither in series with, nor in parallel

25. When a bridge is "balanced" ________.
    a) the voltage across the bridge component is zero
    b) the current through the bridge component is zero
    c) both (a) and (b)
    d) neither (a) nor (b)

26. A bridge employing a decade box is used as a precision ohmmeter. The *only* condition where the value of the decade box is equal to the value of the unknown resistor is when ________.

    a) $R_1 > R_2$

    b) $R_1 < R_2$

    c) $R_1 = R_2$

    d) $R_1 \leqslant R_2$

    e) $R_1 > R_2$

PRACTICE PROBLEMS

1. Calculate the voltage, current, and power values for the circuit in Figure 7.39.

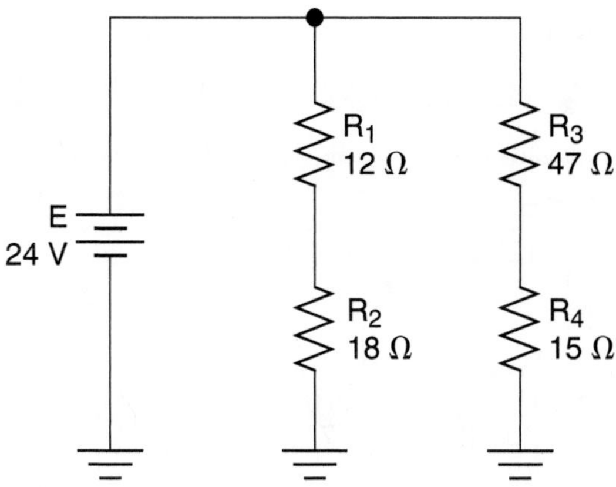

FIGURE 7.39

2. Calculate the voltage, current, and power values for the circuit in Figure 7.40.

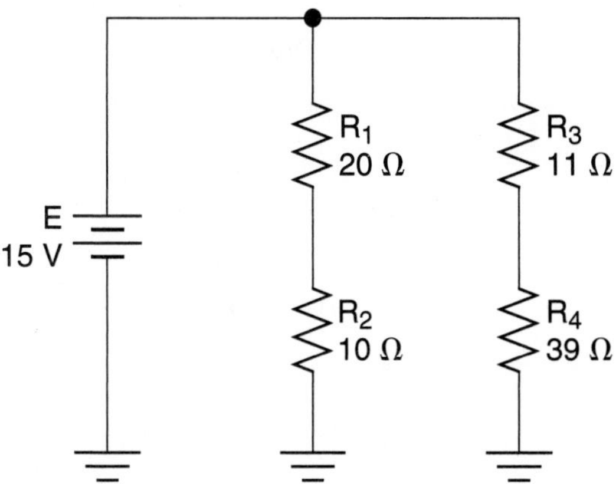

FIGURE 7.40

3. Calculate the voltage, current, and power values for the circuit in Figure 7.41.
4. Calculate the voltage, current, and power values for the circuit in Figure 7.42.
5. Calculate the voltage, current, and power values for the circuit in Figure 7.43.

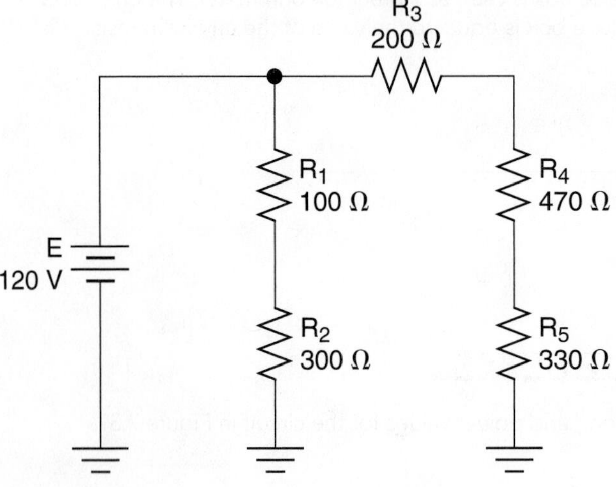

FIGURE 7.41

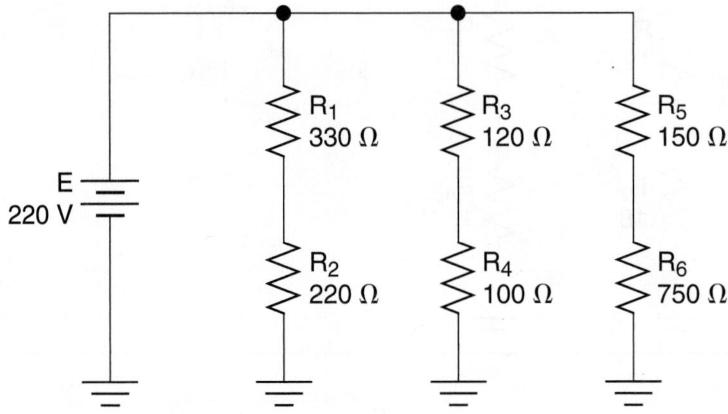

FIGURE 7.42

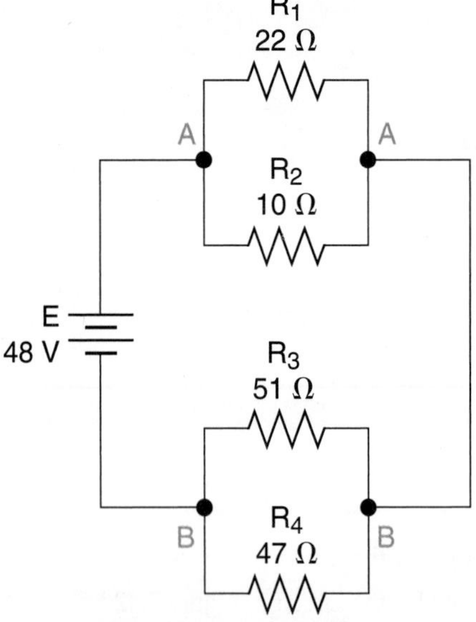

FIGURE 7.43

6. Calculate the voltage, current, and power values for the circuit in Figure 7.44.

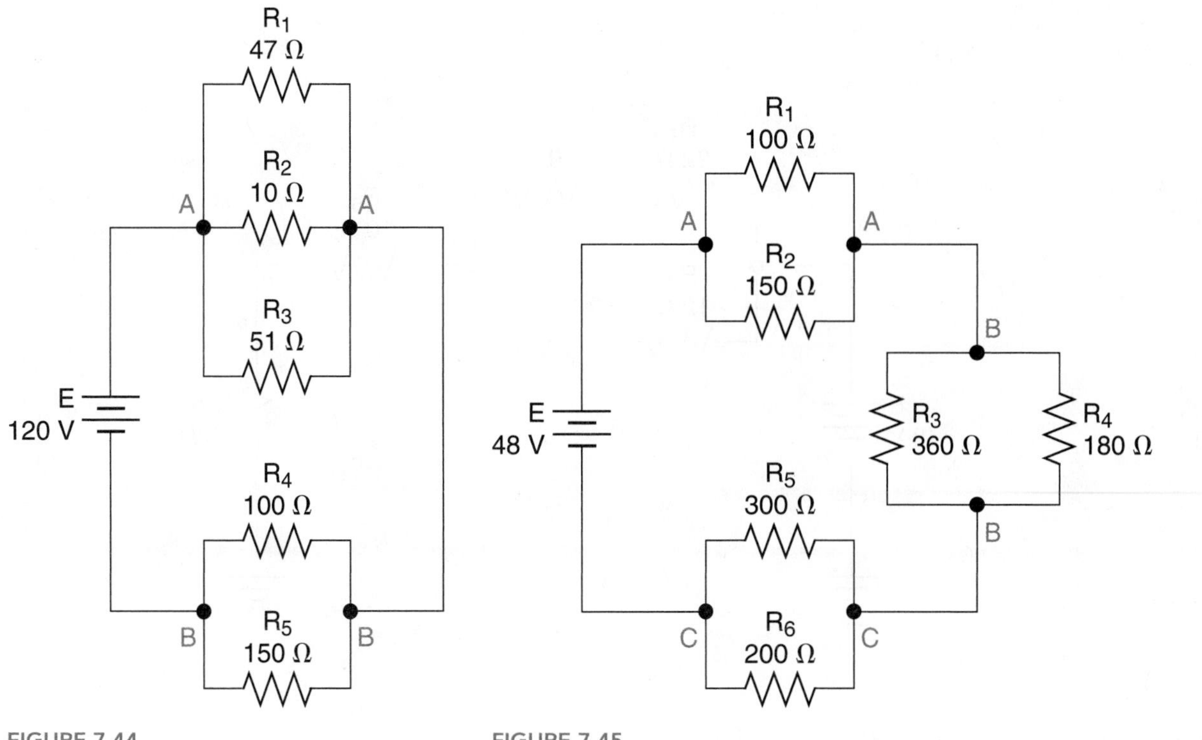

FIGURE 7.44

FIGURE 7.45

7. Calculate the voltage, current, and power values for the circuit in Figure 7.45.
8. Calculate the voltage, current, and power values for the circuit in Figure 7.46.

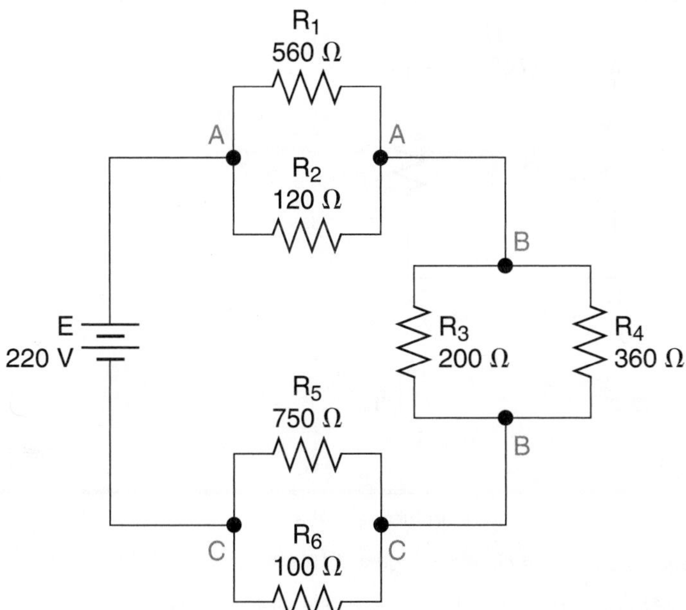

FIGURE 7.46

9. Derive the series-equivalent of the circuit in Figure 7.47a.
10. Derive the series-equivalent of the circuit in Figure 7.47b.

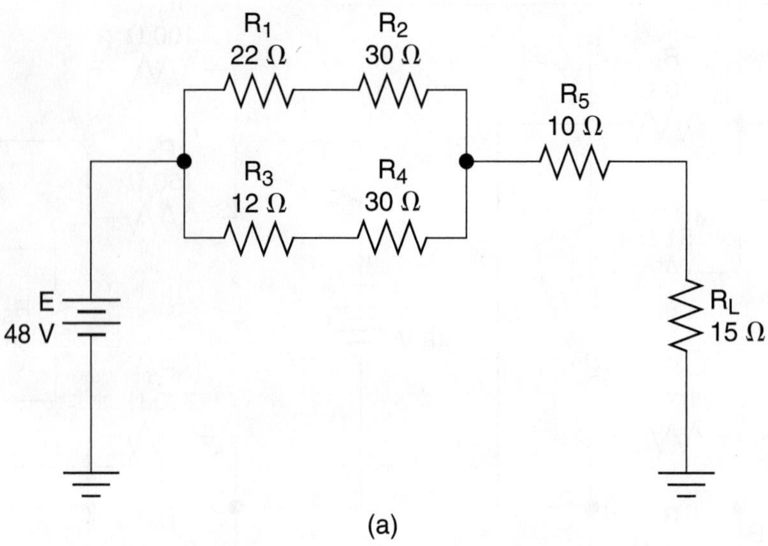

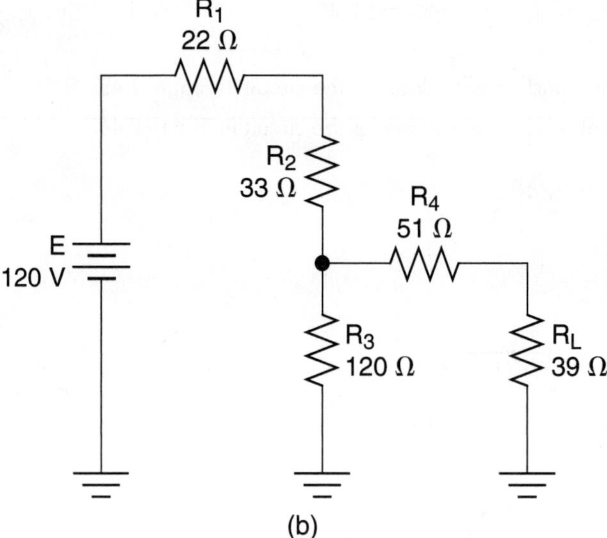

FIGURE 7.47

11. Derive the parallel-equivalent of the circuit in Figure 7.48a.
12. Derive the parallel-equivalent of the circuit in Figure 7.48b.
13. Determine the value of load voltage (V_{RL}) for the circuit in Figure 7.49a.
14. Determine the value of load voltage (V_{RL}) for the circuit in Figure 7.49b.

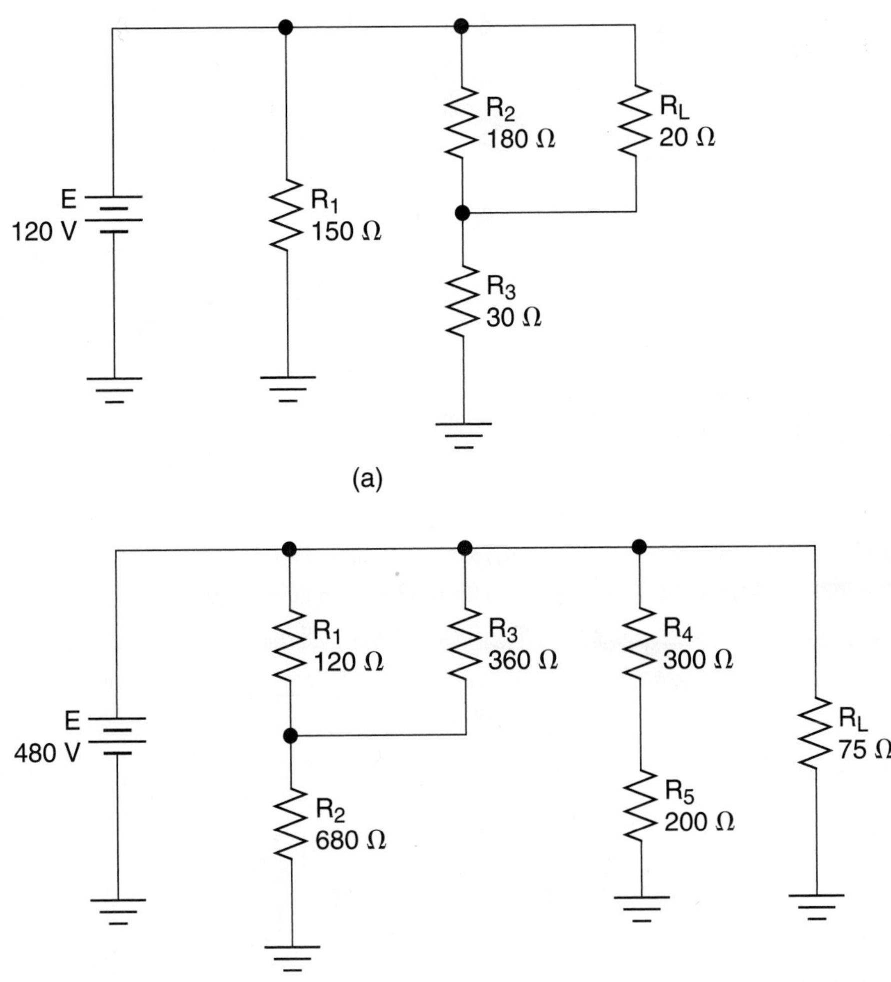

FIGURE 7.48

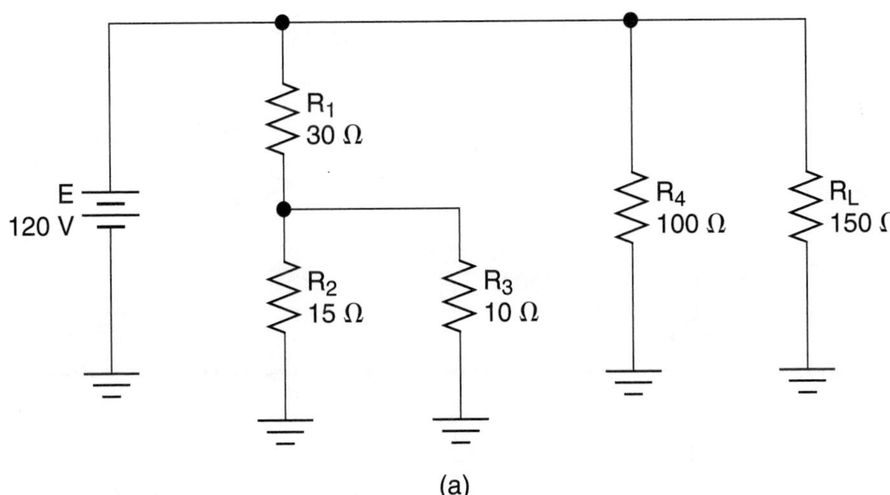

FIGURE 7.49

R1
120 Ω
R3
100 Ω
RL
150 Ω
E
120 V
R2
200 Ω

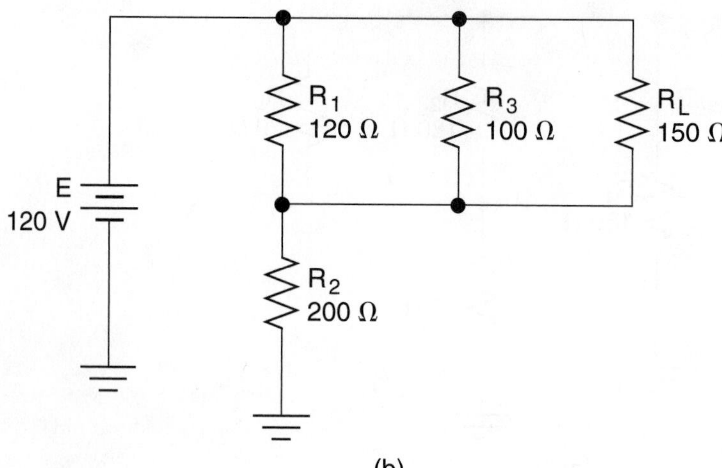

FIGURE 7.49 *(Continued)* (b)

15. Determine the value of load current (I_{RL}) for the circuit in Figure 7.50a.
16. Determine the value of load current (I_{RL}) for the circuit in Figure 7.50b.
17. Determine the value of load power (P_{RL}) for the circuit in Figure 7.51a.

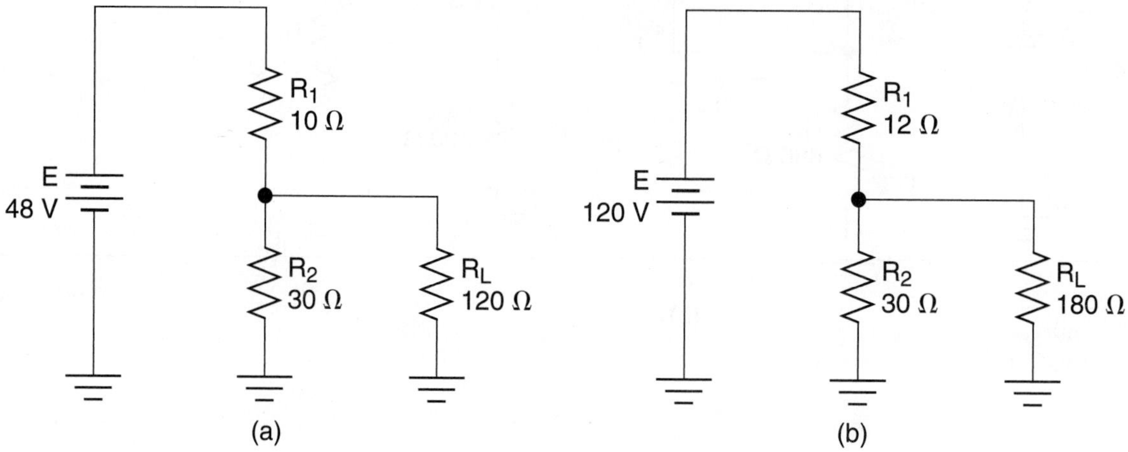

R1
10 Ω
E
48 V
R2
30 Ω
RL
120 Ω
(a)
R1
12 Ω
E
120 V
R2
30 Ω
RL
180 Ω
(b)

FIGURE 7.50

R1
12 Ω
E
120 V
R2
36 Ω
RL
360 Ω
(a)
R1
120 Ω
E
200 V
R2
36 Ω
RL
750 Ω
(b)

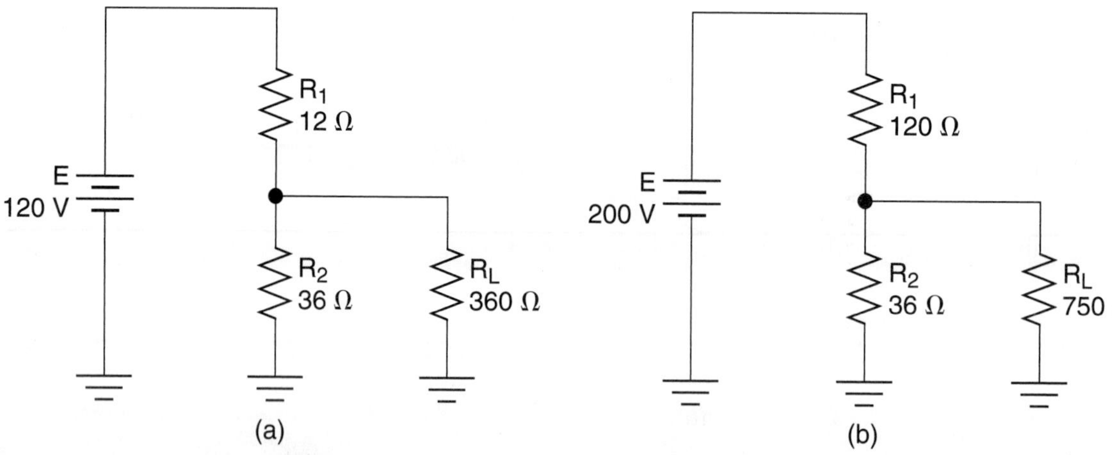

FIGURE 7.51

18. Determine the value of load power (P_{RL}) for the circuit in Figure 7.51b.
19. Determine the values of V_L and V_{NL} for the circuit in Figure 7.52a.
20. Determine the values of V_L and V_{NL} for the circuit in Figure 7.52b.

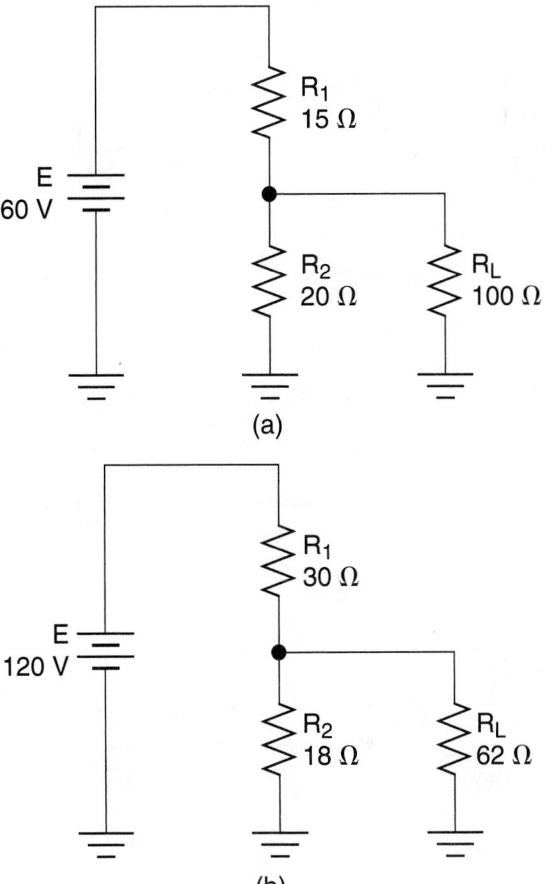

FIGURE 7.52

21. Determine the value of V_{R2} for the circuit in Figure 7.53a. Then, determine the reading that will be produced when the voltmeter is connected as shown in Figure 7.53b.

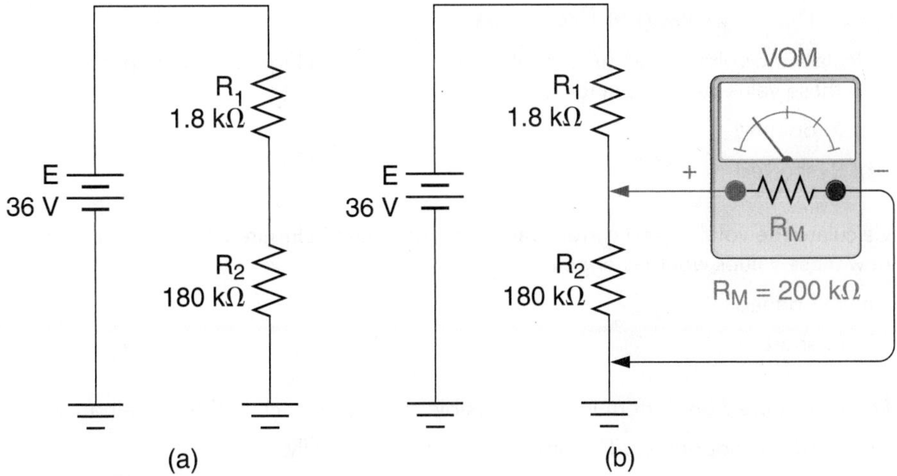

FIGURE 7.53

22. Repeat problem 21 for the circuit shown in Figure 7.54.
23. Determine the value of R_X for the circuit in Figure 7.55.

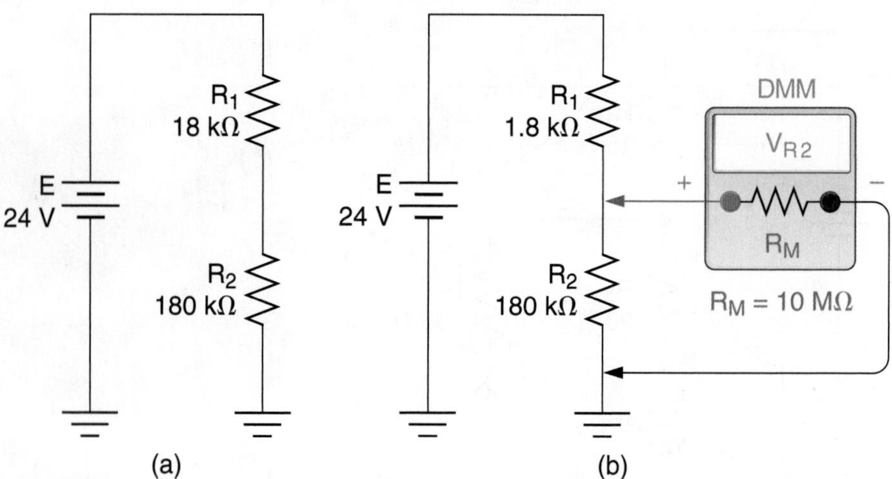

FIGURE 7.54

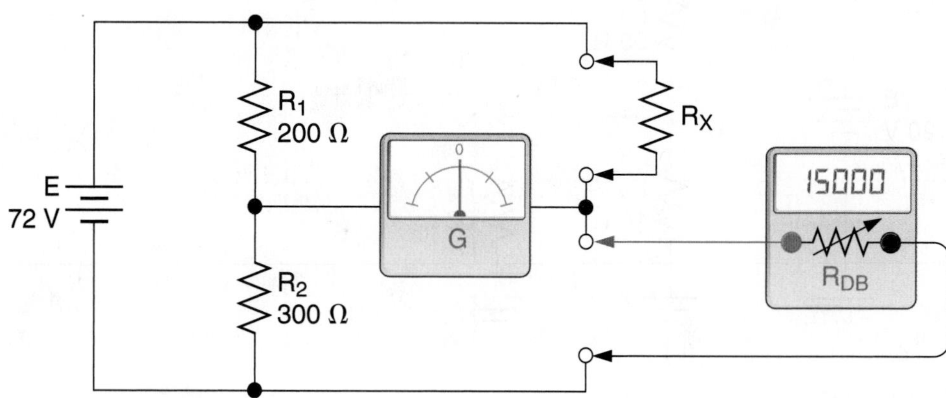

FIGURE 7.55

Troubleshooting Practice Problems

24. Calculate the voltage and current values for the circuit in Figure 7.56. Then, predict how these values would change if:
    a) R_1 opened
    b) R_3 shorted

25. Calculate the voltage and current values for the circuit in Figure 7.57. Then, predict how these values would change if
    a) R_2 opened
    b) R_4 shorted

26. Refer to Figure 7.58. Determine the probable cause(s) of each of the following:
    a) Load 1 is inoperative. The others are working normally.
    b) Loads 2 and 3 are inoperative. Load 1 is working normally.

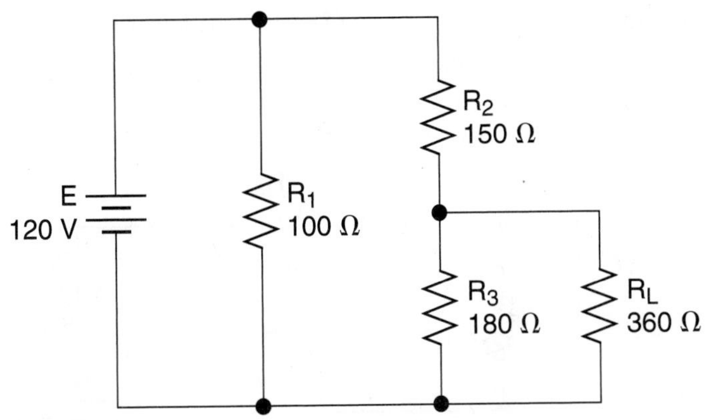

FIGURE 7.56

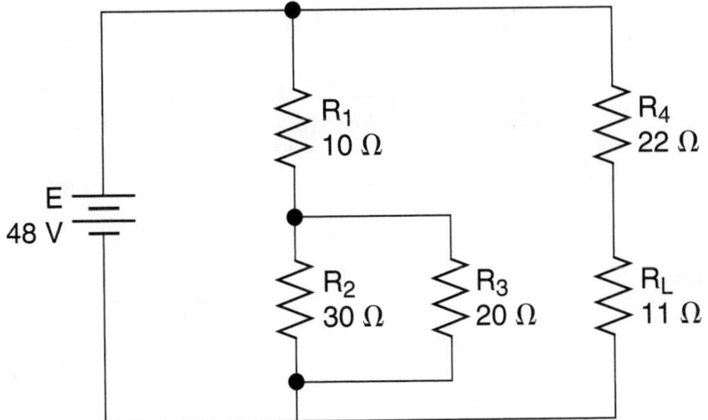

FIGURE 7.57

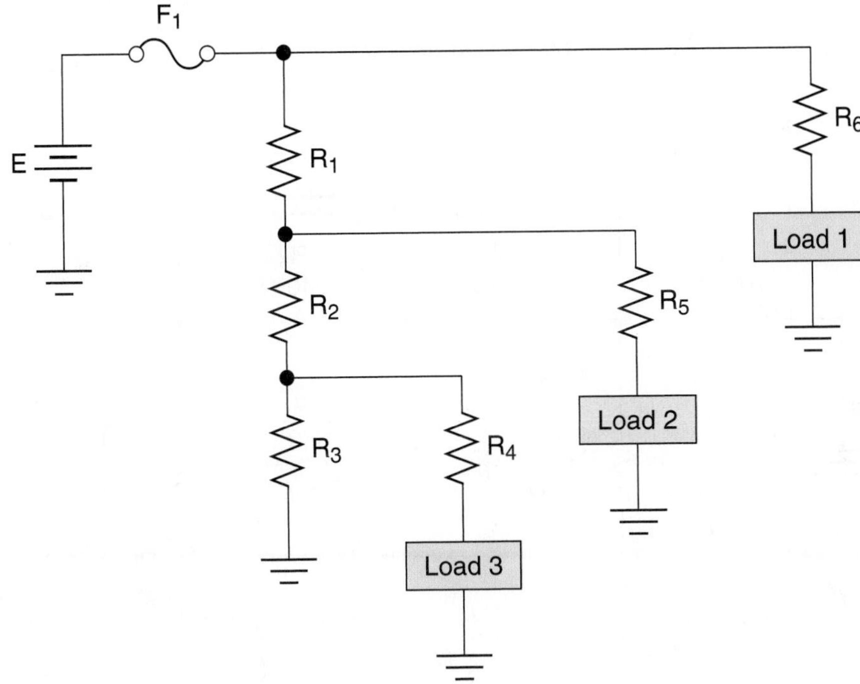

FIGURE 7.58

27. Refer to Figure 7.59. Determine the probable cause(s) of each of the following:
    a) Load 2 is inoperative and L_2 is not lit. The others are working normally.
    b) None of the loads are working. None of the lamps are lit.

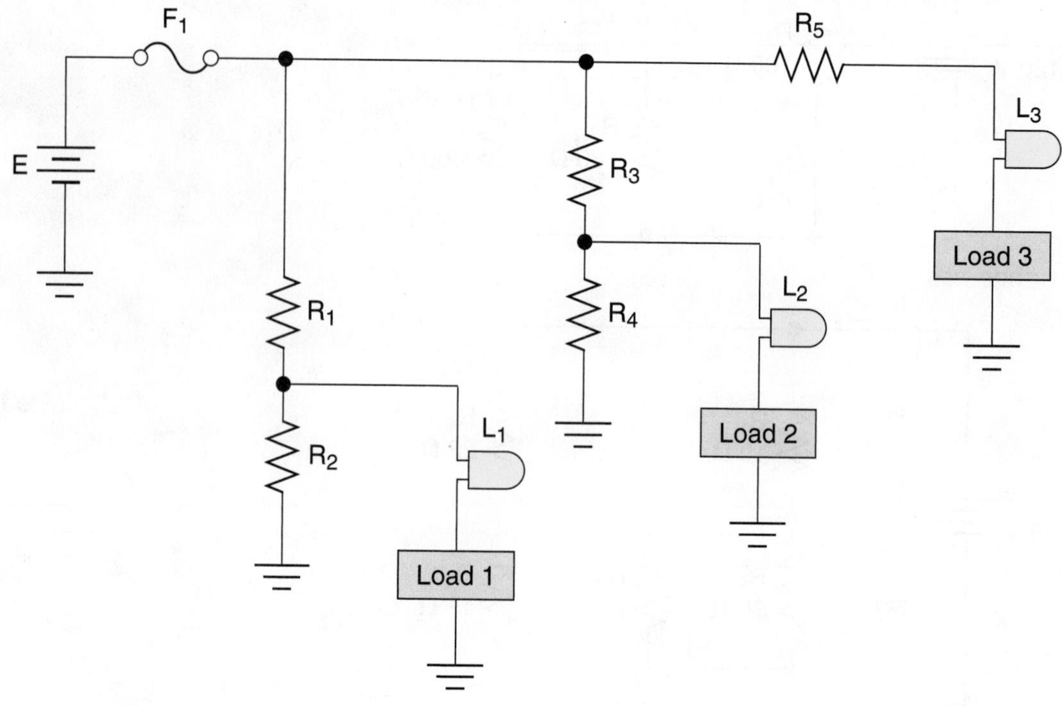

FIGURE 7.59

Challenger

28. Determine the value of V_{R6} for the circuit in Figure 7.60.

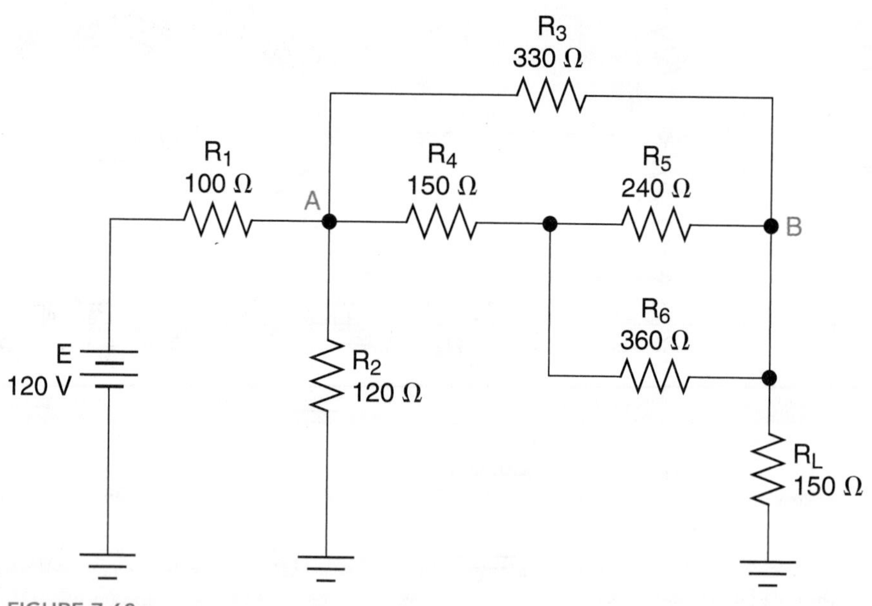

FIGURE 7.60

ANSWERS TO THE EXAMPLE PRACTICE PROBLEMS

7.1 R_A = 55 Ω, I_A = 2.18 A, V_{R1} = 47.96 V, V_{R2} = 71.94 V, P_{R1} = 105 W, P_{R2} = 157 W, R_B = 22 Ω, I_B = 5.45 A, V_{R3} = 65.4 V, V_{R4} = 54.5 V, P_{R3} = 356 W, P_{R4} = 297 W, P_B = 655 W, I_T = 7.63 A, R_T = 15.7 Ω, P_T = 916 W

7.2 R_A = 6 Ω, R_B = 12 Ω, V_A = 8 V, V_B = 16 V, I_{R1} = 800 mA, I_{R2} = 533 mA, I_{R3} = 889 mA, I_{R4} = 444 mA, P_{R1} = 6.4 W, P_{R2} = 4.27 W, P_{R3} = 14.2 W, P_{R4} = 7.1 W, I_T = 1.33 A, P_T = 31.9 W.

7.3 13 V

7.4 The circuit contains the voltage source (E) in series with a 19 Ω equivalent resistance (for R_1 through R_4) and the load resistance (10 Ω).

7.5 The circuit contains 3 branches: Branch 1 contains a 48 Ω resistor (for $R_1 + R_2$); branch 2 contains a 29.5 Ω resistor (for $R_3 \parallel R_4 + R_2$); and branch 3 contains R_6 (75 Ω).

7.6 6.38 W

7.7 V_L = 60 V, V_{NL} = 72 V

7.8 15.9 V

7.9 7.5 kΩ

7.10 The values in the R_1–R_2 branch are normal: I = 2.4 mA, V_{R1} = 4.8 V, and V_{R2} = 7.2 V. The R_3–R_4 branch is open, so I_B = 0 A, V_{R3} = 0 V, and V_{R4} = 12 V.

7.11 The total resistance decreases to 120 Ω, which allows the total current to increase to 275 mA. The voltage across the R_3-R_4 loop drops to 0 V, while the voltage across the R_1–R_2 loop increases to 33 V. I_{R1} now equals 165 mA and I_{R2} now equals 110 mA.

CHAPTER 8

INTRODUCTION TO RESIDENTIAL CIRCUITS: A PRACTICAL SERIES-PARALLEL APPLICATION

PURPOSE

The focus of this chapter is twofold. The first is to introduce some of the basic concepts and terminology associated with residential wiring. The second is to demonstrate a real-world series-parallel circuit that many of you will encounter throughout your professional careers. As you may already know, the voltages that are supplied by the utility company to a residence are AC (alternating current) voltages, not DC voltages. Although we will not cover AC circuits until Chapter 13, this should not affect our discussion. The basic rules of series, parallel, and series-parallel circuits apply to both DC and AC circuits.

It should be noted that this chapter is intended only to provide an introduction to residential circuits. It is NOT intended to provide a guide for installing or repairing any residential circuits or circuit elements. You should work on residential electrical circuits only after being fully trained by a qualified electrical instructor and participating in an apprenticeship program.

KEY TERMS

The following terms are introduced and defined in this chapter on the pages indicated:

OBJECTIVES

After completing this chapter, you should be able to:

1. Identify the elements that make up a residential service entrance.
2. Contrast the two types of service drops.
3. Describe a service lateral.
4. Calculate energy in *kilowatt hours* (kWh).
5. State the function of the main disconnect switch in a service entrance.
6. Describe the physical layout of a typical residential service panel.
7. Discuss the need for grounding and common grounding techniques.
8. State the functions of switches and receptacles.
9. Describe the construction and operation of a single-pole switch.
10. Describe the construction and operation of three-way switches.
11. Describe the construction and operation of four-way switches.
12. Describe the receptacles that are commonly found in residential AC circuits.
13. Identify the two types of residential AC circuits and list the applications for each.
14. Identify the number and size of the conductors in a nonmetallic cable using the label on its jacket.

8.1 THE SERVICE ENTRANCE AND SERVICE DROP

Power enters a residence through the **SERVICE ENTRANCE**. The service entrance includes the **ELECTRIC METER** that measures the amount of energy delivered to the home, and the **SERVICE PANEL** that houses the circuit breakers or fuses. The service panel also distributes power to the various circuits in the house.

SERVICE ENTRANCE The place where power from the utility company enters a residence.

ELECTRIC METER A meter that measures the amount of energy that is provided to a residence.

SERVICE PANEL The box that houses circuit breakers and/or fuses and distributes power to the various circuits in a residence.

SERVICE DROP An overhead power connection from the utility lines to the service entrance.

The Service Drop

An overhead power connection from the utility lines to the service entrance is called the **SERVICE DROP**. The service drop illustrated in Figure 8.1 has two 120 V lines and a neutral conductor. The three lines may be independent conductors or housed in a three-conductor cable called a *triplex cable* (as show in the figure).

NOTE: The representation of the utility pole in Figure 8.1 is greatly simplified. In practice, the service drop would originate at a transformer or a set of three conductors.

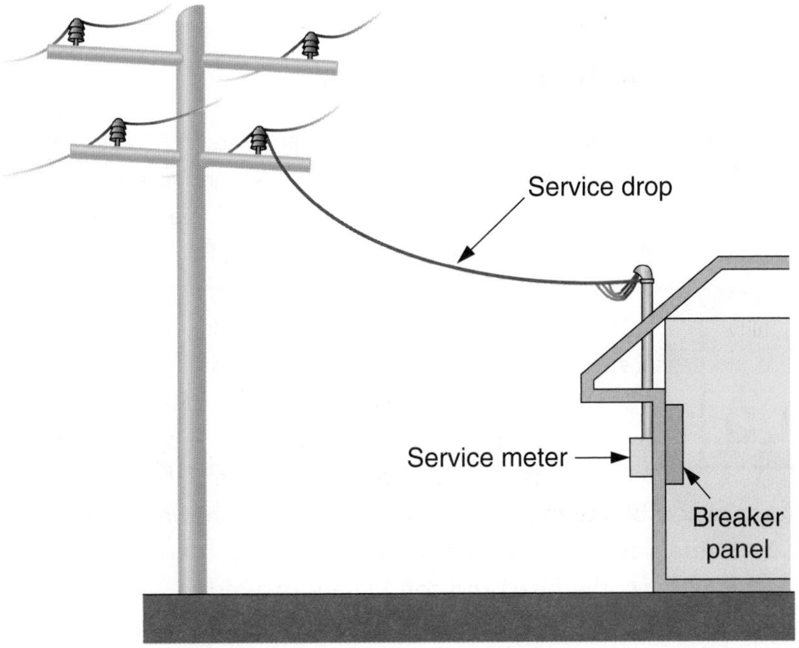

FIGURE 8.1 A service drop.

The three conductors comprising the service drop consist of two hot lines and one neutral. Each hot line has a potential of 120 V to the neutral line. Between the two hot lines there is a difference of potential of 240 V. This is why the service is identified as a 120/240 V service. This concept is illustrated in Figure 8.2.

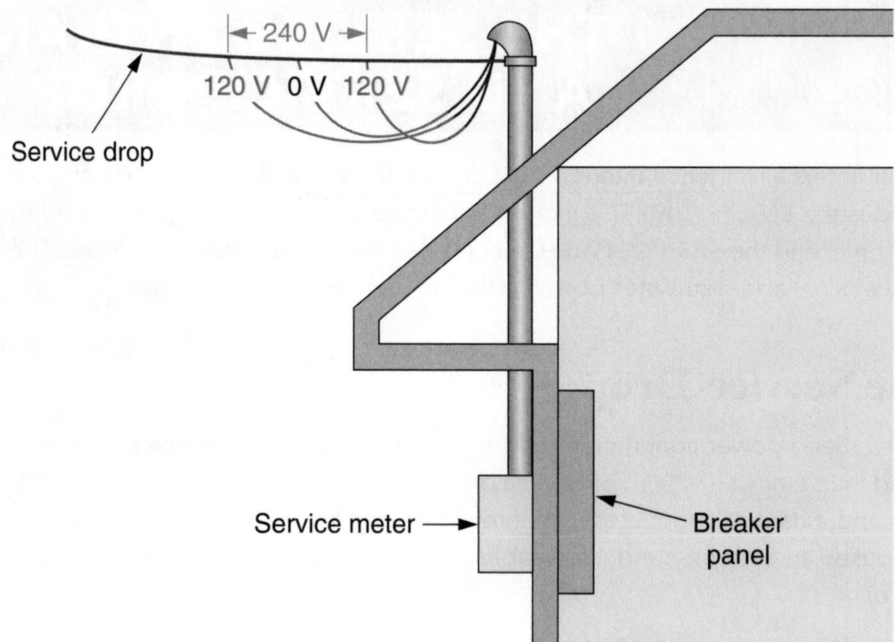

FIGURE 8.2 Service drop voltages.

MAST In a mast-type service drop, the conduit and weatherhead that extend upward from the roof.

MAST KNOB On a mast-type service conduit, the structure on the mast that the service lines are attached to.

TAILS The wire loops at the service entrance that prevent any mechanical stresses on the power lines and prevent any water (due to rain) from traveling along the lines into the service drop conduit. Also known as *drip loops.*

Mast and Clevis Service Drops

There are two types of overhead service drops: *mast* and *clevis.* A mast service drop is illustrated in Figure 8.3. The term **MAST** refers to the conduit and weatherhead that extend upward from the roof. The service drop is attached to the mast at the **MAST KNOB**.

In overhead residential power drops, the neutral conductor is uninsulated and is used as a means of support for the triplex cable between the utility distribution point and the residence.

IN THE FIELD

The conductor **TAILS** (or *drip loops*) serve two purposes. First, they provide slack that reduces any mechanical stresses on the power lines. This is referred to as *strain relief.* Second, they prevent any water (due to rain) from traveling along the lines into the service drop conduit.

A clevis service drop has fasteners that secure the power lines (or triplex cable) to the side of the residence, as shown in Figure 8.4. The term **CLEVIS** refers to the connectors that fasten the conductors to the building. Note that the weatherhead and conduit are secured to the side of the residence below the roof line. This distinguishes the clevis service drop from the mast service drop.

CLEVIS The mechanical connectors that fasten utility conductors to the side of a building.

Service Lateral

SERVICE LATERAL An underground service entrance.

TRANSFORMERS Electrical devices that are used to raise or lower a voltage level.

An underground service connection, or **SERVICE LATERAL**, is illustrated in Figure 8.5. The primary power lines pass through the conduit to the pad transformer input. The secondary power lines connect the transformer output to the electrical service meter. **TRANSFORMERS** are devices that are used to raise or

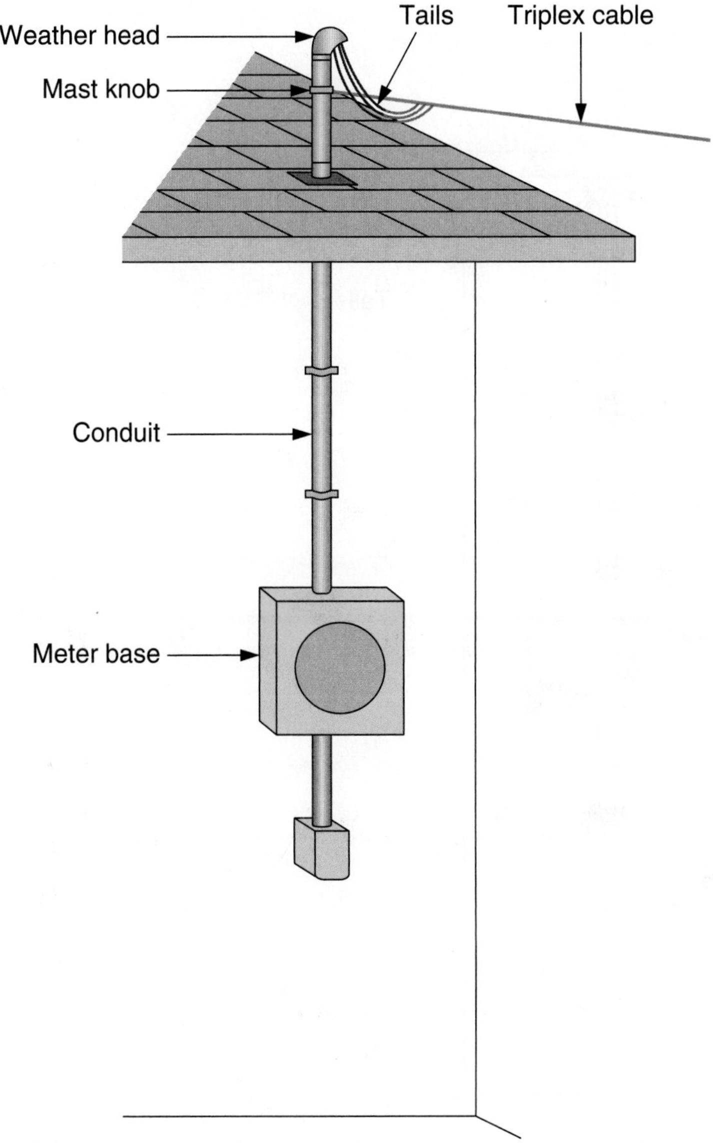

FIGURE 8.3 Mast-type service drop.

lower a voltage level. In this case, the transformer is used to lower the voltage from the utility to the 120/240 V residential level. We will cover transformers in Chapter 21.

The Service Meter

As stated earlier, the *service meter* measures the amount of energy that is used by the customer. Two electrical service meters are shown in Figure 8.6, one providing an *analog* display (rotary dials) and the other providing a *digital* display.

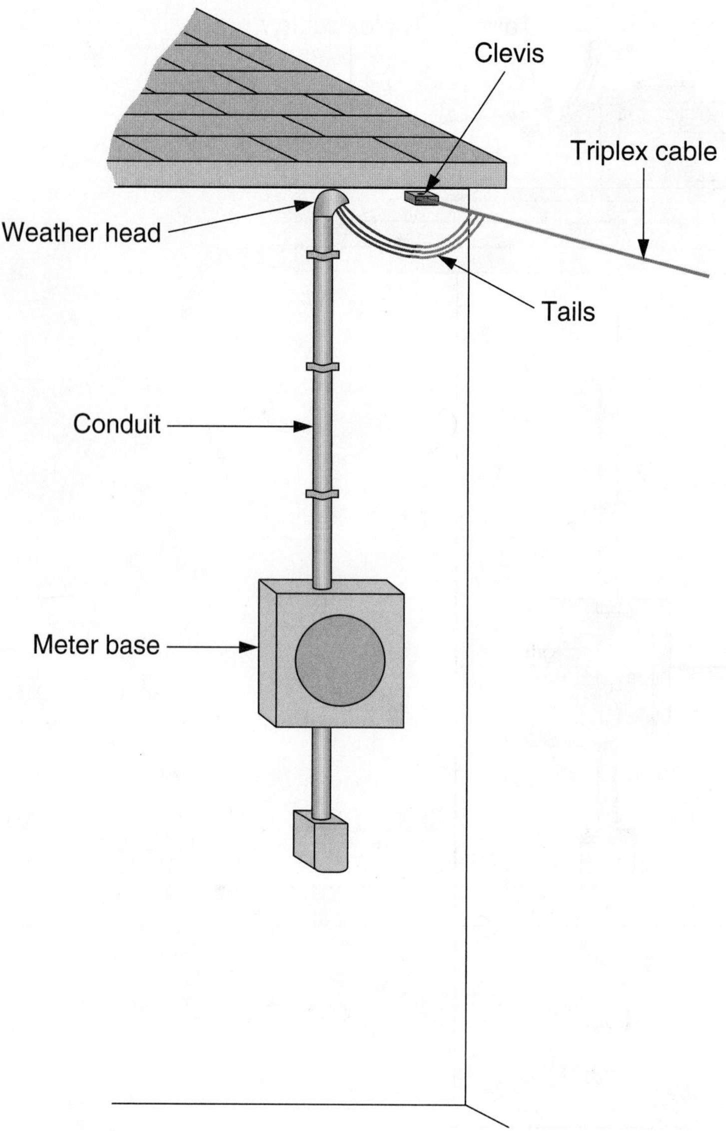

FIGURE 8.4 Clevis-type service drop.

KILOWATT HOUR (KWH) The amount of energy used in one hour.

The electrical service meter measures energy in **KILOWATT-HOURS (KWH)**. As you learned in Chapter 4, a kilowatt-hour is the amount of energy used in one hour, found as

(8.1)

$$E_{kWh} = \frac{P \times t}{1000}$$

where

P = power, in watts (W)
t = time, in hours (h)

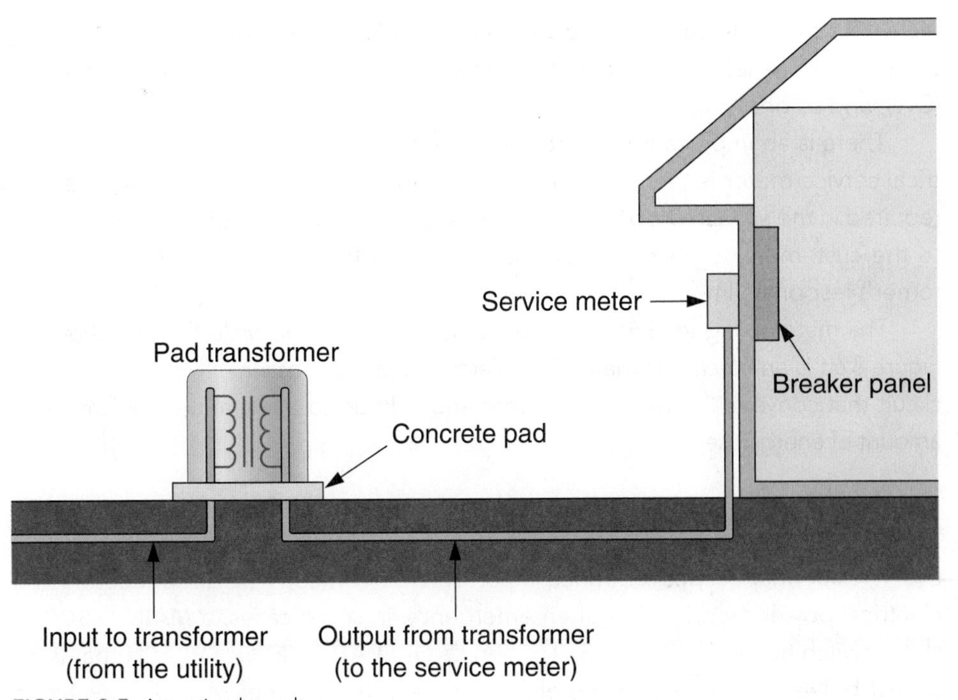

FIGURE 8.5 A service lateral.

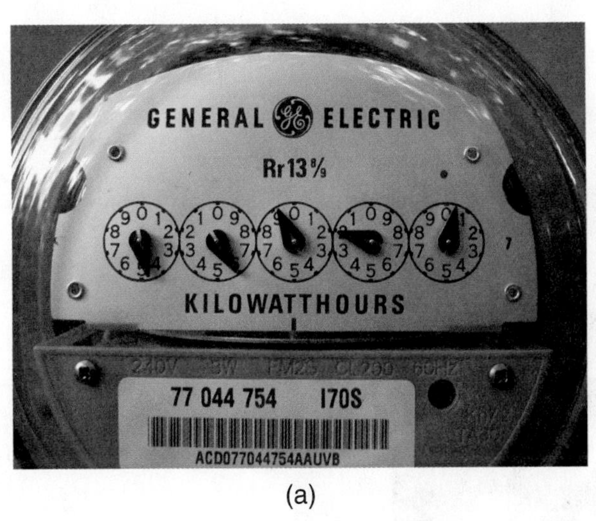

(a)

(b)

FIGURE 8.6 Analog vs. digital service meters.
(a) General Electric; (b) Ontario's Independent Electricity System Operator

As stated in Chapter 4, the 1000 in the denominator of the equation converts the power to kilowatts (kW).

The dials on the meter in Figure 8.6a are read from left to right. The weights of the dials (from left to right) are 10000, 1000, 100, 10, and 1. The value indicated by each dial equals the reading times its weight (the value printed above the dial). For example, the value indicated by the meter in Figure 8.6a is found as

$$4 \times 10000 + 5 \times 1000 + 9 \times 100 + 2 \times 10 + 0$$
$$= 45{,}920 \text{ kWh}$$

Note that the dials on the analog meter rotate in alternating directions. That is, the dial on the left rotates counterclockwise (CCW), the next rotates clockwise (CW), and so on.

There is an important point to be made at this time. In most cases, the electrical service meter is the point where the utility's responsibility ends. Any repairs required in the service drop or service meter are made by the utility with no charge to the customer. However, any problems beyond the service meter are the customer's responsibility.

The meter in Figure 8.6a is an *electromechanical* meter, while the one shown in Figure 8.6b is an *electronic* meter. The electronic meter is essentially a series-parallel circuit that converts the incoming current and voltage to a signal that indicates the amount of energy used.

MAIN DISCONNECT A switch that is used to disconnect power from a residence in case of an emergency.

EXTERNALLY OPERATED (EXO) SWITCH A main disconnect switch that is positioned between the service entrance and the electrical panel.

Main Disconnect

Every residential service entrance must provide a means of disconnecting the electrical power feed in case of an emergency. In some cases, a **MAIN DISCONNECT** switch (or breaker) is an **EXTERNALLY OPERATED (EXO) SWITCH** that is inserted between the service meter and the electrical panel. An EXO switch is shown in Figure 8.7. In other cases, one or more circuit breakers are housed in the electrical panel that provides the required main disconnect capability.

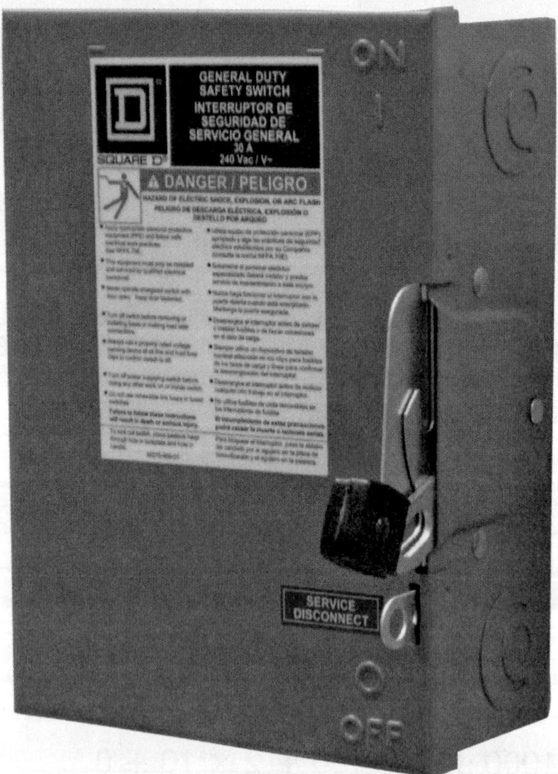

FIGURE 8.7 An externally operated (EXO) main disconnect.

Regardless of the type of main disconnect used, the breaker(s) must be in series with both hot lines. The main disconnect must cut power (interrupt current) to all circuits. This can be accomplished only if it is connected in series with them.

The Service Panel

The final element in the service entrance is the *service panel*. As stated earlier, the service panel performs two important functions:

1. It houses the overload protection components (circuit breakers in most cases, and fuses in older panels).
2. It serves as the connection point between the main **FEEDER** and **BRANCH CIRCUITS**.

A typical residential service panel is illustrated in Figure 8.8.

The power and neutral lines enter the service panel in Figure 8.8 through the conduit that is secured to the panel. The two hot lines (black and red) are connected to a **MAIN BREAKER**, a double-pole, double-throw (DPDT) switch that serves as the main disconnect for the service entrance. As is the case with the main disconnect, the main breaker is in series with both hot lines. The neutral conductor (white) is connected to the **NEUTRAL BAR** (or *ground bar*), which is the common connection point for all the neutral, or **RETURN**, lines in the residence. You can think of this as a *common ground.*

When the main breaker is closed, it connects the feeder conductors to two metal structures called **POWER BUSES**, with the black wire being connected to one power bus and the red wire connected to the other. The power buses provide the 120 V connections and mounting structures for the remaining circuit breakers. The power buses are configured so that adjacent circuit breakers are connected to alternating buses. For example, in Figure 8.8, the top circuit breaker is connected to bus A, then next to bus B, and so on. This arrangement serves two purposes:

1. By alternating power buses, the loads (residential branch circuits) are balanced between the two service inputs. That is, a nearly equal number of circuits draw power from the black power line and the red power line.
2. A 240 V connection can be made using breakers that are connected to adjacent 120 V bus connections.

For example, a 240 V connection is formed by the top two breakers in Figure 8.8.

FEEDER CIRCUITS The lines that bring power to the electrical panel.

BRANCH CIRCUITS The circuits that distribute power throughout a residence or other building.

MAIN BREAKER A double-pole, double-throw (DPDT) switch that serves as the main disconnect for an electrical panel.

NEUTRAL BAR The common connection point for all the neutral lines in a residence.

RETURN Another name for a neutral wire.

POWER BUSES Two metal structures that provide the 120 V connections and mounting structures for branch circuit breakers.

The ground path also carries ground-fault current from the point of where the fault occurs to the circuit power source. This current then activates the overcurrent protective device to de-energize the faulty circuit.

IN THE FIELD

Grounding

A continuous low-resistance ground path is a part of any residential electrical system. This ground path ensures that the voltage levels in a residential wiring system are stable, and that the overload protection elements are as effective as possible.

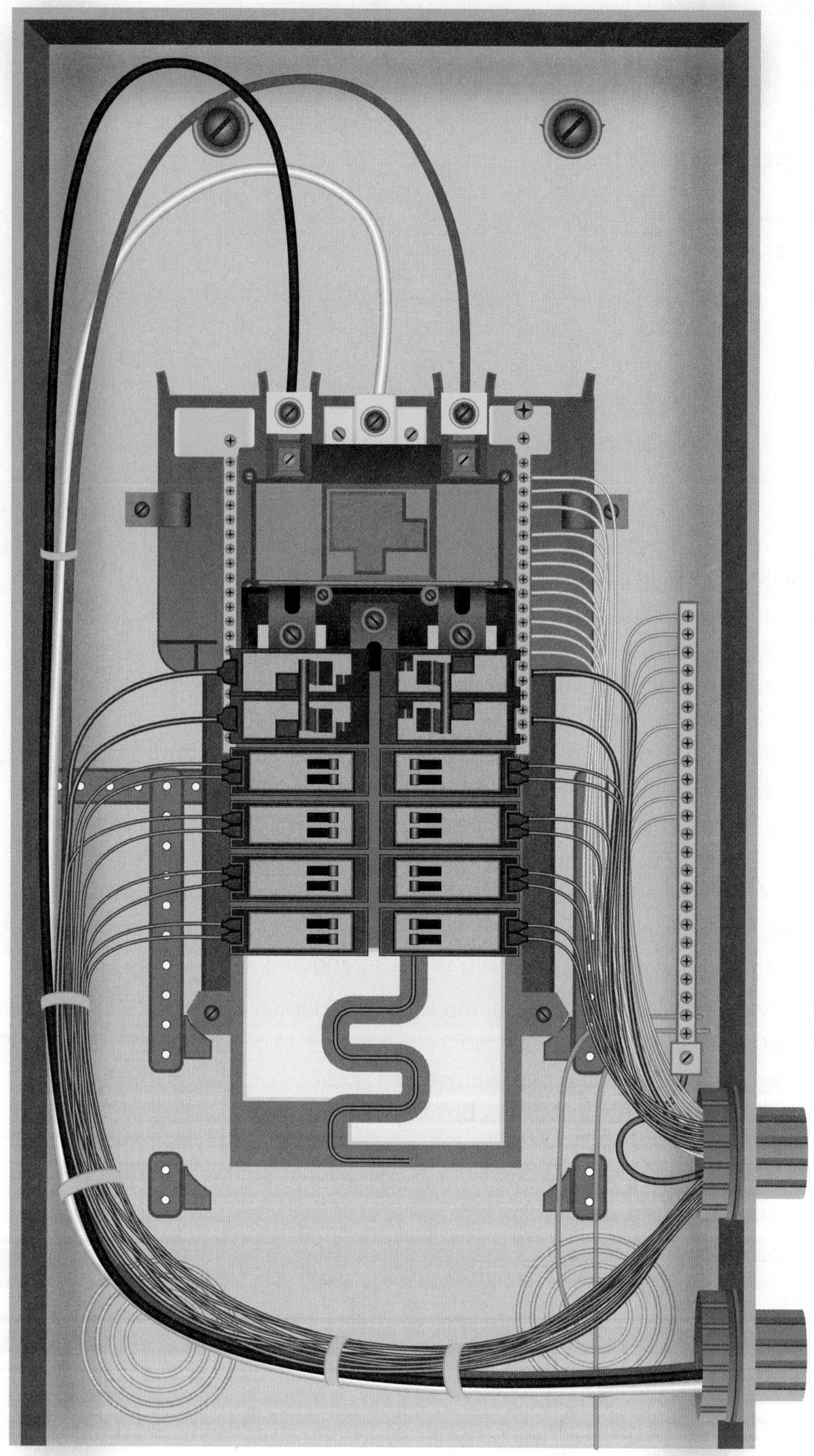

FIGURE 8.8 Typical residential service panel.

A grounding circuit is illustrated in Figure 8.9. The incoming neutral line is connected to the neutral bus. The neutral bus, in turn, is connected directly to earth ground using one of two methods:

1. The ground conductor leads from the neutral bar to the water pipes or gas pipes.
2. If neither of the connections in (1) will work, a ground conductor is run from the neutral bar to a **GROUNDING ROD** outside of the residence.

GROUNDING ROD A metal rod, driven into the ground outside a residence, that provides an electrical ground connection for the neutral bar in the service panel.

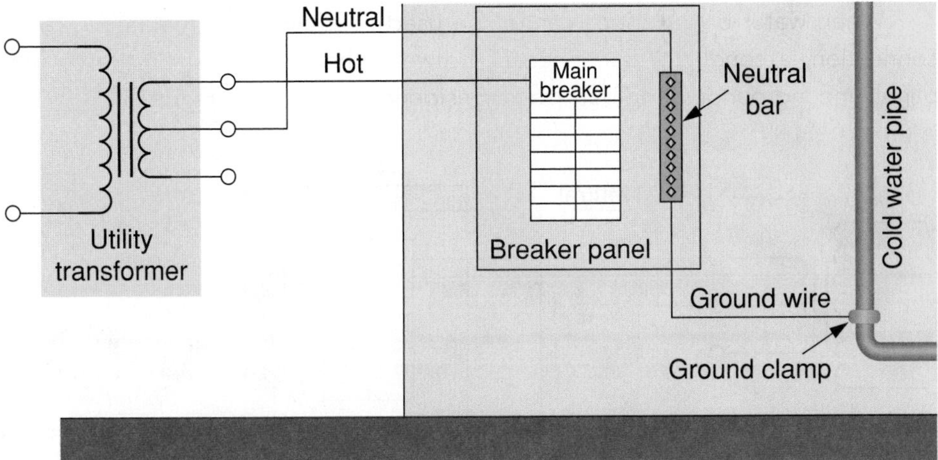

FIGURE 8.9 A grounding circuit.

The ground conductor in Figure 8.9 leads from the neutral bar to a **GROUND CLAMP** on the water pipe. Because the water pipes go underground the grounding conductor establishes a connection between earth ground and the neutral bar in the service panel. When the neutral wire in a branch circuit terminates at the neutral bar, it becomes the ground conductor for that circuit.

GROUND CLAMP A clamp used to connect a ground wire to a water pipe, establishing a connection between earth ground and the neutral bar in the service panel.

To have a *continuous* ground path, steps must be taken to make sure that nothing breaks (interrupts) that path. Consequently, a connection must be made *around* (in parallel with) anything that breaks the ground path. For example, the water meter in Figure 8.10 breaks the electrical continuity in the water pipe. To ensure

FIGURE 8.10 A grounding conductor bypasses a water meter.

Water filtering devices also break the ground path and therefore need to have a copper jumper connected from the filter input pipe to the filter output pipe.

IN THE FIELD

that the water pipe provides a continuous ground path, a conductor is connected between the water pipes on either side of the meter. This provides a ground connection that *bypasses* (goes around) the meter.

It should be noted that *gas pipes* can be used to provide the required grounding connections. In this case, it is the gas meter that will require a bypass conductor like the one shown in Figure 8.10.

When water or gas pipes cannot be used to provide the required ground connection, a conductor is connected between the neutral bar (in the service panel) and a grounding rod outside the residence, as shown in Figure 8.11.

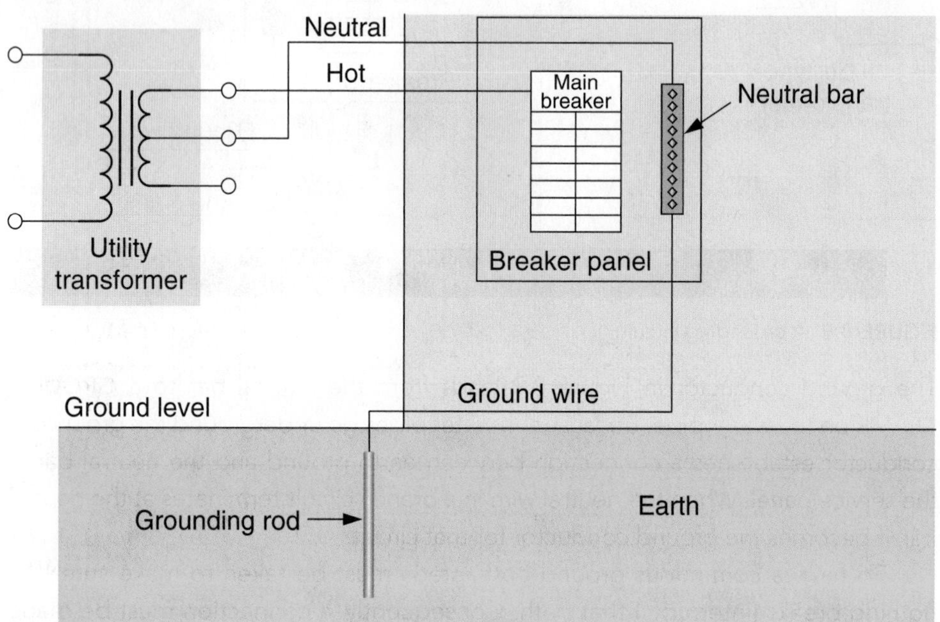

FIGURE 8.11 Grounding rod connection.

PROGRESS CHECK

1. Describe a service entrance and list the components that it contains.
2. Describe the differences between a mast service drop, a clevis service drop, and a service lateral.
3. What are drip loops and what purposes do they serve?
4. Describe the dials on a service meter, how they are read, and how they rotate.
5. What is a main disconnect? What is an EXO switch?
6. What are the two main functions of a service panel?
7. What color are the conductors that enter the service panel from the service entrance? Where are they connected in the service panel?
8. Describe the two ways that a ground is established in a residential electrical system.

8.2 SWITCHES AND RECEPTACLES

In the previous section, we examined the service entrance, which serves as the power source for a residential electrical system. In this section, we shift our focus to the control and load elements: switches, receptacles, and fixtures. **SWITCHES** are the control elements in the residential electrical system. Because switches are always used to make or break a connection, they are always wired in series with the circuit or device they are meant to control. **RECEPTACLES** provide the connection points for such items as lamps, home electronics, and appliances; that is, for various loads on the system. All the receptacles on a given branch circuit are wired in parallel.

SWITCHES The control elements in the residential electrical system.

RECEPTACLES The connection points for such items as lamps, home electronics, and appliances; that is, for various loads on the system.

Single-Pole Switches

Single-pole, single-throw (SPST) switches are the most commonly encountered switches in residential circuits. They are most often used to switch light fixtures on and off.

The single-pole switch is used to either close (energize) or open (deenergize) a circuit. For example, the switch in a lighting circuit energizes the lights (turns them on) when switched to the "up" position and de-energizes the lights (turns them off) when switched to the "down" position—assuming that the switch is installed correctly.

A single-pole switch is shown in Figure 8.12. As you can see, the switch has two screws on its right side. These screws are used to wire the switch into a circuit as shown in Figure 8.13a. When wired as shown, the switch can be used to make or break the connection between the two "hot" conductors as indicated in Figure 8.13b. The switch is always wired in series with the black hot conductor, not in series with the white neutral wire. If the switch was part of the neutral circuit, it would be bypassed if there were a short to ground somewhere in the circuit, and the circuit would be energized even if the switch was open.

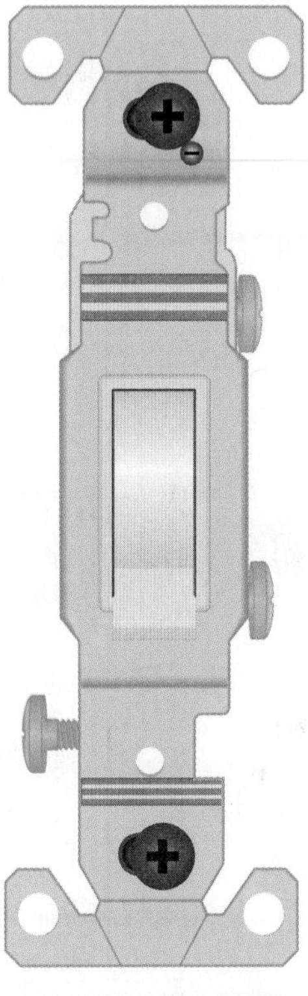

FIGURE 8.12 A SPST light switch.

Three-Way Switches

A three-way switch is used to provide two on/off controls, in two different locations, for a common circuit. The most common application for three-way switches is a lighting fixture that has switches at the bottom and top of a staircase, where either switch can be used to turn the fixture on or off.

The three-way switch is a single-pole, double-throw (SPDT) switch, as illustrated in Figure 8.14a. When the toggle is down, the *common* terminal is connected to the lower *traveler* terminal. When the toggle is up, the common terminal is connected to the upper traveler terminal.

Two three-way switches are wired to control a common light fixture as illustrated in Figure 8.15. With both switches toggled down, current passes through the lower traveler of each switch as shown in Figure 8.15a, lighting the light. The same holds true when both switches are toggled up. The current, in this case, passes

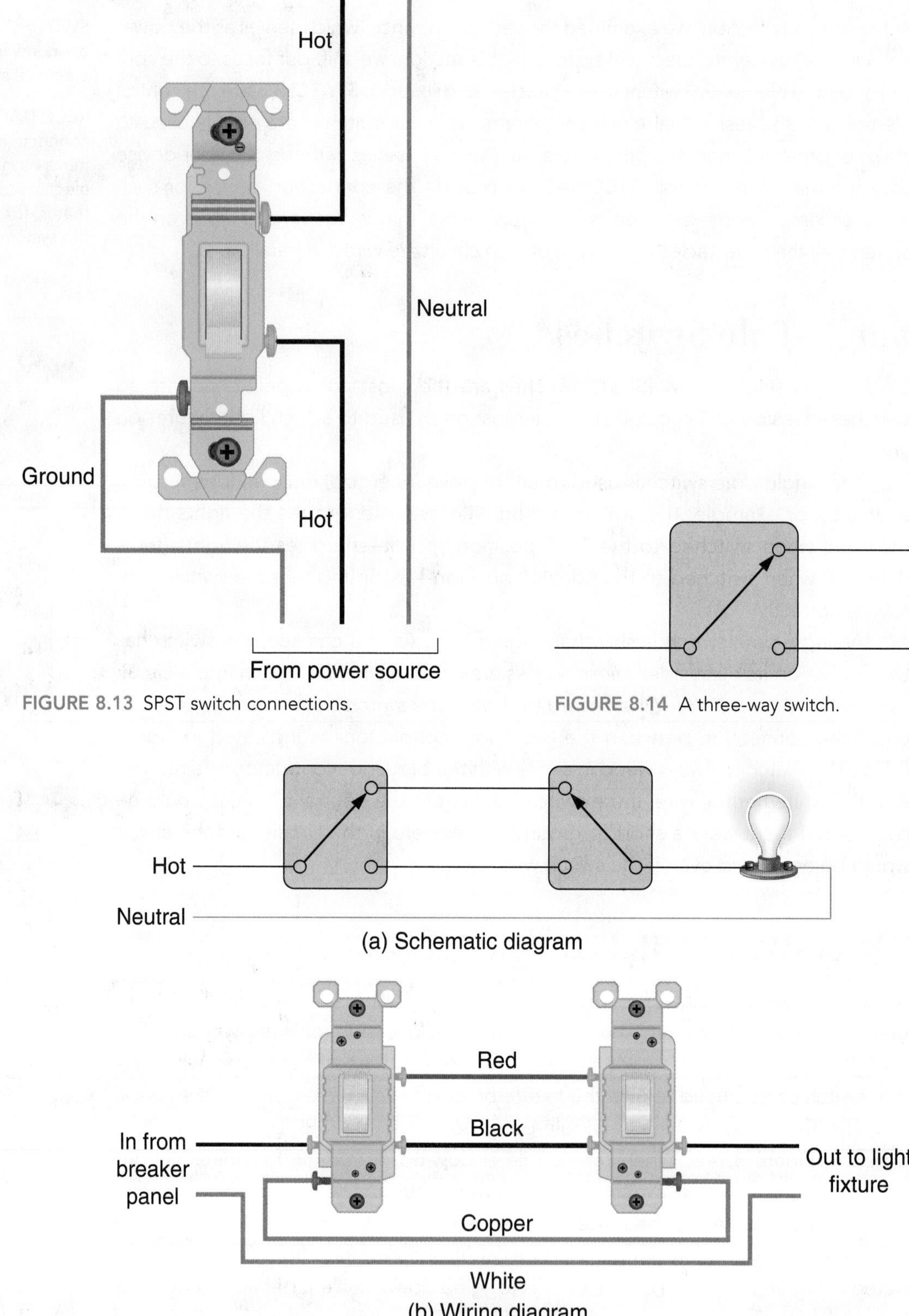

FIGURE 8.13 SPST switch connections.

FIGURE 8.14 A three-way switch.

(a) Schematic diagram

(b) Wiring diagram

FIGURE 8.15 A three-way switch circuit for a light fixture.

through the upper traveler as shown in Figure 8.15b, lighting the light. When the light is on, toggling either switch breaks the current path, turning the light off.

Note that the wiring scheme illustrated in Figure 8.15 is only one of several that can be implemented using three-way switches. *Note also that this wiring scheme is a practical example of a series-parallel circuit.*

Four-Way Switches

A four-way switch is a double-pole, double-throw (DPDT) switch that is used in conjunction with three-way switches to provide control of a light fixture from multiple locations. The internal connections made by a four-way switch are illustrated in Figure 8.16.

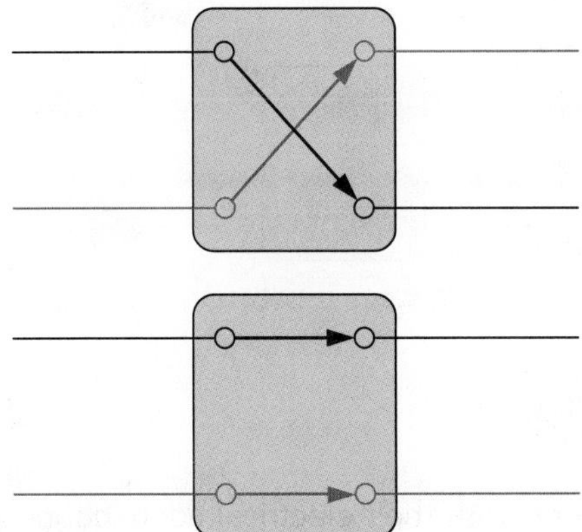

FIGURE 8.16 A four-way switch.

A four-way switch allows a light fixture to be controlled independently by three switches. For example, adding a four-way switch to the circuit in Figure 8.14 creates the three-switch circuit illustrated in Figure 8.17. With the added four-way switch, the light can be controlled from three locations.

When the switches are in the positions shown in Figure 8.17a, current passes through all three switches and the light is on. When the four-way switch is toggled as shown in Figure 8.17b, the current path is broken and the light is off. When the three-way switch SW1 is toggled, the current path is reestablished as shown in Figure 8.17c, and the light turns on again. If you keep in mind that toggling any switch turns an "on" light "off" (or vice versa), you can identify the current path (or break) for any combination of switch positions in Figure 8.17. Once again, we have a practical example of a series-parallel circuit.

Receptacles

A *receptacle* is a device that provides a means of connecting various appliances and electronic devices to the power distribution circuits. Receptacles are often referred to as **OUTLETS**, though the term *outlet* more correctly refers to the receptacle, its mounting box, and wiring.

OUTLET A term that is often used in reference to a receptacle even though it more properly refers to the receptacle, its mounting box, and wiring.

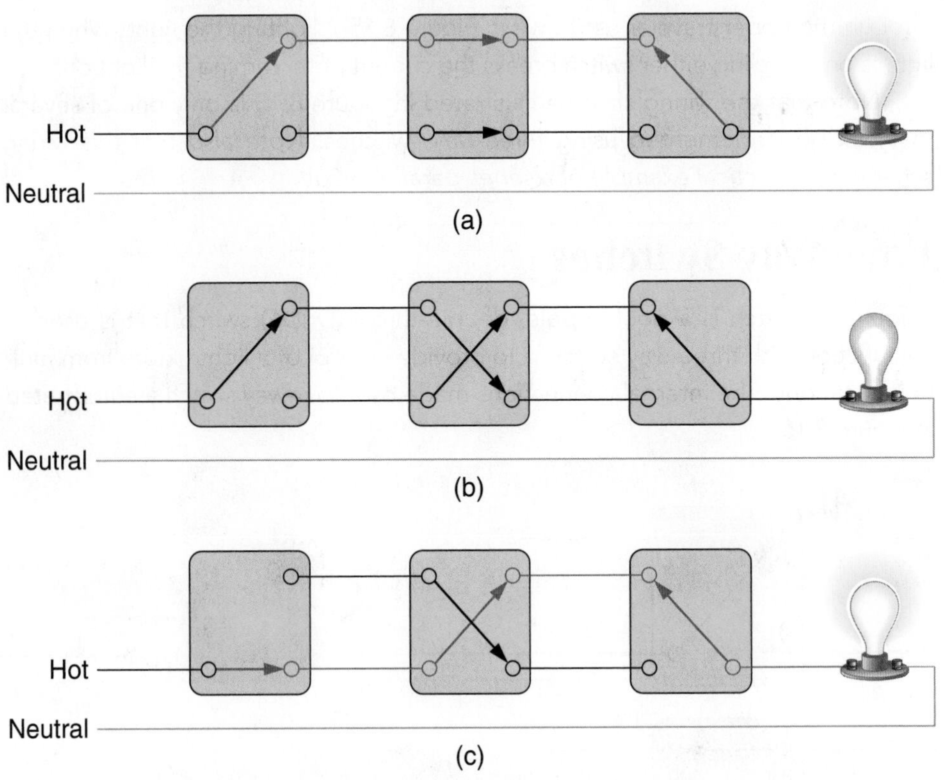

FIGURE 8.17 A four-way switch circuit for a light fixture.

Some common receptacles are illustrated (along with their electrical connections) in Figure 8.18. The receptacle *poles* are electrical connections. A two-pole receptacle connects to one hot conductor and the neutral (ground) conductor, providing 120 V connections. A three-pole receptacle connects to the two hot conductors, providing a 240 V connection. It also provides a ground connection. The applications for each of the receptacles in Figure 8.18 are included in the figure.

Ground Fault Circuit Interrupters (GFCIs)

GROUND FAULT CIRCUIT INTERRUPTER (GFCI) A special type of receptacle designed to switch off in the event that there is a measurable difference between the current levels on the hot and neutral lines.

A **GROUND FAULT CIRCUIT INTERRUPTER (GFCI)**, sometimes referred to as a *GFI*, is a special type of receptacle. Its job is to prevent electric shock. Here is how it works: The internal circuitry of the GFCI monitors the current in both the hot and neutral lines. If there is no ground fault, the current in the hot line should equal the current in the neutral line. If there is another path to ground (a ground fault), then the current in the neutral will be lower than the current in the hot line and the GFCI will break the circuit. A differential of 4 mA to 5 mA between the two currents is all it takes to cause the GFCI to open the circuit. It can respond in about 1/30 of a second.

Certain GFCI receptacles have feed-through capabilities so that receptacles connected downstream of them are also GFCI protected.

IN THE FIELD

A GFCI can be reset after it has broken a circuit. It also has a test button that allows you to make sure it is working

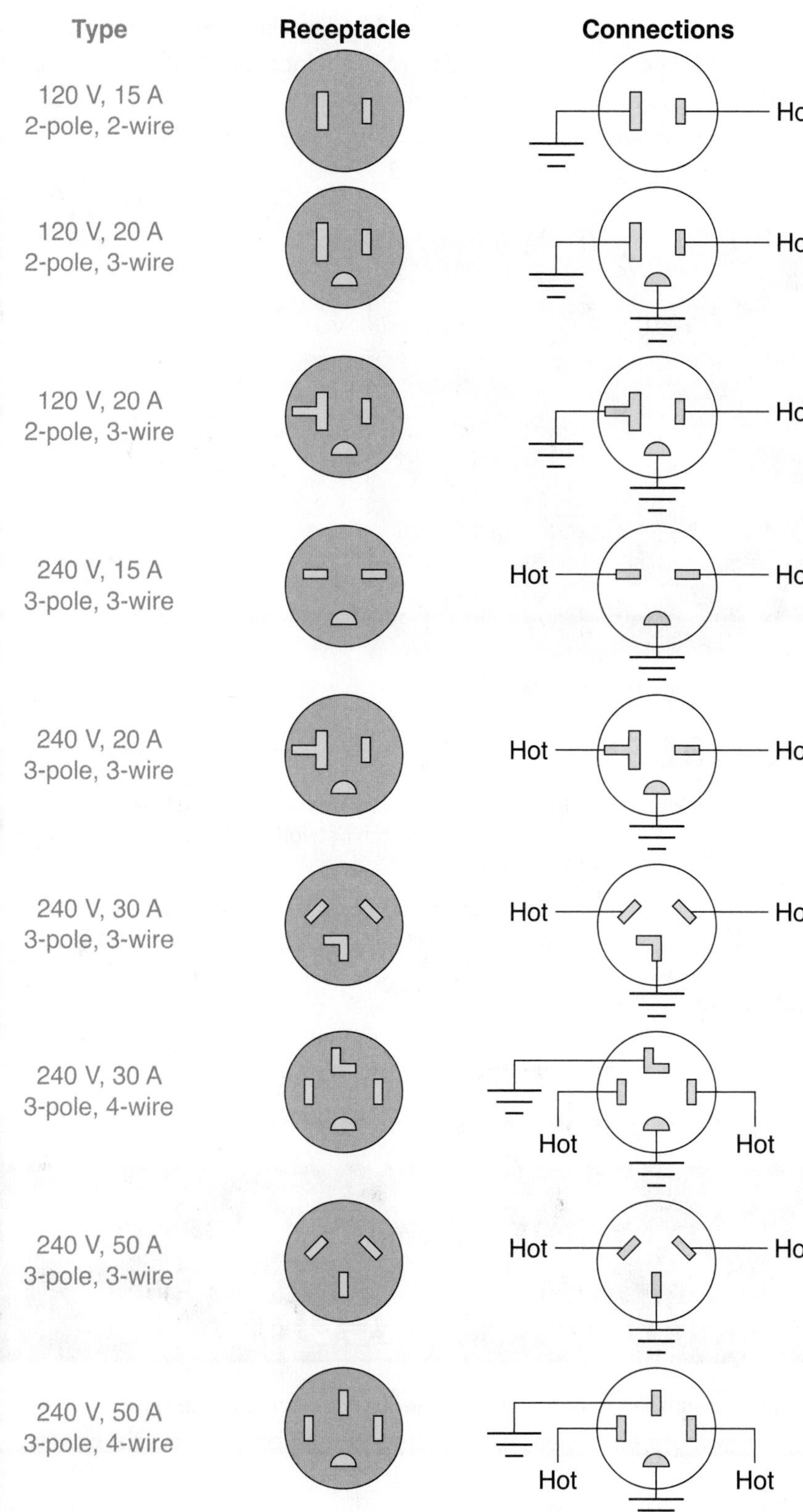

FIGURE 8.18 Receptacle configurations.

properly. Like any device, GFCIs do wear out, so regular testing is recommended. GFCIs are mandated by code to be installed in certain locations in the home, but a good rule of thumb is that any location where water is present, such as bathrooms and kitchen counters, requires a GFCI. There are also GFCI-type circuit breakers. A GFCI receptacle is shown in Figure 8.19.

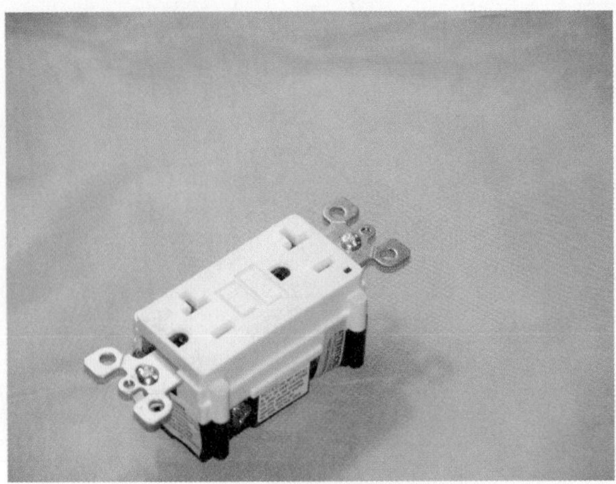

FIGURE 8.19 Ground-fault circuit interrupter (GFCI).

PROGRESS CHECK

1. What is the most common use for a single-pole switch?
2. Which conductor is connected to the terminals on a single-pole switch? Why is this important?
3. What is the most common use for a three-way switch?
4. Describe one common wiring scheme for a three-way switch. Explain how this is an example of a series-parallel circuit.
5. Explain what components you would use in a three-location switch network and how you would wire them.
6. What is a receptacle pole? How many poles does a 120 V receptacle have? How many poles does a 240 V receptacle have?

8.3 RESIDENTIAL AC CIRCUITS

Now that we have discussed residential service entrances, switches, and receptacles, we will look at how these elements are connected to form basic residential circuits.

Conductors

Most residential circuits utilize one of the conductor combinations identified in Table 8.1. The conductor combinations listed are illustrated in Figure 8.20.

TABLE 8.1 • Circuits, applications, and conductor combinations.

CIRCUIT	PROVIDES POWER FOR...	CONDUCTOR COMBINATION
120 V	Receptacles and light fixtures	One hot, one neutral, and ground
240 V	Dryers, stoves, central air conditioning units	Two hot, one neutral, and ground

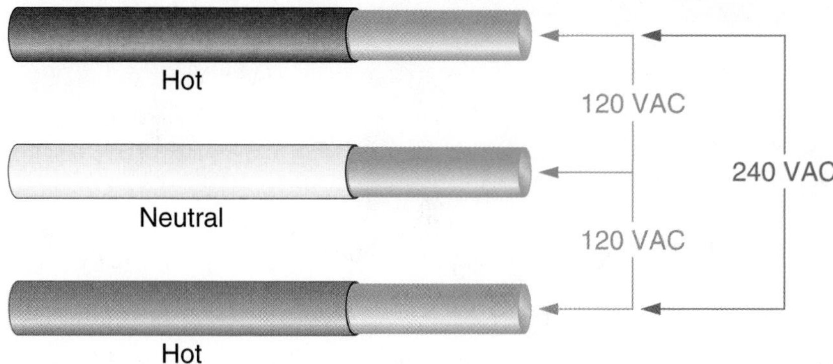

FIGURE 8.20 Conductor voltages in a 240-V circuit.

The current level that each conductor in Figure 8.20 can handle depends on its gauge. For example,

- 14 gauge wires can handle currents up to 15 A.
- 12 gauge wires can handle currents up to 20 A.
- 10 gauge wires can handle currents up to 30 A.
- 8 gauge wires can handle currents up to 40 A.

Note that these statements indicate current handling capabilities only. They do not take into consideration any local building code restrictions. Note also that a wire can be substituted for a smaller-diameter (higher gauge) wire, but not for a larger-diameter (lower gauge) wire. For example, 12 gauge wire can be used in place of 14 gauge wire in a 15 A circuit. However, 12 gauge wire cannot be used in place of 10 gauge wire in a 30 A circuit.

Nonmetallic Cables

A **NONMETALLIC (NM) CABLE** (also called *Romex* cable) is a group of insulated conductors housed in a protective plastic jacket. The lower nonmetallic cable shown in Figure 8.21 contains a hot conductor, a neutral conductor, and a bare ground wire. As indicated in Table 8.1, this cable is well suited for use in a 120 V circuit. The upper NM cable in Figure 8.21 has two hot conductors, a neutral conductor, and a bare ground wire. As such, it is well suited for use in a 240 V circuit.

NONMETALLIC (NM) CABLE A group of insulated conductors housed in a protective plastic jacket; also called *Romex*.

The label on an NM cable indicates the size (gauge) and number of conductors in the cable. For example, the label

14/2 w/GR

("*fourteen/two with ground*") indicates that the cable contains two 14 gauge conductors and a ground wire. By the same token, the label

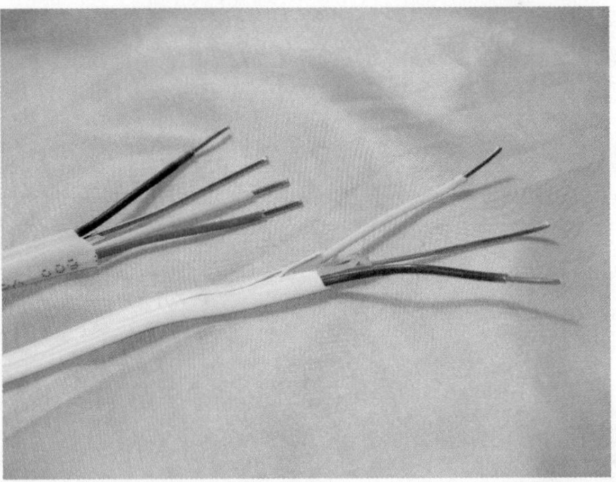

FIGURE 8.21 Nonmetalic cables.

12/3 w/GR

(*"twelve/three with ground"*) indicates that the cable contains three 12 gauge conductors and a ground wire.

The cable type—either NM or NMC—indicates the suitability of the cable cover for use in different environments. For example, type NMC cable is better suited than type NM cable for use in wet environments. Consequently, type NM cable is installed in dry locations (such as interior walls), while NMC cable is typically installed in masonry and other damp locations.

Another type of cable that can be used in place of NM cable is housed in a flexible metal cover. This type of cable is often referred to as *BX cable* or *armored cable*, but is properly called *Type AC cable*. Type AC cable with an additional external plastic sheath is called *TECK cable*. It is also common practice to run NM cable in a rigid metal conduit if it is not inside a wall or otherwise protected.

A Typical 120 V Residential Circuit

As indicated earlier, most 120 V circuits are used to provide power to one or more receptacles and/or light fixtures. A circuit containing two receptacles, a switch, and a light fixture is illustrated in Figure 8.22. The circuit originates with a 15 A breaker in the service panel (not shown). The breaker is in series with both receptacles, the switch, and the light fixture to enable it to cut power to the entire circuit if an overcurrent event occurs. An NM cable extends from the breaker to one receptacle, and the second receptacle is connected in parallel with the first. From there, the cable continues to a single-pole switch that controls the light fixture. Note that both receptacles and the light fixture are wired in parallel with each other and with the rest of the circuit.

This is a very important point. As you know, one of the characteristics of a parallel circuit is that all branches in the circuit have the same voltage. Think what would happen if all the receptacles in a house were *not* wired in parallel. If you were to turn on the TV, the voltage drop across the TV would affect every device connected in series with it. Every device that we connect to a 120 V receptacle

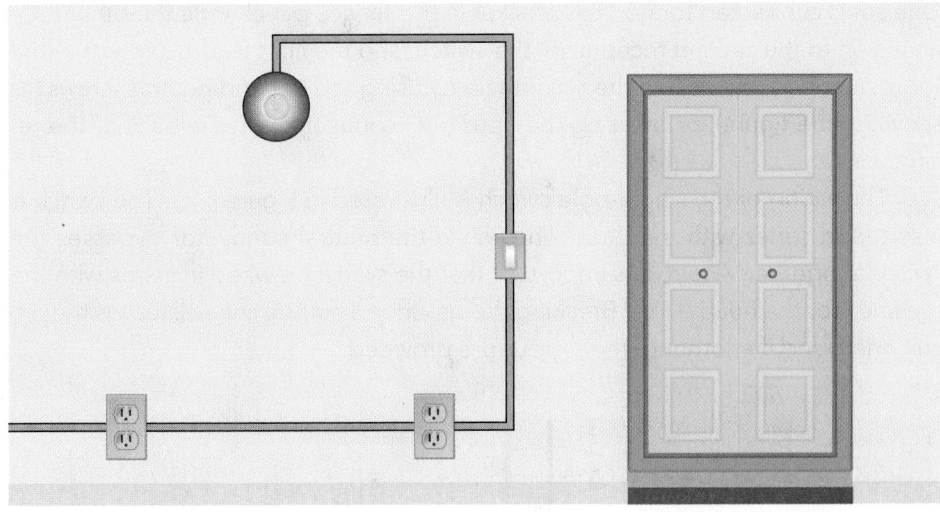

FIGURE 8.22 A typical 120-V circuit.

expects to get 120 V, so all 120 V receptacles are wired in parallel with each other. The switch, on the other hand, is connected in series with the light fixture, as it must make or break a connection between the light and the receptacle. It cannot be in parallel with the light fixture, and it cannot be in series with any of the outlets, unless you want the outlets to be turned off and on with the light.

The wiring of each receptacle in Figure 8.22 is illustrated in Figure 8.23. As you can see, two sets of hot and neutral conductors are connected to the receptacle.

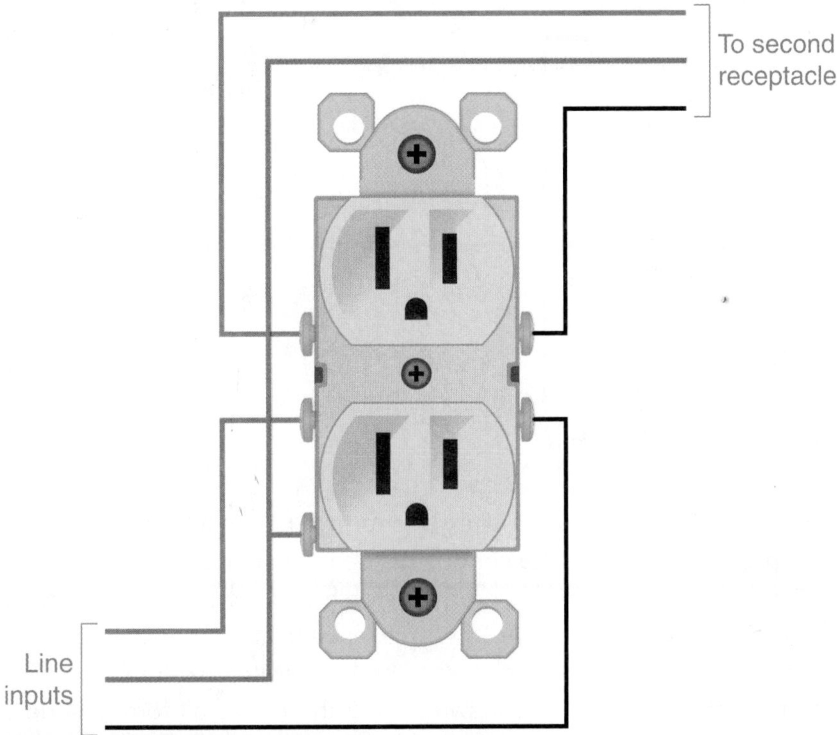

FIGURE 8.23 Receptacle connections.

One set is connected to the 15 A breaker in the service panel while the other continues on to the second receptacle, the switch, and the light fixture. Note that the wires can be connected to the receptacle by using the hot and neutral screws (as shown in the figure), or by using the "push-in" connectors on the back of the receptacle.

The wiring of the single-pole switch is illustrated in Figure 8.24. The switch is inserted in series with the "hot" line, while the neutral conductor bypasses the switch altogether. Again, it is important that the switch be wired in series with the hot line, not the neutral line. Breaking the grounded neutral line still leaves the circuit energized if another path to ground is provided.

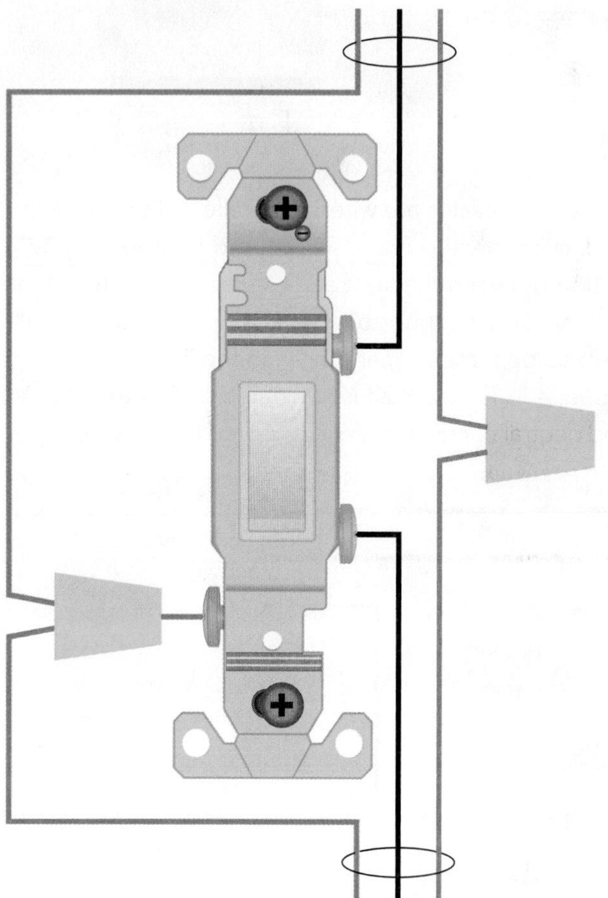

FIGURE 8.24 Light switch connections.

The light fixture in Figure 8.22 is wired as shown in Figure 8.25. Because this is the final element in the circuit, the conductors do not continue beyond the fixture. Note that the series combination of the switch and the light fixture are in parallel with the two receptacles. The 15 A breaker back at the panel is in series with the parallel-connected receptacles and the light switch/fixture circuit. So, the breaker, the receptacles, the light switch, and the fixture all form a series-parallel circuit.

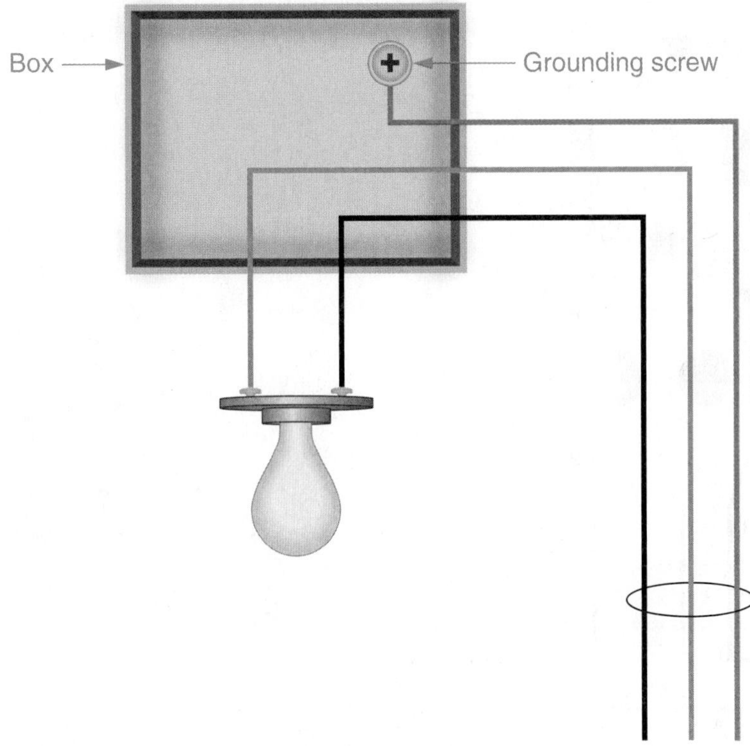

FIGURE 8.25 Light-fixture connections.

A Typical 240 V Circuit

A typical 240 V circuit is illustrated in Figure 8.26. The circuit shown provides 240 V to the receptacle for an electric dryer. The nonmetallic (NM) cable—which is labeled "10/3 w/GR"—originates at a 30 A double-throw breaker pair and provides 240 V for the dryer receptacle. As is always the case for a breaker or fuse, the circuit protector (breaker or fuse) must be in series with the circuit if it is to break the circuit whenever an overcurrent situation develops. One side of the breaker is in series with one hot conductor and the other side of the breaker is in series with the other hot conductor. The cable is connected to the receptacle as shown in Figure 8.27.

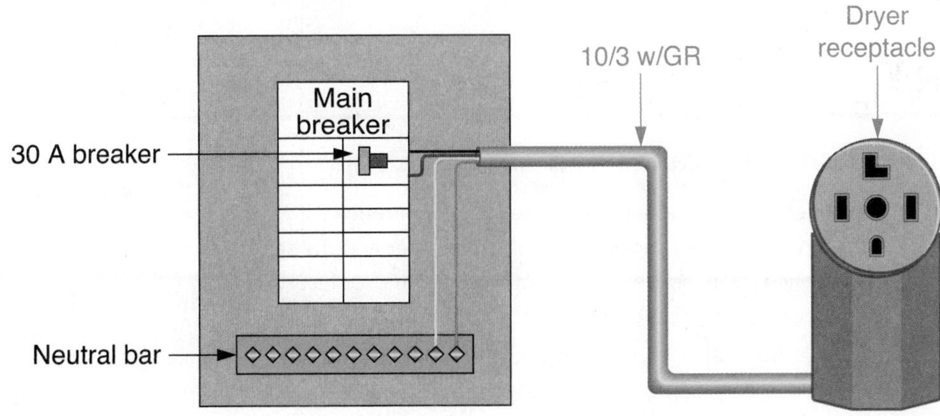

FIGURE 8.26 A 240-V circuit.

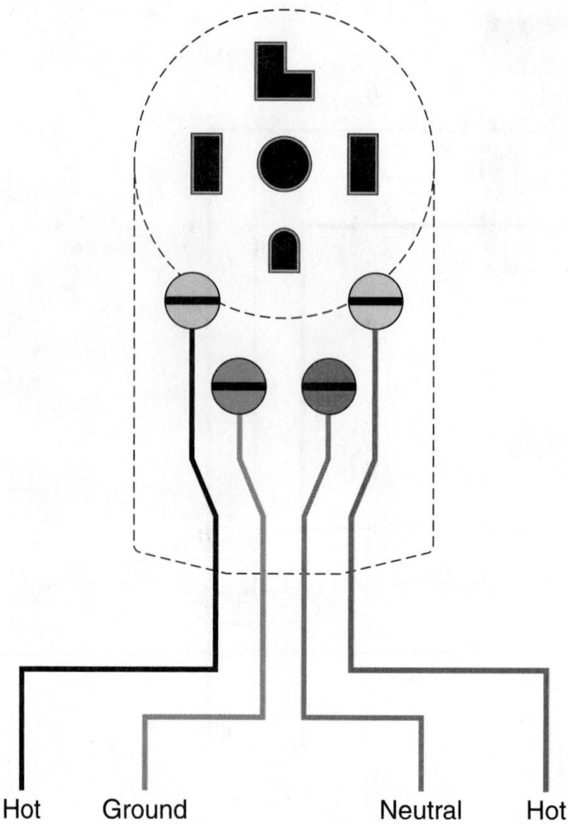

FIGURE 8.27 Internal connections for the receptacle in Figure 8.26.

PROGRESS CHECK

1. Describe the conductor combinations for 120 V and 240 V circuits.
2. What is the relationship between wire gauge and current capacity?
3. What is the basic rule for wire substitution?
4. Describe a nonmetallic cable. What information does the label on a nonmetallic conductor tell you?
5. Describe a typical 120 V installation.
6. Describe a typical 240 V installation.

8.4 SUMMARY

The Service Entrance and Service Drop

- AC power enters the home through the service entrance.
  - The service entrance includes the electric meter and the service panel.
  - The service panel houses the circuit breakers and/or fuses.
  - The service panel distributes power throughout the home.
- The service drop is the connection from the utility power lines to the service entrance.
  - The service drop contains two 120 V hot lines and one neutral.
  - These three lines may be independent conductors, or they may be contained in a single cable called a *triplex cable*.

- There are two types of overhead service drops: mast and clevis.
- A mast service drop includes a conduit and weatherhead that extends above the roof. (See Figure 8.3.)
  - The conductor(s) tail is the point where the utility lines enter the mast. This is also known as the *drip loop*. It prevents mechanical stress on the power lines and prevents rain water from entering the mast conduit.
- A clevis service drop is fastened to the side of the residence. (See Figure 8.4.)
  - The term clevis refers to the fasteners used to secure the cable(s).
  - The weatherhead and conduit are below the roof of the residence.
- An underground service lateral connects the utility lines to the meter through a pad transformer. (See Figure 8.5.)
- The electrical service meter measures the amount of energy used by the customer.
  - Meters can be analog (electromechanical) or digital (electronic).
- Electrical meters measure energy in kilowatt-hours (kWh).
  - The dials on a meter are read from left to right. (See Figure 8.6.)
  - Note that the dials rotate in alternating directions.
- The utility is not responsible for anything beyond the electrical service meter.
  - Anything beyond the meter is the responsibility of the customer (homeowner).
- Every residential service entrance must have a main disconnect switch (or breaker).
  - The disconnect may be housed in the service panel, or it may be an external unit between the meter and the panel.
  - If it is external to the panel, it is referred to as an externally operated (EXO) switch.
- The final element in the service entrance is the service panel. (See Figure 8.8.)
  - It houses the overload protection components (circuit breakers or fuses in older homes).
  - It is the connection point between the main feeder and the branch circuits.
- The two hot lines and one neutral line enter the service panel.
  - The two hot lines (black and red) are connected to the main breaker.
  - The main breaker is a double-pole, double-throw (DPDT) switch.
  - The neutral line (white) is connected to the neutral or ground bar. This is the common point for all the return lines in the residence.
- When the main breaker is closed, the feeder conductors are connected to the power buses.
  - The black conductor connects to one bus, and the red to the other.
  - Each power bus provides 120 V (with respect to the neutral line). It also provides the mounting structure for the individual circuit breakers.
- The power buses are configured so that each of the adjacent circuit breakers connects to alternating buses.
  - This is to keep the residential load balanced between the two hot lines.
  - It also allows a 240 V connection by means of a double breaker. (See Figure 8.8.)
- Every residential electrical system must have a low-resistance path to ground.
  - The ground path keeps the voltage levels stable and assures that the overvoltage protection elements work effectively.
- The neutral line is connected to the neutral bus, which in turn is connected to earth ground using one of two methods.
  - The neutral bus is connected to a metal water or gas pipe. Since the water and gas pipes go underground, they provide a low-resistance path to earth ground.
  - The neutral bus is connected to a grounding rod outside the residence.

- In order to have a continuous ground path, steps must be taken to make sure nothing breaks the ground path.
  - For instance, if a water or gas pipe is used, the ground line must bypass (go around) the gas or water meter.

Switches and Receptacles

- Switches are the control elements that connect or disconnect power to various fixtures, like lights.
- Receptacles are the connection points that power the various loads on the system.
- Single-pole switches are the most common residential switches. A common example is a standard light switch.
  - The single-pole switch is used to energize or de-energize a circuit.
  - If the switch is properly installed, it energizes a circuit when in the up position, and de-energizes the circuit in the down position.
- The two hot (black) conductors are connected to either side of the switch by means of two screws on the right side of the switch. The neutral line is not connected to the switch. (See Figure 8.13b.)
- A three-way switch is used to provide on/off control for a circuit from two different locations. It is a single-pole, double-throw (SPDT) switch.
  - The most common application for this type of switch is to control an overhead light from the top and bottom of a staircase.
  - One possible three-way switch wiring configuration is illustrated in Figure 8.15.
- A four-way switch is a double-pole, double-throw (DPDT) switch. See Figure 8.16 for its internal wiring.
- A four-way switch is used in conjunction with three-way switches to provide control of a circuit from multiple locations. (See Figure 8.17.)
- Receptacles provide a means of connecting various appliances or other devices to the power distribution circuits.
  - Receptacles are also known as outlets, though the term outlet really refers to the receptacle, its mounting box, and wiring.
- The receptacle connections are referred to as poles.
  - A two-pole receptacle connects the neutral to one of the hot lines providing a 120 V connection.
  - A three-pole receptacle provides a connection between both 120 V lines for a 240 V connection. It also provides a ground connection.
- A ground fault circuit interrupter (GFCI) is a receptacle designed to open the circuit if there is a difference between the "hot" and "neutral" line currents.
  - A difference of as little as 4–5 mA is all it takes to trip a GFCI.
  - A GFCI has a "reset" button and a "test" button. (See Figure 8.19.)

Residential AC Circuits

- Most residential circuits utilize one of the following conductor combinations:
  - One hot, the neutral, and a ground, used for 120 V receptacles and light fixtures.
  - Two hot lines, one neutral, and ground, used for 240 V appliances like stoves, dryers, and air conditioners.
- The current level that a conductor can handle is determined by its gauge.
  - 14 gauge wire handles currents up to 15 A.
  - 12 gauge wire handles currents up to 20 A.

  - 10 gauge wire handles currents up to 30 A.
  - 8 gauge wire handles currents up to 40 A.
- It should be noted that these ratings indicate current handling capabilities only, and do not take into consideration any local building codes.
- Any lower-gauge (larger-diameter) wire can be substituted for any higher-gauge wire.
  - You must never substitute a higher-gauge wire (smaller diameter) for a lower-gauge wire.
- Nonmetallic (NM) cable (commonly referred to as *Romex*) is a group of insulated conductors housed in a protective plastic jacket. (See Figure 8.21.)
- The label on the outside of the cable indicates the size (gauge) and number of conductors.
  - For example, the label 14/2 w/GR (*fourteen/two with ground*) indicates that the jacket contains two 14 gauge wires and a ground wire.
  - A label of 12/3 w/GR (twelve/three with ground) indicates the jacket contains three 12 gauge conductors and a ground.
- The cable type is labeled as NM or NMC.
  - NM is installed in dry areas like internal walls.
  - NMC can be used in masonry and other damp locations.
- A typical 120 V circuit is illustrated in Figure 8.22.
  - The circuit originates at a 15 A breaker in the service panel.
  - An NM cable connects the breaker to a receptacle.
  - From there the cable continues to a switch, and then to a light fixture.
  - Note that with the exception of the switch and the breaker, all the components are connected in parallel so that voltage remains constant and one load cannot affect any other.
- The wiring in a receptacle is illustrated in Figure 8.23.
  - There are two sets of hot (brass) and two sets of neutral (silver) screws on each receptacle.
  - Some receptacles have push-in connectors.
- The switch is connected in series with the light fixture.
  - It is inserted in the hot line, while the neutral bypasses the switch.
  - The light fixture is wired as shown in Figure 8.25.
- A typical 240 V circuit is illustrated in Figure 8.26.
  - It originates at a double 30 A breaker and continues to a 240 V dryer receptacle.
  - The cable is connected to the receptacle as shown in Figure 8.27.

CHAPTER REVIEW

1. AC power enters the home through the ________.
   a) mast
   b) clevis
   c) service entrance
   d) service drop

2. The utility connection to the residence is called the ________.
   a) mast
   b) clevis
   c) service entrance
   d) overhead service drop or service lateral

3. If the wires from the utility are contained in a single jacket, it is referred to as ________.
   a) MN cable
   b) BX cable
   c) 10/3 cable
   d) triplex cable

4. The difference of potential between the two hot conductors from the utility is ________.
   a) 240 V
   b) 120 V
   c) 0 V
   d) None of the above

5. The conduit and weatherhead located above the roof is referred to as a ________.
   a) mast
   b) mast knob
   c) clevis
   d) clevis loop

6. Drip loops are also known as ________.
   a) conductor tails
   b) rain loops
   c) clevis loops
   d) mast loops

7. The term clevis refers to ________.
   a) the mast head
   b) wall-mounted conductor connectors
   c) the mast conduit
   d) the mast service entrance

8. A residential electrical service meter measures energy in ________.
   a) watts
   b) kilowatt-hours
   c) kVA
   d) watt-hours

9. The dials on an analog service are read ________.
   a) in reverse sequence
   b) left to right
   c) counterclockwise
   d) right to left

10. The customer is responsible for any problems in the residential electrical system beyond the ________.
    a) service meter
    b) service entrance

c) mast conduit
d) main disconnect

11. A main disconnect that is external to the service panel is referred to as a(n) ________.
    a) clevis switch
    b) EXO switch
    c) outboard switch
    d) slave switch

12. The main service panel serves as a connection point between the ________ and the ________.
    a) EXO switch, NM wiring
    b) breakers, meter
    c) feeder, branch circuits
    d) 120 V lines, 240 V lines

13. The neutral is also known as the ________.
    a) return line
    b) power bus
    c) neutral bus
    d) cool line

14. The ________ connect to the two power buses.
    a) common lines
    b) NM lines
    c) hot lines
    d) 240 V lines

15. The colors of the lines that connect to the power buses are ________.
    a) black and red
    b) black and white
    c) red and white
    d) red and green

16. Adjacent circuit breakers connect to alternate power buses so that ________.
    a) you can get 120 V power
    b) you can get 240 V power
    c) the voltage is inverted
    d) the load is balanced
    e) both (b) and (d)

17. The common bus may be connected to ________ to establish an earth ground.
    a) a water pipe
    b) a grounding rod
    c) a gas line
    d) all of the above

18. A single bypass line across a gas or water meter could be used to ________.
    a) maintain a continuous ground path
    b) steal water
    c) neutralize the meter
    d) prevent static electricity buildup

19. If a switch is correctly installed, it makes a connection when it is ________.
    a) normally closed
    b) in the up position
    c) normally open
    d) down position

20. A standard light switch is usually a(n) ________ switch.
    a) SPST
    b) SPDT
    c) DPST
    d) DPDT

21. A three-way switch is usually a(n) ________.
    a) SPST
    b) SPDT
    c) DPST
    d) DPDT

22. A four-way switch is usually a(n) ________.
    a) SPST
    b) SPDT
    c) DPST
    d) DPDT

23. A four-way switch is used in conjunction with ________ to provide multilocation control.
    a) three-way switches
    b) a dimmer
    c) a DPST switch
    d) an SPST switch

24. Although used to describe a receptacle, an outlet correctly includes ________.
    a) the mounting box
    b) the strain relief
    c) the socket
    d) the mounting box, wiring, and receptacle

25. A 240 V conductor combination contains ________.
    a) two hots and a ground
    b) two hots, a neutral, and a ground

c) three hots and a ground

d) two hots, a neutral, and no ground

26. A 10 gauge wire can handle currents up to ________.
    a) 15 A
    b) 20 A
    c) 30 A
    d) 40 A

27. A label on a cable that reads 10/3 w/GR means that the cable contains ________.
    a) three 10-gauge conductors and a ground wire
    b) three 10-gauge conductors, a neutral, and a ground wire
    c) three 10-gauge conductors, but no ground wire
    d) three 10-gauge conductors, a neutral, but no ground wire

28. All the receptacles on a single circuit are wired ________.
    a) for 15 A service
    b) in parallel
    c) for 30 A service
    d) in series

29. From the service panel to the load, the only components wired in series are ________.
    a) the breakers and the receptacles
    b) the breakers and the light fixtures
    c) the light fixtures and the switches
    d) the breakers and the switches

1
5
10
15
20
25
30
35
40
45
50
55
60
65
70
75
80
85
90
95
100
105
110
115
180
PA
VDD
GND

CHAPTER 9

SOURCE AND LOAD ANALYSIS

PURPOSE

In Chapter 7, you were introduced to the fundamentals of series-parallel circuit operation and analysis. In this chapter, we will look at some tools that are used to address some specific types of circuit analysis problems. These tools are generally referred to as **NETWORK THEOREMS**.

In this chapter, you will be introduced to the most commonly used network theorems. They are generally used to predict how the output from a voltage source (or current source) reacts to a change in load demand. Analyzing the effect that a change in load has on the output from a voltage or current source is referred to as **LOAD ANALYSIS**.

KEY TERMS

The following terms are introduced and defined in this chapter on the pages indicated:

Ideal current source, 365
Load analysis, 362
Multisource circuit, 411
Network theorem, 362
Norton current (I_N), 400
Norton equivalent circuit, 400
Norton resistance (R_N), 400
Norton's theorem, 400
Shorted-load output current (I_{SL}), 369
Superposition theorem, 412
Thevenin equivalent circuit, 377
Thevenin resistance (R_{TH}), 377
Thevenin's theorem, 377
Thevenin voltage (E_{TH}), 377

OBJECTIVES

After completing this chapter, you should be able to:

1. Discuss the characteristics of ideal and practical *voltage and current sources*.
2. Convert a voltage source to a current source and vice versa.
3. Describe *load analysis*.
4. Discuss *Thevenin's theorem* and the purpose it serves.
5. Describe the processes used to determine the values of *Thevenin voltage* (V_{th}) and *Thevenin resistance* (R_{th}) for a series-parallel circuit.
6. Derive the *Thevenin equivalent* of a given series-parallel circuit.
7. Use Thevenin's theorem to determine the range of output voltages for a series-parallel circuit with a variable load.

8. Use Thevenin's theorem to determine the maximum possible value of load power for a series-parallel circuit with a variable load.
9. Discuss *Norton's theorem* and the purpose it serves.
10. Describe the processes used to determine the values of *Norton current* (I_N) and *Norton resistance* (R_N) for a series-parallel circuit.
12. Analyze the operation of a series-parallel circuit using a *Norton equivalent circuit.*
13. Describe and perform *Thevenin-to-Norton* and *Norton-to-Thevenin* conversions.
14. Analyze multisource circuits using the *superposition* theorem.

NETWORK THEOREMS Analysis techniques that are used to solve specific types of circuit problems.

LOAD ANALYSIS Analyzing the effect that a change in load has on the output from a voltage or current source.

9.1 VOLTAGE AND CURRENT SOURCES

In Chapter 7, you learned that circuit output voltage varies directly with load resistance. That is,

- An increase in load resistance causes an increase in circuit output voltage
- A decrease in load resistance causes a decrease in circuit output voltage.

In this section, we begin our study of nework theorems by reviewing the operating characteristics of voltage and current sources and the methods used to convert each type of source to the other.

Practical Voltage Sources: The Effects of Source Resistance

An *ideal voltage source* maintains a constant output voltage regardless of its load resistance. For example, let's say that the voltage source in Figure 9.1a is an ideal

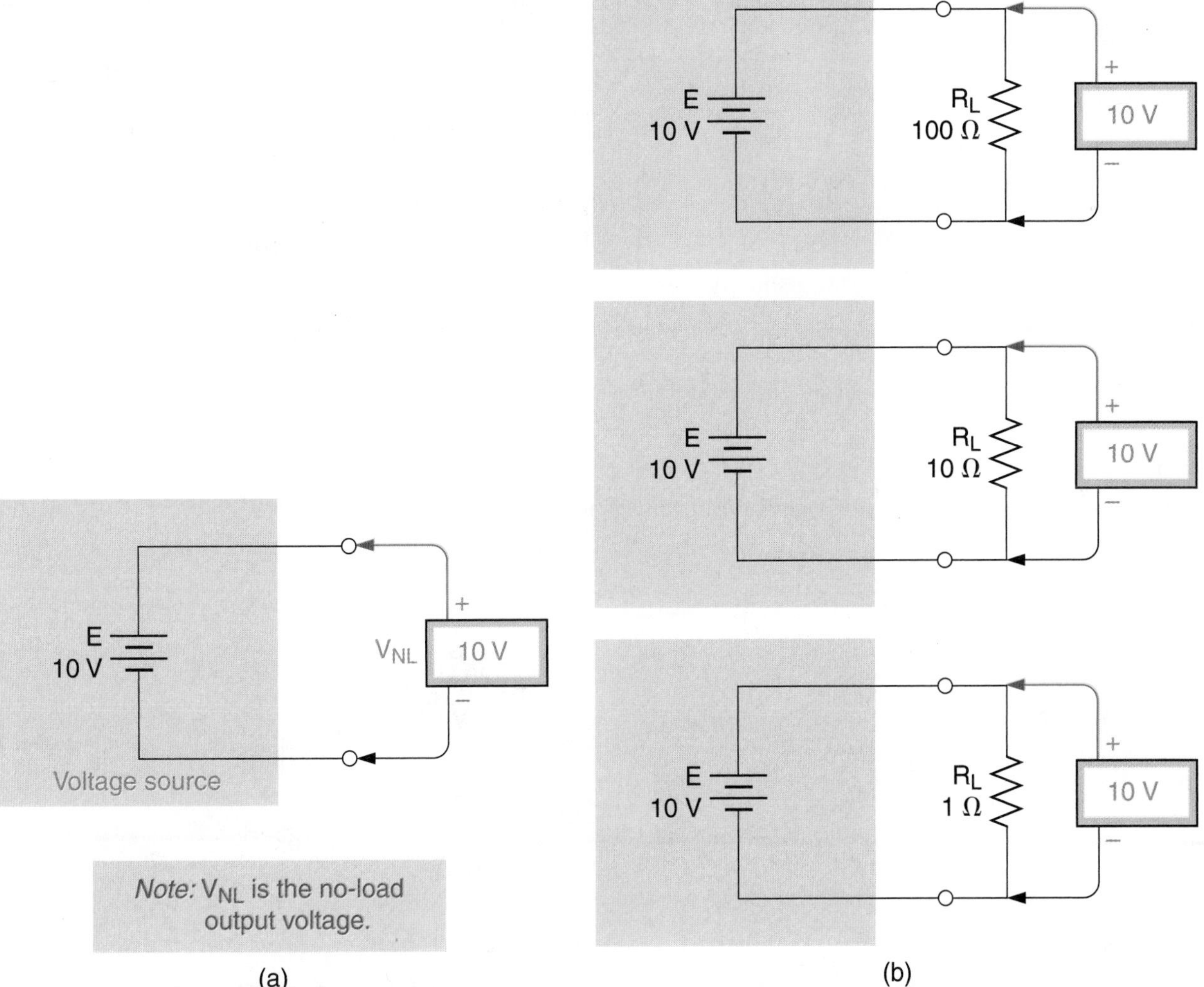

FIGURE 9.1 Output voltages from an ideal voltage source.

voltage source. As shown in the figure, the voltage across the open terminals of the source is 10 V. In Chapter 7, you learned that this "open terminal" voltage is referred to as the no-load output voltage (V_{NL}). When the various load resistances shown in Figure 9.1b are connected to the source, it maintains the same 10 V output. Thus, for an ideal voltage source,

$$V_{NL} = V_{RL}$$

regardless of the value of (R_L).

Unfortunately, ideal voltage sources do not exist at this time. For any *practical voltage source*, a decrease in load resistance results in a decrease in the source voltage. The reason for this lies in the fact that every voltage source has some amount of internal *source resistance*, as represented by the resistor (R_S) in Figure 9.2a. When a load is connected to the source as shown in Figure 9.2b, it forms a voltage divider with the internal resistance of the source. This causes V_{RL} to be lower than the no-load output voltage (V_{NL}), as demonstrated in Example 9.1.

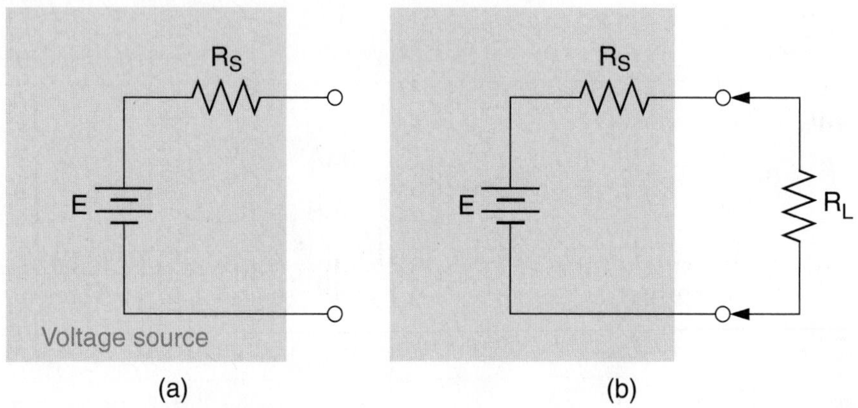

FIGURE 9.2 A practical voltage source.

EXAMPLE 9.1

The no-load output voltage of the source in Figure 9.3 is 12 V. Calculate the values of V_{RL} for $R_L = 100\ \Omega$ and $R_L = 20\ \Omega$.

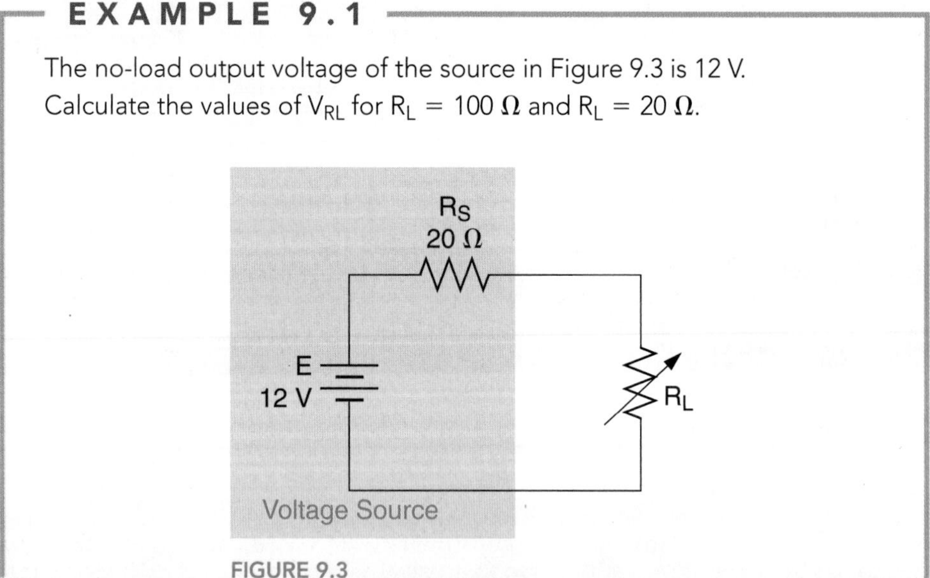

FIGURE 9.3

SOLUTION

The load resistance forms a voltage divider with the internal resistance of the source (R_S). When $R_L = 100\ \Omega$, V_{RL} is found as

$$V_{RL} = E \times \frac{R_L}{R_S + R_L}$$

$$= 12\text{ V} \times \frac{100\ \Omega}{120\ \Omega}$$

$$= 10\text{ V}$$

When $R_L = 20\ \Omega$, V_{RL} is found as

$$V_{RL} = E \times \frac{R_L}{R_S + R_L}$$

$$= 12\text{ V} \times \frac{20\ \Omega}{40\ \Omega}$$

$$= 6\text{ V}$$

As you can see, the decrease in load resistance caused a drastic decrease in V_{RL}.

PRACTICE PROBLEM 9.1

A voltage source with 40 Ω of internal resistance has a no-load output of 14 V. Determine the values of V_{RL} when $R_L = 240\ \Omega$ and $R_L = 120\ \Omega$.

The internal resistance of most DC voltage sources is 50 Ω or less. As such, it does not present a major problem for loads in the low kΩ range or higher. However, it can cause a significant drop in output voltage when a low-resistance load is present. This is why low internal resistance is considered desirable for a DC voltage source.

The Bottom Line

An ideal voltage source provides a constant output, regardless of the value of R_L. A practical voltage source, on the other hand, has an output voltage that varies directly with the value of R_L. This means that a change in load resistance will cause a similar change in load voltage. The characteristics of ideal and practical voltage sources are summarized in Figure 9.4.

Practical Current Sources: The Effects of Source Resistance

An **IDEAL CURRENT SOURCE** maintains a constant output current regardless of its load resistance. For example, let's say the current source in Figure 9.5a is an ideal source. As shown in the figure, the current source is rated at 100 mA. When the

IDEAL CURRENT SOURCE A current source that maintains a constant output current regardless of its load resistance.

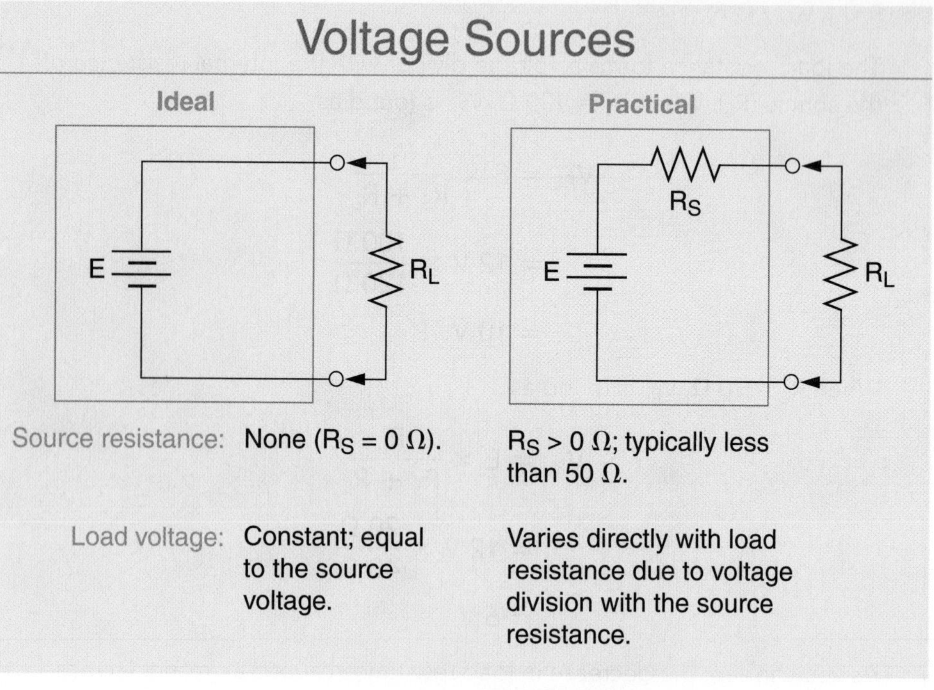

FIGURE 9.4 Ideal vs. practical voltage sources.

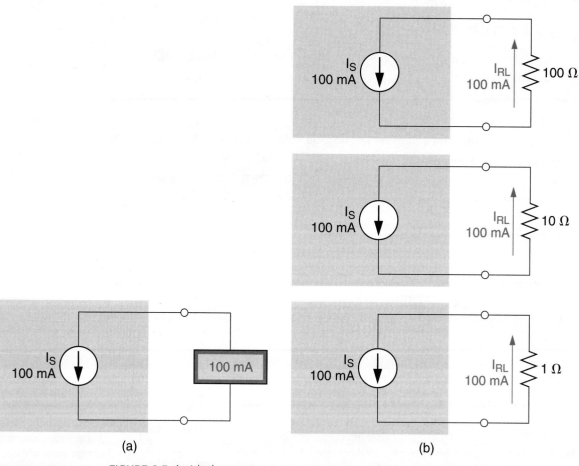

FIGURE 9.5 An ideal current source.

various load resistances shown in Figure 9.5b are connected to the source, it maintains the same 100 mA output. Thus, for an ideal current source,

$$I_S = I_{RL}$$

regardless of the value of R_L.

An ideal current source maintains a constant output because its internal resistance is infinite. For example, take a look at Figure 9.6a. The *source resistance* (R_S) is represented by a resistor that is connected in parallel with the source. With the source resistance being infinite, all of the circuit current passes through the load, regardless of its resistance (R_L).

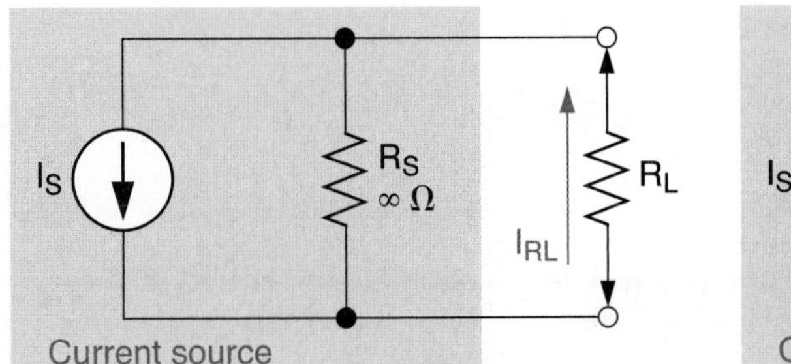

(a) Ideal current source

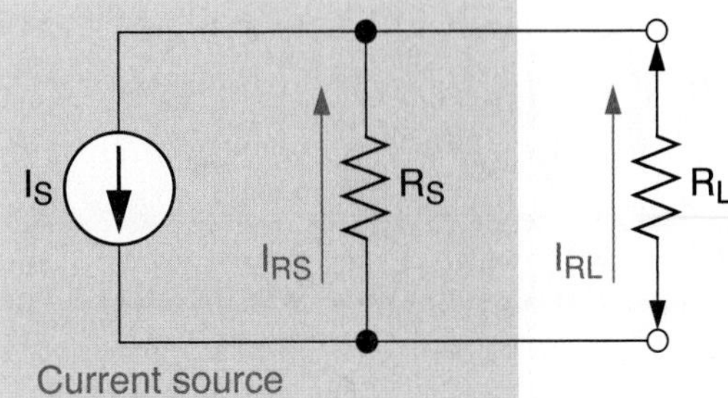

(b) Practical current source

FIGURE 9.6 A practical current source.

In practice, the output from a current source does vary when there is a change in load resistance because R_S is not infinite. When a load is connected to a practical current source, it forms a current divider with the source resistance as shown in Figure 9.6b. As a result, there is a change in the output from the current source when the load resistance changes, as demonstrated in Example 9.2.

EXAMPLE 9.2

The current source in Figure 9.7 has a variable load resistance.

Calculate the change in load current that occurs if the load resistance is changed from 200 Ω to 500 Ω.

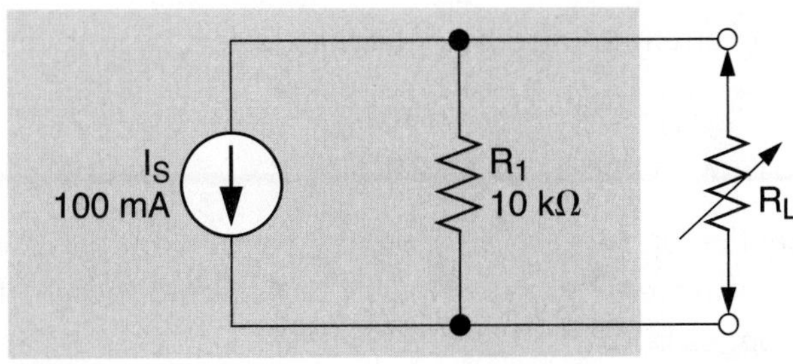

FIGURE 9.7

SOLUTION

As shown in Figure 9.7, the load and source resistances are in parallel. When $R_L = 200\ \Omega$,

$$R_T = \frac{1}{\dfrac{1}{R_S} + \dfrac{1}{R_L}}$$

$$= \frac{1}{\dfrac{1}{10{,}000\ \Omega} + \dfrac{1}{200\ \Omega}}$$

$$= 196\ \Omega$$

and

$$I_{RL} = I_S \times \frac{R_T}{R_L}$$

$$= 100\text{ mA} \times \frac{196\ \Omega}{200\ \Omega}$$

$$= 98\text{ mA}$$

When $R_L = 500\ \Omega$,

$$R_T = \frac{1}{\dfrac{1}{R_S} + \dfrac{1}{R_L}}$$

$$= \frac{1}{\dfrac{1}{10{,}000\ \Omega} + \dfrac{1}{500\ \Omega}}$$

$$= 476\ \Omega$$

and

$$I_{RL} = I_S \times \frac{R_T}{R_L}$$

$$= 100\text{ mA} \times \frac{476\ \Omega}{500\ \Omega}$$

$$= 95.2\text{ mA}$$

The change in current equals the difference between our two I_{RL} values, 2.8 mA.

PRACTICE PROBLEM 9.2

A circuit like the one in Figure 9.7 has a value of $R_S = 200\ \text{k}\Omega$ and $I_S = 50$ mA. Calculate the change in load current that occurs if the load resistance changes from 1 kΩ to 5 kΩ.

The change in load current calculated in the example (from 98 mA to 95.2 mA) is very slight when you consider the fact that load resistance *doubled.*

The maximum output from a current source is referred to as the **SHORTED-LOAD OUTPUT CURRENT (ISL)**. The value of I_{SL} can be measured as shown in Figure 9.8. Assuming that the internal resistance of the current meter is 0 Ω, all of the source current passes through the meter when it is connected as shown in the figure (because the meter shorts out R_S). Thus, I_{SL} equals the rated value of the source current (I_S). At the same time, the current through the load (I_{RL}) is less than I_{SL} when $R_L > 0\ \Omega$. This relationship is demonstrated in Example 9.3.

SHORTED-LOAD OUTPUT CURRENT (I_{SL}) The maximum output from a current source, measured with its output terminals shorted.

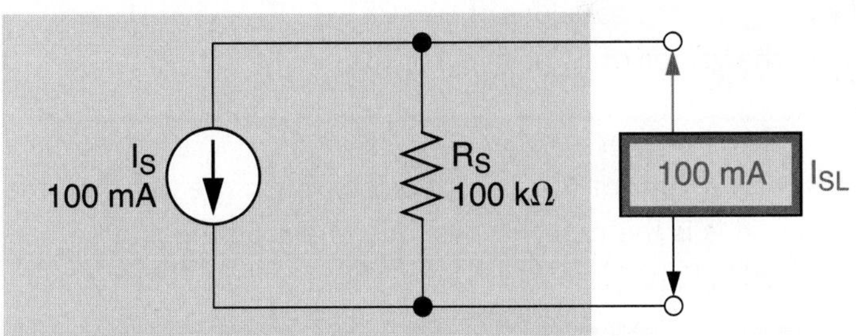

FIGURE 9.8 Measuring short circuit output current.

EXAMPLE 9.3

Calculate the values of I_{SL} and I_{RL} for the circuit in Figure 9.9.

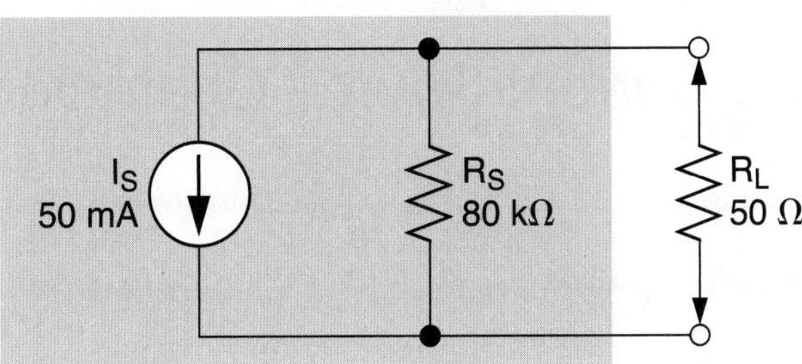

FIGURE 9.9

SOLUTION

When the load resistance is shorted, the current through the short equals the rated source current, as follows:

$$I_{SL} = I_S = 50\text{ mA}$$

When the load is connected as shown, the total circuit resistance is found as

$$R_T = \frac{1}{\dfrac{1}{R_S} + \dfrac{1}{R_L}}$$

$$= \frac{1}{\dfrac{1}{80{,}000\ \Omega} + \dfrac{1}{50\ \Omega}}$$

$$= 49.97\ \Omega$$

and

$$I_{RL} = I_S \times \frac{R_T}{R_L}$$

$$= 50 \text{ mA} \times \frac{49.97 \ \Omega}{50 \ \Omega}$$

$$= 49.97 \text{ mA}$$

PRACTICE PROBLEM 9.3

A circuit like the one in Figure 9.9 has values of $I_S = 25$ mA, $R_S = 50$ kΩ, and $R_L = 5$ kΩ. Calculate the values of I_{SL} and I_{RL} for the circuit.

NOTE: The higher the load resistance resistance, the greater the difference between the calculated and measured values of load current. This relationship is the opposite of that between load resistance the the internal resistance of a voltage source.

The internal resistance of a current source is typically much greater than the resistance of its load. When this is the case, R_S has little effect on the source output current (as demonstrated in Example 9.3).

Voltage and Current Sources: A Comparison

The characteristics of voltage and current sources are compared in Figure 9.10. As you can see, the internal resistance of the voltage source causes the circuit to act

Voltage and Current Source Characteristics

	Voltage Source	Current Source
	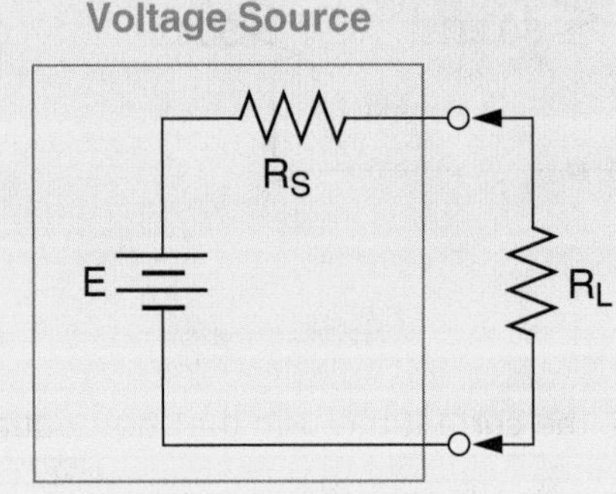	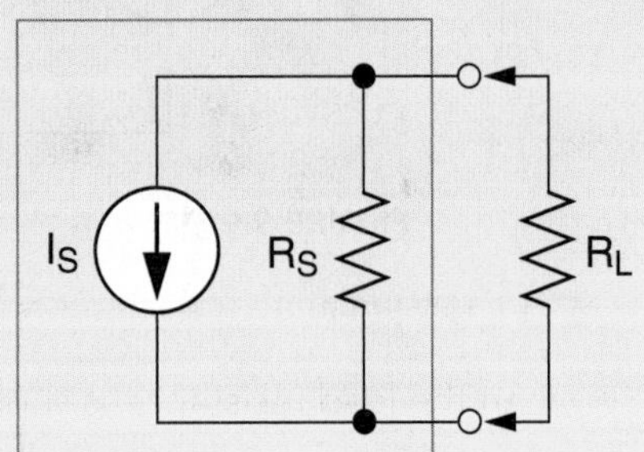
Source resistance:	Ideally zero; typically less than 50 Ω.	Ideally infinite; typically in the high-kΩ range.
Output:	The output voltage is determined by the source voltage rating, the source resistance, and the load resistance. Varies directly with load resistance.	The output current is determined by the source current rating, the source resistance, and the load resistance. Varies inversely with load resistance.

FIGURE 9.10 Voltage vs. current sources.

as a voltage divider. The internal resistance of the current source causes that circuit to act as a current divider. In each case, the appropriate "divider" relationship is used to determine the output value when a load is connected.

Equivalent Voltage and Current Sources

For every voltage source, there exists an equivalent current source (and vice versa). For any value of load resistance, a voltage source and its current source equivalent will provide the same output values. This point is demonstrated in Example 9.4.

EXAMPLE 9.4

Figure 9.11 shows a voltage source and its current source equivalent. Calculate the value of V_{RL} provided by each of the circuits for values of $R_L = 400\ \Omega$ and $R_L = 100\ \Omega$.

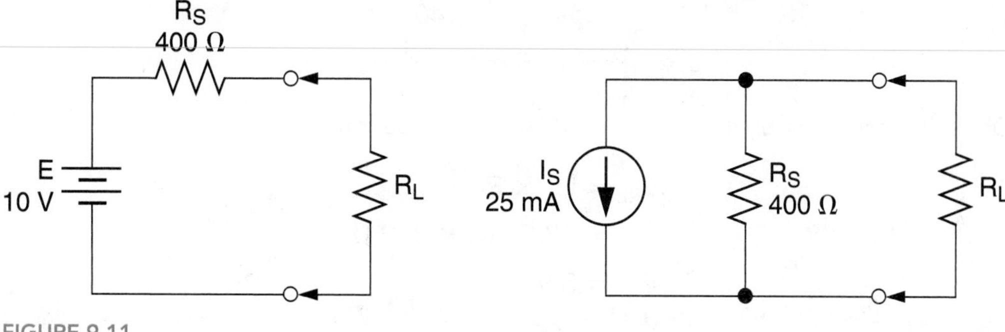

FIGURE 9.11

SOLUTION

First, we'll calculate the voltage source values. When $R_L = 400\ \Omega$, the output from the voltage source can be found as

$$V_{RL} = E \times \frac{R_L}{R_L + R_S}$$

$$= 10\text{ V} \times \frac{400\ \Omega}{800\ \Omega}$$

$$= 5\text{ V}$$

When $R_L = 100\ \Omega$, the output from the voltage source can be found as

$$V_{RL} = E \times \frac{R_L}{R_L + R_S}$$

$$= 10\text{ V} \times \frac{100\ \Omega}{500\ \Omega}$$

$$= 2\text{ V}$$

Now, we'll calculate the current source values. Since it provides an output current, an extra step will be needed to determine the output voltage for each value of R_L. When $R_L = 400\ \Omega$, the output from the current source can be found as

$$R_T = \frac{1}{\frac{1}{R_S} + \frac{1}{R_L}}$$

$$= \frac{1}{\frac{1}{400\ \Omega} + \frac{1}{400\ \Omega}}$$

$$= 200\ \Omega$$

and

$$I_{RL} = I_S \times \frac{R_T}{R_L}$$

$$= 25\text{ mA} \times \frac{200\ \Omega}{400\ \Omega}$$

$$= 12.5\text{ mA}$$

Using Ohm's law, the value of V_{RL} (for $R_L = 400\ \Omega$) can now be found as

$$V_{RL} = I_{RL} \times R_L$$

$$= 12.5\text{ mA} \times 400\ \Omega$$

$$= 5\text{ V}$$

When $R_L = 100\ \Omega$, the value of I_{RL} can be found as

$$R_T = \frac{1}{\frac{1}{R_S} + \frac{1}{R_L}}$$

$$= \frac{1}{\frac{1}{400\ \Omega} + \frac{1}{100\ \Omega}}$$

$$= 80\ \Omega$$

and

$$I_{RL} = I_S \times \frac{R_T}{R_L}$$

$$= 25\text{ mA} \times \frac{80\ \Omega}{100\ \Omega}$$

$$= 20\text{ mA}$$

Using Ohm's law, the value of V_L (for $R_L = 100\ \Omega$) can now be found as

$$V_{RL} = I_{RL} \times R_L$$
$$= 20\text{ mA} \times 100\ \Omega$$
$$= 2\text{ V}$$

PRACTICE PROBLEM 9.4

Change the value of R_S for each circuit in Figure 9.11 to 100 Ω. Then, show that the circuits will produce the same load *current* for values of $R_L = 100\ \Omega$ and $R_L = 25\ \Omega$.

The results from Example 9.4 can be used to demonstrate the relationship between the sources. As a reference, the results from the example are summarized as follows:

Voltage Source		*Current Source*	
LOAD RESISTANCE (R_L)	LOAD VOLTAGE (V_{RL})	LOAD CURRENT (I_{RL})	LOAD VOLTAGE (V_{RL})
400 Ω	5 V	12.5 mA	5 V
100 Ω	2 V	20 mA	2 V

The fact that the two circuits produced the same output voltage for each value of load resistance demonstrates that they *are* equivalent circuits.

Source Conversions

Converting a voltage source to a current source (or vice versa) is fairly simple. The relationships used to convert one type of source to the other are shown in Figure 9.12. As you can see, the value of R_S does not change from one circuit to the other. When converting a voltage source to a current source (as shown in Figure 9.12a), the source current (I_S) is found as

(9.1)
$$I_S = \frac{E}{R_S}$$

Once the value of I_S has been calculated, the current source is drawn as shown in Figure 9.12a. Note that the direction of the arrow in the current source must match the direction of the current from the voltage source. The process for converting a voltage source to a current source is demonstrated further in Example 9.5.

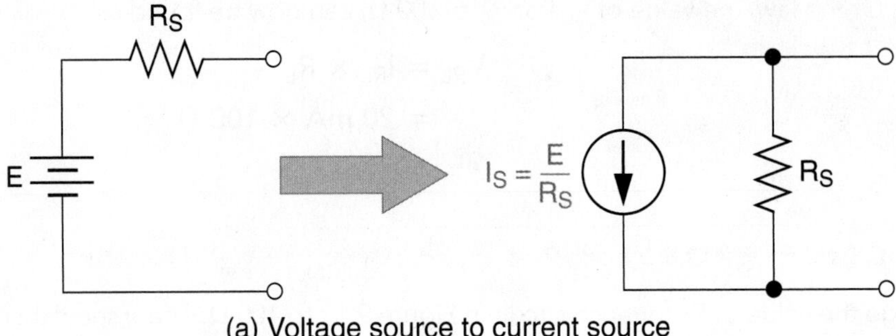

(a) Voltage source to current source

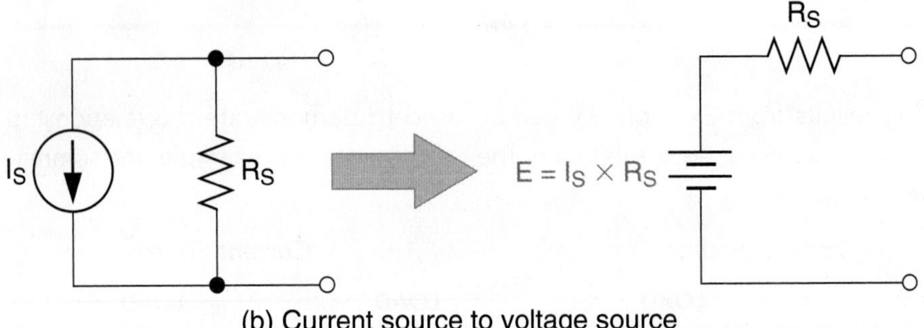

(b) Current source to voltage source

FIGURE 9.12 Source conversions.

EXAMPLE 9.5

Convert the voltage source in Figure 9.13a to an equivalent current source.

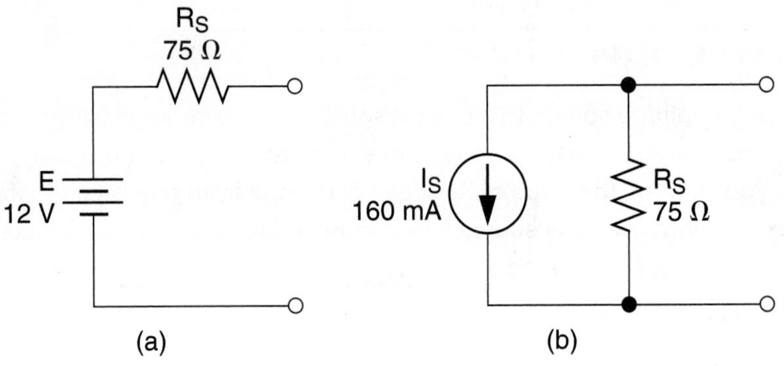

FIGURE 9.13

SOLUTION

The value of source resistance remains the same as shown in the figure, 75 Ω. The value of I_S is found using equation (9.1), as follows:

$$
\begin{aligned}
I_S &= \frac{E}{R_S} \\
&= \frac{12\ \text{V}}{75\ \Omega} \\
&= 160\ \text{mA}
\end{aligned}
$$

Using the values of $I_S = 160$ mA and $R_S = 75\ \Omega$, the current source is drawn as shown in Figure 9.13b.

PRACTICE PROBLEM 9.5

A voltage source like the one in Figure 9.13a has values of $E = 8$ V and $R_S = 16\ \Omega$. Convert the voltage source to an equivalent current source. Then, determine the output *current* provided by both circuits for a 32 Ω load.

When converting a current source to a voltage source (as shown in Figure 9.12), the value of source voltage is found as

$$E = I_S \times R_S \tag{9.2}$$

Once the value of E is known, the voltage source is drawn as shown in Figure 9.12b. Again, care must be taken to make sure the polarity of the voltage source matches the direction of current indicated by the current source. The process for converting a current source to a voltage source is demonstrated in Example 9.6.

EXAMPLE 9.6

Convert the current source in Figure 9.14a to an equivalent voltage source.

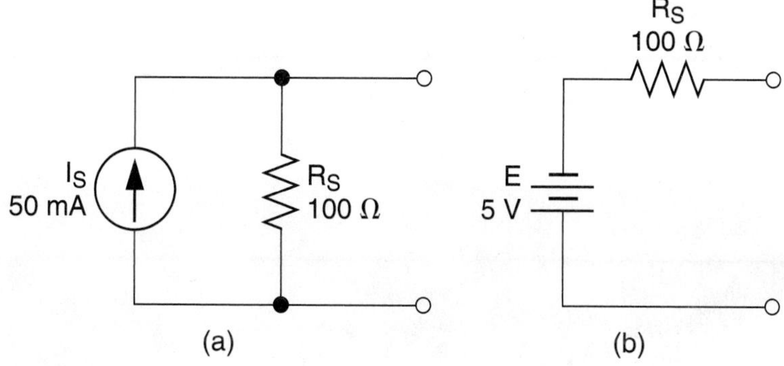

FIGURE 9.14

SOLUTION

The value of the source resistance remains the same as shown in the figure, 100 Ω.

The value of E is found using equation (9.2), as follows:

$$\begin{aligned} E &= I_S \times R_S \\ &= 50\text{ mA} \times 100\ \Omega \\ &= 5\text{ V} \end{aligned}$$

Using the values of E = 5 V and R_S = 100 Ω, the voltage source is drawn as shown in Figure 9.15b. Note that the polarity of the voltage source matches that indicated by the current source.

PRACTICE PROBLEM 9.6

A current source like the one in Figure 9.14a has values of I_S = 40 mA and R_S = 150 Ω. Convert the current source to an equivalent voltage source. Then, determine the output voltage provided by both circuits for a 100 Ω load.

Converting a voltage source to a current source (or vice versa) is not something that you will do on a regular basis. However, later in this chapter you will see how source conversions can simplify certain circuit analysis problems.

PROGRESS CHECK

1. In terms of source resistance, how does the ideal voltage source differ from the practical voltage source?
2. In terms of *load voltage*, how does the ideal voltage source differ from the practical voltage source?
3. In terms of *source resistance*, how does the ideal current source differ from the practical current source?
4. In terms of *load current*, how does the ideal current source differ from the practical current source?
5. Compare and contrast *voltage sources* and *current sources*.
6. Describe the process for converting a voltage source to a current source.
7. Describe the process for converting a current source to a voltage source.

9.2 THEVENIN'S THEOREM

Thevenin's theorem is especially useful in analyzing power systems and other circuits where the load resistance varies. In such cases, Thevenin's theorem makes it easier to recalculate load voltages and currents when a change in load resistance occurs.

In the last section, you saw how many electronic circuits experience a change in output voltage and current when there is a change in load resistance.

The load analysis of a series or parallel circuit is relatively simple and straightforward. For example, if the load in Figure 9.15a changes, we can easily predict the change in load voltage using the voltage-divider relationship. By the same token, the current-divider relationship can be used to predict how the circuit in Figure 9.15b will respond to a change in load resistance.

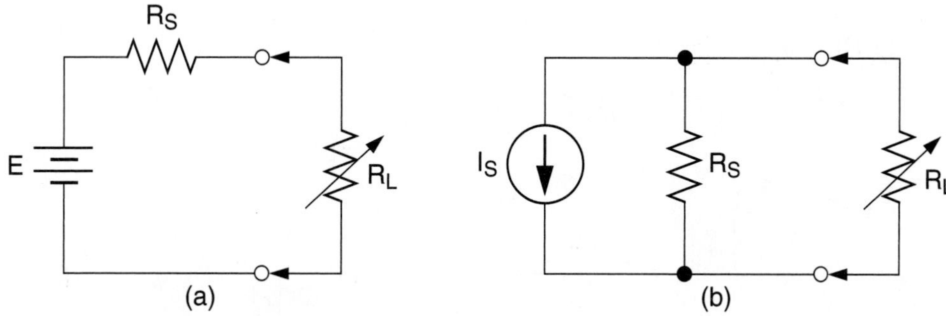

FIGURE 9.15 Voltage and current sources.

Using only the analysis techniques we have established so far, the load analysis of a series-parallel circuit could be a long and tedious process. For example, consider the circuit shown in Figure 9.16. Imagine trying to predict the value of load voltage produced by this circuit for the load resistance values shown. For each value of load resistance, we would have to:

1. Derive a new series-equivalent for the circuit,
2. Solve for the values in the equivalent circuit, and
3. Use those values to calculate the actual load voltage.

Solving this problem could take a *long* time. Fortunately, the type of problem represented in Figure 9.16 can be simplified by using **THEVENIN'S THEOREM**.

THEVENIN'S THEOREM A theorem stating that any resistive circuit or network, no matter how complex, can be represented as a voltage source in series with a source resistance.

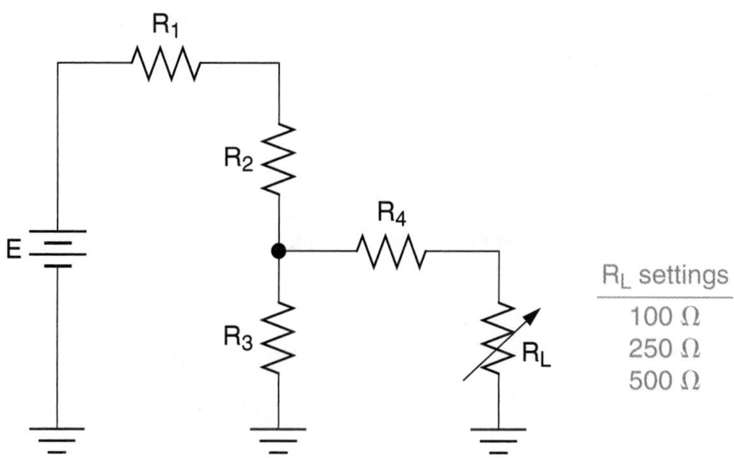

FIGURE 9.16 A loaded voltage divider.

Thevenin's theorem states that any resistive circuit or network, no matter how complex, can be represented as a voltage source in series with a source resistance. For example, the series-parallel circuit in Figure 9.17 can be represented by the **THEVENIN EQUIVALENT CIRCUIT** shown in the figure. Note that the source voltage and resistance in Figure 9.17b are referred to as the **THEVENIN VOLTAGE (E_{TH})** and the **THEVENIN RESISTANCE (R_{TH})**.

Later in this section, you are shown how to derive the Thevenin equivalent of a given series-parallel circuit. First, let's take a look at how this theorem can be used to simplify the load analysis of a series-parallel circuit.

THEVENIN EQUIVALENT CIRCUIT For any circuit, an equivalent that contains a voltage source (E_{TH}) in series with a source resistance (R_{TH}).

THEVENIN VOLTAGE (E_{TH}) The voltage source in a Thevenin equivalent circuit.

THEVENIN RESISTANCE (R_{TH}) The source resistance in a Thevenin equivalent circuit.

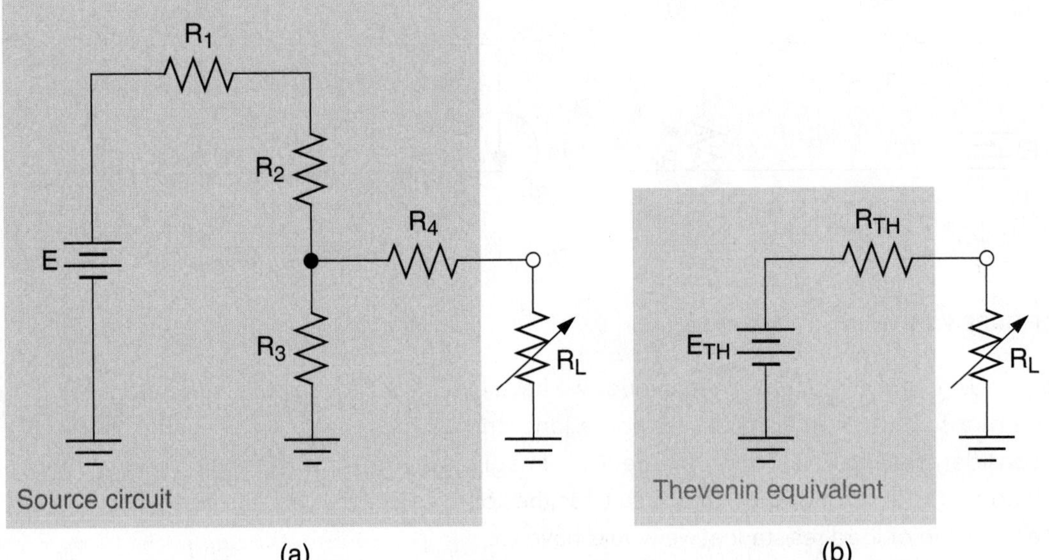

FIGURE 9.17 A circuit and its Thevenin equivalent.

Applications of Thevenin's Theorem

The strength of the Thevenin equivalent circuit lies in the fact that it produces the same output values as the original circuit for any given value of load resistance. For example, the circuits in Figures 9.17a and 9.17b will produce the same values of load voltage, current, and power when their load resistance values are equal. This relationship is demonstrated in Example 9.7.

EXAMPLE 9.7

Figure 9.18 contains a loaded voltage divider and its Thevenin equivalent circuit. Determine the load voltage produced by each circuit for load resistance values of 10 Ω and 18 Ω.

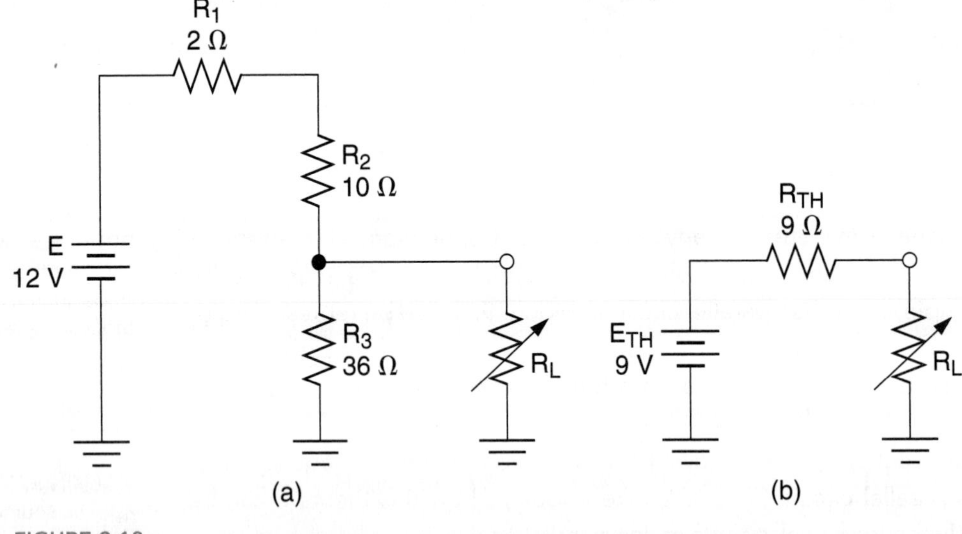

FIGURE 9.18

SOLUTION

When $R_L = 10\ \Omega$, the loaded voltage divider in Figure 9.18a has values of

$$R_T = \frac{1}{\frac{1}{R_3} + \frac{1}{R_L}}$$

$$= \frac{1}{\frac{1}{36\ \Omega} + \frac{1}{10\ \Omega}}$$

$$= 7.83\ \Omega$$

and

$$V_{EQ} = E \times \frac{R_{EQ}}{R_T}$$

$$= 12\ V \times \frac{7.83\ \Omega}{19.83\ \Omega}$$

$$= 4.74\ V$$

Therefore, $V_L = 4.74$ V when $R_L = 10\ \Omega$. When $R_L = 18\ \Omega$,

$$R_T = \frac{1}{\frac{1}{R_3} + \frac{1}{R_L}}$$

$$= \frac{1}{\frac{1}{36\ \Omega} + \frac{1}{18\ \Omega}}$$

$$= 12\ \Omega$$

and

$$V_{EQ} = E \times \frac{R_{EQ}}{R_T}$$

$$= 12\ V \times \frac{12\ \Omega}{24\ \Omega}$$

$$= 6\ V$$

Therefore, $V_L = 6$ V when $R_L = 18\ \Omega$.

Using the Thevenin equivalent circuit, we could solve the same problem as follows: When $R_L = 10\ \Omega$,

$$V_{RL} = E_{TH} \times \frac{R_L}{R_L + R_{TH}}$$

$$= 9\ V \times \frac{10\ \Omega}{19\ \Omega}$$

$$= 4.74\ V$$

When $R_L = 18\ \Omega$,

$$V_{RL} = E_{TH} \times \frac{R_L}{R_L + R_{TH}}$$

$$= 9\text{ V} \times \frac{18\ \Omega}{27\ \Omega}$$

$$= 6\text{ V}$$

PRACTICE PROBLEM 9.7

Figure 9.19 shows a series-parallel circuit and its Thevenin equivalent. Determine the output voltage produced by each circuit for values of $R_L = 150\ \Omega$ and $R_L = 30\ \Omega$.

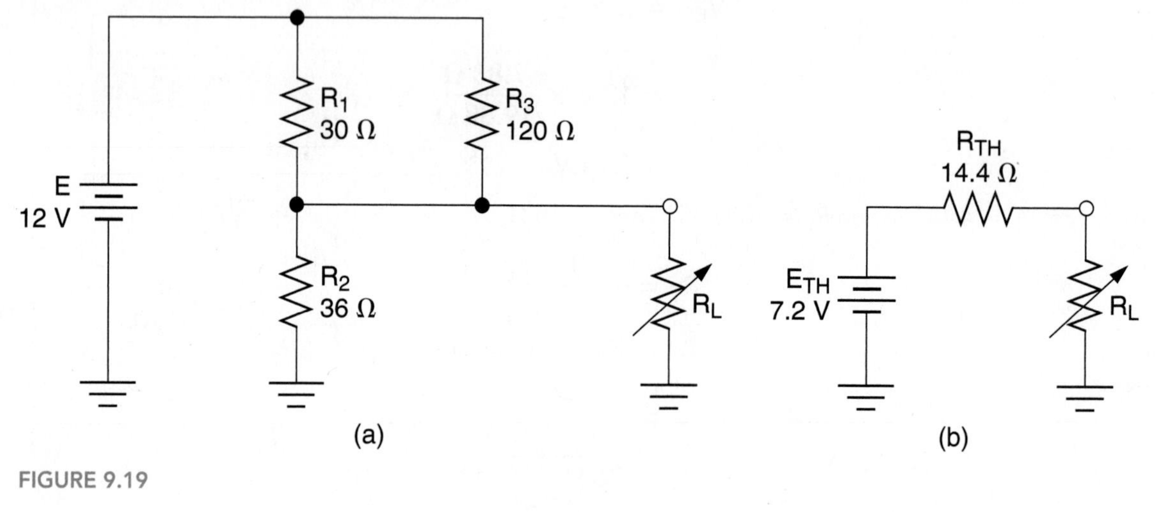

FIGURE 9.19

As you can see, the Thevenin equivalent for a series-parallel circuit produces the same output for any given value of load resistance. This means that the load analysis of a series-parallel circuit (like the one in Figure 9.17) can be performed as follows:

1. Derive the Thevenin equivalent for the series-parallel circuit.
2. For each value of load resistance, determine the output from the Thevenin equivalent circuit using the voltage-divider relationship.

The values obtained using this simple process are the same as those that are produced by the original circuit.

Calculating the Value of E_{TH}

Deriving the Thevenin equivalent of a series-parallel circuit begins with determining the value of Thevenin voltage (E_{TH}). Though we've given it a new name, the

Thevenin voltage for a series-parallel circuit is nothing more than its no-load output voltage (V_{NL}). That is, the voltage that is present at the output terminals of the circuit with the load removed. The measurement of E_{TH} for the circuit in Figure 9.20a is illustrated in Figure 9.20b. The process for calculating the value of E_{TH} is demonstrated in Example 9.8.

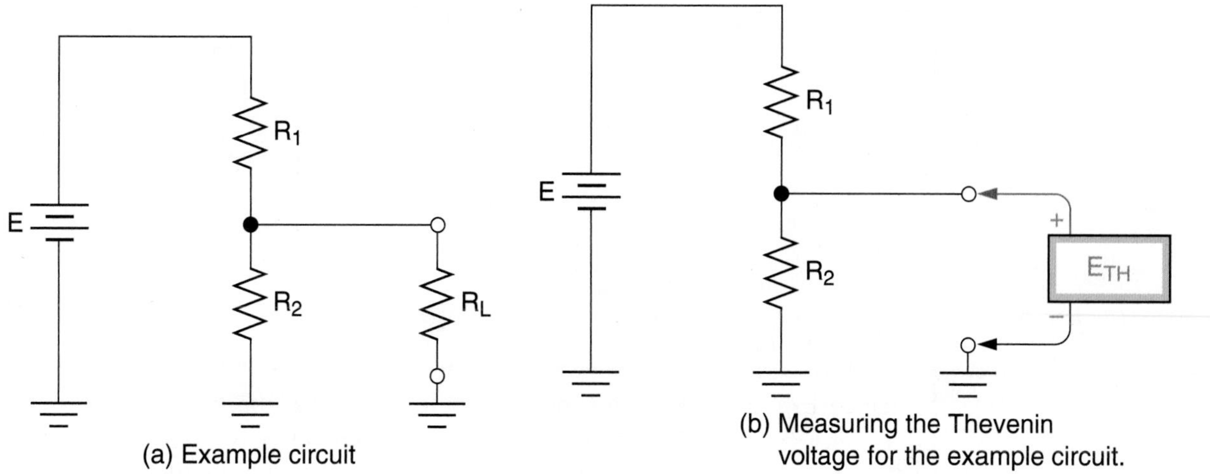

FIGURE 9.20 Measuring Thevenin voltage (E_{TH}).

EXAMPLE 9.8

Calculate the value of E_{TH} for the series-parallel circuit shown in Figure 9.21a.

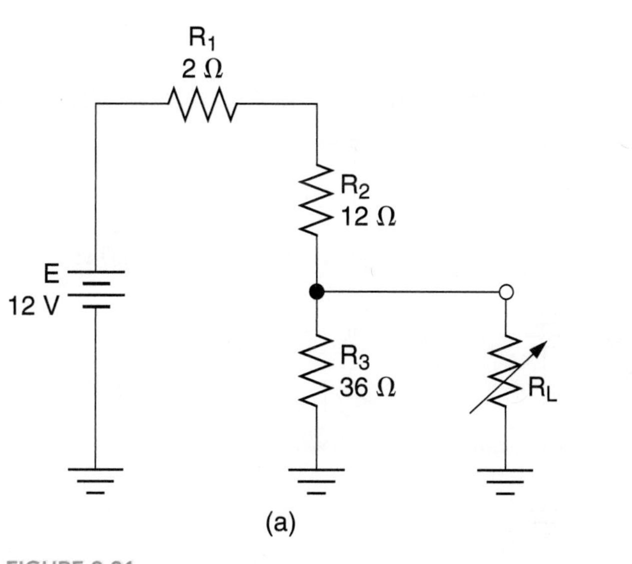

FIGURE 9.21

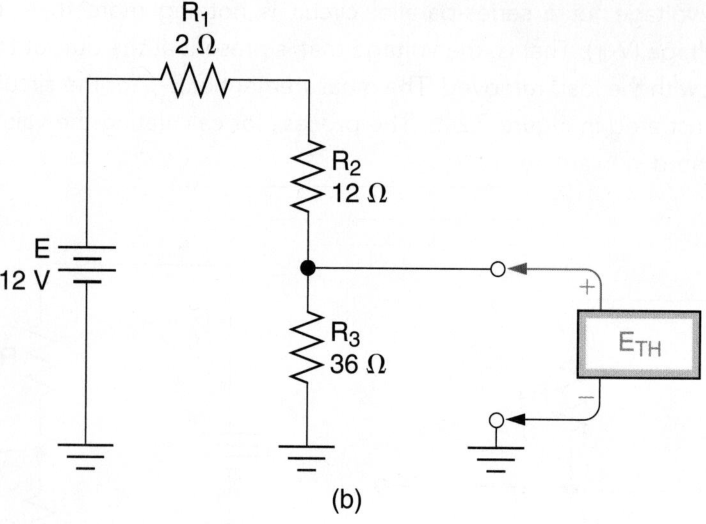

FIGURE 9.21 *(Continued)*

SOLUTION

Since E_{TH} is equal to the no-load output voltage, the first step is to remove the load. This gives us the circuit shown in Figure 9.21b. For this circuit, the voltage across the open load terminals is equal to the voltage across R_3. Therefore, E_{TH} can be found as

$$E_{TH} = E \times \frac{R_3}{R_T}$$

$$= 12\text{ V} \times \frac{36\ \Omega}{48\ \Omega}$$

$$= 9\text{ V}$$

PRACTICE PROBLEM 9.8

Determine the value of E_{TH} for the circuit shown in Figure 9.22.

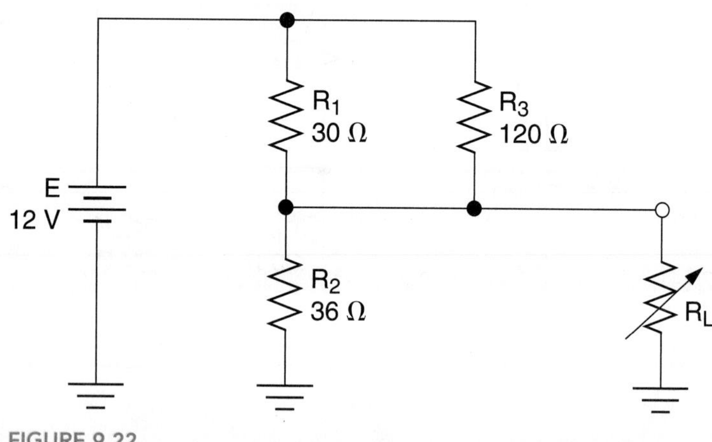

FIGURE 9.22

Thevenin Resistance (R_{TH})

The next step in deriving a Thevenin equivalent circuit is determining the value of Thevenin resistance (R_{TH}). In essence, R_{TH} can be described as the no-load output resistance (R_{NL}) of a circuit; that is, the resistance measured across the output terminals with the load removed. Figure 9.23 shows how this resistance can be measured. Note that:

1. The load has been removed.
2. The open source (E) terminals have been shorted together.

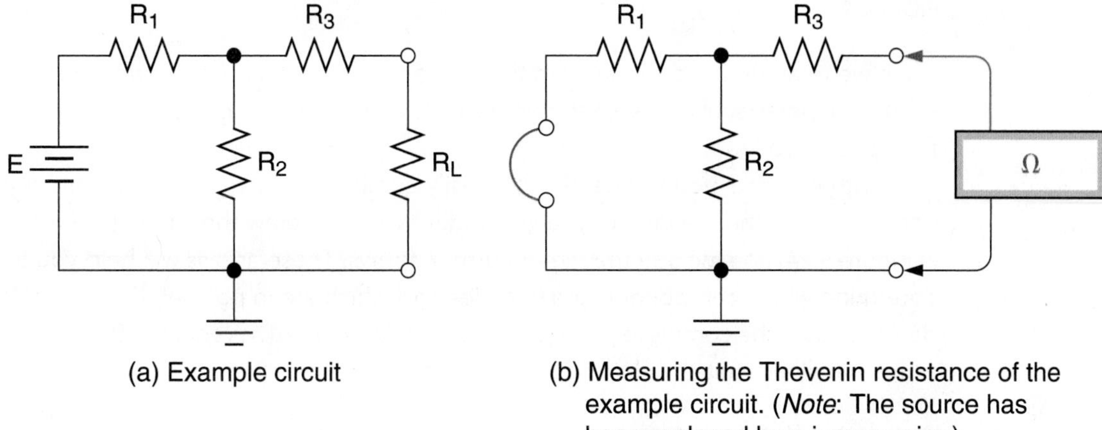

(a) Example circuit

(b) Measuring the Thevenin resistance of the example circuit. (*Note*: The source has been replaced by a jumper wire.)

FIGURE 9.23 Measuring Thevenin resistance (R_{TH}).

Why is the voltage source replaced with a wire? First, you know that the voltage source must be removed from the circuit before making any kind of resistance measurement. If we want an accurate resistance measurement, we must replace the source with a resistance that is equal to the source resistance (R_S). For example, if we assume that E is an ideal voltage source with an internal resistance of 0 Ω, then we must replace it with a resistance of approximately 0 Ω, in this case a wire. Otherwise, the measured output resistance will not be accurate. In most cases, the assumption that a DC voltage source has an internal resistance of close to 0 Ω is a valid one. For example, most modern batteries have internal resistance values that are less than 1 Ω!

The currents shown in Figure 9.24 represent the circuit currents generated by the meter. These currents indicate that (from the meter's perspective):

1. R_1 is in parallel with R_2.
2. R_3 is in series with the other two resistors.

Therefore, the Thevenin resistance of the circuit in Figure 9.24 is found as

$$R_{TH} = (R_1 \| R_2) + R_3$$

Calculating circuit resistance from the viewpoint of the load is a process that usually takes a while to grow accustomed to. In our previous experiences, we simplified circuits by combining resistor values from load to source. However,

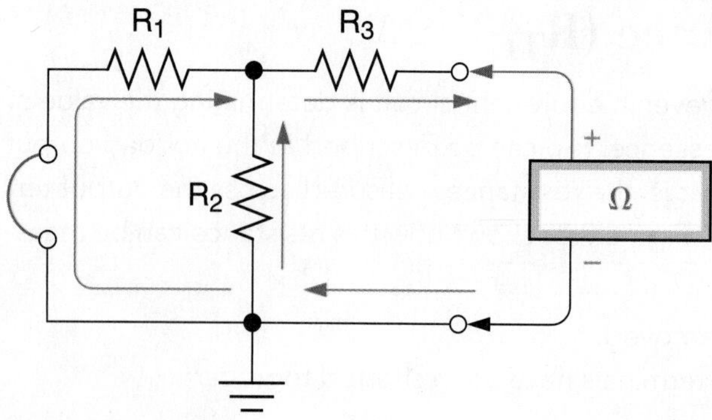

FIGURE 9.24

when we're analyzing a circuit from the viewpoint of the load (as shown in Figure 9.24), our perspective has changed. We are now combining resistance values *from source to load.*

With practice, you will be able to easily combine resistance values from any perspective. In the meantime, you will find it helpful to draw the circuit (as shown in Figure 9.24) and include the meter current arrows. These arrows will help you to determine which components are in series and which are in parallel. Example 9.9 demonstrates the complete procedure for calculating the Thevenin resistance of a series-parallel circuit.

EXAMPLE 9.9

Calculate the Thevenin resistance for the circuit in Figure 9.25a.

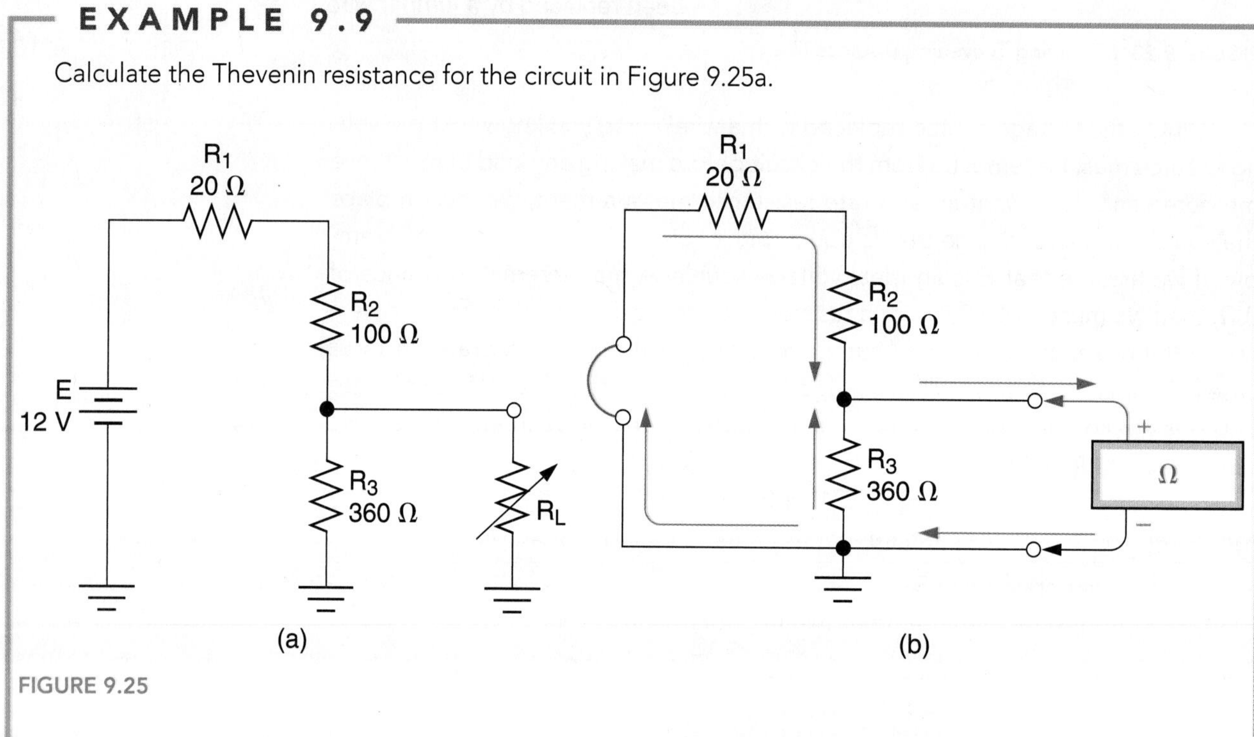

FIGURE 9.25

SOLUTION

First, the circuit is drawn as shown in Figure 9.25b. Note that the load has been removed and the source has been replaced by a wire. The current arrows in the figure represent the current that would be produced by an ohmmeter connected to the open load terminals. These currents indicate that:

1. R_1 is in series with R_2.
2. R_3 is in parallel with the other two resistors.

Therefore, the Thevenin resistance for the circuit can be found as

$$R_{TH} = (R_1 + R_2) \| R_3$$

$$= \frac{(R_1 + R_2) \times R_3}{(R_1 + R_2) + R_3}$$

$$= \frac{120 \times 360}{120 + 360}\ \Omega$$

$$= 90\ \Omega$$

PRACTICE PROBLEM 9.9

Calculate the Thevenin resistance for the circuit in Figure 9.22 (Example 9.8).

The results from the last two examples are illustrated in Figure 9.26. As you can see, the values of E_{TH} and R_{TH} were combined into a single Thevenin equivalent circuit. The relationship between the series-parallel circuit and its Thevenin equivalent was demonstrated in Example 9.7.

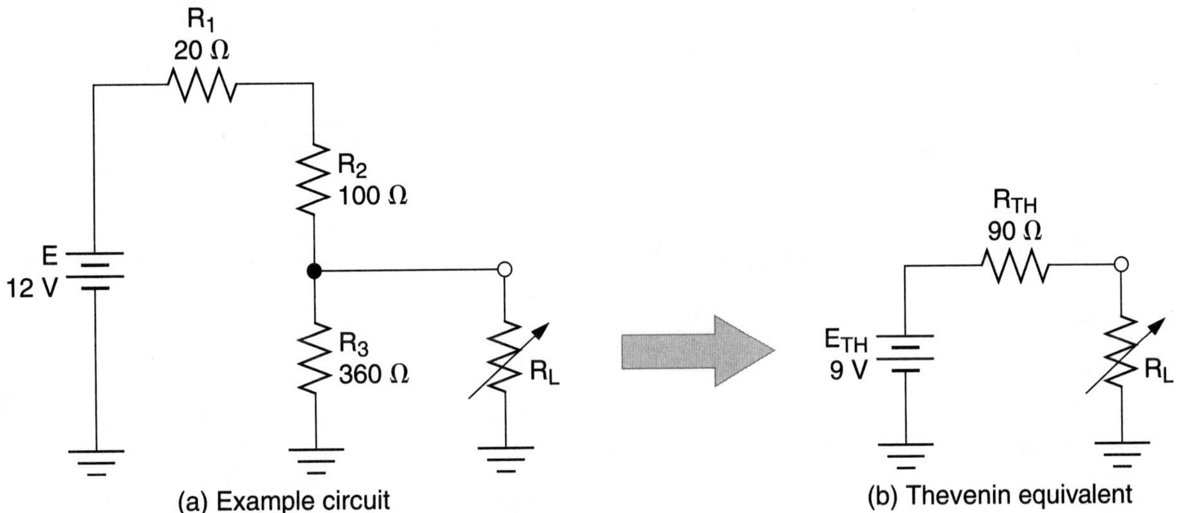

FIGURE 9.26 A series-parallel circuit and its Thevenin equivalent.

Deriving Thevenin Equivalent Circuits

You have been shown how to determine the Thevenin equivalent voltage and resistance for a series-parallel circuit. At this point, we'll put it all together by deriving the Thevenin equivalents for two example circuits.

EXAMPLE 9.10

Derive the Thevenin equivalent of the circuit shown in Figure 9.27a.

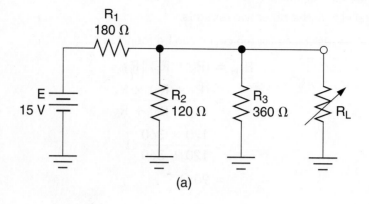

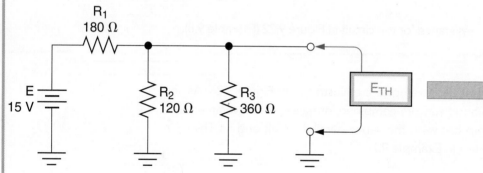

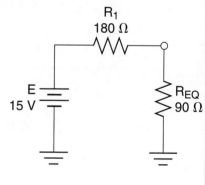

R_1 180 Ω
E 15 V
R_2 120 Ω
R_3 360 Ω
R_L

(a)

R_1 180 Ω
E 15 V
R_2 120 Ω
R_3 360 Ω
E_{TH}

R_1 180 Ω
E 15 V
R_{EQ} 90 Ω

(b)

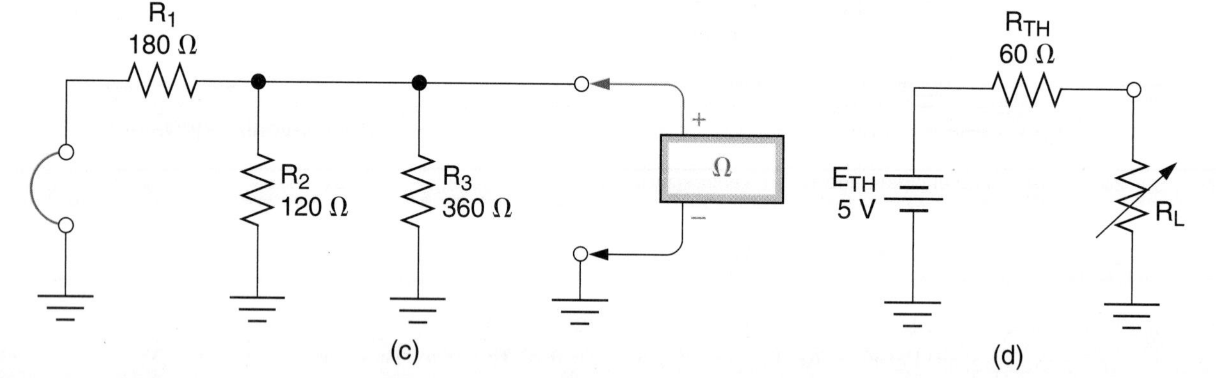

FIGURE 9.27

SOLUTION

We start by drawing the circuit with the load removed, as shown in Figure 9.27b. As you can see, the voltage across the open load terminals (E_{TH}) is equal to the voltage across the parallel combination of R_2 and R_3. Combining these two resistors into a single equivalent resistance, we get the series-equivalent circuit shown. The value of R_{EQ} for this circuit is found as

$$R_{EQ} = \frac{1}{\frac{1}{R_2} + \frac{1}{R_3}}$$

$$= \frac{1}{\frac{1}{120\ \Omega} + \frac{1}{360\ \Omega}}$$

$$= 90\ \Omega$$

Since the Thevenin voltage is equal to the voltage across R_{EQ}, its value can be found as

$$E_{TH} = E \times \frac{R_{EQ}}{R_1 + R_{EQ}}$$

$$= 15\ \text{V} \times \frac{90\ \Omega}{270\ \Omega}$$

$$= 5\ \text{V}$$

To calculate the value of R_{TH}, we need to replace the voltage source with a wire, as shown in Figure 9.27c. The resistance between the open output terminals (R_{TH}) equals the parallel combination of R_1, R_2, and R_3. By formula,

$$R_{TH} = R_1 \| R_2 \| R_3$$

Because we have three resistors in parallel, we'll use the reciprocal approach to calculating R_{TH}, as follows:

$$R_{TH} = \frac{1}{\frac{1}{R_1} + \frac{1}{R_2} + \frac{1}{R_3}}$$

$$= \frac{1}{\frac{1}{180\ \Omega} + \frac{1}{120\ \Omega} + \frac{1}{360\ \Omega}}$$

$$= 60\ \Omega$$

Using our calculated values of E_{TH} and R_{TH}, the Thevenin equivalent of the original circuit is drawn as shown in Figure 9.27d.

PRACTICE PROBLEM 9.10

Derive the Thevenin equivalent of the circuit in Figure 9.28.

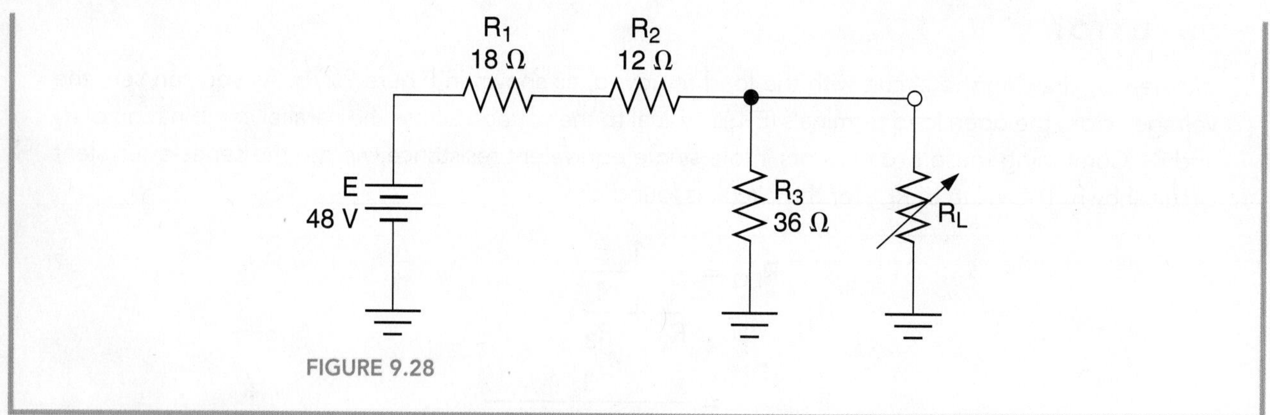

FIGURE 9.28

When a resistor is in series with the load, it has no effect on the value of E_{TH}. This point is demonstrated in Example 9.11.

EXAMPLE 9.11

Derive the Thevenin equivalent of the circuit in Figure 9.29a.

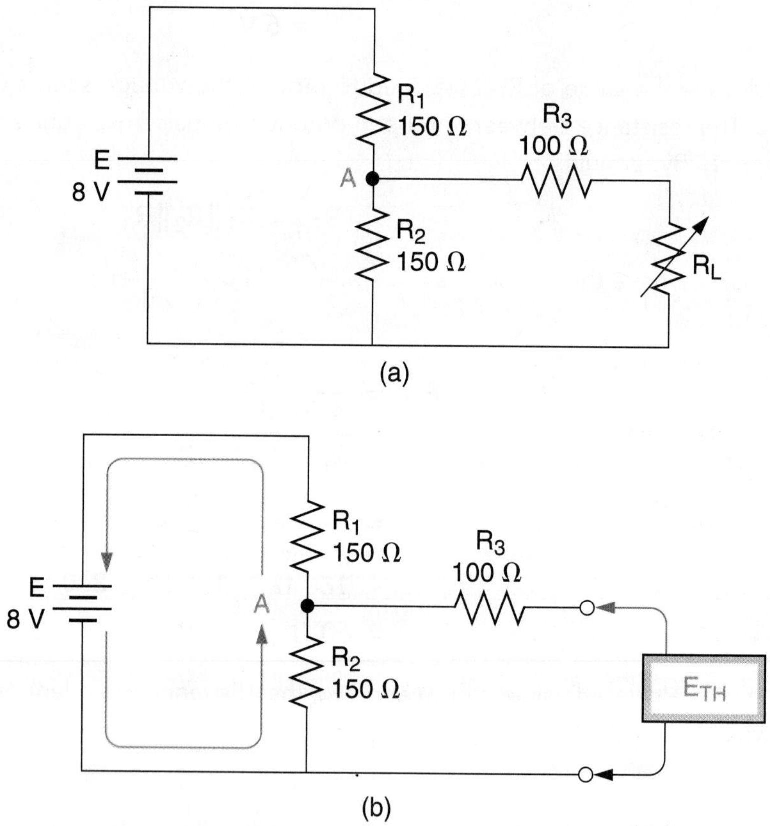

(a)

(b)

FIGURE 9.29

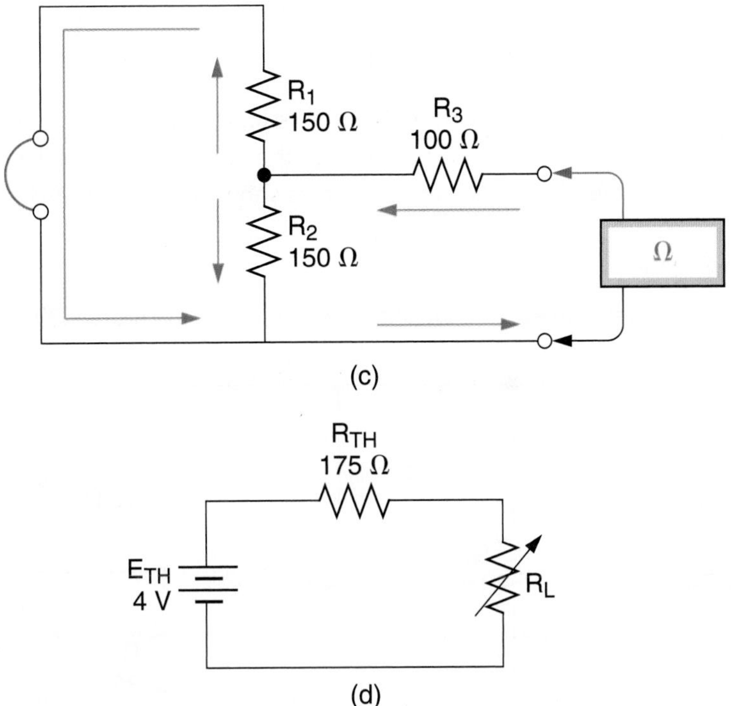

FIGURE 9.29 *(Continued)*

SOLUTION

First, the load is removed, giving us the circuit shown in Figure 9.29b. Note that the source current passes only through R_1 and R_2, as R_3 is not a part of any complete current path. Since $I_{R3} = 0$ A, there is no voltage across R_3. Therefore, the voltage at the load terminal equals the voltage at point (A); that is $E_{TH} = V_{R2}$. The value of E_{TH} for Figure 9.29a can be found as

$$E_{TH} = E \times \frac{R_2}{R_1 + R_2}$$

$$= 8\text{ V} \times \frac{150\ \Omega}{300\ \Omega}$$

$$= 4\text{ V}$$

The value of R_{TH} for the circuit can be calculated with the help of Figure 9.29c. The current arrows represent the current produced by an ohmmeter connected to the output terminals of the circuit. As these arrows indicate, the parallel combination of R_1 and R_2 is in series with R_3. Combining these combinations into a single equation, we can calculate the value of R_{TH} as follows:

$$R_{TH} = \frac{R_1 \times R_2}{R_1 + R_2} + R_3$$

$$= \frac{150 \times 150}{150 + 150}\ \Omega + 100\ \Omega$$

$= 75\ \Omega + 100\ \Omega$

$= 175\ \Omega$

As you can see, R_3 (which is in series with the load) did not affect the value of E_{TH}. At the same time, it *did* weigh into the value of R_{TH}. The Thevenin equivalent of the original circuit is drawn as shown in Figure 9.29d.

PRACTICE PROBLEM 9.11

Derive the Thevenin equivalent of the circuit shown in Figure 9.30.

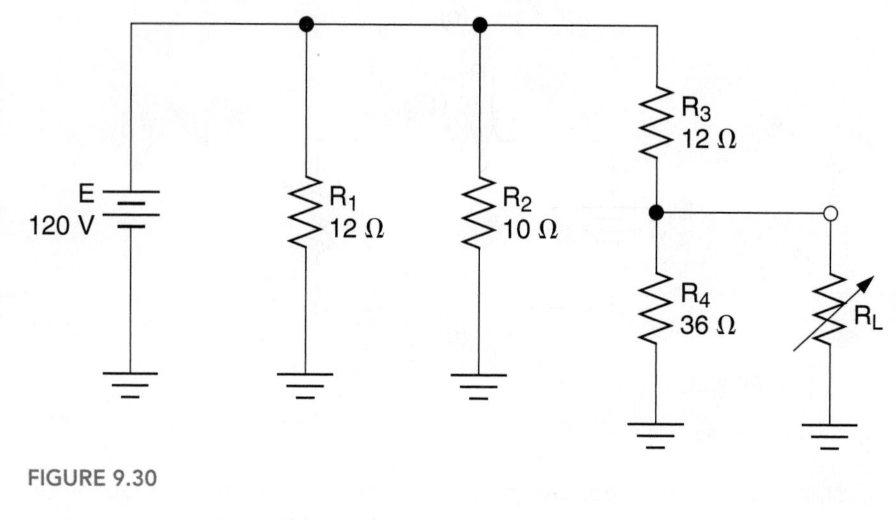

FIGURE 9.30

Though it may seem complicated at first, deriving the Thevenin equivalents of series-parallel circuits becomes relatively simple with practice. Once you have become comfortable with using it, Thevenin's theorem can be used to solve circuit problems that would be difficult to solve otherwise.

PROGRESS CHECK

1. What is *Thevenin's theorem*?
2. What purpose is served by Thevenin's theorem?
3. Using Thevenin's theorem, what is the procedure for performing the load analysis of a series-parallel circuit?
4. What is *Thevenin voltage*? How is it measured?
5. What is *Thevenin resistance*? How is it measured?
6. When measuring Thevenin resistance, why is the circuit voltage source replaced by a wire?

9.3 MAXIMUM POWER TRANSFER

In Chapter 4, you were told that electrical systems are designed for the purpose of delivering power (in one form or another) to a load. It would follow that we would be interested in knowing the maximum possible output power for a given source.

The *maximum power transfer theorem* establishes a relationship among source resistance (R_S), load resistance (R_L), and maximum load power. According to this theorem, maximum power transfer from a voltage source to its load occurs when the load resistance is equal to the source resistance. This means load power reaches its maximum possible value when $R_L = R_S$. For example, consider the circuit shown in Figure 9.31a. According to the maximum power transfer theorem, the voltage source transfers maximum power to its load when the rheostat is set to 10 Ω (the value of R_S).

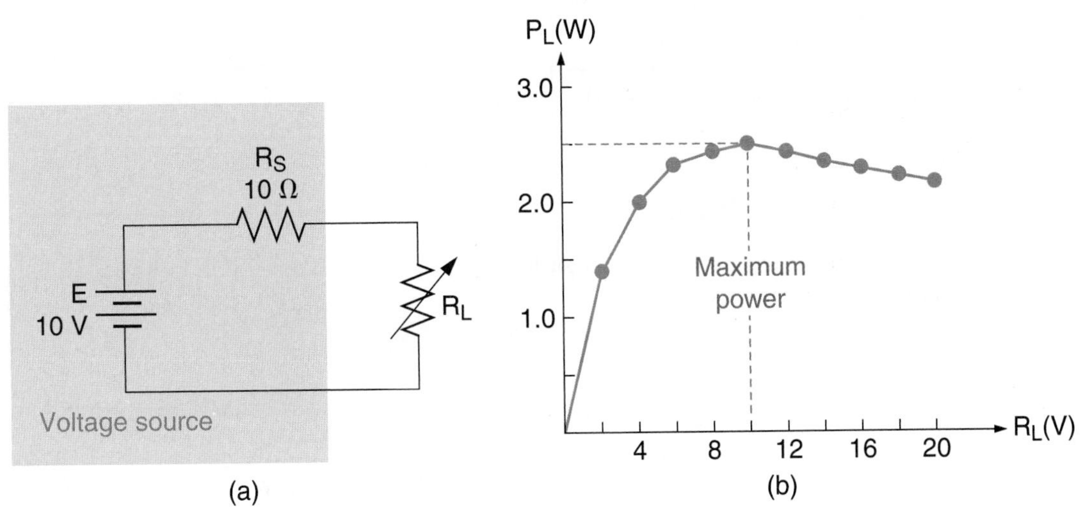

FIGURE 9.31 Load power vs. load resistance.

While the maximum power transfer theorem is not easy to derive mathematically, it can be demonstrated by performing a series of power calculations on the circuit in Figure 9.31a. In Table 9.1, we see the values of load voltage, current, and power that occur when the load in Figure 9.31a is varied from 0 Ω to 20 Ω (at 2 Ω increments). If you look at the P_{RL} column of values, you'll see that load power increases as the value of R_L increases from 0 Ω to 10 Ω. Then, as the value of R_L progresses above 10 Ω, load power starts to decrease again. Note that the maximum value of load power (2.50 W) occurs when $R_L = 10\ \Omega$, the value of the source resistance.

If we plot a graph of P_{RL} versus R_L, we get the curve shown in Figure 9.31b. Again, you can see that maximum load power in this circuit occurs when $R_L = R_S$.

The maximum power transfer theorem allows us to easily calculate the maximum possible load power for a circuit when the source resistance is known. This is demonstrated in Example 9.12.

TABLE 9.1 • Load Value Combinations for Figure 9.31.

R_L	V_{RL}	I_{RL}	P_{RL}
0 Ω	0 V	1 A	0 W
2 Ω	1.66 V	833 mA	1.38 W
4 Ω	2.86 V	714 mA	2.04 W
6 Ω	3.75 V	625 mA	2.34 W
8 Ω	4.44 V	556 mA	2.47 W
10 Ω	5.00 V	500 mA	2.50 W
12 Ω	5.45 V	456 mA	2.48 W
14 Ω	5.83 V	417 mA	2.43 W
16 Ω	6.15 V	385 mA	2.37 W
18 Ω	6.43 V	357 mA	2.30 W
20 Ω	6.67 V	334 mA	2.23 W

EXAMPLE 9.12

Calculate the maximum possible load power for the circuit in Figure 9.32.

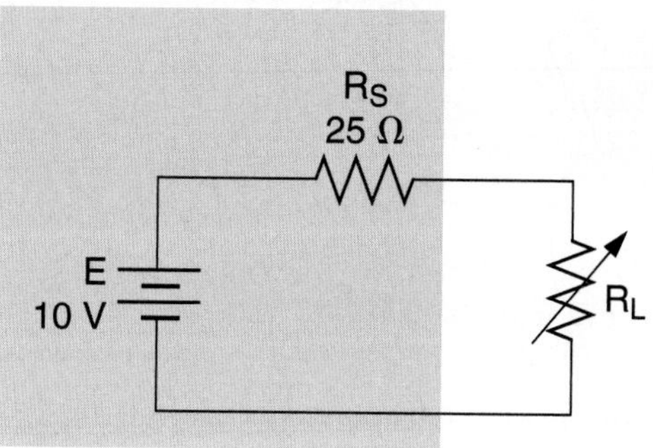

FIGURE 9.32

SOLUTION

Maximum power is transferred to the load when R_L equals the value of R_S, 25 Ω. When the rheostat is set to this value, the load voltage is one-half the source voltage, and

$$P_{RL} = \frac{V_{RL}^2}{R_L}$$

$$= \frac{(5\text{ V})^2}{25\ \Omega}$$

$$= 1\text{ W} \qquad \text{(maximum)}$$

PRACTICE PROBLEM 9.12

A circuit like the one in Figure 9.32 has values of $E = 32$ V and $R_S = 40\ \Omega$. Calculate the maximum possible load power for the circuit.

There are several points that should be mentioned at this time:

1. Maximum load power does not necessarily correspond to maximum load voltage. This point is verified by the values listed in Table 9.1.
2. In Chapter 4, we discussed the concept of source efficiency. You may recall that the efficiency of a power supply is a ratio of its output power to its input power. Maximum power transfer does not mean that the power supply is 100% efficient. In fact, when maximum power transfer occurs in a circuit with a fixed source resistance, source efficiency is only 50%. This is due to the fact that R_L and R_S are equal in value. Therefore, each is using half of the power being generated by the source.

Thevenin's Theorem and Maximum Power Transfer

For a series-parallel circuit with a variable load, maximum power transfer occurs when the load resistance equals the Thevenin resistance of the circuit. In other words, load power reaches its maximum possible value when $R_L = R_{TH}$. Example 9.13 demonstrates the calculation of maximum load power for a series-parallel circuit.

There are some practical limitations to using a Thevenin equivalent circuit. For example, most circuits are only linear over a certain range of values, and Thevenin equivalent circuits are valid only within that linear range. Also, while Thevenin equivalent circuits can be used to predict load power values, they cannot be used to calculate the power dissipated by the remaining components in the circuit.

KEY CONCEPT

EXAMPLE 9.13

Calculate the maximum possible load power for the circuit in Figure 9.33a.

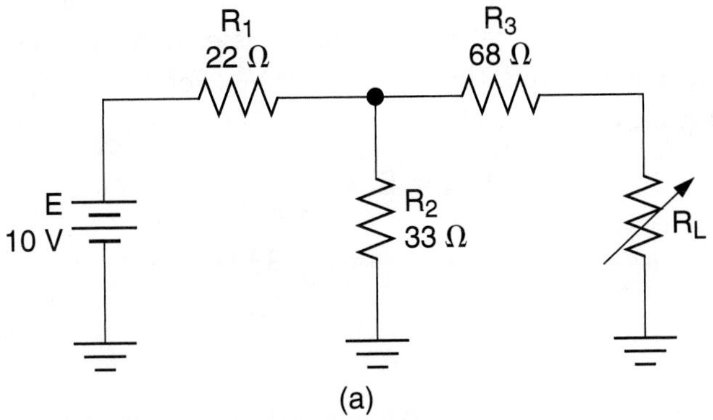

R1 22 Ω
R3 68 Ω
E 10 V
R2 33 Ω
RL
(a)

FIGURE 9.33

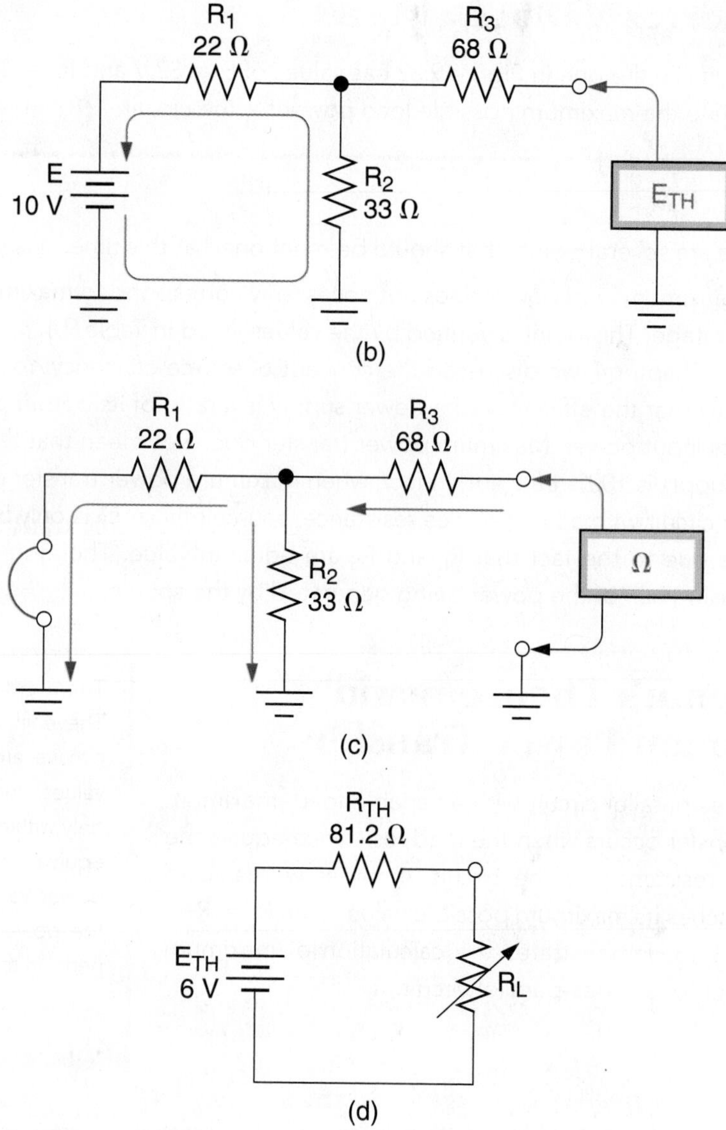

FIGURE 9.33 *(Continued)*

SOLUTION

The solution starts with deriving the Thevenin equivalent of the circuit. As shown in Figure 9.33b, the circuit current passes only through R_1 and R_2 when the load is removed. Therefore, $E_{TH} = V_{R2}$. For this circuit, the value of E_{TH} is found as:

$$E_{TH} = E \times \frac{R_2}{R_1 + R_2}$$

$$= 10\text{ V} \times \frac{33\ \Omega}{55\ \Omega}$$

$$= 6\text{ V}$$

If an ohmmeter were connected to the open load terminal, it would generate the currents shown in Figure 9.33c. As these currents indicate, the parallel combination of R_1 and R_2 is in series with R_3. Combining these relationships into a single equation, we can calculate the value of R_{TH} as follows:

$$R_{TH} = \frac{R_1 \times R_2}{R_1 + R_2} + R_3$$
$$= \frac{22 \times 33}{22 + 33}\ \Omega + 68\ \Omega$$
$$= 13.2\ \Omega + 68\ \Omega$$
$$= 81.2\ \Omega$$

Using the calculated values of E_{TH} and R_{TH}, the Thevenin equivalent circuit is drawn as shown in Figure 9.33d. Since $R_{TH} = 81.2\ \Omega$, maximum load power occurs when the load is also set to 81.2 Ω (as shown in the figure). When $R_L = 81.2\ \Omega$, the load voltage is found as

$$V_{RL} = E_{TH} \times \frac{R_L}{R_L + R_{TH}}$$
$$= 6\text{ V} \times \frac{81.2\ \Omega}{281.2\ \Omega}$$
$$= 1.73\text{ V}$$

and the load power is found as

$$P_{RL} = \frac{V_{RL}^2}{R_L}$$
$$= \frac{(1.73\text{ V})^2}{81.2\ \Omega}$$
$$= 111\text{ mW}$$

PRACTICE PROBLEM 9.13

Calculate the maximum load power for the circuit in Figure 9.34.

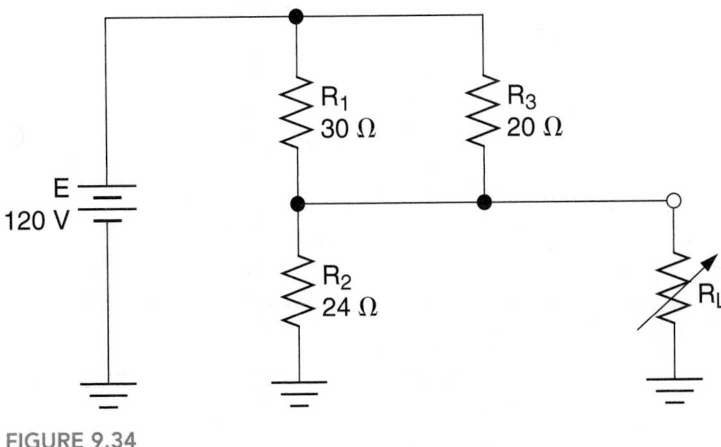

FIGURE 9.34

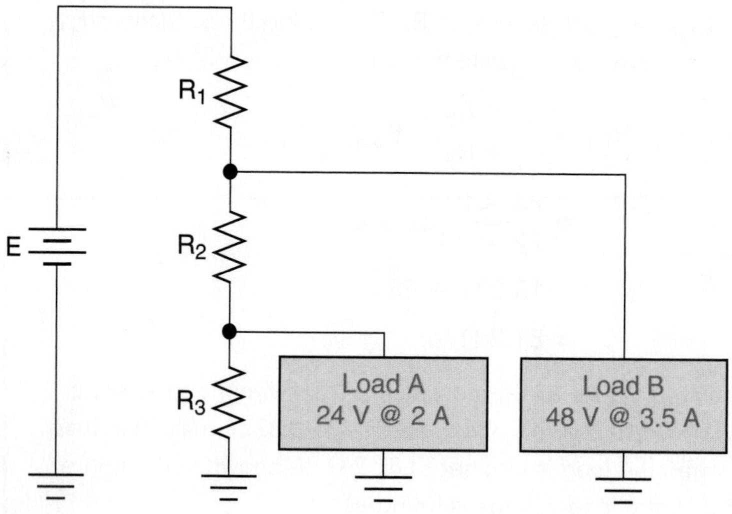

FIGURE 9.35

Multiload Circuits

As you learned in Chapter 7, loaded voltage dividers often have more than one load. For example, the voltage divider in Figure 9.35 is being used to supply the operating voltages for two loads.

When using Thevenin's theorem to analyze a circuit like the one in Figure 9.35, there are several points you need to keep in mind:

1. Because they are connected to the circuit at different points, each load in Figure 9.35 has its unique Thevenin equivalent of the original circuit.
2. When deriving the Thevenin equivalent for any load, the resistance of the other load(s) must be taken into account.

These points are illustrated in Figure 9.36. Here we have a loaded voltage divider and two equivalent circuits. Figure 9.36b shows the circuit with an open Load 1

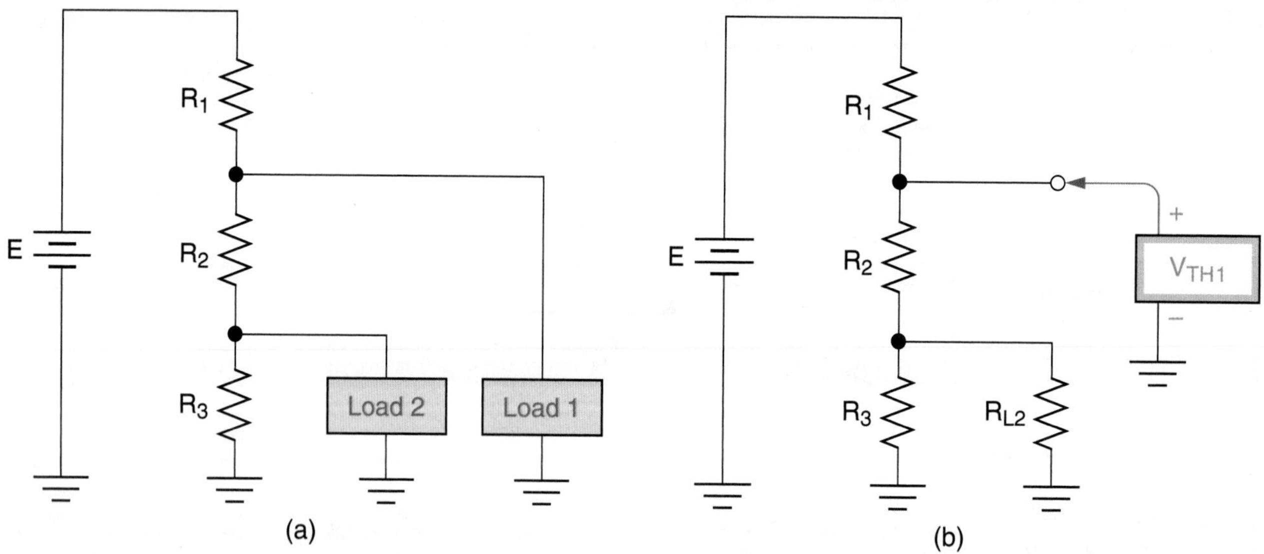

FIGURE 9.36

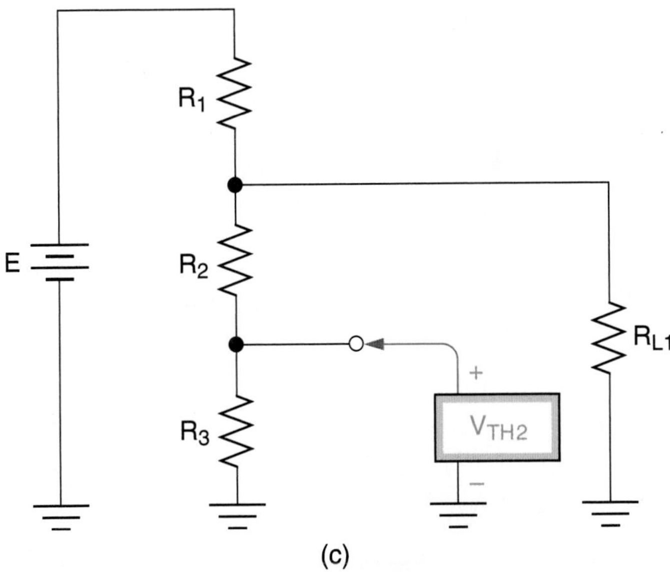

FIGURE 9.36 *(Continued)*

terminal. Figure 9.36c shows the circuit with an open Load 2 terminal. As you can see, each load "sees" a unique source circuit. That is, the source circuit for Load 1 is different from the one for Load 2.

When analyzing the circuit in Figure 9.36a, a Thevenin equivalent is derived for each load. Then, these circuits are used separately to perform each load analysis. The analysis of one load in a multiload circuit is demonstrated in Example 9.14.

EXAMPLE 9.14

Determine the maximum value of load power for Load 1 in Figure 9.37a. Assume that Load 2 is fixed at a value of $R_{L2} = 100\ \Omega$.

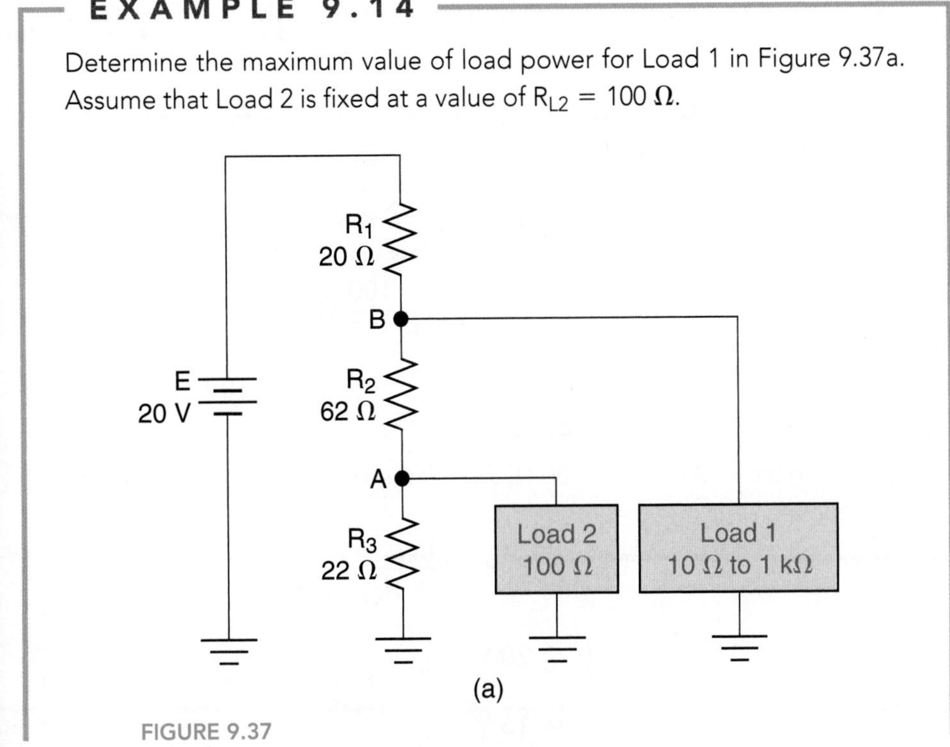

FIGURE 9.37

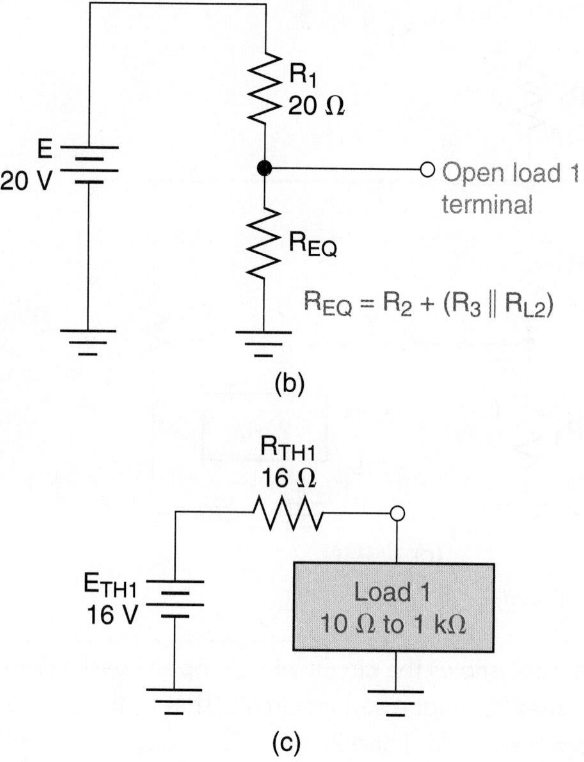

FIGURE 9.37 *(Continued)*

SOLUTION

The simplest approach to this problem is to combine R_2, R_3, and the resistance of Load 2 (R_{L2}) into a single equivalent resistance (as shown in Figure 9.37b).

The value of R_{EQ} (which represents the total resistance from point B to ground) can be found as

$$R_{EQ} = R_2 + \frac{R_3 \times R_{L2}}{R_3 + R_{L2}}$$

$$= 62\ \Omega + \frac{22 \times 100}{22 + 100}\ \Omega$$

$$= 62\ \Omega + 18\ \Omega$$

$$= 80\ \Omega$$

Now, using the value of $R_{EQ} = 80\ \Omega$, the voltage at the open Load 1 terminal can be found as

$$E_{TH1} = E \times \frac{R_{EQ}}{R_1 + R_{EQ}}$$

$$= 20\text{ V} \times \frac{80\ \Omega}{100\ \Omega}$$

$$= 16\text{ V}$$

Measured from the open Load 1 terminal, the value of R_{TH} in Figure 9.37b is equal to the parallel combination of R_1 and R_{EQ}. Therefore:

$$R_{TH1} = \frac{1}{\frac{1}{R_1} + \frac{1}{R_{EQ}}}$$

$$= \frac{1}{\frac{1}{20\ \Omega} + \frac{1}{80\ \Omega}}$$

$$= 16\ \Omega$$

This value indicates that maximum load power is generated when R_{L1} is 16 Ω. When $R_{L1} = 16\ \Omega$:

$$V_{RL} = E_{TH} \times \frac{R_L}{R_L + R_{th}}$$

$$= 16\ V \times \frac{16\ \Omega}{32\ \Omega}$$

$$= 8\ V$$

and

$$P_{L(max)} = \frac{V_{RL}^2}{R_L}$$

$$= \frac{(8\ V)^2}{16\ \Omega}$$

$$= 4\ W$$

PRACTICE PROBLEM 9.14

Assume that the positions of Load 1 and Load 2 in Figure 9.37a have been reversed. Calculate the maximum load power for Load 1.

The analysis technique demonstrated in Example 9.14 can be used on any number of loads. However, there is one restriction: Each load (other than the load of interest) must be assumed to have a fixed resistance.

PROGRESS CHECK

1. Describe how Thevenin's theorem can be used to perform the load analyses of a multiload circuit.
2. What restriction applies to the Thevenin analysis of a multiload circuit?

9.4 NORTON'S THEOREM

NORTON'S THEOREM A theorem stating that any resistive circuit or network, no matter how complex, can be represented as a current source in parallel with a source resistance.

NORTON EQUIVALENT CIRCUIT For any circuit, an equivalent circuit that contains a current source (I_N) in parallel with a source resistance (R_N).

NORTON CURRENT (I_N) The current source in a Norton equivalent circuit.

NORTON RESISTANCE (R_N) The source resistance in a Norton equivalent circuit.

In this section, we will discuss another network theorem that can be used in the load analysis of a given series-parallel circuit. As you will see, this theorem is very closely related to Thevenin's theorem.

NORTON'S THEOREM states that any resistive circuit or network, no matter how complex, can be represented as a current source in parallel with a source resistance. A series-parallel circuit is shown with its **NORTON EQUIVALENT CIRCUIT** in Figure 9.38. Note that the current source and resistance are referred to as the **NORTON CURRENT (I_N)** and the **NORTON RESISTANCE (R_N)**.

Norton's theorem reduces a circuit down to a single resistance in parallel with a constant current source while Thevenin's theorem reduces the circuit down to a single resistance in series with a single voltage. As such, Norton's theorem can be thought of as the current counterpart of Thevenin's theorem.

KEY CONCEPT

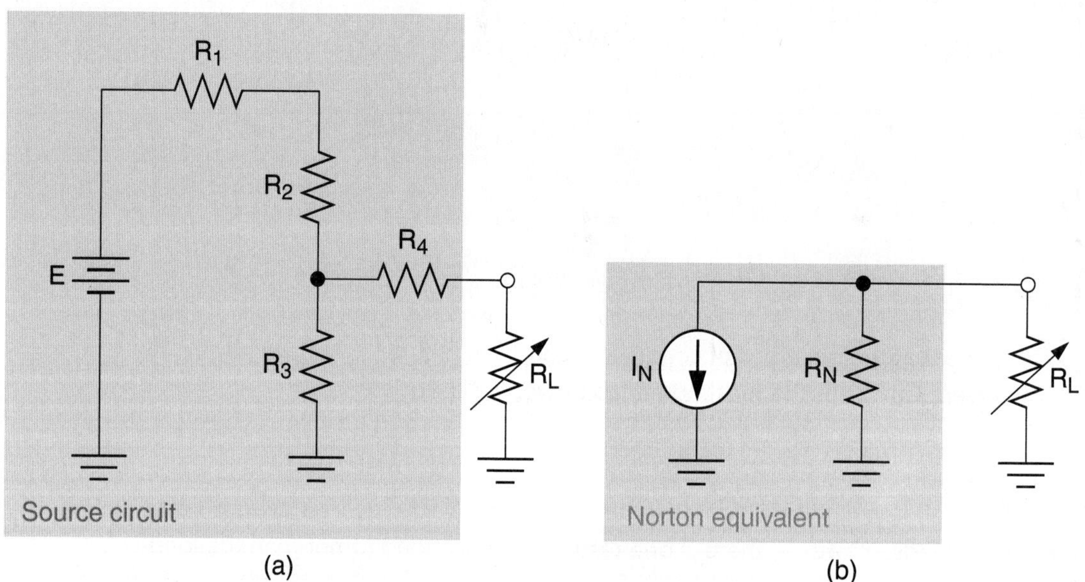

FIGURE 9.38 A series-parallel circuit and its Norton equivalent.

Obviously, Norton's theorem is very similar to Thevenin's theorem. The primary difference is that Norton represents a complex circuit as a practical current source while Thevenin represents the same circuit as a practical voltage source.

Determining the Value of I_N

By definition, Norton current (I_N) is the current through the shorted load terminals. The measurement of I_N is illustrated in Figure 9.39a. As you can see:

1. The load has been removed.
2. An ammeter has been connected across the load terminals.

Since the internal resistance of a current meter is (ideally) 0 Ω, the current being measured in Figure 9.39a is equal to the Norton current for the circuit.

To calculate the value of I_N for a series-parallel circuit, you short the load terminals as shown in Figure 9.39b. Then, you calculate the current through the shorted load terminals. This procedure is demonstrated in Example 9.15.

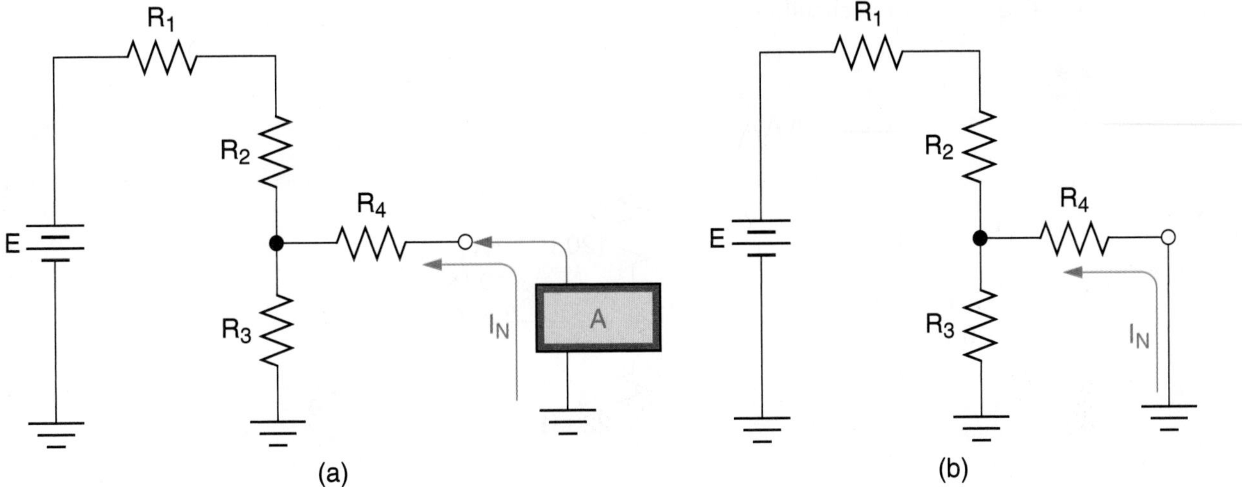

FIGURE 9.39 Measuring Norton current (I_N).

EXAMPLE 9.15

Calculate the Norton current (I_N) for the circuit in Figure 9.40a.

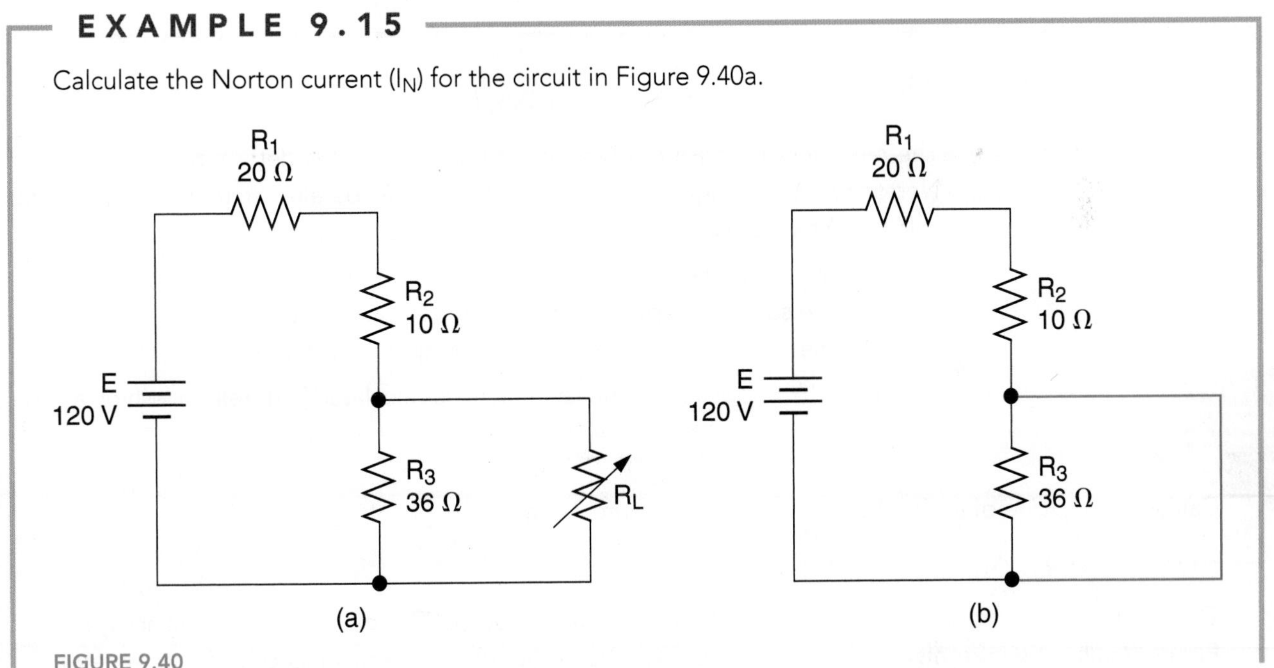

FIGURE 9.40

SOLUTION

The first step is to short the load as shown in Figure 9.40b. When the load is shorted, it also shorts out R_3. Therefore, the value of I_N is determined by the source voltage (E) and the series combination of R_1 and R_2, as follows:

$$I_N = \frac{E}{R_1 + R_2}$$

$$= \frac{120\text{ V}}{30\ \Omega}$$

$$= 4\text{ A}$$

PRACTICE PROBLEM 9.15

Calculate the value of I_N for the circuit in Figure 9.41.

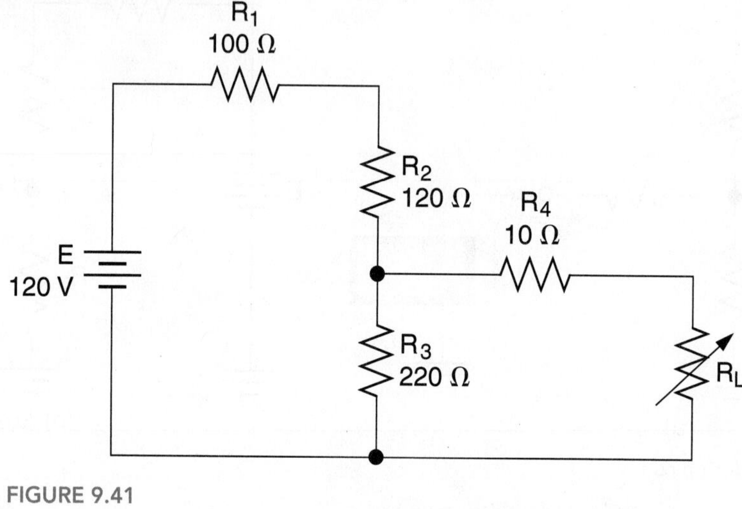

FIGURE 9.41

Norton Resistance (R_N)

The next step in deriving a Norton equivalent circuit is determining the value of Norton resistance (R_N). The value of R_N for a series-parallel circuit is measured and calculated exactly like Thevenin resistance. That is:

1. The load is removed.
2. The source is replaced by its resistive equivalent.
3. The resistance (as measured at the open load terminals) is calculated.

Example 9.16 demonstrates the fact that R_N is calculated in the same fashion as R_{TH}.

EXAMPLE 9.16

Calculate the value of R_N for the circuit shown in Figure 9.42a.

SOLUTION

The first step is to remove the load and short out the voltage source. This gives us the circuit shown in Figure 9.42b. The currents shown in the circuit are those that would be produced by an ohmmeter

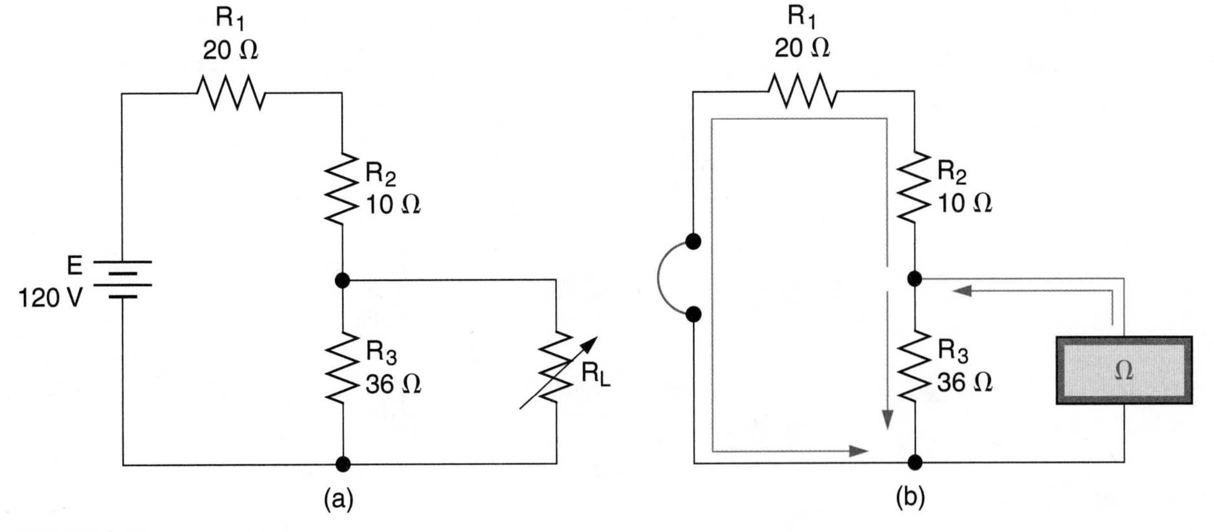

FIGURE 9.42

connected to the load terminals. As these arrows indicate, the series combination of R_1 and R_2 is in parallel with R_3. Therefore, the value of R_N can be calculated as follows:

$$R_N = \frac{(R_1 + R_2) \times R_3}{(R_1 + R_2) + R_3}$$

$$= \frac{30 \times 36}{30 + 36}\ \Omega$$

$$= 16.4\ \Omega$$

PRACTICE PROBLEM 9.16

Calculate the value of R_N for the circuit shown in Figure 9.41.

As you can see, there is no difference between R_N and R_{TH}. They are calculated (and measured) in exactly the same fashion.

Load Analysis Using Norton Equivalent Circuits

Norton equivalent circuits are used in the same fashion as Thevenin equivalent circuits. Once the Norton equivalent of a circuit is derived, the load is placed in the equivalent circuit. Then, the current divider relationship is used to determine the value of load current.

The following series of examples demonstrate how load analysis problems can be solved using Norton equivalent circuits.

EXAMPLE 9.17

Predict the values of I_{RL} that will be produced by the circuit in Figure 9.43a for values of $R_L = 25\ \Omega$ and $R_L = 75\ \Omega$.

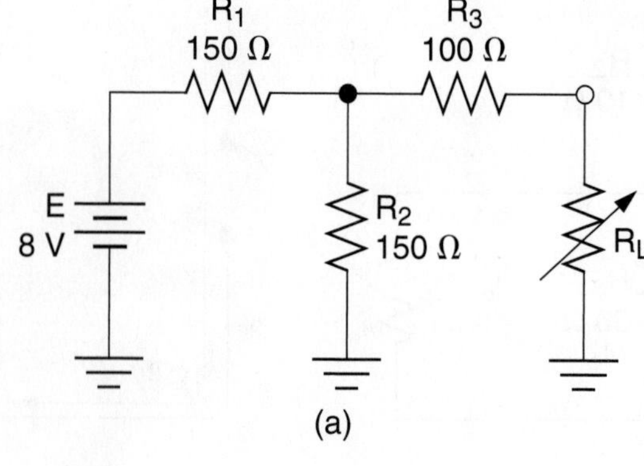

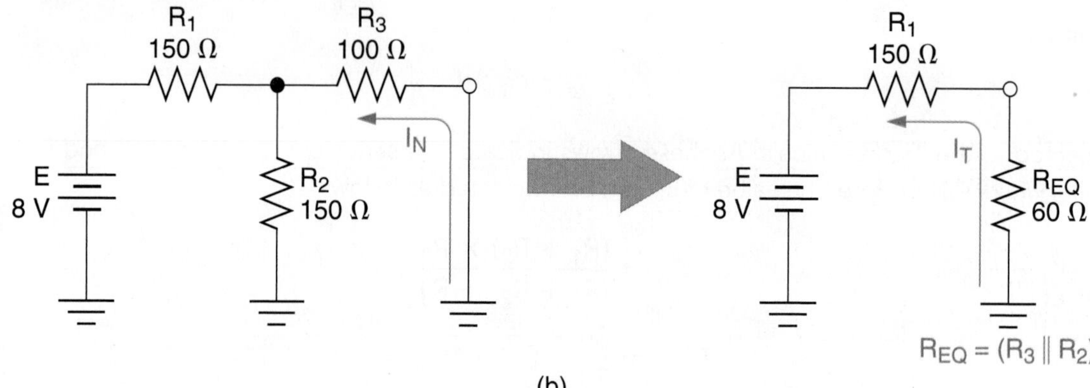

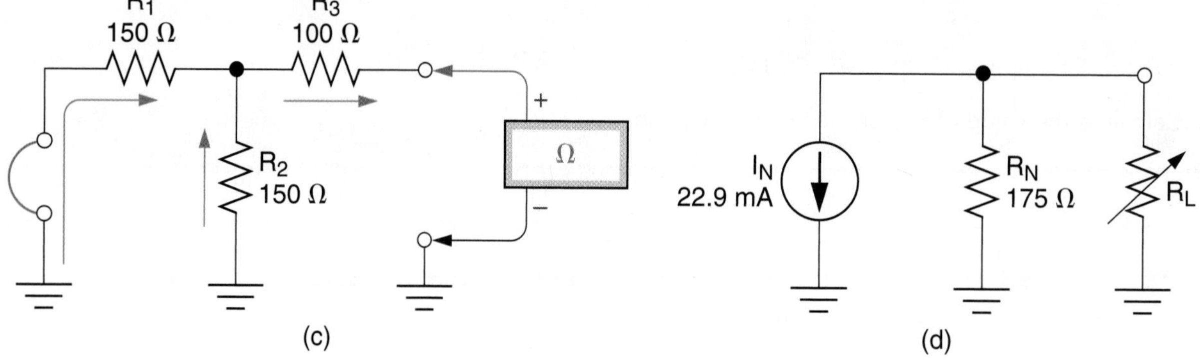

FIGURE 9.43

SOLUTION

The Norton current (I_N) equals the current through the shorted load terminals (as shown in Figure 9.43b). With the load shorted, the value of I_N equals the current through R_3. Calculating the value of I_N begins with combining R_2 and R_3 into a single parallel-equivalent resistance, as follows:

$$R_{EQ} = \frac{1}{\dfrac{1}{R_2} + \dfrac{1}{R_3}}$$

$$= \frac{1}{\dfrac{1}{150\ \Omega} + \dfrac{1}{100\ \Omega}}$$

$$= 60\ \Omega$$

Now, the voltage across R_{EQ} is found as

$$V_{EQ} = E \times \frac{R_{EQ}}{R_1 + R_{EQ}}$$

$$= 8\text{ V} \times \frac{60\ \Omega}{210\ \Omega}$$

$$= 2.29\text{ V}$$

This result indicates that there is 2.29 V across R_3 (as well as R_2). Therefore, the value of I_{R3} can be found using

$$I_{R3} = \frac{V_{EQ}}{R_3}$$

$$= \frac{2.29\text{ V}}{100\ \Omega}$$

$$= 22.9\text{ mA}$$

Since this current passes through the shorted load terminals, the value of I_N for the circuit is 22.9 mA.

If we remove the load and replace the source in our original circuit with a wire, we get the circuit shown in Figure 9.43c. The currents shown in the figure are those that would be produced by an ohmmeter connected across the open load terminals. As these arrows indicate, R_N equals the parallel combination of R_1 and R_2 in series with R_3. Therefore,

$$R_N = \frac{R_1 \times R_2}{R_1 + R_2} + R_3$$

$$= \frac{150 \times 150}{150 + 150}\ \Omega + 100\ \Omega$$

$$= 75\ \Omega + 100\ \Omega$$

$$= 175\ \Omega$$

Now, using the calculated values of I_N and R_N, the Norton equivalent circuit is constructed as shown in Figure 9.43d. When $R_L = 25\ \Omega$, the value of I_{RL} can be found using the current-divider relationship, as follows: The parallel combination of R_L and R_N is found as

$$R_T = \frac{1}{\dfrac{1}{R_L} + \dfrac{1}{R_N}}$$

$$= \frac{1}{\dfrac{1}{25\ \Omega} + \dfrac{1}{175\ \Omega}}$$

$$= 21.9\ \Omega$$

and

$$I_{RL} = I_N \times \frac{R_T}{R_L}$$
$$= (22.9 \text{ mA})\frac{21.9\ \Omega}{25\ \Omega}$$
$$= 20.1 \text{ mA}$$

When $R_L = 75\ \Omega$, the total circuit resistance is found as

$$R_T = \frac{R_L \times R_N}{R_L + R_N}$$
$$= \frac{75 \times 175}{75 + 175}\ \Omega$$
$$= 52.5\ \Omega$$

and the value of I_{RL} can be found as

$$I_{RL} = I_N \times \frac{R_T}{R_L}$$
$$= (22.9 \text{ mA})\frac{52.5\ \Omega}{75\ \Omega}$$
$$= 16 \text{ mA}$$

PRACTICE PROBLEM 9.17

Predict the values of I_{RL} that will be produced by the circuit in Figure 9.44 for values of $R_L = 100\ \Omega$ and $R_L = 150\ \Omega$.

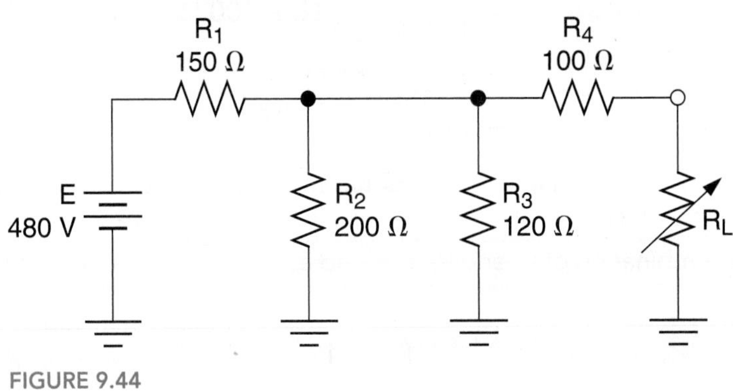

FIGURE 9.44

Now, let's see how Norton's theorem can be used to calculate the maximum possible load power for a series-parallel circuit with a variable load.

EXAMPLE 9.18

Determine the maximum load power for the circuit in Figure 9.45a.

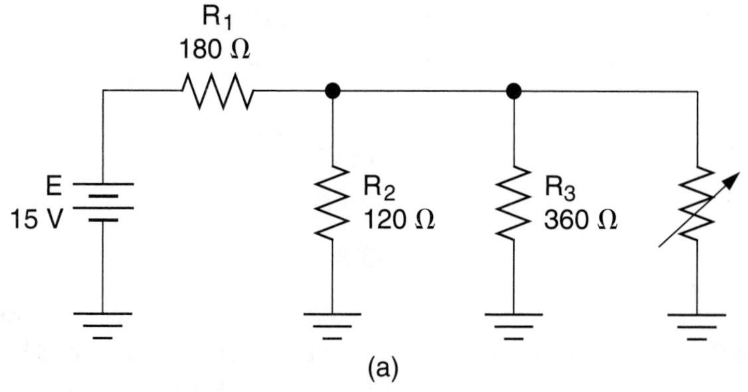

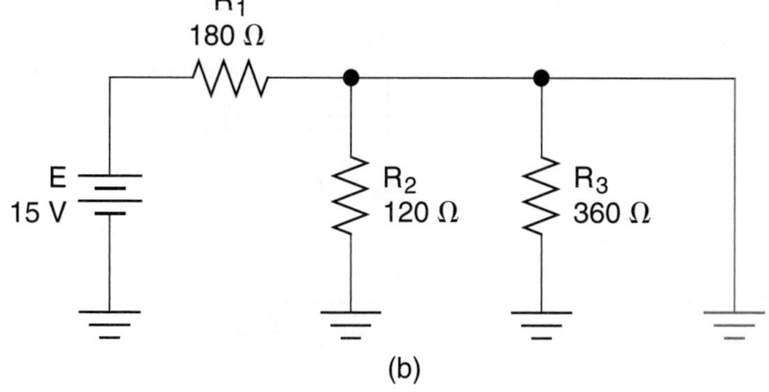

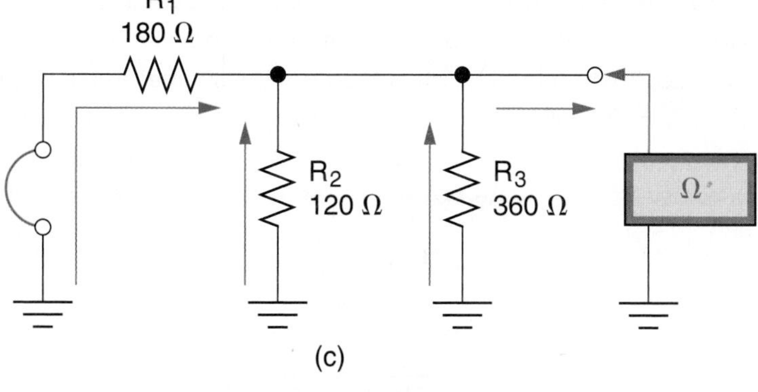

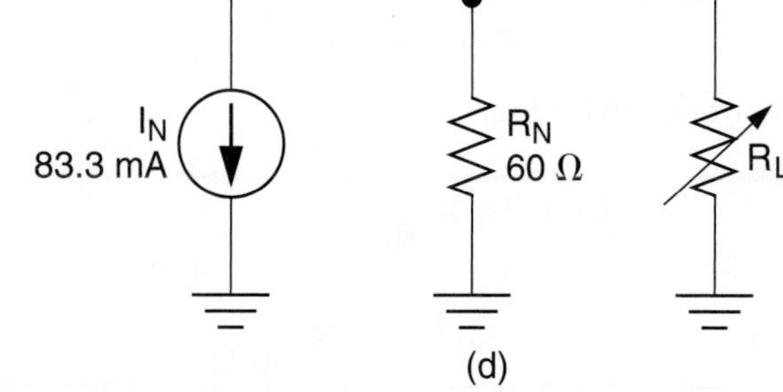

FIGURE 9.45

SOLUTION

The first step is to short out the load. In this case, shorting out the load also shorts out both R_2 and R_3 (as shown in Figure 9.45b). As a result, the current through the shorted load can be found simply as

$$I_N = \frac{E}{R_1} = \frac{15\text{ V}}{180\ \Omega} = 83.3\text{ mA}$$

If we remove the load and short the source, we get the circuit shown in Figure 9.45c. The current arrows represent the current that would be generated by an ohmmeter connected to the open load terminals. As these arrows indicate, the Norton resistance equals the parallel combination of R_1, R_2, and R_3. Using the reciprocal method,

$$R_N = \frac{1}{\frac{1}{R_1} + \frac{1}{R_2} + \frac{1}{R_3}} = \frac{1}{\frac{1}{180} + \frac{1}{120} + \frac{1}{360}}\ \Omega = 60\ \Omega$$

Using the values of I_N and R_N, the Norton equivalent circuit is constructed as shown in Figure 9.45d. As you know, maximum power transfer to a variable load occurs when load resistance equals the source resistance, in this case, when $R_L = R_N$. When $R_L = 60\ \Omega$, the total circuit resistance is found as

$$R_T = \frac{1}{\frac{1}{R_L} + \frac{1}{R_N}} = \frac{1}{\frac{1}{60\ \Omega} + \frac{1}{60\ \Omega}} = 30\ \Omega$$

and the load current can be found as

$$I_{RL} = I_N \times \frac{R_T}{R_L} = 83.3\text{ mA} \times \frac{30\ \Omega}{60\ \Omega} = 41.7\text{ mA}$$

Finally, the maximum load power can be found as

$$P_{RL} = I_{RL}^2 \times R_L$$
$$= (41.7\text{ mA})^2(60\ \Omega)$$
$$= 104\text{ mW}$$

PRACTICE PROBLEM 9.18

Calculate the maximum load power for the circuit shown in Figure 9.46.

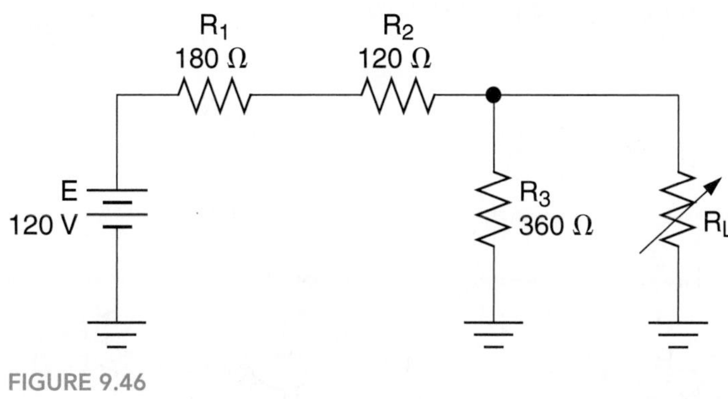

FIGURE 9.46

Norton-to-Thevenin and Thevenin-to-Norton Conversions

Every Thevenin equivalent circuit has a matching Norton equivalent circuit, and vice versa. For example, Figure 9.47 contains a series-parallel circuit along with its Thevenin and Norton equivalent circuits. The load calculations shown for each equivalent circuit were made using an assumed value of $R_L = 60\ \Omega$. As you can see, the equivalent circuits provide identical values of load voltage and current.

Using the source conversion techniques introduced earlier in the chapter, you can convert either type of equivalent circuit to the other. The relationships used to perform both conversions are shown in Figure 9.49.

If we apply the Thevenin-to-Norton relationships to the Thevenin circuit in Figure 9.48, we get:

$$R_N = R_{TH} = 60\ \Omega \qquad \text{and} \qquad I_N = \frac{V_{TH}}{R_{TH}} = \frac{5\text{ V}}{60\ \Omega} = 83.3\text{mA}$$

These values match those in the Norton equivalent circuit in Figure 9.49.

One Final Note

As your study of electricity continues, you'll find that there are situations that lend themselves more toward one theorem than the other. In most cases, the

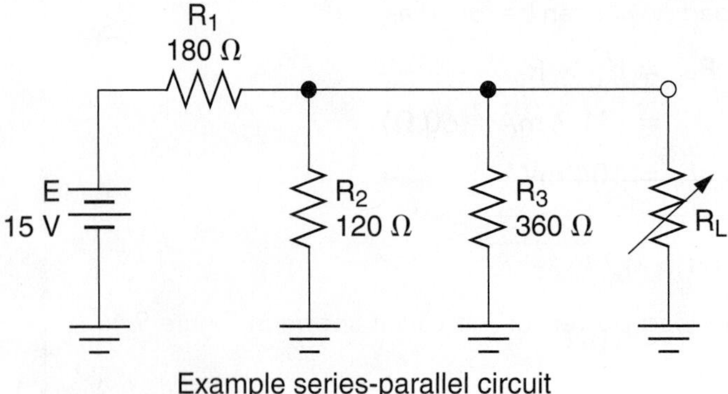

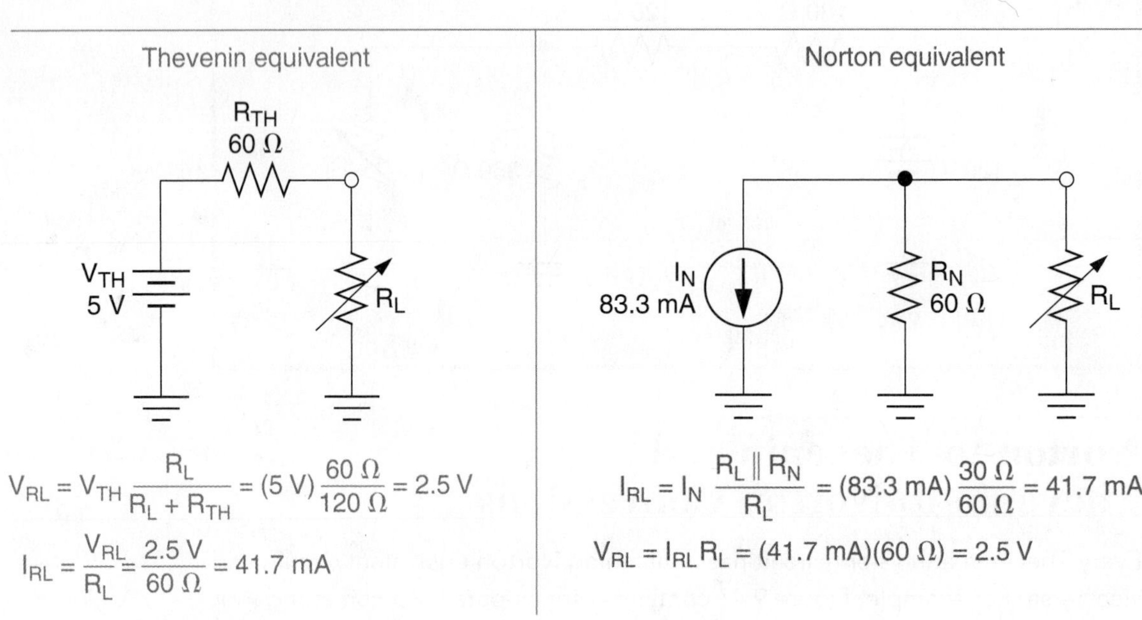

$$V_{RL} = V_{TH}\frac{R_L}{R_L + R_{TH}} = (5\text{ V})\frac{60\ \Omega}{120\ \Omega} = 2.5\text{ V}$$

$$I_{RL} = \frac{V_{RL}}{R_L} = \frac{2.5\text{ V}}{60\ \Omega} = 41.7\text{ mA}$$

$$I_{RL} = I_N\frac{R_L \parallel R_N}{R_L} = (83.3\text{ mA})\frac{30\ \Omega}{60\ \Omega} = 41.7\text{ mA}$$

$$V_{RL} = I_{RL}R_L = (41.7\text{ mA})(60\ \Omega) = 2.5\text{ V}$$

FIGURE 9.47 Thevenin and Norton equivalents for a series-parallel circuit.

approach you take to load analysis is up to you. Many technicians find it easier to become proficient at using one type of theorem and then using source conversions to derive the other (if the need arises). Others find it beneficial to become proficient at both Norton's and Thevenin's theorems. Again, the choice is yours.

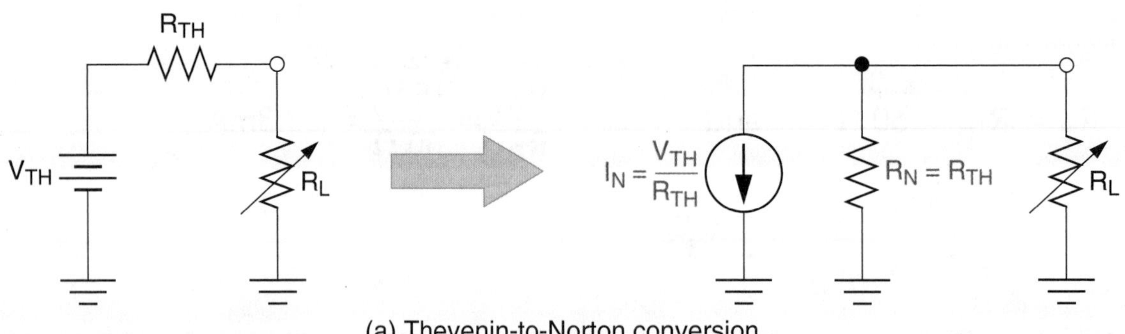

(a) Thevenin-to-Norton conversion

FIGURE 9.48

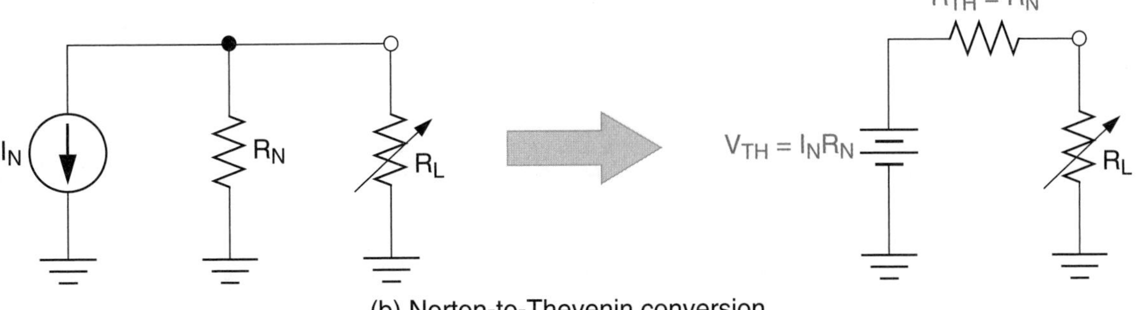

(b) Norton-to-Thevenin conversion

FIGURE 9.48 *(Continued)*

PROGRESS CHECK

1. Describe the process for determining the value of Norton current for a series-parallel circuit.
2. Describe the process for determining the value of Norton resistance for a series-parallel circuit.
3. In terms of load analysis, how does Norton's theorem differ from Thevenin's theorem?
4. Describe the relationship between the Norton and Thevenin equivalents of a given circuit.
5. How do you convert from one equivalent circuit (Norton or Thevenin) to the other?

9.5 SUPERPOSITION

All of the series-parallel circuits we've discussed up to this point have had only one voltage source. There are circuits, however, that contain more than one voltage (or current) source. A circuit of this type is referred to as a **MULTISOURCE CIRCUIT**. Several examples of multisource circuits are shown in Figure 9.49.

MULTISOURCE CIRCUIT A circuit with multiple voltage and/or current sources.

When analyzing multisource circuits, the effects of each voltage (or current) source must be taken into account. In this section, we will cover one of several methods used to analyze multisource circuits.

The Superposition Theorem

Superposition theorem cannot be applied to circuits with voltage-controlled or current-controlled resistance values (such as some solid-state electronic circuits).

KEY CONCEPT

The **SUPERPOSITION THEOREM** states that the response of a circuit to more than one source can be determined by analyzing the circuit's response to each source (alone) and combining the results. For example, consider the circuit

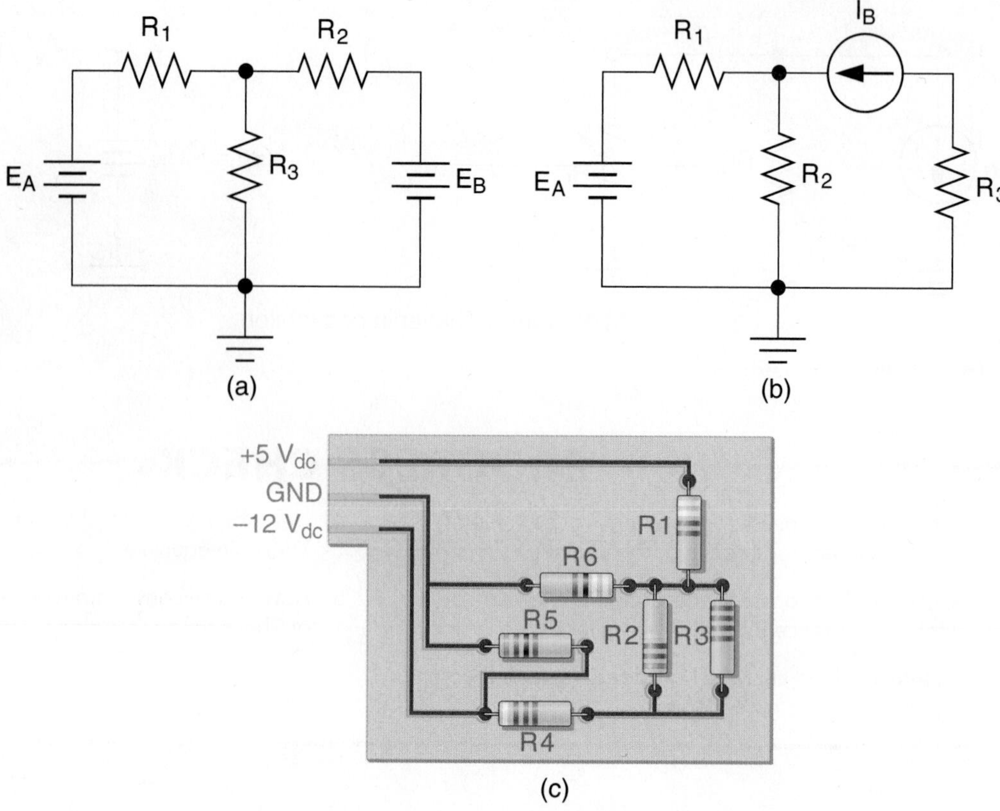

FIGURE 9.49 Multisource circuits.

SUPERPOSITION THEOREM
A theorem stating that the response of a circuit to more than one source can be determined by analyzing the circuit's response to each source and combining the results.

shown in Figure 9.50. According to the superposition theorem, we can determine the component voltages in the circuit by:

1. Analyzing the circuit as if E_A were the only voltage source.
2. Analyzing the circuit as if E_B were the only voltage source.
3. Combining the results from steps 1 and 2.

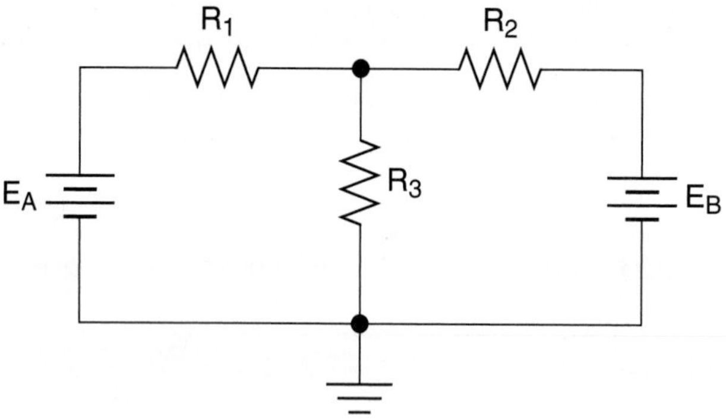

FIGURE 9.50

When using the superposition theorem to solve a multisource circuit, it is very helpful to draw the original circuit as two single-source circuits. For example, the circuit in Figure 9.50 would be redrawn as shown in Figure 9.51. Note that each

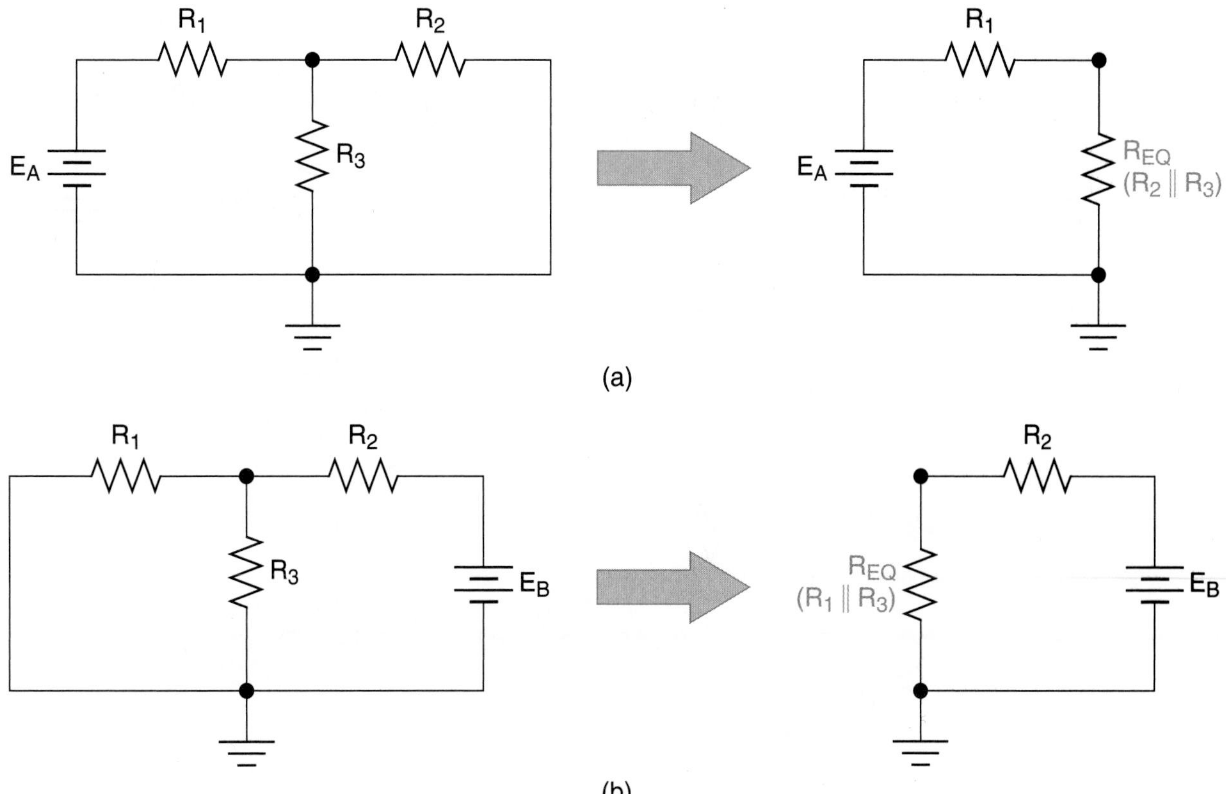

FIGURE 9.51

circuit is identical to the original, except that one source has been removed. When drawn as shown, the analysis of the circuit for each voltage source becomes a problem like those we covered in Chapter 7. For example, the circuit in Figure 9.51a can be simplified to a series-equivalent by combining R_2 and R_3 into a parallel equivalent resistance. At that point, determining the component voltages is simple. The circuit in Figure 9.51b can be simplified in the same manner: R_1 and R_2 are combined into a parallel equivalent resistance, leaving us with a simple series circuit.

After calculating the voltage values for the circuits in Figure 9.51, we combine the values to obtain the actual circuit voltages. However, care must be taken to ensure that the voltages are combined properly. This point is illustrated in Figure 9.52. Figures 9.52a and 9.52b show the polarities of the voltages produced by their respective sources. If we combine these voltages as shown in Figure 9.52c, you can see that:

1. $V_{R1(A)}$ and $V_{R1(B)}$ are *opposing voltages.* Therefore, the actual component voltage equals *the difference between the two.*
2. $V_{R2(A)}$ and $V_{R2(B)}$ are also opposing voltages. Again, the actual component voltage equals the difference between the two.
3. $V_{R3(A)}$ and $V_{R3(B)}$ are *aiding voltages.* Therefore, the actual component voltage equals *the sum of the two.*

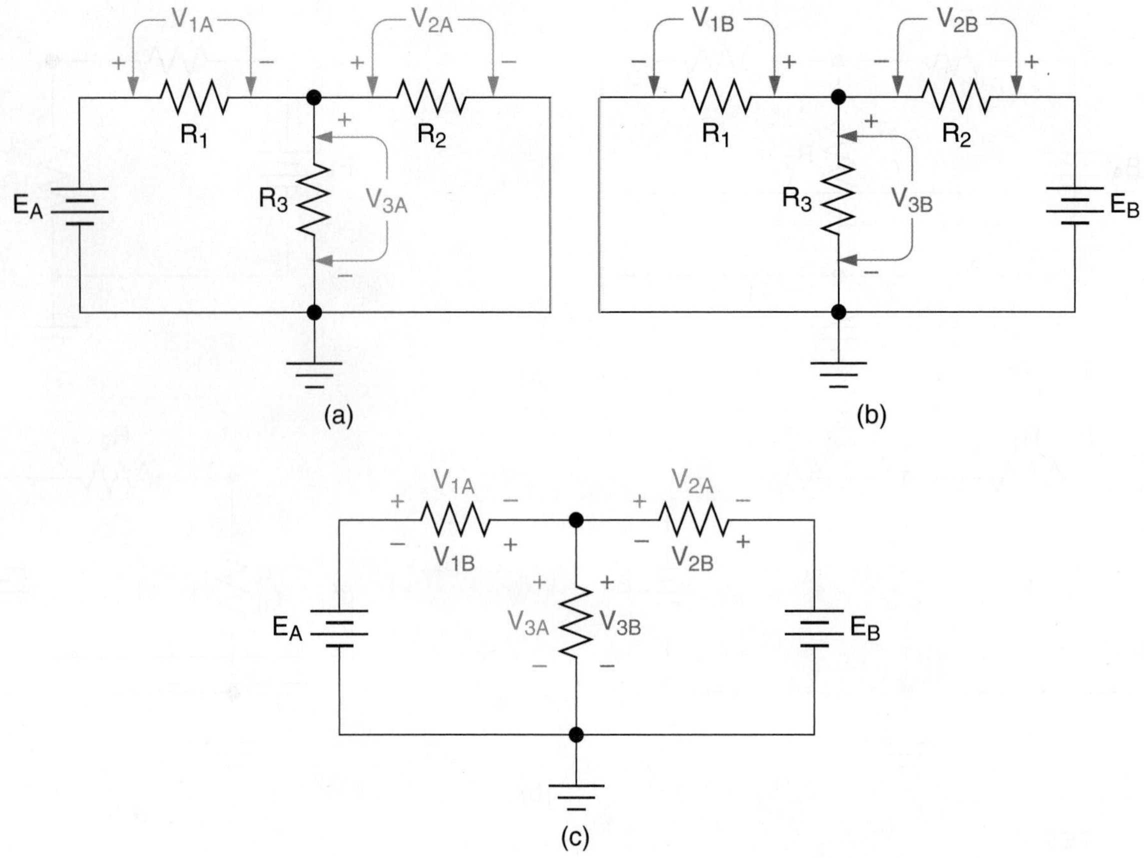

FIGURE 9.52 Voltages in a multisource circuit.

The superposition analysis of a simple multisource circuit is demonstrated in Example 9.19. As you will see, the polarities of the various voltages determine how they are combined in the final steps of the analysis.

EXAMPLE 9.19

Determine the values of V_{R1}, V_{R2}, and V_{R3} for the circuit shown in Figure 9.53a.

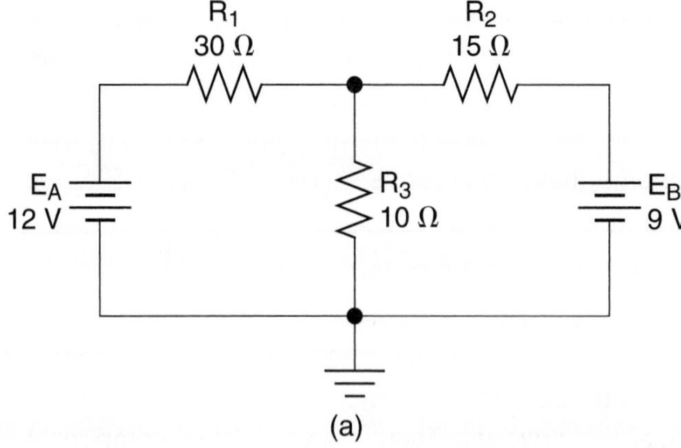

FIGURE 9.53

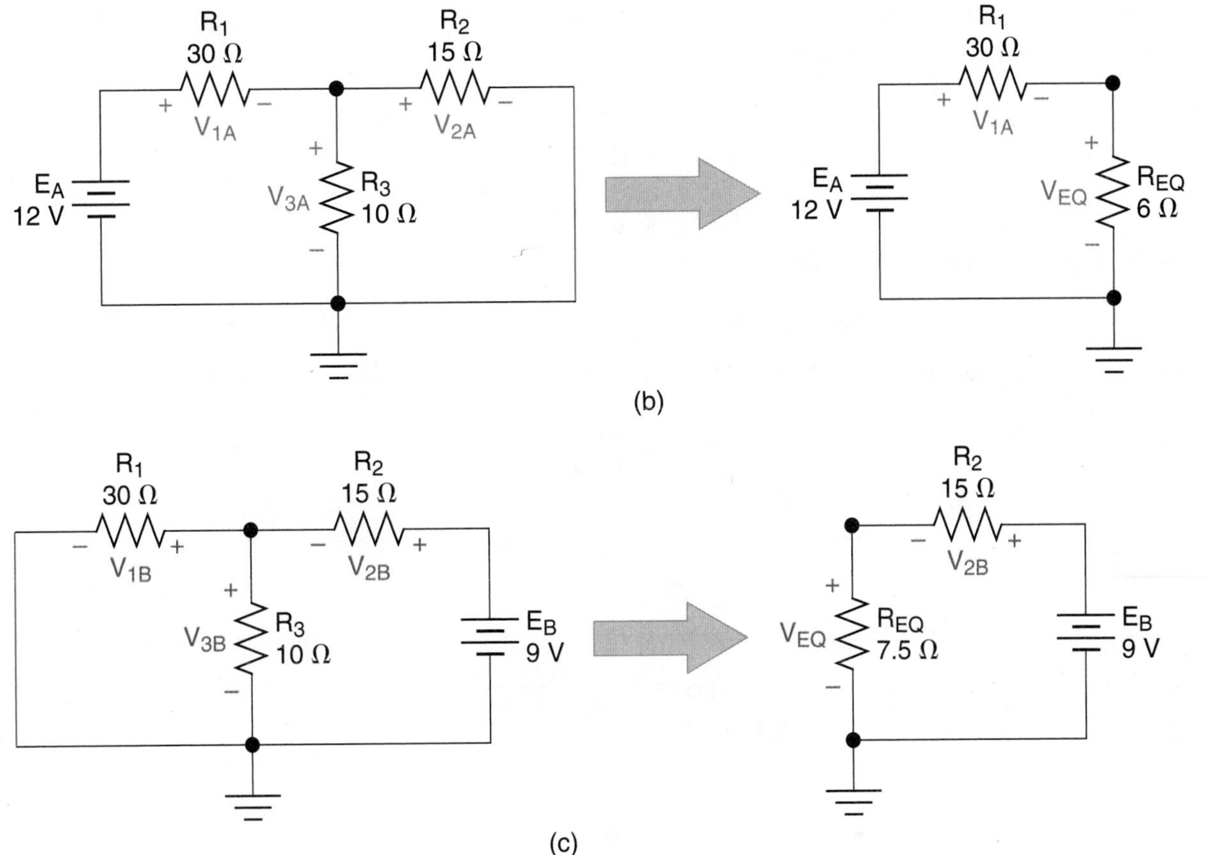

FIGURE 9.53 *(Continued)*

SOLUTION

The first step is to split the circuit into two single-source circuits. These circuits (along with their series-equivalents) are shown in Figures 9.53b and 9.53c. For the circuit in Figure 9.53b, R_2 and R_3 are combined into an equivalent resistance (R_{EQ}) as follows:

$$R_{EQ} = \frac{R_2 \times R_3}{R_2 + R_3}$$

$$= \frac{15 \times 10}{15 + 10}\,\Omega$$

$$= 6\,\Omega$$

Using this value in the voltage-divider equation, the value of V_{EQ} can be found as

$$V_{EQ} = E_A \times \frac{R_{EQ}}{R_{EQ} + R_1}$$

$$= 12\text{ V} \times \frac{6\,\Omega}{36\,\Omega}$$

$$= 2\text{ V}$$

Since this voltage is across the parallel combination of R_2 and R_3, $V_{R2(A)}$ and $V_{R3(A)}$ both equal 2 V. The voltage across R_1 can now be found as

$$\begin{aligned} V_1 &= E_A - V_{EQ} \\ &= 12\text{ V} - 2\text{ V} \\ &= 10\text{ V} \end{aligned}$$

At this point, we have calculated the following values:

$$V_{1(A)} = 10\text{ V} \qquad V_{2(A)} = 2\text{ V} \qquad V_{3(A)} = 2\text{ V}$$

Performing a similar analysis of the circuit in Figure 9.53c, we get the following values:

$$\begin{aligned} R_{EQ} &= \frac{R_1 \times R_3}{R_1 + R_3} \\ &= \frac{30 \times 10}{30 + 10}\ \Omega \\ &= 7.5\ \Omega \end{aligned}$$

$$\begin{aligned} V_{EQ} &= E_B \frac{R_{EQ}}{R_{EQ} + R_1} \\ &= 9\text{ V} \times \frac{7.5\ \Omega}{22.5\ \Omega} \\ &= 3\text{ V} \end{aligned}$$

and

$$\begin{aligned} V_{R2} &= E_B - V_{EQ} \\ &= 9\text{ V} - 3\text{ V} \\ &= 6\text{ V} \end{aligned}$$

For the circuit in Figure 9.53c, we have calculated the following values:

$$V_{1(B)} = 3\text{ V} \qquad V_{2(B)} = 6\text{ V} \qquad V_{3(B)} = 3\text{ V}$$

Finally, we can determine the actual component voltages by combining the (A) and (B) results. If you compare the polarities of $V_{R1(A)}$ and $V_{R1(B)}$ in Figure 9.53, you can see that they are opposing. Therefore,

$$\begin{aligned} V_{R1} &= |V_{R1(A)} - V_{R1(B)}| \\ &= 10\text{ V} - 3\text{ V} \\ &= 7\text{ V} \end{aligned}$$

Figure 9.53 also shows $V_{R2(A)}$ and $V_{R2(B)}$ to be opposing. Therefore,

$$\begin{aligned} V_2 &= |V_{2(B)} - V_{2(A)}| \\ &= 6\text{ V} - 2\text{ V} \\ &= 4\text{ V} \end{aligned}$$

Since $V_{R2(A)}$ and $V_{R2(B)}$ are aiding voltages, the actual value of V_{R3} is found as

$$V_{R3} = |V_{R3(A)} + V_{R3(B)}|$$
$$= 2\text{ V} + 3\text{ V}$$
$$= 5\text{ V}$$

For the circuit in Figure 9.53a, the actual component voltages are as follows:

$$V_1 = 7\text{ V} \qquad V_2 = 4\text{ V} \qquad V_3 = 5\text{ V}$$

PRACTICE PROBLEM 9.19

A circuit like the one in Figure 9.53a has the following values: $E_A = 14\text{ V}$, $E_B = 7\text{ V}$, $R_1 = 100\ \Omega$, $R_2 = 150\ \Omega$, and $R_3 = 150\ \Omega$. Calculate the values of V_{R1}, V_{R2}, and V_{R3} for the circuit.

The results from Example 9.1 can be used to make several important points. Figure 9.6a shows the original circuit with the calculated component voltages included. First, note the polarities of V_{R1} and V_{R2}. In both cases, the assumed polarity matches that of the greater voltage. Since $V_{R1(A)}$ is greater than $V_{R1(B)}$, the assumed polarity of V_{R1} matches that of $V_{R1(A)}$. By comparing Figure 9.54 to Figure 9.53, you

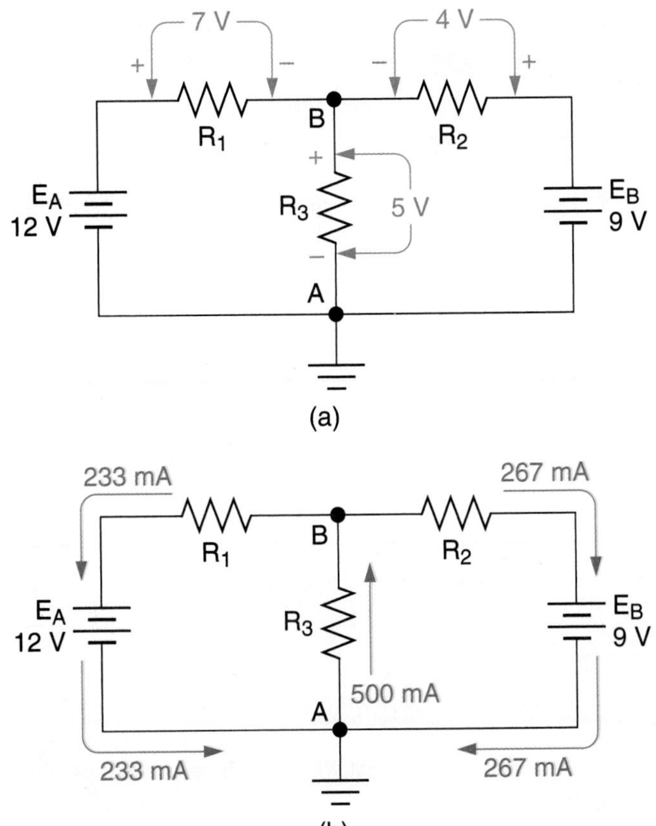

FIGURE 9.54 Calculated currents and voltages for the circuit in Example 9.19.

can see that the same holds true in the case of V_{R2}. The polarity of V_{R2} matches the polarity of $V_{R2(B)}$, which is greater than $V_{R2(A)}$.

Figure 9.54a demonstrates another interesting point. As you know, Kirchhoff's voltage law states that series voltages must add up to equal the source voltage. If you look at the values in Figure 9.54a, you can see that

$$E_A = V_{R1} + V_{R3}$$

and

$$E_B = V_{R2} + V_{R3}$$

Figure 9.54b shows the circuit currents and their values. Note that the value of each current was found using Ohm's law and the values shown in Figure 9.54a. Using the current values shown, the overall operation of the circuit can be described as follows:

1. E_A is generating a total current of 233 mA.
2. E_B is generating a total current of 267 mA.
3. At point (A), the source currents combine. The sum of these currents (500 mA) passes through R_3.
4. At point (B), the currents split. Each current returns to its source via the series resistor.

Figure 9.54b also demonstrates the fact that the circuit operates according to Kirchhoff's current law. As you can see, the total current entering a given point or component equals the total current leaving that point or component.

PROGRESS CHECK

1. What is the *superposition* theorem? To what type of circuit does it apply?
2. Why is it helpful to redraw a multisource circuit as two (or more) single-source circuits when using the superposition theorem?
3. How can you perform a quick check of your analysis on a given multisource circuit?

9.6 SUMMARY

Here is a summary of the major points that were made in this chapter:

Voltage and Current Sources

- Analyzing the effect that a load has on the output from a given circuit is referred to as load analysis.
- The primary difference between the ideal and practical voltage source is the effect that source resistance has on load voltage.
  - The output from the ideal voltage source remains constant, despite variations in load resistance.

  - The output from a practical voltage source varies directly with changes in load resistance.
- The primary difference between the ideal and practical current source is the effect that source resistance has on load current.
  - The output from the ideal current source remains constant, despite variations in load resistance.
  - The output from a practical current source varies inversely with changes in load resistance.
- For every voltage source, there exists an equivalent current source (and vice versa).
  - Two sources are considered to be equivalent when they provide the same output values for any given value of load resistance.

Thevenin's Theorem

- *Thevenin's theorem* states that any resistive circuit or network, no matter how complex, can be represented as a voltage source in series with a source resistance.
  - Any series-parallel circuit can be represented as a Thevenin equivalent circuit.
  - The components of this circuit are called the *Thevenin voltage* (E_{TH}) and the *Thevenin resistance* (R_{TH}).
- The strength of the Thevenin equivalent circuit lies in the fact that it produces the same output values as the original circuit for any given value of load resistance.
- When using Thevenin's theorem, the load analysis of a series-parallel circuit is performed as follows:
  - Derive the Thevenin equivalent for the series-parallel circuit.
  - For each value of load resistance, determine the output from the Thevenin equivalent circuit using the voltage-divider relationship.
- The values obtained will equal the actual outputs from the original circuit.
- To determine the value of E_{TH}:
  - Remove the load.
  - Calculate the voltage across the open load terminals. This voltage is E_{TH}.
- Thevenin voltage (E_{TH}) is the no-load output voltage (V_{NL}) of the original circuit.
- To determine the value of R_{TH}:
  - Remove the load and replace the voltage source with a wire (to represent the ideal source resistance of 0 Ω).
  - Calculate the resistance that would be measured across the open load terminals. This resistance is R_{TH}.
- Thevenin resistance (R_{TH}) is the no-load output resistance (R_{NL}) of the original circuit.
- Thevenin's theorem is commonly used to predict the change in load voltage that results from a change in load resistance.

Maximum Power Transfer

- Maximum power transfer occurs when load resistance equals source resistance.
- For a series-parallel circuit with a variable load, maximum power transfer occurs when R_L is set to the value of R_{TH}.
- In a multiload circuit, no two loads have the same Thevenin equivalent circuit.
  - To derive the Thevenin equivalent circuit for each load, you must assume fixed values of resistance for the other loads.

Norton's Theorem

- Norton's theorem states that any resistive circuit or network, no matter how complex, can be represented as a current source in parallel with a source resistance.
  - Any series-parallel circuit can be represented as a Norton equivalent circuit. The components of this circuit are called the *Norton current* (I_N) and the *Norton resistance* (R_N).
  - In essence, Norton's theorem is a "current-source version" of Thevenin's theorem.
- To determine the value of I_N:
  - Short the load terminals.
  - Calculate the value of current through the shorted load terminals. This current is I_N.
- The value of R_N is found using the same procedure as the one given for finding R_{TH}.
- For every Norton equivalent circuit, there is a matching Thevenin equivalent circuit (and vice versa).
- Using the appropriate source conversion technique, a Thevenin equivalent circuit can be converted to a Norton equivalent circuit (and vice versa).

Superposition Theorem

- A *multisource circuit* is one that contains more than one voltage (or current) source.
  - When analyzing a multisource circuit, the effects of all the sources must be taken into account.
- The *superposition theorem* states that the response of a circuit to more than one source can be determined by analyzing the circuit's response to each source (alone) and combining the results.

CHAPTER REVIEW

1. Problem solving techniques that are applied to specific types of circuits are often referred to as ________.
   a) bridge analysis
   b) complex circuit analysis
   c) network theorems
   d) network analysis

2. Load analysis is used to determine how voltage and current sources respond to ________.
   a) an increase in load resistance
   b) a decrease in load resistance
   c) an increase in load demand
   d) all of the above

3. An ideal voltage source ________.
   a) maintains a constant output voltage regardless of load resistance
   b) has zero source resistance
   c) has a V_{NL} value that is equal to V_{RL}
   d) all of the above

4. In any practical voltage source, if the load current *decreases*, the difference between V_{RL} and V_{NL} will ________ .
   a) increase
   b) decrease
   c) remain the same
   d) it depends on the value of R_S

5. A voltage source with a V_{NL} value of 60 V produces 40 V across a 20 Ω load. This means that the value of R_S equals ________.
   a) 10 Ω
   b) 13.3 Ω
   c) 30 Ω
   d) 26.6 Ω

6. A voltage source develops 110 volts across a 15 Ω load. It has an internal resistance of 10 Ω. This means that the value of V_{NL} equals ________.
   a) 165 V
   b) 275
   c) 183 V
   d) 168 V

7. Most practical DC voltage sources have values of R_S ________.
   a) that are less than 50 Ω
   b) that can have a significant effect on high value loads
   c) that are greater than 50 Ω
   d) none of the above

8. Source resistance is a ________ component in a practical voltage source, and a ________ component in a practical current source.
   a) series, series
   b) series, parallel
   c) parallel, series
   d) parallel, parallel

9. Source resistance is ________ in a practical voltage source and ________ in a practical current source.
   a) high, low
   b) high, high
   c) low, low
   d) low, high

10. The ________ provided by a practical current source is lower than the value provided by an ideal current source having the same output rating.
    a) load voltage
    b) load current
    c) load power
    d) all of the above

11. In a practical current source, the ________ the value of R_S, the ________ the value of I_{RL}.
    a) higher, lower
    b) higher, higher
    c) lower, higher
    d) lower, lower

12. A 100 mA current source delivers 90.9 mA to a 10 kΩ load. The source resistance of this current source is ________.
    a) 100 kΩ
    b) 110 kΩ
    c) 90 kΩ
    d) 90.9 kΩ

13. A current source with a 330 kΩ source resistance is rated at 2.4 A. A ________ load would pass 1.8 A if connected to this current source.
    a) 100 kΩ
    b) 110 kΩ
    c) 90 kΩ
    d) 90.9 kΩ

14. The maximum output from a current source is referred to as ________.
    a) I_{SL}
    b) I_{Lmax}
    c) I_{Omax}
    d) I_N

15. A current source has values of R_S = 10 kΩ and I_S = 6 A. The equivalent voltage source would have an R_S value of ________.
    a) 60 kΩ
    b) 1.67 kΩ
    c) 16.7 kΩ
    d) 10 kΩ

16. The no-load voltage of an equivalent voltage source to the current source described in question 15 would be ________.
    a) 60 kV
    b) 60 V
    c) 6 V
    d) 600 mV

17. A 24 V voltage source has 20 Ω of source resistance. The equivalent current source would contain an ideal current source rated at ________ in parallel with ________ of resistance.
    a) 4.8 A, 20 Ω
    b) 1.2 A, 4.8 kΩ
    c) 1.2 A, 20 Ω
    d) 4.8 A, 4.8 kΩ

18. R_{TH} is measured from the perspective of ________.
    a) the source resistance
    b) the source
    c) the output resistance
    d) the load

19. E_{TH} is equal to ________.
    a) V_{NL}
    b) V_{EQ}
    c) V_L
    d) all of the above
    e) none of the above

20. When solving for R_{TH} the voltage source is replaced with a short because ________.
    a) of its high impedance
    b) it has a very low R_S value
    c) sources are always shorted by ohmmeters
    d) the load is removed

21. The amp for a night club sound system is designed to deliver maximum power to two 8 Ω speakers connected in parallel. This means that the equivalent output resistance of this amplifier must be ________.
    a) very high
    b) 4 Ω
    c) very low
    d) 8 Ω

22. R_N is ________ R_{TH} for equivalent circuits.
    a) usually higher than
    b) always higher than
    c) usually lower than
    d) always lower than
    e) equal to

23. In order to produce a Thevenin equivalent circuit for a multiload circuit ________.
    a) all but one of the loads must be fixed
    b) all loads must be fixed
    c) all but one of the loads must be variable
    d) all loads must be variable

24. When measuring I_N you must ________.
    a) open the load
    b) short out the load
    c) replace the source with an open
    d) replace the source with a short

25. A Thevenin circuit and its Norton equivalent are found to have the following values: $R_{TH} = 2.2\ k\Omega$ and $I_N = 250$ mA. This means $E_{TH} =$ ________.
    a) 250 V
    b) 9.8 kV
    c) 550 V
    d) 114 V

26. Multisource circuits have ________.
    a) two voltage sources
    b) two or more voltage sources
    c) two or more voltage and/or current sources
    d) two or more current sources

27. In order to solve a network with multiple voltage sources using the superposition theorem, the first step is to ________.
    a) replace all voltage sources with an open
    b) replace one of the voltage sources with an open
    c) replace all voltage sources with a short
    d) replace all but one of the voltage sources with a short

28. Two voltages with opposite polarity are called ________.
    a) aiding voltages
    b) opposing voltages
    c) reverse voltages
    d) either aiding or opposing, depending on current direction

PRACTICE PROBLEMS

1. Using a value of $R_L = 100\ \Omega$, determine whether or not the circuits in Figure 9.55a are equivalent circuits.
2. Using a value of $R_L = 2.2\ k\Omega$, determine whether or not the circuits in Figure 9.55b are equivalent circuits.

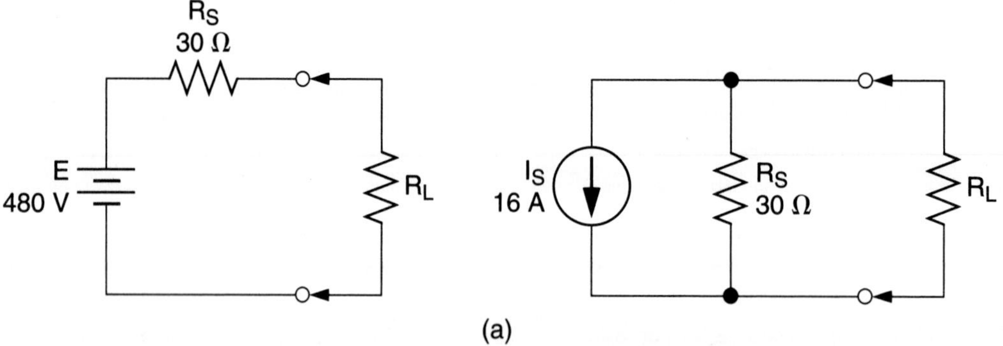

FIGURE 9.55

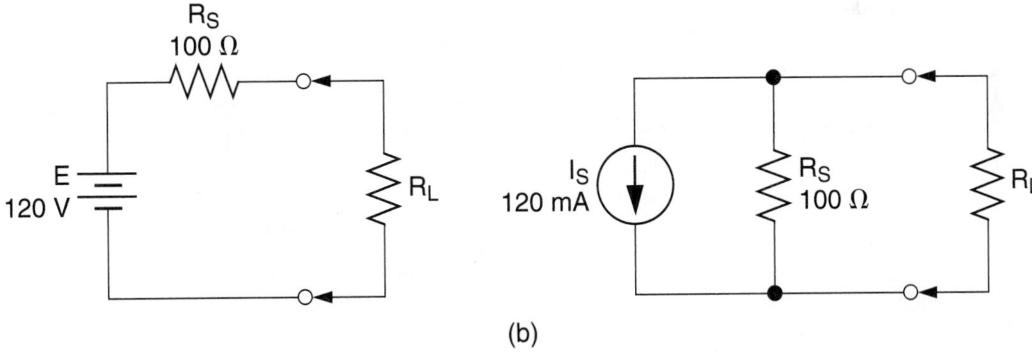

FIGURE 9.55 *(Continued)*

3. Convert the voltage source in Figure 9.56a to an equivalent current source.
4. Convert the voltage source in Figure 9.56b to an equivalent current source.
5. Derive the current source equivalent of the voltage source in Figure 9.56c. Then calculate the values of V_{RL} produced by each circuit for values of $R_L = 100\ \Omega$ and $R_L = 330\ \Omega$.
6. Derive the current source equivalent of the voltage source in Figure 9.56d. Then calculate the values of I_L produced by each circuit for values of $R_L = 180\ \Omega$ and $R_L = 1\ \text{k}\ \Omega$.

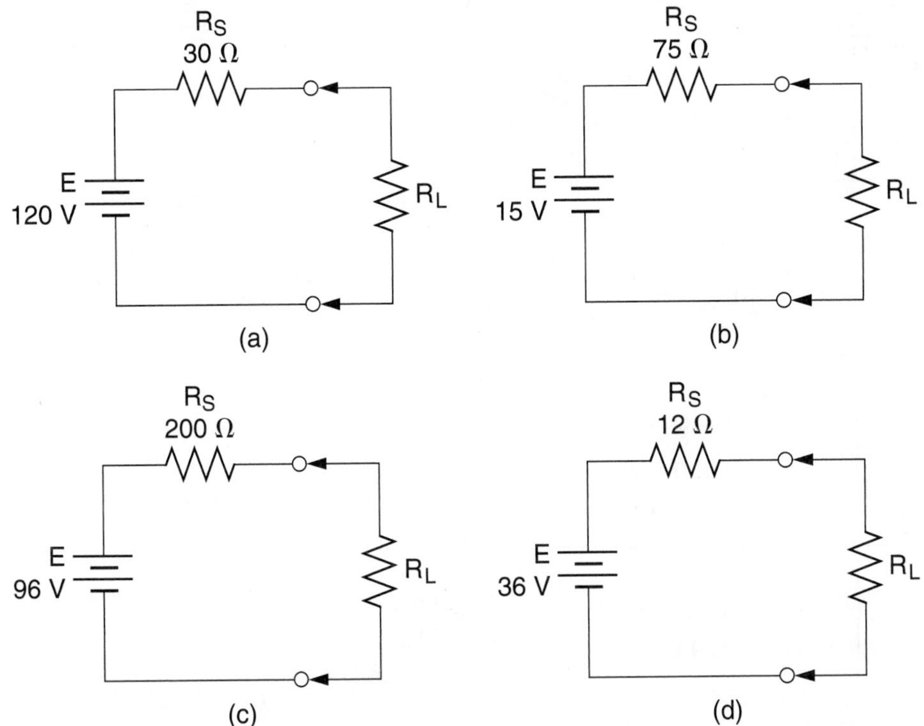

FIGURE 9.56

7. Convert the current source in Figure 9.57a to an equivalent voltage source.
8. Derive the voltage source equivalent of the current source in Figure 9.57b. Then calculate the values of V_{RL} produced by each circuit for values of $R_L = 75\ \Omega$ and $R_L = 150\ \Omega$.

I_S 4 A
R_S 7.5 Ω
R_L

(a)

I_S 600 mA
R_S 20 Ω
R_L

(b)

FIGURE 9.57

9. Derive the Thevenin equivalent of the circuit in Figure 9.58a.
10. Derive the Thevenin equivalent of the circuit in Figure 9.58b.

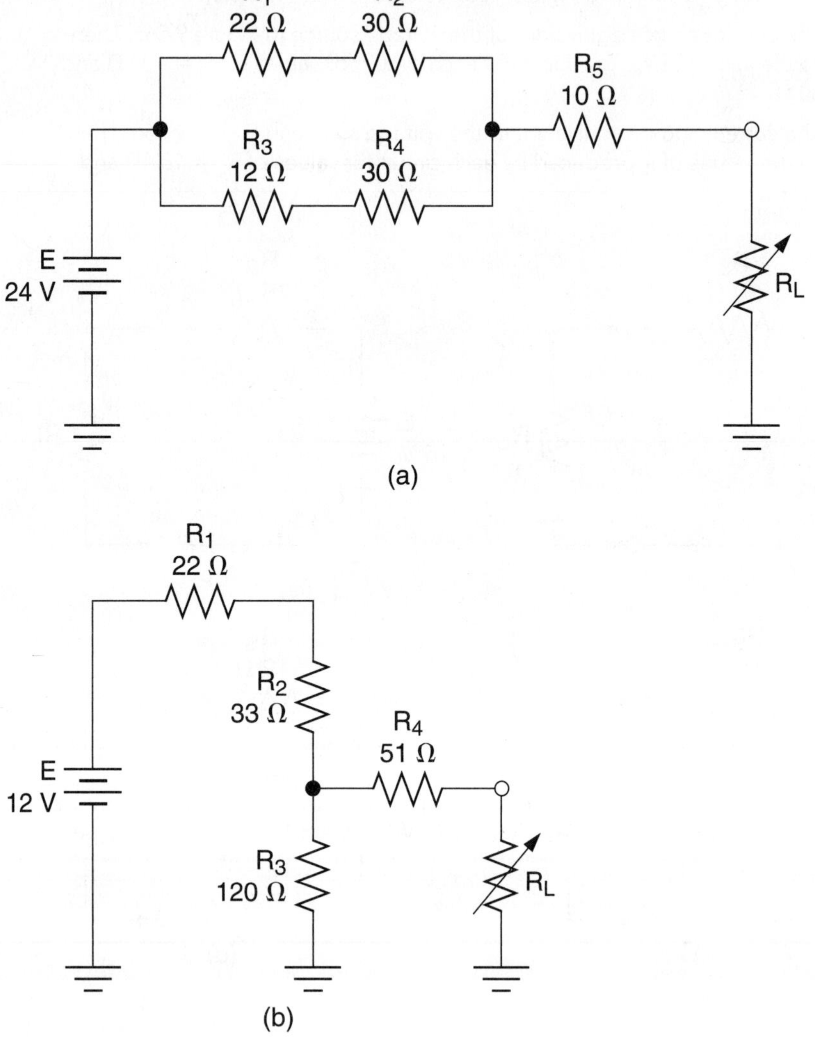

FIGURE 9.58

11. Derive the Thevenin equivalent of the circuit in Figure 9.59a.
12. Derive the Thevenin equivalent of the circuit in Figure 9.59b.

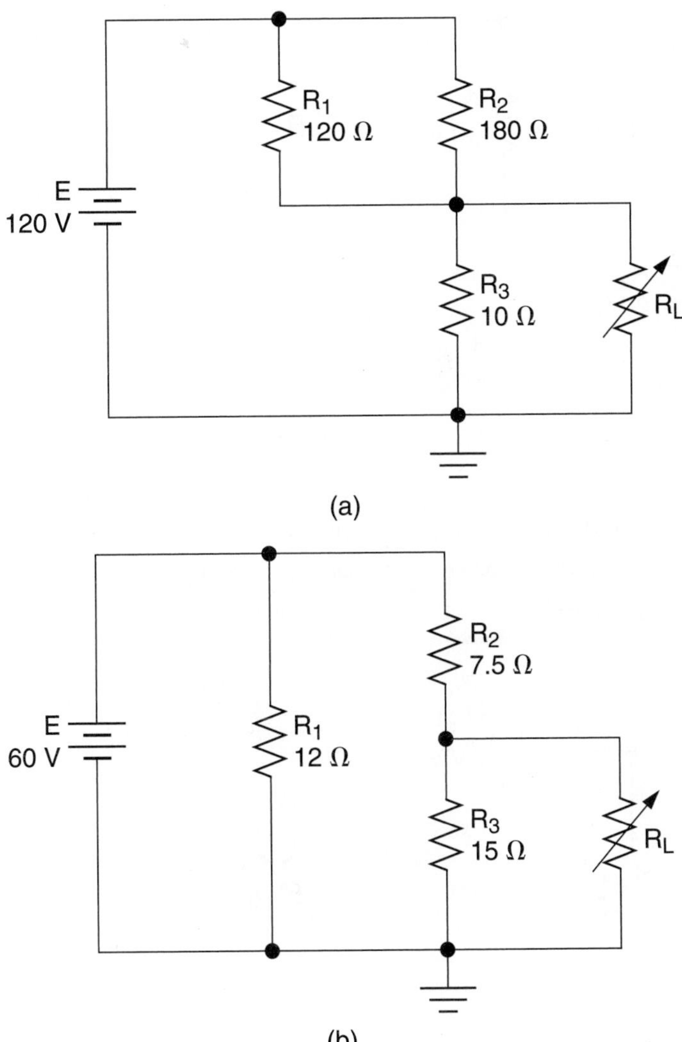

FIGURE 9.59

13. Calculate the maximum possible load power for the circuit in Figure 9.60a.

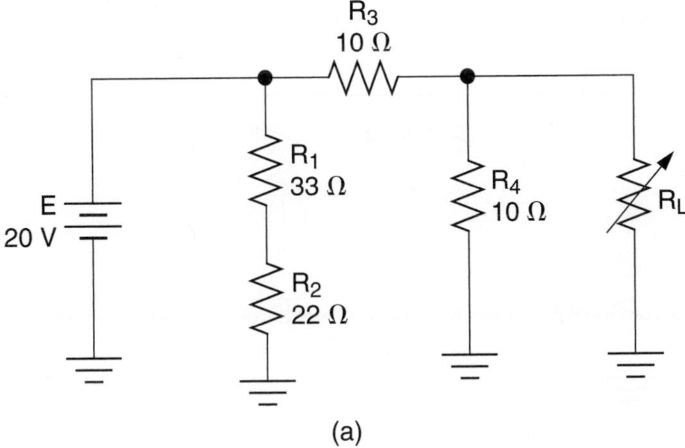

FIGURE 9.60

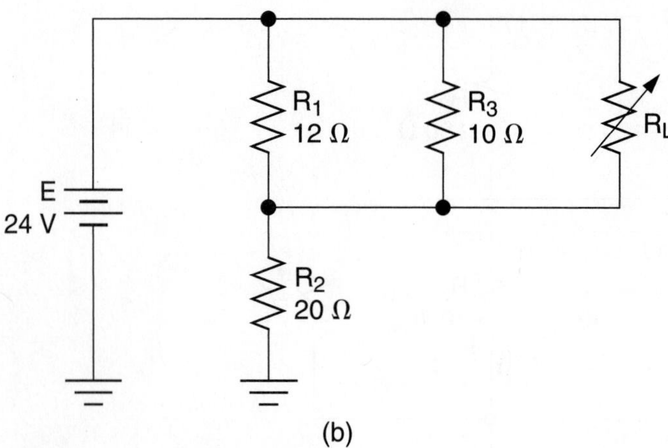

FIGURE 9.60 *(Continued)*

14. Calculate the maximum possible load power for the circuit in Figure 9.60b.
15. Calculate the maximum possible load power for the circuit in Figure 9.61a.
16. Calculate the maximum possible load power for the circuit in Figure 9.61b.

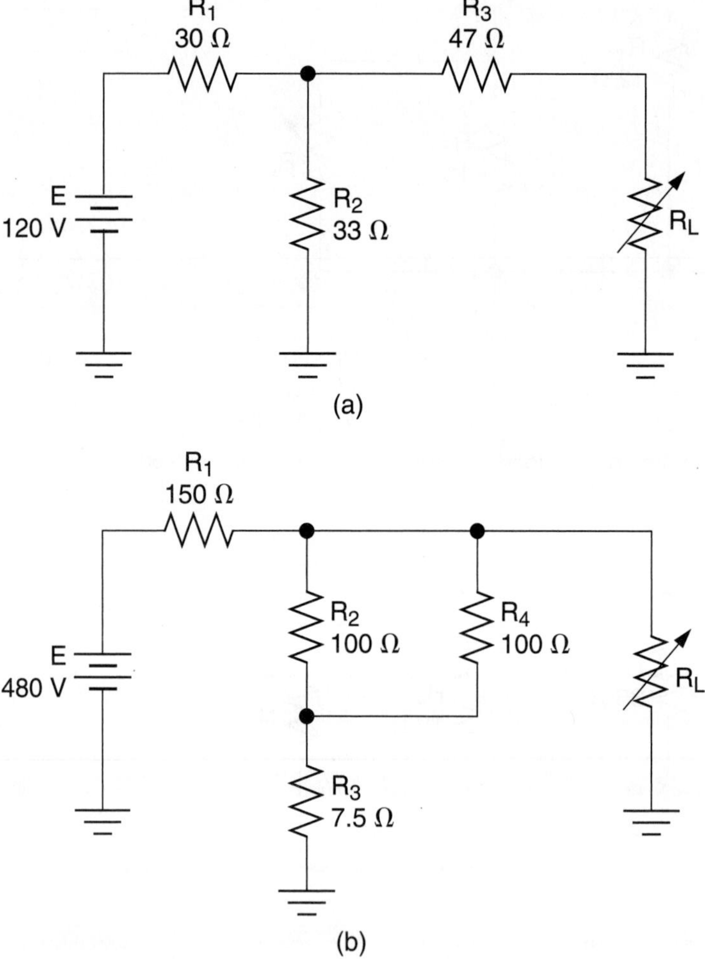

FIGURE 9.61

17. Derive the Thevenin equivalent circuit for Load 1 in Figure 9.62.

18. Derive the Thevenin equivalent circuit for Load 2 in Figure 9.62.

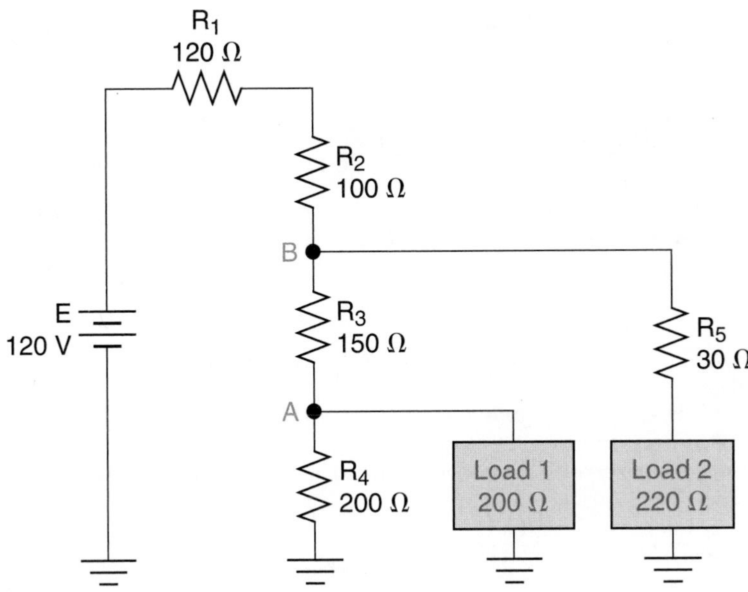

FIGURE 9.62

19. Determine the maximum possible load power for the bridge circuit in Figure 9.63a.

20. Determine the maximum possible load power for the bridge circuit in Figure 9.63b.

21. Derive the Norton equivalent for the circuit in Figure 9.64a.

22. Derive the Norton equivalent for the circuit in Figure 9.64b.

23. Derive the Norton equivalent for the circuit in Figure 9.65a.

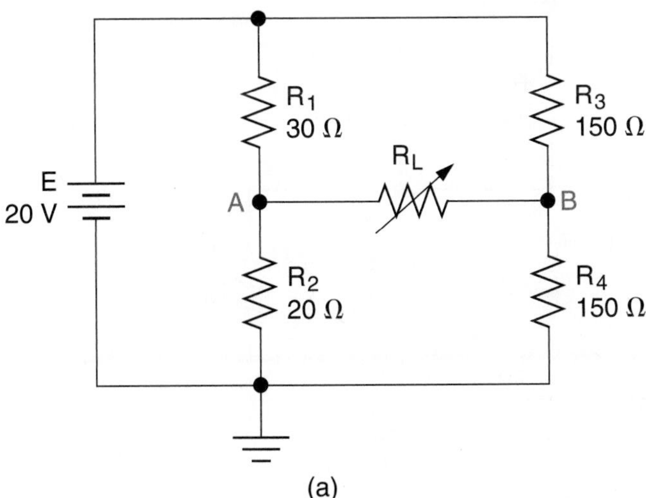

FIGURE 9.63

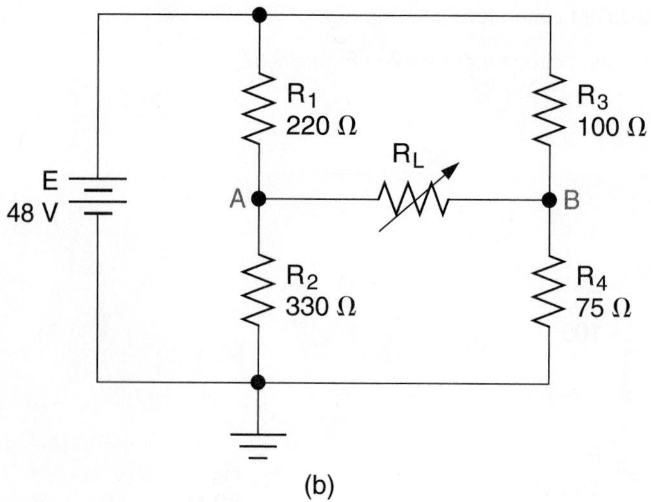

FIGURE 9.63 (*Continued*)

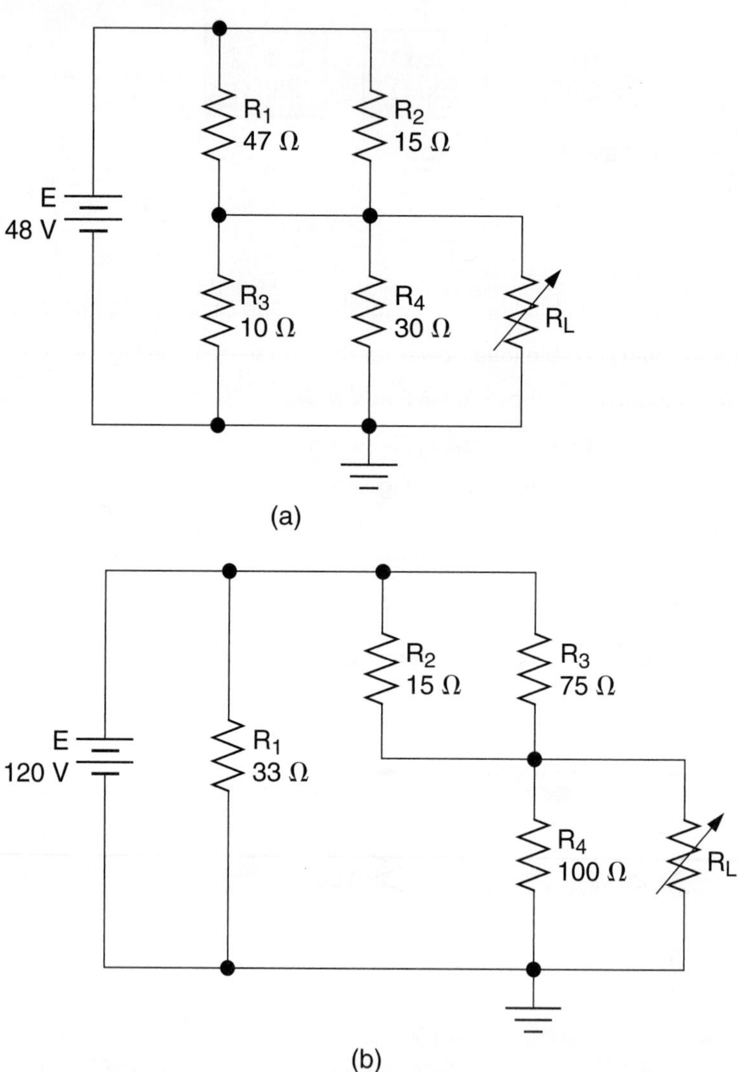

FIGURE 9.64

24. Derive the Norton equivalent for the circuit in Figure 9.65b.
25. Determine the values of V_{R1}, V_{R2}, and V_{R3} for the circuit in Figure 9.66a.

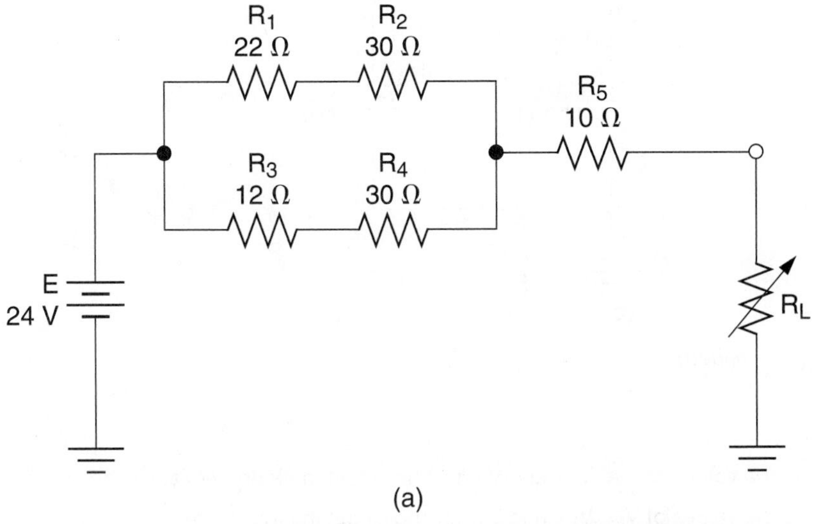

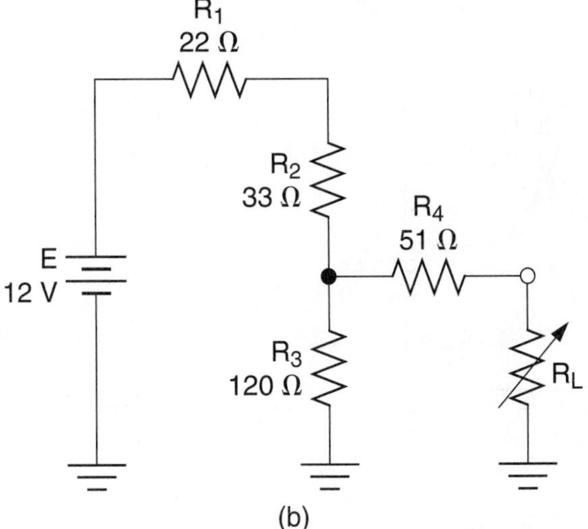

FIGURE 9.65

26. Determine the values of V_{R1}, V_{R2}, and V_{R3} for the circuit in Figure 9.66b.

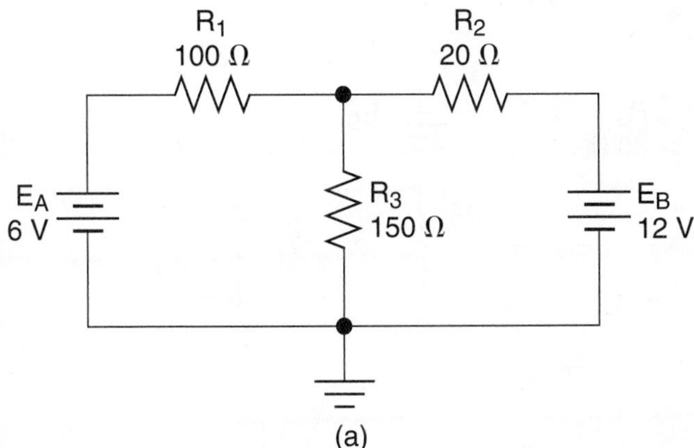

FIGURE 9.66

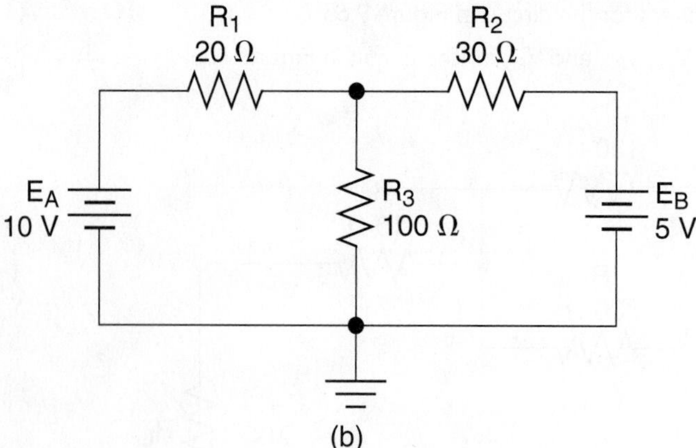

FIGURE 9.66 *(Continued)*

27. Determine the values of V_{R1}, V_{R2}, and V_{R3} for the circuit in Figure 9.67a.
28. Determine the values of V_{R1}, V_{R2}, and V_{R3} for the circuit in Figure 9.67b.

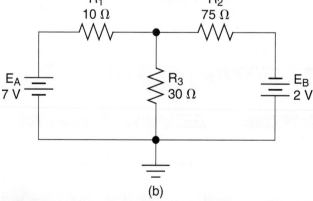

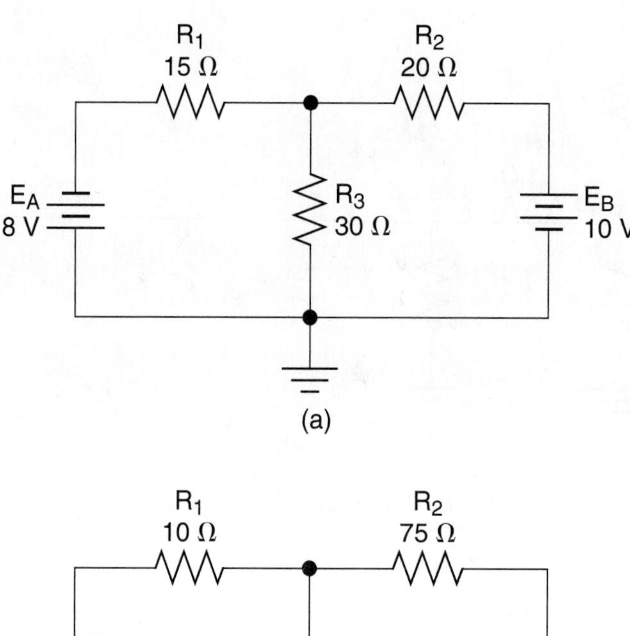

FIGURE 9.67

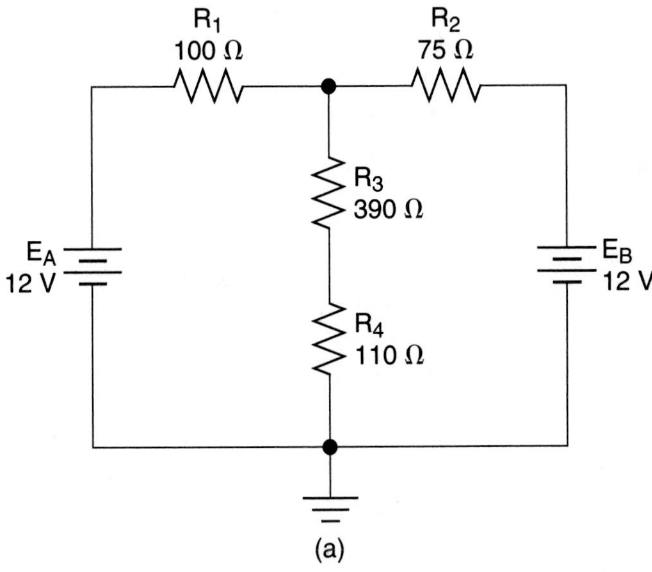

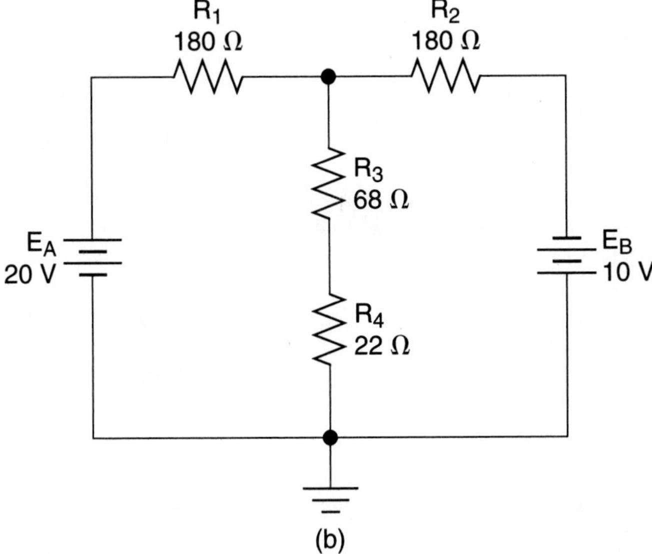

FIGURE 9.68

29. Determine the values of V_{R1} through V_{R4} for the circuit in Figure 9.68a.
30. Determine the values of V_{R1} through V_{R4} for the circuit in Figure 9.68b.

CHALLENGERS

31. Calculate the range of V_{RL} values for each load in Figure 9.69.
32. Calculate the range of load voltage values for the circuit in Figure 9.70.
33. Determine the maximum possible load power for the circuit shown in Figure 9.71.

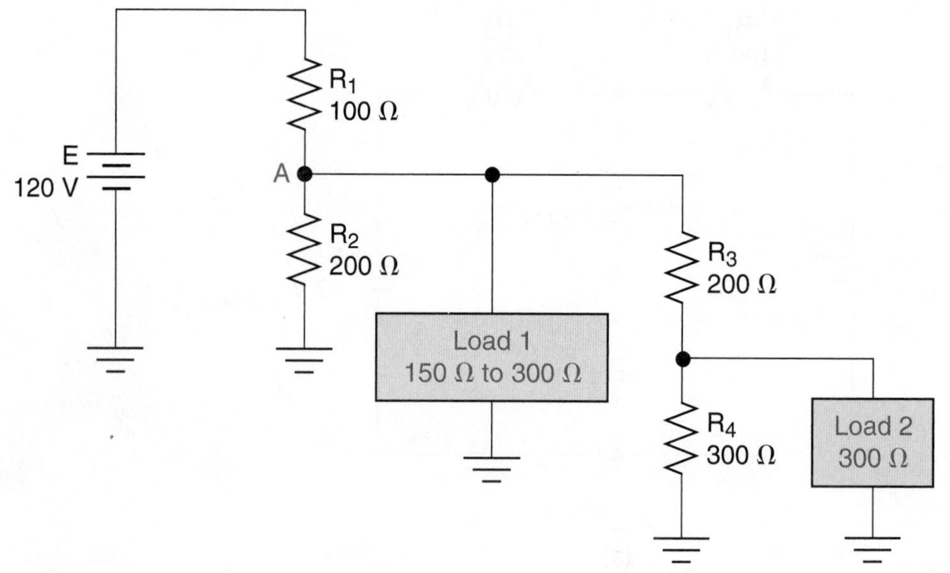

FIGURE 9.69

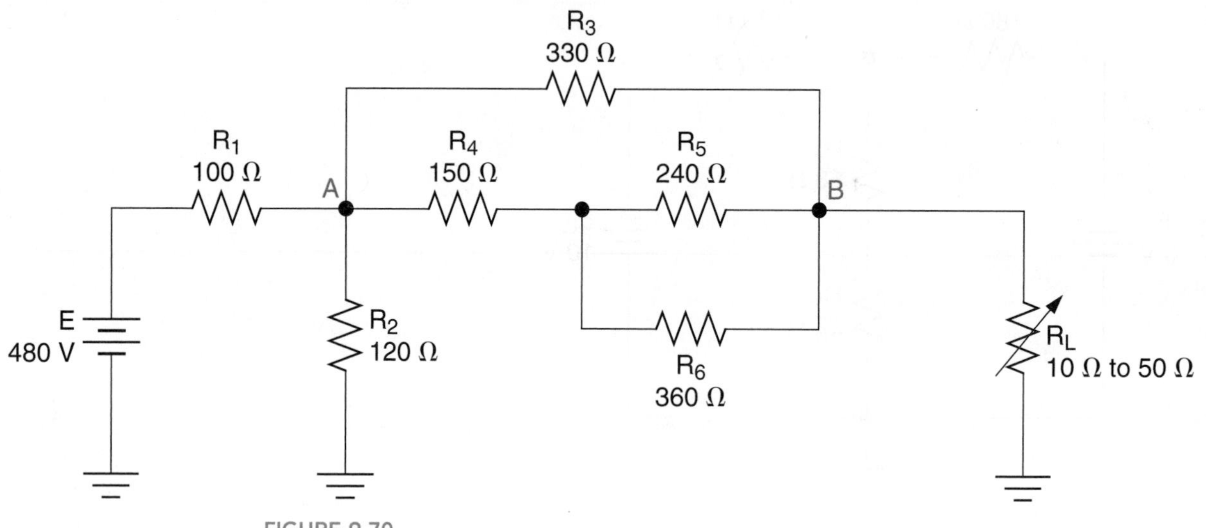

FIGURE 9.70

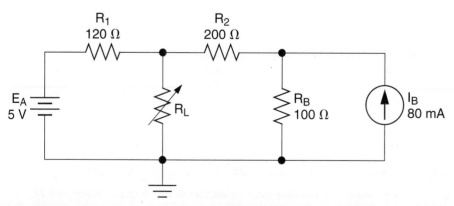

R1
120 Ω
R2
200 Ω
EA
5 V
RL
RB
100 Ω
IB
80 mA

FIGURE 9.71

ANSWERS TO THE EXAMPLE PRACTICE PROBLEMS

9.1	12 V, 10.5 V
9.2	49.8 mA to 48.8 mA
9.3	I_{SL} = 25 mA, I_{RL} = 22.7 mA
9.4	I_L = 50 mA @ R_L = 100 Ω, I_L = 80 mA @ R_L = 25 Ω
9.5	I_S = 500 mA, R_S = 16 Ω, I_{RL} = 167 mA
9.6	E = 6 V, R_S = 150 Ω, V_{RL} = 2.4 V
9.7	V_{RL} = 6.57 V @ R_L = 150 Ω, V_{RL} = 4.86 V @ R_L = 30 Ω
9.8	7.2 V
9.9	14.4 Ω
9.10	E_{TH} = 26.2 V, R_{TH} = 16.4 Ω
9.11	E_{TH} = 90 V, R_{TH} = 9 Ω
9.12	6.4 W
9.13	200 W
9.14	193 mW
9.15	500 mA
9.16	120 Ω
9.17	640 mA @ R_L = 100 Ω, 533 mA @ R_L = 150 Ω
9.18	6.54 W
9.19	V_{R1} = 6 V, V_{R2} = 1 V, V_{R3} = 8 V

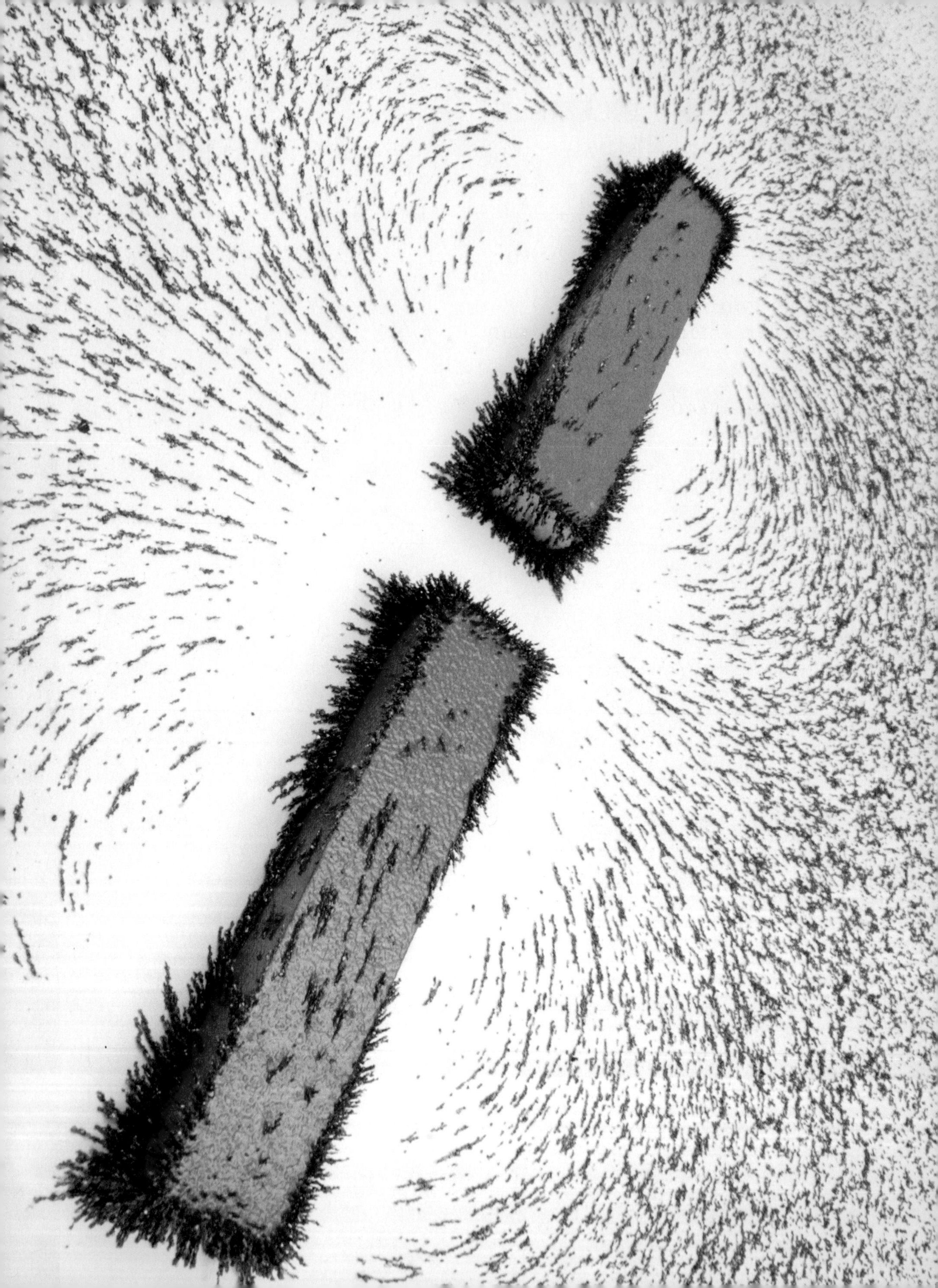

CHAPTER 10

MAGNETISM

PURPOSE

Magnetism plays an important role in the operation of many electric components and systems. For example, electric motors and generators, analog meters, and electric relays utilize the properties of magnetism.

In this chapter, we'll examine many of the basic properties of magnetism. A complete study of the subject would require several volumes, so our discussions will be limited to those principles that relate directly to the study of electricity.

KEY TERMS

The following terms are introduced and defined in this chapter on the pages indicated:

Air gap, 465
Ampere-turns ($A \cdot t$), 457
Coercive force, 461
Coil, 454
Core, 454
Domain theory, 448
Ferromagnetic, 451
Field strength, 442
Flux density, 441
Hysteresis, 460
Induction, 449
Left-hand rule, 453
Magnetic field, 441
Magnetic flux, 441
Magnetic induction, 450
Magnetic north, 439
Magnetic shielding, 465
Magnetic south, 439
Magnetization curve, 460
Magnetomotive force (mmf), 455
Maxwell (Mx), 441
Meter movement, 466
Microweber (μWb), 441
North-seeking pole, 439
Permeability (μ), 445
Poles, 439
Relative permeability (μ_r), 445
Relay, 461
Reluctance, 450
Residual flux density, 460
Retentivity, 450
Ring magnet, 466
Rowland's law, 456
South-seeking pole, 439
Tesla (T), 442
Weber (Wb), 441

OBJECTIVES

After studying the material in this chapter, you should be able to:

1. Describe magnetic force.
2. Identify the poles of a magnet and state the relationships between them.
3. List and define the common units of magnetic flux.
4. Contrast the following: magnetic fields, magnetic flux, and flux density.
5. List and describe the common units of flux density.

6. Calculate the flux density at any point in a magnetic field, given the cross-sectional area and the total flux.
7. Compare and contrast permeability with relative permeability.
8. Given the permeability of a material, calculate its relative permeability.
9. Discuss the domain theory of the source of magnetism.
10. Describe the magnetic induction methods of producing a magnet.
11. Define retentivity and discuss its relationship to permeability.
12. Define reluctance and discuss its relationship to permeability.
13. List and describe the magnetic classifications of materials.
14. Describe the relationship between current and a resulting magnetic field.
15. Compare the basic magnetic and electric quantities.
16. Discuss the relationships that exist among the following: magnetomotive force, ampere-turns, coil current, core permeability, and coil flux density.
17. Define and discuss hysteresis.
18. Describe the operation of electromagnetic relays.
19. Describe magnetic shielding.
20. List and describe the various shapes of magnets.
21. Describe the operation of a D'Arsonval meter movement.
22. Describe the proper care and handling of magnets.

10.1 MAGNETISM: AN OVERVIEW

As you know, a magnet can attract (or repel) another magnet or any *ferrous* metal (metal containing iron). The force that a magnet exerts on the objects around it is referred to as *magnetic force*.

When exposed to magnetic force, iron filings line up as shown in Figure 10.1a. For this reason, magnetic force is commonly represented as a series of lines. These lines form closed loops around (and through) the magnet, as illustrated in Figure 10.1b.

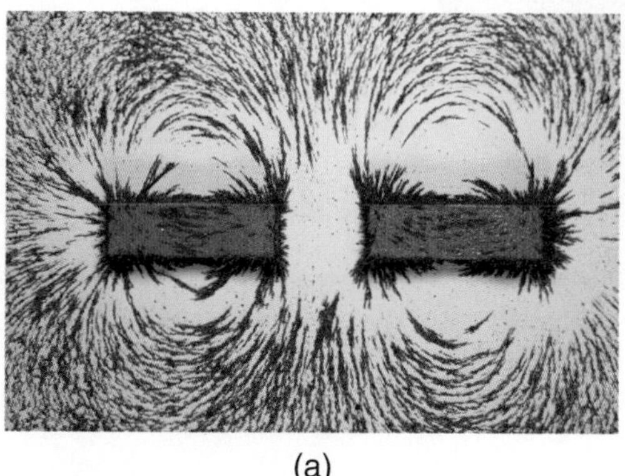

(a)

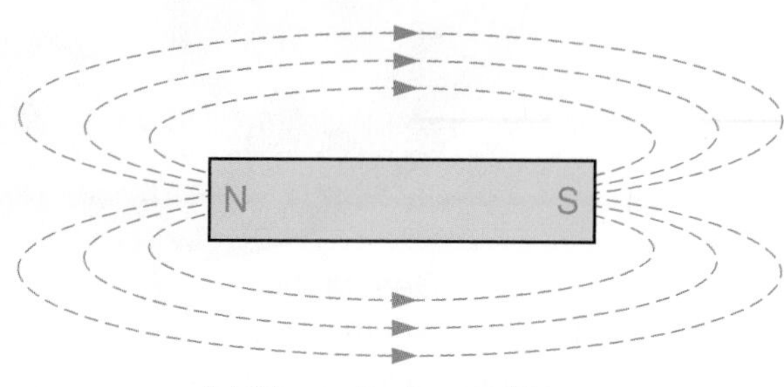

(b) Magnetic lines of force

FIGURE 10.1 Iron filings align with magnetic lines of force.
(a) Istock.com

Magnetic Poles

The points where magnetic lines of force leave (and return to) a magnet are referred to as the **POLES**. By convention, the lines of force are assumed to emanate from the **NORTH-SEEKING POLE (N)** and return to the magnet via its **SOUTH-SEEKING POLE (S)**. Within the magnet, the lines of force continue from (S) to (N).

The terms used to describe the poles of a magnet originate as shown in Figure 10.2. The earth itself is one huge magnet, with **MAGNETIC NORTH** and **MAGNETIC SOUTH** poles. If we suspend a magnet as shown, the magnet turns so that one pole points to magnetic north and the other points to magnetic south. The poles are then identified as north-seeking and south-seeking.

POLES The points where magnetic lines of force leave (and return to) a magnet.

NORTH-SEEKING POLE The pole where lines of force leave a magnet.

SOUTH-SEEKING POLE The pole where lines of force return to a magnet.

MAGNETIC NORTH The earth's magnetic north pole.

MAGNETIC SOUTH The earth's magnetic south pole.

Like Poles and Unlike Poles

You may recall that like charges repel (and unlike charges attract) each other. Magnetic poles have similar effects on each other. That is:

1. *Like poles repel* each other.
2. *Unlike poles attract* each other.

These interactions between the poles are illustrated in Figure 10.3.

FIGURE 10.2 The earth's magnetic poles.

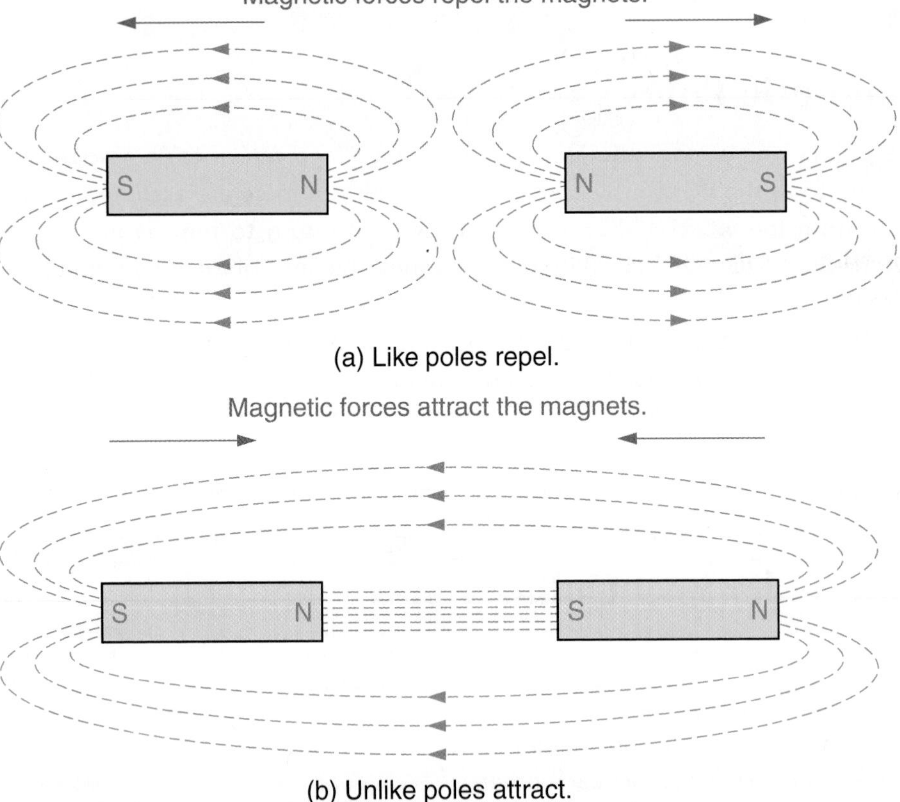

FIGURE 10.3 Like and unlike magnetic poles.

Magnetic Flux

The lines of force produced by a magnet are collectively referred to as **MAGNETIC FLUX**. There are two fundamental units of measure for magnetic flux. These units are identified and defined as follows:

MAGNETIC FLUX The lines of force produced by a magnet.

UNIT OF MEASURE	UNIT SYSTEM	VALUE (DEFINED)
Maxwell (Mx)	cgs	1 Mx = 1 line of flux
Weber (Wb)	SI	1 Wb = 10^8 Mx

The **MAXWELL (Mx)** is easy to visualize. The magnet represented in Figure 10.1b has six lines of force. Therefore, the magnetic flux (in maxwells) is given as

MAXWELL (Mx) A unit of measure of magnetic flux, equal to 1 line of flux.

$$\Phi = 6 \text{ Mx}$$

where Φ (the Greek letter *phi*) is used to represent magnetic flux.

Unfortunately, the maxwell is not a practical unit of measure. Even small magnets produce flux in the thousands of maxwells (and higher). On the other hand, the **WEBER (Wb)** is often too large a unit of measure to be useful. In such cases, the **MICROWEBER (μWb)** is the preferred unit of measure. Keeping in mind that 1 Wb = 10^8 Mx, the amount of flux contained in one microweber can be calculated as follows

WEBER (Wb) A unit of measure of magnetic flux, equal to 10^8 Mx (lines of flux).

MICROWEBER (μWb) A practical unit of measure of magnetic flux, equal to 100 Mx (lines of flux).

$$\begin{aligned} 1\ \mu\text{Wb} &= 1 \times 10^{-6}\ \text{Wb} \\ &= (1 \times 10^{-6})(1 \times 10^{8}\ \text{Mx}) \\ &= 1 \times 10^{2}\ \text{Mx} \\ &= 100\ \text{Mx} \end{aligned}$$

Therefore, one microweber contains 100 lines of flux.

Magnetic characteristics are typically measured in both *cgs* (centimeter-gram-second) units and *SI* (Systeme International) units. Throughout the remainder of this chapter, we will focus on the SI units of measure for magnetic quantities, as they are used more often than *cgs* units.

Flux Density

The area of space surrounding a magnet contains magnetic flux. This area of space is referred to as a **MAGNETIC FIELD**. A magnetic field is illustrated in Figure 10.4. Note that the field is a three-dimensional space (similar to a cylinder) that surrounds the magnet.

MAGNETIC FIELD The area of space surrounding a magnet that contains magnetic flux.

The "strength" of a magnet depends on:

- The field size.
- The amount of flux it contains.

The greater the amount of flux per unit area, the stronger the magnetic force. **FLUX DENSITY** is a measure of flux per unit area, and therefore, an indicator of

FLUX DENSITY A measure of *flux per unit area* and, therefore, an indicator of magnetic strength.

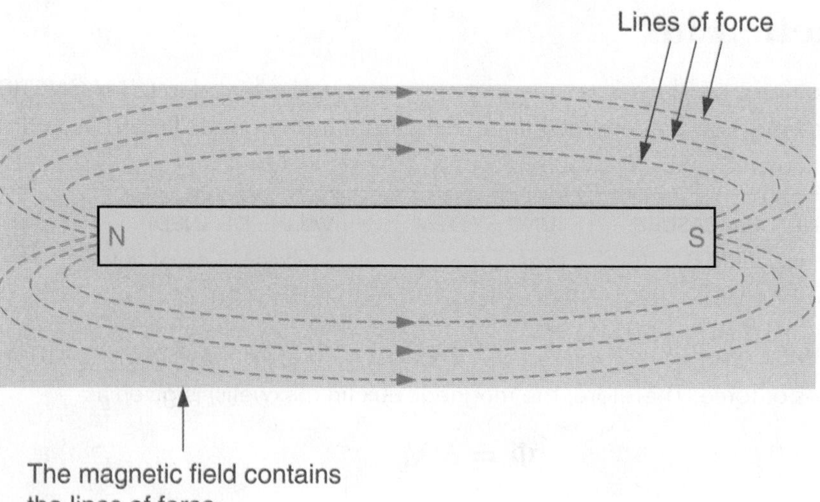
Lines of force
N
S
The magnetic field contains
the lines of force.

FIGURE 10.4 A magnetic field.

FIELD STRENGTH Another name for flux density.

magnetic strength. In fact, flux density is also referred to as **FIELD STRENGTH**. By formula, flux density can be expressed as

$$\text{Flux density} = \frac{\text{magnetic flux}}{\text{area}}$$

or

(10.1)

$$B = \frac{\Phi}{A}$$

where

B = the flux density
Φ = the amount of flux
A = the cross-sectional area containing the flux

The concept of flux density is illustrated in Figure 10.5. The area in Figure 10.5a has six lines of flux passing through it. An area of equal size (shown in Figure 10.5b) has only three lines of flux. Since flux density is a measure of flux per unit area, the flux density in (a) is twice the flux density in (b).

TESLA (T) A unit of measure of flux density, equal to one weber per square meter (Wb/m^2).

The SI unit of measure for flux density is the **TESLA (T)**. One tesla of flux density is equal to one weber per square meter (Wb/m^2). Written mathematically,

$$1\ T = 1\ Wb/m^2$$

Earlier, you learned that one weber equals 10^8 lines of magnetic flux. Therefore, one tesla is equal to 10^8 lines of flux per square meter. The calculation of flux density (in teslas) is demonstrated in Example 10.1.

N S

Area

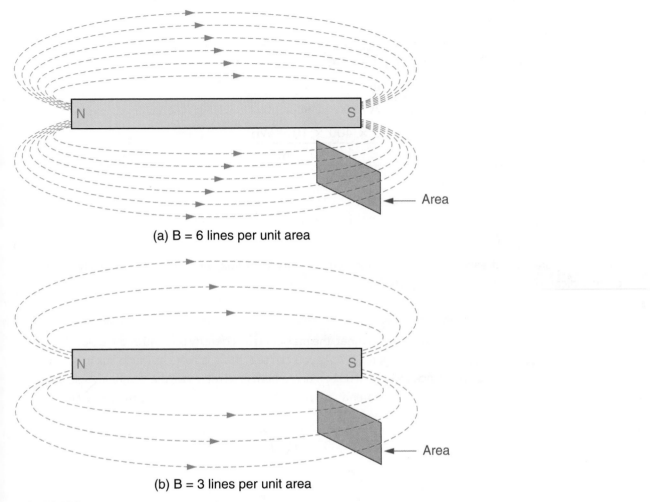

(a) B = 6 lines per unit area

N S

Area

(b) B = 3 lines per unit area

FIGURE 10.5

EXAMPLE 10.1

Calculate the flux density for the shaded area in Figure 10.6. Assume that each line of force represents 400 μWb of flux.

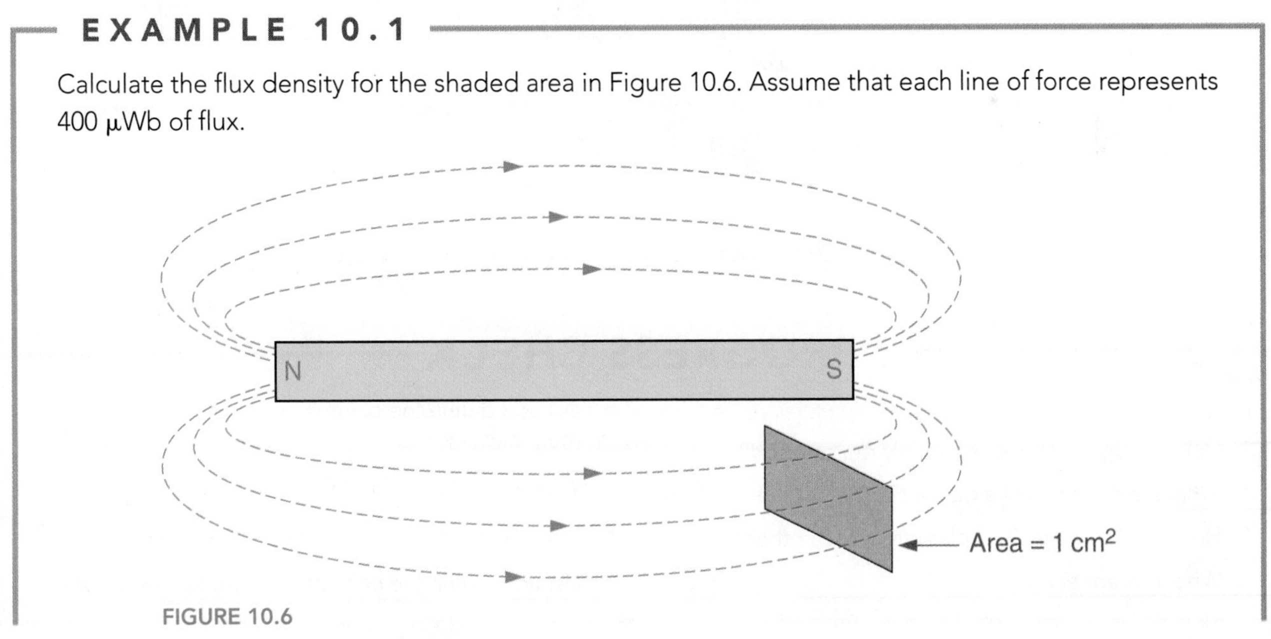

FIGURE 10.6

SOLUTION

The shaded area is one square centimeter (1 cm^2), which is equal to 1×10^{-4} square meters (m^2). Therefore, the flux density in Figure 10.6 is found as

$$B = \frac{\Phi}{A}$$

$$= \frac{3 \times 400 \times 10^{-6}\,\text{Wb}}{1 \times 10^{-4}\,\text{m}}$$

$$= 12\ \text{Wb/m}^2$$

$$= 12\ \text{T}$$

PRACTICE PROBLEM 10.1

Refer to Figure 10.6. If each line of force represents 150 μWb of flux, what is the flux density at the point indicated? Assume the area equals 3 cm^2.

For any given magnet, the maximum flux density is found at the poles. This point is illustrated in Figure 10.7. As you can see, there is twice as much flux in area A as in area B. Since all lines of flux emanate from (and return to) the magnetic poles, the flux density must be greater at the poles than anywhere else in the magnetic field.

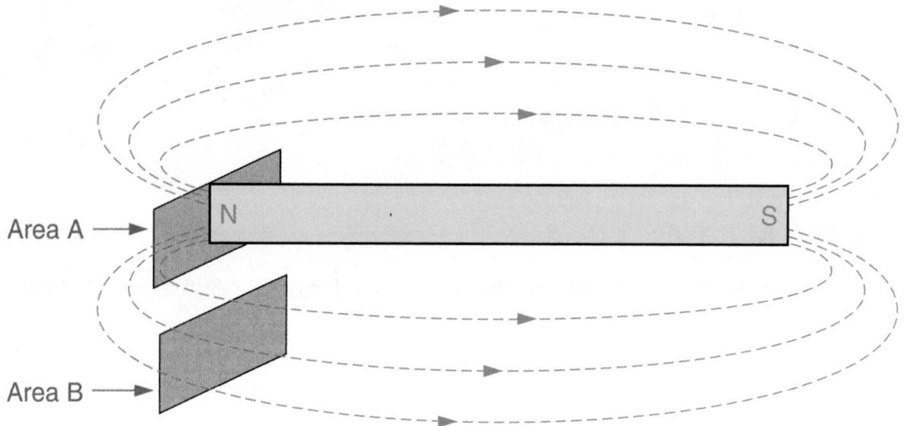

FIGURE 10.7 Flux density is greater at the poles of a magnet.

PROGRESS CHECK

1. What is *magnetic force*?
2. How is magnetic force commonly represented?
3. What are the *poles* of a magnet? What are they called?
4. How do like and unlike poles affect each other?
5. What is *magnetic flux*?
6. List and define the common units of measure for magnetic flux.
7. What is a *magnetic field*?
8. What is *flux density*? What does it indicate?
9. List and define the units of measure for flux density.

10.2 MAGNETIC CHARACTERISTICS

Some materials respond to the presence of a magnetic force, while others do not. In this section, we will discuss the magnetic characteristics and classifications of materials.

Permeability

PERMEABILITY (μ) A measure of the ease with which lines of magnetic force are established within a material.

PERMEABILITY (μ) is a measure of the ease with which lines of magnetic force are established within a material. The higher the permeability of a given material, the easier it is to produce magnetic lines of force within the material.

Increasing permeability has always been one of the goals of transformer design. A variety of core materials, including silicon iron in various forms, have been used to enhance performance.

IN THE FIELD

Materials with high permeability have the ability to concentrate magnetic lines of force. This point is illustrated in Figure 10.8. The bar shown is made of soft iron, a material with a permeability that is much higher than that of the surrounding air. Because the iron bar more readily accepts the lines of force, they align themselves in the iron bar as shown.

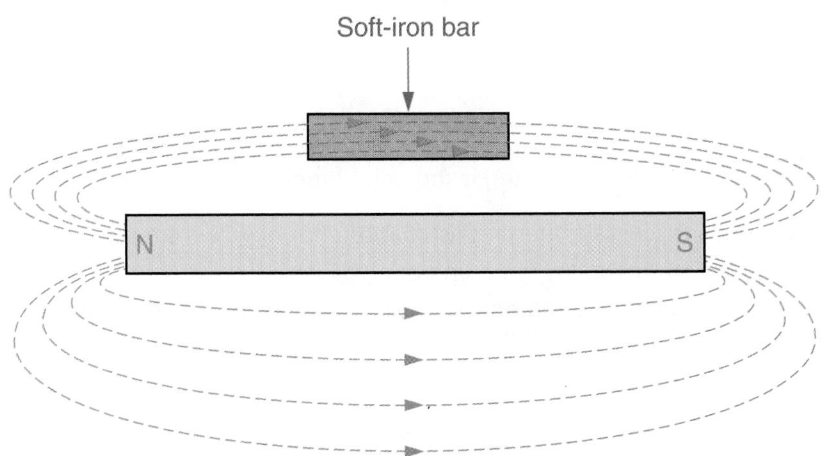

FIGURE 10.8 A soft-iron bar concentrates magnetic flux.

Since the lines of force in Figure 10.8 are drawn to the iron bar, the flux density at any point in the bar is greater than in the surrounding air. For example, assume that the cross-sectional areas shown in Figure 10.9 represent one square centimeter (1 cm^2). As you can see, the flux density in the iron bar (area A) is twice that of area B. Thus, high-permeability materials can be used to increase flux density.

Relative Permeability

RELATIVE PERMEABILITY (μ_r) The ratio of the material's permeability to that of free space.

In general, materials are rated according to their **RELATIVE PERMEABILITY** (μ_r). The relative permeability of a material is the ratio of the material's

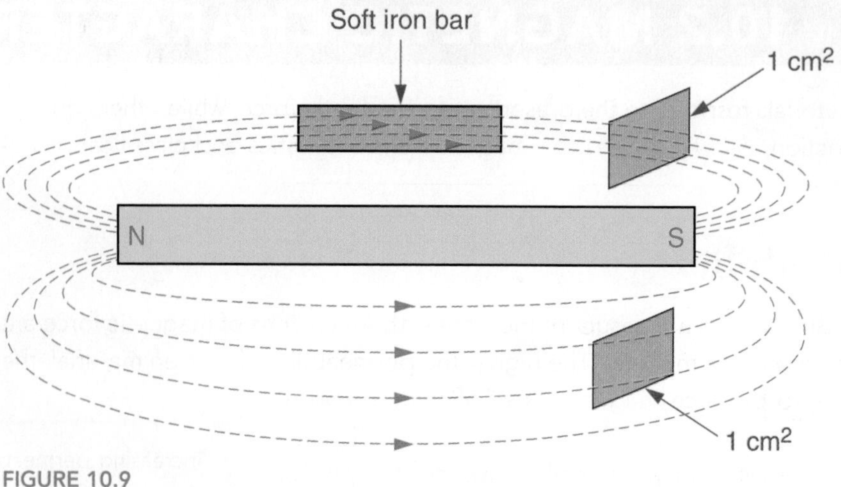

FIGURE 10.9

permeability to that of free space. In SI units, the permeability of free space is known to be

$$\mu_0 = 4\pi \times 10^{-7} \frac{\text{Wb}}{\text{A} \cdot \text{m}}$$

where

μ_0 is used to represent the permeability of free space

$\frac{\text{Wb}}{\text{A} \cdot \text{m}}$ (webers per ampere-meter) is the unit of measure for permeability

The ampere-meter is discussed later in this chapter. For now, we simply want to establish the fact that permeability has a unit of measure.

When the permeability of a material is known, its relative permeability is calculated using

(10.2)

$$\mu_r = \frac{\mu_m}{\mu_0}$$

where

μ_r = the relative permeability of the material

μ_m = the permeability of the material, in $\frac{\text{Wb}}{\text{A} \cdot \text{m}}$

μ_0 = the permeability of free space

Since relative permeability is a ratio of one permeability value to another, it has no unit of measure. The concept of relative permeability is illustrated further in Example 10.2.

EXAMPLE 10.2

The permeability of iron is approximately $2.5 \times 10^{-4} \frac{\text{Wb}}{\text{A} \cdot \text{m}}$. What is its relative permeability?

SOLUTION

Using the values given for iron and free space, the relative permeability of iron is found as

$$\mu_r = \frac{\mu_m}{\mu_0}$$

$$= \frac{2.5 \times 10^{-4}\,\text{Wb/A} \cdot \text{m}}{4\pi \times 10^{-7}\,\text{Wb/A} \cdot \text{m}}$$

$$= 199$$

This result indicates that the permeability of iron is approximately 200 times greater than an equal volume of free space.

Relative permeability is important because it equals the ratio of flux density within a material to flux density in free space. In other words, it tells us how much a material increases flux density, relative to the space surrounding it. For example, if we assume that the flux density in the space surrounding an iron bar is 2×10^{-3} teslas, the flux density in the iron bar is 200 times as great, or 4×10^{-1} teslas.

The Source of Magnetism: Domain Theory

As you know, all atoms contain electrons that orbit a nucleus. It is believed that every electron in a given atom has both an electric charge and a magnetic field. The electric charge, as you know, is always *negative* (and is offset by a matching positive charge in the nucleus). The polarity of the magnetic field, however, depends on the direction the electron spins on its own axis.

Each electron in an atom spins in a clockwise or counterclockwise direction. For example, let's assume that the electrons in Figure 10.10 are spinning in the directions indicated.

When an atom contains an equal number of electrons spinning in each direction, their magnetic fields cancel each other and the atom does not generate magnetic force. When the electrons spinning in one direction outnumber the ones spinning in the other direction, the atom acts as a source of magnetic force. The polarity of the magnetic force depends on the net spin direction of the electrons.

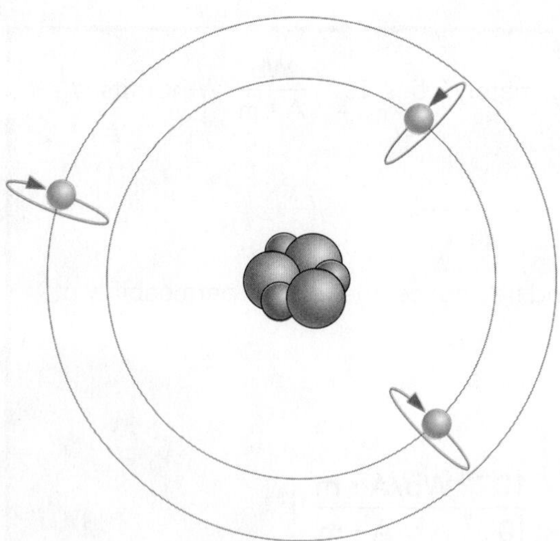

FIGURE 10.10 Electron spin.

DOMAIN THEORY The theory that atoms with like magnetic fields join to form *magnetic domains.*

DOMAIN THEORY states that atoms with like magnetic fields will join together to form *magnetic domains.* Each domain acts as a magnet, with two poles and a magnetic field. In a nonmagnetized material, the domains are randomly positioned as shown in Figure 10.11a. As a result, the magnetic fields produced by the domains cancel each other, and the material has no net magnetic force. However, if the domains can be aligned (as shown in Figure 10.11b), the material generates magnetic force.

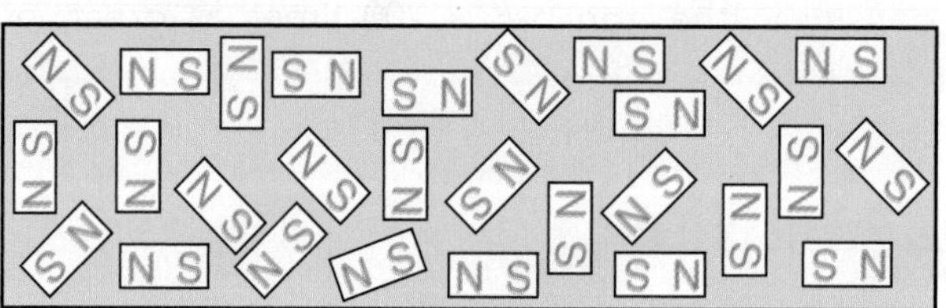

(a) Random poles

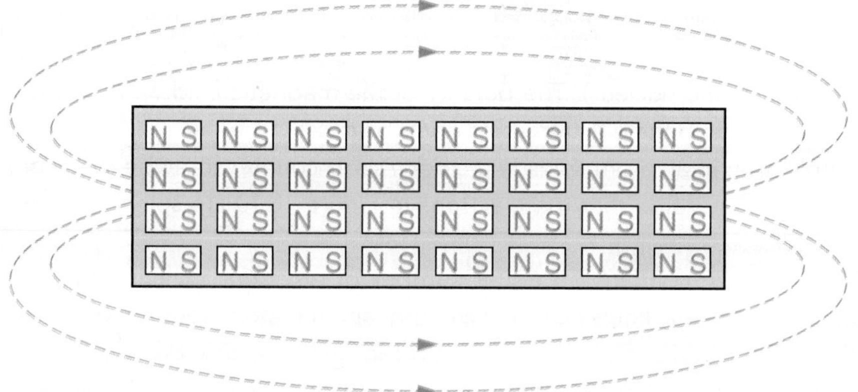

(b) Aligned poles and resulting flux

FIGURE 10.11 Magnetic domains.

Magnetic Induction

Most magnets are artificially produced. That is, their magnetic domains are aligned using an external force. The process of producing a magnet in this fashion is called **INDUCTION**. Three methods of producing a magnet by induction are illustrated in Figure 10.12.

The combination of mutual induction between coils and self-induction in each coil form the basis of transformer action. For a transformer to do its job effectively, its coils must be coupled tightly and must have high self-induction.

IN THE FIELD

INDUCTION Producing a magnet by aligning magnetic domains using an external force.

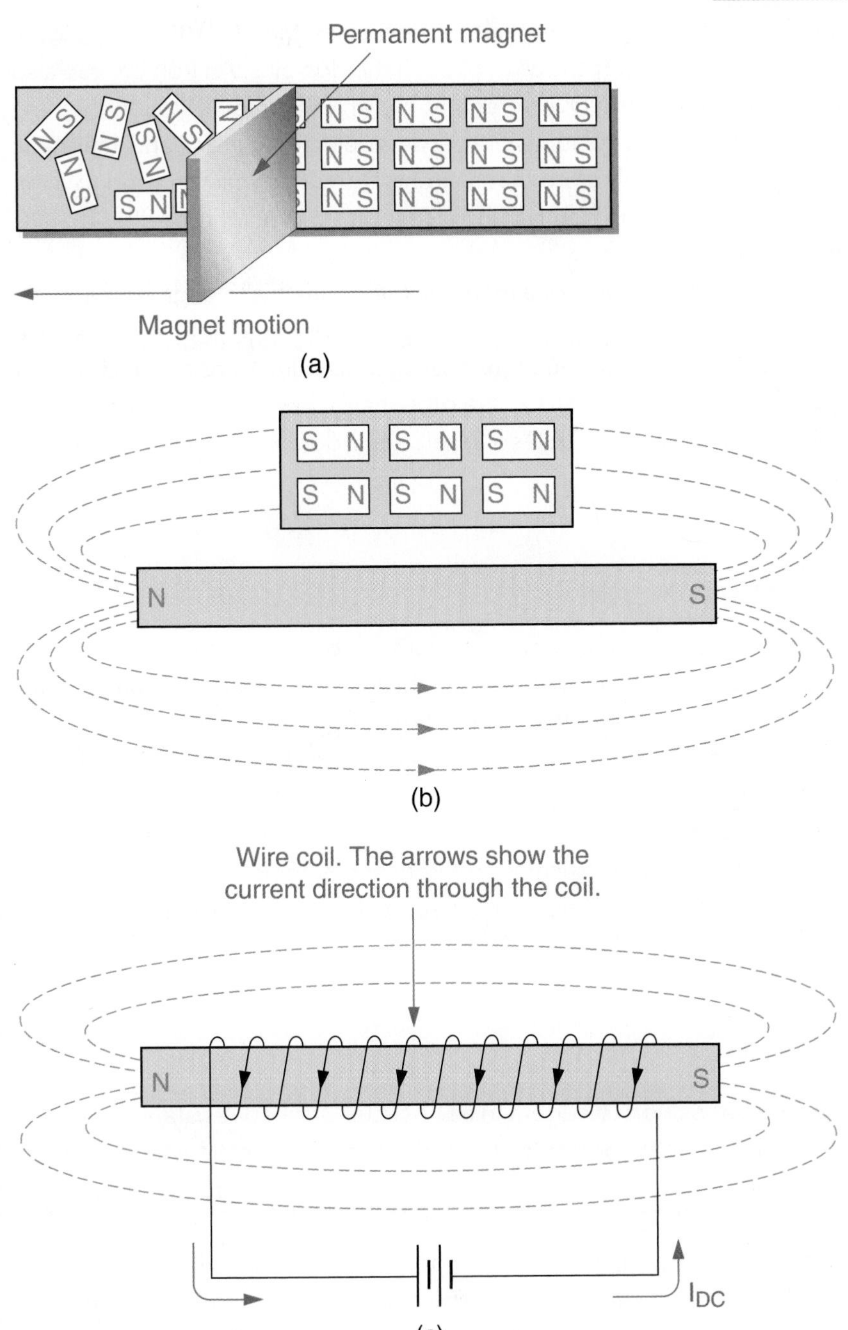

FIGURE 10.12 Magnetic induction.

In Figure 10.12a, a bar of steel is stroked with a magnet. The magnetic field generated by the magnet aligns the magnetic domains in the steel. As a result, the steel becomes magnetized. It is also possible to magnetize an object simply by placing it within the field of a magnet. This method of producing a magnet is illustrated in Figure 10.12b. In this case, the flux produced by the magnet aligns the domains in the iron bar. As a result, the iron bar becomes magnetized. Note that the poles in the iron bar are opposite those in the magnet.

MAGNETIC INDUCTION Using a magnetic field to align magnetic domains, thereby producing an artificial magnet.

The methods described above are referred to as **MAGNETIC INDUCTION**, since an external magnetic force is used to align the domains. An iron bar can also be magnetized using a strong DC current, as shown in Figure 10.12c. In the next section, we will discuss the relationship between electric current and magnetic fields. For now, we will simply state that a strong DC current can be used to align magnetic domains. Thus, the iron bar in Figure 10.12c is magnetized by the DC current through the coil.

The higher the permeability of a material, the easier it is to align its magnetic domains, and therefore, to magnetize. Because high-permeability materials have the ability to concentrate magnetic flux, they increase flux density, which better aligns the domains in the material. On the other hand, low-permeability materials do not concentrate magnetic flux. As a result, their domains do not align themselves as well.

Retentivity

RETENTIVITY The ability of an artificial magnet to retain its magnetic characteristics.

High-permeability materials are easily magnetized, but their magnetic strength quickly fades. On the other hand, low-permeability materials are difficult to magnetize, but tend to retain most of their magnetic strength for a long period of time. The ability of an artificial magnet to retain its magnetic characteristics is referred to as **RETENTIVITY**. High-retentivity materials, such as hardened steel, are used to produce *permanent magnets*. Once magnetized, they remain magnetic for a long time. Low-retentivity materials, such as soft iron, act as *temporary magnets*. That is, they lose most of their magnetic strength soon after being isolated from an external source of magnetic force.

Reluctance

RELUCTANCE The opposition that a material presents to magnetic lines of force.

As you know, the opposition to current in a DC circuit is called *resistance*. The opposition that a material presents to magnetic lines of force is called **RELUCTANCE**. In essence, reluctance can be viewed as the *magnetic resistance* of a given material.

Reluctance is typically given as a *ratio*. For example, the SI unit of measure for reluctance is *amperes per weber* (A/Wb). The reluctance of a material can be calculated using

(10.3)

$$\Re = \frac{\ell}{\mu A}$$

where

$\Re$ = the reluctance of the material
ℓ = the length of the material
μ = the permeability of the material
A = the cross-sectional area of the material

If you look back on page 59, you'll see that equation (10.3) is very similar to equation (2.2), which is used to find the resistance of a given material. This supports the concept of reluctance as magnetic resistance.

Equation (10.3) also demonstrates that reluctance varies inversely with permeability. That is, high-permeability materials have low reluctance values, and vice versa. This makes sense when you consider that:

- Permeability is a measure of how easily a material passes magnetic flux.
- Reluctance is a measure of opposition to magnetic flux.

Magnetic Classifications of Materials

All materials can be classified according to their magnetic characteristics. The magnetic classifications of materials are identified in Table 10.1, along with their characteristics. As Table 10.1 illustrates:

1. High-permeability (**FERROMAGNETIC**) materials are considered to be magnetic, while all others are generally considered to be nonmagnetic.

FERROMAGNETIC High-permeability materials, considered to be magnetic.

TABLE 10.1 • Magnetic Classifications of Materials

CLASSIFICATION:	FERROMAGNETIC	PARAMAGNETIC	DIAMAGNETIC
Relative permeability (μ_r):	Much greater than 1	Slightly greater than 1	Less than 1
Retentivity:	Low	Moderate	High
Reluctance:	Low	Moderate	High
Considered to be . . .	Magnetic	Slightly magnetic	Nonmagnetic
Examples:	Iron	Aluminum	Copper
	Steel	Air	Silver
	Nickel	Platinum	Gold
	Cobalt		Zinc
	*Alnico		
	*Permalloy		

*These are *alloys*. Alnico is made of aluminum, nickel, and cobalt. Permalloy is made of iron and nickel.

2. Retentivity varies inversely with permeability and directly with reluctance. That is, the more difficult it is to induce magnetism in a material, the longer the magnetic characteristics are retained. By the same token, the easier it is to induce magnetism, the faster the magnetic characteristics fade.

As you will learn, high retentivity is a desirable characteristic in certain applications and a liability in others.

PROGRESS CHECK

1. What is permeability?
2. What effect does a high-permeability material have on magnetic flux?
3. What is relative permeability?
4. According to domain theory, what determines the magnetic characteristics of an atom?
5. What is a magnetic domain?
6. List and describe the methods of magnetic induction.
7. What is retentivity?
8. What types of materials are used to produce permanent magnets? Temporary magnets?
9. What is reluctance? What electrical property is it similar to?
10. What is the relationship between reluctance and permeability? Explain your answer.
11. Using the information contained in Table 10.1, compare and contrast the magnetic characteristics of aluminum and iron.

10.3 ELECTROMAGNETISM

In 1820, a Danish physicist named *Hans Christian Orstead* discovered that an electric current produces a magnetic field. Using a structure similar to the one shown in Figure 10.13a, he found that the compass always aligned itself at 90° angles to a current-carrying wire. When the current was in the direction shown in Figure 10.13b, the compass aligned itself as shown. When the current was reversed (as shown in Figure 10.13c), the compass reversed direction, as shown in the figure.

As a result of Orstead's work, it is known that current passing through a wire generates lines of magnetic force that circle the wire as illustrated in Figure 10.14. The figure also shows that reversing the direction of the current causes the polarity of the flux to reverse.

STAYING SAFE Warning: To avoid heating a surrounding ferrous metal raceway by induction, all AC phase conductors, along with any neutral and grounding conductors, must be grouped together within the raceway. When single AC conductors penetrate a ferrous metal, slots must be cut between holes through which the conductors pass or insulating rings must be installed in the holes to minimize the inductive effect.

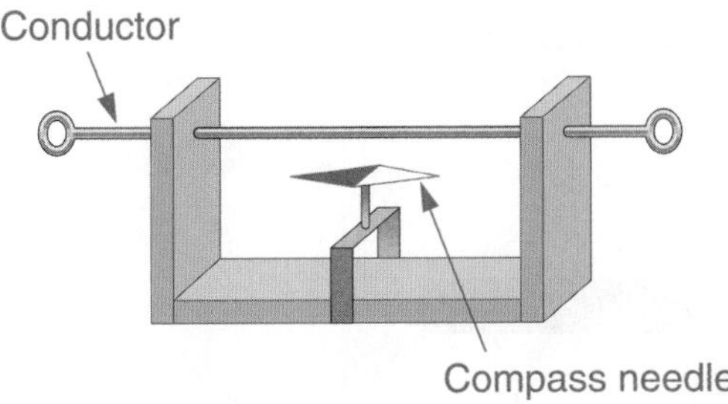

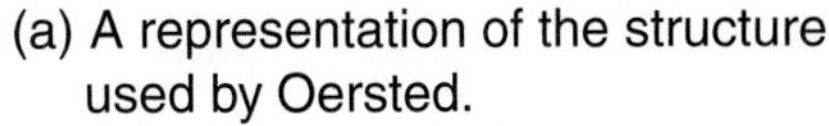

(a) A representation of the structure used by Oersted.

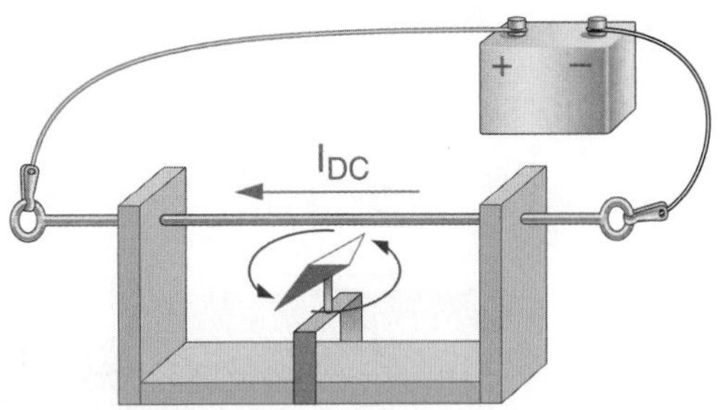

(b) The current through the conductor produces a magnetic field that moves the needle.

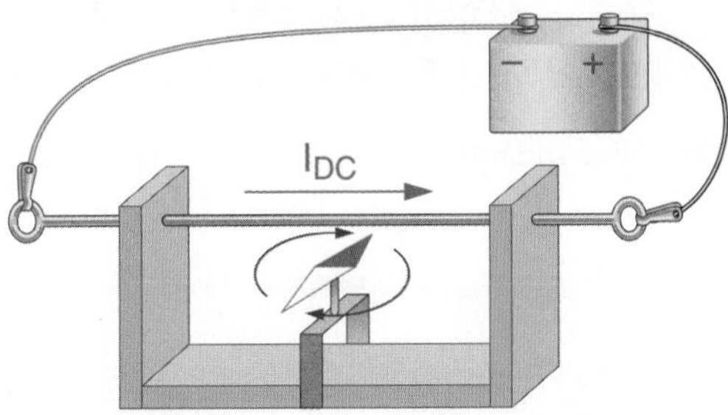

(c) Reversing the current reverses the magnetic field, causing the needle to move in the opposite direction.

FIGURE 10.13 Current produces a magnetic field.

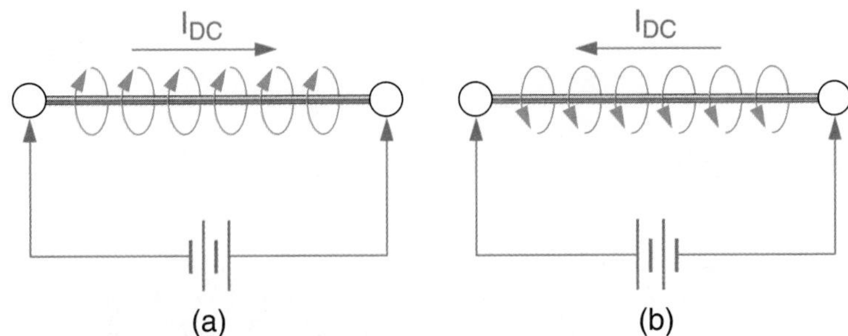

FIGURE 10.14 Current direction vs. magnetic flux direction.

The Left-Hand Rule

The **LEFT-HAND RULE** is a memory aid designed to help you remember the relationship between the direction of electron flow and the direction of the resulting magnetic field. The left-hand rule is illustrated in Figure 10.15.

LEFT-HAND RULE A memory aid that helps you remember the relationship between the direction of electron flow and the direction of the resulting magnetic field.

As you can see, the fingers indicate the direction taken by the flux when the thumb points in the direction of the current. If the current is reversed, the left hand is turned the other way. When repositioned, the fingers indicate that the flux direction has also changed.

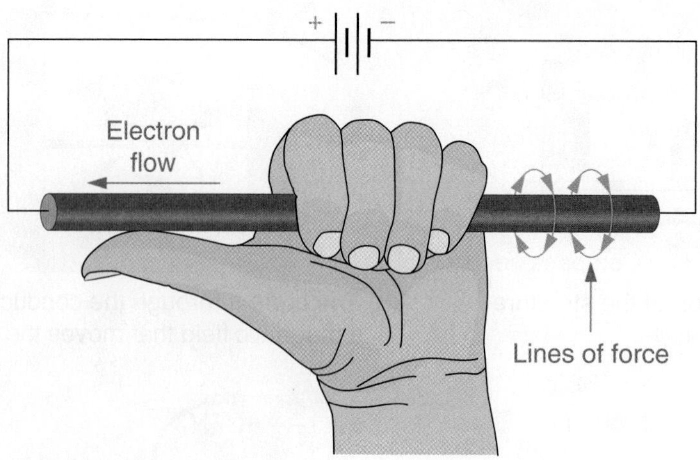

FIGURE 10.15 The left-hand rule.

Coils

COIL A wire wrapped into a series of loops for the purpose of concentrating magnetic flux.

If we take a wire and wrap it into a series of loops, called a **COIL**, the lines of force generated by the current add together to form a magnetic field. The magnetic field produced by a coil is illustrated in Figure 10.16.

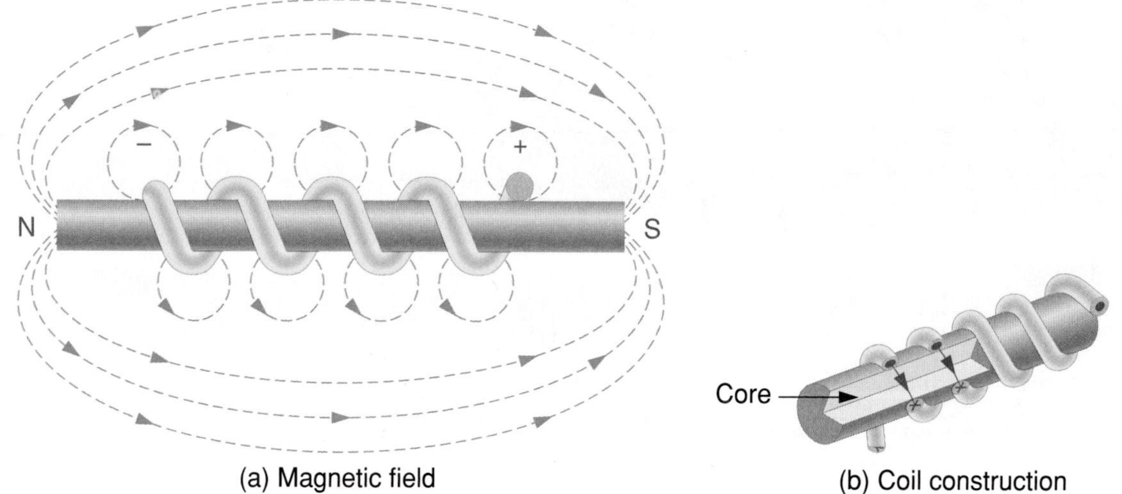

FIGURE 10.16 Coil current and magnetic lines of force.

CORE The space in the center of a coil.

The space inside the coil is called the **CORE**. If the core is made using an iron bar, the lines of flux are concentrated in the bar. As described in the last section, the iron bar becomes magnetic by induction. However, the magnetic characteristics quickly fade after power is removed from the coil because of iron's low retentivity.

The left-hand rule can be applied to a coil as shown in Figure 10.17. In this case, the thumb indicates the direction of the magnetic flux when the fingers point in the direction of the coil current. As shown, the direction of the flux reverses when the current direction through the coil reverses.

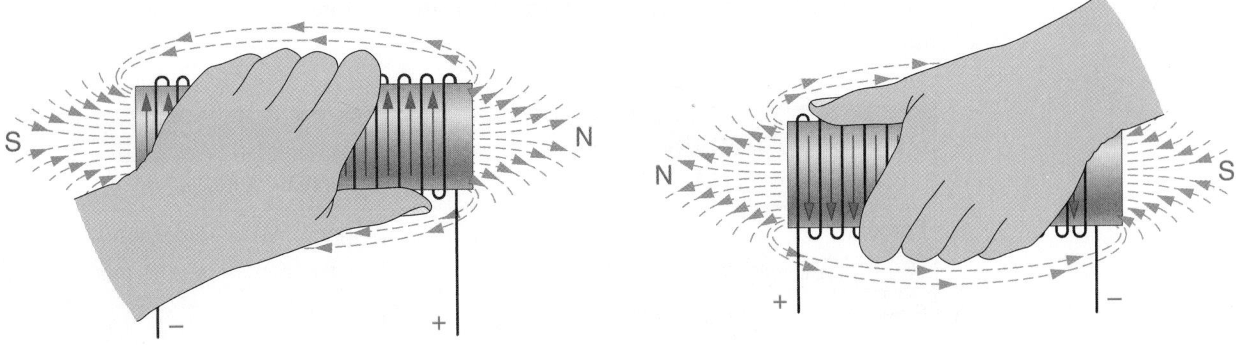

FIGURE 10.17 The left-hand rule for coils.

Figure 10.18 shows how a coil can be used to produce a magnet. When current passes through the coil of wire, the iron bar (nail) in the core becomes magnetic and lifts the pins off the table. The strength of the magnet depends (in part) on the amount of **MAGNETOMOTIVE FORCE (mmf)** produced by the current through the coil.

MAGNETOMOTIVE FORCE (mmf) The force that generates magnetic flux in a material that has reluctance; the magnetic counterpart to electromotive force (emf).

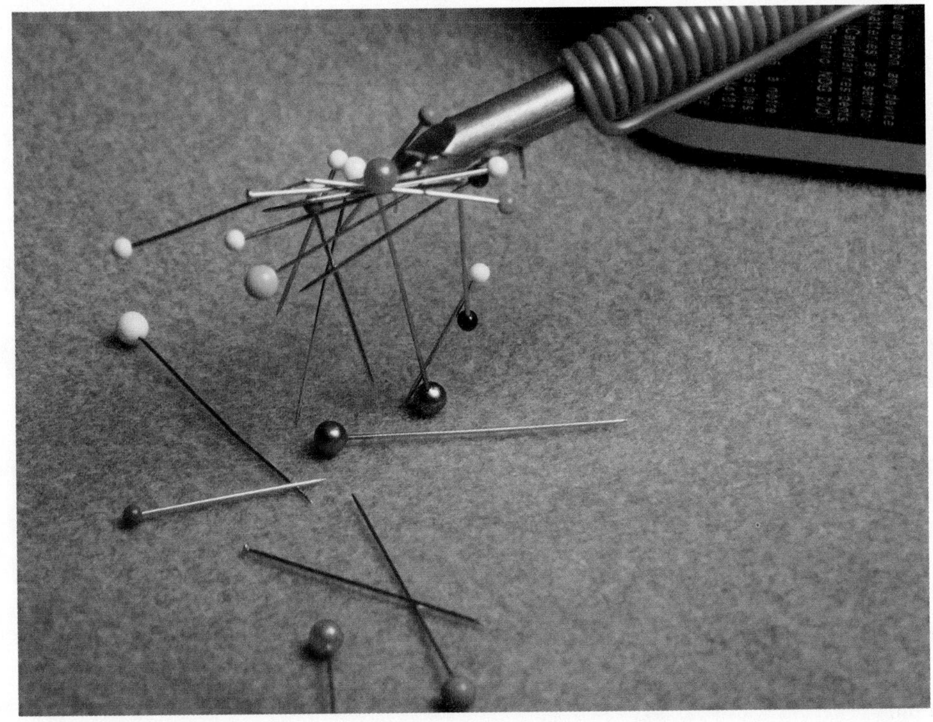

FIGURE 10.18 The magnetic field produced by a coil.

Magnetomotive Force (mmf)

In Figure 10.18, the current through the coil is generating magnetic flux through the reluctance of the iron bar. Generating magnetic flux through any value of reluctance requires a certain amount of magnetomotive force (mmf).

As Table 10.2 indicates, mmf is to magnetic circuits what electromotive force (emf) is to electric circuits.

TABLE 10.2 • Electric vs. Magnetic Circuits

	ELECTRIC CIRCUIT	MAGNETIC CIRCUIT
Force:	Electromotive force (emf) (measured in *volts*)	Magnetomotive force (mmf) (measured in *ampere-turns*)
Flow:	Current (I) (measured in *amperes*)	Flux (ϕ) (measured in *webers*)
Opposition:	Resistance (R) (measured in *ohms*)	Reluctance ($\Re$) (measured in *ampere-turns/weber*)

ROWLAND'S LAW The magnetic counterpart of Ohm's law, it defines the relationship between mmf, flux, and reluctance.

Just as Ohm's law defines the relationship between emf, current, and resistance, **ROWLAND'S LAW** defines the relationship between mmf, flux, and reluctance. Rowland's law states that magnetic flux is directly proportional to magnetomotive force and inversely proportional to reluctance. By formula,

(10.4)

$$\Phi = \frac{F}{\Re}$$

where

Φ = the magnetic flux, in *webers* (Wb)
F = the magnetomotive force, in *ampere-turns* (A · t)
$\Re$ = the reluctance, in *ampere-turns per weber* (A · t/Wb)

Equation (10.4) is useful because it shows that there are two ways to increase the amount of flux produced by a magnet like the one in Figure 10.16. The amount of flux produced by the magnet can be increased by either:

- Increasing the amount of magnetomotive force, or
- Using a lower-reluctance material as the core.

For example, a given value of mmf will generate more flux in an iron-core coil than it will in a comparable air-core coil because the reluctance of iron is lower than that of air.

Ampere-Turns

The magnetomotive force produced by a coil is proportional to

- The current through the coil, and
- The number of turns.

These two factors are combined into a single value called **AMPERE TURNS (A · t)**. The ampere turns value for a given coil equals the product of the current and the number of turns, as illustrated in Figure 10.19. As you can see, the value of (A · t) for a coil can be varied by changing either the amount of current or the number of turns in the coil. Note that ampere-turns are commonly represented using NI (*N* for *number of turns*, and *I* for current).

AMPERE TURNS (A · t) For a given coil, the product of coil current and the number of turns.

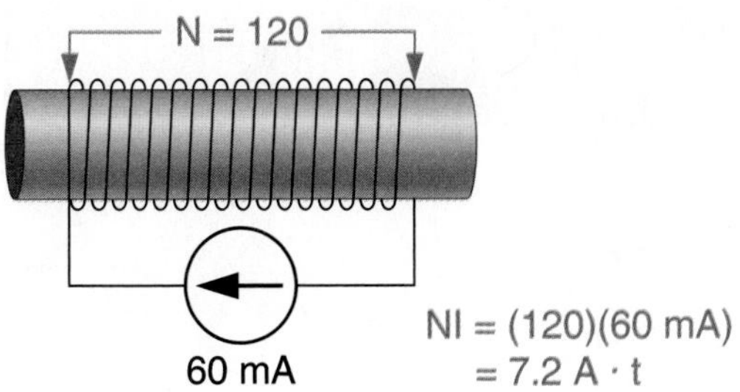

FIGURE 10.19 Calculating ampere-turns.

The magnetomotive force produced by a coil is proportional to the ampere-turns value of the coil. By formula,

(10.5)

$$\text{mmf} = N \times I$$

The flux density produced by a coil depends on three factors:

1. The permeability of the core,
2. The ampere-turns of the coil, and
3. The length of the coil.

When these values are known, the flux density produced by a coil (in teslas) is found as

(10.6)

$$B = \frac{\mu_m NI}{\ell}$$

where

B = the flux density, in teslas
μ_m = the permeability of the core material, in Wb/A · m
NI = the ampere-turns value of the coil
ℓ = the length of the coil (in meters)

The use of this equation is demonstrated in Example 10.3.

EXAMPLE 10.3

The coil in Figure 10.20 has an *air* core. Calculate the flux density (in teslas) for the coil.

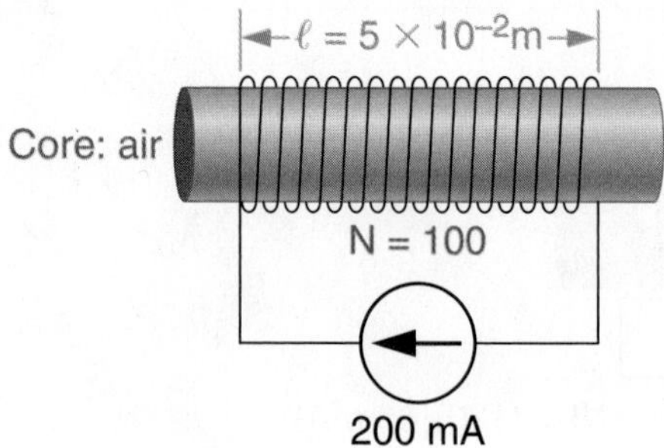

FIGURE 10.20

SOLUTION

As stated earlier, the permeability of air is approximately $4\pi \times 10^{-7} \frac{\text{Wb}}{\text{A} \cdot \text{m}}$. The length of the coil is shown to be 5×10^{-2}m. Using these values in equation (10.6), the flux density produced by the coil is found as

$$B = \frac{\mu_m N I}{\ell}$$

$$= \frac{\left(4\pi \times 10^{-7} \frac{\text{Wb}}{\text{A} \cdot \text{m}}\right) \times (100) \times (200 \times 10^{-3}\,\text{A})}{5 \times 10^{-2}\,\text{m}}$$

$$= 503 \times 10^{-6}\,\text{Wb/m}^2$$

$$= 503\,\mu\text{T}$$

PRACTICE PROBLEM 10.3

The air-core coil in Figure 10.20 is replaced with an iron-core coil having the same dimensions. Calculate the flux density produced by the coil.

If you take another look at equation (10.6), you'll see that:

1. The values of N and ℓ are determined by the physical characteristics of the coil.
2. The value of μ_m is determined by the type of material found in the core.

As such, these values cannot easily be changed for a given coil. The only value that can easily be changed to vary flux density is coil current (I). The effect of varying current on the value of flux density is demonstrated in the following example.

EXAMPLE 10.4

The current through the coil in Figure 10.20 is *increased* to 250 mA. Calculate the resulting change in flux density for the coil.

SOLUTION

Using a value of I = 250 mA, the flux density for the coil is found as

$$B = \frac{\mu_m NI}{\ell}$$

$$= \frac{\left(4\pi \times 10^{-7}\,\frac{\text{Wb}}{\text{A}\cdot\text{m}}\right) \times (100) \times (250 \times 10^{-3}\,\text{A})}{5 \times 10^{-2}\,\text{m}}$$

$$= 628 \times 10^{-6}\,\text{Wb/m}^2$$

$$= 628\ \mu\text{T}$$

In Example 10.3, we calculated a value of B = 503 μT for I = 200 mA. Using this value, the change in flux density is found as

$$\Delta B = 628\ \mu\text{T} - 503\ \mu\text{T}$$

$$= 125\ \mu\text{T}$$

PRACTICE PROBLEM 10.4

The current through the coil in Practice Problem 10–3 is decreased to 150 mA. Calculate the change in flux density that results from the change in current. (Compare the percent of change in current to the percent of change in flux density.)

Hysteresis

As you know, most materials lose much of their magnetic strength (flux density) when a magnetizing force is removed. For example, the iron core in Figure 10.21 experiences a significant drop in flux density when current is removed from the coil.

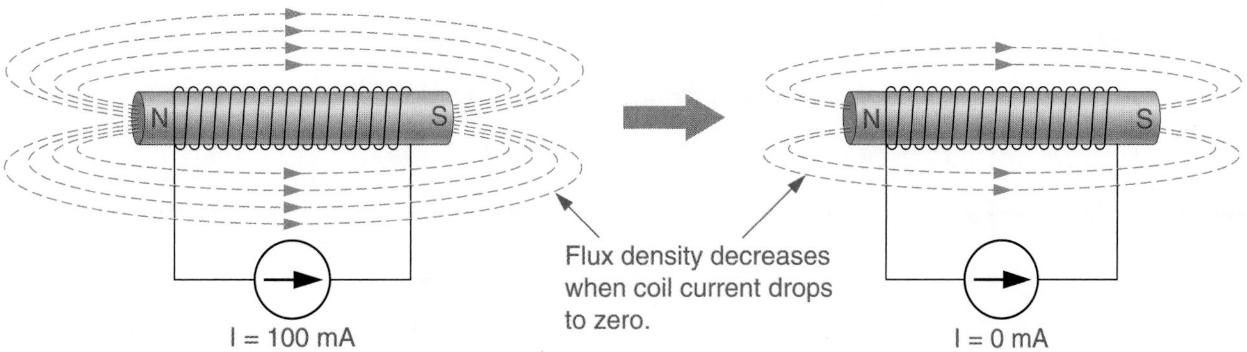

FIGURE 10.21 Flux density vs. coil current.

HYSTERESIS The time lag between the removal of a magnetizing force and the resulting drop in flux density.

MAGNETIZATION CURVE A curve that plots flux density (B) as a function of magnetic field strength (H).

RESIDUAL FLUX DENSITY The flux density that remains in a material after a magnetizing force is removed.

The time required for the flux density to drop after the magnetizing force is removed depends on the rententivity of the material. The time lag between the removal of a magnetizing force and the drop in flux density is referred to as **HYSTERESIS**.

The retentivity of a given material is commonly represented using a **MAGNETIZATION CURVE**. As shown in Figure 10.22, a magnetization curve represents flux density (B) as a function of magnetic field strength (H). For this reason, a given magnetization curve is often referred to as a *B-H curve*.

Figure 10.22b shows what happens when a current is applied to a coil like the one shown in Figure 10.21 where the iron core is not initially magnetized. When the magnetic field strength produced by the current reaches its maximum value ($+H_M$), the flux density in the core also reaches its maximum value ($+B_M$). When a current is no longer applied to the coil, the field strength (H) drops to zero. However, some **RESIDUAL FLUX DENSITY** remains in the core. This residual flux density is represented by the positive value ($+B_0$) in Figure 10.22c. The retentivity of the material determines:

- The *time* required for the flux density to drop from $+B_M$ to $+B_0$.
- The *value* of $+B_0$, in teslas (T).

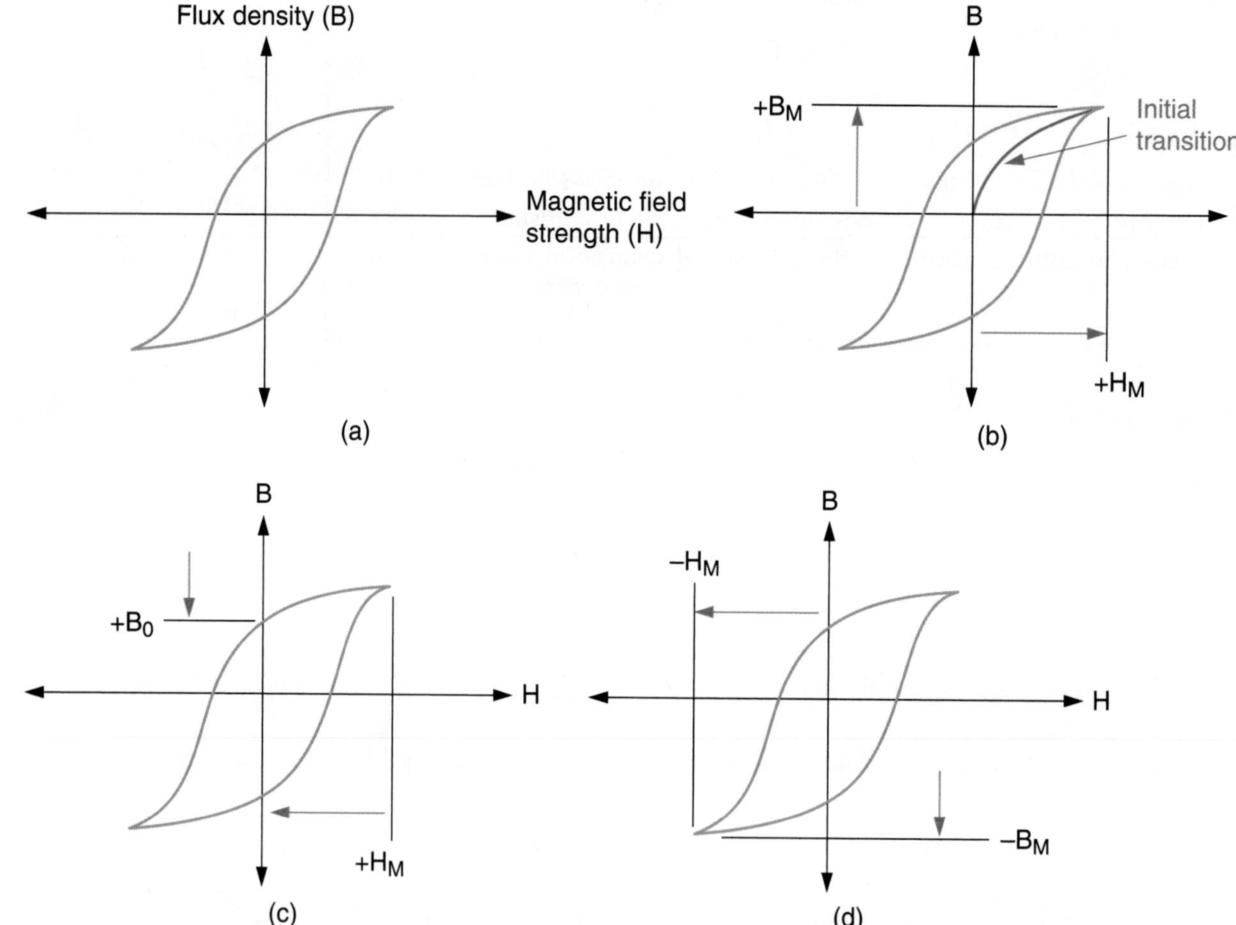

FIGURE 10.22 Magnetization curves.

In other words, the higher the retentivity of a given material:

- The longer it takes for the flux density to drop from $+B_M$ to $+B_0$.
- The greater the value of $+B_0$.

If the coil current is reversed, the polarity of the magnetizing force and the resulting magnetic flux also reverse. This condition is represented by Figure 10.22d. Again, when current is no longer applied to the coil, magnetic field strength drops to zero and the flux density drops to the value shown on the curve ($-B_0$). Note that the magnetizing force required to return the flux density (B) to zero is referred to as the **COERCIVE FORCE**.

COERCIVE FORCE The magnetizing force required to return the residual flux density (B) in a material to zero.

It is important to note that the two halves of the magnetization curve are simply mirror images. One half represents the response of the core to a current generated in one direction. The other half shows the response of the core when the coil current reverses direction.

PROGRESS CHECK

1. What relationship did Orstead discover between magnetism and current?
2. What happens to the polarity of a magnetic field when current is reversed?
3. What is a coil? What is the core of a coil?
4. What is magnetomotive force (mmf)?
5. What is Rowland's law? Which electrical law does it resemble?
6. What is an ampere-turn?
7. What is the relationship between ampere-turns and flux density?
8. In a practical sense, how is the amount of magnetomotive force produced by a given coil usually varied?
9. What is hysteresis?
10. What is residual flux density?
11. Using the B-H curve, describe what happens when power is applied to and removed from a coil.

10.4 ELECTROMAGNETIC RELAYS: AN APPLICATION OF ELECTROMAGNETIC PRINCIPLES

In the last section, you were introduced to the relationship between magnetism and current. In this section, you will be shown how one common component makes use of this relationship.

Electromagnetic Relays

A **RELAY** is a switch that is opened or closed by an input signal. The construction of an *electromagnetic relay* is illustrated in Figure 10.23. The primary components

RELAY A switch that is opened or closed by an input signal.

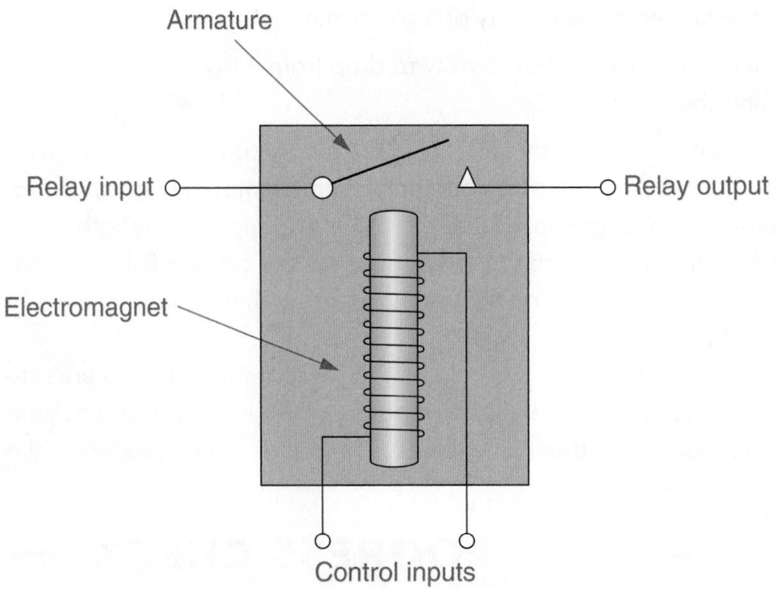

FIGURE 10.23 An electromagnetic relay.

in the relay are the coil and the *armature*. The armature is held open (or closed) by a spring (not shown). When a control input voltage generates a current through the coil in Figure 10.23, it produces a magnetic field that attracts the armature, closing the switch. When the switch closes, the relay input is connected to the relay output. As long as the control input is present, the switch remains closed. Once the control input is removed, the magnetic field collapses and the switch opens, breaking the connection between the relay input and output terminals.

An electromagnetic relay allows you to use a low-current input to control a higher-current circuit. For example, the bell control circuit in Figure 10.24 uses a

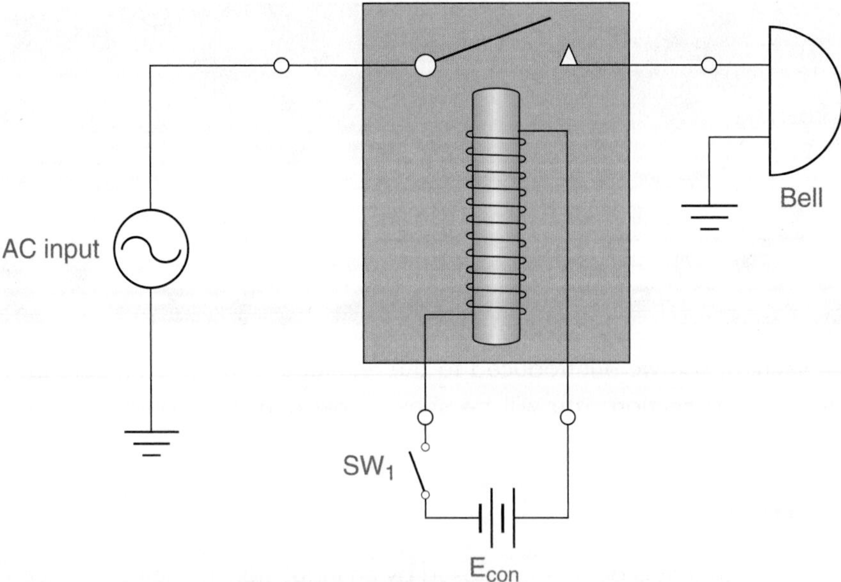

FIGURE 10.24 An electromagnetic relay application.

simple battery (E_{CON}) to activate a bell. When the switch (SW1) is closed, the battery energizes the electromagnet, causing the relay armature to move to the closed position. When the relay switch is closed, the AC source is connected to the bell, and the bell rings.

Currents in the milliamp range are sufficient to activate the armature in some electromagnetic relays. At the same time, the current passing from the relay input to its output can be in the tens of amperes or higher, depending on the relay. For example, one Allen Bradley® relay is rated to handle 600 V @ 35 A. Note that electromagnetic relays come in *normally-open* (NO) and *normally-closed* (NC) configurations. Figure 10.25 shows an electromagnetic relay that employs a moveable plunger to connect the relay input and output terminals.

There are many electromagnetic relay configurations that are manufactured for various applications. For example, certain configurations allow you to control more than one AC signal or control motor direction with a single relay.

Like any other component, electromagnetic relays have their strengths and weaknesses. Among their strengths, electromagnetic relays:

- Allow you to control a high-value current with a low-current control signal.
- Provide complete electrical isolation between their source and load circuits when open.
- Can be heard switching on and off, which makes for relatively simple troubleshooting.

Among their weaknesses:

- They are mechanical components, which makes their moving parts susceptible to physical breakdown. Dirty contacts can also cause electromechanical relays to operate intermittently (which is often indicated by a chattering sound).
- Electromagnetic relays are slow. They cannot be opened and closed as quickly as some other types of electrical switches.
- Electromagnetic relays produce a reverse voltage (called a *counter emf* or *kick emf*) that can damage other components when they switch from on to off. The source of this counter emf is discussed in Chapter 11.

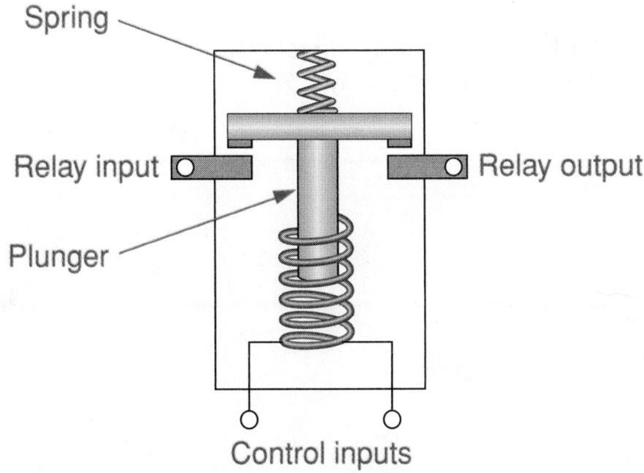

FIGURE 10.25 An electromagnetic relay.

Reed Relays

A *reed relay* is a variation on the electromagnetic relay in Figure 10.23. The construction of a reed relay is illustrated in Figure 10.26. The relay has two metal strips that are housed in a glass case. An electromagnetic coil is wound directly around the glass case, as shown in Figure 10.27. When current passes through the coil, the resulting magnetic field causes the reeds to make contact, allowing current to pass through the relay. When the control current is removed, the magnetic field collapses, and the reeds spring back into their original positions. This opens the current path between the relay input and output terminals. The advantage of this type of magnetic relay is its relatively small size (as compared to a conventional electromagnetic relay).

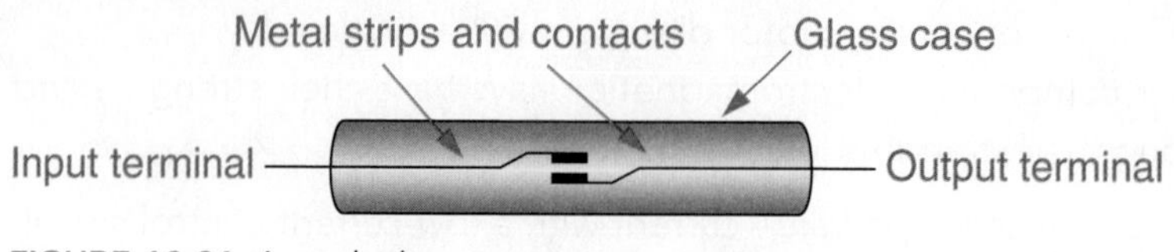

FIGURE 10.26 A reed relay.

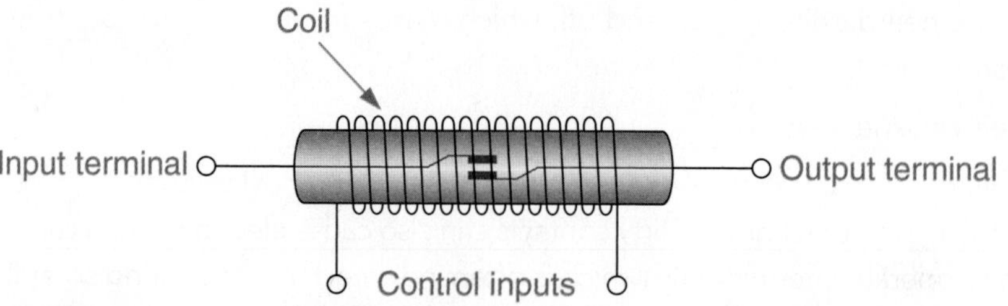

FIGURE 10.27 A reed relay control circuit.

PROGRESS CHECK

1. Describe the construction of a conventional electromagnetic relay.
2. Describe how the control input to a conventional electromagnetic relay controls the relay armature.
3. List the strengths and weaknesses of conventional electromagnetic relays.
4. Describe the physical construction of a reed relay.
5. How are the metal strips (armatures) in a reed relay activated?
6. What advantage does a reed relay have over a conventional electromagnetic relay?

10.5 RELATED TOPICS

In this section, we briefly discuss several topics that relate to magnetism. Among the topics we will cover are magnetic shielding, common shapes of magnets, and their proper care and storage.

Magnetic Shielding

Some instruments, like analog meter movements, are sensitive to magnetic flux. That is, exposure to magnetic flux causes them to operate improperly. When such an instrument must be placed in a magnetic field, **MAGNETIC SHIELDING** is used to insulate the instrument from the effects of the magnetic flux.

MAGNETIC SHIELDING
Insulating an instrument from the influence of magnetic flux.

It should be noted that there is no known magnetic insulating material. Flux passes through any material, regardless of its reluctance. Because of this, magnetic shielding is accomplished by diverting lines of flux around the object to be shielded. The means by which this is accomplished is illustrated in Figure 10.28.

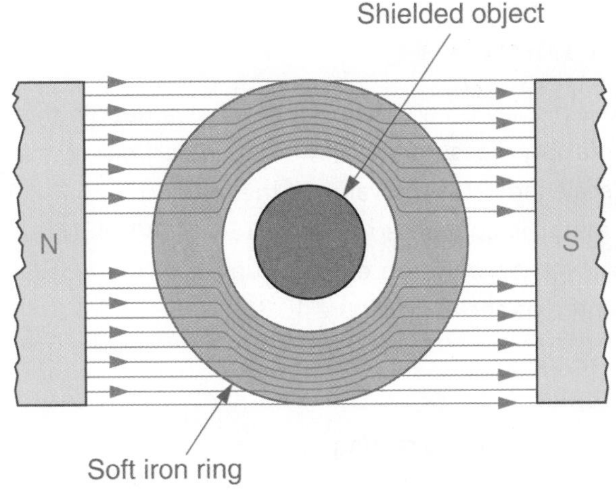

FIGURE 10.28 Magnetic shielding.

Figure 10.28 shows the lines of force between two magnetic poles. As you can see, a soft iron ring has been placed between the poles. Rather than follow a straight-line path between the poles, the lines of force bend and pass through the high-permeability material. If a magnetically sensitive device is placed in the middle of the iron ring, the lines of force are diverted around the device. As a result, the device is shielded from the flux.

Air Gaps

The space directly between two magnetic poles is called an *air gap*. The air gaps for several types of magnets are illustrated in Figure 10.29.

The narrower the air gap between two poles, the stronger the force of attraction between those two poles. With a narrow air gap, there is little room for

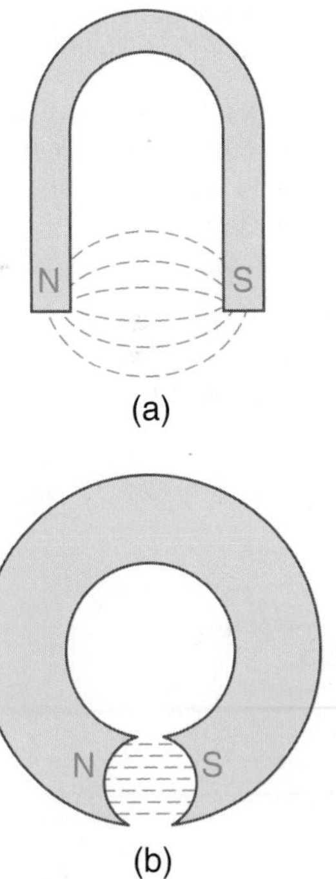

FIGURE 10.29 Air gaps.

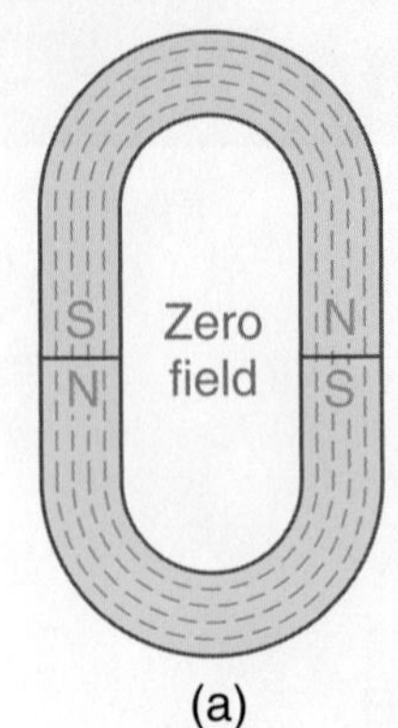

(a)

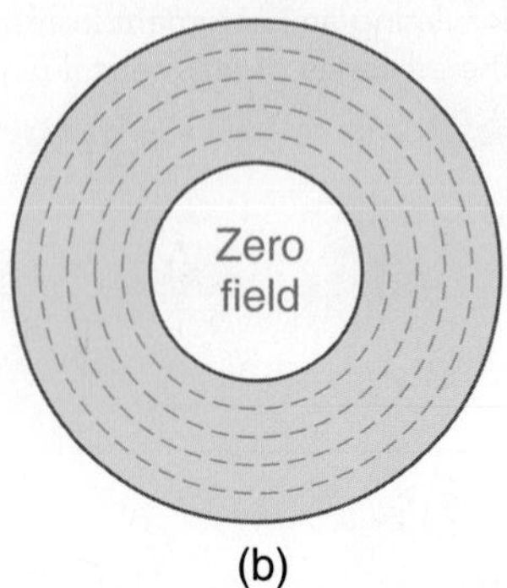

(b)

FIGURE 10.30 Ring magnets.

RING MAGNET A magnet that does not have any identifiable poles or air gaps, and does not generate an external magnetic field.

METER MOVEMENT A structure that moves a meter pointer in response to an electric current.

the lines of force between the poles to spread out. As a result, a narrow air gap produces a magnetic field with relatively high flux density. This is why the force of attraction between two magnets increases when you move them closer together.

Ring Magnets

RING MAGNETS are unique because they do not have any identifiable poles or air gaps, and do not generate an external magnetic field. A ring magnet can be produced in one of several fashions, as shown in Figure 10.30.

Figure 10.30a shows a ring magnet that is actually made up of two horseshoe magnets. When connected as shown, the lines of flux remain in the ring magnet, and there is no magnetic field in the center area of the magnet.

When constructed as shown in Figure 10.30b, a ring magnet has no identifiable poles. Rather, a continuous flow of magnetic flux circles within the magnet. Among other applications, ring magnets are commonly used in analog meter movements.

Analog Meter Movements

In Chapter 3, you learned that the display on an analog meter has a needle that moves over a fixed scale. When taking measurements with an analog meter, the position of the pointer over the scale indicates the value of the reading.

The pointer on an analog meter is connected to a **METER MOVEMENT**, a structure that moves the pointer in response to an electric current. A *D'Arsonval* meter movement can be represented as shown in Figure 10.31.

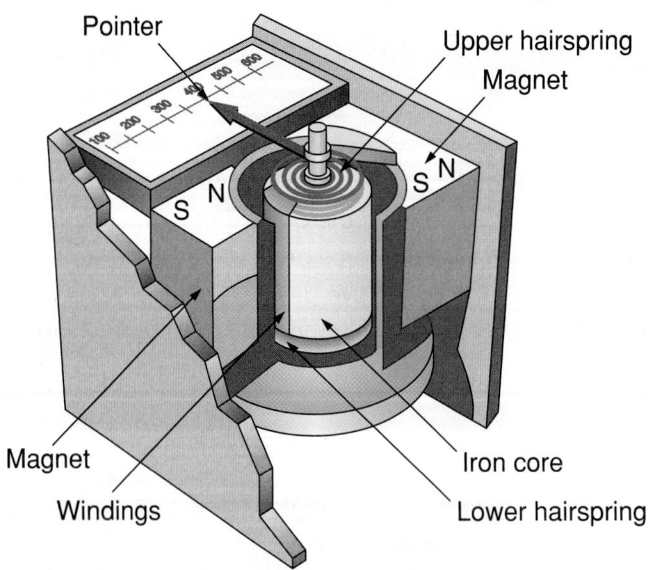

FIGURE 10.31 A D'Arsonval meter movement.
Redrawn from DexterMagnetic Tech.

The D'Arsonval meter movement has four primary parts:

1. A permanent magnet.
2. An iron core suspended between the poles of the magnet.
3. A coil wrapped around the core.
4. A pair of hairsprings attached to the core mounting pin above and below the iron core.

When current passes through the coil in Figure 10.31, it generates a magnetic field around the iron core. This magnetic field interacts with the field produced by the permanent magnet, causing the iron core to rotate clockwise. When the iron core in the movement rotates, the pointer moves over the fixed scale. The position of the pointer over the scale indicates the value of the reading.

The hairsprings in the D'Arsonval movement apply pressure that opposes the core rotation. That is, while the input current causes the iron core to rotate, the hairsprings push the other way, limiting the degree of rotation. As a result, the degree of core rotation varies directly with the coil current; the greater the coil current, the greater the degree of rotation. When the coil current returns to zero, the hairsprings push the core (and pointer) back to its resting position.

Care and Storage

Over time, even a "permanent" magnet can lose most of its magnetic strength if not properly cared for. The most common causes of loss of magnetic strength for a permanent magnet are:

1. Allowing the magnet to be jarred or dropped.
2. Exposing the magnet to sufficiently high temperatures.
3. Improperly storing a magnet.

When a magnet is jarred or dropped, the impact can knock the magnetic domains out of alignment. As a result, magnetic strength is lost. Excessive heat can also cause a misalignment of the domains, resulting in a loss of magnetic strength.

When storing magnets, care must be taken to ensure that the flux produced by the magnet is not lost externally. This is accomplished by storing one or more magnets as shown in Figure 10.32.

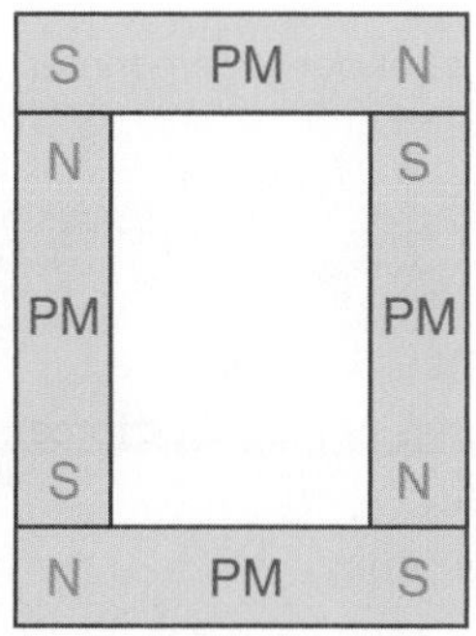

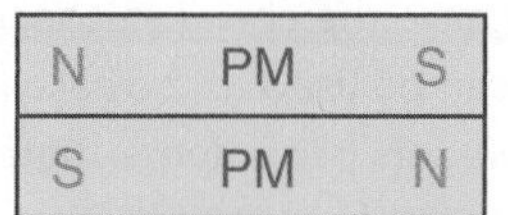

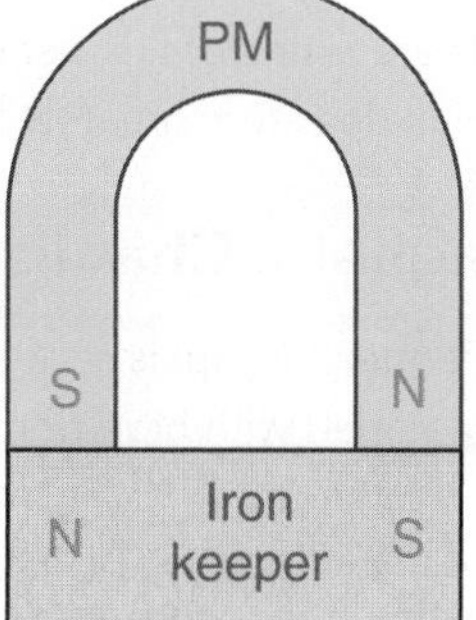

PM: permanent magnet

FIGURE 10.32 Magnet storage configurations.

Each of the storing methods shown has the same effect as constructing a ring magnet. That is, the flux generated by the magnet(s) stays within a loop, rather than being lost externally.

PROGRESS CHECK

1. What is *magnetic shielding*? How is it accomplished?
2. What is an *air gap*?
3. What is the relationship between air gap width and magnetic strength?
4. What characteristics distinguish a *ring magnet* from other types of magnets?
5. What is a common application for ring magnets?
6. List the primary components of the D'Arsonval meter movement.
7. Describe the operation of the D'Arsonval meter movement.
8. List the most common causes of loss of magnetic strength for a permanent magnet.
9. Describe the proper storage of a magnet.

10.6 SUMMARY

Here is a summary of the major points that were made in this chapter:

Magnetism: An Overview

- The force that a magnet exerts on the objects around it is referred to as *magnetic force*.
- The points where magnetic lines of force leave (and return to) a magnet are referred to as the *poles*.
  - The two poles are the north-seeking (N) and south-seeking (S) poles.
  - Like poles repel each other. Unlike poles attract.
- The lines of force produced by a magnet are collectively referred to as *magnetic flux*.
- There are two units of measure for flux.
  - The maxwell (Mx): 1 Mx = 1 line of flux
  - The weber (Wb): 1 Wb $= 1 \times 10^8$ Mx
- The microweber (μWb), which equals 100 Mx, is a commonly used unit of measure for flux.
- The area of space that surrounds a magnet and contains magnetic flux is called the *magnetic field.*
- *Flux density* is a measure of flux per unit area.
  - Flux density is an indicator of magnetic strength.
- There SI unit of flux density is the *tesla (T)*. One tesla equals one weber per square meter.
- Flux density is greatest at the poles of a magnet.

Magnetic Characteristics

- *Permeability* (μ) is a measure of how easily a material passes magnetic lines of force.
- Materials with high permeability have the ability to concentrate lines of magnetic force.
  - High-permeability materials can be used to increase flux density.
  - Permeability is measured in webers per ampere-meter (Wb/A · m).
- *Relative permeability* (μ_r) is the ratio of a material's permeability to the permeability of free space (μ_0)
  - Since μ_r is a ratio, it has no units.

- Relative permeability is important because it equals the ratio of flux density within a material to flux density in free space.
- Domain theory states that atoms with like magnetic fields join together to form magnetic domains.
  - Every atom has both an electric field and a magnetic field. Atoms with like magnetic fields join to form domains.
  - Each domain acts as a magnet, with two poles and a magnetic field.
  - In nonmagnetized materials, the domains are randomly positioned, so their magnetic fields cancel.
  - In magnetic materials, the domains can be aligned to produce a magnet.
- *Induction* is the process of producing an artificial magnet.
  - An external source of magnetism is used to align the magnetic domains in a material.
- There are *three* methods of induction, as shown in Figure 10.12 (page 449).
  - Induction can be accomplished using the force of an external magnet, or using an electric current.
  - The domains in high-permeability materials are relatively easy to align, since these materials concentrate magnetic flux.
- The ability of an artificial magnet to retain its magnetic characteristics is called *retentivity*.
  - High-permeability materials have *low* retentivity. Low-permeability materials have *high* retentivity.
  - High-retentivity (low-permeability) materials are used to produce permanent magnets.
  - Low-retentivity (high-permeability) materials are used to produce temporary magnets.
- The opposition that a material presents to magnetic lines of force is called *reluctance*.
  - Reluctance is (essentially) the magnetic equivalent of resistance.
  - Reluctance varies inversely with permeability.
  - In the SI, reluctance is measured in ampere-turns per weber (A · t/Wb).
- All materials fall within one of three magnetic classifications.
  - *Ferromagnetic* materials have high permeability, and are considered to be magnetic.
  - *Paramagnetic* materials have relatively low permeability, and are not considered to be magnetic.
  - *Diamagnetic* materials have very low permeability, and are not considered to be magnetic.

Electromagnetism

- In 1820, Orstead discovered that an electric current produces a magnetic field.
  - The magnetic field resulting from a current exists at 90° angles to the current.
- The *left-hand rule* is a memory aid designed to help you remember the relationship between current direction and the direction of the resulting magnetic field.
- A wire that is wrapped into a series of loops is called a *coil*.
  - The area in the center of the coil is called the *core*.
- The lines of force produced by the current through a coil join to form a magnetic field with relatively high flux density.
  - If the core contains an iron bar, the bar becomes magnetic by induction.
- The *left-hand rule* can be applied to a coil as illustrated in Figure 10.17 (page 455). Generating magnetic flux through any value of reluctance requires *magnetomotive force (mmf)*.
  - Magnetomotive force is to magnetic circuits what electromotive force (emf) is to electric circuits.

- *Rowland's law* defines the relationship between mmf, reluctance, and magnetic flux.
  - Magnetic flux is directly proportional to mmf and inversely proportional to reluctance.
  - Rowland's law is similar in form to Ohm's law.
- The magnetomotive force produced by a coil is proportional to both the current through the coil and the number of turns (loops) contained in the coil.
  - These two factors are combined in the *ampere-turn*, which equals the product of the two.
  - Ampere-turns are commonly expressed as NI (N for the number of turns and I for the current).
- The flux density produced by a coil depends on:
  - The permeability of the core
  - The number of coil turns
  - The current through the coil
  - The length of the coil
- Of the values listed (above), coil current is the value that is normally varied to control the flux density produced by a coil.
- Many materials lose much of their magnetic strength (flux density) when a magnetizing force is removed.
  - The time lag between removing a mmf and the resulting loss in flux density is referred to as *hysteresis*.
  - The principle of hysteresis is commonly represented using the magnetization curve (Figure 10.22, page 460).
  - The flux density that remains when a magnetizing force is removed is called *residual flux density*.
  - The amount of residual flux density depends on the retentivity of the material.

Electromagnetic Relays: An Application of Electromagnetic Properties

- A relay is a switch that is opened or closed by an input signal.
  - An electromagnetic relay contains a coil and an armature.
    - When current passes through the coil, a magnetic field is generated. The magnetic field attracts the armature, closing the switch.
    - When the coil current drops to zero, the magnetic field collapses and the relay opens.
- Relays come in a variety of configurations that are used for different purposes.
- Electromagnetic relays:
  - Allow you to control a high-value current with a low-value control signal.
  - Provide complete electrical isolation between their source and load circuits when open.
  - Are relatively easy to troubleshoot
  - Are susceptible to mechanical breakdown
  - Are relatively slow (when compared to other types of relays)
  - Produce a reverse ("kick") emf when turned off that must be accounted for in circuit design
- A reed relay is an electromagnetic relay that is smaller than its conventional counterpart.

Related Topics

- *Magnetic shielding* is used to insulate sensitive instruments from magnetic flux.
  - There is no known magnetic insulating material.
  - Shielding is accomplished by diverting magnetic flux around the sensitive instrument.

- The space directly between two magnetic poles is called an *air gap*.
  - The narrower the air gap, the stronger the force of attraction (or repulsion) between the poles.
  - *Ring magnets* are unique because they do not have any identifiable poles or air gaps, and do not generate an external magnetic field.
- An *analog meter movement* moves the pointer in an analog meter.
- The *D'Arsonval* meter movement has four primary parts.
  - A permanent magnet
  - An iron core suspended between the poles of the magnet
  - A coil wrapped around the core
  - A pair or hairsprings attached to the core mounting pin above and below the iron core (see Figure 10.31)
- When current passes through the coil, the iron core rotates on its axis.
  - The hairsprings provide a counter force that limits how far the core rotates.
- When coil current is cut off, the hairsprings return the pointer to its resting position.
- The most common causes of loss of magnetic strength for a permanent magnet are:
  - Allowing the magnet to be jarred or dropped
  - Exposing the magnet to sufficiently high temperatures
  - Improperly storing a magnet
- When storing a magnet, care must be taken to ensure that the flux produced by the magnet is not lost externally.
- Magnets should be stored as shown in Figure 10.32 (page 467).

CHAPTER REVIEW

1. The points from which magnetic lines of force emanate from a magnet are referred to as ________.
   a) nodes
   b) poles
   c) junctions
   d) plates

2. The magnetic lines of force produced by a magnet are collectively referred to as ________.
   a) north and south poles
   b) magnetic flow
   c) magnetic flux
   d) electromagnetic flux

3. One maxwell equals ________ line(s) of magnetic force while 1 weber equals ________ line(s).
   a) 10^8, 1
   b) 1, 10^8
   c) 1, 10^2
   d) 10^2, 1

4. The most common unit of measure for magnetic flux is ________.
   a) the weber
   b) the maxwell
   c) the μWb
   d) the μMx

5. The symbol for flux density is ________ and the symbol for magnetic flux is ________.
   a) B, Φ
   b) Φ, B
   c) μ, B
   d) Φ, μ

6. The unit of measure for flux density is ________.
   a) the weber
   b) the maxwell
   c) the microweber
   d) the tesla

7. 10 microwebers equal ________ maxwells.
   a) 100
   b) 10
   c) 1000
   d) 1

8. The *maximum* flux density of a magnet is ________.
   a) 1×10^8 μWb
   b) at the poles
   c) proportional to area
   d) is inversely proportional to Φ

9. The ease with which a material establishes lines of magnetic flux within itself is referred to as ________.
   a) flux density
   b) permittivity
   c) permeability
   d) flux conductance

10. The μ_m rating of a magnetic material is ________.
    a) much less than that of free space
    b) much greater than that of free space
    c) equal to that of free space
    d) much greater than 1

11. A permanent magnet is placed near a piece of iron. This flux density of the magnetic field in the iron will be ________ in the air around the magnet.
    a) much greater than
    b) much lower than
    c) about the same as
    d) slightly lower than

12. The domain theory of magnetism is based upon the theory that ________.
    a) all electrons have a magnetic field
    b) electron field polarity depends upon the direction of spin
    c) atoms with like magnetic fields can join together
    d) all of the above

13. The magnetization of a material through the use of some external force is referred to as ________.
    a) magnetic coercion
    b) magnetic susceptance
    c) magnetic induction
    d) magnetic conduction

14. Low-permeability materials are ________ to magnetize, while they tend to ________ their magnetic properties.
    a) difficult, retain
    b) difficult, quickly lose
    c) easy, retain
    d) easy, quickly lose

15. The ability of a magnet to remain magnetized after the magnetizing force that produced it is removed is referred to as ________.
    a) permittivity
    b) permeability
    c) reluctance
    d) retentivity

16. Reluctance is the magnetic equivalent of electrical ________.
    a) permittivity
    b) impedance
    c) resistance
    d) susceptance

17. Temporary magnets ________.
    a) are easy to magnetize
    b) lose their magnetic characteristics easily
    c) have high permeability

d) have low retentivity
e) all of the above

18. Reluctance is directly proportional to the _______ of a given material.
    a) permeability
    b) retentivity
    c) length
    d) area

19. Materials that are only slightly magnetic are called _______ materials while those that are nonmagnetic are called _______ materials.
    a) paramagnetic, diamagnetic
    b) diamagnetic, paramagnetic
    c) ferromagnetic, paramagnetic
    d) paramagnetic, ferromagnetic

20. Hans Orstead discovered _______.
    a) the lines of magnetic flux
    b) the laws of magnetic repulsion and attraction
    c) the left-hand rule
    d) that any conductor carrying an electric current produces a magnetic field

21. Suppose that you are looking directly into a conductor and the current is flowing toward you. The left-hand rule tells you that the direction of the magnetic field is _______.
    a) to the right
    b) clockwise
    c) to the left
    d) counterclockwise

22. The space inside a coil is called _______.
    a) the inductor
    b) the pole piece
    c) the armature
    d) the core

23. Rowland's law states that magnetic flux is directly proportional to _______ and inversely proportional to _______.
    a) reluctance, mmf
    b) mmf, reluctance
    c) reluctance, ampere-turns
    d) mmf, retentivity

24. Ampere-turns is usually represented using _______.
    a) A · t
    b) NI

c) IT

d) IN

25. The flux density of a coil depends on three factors. They are ________.

a) length, area, and μ_m

b) length, reluctance, and μ_m

c) length, ampere-turns, and μ_m

d) area, length, and μ_m

26. The time lag between the removal of a magnetizing force and the drop in flux density for a given material is called ________.

a) reluctance

b) residual flux density

c) hysteresis

d) retentivity

27. The magnetization curve for a given material is often referred to as a(n) ________.

a) retentivity curve

b) B-H curve

c) reluctance curve

d) NI curve

28. Electromagnetic relays ________.

a) come as both NC and NO switches

b) electrically isolate the input from the output

c) allow a low-current input to control a high-current output

d) all of the above

29. Electromagnetic relays produce a reverse voltage that is referred to as ________.

a) counter voltage

b) kick emf

c) inverse voltage

d) breakover voltage

e) all of the above

30. Electromagnetic relays have two primary components called ________.

a) the pole and the throw

b) the pole piece and the coil

c) the coil and the armature

d) the armature and the pole piece

31. Ring magnets ________.

a) have no poles

b) retain all magnetic flux within themselves

c) are often used in analog meter movements
d) all of the above

32. The hairsprings in a D'Arsonval meter movement ________.
a) counteract the magnetically induced core rotation
b) support the permanent magnet
c) support the coil
d) all of the above

33. Permanent magnets can lose their magnetic strength through ________.
a) being heated to a high temperature
b) being stored improperly
c) being dropped
d) all of the above

PRACTICE PROBLEMS

1. Calculate the flux density for the shaded area in Figure 10.33a. Assume that each line of force represents 120 μWb of flux.

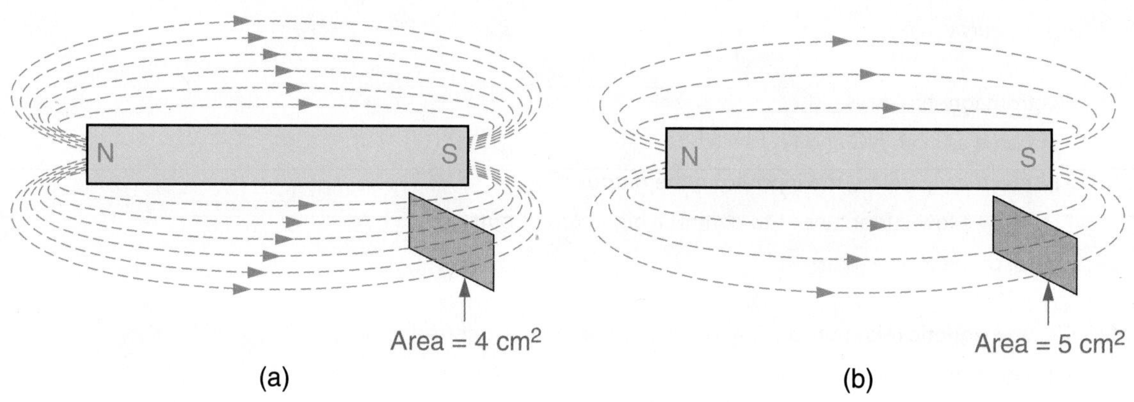

FIGURE 10.33

2. Calculate the flux density for the shaded area in Figure 10.33b. Assume that each line of force represents 200 μWb of flux.
3. A magnet with a cross-sectional area of 1.5 cm^2 generates 300 μWb of flux at its poles. Calculate the flux density at the poles of the magnet.
4. A magnet with a cross-sectional area of 3 cm^2 generates 1200 μWb of flux at its poles. Calculate the flux density at the poles of the magnet.
5. The permeability of a piece of low-carbon steel is approximately 7.54×10^{-4} Wb/A · m. Calculate the relative permeability of this material.
6. Calculate the relative permeability of free space.
7. The permeability of a piece of permalloy (an alloy made primarily of iron and nickel) is approximately 0.1 Wb/A · m. Calculate the relative permeability of this material.
8. The permeability of a piece of iron ingot is approximately 3.77×10^{-3} Wb/A · m. Calculate the relative permeability of this material.

9. Calculate the flux density produced by the coil in Figure 10.34a.
10. Calculate the flux density produced by the coil in Figure 10.34b.

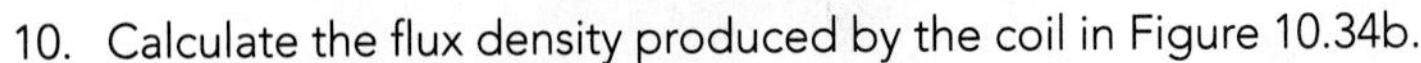

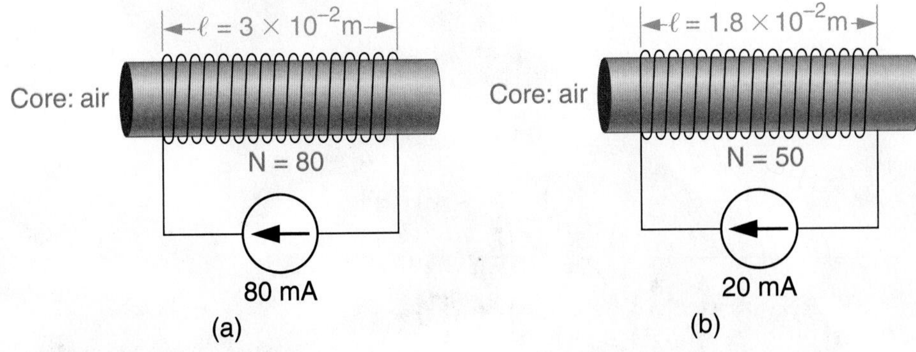

11. Calculate the flux density produced by the coil in Figure 10.35a.
12. Calculate the flux density produced by the coil in Figure 10.35b.

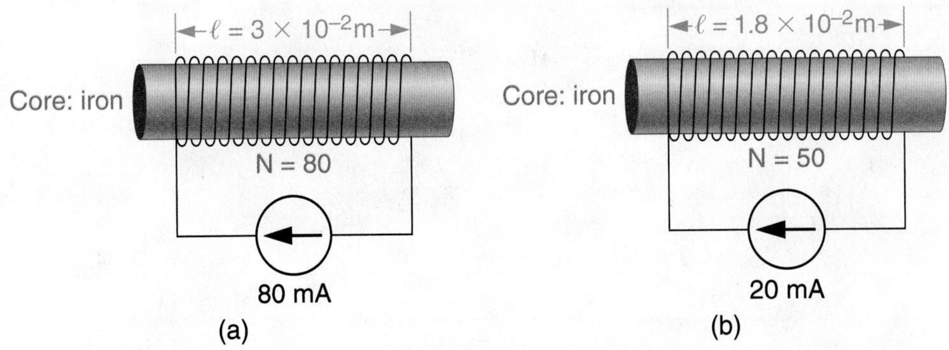

FIGURE 10.35

ANSWERS TO THE EXAMPLE PRACTICE PROBLEMS

10.1	1.5 T
10.3	1×10^{-1} T
10.4	2.5×10^{-2} T; ΔI = 25%; ΔB = 75%

CHAPTER 11

INDUCTANCE AND INDUCTORS

PURPOSE

You have seen that the current through a conductor produces a magnetic field. In this chapter, we're going to take the relationship between current and magnetism a step further by discussing a property called **INDUCTANCE**.

Technically defined, *inductance* is the ability of a component to induce a voltage across itself or a nearby circuit by generating a changing magnetic field. Based on its overall effect, inductance is generally described as the electrical property that:

- Opposes any change in current.
- Stores energy in an electromagnetic field.

As you will learn, both of these descriptions are valid.

An **INDUCTOR** is a component designed to provide a specific measure of inductance. Inductors are capable of storing energy in an electromagnetic field, as well as oppose any change in current. These characteristics make the inductor useful in a variety of applications, some of which will be introduced later in this chapter.

It should be noted that inductors are not commonly found in residential and commercial electrical systems. However, the principles of inductance play a crucial role in the operation of many common electrical system components, such as transformers, motors, generators, and relays (among others).

KEY TERMS

The following terms are introduced and defined in this chapter on the pages indicated:

Coefficient of coupling (k), 490
Counter emf, 485
Coupled, 490
Decay curve, 500
Ferrite core, 504
Henry (H), 485
Inductance, 480
Inductor, 480
Lenz's law, 484
Mutual inductance, 490
Rise curve, 498
Self-inductance, 484
Time constant (τ), 499
Toroid, 505
Transition, 497
Unity coupling, 491

OBJECTIVES

After completing this chapter, you should be able to:

1. Describe the effect of varying current on a magnetic field.
2. List and discuss the first three of Faraday's laws of induction.
3. Describe the concept of *self-induction*.
4. Explain the concept of counter emf.

5. Define the henry (H) unit of inductance.
6. Calculate the value of an inductor (in henries), given its physical dimensions and core permeability.
7. Discuss mutual inductance and list the factors that affect its value.
8. Discuss coefficient of coupling, its normal range of values, and the factors that affect its values.
9. Calculate the total inductance for any group of inductors connected in series.
10. Describe the possible effects of mutual inductance on total series inductance.
11. Calculate the total inductance for any group of inductors connected in parallel.
12. Describe the current transitions in a resistive-inductive DC circuit.
13. Describe the universal current curve for a resistive-inductive DC circuit.
14. Calculate the current through a DC resistive-inductive circuit at any time constant interval.
15. Calculate the time constant for a DC resistive-inductive circuit.
16. Compare and contrast iron-core, air-core, and ferrite-core inductors.
17. List and describe the various types of inductors.

INDUCTANCE The ability of a component to induce a voltage across itself or a nearby circuit by generating a changing magnetic field.

INDUCTOR A component designed to provide a specific measure of inductance.

An inductor is a conducting coil of wire that generates an electromagnetic field. Inductors are one of the more common electrical components because of their ability to delay and reshape changing currents.

IN THE FIELD

11.1 INTRODUCTION TO INDUCTANCE

As indicated in the chapter opening, inductance makes use of the relationship between current and magnetism. As such, we will start our discussion on inductance by reviewing some principles of magnetism.

Review: Current and Magnetism

In Chapter 10, you were shown that a current-carrying coil generates magnetic flux, as shown in Figure 11.1. Here are some of the principles that were established on the relationship between current and the resulting flux:

- Current through a wire generates magnetic lines of force (flux).
- When a wire is wrapped in a series of loops, called a coil, the lines of force generated by the current add together to form a magnetic field.
- The strength of the magnetic field is determined in part by the magnitude of the current (I) and number of turns (N) in the coil. The product of current and coil turns (NI) is measured in ampere-turns (A • t).
- The polarity of the flux generated by a current is determined by the current direction.

As you will see, all of these factors play a role in the principles of inductance.

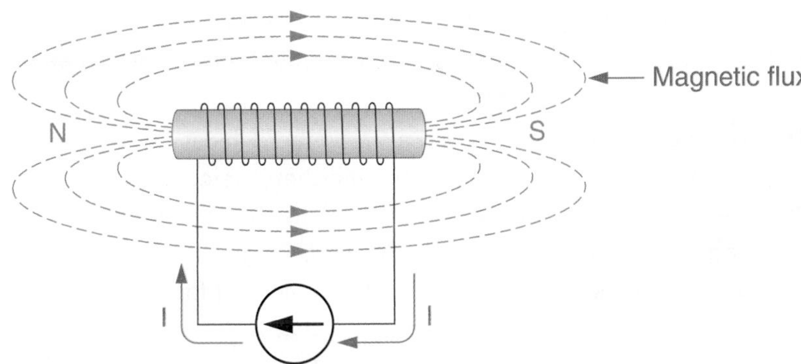

FIGURE 11.1 Current through a coil generates a magnetic field.

The Effect of Varying Current on a Magnetic Field

An inductor is simply a coil like the one in Figure 11.1. When current passes through an inductor, magnetic flux is generated as shown in the figure. The flux density, or *field strength*, can be calculated using

(11.1)

$$B = \frac{\mu_m NI}{\ell}$$

where

B = the flux density, in webers per square meter (Wb/m^2)
μ_m = the permeability of the core material
NI = the ampere-turns product for the component
ℓ = the length of the coil, in meters

Equation (11.1) was first introduced in Chapter 10. According to this relationship:

- An increase in inductor current causes an increase in flux density.
- A decrease in inductor current causes a decrease in flux density.

In other words, flux density varies directly with inductor current.

The effect of varying inductor current on its magnetic field is illustrated in Figure 11.2. The circuit shown contains an AC source that generates a constantly changing current. As the inductor current increases (during t_1), flux density increases and the magnetic field expands outward from the component. As the inductor current decreases (during t_2), flux density decreases and the magnetic field collapses back into the component. The impact of this changing magnetic field is described by three of Faraday's laws of induction.

Faraday's Laws of Induction

In 1831, an English scientist named Michael Faraday discovered that a magnetic field could be used to induce (generate) a voltage across a coil through a process called *electromagnetic induction*. Faraday's observations on the results of his experiments have come to be known as *Faraday's laws of induction*. The first three of these laws can be paraphrased as follows:

Law 1: To induce a voltage across a wire, there must be a relative motion between the wire and the magnetic field.

Law 2: The voltage induced is proportional to the rate of change in magnetic flux encountered by the wire.

Law 3: When a wire is cut by 10^8 perpendicular lines of force per second (1 Wb/sec), one volt is induced across that wire.

Faraday's first law states that induction takes place when there is relative motion between the conductor and the magnetic field. As shown in Figure 11.2, a varying coil current generates an expanding and collapsing magnetic field that passes through the stationary inductor. Therefore, there is relative motion between the inductor and the magnetic field, and Faraday's first law of induction applies. The changing magnetic field induces a voltage across the inductor.

Faraday's second law states that induced voltage is proportional to the rate of change in magnetic flux. That is:

- The greater the rate of change in current, the greater the voltage induced across the component.
- The lower the rate of change in current, the lower the voltage induced across the component.

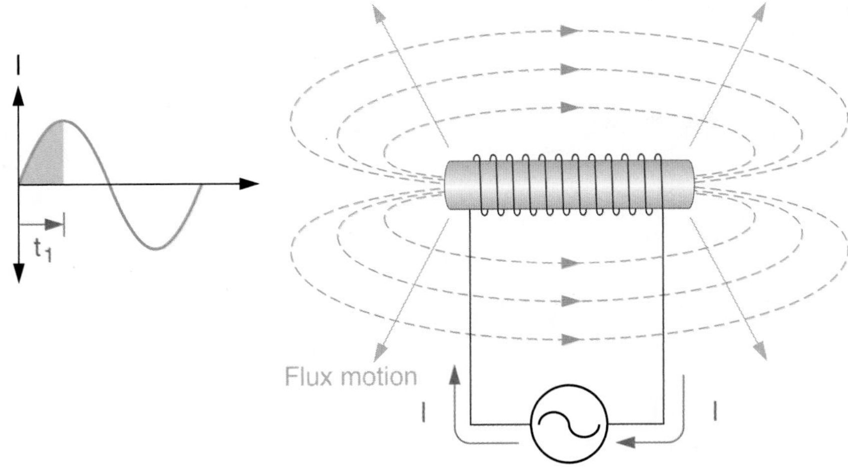

(a) During t_1, the magnetic field produced by the coil expands outward as current increases.

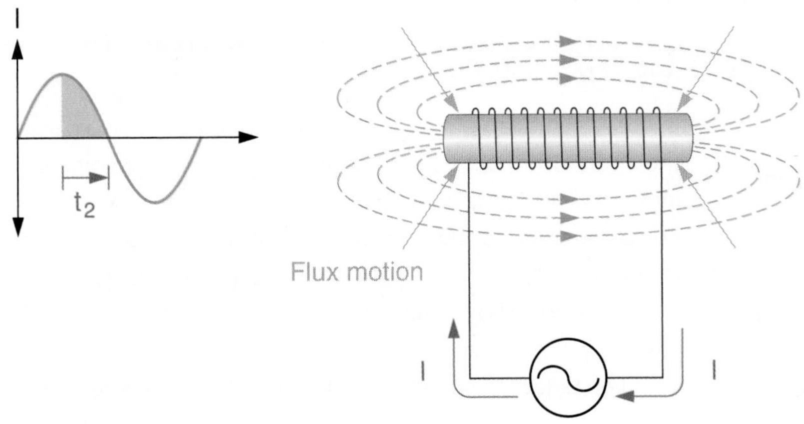

(b) During t_2, the magnetic field produced by the coil collapses inward as current decreases.

FIGURE 11.2 Magnetic field strength varies directly with coil current.

The relationship given in Faraday's third law is illustrated in Figure 11.3. The magnetic field is expanding from the core of the inductor. When the magnetic flux passes through the turns of the inductor at a rate of one weber per second, 1 V is induced across each turn in the component. Thus, 12 V is induced across the inductor in Figure 11.3, one volt for each turn.

We have now established the following principles of operation for a current-carrying inductor:

- The current through an inductor generates magnetic flux.
- As inductor current increases, the magnetic field expands. As inductor current decreases, the magnetic field collapses.
- As the magnetic field expands and collapses, the flux cuts through the stationary inductor.
- The flux cutting through the coils induces a voltage across the inductor.

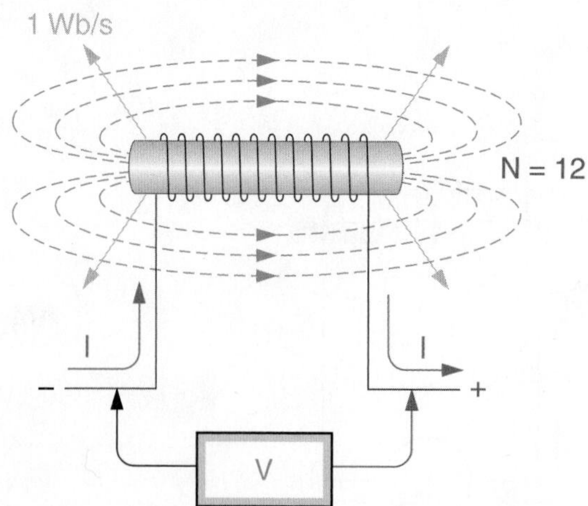

FIGURE 11.3

SELF-INDUCTANCE The process whereby a coil induces a voltage across itself by generating a changing magnetic field.

The process just described is referred to as **SELF-INDUCTANCE**. To understand the full impact of self-inductance, we have to take a look at another law of magnetic induction, called **LENZ'S LAW**.

An example of Lenz's Law can be seen in the following example: The speed of the armature of a 100% efficient motor having no friction or winding losses will increase until the counter emf is equal to the applied emf. In other words, there will be no net emf, no current, and hence, no net mechanical force. As such, the armature will spin at a constant rate of its own accord.

IN THE FIELD

Lenz's Law

In 1834, a Russian physicist named Heinrich Lenz discovered the relationship between a magnetic field and the voltage it induces. This relationship, which has come to be known as *Lenz's law*, can be paraphrased as follows: *An induced voltage always opposes its source.* That is, the polarity of the voltage induced by a magnetic field opposes the change in current that produced the magnetic field in the first place.

Lenz's law is illustrated in Figure 11.4. In Figure 11.4a:

LENZ'S LAW A law stating that an induced voltage always opposes its source.

- An increase in inductor current causes the magnetic field to expand.
- As the magnetic field expands, it cuts through the coil, inducing a voltage.
- The polarity of the voltage opposes the increase in current.

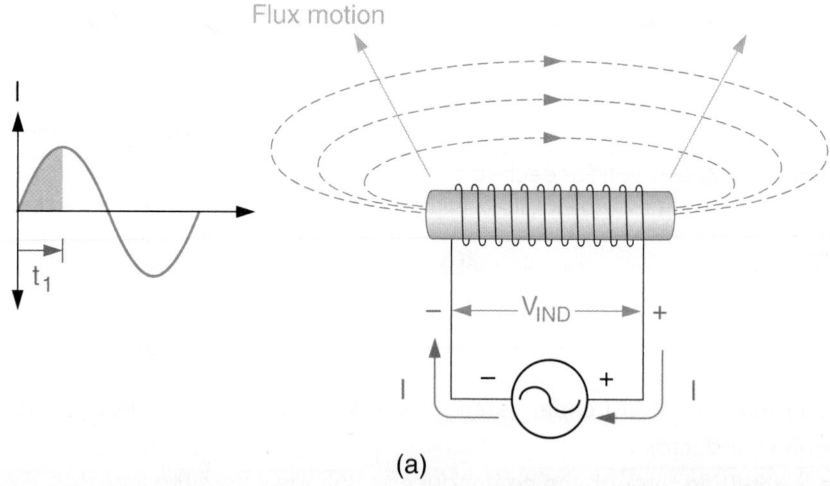

FIGURE 11.4 The voltage induced across an inductor opposes inductor current.

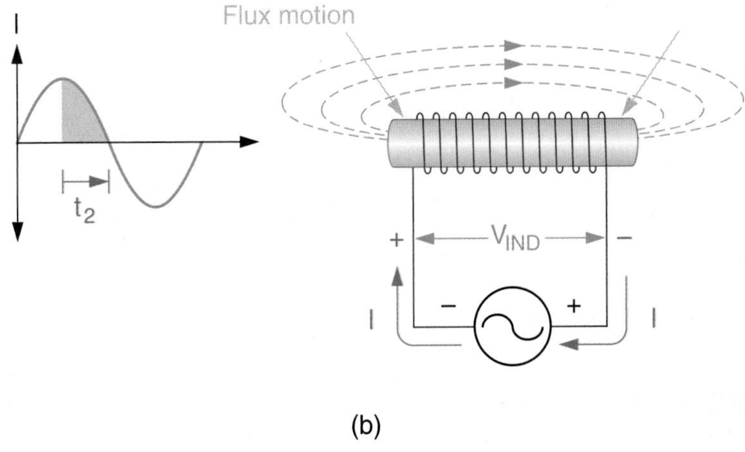

FIGURE 11.4 (*Continued*)

In Figure 11.4b:

- A decrease in inductor current causes the magnetic field to collapse.
- As the magnetic field collapses, it cuts through the coil, inducing a voltage across the component. (Note that the polarity of the voltage has reversed.)
- The polarity of the induced voltage opposes the decrease in current.

As you can see, the polarity of the induced voltage always opposes the change in coil current. Because it always opposes the change in coil current, the induced voltage is referred to as **COUNTER EMF**.

COUNTER EMF The induced voltage across an inductor that opposes any change in coil current.

Induced Voltage

As you know, a voltage is induced across an inductor when the current through the component changes. This relationship can be expressed as

(11.2)

$$V_{IND} = L\frac{\Delta I}{\Delta t}$$

where

V_{IND} = the value of induced voltage

L = the inductance of the coil, measured in henries (H)

$\frac{\Delta I}{\Delta t}$ = the rate of change in inductor current

The unit of measure of inductance is the **HENRY (H)**. The henry can be explained with the help of Figure 11.5. The current through the inductor is changing at a rate of one ampere per second (1 A/s). This change in current is inducing 1 V across the inductor. When a change of 1 A/s induces 1 V across an inductor, the value of the inductance is said to be one henry (1 H). The relationship among inductance, inductor current, and inductor voltage is demonstrated further in Example 11.1.

HENRY (H) The value of inductance that generates a 1-V counter emf when inductor current changes by one ampere per second (A/s).

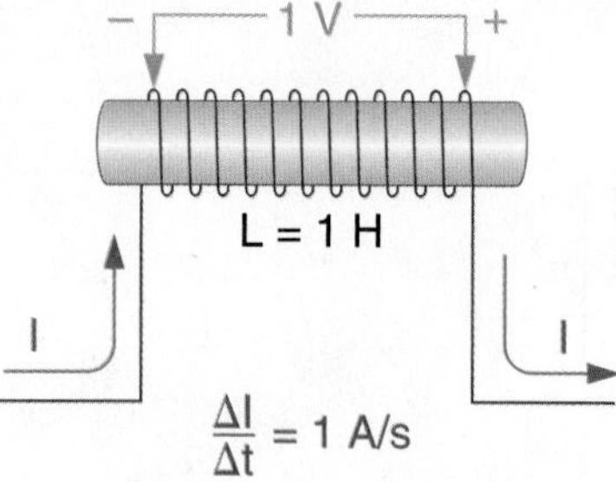

FIGURE 11.5 One henry (H) of inductance.

EXAMPLE 11.1

The schematic symbol for a 3.3 H air-core inductor is shown in Figure 11.6. If the current through the component changes at a rate of 2 A per second, what is the value of the induced voltage?

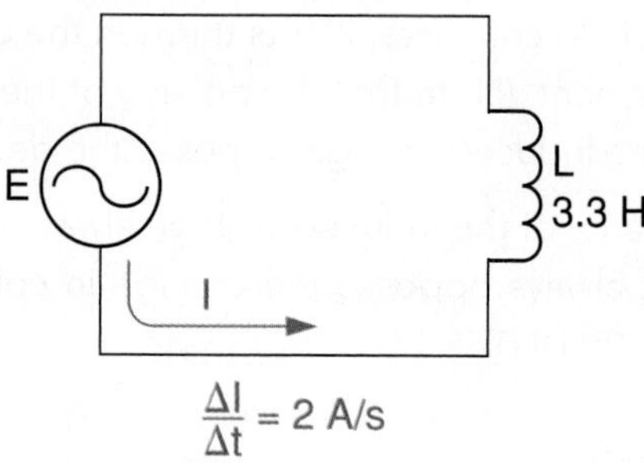

FIGURE 11.6

SOLUTION

Using the information provided above, the value of the induced voltage can be found as

$$V_{IND} = L\frac{\Delta I}{\Delta T}$$

$$= (3.3\text{ H}) \times (2\text{ A/s})$$

$$= 6.6\text{ V}$$

PRACTICE PROBLEM 11.1

The current through a 1.5 H inductor changes at a rate of 5 A/s. Calculate the value of the induced voltage.

Equation (11.2) defines the henry in terms of induced voltage and rate of change in current. You can also approximate the value of an inductor based on its physical characteristics, as follows:

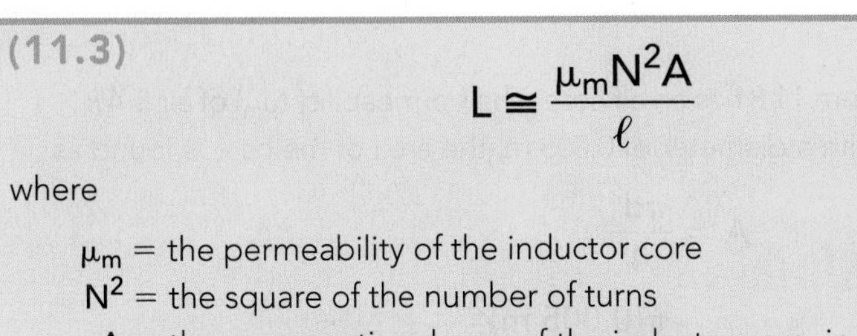

(11.3)

$$L \cong \frac{\mu_m N^2 A}{\ell}$$

where

μ_m = the permeability of the inductor core
N^2 = the square of the number of turns
A = the cross-sectional area of the inductor core, in square meters (m^2)
ℓ = the length of the coil, in meters (m)

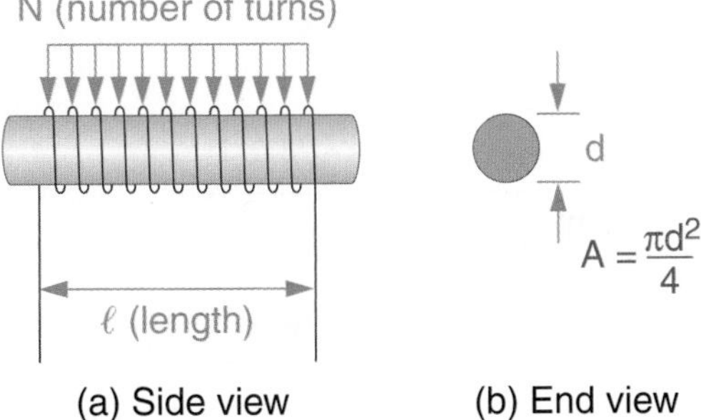

FIGURE 11.7 Inductor physical characteristics.

The physical characteristics in equation (11.3) are illustrated in Figure 11.7. Note that the area (A) of the core can be found using

(11.4)

$$A = \frac{\pi d^2}{4}$$

where **d** = the diameter of the inductor core, in meters (m)

Example 11.2 shows how the value of an inductor can be approximated when its physical dimensions are known.

EXAMPLE 11.2

What is the value of the inductor in Figure 11.8?

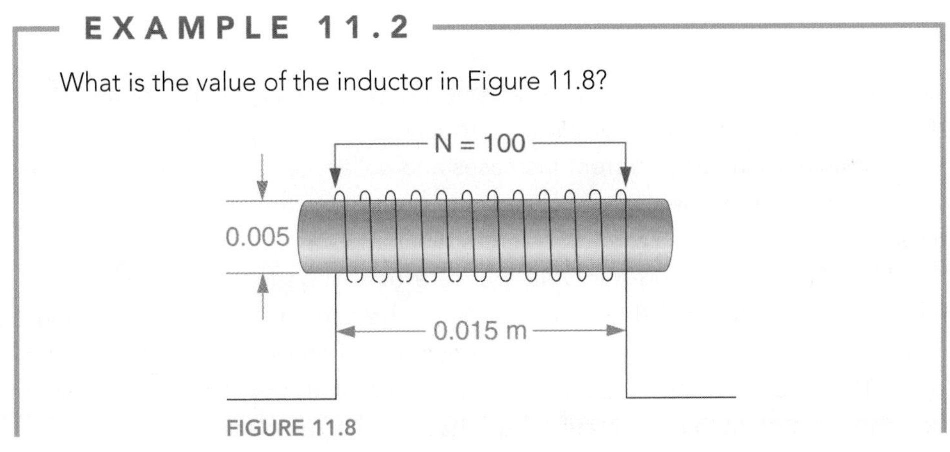

FIGURE 11.8

SOLUTION

The inductor in Figure 11.8 has an air core. The permeability (μ_m) of air is $4\pi \times 10^{-7}$ Wb/A · m. With a diameter of 0.005 m, the area of the core is found as

$$A = \frac{\pi d^2}{4}$$

$$= \frac{\pi(0.005\ \text{m})^2}{4}$$

$$= 19.6 \times 10^{-6}\ \text{m}^2$$

and the value of the inductor can be found as

$$L \cong \frac{\mu_m N^2 A}{\ell}$$

$$= \frac{(4\pi \times 10^{-7}\ \text{Wb/A} \cdot \text{m})(100)^2(19.6 \times 10^{-3}\ \text{m}^2)}{0.015\ \text{m}}$$

$$\cong 16.4\ \mu\text{H}$$

PRACTICE PROBLEM 11.2

An air-core inductor has the following dimensions: $d = 0.0025$ m, $N = 150$, and $\ell = 0.02$ m. Calculate the value of the component.

Though you are shown how to solve equation (11.3), it is rarely used in practice for several reasons: First, the value of any given inductor is typically known, making the calculation unnecessary. Second, it works only for specific types of inductors. There are inductors that require the use of a different equation.

The true value of equation (11.3) lies in what it tells us about the factors that affect inductance. The equation indicates that the value of an inductor is:

- Directly proportional to the permeability of its core
- Directly proportional to the number of turns it has
- Inversely proportional to its length

Putting It All Together

An inductor is a component that is designed to provide a specific amount of inductance. A changing inductor current produces a changing magnetic field. This field expands as inductor current increases and collapses as inductor current decreases. In the process of expanding and collapsing, the magnetic field cuts through the coil, inducing a voltage across the component.

Inductance is measured in volts per rate of change in current. The unit of measure of inductance is the henry (H). When the current through 1 H of inductance changes at a rate of 1 A/s, the difference of potential induced across the coil is 1 V. The polarity of this voltage is such that it opposes the change (increase or decrease) in current that generated the voltage in the first place.

The value of an inductor is directly proportional to:

- The permeability of its core
- The cross-sectional area of its core
- The square of the number of turns

and inversely proportional to its length. The basic characteristics of inductors are summarized for you in Figure 11.9.

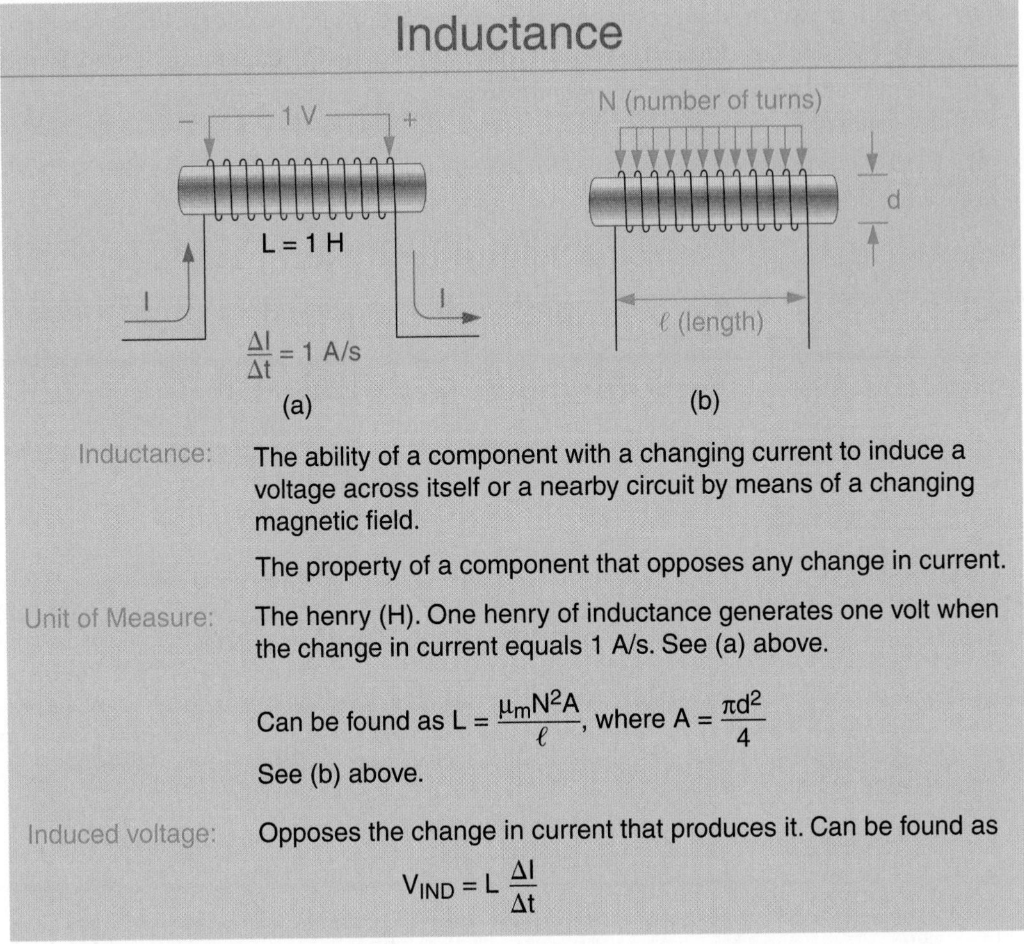

Inductance
1 V
L = 1 H
I
I
$\frac{\Delta I}{\Delta t} = 1$ A/s
(a)
N (number of turns)
d
ℓ (length)
(b)
Inductance: The ability of a component with a changing current to induce a voltage across itself or a nearby circuit by means of a changing magnetic field.
The property of a component that opposes any change in current.
Unit of Measure: The henry (H). One henry of inductance generates one volt when the change in current equals 1 A/s. See (a) above.
Can be found as $L = \frac{\mu_m N^2 A}{\ell}$, where $A = \frac{\pi d^2}{4}$
See (b) above.
Induced voltage: Opposes the change in current that produces it. Can be found as
$V_{IND} = L \frac{\Delta I}{\Delta t}$

FIGURE 11.9 Summary of basic inductance characteristics.

PROGRESS CHECK

1. What is inductance?
2. What is an *inductor*?
3. What is the relationship between flux density and:
   a. core permeability?
   b. the ampere-turns of a coil?
   c. coil length?
4. How does a change in coil current affect the magnetic field generated by the component?
5. How is Faraday's first law of induction fulfilled by a coil and its magnetic field?
6. What is Lenz's law?
7. What is the unit of measure for inductance? In terms of volts per change in current, what does this unit equal?
8. Describe the relationship between the value of an inductor and its physical characteristics.

11.2 MUTUAL INDUCTANCE

MUTUAL INDUCTANCE When the expanding and collapsing flux produced by an inductor induces a voltage across another inductor that is in close proximity.

COUPLED A term used to describe two components (or circuits) that are connected so that energy is transferred from one to the other.

When one inductor is placed in close proximity to another, the expanding and collapsing flux produced by either coil can induce a voltage across the other. This phenomenon, called **MUTUAL INDUCTANCE**, is illustrated in Figure 11.10. When positioned as shown in Figure 11.10a, the flux produced by the changing current through L_1 cuts through the turns of L_2. As a result, a voltage is induced across L_2. Note that two or more components (or circuits) are said to be **COUPLED** when energy is transferred from one to the other. In Figure 11.10a, energy is transferred from L_1 to L_2 via the lines of magnetic flux.

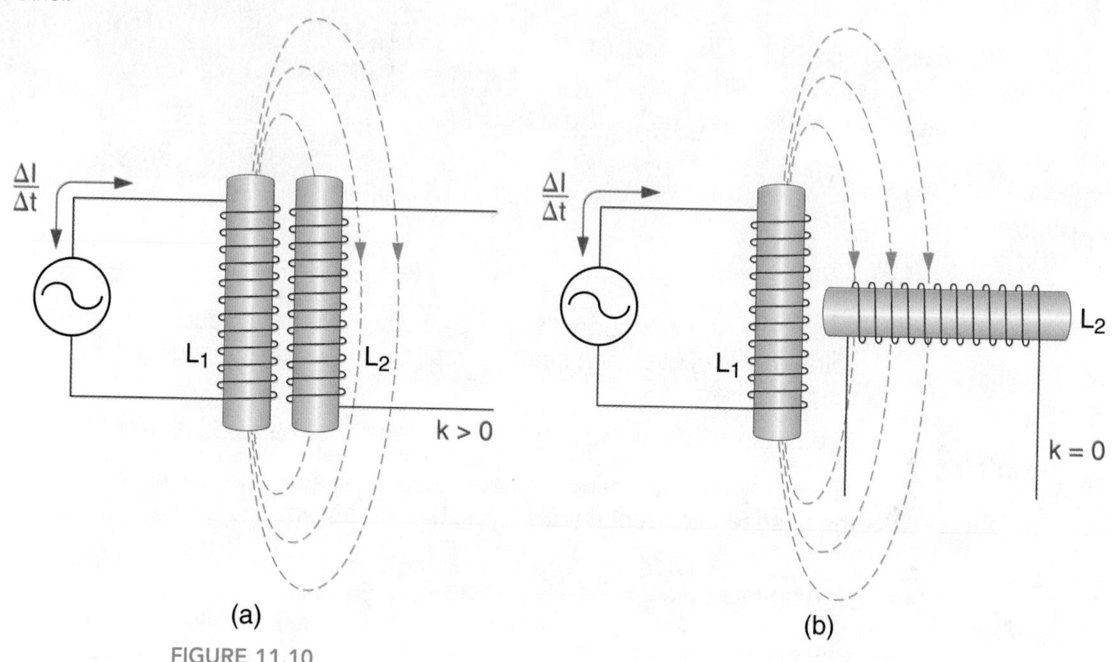

FIGURE 11.10

The amount of mutual inductance between two coils can be found as

(11.5)

$$L_M = k\sqrt{L_1 L_2}$$

COEFFICIENT OF COUPLING (k) The degree of coupling that takes place between two or more coils.

where k represents the coefficient of coupling between the coils. The **COEFFICIENT OF COUPLING (k)** between two or more coils is a measure of the degree of coupling that takes place between the coils. The higher the coefficient of coupling, the greater the mutual inductance between the coils, and the greater the energy transfer from one component to the other(s).

The value of k for any pair of inductors has a range of 0 to 1. When $k = 0$, there is no coupling between the two inductors. When $k = 0.5$, half the flux from one inductor cuts through the turns of the second coil. When $k = 1$, all of the flux from the first coil cuts through the turns of the second coil.

It should be noted that a value of $k = 1$ represents an ideal condition called **UNITY COUPLING**. Unity coupling cannot be achieved in practice because the flux produced by a coil expands outward in all directions from the component. As such, it is not possible for all the generated flux to pass through a second inductor. However, values of $k > 0.9$ can be achieved in practice.

UNITY COUPLING An ideal condition where all of the flux produced by one coil passes through another.

The coefficient of coupling between two coils depends on several factors, including

- The physical distance between the components
- The relative angle between them

The first of these is easy to visualize. To have mutual inductance, the coils must be close enough for the flux from one to pass through the turns of the other. If the coils are far enough apart, $k = 0$ and there is no mutual inductance between the coils. The second factor is illustrated in Figure 11.10b. In this case, the coils are positioned so that the flux from L_1 passes through L_2. However, the coils are perpendicular, so the flux produced by L_1 does not cut through the turns of L_2. Therefore, $k \cong 0$ and little (if any) voltage is induced across L_2.

Series-Connected Coils

When inductors are connected in series, care is taken to ensure that there is no mutual inductance between the components. When this is the case, the total inductance of the circuit can be found as

(11.6)

$$L_T = L_1 + L_2 + \cdots + L_n$$

where L_n is the highest numbered inductor in the circuit.

As you can see, total inductance is found in the same fashion as total resistance in a series circuit. This is demonstrated further in Example 11.3.

EXAMPLE 11.3

What is the total circuit inductance in Figure 11.11?

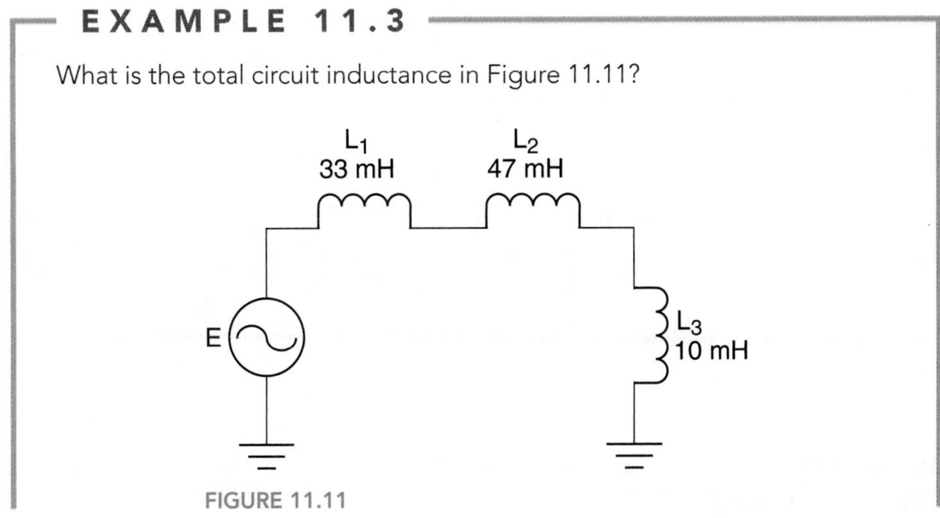

FIGURE 11.11

SOLUTION

The circuit has a total inductance of

$$\begin{aligned} L_T &= L_1 + L_2 + L_3 \\ &= 33 \text{ mH} + 47 \text{ mH} + 10 \text{ mH} \\ &= 90 \text{ mH} \end{aligned}$$

PRACTICE PROBLEM 11.3

A circuit like the one in Figure 11.11 has the following values: $L_1 = 100\ \mu\text{H}$, $L_2 = 3.3\text{ mH}$, and $L_3 = 330\ \mu\text{H}$. Calculate the value of L_T.

When two series-connected inductors have a value of $k > 0$, their mutual inductance affects the total circuit inductance. The total inductance of two coils that have some mutual inductance between them is found as

(11.7)
$$L_T = L_1 + L_2 \pm 2L_M$$

where L_M is the mutual inductance between the two inductors (L_1 and L_2). If the inductors are connected so that their magnetic fields have the same polarity, the mutual inductance adds to the total inductance. In this case,

$$L_T = L_1 + L_2 + 2L_M$$

If they are connected in such a way that their magnetic fields have opposing polarities, the mutual inductance subtracts from the total inductance. In this case,

$$L_T = L_1 + L_2 - 2L_M$$

Parallel-Connected Coils

When inductors are connected in parallel, care is taken to ensure that there is no mutual inductance between the components. When this is the case, the total inductance of the circuit can be found as

(11.8)
$$L_T = \frac{1}{\frac{1}{L_1} + \frac{1}{L_2} + \cdots + \frac{1}{L_n}}$$

where L_n is the value of the highest numbered inductor in the circuit.

As you can see, total inductance is calculated in the same fashion as total resistance in a parallel circuit. This means that we can also calculate total parallel inductance using

(11.9)

$$L_T = \frac{L_1 \times L_2}{L_1 + L_2}$$

for an inductive circuit containing only two branches. Finally, for equal-value inductors in parallel, the total inductance can be found using

(11.10)

$$L_T = \frac{L}{n}$$

where

L = the value of the inductors
n = the number of parallel-connected components

If you compare the parallel inductance equations with the parallel resistance equations in Chapter 6, you'll see that they are identical formats used under the same conditions. The following examples demonstrate the use of these equations.

EXAMPLE 11.4

What is the total circuit inductance in Figure 11.12?

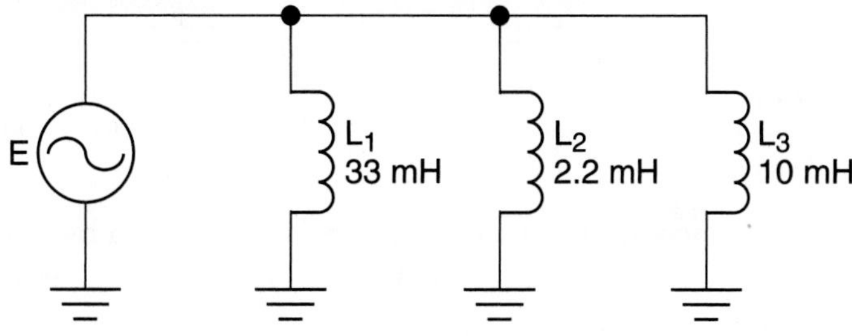

FIGURE 11.12

SOLUTION

The total inductance in the circuit in Figure 11.12 can be found as follows:

$$L_T = \frac{1}{\frac{1}{L_1} + \frac{1}{L_2} + \frac{1}{L_3}}$$

$$= \frac{1}{\frac{1}{33} + \frac{1}{2.2} + \frac{1}{10}} \text{ mH}$$

$$= \frac{1}{0.5848} \text{ mH}$$

$$= 1.71 \text{ mH}$$

PRACTICE PROBLEM 11.4

A circuit like the one in Figure 11.12 has the following values: $L_1 = 330\ \mu H$, $L_2 = 1\ mH$, and $L_3 = 470\ \mu H$. Calculate the total inductance in the circuit.

When you compare the result in Example 11.4 with the given branch values, you see another similarity between parallel inductive and resistive circuits: In any parallel circuit, the total inductance is lower than any of the branch values.

EXAMPLE 11.5

What is the total circuit inductance in Figure 11.13?

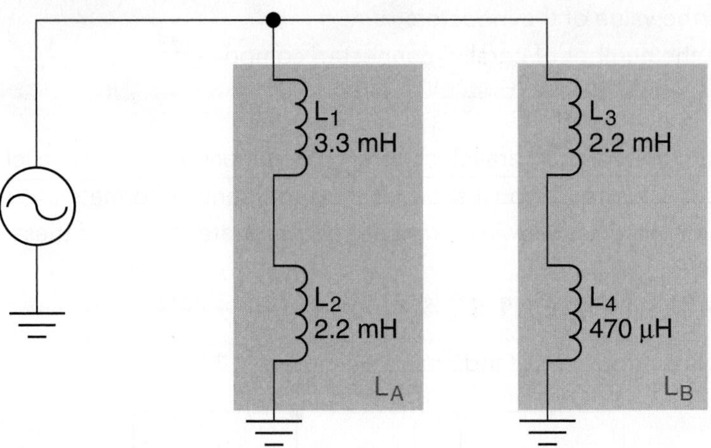

FIGURE 11.13

SOLUTION

The total inductance in the circuit in Figure 11.13 is found by calculating the branch inductance values and combining the results. First, the total inductance in the first branch (L_A) has a value of

$$\begin{aligned} L_A &= L_1 + L_2 \\ &= 3.3\ mH + 2.2\ mH \\ &= 5.5\ mH \end{aligned}$$

The total inductance in the second branch (L_B) has a value of

$$\begin{aligned} L_B &= L_3 + L_4 \\ &= 2.2\ mH + 470\ \mu H \\ &= 2.67\ mH \end{aligned}$$

Finally, the total inductance in the circuit has a value of

$$L_T = \frac{1}{\frac{1}{L_A} + \frac{1}{L_B}}$$

$$= \frac{1}{\frac{1}{5.5} + \frac{1}{2.67}} \text{ mH}$$

$$= \frac{1}{0.5564} \text{ mH}$$

$$= 1.8 \text{ mH}$$

As is always the case, L_T is lower than the total inductance of either branch.

PRACTICE PROBLEM 11.5

A circuit like the one in Figure 11.13 has the following values: $L_1 = 10$ mH, $L_2 = 1$ mH, $L_3 = 4.7$ mH, and $L_4 = 3.3$ mH. Calculate the total inductance in the circuit using equation (11.9).

In Example 11.5, the total circuit inductance is lower than either of the total branch inductance values (L_A or L_B).

EXAMPLE 11.6

Four 1.5 H inductors are connected in parallel. What is the total inductance of the circuit?

SOLUTION

The total inductance of the circuit can be found as

$$L_T = \frac{L}{N}$$

$$= \frac{1.5 \text{ H}}{4}$$

$$= 375 \text{ mH}$$

PRACTICE PROBLEM 11.6

A circuit contains five 3.3 H inductors connected in parallel. Calculate the total inductance in the circuit.

If two parallel inductors are wired in close proximity to each other, there may be some measurable amount of mutual inductance between them. In this case, the total circuit inductance is found as

(11.11)

$$L_T = \frac{(L_1 \pm L_M)(L_2 \pm L_M)}{(L_1 \pm L_M) + (L_2 \pm L_M)}$$

where L_M is the mutual inductance between the two inductors (L_1 and L_2). As with the series inductors, the L_M is added if the inductors are connected so that their magnetic fields have the same polarity, and subtracted if the inductors are connected in such a way that their magnetic fields have opposing polarities.

A Practical Consideration

The potential effects of mutual inductance are usually taken into account in circuit design. That is, coils are normally positioned to avoid generating any mutual inductance. Therefore, you will rarely be required to account for mutual inductance in the analysis of any circuit.

So, why introduce equations showing the effects of mutual inductance on total inductance? At times, there is a discrepancy between actual component values and total inductance. When this is the case, the probable cause is some amount of mutual inductance that is increasing (or reducing) the total circuit inductance.

PROGRESS CHECK

1. What is mutual inductance?
2. What is *unity coupling*? What do we mean when we say that two (or more) components are electrically *coupled*?
3. What is indicated by the *coefficient of coupling (k)* between two coils?
4. How do you determine the total inductance in a series circuit when
   a. There is no mutual inductance?
   b. There is some measurable amount of mutual inductance?
5. How do you determine the total inductance in a parallel circuit when
   a. There is no mutual inductance?
   b. There is some measurable amount of mutual inductance?

11.3 INDUCTORS IN DC CIRCUITS

The voltage across an ideal inductance with constant DC current is zero because the rate of change of current is zero. Since the inductor has current through it but no voltage across it, it is a short circuit.

KEY CONCEPT

Earlier in the chapter, inductance was described as a property that:

- Opposes any increase or decrease in current
- Stores energy in an electromagnetic field

These characteristics of inductance can be seen clearly when we discuss the effects that inductance has on the operation of a DC circuit.

Current Transitions in a Resistive DC Circuit

When power is applied to a resistive DC circuit, the current instantly jumps to its maximum value. This point is illustrated in Figure 11.14. The circuit in Figure 11.14a contains a DC source, a switch, and a load. As shown in Figure 11.14b, the load current equals zero prior to closing the switch (as you would expect). When the switch is closed, the load current instantly increases to an assumed maximum value of 2 A. When the switch re-opens, the current level drops back to its initial level (0 A). The changes in current that occur when the switch is opened and closed are called **TRANSITIONS**. The transition from 0 A to 2 A at t_1 is called a *positive-going transition*. The transition from 2 A to 0 A at t_2 is called a *negative-going transition*.

TRANSITION A sudden change in current from one level to another.

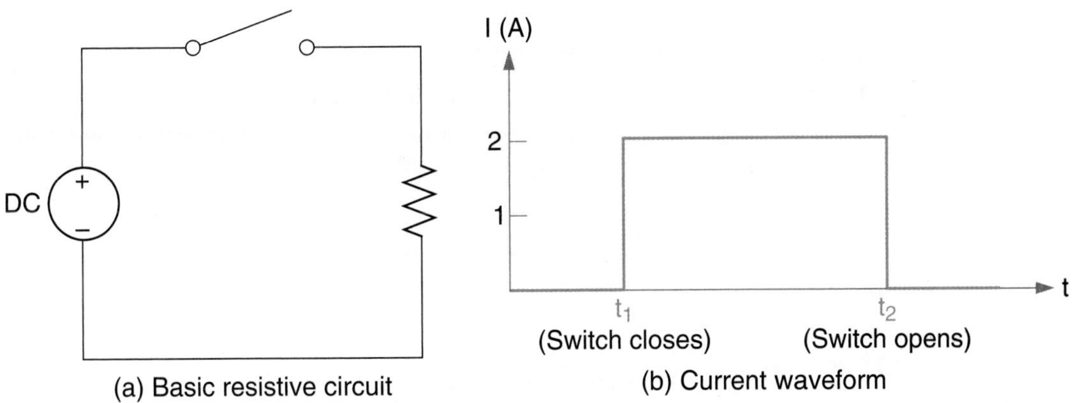

FIGURE 11.14 Current transitions in a resistive circuit.

Current Transitions in a Resistive-Inductive DC Circuit

The current transitions in a resistive-inductive (RL) circuit are more gradual than those in a purely resistive circuit, as illustrated in Figure 11.15.

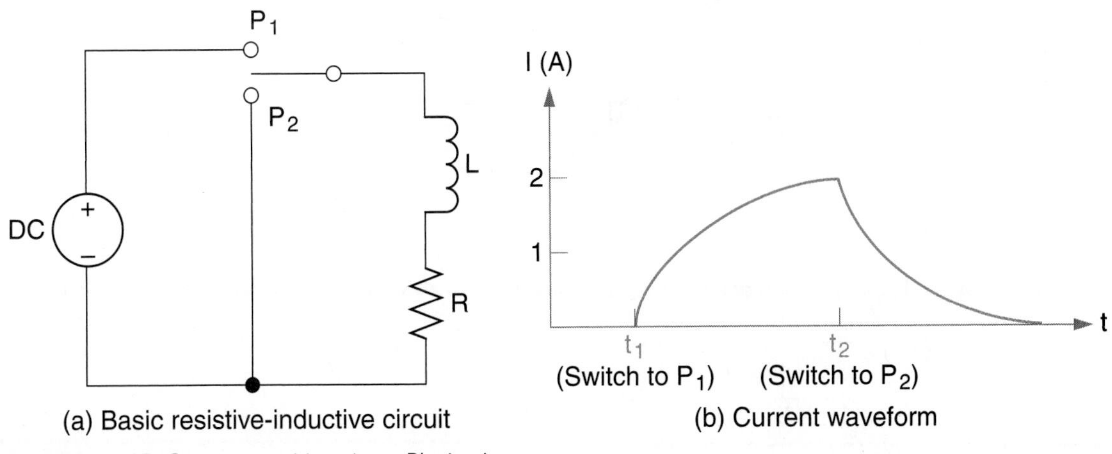

FIGURE 11.15 Current transitions in an RL circuit.

The circuit in Figure 11.15a has a DC source, a SPDT switch, an inductor, and a resistor. Here is what happens when the switch in the RL circuit is set to position (1): With the source connected to the circuit, the current tries to make a positive-going transition like the one shown in Figure 11.14b. However, the inductor opposes any change in current, so the circuit current cannot instantly rise to its maximum value (as it did in the purely resistive circuit). Rather, the current rises to its maximum value over the time period between t_1 and t_2. Over the same time period, the increasing inductor current generates an expanding electromagnetic field around the inductor. At t_2, the inductor current and the electromagnetic field strength both reach their maximum values.

Here is what happens when the switch in the RL circuit is set to position (2): With the source disconnected from the circuit, the current tries to drop to 0 A instantly as it does in a purely resistive circuit. However, the flux generated by the inductor begins to collapse back through the component. As a result, the following occurs when the switch is set to position (2):

1. The magnetic field surrounding the inductor collapses.
2. The collapsing magnetic flux generates the current shown in the figure.
3. The current generated by the inductor decays as its magnetic field decays.

This sequence of events produces the current waveform that follows t_2 in Figure 11.15b.

The Universal Rise Curve

The shape of the curve in Figure 11.15b is a result of the inductor's opposition to any increase in current. This means that the shape of the curve is the same regardless of the value of the circuit current. For example, if we double the source voltage in Figure 11.15a, the maximum inductor current also doubles. However, the shape of the curve remains unchanged.

RISE CURVE The portion of the RL curve that represents the increase in current that occurs after power is applied to the circuit.

The **RISE CURVE** for any resistive-inductive DC circuit can be drawn as shown in Figure 11.16. The curve is described as a *universal curve* because it applies to all resistive-inductive DC circuits.

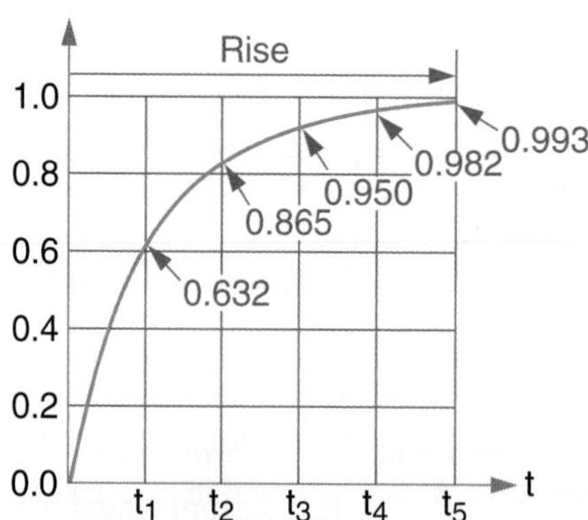

FIGURE 11.16 Current rise curve.

The curve in Figure 11.16 is divided into five equal time intervals; the first from t_0 to t_1, the second from t_1 to t_2, and so on. The decimal value at the end of each interval is used, along with the maximum circuit current, to calculate the current at that point on the curve. The current at any point on the curve is calculated as shown in Example 11.7.

EXAMPLE 11.7

The current through a series RL circuit has a maximum value of 10 A. What are the values at t_1 and t_2 on the current rise curve?

SOLUTION

At t_1, the inductor current (I_L) has a value of

$$I_L = 0.632 \times 10 \text{ A}$$
$$= 6.32 \text{ A}$$

At t_2, the inductor current has a value of

$$I_L = 0.865 \times 10 \text{ A}$$
$$= 8.65 \text{ A}$$

PRACTICE PROBLEM 11.7

For the circuit described above, calculate the current values at t_3, t_4, and t_5.

The multipliers in Figure 11.16 work with all series RL circuits. For example, if the circuit described in Example 11.7 had a maximum current of 30 A, the inductor current values at t_1 and t_2 would be found as

$$I_L = 0.632 \times 30 \text{ A}$$
$$= 18.96 \text{ A}$$

and

$$I_L = 0.865 \times 30 \text{ A}$$
$$= 29.95 \text{ A}$$

The RL Time Constant (τ)

Each time interval in the rise curve represents a real-time value called a **TIME CONSTANT (τ)**. The term *constant* is used because τ is determined by the circuit inductance and resistance. Therefore, its value is constant for a given circuit.

TIME CONSTANT (τ) One of five equal time periods in the rise and decay curves for an RL circuit.

For a series resistive-inductive circuit, the value of the RL time constant is calculated using

(11.12)

$$\tau = \frac{L}{R}$$

where

τ = the duration of each time interval, in seconds
L = the total series inductance
R = the total series resistance

Example 11.8 demonstrates the use of this equation.

EXAMPLE 11.8

A series RL circuit has a 10 mH inductor and a 100 Ω resistor. What is the time constant for the circuit?

SOLUTION

The circuit time constant has a value of

$$\tau = \frac{L}{R}$$

$$= \frac{10 \text{ mH}}{100 \ \Omega}$$

$$= 100 \ \mu\text{s}$$

PRACTICE PROBLEM 11.8

Calculate the time constant for a series RL circuit that contains a 330 mH inductor and a 750 Ω resistor.

Figure 11.17 shows the rise curve for the circuit in Example 11.8. As you can see, each time interval represents the value of the circuit time constant (100 μs). The total time required for the circuit current to complete its rise is approximately equal to five time constants (5τ); in this case, 500 μs.

The Universal Decay Curve

DECAY CURVE The portion of the RL curve that represents the decrease in current that occurs after power is disconnected from the circuit.

The portion of the curve in Figure 11.15b that represents the decrease in current (after t_2) is called a **DECAY CURVE**. The decay curve is a result of the magnetic field collapsing back into the inductor. Like the rise curve, the shape of the decay curve is not affected by the circuit values. For example, if we double the source voltage in Figure 11.15a, the current at the start of the decay cycle is twice the value shown in Figure 11.15b. Even so, the shape of the decay curve remains unchanged.

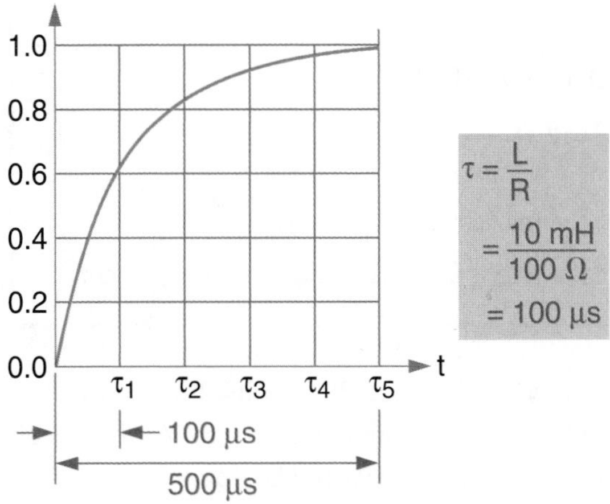

FIGURE 11.17 Current rise curve.

The current decay is represented by the *universal decay curve* shown in Figure 11.18. Like the rise curve described earlier, the curve in Figure 11.18 is called a universal curve because it applies to all resistive-inductive DC circuits.

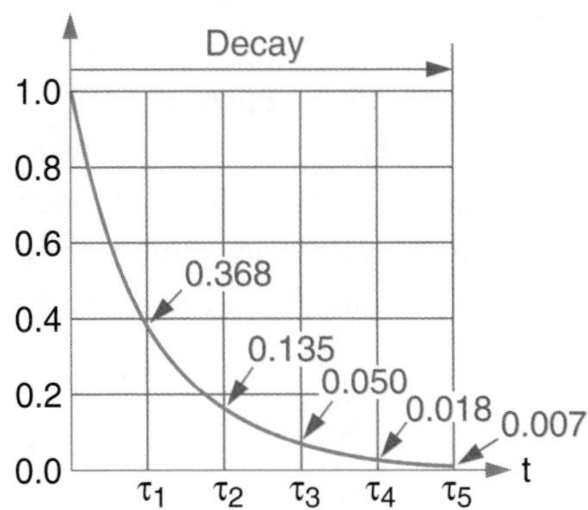

FIGURE 11.18 Current decay curve.

The curve in Figure 11.18 is divided into five equal time intervals; the first ends at τ_1, the second ends at τ_2, and so on. The decimal value at the end of each interval is used, along with the maximum circuit current, to calculate the current at that point on the curve. The current at each point on the curve is calculated as shown in Example 11.9.

EXAMPLE 11.9

The current through a series RL circuit has a maximum value of 10 A. What are the values at τ_1 and τ_2 on the current decay curve?

SOLUTION

At τ_1 on the decay curve, the inductor current has a value of

$$
\begin{aligned}
I_L &= 0.368 \times 10 \text{ A} \\
&= 3.68 \text{ A}
\end{aligned}
$$

At τ_2, the inductor current has a value of

$$I_L = 0.135 \times 10\text{ A}$$
$$= 1.35\text{ A}$$

PRACTICE PROBLEM 11.9

For the circuit described above, calculate the current values at τ_3, τ_4, and τ_5.

The multipliers in Figure 11.18 apply to all series RL circuits. For example, if the circuit described in Example 11.9 had a maximum current of 30 A, the circuit current values at τ_1 and τ_2 on the current decay curve would be found as

$$I_L = 0.368 \times 30\text{ A}$$
$$= 11.04\text{ A}$$

and

$$I_L = 0.135 \times 30\text{ A}$$
$$= 4.05\text{ A}$$

As shown in Figure 11.18, each time interval in the decay curve equals the time constant (τ) for the circuit. In fact, the rise curve and decay curve are essentially mirror images of each other. For example, Figure 11.19 shows the combined rise and decay curves for the current in a series RL circuit with a maximum current of 100 mA. Assuming the circuit has values of L = 300 mH and R = 75 Ω, the time constant for the circuit is calculated as follows:

$$\tau = \frac{L}{R}$$
$$= \frac{300\text{ mH}}{75\ \Omega}$$
$$= 4\text{ ms}$$

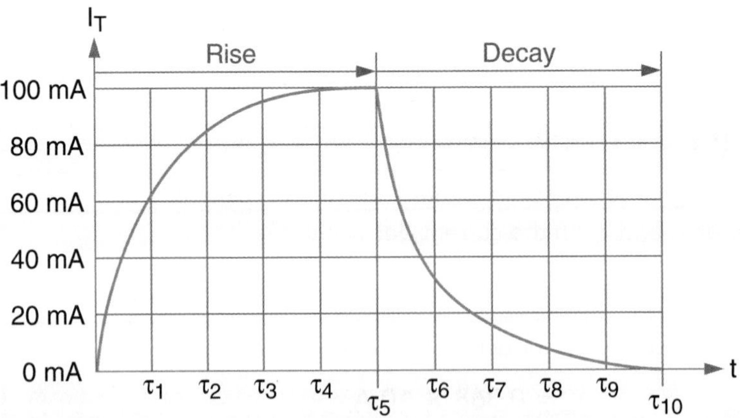

FIGURE 11.19 The combined rise and decay curves for an RL circuit.

The current rise and decay curves each take a total of 5×4 ms = 20 ms. Thus, the two halves of the curve have equal time spans and peak values.

One Final Note

The purpose of this section has been to demonstrate the characteristics of inductance. Using a DC circuit, we were able to see how the component opposes a change in current and stores energy in an electromagnetic field. As your education continues, you'll learn that inductance plays a role in a wide variety of applications.

Measuring inductor resistance to detect a short is often of little value, especially if only a few turns are shorted. This is because the few shorted turns will not produce a measurable change in the total component resistance.

IN THE FIELD

The principles of electromagnetic induction are applied in many electrical devices and systems, including induction motors, generators, transformers, current transformer (CT) meters, Hall-effect meters, and clamp meters.

IN THE FIELD

PROGRESS CHECK

1. What is a *transition*? What two types of transitions occur in a DC circuit when the switch is closed and opened?
2. Refer to Figure 11.15. What causes the shape of the current transitions?
3. What is the relationship between the shape of a current rise (or decay) curve and the maximum circuit current?
4. What is a universal curve? What purpose does it serve?
5. What is implied by the word "constant" in the term *time constant*?
6. How is the value of the time constant in a series RL circuit found?
7. How many time constants are required to complete a rise (or decay) curve?

11.4 TYPES OF INDUCTORS

There are many types of inductors that are designed for specific applications. Though these inductors each have unique characteristics, they all work according to the principles covered in this chapter.

Iron-Core, Air-Core, and Ferrite-core Inductors

Inductors have iron, ferrite, or air cores. In schematic diagrams, the cores are identified as shown in Figure 11.20.

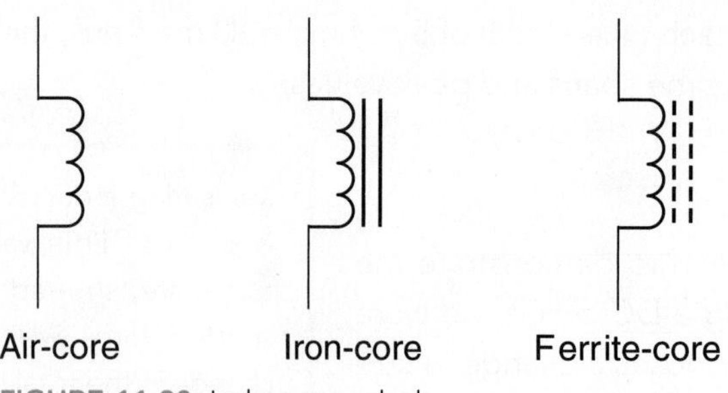

FIGURE 11.20 Inductor symbols.

Iron-core inductors are better suited for use in low-frequency applications, such as DC and AC power circuits, because the DC winding resistance of an iron-core inductor is much lower than that of an equal-value air-core inductor.

In Chapter 10, you learned that *relative permeability* (μ_r) is a ratio of the permeability of a material to that of air. The relative permeability of magnetic iron is approximately 200. Because magnetic iron has 200 times the permeability of air, it takes far fewer turns (and therefore, a shorter length of wire) to produce a given value of inductance using an iron core. Because of its shorter length, the winding resistance of an iron-core inductor is always significantly less than that of an equal-value air-core inductor.

The relatively low winding resistance of an iron-core inductor makes it better suited for use in high-current DC and AC applications. With its low winding resistance, a given iron-core inductor dissipates much less power than an equal-value air-core inductor. (Power dissipation is always an important consideration in high-current circuitry.) At the same time, iron cores can experience significant power losses when operated at higher frequencies, making air-core inductors more suitable for such applications.

Air Cores

The wire turns of an air-core inductor are either self-supporting or physically supported by a nonmagnetic form, such as a ceramic tube. An inductor formed around a ceramic tube is sometimes called a *ceramic core inductor.* A ceramic core holds its shape when heated and has no magnetic properties. As such, it provides physical support for the coil without affecting the component's value.

Ferrite-core Inductors

FERRITE CORE A magnetic core that consists primarily of iron oxide.

A **FERRITE CORE** is a magnetic core that consists primarily of iron oxide. A ferrite-core inductor typically has much higher permeability than does a comparable magnetic-iron core inductor. As such, higher values of inductance can be obtained with ferrite cores than with comparable magnetic-iron cores. Ferrite cores can be solid or powdered.

Toroids

A **TOROID** is a doughnut-shaped magnetic core, as shown in Figure 11.21. Because of its shape, nearly all of the flux produced by the coil remains in the core. That is, very little of the flux is lost to the air surrounding the coil. As a result, toroids have:

- Inductance values that are much greater than their physical size would indicate.
- Extremely accurate rated values.

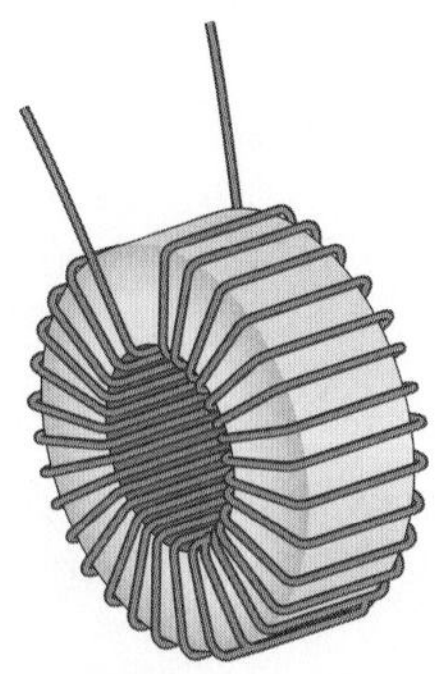

FIGURE 11.21 A toroid.

TOROID A doughnut-shaped magnetic core.

Toroids are smaller, by volume and weight, and produce less electromagnetic interference (EMI) than other types of inductors. Also, their windings cool better because their larger surface area and shape lead to near complete magnetic field cancellation outside of the coil.

IN THE FIELD

PROGRESS CHECK

1. Why are iron-core inductors better suited for low-frequency applications than air-core inductors?
2. What are the two basic types of air-core inductors?
3. What is a toroid? What results from the shape of the toroid core?

11.5 SUMMARY

Here is a summary of the major points that were covered in this chapter:

Introduction to Inductance

- *Inductance* is the electrical property that opposes any change in current.
  - The current through an inductive component produces magnetic flux.
  - The magnetic flux induces a voltage across the component that opposes any change in the current level.
- An *inductor* is a component designed to provide a specific amount of inductance. An inductor can
  - Store energy in an electromagnetic field
  - Oppose any change in current
- The relationship between current and any resulting magnetic flux can be summarized as follows:
  - The current through a wire produces magnetic lines of force (flux).
  - When a wire is wrapped into a coil, the lines of force generated by the current join together to form a magnetic field. The strength of the magnetic field is determined in part by the level of current and the number of turns (loops) in the coil.
    - These two factors are combined into a single value called the *ampere-turn (NI)*.

    - The polarity of the flux generated by a current is determined by the current direction. That is, if the current through a coil reverses direction, so do the poles of the resulting magnetic field.
- Flux density varies directly with inductor current.
    - Flux density increases when inductor current increases.
    - Flux density decreases when inductor current decreases.
- The first three of Faraday's laws of induction can be summarized as follows:
    - Law 1: To induce a voltage across a wire, there must be a relative motion between the wire and the magnetic field.
    - Law 2: The voltage induced is proportional to the rate of change in magnetic flux encountered by the wire.
        - The greater the rate of change in current, the greater the voltage induced across the component.
    - Law 3: When a wire is cut by 10^8 perpendicular lines of force per second, one volt is induced across that wire.
- The principles of inductor operation can be summarized as follows:
    - The current through an inductor generates magnetic flux. The amount of flux varies directly with inductor current.
    - As inductor current increases, the magnetic field expands. As inductor current decreases, the magnetic field collapses.
    - The expanding and collapsing magnetic flux cuts through the stationary inductor.
    - The flux cutting through the coils induces a voltage across the inductor.
- According to Lenz's law, an induced voltage always opposes its source.
    - Because it always opposes a change in coil current, induced voltage is referred to as *counter emf.*
- The henry (H) is the unit of measure of inductance.
    - A change in current equal to 1 A/s induces 1 V across 1 H of inductance.
- The value of an inductor is:
    - Directly proportional to the permeability of its core
    - Directly proportional to the number of turns it has
    - Inversely proportional to its length

Mutual Inductance

- When the magnetic field produced by one coil induces a voltage across another coil, there is mutual inductance between the coils.
- The amount of mutual inductance between two coils depends on
    - The distance between the coils
    - The values of the inductors
    - The physical orientation of the coils.
- The coefficient of coupling (k) between two coils indicates the degree of coupling that occurs between the two.
    - The value of k has a range of 0 to 1.
        - When none of the flux produced by one coil cuts through the other, $k = 0$.
        - When all of the flux produced by one coil cuts through the other, $k = 1$. This condition is called *unity coupling.*
- When inductors are connected in series, the total inductance equals the sum of the individual component values (assuming no mutual inductance is present).
    - Total series inductance is calculated in the same fashion as total series resistance.

- When inductors are connected in parallel, the total inductance is lower than any of the branch inductance values.
  - Total parallel inductance is calculated in the same fashion as total parallel resistance.

Inductors in DC Circuits

- When the power switch in a purely resistive series circuit is:
  - Closed (i.e., power is connected), the voltage and current levels instantly increase to their maximum levels.
  - Re-opened (i.e., power is disconnected), the voltage and current levels instantly drop to 0 V and 0 A.
- Abrupt changes in voltage and current levels are called *transitions.*
  - Transitions are described as being positive-going or negative-going.
- When the power switch in a resistive-inductive (RL) series circuit is closed:
  - The current begins to make a positive-going transition.
  - The inductor slows the increase in current.
  - The increasing current generates an electromagnetic field around the inductor.
- The graphical representation of the rise in inductor current is called a *rise curve.*
  - Because the curve has the same shape regardless of circuit component values and maximum current levels, it is also called a *universal curve.*
- The universal rise curve is divided into five equal time intervals.
  - During the five time intervals, the circuit current increases from 0 A to its maximum value.
  - The curve contains constants (multipliers) that are used to calculate the total circuit current at the end of each time interval.
- Each time interval on the rise curve represents a real-time value called a *time constant (τ).*
  - The time constant for a series RL circuit is calculated using $\tau = L/R$.
  - It takes 5 time constants for the current in a series RL circuit to reach its maximum value.
- When the power switch in a resistive-inductive (RL) series circuit is re-opened and a discharge current path is provided:
  - The voltage across the inductor reverses.
  - The electromagnetic field collapses back into the inductor.
  - The inductor generates a decreasing current through the series resistance.
- The graphical representation of the decrease in inductor current is called a *decay curve.*
  - Because the curve has the same shape regardless of circuit component values and maximum current levels, it is also called a *universal curve.*
- The universal decay curve is divided into five equal time intervals.
  - During the five time intervals, the circuit current decreases from its maximum value to 0 A.
  - The curve contains constants (multipliers) that are used to calculate the total circuit current at the end of each time interval.
- Each time interval on the decay curve represents a real-time value called a *time constant (τ).*
  - The time constant for a series RL circuit is calculated using $\tau = L/R$.
  - It takes 5 time constants for the current in a series RL circuit to drop from its maximum value to 0 A.

Types of Inductors

- Iron-core inductors are better suited than air-core inductors for *low-frequency* applications.
  - Compared to equal-value air-core inductors, iron-core inductors tend to have:
    - Higher inductance values
    - Lower winding resistance
- Air-core inductors are better suited to high-frequency applications than are iron-core inductors.
- The coil of an air-core inductor is self-supporting or supported by a nonmetallic form like a ceramic tube.
  - Air-core inductors built with ceramic tubes are often called *ceramic-core inductors.*
- A ferrite core is a magnetic core that consists primarily of iron oxide.
  - Ferrite cores have much higher permeability (and therefore, inductance) than comparable magnetic-iron cores.
- A toroid is a doughnut-shaped magnetic core. Toroid coils have:
  - High inductance values for their physical size.
  - Extremely accurate rated values.

CHAPTER REVIEW

1. An inductor ________.
   a) provides a specific value of inductance
   b) stores energy in a magnetic field
   c) opposes any change in current
   d) all of the above

2. Inductors are *not* found in ________.
   a) generators
   b) optocouplers
   c) motors
   d) relays

3. The strength of the magnetic field around a coil is determined by ________ and ________.
   a) the magnitude of current, its direction
   b) the magnitude of current, the number of turns
   c) the direction of current, the number of turns
   d) the rate of change of current, its magnitude

4. Flux density is *inversely* proportional to ________.
   a) the number of turns
   b) the permeability of the core
   c) the length of the inductor
   d) the magnitude of the current

5. In order to induce a voltage across a coil ________.
   a) the coil must be cut by at least 10^8 lines of force per second
   b) the coil must be in motion

c) current must be increasing

d) there must be relative motion between the coil and the magnetic field

6. A coil with 200 turns is cut perpendicularly by 200 mWb/second of magnetic flux. The voltage induced across this inductor will be ________.

   a) 40 V

   b) 400 V

   c) 400 mV

   d) 0 V

7. The polarity of the voltage induced in the inductor described in question 6 is determined by ________.

   a) the magnitude of the current

   b) the permeability of the coil

   c) whether the magnetic field is expanding or collapsing

   d) the direction of the current

8. The production of a voltage across an inductor as a result of its own expanding or contracting magnetic field is called ________.

   a) counter emf

   b) self-inductance

   c) flux density

   d) Faraday's second law

9. Lenz's law states that the induced voltage across an inductor ________.

   a) always opposes the applied voltage

   b) always opposes the current through the inductor

   c) always opposes any change in current

   d) changes polarity as current changes direction

10. The current through a 470 mH inductor is changing at a rate of 2.6 A/s. The voltage induced across the inductor is ________.

    a) 181 mV

    b) 1.22 V

    c) 5.53 V

    d) 122 mV

11. Two coils are identical, but coil 1 has 79% as many turns as coil 2. If coil 1 is known to be exactly 21.9 mH then coil 2 is ________.

    a) 35.1 mH

    b) 27.7 mH

    c) 17.3 mH

    d) 13.7 mH

12. Two coils are identical but one has a diameter that is 1.2 times that of the other. The inductance of the coil with the larger diameter is ________ times ________ than the other.

    a) 1.44, smaller

    b) 1.44, larger

c) 1.2, larger
d) 0.83, larger

13. The voltage that is produced in an inductor as a result of the changing magnetic field of another inductor is referred to as ________.
    a) self-inductance
    b) common inductance
    c) mutual inductance
    d) proximity inductance

14. The coefficient of coupling between two inductors is ________.
    a) at right angles
    b) always less than one
    c) always greater than one
    d) dependent on polarity

15. When two inductors are at right angles to each other, the value of k is ________.
    a) close to unity
    b) close to zero
    c) very high
    d) negative

16. A 10 mH inductor and a 4.7 mH inductor are connected in series. The mutual inductance between them is +0.4 mH. The total inductance of this circuit is ________.
    a) 15.1 mH
    b) 14.3 mH
    c) 15.5 mH
    d) 13.9 mH

17. The two inductors listed in question 16 are connected in parallel and have the same mutual inductance between them. The total inductance of this circuit is ________.
    a) 2.22 mH
    b) 3.64 mH
    c) 3.42 mH
    d) 2.66 mH

18. Unity coupling occurs when ________.
    a) there is very little coupling between two inductors
    b) the inductors are of equal value
    c) $k = 1$
    d) $L_M = 1$

19. The current in an RL circuit changes from 2 A to 1.5 A. This would result in a ________.
    a) current surge
    b) current spike

c) positive-going transition

d) negative-going transition

20. The voltage induced across an inductor as its magnetic field collapses is called ________.

    a) reverse voltage
    b) counter voltage
    c) counter emf
    d) transition voltage

21. The voltage induced across an inductor as its magnetic field expands is called ________.

    a) forward voltage
    b) aiding voltage
    c) counter emf
    d) transition voltage

22. 10 V is applied to a given RL circuit producing a rise curve. If 20 V had been applied to the same circuit, the shape of the rise curve would ________.

    a) be the same
    b) be more extended
    c) be more compressed
    d) take longer to occur

23. A circuit is composed of a 10 mH inductor in series with a 33 kΩ resistor. The time constant for this circuit is ________.

    a) 330 s
    b) 3.03 μs
    c) 3.3 μs
    d) 0.303 μs

24. The resistor in question 23 is changed to 2.2 kΩ. This means that the time required for circuit current to reach its maximum value has ________.

    a) increased
    b) decreased
    c) remained the same
    d) remained the same, but the maximum current has increased

25. Assume that the current in the circuit described in question 24 starts at 0 A. It would take ________ for circuit current to reach its maximum if 10 V is applied.

    a) 4.55 μs
    b) 22.7 μs
    c) 2.27 μs
    d) 45.5 μs

26. A series RL circuit contains a 200 mH coil in series with a 470 Ω resistor. A switch is closed, applying 24 V to the circuit. Ignoring any winding resistance of the coil, the value of circuit current after 2 time constants is ________.
    a) 4.42 mA
    b) 32.3 mA
    c) 44.2 mA
    d) 323 mA

27. Assume that a circuit has a time constant of 20 ms. The circuit current is allowed to rise for only 2 time constants. This means that it will take ________ for the current to decay and reach 0 A.
    a) 3 time constants
    b) 5 time constants
    c) 60 ms
    d) 100 ms
    e) a and c
    f) b and d

28. If an iron-core inductor and an air-core inductor are identical in every other way, ________ has the most inductance.
    a) the air-core inductor
    b) the iron-core inductor
    c) neither inductor

29. For a given inductance a(n) ________-core inductor has more resistance than an ________-core.
    a) air, iron
    b) iron, air
    c) ferrite, iron
    d) ferrite, air

30. A ceramic-core inductor ________.
    a) is the same as an air-core inductor
    b) is not a good choice for high-power applications
    c) has relatively high winding resistance
    d) all of the above

31. Assume that an inductor is to be used in an environment that is very sensitive to magnetic fields. A(n) ________ would be a good choice for this application.
    a) permalloy core inductor
    b) ferrite-core inductor
    c) inductor wound on a toroid core
    d) ceramic-core inductor

PRACTICE PROBLEMS

1. Calculate the value of V_{IND} developed across a 33 mH coil when the rate of change in component current equals 200 mA/s.
2. Calculate the value of V_{IND} developed across a 4.7 H coil when the rate of change in component current equals 1.5 A/s.
3. Calculate the value of V_{IND} developed across a 100 mH coil when the rate of change in component current equals 25 mA/ms. (Hint: Convert the rate of change in current into A/s.)
4. Calculate the value of V_{IND} developed across a 47 mH coil when the rate of change in component current equals 4 A/s.
5. Calculate the value of the air-core inductor in Figure 11.22a.
6. Calculate the value of the air-core inductor in Figure 11.22b.

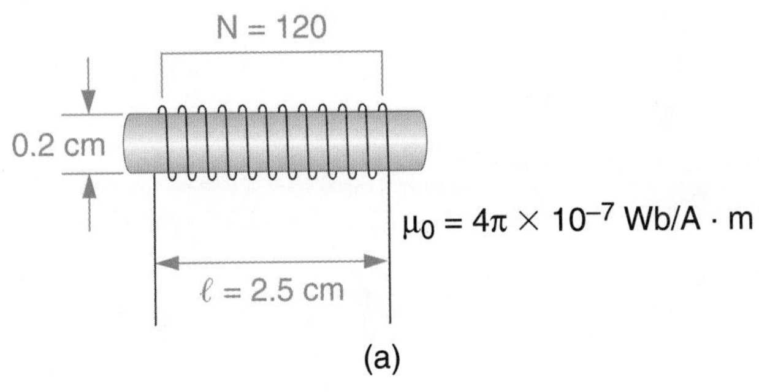

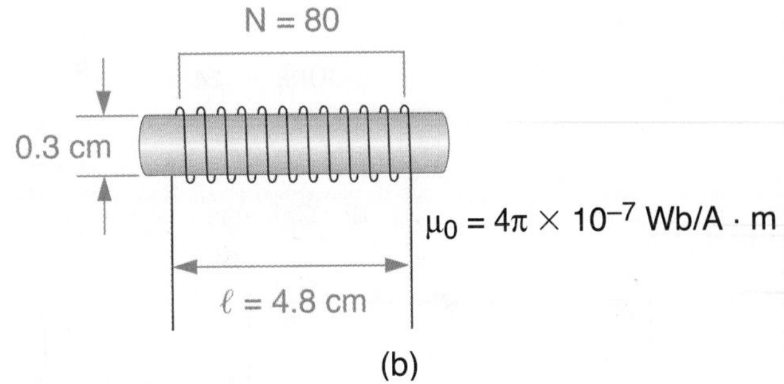

FIGURE 11.22

7. Calculate the value of the inductor shown in Figure 11.23a.
8. Calculate the value of the inductor shown in Figure 11.23b.

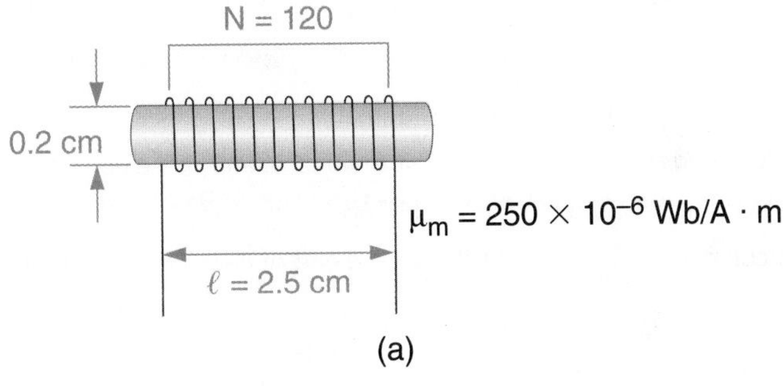

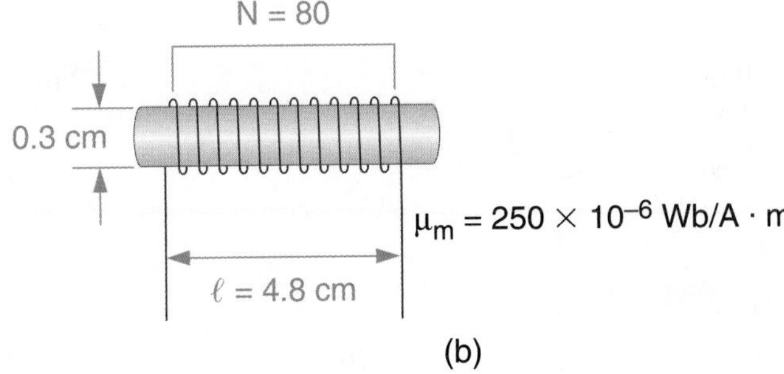

FIGURE 11.23

9. Calculate the total inductance in Figure 11.24a.
10. Calculate the total inductance in Figure 11.24b.

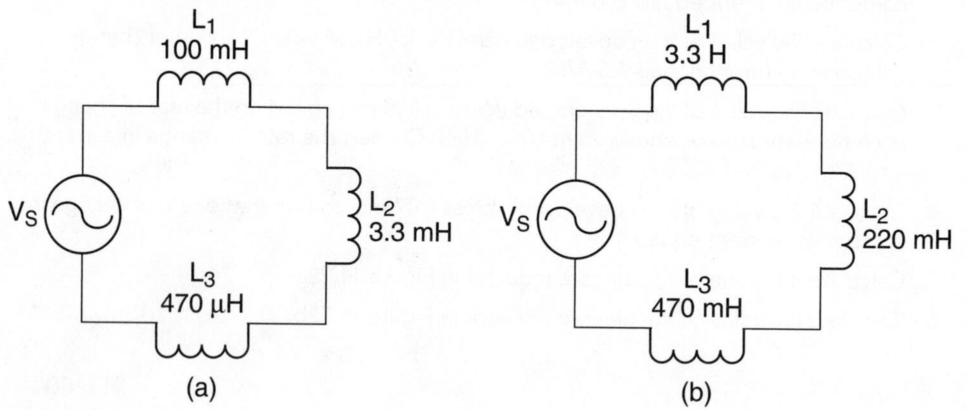

FIGURE 11.24

11. Calculate the total inductance in Figure 11.25a.
12. Calculate the total inductance in Figure 11.25b.

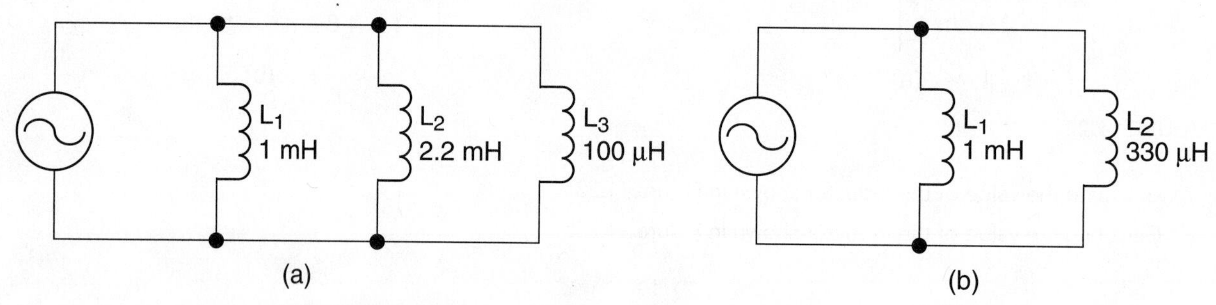

FIGURE 11.25

13. Determine the time constant for the circuit in Figure 11.26a.
14. Determine the value of circuit current in Figure 11.26a at the end of each time constant.
15. Determine the time constant for the circuit in Figure 11.26b.
16. Determine the value of circuit current in Figure 11.26b at the end of each time constant.

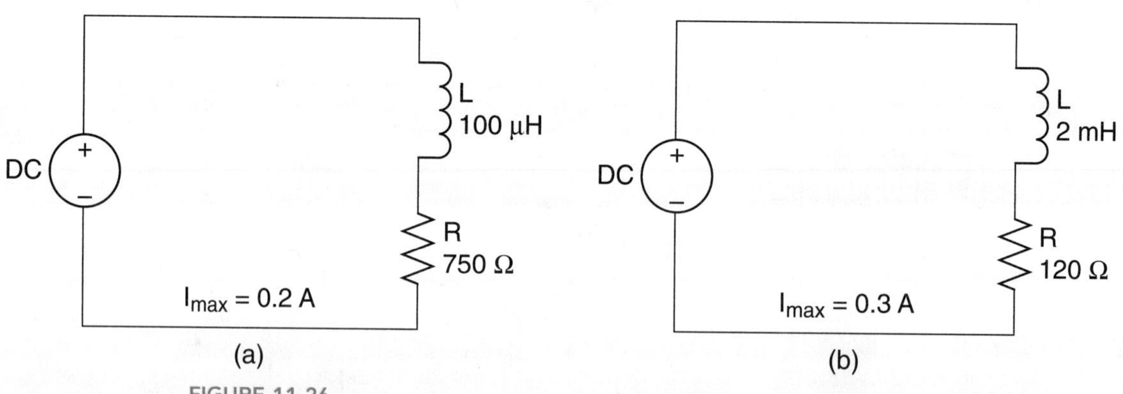

FIGURE 11.26

ANSWERS TO THE EXAMPLE PRACTICE PROBLEMS

11.1	7.5 V
11.2	6.94 μH
11.3	3.73 mH
11.4	162 μH
11.5	4.63 mH
11.6	660 mH
11.7	9.5 A, 9.82 A, 9.93 A
11.8	440 μs
11.9	50 mA, 18 mA, 7 mA

DA3

CHAPTER 12

CAPACITANCE AND CAPACITORS

PURPOSE

In Chapter 11, you were introduced to inductance, a property that opposes any change in current. In this chapter, we are going to examine capacitance, an electrical property that is the opposite of inductance in many ways.

Technically defined, **CAPACITANCE** is the ability of a component to store energy in the form of an electrostatic charge. Based on its operating characteristics, capacitance is also described as an electrical property that opposes any change in voltage. As you will learn, both of these descriptions accurately describe the property of capacitance.

A **CAPACITOR** is a component designed to provide a specific value of capacitance. Capacitors:

- Store energy in an electrostatic field
- Oppose any change in voltage

These characteristics make the capacitor useful in a variety of applications.

KEY TERMS

The following terms are introduced and defined in this chapter on the pages indicated:

OBJECTIVES

After completing this chapter, you should be able to:

1. Define capacitance.
2. Describe the physical makeup of a capacitor.
3. Describe the charge and discharge characteristics of capacitors.
4. Discuss capacity and its unit of measure.

5. Describe the relationship between the physical construction of a capacitor and its value.
6. Calculate the total capacitance of any number of capacitors connected in series or parallel.
7. Describe the voltage transitions in a resistive-capacitive DC circuit.
8. Describe the universal voltage curve for a resistive-capacitive DC circuit.
9. Calculate the capacitor voltage in a resistive-capacitive DC circuit at any interval on the charge or discharge curve.
10. Calculate the time constant for a resistive-capacitive DC circuit.
11. Compare and contrast the commonly used types of capacitors.

CAPACITANCE The ability of a component to store energy in the form of an electrostatic charge.

CAPACITOR A component that provides a specific value of capacitance.

12.1 CAPACITANCE AND CAPACITORS: AN OVERVIEW

The simplest approach to understanding capacitance begins with a description of the capacitor. In this section, we will look at the construction of a basic capacitor and what happens when it is connected to a DC voltage source. Using this circuit as an example, we will then discuss the property of capacitance.

Capacitor Construction

PLATES The conductive surfaces of a capacitor.

DIELECTRIC The insulating layer in a capacitor.

A capacitor is made up of two conductive surfaces that are separated by an insulating layer, as shown in Figure 12.1a. The conductive surfaces are referred to as **PLATES** and the insulating layer is referred to as the **DIELECTRIC**.

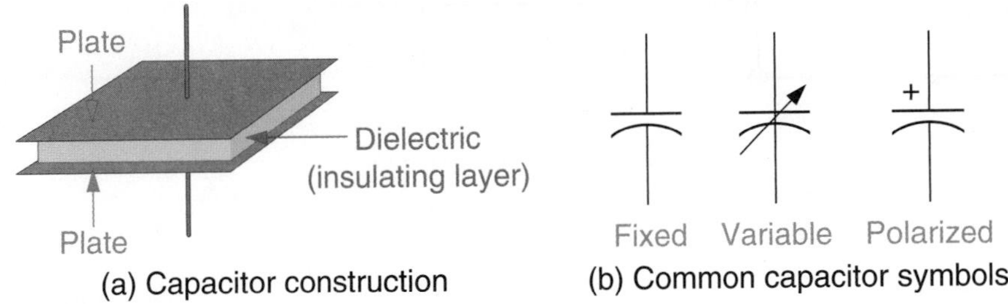

FIGURE 12.1 Capacitor construction and symbols.

Several capacitor schematic symbols are shown in Figure 12.1b. Note the similarity between the capacitor symbol and its physical construction.

Capacitor Charge

ELECTROSTATIC CHARGE The attraction between charges that are stationary.

When a capacitor is connected to a DC voltage source, an **ELECTROSTATIC CHARGE** develops across the plates of the component. This concept is illustrated in Figure 12.2. In Figure 12.2a, the capacitor is connected to a DC voltage source via a switch (SW1). Initially, the capacitor has a 0 V difference of potential across its plates. When SW1 is closed (as shown in Figure 12.2b), the source voltage is applied to the component. This connection causes several things to happen simultaneously:

- The positive terminal of the source draws electrons away from the positive plate of the capacitor, leaving an excess of positive charges on this plate.
- The negative terminal of the source forces electrons toward the negative plate of the capacitor, producing an excess of negative charges on this plate.

The negative and positive charges on the plates of the capacitor exert a force of attraction on each other. This force of attraction extends through the dielectric, holding the positive and negative charges at the plates.

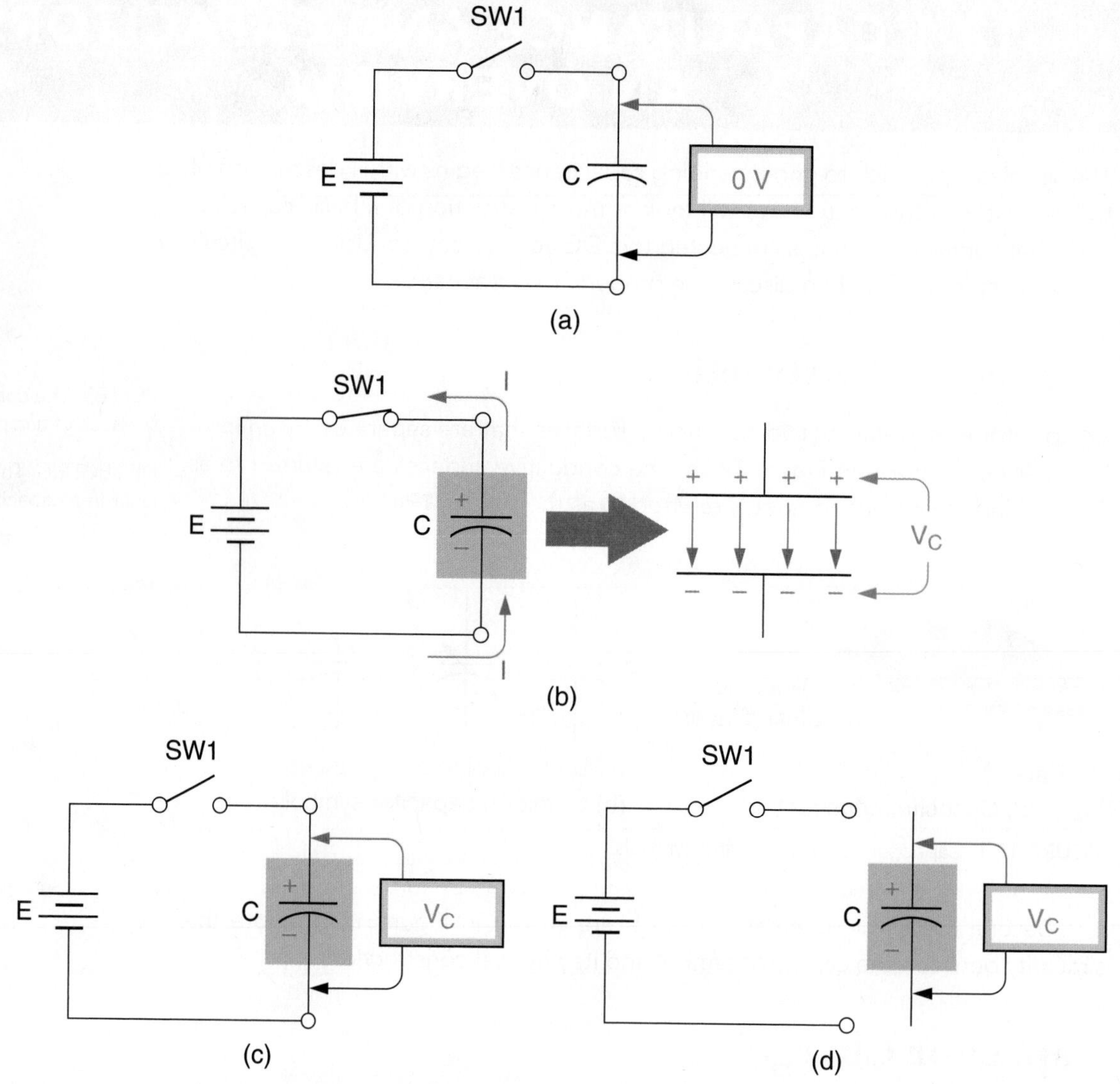

FIGURE 12.2 Charging a capacitor.

CAPACITY The ability of a capacitor to store a specific amount of charge (per volt applied).

The total charge that a capacitor can store is determined by its **CAPACITY**. Once the capacitor reaches its capacity, the flow of charge shown in Figure 12.2b ceases. That is, the charged capacitor blocks the flow of charge (current) in the circuit. Because the charged capacitor blocks any current through the circuit, it can be viewed as a DC open. (Remember, an open is a break in a current path that blocks current and drops the applied voltage.)

If the source is disconnected from the capacitor (as shown in Figure 12.2c), the charges remain on the plates of the component. Since current cannot pass through an open switch, the positive and negative charges remain on the plates of the capacitor. Thus, the capacitor retains its "plate-to-plate" charge even after the voltage source has been disconnected. In effect, the capacitor is storing energy in the electrostatic field that is permeating the dielectric. This energy remains in the capacitor as long as there is no discharge path for the component. In fact, if the capacitor is removed from the circuit (as shown in Figure 12.2d), there are still charges being held at its plates. Note that these charges maintain a difference of potential (voltage) across the capacitor, even though it is no longer connected to a voltage source.

Capacitor Discharge

To discharge a capacitor, an external path must be provided for the flow of charge (current) between the plates of the component (like the one shown in Figure 12.3). In Figure 12.3a, the switch is set so that the voltage source is connected to the capacitor. As described earlier, the capacitor charges until it has reached its full storage capacity. At that point, the flow of charge (DC current) ceases in the circuit.

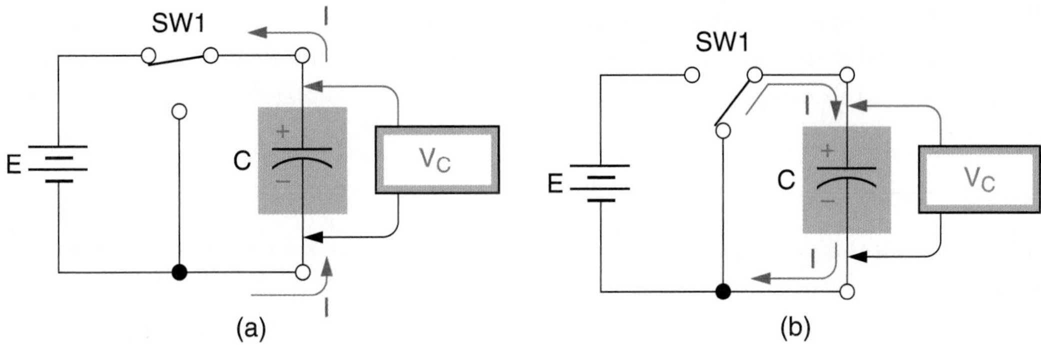

FIGURE 12.3 Discharging a capacitor.

When the switch is set to the lower position, an external current path is provided between the plates of the capacitor. As a result, the capacitor discharges as shown in Figure 12.3b. Note that the charge flows from one plate to the other until there is no longer any difference of potential across the plates. That is, the current ceases when

$$V_C = 0\text{ V}$$

Also note that the capacitor discharge current is in the opposite direction of the capacitor charge current. After the capacitor has fully discharged, the difference of potential across its plates remains at 0 V until it is reconnected to a voltage source.

There is an important point to be made: In the descriptions of capacitor charge and discharge cycles, the flow of charge is to and from the capacitor plates. Nowhere in our discussion was current described as passing *through* the capacitor. The dielectric of a capacitor is an insulator that (ideally) does not allow current to pass through the component. Rather, charge flows to and from the plates as described in this section.

A Practical Consideration

Since an external current path must be provided to discharge a capacitor, you must ensure that all capacitors have been discharged whenever working with capacitive circuits. You see, a break in the current path in a capacitive circuit could prevent the capacitor(s) from discharging. For example, refer back to Figure 12.3. An open contact in the switch could prevent the capacitor from discharging when SW1 is set to position (2). In this case, the capacitor could be storing sufficient charge to injure

STAYING SAFE Never assume that a capacitor is not holding a charge, even when power is disconnected from a circuit. Capacitors in high-power circuits can store enough charge to cause serious injury.

you (or your test equipment) should you come into contact with the component leads.

When working with a capacitive circuit, a **SHORTING TOOL** is often used to ensure that all of the capacitors are discharged. The shorting tool, which is simply a conductor with a double-insulated handle, is placed across the terminals of the component. If any residual charge remains in the component, it is shorted out by the conductor.

SHORTING TOOL A conductor with a double-insulated handle that is used to discharge a capacitor.

Capacity

Every capacitor has the ability to store a specific amount of charge per volt applied. For example, look at the capacitor shown in Figure 12.4a. The component is storing six coulombs (6 C) of charge when its plate-to-plate voltage is 2 V.

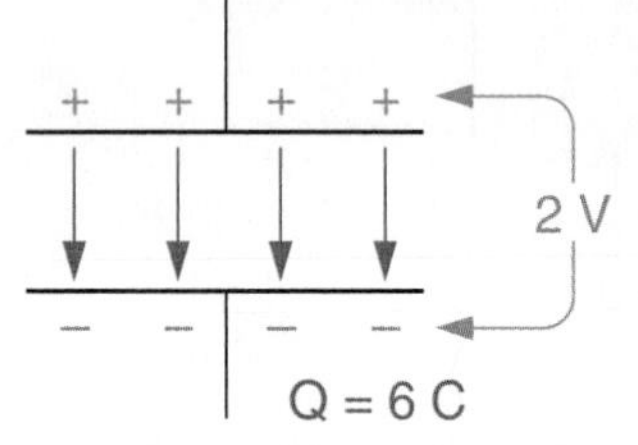

$$\text{Capacity} = \frac{Q}{V_C} = \frac{6\text{ C}}{2\text{ V}} = 3\text{ C/V}$$

(a)

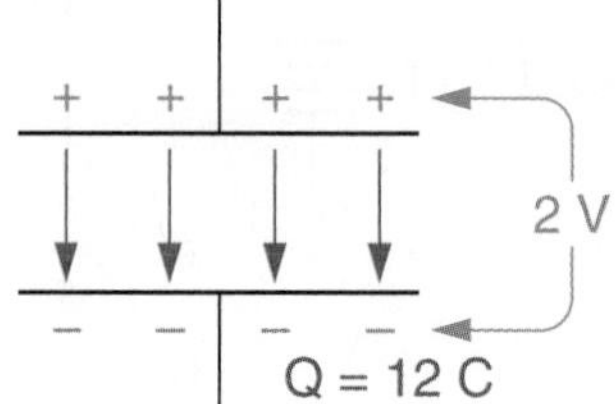

$$\text{Capacity} = \frac{Q}{V_C} = \frac{12\text{ C}}{2\text{ V}} = 6\text{ C/V}$$

(b)

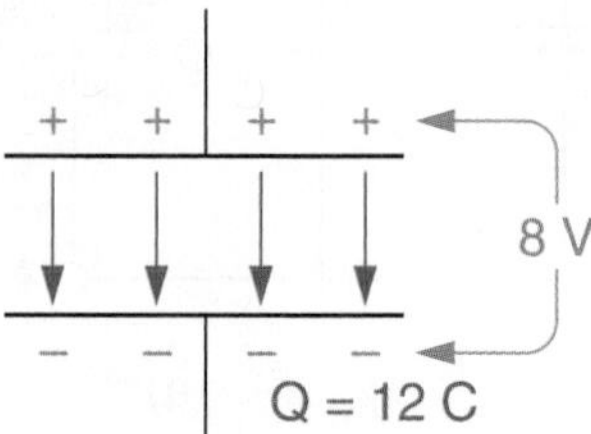

$$\text{Capacity} = \frac{Q}{V_C} = \frac{12\text{ C}}{8\text{ V}} = 1.5\text{ C/V}$$

(c)

FIGURE 12.4 Capacity.

The maximum charge a capacitor can store is called its *capacity.* (In fact, the term *capacitance* comes from the word *capacity.*) Capacity is measured in *coulombs per volt* (C/V). For example, the capacity of the component in Figure 12.4a is expressed as

$$\text{Capacity (C)} = 3\text{ C/V} \qquad \text{(coulombs per volt)}$$

Capacity is directly proportional to charge and inversely proportional to voltage. By formula,

(12.1)

$$C = \frac{Q}{V}$$

where

C = the capacity (or capacitance) of the component, in coulombs per volt (C/V)
Q = the total charge stored by the component, in coulombs (C)
V = the voltage across the capacitor, in volts (V)

The use of this equation is demonstrated under each capacitor in Figure 12.4.

The Unit of Measure for Capacitance

Capacitance is measured in **FARADS (F)**. One farad is the amount of capacitance that stores one coulomb of charge for each 1-V difference of potential across a capacitor's plates. That is,

$$1 \text{ farad} = 1 \text{ coulomb/volt}$$

FARAD The amount of capacitance that stores one coulomb of charge for each 1-V difference of potential across a capacitor's plates.

Stated mathematically,

(12.2)

$$1\text{F} = 1 \text{ C/V}$$

Example 12.1 demonstrates the relationship among C, Q, and V.

EXAMPLE 12.1

A capacitor stores 0.1 C of charge when there is 10 V across its plates. What is the value of the component?

SOLUTION

The capacitor has a value of

$$C = \frac{Q}{V}$$

$$= \frac{0.1\text{ C}}{10\text{ V}}$$

$$= 0.01\text{ C/V}$$

$$= 0.01\text{ F}$$

PRACTICE PROBLEM 12.1

A capacitor stores 0.03 C of charge when there is 20 V across its plates. Calculate the capacity (in farads) of the component.

Capacitor Ratings

Most capacitors are rated in the *picofarad* (pF) to *microfarad* (μF) range. For some reason, capacitor values in the *millifarad* (mF) range are typically expressed in thousands of microfarads. For example, one well-known U.S. manufacturer uses a 68 mF capacitor in one of its electrical products. In the product's technical diagrams, the component is identified as a 68,000 μF capacitor.

It should be noted that most capacitors have fairly poor tolerance ratings. That is, the actual value of a capacitor may vary significantly from its rated value. Even so, the poor tolerance ratings of capacitors do not present a significant problem in most electrical circuits and applications.

Capacitor Physical Characteristics

Earlier, you were told that the value of a capacitor depends on its physical characteristics. The relationship between a capacitor's physical makeup and its capacity is given as

(12.3)

$$C = (8.85 \times 10^{-12})\epsilon_r \frac{A}{d}$$

where

C = the capacity of the component, in farads (F)
(8.85×10^{-12}) = the permittivity of a vacuum, in farads per meter (F/m)
ϵ_r = the relative permittivity of the dielectric
A = the area of either plate, in square meters (m^2)
d = the distance between the plates, in meters (m)

Equation (12.3) is introduced here because it identifies the capacitor characteristics that affect the overall value of the component. Let's take a brief look at the properties contained in the equation.

Permittivity

PERMITTIVITY A measure of the ease with which lines of electrical force are established within a given material.

RELATIVE PERMITTIVITY The ratio of a material's permittivity to that of a vacuum.

PERMITTIVITY is a measure of the ease with which lines of electrical force are established within a given material. For example, in Figure 12.4, the arrows between the plates represent the force of attraction between the charges. The permittivity of the dielectric is a measure of how easily these lines of force are established within the material. In a sense, permittivity can be viewed as the electrical equivalent of magnetic permeability. **RELATIVE PERMITTIVITY** (ϵ_r) is the ratio of a material's permittivity to that of a vacuum. The product of ϵ_r and the constant in equation (12.3) equals the actual permittivity of the dielectric material.

Plate Area

Every ion takes up some amount of physical space. With this in mind, take a look at the capacitors represented in Figure 12.5. The capacitor on the right would be able to store a greater amount of charge than the one on the left, since there is

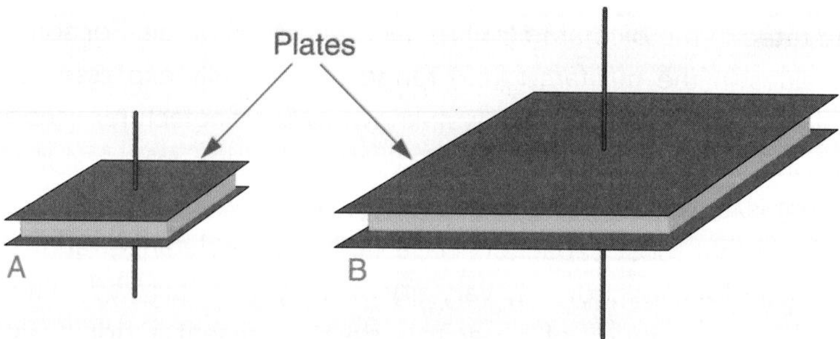

FIGURE 12.5 Capacity varies directly with plate area.

more room for charges to accumulate on each of the plates. As equation (12.3) indicates, capacity is directly proportional to the area of either plate.

Dielectric Thickness

The dielectric thickness of a capacitor determines the distance between its plates. The greater the distance between the plates, the greater the distance between the charges that accumulate on the plates and the lower the force of attraction between them. Therefore, capacity varies inversely with the dielectric thickness of a capacitor, as given in equation (12.3).

A Practical Consideration

The relationships among capacity, plate area, and dielectric thickness may seem purely theoretical at first, given the fact that we cannot physically alter the dimensions of a fixed capacitor. However, you will see that these relationships can be used to explain the relationship between capacitor values and total capacitance in both series and parallel circuits.

PROGRESS CHECK

1. What is capacitance? What does capacitance oppose?
2. What is a *capacitor*?
3. Draw a capacitor and label its parts.
4. What happens to the charge on a capacitor when the component is disconnected from its power source?
5. What is required to discharge a capacitor?
6. What precaution should be taken when working on capacitive circuits?
7. What is *capacity*?
8. What is the relationship between capacity and charge? Between capacity and voltage?
9. What is the unit of measure for capacity?
10. What is *permittivity*?
11. What is the relationship between capacity and plate area?
12. What is the relationship between capacity and dielectric thickness?

12.2 SERIES AND PARALLEL CAPACITORS

In this section, we will establish the methods used to calculate total capacitance in series and parallel circuits. Among other things, we will take a look at each of the following relationships:

- Total series capacitance is lower than the lowest value component in the circuit.
- Total parallel capacitance equals the sum of the individual component values.

As these statements imply:

- Total series capacitance is calculated in the same fashion as total parallel inductance or resistance.
- Total parallel capacitance is calculated in the same fashion as total series inductance or resistance.

Series Capacitors

When two or more capacitors are connected in series, the total capacitance is lower than any individual component value. By formula,

(12.4)

$$C_T = \frac{1}{\frac{1}{C_1} + \frac{1}{C_2} + \cdots + \frac{1}{C_n}}$$

where

C_T = the total series capacitance
C_n = the highest numbered capacitor in the string

Example 12.1 demonstrates the relationship between total series capacitance and individual capacitor values.

EXAMPLE 12.2

What is the total capacitance in Figure 12.6?

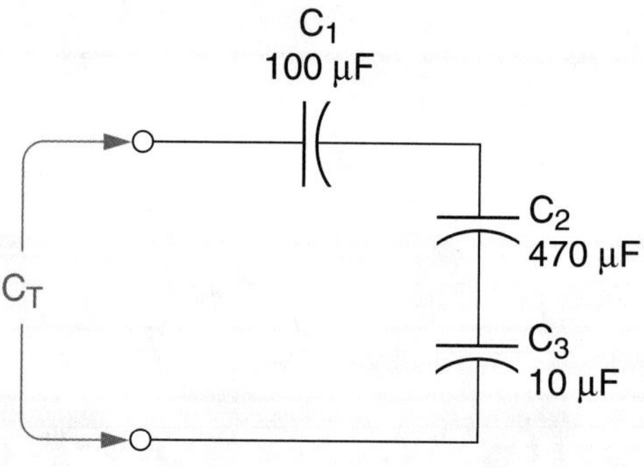

FIGURE 12.6

SOLUTION

The total circuit capacitance has a value of

$$C_T = \frac{1}{\frac{1}{C_1} + \frac{1}{C_2} + \frac{1}{C_3}}$$

$$= \frac{1}{\frac{1}{100} + \frac{1}{470} + \frac{1}{10}} \mu F$$

$$= 8.92 \mu F$$

PRACTICE PROBLEM 12.2

A circuit like the one in Figure 12.6 has the following values: $C_1 = 1500\ \mu F$, $C_2 = 2200\ \mu F$, and $C_3 = 3300\ \mu F$. Calculate the total circuit capacitance.

When you compare the total capacitance (C_T) in Example 12.2 to the individual capacitor values, you can see that the total capacitance is lower than any individual component value. This is always the case when capacitors are connected in series.

Alternate Approaches to Calculating Total Series Capacitance

You may have noticed the similarity between equation (12.4) and the relationships introduced earlier for calculating total parallel resistance and inductance. For example, the reciprocal equations for two-component resistive, inductive, and capacitive circuits are provided as follows:

PARALLEL RESISTANCE	PARALLEL INDUCTANCE	SERIES CAPACITANCE
$R_T = \frac{1}{\frac{1}{R_1} + \frac{1}{R_2}}$	$L_T = \frac{1}{\frac{1}{L_1} + \frac{1}{L_2}}$	$C_T = \frac{1}{\frac{1}{C_1} + \frac{1}{C_2}}$

Do you recall the "special-case" approaches to calculating total parallel resistance or inductance? The same approaches can be used to calculate the total capacitance in a series circuit under specific conditions. For example, the total capacitance of two series-connected capacitors can be found using the product-over-sum approach that follows:

$$C_T = \frac{C_1 \times C_2}{C_1 + C_2} \tag{12.5}$$

For example, the total capacitance for a series circuit with values of $C_1 = 1000\ \mu F$ and $C_2 = 4700\ \mu F$ can be found as

$$C_T = \frac{C_1 \times C_2}{C_1 + C_2}$$

$$= \frac{1000 \times 4700}{1000 + 4700} \mu F$$

$$= 825 \mu F$$

When equal-value capacitors are connected in series, the total circuit capacitance can be found as

(12.6)

$$C_T = \frac{C}{n}$$

where

C = the value of any one capacitor in the series circuit
n = the total number of capacitors in the series circuit

For example, the total capacitance of five 3300 μF capacitors connected in series can be found as

$$\begin{aligned} C_T &= \frac{C}{n} \\ &= \frac{3300\ \mu F}{5} \\ &= 660\ \mu F \end{aligned}$$

As you can see, we calculate series capacitance using the same relationships we use to calculate parallel resistance and inductance.

Why Connecting Capacitors in Series Reduces Overall Capacitance

You have seen the equations used to calculate total series capacitance, but they don't really help you to understand *why* capacitance is reduced when components are connected in series. The effect of connecting two identical capacitors in series is illustrated in Figure 12.7.

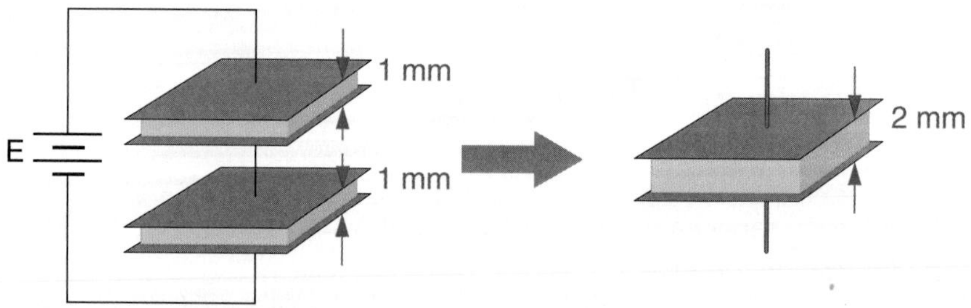

FIGURE 12.7 Connecting capacitors in series.

As stated earlier, the value of a capacitor is inversely proportional to the thickness of its dielectric. With this in mind, assume that each capacitor in Figure 12.7 has a dielectric thickness of 1 mm. By connecting the capacitors in series, the total dielectric thickness between the positive and negative terminals of the source increases to 2 mm. Since the dielectric thickness has effectively doubled, the total capacitance is one-half the value of either capacitor alone.

Connecting Capacitors in Parallel

When capacitors are connected in parallel, the total capacitance equals the sum of the individual component values. By formula,

(12.7)

$$C_T = C_1 + C_2 + \cdots + C_n$$

where

C_n = the highest numbered capacitor in the parallel circuit

The use of equation (12.7) is demonstrated in Example 12.3.

EXAMPLE 12.3

What is the total circuit capacitance in Figure 12.8?

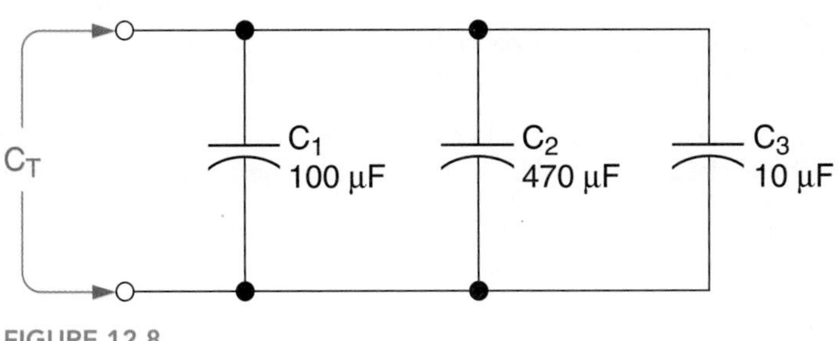

FIGURE 12.8

SOLUTION

The total circuit capacitance has a value of

$$C_T = C_1 + C_2 + C_3$$
$$= 100\ \mu F + 470\ \mu F + 10\ \mu F$$
$$= 580\ \mu F$$

PRACTICE PROBLEM 12.3

A circuit like the one in Figure 12.8 has the following values: $C_1 = 330\ \mu F$, $C_2 = 220\ \mu F$, and $C_3 = 1000\ \mu F$. Calculate the total circuit capacitance.

Why Connecting Capacitors in Parallel Increases Overall Capacitance

The effect of connecting two identical capacitors in parallel is illustrated in Figure 12.9.

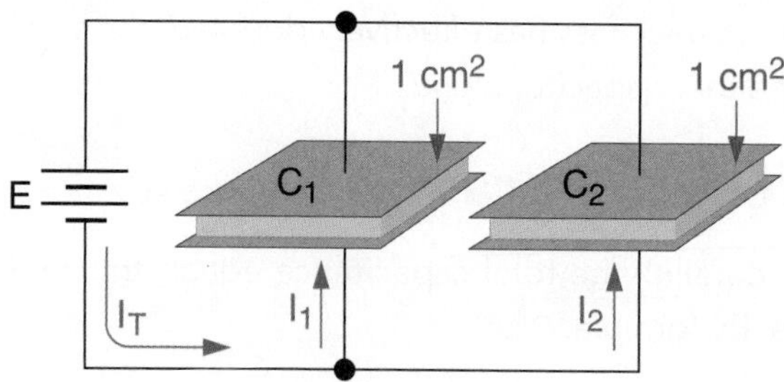

FIGURE 12.9 Connecting capacitors in parallel.

As stated earlier, the value of a capacitor is directly proportional to plate area. With this in mind, assume that each capacitor in Figure 12.9 has a plate area of 1 cm^2. When the capacitors are connected in parallel, the source provides equal currents to the two components. As such, the source current "sees" a total plate area of 2 cm^2. Since the plate area has effectively doubled, so has the total capacitance in the circuit.

PROGRESS CHECK

1. What is the relationship between total capacitance and the individual capacitor values in a series circuit?
2. What is the relationship between total capacitance and the individual capacitor values in a parallel circuit?
3. List the equations that are commonly used to calculate the total capacitance in a series circuit. Identify the special conditions (if any) of each.
4. Explain the effect that connecting capacitors in series has on the overall value of circuit capacitance.
5. Explain the effect that connecting capacitors in parallel has on the overall value of circuit capacitance.

12.3 CAPACITORS IN DC CIRCUITS

If a capacitor is placed in a DC circuit, it effectively behaves like a break in the circuit after it charges. This is due to the extremely high resistance of the dielectric, which effectively acts as a DC open circuit once the component has charged.

KEY CONCEPT

Earlier in the chapter, capacitance was described as a property that:

- Opposes an increase or decrease in voltage
- Stores energy in an electrostatic field

These characteristics of capacitance can be seen clearly when we discuss the effects that capacitance has on the operation of a DC circuit.

Voltage Transitions in a Resistive DC Circuit

When power is applied to a resistive DC circuit, the voltage and current instantly jump to their maximum values. This point is illustrated in Figure 12.10.

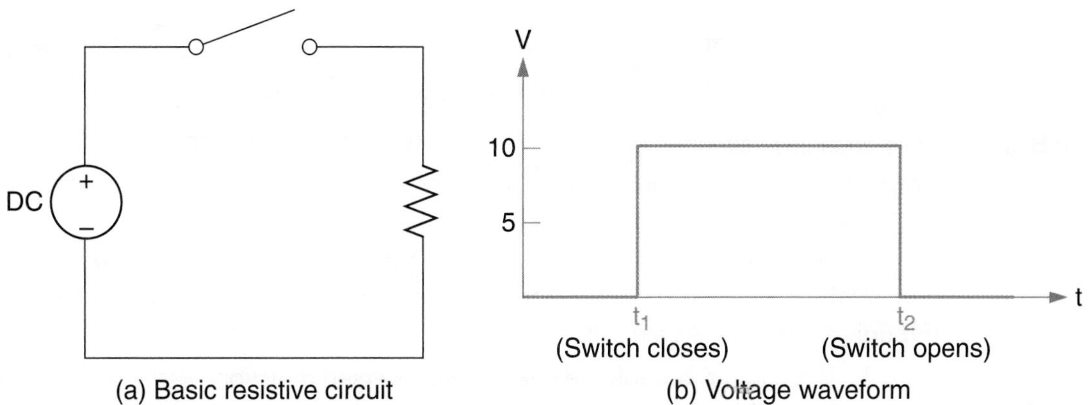

FIGURE 12.10 Current transitions in a resistive DC circuit.

The circuit in Figure 12.10a contains a power source, a switch, and a load. As shown in Figure 12.10b, the load voltage equals zero prior to closing the switch (as you would expect). When the switch is closed, the load voltage instantly increases to its maximum value. When the switch is re-opened, the voltage drops back to its initial level (0 V). As you learned in Chapter 11, the changes in voltage that occur when the switch is opened and closed are called *transitions.* The transition from 0 V to 10 V between t_0 and t_1 is a positive-going transition. The transition from 10 V to 0 V (after t_2) is a negative-going transition.

Voltage Transitions in a Resistive-Capacitive DC Circuit

The voltage transitions in a resistive-capacitive (RC) circuit require more time than those in a purely resistive circuit, as illustrated in Figure 12.11.

The circuit in Figure 12.11a contains a DC source, an SPDT switch, a capacitor and a resistor. Here is what happens when the switch in the RC circuit is set to its upper position: With the source connected to the circuit, the capacitor charges as shown in Figure 12.11b. As the capacitor charges:

1. The voltage across the component rises to its maximum value (which equals the source voltage, E).
2. Energy is stored in the electrostatic field between the capacitor plates.

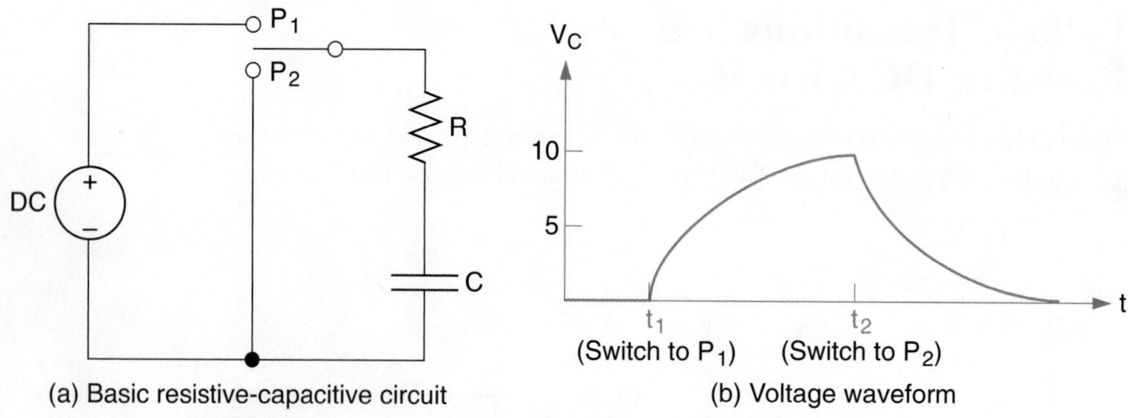

FIGURE 12.11 Capacitor voltage transitions in an RC circuit.

Here is what happens when the switch in the RC circuit is set to its lower position: With the source disconnected from the circuit, the capacitor discharges through the resistor. As a result,

1. The current through the resistor has reversed direction.
2. The capacitor discharges to 0 V.

As you can see, the capacitor accepted an electrical charge during its charge cycle, stored that charge, and returned it to the circuit when the source was disconnected.

The Rise Curve

RISE CURVE The curve representing capacitor voltage during the component's DC charge cycle.

The curve in Figure 12.11b that represents the increase in capacitor voltage between t_1 and t_2 is called a **RISE CURVE**. The shape of the rise curve is a result of the charging action of the capacitor. This means that the shape of the curve is not affected by the circuit voltage. For example, if we double the source voltage in Figure 12.11, the capacitor charging current also doubles. However, the shape of the curve remains the same.

The *universal rise curve* for any resistive-capacitive DC circuit can be drawn as shown in Figure 12.12. The curve is called a *universal curve* because it applies to all resistive-capacitive DC circuits.

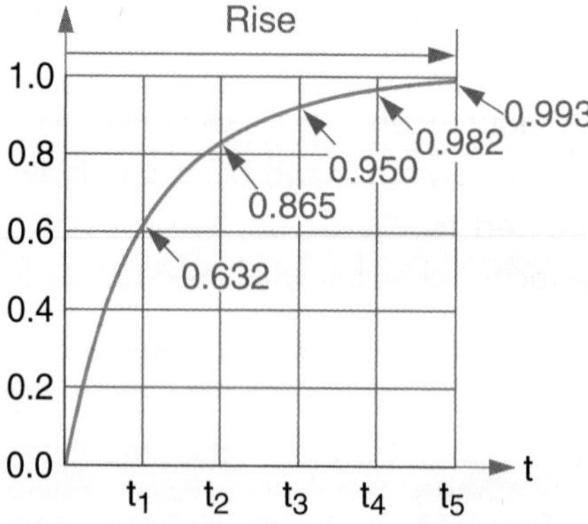

FIGURE 12.12 Capacitor voltage rise curve.

The rise curve in Figure 12.12 is divided into five equal time intervals; the first from t_0 to t_1, the second from t_1 to t_2, and so on. The decimal value at the end of each interval is used, along with the applied voltage, to calculate the voltage at that point on the curve. The voltage at each point on the rise curve is calculated as shown in Example 12.4.

EXAMPLE 12.4

An RC circuit has a value of E = 10 V. Calculate the capacitor voltage values at t_1 and t_2 on the voltage curve.

SOLUTION

At t_1, the capacitor voltage has a value of

$$V_C = 0.632 \times 10\text{ V}$$
$$= 6.32\text{ V}$$

At t_2, the capacitor voltage has a value of

$$V_C = 0.865 \times 10\text{ V}$$
$$= 8.65\text{ V}$$

PRACTICE PROBLEM 12.4

For the circuit described above, calculate the capacitor voltage at t_3, t_4, and t_5.

The multipliers in Figure 12.12 work with all series RC circuits. For example, if the applied voltage in Example 12.4 had a value of 24 V, the capacitor voltage values at t_1 and t_2 would be found as

$$V_C = 0.632 \times 24\text{ V}$$
$$= 15.2\text{ V}$$

and

$$V_C = 0.865 \times 24\text{ V}$$
$$= 20.8\text{ V}$$

The RC Time Constant (τ)

Each time interval in the voltage curve represents a real-time value called a *time constant*, represented by the Greek letter *tau* (τ). The term *constant* is used because τ is determined by the circuit capacitance and resistance. Therefore, its value is constant for a given circuit.

For a series resistive-capacitive circuit, the value of the RC time constant is calculated using

(12.8)

$$\tau = R \times C$$

where

τ = the duration of each time interval, in seconds
R = the total series resistance
C = the total series capacitance

Example 12.5 demonstrates the use of this equation.

EXAMPLE 12.5

A resistive-capacitive DC circuit has a 1000 μF capacitor and a 100 Ω resistor. Calculate the circuit time constant.

SOLUTION

The circuit time constant has a value of

$$\begin{aligned}\tau &= R \times C \\ &= 100\ \Omega \times 1000\ \mu F \\ &= 100{,}000\ \mu s \\ &= 100\ ms\end{aligned}$$

PRACTICE PROBLEM 12.5

Calculate the time constant for a series RC circuit that contains a 3300 μF capacitor and a 750 Ω resistor.

Figure 12.13 shows the rise curve for the circuit in Example 12.5. As you can see, each time interval represents the value of the circuit time constant (100 ms). The total time required for the capacitor voltage to complete its rise is equal to five time constants (5τ), in this case, 500 ms.

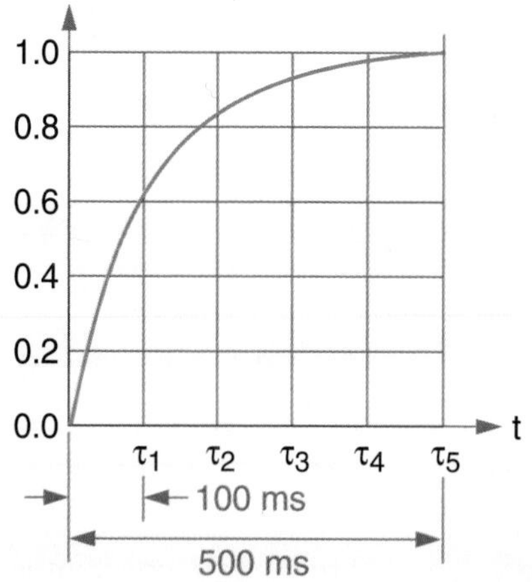

FIGURE 12.13 Capacitor voltage rise curve.

The Decay Curve

The curve in Figure 12.11b that represents the decrease in capacitor voltage (to the right of t_2) is called a **DECAY CURVE**. The shape of the decay curve is a result of the capacitor discharge action. This means that the shape of the curve is not affected by the circuit values. For example, if we double the source voltage in Figure 12.11, the capacitor voltage at the start of the decay cycle is twice the value shown. However, the shape of the decay curve remains unchanged.

DECAY CURVE The curve representing capacitor voltage during the component's DC discharge cycle.

The *universal decay curve* for any resistive-capacitive DC circuit can be drawn as shown in Figure 12.14. Like the rise curve described earlier, the curve in Figure 12.14 is called a *universal* curve because it applies to all resistive-capacitive DC circuits.

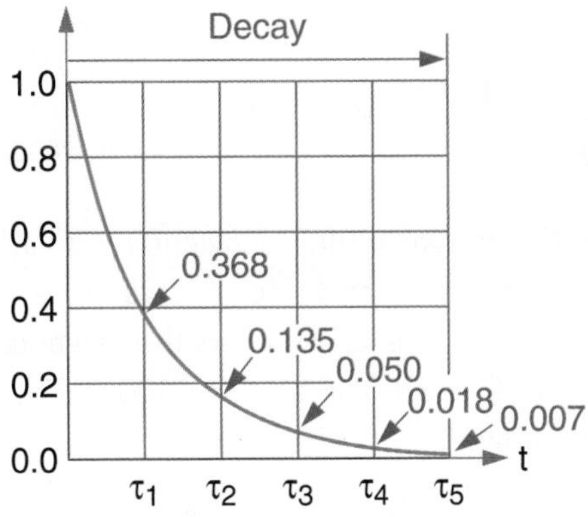

FIGURE 12.14 Capacitor voltage discharge curve.

The curve in Figure 12.14 is divided into five equal time intervals; the first from τ_0 to τ_1, the second from τ_1 to τ_2, and so on. The decimal value at the end of each interval is used, along with the circuit input voltage, to calculate the capacitor voltage at that point on the curve. The capacitor voltage at each point on the curve is calculated as shown in Example 12.6.

EXAMPLE 12.6

The voltage applied to a series RC circuit has a value of 10 V. What is the value of the capacitor voltage at τ_1 and τ_2 on the circuit decay curve?

SOLUTION

At τ_1 on the decay curve, the capacitor voltage has a value of

$$V_C = 0.368 \times 10\text{ V}$$
$$= 3.68\text{ V}$$

At τ_2, the capacitor voltage has a value of

$$V_C = 0.135 \times 10\text{ V}$$
$$= 1.35\text{ V}$$

PRACTICE PROBLEM 12.6

For the circuit described, calculate the capacitor voltage at τ_3, τ_4, and τ_5.

The multipliers in Figure 12.14 apply to all series RC circuits. For example, if the applied voltage in Example 12.6 had a value of 24 V, the circuit current values at τ_1 and τ_2 on the current decay curve would be found as

$$V_C = 0.368 \times 24\text{ V}$$
$$= 8.83\text{ V}$$

and

$$V_C = 0.135 \times 24\text{ V}$$
$$= 3.24\text{ V}$$

As is the case with the rise curve, each time interval in the decay curve equals the time constant (τ) for the circuit. In fact, the rise curve and decay curve are essentially mirror images of each other. For example, Figure 12.15 shows the rise and decay curves for a series RC circuit with values of C = 3300 μF and R = 100 Ω. The time constant for each curve is calculated as follows:

$$\tau = R \times C$$
$$= 100\ \Omega \times 3300\ \mu\text{F}$$
$$= 330\text{ ms}$$

As shown in the figure, the rise in capacitor voltage takes a total of 5 × 330 ms = 1650 ms. The same holds true for the capacitor voltage decay curve.

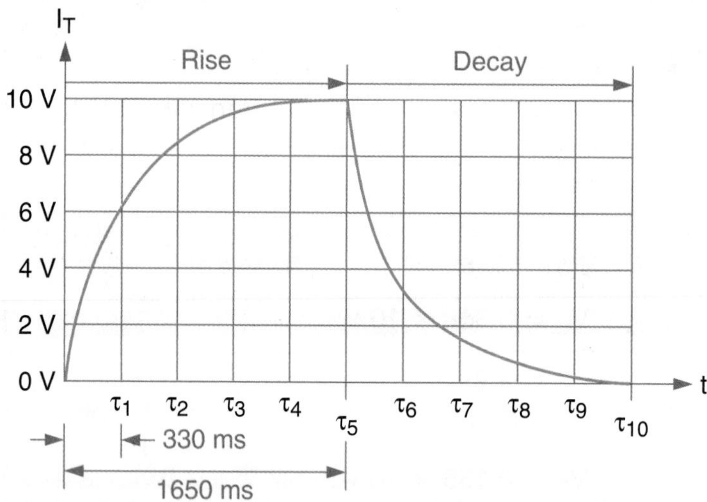

FIGURE 12.15

One Final Note

The purpose of this section has been to demonstrate the characteristics of capacitance. Using a DC circuit, we were able to see how the component opposes a change in voltage and stores energy in an electrostatic field.

As your education continues, you'll learn that capacitance plays a role in a wide variety of applications. For future reference, many of the principles and relationships introduced in this section are summarized in Figure 12.16.

Banks (groups) of capacitors are used in AC power systems to correct for poor power factor. However, harmonic currents generated by nonlinear loads can result in overheating, and insulation breakdown.

IN THE FIELD

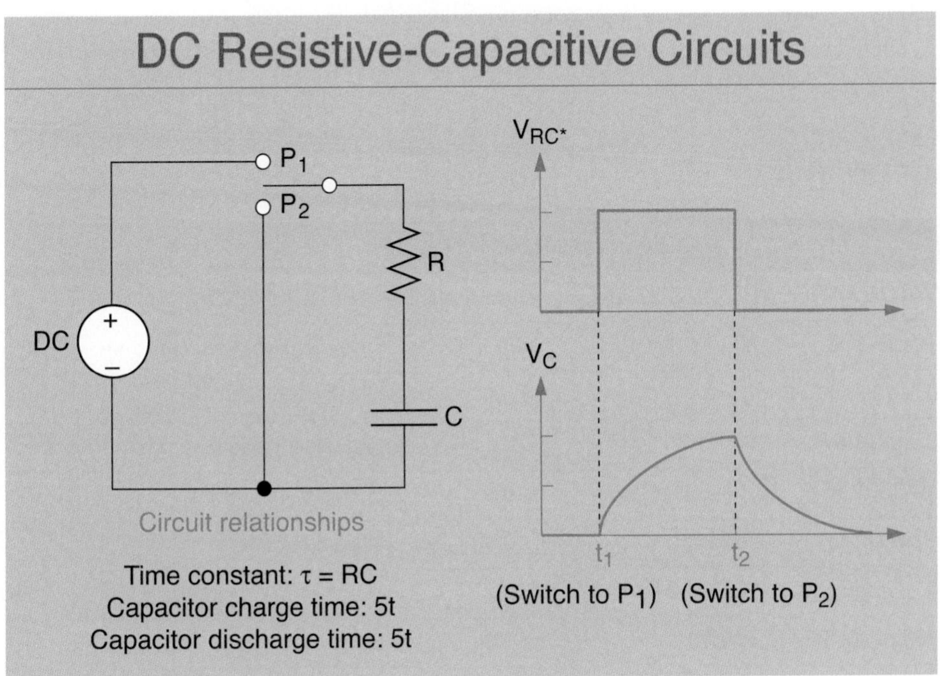

*Note: V_{RC} represents the voltage at the right-hand switch terminal.

FIGURE 12.16 DC resistive-capacitive circuit characteristics.

PROGRESS CHECK

1. What is a *transition*? What two types of transitions occur in a DC circuit when the switch is closed and opened?
2. Refer to Figure 12.12. What causes the shape of the V_C transitions?
3. What is the relationship between the shape of the capacitor rise (or decay) curve and the circuit current?
4. What is a universal curve? What purpose does it serve?
5. What is implied by the word "constant" in the term *time constant*?
6. How is the value of the time constant in a series RC circuit found?
7. How many time constants are required to complete a rise (or decay) curve?

12.4 CAPACITOR TYPES

Most capacitors are named for their dielectric material. Table 12.1 contains a summary of some common types of capacitors and their characteristics. The values and applications given in the table are *typical*; that is, you may come across uses and component values that are not listed here.

Super-capacitors, used as energy-storage devices for AC power systems, are combined in parallel strings and can achieve hundreds of volts. They show great promise for industrial applications because they contain no hazardous chemicals (and therefore have limited environmental impact) and their expected performance vs. reasonable cost.

IN THE FIELD

TABLE 12.1 • Common Capacitors[a]

TYPE	VALUE RANGE	VOLTAGE RANGE	TOLERANCE	APPLICATIONS/COMMENTS
Ceramic	10 pF–1 μF	50 V–30 kV	±(20+)%	Inexpensive, very common
Double layer	0.1–10 F	1.5–6 V	±20%	Power backup
Electrolytic	0.1 μF–1.6 F	3–600 V	−20 to +100%	Power supplies, normally polarized
Glass	10–1000 pF	100–600 V	±10%	Long-term stability
Mica	1 pF–0.1 μF	100–600 V	±10%	Radio circuits
Oil	0.1–20 μF	200 V–10 kV	±10%	High-voltage circuits
Polycarbonate[b]	100 pF–30 μF	50–800 V	±10%	High quality
Polypropylene[b]	100 pF–50 μF	100–800 V	±5%	High quality
Polystyrene[b]	10 pF–2.7 μF	100–600 V	±10%	Signal filters
Porcelain	100 pF–0.1 μF	50–400 V	±10%	Good long-term stability
Tantalum	0.1–500 μF	6–100 V	±20%	High capacity, polarized
Teflon	1 nF–2 μF	50–200 V	±5%	Highest quality
Vacuum	1–5000 pF	2–36 kV	±10%	Radio and TV transmitters

[a]All values and ranges are typical.
[b]These are collectively referred to as *plastic film resistors*.

Electrolytic Capacitors

ELECTROLYTIC CAPACITOR
A capacitor that contains a conducting liquid that serves as one of its plates.

ELECTROLYTIC CAPACITORS are common components that are used in low-frequency circuits, such as DC power supplies. These capacitors are housed in metal cans that make them relatively easy to identify. Several electrolytic capacitors are shown in Figure 12.17.

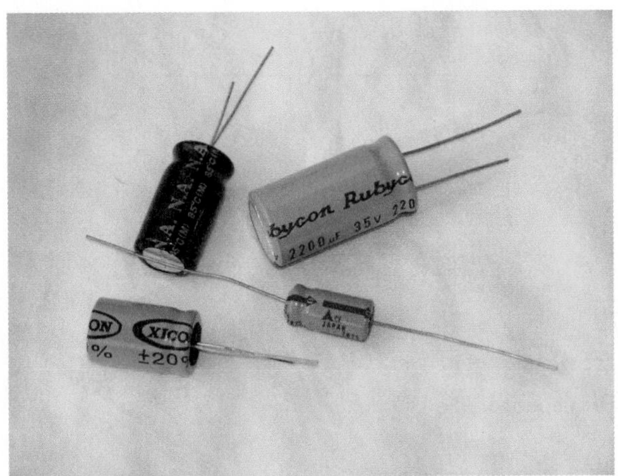

FIGURE 12.17 Electrolytic capacitors.

Aluminum electrolytic capacitors are subject to failures induced by thermal cycling (changes in temperature). A common rule-of-thumb is that stress and failure rates double for each 10°C increase in operating temperature between +75°C and the full-rated temperature of the component.

Electrolytic capacitors contain an **ELECTROLYTE** that makes it possible to produce high-capacity components that are relatively small.

ELECTROLYTE A chemical that contains free ions and conducts electricity.

Polarized Electrolytic Capacitors

Most electrolytic capacitors are *polarized,* meaning that one component lead must always be more negative than the other. The schematic symbol for a *polarized electrolytic capacitor* is shown in Figure 12.18a. The positive (or negative) terminal of a polarized electrolytic capacitor is usually identified as shown in Figure 12.18b.

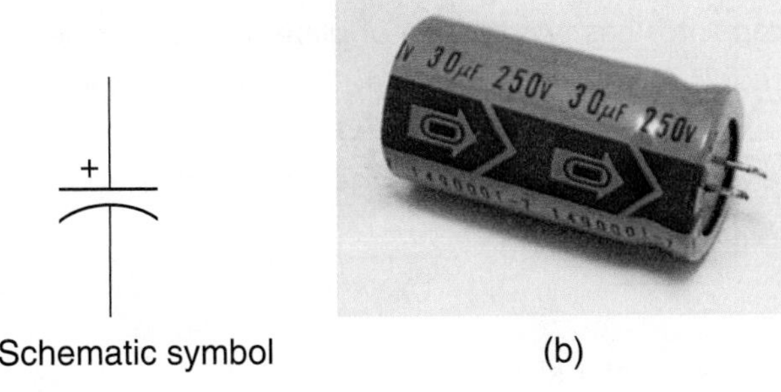

(a) Schematic symbol (b)

FIGURE 12.18

When replacing a polarized capacitor, care must be taken to match the polarity of the capacitor to that of any DC voltage in the circuit. This point is illustrated in Figure 12.19. When the capacitor is inserted correctly (as shown in Figure 12.19a), its polarization matches the polarity of the voltage source. If a polarized electrolytic

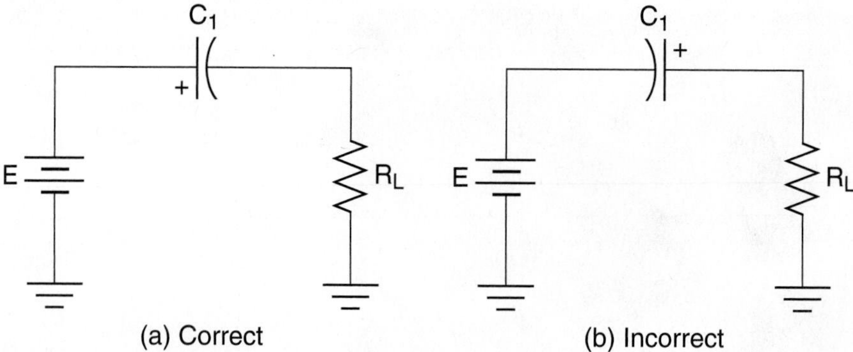

FIGURE 12.19 Connecting a polarized capacitor in a DC circuit.

capacitor is inserted incorrectly (as shown in Figure 12.19b), the component will become extremely hot. This condition can be dangerous for two reasons:

1. The component may become hot enough to cause severe burns if touched.
2. As it heats, the electrolyte of the capacitor may break down and burn, producing gases that may be trapped within the metal can. As the electrolyte continues to burn, gas pressure builds within the component, and may eventually cause it to explode.

As a polarized electrolytic capacitor heats up, it produces a strong burning odor. If you smell such an odor, immediately disconnect the power source. Be sure to allow the component to cool for several minutes before attempting to handle it.

Variable Capacitors

There are a variety of variable capacitors that allow the user to adjust for specific values of capacitance. At this point, we're going to briefly discuss two types of variable capacitors.

Interleaved-Plate Capacitors

One common variable capacitor is the *interleaved-plate capacitor*. Interleaved-plate capacitors are found primarily in high-power systems.

STATOR The stationary plates in an interleaved capacitor.

ROTOR The rotating plates in an interleaved capacitor.

An interleaved-plate capacitor has two groups of plates and an air dielectric. An interleaved-plate capacitor is illustrated in Figure 12.20. The capacitor shown is divided into two parts, called the **STATOR** and the **ROTOR**. Each of these parts contains a series of plates that are separated by air. When the rotor is adjusted, the rotor

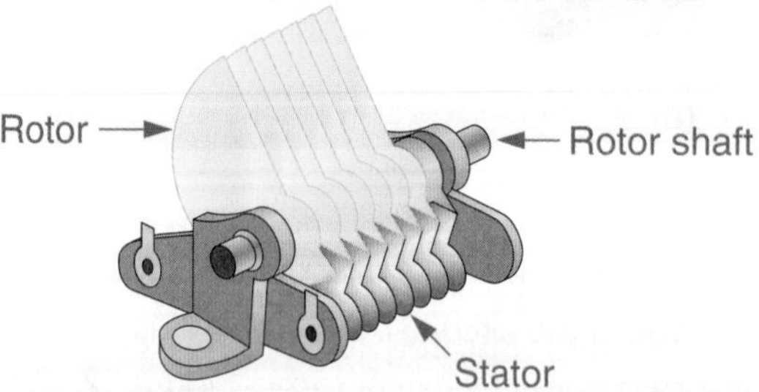

FIGURE 12.20 An interleaved-plate capacitor.

plates cut into the spaces between the stator plates. When the rotor is adjusted so that its plates are between those of the stator, the plates are said to be interleaved.

Transmitter-Type Interleaved Capacitors

These components are designed specifically for high-power applications. The *vanes* (metal plates) are spaced farther apart than they are for a common interleaved plate capacitor. Oil dielectrics may be used to provide higher capacity and breakdown voltage ratings.

PROGRESS CHECK

1. Why are electrolytic capacitors easy to identify?
2. What is the advantage that electrolytic capacitors have over other types of capacitors?
3. List the drawbacks of using electrolytic capacitors.
4. What precautions must be taken when working with polarized electrolytic capacitors?
5. What are interleaved-plate capacitors?

12.5 SUMMARY

Here is a summary of the major points that were covered in this chapter:

Capacitance and Capacitors: An Overview

- Capacitance is the ability of a component to store energy in the form of an electrostatic charge. It is often described as the ability of a component to oppose a change in voltage.
- A capacitor is a component designed to provide a specific measure of capacitance.
- A capacitor is made up of two conducting surfaces (called *plates*) that are separated by an insulating layer (called a *dielectric*).
- When a capacitor is connected to a DC voltage source, an electrostatic charge develops between the plates of the component.
- When a DC voltage source is connected across a capacitor:
    - An excess of positive charge accumulates on the plate that is connected to the positive terminal of the battery.
    - An excess of negative charge accumulates on the plate that is connected to the negative terminal of the battery.
- The positive and negative charges on the plates exert a force of attraction on each other through the dielectric. This force of attraction keeps the charges at the plates.
- The number of charges on the plates is determined by the capacity of the component.
- Once charged:
    - A capacitor blocks the flow of charge (current) in a DC circuit.
    - The difference of potential across the capacitor plates is approximately equal to the source voltage.
- To discharge a capacitor, an external path must be provided for the flow of charge (current) between the plates of the component.

- An open in the discharge path for a capacitor can prevent the component from discharging.
  - Never assume that a capacitor is not holding a charge.
  - When working with a capacitive circuit, a shorting tool is often used to ensure that all of the capacitors have discharged.
- The capacity of a capacitor is the amount of charge the component can store per volt applied. Capacity is measured in coulombs per volt (C/V).
- The amount of charge stored by a given capacitor depends on its capacity and the applied capacitor voltage.
- Capacitance is measured in farads (F). One farad is a capacity of one coulomb of charge per volt applied.
- Capacity is determined by the physical characteristics of a given capacitor.
  - It is directly proportional to the permittivity of the dielectric.
  - It is directly proportional to the cross-sectional area of the plates.
  - It is inversely proportional to the distance between the plates (i.e., the thickness of the dielectric).
- Permittivity is a measure of the ease with which lines of electrical force are established within a given material, measured in farads per meter.
- The higher the permittivity of a material, the easier it is to establish lines of electrical force within that material.

Series and Parallel Capacitors

- The total capacitance in a series circuit is lower than the lowest value component in the circuit.
- Total series capacitance is calculated in the same fashion as parallel resistance or inductance.
- Connecting capacitors in series effectively increases dielectric thickness. This is why overall capacitance is reduced by the series connection.
- The total capacitance in a parallel circuit equals the sum of the branch capacitances.
- Connecting capacitors in parallel effectively increases plate area. This is why overall capacitance is increased by the parallel connection.

Capacitors in DC Circuits

- When the power switch in a purely resistive series circuit is:
  - Closed (i.e., power is connected), the voltage level instantly increases to its maximum level.
  - Opened (i.e., power is disconnected), the voltage level instantly drops to its initial value (0 V).
- Abrupt changes in voltage levels are called *transitions.*
  - Transitions are described as positive-going or negative-going.
- When the power switch in a resistive-capacitive (RC) series circuit is closed:
  - Current is generated in the circuit, charging the capacitor.
  - The capacitor voltage increases as the capacitor charges.
  - Energy is stored in the electrostatic field between the plates.
- A graphical representation of the rise in capacitor voltage is called a *rise curve.*
  - Because the curve has the same shape regardless of circuit component values and maximum voltage levels, it is also called a universal curve.

- The universal rise curve is divided into five equal time intervals.
  - During the five time intervals, the capacitor voltage increases from 0 V to its maximum value.
  - The curve contains constants (multipliers) that are used to calculate the capacitor voltage at the end of each time interval.
- Each time interval on the rise curve represents a real-time value called a *time constant* (τ).
  - The time constant for a series RC circuit is calculated using $\tau = R \times C$.
  - It takes 5 time constants for the voltage in a series RC circuit to reach its maximum value.
- When the power switch in a resistive-capacitive (RC) series circuit is re-opened and a discharge current path is provided, the capacitor generates a decreasing current through the series resistance.
  - The discharge current is in the opposite direction of the charging current.
- The graphical representation of the decrease in capacitor voltage is called a *decay curve*.
  - Because the curve has the same shape regardless of circuit component values and maximum voltage levels, it is also called a *universal curve*.
- The universal decay curve is divided into five equal time intervals.
  - During the five time intervals, the capacitor voltage decreases from its maximum value to 0 V.
  - The curve contains constants (multipliers) that are used to calculate the capacitor voltage at the end of each time interval.
- Each time interval on the decay curve represents a real-time value called a *time constant* (τ).
  - The time constant for a series RC circuit is calculated using $\tau = R \times C$.
  - It takes 5 time constants for the capacitor voltage in a series RC circuit to drop from its maximum value to 0 V.
- Most capacitors are named for their dielectric materials.
- Electrolytic capacitors are common components that are used in low-frequency applications. Electrolytic capacitors:
  - Are housed in metal cans.
  - Contain an electrolyte that makes it possible to produce high-capacity components that are relatively small.
- Most electrolytic capacitors are polarized.
- When replacing a polarized electrolytic capacitor, care must be taken to match the polarity of the capacitor to that of any DC voltage in the circuit.
- If a polarized electrolytic capacitor is inserted in a circuit backwards:
  - It may become hot enough to cause severe burns if touched.
  - The burning electrolyte may produce gases, eventually causing the component to explode.
  - The burning electrolyte generally produces a strong burning odor.
- If a polarized electrolytic capacitor smells like it's burning:
  - Immediately remove power from the circuit.
  - Allow the component to cool for several moments before attempting to touch it.
- The interleaved-plate capacitor is a variable component that is used primarily in high-power systems and older consumer electronics systems.
- An interleaved-plate capacitor has two groups of plates and an air dielectric.

CHAPTER REVIEW

1. A capacitor ________.
   a) provides a specific value of capacitance
   b) stores energy in an electrostatic field
   c) opposes any change in voltage
   d) all of the above

2. The two leads of a capacitor are connected to the ________.
   a) plates
   b) conductors
   c) dielectric
   d) terminals

3. The insulating layer of a capacitor is called the ________.
   a) plate
   b) conductor
   c) dielectric
   d) terminal

4. The side of a charged capacitor that is connected to the positive side of the supply voltage has an excess of ________ charge.
   a) capacitive
   b) inductive
   c) negative
   d) positive

5. The amount of charge that a capacitor can accept is determined by ________.
   a) its capacity
   b) the plate area
   c) the dielectric thickness
   d) all of the above

6. A(n) ________ is developed across the plates of a charged capacitor.
   a) electrostatic charge
   b) potential difference
   c) force of attraction
   d) all of the above

7. A charged capacitor acts like ________.
   a) a DC open
   b) a momentary short
   c) an inverse voltage
   d) a current source

8. When a capacitor discharges, ________.
   a) current reverses direction compared to the charge current
   b) it acts like a DC open
   c) the electrostatic charge remains
   d) plate voltage increases

9. Capacitors will hold a charge until ________.
   a) the applied voltage drops to zero
   b) 5 time constants have occurred
   c) an external discharge path is provided
   d) 1 time constant has occurred

10. A 470 μF capacitor will hold ________ of charge if 120 V is applied.
    a) 2.55×10^5 C
    b) 3.92 μC
    c) 56.4 mC
    d) 56.4 C

11. A 2200 μF capacitor when fully charged stores 0.45 coulombs of charge. This means that the applied voltage is ________.
    a) 205 V
    b) 4.89 V
    c) 489 mV
    d) 20.5 V

12. Capacitance is inversely proportional to ________.
    a) charge
    b) dielectric thickness
    c) plate area
    d) applied voltage

13. The ease with which electrical force is established in a given material is referred to as ________.
    a) capacity
    b) dielectric strength
    c) permittivity
    d) permeability

14. Three capacitors are connected together in a circuit. The total capacitance is found to be 83 μF. The smallest capacitor in the circuit is 100 μF. This means that the capacitors are most likely ________.
    a) connected in series
    b) equal in value
    c) connected in parallel
    d) not polarized

15. Three identical capacitors are connected in series. The total capacitance is found to be 227 μF. The value of each capacitor is most likely ________.
    a) 680 μF
    b) 75 μF
    c) 114 μF
    d) 228 μF

16. A 470 μF capacitor is connected in parallel with two 10 μF capacitors, also connected in parallel. The total capacitance of this circuit is ________.
    a) 45 μF
    b) 450 μF
    c) 490 μF
    d) 49 μF

17. Two capacitors are connected in series. A third is then connected in parallel with the first two. What happens to the total circuit capacitance as a result of adding the third capacitor?
    a) It increases.
    b) It decreases.
    c) Can't tell, depends on component values.
    d) It remains the same.

18. Connecting capacitors in parallel effectively causes the ________ to ________.
    a) distance between the plates, increase
    b) distance between the plates, decrease
    c) plate area, increase
    d) plate area, decrease

19. Connecting capacitors in series effectively causes the ________ to ________.
    a) distance between the plates, increase
    b) distance between the plates, decrease
    c) plate area, increase
    d) plate area, decrease

20. There is (are) ________ equation(s) that can be used to solve for total capacitance in a series circuit.
    a) one
    b) two
    c) three
    d) four

21. Compared to a purely resistive circuit, voltage transitions ________ in an RC circuit.
    a) take longer
    b) are more linear
    c) are inverted
    d) have a lower maximum value

22. The difference between the universal rise curve in an RL circuit and the universal rise curve in an RC circuit is ________.
    a) the RC curve plots voltage, while the RL curve plots current
    b) the RC curve is the inverse of the RL curve
    c) the RC rise curve is the same as the RL discharge curve
    d) there is no difference

23. Two 100 Ω resistors are in series with a 47 μF capacitor. The time constant for this circuit is ________.
    a) 0.47 μs
    b) 0.24 μs
    c) 4.7 ms
    d) 9.4 ms

24. The circuit described in question 23 has a 33 μF capacitor added in parallel with the 47 μF capacitor. Assuming the applied voltage is 40 V, and the capacitors are fully discharged when the voltage is applied, the time it would take for the voltage across the capacitors to reach its maximum value would be ________.
    a) 40 ms
    b) 60 ms
    c) 80 ms
    d) 80 μs

25. Refer to question 24. Assume that a discharge path is available. The time it would take for circuit current to drop to zero once the circuit is fully charged would be ________.
    a) 40 ms
    b) 60 ms
    c) 80 ms
    d) 80 μs

26. Refer to question 24. When fully charged, the voltage across the 47 μF capacitor would be ________ while the voltage across the 33 μF capacitor would be ________.
    a) 20 V, 20 V
    b) 40 V, 40 V
    c) 23.5 V, 16.5 V
    d) 16.5 V, 23.5 V

27. Assume that an RC circuit has a time constant of 40 ms. The applied voltage is disconnected after only 3 time constants. This means the voltage across the capacitor would be ________ if the applied voltage is 45 V.
    a) 27 V
    b) 18 V
    c) 42.8 V
    d) 2.25 V

28. In an RC circuit, discharge current is ________ charge current.
    a) less than
    b) equal to
    c) in the opposite direction of
    d) greater than

29. Capacitors are usually named based upon ________.
    a) the dielectric material
    b) the maximum voltage rating
    c) the type of container
    d) their application

30. Electrolytic capacitors ________.
    a) are housed in metal cans
    b) have a lot of capacitance for their size compared to other capacitors
    c) are used in low-frequency applications
    d) all of the above

31. When replacing an electrolytic capacitor, you must be careful to determine ________ or the capacitor might explode.
    a) the correct frequency
    b) the correct time constant
    c) the correct polarity
    d) if a mica capacitor should be used

32. The dielectric in most interleaved capacitors is ________.
    a) mica
    b) polypropylene
    c) air
    d) polystyrene

33. The two primary parts of an interleaved capacitor are ________.
    a) the pole and the throw
    b) the plate and the gap
    c) the rotor and the stator
    d) the hub and the rotor

PRACTICE PROBLEMS

1. A capacitor is storing 200 mC of charge when the difference of potential across its plates is 25 V. Calculate the value of the component.
2. A capacitor is storing 100 μC of charge when the difference of potential across its plates is 40 V. Calculate the value of the component.
3. Determine the charge that a 22 μF capacitor stores when the difference of potential across its plates is 18 V.
4. Determine the charge that a 100 pF capacitor stores when the difference of potential across its plates is 25 V.

5. Calculate the total capacitance in the series circuit shown in Figure 12.21a.
6. Calculate the total capacitance in the series circuit shown in Figure 12.21b.

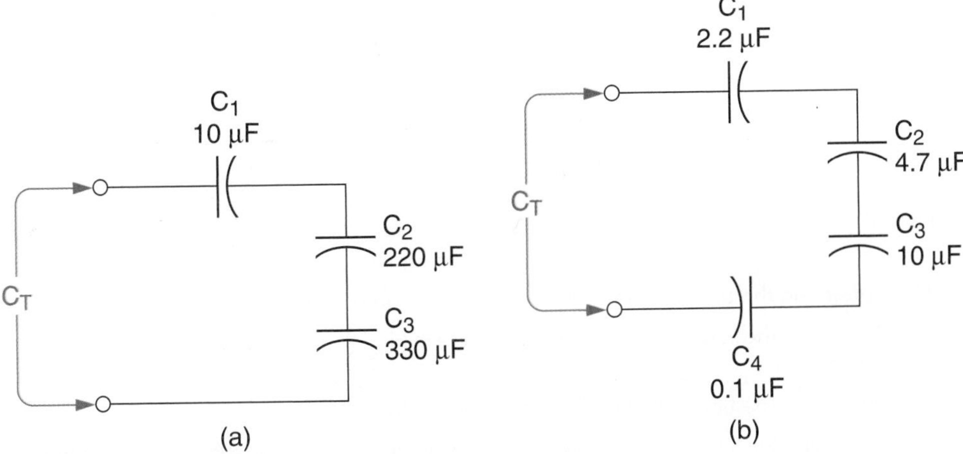

FIGURE 12.21

7. Calculate the total capacitance in the series circuit shown in Figure 12.22a.
8. Calculate the total capacitance in the series circuit shown in Figure 12.22b.

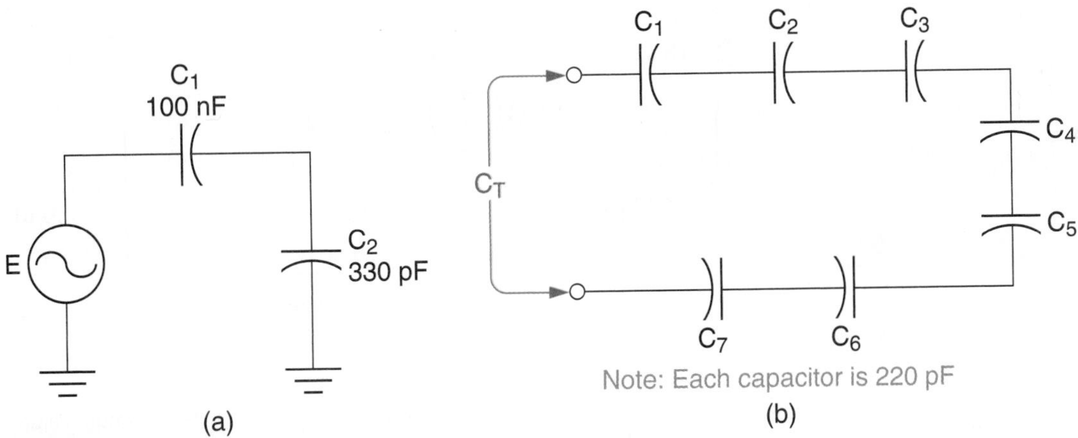

FIGURE 12.22

9. Calculate the total capacitance in the parallel circuit shown in Figure 12.23a.
10. Calculate the total capacitance in the parallel circuit shown in Figure 12.23b.

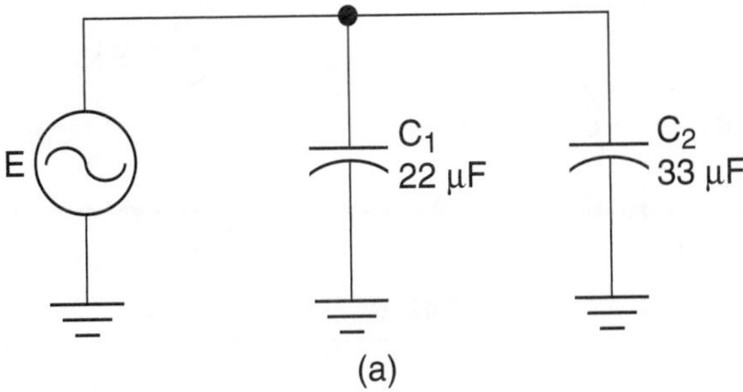

FIGURE 12.23

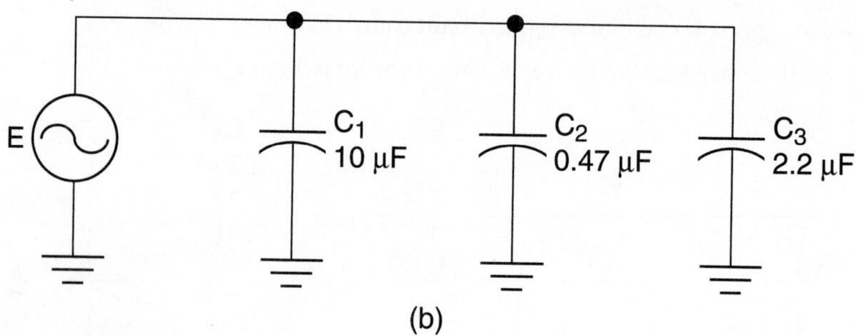

FIGURE 12.23 *(Continued)*

11. Determine the time constant for the circuit in Figure 12.24a.
12. Determine the value of the capacitor voltage in Figure 12.24a at the end of each time constant.
13. Determine the time constant for the circuit in Figure 12.24b.
14. Determine the value of the capacitor voltage in Figure 12.24b at the end of each time constant.

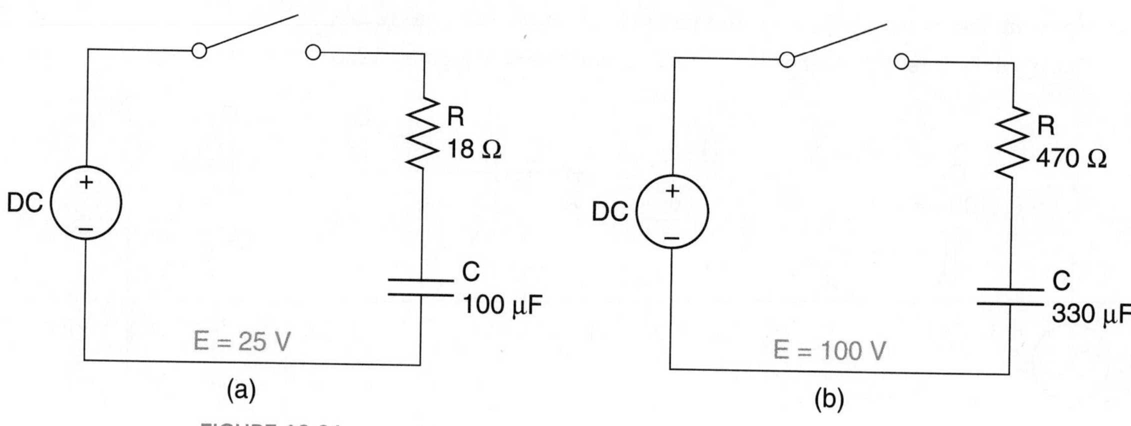

FIGURE 12.24

15. Determine the time required for the current in Figure 12.25a to reach its maximum value.
16. Determine the time required for the current in Figure 12.25b to reach its maximum value.

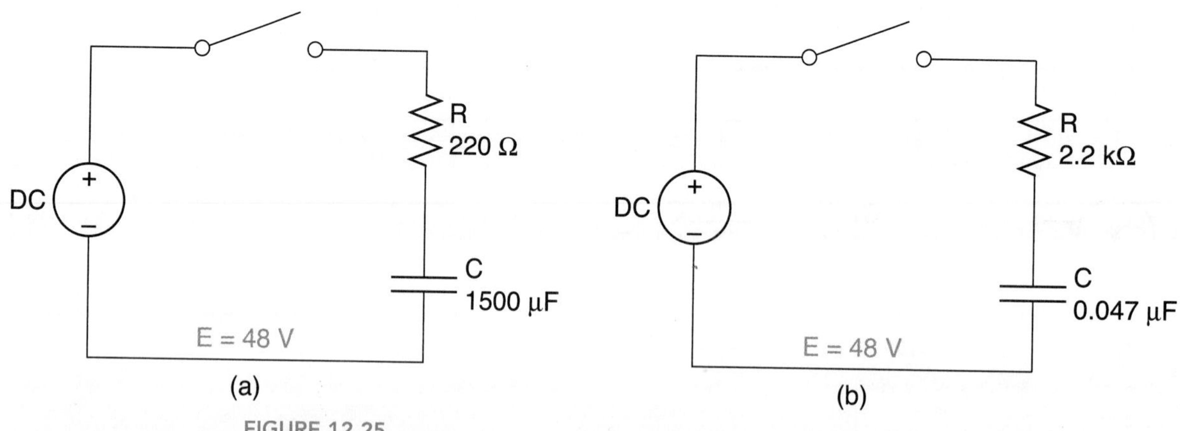

FIGURE 12.25

ANSWERS TO THE EXAMPLE PRACTICE PROBLEMS

12.1	0.0015 F
12.2	702 μF
12.3	1550 μF
12.4	9.5 V, 9.82 V, 9.93 V
12.5	2.48 s
12.6	0.5 V, 0.18 V, 0.07 V

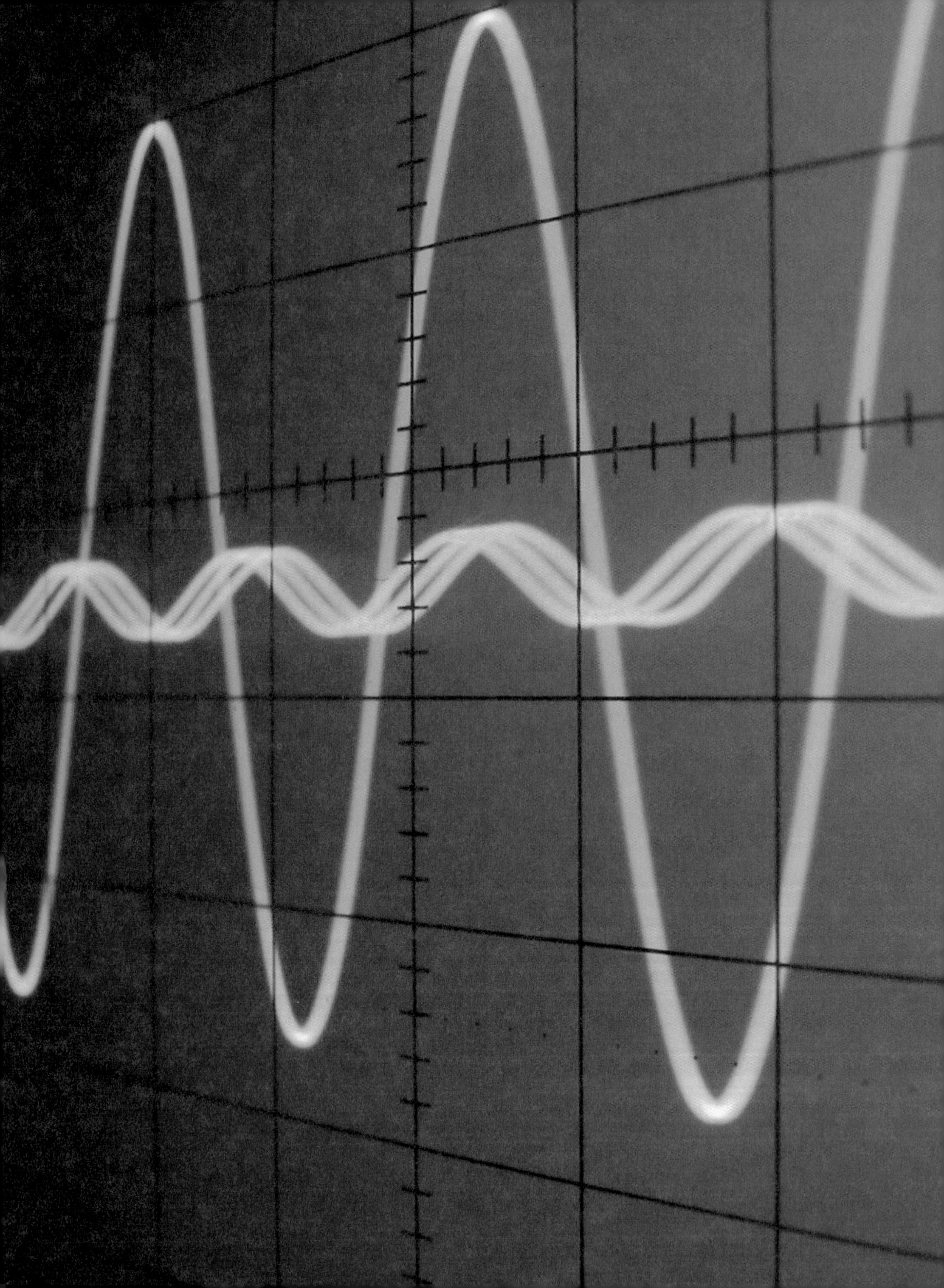

CHAPTER 13

ALTERNATING CURRENT (AC)

PURPOSE

In Chapter 2, you were told that there are two types of current: *direct current* (DC) and *alternating current* (AC). In DC circuits, current is always in the same direction. In AC circuits, current periodically changes direction. Up to this point, we have limited our discussions to DC circuits and characteristics.

The power provided to homes, businesses, and industries is in the form of alternating current and voltage. As such, you must be thoroughly familiar with the characteristics of alternating current and voltage if you hope to work on common electrical circuits and systems.

In this chapter, we will focus on AC measurements and characteristics. In future chapters, we will analyze the operation of a variety of AC components and circuits.

KEY TERMS

The following terms are introduced and defined in this chapter on the pages indicated:

OBJECTIVES

After completing this chapter, you should be able to:

1. Describe *alternating current (AC)*.
2. Describe the makeup of a sinusoidal waveform.
3. Describe the relationship between cycle time and frequency.
4. Determine the cycle time and frequency of a waveform that is displayed on an oscilloscope.

5. Define and identify each of the following magnitude-related values: *peak, peak-to-peak, instantaneous, full-cycle average,* and *half-cycle average.*
6. Describe the relationship between *average AC power* and *rms values.*
7. Describe the relationship between *peak* and *rms* values.
8. Describe how each of the magnitude-related values is measured.
9. Describe the magnetic induction of current.
10. Compare and contrast *phase* and *time* measurements.
11. Discuss *phase angles* and the means by which they are measured.
12. Calculate the instantaneous value at any point on a sine wave using the *degree* approach.
13. Discuss the effects of a *DC offset* on sine wave measurements.
14. Describe *wavelength* and calculate its value for any waveform.
15. Discuss the relationship between a waveform and its *harmonics.*
16. Describe and analyze *rectangular waveforms.*

13.1 ALTERNATING CURRENT (AC): OVERVIEW AND TIME MEASUREMENTS

Generally, the term **ALTERNATING CURRENT (AC)** is used to describe any current that periodically changes direction. For example, take a look at the *current versus time* relationship shown in Figure 13.1. The horizontal axis of the graph is used to represent time (t). The vertical axis of the graph is used to represent both the **MAGNITUDE** (value) and direction of the current. As the figure shows, the current builds to a peak value in one direction and returns to zero (t_0 to t_1), and then builds to a peak value in the other direction and returns to zero (t_1 to t_2). Note that the current not only changes direction, but constantly changes in magnitude.

ALTERNATING CURRENT (AC) Any current that periodically changes direction.

MAGNITUDE The value of a voltage or current.

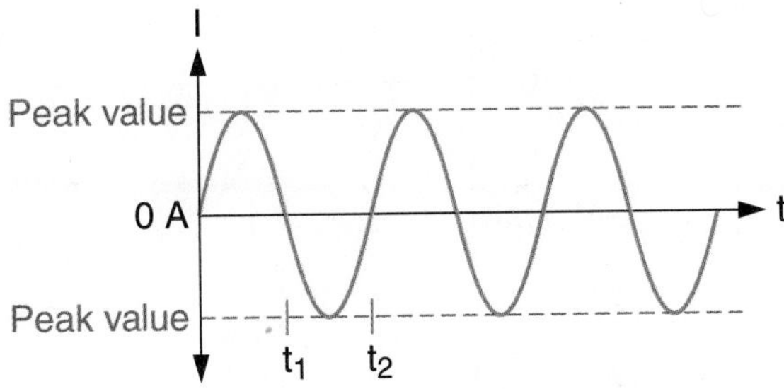

FIGURE 13.1 Alternating current (AC).

A graph of the relationship between magnitude and time is referred to as a *waveform*. For reasons that will be given later, the waveform in Figure 13.1 is called a **SINUSOIDAL WAVEFORM**, or **SINE WAVE**. It should be noted that the sine wave is not the only type of AC waveform. Some others will be introduced briefly later in this chapter.

As you will learn, the magnitude of a sine wave voltage or current can be expressed in several ways. The most commonly used in practice are *RMS*, *peak*, and *average*.

IN THE FIELD

SINUSOIDAL WAVEFORM A waveform whose magnitude varies with the sine of its phase.

SINE WAVE Another name for a sinusoidal waveform.

Basic AC Operation

The operation of a simple AC circuit is illustrated in Figure 13.2. Figure 13.2a highlights the operation of the circuit during the positive half of the AC input. Figure 13.2b highlights the operation of the circuit during the negative half of the AC input. Using the time periods shown in the figure, the operation of the circuit can be described as follows:

TIME PERIOD	CIRCUIT OPERATION
t_0 to t_1	The output from the voltage source increases from 0 V to its positive peak value (+V peak). As the voltage increases, the current also increases from 0 A to its peak value. Note the voltage polarity and current direction shown in the figure.
t_1 to t_2	The output from the voltage source returns to 0 V and the current returns to 0 A.

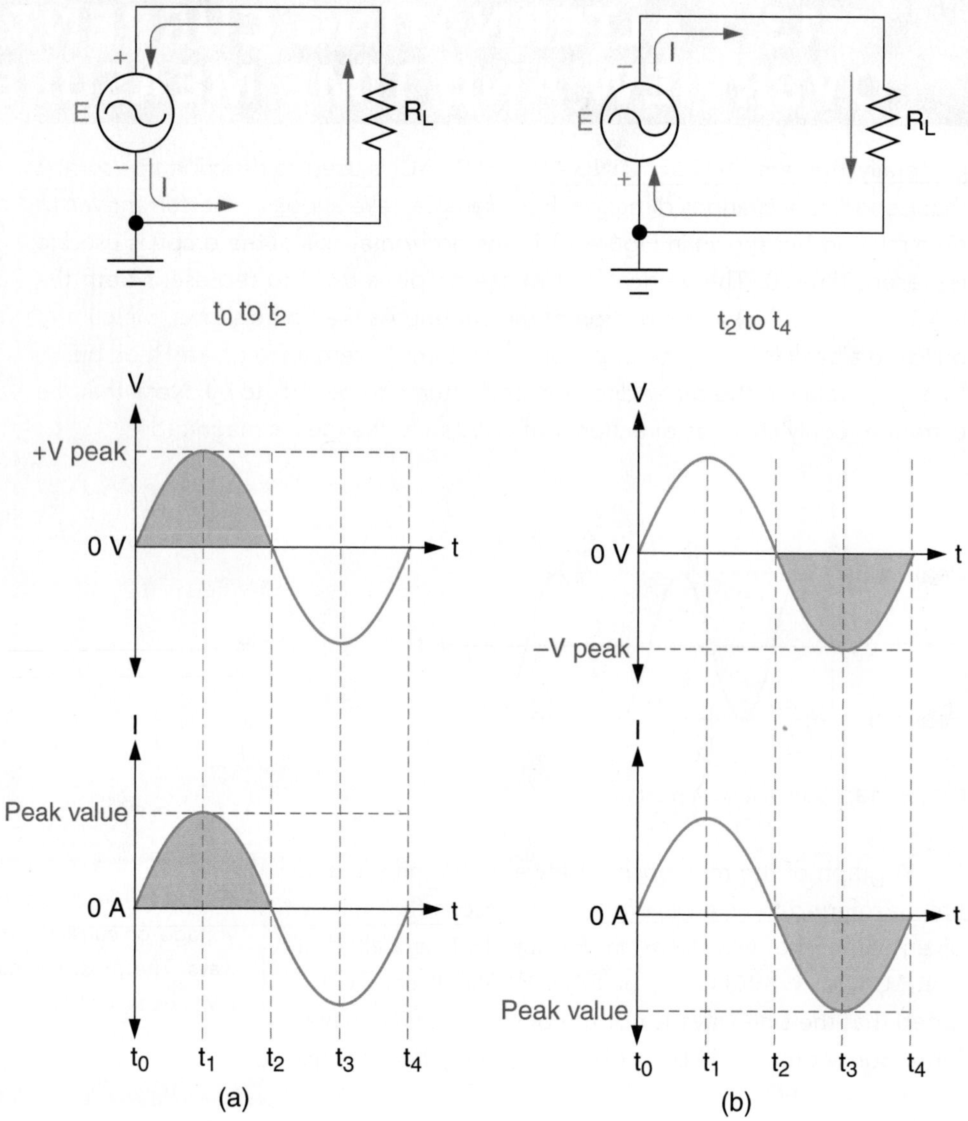

FIGURE 13.2 AC circuit operation.

t_2 to t_3 The output from the source increases to its negative peak value (−V peak), as does the circuit current. Note that both the voltage polarity and the current direction have reversed.

t_3 to t_4 The output from the voltage source returns to 0 V and the current returns to 0 A.

Crest factor (the ratio of a waveform's peak value to its RMS value) can be used to verify the amount of distortion in that waveform. For a purely sinusoidal AC waveform, the crest factor is 1.414 (1/0.707). Some DMMs have a *crest factor rating* that describes their ability to measure distortion.

IN THE FIELD

Alternations and Cycles

The positive and negative halves of the waveform in Figure 13.2 are referred to as **ALTERNATIONS**. The complete transition through one *positive alternation* and one *negative alternation* is referred to as a **CYCLE**. These terms are used to identify the parts of the waveform in Figure 13.3a. Since two alternations make up a cycle, each alternation is commonly referred to as a **HALF-CYCLE**. As shown in Figure 13.3b, a sine wave is made of a continuous stream of cycles. A waveform that repeats itself continuously (like the one shown in Figure 13.3b) is called a **PERIODIC WAVEFORM**.

ALTERNATIONS The positive and negative halves of a waveform.

CYCLE The complete transition through one positive and one negative alternation.

HALF-CYCLE Another way of referring to either alternation of a waveform.

PERIODIC WAVEFORM A waveform that repeats itself continuously.

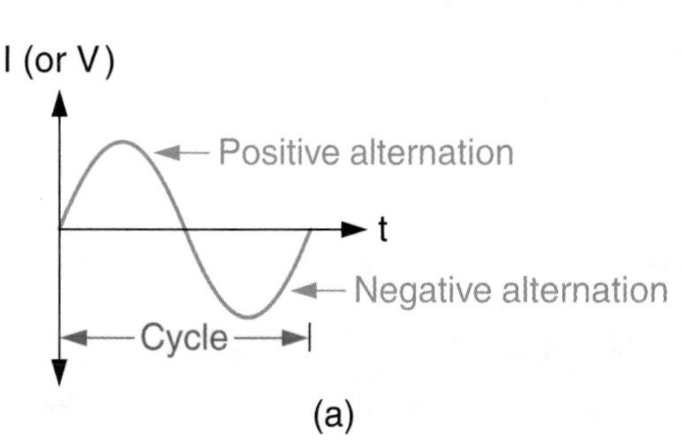

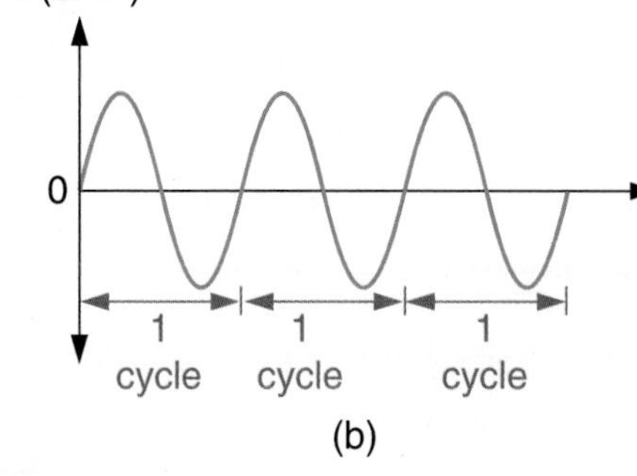

FIGURE 13.3 Sine wave alternations and cycles.

Waveform Period and Frequency

The time required to complete one cycle of an AC waveform is referred to as its **PERIOD**. Example 13.1 demonstrates the concept of waveform period.

PERIOD The time required to complete one cycle of an AC waveform.

EXAMPLE 13.1

What is the period of the waveform represented in Figure 13.4?

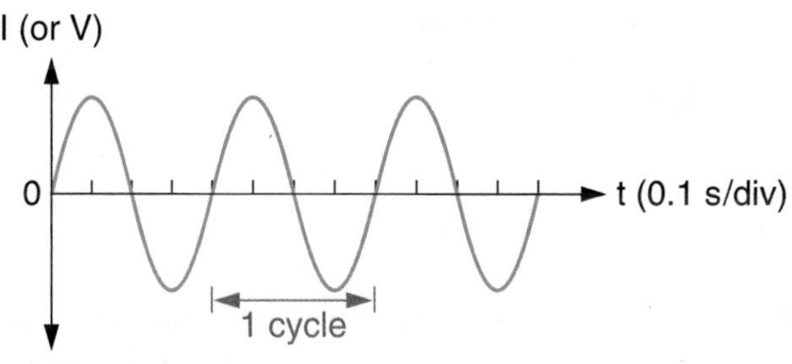

FIGURE 13.4

SOLUTION

Each division on the horizontal axis of the graph represents a time period of 0.1 s. One cycle of the waveform is four divisions in length. Therefore, the period (T) of the waveform is found as

$$T = (4\text{ divisions}) \times 0.1\frac{\text{s}}{\text{div}}$$

$$= 0.4 \text{ s}$$
$$= 400 \text{ ms}$$

PRACTICE PROBLEM 13.1

Calculate the period of the waveform shown in Figure 13.5.

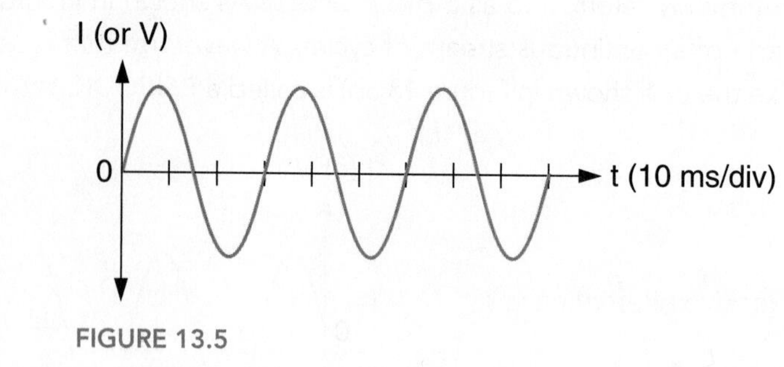

FIGURE 13.5

The period of a waveform can be measured from any given point on the waveform to the identical point in the next cycle, as illustrated in Figure 13.6. With a horizontal scale of 20 ms/div, the total cycle time (regardless of where it is measured) works out to be

$$T_C = (4 \text{ div}) \times 20\frac{\text{ms}}{\text{div}}$$
$$= 80 \text{ ms}$$

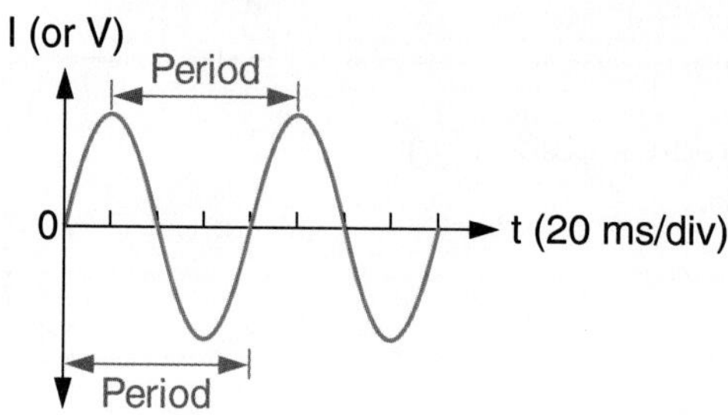

FIGURE 13.6 Waveform period measurements.

FREQUENCY The rate at which waveform cycles repeat themselves, in cycles per second.

The **FREQUENCY (f)** of a waveform is the rate at which the cycles repeat themselves, in cycles per second. The concept of frequency is illustrated in Figure 13.7. The waveform shown has a cycle time of 200 ms. Therefore, the cycle repeats itself five times every second, and the frequency (f) of the waveform is expressed as

$$f = 5 \text{ cycles per second}$$

HERTZ (Hz) The unit of measure for frequency, equal to 1 cycle per second.

The unit of measure for frequency is the **HERTZ (Hz)**. One hertz is equal to one cycle per second. Therefore, the frequency of the waveform in Figure 13.7 would be written as

$$f = 5 \text{ Hz}$$

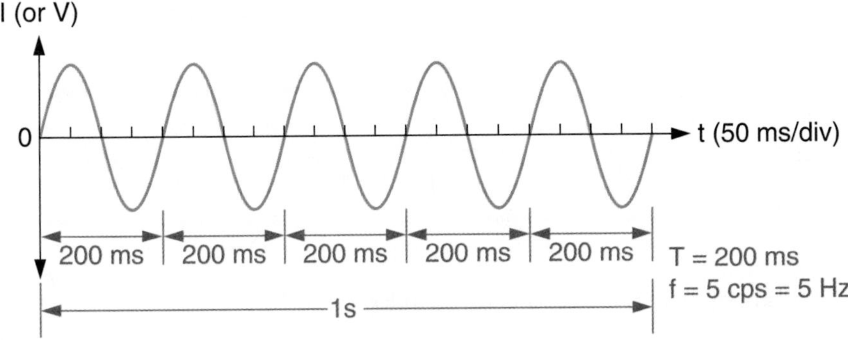

FIGURE 13.7

When the cycle time of a waveform is known, the frequency of the waveform can be found as

(13.1)

$$f = \frac{1}{T}$$

where

T = the period of the waveform in seconds

The use of this equation is demonstrated in Example 13.2.

EXAMPLE 13.2

What is the frequency of a sine wave that has a period of 100 ms?

SOLUTION

With a period of 100 ms, the sine wave has a frequency of

$$f = \frac{1}{T}$$

$$= \frac{1}{100 \times 10^{-3} s}$$

$$= 10 \text{ cycles per second}$$

$$= 10 \text{ Hz}$$

PRACTICE PROBLEM 13.2

A sine wave has a period of 40 ms. Calculate the frequency of the waveform.

When the frequency of a waveform is known, its period can be found as

(13.2)
$$T = \frac{1}{f}$$

The use of this relationship is demonstrated in Example 13.3.

EXAMPLE 13.3

What is the period of a 400 Hz sine wave?

SOLUTION

With a frequency of 400 Hz, the sine wave has a period of

$$T = \frac{1}{f}$$
$$= \frac{1}{400\ \text{Hz}}$$
$$= 2.5\ \text{ms}$$

This means that a 400 Hz sine wave completes one cycle every 2.5 ms.

PRACTICE PROBLEM 13.3

Calculate the period of a 60 Hz sine wave.

OSCILLOSCOPE A piece of test equipment that provides a visual representation of a voltage waveform.

Oscilloscopes can also be used to observe distortion in an electrical signal and measure the time span between two events, such as the relative timing of two related signals.

IN THE FIELD

STAYING SAFE The ground clip of an oscilloscope probe should be connected only to earth ground or an isolated common in the equipment being tested to avoid electrical shock and/or equipment damage due to the presence of differing ground planes.

Oscilloscope Time and Frequency Measurements

An **OSCILLOSCOPE** is a piece of test equipment that provides a visual representation of a voltage waveform. This visual display can be used to make a variety of magnitude and time-related measurements. An oscilloscope is shown in Figure 13.8, along with the display of a sine wave.

The oscilloscope screen is divided into a series of *major divisions* and *minor divisions*, as shown in Figure 13.9. The divisions along the *x*-axis are used to measure time, while the divisions along the *y*-axis are used to measure voltage. For now, we are interested only in the time-related measurements. (Voltage measurements are addressed in the next section.)

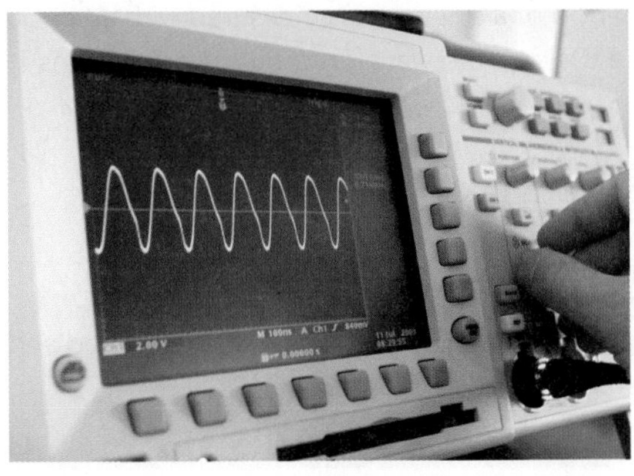

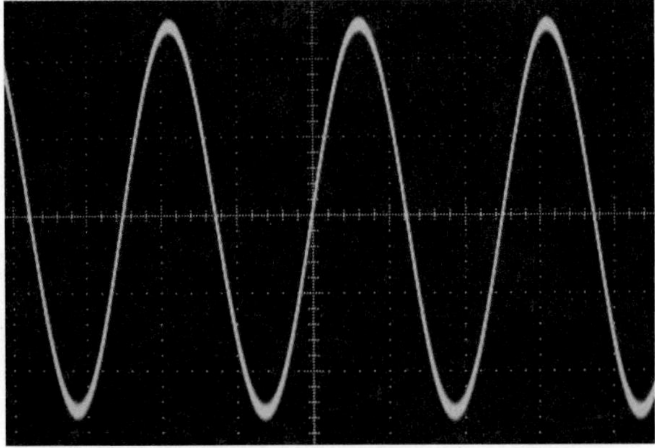

FIGURE 13.8 An oscilloscope and sine wave display.
Istock.com

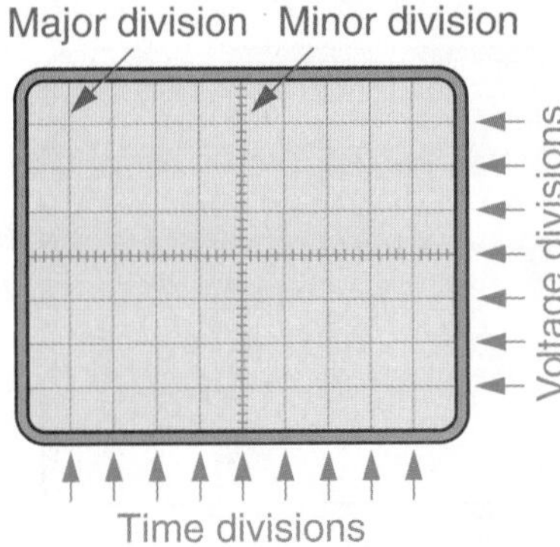

FIGURE 13.9 An oscilloscope display grid.

The setting of the **TIME BASE** control on the oscilloscope determines the time represented by the interval between any two adjacent major divisions along the *x*-axis. This point is illustrated in Figure 13.10a. If the time base control is set to 5 ms/div, the time intervals have the values shown. In each case, the time interval represented on the screen equals the time base setting times the number of major divisions.

TIME BASE The oscilloscope control that determines the time represented by the interval between any two adjacent major divisions along the *x*-axis.

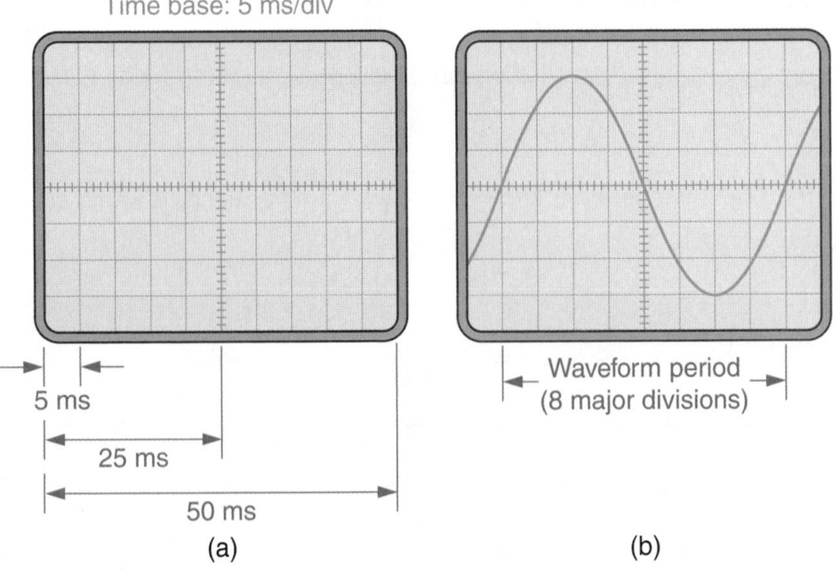

FIGURE 13.10 Oscilloscope time measurements.

Figure 13.10b shows how the period of a sine wave is measured with an oscilloscope. As shown in the figure, one cycle of the waveform is 8 major divisions in length. If the time base control is set to 100 ms/div, the sine wave period is

$$T = (8 \text{ divisions}) \times \frac{100 \text{ ms}}{\text{div}}$$
$$= 800 \text{ ms}$$

The frequency of a waveform can be determined by measuring its period and then calculating its frequency, as demonstrated in Example 13.4.

EXAMPLE 13.4

What is the frequency of the sine wave in Figure 13.11a?

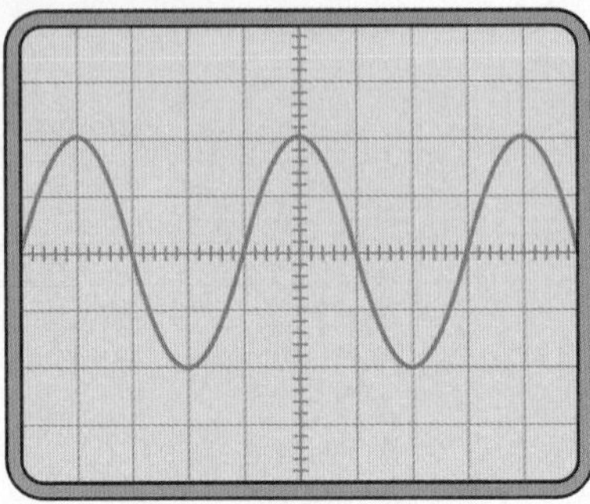

Time base: 5 ms/div

(a)

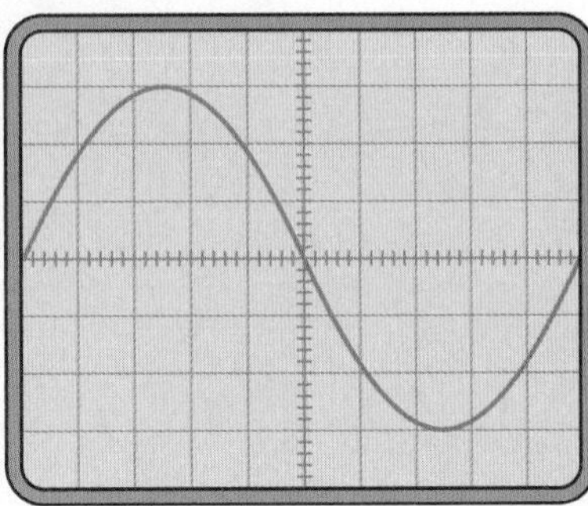

Time base: 2 ms/div

(b)

FIGURE 13.11

SOLUTION

The display shows $2\frac{1}{2}$ cycles of a sine wave. Each cycle is 4 divisions in length. With a time base of 5 ms/div, the sine wave period is

$$T = (4 \text{ divisions}) \times \frac{5 \text{ ms}}{\text{div}}$$
$$= 20 \text{ ms}$$

Once the sine wave period is known, the waveform frequency is found as

$$f = \frac{1}{T}$$
$$= \frac{1}{20 \text{ ms}}$$
$$= 50 \text{ Hz}$$

PRACTICE PROBLEM 13.4

Calculate the frequency of the waveform shown in Figure 13.11b.

When you solve Practice Problem 13.4, you get a result of 50 Hz; the same value that was calculated for the waveform in Figure 13.11a. The fact that these waveforms have the same frequency demonstrates an important characteristic of the time base control on the oscilloscope: When you vary the time base setting, you do not change the frequency of the waveform, only the time represented by each division along the horizontal axis of the display. In other words, changing the time base setting changes the waveform display, not the waveform itself.

Summary of Time-Related Characteristics

You have now been introduced to the most common time-related AC characteristics. These characteristics (and terms) are summarized in Figure 13.12. In the next section, we'll move on to look at the common magnitude-related characteristics.

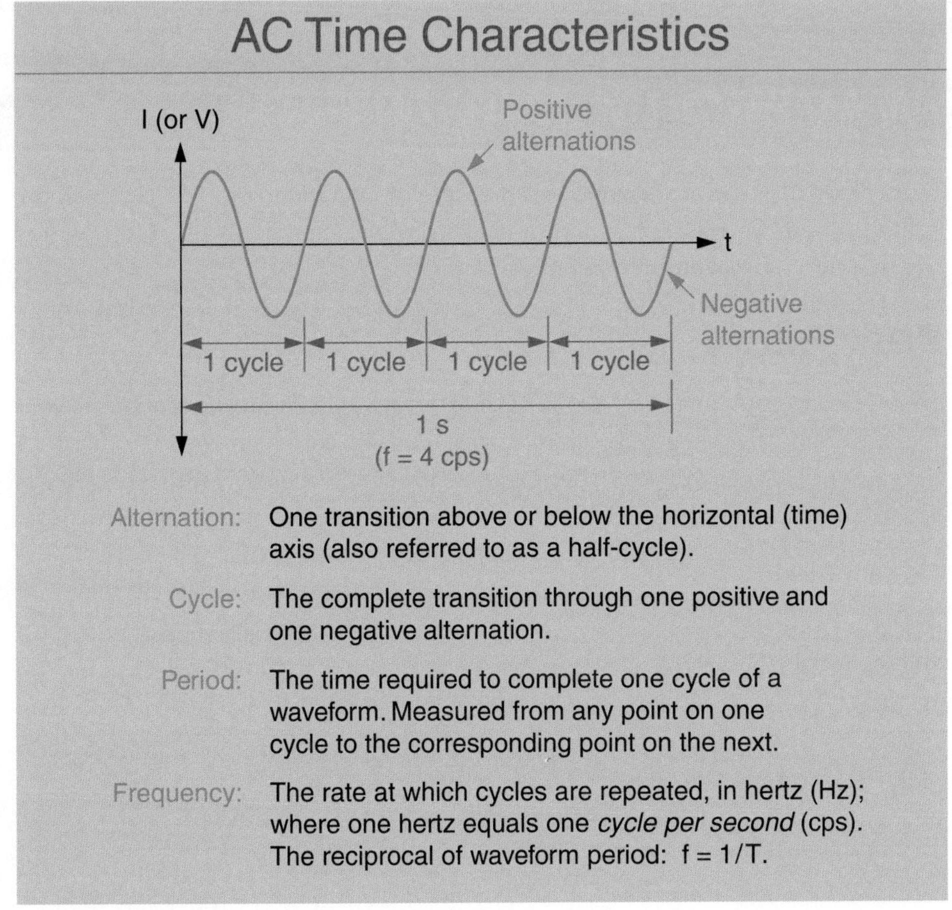

AC Time Characteristics
I (or V)
Positive alternations
t
Negative alternations
1 cycle
1 cycle
1 cycle
1 cycle
1 s
(f = 4 cps)
Alternation: One transition above or below the horizontal (time) axis (also referred to as a half-cycle).
Cycle: The complete transition through one positive and one negative alternation.
Period: The time required to complete one cycle of a waveform. Measured from any point on one cycle to the corresponding point on the next.
Frequency: The rate at which cycles are repeated, in hertz (Hz); where one hertz equals one *cycle per second* (cps). The reciprocal of waveform period: f = 1/T.

FIGURE 13.12 AC time-related characteristics.

It should be noted that many electrical professionals do not use oscilloscopes on the job. Even so, the oscilloscope display provides a convenient means of illustrating many waveform characteristics. That is why we focus on oscilloscope displays and measurements in this chapter.

PROGRESS CHECK

1. What is the term *alternating current* used to describe?
2. What is a waveform?
3. What is the relationship between alternations and cycles?
4. What is the period of a waveform? By what other name is it known?
5. Where on a sine wave can its period be measured?
6. What is frequency? What is its unit of measure?
7. Describe the relationship between waveform period and frequency.
8. What is an oscilloscope?
9. What changes when you adjust the time base control on an oscilloscope?
10. How do you determine waveform period and frequency with an oscilloscope?

13.2 WAVEFORM MAGNITUDE

There are many ways to describe and measure the magnitude of an AC waveform. In this section, we will discuss the methods that are commonly used to describe and measure AC waveform magnitude.

Peak Values

PEAK VALUES The maximum value reached on each alternation of a waveform.

Every AC waveform has two **PEAK VALUES**, as shown in Figure 13.13. For the voltage waveform, the peak values are ±10 V. For the current waveform, the peak values are ±1 A. As you can see, the waveforms in Figure 13.13 have identical positive and negative alternations. AC waveforms that have identical positive and negative alternations are sometimes called *true AC* waveforms.

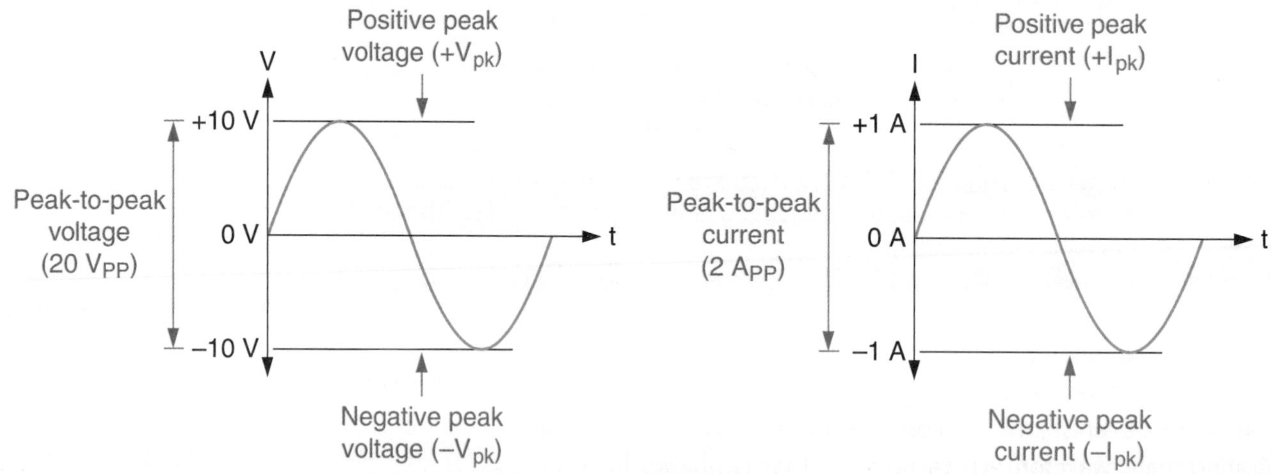

FIGURE 13.13 Peak and peak-to-peak values.

Peak-to-Peak Values

The **PEAK-TO-PEAK VALUE** of a waveform is the difference between its positive and negative peak values. For example, the voltage waveform in Figure 13.13 has a peak-to-peak value of

PEAK-TO-PEAK VALUE The difference between the positive and negative peak values of a waveform.

$$\begin{aligned} V_{PP} &= +10\text{ V} - (-10\text{ V}) \\ &= 10\text{ V} + 10\text{ V} \\ &= 20\text{ V} \end{aligned}$$

which equals the value indicated in the figure. For the current waveform, the difference between the peak values is shown to be 2 A.

The peak-to-peak value of a pure AC waveform is also equal to twice its peak value. By formula,

(13.3)

$$V_{PP} = 2V_{pk}$$

where

V_{PP} = the peak-to-peak voltage
V_{pk} = the peak voltage

and

(13.4)

$$I_{PP} = 2I_{pk}$$

where

I_{PP} = the peak-to-peak current
I_{pk} = the peak current

These relationships can be verified by comparing the peak and peak-to-peak values shown in Figure 13.13. In each case, the peak-to-peak value is twice the peak value.

Measuring Peak and Peak-to-Peak Voltages

In the last section, you were shown how the oscilloscope is used to measure cycle time and frequency. You were also told that the oscilloscope can be used to make magnitude-related measurements.

Both peak and peak-to-peak voltages can be measured using an oscilloscope. These measurements are illustrated in Figure 13.14.

Voltage values are measured along the vertical axis (y-axis) of the display. The value represented by each major division is determined by the **VERTICAL SENSITIVITY** control. Note that the vertical sensitivity control is labeled *volts/division* on

VERTICAL SENSITIVITY The value represented by each major vertical division on an oscilloscope, expressed in *volts/division*.

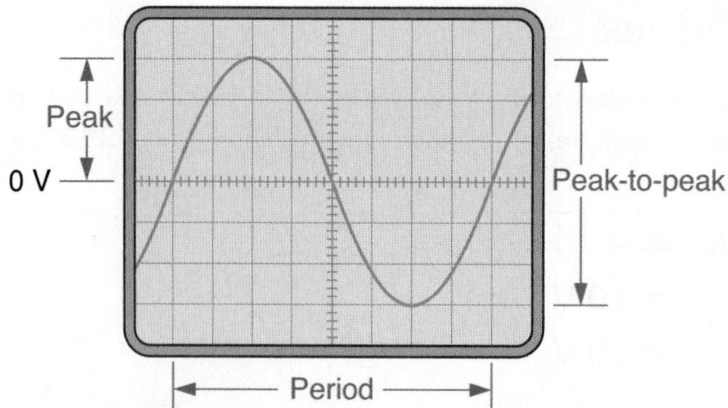

FIGURE 13.14 Sine wave period, peak, and peak-to-peak measurements.

STAYING SAFE To prevent personal injury and equipment damage, do not apply excessive voltage to an oscilloscope's input jacks. Voltage limits can be found in the test unit's operating manual or printed on the oscilloscope itself.

most oscilloscopes. With a vertical sensitivity of 5 volts/div, the waveform has a peak value of 15 V. If the volts/div setting is changed, the voltage represented by each major division changes. However, the waveform values do not.

When the center of the grid is established as 0 V, the major divisions above the center line represent positive voltages and the major divisions below the center line represent negative voltages.

Instantaneous Values

INSTANTANEOUS VALUE The voltage or current magnitude at a specific point on a waveform.

An **INSTANTANEOUS VALUE** is the voltage or current magnitude at a specific point on a waveform. For example, two times (t_1 and t_2) are identified on the voltage sine wave in Figure 13.15. Each of these times has a corresponding instantaneous voltage. These voltages are labeled v_1 and v_2, respectively. Note that instantaneous values are identified using lowercase letters (rather than uppercase letters).

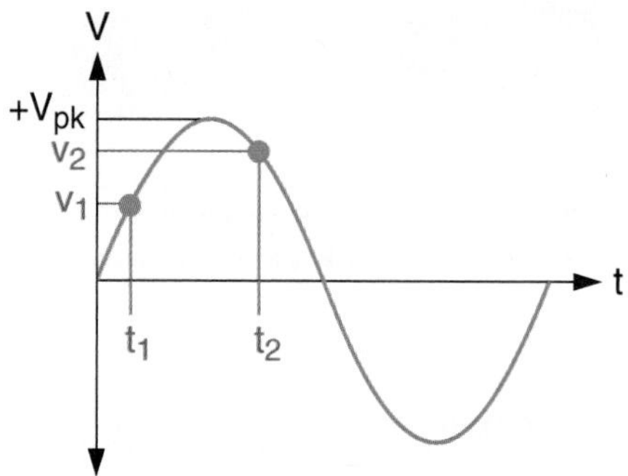

FIGURE 13.15 Instantaneous values.

Half-Cycle Average

HALF-CYCLE AVERAGE The average of all the instantaneous voltage or current values in either of a waveform's alternations.

In some applications, it is important to know the **HALF-CYCLE AVERAGE** of a sine wave. The half-cycle average of a sine wave is the average of all the instantaneous voltage or current values in either of its alternations, as illustrated in Figure 13.16a. As you can see, all of the instantaneous values are taken from the positive alternation.

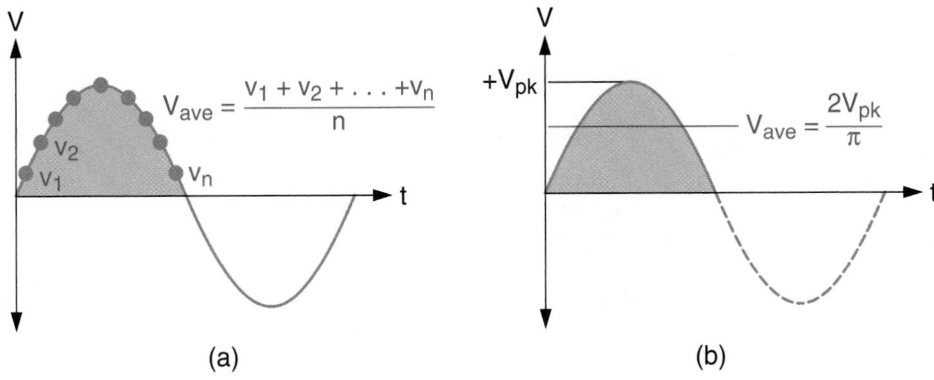

FIGURE 13.16 Half-cycle average values.

It is not necessary to measure and average a series of instantaneous values to determine the value of V_{AVE} for a given sine wave. In fact, the average value of a pure AC sine wave can be calculated using

(13.5)

$$V_{AVE} = \frac{2 \times V_{pk}}{\pi}$$

where V_{pk} is the peak value of the sine wave. Since $\frac{2}{\pi} \cong 0.637$, the above equation can be rewritten as

(13.6)

$$V_{AVE} \cong 0.637 \times V_{pk}$$

The value of V_{AVE} for a sine wave is calculated as shown in Example 13.5.

EXAMPLE 13.5

What is the half-cycle average of the sine wave in Figure 13.17a?

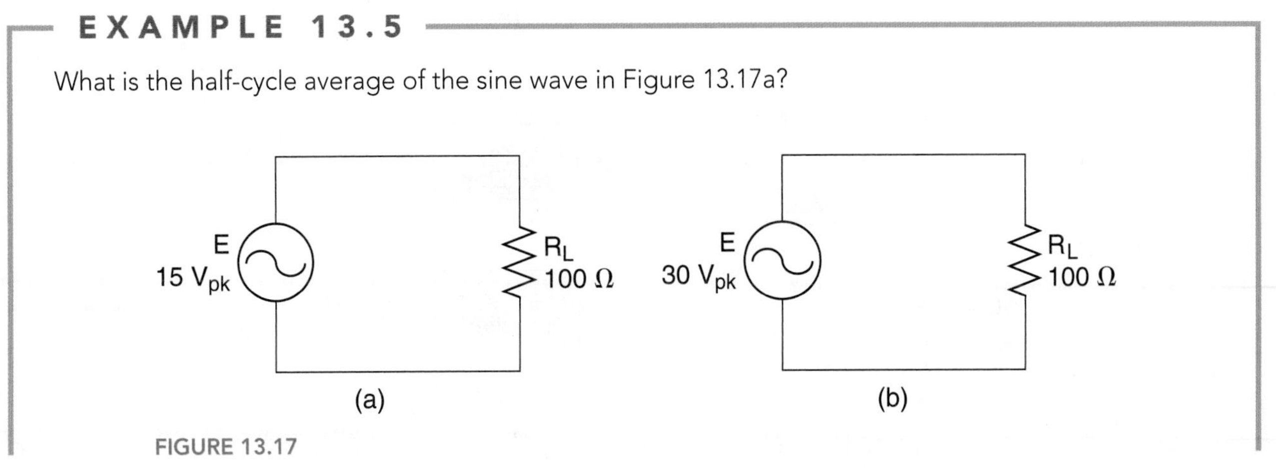

FIGURE 13.17

SOLUTION

The half-cycle average voltage (V_{AVE}) for the sine wave is calculated as follows:

$$V_{AVE} = \frac{2 \times V_{pk}}{\pi}$$
$$= \frac{2 \times 15\text{ V}}{\pi}$$
$$= \frac{30\text{ V}}{\pi}$$
$$= 9.55\text{ V}$$

PRACTICE PROBLEM 13.5

Calculate the value of V_{AVE} for the waveform in Figure 13.17b.

Measuring Half-Cycle Average Values

The average value of any AC voltage can be measured using a DC voltmeter. There is, however, one requirement: *The waveform must be altered so its alternations are all positive or all negative.* The reason for this is illustrated in Figure 13.18a. If we attempt to read the value of V_{AVE} as shown, both alternations of the source voltage are applied to the meter. As a result, the meter reads the average of the entire cycle, which is 0 V. So, if we want to read the average of either alternation, we have

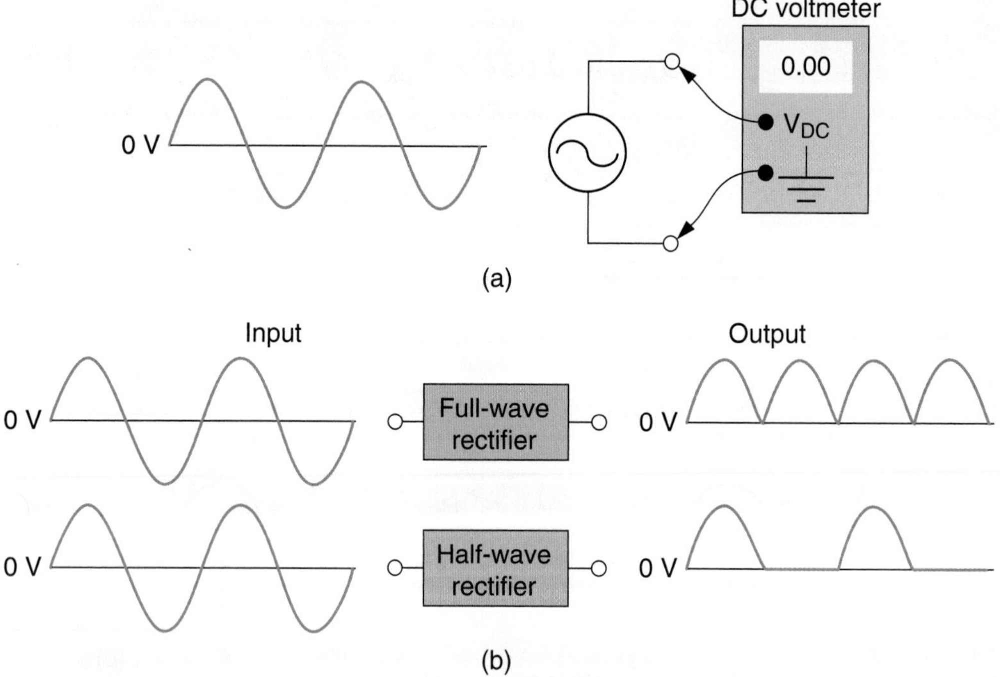

FIGURE 13.18 Rectifier circuits.

to make sure that the DC voltmeter "sees" only positive alternations or negative alternations—not both. This is accomplished using a circuit called a **RECTIFIER**.

RECTIFIER A circuit that converts AC to pulsating DC.

A rectifier is a circuit that converts AC to pulsating DC. That is, it converts the AC to a series of alternations that are all positive or all negative. The *full-wave rectifier* in Figure 13.18b essentially converts the negative input alternations to positive alternations.

Once an AC voltage has been rectified, a DC voltmeter can be used to measure its half-cycle average. For example, if a DC voltmeter is connected to the output of the full-wave rectifier in Figure 13.18b, the meter reading will equal the average of the positive alternation of the input waveform. As shown in the figure,

$$V_{DC} = \frac{2 \times V_{pk}}{\pi} = 0.637 \times V_{pk}$$

If you compare this relationship to equations (13.5) and (13.6), you'll see that they all define the DC average of one alternation of a voltage sine wave. Once a sine wave is rectified, its half-cycle average current can be measured using a DC ammeter.

Average AC Power and RMS Values

In Chapter 4, you were shown that load power can be found as

$$P_{RL} = \frac{V_{RL}^2}{R_L}$$

When dealing with AC circuits like the one in Figure 13.19a, we are interested in the **AVERAGE AC POWER** (P_{AC}) generated throughout each cycle. The concept of average AC power is illustrated in Figure 13.19b. If we assume that the curve shown represents the variations in load power over one cycle of load voltage, average AC power for the load is the value that falls midway between 0 W and the peak power value (P_{pk}).

AVERAGE AC POWER The value that falls midway between 0 W and the peak power value (P_{Pk}) for a sine wave.

As you know, power is proportional to the square of voltage. If we square the instantaneous voltage values in a sine wave, we end up with a V^2 wave like the one shown in Figure 13.19c. Comparing this waveform to the power waveform shows that the shapes of the two are identical.

Because the V^2 and power waveforms have identical shapes, we can combine their traits as shown in Figure 13.19d. As the combined waveform shows,

$$P = P_{AVE} \quad \text{when} \quad V^2 = \frac{V_{pk}^2}{2}$$

which can be rewritten as

$$P = P_{AVE} \quad \text{when} \quad V = \sqrt{\frac{V_{pk}^2}{2}}$$

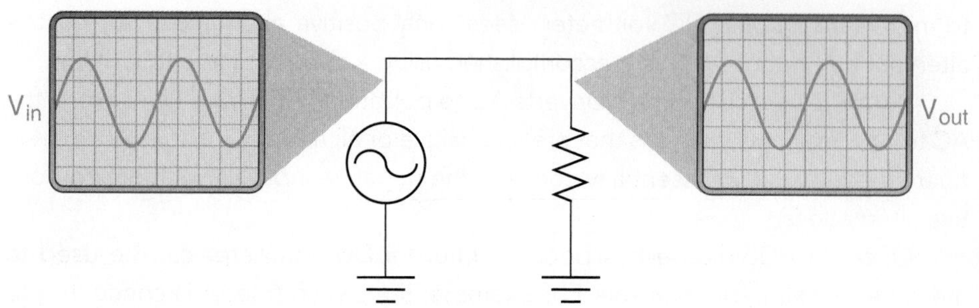

(a) Voltage changes in an ac circuit.

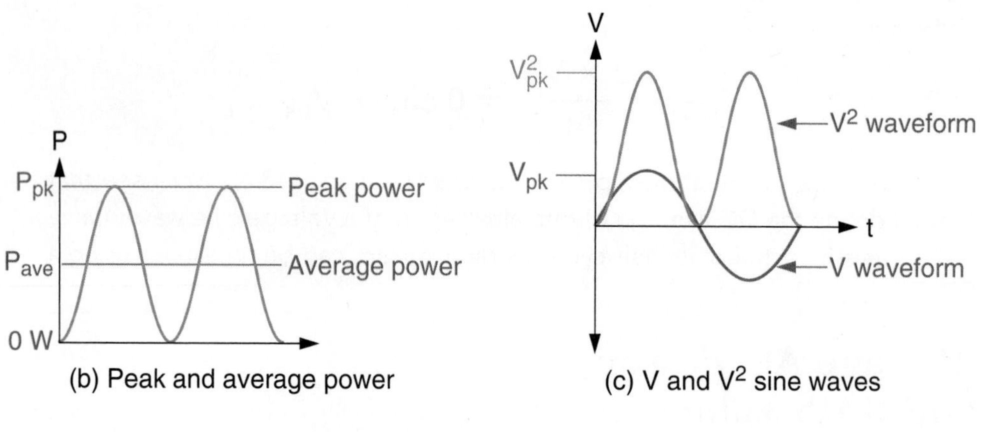

(b) Peak and average power

(c) V and V^2 sine waves

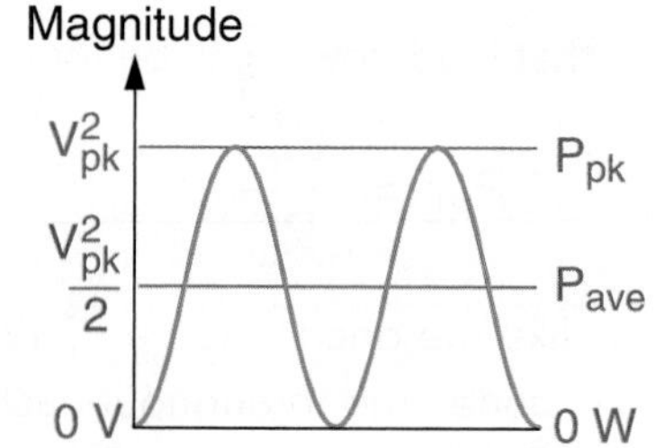

(d) Corresponding V^2 and P waveform values

FIGURE 13.19 Sine wave power and voltage.

The voltage relationship that precedes Figure 13.19 defines what is called the **ROOT-MEAN-SQUARE (RMS) VALUE** of voltage for the circuit in Figure 13.19a. The RMS value of voltage or current, when used in the appropriate power equations, gives you the average AC power of a waveform. By formula,

$$V_{RMS} = \sqrt{\frac{V^2_{pk}}{2}}$$

The equation above can be rewritten as

(13.7)

$$V_{RMS} = \frac{V_{pk}}{\sqrt{2}}$$

or

(13.8)
$$V_{RMS} = 0.707 \times V_{pk}$$

where $0.707 = \frac{1}{\sqrt{2}}$. By the same token, the RMS value of a current sine wave can be calculated using

(13.9)
$$I_{RMS} = \frac{I_{pk}}{\sqrt{2}}$$

or

(13.10)
$$I_{RMS} = 0.707 \times I_{pk}$$

Example 13.6 demonstrates one approach to calculating RMS values.

EXAMPLE 13.6

What are the RMS values of voltage and current for the circuit in Figure 13.20a?

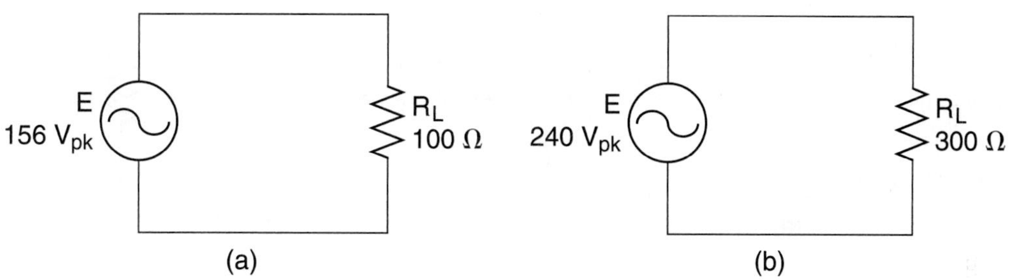

FIGURE 13.20

SOLUTION

The RMS voltage is found as

$$V_{RMS} = 0.707 \times V_{pk}$$
$$= 0.707 \times 156\text{ V}$$
$$= 110\text{ V}$$

The peak circuit current can be calculated using Ohm's law, as follows:

$$I_{Pk} = \frac{V_{pk}}{R_L}$$

$$= \frac{156 \text{ V}}{100 \ \Omega}$$
$$= 1.56 \text{ A}$$

The RMS current can now be found as

$$I_{RMS} = 0.707 \times I_{pk}$$
$$= 0.707 \times 1.56 \text{ A}$$
$$= 1.1 \text{ A}$$

Of course, we could have simply divided the value of V_{RMS} by the value of R_L to calculate the RMS current. The result would have been the same.

PRACTICE PROBLEM 13.6

Calculate the RMS voltage and current for the circuit in Figure 13.20b.

As stated earlier, AC power calculations are always made using RMS values. Example 13.7 demonstrates the process for determining average load power when the peak load voltage is known.

EXAMPLE 13.7

What is the average load power being generated by the circuit in Figure 13.21a?

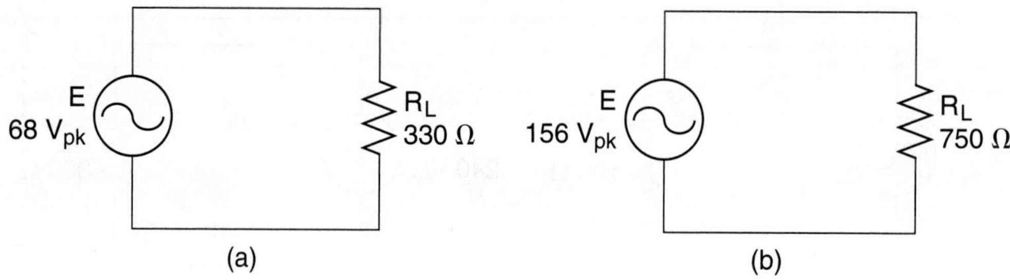

FIGURE 13.21

SOLUTION

Calculating the average load power for the circuit in Figure 13.21a begins with calculating the RMS load voltage, as follows:

$$V_{RMS} = 0.707 \times V_{pk}$$
$$= 0.707 \times 68 \text{ V}$$
$$= 48.1 \text{ V}$$

Once the peak voltage is converted to an RMS value, the load power can be found as

$$P_L = \frac{V_{RMS}^2}{R_L}$$

$$= \frac{(48.1\ V)^2}{330\ \Omega}$$

$$= 7.01\ W$$

PRACTICE PROBLEM 13.7

Calculate the average load power for the circuit in Figure 13.21b.

There are several more points regarding RMS values that need to be made:

- Average AC power and DC power are equivalents. That is, a 48-V RMS sine wave provides the same amount of power to a resistive load as a 48-V DC source.
- AC power calculations are always performed using RMS values. For example, RMS values of voltage and current must be used in the P = I × E equation to obtain the correct result.
- Because power is a primary concern with any electrical system, RMS values are the primary values of interest in electrical circuits.
- Any magnitude-related measurement that is not specifically identified as peak, peak-to-peak, or average is an RMS value.

Measuring RMS Values Using a DMM

RMS values are commonly referred to as **EFFECTIVE VALUES**. The name stems from the fact that RMS values produce the same heating effect as their equivalent DC values.

EFFECTIVE VALUES RMS values, which produce the same heating effect as their equivalent DC values.

Effective (RMS) voltage is measured using an *AC voltmeter*, while effective current is measured using an *AC ammeter*.

Figure 13.22 shows a DMM function *selector switch*. With most DMMs, switching from a DC scale to an AC scale is simply a matter of turning the selector

NOTE: The DC voltage scales in Figure 13.22 are identified by the V⎓ label and the AC voltage scales are identified by the V~ label. This particular meter does not measure alternating current.

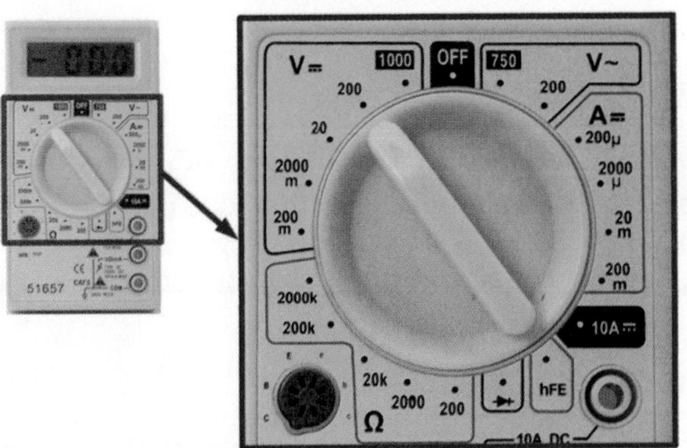

FIGURE 13.22 A DMM function selector switch.

switch to the desired measurement. On others, you may also need to switch your probe to a specially designated input.

Averaging Meters vs. "True-RMS" Meters

There are actually two types of AC meters. The first of these, which is the less expensive and more common of the two, is the averaging meter.

AVERAGING METER An AC meter that measures the half-cycle average of a waveform and then converts that value to an RMS readout.

An **AVERAGING METER** does not directly measure RMS values. Rather, it measures the half-cycle average of an AC waveform and then converts that average to an RMS readout. The meter contains a rectifier that converts the AC waveform input to pulsating DC (as described earlier). The meter then measures the rectifier output and multiplies the measured value by 1.11 to produce an RMS reading. Why multiply the reading by 1.11? According to equation (13.6) the half-cycle average of a sine wave has a value of

$$V_{AVE} = 0.637 \times V_{pk}$$

If we multiply this equation by 1.11, we get

$$\begin{aligned} 1.11 \times V_{AVE} &= 1.11 \times 0.637 \times V_{pk} \\ &= 0.707 \times V_{pk} \end{aligned}$$

By definition, $V_{RMS} = 0.707 \times V_{pk}$, so multiplying V_{AVE} by 1.11 gives us V_{RMS}.

Averaging meters are limited in their applications. They can only be used to measure sinusoidal (sine wave) voltages and currents. They cannot be used on other waveforms, like those shown in Figure 13.23.

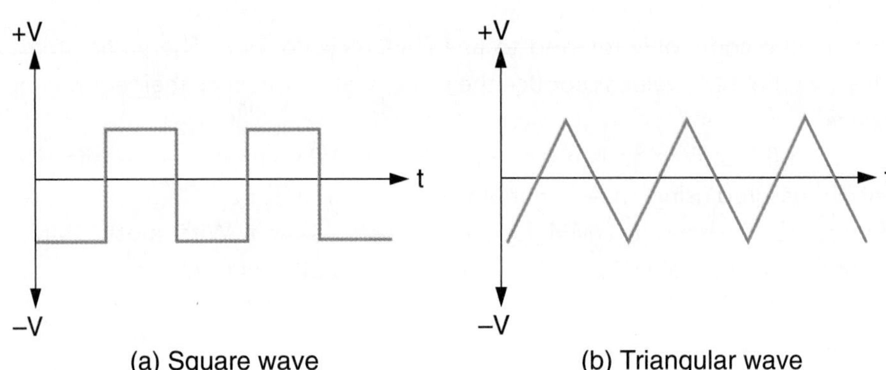

FIGURE 13.23 Square waves and triangular waves.

TRUE-RMS METER An AC meter that uses relatively complex circuitry to directly measure the RMS value of any true AC waveform.

A **TRUE-RMS METER** uses relatively complex circuitry to directly measure the RMS value of any true AC waveform. You may recall that a true AC waveform is one that has identical positive and negative alternations. A true-RMS meter can be used to accurately measure either of the waveforms in Figure 13.23, while an averaging RMS meter cannot.

Summary

In this section, you have been introduced to peak, peak-to-peak, average, and effective (RMS) values and the means by which each is measured. The various sine

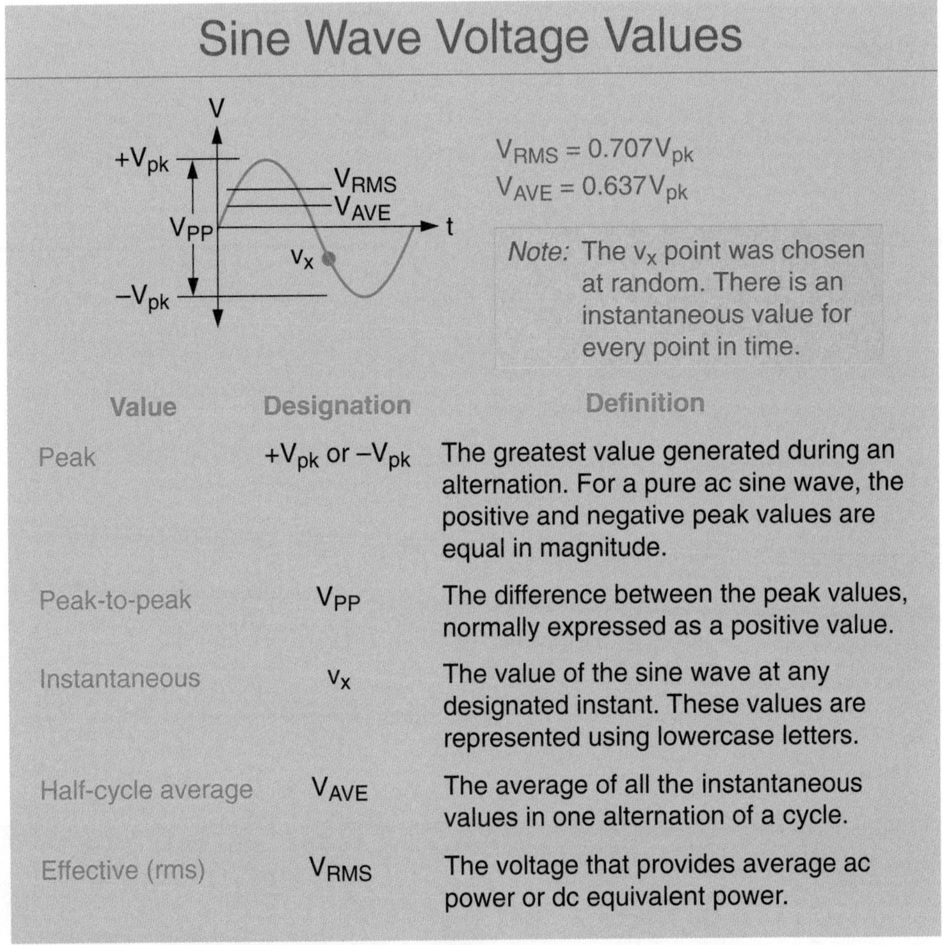

Sine Wave Voltage Values
V
$+V_{pk}$
V_{RMS}
V_{AVE}
V_{PP}
t
v_x
$-V_{pk}$
$V_{RMS} = 0.707V_{pk}$
$V_{AVE} = 0.637V_{pk}$
Note: The v_x point was chosen at random. There is an instantaneous value for every point in time.
Value
Designation
Definition
Peak
$+V_{pk}$ or $-V_{pk}$
The greatest value generated during an alternation. For a pure ac sine wave, the positive and negative peak values are equal in magnitude.
Peak-to-peak
V_{PP}
The difference between the peak values, normally expressed as a positive value.
Instantaneous
v_x
The value of the sine wave at any designated instant. These values are represented using lowercase letters.
Half-cycle average
V_{AVE}
The average of all the instantaneous values in one alternation of a cycle.
Effective (rms)
V_{RMS}
The voltage that provides average ac power or dc equivalent power.

FIGURE 13.24 Sine wave voltage measurements.

wave voltage measurements are summarized in Figure 13.24. Even though voltages are emphasized in the figure, sine wave currents are labeled and defined in a similar fashion.

Sine wave voltages and currents are measured using various pieces of test equipment. Voltage and current measurements are summarized in Table 13.1.

TABLE 13.1 • AC Measurements

VALUE	...IS MEASURED USING A/AN...
Effective (RMS) voltage	AC voltmeter
Effective (RMS) current	AC ammeter
Half-cycle average voltage	DC voltmeter
Half-cycle average current	DC ammeter
Peak voltage	oscilloscope
Peak-to-peak voltage	oscilloscope

Note that peak current and peak-to-peak current cannot be measured using any standard piece of test equipment.

You have now been introduced to the time, voltage, and current characteristics of sinusoidal waveforms. In the next section, we are going to take a closer look at sine waves and the methods used to generate them.

PROGRESS CHECK

1. What is the relationship between the two alternations of a pure AC waveform?
2. How do you calculate the peak-to-peak value of a waveform?
3. What are instantaneous values? How are they designated?
4. What is the half-cycle average of a sine wave?
5. Why is average AC power an important consideration?
6. What is the root-mean-square (RMS) value of a sine wave?
7. What is the relationship between RMS voltage (and current) and DC equivalent power?
8. In terms of voltage (or current) designations, how do you know when a value is an RMS value?
9. Describe the use of the oscilloscope to measure peak voltages and peak-to-peak voltages.
10. What is used to measure half-cycle average voltage? Half-cycle average current?
11. What is used to measure effective (RMS) voltage? Effective (RMS) current?
12. What is the difference between an averaging meter and a true-RMS meter?

13.3 SINE WAVES: PHASE AND INSTANTANEOUS VALUES

In Chapter 10, you were shown how the current through a coil can be used to generate a magnetic field. As a review, this principle of magnetic induction is illustrated in Figure 13.25.

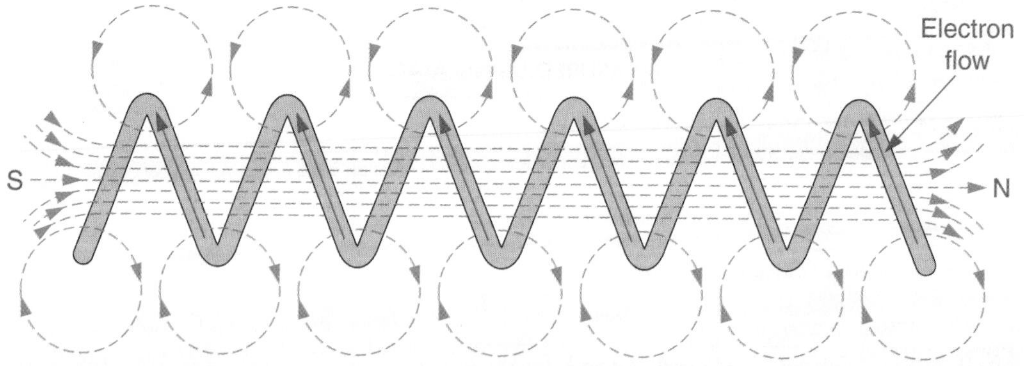

FIGURE 13.25 Coil current and magnetic lines of flux.

Just as current can be used to produce a magnetic field, the field produced by a magnet can be used to generate current through a conductor. In this section, you will be shown how a magnetic field can be used to generate a sinusoidal alternating current. You will also be shown how sine wave characteristics affect instantaneous values, and the methods used to calculate those values.

Magnetic Induction of Current

If a conductor is cut by perpendicular lines of magnetic force, maximum current is induced in the conductor. This principle is illustrated in Figure 13.26. In Figure 13.26a, a conductor is at a 90° angle to the lines of force. When the conductor is moved downward through the lines of force, a current is generated through the conductor in the direction shown.

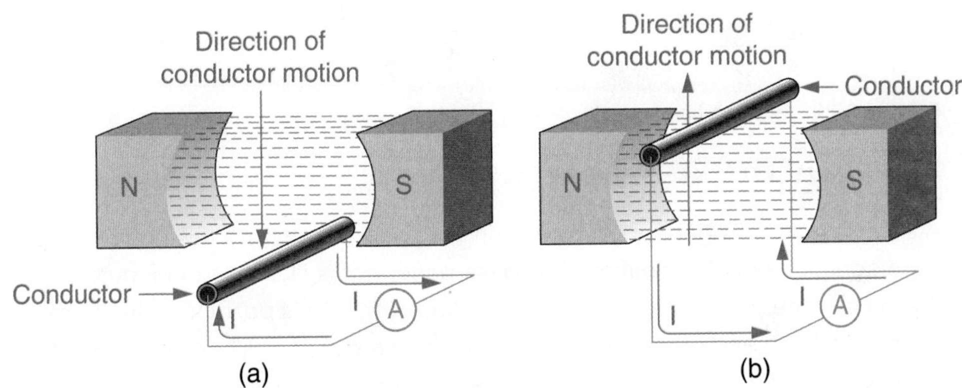

FIGURE 13.26 Magnetic induction of current.

In Figure 13.26b, the direction of conductor motion is reversed. As the conductor moves upward through the lines of force, a current is generated in the direction shown. As you can see, reversing the direction of motion reverses the direction of the conductor current.

In Figures 13.26a and 13.26b, the conductor, lines of force, and direction of conductor motion are all at 90° angles to each other. As long as the angles are maintained at 90°, maximum current is generated in the wire. However, if any one of the angles changes, the amount of current through the conductor decreases. For example, the current generated in the conductor drops to 0 A when the conductor motion is parallel to the direction of the flux. This drop in current happens because the conductor is no longer cutting the lines of force.

Applying the Left-Hand Rule

In Chapter 10, you were introduced to the left-hand rule. You may recall that the left-hand rule is a memory aid designed to help you remember the relationship between a current and its associated magnetic field.

We can use the left-hand rule to help determine the direction of the current that is generated in a conductor passing through a magnetic field. The application of the left-hand rule in this instance is illustrated in Figure 13.27.

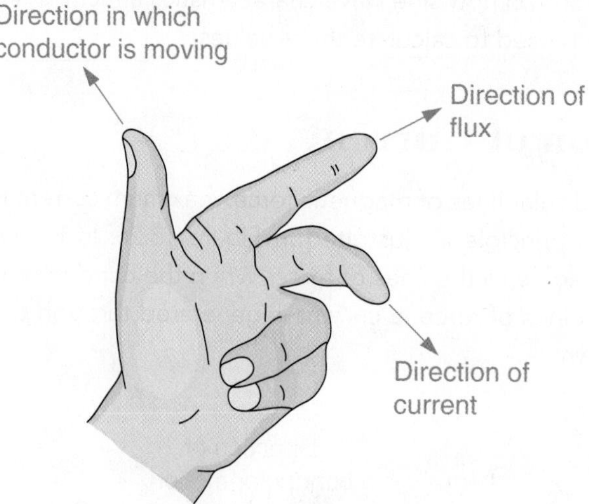

FIGURE 13.27 The left-hand rule.

Generating a Sine Wave

A sine wave can be generated by rotating a "loop conductor" through a stationary magnetic field. A simplified sine wave generator is illustrated in Figure 13.28. As you can see, there is a *loop conductor* (or *rotor*) that is positioned in the magnetic field. Each end of the rotor is connected to a *slip ring*. When an ammeter (or some other load) is connected across the slip rings, the circuit is completed, providing a path for any current produced by the generator.

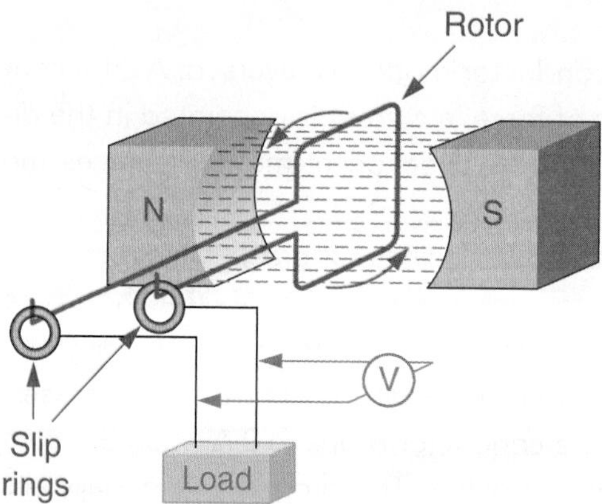

FIGURE 13.28 A simplified sine wave generator.

As it rotates through the magnetic field, a current is induced in the rotor. The magnitude of the current and its direction are determined by the direction and angle of rotor movement relative to the magnetic lines of force. This point is illustrated in Figure 13.29.

Figure 13.29 illustrates the operation of the simplified generator shown in Figure 13.28. The voltage curves represent the load voltage that would be produced

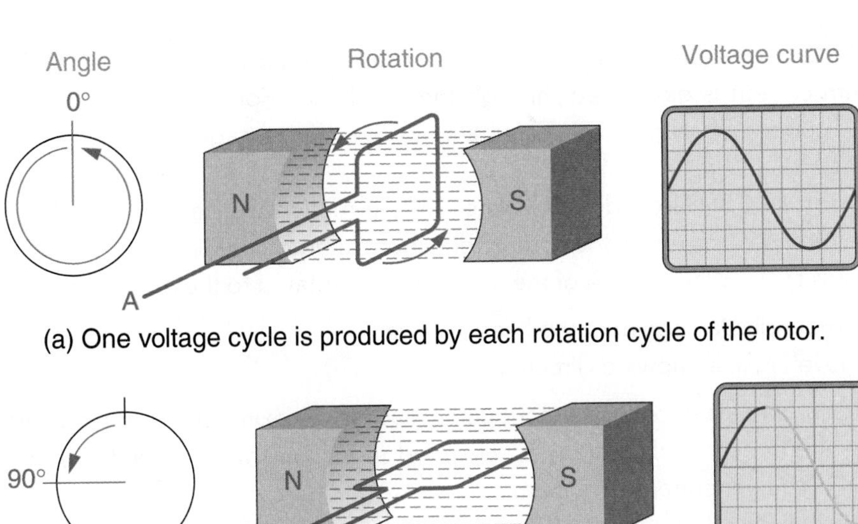

(a) One voltage cycle is produced by each rotation cycle of the rotor.

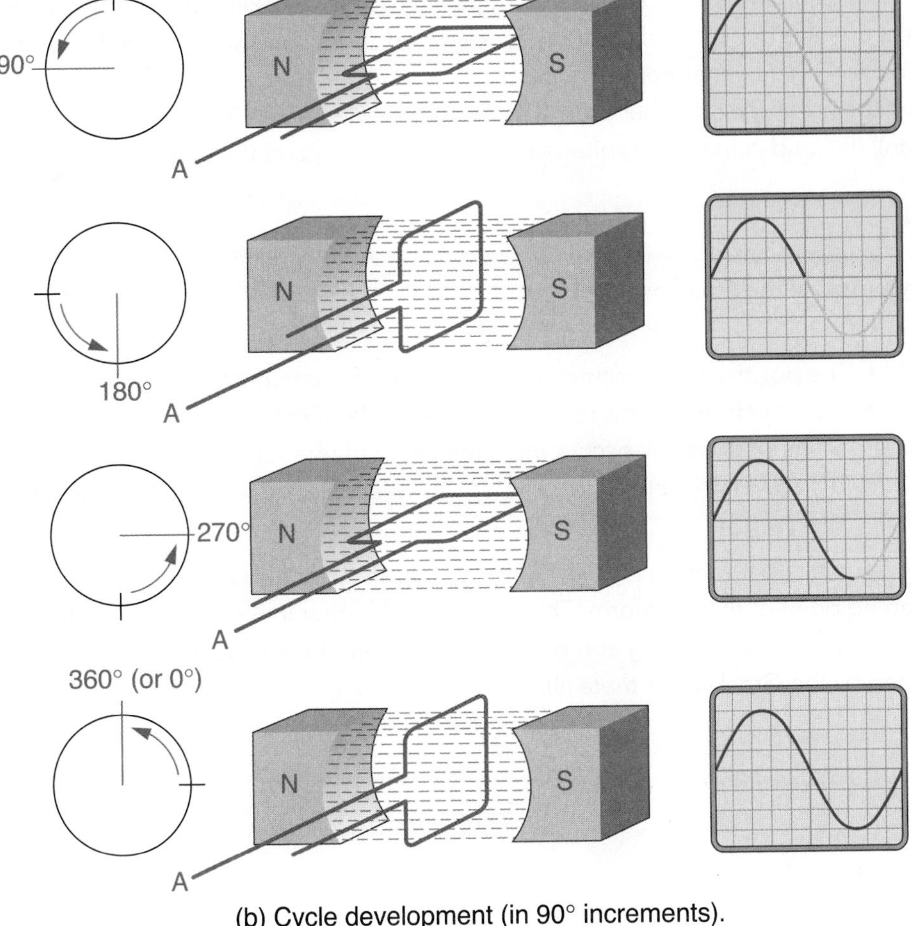

(b) Cycle development (in 90° increments).

FIGURE 13.29 Generating a sine wave.

by any generated current. (To simplify this illustration, the slip rings and load have been omitted.)

When side A of the conductor is at 0°, its motion is parallel to the lines of force. Since it is not cutting the lines of force, the current through the conductor is 0 A and there is no voltage across the load. When A rotates through the 90° point, it is:

- Perpendicular to the lines of force
- Traveling in a downward direction

Since the angle of the conductor motion (relative to the lines of force) is 90°, maximum current is generated through the conductor. For the sake of discussion, this current is assumed to generate a positive load voltage, as shown in the voltage curve.

When side A of the conductor reaches 180°, its motion is once again parallel to the lines of force. As a result, the conductor current returns to 0 A and the load voltage returns to 0 V. When side A of the conductor has rotated to the 270° point, it is:

- Perpendicular (again) to the lines of force
- Traveling in an upward direction

Once again, current through the conductor reaches a maximum value. However, because the direction of motion has changed (from downward to upward), the direction of the generated current has reversed. As a result, the polarity of the load voltage has reversed. When side A of the conductor reaches 360° (its starting position), the load current and voltage both return to zero, and the cycle begins again.

Phase

If you compare each curve in Figure 13.29 with its corresponding angle of rotation, you'll see that:

- The positive peak occurs when the angle of rotation is 90°.
- The waveform returns to 0 V when the angle of rotation reaches 180°.
- The negative peak occurs when the angle of rotation reaches 270°.
- The waveform returns to 0 V (again) when the angle of rotation reaches 360° (or 0°).

As the angle diagrams in Figure 13.29 indicate, these angles are all relative to the starting point of the waveform. This is illustrated further in Figure 13.30. Note that the horizontal axis of the graph is used to represent the angles of rotation, designated by the Greek letter *theta* (θ).

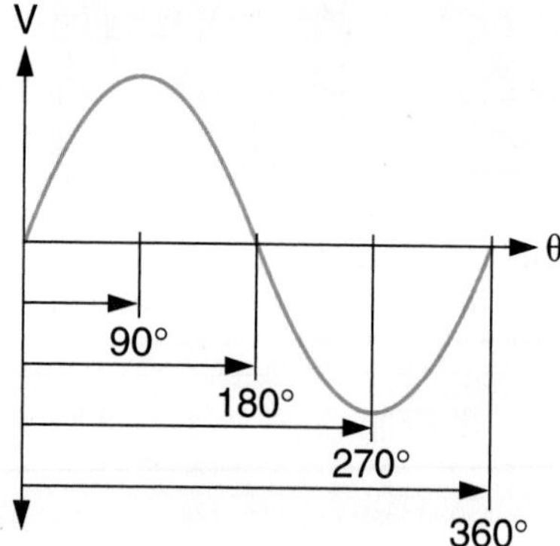

FIGURE 13.30 A sine wave measured in degrees.

PHASE The position of a point on a waveform relative to the start of the waveform, usually given in degrees.

Any point on a waveform can be identified by its phase. The **PHASE** of a given point is its position relative to the start of the waveform, usually measured in degrees. Several points are identified by phase in Figure 13.31.

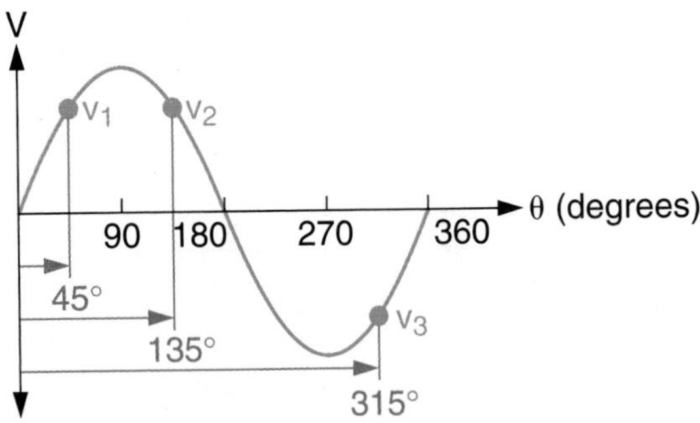

FIGURE 13.31 More sine wave degree measurements.

Sine waves are often described in terms of phase for a couple of reasons:

1. Phase measurements allow us to identify points on a waveform independent of waveform period and frequency.
2. Phase plays a role in many RC and RL circuit voltage, current, and power dissipation values.

The first of these points is illustrated in Figure 13.32. The two waveforms shown have frequencies of 250 Hz and 10 kHz. The 250 Hz waveform reaches its positive peak after 1 ms. The 10 kHz waveform reaches the same point after only 25 μs. Even so, both waveforms reach a positive peak when $\theta = 90°$. As you can see, the phase allows us to identify corresponding points on two waveforms independent of time and frequency.

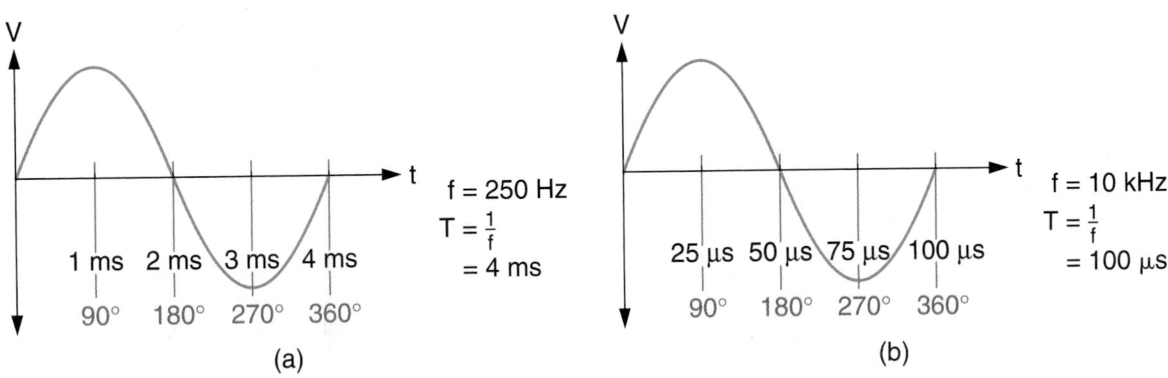

FIGURE 13.32 Sine wave phase and time measurements.

Phase Angles

There are many cases where a circuit will contain two (or more) sine waves of equal frequency that do not reach their peaks at the same time. Waveforms of this nature are said to be *out of phase*. Figure 13.33a shows an example of two waveforms that are out of phase. As the figure implies, the two waveforms have the same period, but they do not reach their peak values and zero-crossings at the same time. The phase difference between two waveforms is referred to as their **PHASE ANGLE**. The phase angle between two waveforms is identified in Figure 13.33b.

PHASE ANGLE The phase difference between two waveforms.

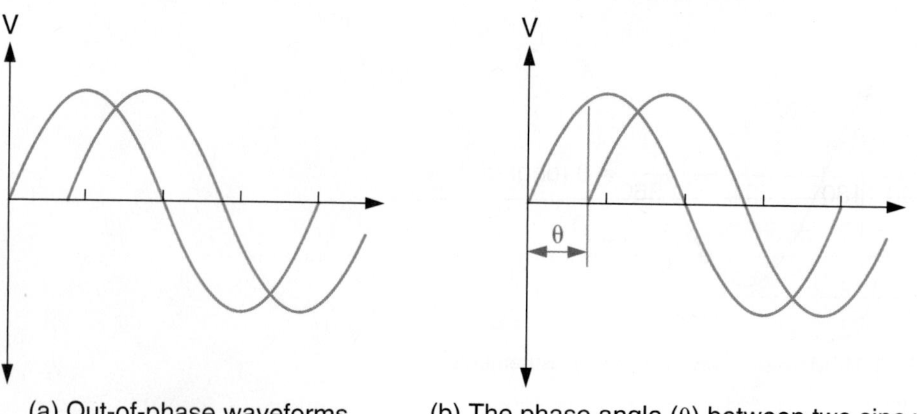

FIGURE 13.33 Phase angle.

Sometimes it is easy to determine the phase relationship between a pair of waveforms. For example, the waveforms in Figure 13.34a are in phase. This is indicated by the fact that their peaks and zero crossings occur at the same time. The waveforms in Figure 13.34b are 90 degrees out of phase. Two waveforms are 90 degrees out of phase when the 0° point of one occurs at the same time as the 90° point of the other. The waveforms in Figure 13.34c are 180° out of phase. This is indicated by the fact that one waveform reaches its positive peak at the same time the other reaches its negative peak.

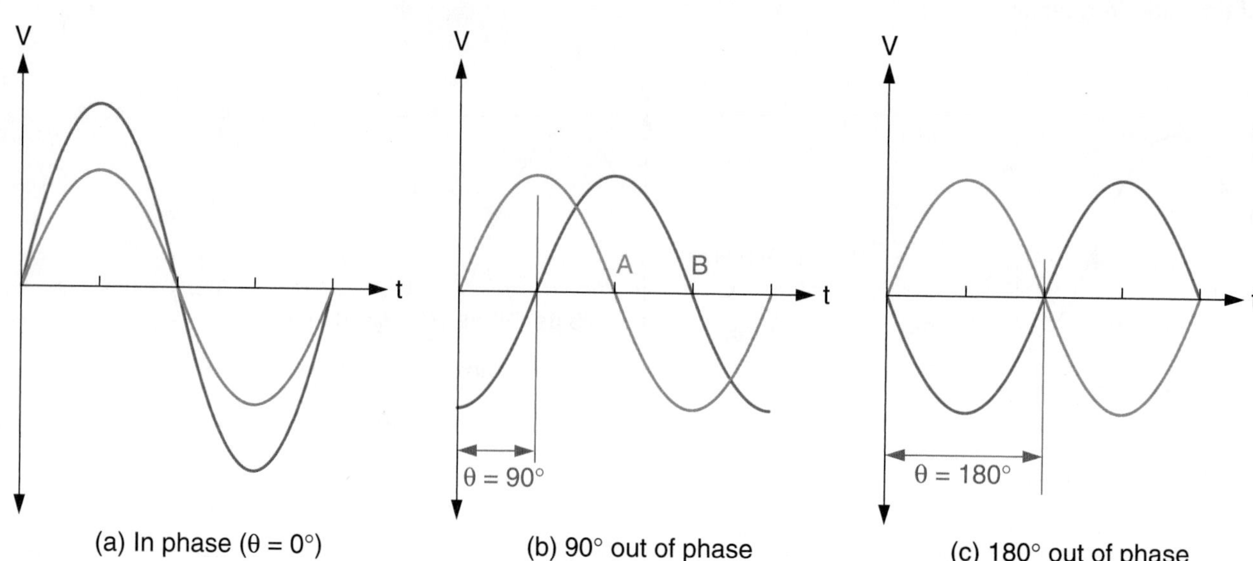

FIGURE 13.34 Phase angles that are easy to recognize.

Instantaneous Values

Earlier in this chapter, you were told that an instantaneous value is the magnitude of a voltage (or current) at a specified point in time. For example, v_1 in Figure 13.35a is an instantaneous value. Now that we have discussed phase relationships, we are ready to look at the means used to determine any instantaneous value.

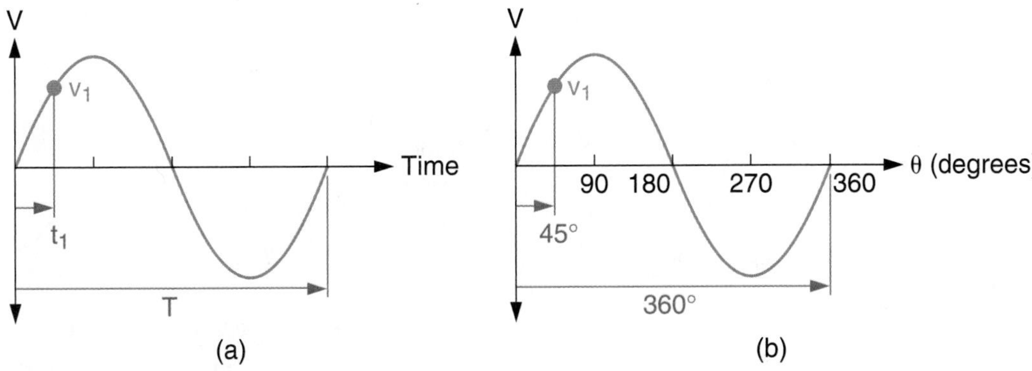

FIGURE 13.35 Time and phase measurements of an instantaneous value.

The instantaneous value of a sine wave voltage (or current) is proportional to the peak value of the waveform and the sine of the phase angle. By formula,

(13.11)
$$v = V_{pk}\sin\theta$$

Example 13.8 demonstrates the use of this equation.

EXAMPLE 13.8

Assuming that the waveform in Figure 13.35b has a peak value of 10 V, what is the value of v_1?

SOLUTION

With a peak value of 10 V, the value of v_1 can be found as

$$\begin{aligned} v_1 &= V_{pk}\sin\theta \\ &= (10\text{ V}) \times (\sin 45°) \\ &= (10\text{ V}) \times (0.707) \\ &= 7.07\text{ V} \end{aligned}$$

This result indicates that $v_1 = 7.07$ V when $\theta = 45°$.

PRACTICE PROBLEM 13.8

A sine wave has a peak value of 12 V. Determine the instantaneous value of the waveform when $\theta = 60°$.

The peak value of a sine wave (like those in Figure 13.35) is a constant. That is, the peak value does not change over the course of a cycle. However, the phase angle is always changing. Since the instantaneous values are proportional to the sine of the phase angle (sin θ), the waveform is said to vary at a "sinusoidal" rate. This is where the name *sine wave* comes from.

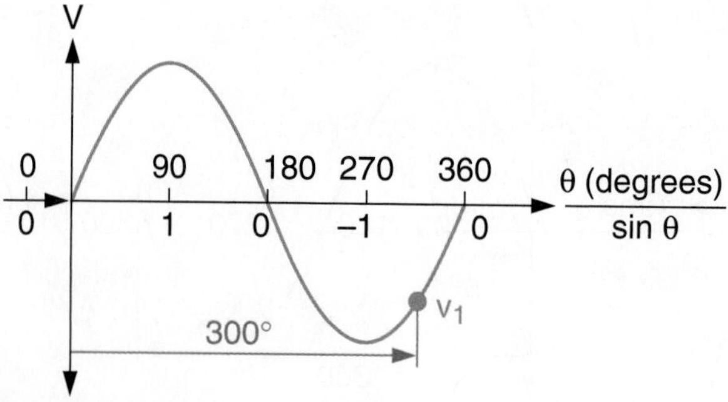

FIGURE 13.36

Figure 13.36 shows a sine wave divided into 90° increments. At each increment, the value of sin θ is shown. From the sin θ values shown, we can draw the following conclusions:

1. During the positive alternation of the waveform, sin θ is positive. Therefore, the value of $V_{pk}\sin\theta$ is also positive.
2. During the negative alternation of the waveform, sin θ is negative. Therefore, the value of $V_{pk}\sin\theta$ is also negative.

This means that equation (13.11) provides the correct polarity for any instantaneous value in either alternation. This is demonstrated further in Example 13.9.

EXAMPLE 13.9

Assuming the waveform in Figure 13.36 has a peak value of 15 V, what is the value of v_1?

SOLUTION

Assuming a peak value of 15 V, the value of v_1 can be found as

$$\begin{aligned} v_1 &= V_{pk}\sin\theta \\ &= (15\text{ V}) \times (\sin 300°) \\ &= (15\text{ V}) \times (-0.866) \\ &= -12.99\text{ V} \end{aligned}$$

PRACTICE PROBLEM 13.9

A sine wave has a peak value of 30 V. Determine the instantaneous value of the waveform when $\theta = 200°$.

Putting It All Together

A sine wave can be generated through magnetic induction. When a conductor cuts through a stationary magnetic field, a current is induced in the conductor. The value of the current is proportional to the angle of the conductor relative to the magnetic field:

- When the conductor cuts through the magnetic field at a 90° angle, maximum current is induced in the conductor.
- When the conductor is parallel to the magnetic field, minimum current is generated through the conductor.

A sine wave can be generated by rotating a loop conductor, or *rotor*, through a magnetic field (as illustrated in Figure 13.29):

- The positive peak of the waveform occurs when the angle of rotation is 90°.
- The waveform returns to zero when the angle of rotation equals 180°.
- The negative peak of the waveform occurs when the angle of rotation is 270°.
- The waveform returns to 0 V (again) when the angle of rotation is 360° (or 0°).

The phase of a point on a sine wave indicates its position relative to the start of the waveform, usually expressed in degrees. Sine waves are often described in terms of phase.

When two sine waves of equal frequency do not peak at the same time, they are said to be *out of phase*. The phase angle between two such waves is measured in degrees (as shown in Figure 13.34).

The instantaneous value of voltage or current at any point on a sine wave equals the product of its peak value and sin θ.

One More Point

You have now been introduced to the basic concepts of sine waves and sine wave measurements. In the next section, we will close out the chapter by briefly discussing another type of waveform, called a *rectangular wave*. Though not encountered by many electrical workers on a regular basis, rectangular waves play an important role in industrial control systems and in some telecommunications cabling applications.

PROGRESS CHECK

1. Briefly describe how current can be generated by magnetic induction.
2. Briefly describe the operation of the simple AC generator in Figure 13.29.
3. What is the phase of a point on a waveform?
4. Why are sine waves commonly described in terms of phase?
5. What is meant by the term *phase angle*?
6. What factors determine the instantaneous value of a sine wave? Which of these values is a constant for a given waveform? Which varies from one point on the waveform to the next?

13.4 RECTANGULAR WAVES

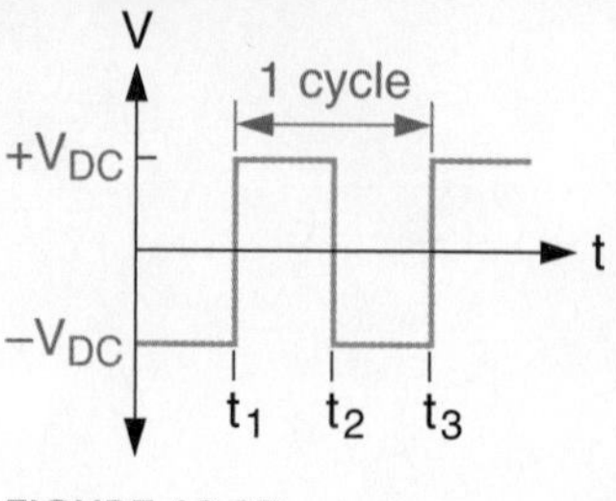

FIGURE 13.37

Rectangular waves are waveforms that alternate between two DC levels. An example of a rectangular waveform is shown in Figure 13.37. The waveform shown alternates between two levels, labeled $-V_{DC}$ and $+V_{DC}$.

Figure 13.38 shows the response of a circuit to a rectangular-wave input. From t_1 to t_2, the source has the polarity and current direction shown in Figure 13.38b. From t_2 to t_3, the source polarity and current direction are reversed as shown in Figure 13.38c. Even though the rectangular wave is made up of alternating DC levels, it produces current that alternates direction. Therefore, it is correctly classified as being an AC waveform.

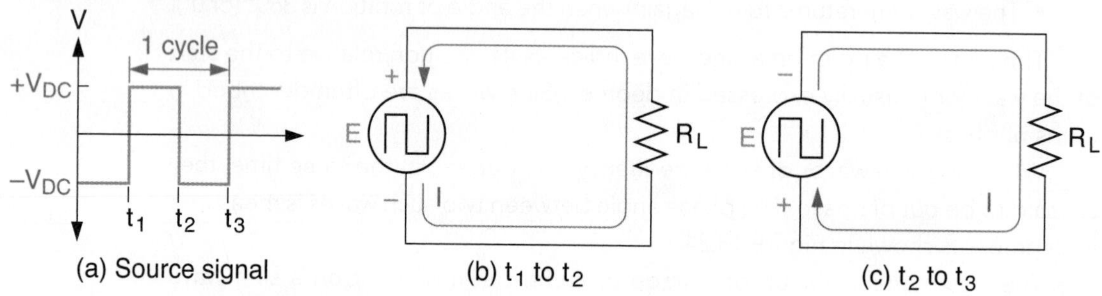

FIGURE 13.38 Circuit currents generated by a rectangular wave.

Terminology and Time Measurements

PULSE WIDTH (PW) The positive alternation of a rectangular waveform.

SPACE WIDTH (SW) The negative alternation of a rectangular waveform.

The parts of a rectangular waveform are identified as shown in Figure 13.39. The positive half-cycle is referred to as the **PULSE WIDTH (PW)**. The negative half-cycle is referred to as the **SPACE WIDTH (SW)**. The sum of the pulse width and space width equals the period of the waveform.

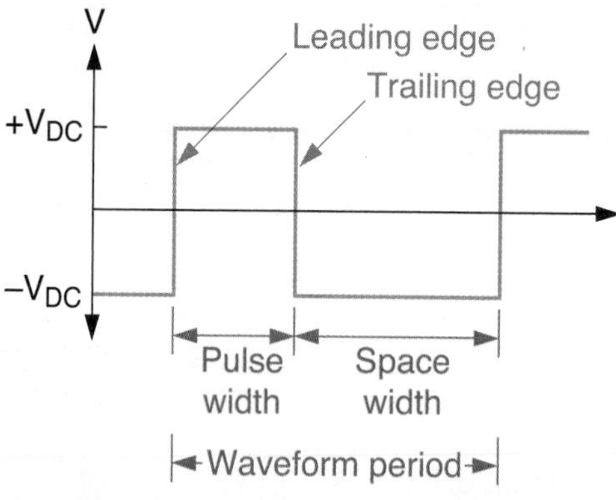

FIGURE 13.39 Parts of a rectangular waveform.

Duty Cycle

DUTY CYCLE The ratio of pulse width to waveform period, expressed as a percent.

The **DUTY CYCLE** of a rectangular waveform is the ratio of pulse width to the waveform period, expressed as a percent. By formula,

(13.12)

$$\text{Duty cycle}(\%) = \frac{PW}{T} \times 100$$

where

PW = the pulse width of the circuit input
T = the period of the circuit input

Example 13.10 demonstrates the calculation of duty cycle for a rectangular waveform.

EXAMPLE 13.10

What is the duty cycle of the waveform shown in Figure 13.40a?

SOLUTION

The waveform duty cycle has a value of

$$\text{Duty cycle}(\%) = \frac{PW}{T} \times 100$$
$$= \frac{50\ \mu s}{150\ \mu s} \times 100$$
$$= 33.3\%$$

This value indicates that the pulse width makes up 33.3% of the total cycle.

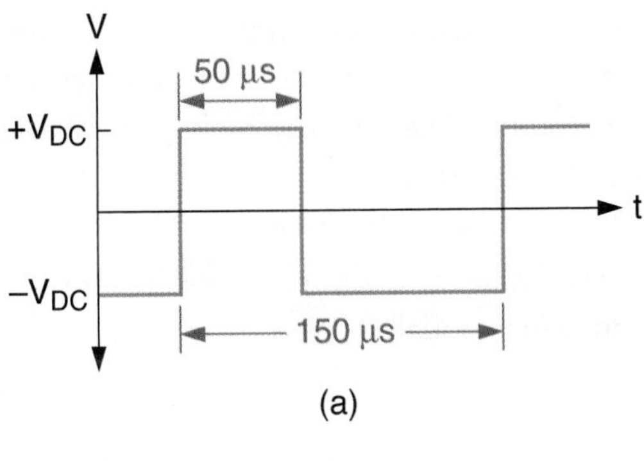

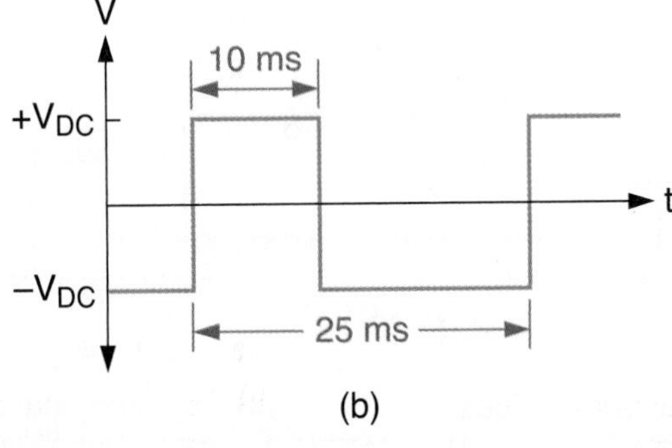

FIGURE 13.40

PRACTICE PROBLEM 13.10

Calculate the duty cycle of the waveform shown in Figure 13.40b.

Square Waves

SQUARE WAVE A special-case rectangular waveform that has equal pulse width and space width values.

The **SQUARE WAVE** is a special-case rectangular waveform that has equal pulse width and space width values. A square wave is shown in Figure 13.41. Since the values of PW and SW are equal, the duty cycle of a square wave is always 50%.

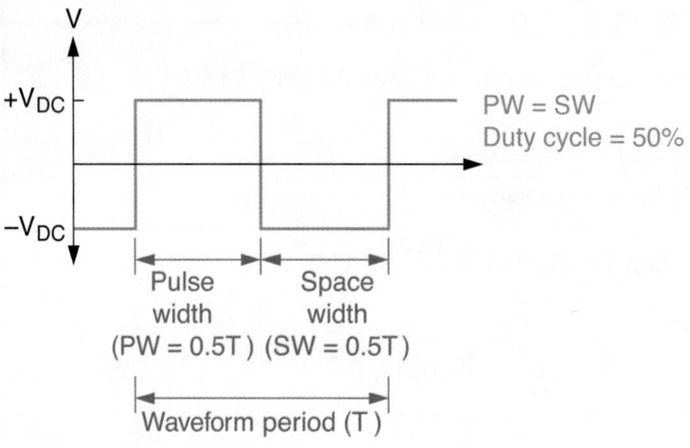

FIGURE 13.41 A square wave.

SYMMETRICAL WAVEFORM A waveform that has identical alternations.

A **SYMMETRICAL WAVEFORM** is one with cycles that are made up of identical halves. When the two half-cycles of a waveform have equal time spans, they are said to be *symmetrical in time*. When the two half-cycles of a waveform have equal peak values, they are said to be *symmetrical in amplitude*. For example, look at the square wave in Figure 13.42a. In this case:

1. PW = SW, so the waveform is symmetrical in time.
2. The waveform varies equally above and below 0 V, so it is symmetrical in amplitude.

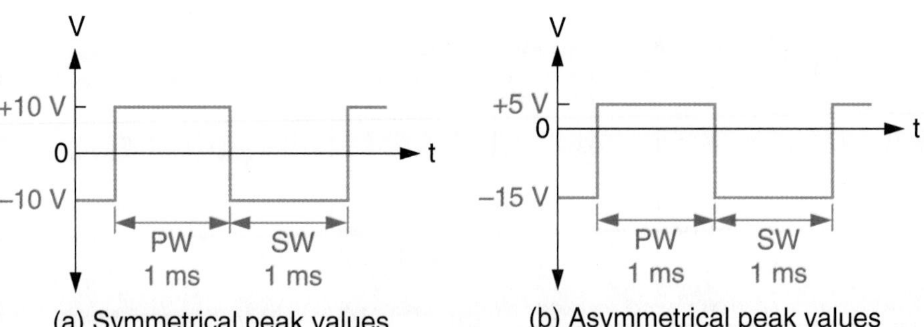

FIGURE 13.42 Symmetrical and asymmetrical amplitudes.

A square wave with equal positive and negative peak values has an average value of 0 V. For a square wave with asymmetrical peak values, the average voltage can be calculated using

(13.13)
$$V_{AVE} = \frac{+V_{pk} + (-V_{pk})}{2}$$

For the waveform in Figure 13.42b, the average value can be found as

$$\begin{aligned} V_{AVE} &= \frac{+V_{pk} + (-V_{pk})}{2} \\ &= \frac{+5\text{ V} + (-15\text{ V})}{2} \\ &= \frac{-10\text{ V}}{2} \\ &= -5\text{ V} \end{aligned}$$

Solid-State Relays: A Rectangular Wave Application

Rectangular waves can be found in a variety of electrical applications. One common application is illustrated in Figure 13.43. In Chapter 10, you were introduced to the electromagnetic relay, a component that is used to control an AC signal. A **SOLID-STATE RELAY** is a component that is used for the same purpose.

SOLID-STATE RELAY A semiconductor relay that is used to control an AC signal.

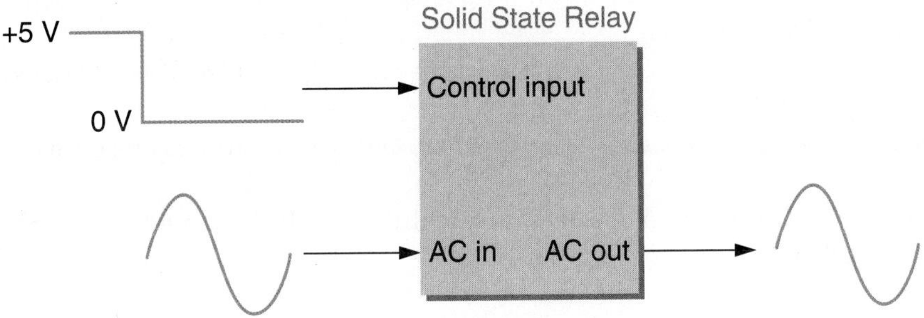

FIGURE 13.43

The solid-state relay in Figure 13.43 has two inputs and one output. The signal at the control input determines whether the AC input passes through the relay to the AC output terminal. For example, when the control input to the relay in Figure 13.43 is $+5\ V_{DC}$, the relay blocks the AC input signal. When the control input goes to ground, the AC input signal passes through the relay to its load, such as a motor. As long as the control input remains low, the AC signal is applied to the load, and it operates normally. As soon as the control input returns to $+5\ V_{DC}$, the AC input is blocked again, and the load ceases to operate.

PROGRESS CHECK

1. What is a rectangular waveform?
2. Draw a rectangular waveform and identify each of the following: the pulse width, the space width, and the waveform period.
3. What is duty cycle?
4. What is a square wave?
5. What is a symmetrical waveform?
6. Describe the operation of a solid-state relay.

13.5 SUMMARY

Here is a summary of the major points that were covered in this chapter:

Alternating Current (AC): Overview and Time Measurements

- The term *alternating current* is used to describe any current that periodically changes direction.
- A graph of the relationship between magnitude and time is referred to as a waveform. On a waveform display, magnitude is measured along the vertical (*y*) axis. Time is measured along the horizontal (*x*) axis.
- The positive and negative transitions of a waveform are called *alternations*.
- The complete transition through one positive alternation and one negative alternation is referred to as a *cycle*. Because two alternations make up one cycle, each alternation is commonly referred to as a half-cycle.
- The period of an AC waveform is the time required to complete one cycle of the waveform.
- Waveform period can be measured from any given point on the waveform to the identical point on the next cycle.
- The frequency of a waveform is the rate at which the cycles repeat themselves, measured in cycles per second.
- The unit of measure for frequency is the hertz (Hz). One hertz is equal to one cycle per second.
- Frequency and cycle time are inversely proportional. Each can be calculated as the reciprocal of the other. That is:

$$T = \frac{1}{f} \quad \text{and} \quad f = \frac{1}{T}$$

- An oscilloscope is a piece of test equipment that provides a visual representation of a voltage waveform. The visual display can be used to make a variety of magnitude-related and time-related measurements.
    - The setting of the time base control determines the amount of time represented by the space between each pair of major divisions along the x-axis.
    - When you vary the time base setting on the oscilloscope, you change only the waveform display, not the waveform itself.

Waveform Magnitude

- The maximum value that a waveform reaches during each transition is the peak value of that transition.
- A pure AC waveform is one with identical alternations having the same peak value.
- The peak-to-peak value of a waveform is equal to the difference between its peak values.
  - For a pure AC waveform, the peak-to-peak value is twice its peak value.
- An instantaneous value is the magnitude of a waveform at a specified point in time.
  - Instantaneous values are represented using lowercase letters (as opposed to constant values, which are represented using capital letters).
- The half-cycle average of a waveform is the average of all its instantaneous values of voltage (or current) through one alternation of the input cycle.
- When dealing with AC circuits, we are interested in the average AC power generated throughout each cycle.
- The RMS (root-mean-square) value of a voltage (or current) waveform is the value that, when used in the appropriate power equation, gives us the average AC power of the waveform.
- The RMS value of any sine wave is equal to 70.7% of its peak value.
- Power has always been used as the basis of comparison for AC and DC.
  - Two sources, one DC and one AC, are considered equivalents when they deliver the same amount of power to a given load.
  - Average AC power is the equivalent of DC power.
  - Since RMS values are used to calculate average AC power, they also provide equivalent DC power values.
- When a magnitude-related value is not specifically identified (as peak, peak-to- peak, or average), it is an RMS value.
- Peak and peak-to-peak voltages are measured with an oscilloscope.
- Peak and peak-to-peak currents cannot be measured by any standard piece of test equipment.
- Half-cycle average voltage can be measured using a DC voltmeter and a circuit called a rectifier.
  - A rectifier is a circuit that converts AC to pulsating DC.
- Half-cycle average current can be measured using a DC ammeter and a rectifier.
- The RMS values of a sine wave are also referred to as effective values.
- Effective values are measured using AC voltmeters and AC ammeters.
- The tools used to measure the various AC values are summarized in Table 13.1.

Sine Waves: Phase and Instantaneous Values

- Just as current can be used to produce a magnetic field, a magnetic field can be used to generate current through a conductor.
  - If a conductor cuts through lines of magnetic force, current is induced in the wire.
  - The direction of current depends on the direction of motion that the conductor takes through the magnetic field.
- A sine wave can be generated by rotating a loop conductor through a stationary magnetic field.

- The phase of a given point is its position relative to the start of the waveform, usually expressed in degrees.
  - Phase measurements allow us to identify points on a waveform independent of cycle time and frequency.
  - Phase plays an important role in many RC and RL circuit voltage, current, and power dissipation values.
- Two (or more) sine waves of equal frequency that do not reach their peaks at the same time are said to be *out of phase*.
- The phase difference between two waveforms is referred to as their phase angle.
- The instantaneous value of a sine wave voltage (or current) is proportional to the peak value of the waveform and the sine of the phase angle (sin θ).
- Since instantaneous values are proportional to the sine of the phase angle (sin θ), the waveform is said to vary at a "sinusoidal" rate. This is where the term *sine wave* comes from.

Rectangular Waves

- A rectangular wave is one that alternates between two DC levels.
- The alternations of a rectangular wave are referred to as the pulse width (PW) and space width (SW).
  - The sum of PW and SW equals the waveform period.
- The duty cycle of a rectangular wave is the ratio of pulse width to waveform period, expressed as a percent.
- A square wave is a special-case rectangular wave that has equal PW and SW values.
  - The duty cycle of a square wave is always 50%.
- A symmetrical waveform is one with cycles that are made up of identical half-cycles.
- There are two types of symmetry: time and amplitude.

CHAPTER REVIEW

1. The main difference between AC and DC is ________.
   a) DC is more efficient
   b) current changes direction in AC and not in DC
   c) AC is more efficient
   d) current changes direction in DC and not in AC

2. In a current-versus-time graph, the vertical axis represents ________ and the horizontal axis represents ________.
   a) time, phase
   b) magnitude, time
   c) time, magnitude
   d) magnitude, frequency

3. The graph of the relationship between magnitude and time is referred to as a(n) ________.
   a) cycle
   b) schematic
   c) alternation
   d) waveform

4. When voltage polarity changes, ________ changes in an AC circuit.
   a) current direction
   b) current polarity
   c) current phase
   d) current magnitude

5. Each half-cycle of an AC waveform is referred to as a(n) ________.
   a) period
   b) demi-cycle
   c) alternation
   d) transition

6. The horizontal axis of a waveform display is set so that each major division represents 20 ms. The positive half-cycle occupies 2.5 divisions of the display. This means that the period of this waveform is ________.
   a) 50 ms
   b) 20 ms
   c) 100 ms
   d) 8 ms

7. A sine wave with a period of 0.4 ms has a frequency of ________.
   a) 2.5 kHz
   b) 250 Hz
   c) 25 kHz
   d) 25 Hz

8. A sine wave with a frequency of 120 Hz has a period of ________.
   a) 83.3 ms
   b) 8.33 ms
   c) 833 ms
   d) 833 μs

9. Suppose that you are viewing a 1 kHz sine wave on an oscilloscope and 5 cycles occupy the whole display. Now suppose that you need to view a 60 Hz sine wave. In order to see only 5 cycles of the waveform you will need to adjust the ________ control to a ________ setting.
   a) major division, higher
   b) volts/division, lower
   c) time-base, higher
   d) time-base, lower

10. If frequency increases, the ________ of an AC waveform decreases.
    a) magnitude
    b) amplitude
    c) phase shift
    d) period

11. A pure AC sine wave has a peak value of 18.5 V. This means that V_{PP} = ________.
    a) 37 V
    b) 18.5 V
    c) 11.8 V
    d) 9.25 V
12. In order to measure the peak voltage output from a power supply you would have to use a(n) ________.
    a) DC voltmeter
    b) AC voltmeter
    c) oscilloscope
    d) DMM
13. The vertical sensitivity control on an oscilloscope controls ________.
    a) the value of each major division of the y-axis
    b) the value of each major division of the x-axis
    c) the period of the waveform
    d) the amplitude of the waveform
14. Any specific point on a waveform is referred to as a(n) ________.
    a) point value
    b) unique value
    c) phase value
    d) instantaneous value
15. The half-cycle average of a 170 V_{pk} pure sine wave is ________.
    a) 85 V
    b) 108 V
    c) 54 V
    d) 0 V
16. The full-cycle average of a 170 V_{pk} pure sine wave is ________.
    a) 85 V
    b) 108 V
    c) 54 V
    d) 0 V
17. A full-wave rectifier is used to ________.
    a) convert AC to pulsating DC
    b) convert negative alternations to positive alternations
    c) allow a DC voltmeter to measure the half-cycle average voltage
    d) all of the above
18. When solving for power in an AC circuit, you must use ________ in the V × I equation.
    a) peak values
    b) peak-to-peak values
    c) RMS values
    d) any of the above so long as both values are the same

19. The power waveform in an AC circuit has the same shape as a ________.
    a) V^2 waveform
    b) $\sqrt{2}$ waveform
    c) $\frac{1}{\sqrt{2}}$ waveform
    d) rectified waveform

20. The equation used to solve for an RMS voltage is V_{RMS} = ________.
    a) $\frac{V_{pk}}{\sqrt{2}}$
    b) $\sqrt{\frac{V_{pk}^2}{2}}$
    c) $V_{Pk} \times 0.707$
    d) all of the above

21. The effective value of a 680 V_{PP} sine wave is ________.
    a) 481 V
    b) 433 V
    c) 240 V
    d) 217 V

22. The power dissipated by a light bulb is 120 W. The voltage across the light bulb is measured, using an AC voltmeter, at 120 V. This means that the peak-to-peak current through the light bulb is ________.
    a) 2.83 A
    b) 2 A
    c) 1 A
    d) 707 mA

23. A 679 V_{PP} sine wave is applied to a 1500 W heater. The resistance of this heater is ________.
    a) 307 Ω
    b) 2.2 Ω
    c) 38.4 Ω
    d) 154 Ω

24. The RMS current through the heater described in question 23 is ________.
    a) 453 mA
    b) 3.12 A
    c) 6.25 A
    d) 625 mA

25. In order to measure the effective value of a waveform that is *not* a sine wave, you must use a(n) ________.
    a) DMM
    b) oscilloscope
    c) averaging meter
    d) true-RMS meter

26. In order to measure the peak current in an AC circuit, you must use a(n) ________.
    a) true-RMS ammeter
    b) averaging ammeter
    c) oscilloscope
    d) none of the above

27. Maximum current is generated in a conductor when it cuts magnetic lines of force at an angle of ________.
    a) 0°
    b) 90°
    c) 45°
    d) 180°

28. A loop conductor is rotating in a magnetic field in order to generate current. The ends of the loop are connected to ________.
    a) an ammeter
    b) a voltmeter
    c) slip rings
    d) an alternator

29. The position of any point on a waveform, in relation to the start of that waveform, is identified as its ________.
    a) phase
    b) amplitude
    c) rotation
    d) time shift

30. A sine wave reaches its negative peak value at ________.
    a) 0°
    b) 90°
    c) 180°
    d) 270°

31. Two sine waves have the same frequency, but one of them is at its positive peak when the other is at zero going negative. These two waveforms are ________ out of phase.
    a) 90°
    b) 180°
    c) 45°
    d) 270°
    e) 90° or 270°

32. A 220 V_{pk} sine wave has an instantaneous value of ________ when the phase angle is 195°.
    a) +56.9 V
    b) −56.9 V
    c) −212 V
    d) +212 V

33. The sum of the pulse width and the space width of a rectangular waveform defines the ________ of the waveform.
    a) period
    b) phase shift
    c) frequency
    d) duty cycle

34. A rectangular wave has a duty cycle of 66%. If the space width is 400 ms then the pulse width must be ________.
    a) 406 ms
    b) 264 ms
    c) 777 ms
    d) 606 ms

35. A square wave has values of $+V_{pk} = 9$ V and $-V_{pk} = -20$ V. The average value of this waveform is ________.
    a) 0 V
    b) −11 V
    c) −5.5 V
    d) −3.89 V

PRACTICE PROBLEMS

1. Calculate the period of the waveform in Figure 13.44a.
2. Calculate the period of the waveform in Figure 13.44b.

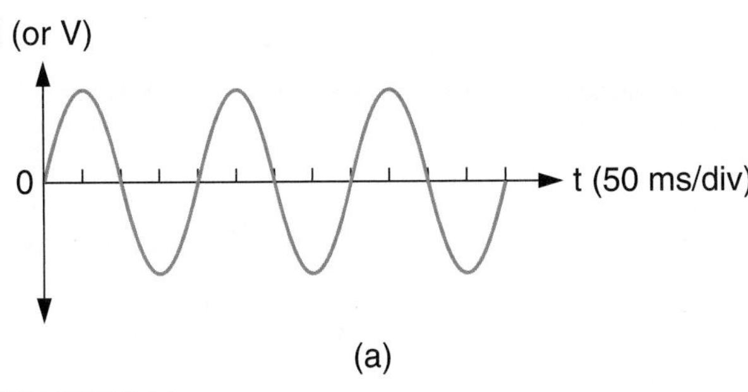

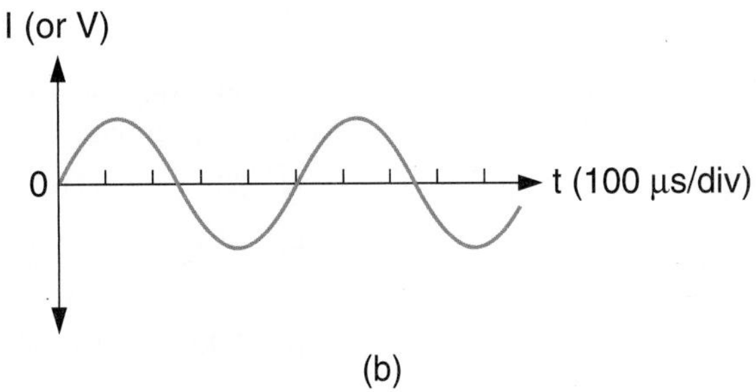

FIGURE 13.44

3. Calculate the period of the waveform in Figure 13.45a.
4. Calculate the period of the waveform in Figure 13.45b.

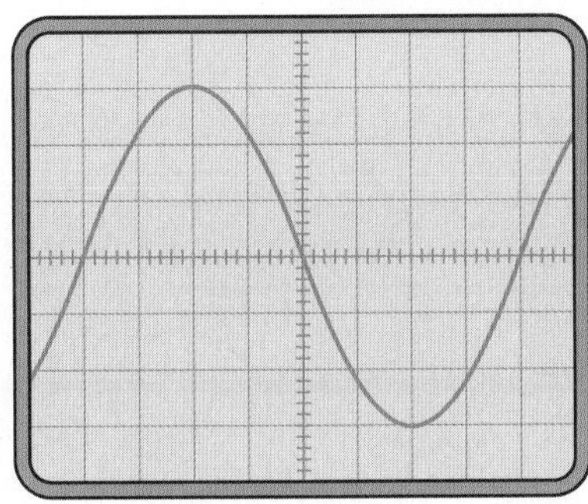

Time base: 10 μs/div
(a)

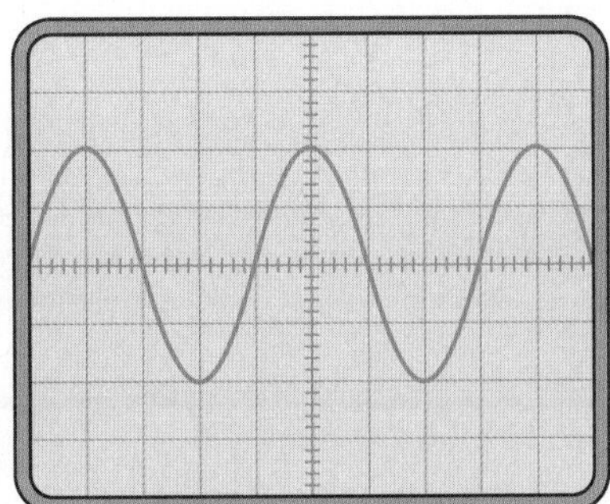

Time base: 20 ms/div
(b)

FIGURE 13.45

5. Calculate the frequency of each waveform in Figure 13.44.
6. Calculate the frequency of each waveform in Figure 13.45.
7. Calculate the frequency of the waveform in Figure 13.46a.
8. Calculate the frequency of the waveform in Figure 13.46b.

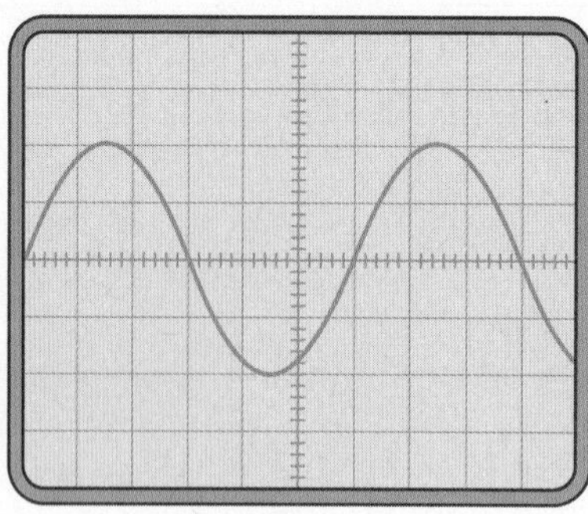
Time base: 5 μs/div
(a)

Time base: 10 ms/div
(b)

FIGURE 13.46

9. Calculate the waveform period that corresponds to each of the following frequencies:
   a) 500 Hz
   b) 2.5 kHz
   c) 1 MHz
   d) 75 kHz
10. Calculate the range of waveform periods that corresponds to each of the following frequency ranges:
    a) 100 Hz to 300 Hz
    b) 1 kHz to 5 kHz
    c) 20 Hz to 20 kHz
    d) 1.5 MHz to 10 MHz
11. A sine wave has peak values of ±12 V. Calculate the waveform's half-cycle average voltage.
12. A sine wave has peak values of ±20 V. Calculate the waveform's half-cycle average voltage.
13. Determine the peak-to-peak and half-cycle average voltages for the waveform in Figure 13.47a.

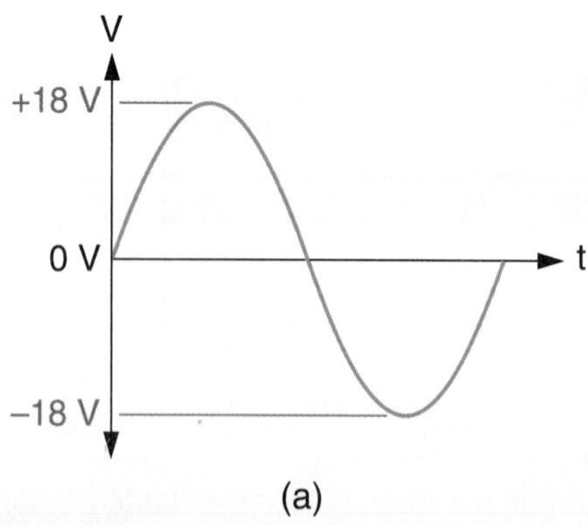

(a)

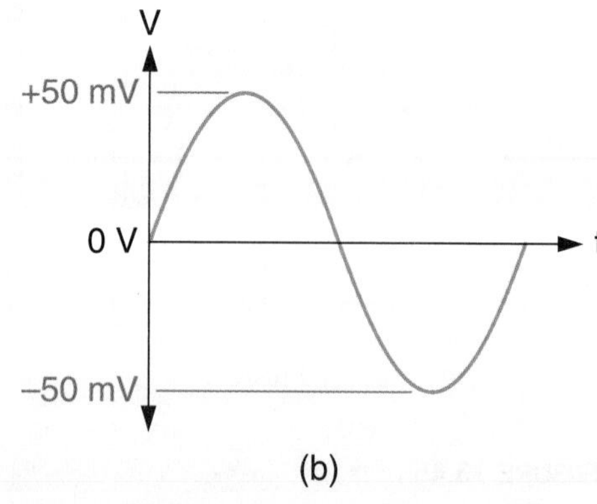

(b)

FIGURE 13.47

14. Determine the peak-to-peak and half-cycle average voltages for the waveform in Figure 13.47b.
15. Calculate the RMS values of load voltage and current for the circuit in Figure 13.48a.
16. Calculate the RMS values of load voltage and current for the circuit in Figure 13.48b.

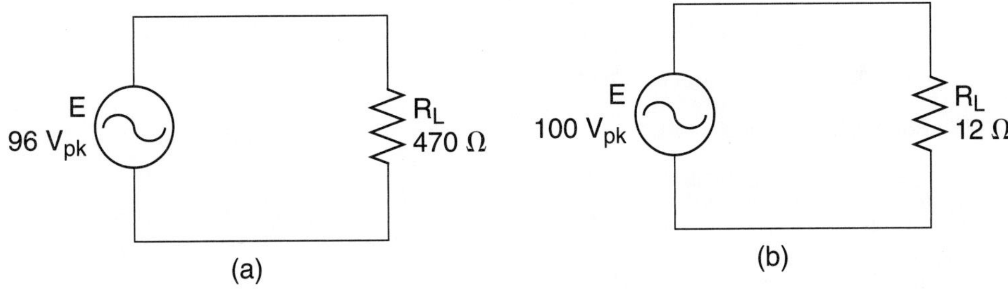

FIGURE 13.48

17. Calculate the RMS values of load voltage and current for the circuit in Figure 13.49a.
18. Calculate the RMS values of load voltage and current for the circuit in Figure 13.49b.

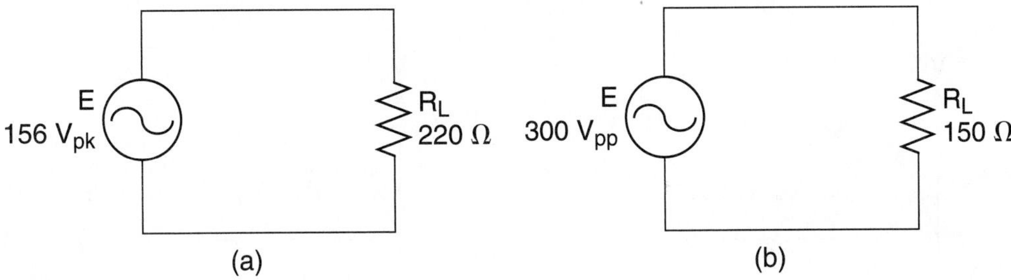

FIGURE 13.49

19. Calculate the load power for the circuit in Figure 13.50a.
20. Calculate the load power for the circuit in Figure 13.50b.

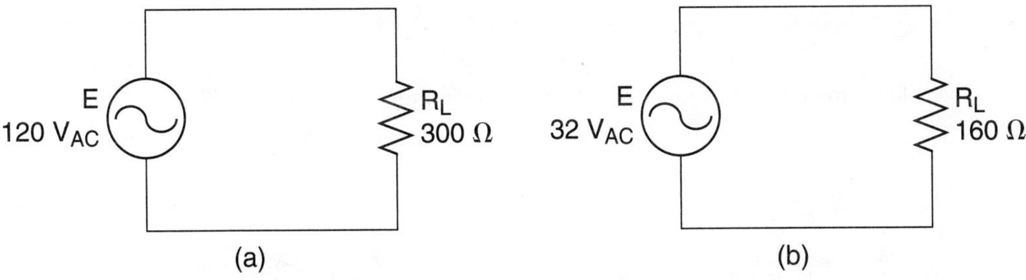

FIGURE 13.50

21. Determine the peak and peak-to-peak values of the waveform in Figure 13.45a. Assume that the vertical sensitivity setting of the oscilloscope is 50 mV/div.
22. Determine the peak and peak-to-peak values of the waveform in Figure 13.45b. Assume that the vertical sensitivity setting of the oscilloscope is 100 μV/div.
23. Determine the peak and peak-to-peak values of the waveform in Figure 13.46a. Assume that the vertical sensitivity setting of the oscilloscope is 10 V/div.

24. Determine the peak and peak-to-peak values of the waveform in Figure 13.46b. Assume that the vertical sensitivity setting of the oscilloscope is 200 mV/div.
25. For a $\pm 15\ V_{pk}$ sine wave,
    a) v = ________ when $\theta = 60°$
    b) v = ________ when $\theta = 150°$
    c) v = ________ when $\theta = 240°$
    d) v = ________ when $\theta = 350°$
26. For a $\pm 170\ V_{pk}$ sine wave,
    a) v = ________ when $\theta = 45°$
    b) v = ________ when $\theta = 190°$
    c) v = ________ when $\theta = 10°$
    d) v = ________ when $\theta = 180°$
27. Refer to the waveforms in Figure 13.51a. Determine the instantaneous value of waveform A when waveform B is at 0 degrees.
28. Refer to the waveforms in Figure 13.51b. Determine the instantaneous value of waveform A when waveform B is at 20 degrees.

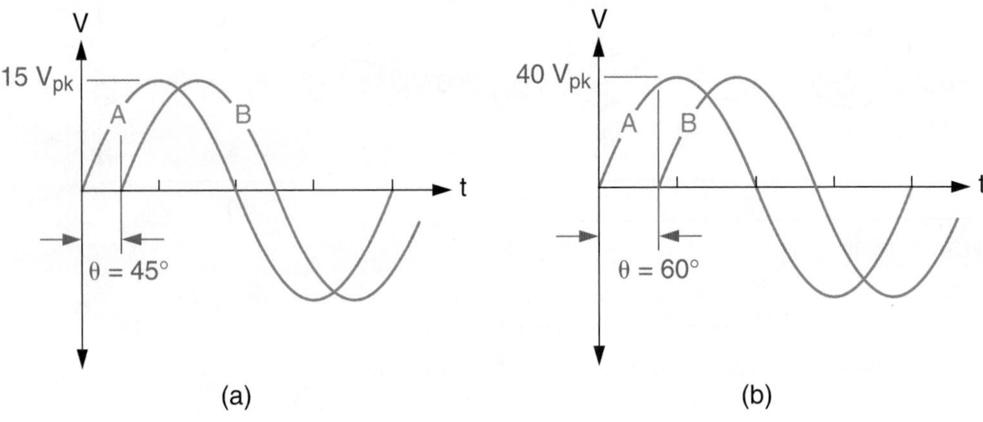

FIGURE 13.51

29. Determine the duty cycle of the waveform in Figure 13.52a.
30. Determine the duty cycle of the waveform in Figure 13.52b.

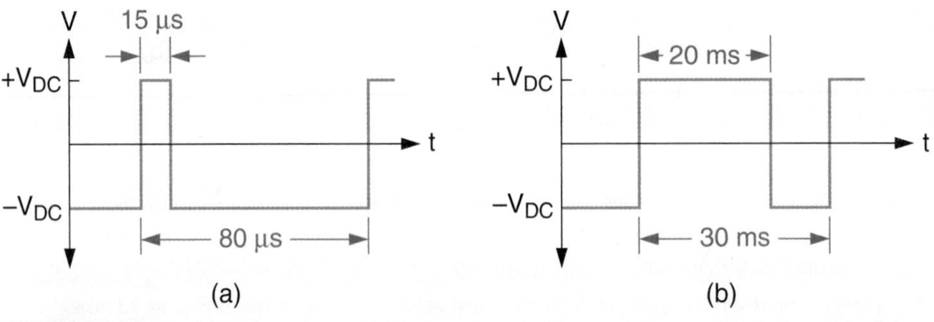

FIGURE 13.52

31. Determine the duty cycle of the waveform in Figure 13.53a.
32. Determine the duty cycle of the waveform in Figure 13.53b.

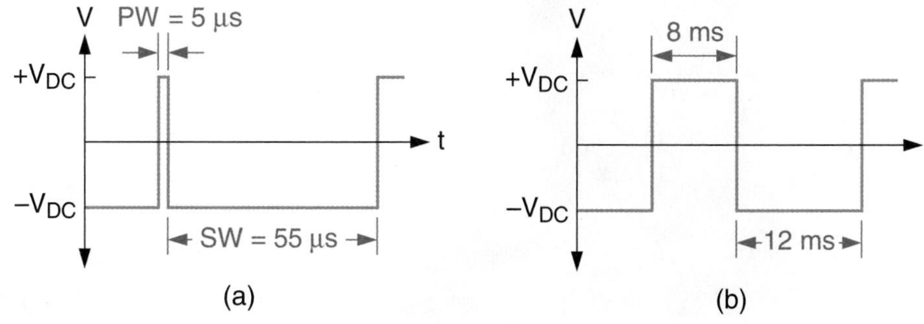

FIGURE 13.53

33. Determine the average value of the waveform in Figure 13.54a.
34. Determine the average value of the waveform in Figure 13.54b.

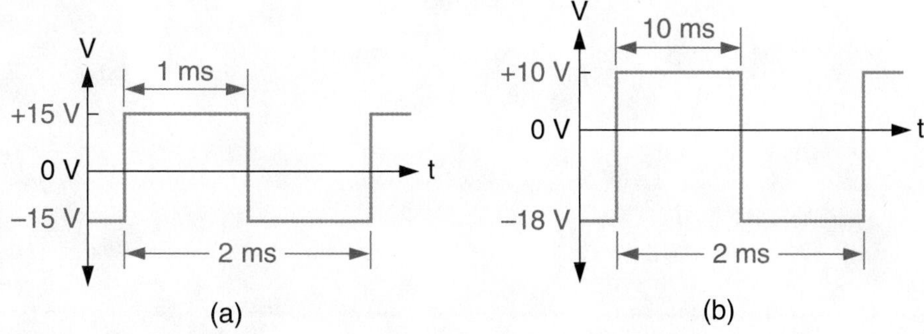

FIGURE 13.54

ANSWERS TO THE EXAMPLE PRACTICE PROBLEMS

13.1	30 ms
13.2	25 Hz
13.3	16.7 ms
13.4	50 Hz
13.5	19.1 V
13.6	170 V, 567 mA
13.7	16.2 W
13.8	10.4 V
13.9	−10.3 V
13.10	40%

CHAPTER 14

SERIES RESISTIVE-INDUCTIVE (RL) CIRCUITS

PURPOSE

As you may recall, inductance is the ability of a component with a changing current to induce a voltage across itself or a nearby circuit by generating a changing magnetic field. Based on its overall effect, inductance is often described as the ability of a component to:

1. Oppose any change in current.
2. Store energy in an electromagnetic field.

In Chapter 11, we discussed the physical characteristics of inductors, series-connected and parallel-connected inductors, and the response of an inductor to a DC source. In this chapter, we will examine the AC characteristics of series resistive-inductive (RL) circuits, starting with the effect of inductance on AC phase relationships.

KEY TERMS

The following terms are introduced and defined in this chapter on the pages indicated:

Apparent power (P_{APP}), 613
Dynamic, 612
Imaginary power, 615
Impedance (Z), 612
Inductive reactance (X_L), 608
Power factor, 635
Quality (Q), 616
Reactive power, 615
Resistive power, 615
Static, 612
True power, 615
Volt-amperes (VA), 615
Volt-amperes-reactive (VAR), 615
Winding resistance (R_W), 614

OBJECTIVES

After completing this chapter, you should be able to:

1. Describe the phase relationship between inductor current and voltage.
2. Discuss inductive reactance (X_L) and its source.
3. Calculate the reactance of an inductor operated at a specific frequency.
4. Compare and contrast series resistive, inductive, and resistive-inductive (RL) circuits.
5. Compare and contrast apparent power, reactive power, and resistive (or *true*) power.
6. Describe inductor quality (Q) and calculate its value for an inductor.
7. Describe the voltage and phase characteristics of series RL circuits.

8. Describe the impedance and current characteristics of series RL circuits.
9. Calculate the impedance, voltage, and current values for any series RL circuit.
10. Perform all relevant power calculations for a series RL circuit.
11. Describe the concept of power factor.
12. Calculate the power factor for a series RL circuit.

14.1 THE PHASE RELATIONSHIP BETWEEN INDUCTOR CURRENT AND VOLTAGE

In any purely resistive circuit, current and voltage are directly proportional (as described by Ohm's law). In a purely inductive circuit, the relationship between current and voltage gets a bit more complicated because inductor voltage is directly proportional to *the rate of change in current.* The relationship between inductor voltage and current was given in Chapter 11 as

A *linear* load is one that is purely resistive, inductive, or capacitive. A *nonlinear* load is one that has two or more of these properties. In practice, most loads (such as motors and industrial power distribution circuits) are nonlinear loads.

IN THE FIELD

(14.1)

$$V_{IND} = L\frac{\Delta I}{\Delta t}$$

With this relationship in mind, look at the current waveform shown in Figure 14.1. From 0° to 90° on the waveform, the current increases. From 90° to 180°, the current decreases. The same holds true for the negative alternation. That is, current increases between 180° and 270°, and decreases between 270° and 360°. The only difference is the direction of the current.

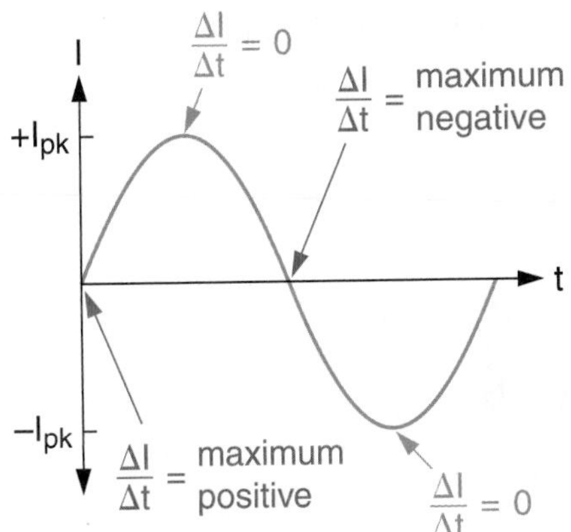

FIGURE 14.1 Rates of change in current.

Now, here's what happens at the 90° and 270° points on the current waveform: At the instant the waveform reaches a peak value, the rate of change in current is zero amperes per second (A/s). (As an analogy, picture a baseball that is thrown straight up into the air. When it reaches its highest point, it stops for just an instant before falling. At that instant, its rate of change in position is zero.) When the current waveform is halfway between the peaks (at 180° and 360°), the rate of change in current reaches its maximum value. These relationships are summarized in the first two columns of Table 14.1.

TABLE 14.1 • Rates of Change in Sine Wave Current.

WAVEFORM PHASE	RATE OF CHANGE IN CURRENT	INDUCTOR VOLTAGE
90°	Zero	Zero
180°	Maximum	Maximum
270°	Zero	Zero
360°	Maximum	Maximum

The third column in Table 14.1 is based on the fact that inductor voltage is directly proportional to the rate of change in inductor current. That is,

- V_L reaches its minimum value (0 V) when the rate of change in inductor current is zero.
- V_L reaches its peak value (V_{Pk}) when the rate of change in current reaches its maximum value.

Now, let's focus on the relationship between inductor voltage and the phase of the current waveform. According to Table 14.1,

- V_L reaches its minimum value (0 V) at the 90° and 270° points on the current waveform.
- V_L reaches its peak value (V_{Pk}) at the 180° and 360° points on the current waveform.

Keeping these statements in mind, look at the waveforms in Figure 14.2.

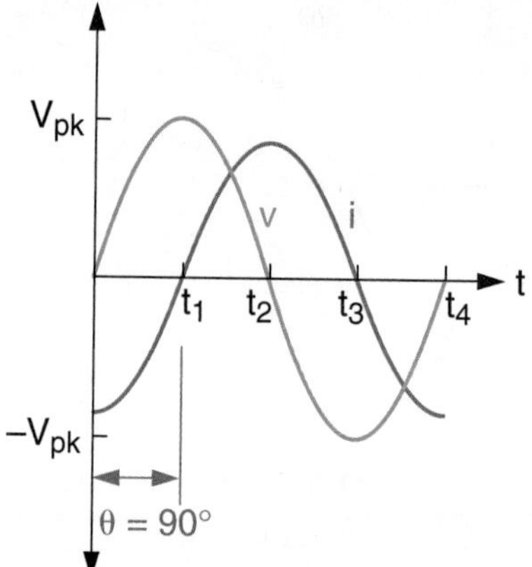

FIGURE 14.2 Voltage and current in a purely inductive circuit.

As shown in Figure 14.2, voltage reaches its maximum value when current is at its minimum value, and vice versa. In other words, there is a 90° phase shift between the voltage and current waveforms. If we use the voltage waveform as a reference, the inductor current begins its positive half-cycle 90° *after* the inductor voltage. Note that this relationship is normally described in one of two ways:

1. "Voltage *leads* current by 90 degrees."
2. "Current *lags* voltage by 90 degrees."

You will see that the terms *lead* and *lag* are commonly used when describing the phase relationships in AC circuits.

An Important Point

The current-voltage phase relationship described in this section applies only to a purely inductive circuit; that is, a circuit that does not contain any significant amount of resistance. As you will learn later in this chapter, the phase relationship between circuit current and source voltage changes when any significant amount of resistance is added to an inductive circuit.

PROGRESS CHECK

1. What is the relationship between inductor voltage and the rate of change in inductor current?
2. Explain the phase relationship shown in Figure 14.2.
3. What is meant by the terms *lead* and *lag*?

14.2 INDUCTIVE REACTANCE (X_L)

According to Ohm's law, any component that has a voltage across it and a current through it is providing some measurable opposition to that current. For example, the ammeter in Figure 14.3a indicates that the resistor is limiting the circuit current to 1 A. With 24 V applied to the circuit, the opposition to current provided by the resistor is found as $R = E/I$; in this case, 24 Ω.

Like a resistor, an inductor with a sine wave input provides a measurable amount of opposition to current. For example, consider the inductive waveforms and circuit shown in Figure 14.3b. As the waveforms indicate, there is a 90° phase difference between inductor current and voltage. At the same time, each waveform has a measurable RMS (or *effective*) value. If we were to measure these RMS values, we could determine the total opposition to current provided by the inductor. Using the RMS values shown in Figure 14.3, this opposition can be found as

$$\text{opposition} = \frac{V_{RMS}}{I_{RMS}}$$

$$= \frac{120\text{ V}}{1\text{ A}}$$

$$= 120\ \Omega$$

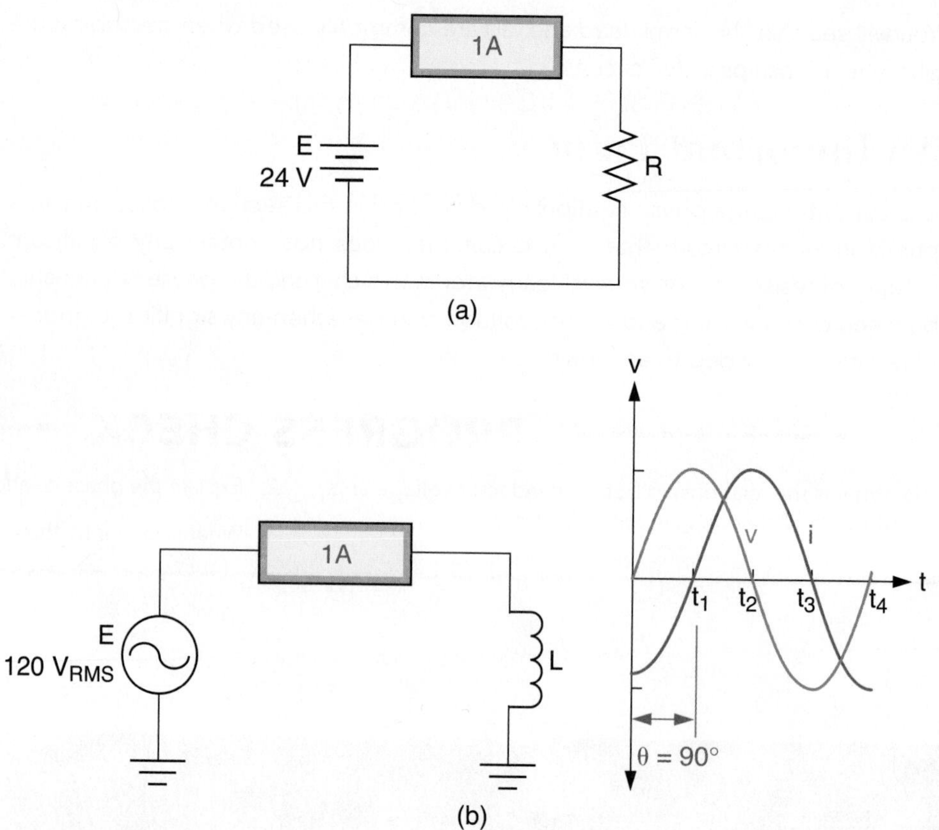

FIGURE 14.3 Resistors and inductors oppose alternating current.

As you know, an inductor is basically a wire that is wrapped into series of loops. If we were to measure the resistance of a coil, we would obtain an extremely low reading because of the relatively low resistance of copper wire. For example, if the coil in Figure 14.3 was made up of 2 ft of 30 gauge wire, its resistance could be approximated as

$$R = (\text{length}) \times (\text{ohms/unit length})$$
$$= 2\text{ ft} \times \frac{103.2\ \Omega}{1000\text{ ft}}$$
$$\cong 0.21\ \Omega$$

INDUCTIVE REACTANCE (X_L) The opposition that an inductor presents to an alternating current, measured in ohms (Ω).

This is significantly lower than the 120 Ω of opposition that we calculated using Ohm's law. Assuming that all the calculations are correct, there is only one way to account for the difference between our calculated values of 0.21 Ω and 120 Ω: *The inductor is providing some type of opposition to current other than resistance.* This "other" opposition to current is called **INDUCTIVE REACTANCE (X_L)**.

Reactance (X)

You have been told that inductance can be defined as *the ability of a component to oppose a change in current.* In an inductive circuit with a sine wave input (like the one in Figure 14.3b), the source current is constantly changing. The opposition (in ohms) that an inductor presents to this changing current is inductive reactance (X_L). Note that the letter *X* is used to designate reactance, and the subscript *L* is used to identify the reactance as being inductive. (In Chapter 16, you will be introduced to another type of reactance.)

A current-carrying wire generates a circular magnetic field around itself. When the current changes direction—as it does dozens of times a second with AC—the magnetic field changes polarity. This changing polarity generates a voltage that opposes the original current.

KEY CONCEPT

Calculating the Value of (X_L)

You have already been shown one way to calculate the value of X_L for an inductor. When the RMS values of inductor current and voltage are known, the reactance of the component can be found as

$$X_L = \frac{V_{RMS}}{I_{RMS}} \tag{14.2}$$

which is just a version of Ohm's law. We used this equation to calculate the reactance for the circuit in Figure 14.3b.

There is another (and more commonly used) method for calculating the reactance of an inductor. When the value of an inductor and the frequency of operation are known, the reactance of the inductor can be found using

$$X_L = 2\pi fL \tag{14.3}$$

Inductive reactance (X_L) is directly proportional to signal frequency (f) and inductance (L). The greater the frequency or inductance, the greater the inductive reactance.

KEY CONCEPT

The use of this equation is demonstrated in Example 14.1.

EXAMPLE 14.1

What is the reactance of the inductor in Figure 14.4?

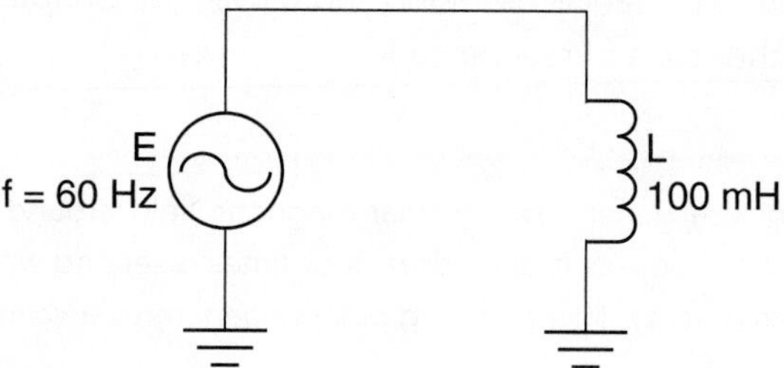

FIGURE 14.4

SOLUTION

Using the values shown, the reactance of the inductor can be found as

$$\begin{aligned} X_L &= 2\pi fL \\ &= 2\pi \times 60\text{ Hz} \times 100\text{ mH} \\ &\cong 37.7\ \Omega \end{aligned}$$

PRACTICE PROBLEM 14.1

A 470 mH inductor is operated at a frequency of 120 Hz. Calculate the reactance provided by the component.

X_L and Ohm's Law

In an inductive circuit, inductive reactance (X_L) can be used with Ohm's law to determine the total circuit current (just like resistance). The analysis of a simple inductive circuit is demonstrated in Example 14.2.

EXAMPLE 14.2

Calculate the inductor current in Figure 14.5.

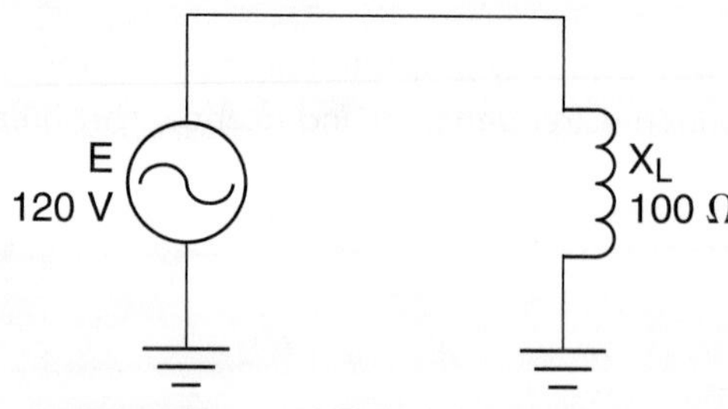

FIGURE 14.5

SOLUTION

The current through the circuit has a value of

$$I = \frac{E}{X_L}$$

$$= \frac{120\text{ V}}{100\ \Omega}$$

$$= 1.2\text{ A}$$

PRACTICE PROBLEM 14.2

An inductive circuit has the following values: E = 120 V and $X_L = 80\ \Omega$. Calculate the total circuit current.

When the values of inductance, source voltage, and operating frequency for a circuit are known, the circuit current can be calculated as demonstrated in Example 14.3.

EXAMPLE 14.3

Calculate the current through the circuit in Figure 14.6a.

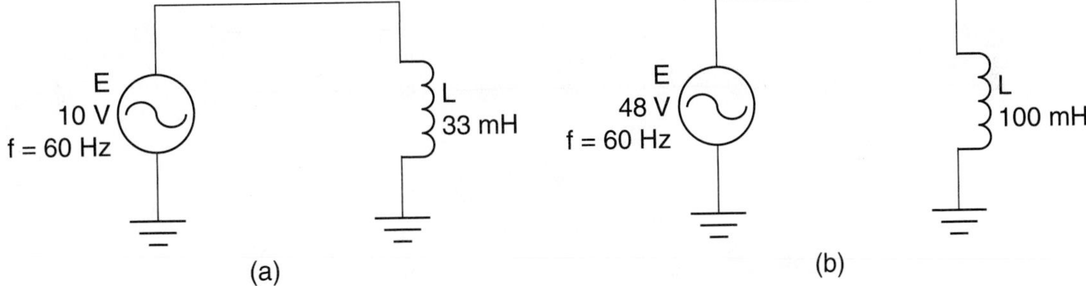

FIGURE 14.6

SOLUTION

To calculate the circuit current, you begin by calculating the reactance of the coil, as follows:

$$X_L = 2\pi fL$$

$$= 2\pi \times 60\text{ Hz} \times 33\text{ mH}$$

$$= 12.4\ \Omega$$

Next, using X_L and the applied voltage, the value of the circuit current is found as

$$I = \frac{E}{X_L}$$

$$= \frac{10\text{ V}}{12.4\ \Omega}$$

$$= 806\text{ mA}$$

PRACTICE PROBLEM 14.3

Calculate the value of current for the circuit in Figure 14.6b.

Series and Parallel Reactances

When reactive components are connected in series or parallel, the total reactance is found in the same manner as total resistance. This point is illustrated in Figure 14.7. (We have used similar equations for resistance calculations, so we do not need to rework them here.)

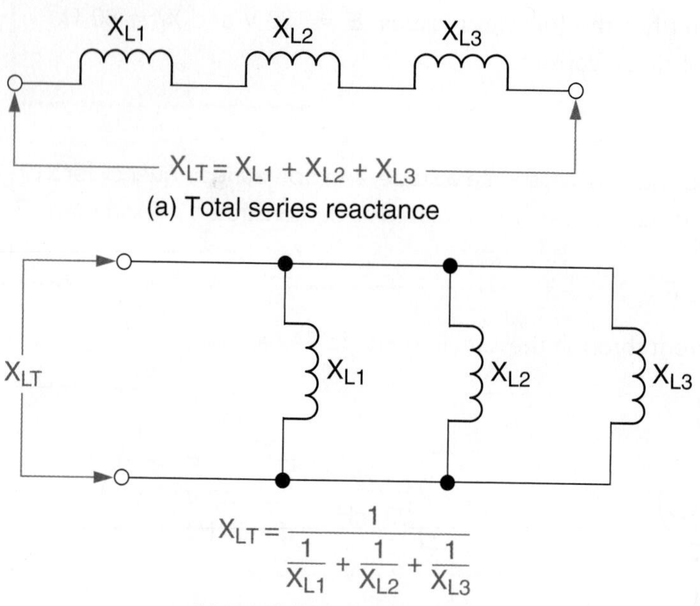

FIGURE 14.7 Calculating total series and parallel reactance.

Resistance, Reactance, and Impedance (Z)

STATIC A value that does not change in response to circuit conditions.

DYNAMIC A value that changes when specific circuit conditions change.

IMPEDANCE (Z) The total opposition to current in an AC circuit, consisting of resistance and/or reactance.

Resistance is usually a **STATIC** value. That is, the resistance in a circuit usually does not vary when circuit conditions change. Reactance, on the other hand, is a **DYNAMIC** value. That is, the reactance of an inductor changes when the circuit operating frequency changes.

Though they are both oppositions to current (measured in ohms), resistance and reactance cannot be added algebraically to determine the total opposition to current in a circuit.

Later in this chapter, you will learn how to combine resistance and reactance into a single value called **IMPEDANCE (Z)**. Technically defined, impedance is the total opposition to current in an AC circuit, consisting of resistance and/or reactance. As you will learn, impedance equals the geometric sum of resistance and reactance, expressed in ohms.

PROGRESS CHECK

1. How is the opposition to current provided by an inductor measured?
2. Why do many inductors have extremely low resistance values?
3. What is *inductive reactance* (X_L)?
4. What two relationships are commonly used to calculate the value of X_L?
5. Which of the relationships in question four is used most often? Why?
6. How is total current determined in a basic inductive circuit?
7. How is total reactance calculated in a series circuit? A parallel circuit?
8. What is impedance (Z)?

14.3 INDUCTOR POWER

Later in the chapter, we will discuss the power values that are found in series RL circuits. To prepare for that discussion, we are going to examine several inductor power values and measurements.

Apparent Power (P_{APP})

When used in an AC circuit, any inductor has measurable values of voltage and current. If these values are plugged into the $P = I \times E$ equation, they provide a numeric result (as does any combination of voltage and current). For the circuit in Figure 14.8, this power equation yields a result of

The combination of reactive power and resistive (true) power is called *apparent power*. Commercial utility customers are charged for apparent power, measured in kilo-volt-amperes (kVA), unlike residential customers who are charged for energy in kilo-watt-hours (kWh).

IN THE FIELD

$$
\begin{aligned}
P &= I \times E \\
&= 318\text{ mA} \times 120\text{ V} \\
&= 38.2\text{ W}
\end{aligned}
$$

The only problem is that the result of this calculation is not a true indicator of energy used. As such, it is referred to as **APPARENT POWER (P_{APP})**. The term *apparent power* is used because most of the energy is actually stored in the electromagnetic field generated by the inductor. Only a small portion of the value obtained is actually dissipated by the component. This point can be explained with the help of Figure 14.9.

APPARENT POWER (P_{APP}) The value of $V \times I$ in an AC circuit, where the result does not indicate the true value of power dissipation in the circuit; the combination of resistive and reactive power, measured in volt-amperes (VA).

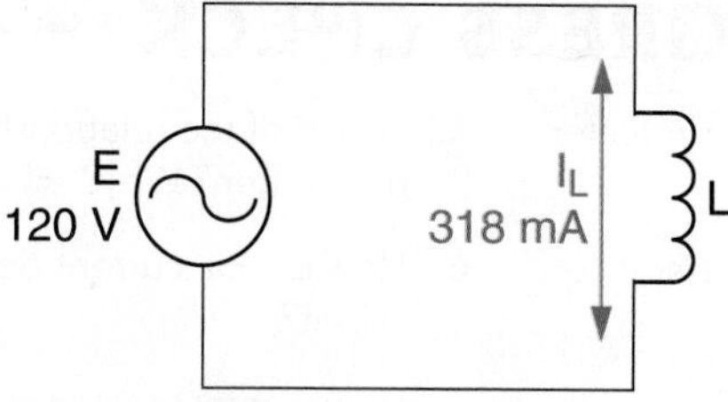

FIGURE 14.8

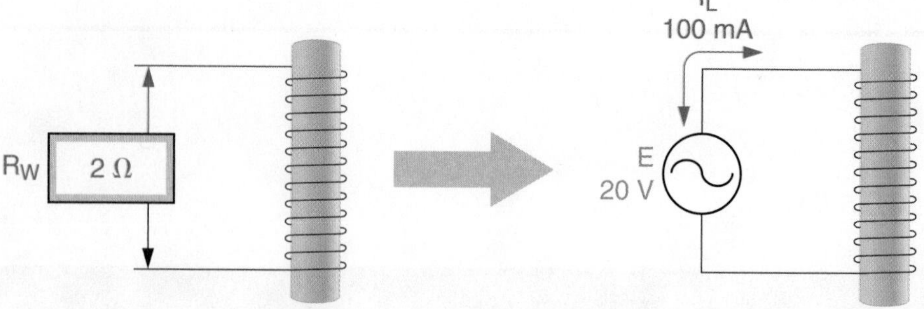

FIGURE 14.9 Inductor reactance and winding resistance.

WINDING RESISTANCE (R_W) The resistance of a coil (inductor).

As you know, power is dissipated whenever current passes through any measurable amount of resistance. With this in mind, consider the inductor shown in Figure 14.9. If we assume that the **WINDING RESISTANCE (R_W)** of the coil has a value of $R_w = 2\ \Omega$, then the power dissipated by the component can be found as

$$\begin{aligned} P &= I^2 \times R_W \\ &= (100\text{ mA})^2 \times 2\ \Omega \\ &= 20\text{ mW} \end{aligned}$$

At the same time, the inductor voltage and current values shown in the figure would indicate that the component is dissipating

$$\begin{aligned} P &= I_L \times V_L \\ &= 100\text{ mA} \times 20\text{ V} \\ &= 2\text{ W} \end{aligned}$$

As you can see, there is a discrepancy between the two calculations. The discrepancy is not in our calculations, but rather, in our interpretation of the results. We assume that the value of 2 W represents an amount of energy used each second. However, only a small portion (20 mW) represents the amount of energy used. The rest represents the rate at which the inductor is transferring energy to its magnetic field. This transfer of energy can be explained with the help of Figure 14.10. The field generated by the inductor consists of magnetic energy. Since energy cannot be created (or destroyed), this magnetic energy had to come from somewhere. The energy in the magnetic field comes from the AC source, via the inductor. In a sense, the inductor can be viewed as an energy converter because it converts electrical energy into magnetic energy.

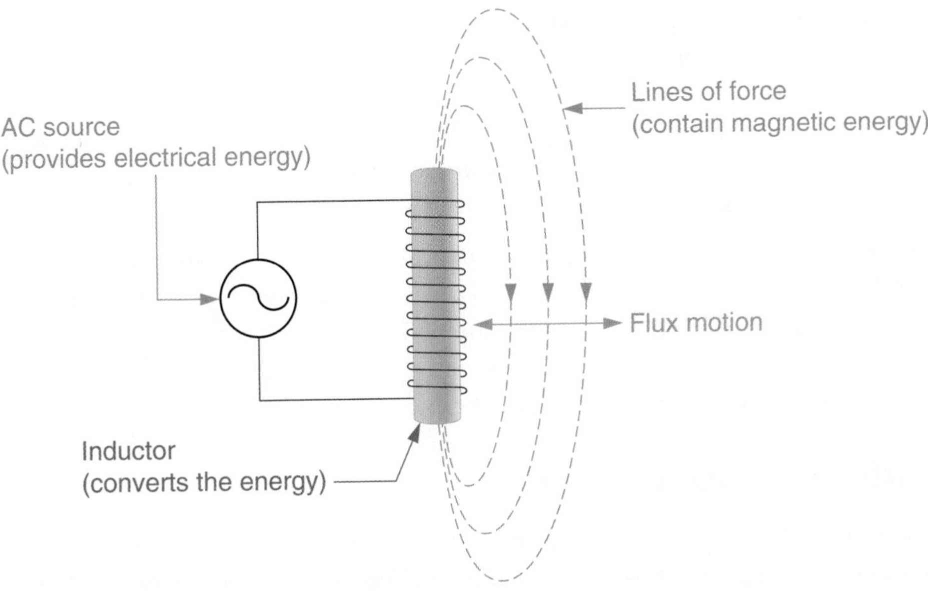

FIGURE 14.10 An inductor converting energy.

We have now established three "power" values that can be calculated for any inductive circuit. The terms used to identify them are listed and defined in Table 14.2.

TABLE 14.2 • Inductor "Power" Values

TERM	DEFINITION
RESISTIVE POWER (P_R)	The power actually dissipated by the winding resistance of an inductor. Also referred to as **TRUE POWER**. Resistive power is measured in watts (W).
REACTIVE POWER (P_X)	A value that indicates the rate at which energy is transferred to and from an inductor's magnetic field. Also referred to as **IMAGINARY POWER**. Reactive power is measured in **VOLT-AMPERES-REACTIVE (VAR)**.
Apparent power (P_{APP})	The combination of resistive and reactive power. Only a small portion of P_{APP} is actually dissipated by the inductor. The rest is transferred to and from the magnetic field by the inductor. Apparent power is measured in **VOLT-AMPERES (VA)**.

RESISTIVE POWER (P_R) The power actually dissipated by the winding resistance of an inductor, measured in watts (W).

TRUE POWER Another term for *resistive power*.

REACTIVE POWER (P_X) A value that indicates the rate at which energy is transferred to and from an inductor's magnetic field, measured in volt-amperes-reactive (VAR); also called *imaginary power*.

IMAGINARY POWER Another term for *reactive power*.

VOLT-AMPERES REACTIVE (VAR) The unit of measure for reactive power.

VOLT-AMPERES (VA) The unit of measure for apparent power.

There are two important points that need to be made at this time:

1. Because reactive (or "imaginary") power doesn't represent a true power value, it has its own unit of measure. To distinguish reactive power from true power, it is measured in *volt-amperes-reactive*, or *VARs*. By the same token, only a portion of apparent power is actually dissipated, so it also has its own unit of measure. As shown in the table, apparent power is measured in *volt-amperes (VA)*. You will see calculations involving both VAR and VA values later in the chapter.

2. When the magnetic field generated around an inductor collapses (due to a polarity change in the AC source), the energy that it contains is returned to the circuit, and therefore, to the AC source. As a result, the net reactive power drawn from the source is 0 VAR. The only power loss experienced by the AC source is the power used by any resistance in the circuit. That is why this power loss is referred to as *true* power.

In practice, every inductor has some measurable amount of winding resistance, and therefore, dissipates some amount of power. The amount of winding resistance determines (in part) the quality of an inductor. That is, the lower the winding resistance, the higher the quality of the component. This is because the *ideal* inductor would have no winding resistance, and therefore, would dissipate no power.

Inductor Quality (Q)

The *quality (Q)* rating of an inductor indicates how close the inductor comes to the power characteristics of the ideal component. The Q of an inductor can be found as the ratio of reactive power to true power for the component. By formula,

(14.4)

$$Q = \frac{P_X}{P_{RW}}$$

where

P_X = the *reactive power* of the inductor, measured in VARs

P_{RW} = the *true power* dissipated by the winding resistance of the inductor, measured in watts

As stated earlier, the ideal inductor has no winding resistance ($R_W = 0\ \Omega$) and dissipates no power ($P_{RW} = 0$ W). Therefore, the ideal inductor has a value of

$$Q = \frac{P_X}{P_{RW}} = \frac{P_X}{0\text{ W}} = \infty$$

Calculating Q

REMINDER: The dot shown in the P_X and P_{RW} equations is commonly used in place of a times sign (×) to indicate multiplication.

The values of reactive and resistive power can be calculated using

$$P_X = I^2 \cdot X_L$$

and

$$P_{RW} = I^2 \cdot R_W$$

If we substitute these relationships into equation (14.4), we get

$$Q = \frac{I^2 \cdot X_L}{I^2 \cdot R_W}$$

or

(14.5)
$$Q = \frac{X_L}{R_W}$$

The use of this equation is demonstrated in Example 14.4.

EXAMPLE 14.4

An inductor has the following values: $X_L = 250\ \Omega$ and $R_W = 4\ \Omega$. Calculate the Q of the inductor.

SOLUTION

Using equation (14.5), the Q of the inductor can be found as

$$Q = \frac{X_L}{R_W}$$
$$= \frac{250\ \Omega}{4\ \Omega}$$
$$= 62.5$$

PRACTICE PROBLEM 14.4

An inductor has the following values: $X_L = 140\ \Omega$ and $R_W = 1.8\ \Omega$. Calculate the Q of the inductor.

In Chapter 19, you will be shown how the Q of the inductor determines (in part) the response that any reactive circuit has to a change in operating frequency.

PROGRESS CHECK

1. What is apparent power? How is it measured?
2. What is reactive power?
3. Why is reactive power sometimes called *imaginary* power?
4. Why is resistive power sometimes called *true* power?
5. What happens to the energy stored in an inductor's magnetic field when the field collapses?
6. What is the relationship between inductor resistance and quality?
7. What power values can be used to determine the quality (Q) of an inductor?

14.4 SERIES RL CIRCUITS

A *resistive-inductive* (RL) circuit is one that contains any combination of resistors and inductors. Several examples of basic RL circuits are shown in Figure 14.11.

As we progress through the rest of this chapter, we will develop the current, voltage, power, and impedance characteristics of RL circuits. First, we're going to compare RL series circuits to *purely resistive* and *purely inductive* series circuits. The goal is to show the similarities and differences between the three.

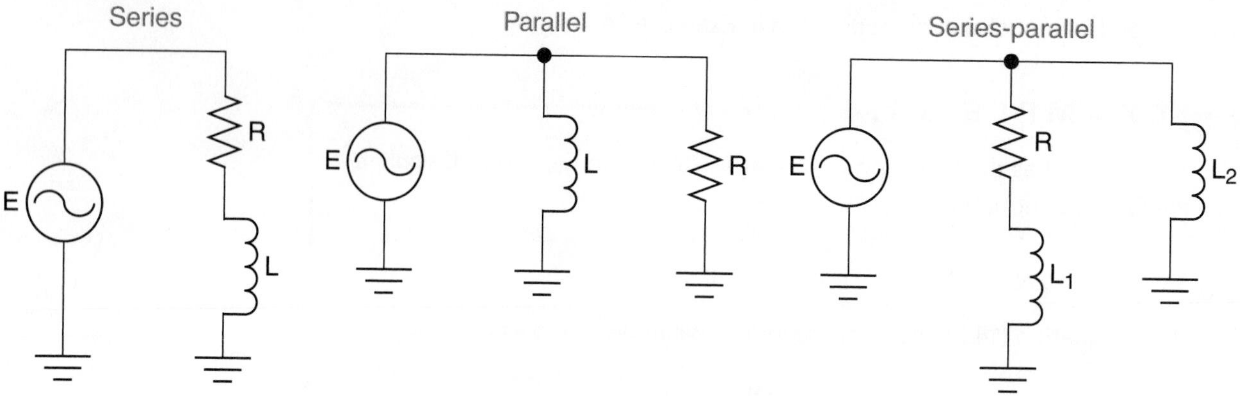

FIGURE 14.11 Basic resistive-inductive (RL) circuits.

Series Circuits: A Comparison

In previous discussions, we established the fundamental characteristics of resistive and inductive series circuits. As a refresher, these characteristics are as follows:

- Current is the same at all points throughout any series circuit.
- The voltage applied to a series circuit equals the sum of the individual component voltages (as given in Kirchhoff's Voltage Law, KVL).
- The total *impedance* (resistance and/or reactance) equals the sum of the individual component impedances.

These relationships are provided (as equations) in Figure 14.12.

The relationships listed for the resistive and inductive circuits should all seem familiar. In each case, the current is constant throughout the circuit, and the component voltages add up to equal the source voltage (E). For the resistive circuit, the total impedance (opposition to current) is equal to the sum of the resistor values. For the inductive circuit, the total impedance is equal to the sum of the reactance (X_L) values.

As we progress through this section, we will discuss each relationship listed for the RL circuit in Figure 14.12. First, let's compare them to those given for the other two circuits. As the I_T equation shows, current is constant throughout the series RL circuit, just as it is in the other series circuits. The voltage and impedance relationships, however, appear to be radically different from those for the resistive and inductive circuits. To understand the change, we need to establish the difference between the *algebraic sum* and the *geometric sum* of two values.

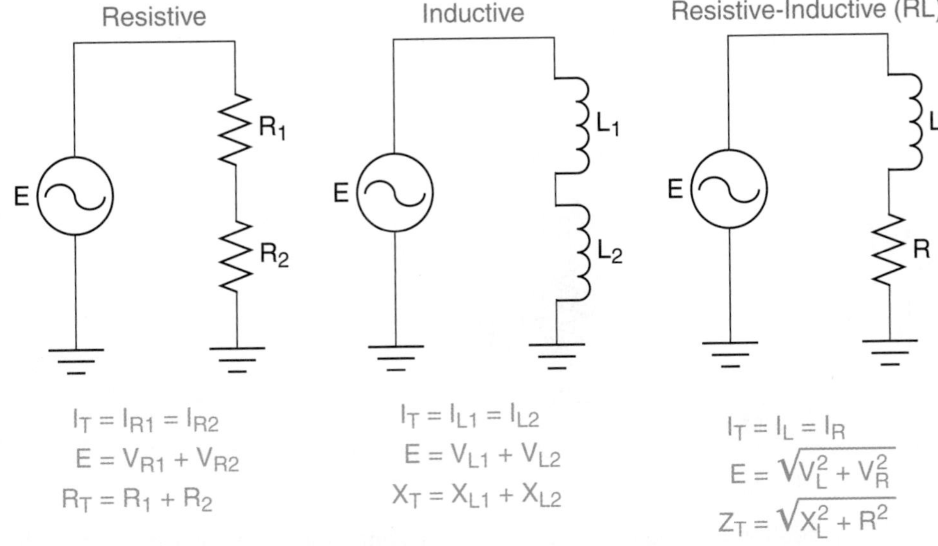

FIGURE 14.12 Resistive, inductive, and resistive-inductive (RL) series circuits.

Algebraic and Geometric Sums

As you begin working with circuits that contain both resistive and reactive components, you'll find that many values must be represented using *vectors*. As such, these values must be added together geometrically rather than algebraically. Given two variables, A and B, the algebraic and geometric sums of the variables are found as follows:

$$\text{Algebraic sum:} \quad C = A + B$$
$$\text{Geometric sum:} \quad C = \sqrt{A^2 + B^2}$$

The reason for using **geometric addition** in resistive-reactive circuits will be made clear as we continue through this chapter. For now, we need only establish that any equation in the form of $C = \sqrt{A^2 + B^2}$ provides a geometric sum of the variables.

RL Circuit Voltage and Impedance

In a series RL circuit, the geometric sum of the component voltages equals the source voltage. By the same token, the geometric sum of resistance and reactance equals the total circuit impedance (opposition to current). Even though we are using a different type of addition, the RL circuit still operates according to the rules we have established for all series circuits. That is:

- Component voltages add (geometrically) to equal the source voltage.
- Resistance and reactance add (geometrically) to equal the total impedance.

Another important characteristic of RL circuits is that the values of source voltage and total impedance each have a phase angle. For example, the circuit in Figure 14.13 was analyzed using the principles and techniques that you will learn

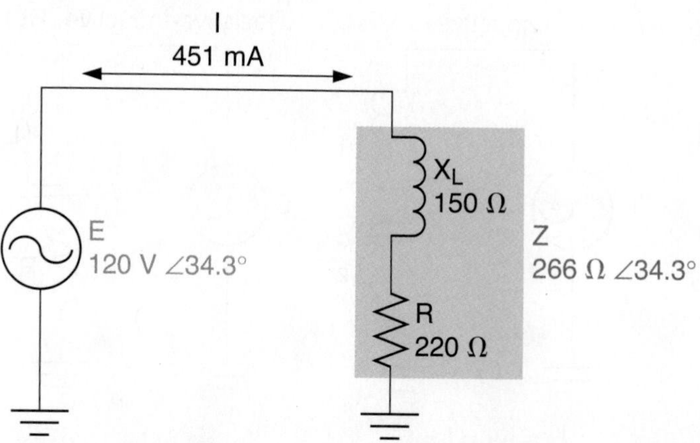

FIGURE 14.13 A series circuit and its current, voltage, and impedance values.

later in this section. As you can see, the source voltage and total impedance each have a phase angle.

Series Voltages

Earlier in the chapter, you were told that voltage leads current by 90 degrees in an inductive circuit. In a resistive circuit, voltage and current are always in phase. These two phase relationships are illustrated in Figure 14.14.

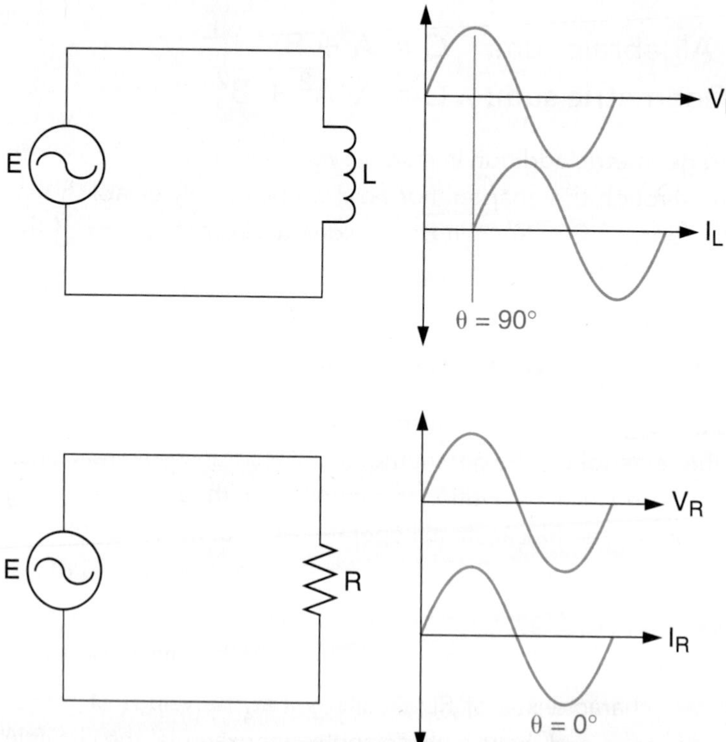

FIGURE 14.14 Phase relationships in series inductive and resistive circuits.

If we combine the circuits in Figure 14.14, we get the series circuit and waveforms shown in Figure 14.15. Since current is constant throughout a series circuit, the inductor and resistor voltage waveforms are referenced to a single current waveform. As you can see:

- The inductor voltage leads the circuit current by 90 degrees.
- The resistor voltage is in phase with the circuit current.

Therefore, *inductor voltage leads resistor voltage by 90 degrees.* This relationship holds true for every series RL circuit with an AC source.

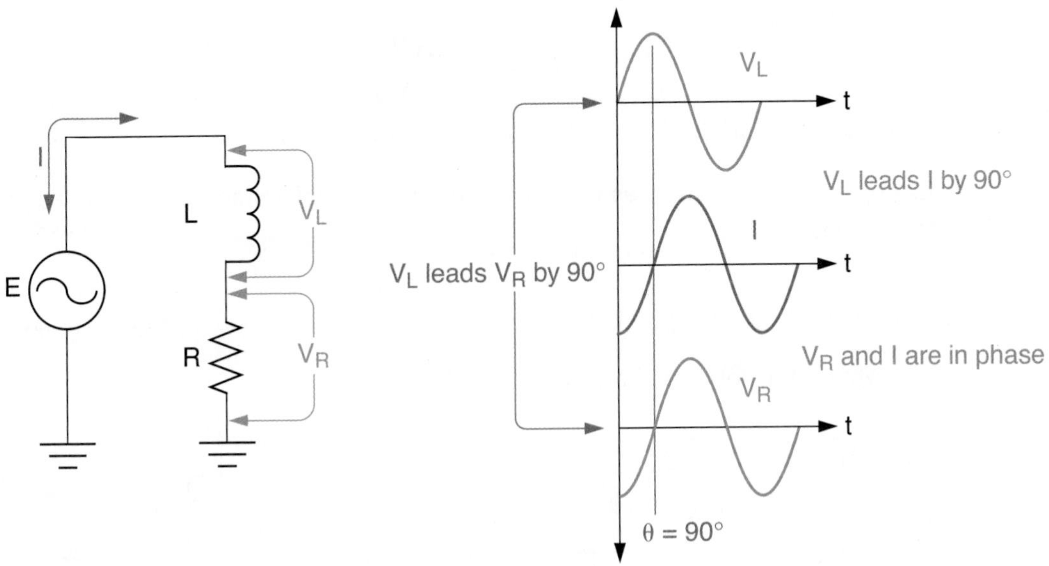

FIGURE 14.15 Phase relationships in a series RL circuit.

Since V_L leads V_R by 90 degrees, the component voltages are represented using vectors as shown in Figure 14.16a. By plotting the value of V_R on the x-axis (0°) and the value of V_L on the y-axis (+90°), the phase relationship between the two is represented in the graph.

(90°)
V_L
90°
V_R
(0°)
(a)

(90°)
V_L
E
θ
V_R
(0°)
(b)

$$E = \sqrt{V_L^2 + V_R^2}$$

$$\theta = \tan^{-1}\left(\frac{V_L}{V_R}\right)$$

FIGURE 14.16 Voltage relationships in a series RL circuit.

Like any other series circuit, the source voltage in an RL circuit must equal the sum of the component voltages. Since the component voltages in an RL circuit are vector values, they must be added geometrically, as follows:

(14.6)
$$E = \sqrt{V_L^2 + V_R^2}$$

Note that the E vector in Figure 14.16b represents the geometric sum of V_L and V_R. The circuit *phase angle* identified in the figure is the phase difference between E and V_R. Because V_R is in phase with the circuit current, θ represents the phase difference between the source voltage (E) and the circuit current. The resistor and inductor voltages can be used to calculate the value of θ as follows:

NOTE: The *inverse tangent* of a value is also called the **arc tangent** (or *arctan*) of that value. The symbol used ($\tan^{-1}$) is a shorthand notation and does not indicate a reciprocal relationship.

(14.7)
$$\theta = \tan^{-1}\frac{V_L}{V_R}$$

Example 14.5 demonstrates the calculation of the total voltage and phase angle for a series RL circuit.

EXAMPLE 14.5

What is the source voltage in Figure 14.17a?

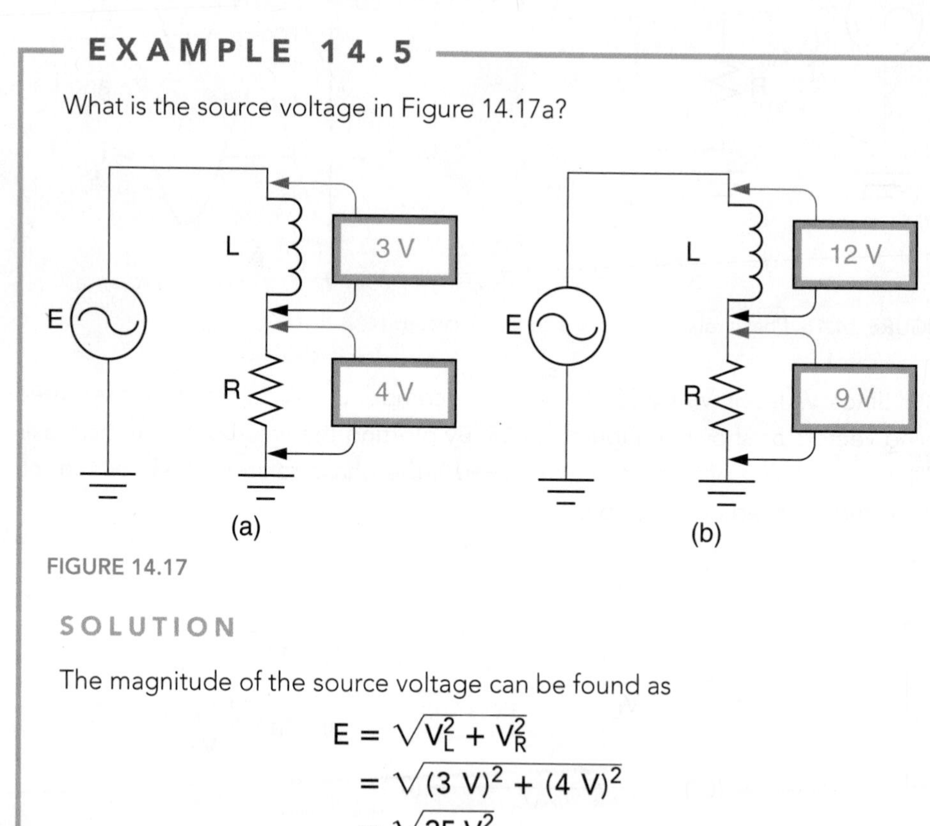

FIGURE 14.17

SOLUTION

The magnitude of the source voltage can be found as

$$\begin{aligned} E &= \sqrt{V_L^2 + V_R^2} \\ &= \sqrt{(3\text{ V})^2 + (4\text{ V})^2} \\ &= \sqrt{25\text{ V}^2} \\ &= 5\text{ V} \end{aligned}$$

The circuit phase angle can be found as

$$\theta = \tan^{-1}\left(\frac{V_L}{V_R}\right)$$

$$= \tan^{-1}\left(\frac{3\text{ V}}{4\text{ V}}\right)$$

$$= \tan^{-1}(0.75)$$

$$\cong 36.9°$$

Therefore, the source voltage for the circuit is 5 V at a phase angle of 36.9°, meaning the source voltage leads the circuit current by 36.9°.

PRACTICE PROBLEM 14.5

Calculate the value of E and the phase angle for the circuit in Figure 14.17b.

The limits on the phase angle between the source voltage and circuit current in any RL circuit are given as follows:

(14.8)
$$0° < \theta < 90°$$

which reads, "Theta is greater than 0° and less than 90°." This range of values makes sense when you consider that:

- The phase angle in a purely resistive circuit is 0°.
- The phase angle in a purely inductive circuit is 90°.

A circuit containing both resistance and inductance is neither purely resistive nor purely inductive. Therefore, it must have a phase angle that is greater than 0 degrees and less than 90 degrees. As you will see, the value of θ for a given RL circuit depends on the ratio of inductive reactance to resistance.

Series Impedance

In an RL circuit, the total impedance (opposition to alternating current) is made up of resistance and inductive reactance. Like series voltages, the values of X_L and R for a series circuit are vector quantities that must be added geometrically. By formula,

(14.9)
$$Z = \sqrt{X_L^2 + R^2}$$

Earlier, you were shown how the voltage phase angle for a series RL circuit can be calculated using the component voltages (V_L and V_R). The phase angle of an RL circuit can also be calculated using X_L and R. By formula,

(14.10)
$$\theta = \tan^{-1}\frac{X_L}{R}$$

Example 14.6 demonstrates the procedure for determining the total impedance in a series RL circuit.

EXAMPLE 14.6

Calculate the values of Z and θ for the circuit in Figure 14.18a.

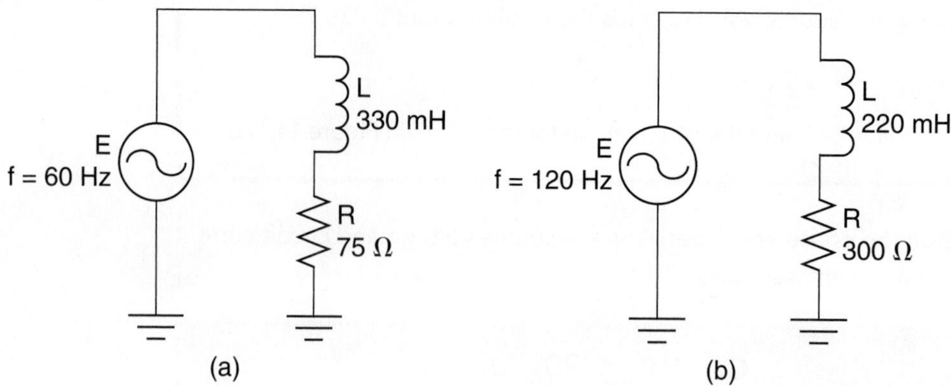

FIGURE 14.18

SOLUTION

The value of X_L for the circuit is found as

$$\begin{aligned} X_L &= 2\pi fL \\ &= 2\pi(60\text{ Hz})(330\text{ mH}) \\ &\cong 124\ \Omega \end{aligned}$$

Now, the magnitude of the circuit impedance can be found as

$$\begin{aligned} Z &= \sqrt{X_L^2 + R^2} \\ &= \sqrt{(124\ \Omega)^2 + (75\ \Omega)^2} \\ &= 145\ \Omega \end{aligned}$$

and the phase angle can be found as

$$\begin{aligned} \theta &= \tan^{-1}\left(\frac{X_L}{R}\right) \\ &= \tan^{-1}\left(\frac{124\ \Omega}{75\ \Omega}\right) \\ &= \tan^{-1}(1.653) \\ &= 58.8^\circ \end{aligned}$$

These results indicate that the circuit impedance is 145 Ω and that the circuit phase angle is 58.8°.

PRACTICE PROBLEM 14.6

Calculate the total impedance and phase angle for the circuit in Figure 14.18b.

Calculating Series Circuit Current

The total current through a series RL circuit is calculated using the source voltage and the circuit impedance. For example, calculating the current through the circuit in Figure 14.19 begins by finding the value of X_L, as follows:

$$\begin{aligned} X_L &= 2\pi fL \\ &= 2\pi(120\text{ Hz})(2\text{ H}) \\ &= 1.51\text{ k}\Omega \end{aligned}$$

The next step is to calculate the circuit impedance, as follows:

$$\begin{aligned} Z_T &= \sqrt{X_L^2 + R^2} \\ &= \sqrt{(1.51\text{ k}\Omega)^2 + (560\ \Omega)^2} \\ &= 1.61\text{ k}\Omega \end{aligned}$$

Now, Ohm's law provides us with a circuit current of

$$\begin{aligned} I &= \frac{E}{Z_T} \\ &= \frac{240\text{ V}}{1.61\text{ k}\Omega} \\ &= 149\text{ mA} \end{aligned}$$

Finally, we can calculate the phase angle between the source voltage and circuit current as follows:

$$\begin{aligned} \theta &= \tan^{-1}\left(\frac{X_L}{R}\right) \\ &= \tan^{-1}\left(\frac{1.51\text{ k}\Omega}{560\ \Omega}\right) \\ &= \tan^{-1}(2.696) \\ &= 69.6^\circ \end{aligned}$$

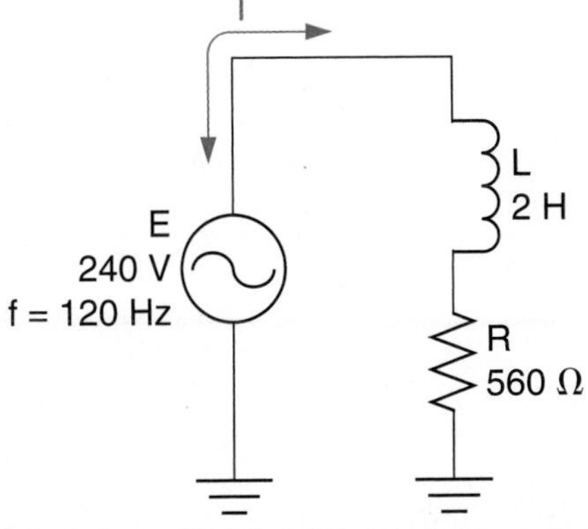

FIGURE 14.19 A series RL circuit.

These calculations indicate that the circuit current is 149 mA, and that the source voltage leads the circuit current by 69.6°.

Series Circuit Analysis

You have been shown how the voltage, current, and impedance values are calculated for a basic series RL circuit. At this point, we are going to work through several circuit analysis problems. The goal is to demonstrate the approaches taken to a variety of series RL circuit analysis problems.

EXAMPLE 14.7

Calculate the voltage, current, and impedance values for the circuit in Figure 14.20a.

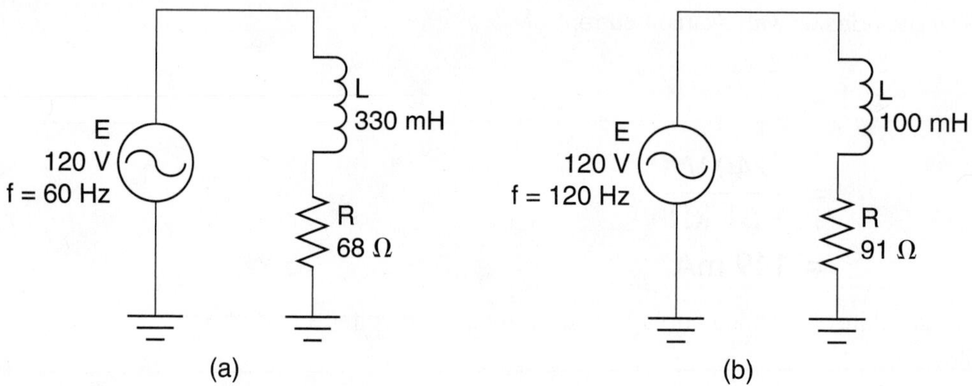

FIGURE 14.20

SOLUTION

Calculating all of the voltage, current, and impedance values for the circuit begins with calculating the value of X_L, as follows:

$$X_L = 2\pi fL$$
$$= 2\pi(60 \text{ Hz})(330 \text{ mH})$$
$$= 124 \ \Omega$$

The magnitude of the circuit impedance can now be found as

$$Z_T = \sqrt{X_L^2 + R^2}$$
$$= \sqrt{(124 \ \Omega)^2 + (68 \ \Omega)^2}$$
$$= 141 \ \Omega$$

Using the values of Z_T and E, the circuit current is found as

$$I = \frac{E}{Z_T}$$

$$= \frac{120 \text{ V}}{141 \ \Omega}$$

$$= 851 \text{ mA}$$

Now, we can calculate the magnitudes of V_L and V_R as follows:

$$V_L = I \cdot X_L$$
$$= 851 \text{ mA} \times 124 \ \Omega$$
$$= 106 \text{ V}$$

and

$$V_R = I \cdot R$$
$$= 851 \text{ mA} \times 68 \ \Omega$$
$$= 57.9 \text{ V}$$

Finally, we can calculate the value of the circuit phase angle as follows:

$$\theta = \tan^{-1}\left(\frac{V_L}{V_R}\right)$$
$$= \tan^{-1}\left(\frac{106 \text{ V}}{57.9 \text{ V}}\right)$$
$$= 61.4°$$

PRACTICE PROBLEM 14.7

Calculate the voltage, current, and impedance values for the circuit in Figure 14.20b.

Using the Voltage-Divider Equation

We can calculate the component voltages in a series RL circuit using the voltage-divider equation (just as we did in DC circuits). However, the voltage-divider equation must be modified as follows:

(14.11)

$$V_n = E \times \frac{Z_n}{Z_T}$$

where

Z_n = the value of R or X_L
Z_T = the geometric sum of R and X_L

Example 14.8 demonstrates the use of the voltage-divider relationship in series RL circuit analysis.

EXAMPLE 14.8

Calculate the component voltages for the circuit in Figure 14.21a.

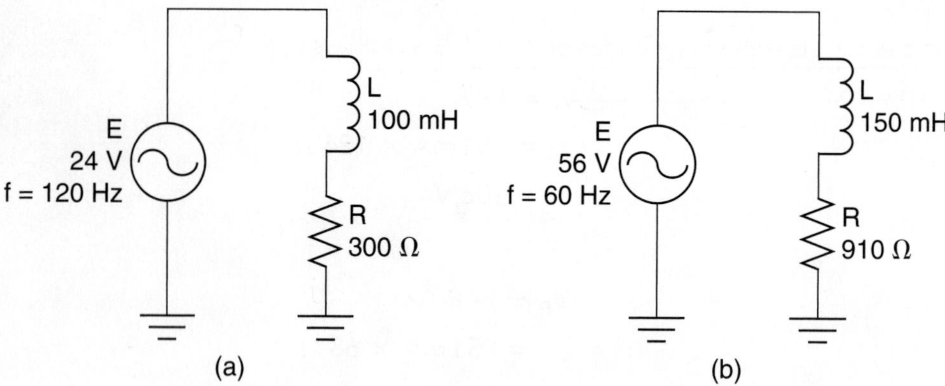

FIGURE 14.21

SOLUTION

Calculating the component voltages begins with calculating the value of X_L as follows:

$$X_L = 2\pi fL$$
$$= 2\pi(120\text{ Hz})(100\text{ mH})$$
$$= 75.4\ \Omega$$

Next, the circuit impedance is found as

$$Z_T = \sqrt{X_L^2 + R^2}$$
$$= \sqrt{(75.4\ \Omega)^2 + (300\ \Omega)^2}$$
$$= 309\ \Omega$$

Once we know the values of X_L and Z_T, the inductor voltage can be found as

$$V_L = E \times \frac{X_L}{Z_T}$$
$$= 24\text{ V} \times \frac{75.4\ \Omega}{309\ \Omega}$$
$$= 5.86\text{ V}$$

and the magnitude of the resistor voltage can be found as

$$V_R = E \times \frac{R}{Z_T}$$
$$= 24\text{ V} \times \frac{300\ \Omega}{309\ \Omega}$$
$$= 23.3\text{ V}$$

PRACTICE PROBLEM 14.8

Calculate the component voltages for the circuit in Figure 14.21b.

If the component voltages calculated in Example 14.8 appear to be incorrect, you may be forgetting that the source voltage equals their *geometric* sum. The sum of the component voltages calculated in the example is approximately equal to the source voltage, as follows:

$$E = \sqrt{V_L^2 + V_R^2}$$
$$= \sqrt{(5.86\ V)^2 + (23.3\ V)^2}$$
$$\cong 24\ V$$

At this point, you have been shown how to calculate most of the basic electrical values for series RL circuits. The only remaining basic property—power—is discussed in the next section.

Our final circuit analysis problem involves a series RL circuit that contains two inductors. As you will see, the total circuit inductance (L_T) is used in place of a single component value in several of the circuit calculations.

EXAMPLE 14.9

Calculate the component voltages for the circuit in Figure 14.22a.

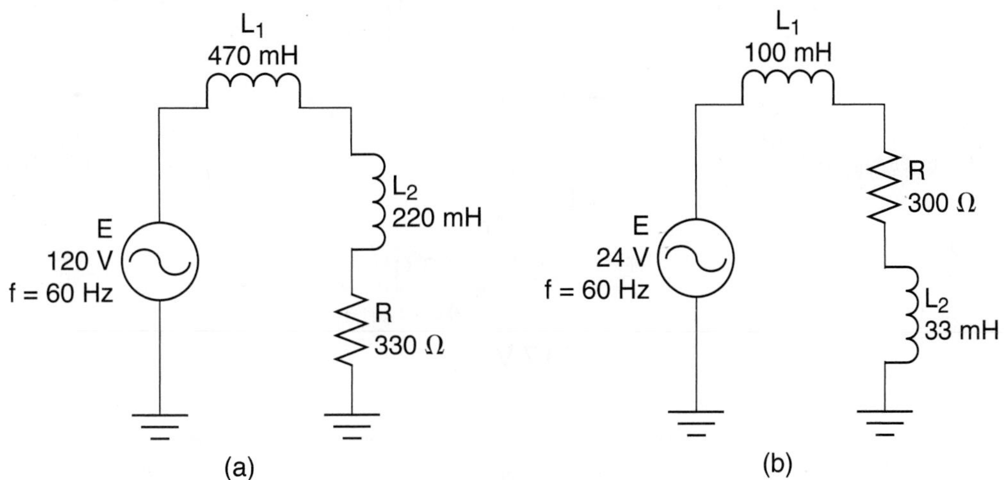

FIGURE 14.22

SOLUTION

Calculating the component voltages begins with determining the reactance of each inductor, as follows:

$$X_{L1} = 2\pi fL_1$$
$$= 2\pi(60\ Hz)(470\ mH)$$
$$= 177\ \Omega$$

and

$$X_{L2} = 2\pi fL_2$$
$$= 2\pi(60\ Hz)(220\ mH)$$
$$= 83\ \Omega$$

To calculate the total circuit impedance, we need to determine the total circuit reactance (X_{LT}). Since X_{L1} and X_{L2} are in phase with each other, the total reactance equals the algebraic sum of their values, as follows:

$$X_{LT} = X_{L1} + X_{L2} = 177\ \Omega + 83\ \Omega = 260\ \Omega$$

Now, using the values of X_{LT} and R, the magnitude of the circuit impedance can be found as

$$Z_T = \sqrt{X_{LT}^2 + R^2} = \sqrt{(260\ \Omega)^2 + (330\ \Omega)^2} = 420\ \Omega$$

Once we know the circuit impedance, we can use the voltage divider equation to calculate the component voltages, as follows:

$$V_{L1} = E \times \frac{X_{L1}}{Z_T} = 120\text{ V} \times \frac{177\ \Omega}{420\ \Omega} = 50.6\text{ V}$$

$$V_{L2} = E \times \frac{X_{L2}}{Z_T} = 120\text{ V} \times \frac{83\ \Omega}{420\ \Omega} = 23.7\text{ V}$$

and

$$V_R = E \times \frac{R}{Z_T} = 120\text{ V} \times \frac{330\ \Omega}{420\ \Omega} = 94.3\text{ V}$$

PRACTICE PROBLEM 14.9

Calculate the values of Z_T, V_{L1}, V_{L2}, and V_R for the circuit in Figure 14.22b. (The placement of the components does not affect the analysis of the circuit.)

When working with circuits containing multiple inductors and/or resistors, just remember the following:

1. Any values that are in phase, such as two or more reactances, can be added algebraically.
2. Any values that are not in phase, such as reactance and resistance, must be added geometrically.
3. Any overall circuit value, such as Z_T or θ, is calculated using the appropriate *total* values. (For example, E would be calculated using V_{LT} and V_{RT}.)

PROGRESS CHECK

1. What are the three fundamental characteristics of every series circuit?
2. What is the geometric sum of two variables? What does it represent?
3. What does the phase angle of a series RL circuit indicate?
4. What electrical values are commonly used to calculate the phase angle of a series RL circuit?
5. How is the *voltage-divider equation* modified for use in series RL circuits?

14.5 POWER CHARACTERISTICS AND CALCULATIONS

Resistive-reactive circuit analysis involves every type of power calculation imaginable. Significant values of apparent power, true power, and reactive power are present in most resistive-reactive circuits. In this section, we examine the fundamental power characteristics and calculations for series RL circuits. In Chapter 15, you'll see that many of the principles covered in this section apply to parallel RL circuits as well.

AC Power Values: A Brief Review

Earlier in the chapter, we discussed apparent power (P_{APP}), resistive power (P_R), and reactive power (P_X) in terms of inductor operation. Each of these values is described in terms of RL circuit values in Table 14.3.

TABLE 14.3 • Power Values in Resistive-Reactive Circuits.

VALUE	DEFINITION
Resistive power (P_R)	The power dissipated by the resistance in an RL circuit. Also called *true power.*
Reactive power (P_X)	The value found using $P = I^2X_L$. The energy (per unit time) found using I^2X_L is actually used by the inductor to build its electromagnetic field, and is returned to the circuit when the field collapses. P_X is measured in *volt-amperes-reactive (VAR)* to distinguish it from true power dissipation values. Also known as *imaginary power.*
Apparent power (P_{APP})	The combination of resistive (true) power and reactive (imaginary) power. Measured in *volt-amperes (VA)* because its value does not represent a true power-dissipation value.

The values of resistive and reactive power for a series RL circuit are calculated as shown in Example 14.10.

EXAMPLE 14.10

Calculate the values of resistive and reactive power for the circuit in Figure 14.23a.

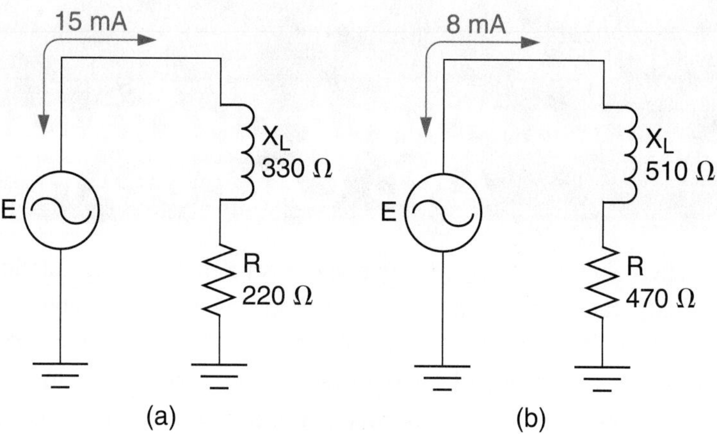

FIGURE 14.23

SOLUTION

Assuming the inductor in Figure 14.23a is an ideal component ($R_W = 0\ \Omega$), its reactive power can be found as

$$P_X = I^2 \cdot X_L$$
$$= (15\text{ mA})^2 \cdot 330\ \Omega$$
$$= 74.3\text{ mVAR}$$

Using the value of R, the circuit resistive power is calculated as follows:

$$P_R = I^2 \cdot R$$
$$= (15\ \text{mA})^2 \cdot 220\ \Omega$$
$$= 49.5\ \text{mW}$$

PRACTICE PROBLEM 14.10

Calculate the values of P_X and P_R for the circuit in Figure 14.23b.

Remember, resistive power is dissipated (as heat) by the resistance in an RL circuit. Reactive power, on the other hand, is stored in the electromagnetic field of the inductor, and returned to the circuit as the field collapses.

Calculating Apparent Power (P_{APP})

Apparent power in a series RL circuit equals the geometric sum of resistive power and reactive power. Like the voltages in a series RL circuit, resistive and reactive power can be represented using vectors, as shown in Figure 14.24a. By plotting the value of P_R on the x-axis (0°) and the value of P_X on the y-axis (90°), the phase relationship between the two is represented in the graph. By formula,

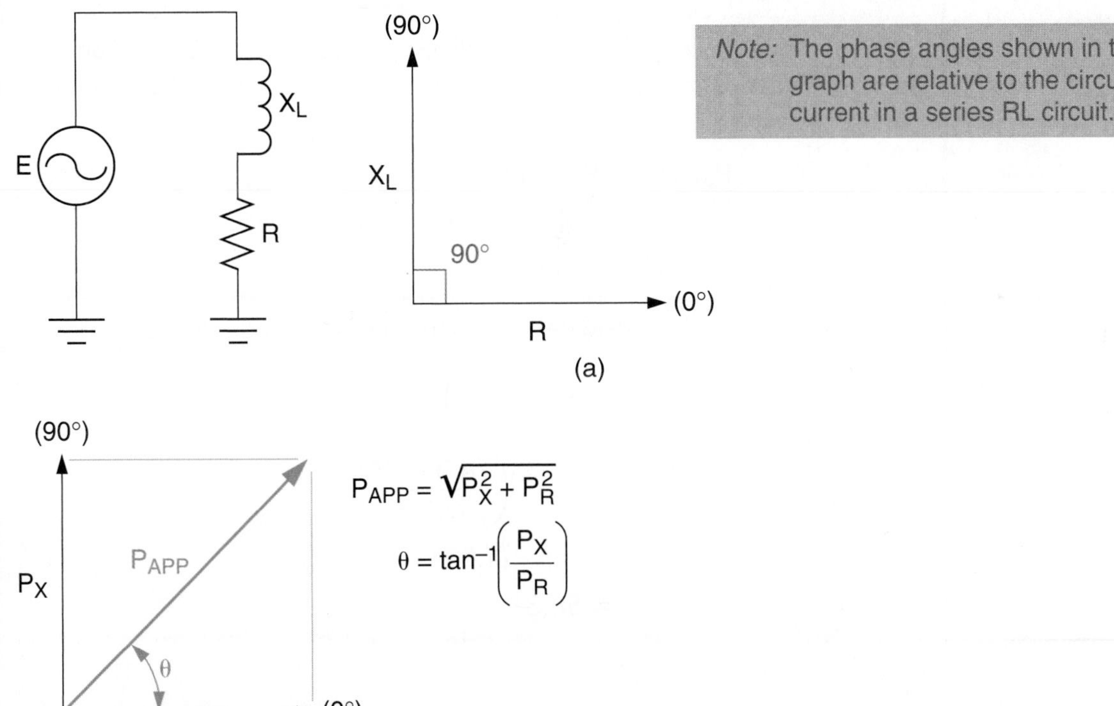

FIGURE 14.24 Series RL circuit power relationships.

(14.12)

$$P_{APP} = \sqrt{P_X^2 + P_R^2}$$

where

P_{APP} = the apparent power, measured in volt-amperes (VA)

Note that the P_{APP} vector in Figure 14.24b represents the geometric sum of P_X and P_R. The circuit *phase angle* identified in the figure is the phase difference between P_{APP} and P_R. Because P_R is in phase with the circuit current, θ represents the phase difference between apparent power and the circuit current. The phase angle of apparent power (relative to the circuit current) is found as

(14.13)

$$\theta = \tan^{-1}\frac{P_X}{P_R}$$

Example 14.11 demonstrates the use of these equations.

EXAMPLE 14.11

What is the value of P_{APP} for the circuit in Example 14.10?

SOLUTION

In Example 14.10, we calculated the following values: $P_X = 74.3$ mVAR and $P_R = 49.5$ mW. Using these values in equation (14.12), the value of apparent power is found as

$$\begin{aligned} P_{APP} &= \sqrt{P_X^2 + P_R^2} \\ &= \sqrt{(74.3 \text{ mVAR})^2 + (49.5 \text{ mW})^2} \\ &= 89.3 \text{ mVA} \end{aligned}$$

and the phase angle of P_{APP} (relative to the circuit current) is found as

$$\begin{aligned} \theta &= \tan^{-1}\left(\frac{P_X}{P_R}\right) \\ &= \tan^{-1}\left(\frac{74.3 \text{ mVAR}}{49.5 \text{ mW}}\right) \\ &= \tan^{-1}(1.501) \\ &= 56.3° \end{aligned}$$

PRACTICE PROBLEM 14.11

Calculate the value and phase angle of P_{APP} for the circuit in Practice Problem 14.10.

When you calculate the apparent power for a circuit, you need to remember that only a portion of its value is actually dissipated by the circuit. (This is why it has its own unit of measure.)

Power Factor

POWER FACTOR is the ratio of resistive power to apparent power. In other words, it is the ratio of actual power dissipation to apparent power. By formula,

POWER FACTOR The ratio of resistive power to apparent power.

(14.14)

$$PF = \frac{P_R}{P_{APP}}$$

where

P_R = the resistive power, in watts

P_{APP} = the apparent power, in volt-amperes (VA)

Example 14.12 demonstrates the use of this relationship.

EXAMPLE 14.12

The circuit in Example 14.11 has the following values: P_R = 49.5 mW and P_{APP} = 89.3 VA. What is the circuit's power factor?

SOLUTION

The circuit power factor has a value of

$$PF = \frac{P_R}{P_{APP}}$$

$$= \frac{49.5 \text{ mW}}{89.3 \text{ mVA}}$$

$$= 0.554$$

PRACTICE PROBLEM 14.12

Calculate the power factor for the circuit in Practice Problem 14.11.

The power factor for a given circuit lies somewhere between 0 and 1. For example, consider the circuits shown in Figure 14.25. The circuit in Figure 14.25a is purely reactive. Therefore,

$$P_R = 0 \text{ W} \qquad P_{APP} = P_X \qquad PF = 0$$

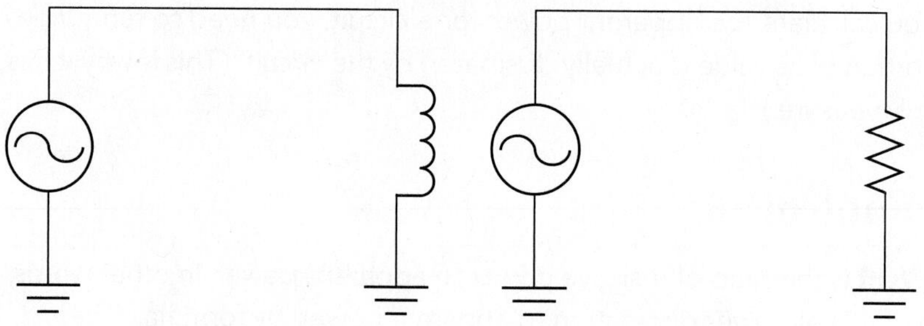

$$PF = \frac{P_R}{P_{APP}} = 0$$
(a)

$$PF = \frac{P_R}{P_{APP}} = 1$$
(b)

FIGURE 14.25 Power factors in purely inductive and purely resistive circuits.

The circuit in Figure 14.25b is purely resistive. Therefore,

$$P_R = 0 \text{ VAR} \qquad P_{APP} = P_R \qquad PF = 1$$

Because an RL circuit is neither purely resistive nor purely reactive, its value must be greater than 0 and less than 1. By formula,

(14.15)
$$0 < PF < 1$$

It should be noted that higher power factors are considered to be desirable in most practical applications. This point is discussed in detail in Chapter 18.

Putting It All Together: Series RL Circuit Analysis

Now that we have covered the power characteristics and calculations associated with series RL circuits, we're going to put it all together in a single circuit analysis problem.

EXAMPLE 14.13

Calculate the voltage, current, impedance, and power values for the circuit in Figure 14.26.

SOLUTION

The reactance in Figure 14.26 is found as

$$X_L = 2\pi fL$$
$$= 2\pi(120 \text{ Hz})(100 \text{ mH})$$
$$= 75.4\ \Omega$$

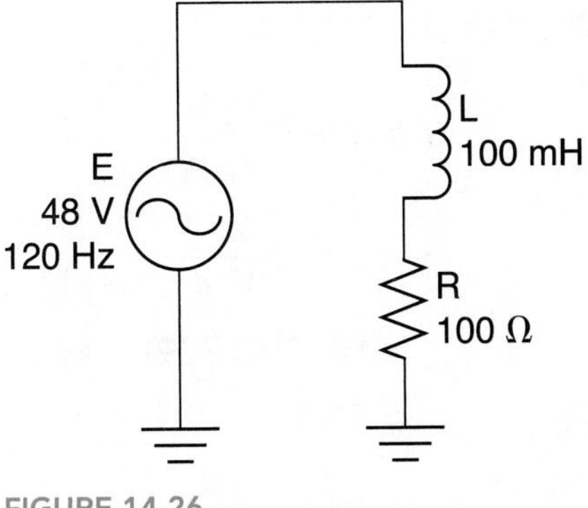

FIGURE 14.26

This, combined with the value of R, gives us a circuit impedance of

$$\begin{aligned} Z_T &= \sqrt{X_L^2 + R^2} \\ &= \sqrt{(75.4\ \Omega)^2 + (100\ \Omega)^2} \\ &= 125\ \Omega \end{aligned}$$

The magnitude of the circuit current can now be found as

$$\begin{aligned} I_T &= \frac{E}{Z_T} \\ &= \frac{48\text{ V}}{125\ \Omega} \\ &= 384\text{ mA} \end{aligned}$$

Using the voltage divider equation, the voltage across the resistor can be found as

$$\begin{aligned} V_R &= E \times \frac{R}{Z_T} \\ &= 48\text{ V} \times \frac{100\ \Omega}{125\ \Omega} \\ &= 38.4\text{ V} \end{aligned}$$

and the voltage across the inductor can be found as

$$\begin{aligned} V_L &= E \times \frac{X_L}{Z_T} \\ &= 48\text{ V} \times \frac{75.4\ \Omega}{125\ \Omega} \\ &= 29\text{ V} \end{aligned}$$

The circuit phase angle can be found as

$$\theta = \tan^{-1}\frac{V_L}{V_R}$$
$$= \tan^{-1}\frac{29\text{ V}}{38.4\text{ V}}$$
$$= \tan^{-1}(0.729)$$
$$= 36.1°$$

The circuit power values can be found as

$$P_R = I \times V_R$$
$$= 384\text{ mA} \times 38.4\text{ V}$$
$$= 14.7\text{ W}$$

$$P_X = I_T \times V_L$$
$$= 384\text{ mA} \times 29\text{ V}$$
$$= 11.1\text{ VAR}$$

and

$$P_{APP} = \sqrt{P_X^2 + P_R^2}$$
$$= \sqrt{(11.1\text{ VAR})^2 + (14.7\text{ W})^2}$$
$$= 18.4\text{ VA}$$

Finally, the circuit power factor is found as

$$PF = \frac{P_R}{P_{APP}}$$
$$= \frac{14.7\text{ W}}{18.4\text{ VA}}$$
$$= 0.799$$

This completes the analysis of the circuit.

PRACTICE PROBLEM 14.13

A circuit like the one in Figure 14.26 has the following values: E = 120 V, f = 60 Hz, L = 470 mH, and R = 150 Ω. Calculate the voltage, current, impedance, and power values for the circuit.

We have concluded our discussion on series RL circuit characteristics. In the next chapter, we will examine the characteristics of parallel RL circuits.

PROGRESS CHECK

1. Compare and contrast the following: apparent power, resistive power, and reactive power.
2. Why is the winding resistance of a coil normally ignored in the resistive power calculations for an RL circuit?
3. State and explain the phase relationship between reactive power (P_X) and resistive power (P_R).
4. What is the power factor for an RL circuit? What does it tell you?
5. How is the power factor of an RL circuit determined?

14.6 SUMMARY

Here is a summary of the major points that were covered in this chapter:

The Phase Relationship Between Inductor Voltage and Current

- Inductor voltage is proportional to the rate of change in inductor current.
  - The rate of change in inductor current is zero amps per second (0 A/s) at the 90° and 270° points on its sine wave input.
  - The rate of change in inductor current reaches its maximum value at the 180° and 360° (0°) points on its sine wave input.
- Inductor voltage is maximum (V_{pk}) at the 180° and 360° points on the current sine wave.
- Inductor voltage is minimum (0 V) at the 90° and 270° points on the current sine wave.
- The relationship between current and voltage in a purely inductive circuit is usually described in either of two ways:
  - Voltage leads current by 90 degrees.
  - Current lags voltage by 90 degrees.

Inductive Reactance (X_L)

- The opposition (in ohms) that an inductor presents to a changing current is referred to as inductive reactance (X_L).
  - X_L is directly proportional to the value of any inductor.
  - X_L is directly proportional to the inductor's operating frequency.
- In any inductive AC circuit, Ohm's law can be used to calculate any one of the following values (given the other two): V_L, I_L, or X_L.
- Total series reactance is found in the same fashion as total series resistance.
- Total parallel reactance is found in the same fashion as total parallel resistance.
- Impedance is the total opposition to current in an AC circuit, consisting of resistance and/or reactance.
  - Resistance is a static value.
  - Reactance is a dynamic value.
  - Resistance and reactance are added *geometrically* to find impedance.

Inductor Power

- The power value calculated using the RMS values of voltage and current in an inductive circuit is referred to as apparent power.
  - The term *apparent power* is used because most of the energy is actually transferred to the electromagnetic field of the inductor. Apparent power is measured in volt-amperes (VA).
  - Apparent power is made up of reactive power and true (resistive) power.
  - Reactive power actually indicates the rate at which energy is transferred to the inductor's electromagnetic field. It is not a power dissipation value, and therefore, is measured in volt-amperes-reactive (VAR) rather than watts (W).
  - True power is the power that is dissipated by the winding resistance of a coil. In most cases, true power makes up a small percentage of the apparent power.
- The ideal inductor (if it could be produced) would have no winding resistance. As a result, all of the energy drawn from the source would be transferred to the component's electromagnetic field.
- The quality (Q) of an inductor is a measure of how close the component comes to having the characteristics of an ideal inductor.

Series RL Circuits

- A resistive-inductive (RL) circuit is one that contains any combination of resistors and inductors.
- Many values in resistive-reactive circuits must be added as vector quantities.
  - Geometric addition must be used to add two or more vector quantities.
  - For two vectors (A and B) at 90° angles, their geometric sum (C) is found as $C = \sqrt{A^2 + B^2}$.
- In a series RL circuit:
  - The source voltage (E) equals the geometric sum of V_R and V_L.
  - The circuit impedance (Z_T) equals the geometric sum of R and X_L.
- Inductor voltage leads resistor voltage by 90° in a series RL circuit.
- The source phase angle ($\angle\theta$) is the phase difference between the source voltage and the circuit current.
  - Voltage leads current in a series RL circuit.
  - The source phase angle falls within the limits of $0° < \theta < 90°$.

Power Characteristics and Calculations

- In any resistive-reactive circuit:
  - Resistive power (or "true power") is measured in watts. P_R is the power that is actually dissipated in the circuit.
  - Reactive power (or "imaginary power") is measured in volt-amperes-reactive (VAR).
  - Apparent power (the geometric sum of resistive and reactive power) is measured in volt-amperes (VA).
- The effect of inductor winding resistance (R_w) on the value of apparent power is usually negligible.
- Apparent power leads the circuit current in a series RL circuit.
- The power factor for a series RL circuit is the ratio of resistive power to apparent power.

CHAPTER REVIEW

1. The voltage induced across an inductor is dependent on ________.
   a) the rate of change of current
   b) the rate of change of current and the inductance of the coil
   c) the rate of change of current and the peak current
   d) the rate of change of current and the polarity of the current

2. A sinusoidal current is flowing through an inductor. When the phase of the current is at 270°, the rate of change of the current is ________, the magnitude of the current is ________, and inductor voltage is ________.
   a) zero, maximum, zero
   b) maximum, zero, maximum
   c) maximum, zero, zero
   d) zero, minimum, zero

3. The voltage across an inductor is at its *minimum* value when inductor current is ________.
   a) at its positive peak value
   b) at its negative peak value
   c) at 90°
   d) at 270°
   e) all of the above

4. The current through an inductor is at zero amps going negative. This means that inductor voltage is ________.
   a) at its positive peak
   b) at its negative peak
   c) lagging the current by 90°
   d) cannot tell with the information given

5. In an inductor, voltage ________ current by 90° and current ________ voltage by 270°.
   a) leads, lags
   b) lags, leads
   c) lags, lags
   d) leads, leads

6. Inductive reactance changes with ________.
   a) the applied voltage
   b) the source resistance
   c) frequency
   d) circuit current

7. The DC resistance of an inductor is usually ________ its reactance.
   a) about the same as
   b) much lower than
   c) much higher than
   d) dependent on

8. The DC resistance of an inductor is usually referred to as its ________.
   a) impedance
   b) DC impedance
   c) winding resistance
   d) true reactance

9. A 0.65 H inductor is installed in an AC circuit with a frequency of 120 Hz. This means that $X_L =$ ________.
   a) 490 Ω
   b) 490 VA
   c) 490 VAR
   d) 1.16 kΩ

10. A 240 V 60 Hz voltage is applied to a 1.15 H coil. The current through the coil is ________.
    a) 208 mA
    b) 5.54 A
    c) 554 mA
    d) 20.8 A

11. A 680 mH coil is added in series to the circuit described in question 10. Assume that the applied voltage is measured and it is found that E is actually 215 V, not 220 V. The total circuit current would be ________.
    a) 312 mA
    b) 614 mA
    c) 3.12 A
    d) 61.4 mA

12. Two 220 mH coils are connected in parallel. They are connected in series with two more 220 mH coils. The applied voltage is 110 V @ 120 Hz. Circuit current in this circuit would be ________.
    a) 2.65 A
    b) 1.66 A
    c) 166 mA
    d) 265 mA

13. If the frequency of the applied voltage in question 12 were 60 Hz rather than 120 Hz, the circuit current would ________.
    a) double
    b) decrease by a factor of 2π
    c) decrease by half
    d) increase by a factor of 2π

14. The opposition to AC current that an inductor provides is considered to be a ________ value.
    a) phase-related
    b) static

c) frequency

d) dynamic

15. Resistive power is also known as ________.

a) reactive power

b) true power

c) apparent power

d) all of the above

16. Imaginary power is also called ________ .

a) reactive power

b) true power

c) impedance power

d) apparent power

17. The power that is dissipated as heat is measured in ________.

a) watts

b) thermal units

c) volt-amperes reactive

d) volt-amperes

18. The combination of resistive and reactive power is called ________ power and its unit of measure is the ________.

a) reactive, VAR

b) apparent, VAR

c) imaginary, VA

d) apparent, VA

19. Two inductors are identical except one has twice the winding resistance of the other. They are both used in identical circuits. The inductor with the higher value of R_W will have ________ real power values and ________ reactive power values.

a) higher, the same

b) higher, higher

c) higher, lower

d) lower, higher

20. The power dissipated by an ideal inductor would be ________.

a) infinite

b) zero

c) very high

d) dependent on frequency

21. If you are adding two resistances together you use ________ addition. If you are adding a resistance to a reactance you use ________ addition.

a) linear, exponential

b) real, imaginary

c) algebraic, geometric

d) geometric, algebraic

22. When working with circuits that can contain both resistive and reactive components, values are often represented as ________.
    a) geometric values
    b) imaginary values
    c) vectors
    d) impedances

23. In a series RL circuit, inductor voltage ________ resistor voltage by ________.
    a) lags, 90°
    b) leads, 90°
    c) lags, 45°
    d) leads, 45°

24. When plotting the values of V_L and V_R, ________ is on the y-axis while ________ is on the x-axis.
    a) V_R, V_L
    b) reactance, resistance
    c) V_L, V_R
    d) resistance, reactance

25. A resistor and an inductor are in series. The value of V_R is 48 V while the value of V_L is 26 V. The magnitude of the source voltage is ________.
    a) 70 V
    b) 29.8 V
    c) 55 V
    d) 58 V

26. The phase angle of the source voltage for the circuit described in question 25 is ________.
    a) 106°
    b) –28°
    c) 10.6°
    d) 28°

27. A 47 Ω resistor is in series with a 470 mH inductor. The circuit frequency is 100 Hz. The total impedance of this circuit is ________ at a phase angle of ________.
    a) 299 Ω, 81°
    b) 299 Ω, 9.04°
    c) 299 Ω, 110 °
    d) 299 Ω, 90.4°

28. If the frequency of the voltage source described in question 27 were to *decrease*, the impedance magnitude would ________ and its phase angle would ________.
    a) increase, increase
    b) decrease, decrease

c) increase, decrease

d) decrease, increase

29. A 550 V @ 60 Hz voltage source is applied to a series circuit containing a 680 Ω resistance and a 1.2 H inductor. Using the voltage divider method, V_R = ________ and V_L = ________.

a) 458 V, 304 V

b) 330 V, 219 V

c) 304 V, 458 V

d) 219 V, 330 V

30. Refer to the circuit described in question 29. The phase angle of the source voltage, relative to circuit current, is ________ while the phase angle of circuit impedance is ________.

a) 33.6°, 56.4°

b) 56.4°, 33.6°

c) 33.6°, −33.6°

d) 33.6°, 33.6°

31. A 2.2 H inductor is connected in series with a 2.2 kΩ resistor. The applied voltage is 220 V @ 60 Hz. Assuming the inductor has negligible winding resistance, the value of I_T = ________.

a) 936 mA

b) 256 mA

c) 93.6 mA

d) 25.6 mA

32. Refer to the circuit described in question 31. In this circuit, the value of P_X = ________ VAR and the value for P_R = ________ W.

a) 7.26, 19.3

b) 19.3, 7.26

c) 77.6, 206

d) 206, 77.6

33. For the circuit described in the previous two questions, the value of P_{APP} = ________ VA at an angle of ________.

a) 20.6, 20.6 °

b) 20.6, 69.4°

c) 69.4, 20.6°

d) 69.4, 69.4°

34. A series RL circuit has power values of P_{APP} = 2.65 VA at an angle of 37.5°. The power factor for this circuit = ________.

a) 60.9%

b) 69.3%

c) 79.3%

d) 70.9%

35. If the frequency of the applied voltage to the circuit described in question 34 were to increase, the power factor would ________.
    a) increase
    b) decrease
    c) remain the same
    d) cannot tell without component values

PRACTICE PROBLEMS

1. Calculate the value of X_L for the circuit in Figure 14.27a.
2. Calculate the value of X_L for the circuit in Figure 14.27b.

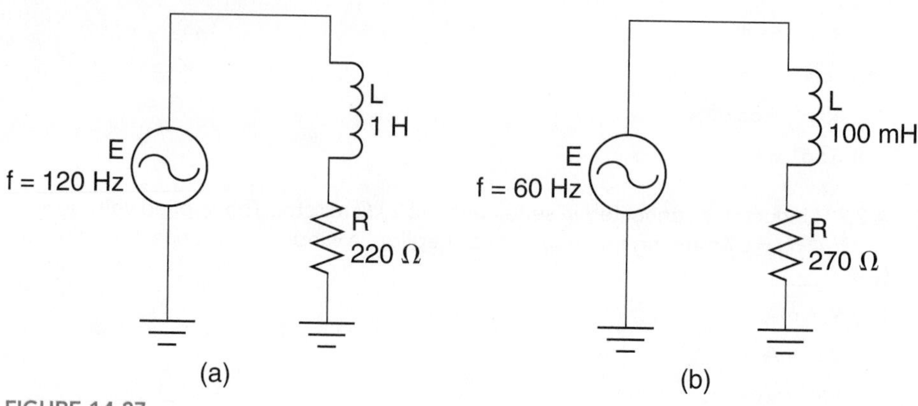

FIGURE 14.27

3. The following table lists values for an inductive circuit. Complete the table.

V_{RMS}	I_{RMS}	X_L
96 V	____	20 Ω
____	3.6 A	10 Ω
20 V	4 A	____

4. The following table lists values for an inductive circuit. Complete the table.

V_{RMS}	I_{RMS}	X_L
480 V	____	200 Ω
____	40 μA	18 MΩ
24 V	0.12 A	____

5. The component voltages shown in Figure 14.28a were measured using an AC voltmeter. Determine the source voltage (E) and phase angle for the circuit.
6. The component voltages shown in Figure 14.28b were measured using an AC voltmeter. Determine the source voltage (E) and phase angle for the circuit.

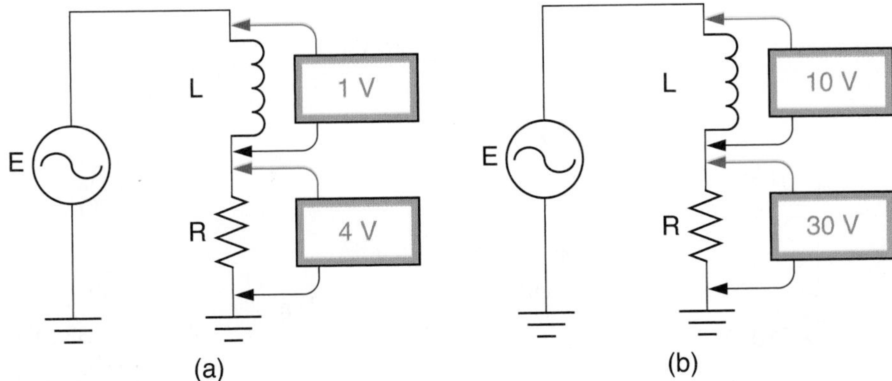

FIGURE 14.28

7. Calculate the circuit impedance and phase angle for the circuit in Figure 14.29a.
8. Calculate the circuit impedance and phase angle for the circuit in Figure 14.29b.

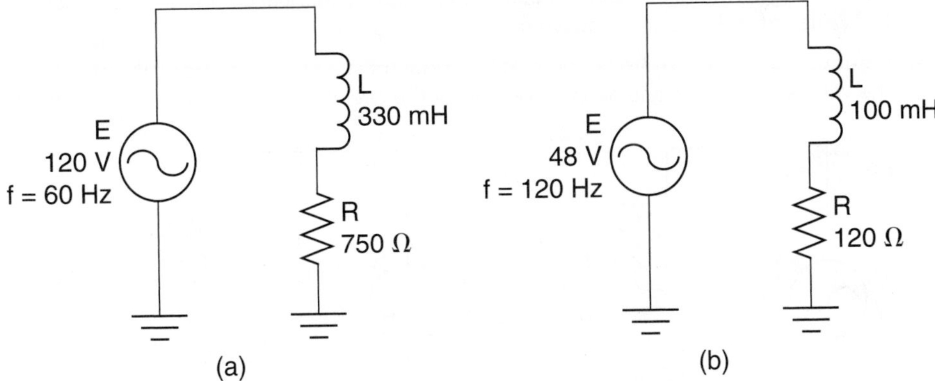

FIGURE 14.29

9. Calculate the values of Z_T, I_T, V_L, V_R, and θ for the circuit in Figure 14.30a.
10. Calculate the values of Z_T, I_T, V_L, V_R, and θ for the circuit in Figure 14.30b.

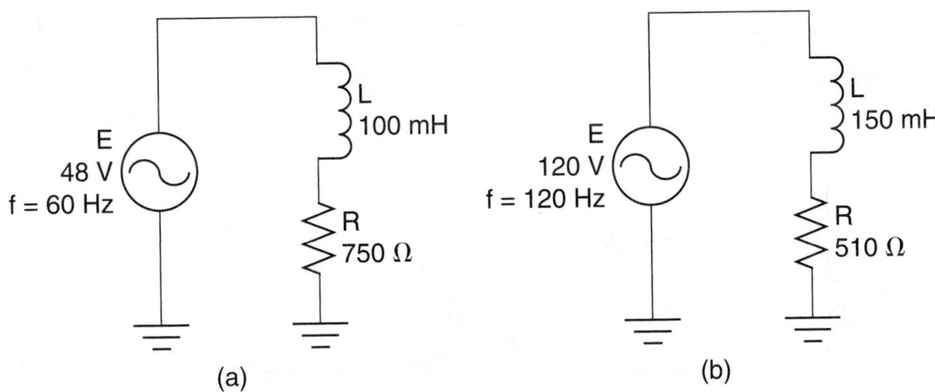

FIGURE 14.30

11. Calculate the values of Z_T, I_T, V_{L1}, V_{L2}, V_R, and θ for the circuit in Figure 14.31a.
12. Calculate the values of Z_T, I_T, V_{L1}, V_{L2}, V_{R1}, V_{R2}, and θ for the circuit in Figure 14.31b.

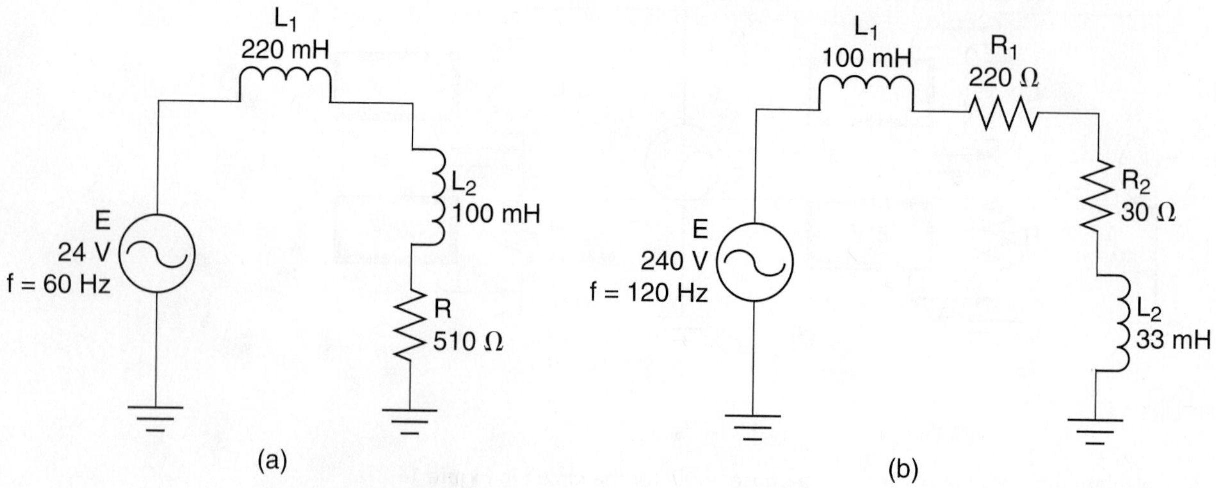

FIGURE 14.31

13. Refer to Figure 14.32a. Assuming that the inductor is an ideal component, calculate the values of resistive and reactive power for the circuit.
14. Refer to Figure 14.32b. Assuming that the inductor is an ideal component, calculate the values of resistive and reactive power for the circuit.

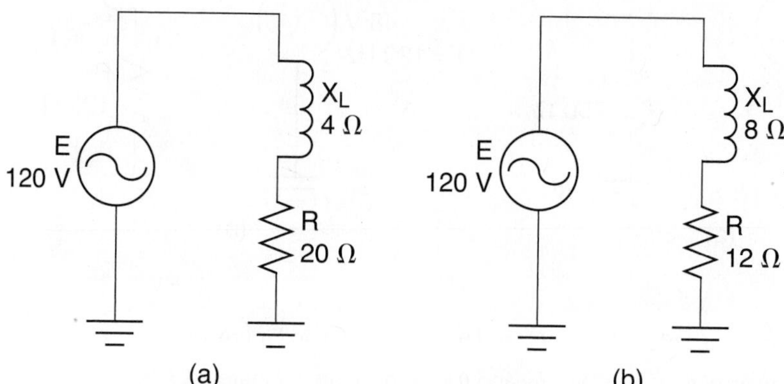

FIGURE 14.32

15. Calculate the magnitude and phase angle of apparent power for the circuit described in problem 13.
16. Calculate the magnitude and phase angle of apparent power for the circuit described in problem 14.

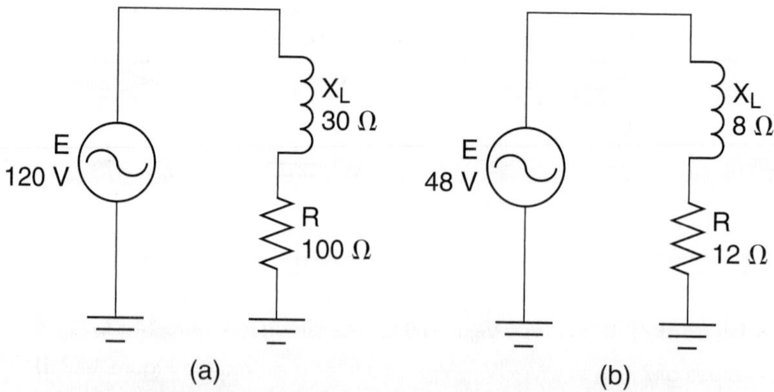

FIGURE 14.33

17. Calculate the component current, voltage, and power values for the circuit in Figure 14.33a. Then calculate the value of apparent power for the circuit.
18. Calculate the component current, voltage, and power values for the circuit in Figure 14.33b. Then calculate the value of apparent power for the circuit.

ANSWERS TO THE EXAMPLE PRACTICE PROBLEMS

14.1 354 Ω

14.2 1.5 A

14.3 1.27 A

14.4 77.8

14.5 15 V ∠53.1°

14.6 343 Ω ∠29°

14.7 X_L = 75.4 Ω, Z = 118 Ω, I = 1.02 A, V_L = 76.9 V, V_R = 92.8 V, θ = 39.6°

14.8 V_L = 3.48 V, V_R = 55.9 V

14.9 Z_T = 304 Ω, V_{L1} = 2.98 V, V_{L2} = 0.979 V, V_R = 23.7 V

14.10 P_X = 32.6 mVAR, P_R = 30.1 mW

14.11 44.4 mVA ∠47.3°

14.12 0.678

14.13 X_L = 177 Ω, Z = 232 Ω, I = 517 mA, V_L = 91.5 V, V_R = 77.6 V, θ = 49.7°, P_L = 47.3 VAR, P_R = 40.1 W, P_{APP} = 62 W, PF = 0.647

CHAPTER 15

PARALLEL RESISTIVE-INDUCTIVE (RL) CIRCUITS

PURPOSE

In Chapter 14, you were introduced to the AC characteristics of inductors and series resistive-inductive (RL) circuits. In this chapter, we conclude our study of RL circuits by discussing the operating characteristics of parallel RL circuits.

OBJECTIVES

After completing this chapter, you should be able to:

1. Compare and contrast parallel resistive, inductive, and resistive-inductive (RL) circuits.
2. Calculate any current or impedance value for a parallel RL circuit.
3. Compare and contrast the values obtained from current-based and impedance-based phase angle calculations.
4. Calculate the power values for a parallel RL circuit.

15.1 PARALLEL RL CIRCUIT CURRENTS

A *parallel RL circuit* is one that contains one or more resistors in parallel with one or more inductors. Several parallel RL circuits are shown in Figure 15.1. As you can see, none of the circuit branches contains more than one component. (If they did, we'd be dealing with a *series-parallel RL circuit*.)

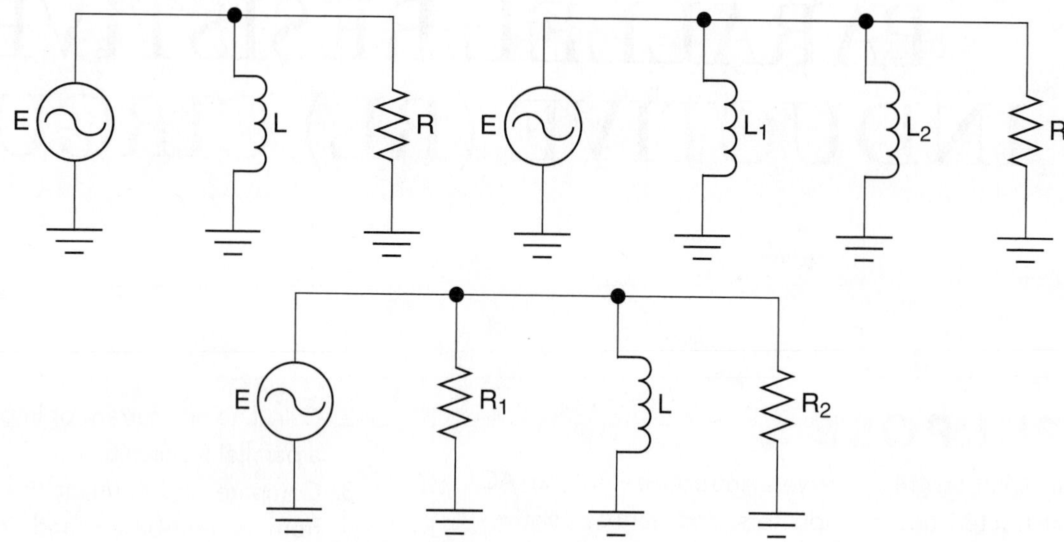

FIGURE 15.1 Parallel resistive-inductive (RL) circuits.

Parallel Circuits: A Comparison

As we did with series circuits, we're going to start by comparing a parallel RL circuit to purely resistive and purely inductive parallel circuits. A comparison of these circuits is provided in Figure 15.2.

In Chapters 6 and 14, we established the fundamental characteristics of parallel-connected resistors and inductors. As a refresher, these characteristics are as follows:

1. All branch voltages equal the source voltage.
2. The circuit current equals the sum of the branch currents.
3. The total impedance (resistance or reactance) is lower than any branch value.

These relationships are provided (as equations) in Figure 15.2 should all seem familiar. In each case, all branch voltages equal the source voltage and the total circuit current equals the sum of the branch currents (in keeping with Kirchhoff's Current Law). The total impedance for the resistive and inductive circuits can be found using the reciprocal or product-over-sum methods shown.

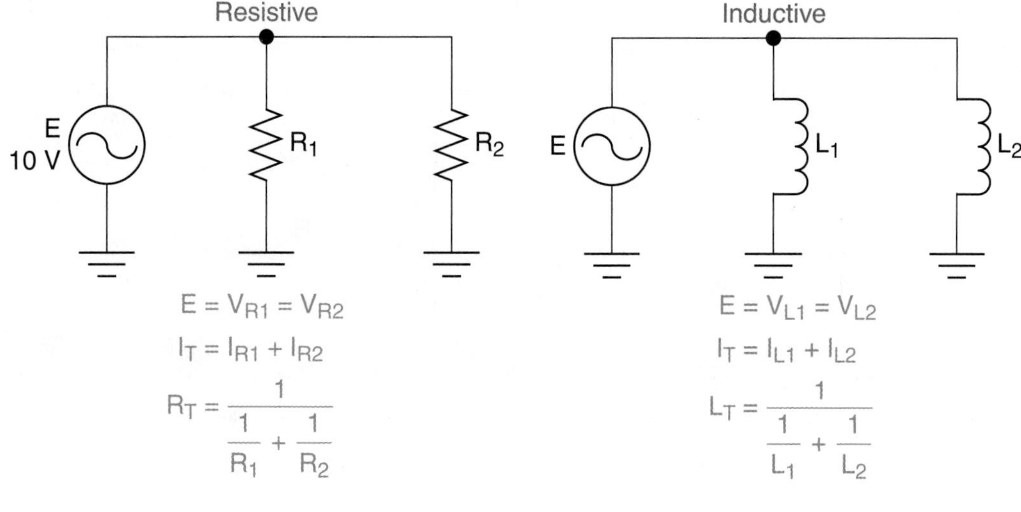

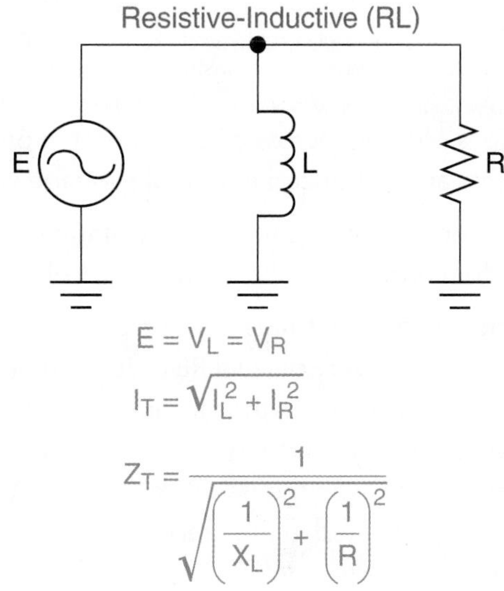

FIGURE 15.2 Parallel resistive, inductive, and resistive-inductive (RL) circuits.

The current and impedance relationships shown for the parallel RL circuit in Figure 15.2 are explained in detail in this chapter. Even so, you can see that the total current in a parallel RL circuit equals the geometric sum of the branch currents. By the same token, calculating total impedance includes the geometric sum of X_L and R in the denominator of the equation. As you may have guessed, these geometric relationships are required because of phase angles within the circuit.

Branch Currents

As you know, current lags voltage by 90 degrees in an inductive circuit. At the same time, current and voltage in a resistive circuit are always in phase. When combined in parallel, a resistor and an inductor produce the waveforms shown in Figure 15.3.

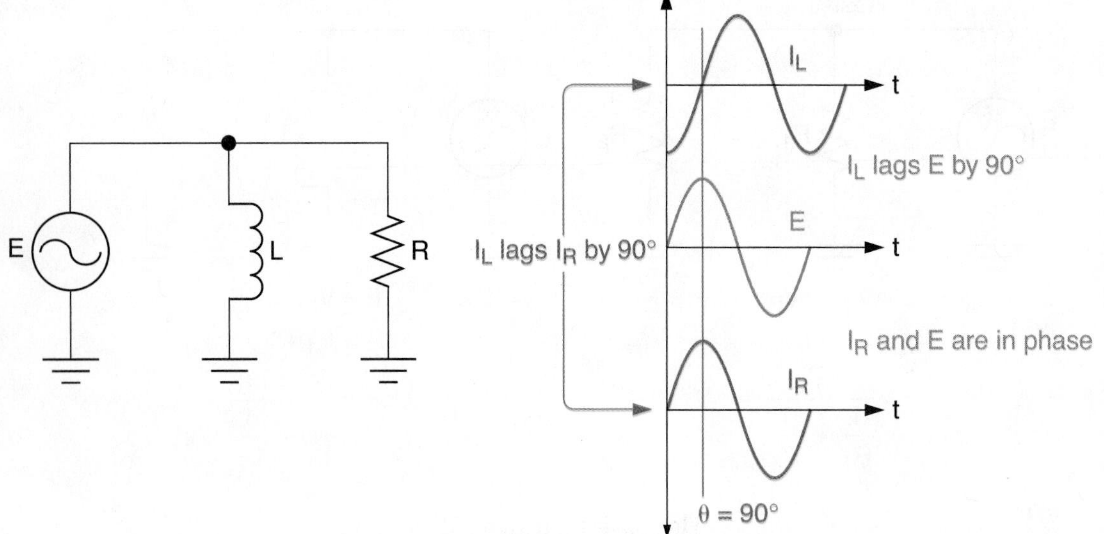

FIGURE 15.3 Parallel RL circuit phase relationships.

Since voltage is constant across all branches in a parallel circuit, the inductor and resistor currents are referenced to a single voltage waveform. As you can see:

1. The inductor current lags the circuit voltage by 90 degrees.
2. The resistor current is in phase with the applied voltage.

Therefore, inductor current lags resistor current by 90 degrees. This phase relationship holds true for every parallel RL circuit with an AC source.

Since I_L lags I_R by 90 degrees, the values of the component currents are represented using *vectors* as shown in Figure 15.4a. By plotting the value of I_R on the x-axis (0°) and the value of I_L on the *negative* y-axis (−90°), the phase relationship between the two is represented in the graph.

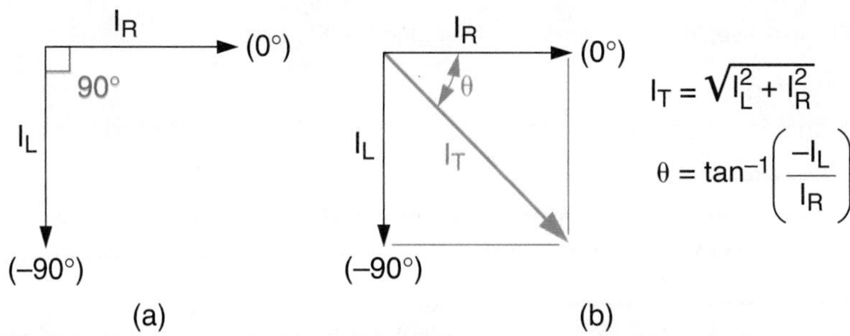

FIGURE 15.4 Parallel RL circuit current vectors.

As in any parallel circuit, the sum of I_L and I_R must equal the total circuit current. In this case, the vector relationship between the two values requires that they be added *geometrically*. By formula,

(15.1)

$$I_T = \sqrt{I_L^2 + I_R^2}$$

A *linear* load is one that is purely resistive, inductive, or capacitive. A *nonlinear* load is one that has two or more of these properties. In practice, most loads (such as motors and industrial power distribution circuits) are nonlinear loads.

IN THE FIELD

The *phase angle* (θ) in a parallel RL circuit is the phase difference between the circuit current and the source voltage. As shown in Figure 15.4b, the value of θ can be found as

(15.2)

$$\theta = \tan^{-1}\frac{-I_L}{I_R}$$

The current phase angle in any parallel RL circuit is always somewhere between 0° (as in a purely resistive circuit) and −90° (as in a purely inductive circuit).

KEY CONCEPT

Since voltage is equal across all branches in a parallel circuit, E is used as the reference for all phase angle measurements. Therefore, the value of θ for the circuit current indicates its phase *relative to the source voltage*. Example 15.1 demonstrates the calculations of total current and θ for a parallel RL circuit.

EXAMPLE 15.1

Calculate the total current and phase angle for the circuit in Figure 15.5a.

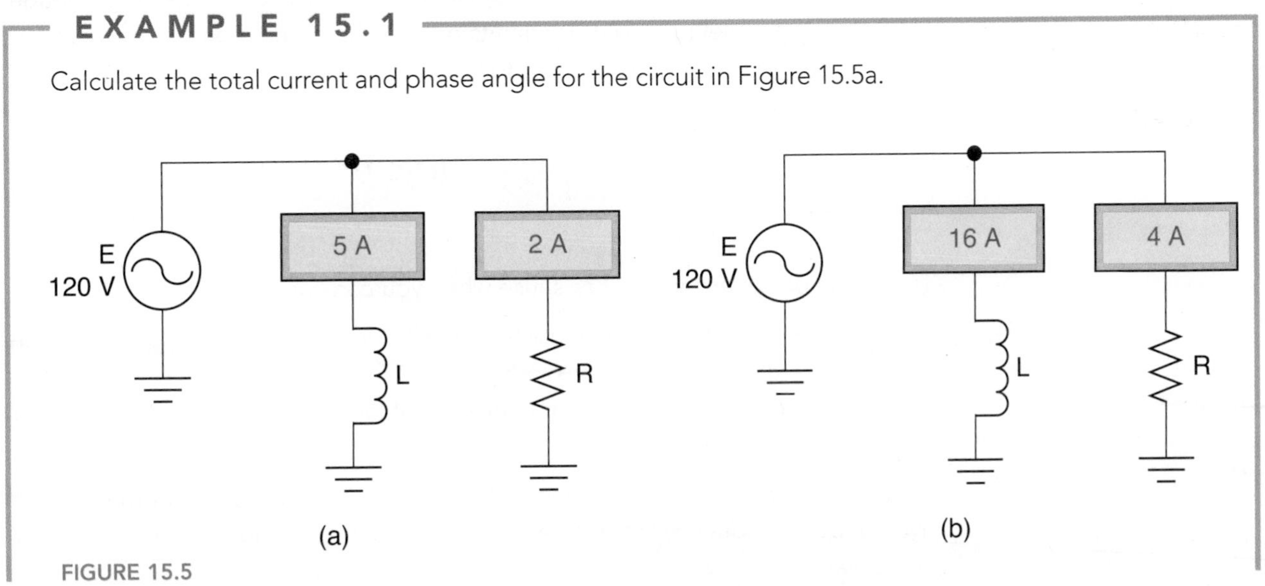

FIGURE 15.5

SOLUTION

The total circuit current is found as

$$\begin{aligned} I_T &= \sqrt{I_L^2 + I_R^2} \\ &= \sqrt{(5\text{ A})^2 + (2\text{ A})^2} \\ &= 5.39\text{ A} \end{aligned}$$

The phase angle of the circuit current (relative to the source voltage) can be found as

$$\begin{aligned} \theta &= \tan^{-1}\left(\frac{-I_L}{I_R}\right) \\ &= \tan^{-1}\left(\frac{-5\text{ A}}{2\text{ A}}\right) \\ &= \tan^{-1}(-2.5) \\ &= -68.2° \end{aligned}$$

The fact that θ is negative indicates that the circuit current lags the source voltage by 68.2 degrees.

PRACTICE PROBLEM 15.1

Calculate the value and phase angle of I_T for the circuit in Figure 15.5b.

You may be wondering why I_L is represented as a negative value in the phase angle equation. If you refer back to Figure 15.4b, you'll see that the vector for I_T lies in the fourth quadrant of the graph (between −90° and 0°). This means that it must have a phase angle that is negative when referenced to the resistor current.

Since resistor current is in phase with the source voltage, the phase angle between the circuit current and the source voltage is also negative. In other words, the negative value of θ indicates that the inductor current lags the source voltage.

For any parallel RL circuit, the limits on the phase angle between circuit current and source voltage are given as follows:

$$-90° < \theta < 0° \tag{15.3}$$

This range of values for θ makes sense when you consider that:

1. The current in a purely resistive circuit is in phase with the source voltage ($\theta = 0°$).
2. The current in a purely inductive circuit lags the source voltage by 90 degrees ($\theta = -90°$).

A parallel circuit containing both inductance and resistance is neither purely resistive nor purely inductive. Therefore, it must have a phase angle that is less than 0° and greater than −90°.

PROGRESS CHECK

1. What are the three fundamental characteristics of any parallel circuit?
2. In a parallel RL circuit, what is the phase relationship between I_L and I_R?
3. In a parallel RL circuit, what does the current phase angle represent?
4. Why does the current in a parallel RL circuit have a negative phase angle?

15.2 PARALLEL RL CIRCUIT IMPEDANCE

As stated earlier, the impedance in an RL circuit is the combined opposition to current provided by the resistance and/or reactance. The easiest way to calculate the total impedance in a parallel RL circuit begins with calculating the circuit current. Once the circuit current is known, Ohm's law states that the circuit impedance can be found as

(15.4)

$$Z_T = \frac{E}{I_T}$$

This approach to calculating the value of Z_T is demonstrated in Example 15.2.

EXAMPLE 15.2

Calculate the total impedance in the circuit in Figure 15.5a.

SOLUTION

In Example 15.1, we calculated a value of $I_T = 5.39$ A. Using this value, the total circuit impedance is found as

$$\begin{aligned} Z_T &= \frac{E}{I_T} \\ &= \frac{120 \text{ V}}{5.39 \text{ A}} \\ &= 22.3\ \Omega \end{aligned}$$

PRACTICE PROBLEM 15.2

Calculate the total impedance for the circuit in Figure 15.5b.

An Observation on the Circuit Impedance in Example 15.2

In every parallel circuit we have discussed, the total impedance has been lower than any branch impedance value. This holds true for resistive-reactive circuits as well. For example, the branch impedance values in Figure 15.5a can be found as

$$X_L = \frac{E}{I_L} = \frac{120 \text{ V}}{5 \text{ A}} = 24\ \Omega$$

and

$$R = \frac{E}{I_R} = \frac{120 \text{ V}}{2 \text{ A}} = 60\ \Omega$$

Each of these values is greater than the value of Z_T calculated in the example (22.3 Ω).

Calculating Parallel-Circuit Impedance

In Chapter 14, you were shown that parallel reactances are calculated in the same way as parallel resistances. The commonly used equations for calculating parallel values of R and X_L are provided (as a reference) in Table 15.1.

TABLE 15.1 • Parallel-Circuit Impedance Equations for Two-Component Resistive and Inductive Circuits.

METHOD	RESISTIVE FORM	INDUCTIVE FORM
Reciprocal:	$R_T = \frac{1}{\frac{1}{R_1} + \frac{1}{R_2}}$	$X_{LT} = \frac{1}{\frac{1}{X_{L1}} + \frac{1}{X_{L2}}}$
Product-over-sum:	$R_T = \frac{R_1 R_2}{R_1 + R_2}$	$X_{LT} = \frac{X_{L1} X_{L2}}{X_{L1} + X_{L2}}$
Equal-value branches:	$R_T = \frac{R}{n}$	$X_{LT} = \frac{X_L}{n}$

(where n = the number of branches in the circuit)

Because there is a phase relationship between X_L and R, the *equal-value branches* approach to calculating total impedance cannot be used on a parallel RL circuit. At the same time, the phase angles of X_L and R make the *reciprocal* method somewhat tedious. For this reason, the *product-over-sum* approach to calculating circuit impedance is preferred for parallel RL circuits. Using the product-over-sum format, the magnitude of parallel RL circuit impedance can be found as

(15.5)
$$Z_T = \frac{X_L R}{\sqrt{X_L^2 + R^2}}$$

Note that the denominator of the fraction contains the geometric sum of the variables. Example 15.3 demonstrates the use of this equation.

EXAMPLE 15.3

Calculate the total impedance in the circuit in Figure 15.6a.

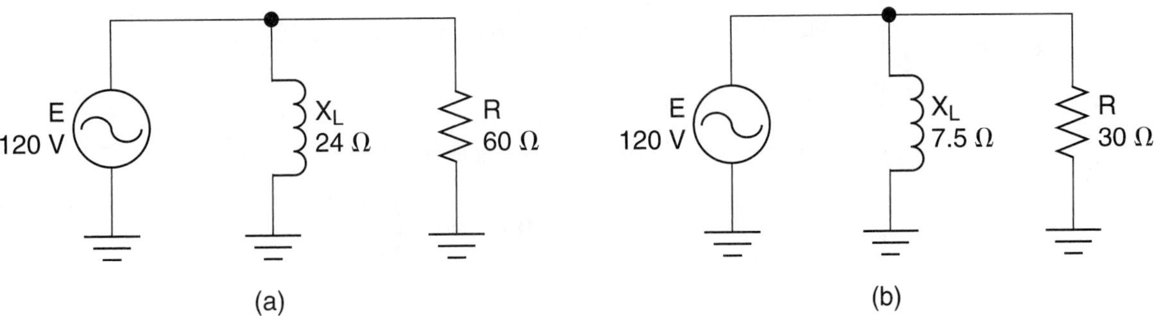

FIGURE 15.6

SOLUTION

The value of Z_T for the circuit is found as

$$Z_T = \frac{X_L R}{\sqrt{X_L^2 + R^2}}$$
$$= \frac{(24\ \Omega)(60\ \Omega)}{\sqrt{(24\ \Omega)^2 + (60\ \Omega)^2}}$$
$$= 22.3\ \Omega$$

To provide a basis for comparison, the values used in this example match the values we calculated for the circuit in Example 15.2. When you compare the results of these two examples, you'll see that they are the same.

PRACTICE PROBLEM 15.3

Calculate the value of Z_T for the circuit in Figure 15.6b. Compare your answer to the one obtained in Practice Problem 15.2.

Calculating the Circuit Phase Angle Using R and X_L

The phase angle of a parallel RL circuit can be found as

(15.6)

$$\theta = \tan^{-1}\frac{R}{X_L}$$

As you can see, the fraction in equation (15.6) is the reciprocal of the one used to calculate the impedance phase angle in a series RL circuit. The use of the equation is demonstrated in Example 15.4.

EXAMPLE 15.4

What is the phase angle in the circuit in Figure 15.6a?

SOLUTION

The phase angle of the circuit is found as

$$\begin{aligned}\theta &= \tan^{-1}\left(\frac{R}{X_L}\right)\\ &= \tan^{-1}\left(\frac{60\ \Omega}{24\ \Omega}\right)\\ &= \tan^{-1}(2.5)\\ &= 68.2^\circ\end{aligned}$$

PRACTICE PROBLEM 15.4

Calculate the phase angle of the circuit impedance for the circuit in Figure 15.6b.

In Example 15.1, we calculated the phase angle of a parallel RL circuit using the values of circuit current. Using equation (15.2), we obtained a value of $\theta = -68.2^\circ$. Then, using the values of R and X_L, we obtained a value of $\theta = 68.2^\circ$

in Example 15.4. While these results may appear to contradict each other, they don't:

1. When calculating θ using I_L and I_R, the result indicates the phase of the circuit current relative to the source voltage. The result is negative because the circuit current lags the source voltage.
2. When calculating θ using R and X_L, the result indicates the phase of the source voltage relative to the circuit current. The result is positive because the source voltage leads the circuit current.

As these statements indicate, the results of the two examples are saying the exact same thing.

PROGRESS CHECK

1. What is the simplest approach to calculating the value of Z_T for a parallel RL circuit?
2. When $X_L = R$ in a parallel RL circuit, can the *equal-value branches* approach to calculating Z_T be used? Explain your answer.
3. Why does equation (15.2) yield a negative value of θ while equation (15.6) yields a positive value of θ?

15.3 PARALLEL CIRCUIT ANALYSIS

In the last section, you were shown how current and impedance are calculated for a parallel RL circuit. Because parallel circuit power calculations (P_X, P_R, and P_{APP}) are identical to those for series circuits, there is no need to revisit them here. At this point, we are going to tie all of these calculations together by working through several circuit analysis problems.

EXAMPLE 15.5

Calculate the current, impedance, and power values for the circuit in Figure 15.7a.

SOLUTION

We begin by calculating the value of X_L, as follows:

$$X_L = 2\pi fL$$

$$= 2\pi(60 \text{ Hz})(470 \text{ mH})$$

$$= 177\ \Omega$$

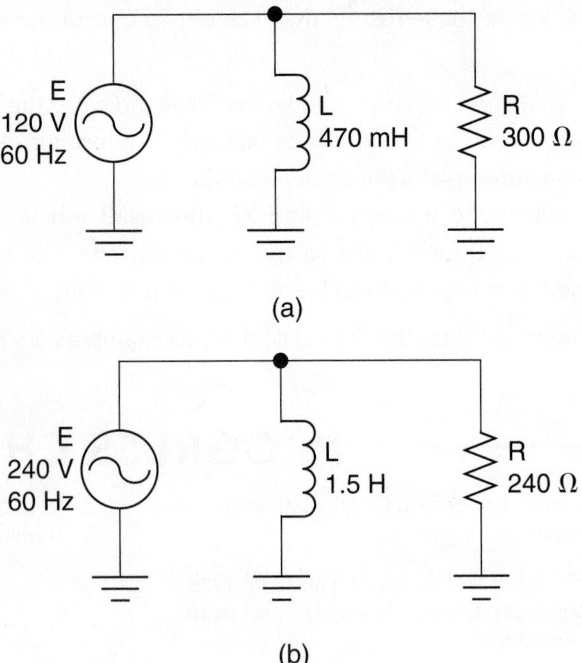

FIGURE 15.7

Now, the branch currents can be found as

$$I_L = \frac{E}{X_L} = \frac{120\text{ V}}{177\ \Omega} = 678\text{ mA}$$

and

$$I_R = \frac{E}{R} = \frac{120\text{ V}}{300\ \Omega} = 400\text{ mA}$$

The magnitude of the circuit current can now be found as

$$I_T = \sqrt{I_L^2 + I_R^2} = \sqrt{(678\text{ mA})^2 + (400\text{ mA})^2} = 787\text{ mA}$$

and the current phase angle (relative to the source voltage) can be found as

$$\theta = \tan^{-1}\left(\frac{-I_L}{I_R}\right)$$

$$= \tan^{-1}\left(\frac{-678 \text{ mA}}{400 \text{ mA}}\right)$$
$$= \tan^{-1}(-1.695)$$
$$= -59.5°$$

Once the total current is known, the circuit impedance can be found as

$$Z_T = \frac{E}{I_T}$$
$$= \frac{120 \text{ V}}{787 \text{ mA}}$$
$$= 152\ \Omega$$

The value of *reactive power* (P_X) for the circuit can be found as

$$P_X = I_L^2 X_L$$
$$= (678 \text{ mA})^2(177\ \Omega)$$
$$= 81.4 \text{ VAR}$$

and the value of *resistive power* (P_R) can be found as

$$P_R = I_R^2 R$$
$$= (400 \text{ mA})^2(300\ \Omega)$$
$$= 48 \text{ W}$$

Once the values of reactive and resistive power are known, the value of *apparent power* (P_{APP}) can be found as

$$P_{APP} = \sqrt{P_X^2 + P_R^2}$$
$$= \sqrt{(81.4 \text{ VAR})^2 + (48 \text{ W})^2}$$
$$= 94.5 \text{ VA}$$

and the circuit power factor can be found as

$$PF = \frac{P_R}{P_{APP}}$$
$$= \frac{48 \text{ W}}{94.5 \text{ VA}}$$
$$= 0.508$$

This completes the circuit calculations.

PRACTICE PROBLEM 15.5

Perform a complete analysis of the circuit shown in Figure 15.7b.

When a parallel RL circuit contains more than one inductive and/or resistive branch, the analysis of the circuit takes a few more steps. This point is demonstrated in Example 15.6.

EXAMPLE 15.6

Calculate the current, impedance, and power values for the circuit in Figure 15.8.

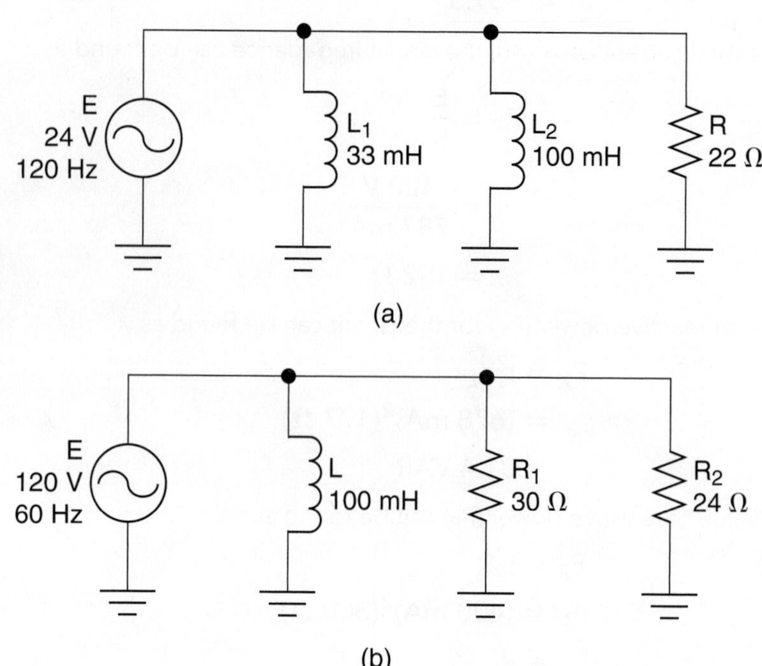

FIGURE 15.8

SOLUTION

We begin by calculating each value of X_L, as follows:

$$X_{L1} = 2\pi f L_1$$
$$= 2\pi(120 \text{ Hz})(33 \text{ mH})$$
$$= 24.9\ \Omega$$

and

$$X_{L2} = 2\pi f L_2$$
$$= 2\pi(120 \text{ Hz})(100 \text{ mH})$$
$$= 75.4\ \Omega$$

Now, the magnitudes of the branch currents can be found as

$$I_{L1} = \frac{E}{X_{L1}}$$
$$= \frac{24 \text{ V}}{24.9\ \Omega}$$
$$= 964 \text{ mA}$$

$$I_{L2} = \frac{E}{X_{L2}} = \frac{24\text{ V}}{75.4\ \Omega} = 318\text{ mA}$$

and

$$I_R = \frac{E}{R} = \frac{24\text{ V}}{22\ \Omega} = 1.09\text{ A}$$

Because I_{L1} and I_{L2} are in phase, their values can be added algebraically, as follows:

$$I_{L(T)} = I_{L1} + I_{L2} = 964\text{ mA} + 318\text{ mA} = 1.28\text{ A}$$

Now, the total inductive current is used (along with I_R) to calculate the circuit current, as follows:

$$I_T = \sqrt{I_{L(T)}^2 + I_R^2} = \sqrt{(1.28\text{ A})^2 + (1.09\text{ A})^2} = 1.68\text{ A}$$

and the circuit phase angle can be found as

$$\theta = \tan^{-1}\left(\frac{-I_{L(T)}}{I_R}\right) = \tan^{-1}\left(\frac{-1.28\text{ A}}{1.09\text{ A}}\right) = \tan^{-1}(-1.17) = -49.6°$$

Once the total circuit current is known, the circuit impedance can be found as:

$$Z_T = \frac{E}{I_T} = \frac{24\text{ V}}{1.68\text{ A}} = 14.3\ \Omega$$

The values of reactive power for the circuit can be found as:

$$\begin{aligned} P_{X(L1)} &= I_{L1}^2 X_{L1} \\ &= (964\text{ mA})^2(24.9\ \Omega) \\ &= 23.1\text{ VAR} \end{aligned}$$

and

$$\begin{aligned} P_{X(L2)} &= I_{L2}^2 X_{L2} \\ &= (318\text{ mA})^2(75.4\ \Omega) \\ &= 7.62\text{ VAR} \end{aligned}$$

Since the inductor values are in phase with each other, the total reactive power can be found as

$$\begin{aligned} P_{X(T)} &= P_{X(L1)} + P_{X(L2)} \\ &= 23.1\text{ VAR} + 7.62\text{ VAR} \\ &= 30.7\text{ VAR} \end{aligned}$$

The resistive power in the circuit can be found as

$$\begin{aligned} P_R &= I_R^2 R \\ &= (1.09\text{ A})^2(22\ \Omega) \\ &= 26.1\text{ W} \end{aligned}$$

Using the resistive power and total reactive power, the value of apparent power for the circuit can be found as

$$\begin{aligned} P_{APP} &= \sqrt{P_{X(T)}^2 + P_R^2} \\ &= \sqrt{(30.7\text{ VAR})^2 + (26.1\text{ W})^2} \\ &= 40.2\text{ VA} \end{aligned}$$

Finally, the circuit power factor can be found as

$$\begin{aligned} PF &= \frac{P_R}{P_{APP}} \\ &= \frac{26.1\text{ W}}{40.2\text{ VA}} \\ &= 0.649 \end{aligned}$$

This completes the analysis of the circuit.

PRACTICE PROBLEM 15.6

Perform a complete analysis on the circuit in Figure 15.8b.

One Final Note

In this chapter, you have been introduced to parallel RL circuit relationships and analysis. Once again, we have a situation where all of the relationships you learned earlier in the text apply. The primary difference is that many circuit values must be added geometrically rather than algebraically (as was the case with series RL circuits).

For future reference, the basic characteristics of parallel RL circuits are summarized in Figure 15.9.

Sample schematic:

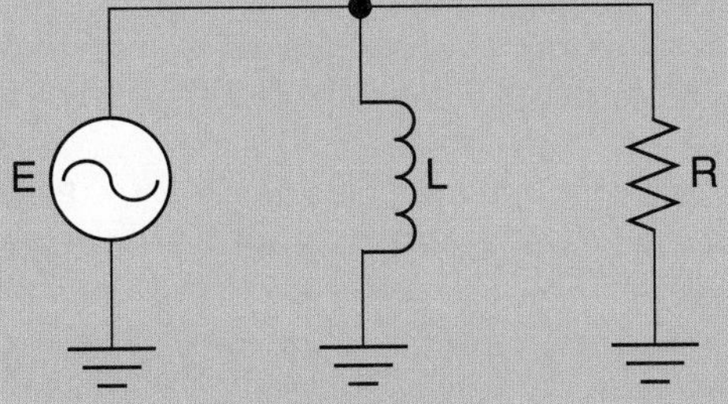

Phase angles: Circuit current lags source voltage. Circuit impedance leads source voltage. The two phase angles have opposite signs and equal magnitudes.

Fundamental relationships:

$$^{*}Z_T = \frac{1}{\sqrt{\frac{1}{X_L^2} + \frac{1}{R^2}}} \quad \text{or} \quad \frac{X_L R}{\sqrt{X_L^2 + R^2}}$$

$$X_L = 2\pi fL$$

$$I_T = \sqrt{I_L^2 + I_R^2}$$

$$\theta = \tan^{-1}\left(\frac{-I_L}{I_R}\right) \text{ or } \theta = \tan^{-1}\left(\frac{R}{X_L}\right)$$

FIGURE 15.9 Parallel RL circuit characteristics.

15.4 SUMMARY

Here is a summary of the major points that were covered in this chapter:

Parallel RL Circuit Currents

- A parallel RL circuit is one that contains one or more resistors in parallel with one or more inductors, each branch containing only one component.
- In a parallel RL circuit:
  - The total circuit current equals the geometric sum of the currents through the resistance and the reactive branches.
  - The inductor current lags the source voltage by 90°.
  - The resistor current is in phase with the circuit voltage.
- The phase angle (θ) in a parallel RL circuit is the phase difference between the total current (I_T) and the source voltage (E).
  - Since voltage is assumed to have an angle of 0° in a parallel circuit, the current phase angle is always negative.
  - For a parallel RL circuit, the current phase angle (relative to the source voltage) has limits of $-90° < \theta < 0°$.

Parallel RL Circuit Impedance

- The impedance phase angle in a parallel RL circuit is always the positive-equivalent of the current phase angle.
- The impedance phase angles for series and parallel RL circuits are calculated using reciprocal fractions, as follows:

Series	Parallel
$\theta = \tan^{-1}\frac{X_L}{R}$	$\theta = \tan^{-1}\frac{R}{X_L}$

- The product-over-sum method for calculating total impedance is preferred for parallel RL circuits. (The geometric sum of X_L and R is used in the denominator.)

CHAPTER REVIEW

1. A parallel circuit must contain only one component in each branch, otherwise it ________.
   a) is too complex to solve
   b) is a susceptive circuit
   c) is a series-parallel circuit
   d) requires calculus to solve

2. Of the basic parallel circuit relationships that we have learned so far, the only one that does *not* apply to parallel RL circuits is ________.
   a) Ohm's law
   b) Kirchhoff's current law
   c) the common voltage in all branches
   d) none of the above

3. When solving for total circuit current, branch currents in a parallel RL circuit must be added geometrically because ________.
   a) resistor current lags inductor current
   b) of the phase difference between branch currents
   c) of the impedance
   d) of the voltage phase angle

4. Current in the inductive branch of a parallel RL circuit always ________ current in the resistive branch.
   a) lags by 90°
   b) lags by more than 0° and less than 90°
   c) leads by 90°
   d) leads by more than 0° and less than 90°

5. The current in the resistive branch always ________ the source voltage while the current in the inductive branch ________.
   a) is in phase with, is in phase with E
   b) lags by 90°, is in phase with I_T
   c) is in phase with, leads I_T by 90°
   d) is in phase with, lags E by 90°

6. The phase angle of I_T in a parallel RL circuit, relative to the applied voltage, is ________.
   a) +90°
   b) −90°
   c) between 0° and −90°
   d) between 0° and 90°

7. The phase angle of Z_T in a parallel RL circuit, relative to the applied voltage, is ________.
   a) +90°
   b) −90°
   c) between 0° and −90°
   d) between 0° and 90°

8. When calculating the phase angle of I_L for a parallel RL circuit, the value of I_T used in the $\tan^{-1}$ equation is ________.
   a) always negative
   b) in the denominator
   c) always positive
   d) always positive and in the numerator

9. A parallel RL circuit has branch currents of $I_R = 2.6$ A and $I_L = 1.85$ A. This means that $I_T =$ ________.
   a) 3.19 A
   b) 3.91 A
   c) 2.11 A
   d) 4.45 A

10. The phase angle of I_T for the circuit described in question 9 is________.
    a) 54.6°
    b) 35.4°
    c) −35.4°
    d) −54.6°

11. The phase angle Z_T for the circuit described in question 9 is ________.
    a) +54.6°
    b) +35.4°
    c) −35.4°
    d) −54.6°

12. The method of solving for circuit impedance that cannot be used in a parallel RL circuit is the ________ method.
    a) product-over-sum
    b) reciprocal
    c) equal-value branches
    d) all of the above
    e) none of the above

13. A 240 V @ 60 Hz supply voltage is applied to a 330 Ω resistor in parallel with a 1.65 H coil. For this circuit, I_T = ________ at a phase angle of ________.
    a) 823 mA, −28°
    b) 1.68 A, −62°
    c) 833 mA, 28°
    d) 833 mA, −62°

14. For the circuit described in question 13, Z_T = ________ at a phase angle of ________.
    a) 292 Ω, −28°
    b) 704 Ω, +28°
    c) 292 Ω, +28°
    d) 704 Ω, −28°

15. If the circuit described in question 13 were connected to a 240 V source voltage in Europe where 50 Hz supplies are used, circuit current would ________ and circuit impedance would ________.
    a) increase, decrease
    b) decrease, increase
    c) decrease, decrease
    d) increase, increase

16. 120 V @ 60 Hz is applied to a parallel RL circuit. One branch has a 1.1 kΩ resistor and the other has a 2.2 H inductor. The value of I_T = ________ at a phase angle of ________.
    a) 322 mA, 53°
    b) 254 mA, −36.9°
    c) 181 mA, −53°
    d) 232 mA, −40°

17. For the circuit described in question 16, P_X = ________ and P_R = ________.
    a) 17.4 VAR, 13.1 W
    b) 13.1 VAR, 17.4 W
    c) 17.4 VA, 13.1 W
    d) 13.1 VA, 17.4 W

18. The value of P_{APP} for the circuit described in question 16 is ________.
    a) 30.5 VA
    b) 30.5 VAR
    c) 21.8 VA
    d) 21.8 VAR

19. The phase angle for the value of P_{APP} solved in question 18 equals ________.
    a) −53°
    b) 40°
    c) 53°
    d) −36.9°

20. The value of P_{APP} for a given parallel RL circuit is known to be 153 VA. It is also known that the value of I_T for this circuit is 1.33 A at a phase angle of −22°. This tells us that the magnitude of the applied voltage is ________ and the power factor of the circuit is ________.
    a) 120 V, 93%
    b) 115 V, 93%
    c) 115 V, 37%
    d) 220 V, 37%

21. For the circuit described in question 20, P_R = ________.
    a) 142 VA
    b) 57 W
    c) 122 W
    d) 142 W

22. Another inductive branch is added to the circuit described in question 20. This means that the power factor for the circuit will ________ and the value of P_R will ________.
    a) increase, increase
    b) decrease, decrease
    c) increase, decrease
    d) decrease, increase
    e) decrease, remain unchanged
    f) increase, remain unchanged

PRACTICE PROBLEMS

1. A parallel RL circuit has values of $I_L = 3$ A and $I_R = 5.6$ A. Calculate the magnitude and phase angle of the circuit current.
2. A parallel RL circuit has values of $I_L = 1.24$ A and $I_R = 980$ mA. Calculate the magnitude and phase angle of the circuit current.

3. Calculate the total impedance for the circuit shown in Figure 15.10a.
4. Calculate the total impedance for the circuit shown in Figure 15.10b.

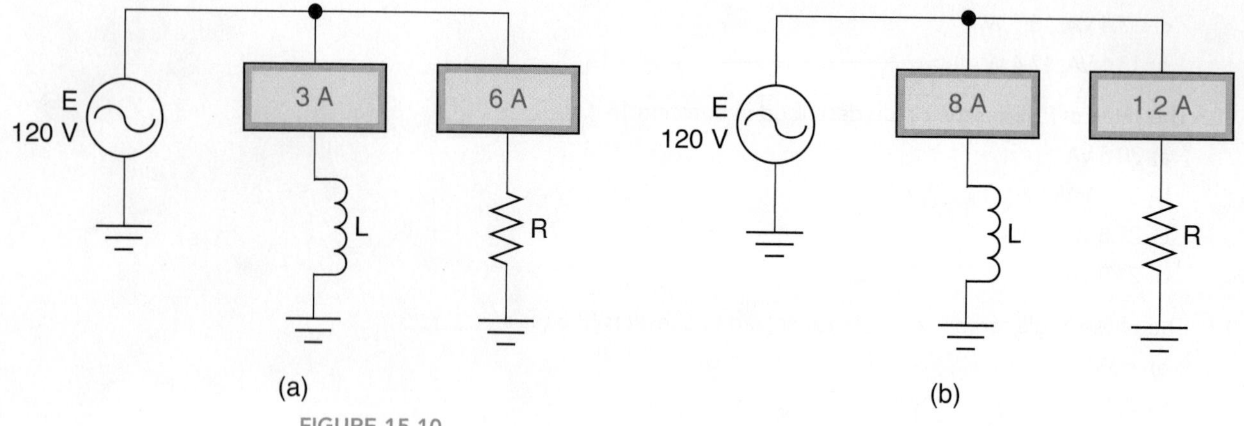

FIGURE 15.10

5. Calculate the circuit impedance and phase angle of the circuit in Figure 15.11a.
6. Calculate the circuit impedance and phase angle of the circuit in Figure 15.11b.

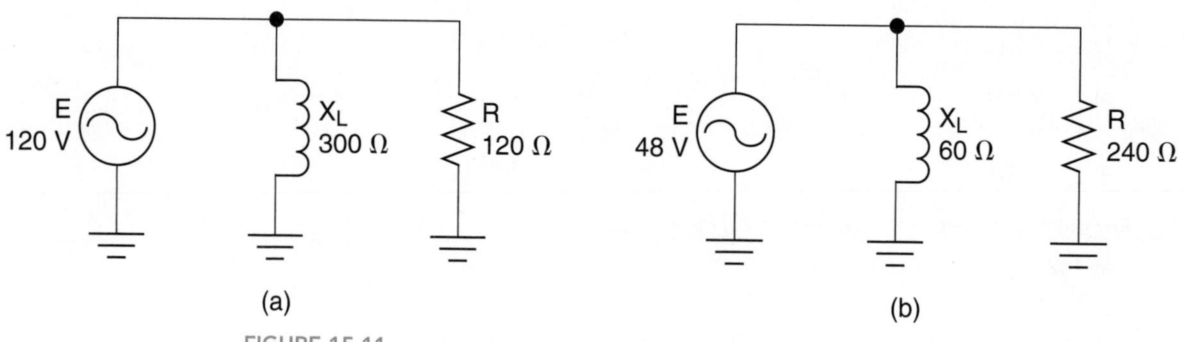

FIGURE 15.11

7. Calculate all the circuit current, impedance, phase, and power values for the circuit in Figure 15.12a.
8. Calculate all the circuit current, impedance, phase, and power values for the circuit in Figure 15.12b.

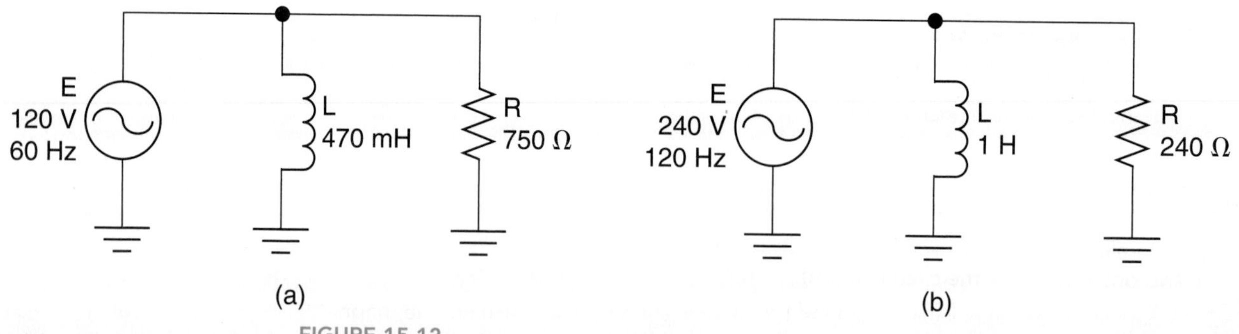

FIGURE 15.12

9. Calculate all the circuit current, impedance, phase, and power values for the circuit in Figure 15.13a.
10. Calculate all the circuit current, impedance, phase, and power values for the circuit in Figure 15.13b.

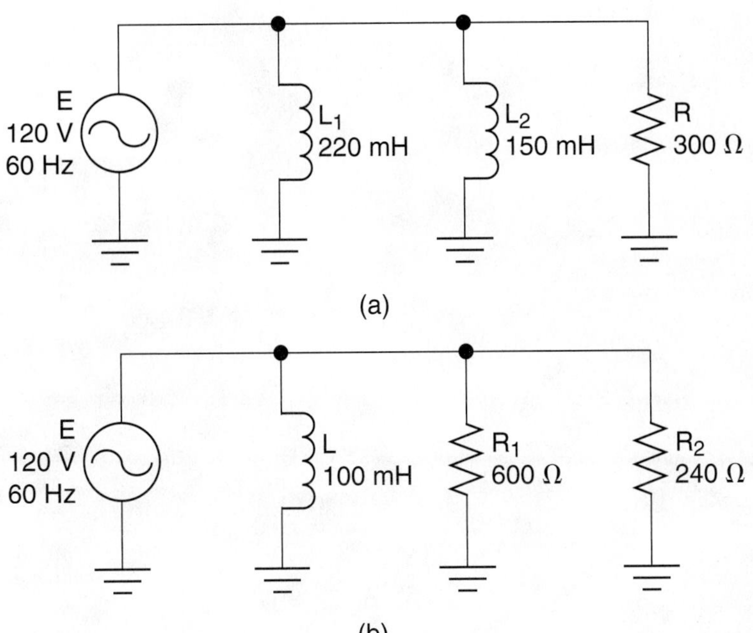

FIGURE 15.13

ANSWERS TO THE EXAMPLE PRACTICE PROBLEMS

15.1 16.5 A

15.2 7.27 Ω

15.3 7.28 Ω

15.4 76°

15.5 $X_L = 565\ \Omega$, $I_R = 1$ A, $I_L = 0.425$ A, $I_T = 1.09$ A, $\theta = -23°$, $Z_T = 220\ \Omega$, $P_X = 102$ VAR, $P_R = 240$ W, $P_{APP} = 261$ VA, PF = 0.92

15.6 $X_L = 37.7\ \Omega$, $I_L = 3.18$ A, $I_{R1} = 4$ A, $I_{R2} = 5$ A, $I_{RT} = 9$ A, $I_T = 9.55$ A, $Z_T = 12.6\ \Omega$, $\theta = -19.5°$, $P_X = 381$ VAR, $P_R = 1.08$ kW, $P_{APP} = 1.15$ kVA, PF = 0.939

CHAPTER 16

SERIES RESISTIVE-CAPACITIVE (RC) CIRCUITS

PURPOSE

The last two chapters dealt with resistive-inductive (RL) circuits. At this point, we are going to turn our attention to resistive-capacitive (RC) circuits.

As you may recall, capacitance is the ability of a component to store energy in an electrostatic field. Based on its overall effect, capacitance is often described as the ability of a component to oppose a change in voltage.

In Chapter 12, we discussed the physical characteristics of capacitors, series-connected and parallel-connected capacitors, and the response of a capacitor to a DC source. In this chapter, we will examine the AC characteristics of series resistive-capacitive (RC) circuits, starting with capacitive AC phase relationships.

KEY TERMS

The following terms are introduced and defined in this chapter on the pages indicated:

Capacitive reactance (X_C), 683
Coupling, 677
DC offset, 679
Isolation, 677
Leakage current, 700

OBJECTIVES

After completing this chapter, you should be able to:

1. Describe the phase relationship between capacitor current and voltage.
2. Discuss capacitive reactance (X_C) and its source.
3. Calculate the reactance of any capacitor (X_C) operated at a specific frequency.

When a capacitor is fully charged, there is no movement of electrons between its plates and its dielectric contains an electrostatic force. Placing too large a charge on a capacitor will result in stressing its dielectric until breakdown occurs. An example of this principle occurs in nature during a thunderstorm. In this case, the dielectric (air) is subjected to electrostatic stress until a breakdown occurs, resulting in a flash of lightning.

KEY CONCEPT

4. Compare and contrast series resistive, capacitive, and resistive-capacitive (RC) circuits.
5. Compare and contrast apparent power, reactive power, and resistive (or *true*) power.
6. Describe the voltage and phase characteristics of series RC circuits.
7. Describe the impedance and current characteristics of series RC circuits.
8. Calculate the impedance, voltage, and current values for any series RC circuit.
9. Perform all relevant power calculations for a series RC circuit.
10. Describe the concept of power factor.
11. Calculate the power factor for a series RC circuit.

16.1 ALTERNATING VOLTAGE AND CURRENT CHARACTERISTICS

In this section, we will examine the response of a capacitor to a sinusoidal input. Many of the principles discussed in this section will seem familiar because they are very similar (in form) to the inductive principles covered in previous chapters. At the same time, you'll see that capacitors and inductors are near-opposites when it comes to many of their operating characteristics.

AC Coupling and DC Isolation

When a component (or circuit) allows electrical energy to pass from one point to another, it is said to provide **COUPLING** between the points. When a component (or circuit) prevents electrical energy from passing between two points, it is said to provide **ISOLATION** between the two.

COUPLING When a component (or circuit) allows electrical energy to pass from one point to another.

ISOLATION When a component (or circuit) prevents electrical energy from passing between two points.

A capacitor provides *AC coupling* and *DC isolation* between its input and output circuits. The means by which a capacitor provides DC isolation is illustrated in Figure 16.1.

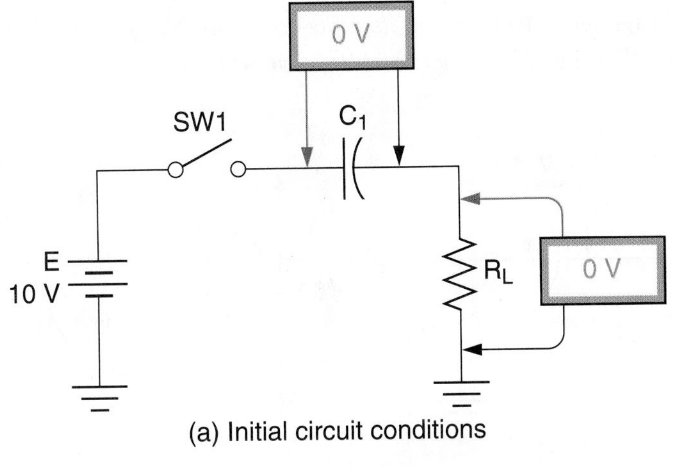

(a) Initial circuit conditions

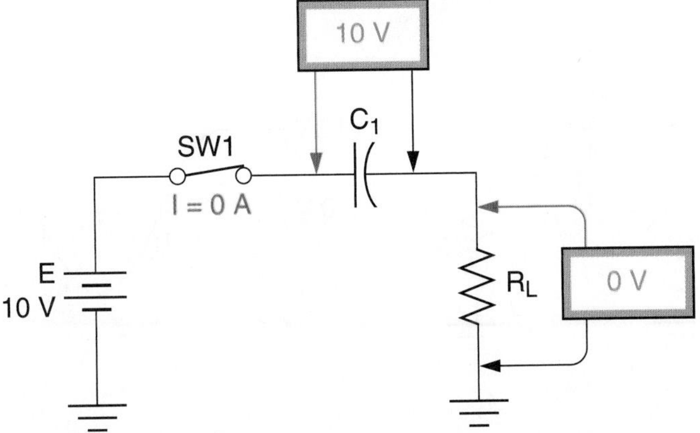

(b) Circuit conditions after the capacitor has charged.

FIGURE 16.1 Voltages before and after a capacitor charges.

Figure 16.1a shows an uncharged capacitor in series with a load. Here's what happens when the switch is closed:

1. A complete path is provided for the current needed to charge the capacitor.
2. The capacitor charges until it reaches its capacity.
3. Once charged, the capacitor prevents any further flow of charge (current) through the circuit.

These statements agree with the description of capacitor charging provided in Chapter 12. The circuit conditions after the capacitor has charged are shown in Figure 16.1b. Note that:

1. The circuit current has dropped to 0 A.
2. The source voltage is dropped across the capacitor.

Since the capacitor is blocking the flow of charge through the circuit, there is no current through the load. As a result, $V_{RL} = 0$ V. In effect, the capacitor is providing DC isolation between E and R_L (once it has charged).

AC Coupling

The AC coupling provided by a capacitor is based on an important component characteristic: *A capacitor always seeks to maintain its plate-to-plate voltage* (V_C). This characteristic is illustrated in Figure 16.2. In Figure 16.2a, a capacitor is shown to have

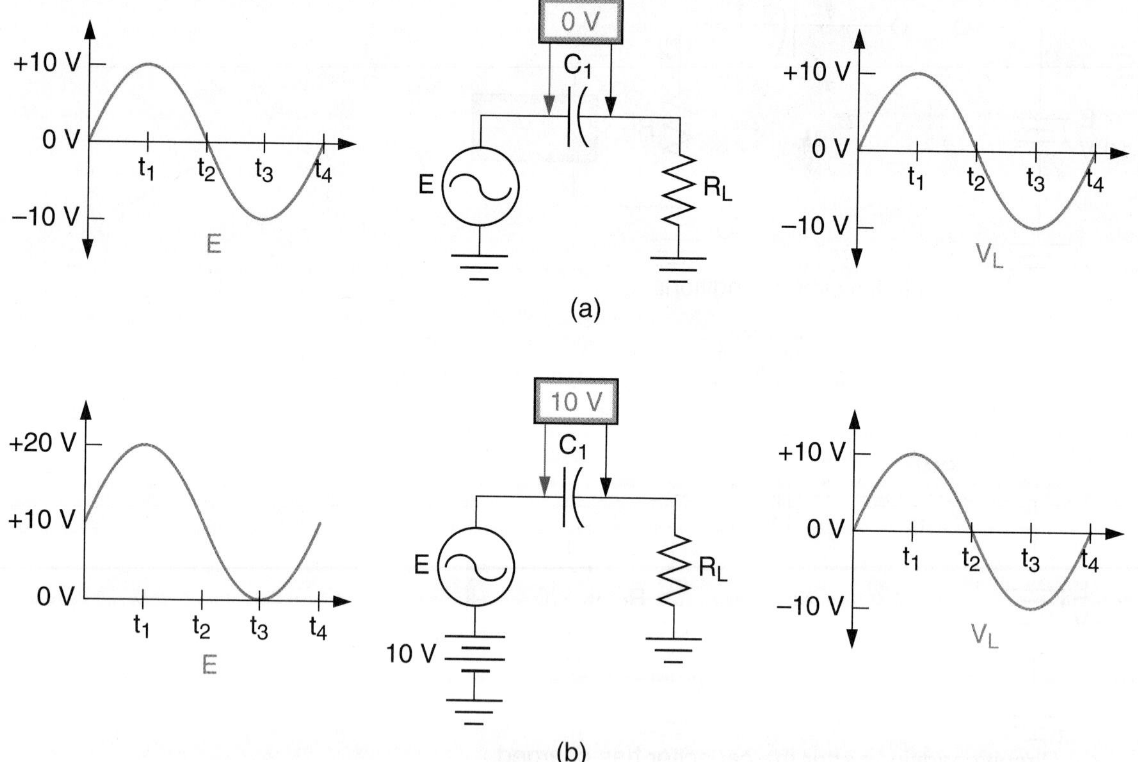

FIGURE 16.2 AC coupling and DC isolation provided by a capacitor.

a 0 V difference of potential across its plates. When the input goes positive (from t_0 to t_1), the capacitor maintains its 0 V plate-to-plate voltage. As a result, the output makes a transition that is identical to the input. As the input makes its negative-going transition (from t_1 to t_3), the capacitor continues to maintain its 0 V plate-to-plate voltage. As a result, the output transition is again identical to the input transition.

If the plate-to-plate voltage (V_C) has some value *other than 0 V*, that voltage shows up as *a difference between the peak input and output voltage values.* For example, the capacitor in Figure 16.2b is shown to have a 10 V difference of potential between its plates. Thus, when the input is at $+20\ V_{pk}$, the output reaches its positive peak value of

$$\begin{aligned} V_{RL} &= E - V_C \\ &= +20\ V_{pk} - 10\ V \\ &= +10\ V_{pk} \end{aligned}$$

When the input reaches its negative peak value, the output reaches a negative peak value of

$$\begin{aligned} V_{RL} &= E - V_C \\ &= 0\ V_{pk} - 10\ V \\ &= -10\ V_{pk} \end{aligned}$$

As you can see, the shape of the waveform has not changed from the input to the output. Only its peak and average values have changed. Note that the difference between the average input and output voltage values equals the difference of potential across the capacitor.

DC Offsets

Unlike the sine waves you saw in previous chapters, the input waveform in Figure 16.2b is centered around +10 V. When a sine wave (or any other waveform) is centered around a voltage other than 0 V, the waveform is said to have a **DC OFFSET**. For example, the input waveform in Figure 16.2b is described as a 20 V_{PP} sine wave with a $+10\ V_{DC}$ offset.

DC OFFSET When a sine wave (or any other waveform) is centered around a voltage other than 0 V.

When the input to a capacitor contains a DC offset, the capacitor isolates the DC offset from the output. As shown in Figure 16.2b, the capacitor couples the 20 V_{PP} sine wave to the load, while blocking the $+10\ V_{DC}$ offset. As a result, the output is a 20 V_{PP} sine wave that is centered around 0 V. In effect, the capacitor has accepted the $+10\ V_{DC}$ offset as a plate-to-plate voltage (as shown in the figure).

Capacitor Current

In a purely capacitive AC circuit, the current is directly proportional to:

- The capacity of the capacitor
- The rate of change in capacitor voltage

These two factors are combined in the following equation:

(16.1)

$$I_C = C\frac{\Delta V}{\Delta t}$$

where

I_C = the instantaneous value of capacitor current
C = the capacity of the component(s), in farads
$\frac{\Delta V}{\Delta t}$ = the rate of change in capacitor voltage

With this relationship in mind, look at the voltage waveform shown in Figure 16.3. From 0° to 90° on the waveform, the voltage increases. From 90° to 180°, the voltage decreases. The same holds true for the negative alternation. That is, voltage increases between 180° and 270°, and decreases between 270° and 360°. The only difference is the polarity of the voltage.

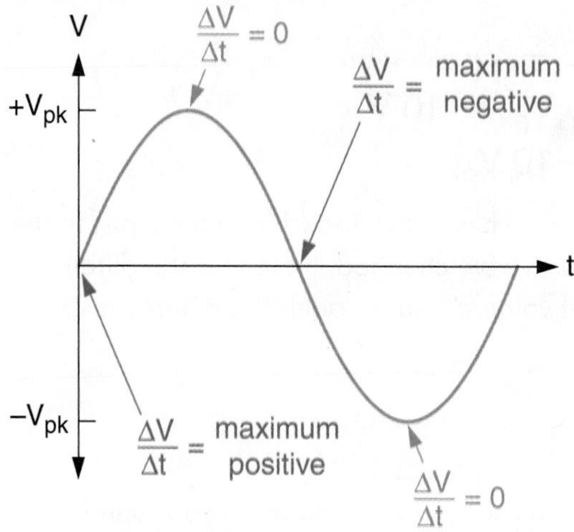

FIGURE 16.3 Rates of change on a voltage sine wave.

Now, here's what happens at the 90° and 270° points on the voltage waveform: At the instant the waveform reaches a peak value, the rate of change in voltage is zero volts per second (V/s). (As an analogy, picture a baseball that is thrown straight up into the air. When it reaches its highest point, it stops for just an instant before falling. At that instant, its rate of change in position is zero.) When the voltage waveform is halfway between the peaks (at 180° and 360°), the rate of change in voltage reaches its maximum value. These relationships are summarized in the first two columns of Table 16.1.

The third column in Table 16.1 is based on the fact that capacitor current is directly proportional to the rate of change in voltage. That is,

- I_C equals its minimum value (0 A) when the rate of change in capacitor voltage is zero.
- I_C equals its maximum value (I_{pk}) when the rate of change in voltage reaches its maximum value.

TABLE 16.1 • Rates of Change in Sine Wave Voltage.

WAVEFORM PHASE	RATE OF CHANGE IN VOLTAGE	CAPACITOR CURRENT
90°	Zero	Zero
180°	Maximum	Maximum
270°	Zero	Zero
360°	Maximum	Maximum

Now, let's focus on the relationship between capacitor current and the phase of the voltage waveform. According to Table 16.1,

- I_C equals its minimum value (0 A) at the 90° and 270° points on the voltage waveform.
- I_C equals its maximum value (I_{Pk}) at the 180° and 360° points on the voltage waveform.

Keeping these statements in mind, look at the waveforms in Figure 16.4.

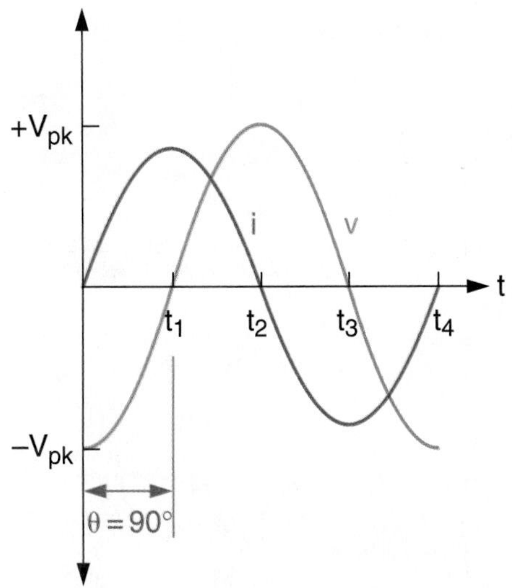

FIGURE 16.4 The phase relationship between capacitor voltage and current.

As shown in Figure 16.4, capacitor current is at its maximum value when capacitor voltage is at its minimum value, and vice versa. In other words, there is a 90° phase shift between the capacitor voltage and current waveforms. If we use the current waveform as a reference, the capacitor voltage begins its positive half-cycle 90° *after* the capacitor current. Note that this relationship is normally described in one of two ways:

1. "Current *leads* voltage by 90 degrees."
2. "Voltage *lags* current by 90 degrees."

Remember, current is used as the phase reference in a series circuit because its value is constant at all points.

An Important Point

The current-voltage phase relationship described in this section applies only to a purely capacitive circuit, that is, a circuit that does not contain any significant amount of resistance. As you will learn later in this chapter, the phase relationship between circuit current and source voltage changes when any significant amount of resistance is added to a capacitive circuit.

Capacitive Versus Inductive Phase Relationships

Earlier, you were told that capacitors and inductors have many similar relationships, but are near-opposites in terms of their operating characteristics. One of the most striking differences between capacitors and inductors are their current versus voltage phase relationships. These phase relationships are contrasted in Figure 16.5.

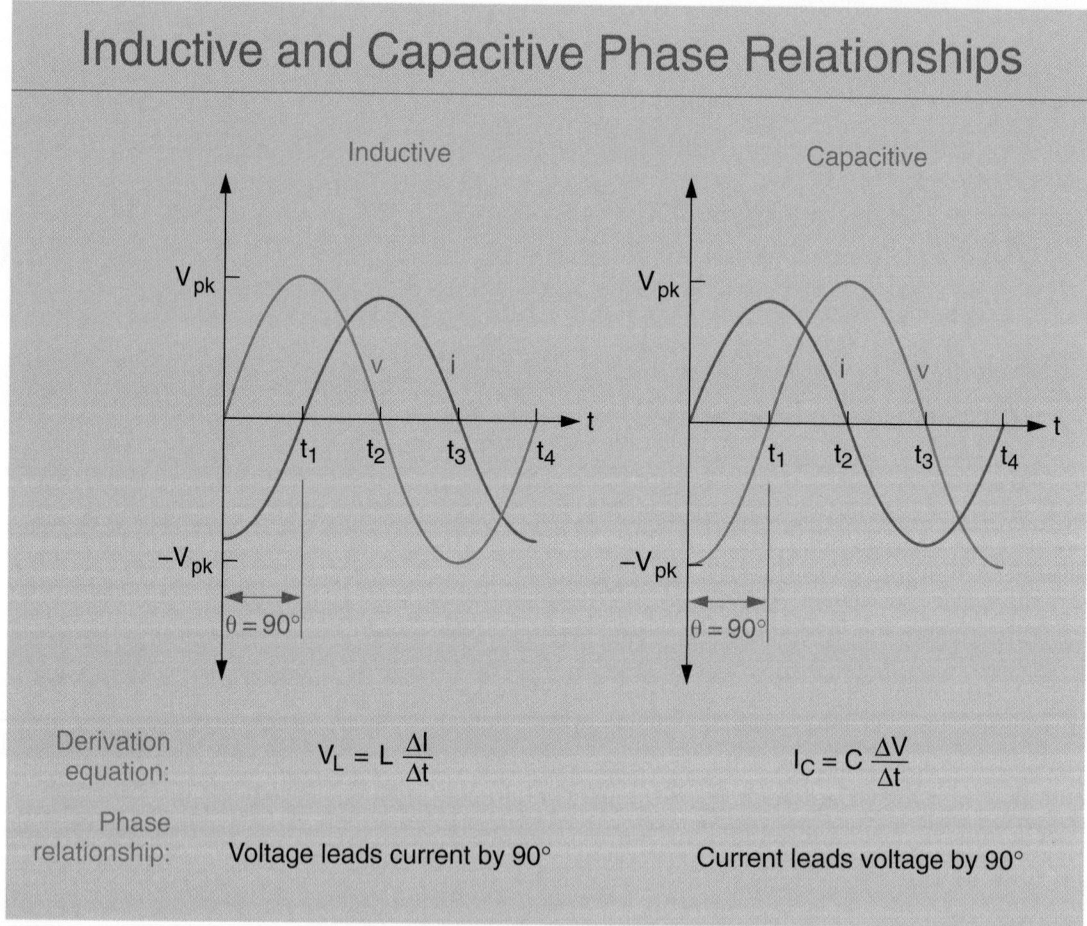

FIGURE 16.5 Phase relationships in inductive and capacitive circuits.

Voltage leads current in an inductive circuit. Current leads voltage in a capacitive circuit. The phrase *ELI the ICEman* can help you remember these phase relationships, as follows:

- Voltage (E) in an inductive (L) circuit leads current (I) by 90° (ELI)
- Current (I) in a capacitive (C) circuit leads voltage (E) by 90° (ICE)

PROGRESS CHECK

1. Describe how a capacitor provides DC isolation between a source and its load.
2. Describe the effect of capacitor voltage (V_C) on input and output peak values.
3. What happens when a capacitor receives an AC input that is centered on a DC offset voltage?
4. What is the relationship between capacitor current and the rate of change in capacitor voltage?
5. State and explain the phase relationship between capacitive current and voltage.
6. Contrast inductive and capacitive current versus voltage phase relationships.

16.2 CAPACITIVE REACTANCE (X_C)

In Chapter 14, you were introduced to inductive reactance (X_L), the opposition to an alternating current that is provided by an inductor. Capacitors also provide a measurable opposition to alternating current, called **CAPACITIVE REACTANCE (X_C)**. In this section, we are going to discuss capacitive reactance, the means by which it is calculated, and its role in AC circuit analysis.

CAPACITIVE REACTANCE (X_C) A capacitor's opposition to alternating current, measured in ohms (Ω).

Capacitor Resistance

When talking about capacitor resistance, we need to distinguish between the *dielectric (true) resistance* and the *effective resistance* of the component. Since the terminals of a capacitor are separated by a dielectric (insulator), the true resistance of the component equals the resistance of its dielectric. This resistance is generally assumed to be, for all practical purposes, infinite. However, the effective resistance of the component can be quite a bit lower. This concept was illustrated in Figure 16.1. In that figure, you were shown that the effective resistance of an uncharged capacitor is extremely low. In fact, at the start of the charge cycle, the component effectively acts as a DC short. Once charged, a capacitor effectively acts as a DC open.

Since the dielectric resistance of a capacitor is extremely high, you would expect the current in a purely capacitive AC circuit to be extremely low. However, this is not necessarily the case. Look at the circuit and waveforms shown in Figure 16.6. The source in Figure 16.6 generates a continual flow of charge (current) to and from the plates of the capacitor. This current is 90° out of phase with the capacitor

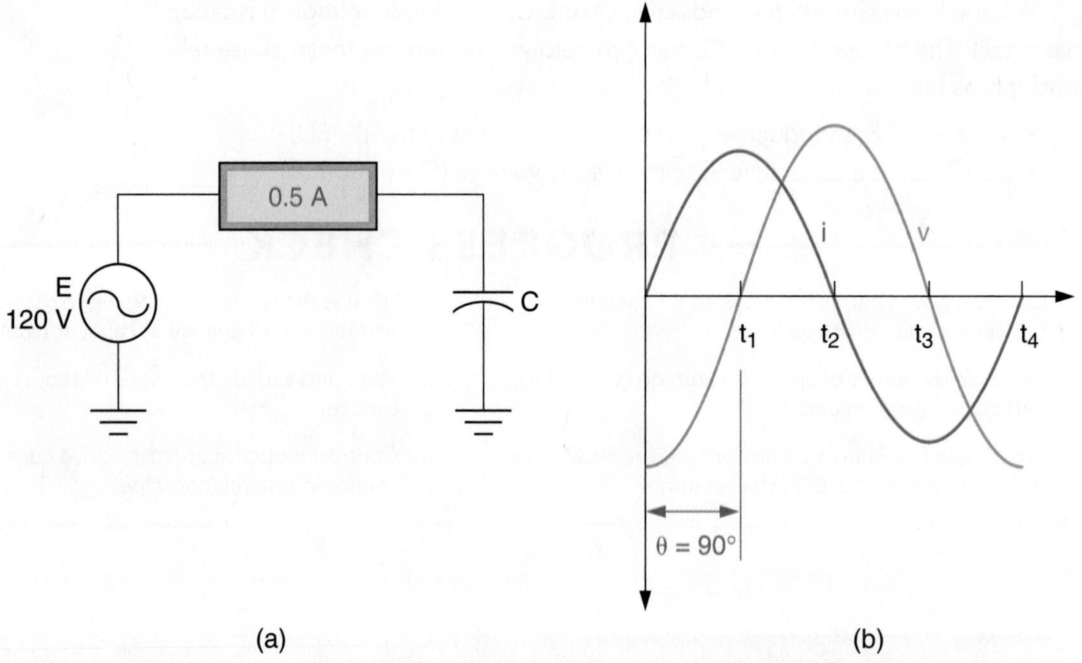

FIGURE 16.6 Capacitor voltage and current.

voltage as shown in the figure. At the same time, each waveform has a measurable *effective* (or RMS) value. If we were to measure these RMS values, we could determine the total opposition to current in the circuit. Using the RMS values shown in the figure, this opposition could be found as

$$\text{opposition} = \frac{V_{RMS}}{I_{RMS}} = \frac{120 \text{ V}}{0.5 \text{ A}} = 240 \ \Omega$$

This is significantly lower than the true resistance of a capacitor. Assuming that the measured values given in the figure are correct, there is only one way to account for the difference between 240 Ω and the true resistance of the capacitor: *The capacitor is providing some type of opposition to current other than resistance.* This "other" opposition to current is called *capacitive reactance (X_C).*

Calculating the Value of X_C

The value of 240 Ω calculated for Figure 16.6 demonstrates that the reactance of a capacitor can be found as

(16.2)

$$X_C = \frac{V_{RMS}}{I_{RMS}}$$

There is another (and more commonly used) method for calculating the reactance of a given capacitor. When the value of a capacitor and the frequency of operation are known, the reactance of a given capacitor can be found as

(16.3)

$$X_C = \frac{1}{2\pi fC}$$

The use of this equation is demonstrated in Example 16.1.

EXAMPLE 16.1

Calculate the reactance of the capacitor in Figure 16.7.

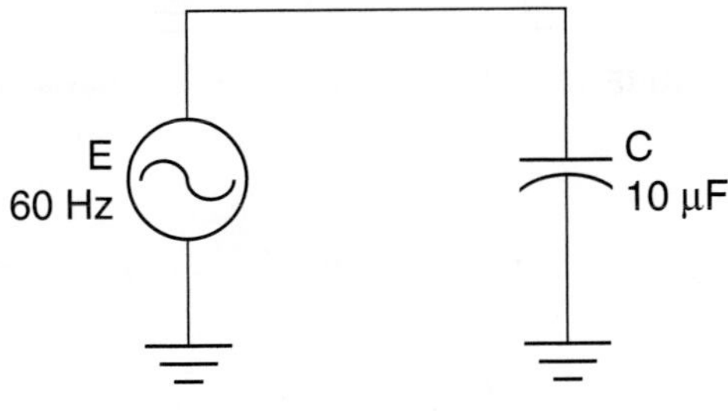

FIGURE 16.7

SOLUTION

The reactance of the capacitor has a value of

$$X_C = \frac{1}{2\pi fC}$$

$$= \frac{1}{2\pi(60\text{ Hz})(10\text{ }\mu\text{F})}$$

$$= 265\ \Omega$$

PRACTICE PROBLEM 16.1

A 22 μF capacitor is operated at a frequency of 60 Hz. Calculate the reactance provided by the component.

In terms of applications, equation (16.3) is usually easier to use than equation (16.2); all you need to know is the value of the capacitor and the frequency of operation. On the other hand, equation (16.2) requires that you calculate (or measure) the circuit values of voltage and current. In many cases, this is easier said than done.

X_C and Ohm's Law

In a purely capacitive circuit, capacitive reactance (X_C) can be used with Ohm's law to determine the total circuit current. The current through a capacitive circuit is calculated as shown in Example 16.2.

EXAMPLE 16.2

Calculate the current through the circuit in Figure 16.8.

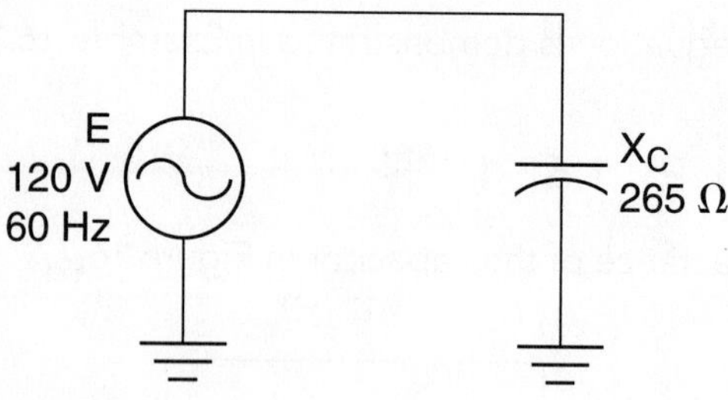

FIGURE 16.8

SOLUTION

The current through the circuit in Figure 16.8 has a value of

$$I_C = \frac{E}{X_C}$$

$$= \frac{120\text{ V}}{265\ \Omega}$$

$$= 453\text{ mA}$$

PRACTICE PROBLEM 16.2

A circuit like the one in Figure 16.8 has the following values: $E = 120$ V and $X_C = 72.5\ \Omega$. Calculate the circuit current.

Series and Parallel Values of X_C

In Chapter 14, you were told that the total reactance in any series or parallel configuration is calculated in the same fashion as total resistance. Figure 16.9 relates this concept to capacitive reactance. (Since you have seen similar equations for both resistance and inductive reactance calculations, we do not need to rework them here.)

There is one point that should be reinforced: Total series capacitance (C_T) and total series capacitive reactance (X_{CT}) are not calculated in the same fashion. X_{CT} is calculated in the same fashion as total series resistance or inductive reactance. On the other hand, total series capacitance (C_T) is calculated in the same fashion as total parallel resistance or inductance (as introduced in Chapter 12).

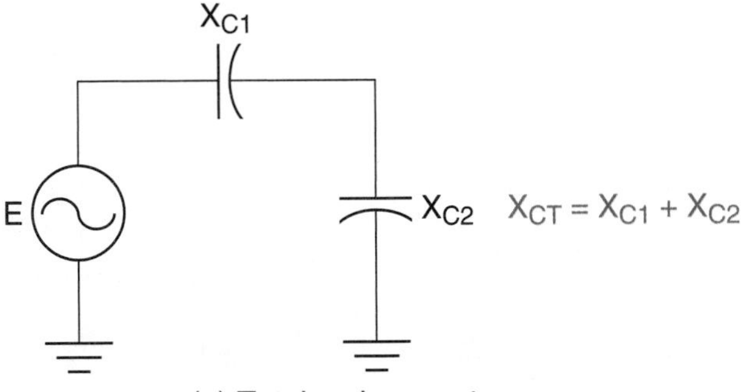

(a) Total series reactance

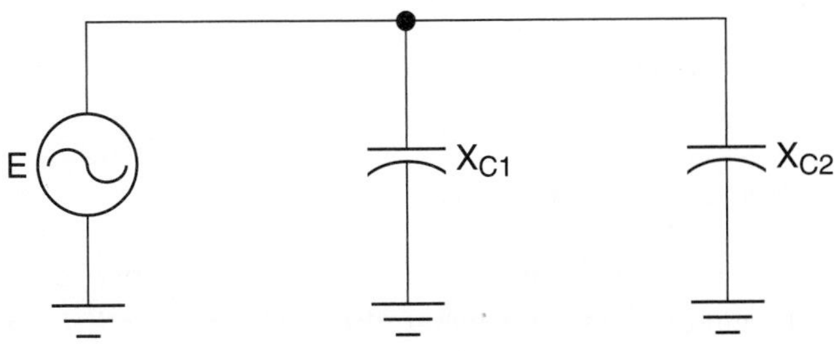

$$X_{CT} = \frac{1}{\frac{1}{X_{C1}} + \frac{1}{X_{C2}}}$$

or

$$X_{CT} = \frac{X_{C1}X_{C2}}{X_{C1} + X_{C2}}$$

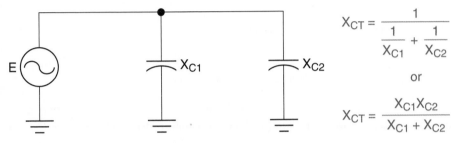

(b) Total parallel reactance

FIGURE 16.9 Calculating total series and parallel circuit reactance.

PROGRESS CHECK

1. Contrast the true and effective resistance of a capacitor.
2. Why do capacitors have extremely high true resistance values?
3. What is capacitive reactance (X_C)?
4. What relationships are commonly used to calculate the value of X_C?
5. How is the magnitude of current in a capacitive circuit calculated?
6. How is the total capacitive reactance calculated in a series circuit? A parallel circuit?

16.3 SERIES RC CIRCUITS

A *resistive-capacitive (RC) circuit* is one that contains any combination of resistors and capacitors. Several examples of basic RC circuits are shown in Figure 16.10.

As we progress through the rest of this chapter, we will develop the current, voltage, power, and impedance characteristics of RC circuits. First, we're going to compare RC series circuits to *purely resistive* and *purely capacitive* series circuits. The goal is to show the similarities and differences among the three.

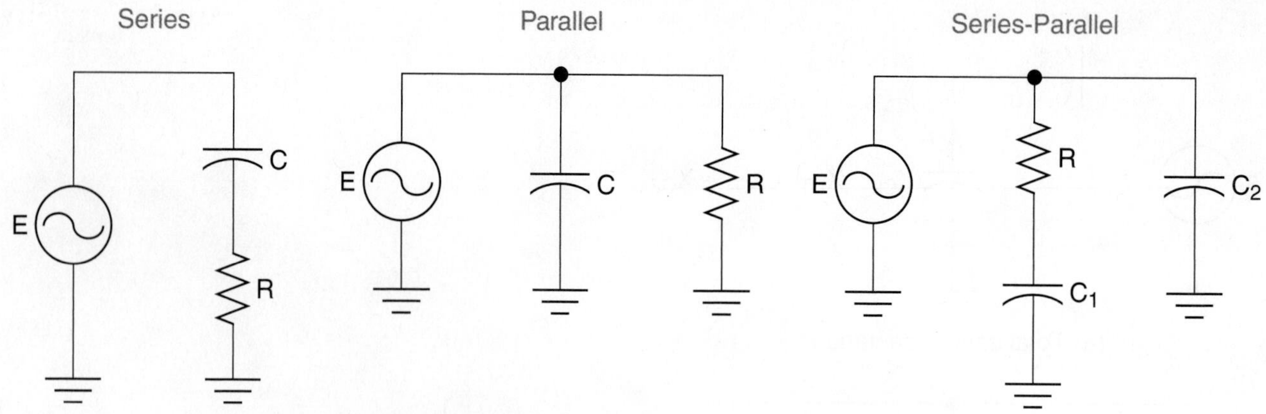

FIGURE 16.10 Series, parallel, and series-parallel RC circuits.

Series Circuits: A Comparison

In previous discussions, we established the fundamental characteristics of resistive and capacitive series circuits. As a refresher, these characteristics are as follows:

- Current is the same at all points throughout the circuit.
- The applied voltage equals the sum of the component voltages (in keeping with Kirchhoff's voltage law).
- The total impedance (resistance and/or reactance) equals the sum of the component impedances.

These relationships are provided (as equations) in Figure 16.11.

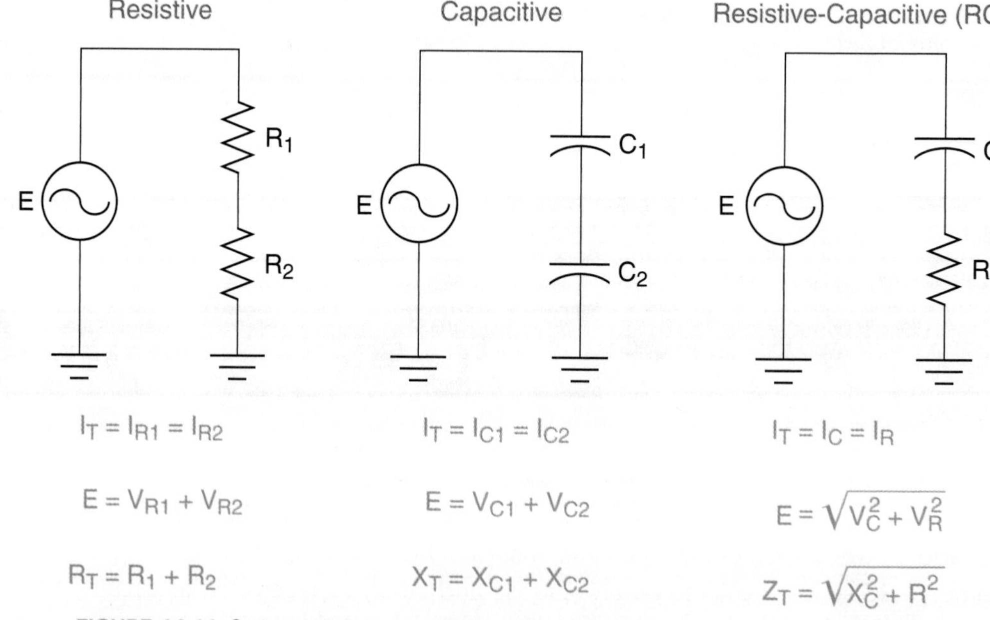

FIGURE 16.11 Series resistive, capacitive, and resistive-capacitive (RC) circuit relationships.

The relationships listed for the resistive and capacitive circuits should seem familiar. In each case, the current is constant throughout the circuit (as is the case for resistive and capacitive circuits). However,

- The applied voltage (E) equals the geometric sum of V_C and V_R.
- The total opposition to current (Z_T) equals the geometric sum of X_C and R.

These series voltage and impedance relationships are nearly identical to those for series RL circuits, as will be demonstrated throughout this section.

STAYING SAFE When discharging large capacitors, a *shorting tool* of sufficient size to handle the discharge current must be used. Also, the shorting procedure must be carried out in a manner such that the operator is "insulated" from the circuit.

RC circuits are commonly used to filter "ripple" voltage from the output of a DC power supply, or to remove harmonics from the output of an AC source (among other applications).

IN THE FIELD

RC Circuit Voltage and Impedance

In a series RC circuit, the geometric sum of the component voltages equals the source voltage. By the same token, the geometric sum of resistance and reactance equals the total impedance (opposition to current) in the circuit. Even though we are using a different type of addition, the RC circuit still operates according to the rules we have established for all series circuits. That is:

- Component voltages add (geometrically) to equal the source voltage.
- Resistance and reactance add (geometrically) to equal the total impedance.

Series Voltages

Earlier in the chapter, you were told that current leads voltage by 90 degrees in a capacitive circuit. In a resistive circuit, voltage and current are always in phase. When we connect a capacitor and a resistor in series, we get the circuit and waveforms shown in Figure 16.12. Since current is constant throughout a series circuit, the capacitor and resistor voltage waveforms are referenced to a single current waveform. As you can see:

- The capacitor voltage lags the circuit current by 90 degrees.
- The resistor voltage is in phase with the circuit current.

Therefore, capacitor voltage lags resistor voltage by 90 degrees. Another way of expressing this relationship is to say that resistor voltage leads capacitor voltage by 90°. This phase relationship holds true for every series RC circuit with an AC source.

Since V_R leads V_C by 90°, the values of the component voltages are represented using vectors as shown in Figure 16.13a. By plotting the value of V_R on the x-axis (0°) and the value of V_C on the *negative* y-axis (−90°), the phase relationship between the two is represented in the graph.

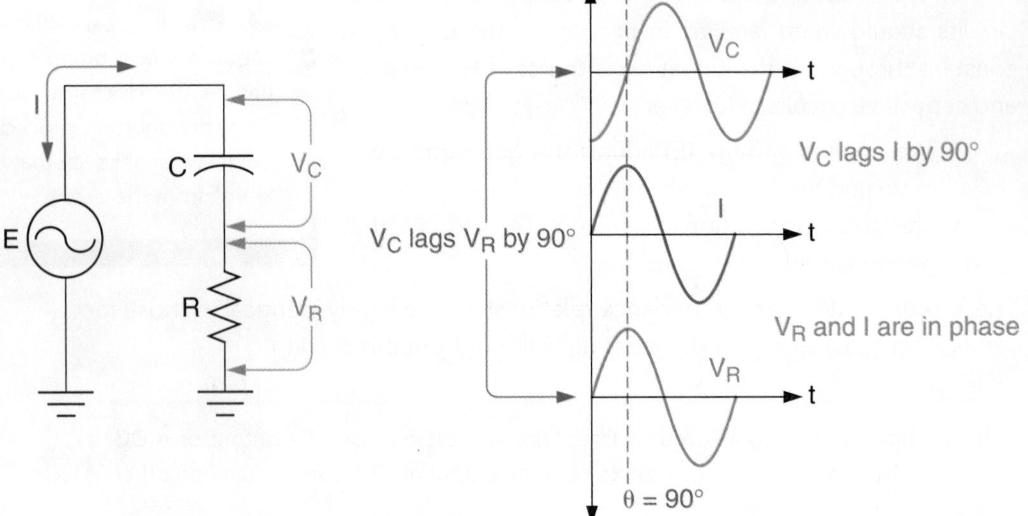

FIGURE 16.12 Series RC circuit voltage and current phase relationships.

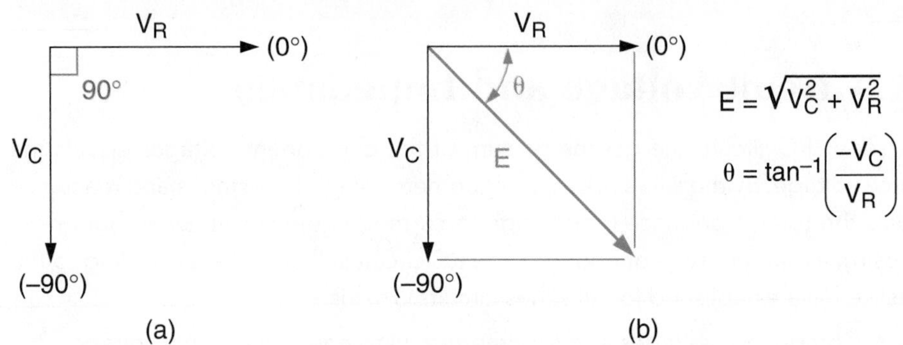

FIGURE 16.13 Series RC circuit voltage vectors.

Like any other series circuit, the source voltage in an RC circuit must equal the sum of the component voltages. Since the component voltages in an RC circuit are vector values, they must be added geometrically. Therefore, the sum of the component voltages represented by the vectors in Figure 16.13b can be found as

(16.4)

$$E = \sqrt{V_C^2 + V_R^2}$$

The phase angle for a series RC circuit represents the phase difference between the source voltage and the circuit current. As shown in Figure 16.13b, the phase angle for a series RC circuit can be found as

(16.5)

$$\theta = \tan^{-1}\frac{-V_C}{V_R}$$

Note that V_C is plotted on the negative y-axis of the graph (as shown in Figure 16.13) because it lags V_R by 90°. The source voltage and phase angle of a series RC circuit are calculated as shown in Example 16.3.

EXAMPLE 16.3

Calculate the source voltage and phase angle for the circuit in Figure 16.14.

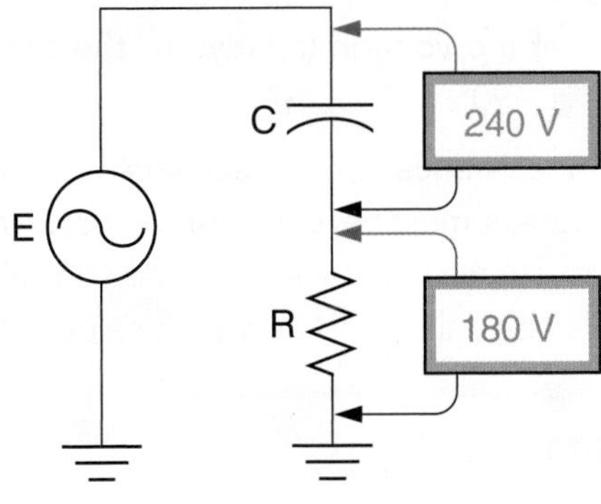

FIGURE 16.14

SOLUTION

The source voltage in Figure 16.14 has a value of

$$E = \sqrt{V_C^2 + V_R^2}$$
$$= \sqrt{(240\text{ V})^2 + (180\text{ V})^2}$$
$$= 300\text{ V}$$

The phase angle of the source voltage (relative to the circuit current) can be found as

$$\theta = \tan^{-1}\left(\frac{-V_C}{V_R}\right)$$
$$= \tan^{-1}\left(\frac{-240\text{ V}}{180\text{ V}}\right)$$
$$= \tan^{-1}(-1.33)$$
$$= -53.1°$$

PRACTICE PROBLEM 16.3

A circuit like the one in Figure 16.14 has values of $V_C = 120$ V and $V_R = 160$ V. Calculate the circuit source voltage and phase angle.

Note that the negative phase angle in Example 16.3 indicates that the source voltage lags the circuit current by 53.1°. For any RC circuit, the voltage phase angle (relative to the circuit current) always falls within the range of

(16.6)

$$-90° < \theta < 0°$$

This range of values for θ makes sense when you consider that:

- The phase angle of a purely resistive circuit is 0°.
- The phase angle of the voltage (relative to the circuit current) in a purely capacitive circuit is −90°.

A circuit containing both resistance and capacitance is neither purely resistive nor purely capacitive. Therefore, it must have a voltage phase angle that is less than 0° and greater than −90° (as shown in Figure 16.15). As you will see, the value of θ for a given RC circuit can also be calculated using the values of X_C and R.

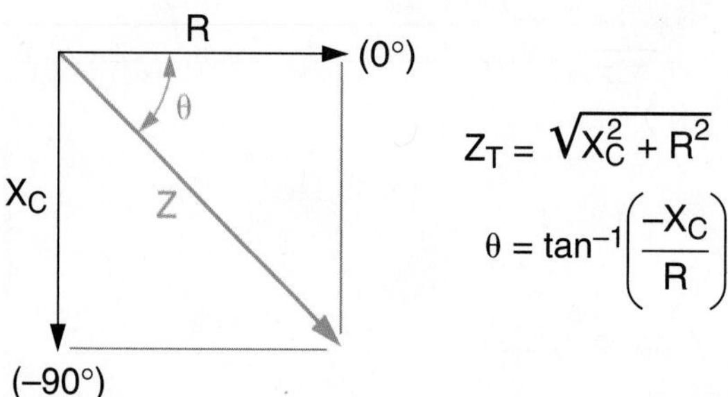

FIGURE 16.15 Series RC circuit reactance and impedance vectors.

Series Impedance

In an RC circuit, the total impedance (opposition to alternating current) is made up of resistance and capacitive reactance. Like series voltages, the values of R and X_C for a series circuit are vector quantities that must be added geometrically. By formula,

(16.7)

$$Z = \sqrt{X_C^2 + R^2}$$

Earlier, you were shown how the phase angle for a series RC circuit can be calculated using the component voltages (V_C and V_R). The phase angle of an RC circuit can also be calculated using X_C and R. By formula,

(16.8)

$$\theta = \tan^{-1}\frac{-X_C}{R}$$

Example 16.5 shows how the values of X_C and R can be used to calculate series RC circuit impedance and phase angle.

EXAMPLE 16.4

Calculate the impedance and phase angle for the circuit in Figure 16.16.

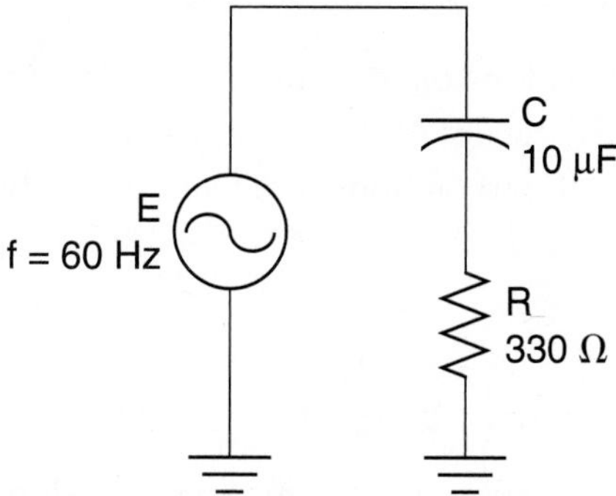

FIGURE 16.16

SOLUTION

Calculating the circuit impedance begins with calculating the value of X_C, as follows:

$$X_C = \frac{1}{2\pi fC}$$
$$= \frac{1}{2\pi(60\text{ Hz})(10\ \mu\text{F})}$$
$$= 265\ \Omega$$

The circuit impedance can now be found as

$$Z = \sqrt{X_C^2 + R^2}$$
$$= \sqrt{(265\ \Omega)^2 + (330\ \Omega)^2}$$
$$= 423\ \Omega$$

and the circuit phase angle is found as

$$\theta = \tan^{-1}\left(\frac{-X_C}{R}\right)$$
$$= \tan^{-1}\left(\frac{-265\ \Omega}{330\ \Omega}\right)$$
$$= \tan^{-1}(-0.803)$$
$$= -38.8°$$

The negative phase angle indicates that the source voltage lags the circuit current by 38.8°.

PRACTICE PROBLEM 16.4

A circuit like the one in Figure 16.16 has the following values: $f = 120$ Hz, $C = 2.2\ \mu F$, and $R = 1\ k\Omega$. Calculate the circuit impedance and phase angle.

Series Circuit Analysis

You have been shown how the voltage, current, and impedance values are calculated for a basic series RC circuit. At this point, we are going to work through two circuit problems. The goal is to demonstrate two approaches to calculating series RC circuit values.

EXAMPLE 16.5

Calculate the voltage, current, and impedance values for the circuit in Figure 16.17a.

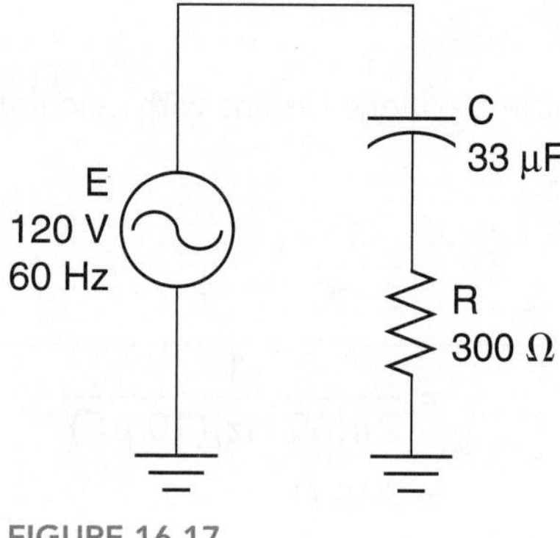

FIGURE 16.17

SOLUTION

We begin with calculating the value of X_C, as follows:

$$X_C = \frac{1}{2\pi fC} = \frac{1}{2\pi(60\text{ Hz})(33\ \mu\text{F})} = 80.4\ \Omega$$

The magnitude and phase angle of the circuit impedance can now be found as

$$Z_T = \sqrt{X_C^2 + R^2} = \sqrt{(80.4\ \Omega)^2 + (300\ \Omega)^2} = 311\ \Omega$$

and

$$\theta = \tan^{-1}\left(\frac{-X_C}{R}\right)$$
$$= \tan^{-1}\left(\frac{-80.4\ \Omega}{300\ \Omega}\right)$$
$$= \tan^{-1}(-0.268)$$
$$= -15^\circ$$

Using the values of Z_T and E, the circuit current is found as

$$I = \frac{E}{Z_T}$$
$$= \frac{120\ \text{V}}{311\ \Omega}$$
$$= 386\ \text{mA}$$

Now, we can calculate the values of V_C and V_R as follows:

$$V_C = I \cdot X_C$$
$$= (386\ \text{mA})(80.4\ \Omega)$$
$$= 31\ \text{V}$$

and

$$V_R = I \cdot R$$
$$= (386\ \text{mA})(300\ \Omega)$$
$$= 116\ \text{V}$$

PRACTICE PROBLEM 16.5

A circuit like the one in Figure 16.17 has the following values: E = 120 V, f = 60 Hz, C = 4.7 μF, and R = 510 Ω. Determine the circuit phase angle, voltage, current, and impedance values for the circuit.

We can calculate the component voltages using the voltage-divider equation, just as we did with series RL circuits. For a series RC circuit, the voltage-divider equation is given as

(16.9)

$$V_n = E \times \frac{Z_n}{Z_T}$$

where

Z_n = the magnitude of R or X_C
Z_T = the geometric sum of R and X_C

Example 16.6 shows how the voltage-divider equation is used in the analysis of a series RC circuit.

EXAMPLE 16.6

Calculate the component voltages for the circuit in Figure 16.18.

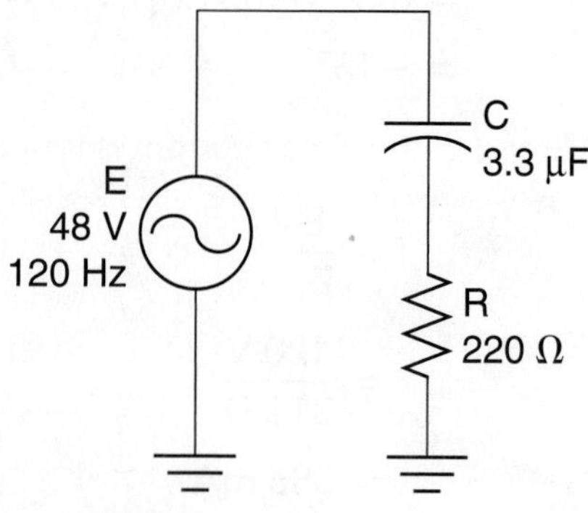

FIGURE 16.18

SOLUTION

Calculating the component voltages for the circuit begins with calculating the value of X_C as follows:

$$X_C = \frac{1}{2\pi fC}$$
$$= \frac{1}{2\pi(120\text{ Hz})(3.3\ \mu\text{F})}$$
$$= 402\ \Omega$$

Then, the magnitude of the circuit impedance is found as

$$Z_T = \sqrt{X_C^2 + R^2}$$
$$= \sqrt{(402\ \Omega)^2 + (220\ \Omega)^2}$$
$$= 458\ \Omega$$

The capacitor voltage can now be found as

$$V_C = E \times \frac{X_C}{Z_T}$$
$$= (48\text{ V})\frac{402\ \Omega}{458\ \Omega}$$
$$= 42.1\text{ V}$$

and the magnitude of the resistor voltage can be found as

$$V_R = E \times \frac{R}{Z_T}$$

$$= (48 \text{ V})\frac{220 \ \Omega}{458 \ \Omega}$$

$$= 23.1 \text{ V}$$

PRACTICE PROBLEM 16.6

A circuit like the one in Figure 16.18 has the following values: E = 120 V, f = 60 Hz, C = 1 μF, and R = 470 Ω. Using the voltage-divider equation, calculate the magnitudes of the component voltages.

If the component voltages calculated in Example 16.6 appear to be incorrect, you may be forgetting that the source voltage equals their geometric sum. The component voltages calculated in the example add up to equal the source voltage, as follows:

$$E = \sqrt{V_C^2 + V_R^2}$$

$$= \sqrt{(42.1 \text{ V})^2 + (23.1 \text{ V})^2}$$

$$= 48 \text{ V}$$

At this point, you have been shown how to calculate most of the basic electrical values for series RC circuits. The only remaining basic property—power—is discussed in the next section.

PROGRESS CHECK

1. What are the three fundamental characteristics of every series circuit?
2. What does the phase angle of a series RC circuit indicate?
3. What electrical values are commonly used to calculate the phase angle of a series RC circuit?
4. How is the voltage-divider equation modified for use in series RC circuits?

16.4 POWER CHARACTERISTICS AND CALCULATIONS

Like RL circuits, RC circuits have significant values of apparent power, reactive power, and true power. In this section, we will look at the fundamental power characteristics and calculations for series RC circuits. As you will learn, many of the principles covered in this section apply to *parallel* RC circuits as well.

AC Power Values: A Brief Review

In Chapters 14 and 15, we discussed apparent power (P_{APP}), resistive power (P_R), and reactive power (P_X) in terms of inductor operation. Each of these values is described in terms of RC circuit values in Table 16.2.

TABLE 16.2 • Power Values in Resistive-Reactive Circuits.

VALUE	DEFINITION
Resistive power (P_R)	The power dissipated by the resistance in an RC circuit. Also known as *true power.*
Reactive power (P_X)	The value found using $P = I^2 \cdot X_C$. The energy (per unit time) that is actually stored in the capacitor's electrostatic field, and is returned to the circuit when the component discharges. P_X is measured in *volt-amperes-reactive* (VAR) to distinguish it from actual power dissipation values. Also known as *imaginary power.*
Apparent power (P_{APP})	The combination of resistive (true) power and reactive (imaginary) power. Measured in *volt-amperes* (VA) because its value does not represent a true power-dissipation value.
Power factor (PF)	The ratio of resistive power to apparent power.

The values of resistive and reactive power for a series RC circuit are calculated as shown in Example 16.7.

EXAMPLE 16.7

Calculate the reactive and resistive power values for the circuit in Figure 16.19.

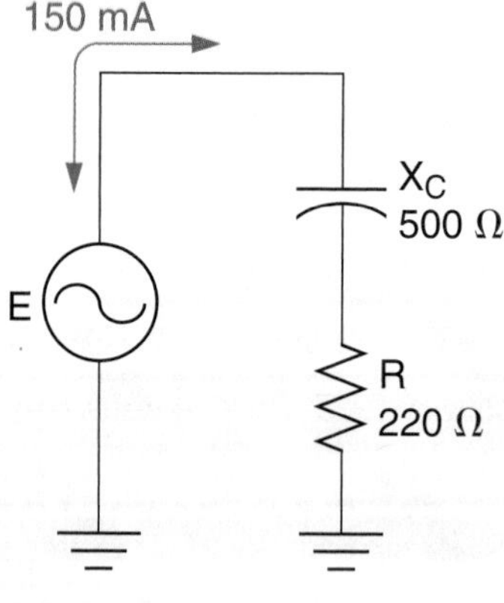

FIGURE 16.19

SOLUTION

The reactive power for the capacitor in Figure 16.19 has a value of

$$P_X = I^2 \cdot X_C$$

$$= (150 \text{ mA})^2 \times (500\ \Omega)$$
$$= 11.3 \text{ VAR}$$

and the value of resistive power is found as

$$P_R = I^2 \cdot R_1$$
$$= (150 \text{ mA})^2 \times (220\ \Omega)$$
$$= 4.95 \text{ W}$$

PRACTICE PROBLEM 16.7

A circuit like the one in Figure 16.19 has the following values: I = 200 mA, X_C = 245 Ω, and R = 180 Ω. Calculate the values of P_X and P_R for the circuit.

You may be wondering why we didn't consider the dielectric resistance of the capacitor in the resistive power (P_R) calculation. The answer lies in the difference between the ideal and practical characteristics of a capacitor.

An ideal capacitor has infinite dielectric resistance. This dielectric resistance is represented as shown in Figure 16.20a. Note that the resistance is drawn as a parallel component (just as it is for a current source). A practical representation of a capacitor must include the true dielectric resistance of the component. For example, let's say that the actual dielectric resistance of the component in Figure 16.20 is

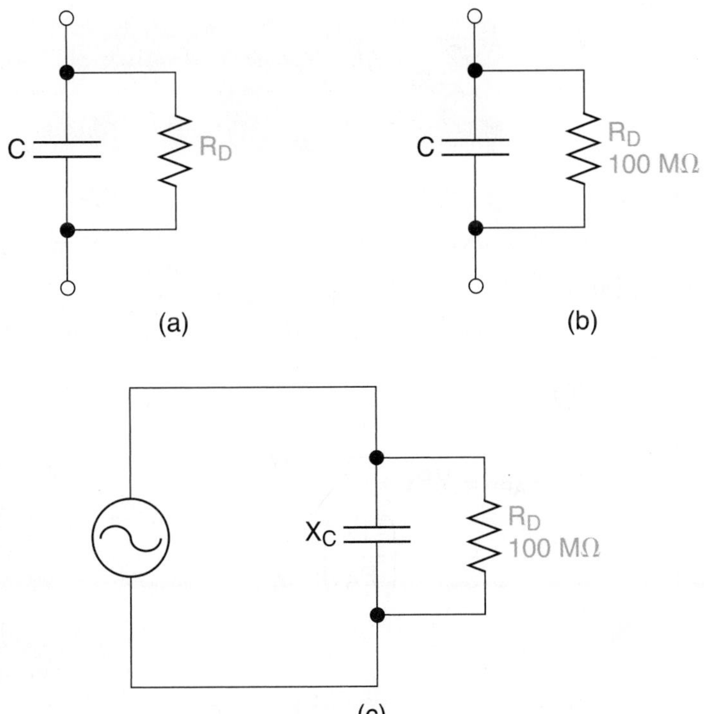

FIGURE 16.20 Capacitor dielectric resistance.

LEAKAGE CURRENT The current that "leaks" through a capacitor's dielectric, typically in the low μA range.

100 MΩ. This value is included as a parallel 100 MΩ resistor (as shown in Figure 16.20b). When connected to a voltage source, the current through the 100 MΩ resistor represents the component **LEAKAGE CURRENT**. That is, the current that "leaks" through the capacitor's dielectric.

When driven by an AC source, the reactance of the capacitor is in parallel with the dielectric resistance of the component (as shown in Figure 16.20c). Normally, $X_C << R_D$. Therefore, the dielectric resistance of the capacitor is (for all practical purposes) shorted out by the component's reactance. For example, assume the capacitor in Example 16.7 has a value of $R_D = 100\ M\Omega$. With a parallel value of $X_C = 500\ \Omega$, the value of R_D is effectively shorted out. This is why the dielectric resistance of a capacitor is not considered in the resistive power calculation for the circuit.

Calculating Apparent Power (P_{APP})

Like voltage and impedance, apparent power in a series RC circuit equals the geometric sum of resistive power and reactive power. Like the voltages in a series RC circuit, resistive and reactive power can be represented using vectors, as shown in Figure 16.21a. By plotting the value of P_R on the x-axis (0°) and P_X on the negative y-axis (−90°), the phase relationship between the two is represented in the graph. By formula,

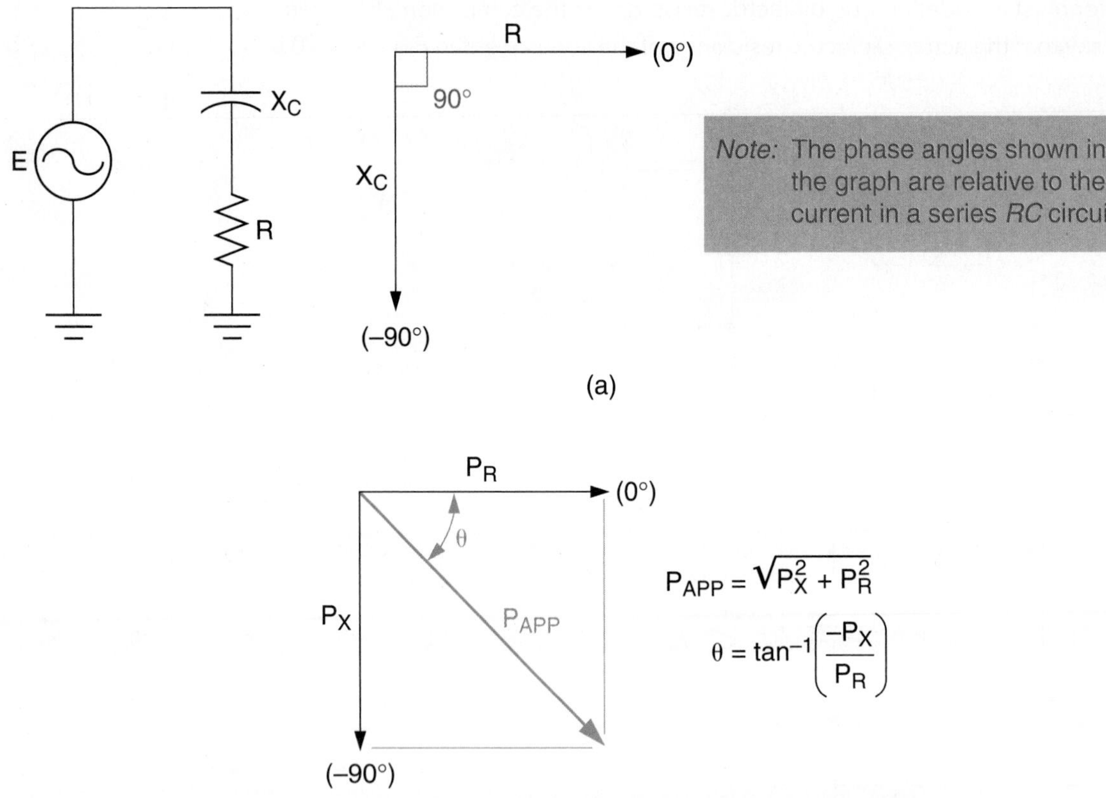

FIGURE 16.21 Parallel relationships in a series RC circuit.

(16.10)

$$P_{APP} = \sqrt{P_X^2 + P_R^2}$$

where

P_{APP} = the apparent power, measured in volt-amperes (VA)

This is the same relationship that we used to determine the value of apparent power for an RL circuit. By the same token, the phase angle equation for apparent power (relative to the circuit current) is very similar to the one used for RL circuits. By formula,

(16.11)

$$\theta = \tan^{-1}\frac{-P_X}{P_R}$$

Example 16.8 demonstrates the use of these equations.

EXAMPLE 16.8

Calculate the apparent power and phase angle values for the circuit in Example 16.7.

SOLUTION

In Example 16.7, we calculated the following values: P_X = 11.3 VAR and P_R = 4.95 W. Using these values, the magnitude of apparent power is found as

$$\begin{aligned} P_{APP} &= \sqrt{P_X^2 + P_R^2} \\ &= \sqrt{(11.3\ \text{VAR})^2 + (4.95\ \text{W})^2} \\ &= 12.3\ \text{VA} \end{aligned}$$

and the phase angle is found as

$$\begin{aligned} \theta &= \tan^{-1}\left(\frac{-P_X}{P_R}\right) \\ &= \tan^{-1}\left(\frac{-11.3\ \text{VAR}}{4.95\ \text{W}}\right) \\ &= \tan^{-1}(-2.28) \\ &= -66.3^\circ \end{aligned}$$

PRACTICE PROBLEM 16.8

Calculate the value of P_{APP} and the phase angle for the circuit described in Practice Problem 16.7.

Don't forget: Only a portion of the apparent power in an RC circuit is actually dissipated by the circuit. This is why it has its own unit of measure.

Power Factor

In Chapter 14, you were told that the *power factor* for a resistive-reactive circuit is the ratio of resistive power to apparent power. By formula,

$$PF = \frac{P_R}{P_{APP}}$$

Example 16.9 shows how this relationship is applied to a series RC circuit.

EXAMPLE 16.9

A series RC circuit has the following values: $P_R = 12$ W and $P_{APP} = 18.6$ VA. What is the circuit power factor?

SOLUTION

Using the given values, the circuit apparent power is found as

$$PF = \frac{P_R}{P_{APP}}$$

$$= \frac{12\text{ W}}{18.6\text{ VA}}$$

$$= 0.645$$

PRACTICE PROBLEM 16.9

A series RC circuit has the following values: $P_R = 33$ W and $P_{APP} = 56$ VA. What is the circuit power factor?

Putting It All Together: Series RC Circuit Analysis

Now that we have covered the power characteristics and calculations associated with series RC circuits, we're going to close out the section with another circuit analysis problem.

EXAMPLE 16.10

Calculate the current, voltage, impedance, and power values for the circuit in Figure 16.22.

SOLUTION

The reactance of the capacitor in Figure 16.22 is found as

$$X_C = \frac{1}{2\pi fC}$$

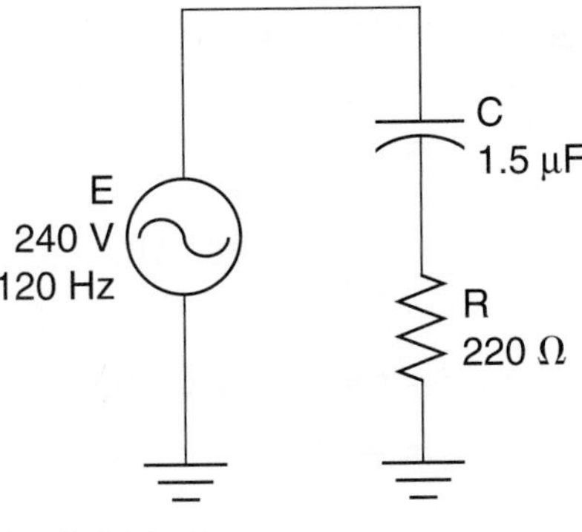

FIGURE 16.22

$$= \frac{1}{2\pi(120\text{ Hz})(1.5\mu\text{F})}$$
$$= 884\ \Omega$$

This, combined with the value of R, gives us a circuit impedance of

$$Z_T = \sqrt{X_C^2 + R^2}$$
$$= \sqrt{(884\ \Omega)^2 + (220\ \Omega)^2}$$
$$= 911\ \Omega$$

The magnitude of the circuit current can now be found as

$$I_T = \frac{E}{Z_T}$$
$$= \frac{240\text{ V}}{911\ \Omega}$$
$$= 263\text{ mA}$$

Using the voltage-divider equation, the voltage across the resistor can be found as

$$V_R = E \times \frac{R}{Z_T}$$
$$= (240\text{ V}) \times \frac{220\ \Omega}{911\ \Omega}$$
$$= 58\text{ V}$$

and the voltage across the capacitor can be found as

$$V_C = E \times \frac{X_C}{Z_T}$$
$$= (240\text{ V})\frac{884\ \Omega}{911\ \Omega}$$
$$= 233\text{ V}$$

The circuit phase angle can be found as

$$\theta = \tan^{-1}\frac{-X_C}{R}$$

$$= \tan^{-1}\frac{-884\ \Omega}{220\ \Omega}$$

$$= \tan^{-1}(-4.018)$$

$$= -76^\circ$$

The circuit power values can be found as

$$P_R = I_T \times V_R$$

$$= 263\text{ mA} \times 58\text{ V}$$

$$= 15.3\text{ W}$$

$$P_X = I_T \times V_C$$

$$= 263\text{ mA} \times 233\text{ V}$$

$$= 61.3\text{ VAR}$$

and

$$P_{APP} = \sqrt{P_X^2 + P_R^2}$$

$$= \sqrt{(61.3\text{ VAR})^2 + (15.3\text{ W})^2}$$

$$= 63.2\text{ VA}$$

Finally, the circuit power factor can be found as

$$PF = \frac{P_R}{P_{APP}}$$

$$= \frac{15.3\text{ W}}{63.2\text{ VA}}$$

$$= 0.242$$

This completes the analysis of the circuit.

PRACTICE PROBLEM 16.10

A circuit like the one in Figure 16.22 has the following values: $E = 120$ V, $f = 60$ Hz, $C = 10\ \mu$F, and $R = 360\ \Omega$. Perform a complete analysis on the circuit.

We have concluded our discussion on series RC circuit characteristics. In the next chapter, we will examine the characteristics of parallel RC circuits.

PROGRESS CHECK

1. Compare and contrast the following: apparent power, resistive power, and reactive power.
2. Why is the dielectric resistance of a capacitor normally ignored in the resistive power calculations for an RC circuit?
3. State and explain the phase relationship between RC circuit reactive power (P_X) and resistive power (P_R).
4. What is the power factor for an RC circuit? What does it tell you?

16.5 SUMMARY

Here is a summary of the major points that were covered in this chapter:

Alternating Voltage and Current Characteristics

- Capacitors and inductors are near-opposites when it comes to many of their operating characteristics.
- A capacitor is often used to provide AC coupling and DC isolation between a source and its load.
- A given capacitor always seeks to maintain its plate-to-plate charge. That is, it opposes any change in its plate-to-plate voltage.
- When the input to a capacitor contains a DC offset voltage, the capacitor isolates the DC offset from the load.
- In a purely capacitive circuit, the current at any instant is directly proportional to the capacity of the capacitor(s) and the rate of change in capacitor voltage.
- In a circuit with a sinusoidal voltage source, capacitor current reaches its maximum value when capacitor voltage reaches its minimum value, and vice versa.
- The current-versus-voltage phase relationships of inductors and capacitors are exact opposites.
  - In a purely inductive circuit, voltage leads current by 90 degrees.
  - In a purely capacitive circuit, current leads voltage by 90 degrees.

Capacitive Reactance (X_C)

- Capacitors provide a measurable opposition to alternating current, called capacitive reactance (X_C).
- Capacitive reactance is inversely proportional to capacitance and operating frequency.
- In a capacitive circuit, X_C can be used with Ohm's law to calculate circuit current.
- Series and parallel values of X_C are combined in the same fashion as series and parallel values of resistance (or inductive reactance).

Series RC Circuits

- A resistive-capacitive (RC) circuit is one that contains any combination of resistors and inductors.
- In a series RC circuit:
  - The source voltage (E) equals the geometric sum of V_R and V_C.
  - The circuit impedance (Z_T) equals the geometric sum of R and X_C.
- Resistor voltage leads capacitor voltage by 90° in a series RC circuit.

- The source phase angle (θ) is the phase difference between the source voltage and the circuit current.
  - Current leads voltage in a series RC circuit.
  - The phase angle falls within the range of $-90° < \theta < 0°$
- The total impedance in a series RC circuit equals the geometric sum of X_C and R.
- The phase angle in a series RC circuit can be calculated using the values of X_C and R.

Power Characteristics and Calculations

- In any resistive-reactive circuit:
  - Resistive power (or "true power") is measured in watts. P_R is the power that is actually dissipated in the circuit.
  - Reactive power (or "imaginary power") is measured in volt-amperes-reactive (VAR).
  - Apparent power (the geometric sum of resistive and reactive power) is measured in volt-amperes (VA).
- Apparent power lags the circuit current in a series RC circuit.
- The *power factor* for a series RC circuit is the ratio of resistive power to apparent power.
  - The power factor for an RC circuit is found as $PF = \frac{P_R}{P_{APP}}$.

CHAPTER REVIEW

1. Capacitors store energy in a(n) ________.
   a) electromagnetic field
   b) electrostatic field
   c) capacitive field
   d) dielectric field

2. The transfer of electrical energy between two points, components, or systems is often referred to as ________.
   a) amplification
   b) transference
   c) coupling
   d) isolation

3. When a component or circuit prevents the transfer of electrical energy between two points, it is said to provide ________.
   a) amplification
   b) transference
   c) coupling
   d) isolation

4. A capacitor is used to connect a DC source to a load. There is a switch connected in series with the circuit. When the switch is closed and the capacitor has fully charged, the voltage across the capacitor is ________ and the voltage across the load is ________.
   a) equal to E, zero volts
   b) zero volts, equal to E

c) 90° out of phase, in phase

d) in phase, 90° out of phase

5. The fact that a capacitor seeks to maintain its plate-to-plate voltage is the basis for the ________ capacitors.

   a) capacity of

   b) phase shift introduced by

   c) reactance of

   d) AC coupling provided by

6. A waveform that is not centered around zero volts is said to have a(n) ________.

   a) DC offset

   b) AC offset

   c) blocking voltage

   d) coupling voltage

7. A sine wave has peak values of +400 V and −100 V. This means that the waveform is centered at ________.

   a) +250 V

   b) +150 V

   c) 0 V

   d) +200 V

8. The two factors that determine capacitor current in an AC circuit are ________.

   a) capacitance and the rate of change of capacitor voltage

   b) capacitance and the rate of change of capacitor current

   c) frequency and the rate of change of capacitor voltage

   d) capacitive reactance and the rate of change of capacitor voltage

9. For a voltage sine wave, the rate of change of voltage is maximum at ________.

   a) 90° and 180 °

   b) 90° and 270°

   c) 0° and 180°

   d) 0° and 270°

10. For a voltage sine wave, the rate of change of voltage is minimum at ________.

    a) 90° and 180°

    b) 90° and 270°

    c) 0° and 180°

    d) 0° and 270°

11. A voltage sine wave is applied to a capacitor. When the waveform is at 360° the rate of change in V_C is ________ and I_C is ________.

    a) maximum, zero

    b) maximum, maximum

c) zero, maximum

d) zero, zero

12. Capacitor current always ________ capacitor voltage by ________.

 a) leads, −90°

 b) lags, 90°

 c) leads, 90°

 d) lags, −90°

13. Capacitor voltage always ________ capacitor current by ________.

 a) leads, −90°

 b) lags, 90°

 c) leads, 90°

 d) lags, −90°

14. The true or DC resistance of a capacitor is equal to ________.

 a) a very low value

 b) the resistance of the plates

 c) the dielectric resistance

 d) the reactance of the capacitor

15. The effective resistance of a capacitor is equal to ________.

 a) a very low value

 b) the resistance of the plates

 c) the dielectric resistance

 d) the reactance of the capacitor

16. The relationship between frequency and X_C is ________.

 a) a direct relationship

 b) an inverse relationship

 c) dependent on the type of capacitor

 d) exponential

17. A 240 V @ 60 Hz sine wave is applied to a 100 μF capacitor. The value of X_C is ________.

 a) 26.5 VAR

 b) 2.65 Ω

 c) 26.5 Ω

 d) 377 Ω

18. 440 V @ 120 Hz is applied to a purely capacitive circuit. If the capacitor is 4.7 μF, capacitor current would be ________.

 a) 1.56 A

 b) 30.5 A

c) 16.6 A

d) 21.8 A

19. Two capacitors are connected in series. You solve for the total capacitance like two resistors in ________ and you solve for total capacitive reactance like two resistances in ________.

   a) series, parallel

   b) series, series

   c) parallel, parallel

   d) parallel, series

20. In an RC circuit, resistor voltage always ________ circuit current.

   a) lags

   b) leads

   c) is in phase with

   d) is 180° out of phase with

21. V_C is always plotted on ________.

   a) the y-axis

   b) the negative y-axis

   c) the x-axis

   d) the negative x-axis

22. A voltage source is connected to a series RC circuit. $V_C = 114$ V and $V_R = 89$ V. This means that E = ________ at a phase angle of ________.

   a) 145 V, −52°

   b) 25 V, 52°

   c) 203 V, −38°

   d) 145 V, 52°

23. A 240 V @ 60 Hz supply is applied to a series RC circuit containing a 330 Ω resistance and a 2.2 μF capacitor. The total impedance of this circuit is ________ at a phase angle of ________.

   a) 1154 Ω, −14.7°

   b) 1554 Ω, −75°

   c) 1554 Ω, −14.7°

   d) 1250 Ω, −75°

24. If the circuit described in question 23 were connected to a European source, where the supply frequency is 50 Hz, this means that circuit current would ________ and the phase angle of the circuit impedance would ________.

   a) increase, become more negative

   b) decrease, become less negative

   c) increase, become less negative

   d) decrease, become more negative

25. A 120 V @ 60 Hz source is applied to a series RC circuit containing a 470 Ω resistance and a 4.7 μF capacitor. For this circuit, V_C = _______ and V_R = _______.
    a) 76.8 V, 92.2 V
    b) 87.6 V, 76.8 V
    c) 92.2 V, 76.8 V
    d) 92.2 V, 87.6 V

26. The phase angle of the source voltage for the circuit described in question 25 is _______.
    a) 39.8°
    b) −50.2°
    c) 50.2°
    d) −39.8°

27. 550 V @ 60 Hz is applied to a 680 Ω resistance in series with a 6.8 μF capacitor. For this circuit, P_X = _______ and P_R = _______.
    a) 192 VAR, 335 W
    b) 335 VAR, 192 W
    c) 103 VAR, 180 W
    d) 180 W, 103 VAR

28. The value of P_{APP} for the circuit described in question 27 is _______ at a phase angle of _______.
    a) 386 VA, 29.8°
    b) 526 VA, −31.3°
    c) 386 VA, −29.8°
    d) 426 VA, 28.9°

29. The power factor for the circuit described in questions 27 and 28 is _______.
    a) 29.8%
    b) 13.2%
    c) 88.6%
    d) 86.8%

30. The true dielectric resistance of a capacitor _______.
    a) is the source of capacitor leakage current
    b) is usually very high
    c) is represented as a parallel resistance
    d) all of the above

PRACTICE PROBLEMS

1. Calculate the value of X_C for the circuit shown in Figure 16.23a.
2. Calculate the value of X_C for the circuit shown in Figure 16.23b.

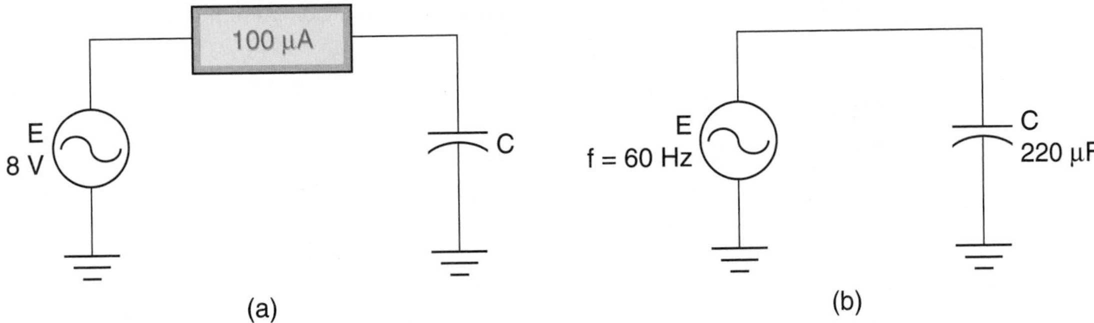

FIGURE 16.23

3. Calculate the value of X_C for the circuit shown in Figure 16.24a.
4. Calculate the value of X_C for the circuit shown in Figure 16.24b.

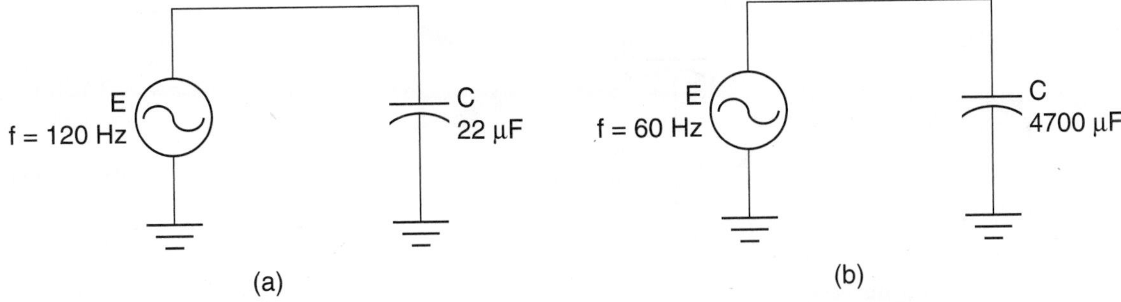

FIGURE 16.24

5. Calculate the value of total current for the circuit in Figure 16.25a.
6. Calculate the value of total current for the circuit in Figure 16.25b.

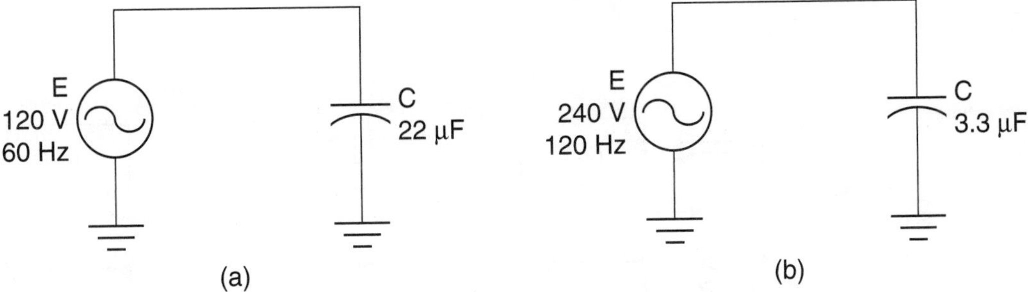

FIGURE 16.25

7. The following table lists values for a capacitive circuit. Complete the table.

V_{RMS}	I_{RMS}	X_C
120 V	______	18 Ω
______	3.6 A	30 Ω
20 V	4.28 A	______

8. The following table lists values for a capacitive circuit. Complete the table.

V_{RMS}	I_{RMS}	X_C
240 V	______	22.5 Ω
______	12 μA	1.98 M Ω
24 V	0.12 A	______

9. Calculate the total reactance in the circuit shown in Figure 16.26a.
10. Calculate the total reactance in the circuit shown in Figure 16.26b.

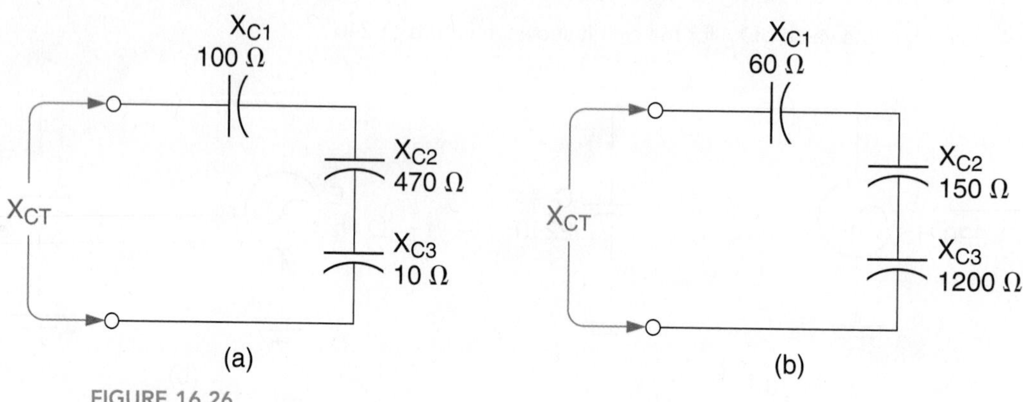

FIGURE 16.26

11. The component voltages shown in Figure 16.27a were measured using an AC voltmeter. Determine the source voltage (E) and phase angle for the circuit.
12. The component voltages shown in Figure 16.27b were measured using an AC voltmeter. Determine the source voltage (E) and phase angle for the circuit.

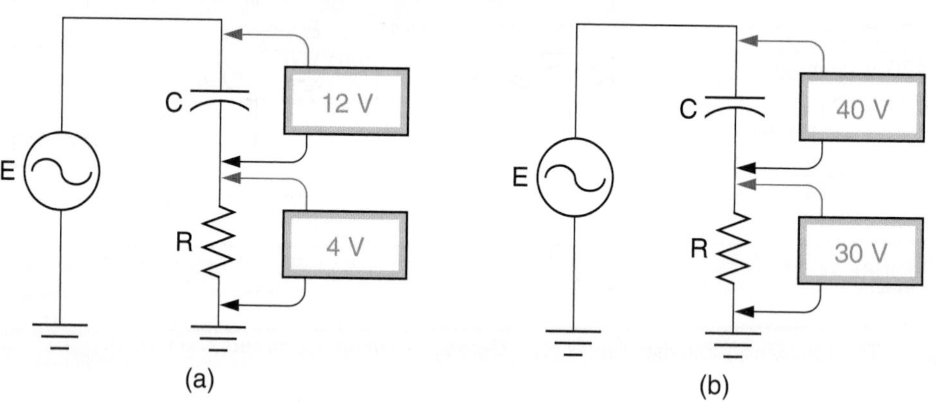

FIGURE 16.27

13. Calculate the circuit impedance and phase angle for the circuit in Figure 16.28a.
14. Calculate the circuit impedance and phase angle for the circuit in Figure 16.28b.

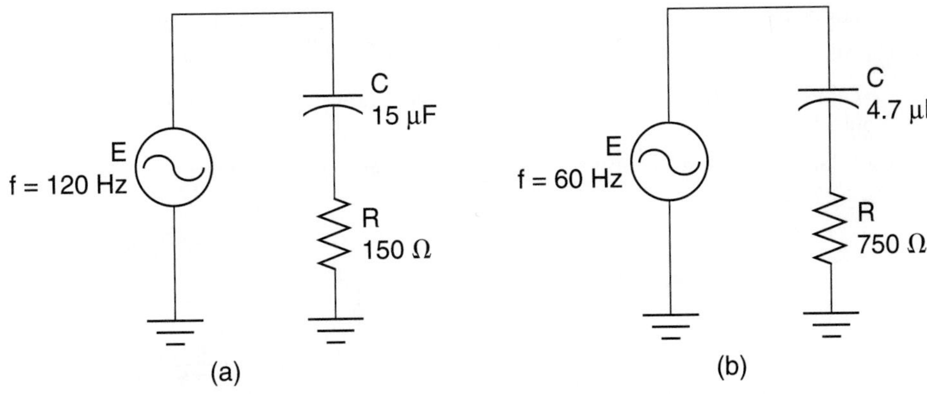

FIGURE 16.28

15. Calculate the values of Z_T, I_T, V_C, V_R, and θ for the circuit in Figure 16.29a.
16. Calculate the values of Z_T, I_T, V_C, V_R, and θ for the circuit in Figure 16.29b.

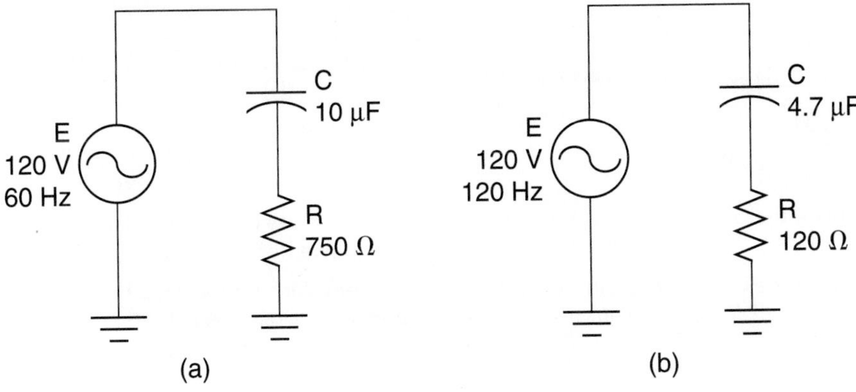

FIGURE 16.29

17. Calculate the values of Z_T, I_T, V_{C1}, V_{C2}, V_R, and θ for the circuit in Figure 16.30a.
18. Calculate the values of Z_T, I_T, V_{C1}, V_{C2}, V_{R1}, V_{R2}, and θ for the circuit in Figure 16.30b.

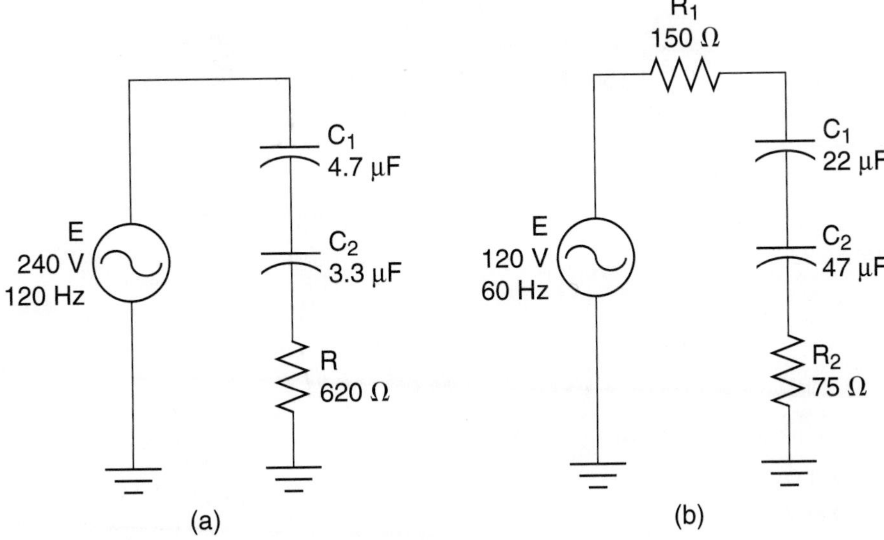

FIGURE 16.30

19. Calculate the values of resistive and reactive power for the circuit in Figure 16.31a.
20. Calculate the values of resistive and reactive power for the circuit in Figure 16.31b.
21. Calculate the magnitude and phase angle of apparent power for the circuit in Figure 16.31a.
22. Calculate the magnitude and phase angle of apparent power for the circuit in Figure 16.31b.

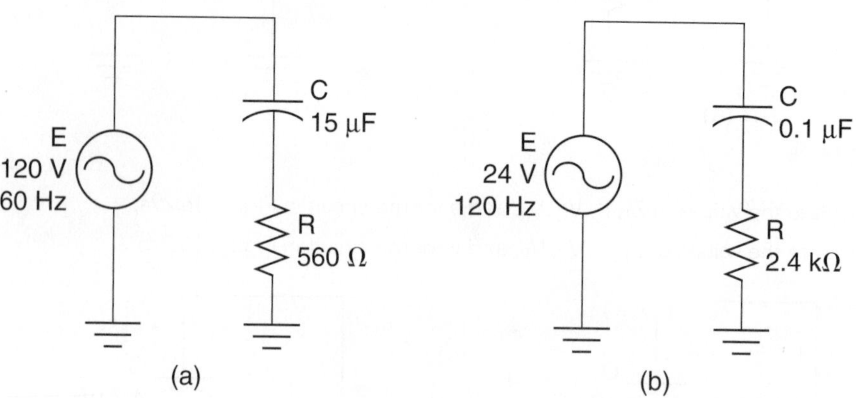

FIGURE 16.31

23. Calculate the component current, voltage, and power values for the circuit in Figure 16.32a. Then calculate the value of apparent power for the circuit.
24. Calculate the component current, voltage, and power values for the circuit in Figure 16.32b. Then calculate the value of apparent power for the circuit.

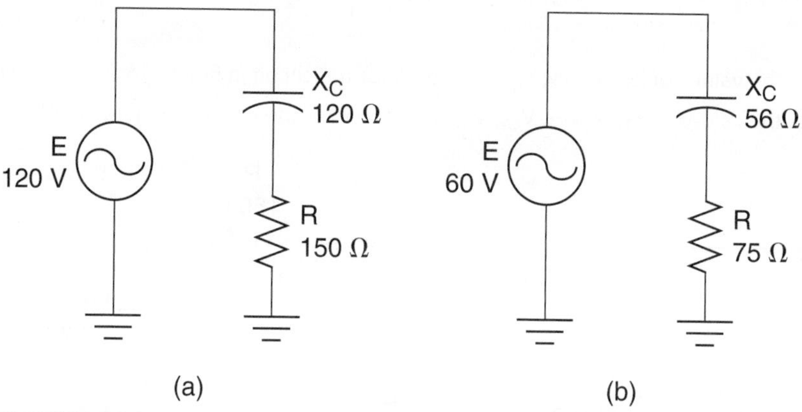

FIGURE 16.32

ANSWERS TO THE EXAMPLE PRACTICE PROBLEMS

16.1 121 Ω

16.2 1.66 A

16.3 200 V $\angle -36.9°$

16.4 $1.17\ k\Omega \angle -31.1°$

16.5 $X_C = 564\ \Omega$, $Z = 760\ \Omega$, $\theta = -47.9°$, $I = 158$ mA, $V_C = 89.1$ V, $V_R = 80.6$ V

16.6 $X_C = 2.65\ k\Omega$, $Z = 2.69$ V, $V_C = 118$ V, $V_R = 21$ V

16.7 $P_X = 9.8$ VAR, $P_R = 7.2$ W

16.8 $12.2\ VA \angle -53.7°$

16.9 0.589

16.10 $X_C = 265\ \Omega$, $Z = 447\ \Omega$, $I = 268$ mA, $V_C = 71$ V, $V_R = 96.5$ V, $\theta = -36.3°$,
$P_X = 19$ VAR, $P_R = 25.9$ W, $P_{APP} = 32.1$ VA, PF = 0.807

CHAPTER 17

PARALLEL RESISTIVE-CAPACITIVE (RC) CIRCUITS

PURPOSE

In Chapter 16, you were introduced to the AC characteristics of capacitors and series resistive-capacitive (RC) circuits. In this chapter, we conclude our study of RC circuits by discussing the operating characteristics of parallel RC circuits. As you will see, the principles of parallel RC circuits are very similar to those of the parallel RL circuits we discussed in Chapter 15.

OBJECTIVES

After completing this chapter, you should be able to:

1. Compare and contrast parallel resistive, capacitive, and resistive-capacitive (RC) circuits.
2. Calculate any current or impedance value for a parallel RC circuit.
3. Compare and contrast the values obtained from current-based and impedance-based phase angle calculations.
4. Calculate the power values for a parallel RC circuit.

17.1 PARALLEL RC CIRCUIT CURRENTS

> Parallel RC circuits are commonly used to filter "ripple" voltage from the output of a DC power supply, or to remove harmonics from the output of an AC source (among other applications).
>
> IN THE FIELD

A *parallel RC circuit* is one that contains one or more resistors in parallel with one or more capacitors. Several parallel RC circuits are shown in Figure 17.1. As you can see, none of the circuit branches contain more than one component. (If they did, we'd be dealing with a *series-parallel* RC circuit.)

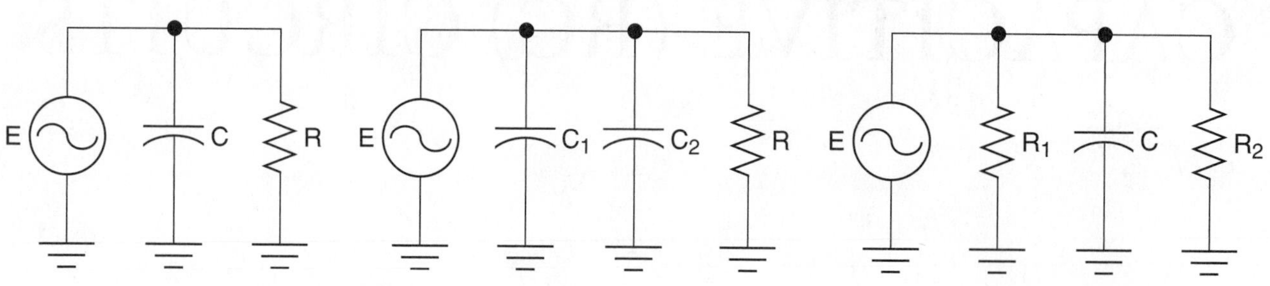

FIGURE 17.1 Parallel RC circuits.

Parallel Circuits: A Comparison

As we did with series circuits, we're going to start by comparing a parallel RC circuit to purely resistive and purely capacitive parallel circuits. A comparison of these circuits is provided in Figure 17.2.

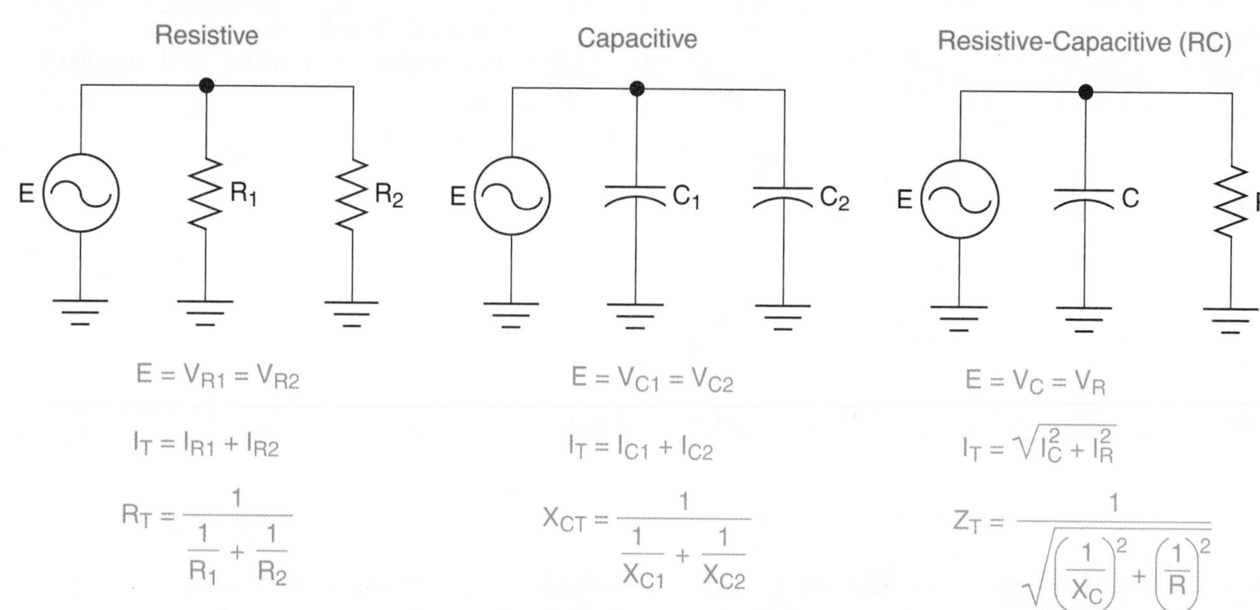

FIGURE 17.2 Parallel resistive, capacitive, and resistive-capacitive (RC) circuits.

In previous chapters, we established some fundamental characteristics of parallel-connected resistances and reactances. Among these characteristics are the following:

1. All branch voltages equal the source voltage.
2. The circuit current equals the sum of the branch currents.
3. The total impedance (resistance and/or reactance) is lower than any branch value, and can be found as shown in Figure 17.2.

The relationships listed for the resistive and capacitive circuits in Figure 17.2 should all seem familiar. In each case, all branch voltages equal the source voltage and the total circuit current equals the sum of the branch currents. The total impedance for the resistive and capacitive circuits can be found using the reciprocal or product-over-sum methods.

The current and impedance relationships shown for the parallel RC circuit in Figure 17.2 are explained in detail in this chapter. Even so, you can see that the total current in a parallel RC circuit equals the geometric sum of the branch currents. By the same token, calculating total impedance using the product over sum method requires the use of the geometric sum of X_C and R in the denominator of the equation. As you may have guessed, these geometric relationships are required because of phase angles within the circuit.

Branch Currents

As you know, current lags voltage by 90 degrees in a capacitive circuit. At the same time, current and voltage in a resistive circuit are always in phase. When combined in parallel, a resistor and a capacitor produce the waveforms shown in Figure 17.3.

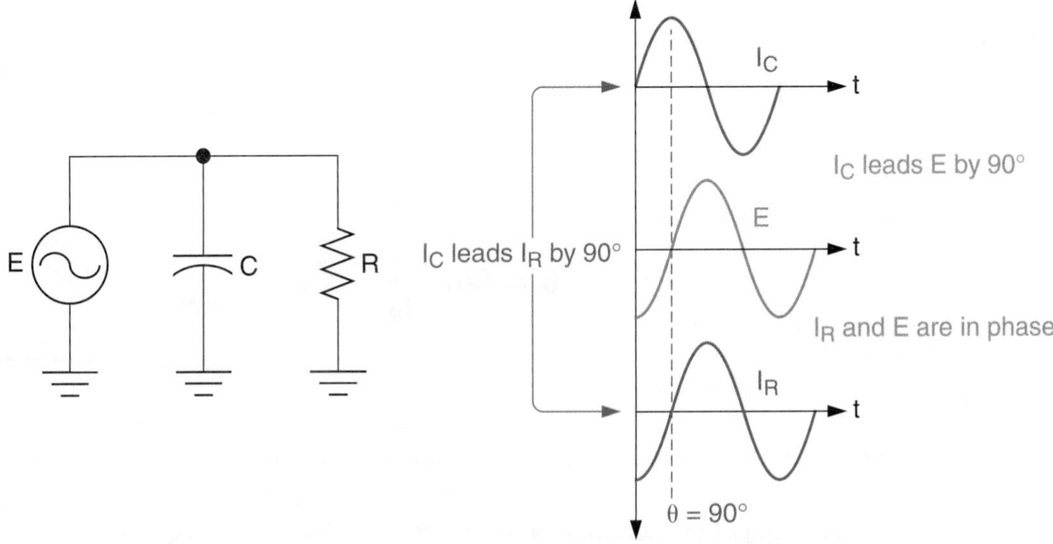

FIGURE 17.3 Parallel RC circuit current and voltage phase relationships.

Since voltage is constant across all branches in a parallel circuit, the capacitor and resistor currents are referenced to a single voltage waveform. As you can see:

1. The capacitor current *leads* the circuit voltage by 90 degrees.
2. The resistor current is in phase with the circuit voltage.

Therefore, *capacitor current leads resistor current by 90 degrees.* This phase relationship holds true for every parallel RC circuit with an AC source.

Since I_C leads I_R by 90 degrees, the values of the component currents are represented using vectors, as shown in Figure 17.4a. By plotting the value of I_R on the *x*-axis (0°) and the value of I_C on the positive *y*-axis (90°), the phase relationship between the two is represented in the graph.

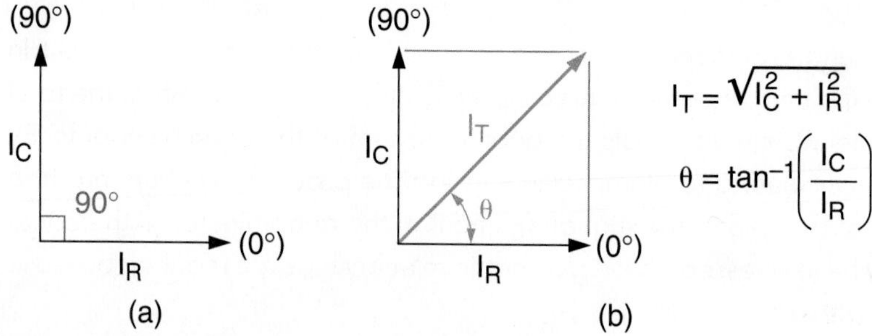

FIGURE 17.4 Current vectors for a parallel RC circuit.

As in any parallel circuit, the sum of I_C and I_R must equal the total circuit current. In this case, the vector relationship between the two values requires that they be added geometrically. By formula,

(17.1)
$$I_T = \sqrt{I_C^2 + I_R^2}$$

The *phase angle* (θ) in a parallel RC circuit is the phase difference between the circuit current and the source voltage. As shown in Figure 17.4b, the value of θ can be found as

(17.2)
$$\theta = \tan^{-1}\frac{I_C}{I_R}$$

Since voltage is equal across all branches in a parallel circuit, E is used as the reference for all phase angle measurements. Therefore, the value of θ for the circuit current indicates its phase relative to the source voltage. The magnitude and phase angle of the current in a parallel RC circuit are calculated as shown in Example 17.1.

EXAMPLE 17.1

Calculate the current through the circuit in Figure 17.5.

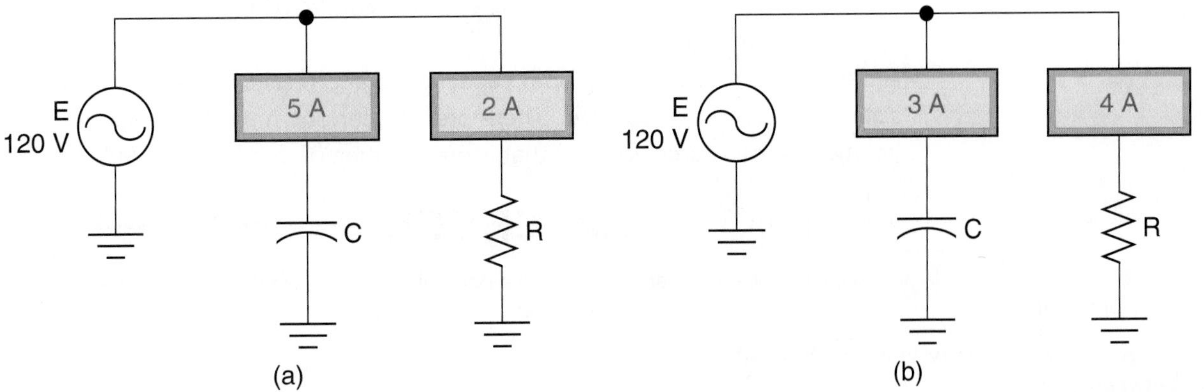

FIGURE 17.5

SOLUTION

The magnitude of the current through the circuit is found as

$$\begin{aligned} I_T &= \sqrt{I_C^2 + I_R^2} \\ &= \sqrt{(5\text{ A})^2 + (2\text{ A})^2} \\ &= 5.39\text{ A} \end{aligned}$$

The current phase angle (relative to the source voltage) can be found as

$$\begin{aligned} \theta &= \tan^{-1}\left(\frac{I_C}{I_R}\right) \\ &= \tan^{-1}\left(\frac{5\text{ A}}{2\text{ A}}\right) \\ &= \tan^{-1}(2.5) \\ &= 68.2° \end{aligned}$$

The value of θ indicates that the circuit current leads the source voltage by 68.2 degrees.

PRACTICE PROBLEM 17.1

Calculate the value of I_T for the circuit in Figure 17.5b.

For any parallel RC circuit, the limits on the phase angle between circuit current and source voltage are given as follows:

$$0° < \theta < 90° \tag{17.3}$$

This range of values for θ makes sense when you consider that:

1. The current in a purely resistive circuit is in phase with the source voltage ($\theta = 0°$).
2. The current in a purely capacitive circuit leads the source voltage by 90 degrees ($\theta = 90°$).

A parallel circuit containing both capacitance and resistance is neither *purely resistive* nor *purely capacitive*. Therefore, its current must have a phase angle (relative to the source voltage) that is greater than 0° and less than 90°.

PROGRESS CHECK

1. What are the three fundamental characteristics of any parallel circuit?
2. In a parallel RC circuit, what is the phase relationship between I_C and I_R?
3. In a parallel RC circuit, what does the current phase angle represent?

17.2 PARALLEL RC CIRCUIT IMPEDANCE

As stated earlier, the impedance in an RC circuit is the combined opposition to current provided by the resistance and the reactance. The easiest way to calculate the total impedance in a parallel RC circuit begins with calculating the circuit current. Once the circuit current is known, Ohm's law states that the circuit impedance can be found as

(17.4)

$$Z_T = \frac{E}{I_T}$$

This approach to calculating the value of Z_T is demonstrated in Example 17.2.

EXAMPLE 17.2

Refer to Example 17.1. What is the circuit impedance in Figure 17.5a?

SOLUTION

The total impedance in the circuit can be found as

$$Z_T = \frac{E}{I_T}$$

$$= \frac{120 \text{ V}}{5.39 \text{ A}}$$
$$= 22.3 \ \Omega$$

PRACTICE PROBLEM 17.2

Calculate the total impedance for the circuit in Figure 17.5b.

Some Observations Based on the Circuit in Example 17.2

In every parallel circuit we have discussed, the total impedance has been lower than any branch impedance value. This holds true for resistive-reactive circuits as well. For example, the branch impedance values in Figure 17.5a can be found as

$$X_C = \frac{E}{I_C}$$
$$= \frac{120 \text{ V}}{5 \text{ A}}$$
$$= 24 \ \Omega$$

and

$$R = \frac{E}{I_R}$$
$$= \frac{120 \text{ V}}{2 \text{ A}}$$
$$= 60 \ \Omega$$

Each of these values is greater than the value of Z_T calculated in the example (22.3 Ω).

Calculating Parallel-Circuit Impedance

In Chapter 14, you were told that parallel reactances are calculated in the same way as parallel resistances. The commonly used equations for calculating parallel values of R and X_C are provided (as a reference) in Table 17.1.

Because there is a phase relationship between X_C and R, the equal-value branches method of calculating total impedance cannot be used on a parallel RC circuit. At the same time, the phase angles of X_C and R make the reciprocal method somewhat complicated. For this reason, the *product-over-sum* approach to calculating circuit impedance is preferred for parallel RC circuits. Using the product-over-sum format, the magnitude of parallel RC circuit impedance can be found as

TABLE 17.1 • Parallel-Circuit Impedance Equations for Two-Component Resistive and Capacitive Circuits.

METHOD	RESISTIVE FORM	CAPACITIVE FORM
Reciprocal:	$R_T = \dfrac{1}{\dfrac{1}{R_1} + \dfrac{1}{R_2}}$	$X_{CT} = \dfrac{1}{\dfrac{1}{X_{C1}} + \dfrac{1}{X_{C2}}}$
Product-over-sum:	$R_T = \dfrac{R_1 R_2}{R_1 + R_2}$	$X_{CT} = \dfrac{X_{C1} X_{C2}}{X_{C1} + X_{C2}}$
Equal-value branches:	$R_T = \dfrac{R}{n}$	$X_{CT} = \dfrac{X_C}{n}$

(where n = the number of branches in the circuit)

(17.5)

$$Z_T = \frac{X_C R}{\sqrt{X_C^2 + R^2}}$$

Note that the denominator of the fraction contains the geometric sum of the variables. Example 17.3 demonstrates the use of this equation.

EXAMPLE 17.3

What is the value of circuit impedance in Figure 17.6a?

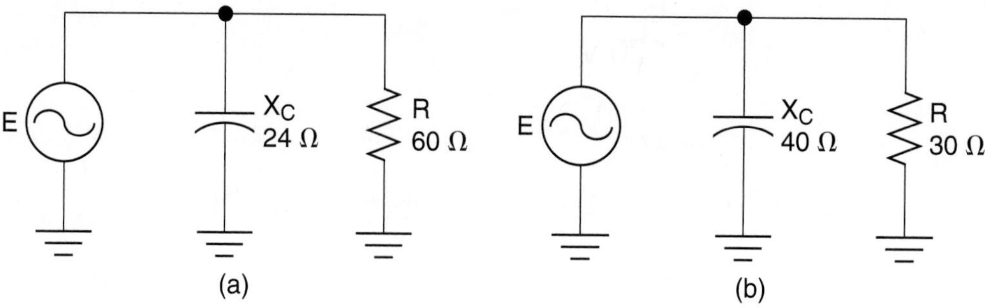

FIGURE 17.6

SOLUTION

The value of Z_T for the circuit is found as

$$Z_T = \frac{X_C R}{\sqrt{X_C^2 + R^2}}$$

$$= \frac{(24\ \Omega)(60\ \Omega)}{\sqrt{(24\ \Omega)^2 + (60\ \Omega)^2}}$$

$$= 22.3\ \Omega$$

To provide a basis for comparison, the values used in this example match the values we calculated for the circuit in Example 17.2. If you compare the results in these two examples, you'll see that they are the same.

PRACTICE PROBLEM 17.3

Calculate the value of Z_T for the circuit in Figure 17.6b. Compare your answer to the one obtained in Practice Problem 17.2.

Calculating the Voltage Phase Angle Using R and X_C

The voltage phase angle of a parallel RC circuit can be found as

(17.6)

$$\theta = \tan^{-1}\frac{R}{-X_C}$$

As you can see, the fraction in equation (17.6) is the *reciprocal* of the one used to calculate the phase angle in a series RC circuit. The use of the equation is demonstrated in Example 17.4.

EXAMPLE 17.4

What is the phase angle for the circuit in Figure 17.6a?

SOLUTION

The circuit phase angle is found as

$$\theta = \tan^{-1}\left(\frac{R}{-X_C}\right)$$

$$= \tan^{-1}\left(\frac{60\ \Omega}{-24\ \Omega}\right)$$

$$= \tan^{-1}(-2.5)$$

$$= -68.2^\circ$$

PRACTICE PROBLEM 17.4

Calculate the phase angle of the circuit impedance for the circuit in Figure 17.6b.

In Example 17.1, we calculated the current phase angle of a parallel RC circuit using the values of circuit current. Using equation (17.2), we obtained a value of $\theta = 68.2°$. Then, we calculated the voltage phase angle for the same circuit in Example 17.4. Using the values of R and X_C, we obtained a value of $\theta = -68.2°$. While these results may appear to contradict each other, they don't:

1. When calculating θ using I_C and I_R, the result indicates the phase of the circuit current relative to the source voltage. The result is positive because the circuit current leads the source voltage.
2. When calculating θ using R and X_C, the result indicates the phase of the source voltage relative to the circuit current. The result is negative because the source voltage lags the circuit current.

As these statements indicate, the results of the two examples are saying the same thing.

PROGRESS CHECK

1. What is the simplest approach to calculating the value of Z_T for a parallel RC circuit?
2. When $X_C = R$ in a parallel RC circuit, can the *equal-value branches* approach to calculating Z_T be used? Explain your answer.
3. Why does equation (17.2) yield a positive value of θ while equation (17.6) yields a negative value of θ?

17.3 PARALLEL CIRCUIT ANALYSIS

In the previous sections, you were shown how current and impedance are calculated for a parallel RC circuit. Parallel circuit power calculations are identical to those for series circuits, so they have already been covered as well. At this point, we are going to tie all of these calculations together by working through two circuit analysis problems.

A *capacitive filter* is a circuit that is used to smooth out any voltage variations in the output from a DC power supply. If the load resistance and capacitance are both large, the ripple voltage (variations in the power supply output voltage) will be small, resulting in a smooth (constant value) output.

KEY CONCEPT

EXAMPLE 17.5

Calculate the current, impedance, and power values for the circuit in Figure 17.7a.

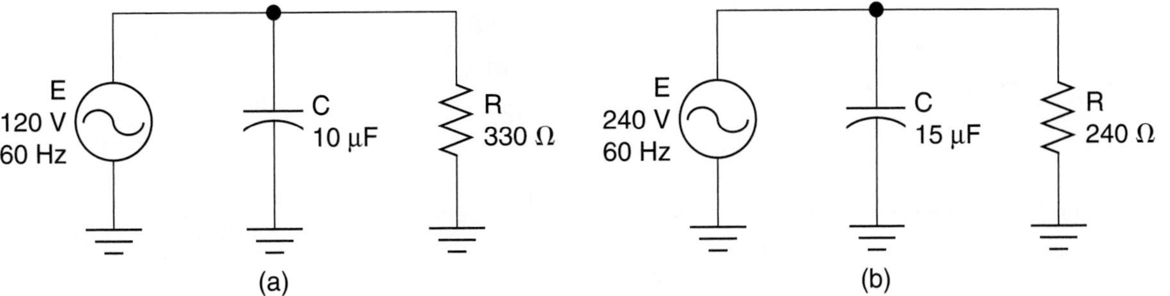

FIGURE 17.7

SOLUTION

An analysis of the circuit begins with calculating the value of X_C, as follows:

$$X_C = \frac{1}{2\pi fC}$$

$$= \frac{1}{2\pi(60\text{ Hz})(10\ \mu\text{F})}$$

$$= 265\ \Omega$$

The branch currents can now be found as

$$I_C = \frac{E}{X_C}$$

$$= \frac{120\text{ V}}{265\ \Omega}$$

$$= 453\text{ mA}$$

and

$$I_R = \frac{E}{R}$$

$$= \frac{120\text{ V}}{300\ \Omega}$$

$$= 400\text{ mA}$$

The magnitude of the circuit current can now be found as

$$I_T = \sqrt{I_L^2 + I_R^2}$$

$$= \sqrt{(453\text{ mA})^2 + (400\text{ mA})^2}$$

$$= 604\text{ mA}$$

and the current phase angle (relative to the source voltage) can be found as

$$\theta = \tan^{-1}\left(\frac{I_C}{I_R}\right)$$
$$= \tan^{-1}\left(\frac{458 \text{ mA}}{400 \text{ mA}}\right)$$
$$= \tan^{-1}(1.133)$$
$$= 48.6°$$

Once the total current is known, the circuit impedance can be found as

$$Z_T = \frac{E}{I_T}$$
$$= \frac{120 \text{ V}}{604 \text{ mA}}$$
$$= 199\ \Omega$$

The value of *reactive power* (P_X) for the circuit can be found as

$$P_X = I_C^2 X_C$$
$$= (453 \text{ mA})^2(265\ \Omega)$$
$$= 54.5 \text{ VAR}$$

and the value of *resistive power* (P_R) can be found as

$$P_R = I_R^2 R$$
$$= (400 \text{ mA})^2\ (300\ \Omega)$$
$$= 48 \text{ W}$$

Once the values of reactive and resistive power are known, the value of *apparent power* (P_{APP}) can be found as

$$P_{APP} = \sqrt{P_X^2 + P_R^2}$$
$$= \sqrt{(54.5 \text{ VAR})^2 + (48 \text{ W})^2}$$
$$= 72.6 \text{ VA}$$

Finally, the power factor for the circuit has a value of

$$PF = \frac{P_R}{P_{APP}}$$
$$= \frac{48 \text{ W}}{72.6 \text{ VA}}$$
$$= 0.6612$$

This completes the analysis of the circuit.

PRACTICE PROBLEM 17.5

Perform a complete analysis of the circuit shown in Figure 17.7b.

When a parallel RC circuit contains more than one capacitive and/or resistive branch, the analysis of the circuit takes a few more steps, as demonstrated in Example 17.6.

EXAMPLE 17.6

Calculate the current, impedance, and power values for the circuit in Figure 17.8.

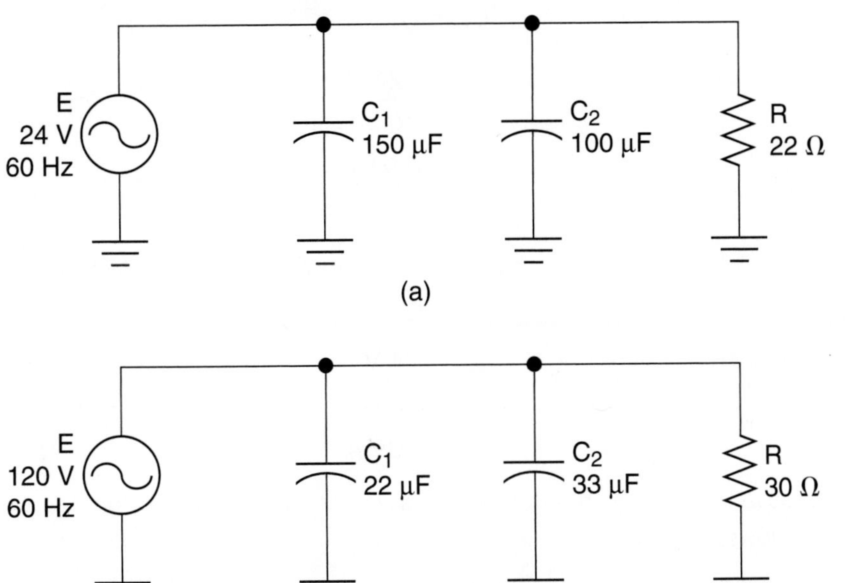

FIGURE 17.8

SOLUTION

An analysis of the circuit in Figure 17.8a begins with calculating the values of X_C as follows:

$$X_{C1} = \frac{1}{2\pi fC}$$

$$= \frac{1}{2\pi(60\text{ Hz})(150\ \mu\text{F})}$$

$$= 17.7\ \Omega$$

and

$$X_{C2} = \frac{1}{2\pi fC}$$

$$= \frac{1}{2\pi(60\text{ Hz})(100\ \mu\text{F})}$$

$$= 26.5\ \Omega$$

Now, the magnitudes of the branch currents can be found as

$$I_{C1} = \frac{E}{X_{C1}} = \frac{24\text{ V}}{17.7\ \Omega} = 1.36\text{ A}$$

$$I_{C2} = \frac{E}{X_{C2}} = \frac{24\text{ V}}{26.5\ \Omega} = 0.91\text{ A}$$

and

$$I_R = \frac{E}{R} = \frac{24\text{ V}}{22\ \Omega} = 1.09\text{ A}$$

Because I_{C1} and I_{C2} are in phase, their values can be added algebraically as follows:

$$I_{C(T)} = I_{C1} + I_{C2} = 1.36\text{ A} + 0.91\text{ A} = 2.27\text{ A}$$

Now, the circuit current is calculated as follows:

$$I_T = \sqrt{I_{C(T)}^2 + I_R^2} = \sqrt{(2.27\text{ A})^2 + (1.09\text{ A})^2} = 2.52\text{ A}$$

and the current phase angle (relative to the applied voltage) can be found as

$$\theta = \tan^{-1}\left(\frac{I_{C(T)}}{I_R}\right) = \tan^{-1}\left(\frac{2.27\text{ A}}{1.09\text{ A}}\right) = \tan^{-1}(2.083) = 64.4^\circ$$

Once the total circuit current is known, the circuit impedance can be found as:

$$Z_T = \frac{E}{I_T}$$

$$= \frac{24\text{ V}}{2.52\text{ A}}$$
$$= 9.5\ \Omega$$

The values of reactive power for the circuit can be found as:

$$P_{X1} = I_{C1}^2 X_{C1}$$
$$= (1.36\text{ A})^2 \times (17.7\ \Omega)$$
$$= 32.7\text{ VAR}$$

and

$$P_{X2} = I_{C2}^2 X_{C2}$$
$$= (0.91\text{ A})^2 \times (26.5\ \Omega)$$
$$= 21.9\text{ VAR}$$

The total reactive power can now be found as

$$P_{X(T)} = P_{X1} + P_{X2}$$
$$= 32.7\text{ VAR} + 21.9\text{ VAR}$$
$$= 54.6\text{ VAR}$$

The resistive power in the circuit can be found as

$$P_R = I_R^2 R$$
$$= (1.09\text{ A})^2(22\ \Omega)$$
$$= 26.1\text{ W}$$

Finally, using resistive power and total reactive power, the value of apparent power for the circuit can be found as

$$P_{APP} = \sqrt{P_{X(T)}^2 + P_R^2}$$
$$= \sqrt{(54.6\text{ VAR})^2 + (26.1\text{ W})^2}$$
$$= 60.5\text{ VA}$$

Finally, the circuit power factor can be calculated as follows:

$$PF = \frac{P_R}{P_{APP}}$$
$$= \frac{26.1\text{ W}}{60.5\text{ VA}}$$
$$= 0.4314$$

This completes the analysis of the circuit.

PRACTICE PROBLEM 17.6

Calculate the current, impedance, and power values for the circuit in Figure 17.8b.

One Final Note

In this chapter, you have been introduced to parallel RC circuit relationships and analysis. Once again, we have a situation where all of the relationships you learned earlier in the text apply. The primary difference is that many circuit values must be added geometrically rather than algebraically (as was the case with parallel RL circuits). For future reference, the basic characteristics of parallel RC circuits are compared to those of parallel RL circuits in Figure 17.9.

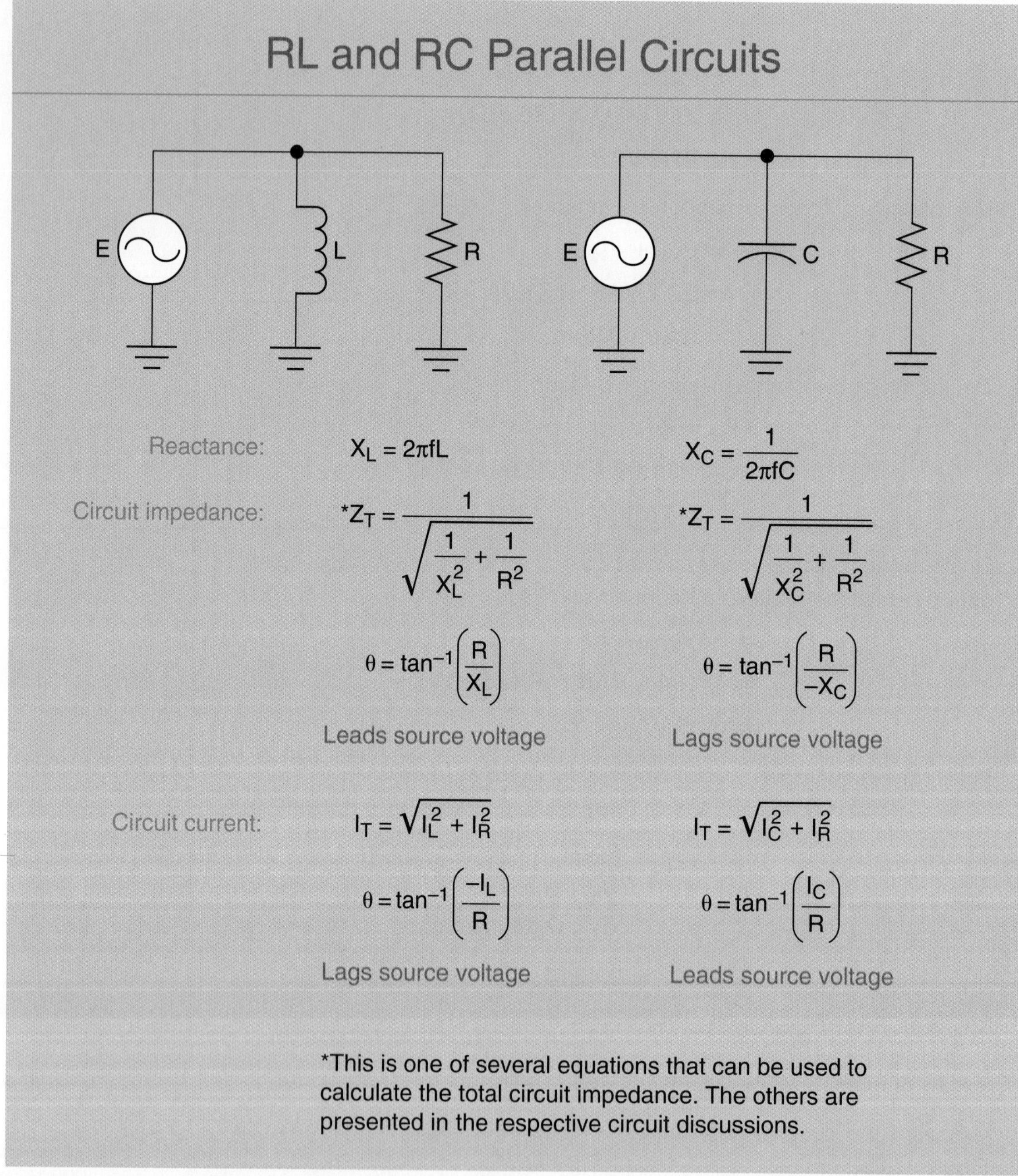

RL and RC Parallel Circuits
E
L
R
E
C
R
Reactance:
$X_L = 2\pi fL$
$X_C = \frac{1}{2\pi fC}$
Circuit impedance:
$*Z_T = \frac{1}{\sqrt{\frac{1}{X_L^2} + \frac{1}{R^2}}}$
$*Z_T = \frac{1}{\sqrt{\frac{1}{X_C^2} + \frac{1}{R^2}}}$
$\theta = \tan^{-1}\left(\frac{R}{X_L}\right)$
$\theta = \tan^{-1}\left(\frac{R}{-X_C}\right)$
Leads source voltage
Lags source voltage
Circuit current:
$I_T = \sqrt{I_L^2 + I_R^2}$
$I_T = \sqrt{I_C^2 + I_R^2}$
$\theta = \tan^{-1}\left(\frac{-I_L}{R}\right)$
$\theta = \tan^{-1}\left(\frac{I_C}{R}\right)$
Lags source voltage
Leads source voltage
*This is one of several equations that can be used to calculate the total circuit impedance. The others are presented in the respective circuit discussions.

FIGURE 17.9

17.4 SUMMARY

Here is a summary of the major points that were covered in this chapter:

Parallel RC Circuit Currents

- A parallel RC circuit is one that contains one or more resistors in parallel with one or more capacitors, each branch containing only one component.
- In a parallel RC circuit:
  - The total circuit current equals the geometric sum of the currents through the resistive and the reactive branches.
  - The capacitor current leads the source voltage by 90°.
  - The resistor current is in phase with the circuit voltage.
- The phase angle (θ) in a parallel RC circuit is the phase difference between the total current (I_T) and the source voltage (E).
  - Since voltage is assumed to have an angle of 0° in a parallel circuit, the current phase angle is always positive.
  - For a parallel RC circuit, the current phase angle (relative to the source voltage) has limits of $0° < \theta < 90°$.

Parallel RC Circuit Impedance

- The product-over-sum method for calculating total impedance is preferred for parallel RC circuits. The geometric sum of X_C and R is used in the denominator of the fraction.
- The reactance(s) and resistance(s) in a parallel RC circuit can be used to calculate the voltage phase angle of the circuit.
- The voltage and current phase angles for series and parallel RC circuits are calculated using reciprocal fractions, as follows:

Series	Parallel
$\theta = \tan^{-1}\frac{-X_C}{R}$	$\theta = \tan^{-1}\frac{R}{-X_C}$

CHAPTER REVIEW

1. A parallel RC circuit must contain only one component in each branch, otherwise it ________.
   a) is too complex to solve
   b) is a susceptive circuit
   c) is a series-parallel circuit
   d) requires calculus to solve
2. The basic rule of parallel resistive circuits that applies to all parallel RC circuits is ________.
   a) Ohm's law
   b) Kirchhoff's current law
   c) the common voltage in all branches relationship
   d) all of the above

3. When solving for total circuit current, branch currents in a parallel RC circuit must be added geometrically because ________ .
   a) resistor current leads capacitor current
   b) of the current phase difference between resistive and reactive branches
   c) of the impedance phase
   d) of the voltage phase difference between resistive and reactive branches
4. Current in the capacitive branch of a parallel RC circuit always ________ current in the resistive branch.
   a) lags by 90°
   b) lags by more than 0° and less than 90°
   c) leads by 90°
   d) leads by more than 0° and less than 90°
5. The current in the resistive branch always ________ the source voltage while the current in the capacitive branch ________.
   a) is in phase with, leads E by 90°
   b) lags by 90°, is in phase with I_T
   c) is in phase with, leads I_T by 90°
   d) is in phase with, lags E by 90°
6. The phase angle of I_T in a parallel RC circuit, relative to the applied voltage, is ________.
   a) +90°
   b) −90°
   c) between 0° and −90°
   d) between 0° and 90°
7. If you use X_C and R to solve for the phase angle, relative to the applied voltage, in a parallel RC circuit, it is ________.
   a) +90°
   b) −90°
   c) between 0° and −90°
   d) between 0° and 90°
8. When calculating the phase angle of I_T for a parallel RC circuit, the value of I_C used in the $\tan^{-1}$ equation is ________.
   a) always negative
   b) in the denominator
   c) always positive
   d) always negative and in the numerator
9. A parallel RC circuit has branch currents of $I_R = 4.1$ A and $I_C = 2.65$ A. This means that $I_T =$ ________.
   a) 4.19 A   b) 3.91 A
   c) 6.75 A   d) 4.88 A
10. The phase angle of I_T for the circuit described in question 9 is ________.
   a) 32.9°   b) 35.4°
   c) −32.9°   d) 54.6°

11. Using X_C and R, the phase angle for the circuit described in question 9 is ________.

    a) +54.6° b) −32.9°

    c) 32.9° d) −54.6°

12. The method of solving for circuit impedance that cannot be used in a parallel RC circuit is the ________ method.

    a) product-over-sum
    b) reciprocal
    c) equal-value branches
    d) all of the above
    e) none of the above

13. A 220 V @ 60 Hz supply voltage is applied to a 47 Ω resistor in parallel with a 47 μF capacitor. For this circuit, I_T = ________ at a phase angle of ________.

    a) 6.09 A, −39.8°
    b) 16.7 A, −62°
    c) 6.09 A, 39.8°
    d) 3 A, 50.2°

14. For the circuit described in question 13, Z_T = ________ at a phase angle of ________.

    a) 73.4 Ω, −50.2°
    b) 73.4 Ω, +50.2°
    c) 36.1 Ω, −39.8°
    d) 36.1 Ω, +39.8°

15. If the circuit described in question 13 were connected to a 220 V source voltage in Europe where 50 Hz supplies are used, circuit current would ________ and the phase angle would be ________ positive.

    a) increase, more
    b) decrease, less
    c) decrease, more
    d) increase, less

16. 240 V @ 60 Hz is applied to a parallel RC circuit. One branch has a 68 Ω resistor and the other has a 100 μF capacitor. The value of I_T = ________ at a phase angle of ________.

    a) 9.72 A, 68.7°
    b) 2.54 A, −36.9°
    c) 73 A, 68.7°
    d) 8.67 A, 40°

17. For the circuit described in question 16, P_X = ________ and P_R = ________.

    a) 2.17 kVAR, 847 W
    b) 13.1 kVAR, 174 W
    c) 174 kVA, 131 W
    d) 214 VA, 847 W

18. The value of P_{APP} for the circuit described in question 16 is ________.

    a) 2.99 kVA b) 2.34 kVAR

    c) 2.34 kVA d) 2.18 kVAR

19. The phase angle for the value of P_{APP} solved in question 18 equals _______.
    a) −68.4°
    b) 40.1°
    c) 53.3°
    d) 68.7°

20. The value of P_{APP} for a given parallel RC circuit is known to be 649 VA. It is also known that the value of I_T for this circuit is 5.5 A at a phase angle of 32°. This tells us that the magnitude of the applied voltage is _______ and the power factor of the circuit is _______.
    a) 120 V, 36.4%
    b) 115 V, 87.2%
    c) 118 V, 84.8%
    d) 236 V, 37.8%

21. For the circuit described in question 20, P_R = _______.
    a) 642 VA
    b) 550 W
    c) 642 W
    d) 570 W

22. Another resistive branch is added to the circuit described in question 20. This means that the power factor for the circuit will _______ and the value of P_X will _______.
    a) increase, increase
    b) decrease, decrease
    c) increase, decrease
    d) decrease, increase
    e) decrease, remain unchanged
    f) increase, remain unchanged

PRACTICE PROBLEMS

1. A parallel RC circuit has values of I_C = 3 A and I_R = 5.6 A. Calculate the magnitude and phase angle of the circuit current.
2. A parallel RC circuit has values of I_C = 1.24 A and I_R = 980 mA. Calculate the magnitude and phase angle of the circuit current.
3. Calculate the total impedance for the circuit shown in Figure 17.10a.
4. Calculate the total impedance for the circuit shown in Figure 17.10b.

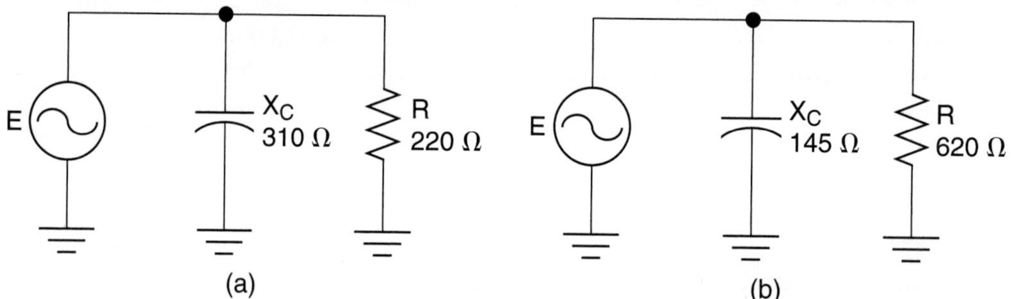

FIGURE 17.10

5. Calculate the circuit impedance and phase angle of the circuit in Figure 17.11a.
6. Calculate the circuit impedance and phase angle of the circuit in Figure 17.11b.
7. Calculate all the circuit current, impedance, phase, and power values for the circuit in Figure 17.12a.
8. Calculate all the circuit current, impedance, phase, and power values for the circuit in Figure 17.12b.

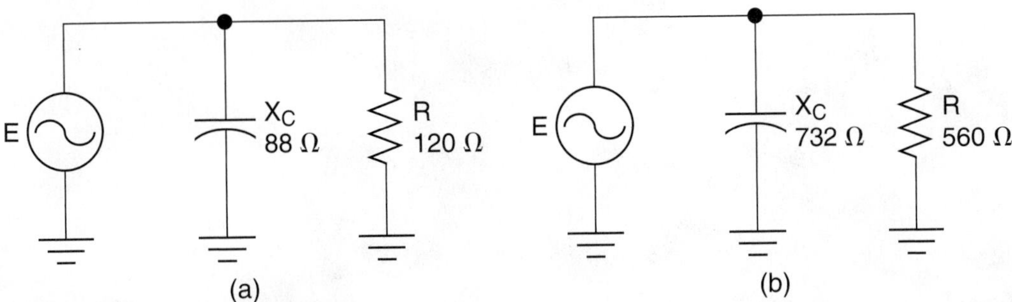

FIGURE 17.11

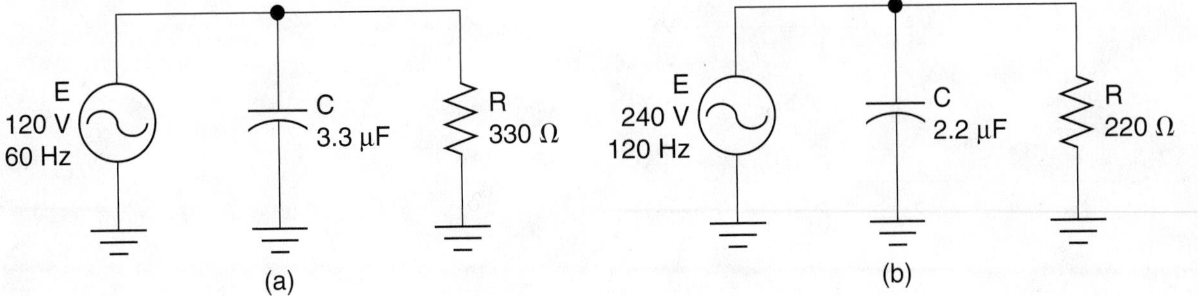

FIGURE 17.12

9. Calculate all the circuit current, impedance, phase, and power values for the circuit in Figure 17.13a.
10. Calculate all the circuit current, impedance, phase, and power values for the circuit in Figure 17.13b.

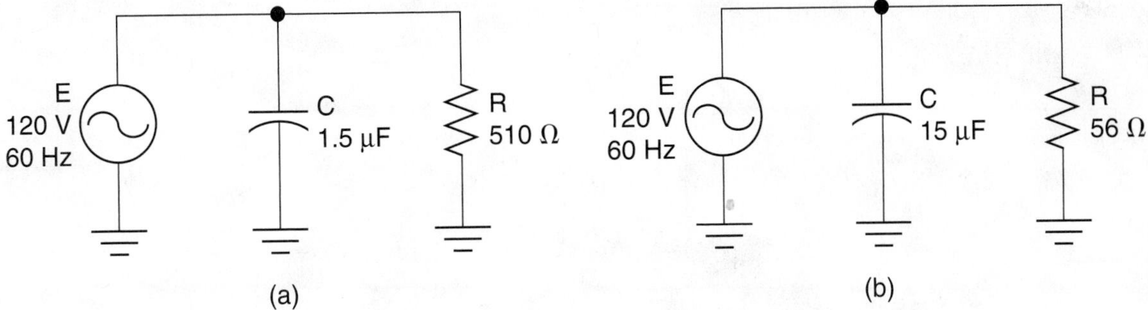

FIGURE 17.13

ANSWERS TO THE EXAMPLE PRACTICE PROBLEMS

17.1 5 A, 36.9°

17.2 24 Ω

17.3 24 Ω. The two values are the same.

17.4 −36.9°

17.5 $X_C = 177\ \Omega$, $I_C = 1.36$ A, $I_R = 1$ A, $I_T = 1.69$ A, $\theta = 53.7°$, $Z = 142\ \Omega$, $P_X = 327$ VAR, $P_R = 240$ W, $P_{APP} = 406$ VA, PF = 0.591

17.6 $X_{C1} = 121\ \Omega$, $X_{C2} = 80.4\ \Omega$, $I_{C1} = 0.992$ A, $I_{C2} = 1.49$ A, $I_R = 4$ A, $I_{C(T)} = 2.48$ A, $I_T = 4.71$ A, $\theta = 31.8°$, $Z = 25.5\ \Omega$, $P_{X1} = 119$ VAR, $P_{X2} = 178$ VAR, $P_{X(T)} = 297$ VAR, $P_R = 480$ W, $P_{APP} = 564$ VA, PF = 0.851

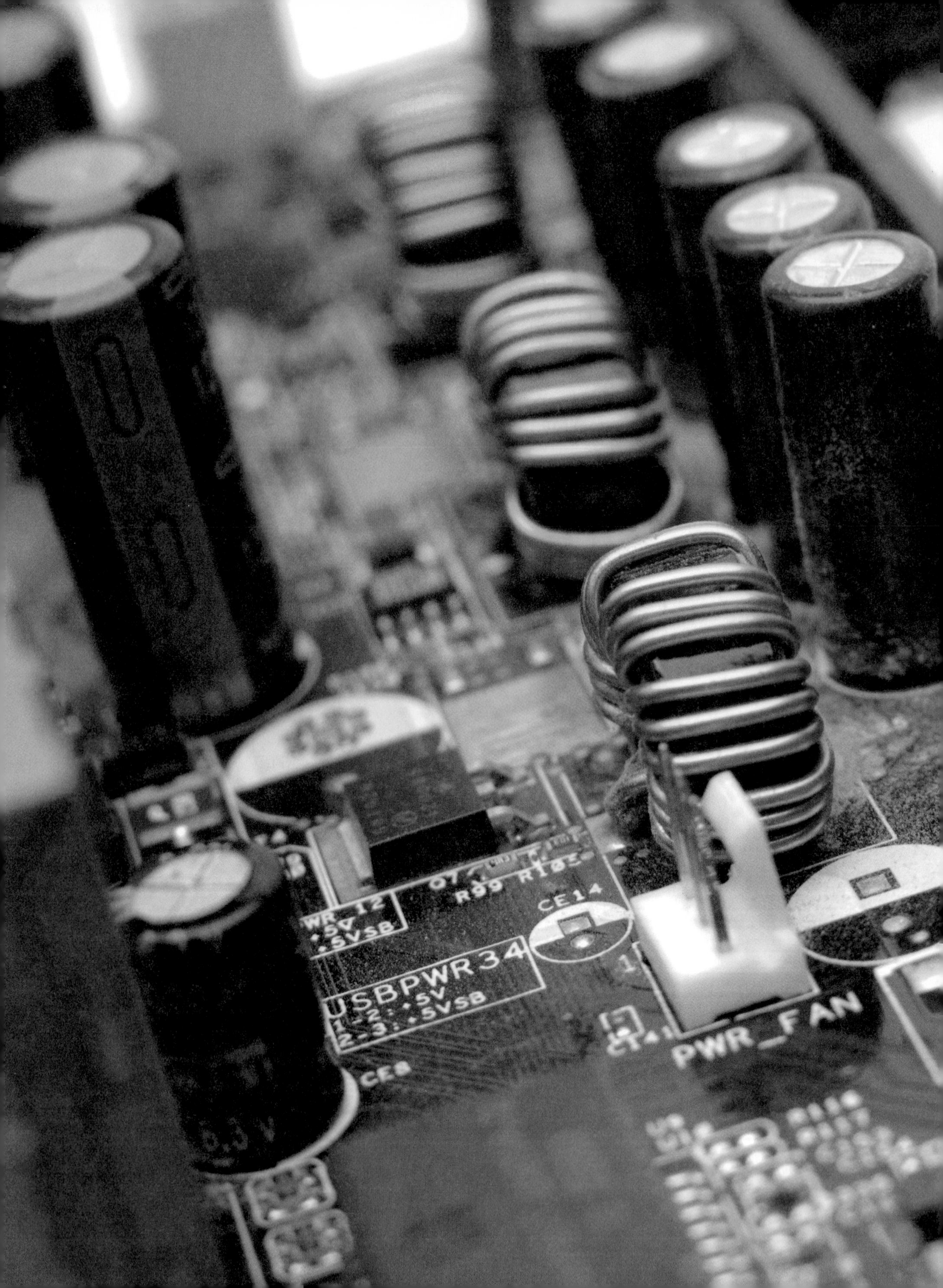

CE14
R99
USBPWR34
1-2: +5V
2-3: +5VSB
PWR_FAN
CE8

CHAPTER 18

RLC CIRCUITS

PURPOSE

Now that we have discussed RL and RC circuits, it is time to combine them into RLC (resistive-inductive-capacitive) circuits. We will start our discussion of RLC circuits with the operating characteristics of LC (inductive-capacitive) circuits.

KEY TERMS

The following terms are introduced and defined in this chapter on the pages indicated:

Bulk correction, 784
Centralized correction, 784
Compensating capacitor, 780
Harmonic, 785
Intelligent power factor controller (IPFC), 784
kVAR-sensitive controller, 784
Lagging VAR, 776
Leading VAR, 777
Power factor correction, 777
Resonance, 756
Resonant frequency (f_r), 756
Static compensation, 780
Static correction 780

OBJECTIVES

After completing this chapter, you should be able to:

1. Discuss the relationship between the component voltages and the operating characteristics of a series LC circuit.
2. Discuss the relationship between the reactance values and operating characteristics of a parallel LC circuit.
3. Calculate the reactance and voltage values for a series LC circuit.
4. Discuss the relationship between the branch currents and the operating characteristics of a parallel LC circuit.
5. Calculate the current and reactance values for a parallel LC circuit.
6. Describe resonance and calculate the resonant frequency of an LC circuit.
7. Describe and analyze the operation of series and parallel resonant LC circuits.
8. Describe and analyze the operation of any series RLC circuit.
9. Describe and analyze the operation of any parallel RLC circuit.
10. Describe power factor correction and the methods commonly used to achieve it.

18.1 SERIES LC CIRCUITS

When an inductor is connected in series with a capacitor, the circuit has the following current and voltage characteristics:

- The inductor and capacitor currents are equal ($I_L = I_C$).
- The inductor and capacitor voltages are 180° out of phase.

The first characteristic needs no clarification. We are dealing with a series circuit, so the component current values are always equal. The second principle can be explained with the help of Figure 18.1.

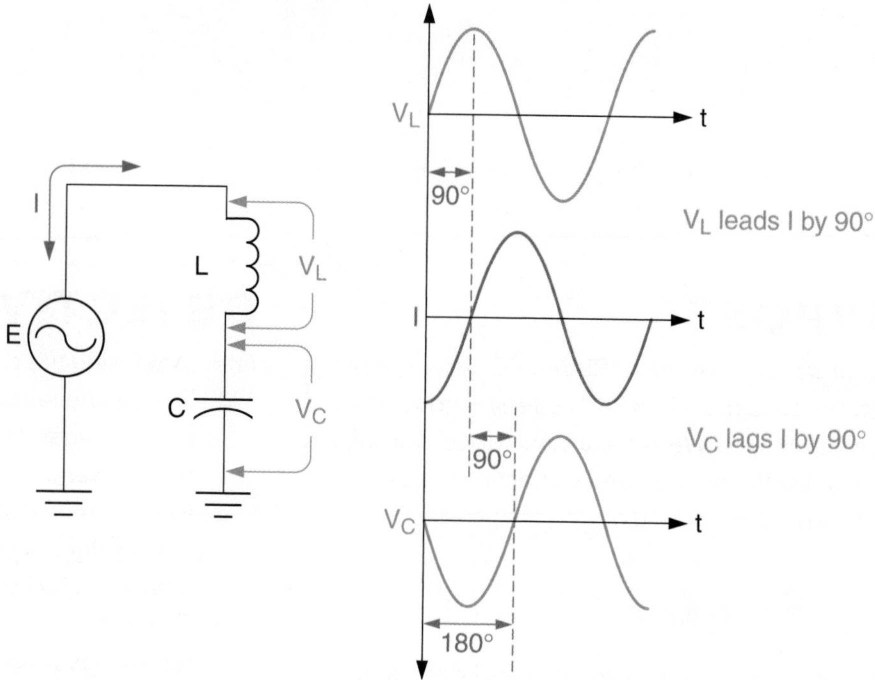

FIGURE 18.1 Phase relationships in a series LC circuit.

When an inductor and capacitor are connected in series, the waveforms in Figure 18.1 are generated. Because the component currents are equal ($I_L = I_C$), the voltage waveforms are referenced to a single current waveform. As shown in the figure:

- The inductor voltage is *leading* the circuit current by 90°.
- The capacitor voltage is *lagging* the circuit current by 90°.

As a result, there is a 180° difference in phase between the two voltage waveforms. Since they are out of phase by 180°, the total voltage equals the difference between V_L and V_C. By formula,

$$E = V_L - V_C \tag{18.1}$$

This equation states that the source voltage must equal the difference between the inductor and capacitor voltages.

Voltage Relationships and Phase Angles

There are three possible relationships between the inductor voltage (V_L) and the capacitor voltage (V_C) in any series LC circuit:

- $V_L > V_C$
- $V_L < V_C$
- $V_L = V_C$

Depending on the relationship between the inductor and capacitor voltages, the circuit has the characteristics of an inductive circuit, a capacitive circuit, or a short (low-resistance) circuit. This principle is illustrated in Figure 18.2. In the graph shown, the circuit current is plotted on the positive *x*-axis. Inductor voltage leads the circuit current by 90°, so V_L is plotted on the positive *y*-axis. By the same token, capacitor voltage lags the circuit current by 90°, so it is plotted on the negative *y*-axis. When plotted as shown in the figure, the voltage vectors are 180° out of phase (as are the component voltages).

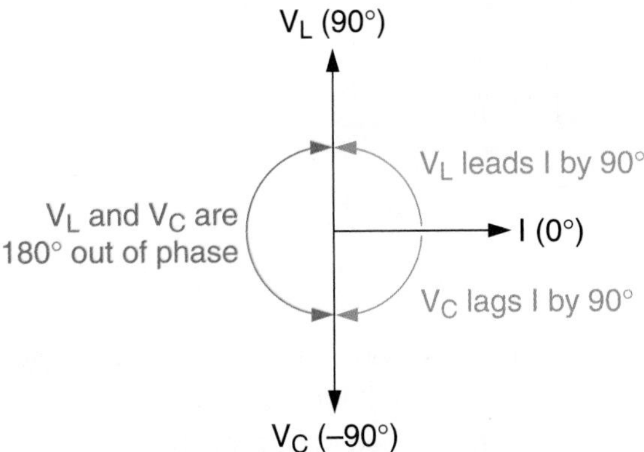

FIGURE 18.2 Vectors for a series LC circuit.

When V_L and V_C are combined, they form a result vector. The direction of the result vector is determined by the relationship between the two values. For example, Figure 18.3a shows the result vector when $V_L > V_C$. In this case, $V_L = 60$ V and $V_C = 20$ V. The result is a 40 V vector that lies on the positive *y*-axis. When the result vector lies on the positive *y*-axis:

- The circuit has inductive characteristics.
- The source voltage leads the circuit current by 90°.

When $V_L < V_C$, the result vector falls on the negative *y*-axis as shown in Figure 18.3b. In this case, $V_L = 10$ V and $V_C = 24$ V. The result is a 14 V vector that falls on the negative *y*-axis. When the result vector lies on the negative *y*-axis:

- The circuit has capacitive characteristics.
- The source voltage lags the circuit current by 90°.

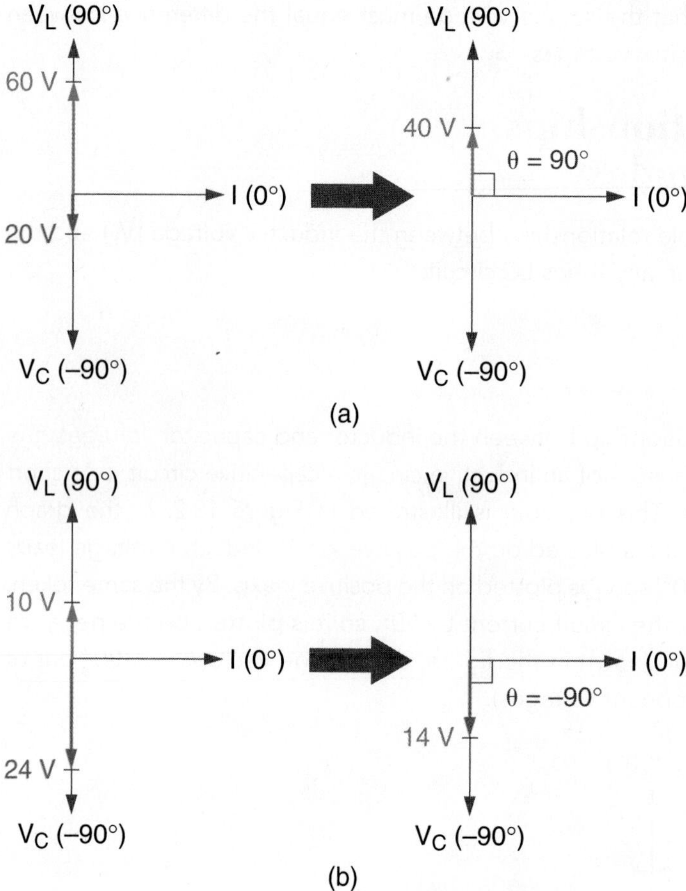

FIGURE 18.3 Example voltage vector sums for a series LC circuit.

When $V_L = V_C$, the 180° phase relationship between the two voltages causes them to cancel out, effectively shorting out the voltage source. This "special case" condition, called *series resonance*, is introduced in Section 18.3. For now, we're going to simplify our discussion by limiting ourselves to cases where $V_L \neq V_C$.

The result of equation (18.1) indicates whether the circuit characteristics are capacitive or inductive. When $V_L > V_C$

- The result of equation (18.1) is *positive*.
- The circuit characteristics are *inductive*.
- The source voltage leads the circuit current by 90°.

When $V_C > V_L$,

- The result of equation (18.1) is *negative*.
- The circuit characteristics are *capacitive*.
- The source voltage lags the circuit current by 90°.

These relationships are illustrated in Examples 18.1 and 18.2.

EXAMPLE 18.1

Are the characteristics of the circuit in Figure 18.4 inductive or capacitive?

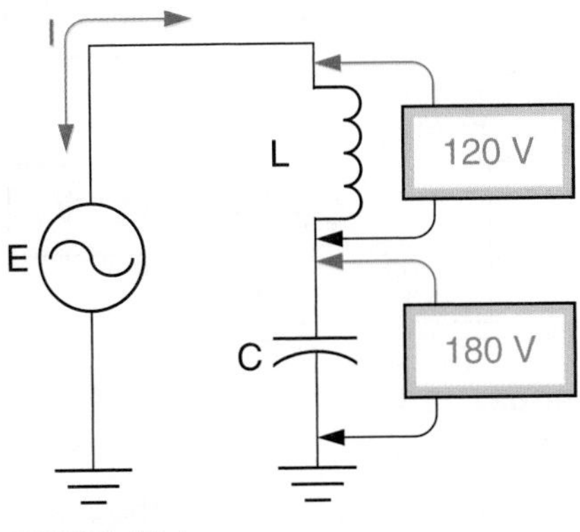

FIGURE 18.4

SOLUTION

The magnitude of the voltage source in Figure 18.4 is found as

$$E = V_L - V_C$$
$$= 120\text{ V} - 180\text{ V}$$
$$= -60\text{ V}$$

We have a negative result, so we know that the result vector lies on the negative *y*-axis and the circuit characteristics are capacitive. That is, the source voltage lags the circuit current by 90°.

PRACTICE PROBLEM 18.1

A circuit like the one in Figure 18.4 has values of $V_L = 120$ V and $V_C = 150$ V. Determine the magnitude and phase angle of the source voltage.

NOTE: The characteristics of a series LC circuit like the one in Figure 18.4 can be determined by simply looking at the component voltages. The component with the highest voltage (the capacitor in this case) determines the overall characteristics of the circuit.

The result in Example 18.1 indicates that the circuit is capacitive in nature. Example 18.2 demonstrates the phase relationship between E and circuit current when $V_L > V_C$.

EXAMPLE 18.2

Are the characteristics of the circuit in Figure 18.5 inductive or capacitive?

SOLUTION

The source voltage in Figure 18.5 is found as

$$E = V_L - V_C$$
$$= 100\text{ V} - 30\text{ V}$$
$$= 70\text{ V}$$

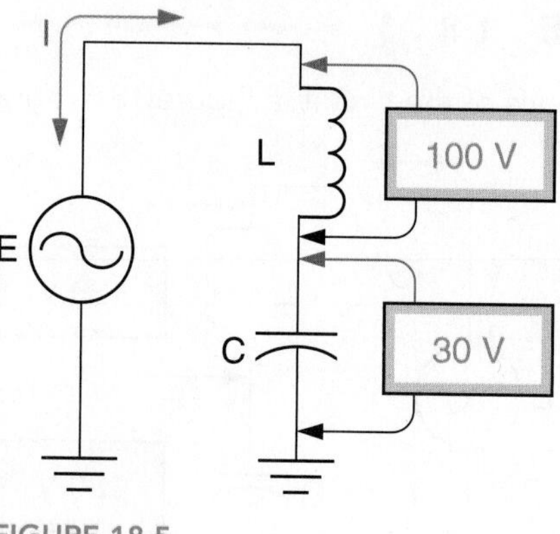

FIGURE 18.5

We have a positive result, so we know that the result vector lies on the positive *y*-axis and the circuit characteristics are inductive. That is, the source voltage leads the circuit current by 90°.

Net Series Reactance (X_S)

When an inductor and capacitor are connected in series, the *net series reactance* (X_S) equals the difference between X_L and X_C. By formula,

(18.2)
$$X_S = X_L - X_C$$

When you compare this relationship to equation (18.1), you see that the component reactances combine in the same fashion as the component voltages.

As was the case with component voltages, the relationship between X_L and X_C indicates whether the circuit characteristics are inductive, capacitive, or those of a short (low-resistance) circuit. When $X_L > X_C$,

- The result of equation (18.2) is *positive.*
- The circuit characteristics are *inductive.*
- The source voltage leads current by 90°.

When $X_C > X_L$,

- The result of equation (18.2) is *negative.*
- The circuit characteristics are *capacitive.*
- The source voltage lags current by 90°.

If the inductive and capacitive reactances are equal, X_S is 0 Ω and the circuit has the characteristics of a short circuit.

The case of $X_L = X_C$ is introduced in Section 18.3. For now, we're going to limit our discussion to cases where $X_L \neq X_C$. Example 18.3 demonstrates the circuit characteristics when $X_L > X_C$.

EXAMPLE 18.3

Are the characteristics of the circuit in Figure 18.6 inductive or capacitive?

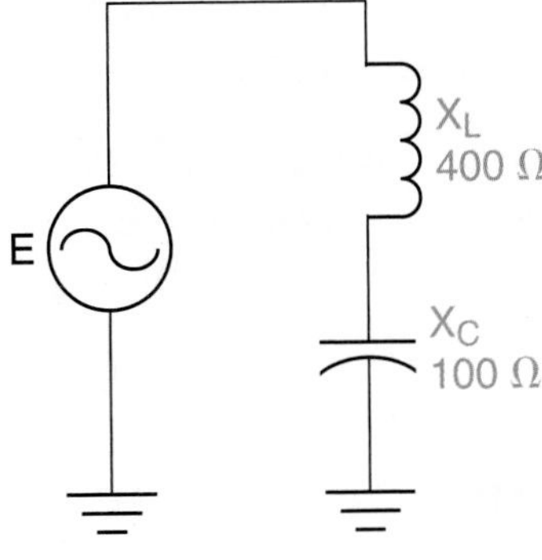

FIGURE 18.6

SOLUTION

The circuit in Figure 18.6 has a net series reactance of

$$X_S = X_L - X_C$$
$$= 400\ \Omega - 100\ \Omega$$
$$= 300\ \Omega$$

Since we obtained a positive result, we know that the result vector lies on the positive *y*-axis and the the circuit characteristics are inductive. That is, the source voltage lags the circuit current by 90°.

PRACTICE PROBLEM 18.3

A circuit like the one in Figure 18.6 has values of $X_L = 80\ \Omega$ and $X_C = 55\ \Omega$. Determine the magnitude of X_S and the voltage phase angle.

The result in Example 18.3 indicates that the circuit characteristics are inductive when $X_L > X_C$. Example 18.4 demonstrates the circuit characteristics when $X_L < X_C$.

EXAMPLE 18.4

Are the characteristics of the circuit in Figure 18.7 inductive or capacitive?

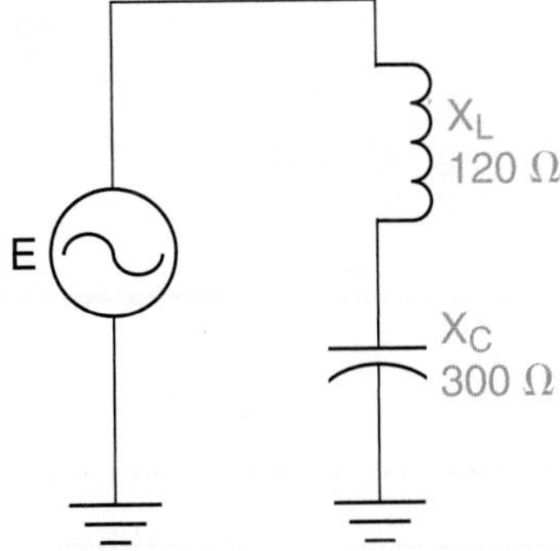

FIGURE 18.7

The circuit in Figure 18.7 has a net series reactance of

$$\begin{aligned} X_S &= X_L - X_C \\ &= 120\ \Omega - 300\ \Omega \\ &= -180\ \Omega \end{aligned}$$

We have a negative result, so we know that the result vector lies on the negative *y*-axis and the circuit characteristics are capacitive. That is, the source voltage lags the circuit current by 90°.

Putting It All Together: Series LC Circuit Characteristics

In this section, we have established the basic voltage and reactance characteristics of series LC circuits. These characteristics are summarized in Table 18.1.

TABLE 18.1 • Series LC Circuit Conditions When $X_L \neq X_C$

REACTANCE RELATIONSHIP	CIRCUIT CHARACTERISTICS
$X_L > X_C$	X_S is positive. The circuit has inductive characteristics. The source voltage leads the circuit current by 90°.
$X_C > X_L$	X_S is negative. The circuit has capacitive characteristics. The source voltage lags the circuit current by 90°.

As you learned in Chapters 13 and 15, the values of X_L and X_C are determined (in part) by the operating frequency of the circuit. In Section 18.3, we will discuss the frequency characteristics of LC circuits. First, we will establish the current and reactance characteristics of *parallel* LC circuits.

PROGRESS CHECK

1. List the current and voltage characteristics of a series LC circuit.
2. Explain the phase relationship between series values of V_L and V_C.
3. Describe the phase angle of source voltage (relative to circuit current) when:
   a. $V_L > V_C$
   b. $V_C > V_L$
4. Describe the phase relationship between source voltage and circuit current in a series LC circuit when:
   a. $X_L > X_C$
   b. $X_C > X_L$

18.2 PARALLEL LC CIRCUITS

When an inductor and a capacitor are connected in parallel (as shown in Figure 18.8), the circuit has the following current and voltage characteristics:

- The inductor and capacitor voltages are equal ($V_L = V_C$).
- The inductor and capacitor currents are 180° out of phase.

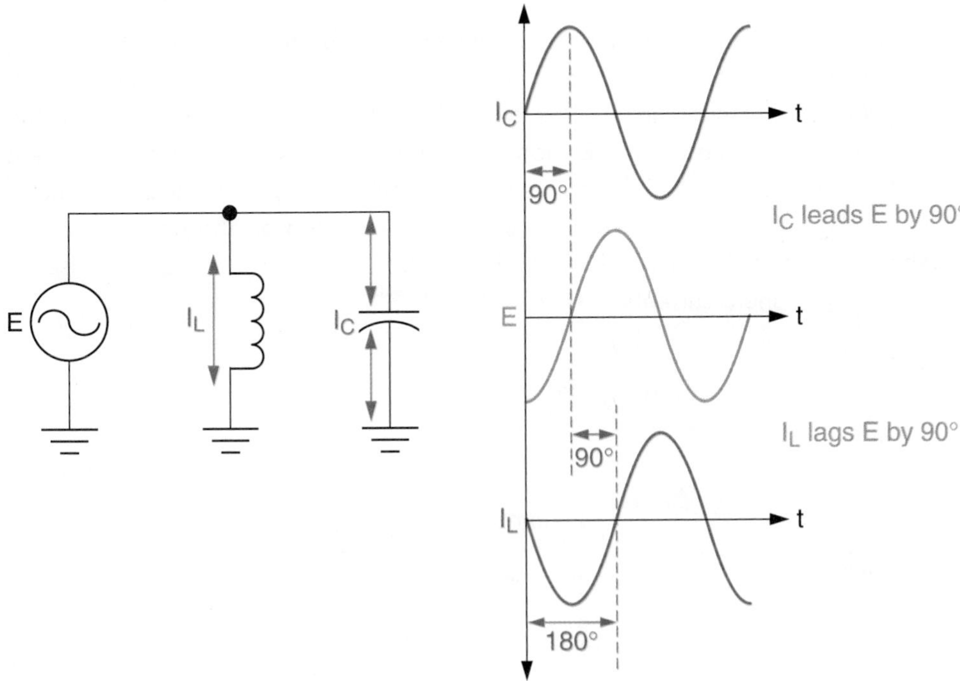

FIGURE 18.8 Current phase relationships in a parallel LC circuit.

The first principle should be familiar by now. We are dealing with a parallel circuit, so the component voltages must all be equal. The second principle can be explained with the help of Figure 18.8.

When an AC source is connected to a parallel LC circuit, the waveforms in Figure 18.8 are generated. As you can see:

- Capacitor current leads capacitor voltage by 90°.
- Inductor current lags inductor voltage by 90°.

As a result, there is a 180° phase difference between the two current waveforms. Because they are out of phase by 180°, the total current is equal to the difference between I_C and I_L. By formula,

(18.3)
$$I_T = I_C - I_L$$

Current Relationships and Phase Angles

There are three possible relationships between capacitor current and inductor current in a parallel LC circuit:

- $I_C > I_L$
- $I_C < I_L$
- $I_C = I_L$

Depending on the relationship between the capacitor and inductor currents, the circuit has the characteristics of an inductive circuit, a capacitive circuit, or an open (high-resistance) circuit. This principle is illustrated in Figure 18.9. In the graph shown, source voltage is plotted on the positive *x*-axis. Capacitor current leads source voltage by 90°, so I_C is represented on the positive *y*-axis. Inductor current lags source voltage by 90°, so it is represented on the negative *y*-axis. When plotted as shown in the figure, the current vectors are 180° out of phase (as are the component currents).

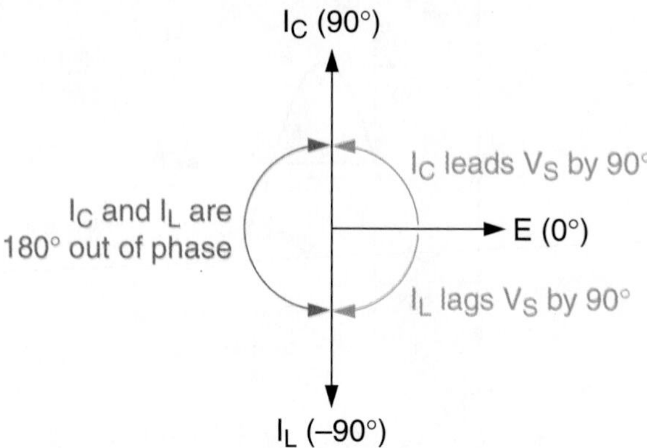

FIGURE 18.9 Current vectors for a parallel LC circuit.

When I_L and I_C are combined, they form a *result vector.* The direction of the result vector is determined by the relationship between the two values. For example, Figure 18.10a shows the result vector when $I_C > I_L$. In this case, $I_C = 8$ A and $I_L = 5$ A. The result is a 3 A vector that lies on the positive *y*-axis. When the result vector lies on the positive *y*-axis:

- The circuit has capacitive characteristics.
- The circuit current leads the source voltage by 90°.

When $I_C > I_L$, the result vector falls on the negative *y*-axis, as shown in Figure 18.10b. In this case, $I_C = 2$ A and $I_L = 6$ A. The result is a 4 A vector that falls on the negative *y*-axis. When the result vector lies on the negative *y*-axis:

- The circuit has inductive characteristics.
- The circuit current lags the source voltage by 90°.

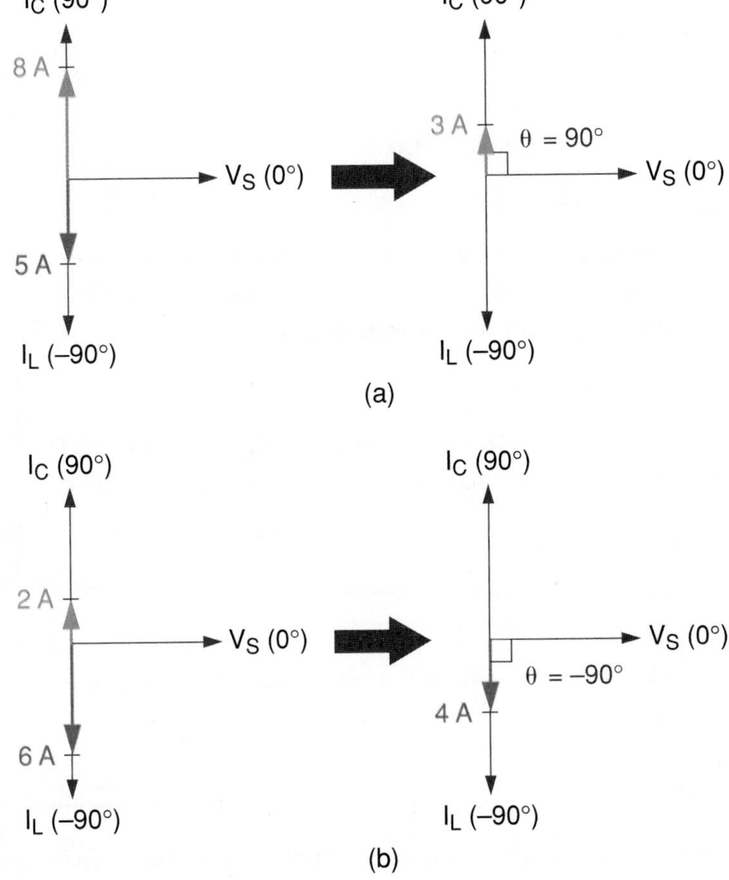

FIGURE 18.10 Example current vector sums for a parallel LC circuit.

When $I_C = I_L$, the 180° phase relationship between the two currents causes them to cancel out. As a result, $I_T = 0$ A and the circuit acts as an open. This "special case" condition, called *parallel resonance*, is introduced in Section 18.3.

The result of equation (18.3) indicates whether the circuit characteristics are capacitive or inductive. When $I_C > I_L$, the result of equation (18.3) is positive and the circuit characteristics are capacitive. When $I_L > I_C$, the result of equation (18.3) is negative and the circuit characteristics are inductive.

EXAMPLE 18.5

Are the characteristics of the circuit in Figure 18.11 inductive or capacitive?

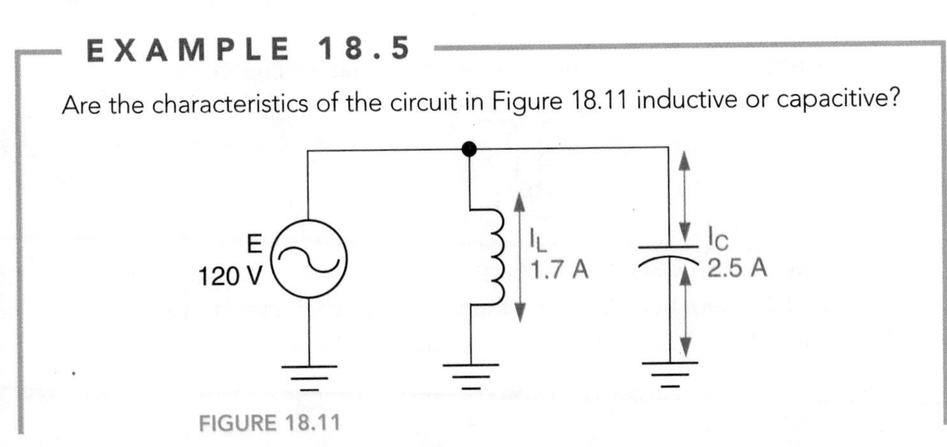

FIGURE 18.11

NOTE: The characteristics of a parallel LC circuit like the one in Figure 18.11 can be determined by simply looking at the branch currents. The component with the highest current (the capacitor in this case) determines the overall characteristics of the circuit.

SOLUTION

The current through the circuit in Figure 18.11 has a value of

$$
\begin{aligned}
I_T &= I_C - I_L \\
&= 2.5\text{ A} - 1.7\text{ A} \\
&= 0.8\text{ A}
\end{aligned}
$$

The result of the I_T calculation is positive, so we know that the result vector lies on the positive *y*-axis and the the circuit characteristics are capacitive. That is, the circuit current leads the source voltage by 90°.

PRACTICE PROBLEM 18.5

A circuit like the one in Figure 18.11 has values of $I_C = 120$ mA and $I_L = 75$ mA. Determine the magnitude and phase angle of the source voltage.

The result in Example 18.5 indicates that the circuit has capacitive characteristics when $I_C > I_L$. Example 18.6 demonstrates the phase relationship between circuit current and E when $I_L > I_C$.

EXAMPLE 18.6

Are the characteristics of the circuit in Figure 18.12 inductive or capacitive?

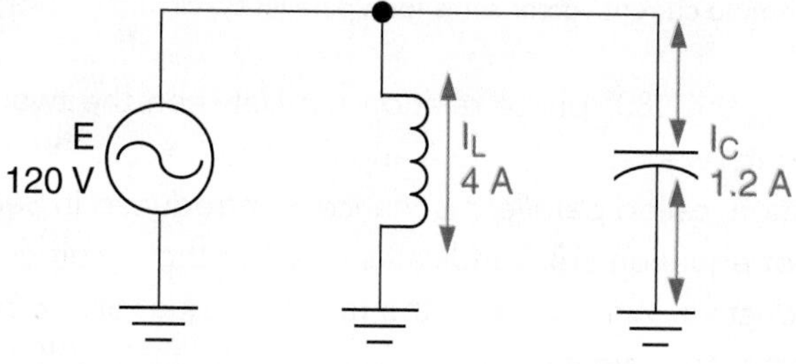

FIGURE 18.12

SOLUTION

The current through the circuit in Figure 18.12 has a value of

$$
\begin{aligned}
I_T &= I_C - I_L \\
&= 1.2\text{ A} - 4\text{ A} \\
&= -2.8\text{ A}
\end{aligned}
$$

The result of the I_T calculation is negative, so we know that the result vector lies on the negative *y*-axis and the the circuit characteristics are inductive. That is, the circuit current lags the source voltage by 90°.

Net Parallel Reactance (X_P)

Because of the unique phase relationship between I_C and I_L, the net reactance of a parallel LC circuit can be greater than either (or both) of the branch reactance values. For example, the circuit in Figure 18.12 has a value of E = 120 V. Using Ohm's law, the reactance values for the circuit can be found as

$$X_L = \frac{E}{I_L} = \frac{120\text{ V}}{4\text{ A}} = 30\ \Omega$$

and

$$X_C = \frac{E}{I_C} = \frac{120\text{ V}}{1.2\text{ A}} = 100\ \Omega$$

If we divide the source voltage by the source current found in Example 18.6, we get a total circuit reactance of

$$X_P = \frac{E}{I_T} = \frac{120\text{ V}}{2.8\text{ A}} = 42.9\ \Omega$$

This value is greater than the value of $X_L = 30\ \Omega$ that was calculated for the circuit. Don't forget: The net reactance of a parallel LC circuit is always greater than the value of one or both of the branch reactances.

When a capacitor and an inductor are connected in parallel, the total parallel reactance (X_P) can be found as

(18.4)

$$X_P = \frac{1}{\frac{1}{X_L} + \frac{1}{X_C}}$$

or

(18.5)
$$X_P = \frac{X_L X_C}{X_L + X_C}$$

The forms of these equations should be familiar to you by now. However, there is one catch: For the equations to work, capacitive reactance must be entered as a *negative* value to account for the difference between the inductive and capacitive phase angles. The X_P equations are solved as shown in Example 18.7.

EXAMPLE 18.7

What is the net parallel reactance (X_P) for the circuit in Figure 18.13?

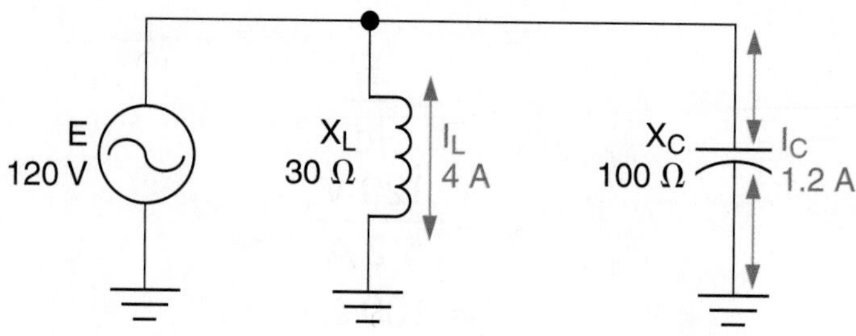

FIGURE 18.13

SOLUTION

Figure 18.13 is Figure 18.12 with added values of source voltage and component reactances. The net parallel reactance for the circuit can be found as

$$X_P = \frac{1}{\frac{1}{X_L} + \frac{1}{X_C}}$$
$$= \frac{1}{\frac{1}{30\ \Omega} + \frac{1}{-100\ \Omega}}$$
$$= \frac{1}{\frac{1}{30\ \Omega} - \frac{1}{100\ \Omega}}$$
$$= 42.9\ \Omega$$

or

$$X_P = \frac{X_L X_C}{X_L + X_C}$$
$$= \frac{(30\ \Omega)(-100\ \Omega)}{(30\ \Omega) + (-100\ \Omega)}$$

$$= \frac{-3{,}000}{-70}\ \Omega$$

$$= 42.9\ \Omega$$

As you can see, both equations provide the same result as the Ohm's law calculations we made earlier.

PRACTICE PROBLEM 18.7

A circuit like the one in Figure 18.13 has the following values: $E = 120$ V, $I_L = 3$ A, and $I_C = 1.8$ A. Calculate the value of X_P using Ohm's law, equation (18.4), and equation (18.5).

When using equation (18.4) or (18.5), the sign of the result indicates the phase relationship between source voltage and current. For example, in Figure 18.13, the value of X_P is positive. This indicates that:

- The circuit characteristics are inductive.
- Circuit current lags the source voltage by 90°.

When equation (18.4) or (18.5) yields a negative result:

- The circuit characteristics are capacitive.
- The circuit current leads the source voltage by 90°.

This phase relationship is demonstrated in Example 18.8.

EXAMPLE 18.8

What is the net parallel reactance (X_P) for the circuit in Figure 18.14?

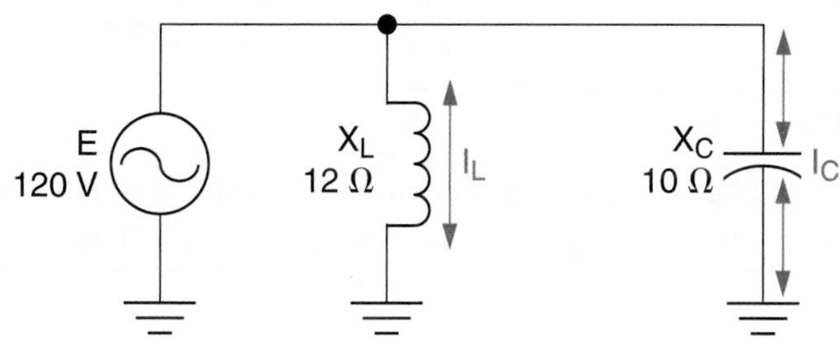

FIGURE 18.14

SOLUTION

The value of X_P for the circuit in Figure 18.14 can be found as

$$X_P = \frac{1}{\frac{1}{X_L} + \frac{1}{X_C}}$$

$$= \frac{1}{\frac{1}{12\ \Omega} + \frac{1}{-10\ \Omega}}$$

$$= \frac{1}{\frac{1}{12\ \Omega} - \frac{1}{10\ \Omega}}$$

$$= -60\ \Omega$$

or

$$X_P = \frac{X_L X_C}{X_L + X_C}$$

$$= \frac{(12\ \Omega)(-10\ \Omega)}{(12\ \Omega) + (-10\ \Omega)}$$

$$= \frac{-120}{2}\ \Omega$$

$$= -60\ \Omega$$

The negative result indicates that the circuit characteristics are capacitive. For this circuit, $X_P = 60\ \Omega$ and the circuit current leads the source voltage by 90°.

Putting It All Together: Basic Parallel LC Circuit Characteristics

The relationships among the currents, voltages, and reactances in a parallel LC circuit can be clarified by comparing the values in Examples 18.7 and 18.8. For convenience, the values of interest are summarized as follows:

EXAMPLE	I_L	I_C	I_T	X_L	X_C	X_P
18.7	4.0 A	1.2 A	Inductive	30 Ω	100 Ω	42.9 Ω
18.8	10 A	12 A	Capacitive	12 Ω	10 Ω	60 Ω

In Example 18.7, we concluded that the circuit is inductive in nature. This conclusion is supported by the fact that the inductor current is greater than the capacitor current. Therefore, the total current (which is equal to the difference between the two) must be inductive. In Example 18.8, we concluded that the circuit was capacitive in nature. This conclusion is supported by the fact that capacitor current is greater than inductor current. Therefore, the total current must be capacitive. The point being made is this: The greater branch current determines the overall nature of the circuit. When $I_L > I_C$, the circuit is inductive, and the circuit current lags the source voltage by 90°. When $I_C > I_L$, the circuit is capacitive, and the circuit current leads the source voltage by 90°.

In terms of reactance relationships, the characteristics of a parallel LC circuit can be summarized as shown in Table 18.2.

TABLE 18.2 • Parallel LC Circuit Conditions When $X_L \neq X_C$

REACTANCE RELATIONSHIP	CIRCUIT CHARACTERISTICS
$X_C < X_L$	$I_C > I_L$ The calculated value of X_P is negative. The circuit is capacitive in nature. Circuit current leads E by 90°.
$X_L < X_C$	$I_L > I_C$ The calculated value of X_P is positive. The circuit is inductive in nature. Circuit current lags E by 90°.

The relationships listed here can be verified by reviewing the calculated values summarized earlier in this section.

PROGRESS CHECK

1. List the current and voltage characteristics of a parallel LC circuit.
2. Explain the phase relationship between parallel values of I_L and I_C.
3. Describe the phase angle of circuit current (relative to source voltage) when:
   a. $I_L > I_C$
   b. $I_C > I_L$
4. State and explain the phase relationship between parallel values of X_L and X_C.
5. Describe the phase angle of X_P (relative to source voltage) when:
   a. $X_L > X_C$
   b. $X_C > X_L$
6. Describe the phase relationship between circuit current and source voltage in a parallel LC circuit when:
   a. $X_L > X_C$
   b. $X_C > X_L$

18.3 RESONANCE

Values of inductive reactance (X_L) and capacitive reactance (X_C) depend in part on circuit operating frequency. As operating frequency increases,

- X_L increases
- X_C decreases

These relationships are illustrated by the reactance curves in Figure 18.15.

RESONANCE For an LC circuit, the operating state where $X_L = X_C$.

RESONANT FREQUENCY (F_R) For an LC circuit, the frequency at which $X_L = X_C$.

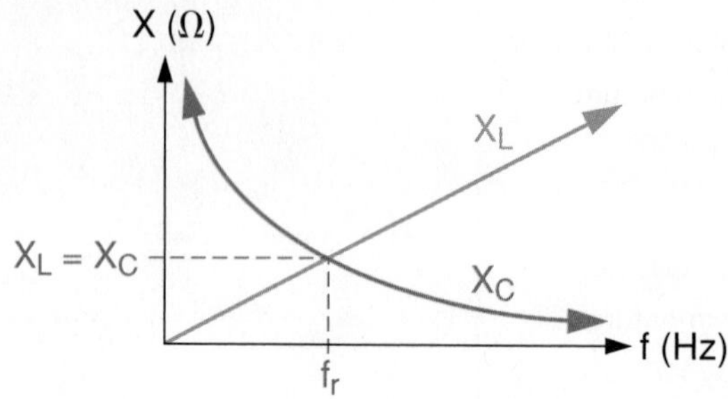

FIGURE 18.15

STAYING SAFE Resonant circuits have many useful functions. At the same time, a resonant circuit may contain relatively high voltages and currents. Always practice caution when working with an LC circuit that is operating at (or near) its resonant frequency.

For every LC circuit, there is a specific frequency where $X_L = X_C$ (as shown in Figure 18.15). When $X_L = X_C$, an LC circuit is said to be operating at **RESONANCE**. The frequency at which resonance occurs for an LC circuit is referred to as its **RESONANT FREQUENCY (f_r)**. The resonant frequency of a given LC circuit is found as

(18.6)

$$f_r = \frac{1}{2\pi\sqrt{LC}}$$

where

f_r = the frequency at which $X_L = X_C$

L, C = the values of inductance and capacitance in the circuit

Resonant circuits are used to respond selectively to signals that fall within a given range of frequencies while rejecting all other frequencies. When a resonant circuit passes a narrower range of frequencies, the circuit is said to have higher *selectivity*.

KEY CONCEPT

The resonant frequency for an LC circuit is calculated as shown in Example 18.9.

EXAMPLE 18.9

What is the resonant frequency (f_r) for the circuit in Figure 18.16?

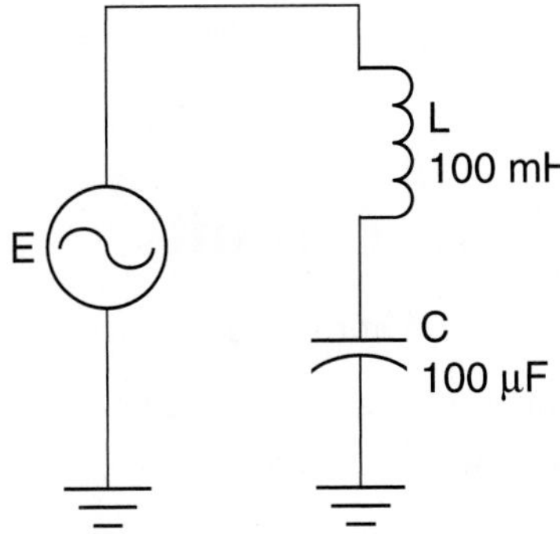

FIGURE 18.16

SOLUTION

The resonant frequency for the circuit is found as

$$f_r = \frac{1}{2\pi\sqrt{LC}}$$

$$= \frac{1}{2\pi\sqrt{(100 \text{ mH})(100 \ \mu\text{F})}}$$

$$= \frac{1}{2\pi\sqrt{10 \times 10^{-6}}}$$

$$= 50.3 \text{ Hz}$$

PRACTICE PROBLEM 18.9

A circuit like the one in Figure 18.16 has values of $L = 330$ mH and $C = 47$ μF. Calculate the resonant frequency of the circuit.

By definition, f_r is the frequency at which $X_L = X_C$. When operated at its resonant frequency, the values of X_L and X_C for the circuit in Figure 18.16 can be found as

$$X_L = 2\pi f L$$

$$= 2\pi(50.3 \text{ Hz})(100 \text{ mH})$$

$$= 31.6 \ \Omega$$

and

$$X_C = \frac{1}{2\pi f C}$$

$$= \frac{1}{2\pi(50.3 \text{ Hz})(100 \ \mu\text{F})}$$

$$= 31.6 \ \Omega$$

As you can see, the circuit values of X_L and X_C are equal at the resonant frequency.

The principles of resonance also apply to parallel LC circuits. That is, every parallel LC circuit

- Has a resonant frequency, found as $f_r = \frac{1}{2\pi\sqrt{LC}}$.
- Has equal values of X_L and X_C when operated at its resonant frequency.

Series Resonant LC Circuits

A series LC circuit that is operating at resonance has the following characteristics:

- The net series reactance of the circuit is 0 Ω.
- The voltage across the series LC circuit is 0 V.
- The circuit voltage and current are in phase; that is, the circuit has *resistive* characteristics.

These characteristics can be explained using the circuit and graph shown in Figure 18.17.

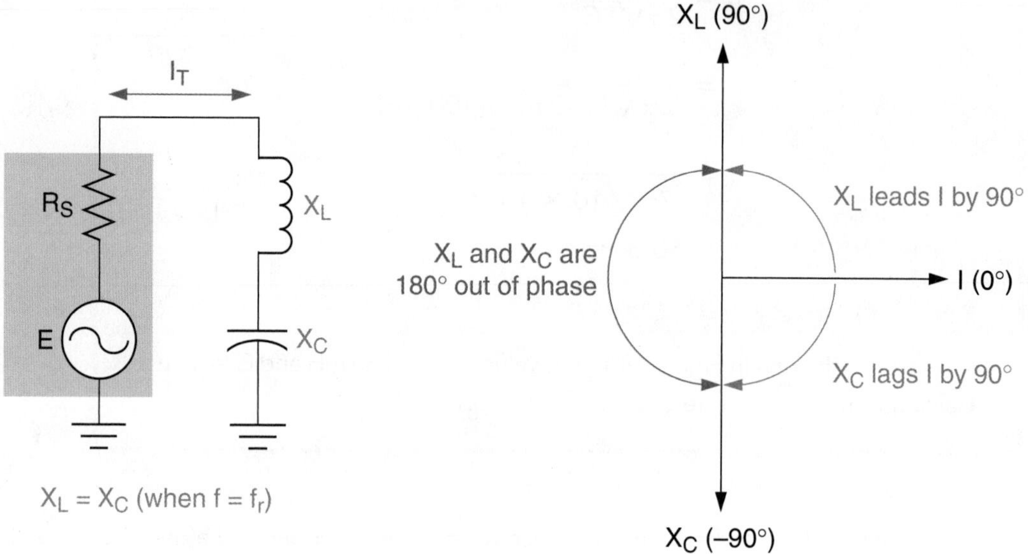

FIGURE 18.17 Series resonant circuit characteristics.

Assuming that the circuit in Figure 18.17 is operating at resonance, the values of X_L and X_C are equal. As shown in the reactance graph, these two values are 180° out of phase. Therefore, the net series reactance (X_S) in the circuit has a value of

$$X_S = X_L - X_C$$
$$= 0\ \Omega$$

Since $X_S = 0\ \Omega$, the voltage across the series combination of L and C must be 0 V. This means the source voltage (E) in Figure 18.17 must be dropped across its own internal resistance (R_S). Therefore, R_S limits the circuit current, and

- The circuit is resistive.
- The source voltage and circuit current are in phase.

The fact that $V_{LC} = 0$ V may lead you to believe that $V_L = 0$ V and $V_C = 0$ V. However, this is not the case. The values of V_L and V_C may be greater than the source voltage. However, because of their phase relationship, the sum of the two voltages is 0 V. This point is demonstrated in Example 18.10.

EXAMPLE 18.10

Calculate the capacitor and inductor voltages for the circuit in Figure 18.18.

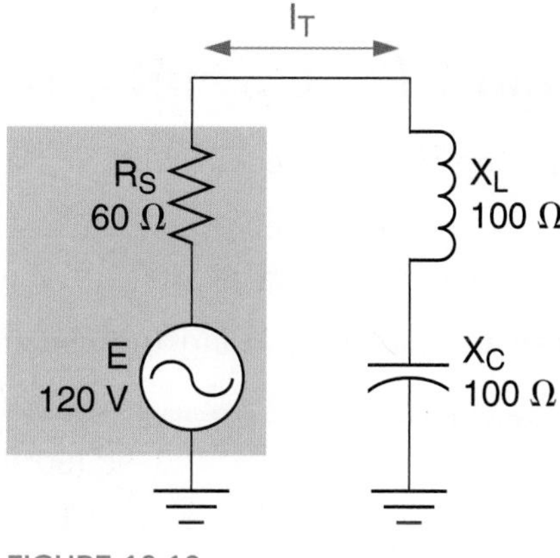

FIGURE 18.18

SOLUTION

The circuit in Figure 18.18 has equal values of X_L and X_C. Therefore, the net series reactance is 0 Ω; the source resistance (60 Ω) is the only opposition to current in the circuit; and the current is found as

$$I = \frac{E}{R_S} = \frac{120\text{ V}}{60\ \Omega} = 2\text{ A}$$

Using I = 2 A, the voltage across the inductor is found as

$$V_L = I \cdot X_L = (2\text{ A})(100\ \Omega) = 200\text{ V}$$

and the voltage across the capacitor is found as

$$V_C = I \cdot X_C = (2\text{ A})(100\ \Omega) = 200\text{ V}$$

Since these voltages are equal and are 180° out of phase, their sum is 0 V (as stated earlier).

PRACTICE PROBLEM 18.10

A circuit like the one in Figure 18.18 has the following values: $E = 120$ V, $R_S = 40\ \Omega$, $X_L = 180\ \Omega$, and $X_C = 180\ \Omega$. Calculate the values of V_L, V_C, and $(V_L + V_C)$ for the circuit.

As you can see, the circuit has values of V_L and V_C that are each greater than the source voltage. However, their sum is 0 V because they are 180° out of phase.

Parallel Resonant LC Circuits

A parallel LC circuit operating at resonance has the following characteristics:

- The net current through the parallel LC circuit is 0 A.
- The circuit has infinite reactance; that is, it acts as an open.

These characteristics can be explained using the circuit and graph shown in Figure 18.19.

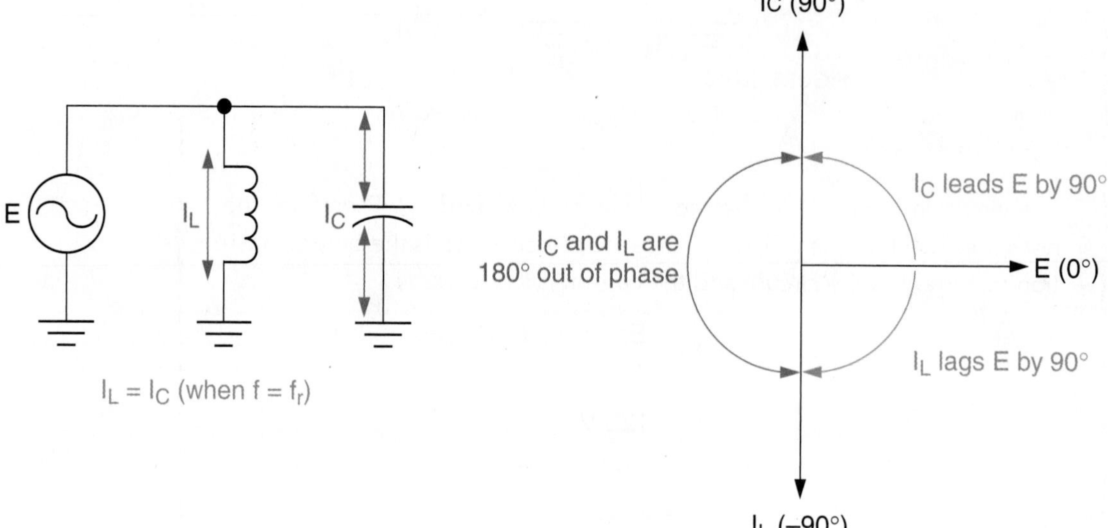

FIGURE 18.19 Current characteristics for a parallel resonant circuit.

Assuming that the circuit in Figure 18.19 is operating at resonance, the values of X_L and X_C are equal. Since the branch reactances and voltages are equal, so are the branch currents. As shown in the graph, I_L and I_C are 180° out of phase. Therefore, the total current in the circuit can be found as

$$I_T = I_C - I_L$$
$$= 0\text{ A}$$

Because the net current through the parallel resonant circuit is 0 A, we can state that the parallel resonant LC circuit is acting as an open. Ideally, the net parallel reactance (X_P) of the parallel resonant circuit is infinite. These parallel resonant circuit characteristics are illustrated further in Example 18.11.

EXAMPLE 18.11

Calculate the net parallel reactance for the circuit in Figure 18.20.

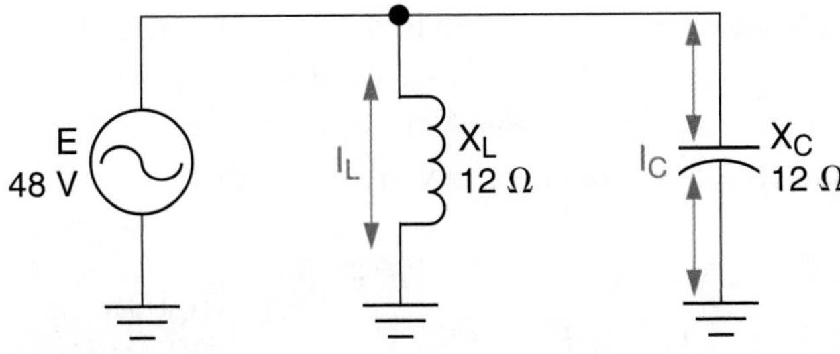

FIGURE 18.20

SOLUTION

The value of I_C for the parallel resonant circuit in Figure 18.20 has a value of

$$I_C = \frac{E}{X_C} = \frac{48\ \text{V}}{12\ \Omega} = 4\ \text{A}$$

The inductor current has a value of

$$I_L = \frac{E}{X_L} = \frac{48\ \text{V}}{12\ \Omega} = 4\ \text{A}$$

The branch currents are equal in magnitude and 180° out of phase, so the net current through the circuit is 0 A. Using this value, the total reactance of the circuit approaches infinity, as follows:

$$X_P = \frac{E}{I_T} = \frac{6\ \text{V}}{0\ \text{A}} \rightarrow \infty\ \Omega$$

The arrow at the end of the calculation indicates that the value of X_P *approaches* infinity (i.e., it becomes too high to measure).

PRACTICE PROBLEM 18.11

A circuit like the one in Figure 18.20 has the following values: $E = 120$ V, $X_L = 40\ \Omega$, and $X_C = 40\ \Omega$. Calculate the values of I_C, I_L, I_T, and X_T for the circuit.

Series Versus Parallel Resonance: A Comparison

In this section, you have been introduced to the characteristics of series and parallel resonant LC circuits. Figure 18.21 provides a summary comparison of the characteristics of these two types of circuits. When you compare the overall reactance, current, and voltage characteristics of the circuits, you can see that they are nearly opposites.

Series-resonant and Parallel-resonant LC Circuits

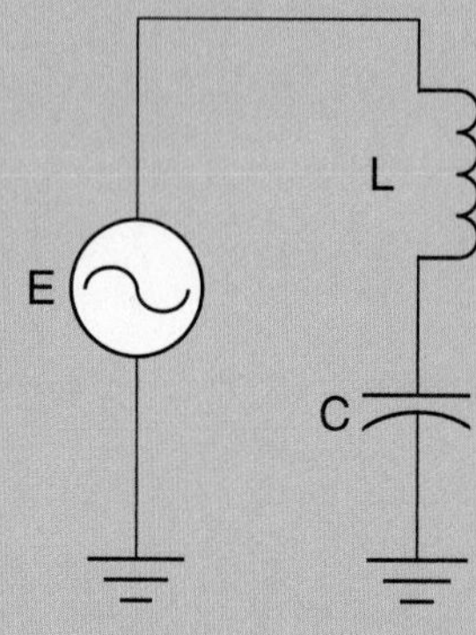

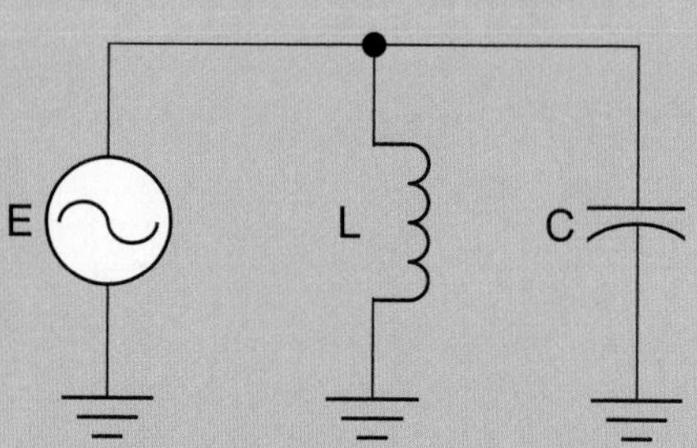

	Series	Parallel
Resonant frequency:	$f_r = \frac{1}{2\pi\sqrt{LC}}$	$f_r = \frac{1}{2\pi\sqrt{LC}}$
Reactance:	$X_L = X_C$ $^*X_S = 0\ \Omega$	$X_L = X_C$ $^{**}X_P$ approaches $\infty\ \Omega$
Other relationships:	V_L and V_C are 180° out of phase. $V_L + V_C = 0$ V I is limited only by source resistance (R_S). I is in phase with E.	I_L and I_C are 180° out of phase. $I_L + I_C = 0$ A

*X_S represents the total series reactance.
**X_P represents the total parallel reactance.

FIGURE 18.21 Series-resonant and parallel-resonant LC circuit characteristics.

PROGRESS CHECK

1. What is the resonant frequency of an LC circuit? How is it calculated?
2. How do changes in component values affect the resonant frequency of an LC circuit?
3. List the basic characteristics of a series resonant LC circuit.
4. List the basic characteristics of a parallel resonant LC circuit.

18.4 SERIES AND PARALLEL RLC CIRCUITS

In this section, we will look at the operation and analysis of basic *resistive-inductive-capacitive* (RLC) circuits. Most of the principles covered in this section will probably seem familiar. For all practical purposes, we are merely combining circuit operating principles that were covered in earlier sections and chapters.

Despite appearing to be complex, series and parallel RLC circuits operate according to the same principles as other series and parallel circuits. When you keep this in mind, RLC circuits are no more difficult to work with than any other series or parallel AC circuits.

IN THE FIELD

Series RLC Circuits

When a resistor, inductor, and capacitor are connected in series, the circuit has the characteristics of one of the following: a purely resistive circuit, an RC circuit, or an RL circuit. The overall characteristics of the circuit are determined by the relationship between X_L and X_C, as given in Table 18.3.

TABLE 18.3 • Reactance Versus Circuit Characteristics.

REACTANCE RELATIONSHIP	RESULTING CIRCUIT CHARACTERISTICS
$X_L > X_C$	The net series reactance is inductive, so the circuit has the characteristics of a series RL circuit: Source voltage leads the circuit current.
$X_L = X_C$	The resonant LC circuit has a net reactance of 0 Ω. Therefore, the circuit is resistive in nature: Source voltage is in phase with the circuit current.
$X_L < X_C$	The net series reactance is capacitive, so the circuit has the characteristics of a series RC circuit: Source voltage lags the circuit current.

The statements in Table 18.3 are based on the phase relationships established in our discussion on LC circuits. They are demonstrated further in the upcoming examples.

Total Series Impedance

In Chapters 14 and 16, you learned that the impedance in a series circuit equals the geometric sum of resistance and reactance. In a series RLC circuit, the total impedance equals the geometric sum of resistance and the *net series reactance* (X_S) in the circuit. By formula,

(18.7)

$$Z_T = \sqrt{X_S^2 + R^2}$$

where

X_S is the net series reactance, equal to $X_L - X_C$

The phase angle of the source voltage (relative to the circuit current) can be found as

(18.8)

$$\theta = \tan^{-1}\frac{X_S}{R}$$

When $X_L > X_C$, the circuit impedance of a series RLC circuit has a positive phase angle, which indicates that the circuit has the characteristics of a series RL circuit. This point is demonstrated in Example 18.12.

EXAMPLE 18.12

Calculate the total impedance and phase angle for the circuit in Figure 18.22.

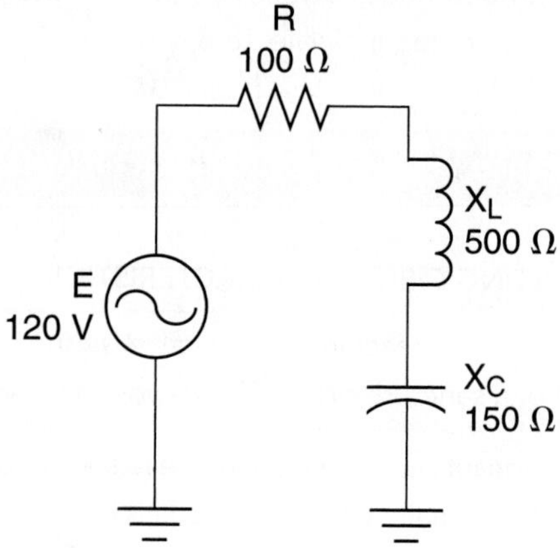

FIGURE 18.22

SOLUTION

First, the net series reactance is found as

$$\begin{aligned} X_S &= X_L - X_C \\ &= 500\ \Omega - 150\ \Omega \\ &= 350\ \Omega \end{aligned}$$

Now, the magnitude of the circuit impedance can be found as

$$\begin{aligned} Z &= \sqrt{X_S^2 + R^2} \\ &= \sqrt{(350\ \Omega)^2 + (100\ \Omega)^2} \\ &= 364\ \Omega \end{aligned}$$

and its phase angle can be found as

$$\theta = \tan^{-1}\frac{X_S}{R}$$
$$= \tan^{-1}\left(\frac{350\ \Omega}{100\ \Omega}\right)$$
$$= 74°$$

The positive phase angle indicates that the circuit has the phase characteristics of a series RL circuit. Therefore, the source voltage leads the circuit current.

PRACTICE PROBLEM 18.12

A circuit like the one in Figure 18.22 has the following values: $R = 300\ \Omega$, $X_L = 840\ \Omega$, and $X_C = 600\ \Omega$. Calculate the total impedance and phase angle for the circuit.

When $X_L < X_C$, the circuit impedance of a series RLC circuit has a negative phase angle, which indicates that the circuit has the characteristics of a series RC circuit. This point is demonstrated in Example 18.13.

EXAMPLE 18.13

Calculate the total impedance and phase angle for the circuit in Figure 18.23.

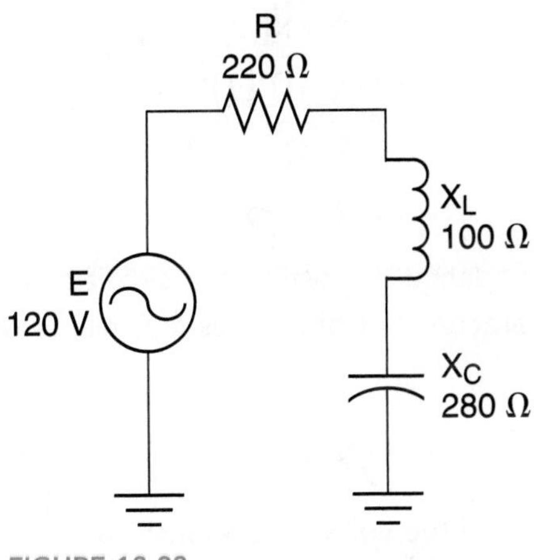

FIGURE 18.23

SOLUTION

First, the net series reactance is found as

$$X_S = X_L - X_C$$
$$= 100\ \Omega - 280\ \Omega$$
$$= -180\ \Omega$$

Now, the magnitude of the circuit impedance can be found as

$$Z = \sqrt{X_S^2 + R^2}$$
$$= \sqrt{(-180\ \Omega)^2 + (220\ \Omega)^2}$$
$$= 284\ \Omega$$

and its phase angle can be found as

$$\theta = \tan^{-1}\frac{X_S}{R}$$
$$= \tan^{-1}\left(\frac{-180\ \Omega}{220\ \Omega}\right)$$
$$= -39.3°$$

The negative phase angle indicates that the circuit has the phase characteristics of a series RC circuit. Therefore, the source voltage lags the circuit current.

PRACTICE PROBLEM 18.13

A circuit like the one in Figure 18.23 has the following values: $R = 120\ \Omega$, $X_L = 300\ \Omega$, and $X_C = 490\ \Omega$. Calculate the total impedance and phase angle for the circuit.

When a series RLC circuit is operating at resonance, the values of inductive and capacitive reactance are equal. When $X_L = X_C$, the net series reactance is $0\ \Omega$. With a value of $X_S = 0\ \Omega$, the circuit impedance has a value of:

$$Z = \sqrt{X_S^2 + R^2}$$
$$= \sqrt{(0\ \Omega)^2 + R^2}$$
$$= \sqrt{R^2}$$
$$= R$$

Because the circuit impedance is resistive, the circuit phase angle is 0°. As these values indicate, the characteristics of a series RLC circuit operating at resonance are purely resistive.

Series Circuit Analysis

Until now, we have ignored the voltage relationships in series RLC circuits. These voltage relationships are illustrated in Figure 18.24.

As indicated in Chapters 14 and 16, current is considered to be the 0° reference in any series circuit. Resistor voltage is always in phase with current, so V_R is used as the reference in Figure 18.24. As you can see:

- Inductor voltage (V_L) leads circuit current by 90°, so it leads resistor voltage by 90° as well.
- Capacitor voltage (V_C) lags circuit current by 90°, so it lags resistor voltage by 90° as well.

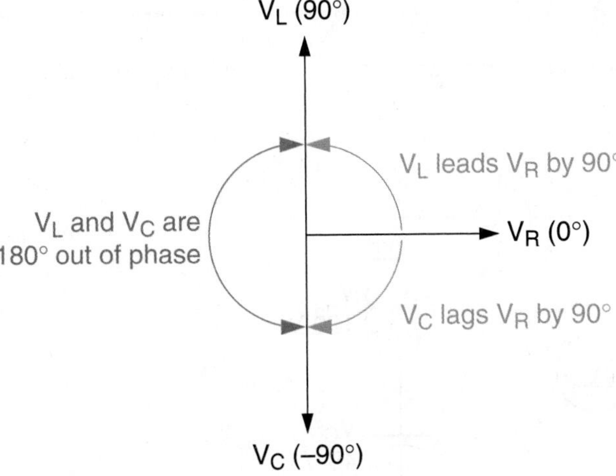

FIGURE 18.24 Voltage vectors for a series RLC circuit.

Therefore, the voltage and capacitor voltages are 180° out of phase, as shown in the figure. With the phase relationships shown, the *net reactive voltage* (V_{LC}) equals the difference between V_L and V_C. By formula,

(18.9)

$$V_{LC} = V_L - V_C$$

As you were shown in Figure 18.3, the net reactive voltage falls on either the positive or negative y-axis. In either case, the source voltage (E) equals the geometric sum of V_{LC} and V_R. By formula:

(18.10)

$$E = \sqrt{V_{LC}^2 + V_R^2}$$

where

E = the source voltage
V_{LC} = the net reactive voltage, found as $V_{LC} = V_L - V_C$
V_R = the voltage across the resistor

As you can see, this equation has the same form as equation (18.7), which was used to calculate the magnitude of series RLC circuit impedance. By the same token, the voltage phase angle (relative to the circuit current) in a series RLC circuit can be found as

(18.11)

$$\theta = \tan^{-1}\frac{V_{LC}}{V_R}$$

The process used to determine the magnitude and phase angle of the voltage source in a series RLC circuit is demonstrated in Example 18.14.

EXAMPLE 18.14

Calculate the source voltage and phase angle for the circuit in Figure 18.25.

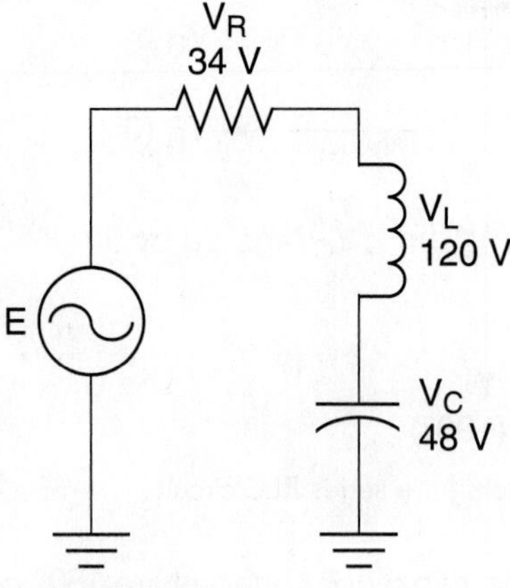

FIGURE 18.25

SOLUTION

First, the net reactive voltage is found as

$$\begin{aligned} V_{LC} &= V_L - V_C \\ &= 120\text{ V} - 48\text{ V} \\ &= 72\text{ V} \end{aligned}$$

Now, the magnitude of the source voltage is found as

$$\begin{aligned} E &= \sqrt{V_{LC}^2 + V_R^2} \\ &= \sqrt{(72\text{ V})^2 + (34\text{ V})^2} \\ &= 80\text{ V} \end{aligned}$$

Finally, the phase angle of the source voltage (relative to the circuit current) is found as

$$\begin{aligned} \theta &= \tan^{-1}\frac{V_{LC}}{V_R} \\ &= \tan^{-1}\left(\frac{72\text{ V}}{34\text{ V}}\right) \\ &= 64.7° \end{aligned}$$

PRACTICE PROBLEM 18.14

A circuit like the one in Figure 18.25 has the following values: $V_R = 20$ V, $V_L = 84$ V, and $V_C = 54$ V. Calculate the source voltage and phase angle for the circuit.

The following observations can be made regarding the circuit in Example 18.14:

- The inductor voltage (V_L) is greater than the capacitor voltage (V_C).
- The positive phase angle (θ) indicates that the circuit is inductive in nature; that is, the source voltage *leads* the circuit current.

Based on these observations, we can say that a series RLC circuit has the characteristics of a series RL circuit when $V_L > V_C$.

When $V_C > V_L$, a series RLC circuit has the characteristics of a series RC circuit. This point is illustrated in Example 18.15.

EXAMPLE 18.15

Calculate the source voltage and phase angle for the circuit in Figure 18.26.

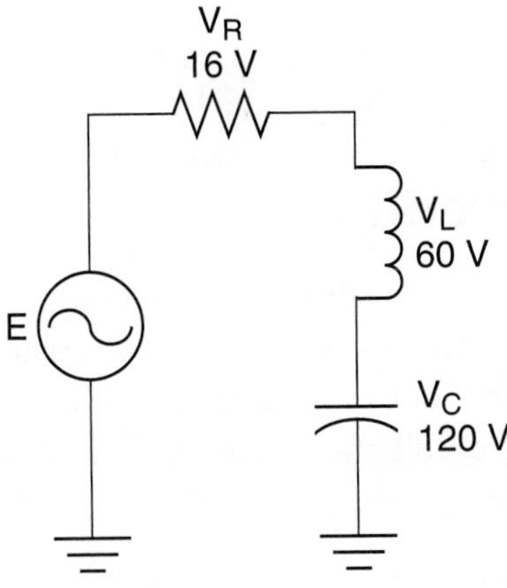

FIGURE 18.26

SOLUTION

First, the net reactive voltage is found as

$$\begin{aligned} V_{LC} &= V_L - V_C \\ &= 60\ \text{V} - 120\ \text{V} \\ &= -60\ \text{V} \end{aligned}$$

Now, the magnitude of the source voltage is found as

$$\begin{aligned} E &= \sqrt{V_{LC}^2 + V_R^2} \\ &= \sqrt{(-60\ \text{V})^2 + (16\ \text{V})^2} \\ &= 62.1\ \text{V} \end{aligned}$$

Finally, the phase angle of the source voltage (relative to the circuit current) is found as

$$\theta = \tan^{-1}\frac{V_{LC}}{V_R}$$

$$= \tan^{-1}\left(\frac{-62.1\text{ V}}{16\text{ V}}\right)$$

$$= -75.6°$$

PRACTICE PROBLEM 18.15

A circuit like the one in Figure 18.26 has the following values: $V_R = 96$ V, $V_L = 120$ V, and $V_C = 240$ V. Calculate the source voltage and phase angle for the circuit.

The following observations can be made regarding the circuit in Example 18.15:

- The capacitor voltage (V_C) is greater than the inductor voltage (V_L).
- The negative phase angle (θ) indicates that the circuit is capacitive in nature; that is, the source voltage *lags* the circuit current.

Based on these observations, we can say that a series RLC circuit has the characteristics of a series RC circuit when $V_C > V_L$.

Parallel RLC Circuits

When a resistor, capacitor, and inductor are connected in parallel, the circuit has the characteristics of one of the following: a purely resistive circuit, a parallel RL circuit, or a parallel RC circuit. For example, look at the circuit in Figure 18.27. The overall characteristics of the circuit are determined by the relationship between I_L and I_C, as given in Table 18.4.

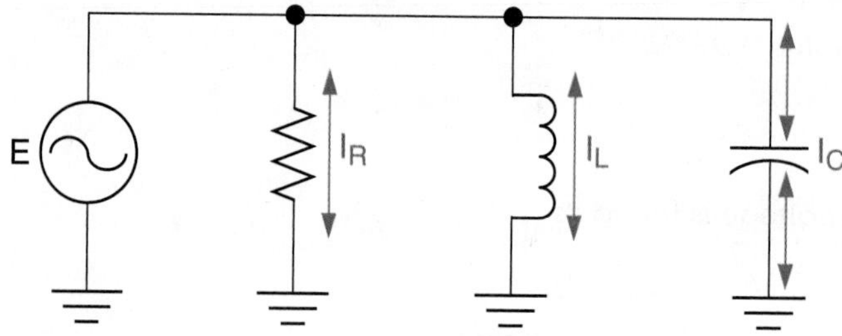

FIGURE 18.27 Parallel RLC circuit currents.

TABLE 18.4 • Reactive Current Versus Circuit Characteristics.

CURRENT RELATIONSHIP	RESULTING CIRCUIT CHARACTERISTICS
$I_L > I_C$	The net reactive current is inductive, so the circuit has the characteristics of a parallel RL circuit: Circuit current lags the source voltage.
$I_L = I_C$	The resonant LC circuit has a net current of 0 A. Therefore, the circuit is resistive in nature: Source voltage and current are in phase.
$I_C > I_L$	The net reactive current is capacitive, so the circuit has the characteristics of a parallel RC circuit: Circuit current leads the source voltage.

The statements in Table 18.4 are based on the phase relationships in Figure 18.19. Since I_L and I_C are 180° out of phase, the *net reactive current* (I_{LC}) is equal to the difference between the two. This means that:

- I_{LC} is *negative* when $I_L > I_C$. In this case, I_{LC} is inductive and the circuit has the characteristics of a parallel RL circuit.
- $I_{LC} = 0$ A when $I_L = I_C$. In this case, the total circuit current equals the resistor current (I_R), meaning that the circuit is purely resistive.
- I_{LC} is *positive* when $I_C > I_L$. In this case, I_{LC} is inductive and the circuit has the characteristics of a parallel RC circuit.

These relationships will be demonstrated in upcoming examples.

Total Parallel Current

Figure 18.28 shows the current vectors for the circuit in Figure 18.27. The I_L and I_C vectors indicate that I_{LC} and I_R are 90° out of phase. As a result of this phase relationship, the circuit current is found as the geometric sum of the two currents. By formula,

(18.12)
$$I_T = \sqrt{I_{LC}^2 + I_R^2}$$

where I_{LC} is the net reactive current, found as $I_{LC} = I_C - I_L$. The phase angle of the circuit current (relative to the source voltage) is found as

(18.13)
$$\theta = \tan^{-1}\frac{I_{LC}}{I_R}$$

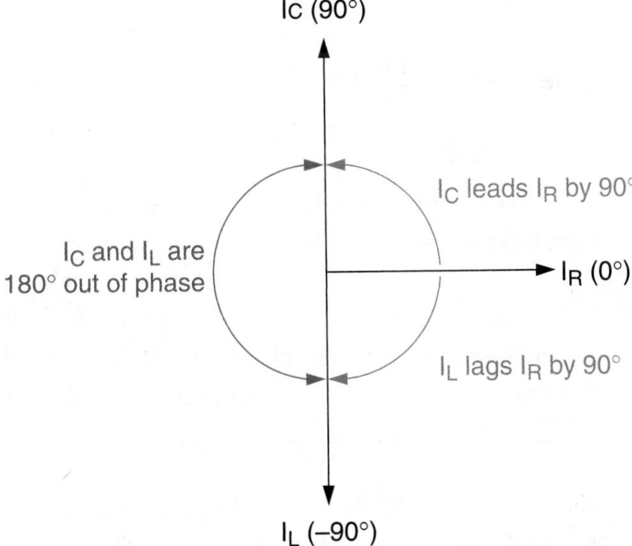

FIGURE 18.28 Current phase relationships.

As stated earlier, a parallel RLC circuit has the characteristics of a parallel RL circuit when $I_L > I_C$. This point is demonstrated in Example 18.16.

EXAMPLE 18.16

Calculate the circuit current and phase angle for the circuit in Figure 18.29.

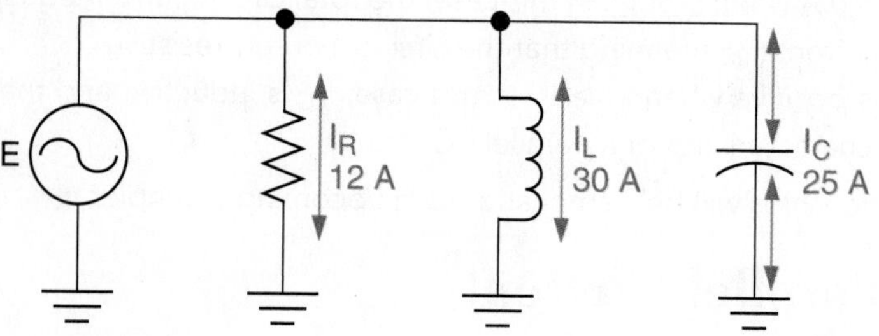

FIGURE 18.29

SOLUTION

First, the net reactive current is found as

$$I_{LC} = I_C - I_L$$
$$= 25\text{ A} - 30\text{ A}$$
$$= -5\text{ A}$$

Now, the magnitude of the circuit current can be found as

$$I_T = \sqrt{I_{LC}^2 + I_R^2}$$
$$= \sqrt{(-5\text{ A})^2 + (12\text{ A})^2}$$
$$= 13\text{ A}$$

and its phase angle can be found as

$$\theta = \tan^{-1}\frac{I_{LC}}{I_R}$$
$$= \tan^{-1}\left(\frac{-5\text{ A}}{12\text{ A}}\right)$$
$$= -22.6°$$

The negative phase angle indicates that the circuit has the characteristics of a parallel RL circuit. Therefore, the source voltage leads the circuit current.

PRACTICE PROBLEM 18.16

A circuit like the one in Figure 18.29 has the following values: $I_R = 6$ A, $I_L = 12$ A, and $I_C = 5$ A. Calculate the circuit current and phase angle for the circuit.

When $I_C > I_L$, the circuit current in a parallel RLC circuit has a positive phase angle, which indicates that the circuit has the characteristics of a parallel RC circuit. This point is demonstrated in Example 18.17.

EXAMPLE 18.17

Calculate the circuit current and phase angle for the circuit in Figure 18.30.

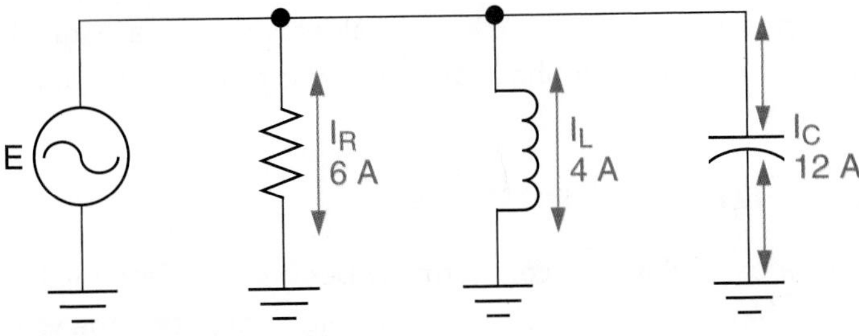

FIGURE 18.30

SOLUTION

First, the net reactive current is found as

$$I_{LC} = I_C - I_L$$
$$= 12\text{ A} - 4\text{ A}$$
$$= 8\text{ A}$$

Now, the total circuit current can be found as

$$I_T = \sqrt{I_{LC}^2 + I_R^2}$$
$$= \sqrt{(8\text{ A})^2 + (6\text{ A})^2}$$
$$= 10\text{ A}$$

and its phase angle can be found as

$$\theta = \tan^{-1}\frac{I_{LC}}{I_R}$$
$$= \tan^{-1}\left(\frac{8\text{ A}}{6\text{ A}}\right)$$
$$= 53.1°$$

The positive phase angle indicates that the circuit has the characteristics of a parallel RC circuit. Therefore, the circuit current leads the source voltage.

PRACTICE PROBLEM 18.17

A circuit like the one in Figure 18.30 has the following values: $I_R = 2$ A, $I_L = 4$ A, and $I_C = 12$ A. Calculate the total current and phase angle of the circuit.

When a parallel RLC circuit is operating at resonance, the values of I_L and I_C are equal. Therefore, the net reactive current is 0 A. With a value of $I_{LC} = 0$ A, the circuit current is found as

$$I_T = \sqrt{I_{LC}^2 + I_R^2}$$
$$= \sqrt{(0\text{ A})^2 + I_R^2}$$

$$= \sqrt{I_R^2}$$
$$= I_R$$

and the circuit phase angle is 0°. As these values indicate, a parallel RLC circuit operating at resonance has no phase angles and is purely resistive in nature.

Parallel Circuit Analysis

The analysis of a parallel RLC circuit normally begins with determining the values of the branch currents and the total circuit current. Then, using the values of E and I_T, the circuit impedance is calculated. The analysis of a parallel RLC circuit is demonstrated in Example 18.18.

EXAMPLE 18.18

Calculate the reactances, currents, impedance, and phase angle for the circuit in Figure 18.31.

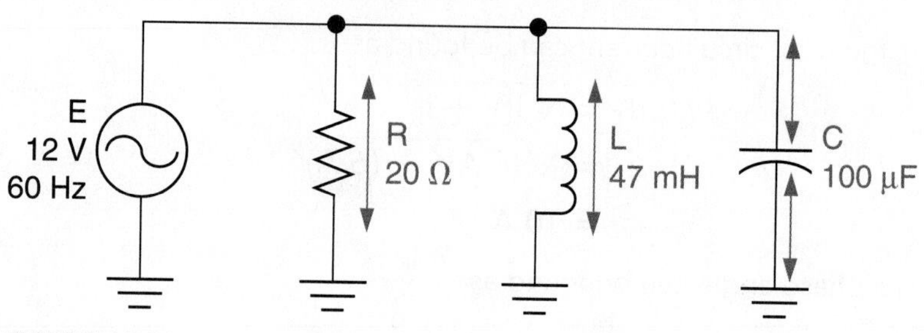

FIGURE 18.31

SOLUTION

First, the circuit reactances are found as

$$X_L = 2\pi fL$$
$$= 2\pi(60\text{ Hz})(47\text{ mH})$$
$$= 17.7\ \Omega$$

and

$$X_C = \frac{1}{2\pi fC}$$
$$= \frac{1}{2\pi(60\text{ Hz})(100\ \mu\text{F})}$$
$$= 26.5\ \Omega$$

The reactive currents can now be found as

$$I_L = \frac{E}{X_L}$$

$$= \frac{12\text{ V}}{17.7\ \Omega}$$

$$= 678\text{ mA}$$

and

$$I_C = \frac{E}{X_C}$$

$$= \frac{12\text{ V}}{26.5\ \Omega}$$

$$= 453\text{ mA}$$

Since the inductive and capacitive currents are 180° out of phase, the net reactive current is found as

$$I_{LC} = I_C - I_L$$

$$= 453\text{ mA} - 678\text{ mA}$$

$$= -225\text{ mA}$$

The value of I_R is found as

$$I_R = \frac{E}{R}$$

$$= \frac{12\text{ V}}{20\ \Omega}$$

$$= 600\text{ mA}$$

Now that we know the values of the branch currents, the magnitude of the total circuit current can be found as

$$I_T = \sqrt{I_{LC}^2 + I_R^2}$$

$$= \sqrt{(-225\text{ mA})^2 + (600\text{ mA})^2}$$

$$= 641\text{ mA}$$

and the phase angle of the circuit current (relative to the source voltage) can be found as

$$\theta = \tan^{-1}\frac{I_{LC}}{I_R}$$

$$= \tan^{-1}\left(\frac{-225\text{ mA}}{600\text{ mA}}\right)$$

$$= -20.6°$$

Once the value of total circuit current is known, the circuit impedance can be found as

$$Z_T = \frac{E}{I_T}$$

$$= \frac{12\text{ V}}{641\text{ mA}}$$

$$= 18.7\ \Omega$$

PRACTICE PROBLEM 18.18

A circuit like the one in Figure 18.31 has the following values: E = 24 V, f = 120 Hz, R = 75 Ω, L = 22 mH, and C = 15 μF. Calculate the reactances, currents, impedance, and phase angle for the circuit.

Leading VAR and Lagging VAR

We have ignored power calculations throughout this section because they haven't changed. Whether you are dealing with a series or parallel RLC circuit, power is calculated as it is for any resistive-reactive circuit. The values of apparent, reactive, and resistive power can be found using the relationships that were introduced earlier in the text.

There is one concept to explore before moving on: the difference between inductive VAR and capacitive VAR. Though both were introduced in earlier chapters, we have not considered the *phase difference* between the two.

LAGGING VAR Another term for inductive VAR.

When a load is inductive, reactive power (VAR) *lags* resistive power (W), as illustrated in Figure 18.32. Based on this relationship, inductive VAR are referred to as **LAGGING VAR**. Since lagging VAR are produced by equipment that has some amount of inductance (like motors and transformers) almost every home or industrial installation has lagging VAR.

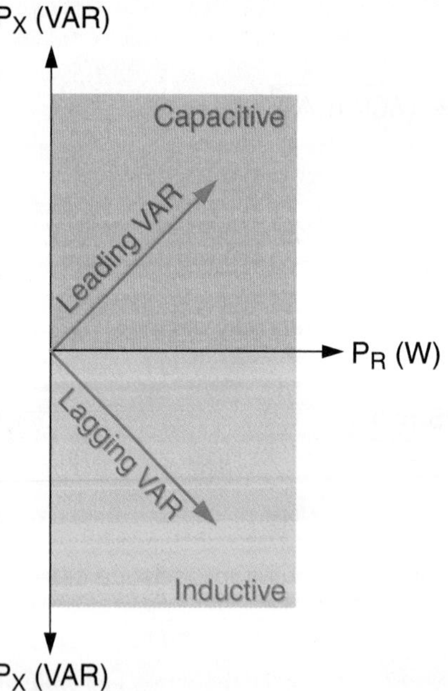

FIGURE 18.32 Leading and lagging VAR.

When a load is capacitive, reactive power *leads* resistive power, as illustrated in Figure 18.32. Because of this relationship, capacitive VAR are referred to as **LEADING VAR**. The fact that capacitance and inductance have leading and lagging VAR (respectively) plays a role in **POWER FACTOR CORRECTION** (as you will see in the next section).

LEADING VAR Another term for capacitive VAR.

POWER FACTOR CORRECTION Using a circuit to improve the power factor of a circuit.

PROGRESS CHECK

1. Describe the phase relationships for a series RLC circuit under the following conditions:
   a. $X_L > X_C$ b. $X_L = X_C$ c. $X_L < X_C$
2. Describe the phase relationships for a parallel RLC circuit under the following conditions:
   a. $I_C > I_L$ b. $I_C = I_L$ c. $I_C < I_L$
3. What are *leading* VAR? What type of circuit has leading VAR?
4. What are *lagging* VAR? What type of circuit has lagging VAR?

18.5 POWER FACTOR CORRECTION

As you learned earlier in the chapter, inductive and capacitive currents are 180° out of phase and cancel each other out when their values are equal. The same is true of inductive VAR and capacitive VAR. This relationship can be used to correct (improve) the power factor of an entire industrial plant, or an individual motor. In this section, we discuss the reason that a low power factor can be a problem to the end user and to the power utility company.

Power Factor (Review)

You may recall that power factor is the ratio of resistive (or *true*) power to apparent power. By formula,

(18.14)
$$PF = \frac{P_R}{P_{APP}}$$

where

PF = power factor

P_R = resistive (or *true*) power, measured in watts (W)

P_{APP} = apparent power, measured in volt-amperes (VA)

When a circuit is *purely reactive*, there is no resistive power ($P_R = 0$ W), and

$$PF = \frac{P_R}{P_{APP}} = \frac{0\text{ W}}{P_{APP}} = 0$$

When a circuit is *purely resistive*, there is no reactive power, $P_R = P_{APP}$, and

$$PF = \frac{P_R}{P_{APP}} = 1$$

As these calculations indicate, power factor has a range of 0 to 1. When a circuit has both resistive and reactive elements, the power factor lies somewhere *between* 0 and 1. That is,

$$0 < PF < 1$$

meaning that the value of PF is greater than 0 and less than 1.

When PF = 1, all of the energy drawn from the power source is converted into a usable form (heat or light). When PF = 0, all of the energy being drawn from the source is stored in the electric and magnetic fields of the reactive components. In other words, none of the energy drawn from the source is converted into a usable form. As such, a value of PF = 1 is considered to be ideal.

The Train Analogy

The lower the power factor of a resistive-reactive circuit, the greater the *volt-ampere* (VA) input required to produce the same amount of heat or light. This principle is often clarified using what has come to be known as *the train analogy.*

Suppose that a locomotive is pulling a boxcar with a chain. When the boxcar is on the same track as the locomotive (as illustrated in Figure 18.33a), the energy expended by the locomotive depends on the mass of the boxcar and the distance travelled. Now, suppose that the boxcar and the locomotive are on parallel sets of tracks (as illustrated in Figure 18.33b). The work required to pull the boxcar (*real power*) is the same, but the locomotive must also provide energy to overcome the lateral (sideways) force that results from being on the other set of tracks. Since the boxcar cannot move sideways, this work (*reactive power*) is wasted—yet it must still be performed. The more offset the rail, the more wasted work. This is analogous to a decreasing power factor.

Costs Associated with a Low Power Factor

Why is a low power factor undesirable? Let's suppose that an industrial plant uses a 240-V supply to power ten 1.5 kW motors. If PF = 1, the total current required to operate these motors is found as

$$I_T = \frac{P_R}{E}$$

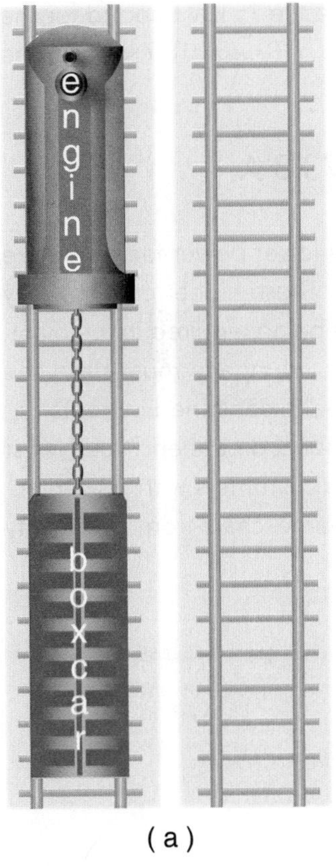

(a)

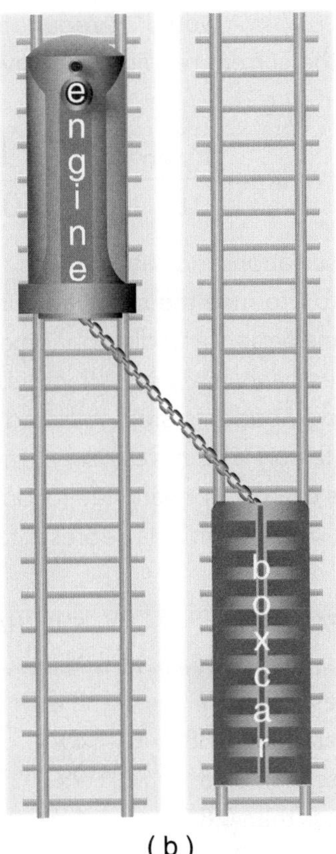

(b)

FIGURE 18.33

$$= \frac{15 \text{ kW}}{240 \text{ V}}$$
$$= 62.5 \text{ A}$$

If the power factor was 0.85, apparent power would have to increase in order to provide the 15 kW required by the motors. This can be demonstrated using equation (18.14), as follows:

$$PF = \frac{P_R}{P_{APP}}$$

For PF = 0.85 and P_R = 15 kW, the equation can be rewritten as

$$0.85 = \frac{15 \text{ kW}}{P_{APP}}$$

Transposing to solve for P_{APP} we get,

$$P_{APP} = \frac{15 \text{ kW}}{0.85}$$
$$= 17.7 \text{ kVA}$$

This means that 17.7 kVA would be required to provide the 15 kW needed for the motors. Finally, the current required to provide a value of P_{APP} = 18.7 kVA can be found as:

$$I_T = \frac{P_{APP}}{E} = \frac{17.7 \text{ kVA}}{240 \text{ V}} = 73.8 \text{ A}$$

As these calculations indicate, circuits that have a lower power factor require higher currents to produce the same amount of work. These higher currents may lead to larger transformers and distribution conductors being required in the plant. If several facilities have low power factors, the power company must have the generating and distribution system required to provide the combined currents.

Low power factors also have a direct monetary impact on the end user. Most utility companies charge a penalty fee (or *surcharge*) to customers with power factors below 0.95. For example, a representative surcharge chart from one utility company is shown below.

POWER FACTOR	SURCHARGE
0.9 to less than 1	Nil
0.88 to less than 0.9	2%
0.85 to less than 0.88	4%
0.8 to less than 0.85	9%
0.75 to less than 0.8	16%
0.7 to less than 0.75	24%
0.65 to less than 0.7	34%
0.6 to less than 0.65	44%
0.55 to less than 0.6	57%
0.5 to less than 0.55	72%
Less than 0.5	80%

Correcting Power Factor

Depending on the size of the installation, power factor correction is accomplished individually at each motor, or centrally using an automatic PF correction controller.

Static PF Correction

STATIC CORRECTION Power factor correction that is implemented for each individual motor.

STATIC COMPENSATION Another term for static correction.

COMPENSATING CAPACITOR A capacitor connected in parallel with a motor for power factor correction.

Correcting the power factor of each individual motor is referred to as **STATIC CORRECTION** or **STATIC COMPENSATION**. A **COMPENSATING CAPACITOR** is connected in parallel with the motor, as shown in Figure 18.34. This means that when the motor is not in use (not adding any inductance to the load) the capacitor is also removed from the circuit. This prevents *overcorrection*, which can be a safety hazard. A leading power factor can result in dangerously high voltages.

An AC motor has both a resistance and inductance, which are represented by the resistor and inductor shown in Figure 18.34a. The power used by the inductive motor windings is reactive power (lagging VAR), but the power required to drive

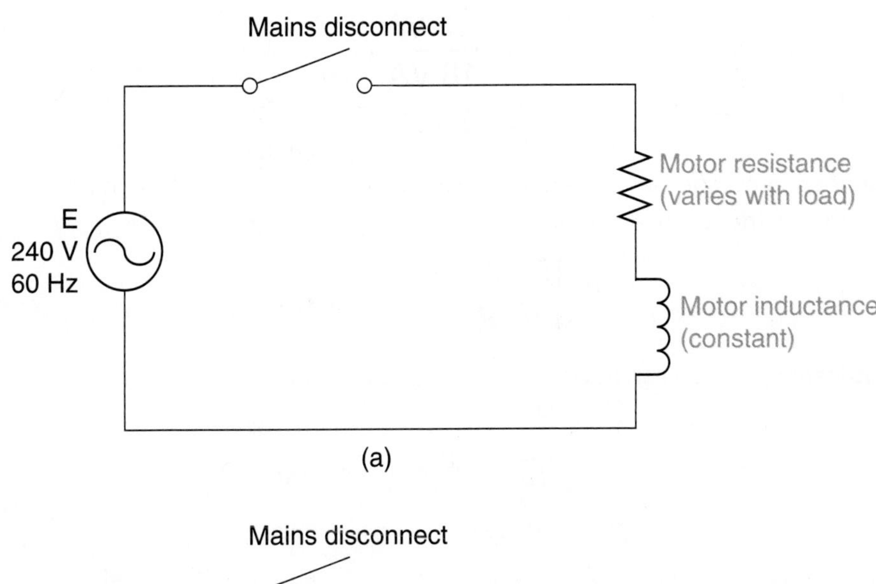

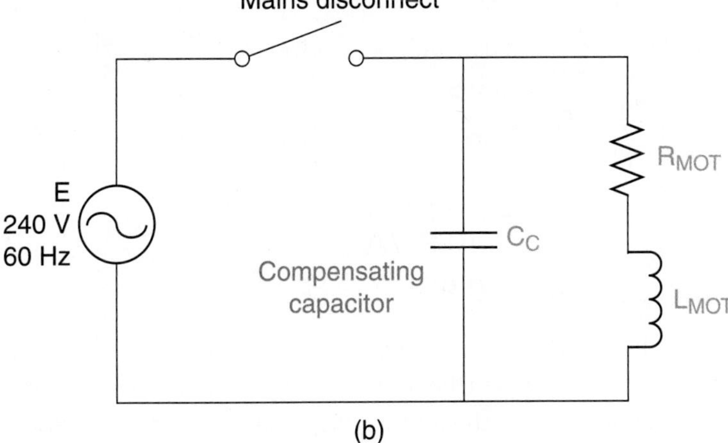

FIGURE 18.34 Compensating capacitor connection.

the load is true power (watts). By placing a compensating capacitor in parallel with the motor (as shown in Figure 18.34b), the *leading* VAR of the capacitor cancel out the *lagging* VAR of the motor. As a result, the motor power factor increases and the apparent power required to drive the motor decreases.

The effect of a compensating capacitor on the power factor of an AC motor can be demonstrated easily. Let's say we have a motor with the following values: $P_R = 10$ W and $P_X = 15$ VAR. Using these values, the apparent power being drawn by the motor is found as

$$P_{APP} = \sqrt{P_X^2 + P_R^2}$$
$$= \sqrt{(15\text{ VAR})^2 + (10\text{ W})^2}$$
$$= 18\text{ VA}$$

and the motor's power factor is found as

$$PF = \frac{P_R}{P_{APP}}$$

$$= \frac{10\ \text{W}}{18\ \text{VA}}$$
$$= 0.556$$

If we connect a compensating capacitor with a value of $P_{XC} = 12$ VAR in parallel with the motor, the reactive power decreases to

$$P_X = 15\ \text{VAR} - 11\ \text{VAR}$$
$$= 4\ \text{VAR}$$

The apparent power being drawn by the motor is now found as

$$P_{APP} = \sqrt{P_X^2 + P_R^2}$$
$$= \sqrt{(4\ \text{VAR})^2 + (10\ \text{W})^2}$$
$$= 10.8\ \text{VA}$$

and the motor's power factor is found as

$$PF = \frac{P_R}{P_{APP}}$$
$$= \frac{10\ \text{W}}{10.8\ \text{VA}}$$
$$= 0.926$$

As these results demonstrate, adding the compensating capacitor to the circuit *decreases* apparent power and *increases* the motor power factor without affecting the true power (10 W) required to drive the motor. This is the basis for using capacitance as a method of power factor correction.

Determining the Value of a Compensating Capacitor

Power factor is usually corrected to between 0.95 and 0.97. While correction to PF = 1 is possible, it can also be dangerous. When PF = 1, the compensating capacitor forms a resonant circuit with the motor and/or other components. This resonant circuit can produce dangerously high voltages. For this reason, power factor correction is usually limited to values below PF = 0.97.

Determining the value required for a compensating capacitor in a specific application is relatively simple. The process is demonstrated in Example 18.19.

EXAMPLE 18.19

A 3 HP motor is connected to a 240 V supply. Using a clamp meter, you determine that the motor is drawing 11.5 A from its power source. Determine the ideal value of a compensating capacitor for the motor.

SOLUTION

The first step is to determine the value of P_{APP}, as follows:

$$P_{APP} = E \times I$$
$$= 240 \text{ V} \times 11.5 \text{ A}$$
$$= 2760 \text{ VA}$$

Since 1 HP = 746 W, the power drawn by a 3 HP motor can be found as

$$P_R = 3 \text{ HP} \times 746 \text{ W/HP}$$
$$= 2238 \text{ W}$$

We can now solve for the power factor as

$$PF = \frac{P_R}{P_{APP}}$$
$$= \frac{2238 \text{ W}}{2760 \text{ VA}}$$
$$= 0.81$$

Next we solve for the motor's reactive power (lagging VAR), as follows:

$$P_X = \sqrt{P_{APP}^2 - P_R^2} = \sqrt{(2760 \text{ VA})^2 - (2238 \text{ W})^2} = 1615 \text{ VAR}$$

The value of P_{APP} that will result in a power factor of 0.95 can be found as

$$P_{APP} = \frac{P_R}{PF} = \frac{2238 \text{ W}}{0.95} = 2356 \text{ VA}$$

For PF = 0.95, the reactive power should have a value of

$$P_X = \sqrt{P_{APP}^2 - P_R^2}$$
$$= \sqrt{(2356 \text{ VA})^2 - (2238 \text{ W})^2} = 736 \text{ VAR}$$

The capacitive (leading) VAR required to reduce the reactive power to 736 VAR equals the difference between the original and desired values of P_X, as follows:

$$P_X = 1615 \text{ VAR} - 736 \text{ VAR}$$
$$= 879 \text{ VAR}$$

Now, using E = 240 V and P_X = 879 VAR, the reactance of the compensating capacitor can be found as

$$X_C = \frac{E^2}{P_X}$$
$$= \frac{240 \text{ V}^2}{879 \text{ VAR}}$$
$$= 65.5 \ \Omega$$

With a line frequency of 60 Hz, we need a capacitor with a value of

$$C = \frac{1}{2\pi f X_C}$$

$$= \frac{1}{2\pi \times 60\text{ Hz} \times 65.5\ \Omega}$$

$$= 40.5\ \mu\text{F}$$

So, to correct the PF to 0.95, a capacitor having a value of approximately 40.5 μF should be connected in parallel between the motor and the mains disconnect, as illustrated in Figure 18.35.

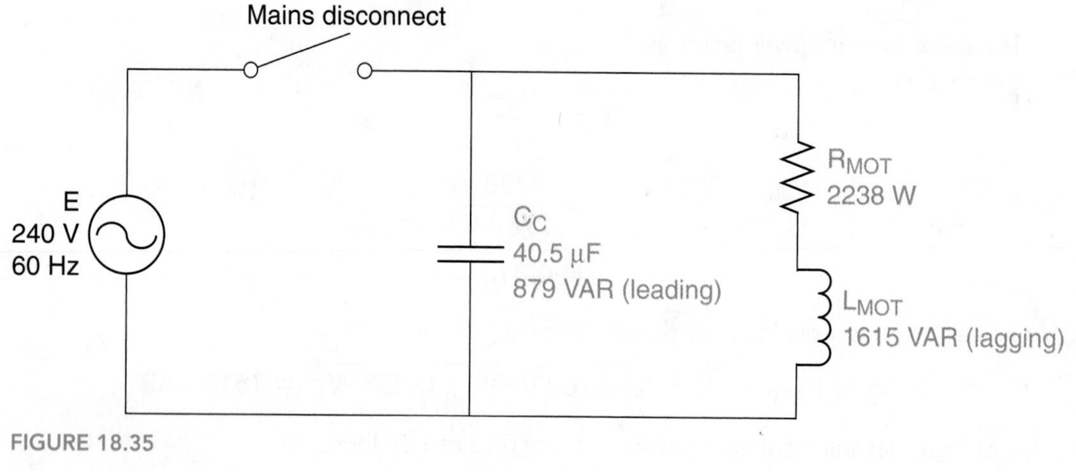

FIGURE 18.35

One final word on static power factor correction (PFC). Extreme care must be taken when employing static PFC on a motor that is driven by a *variable-speed drive* or *inverter*. Connecting a capacitor to the output of a driver can cause serious damage to the driver and the capacitor as the result of a high-frequency voltage that can be produced at the driver output. (See the discussion on *harmonics* later in this section.)

CENTRALIZED CORRECTION A power factor correction technique where compensating capacitor banks are located at the line input to a plant.

BULK CORRECTION Another term for centralized correction.

KVAR-SENSITIVE CONTROLLER A power factor correction circuit that measures changes in the power factor as motors are started or stopped, and switches capacitors in or out as needed to keep the power factor at a preset level.

INTELLIGENT POWER FACTOR CONTROLLER (IPFC) Another name for a kVAR-sensitive controller.

Centralized Correction

CENTRALIZED CORRECTION (also known as **BULK CORRECTION**) is employed in industrial plants where many motors are in use. Rather than correcting motors individually, a centralized control system controls groups (or *banks*) of capacitors that are located at the incoming distribution boards of the facility. These capacitors are connected to the various motors as needed for PF correction.

Most centralized compensation networks are controlled by a **KVAR-SENSITIVE CONTROLLER**. The kVAR-sensitive controller, also known as an **INTELLIGENT POWER FACTOR CONTROLLER (IPFC)**, requires two inputs—one for current and one for voltage. It measures changes in the power factor as motors are started or stopped, and switches capacitors in or out as needed to keep the power factor at a preset level—usually around 95%.

Harmonics

A **HARMONIC** is a whole-number multiple of a given frequency. For example, a 60 Hz sine wave has harmonics at 120 Hz, 180 Hz, 240 Hz, and so on. (These frequencies are referred to as the 2nd, 3rd, and 4th harmonics, respectively.) The higher the frequency of a harmonic, the lower its amplitude.

HARMONIC A whole-number multiple of a given frequency.

The greater the phase angle between AC source current and voltage, the more likely that harmonics will be generated. Since many industries employ semiconductor-controlled switches and drives, harmonics can cause a real problem for power factor correction. The reason has to do with the relationship between frequency and reactance. As frequency increases:

- Capacitive reactance (X_C) decreases.
- The inductive reactance (X_L) of the motor increases.

This can result in very high harmonic currents in the capacitors. To help prevent this, compensating capacitors are often connected to a series inductor. The inductor value is chosen so that it has little reactance at lower harmonic frequencies. However, at the higher harmonics (usually the 5th harmonic and higher) the inductive reactance is high enough to limit harmonically generated currents.

The Future

Today, power factor correction is almost exclusively an industrial application. With today's pressure to conserve energy, PFC at the residential level may not be that far away.

In 2005 a pilot project was initiated to determine the viability of installing automated PFC equipment in residential homes. Here is a brief summary of the findings:

- The uncorrected power factor of the average home was 0.87.
- The cost to install automated PFC equipment to 1000 homes was $450,000.
- The total power saved by the power factor correction was 700 kVA, which had a monetary value of $450,000.

These findings do not take into account the environmental and health costs associated with power generation. However, they do indicate that PFC is economically viable and may be a useful energy conservation tool.

PROGRESS CHECK

1. What type of circuit has a value of PF = 1?
2. What type of circuit has a value of PF = 0?
3. Why is a high power factor desirable?
4. What are the two types of power factor correction?
5. How does a compensating capacitor improve the power factor of an inductive circuit?
6. How is a compensating capacitor connected to a motor?
7. Briefly describe centralized power factor correction.
8. What is a harmonic?
9. Why might a series inductor be connected to a compensating capacitor?

18.6 SUMMARY

Here is a summary of the major points that were covered in this chapter:

Series LC Circuits

- When a capacitor and an inductor are connected in series:
  - the component currents are equal.
  - the component voltages are 180° out of phase.
- The net voltage across a series LC circuit equals the difference between the component voltages.
- A series LC circuit has the characteristics of:
  - an inductive circuit when $V_L > V_C$.
  - a capacitive circuit when $V_C > V_L$.
  - a short circuit when $V_L = V_C$.
- When an inductor and capacitor are connected in series, the total series reactance equals the difference between X_L and X_C.
- For any series LC circuit, the total series reactance is always less than one (or both) of the component values.
- A series LC circuit has the characteristics of:
  - an inductive circuit when $X_L > X_C$.
  - a capacitive circuit when $X_C > X_L$.
  - a short circuit when $X_L = X_C$.

Parallel LC Circuits

- When an inductor and a capacitor are connected in parallel:
  - the component voltages are equal.
  - the component currents are 180° out of phase.
- The net current through a parallel LC circuit equals the difference between I_C and I_L.
- The net current through a parallel LC circuit may be greater than one of the two branch current values.
- A parallel LC circuit has the characteristics of:
  - an inductive circuit when $I_L > I_C$.
  - a capacitive circuit when $I_C > I_L$.
  - an open circuit when $I_L = I_C$.
- A parallel LC circuit has the characteristics of:
  - an inductive circuit when $X_L < X_C$.
  - a capacitive circuit when $X_C < X_L$.
  - an open circuit when $X_L = X_C$.

Resonance

- An LC circuit with equal values of X_L and X_C is said to be operating at resonance.
- The frequency at which resonance occurs for a given LC circuit is referred to as its resonant frequency (f_r).
- Resonant frequency varies inversely with the values of L and C; that is, an increase in either L or C causes f_r to decrease, and vice versa.

- When a series LC circuit is operating at resonance:
  - The total series reactance of the LC circuit is 0 Ω.
  - The voltage across the LC circuit is 0 V.
  - The circuit voltage and current are in phase.
- When a parallel LC circuit is operating at resonance:
  - The sum of the component currents is 0 A.
  - The circuit has infinite reactance; that is, it acts as an open.

Series and Parallel RLC Circuits

- A series RLC circuit has the characteristics of:
  - an RL circuit when $V_L > V_C$.
  - an RC circuit when $V_C > V_L$.
  - a purely resistive circuit when $V_L = V_C$.
- A series RLC circuit has the characteristics of:
  - an RL circuit when $X_L > X_C$.
  - an RC circuit when $X_C > X_L$.
  - a purely resistive circuit when $X_L = X_C$.
- A parallel RLC circuit has the characteristics of:
  - an RL circuit when $I_L > I_C$.
  - an RC circuit when $I_C > I_L$.
  - a purely resistive circuit when $I_L = I_C$.
- Inductive VAR lag resistive power.
  - Inductive VAR are referred to as *lagging* VAR.
- Capactive VAR lead resistive power.
  - Capacitive VAR are referred to as *leading* VAR.

Power Factor Correction (PFC)

- The relationship between leading VAR and lagging VAR can be used to increase the power factor of a resistive-inductive circuit.
- A purely reactive circuit has a value of PF = 0
- A purely resistive circuit has a value of PF = 1
- Power factor values fall within the range of $0 < \text{PF} < 1$
- The lower the power factor of an RL circuit, the greater the volt-ampere (VA) input required to produce the same amount of heat or light.
  - Low-PF circuits require more current than high-PF circuits to accomplish the same amount of work.
  - A plant that has many low-PF circuits may require larger transformers and distribution conductors (to handle the higher currents).
- Most utilities charge a penalty fee (or *surcharge*) to customers with power factors below 0.95.
  - The lower the power factor, the greater the surcharge.
- *Static correction* (or *static compensation*) is a PFC method that is applied to individual motors.
  - A compensating capacitor is connected in parallel with each motor on the output side of the main disconnect.
  - The leading VAR of the capacitor reduces the lagging VAR of the motor, resulting in a higher power factor.

- With the compensating capacitor connected, P_X and P_{APP} are reduced, while true power (which drives the motor) remains unchanged.
- Care must be taken when static correction is applied to a motor driven by a variable-speed drive or an inverter.
  - The capacitor and driver (or inverter) output can be damaged by high voltages that can be generated when the output switches.
- *Centralized correction* (or *bulk correction*) is used when compensation is required for many motors in one location.
  - Most are controlled by a kVAR-sensitive controller.
  - The kVAR-sensitive controller temporarily connects capacitors across motors as needed to provide PFC.
- A harmonic is a whole-number multiple of a given frequency.
- Harmonics are referred to as 2nd, 3rd, etc. harmonics to indicate their relationship to the original frequency.
- At higher harmonic frequencies:
  - Capacitive reactance is reduced.
  - Motor (or transformer) inductive reactance is increased.

CHAPTER REVIEW

1. ________ voltage ________ circuit current by ________ in a series LC circuit.
   a) Capacitor, leads, 180°
   b) Inductor, leads, 90°
   c) Inductor, leads, 180°
   d) Capacitor, leads, 90°

2. ________ voltage ________ circuit current by ________ in a series LC circuit.
   a) Capacitor, leads, 180°
   b) Inductor, lags, 90°
   c) Inductor, leads, 180°
   d) Capacitor, lags, 90°

3. In a series LC circuit ________.
   a) inductor current is equal to capacitor current
   b) the applied voltage equals $V_L - V_C$
   c) there is a 180° phase difference between V_L and V_C
   d) all of the above

4. In a given series LC circuit, $V_L = 85$ V and $V_C = 80$ V. This means that ________.
   a) the result vector equals 5 V
   b) the circuit voltage leads circuit current by 90°
   c) V_L leads V_C by 180°
   d) all of the above

5. It is determined that current leads the applied voltage by 90° in a series LC circuit. This means that ________.
   a) the circuit is net capacitive
   b) X_C is greater than X_L
   c) the circuit will produce leading VAR
   d) all of the above

6. As the frequency applied to a series LC circuit increases, ________.
   a) the phase angle between E and I_T decreases
   b) the circuit becomes more inductive
   c) the phase angle between E and I_T increases
   d) the circuit becomes more capacitive

7. In a series LC circuit, if $X_C > X_L$ ________.
   a) X_S is positive
   b) the resultant vector is positive
   c) circuit current leads E by 90°
   d circuit current lags E by 90°

8. The current in a parallel LC circuit lags the applied voltage. This means that ________.
   a) $I_L > I_C$
   b) $X_L < X_C$
   c) circuit VARs are lagging
   d) all of the above

9. In a parallel LC circuit, if the frequency of the applied voltage increases ________.
   a) the circuit becomes less inductive
   b) the circuit becomes less capacitive
   c) total circuit current increases
   d) capacitor current decreases

10. A 100 mH coil is connected in parallel with a 100 μF capacitor. If the applied voltage has a frequency of 60 Hz, the value of X_P is ________.
    a) −89.2 Ω
    b) 64.2 Ω
    c) 89.2 Ω
    d) 15.6 Ω

11. For the circuit described in question 10, the circuit would have the characteristics of a(n) ________ circuit and E ________ I_T by ________.
    a) capacitive, lags, 90°
    b) capacitive, leads, 90°
    c) inductive, lags, 180°
    d) inductive, leads, 90°

12. In a series resonant circuit ________.
    a) $V_L = V_C$
    b) $X_L = X_C$
    c) $X_S = 0\ \Omega$
    d) all of the above

13. A 10 mH coil is connected in series with a 47 μF capacitor. The frequency at which resonance would occur is ________.
    a) 2.32 kHz
    b) 338 Hz
    c) 232 Hz
    d) 3.38 kHz

14. The components described in question 13 are connected in parallel instead of in series. In this case resonance occurs at ________.
    a) 2.32 kHz
    b) 338 Hz
    c) 232 Hz
    d) 3.38 kHz

15. In a series resonant LC circuit, current is limited only by ________.
    a) X_P
    b) X_S
    c) source resistance
    d) net reactance

16. In a parallel resonant LC circuit, total circuit current ________.
    a) leads E by 90°
    b) is near zero
    c) lags E by 90°
    d) approaches infinity

17. In a parallel resonant LC circuit, component voltages ________.
    a) lag circuit current by 90°
    b) are greater than the applied voltage
    c) are less than the applied voltage
    d) lead circuit current by 90°

18. A 150 mH coil, a 68 μF capacitor, and a 33 Ω resistor are connected in series with a 120 V/60 Hz supply. The value of Z_T for this circuit is ________ at a phase angle of ________.
    a) $37.4\ \Omega \angle 27.9°$
    b) $101\ \Omega \angle 70.9°$
    c) $37.4\ \Omega \angle -27.9°$
    d) $101\ \Omega \angle -70.9°$

19. The circuit described in question 18 is net ________ so the VAR are ________.
    a) capacitive, leading
    b) inductive, leading
    c) capacitive, lagging
    d) inductive, lagging

20. If the circuit described in question 18 was operated at resonance, current would ________.
    a) equal 3.64 A
    b) approach infinity
    c) decrease
    d) lead E by 180°

21. In a series RLC circuit, inductor voltage ________.
    a) is lower than V_C
    b) may be much higher than E
    c) is 90° out of phase with E
    d) lags V_R by 90°

22. A series RLC circuit is operated above resonance. This means that V_L ________ V_C.
    a) is less than
    b) lags
    c) is in phase with
    d) is greater than

23. In a parallel RLC circuit, I_{LC} is positive when ________.
    a) $I_C > I_L$
    b) it is operated below resonance
    c) $X_L > X_C$
    d) the circuit is net capacitive
    e) all of the above

24. A parallel RLC circuit has the following current values: $I_L = 4.6$ A, $I_C = 2.9$ A, $I_R = 5.5$ A. This means that the value of $I_T =$ ________.
    a) 5.76 A $\angle 17.2°$
    b) 9.3 A $\angle 53.7°$
    c) 5.76 A $\angle -17.2°$
    d) 9.3 A $\angle -53.7°$

25. Based upon the information given in question 24, we know that this circuit ________.
    a) has lagging VAR
    b) has leading VAR
    c) is net capacitive
    d) is operating above resonance

26. Power factor is the ratio of ________.
    a) reactive to real power
    b) real to apparent power
    c) real to reactive power
    d) reactive to apparent power

27. If you improve the power factor for an industrial installation ________.
    a) you may lower costs
    b) total current will decrease
    c) you may be able to use lower-capacity conductors in the plant
    d) all of the above

28. Power factor is usually not adjusted to a factor of 1 because ________.
    a) it is too expensive
    b) the technology does not exist
    c) it can result in very high voltages due to resonance
    d) the capacitors would be too large

29. Static power factor correction is ________.
    a) a means of eliminating static electricity
    b) correcting the power factor of individual motors
    c) a means of switching in banks of capacitors
    d) a way of dealing with a leading power factor

30. Harmonics can result in high currents in compensating capacitors because ________.
    a) of their leading power factor
    b) capacitors are nonlinear
    c) inductors are nonlinear
    d) as frequency increases, X_C decreases and X_L increases

PRACTICE PROBLEMS

1. A series LC circuit has values of $V_L = 10$ V and $V_C = 3$ V. Determine the magnitude and phase angle of the voltage source.
2. A series LC circuit has values of $V_L = 6$ V and $V_C = 15$ V. Determine the magnitude and phase angle of the voltage source.
3. Determine the magnitude and phase angle of X_S for the circuit in Figure 18.36a.
4. Determine the magnitude and phase angle of X_S for the circuit in Figure 18.36b.
5. Determine the magnitude and phase angle of X_S for the circuit in Figure 18.37a.
6. Determine the magnitude and phase angle of X_S for the circuit in Figure 18.37b.
7. Determine the values of V_L, V_C, and E for the circuit in Figure 18.38a.
8. Determine the values of V_L, V_C, and E for the circuit in Figure 18.38b.

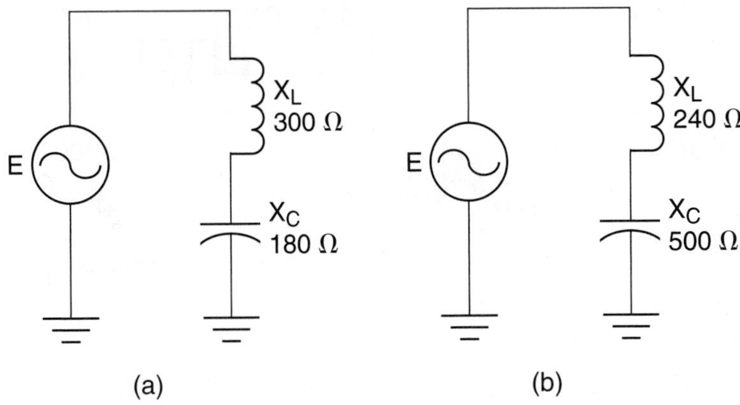

FIGURE 18.36

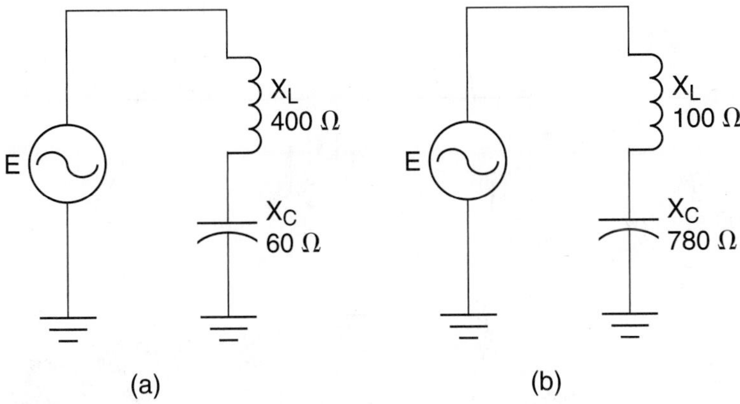

FIGURE 18.37

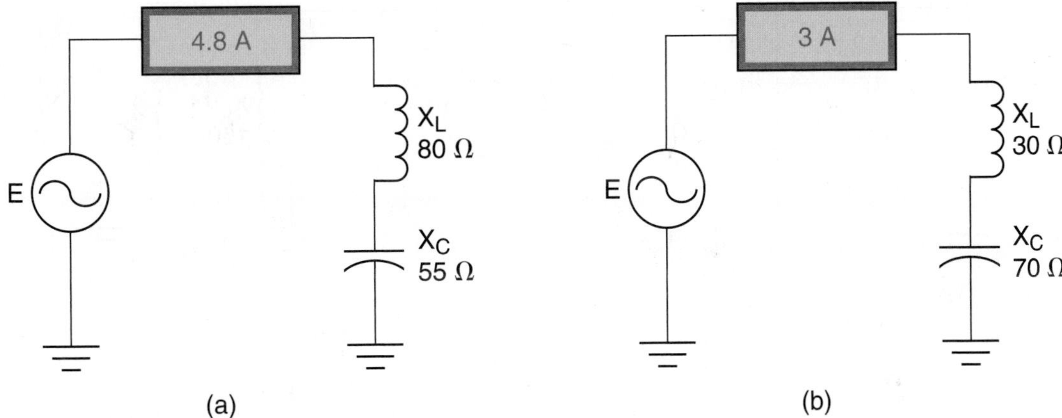

FIGURE 18.38

9. Determine the values of V_L, V_C, and E for the circuit in Figure 18.39a.
10. Determine the values of V_L, V_C, and E for the circuit in Figure 18.39b.
11. Determine the magnitude and phase angle of the circuit current in Figure 18.40a.
12. Determine the magnitude and phase angle of the circuit current in Figure 18.40b.
13. Determine the parallel equivalent reactance for the circuit in Figure 18.41a.
14. Determine the parallel equivalent reactance for the circuit in Figure 18.41b.

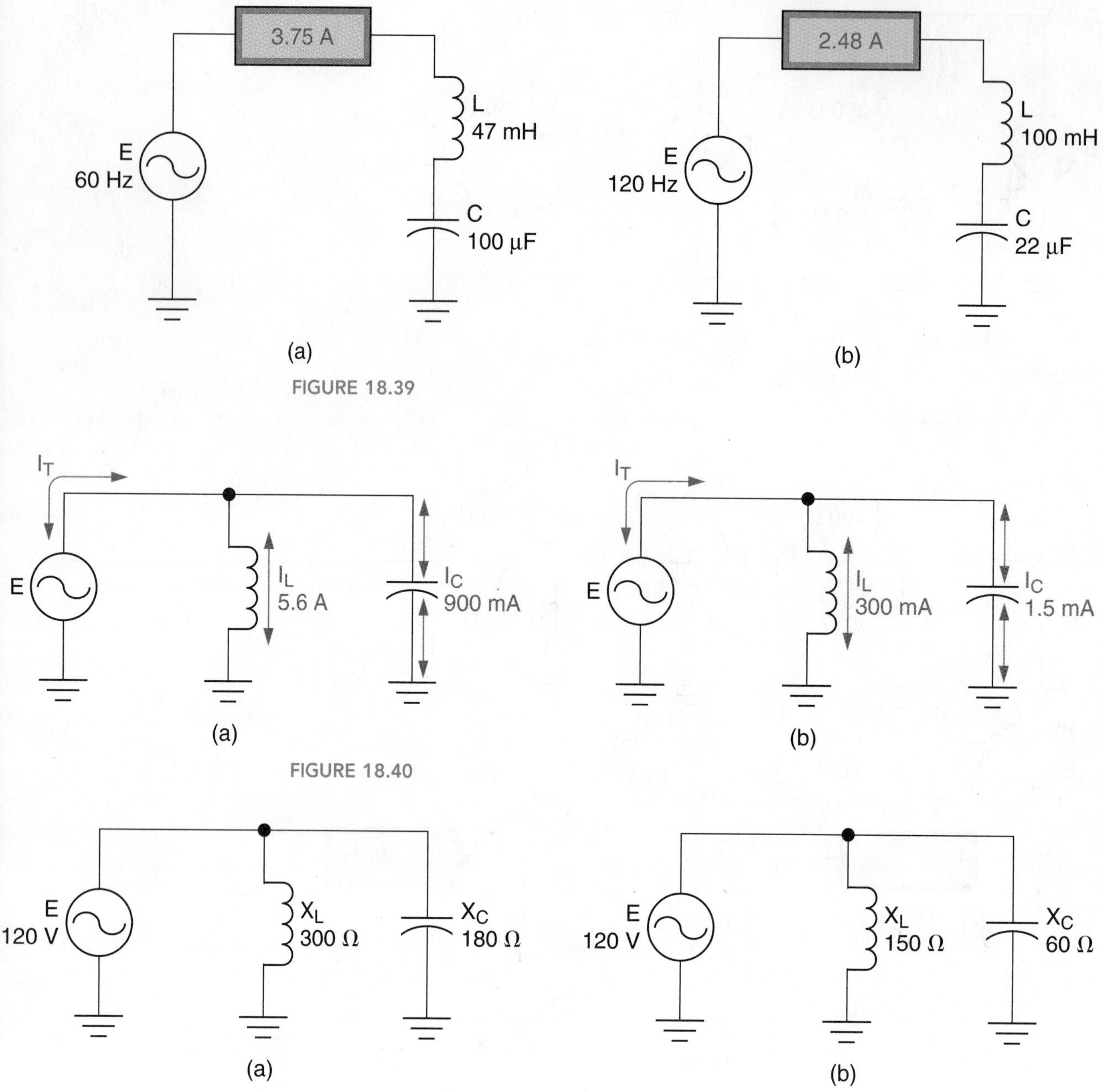

FIGURE 18.39

FIGURE 18.40

FIGURE 18.41

15. Determine the resonant frequency for the circuit in Figure 18.42a.
16. Determine the resonant frequency for the circuit in Figure 18.42b.
17. Calculate the values of X_L and X_C for the circuit in Figure 18.42a when it is operating at its resonant frequency.
18. Calculate the values of X_L and X_C for the circuit in Figure 18.42b when it is operating at its resonant frequency.
19. Calculate the values of V_L, V_C, and V_{LC} for the circuit shown in Figure 18.43a.
20. Calculate the values of V_L, V_C, and V_{LC} for the circuit shown in Figure 18.43b.

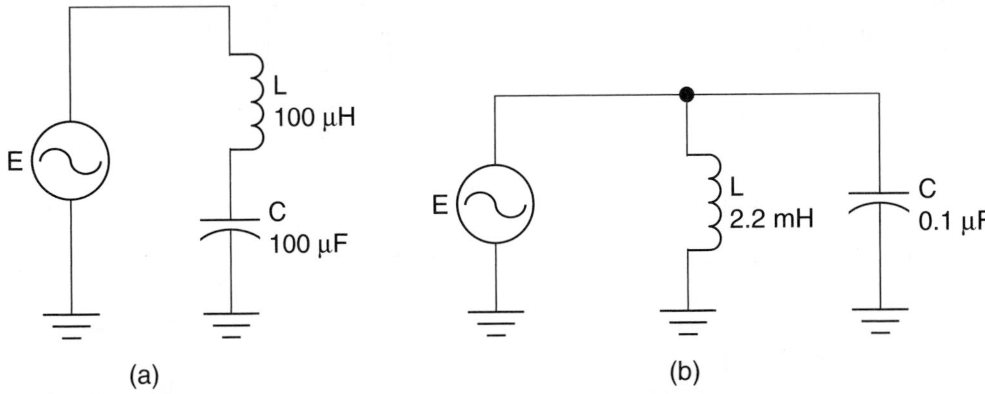

FIGURE 18.42

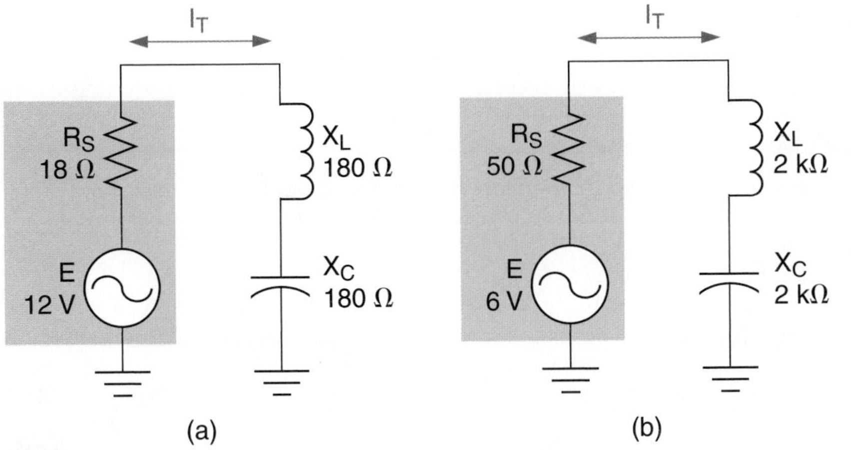

FIGURE 18.43

21. Determine the magnitude and phase angle of the circuit impedance in Figure 18.44a.
22. Determine the magnitude and phase angle of the circuit impedance in Figure 18.44b.

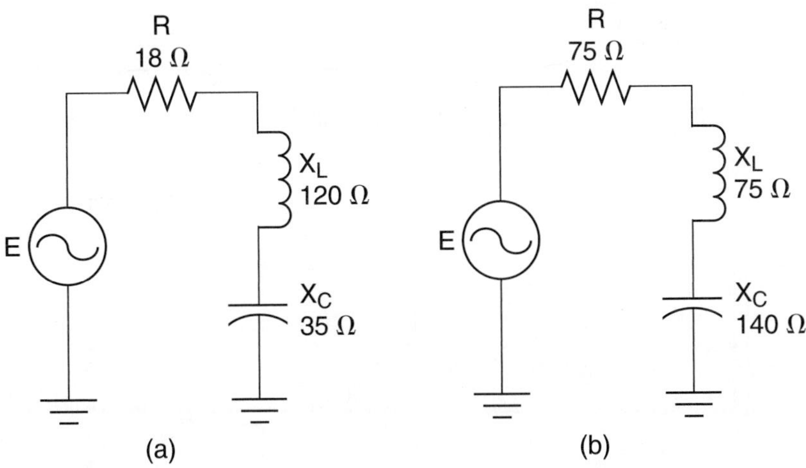

FIGURE 18.44

23. Determine the magnitude and phase angle of the source voltage in Figure 18.45a.
24. Determine the magnitude and phase angle of the source voltage in Figure 18.45b.

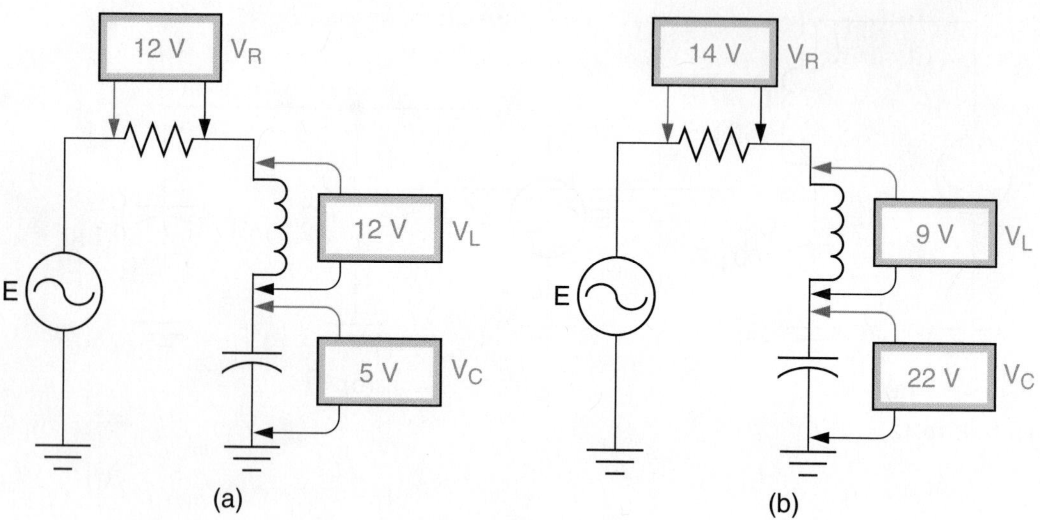

FIGURE 18.45

25. Calculate all current, voltage, impedance, and power values for the circuit shown in Figure 18.46a.
26. Calculate all current, voltage, impedance, and power values for the circuit shown in Figure 18.46b.

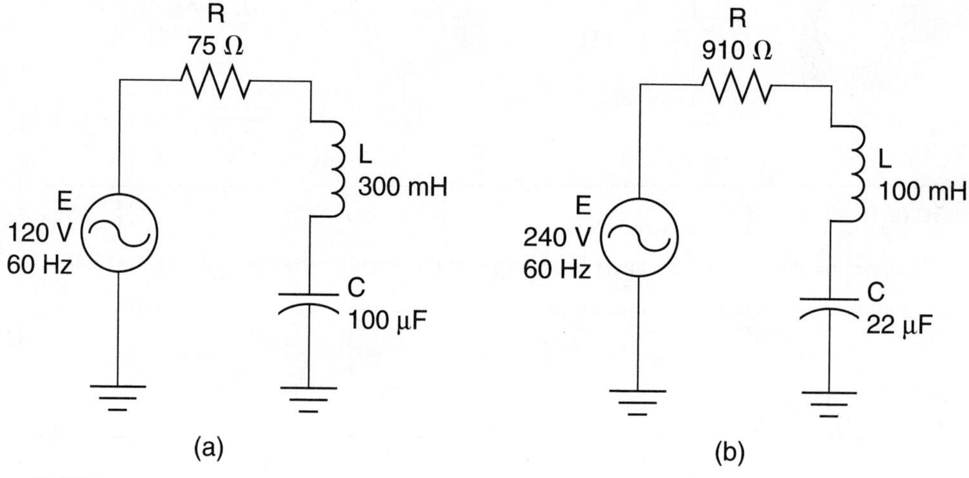

FIGURE 18.46

27. Determine the magnitude and phase angle of the circuit current in Figure 18.47a.

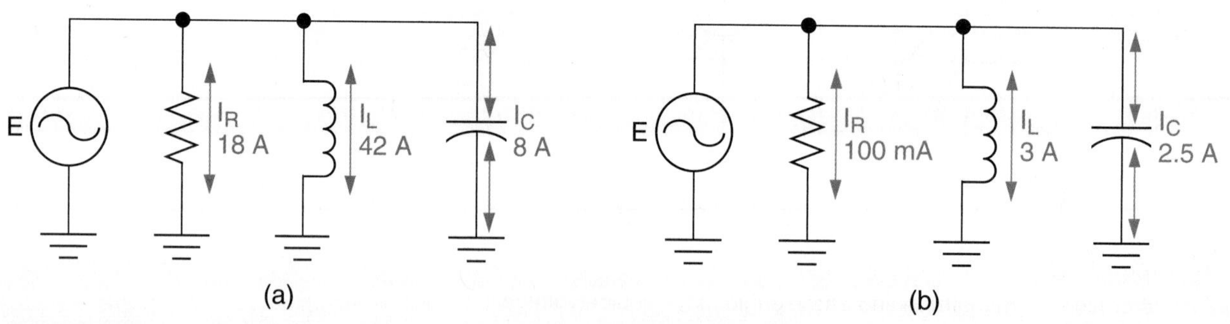

FIGURE 18.47

28. Determine the magnitude and phase angle of the circuit current in Figure 18.47b.
29. Calculate all current, impedance, and power values for the circuit shown in Figure 18.48a.
30. Calculate all current, impedance, and power values for the circuit shown in Figure 18.48b.

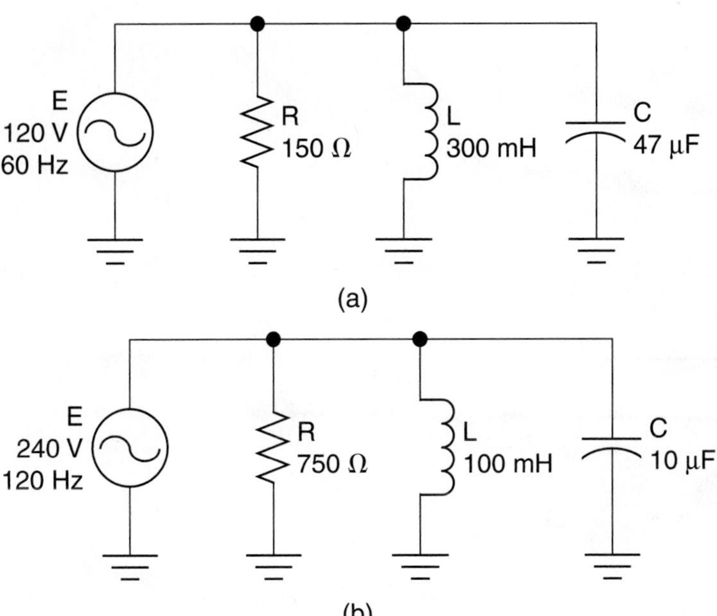

FIGURE 18.48

ANSWERS TO THE EXAMPLE PRACTICE PROBLEMS

18.1 −30 V, capacitive

18.3 25 Ω, inductive

18.5 45 mA, capacitive

18.7 $X_L = 40\ \Omega$, $X_C = 66.7\ \Omega$, $X_P = 100\ \Omega$

18.9 40.4 Hz

18.10 $V_L = 540$ V, $V_C = 540$ V, $V_{LC} = 0$ V

18.11 $I_C = 3$ A, $I_L = 3$ A, $I_T = 0$ A, $Z_T = \infty\ \Omega$

18.12 384 Ω, 38.7°

18.13 225 Ω, -57.7°

18.14 36.1 V, 56.3°

18.15 154 V, −51.3°

18.16 9.22 A, −49.4°

18.17 8.25 A, 76°

18.18 $X_L = 16.6\ \Omega$, $X_C = 88.4\ \Omega$, $I_L = 1.45$ A, $I_C = 271$ mA, $I_{LC} = -1.18$ A, $I_R = 320$ mA, $I_T = 1.22$ A, $\theta = -74.8°$, $Z = 19.7\ \Omega$

CHAPTER 19

FREQUENCY RESPONSE AND FILTERS

PURPOSE

We have touched on the topic of frequency throughout the past five chapters. Now, we will take a more in-depth look at the effect that frequency can have on the characteristics of a group of circuits called *filters*.

KEY TERMS

The following terms are introduced and defined in this chapter on the pages indicated:

Attenuation, 801
Band-stop filter, 804
Bandpass filter, 804
Bandwidth, 805
Center frequency (f_0), 807
Cutoff frequency (f_C), 804
Decade, 811
Decade scale, 812
Filter, 804
Frequency response, 801
Frequency response curve, 803
Gain, 803
High-pass filter, 804
Logarithmic scale, 811
Low-pass filter, 804
Notch filter, 804
Octave, 812
Octave scale, 812
Power gain, 804
Quality (Q), 808
Unity gain, 803
Voltage gain, 803

OBJECTIVES

After completing this chapter, you should be able to:

1. Describe attenuation.
2. Describe the relationship between circuit amplitude and cutoff frequency.
3. List the four primary filters and identify their frequency response curves.
4. Identify (and solve for) each of the following on a frequency response curve: bandwidth, center frequency, lower cutoff frequency, and upper cutoff frequency.
5. Discuss the relationship among filter Q, bandwidth, and center frequency.
6. Describe and calculate the *average frequency* of a bandpass (or notch) filter.
7. Describe the relationship among filter Q, center frequency, and average frequency.
8. List and describe the commonly used logarithmic frequency scales.
9. Describe and analyze the operation of the RC low-pass filter.
10. Describe and analyze the operation of the RL low-pass filter.

11. Describe and analyze the operation of the RC high-pass filter.
12. Describe and analyze the operation of the RL high-pass filter.
13. Describe and analyze the operation of series LC bandpass filters.
14. Describe and analyze the operation of shunt LC bandpass filters.
15. Describe the operation of LC notch filters.

19.1 INTRODUCTION TO FREQUENCY RESPONSE

The reaction a circuit has to a change in frequency is referred to as its **FREQUENCY RESPONSE**. For example, consider the parallel RLC circuit shown in Figure 19.1. If the operating frequency of the circuit *increases*, here's what happens:

FREQUENCY RESPONSE The reaction a circuit has to a change in frequency.

- The reactance of the inductor *increases*, because $X_L = 2\pi fL$. The increase in X_L causes the inductor current (I_L) to *decrease*.
- The reactance of the capacitor *decreases*, because $X_C = \frac{1}{2\pi fC}$. The decrease in X_C causes the capacitor current (I_C) to *increase*.
- The changes in I_L and I_C cause the total circuit current to change, as well as the circuit phase angle (θ).

These changes are collectively referred to as the circuit *frequency response*.

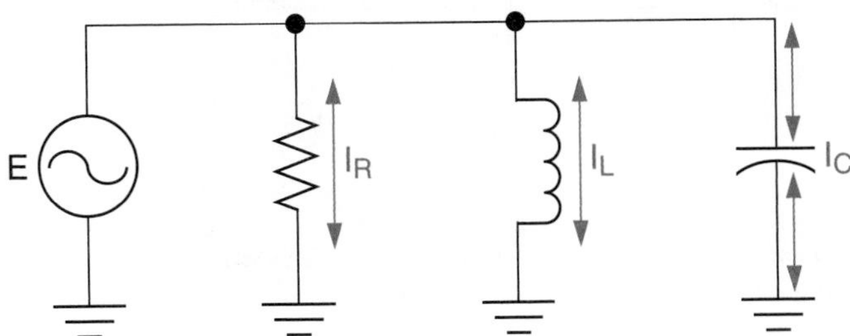

FIGURE 19.1 A parallel RLC circuit.

The subject of frequency response is complex, with many principles and concepts. In this section, we will introduce the most general principles. Many of the topics introduced in this section are discussed in more detail later in the chapter. For now, we are simply establishing a starting point.

Attenuation

Every circuit responds to a change in operating frequency to some extent. The changes may be too subtle to observe, or they may be dramatic. An example of the latter is illustrated in Figure 19.2.

ATTENUATION Any signal loss caused by the frequency response of a circuit.

The circuit represented in Figure 19.2a has input and output waveforms that are nearly identical. When the input frequency is increased by a factor of ten (as shown in Figure 19.2b), the amplitude of the circuit output drops significantly. Any signal loss caused by the frequency response of a circuit is referred to as **ATTENUATION**.

> Decibels are used to express the ratio of one circuit value to another. In electronics, an important consideration is the ratio of circuit output voltage (or power) to circuit input voltage (or power).
>
> KEY CONCEPT

The frequency response just described can be graphed as shown in Figure 19.2c. As you can see, the amplitude of

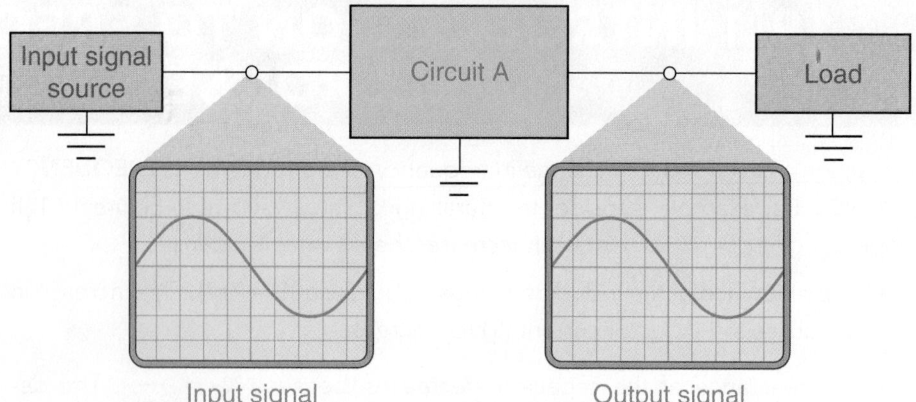

(a) Circuit A input and output signals at a given frequency (f).

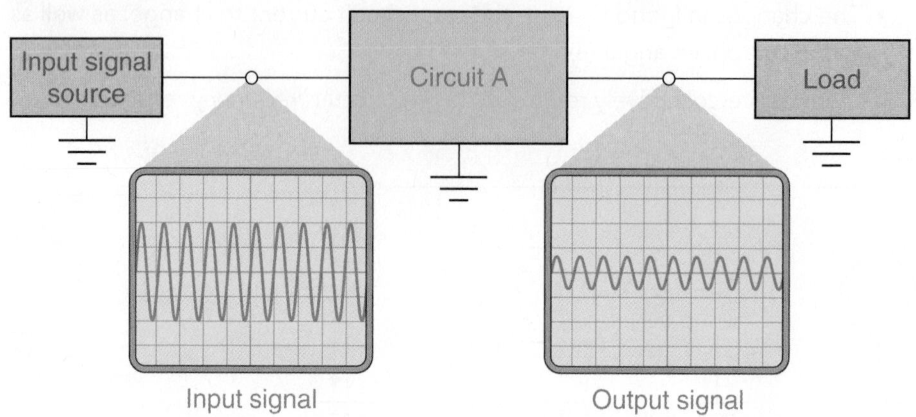

(b) Circuit A input and output signals at 10 times the original frequency (10f).

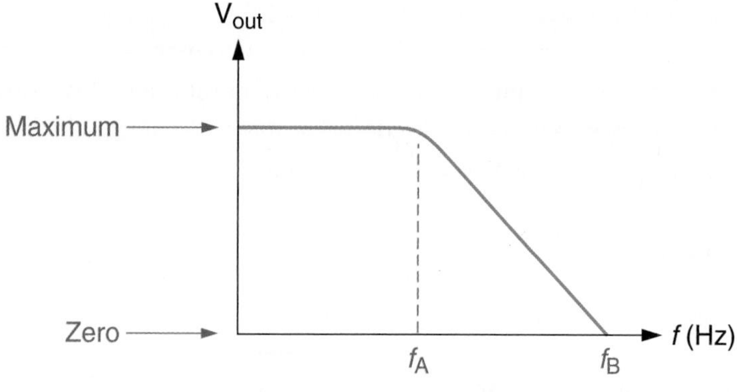

(c) Frequency response curve.

FIGURE 19.2 Attenuation.

the circuit output stays relatively constant over the range of frequencies below the frequency labeled f_A. As operating frequency increases above f_A, the attenuation introduced by the circuit causes the amplitude to drop. If the operating frequency reaches the frequency labeled f_B, the output amplitude decreases to zero. (This means the circuit has no output signal, even though it still has an input signal.)

Frequency Response Curves

The frequency response of a circuit is commonly described in terms of the effect that a change in operating frequency has on *the ratio of output amplitude to input amplitude*. For example, the curve in Figure 19.3 represents the frequency response of a circuit with a maximum value of $\frac{V_{out}}{V_{in}} = 1$. According to the curve, the ratio of V_{out} to V_{in} decreases from

$$\frac{V_{out}}{V_{in}} = 1 \qquad \text{(at } f_A\text{)}$$

to

$$\frac{V_{out}}{V_{in}} = 0 \qquad \text{(at } f_B\text{)}$$

The ratio of a circuit's output amplitude to its input amplitude is referred to as **GAIN**. As such, the circuit represented by the curve in Figure 19.3 is described as having a maximum **VOLTAGE GAIN** of one (1). A gain of one is commonly called **UNITY GAIN**.

GAIN The ratio of a circuit's output amplitude to its input amplitude.

VOLTAGE GAIN The ratio of a circuit's output voltage to its input voltage.

UNITY GAIN A gain of one (1).

> Circuit *gain* has no unit of measure (such as volts or amps) because it is a ratio of two values. For example, the values of current gain, voltage gain, and power gain are ratios of an output value to its corresponding input value (such as output power to input power, etc).
>
> **KEY CONCEPT**

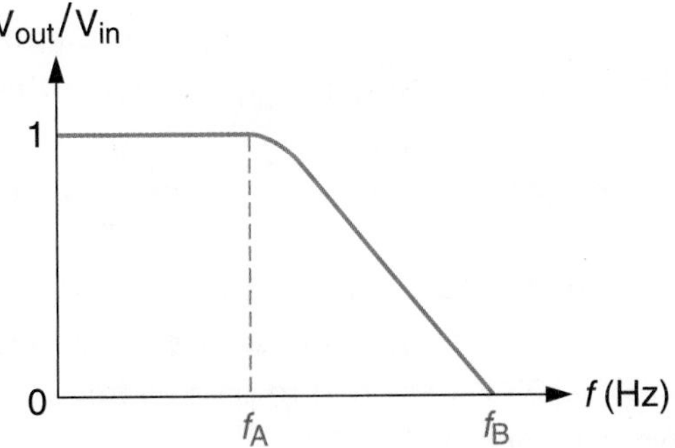

FIGURE 19.3 A graph of voltage gain vs. frequency.

Cutoff Frequency (f_C)

A **FREQUENCY RESPONSE CURVE** is a graph that shows the effect that frequency has on circuit gain. In most cases, these curves use a power ratio as the indicator

FREQUENCY RESPONSE CURVE A graph that shows the effect that frequency has on circuit gain.

POWER GAIN The ratio of a circuit's output power to its input power.

of gain. For example, the curve shown in Figure 19.4 represents the relationship between frequency and **POWER GAIN** (the ratio of output power to input power).

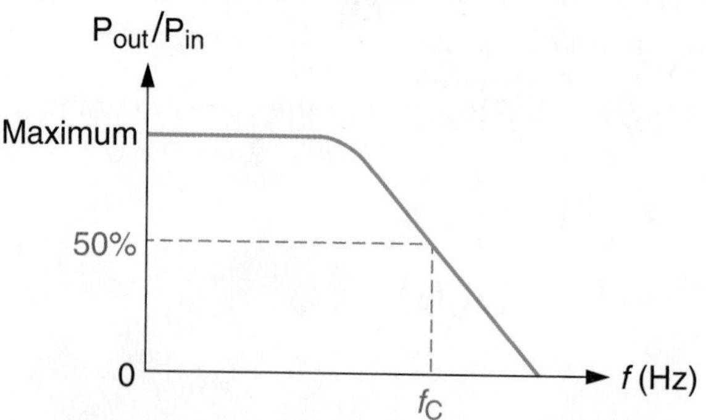

FIGURE 19.4 A graph of power gain versus frequency.

CUTOFF FREQUENCY (f_C) The frequency at which the power gain of a circuit drops to 50% of its maximum value.

FILTER A circuit designed to pass a specific range of frequencies from input to output while blocking others.

The frequency at which the power gain of a circuit drops to 50% of its maximum value is referred to as its **CUTOFF FREQUENCY (f_C)**. The cutoff frequency is identified on the curve in Figure 19.4. By convention, the cutoff frequency on a curve is the dividing line between acceptable and unacceptable output levels from a component or circuit. When the power gain of a circuit is below 50% of its maximum value, the output has been attenuated to the point of being considered unacceptable. It should be noted that the 50% standard for cutoff frequency is an arbitrary value; that is, it is used only because it is agreed upon by professionals in the field. The cutoff frequency of a component or circuit is determined by its resistive and reactive values (as is demonstrated later in this chapter).

Filters

LOW-PASS FILTER A circuit designed to pass all frequencies below its cutoff frequency.

HIGH-PASS FILTER A circuit designed to pass all frequencies above its cutoff frequency.

BANDPASS FILTER A circuit designed to pass the band of frequencies that lies between two cutoff frequencies, labeled f_{C1} and f_{C2}.

BAND-STOP FILTER A circuit designed to block the band of frequencies that lies between its cutoff frequencies, labeled f_{C1} and f_{C2}.

NOTCH FILTER Another name for a band-stop filter.

Many circuits are designed for specific frequency response characteristics; that is, they are designed to pass a specific range of frequencies from input to output while blocking (or *rejecting*) others. These circuits are generally referred to as **FILTERS**.

There are four characteristic filter response curves. These curves are shown in Figure 19.5. Figure 19.5a shows the response curve for a **LOW-PASS FILTER**. A low-pass filter is designed to pass all frequencies *below* its cutoff frequency (f_C). For example, a low-pass filter with a value of $f_C = 1000$ Hz will pass all frequencies below 1000 Hz. In contrast, a **HIGH-PASS FILTER** is designed to pass all frequencies *above* its cutoff frequency (as shown in Figure 19.5b).

A **BANDPASS FILTER** is one is designed to pass the band of frequencies that lies between two cutoff frequencies, labeled f_{C1} and f_{C2}, as shown in Figure 19.5c. In contrast, a **BAND-STOP FILTER**, or **NOTCH FILTER**, is designed to block the band of frequencies that lies between its cutoff frequencies. As shown in Figure 19.5d, the notch filter passes frequencies that are below f_{C1} and above f_{C2}.

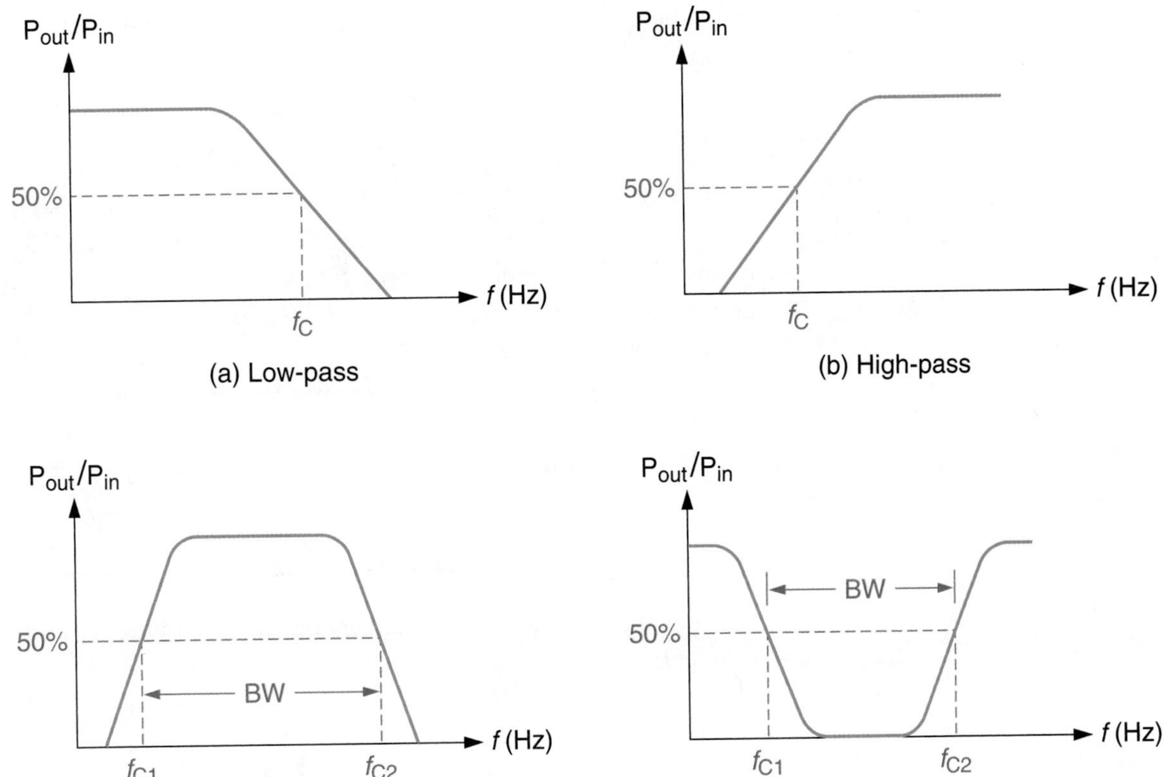

FIGURE 19.5 Filter frequency response curves.

Because bandpass and notch filters have two cutoff frequencies, two values are used to describe their operation that are not commonly applied to low-pass and high-pass filters. These values are called *bandwidth* and *center frequency*.

Bandwidth and Center Frequency

The range (or *band*) of frequencies between the cutoff frequencies of a filter is referred to as its **BANDWIDTH**. The concept of bandwidth is illustrated in Figure 19.6.

BANDWIDTH The range (or *band*) of frequencies between the cutoff frequencies of a filter.

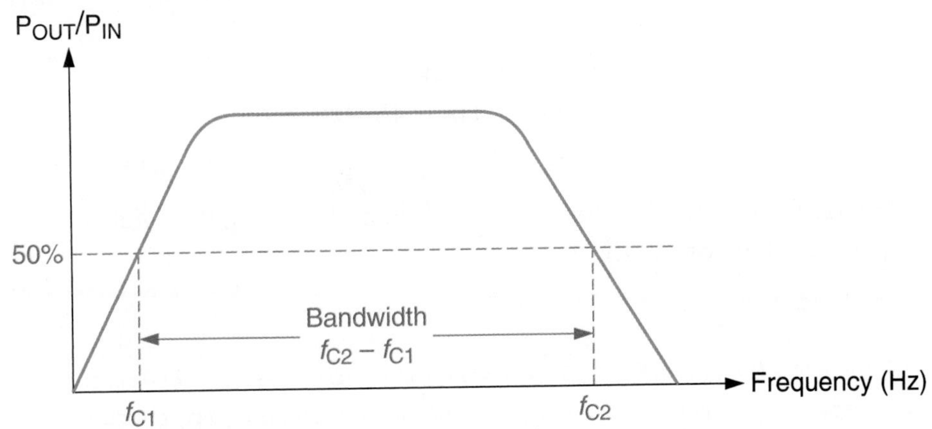

FIGURE 19.6 Bandwidth.

As shown in the figure, bandwidth is equal to the difference between the cutoff frequencies. By formula,

(19.1)

$$BW = f_{C2} - f_{C1}$$

where

BW = the bandwidth of the component or circuit, in Hz
f_{C2} = the upper cutoff frequency
f_{C1} = the lower cutoff frequency

The bandwidth of a filter can be calculated as demonstrated in Example 19.1.

EXAMPLE 19.1

A bandpass filter has the frequency response curve shown in Figure 19.7. Calculate the circuit bandwidth.

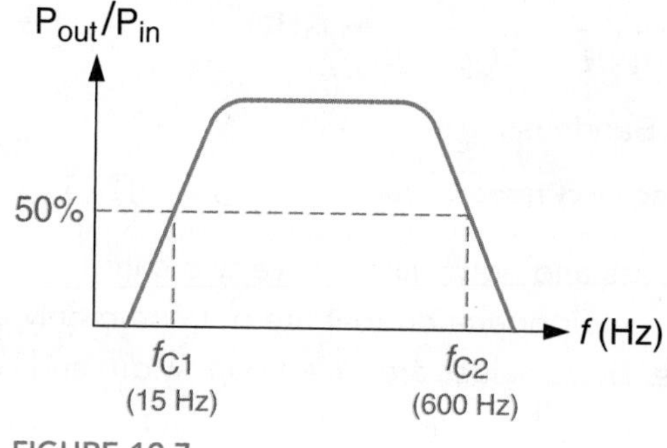

FIGURE 19.7

SOLUTION

Using the cutoff frequencies shown on the curve, the filter bandwidth is found as

$$\begin{aligned} BW &= f_{C2} - f_{C1} \\ &= 600 \text{ Hz} - 15 \text{ Hz} \\ &= 585 \text{ Hz} \end{aligned}$$

This result indicates that there is a 585 Hz spread between the cutoff frequencies of the circuit.

PRACTICE PROBLEM 19.1

A curve like the one in Figure 19.7 has the following values: $f_{C2} = 880$ Hz and $f_{C1} = 62$ Hz. Calculate the bandwidth of the circuit represented by the curve.

Center Frequency (f_0)

To accurately describe the frequency response of a bandpass or notch filter, we need to know more than its bandwidth. We also need to know *where* the frequency response curve is located on the frequency spectrum. For example, consider the bandpass curves shown in Figure 19.8. Each curve has a bandwidth of 100 Hz, yet they have different cutoff frequencies, and therefore are not the same curve. To distinguish between them, we must identify the *center frequency* of each curve.

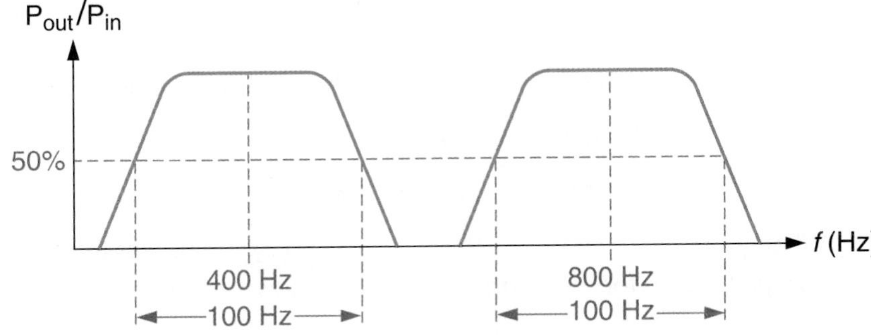

FIGURE 19.8 Center frequency and bandwidth.

Technically defined, **CENTER FREQUENCY (f_0)** is the frequency that equals the geometric average of the cutoff frequencies. The geometric average of any two values equals the square root of their product. Therefore, center frequency is found as

CENTER FREQUENCY (f_0) The frequency that equals the geometric average of the cutoff frequencies.

$$f_0 = \sqrt{f_{C1} f_{C2}} \tag{19.2}$$

Example 19.2 shows how the center frequency is calculated for a filter.

EXAMPLE 19.2

A bandpass filter has the following values: $f_{C1} = 50$ Hz and $f_{C2} = 400$ Hz. Calculate the value of the circuit's center frequency.

SOLUTION

Using the values given, the value of f_0 is found as

$$\begin{aligned} f_0 &= \sqrt{f_{C1} f_{C2}} \\ &= \sqrt{(50 \text{ Hz})(400 \text{ Hz})} \\ &= 141 \text{ Hz} \end{aligned}$$

PRACTICE PROBLEM 19.2

A band-stop filter has the following values: $f_{C1} = 200$ Hz and $f_{C2} = 1200$ Hz. Calculate the value of the circuit's center frequency.

When we first hear the term *center frequency*, we tend to picture a frequency that is halfway between the cutoff frequencies. However, this is not always the case. Consider the results in Example 19.2. Obviously, 141 Hz is not halfway between 50 Hz and 400 Hz. However, it is at the *geometric center* of the two values, meaning that the ratio of f_0 to f_{C1} equals the ratio of f_{C2} to f_0. By formula,

(19.3)
$$\frac{f_0}{f_{C1}} = \frac{f_{C2}}{f_0}$$

If we apply this relationship to the values obtained in the example, we get

$$\frac{141.4\text{ Hz}}{50\text{ Hz}} = \frac{400\text{ Hz}}{141.4\text{ Hz}}$$

which is true. When you solve the ratios with your calculator, you see that they are equal. This is always the case for the geometric average of two values.

Filter Quality (Q)

QUALITY (Q) For a bandpass or notch filter, the ratio of its center frequency to its bandwidth.

The **QUALITY (Q)** of a bandpass or notch filter is the ratio of its center frequency to its bandwidth. By formula,

(19.4)
$$Q = \frac{f_0}{BW}$$

Example 19.3 illustrates this relationship.

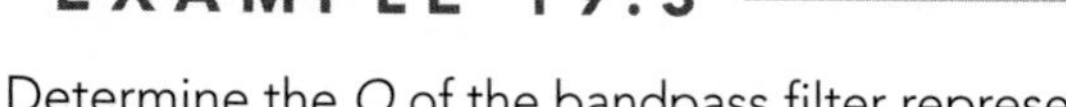

EXAMPLE 19.3

Determine the Q of the bandpass filter represented by the curve in Figure 19.9a.

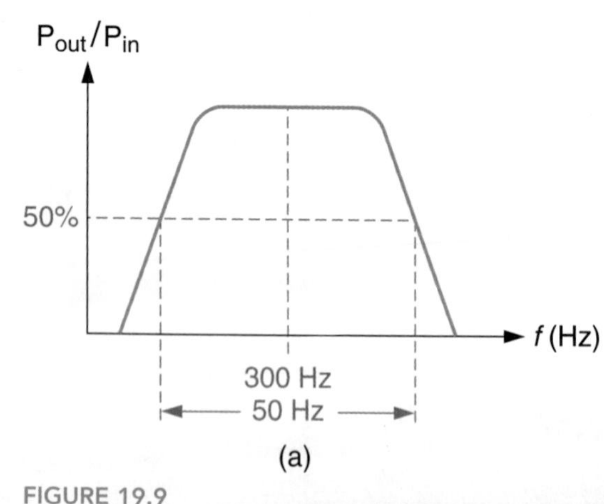

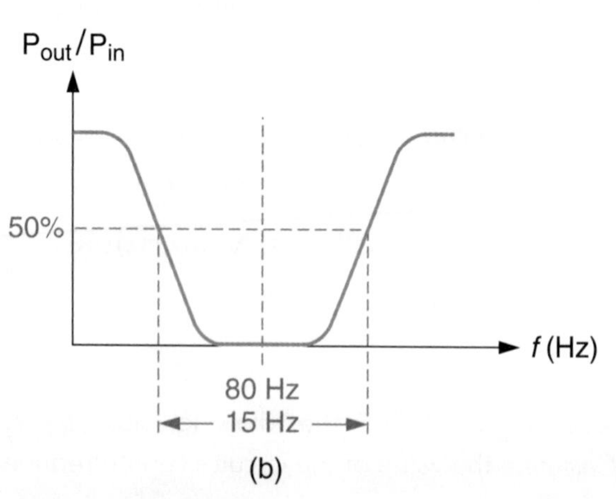

FIGURE 19.9

SOLUTION

The curve has values of $f_0 = 300$ Hz and BW = 50 Hz. Using these two values, the Q of the filter is found as

$$Q = \frac{f_0}{BW} = \frac{300 \text{ Hz}}{50 \text{ Hz}} = 6$$

Note that the result has no unit of measure, since it is a ratio of one frequency to another.

PRACTICE PROBLEM 19.3

Determine the Q of the notch filter represented in Figure 19.9b.

There is an important point to be made at this time. Equation (19.4) is somewhat misleading because it implies that the value of Q depends on the values of center frequency and bandwidth. However, this is not the case. The Q of a filter and its center frequency are both determined by circuit component values (as will be demonstrated later in this chapter). Once their values are known, the filter Q and center frequency are used to calculate the bandwidth of the circuit, as follows:

$$BW = \frac{f_0}{Q} \tag{19.5}$$

Example 19.4 provides a more practical application of the relationship among Q, center frequency, and bandwidth.

EXAMPLE 19.4

Through a series of calculations, a bandpass filter is determined to have values of Q = 4.8 and $f_0 = 960$ Hz. Calculate the circuit bandwidth.

SOLUTION

Using the calculated values of Q and center frequency, the circuit bandwidth is found as

$$BW = \frac{f_0}{Q} = \frac{960 \text{ Hz}}{4.8} = 200 \text{ Hz}$$

PRACTICE PROBLEM 19.4

Using a series of calculations, a notch filter is found to have values of $Q = 3.3$ and $f_0 = 165$ Hz. Determine the circuit bandwidth.

The methods used to calculate the values of Q and f_0 vary from one type of filter to another. For now, we simply want to establish the fact that filter bandwidth is determined by the circuit values of Q and f_0.

Average Frequency (f_{AVE})

Every bandpass and notch filter has both a center frequency (f_0) and an *average frequency* (f_{AVE}). This average frequency is the frequency that lies halfway between the cutoff frequencies. By formula,

(19.6)
$$f_{AVE} = \frac{f_{C1} + f_{C2}}{2}$$

As demonstrated in Example 19.5, there can be a significant difference between the values of f_0 and f_{AVE}.

EXAMPLE 19.5

Determine the values of f_0 and f_{AVE} for the circuit represented by the curve in Figure 19.10a.

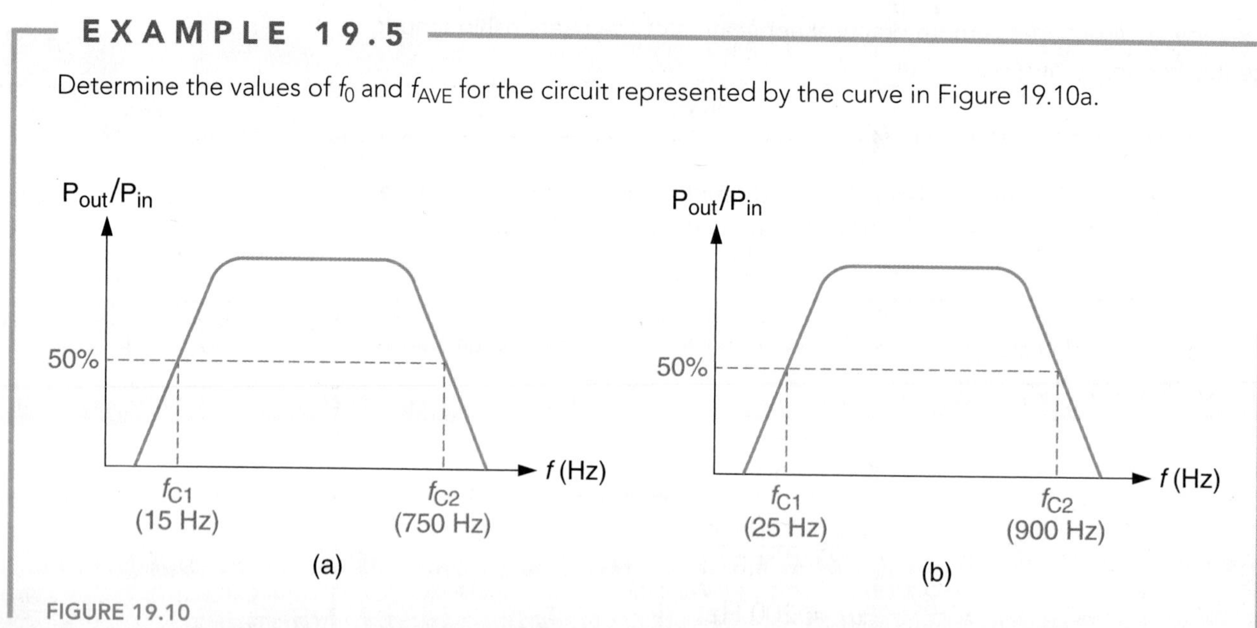

FIGURE 19.10

SOLUTION

Using the values shown on the curve, the *geometric* average of the cutoff frequencies (f_0) is found as

$$f_0 = \sqrt{f_{C1} f_{C2}}$$
$$= \sqrt{(15 \text{ Hz})(750 \text{ Hz})}$$
$$= 106 \text{ Hz}$$

The *algebraic* average of the cutoff frequencies is found as

$$f_{AVE} = \frac{f_{C1} + f_{C2}}{2}$$
$$= \frac{15 \text{ Hz} + 750 \text{ Hz}}{2}$$
$$= 382.5 \text{ Hz}$$

PRACTICE PROBLEM 19.5

Determine the values of f_0 and f_{AVE} for the circuit represented by the curve in Figure 19.10b.

Frequency Scales

Up to this point, we have shown only the given and calculated frequencies on our response curves. In practice, frequency response curves are commonly plotted using frequency scales like those shown in Figure 19.11.

LOGARITHMIC SCALE A scale with a frequency spread between increments that increases at a geometric rate.

DECADE A frequency multiplier of 10.

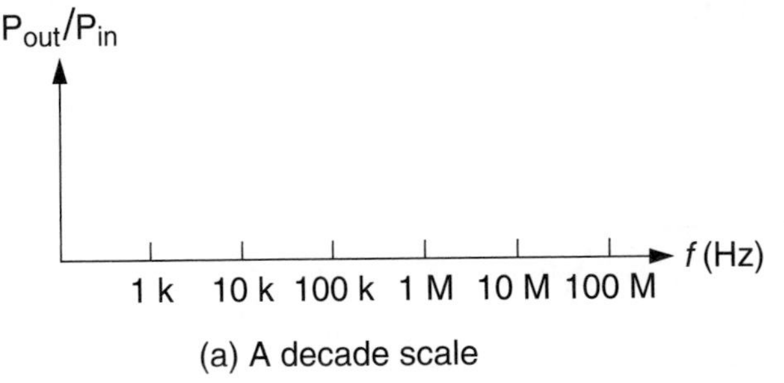

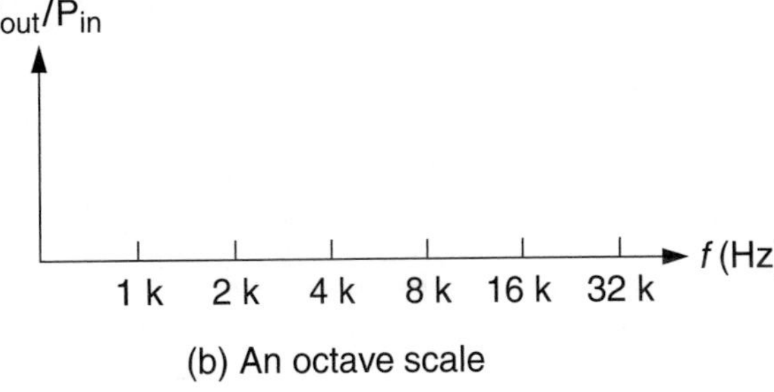

FIGURE 19.11 Logarithmic frequency scales.

The scales shown in the figure are referred to as **LOGARITHMIC SCALES**, meaning that the frequency spread from one increment to the next increases at a geometric rate. In a geometric scale, the value of each increment is a whole-number multiple of the previous increment. For example, the value of each increment in Figure 19.11a is ten times the value of the previous increment.

A frequency multiplier of 10 is referred to as a **DECADE**. For this reason, the scale shown in Figure 19.11a

Commonly encountered frequencies and frequency ranges include 0 Hz (DC), electric power frequencies (60 Hz in U.S.; 50 Hz in Europe), audio frequencies up to 20 kHz, ultrasonic frequencies up to hypersonic frequencies (five times the audio-frequency limit), video frequencies from zero to 4 MHz, and digital equipment frequencies that can extend into the low GHz range.

IN THE FIELD

DECADE SCALE A scale where the value of each increment is 10 times the value of the previous increment.

OCTAVE A frequency multiplier of two (2).

OCTAVE SCALE A scale where the value of each increment is two times the value of the previous increment.

is called a **DECADE SCALE**. In contrast, two (2) is the multiplier used from one increment to the next in Figure 19.11b. A frequency multiplier of 2 is referred to as an **OCTAVE**, so the scale is called an **OCTAVE SCALE**. Note that both decade and octave scales are used in practice. When a large frequency range is needed, a decade scale is used. When a more precise representation over a smaller frequency range is needed, an octave scale may be used.

When a logarithmic scale is used, the center frequency (which is the geometric average of the cutoff frequencies) falls in the physical center of the curve. This point is illustrated in Figure 19.12. The curve shown was plotted using assumed cutoff frequencies of 8 Hz and 128 Hz. Using equation (19.2), the center frequency of the curve can be found as

$$\begin{aligned} f_0 &= \sqrt{f_{C1} f_{C2}} \\ &= \sqrt{(8\text{ Hz})(128\text{ Hz})} \\ &= 32\text{ Hz} \end{aligned}$$

As shown in Figure 19.12a, the geometric center frequency (f_0) falls very close to the physical center of the curve. If an algebraic scale were used, f_0 would fall very close to the left end of the scale, as shown in Figure 19.12b.

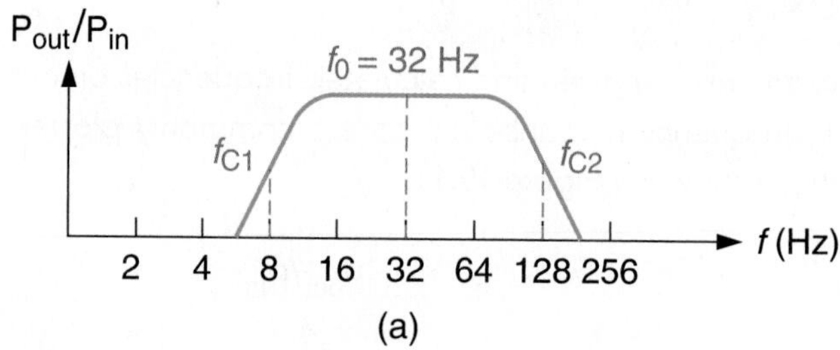

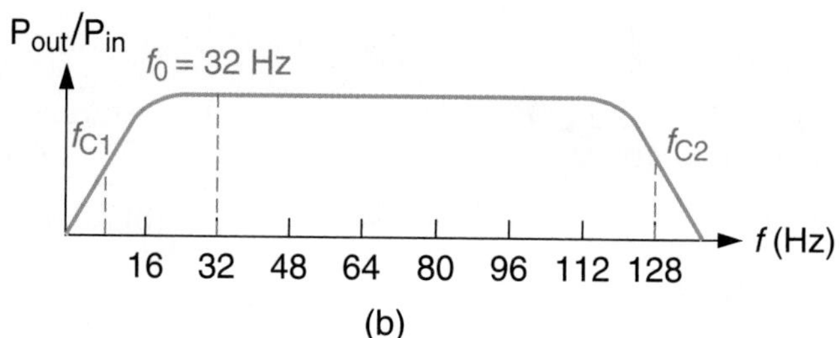

FIGURE 19.12 Plots of f_0 and f_{AVE}.

Critical Frequencies: Putting It All Together

Low-pass and high-pass filters are identified using a single cutoff frequency, as shown in Figure 19.5. Low-pass filters have an upper cutoff frequency (f_{C2}), while high-pass filters have a lower cutoff frequency (f_{C1}). In either case, the cutoff frequency is the operating frequency where the ratio of output power to input power equals 50% of its maximum value. The value of the cutoff frequency for either response curve depends on component values in the circuit itself.

Bandpass and band-stop (or notch) filters each have two cutoff frequencies, as shown in Figure 19.5. A bandpass filter passes the band of frequencies that lies between its cutoff frequencies. A notch filter blocks (attenuates) the band of frequencies that lies between its cutoff frequencies.

Bandpass and notch filters are normally described in terms of their center frequency (f_0) and bandwidth (BW). The center frequency equals the geometric average of the cutoff frequencies, and the bandwidth equals the difference between them. In effect, the center frequency tells us where the curve is located on the frequency spectrum, while the bandwidth tells us the range of frequencies that is passed (or blocked) by the filter.

The average frequency (f_{AVE}) of a bandpass or notch filter equals the algebraic average of its cutoff frequencies. The value of f_{AVE} (not shown in the figure) falls halfway between the cutoff frequencies.

Most frequency response curves are plotted on logarithmic scales. The frequency spread from one increment to the next on a logarithmic scale increases at a geometric rate. That is, the value of each increment is a whole-number multiple of the previous increment.

The two most commonly used logarithmic scales are the decade and octave scales. A decade scale is constructed so that the value of each increment is ten times the value of the previous increment. An octave scale uses two as the multiplier between increments.

PROGRESS CHECK

1. What is attenuation?
2. How is amplitude normally measured on a frequency response curve?
3. What is a cutoff frequency?
4. What is a filter?
5. List the four common types of filters and describe the frequency response of each.
6. What is the bandwidth of a bandpass or notch filter?
7. What two values are required to accurately locate and describe the frequency response curve of a bandpass or notch filter?
8. What is the center frequency of a bandpass or notch filter?
9. What does the quality (Q) value of a filter equal?
10. What is the average frequency (f_{AVE}) of a bandpass or notch filter?
11. What is a logarithmic scale?
12. What is a decade scale? What is an octave scale?

19.2 RC AND RL LOW-PASS FILTERS

Simple RC and RL circuits can be used as low-pass and high-pass filters. In this section, we will look at RC and RL low-pass filters.

RC Low-Pass Filters

An RC circuit acts as a low-pass filter when constructed as shown in Figure 19.13. In the circuit shown, the resistor is positioned directly in the signal path; that is, directly between the source (E) and the load. The capacitor is connected from the signal path to ground, in parallel with the load. Therefore, $V_{RL} = V_C$ (as shown in the figure). When a capacitor (or other component) is connected from a signal path to ground, it is referred to as a *shunt* component.

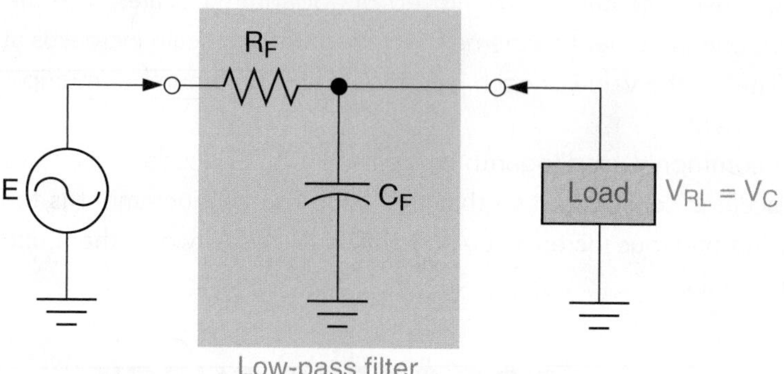

FIGURE 19.13 An RC low-pass filter.

The filtering action of the circuit in Figure 19.13 is a result of the capacitor's response to an increase in frequency. This response can be explained using the curve and circuits in Figure 19.14. The curve in Figure 19.14a shows the relationship between capacitive reactance (X_C) and operating frequency (*f*). As the curve indicates, a capacitor has near-infinite reactance when its operating frequency is 0 Hz. With this in mind, look at the circuit shown in Figure 19.14b. Assuming that the input frequency is 0 Hz, the capacitor effectively acts as an open circuit, so it is left out of the equivalent circuit. In this case, the voltage across the load equals the difference between E and V_{RF}.

At the opposite end of the reactance curve, the value of X_C approaches 0 Ω. When this is the case, the filter has the equivalent circuit shown in Figure 19.14c. As you can see, the capacitor is represented as a short circuit in parallel with the load. In this case, $V_L = 0$ V.

Between the extremes shown in Figure 19.14 lies a range of frequencies over which V_{RL} decreases from its maximum value to approximately 0 V. As V_{RL} decreases, so does the circuit output power. As a result, the circuit has a frequency response curve like the one shown in Figure 19.5a (p 805).

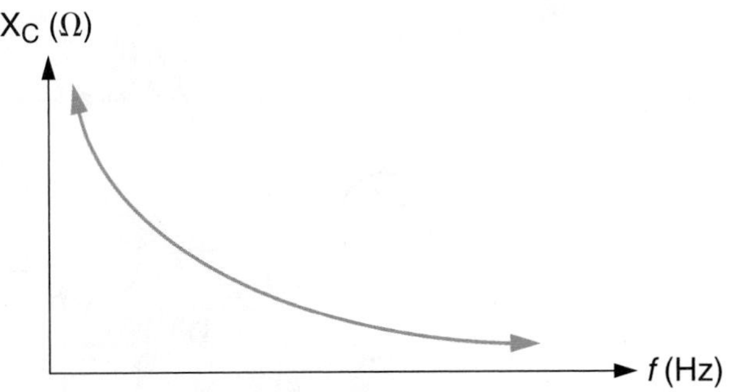

(a) Capacitive reactance vs. frequency

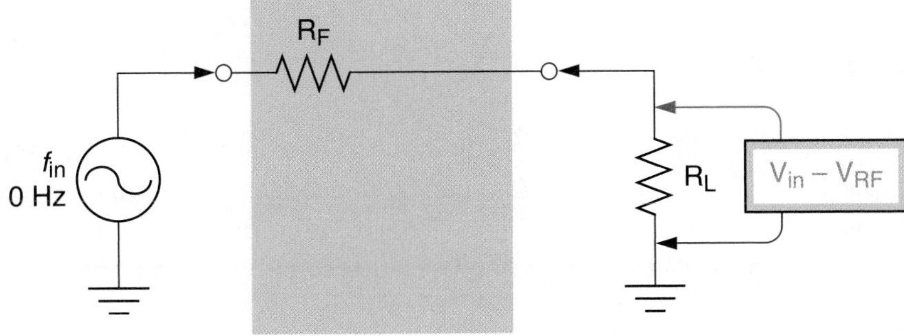

(b) Filter equivalent circuit when $f = 0$ Hz

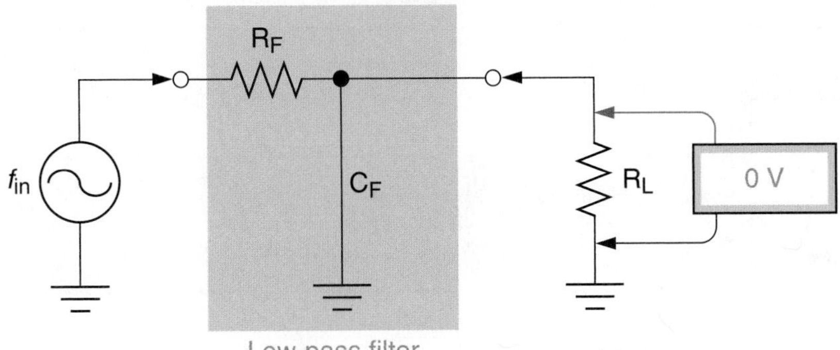

(c) Filter equivalent circuit when $f = \infty$ Hz

FIGURE 19.14 RC low-pass filter operation.

Upper Cutoff Frequency (f_C)

The upper cutoff frequency for an RC low-pass filter is determined by the circuit resistor and capacitor values. By formula,

(19.7)

$$f_C = \frac{1}{2\pi RC}$$

where

f_C = the circuit cutoff frequency
R = the total circuit resistance (as seen by the capacitor)
C = the value of the filter capacitor

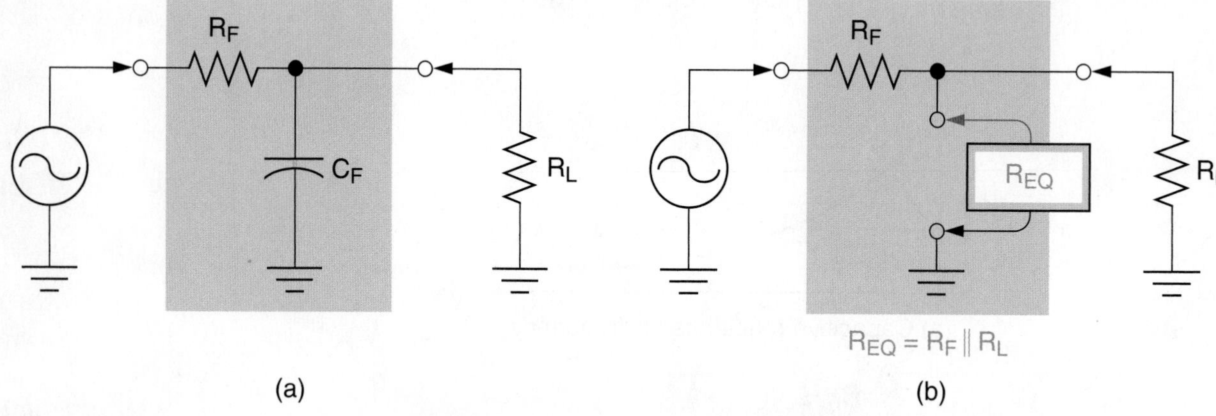

FIGURE 19.15 Resistance in an RC low-pass filter.

The value of R used in the equation is the total resistance as seen by the capacitor. The total resistance (as seen by the capacitor) in Figure 19.15a can be measured by connecting an ohmmeter across the open capacitor terminals as shown in Figure 19.15b. The total resistance seen by the capacitor (R_{EQ}) has a value of

(19.8)

$$R_{EQ} = R_F \parallel R_L$$

and

(19.9)

$$f_C = \frac{1}{2\pi R_{EQ}C}$$

Remember that the ∥ symbol in equation (19.8) indicates that the values are solved as parallel resistors. You can use either the reciprocal approach or the product-over-sum approach to solve for the value of R_{EQ}. Example 19.6 demonstrates the procedure for calculating the cutoff frequency of an RC low-pass filter.

EXAMPLE 19.6

Determine the cutoff frequency for the circuit shown in Figure 19.16a.

SOLUTION

First, the circuit resistance (as seen by the capacitor) is found as

$$R_{EQ} = R_F \parallel R_L$$

$$= \frac{1}{\frac{1}{R_F} + \frac{1}{R_L}}$$

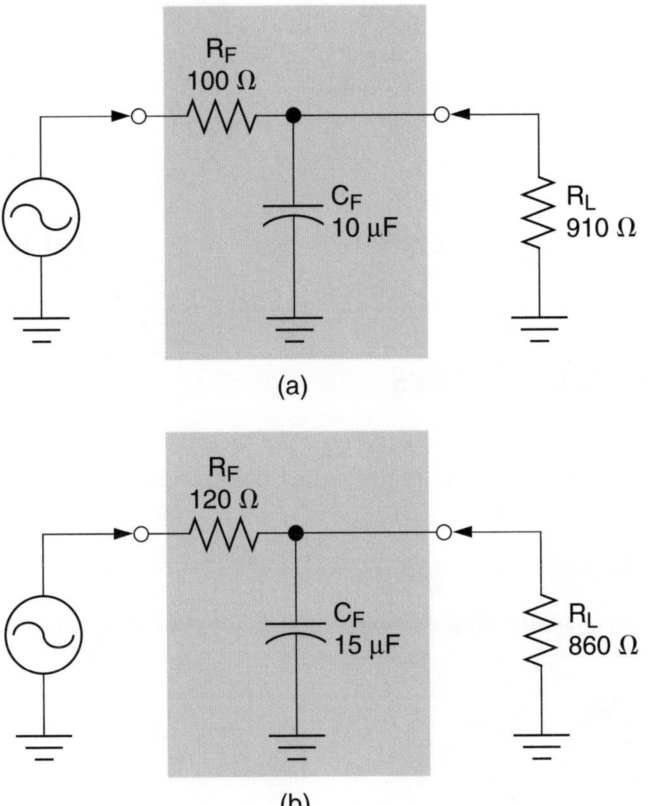

FIGURE 19.16

$$= \frac{1}{\frac{1}{100\ \Omega} + \frac{1}{910\ \Omega}}$$

$$= 90.1\ \Omega$$

Now, the cutoff frequency of the circuit can be found as

$$f_C = \frac{1}{2\pi R_{EQ} C_F}$$

$$= \frac{1}{2\pi(90.1\ \Omega)(10\ \mu\text{F})}$$

$$= 177\ \text{Hz}$$

This result indicates that the power gain of the circuit is reduced to 50% of its maximum value when the operating frequency reaches 177 Hz. The frequency response curve for the circuit is shown in Figure 19.17.

PRACTICE PROBLEM 19.6

Determine the cutoff frequency for the circuit shown in Figure 19.16b.

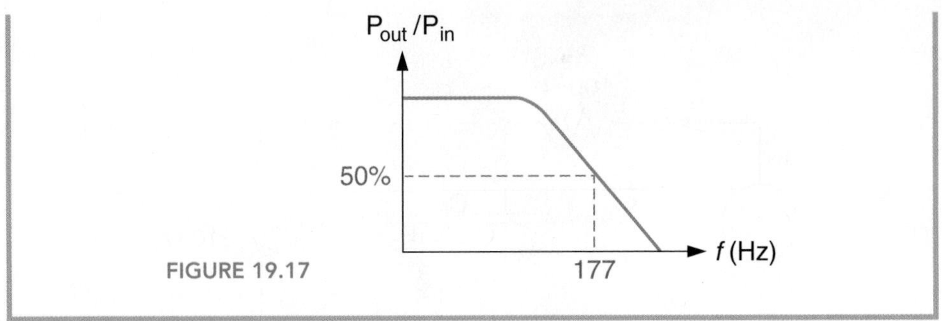

FIGURE 19.17

RL Low-Pass Filters

An RL circuit acts as a low-pass filter when constructed as shown in Figure 19.18. In the circuit shown, the inductor is the series component and the resistor is the shunt component.

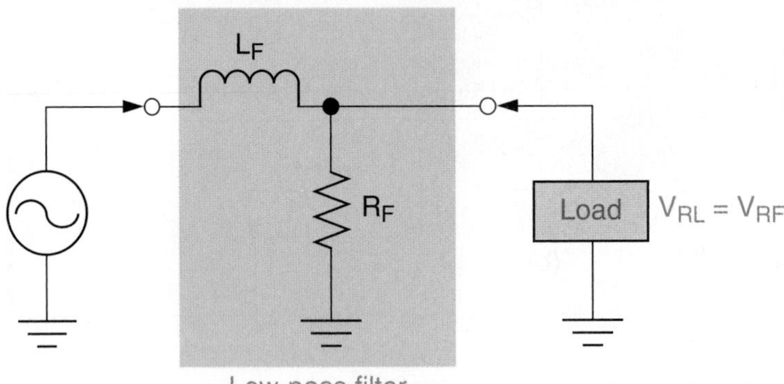

FIGURE 19.18 An RL low-pass filter.

The filtering action of the circuit in Figure 19.18 is a result of the inductor's response to an increase in operating frequency. This response can be explained using the curve and equivalent circuits in Figure 19.19. The curve in Figure 19.19a shows the relationship between inductive reactance (X_L) and operating frequency (f). As the curve indicates, the reactance of an inductor is 0 Ω when the input frequency is 0 Hz. With this in mind, look at the equivalent circuit shown in Figure 19.19b. Assuming that the input frequency is 0 Hz, the inductive reactance is 0 Ω. However, the inductor does have some amount of winding resistance (R_W). Therefore, a resistor representing the winding resistance of the coil has been included in the equivalent circuit. In this case, the load voltage equals the difference between the source voltage (E) and the voltage across R_W.

As shown in the reactance curve, increases in operating frequency cause proportional increases in the value of X_L. Theoretically, the operating frequency can become high enough for the inductor to effectively act as an open. When this is the case, the filter has the equivalent circuit shown in Figure 19.19c. As shown in the circuit, the inductor is represented as a break in the conductor (because of its near-infinite reactance). In this case, $V_{RL} = 0$ V. Between the extremes represented in Figure 19.19 lies a range of frequencies over which V_{RL} decreases from $E - V_{RW}$ to 0 V.

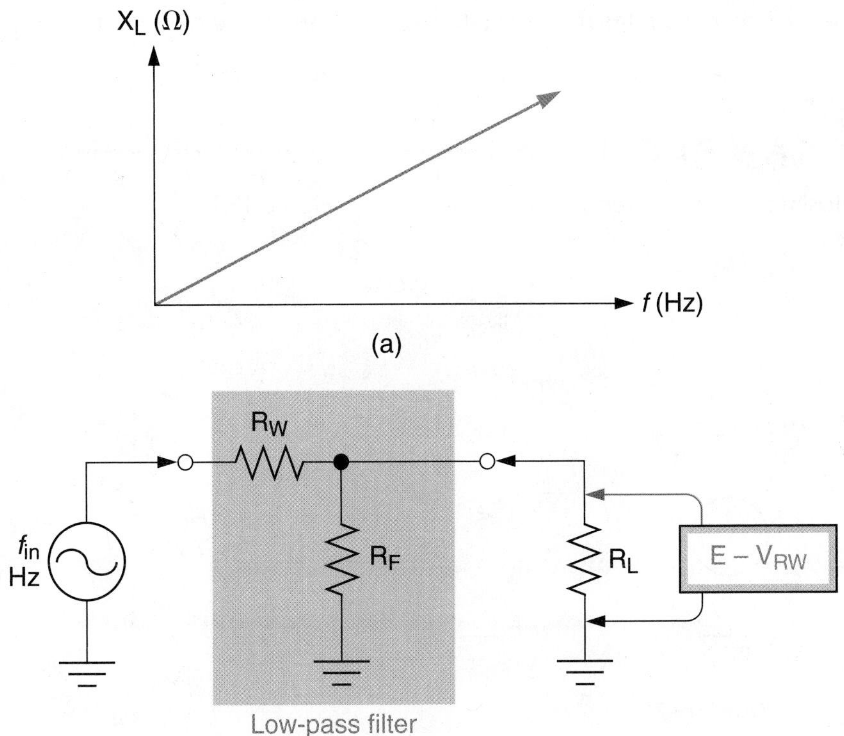

(b) Filter equivalent circuit when $f = 0$ Hz

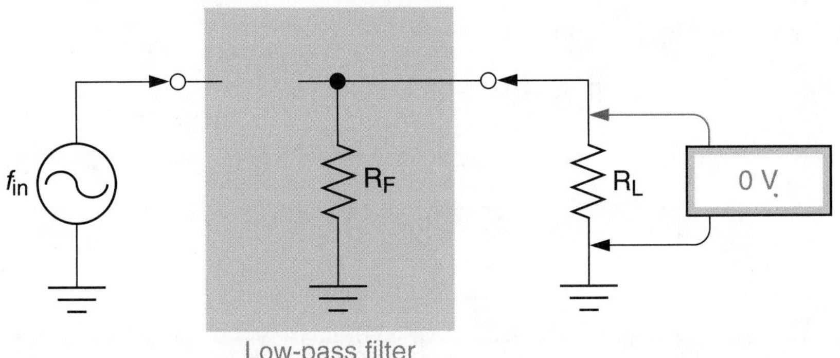

(c) Filter equivalent circuit when $f \rightarrow \infty$ Hz

FIGURE 19.19

Upper Cutoff Frequency (f_C)

The upper cutoff frequency for an RL low-pass filter is determined by the inductor and the parallel combination of R_F and R_L. By formula,

(19.10)

$$f_C = \frac{R_{EQ}}{2\pi L}$$

where

$R_{EQ} = R_F \parallel R_L$

Example 19.7 demonstrates the calculation of f_C for an RL low-pass filter.

EXAMPLE 19.7

Calculate the cutoff frequency for the circuit in Figure 19.20.

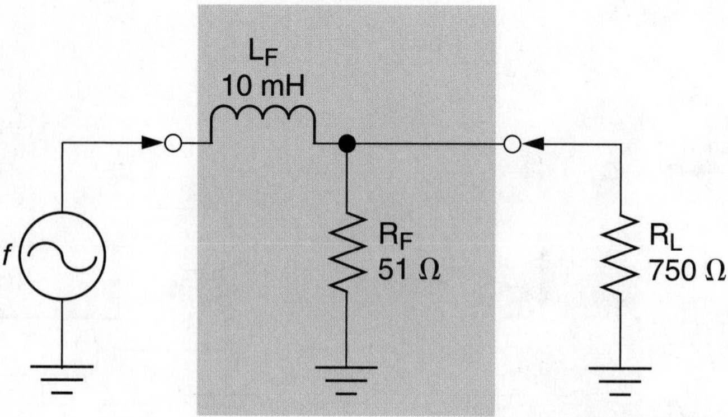

FIGURE 19.20

SOLUTION

First, the value of R_{EQ} is found as

$$R_{EQ} = R_F \parallel R_L$$

$$= \frac{1}{\frac{1}{R_F} + \frac{1}{R_L}}$$

$$= \frac{1}{\frac{1}{51\ \Omega} + \frac{1}{750\ \Omega}}$$

$$= 47.8\ \Omega$$

Now, the filter cutoff frequency is found as

$$f_C = \frac{R_{EQ}}{2\pi L}$$

$$= \frac{47.8\ \Omega}{2\pi(10\ \text{mH})}$$

$$= 761\ \text{Hz}$$

This result indicates that the power gain of the circuit is reduced to 50% of its maximum value when the operating frequency reaches 761 Hz. The frequency response curve for the circuit is shown in Figure 19.21.

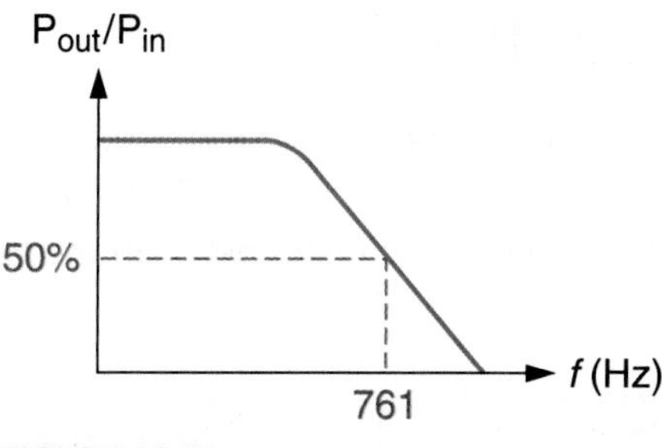

FIGURE 19.21

PRACTICE PROBLEM 19.7

A circuit like the one in Figure 19.20 has the following values: $R_L = 820\ \Omega$, $R_F = 68\ \Omega$, and $L = 33$ mH. Calculate the circuit's cutoff frequency.

A Look Ahead

We have now discussed the operation and analysis of RC and RL low-pass filters. In the following section, we will take a similar look at RC and RL high-pass filters. As you will see, most of the principles discussed in this section apply to those circuits as well.

PROGRESS CHECK

1. What is a *shunt* component?
2. Explain the filtering action of the RC low-pass filter in Figure 19.13.
3. Explain the filtering action of the RL low-pass filter in Figure 19.19.

19.3 RC AND RL HIGH-PASS FILTERS

As you know, a high-pass filter is one that is designed to pass all frequencies above its cutoff frequency. High-pass filters are formed by reversing the positions of the resistive and reactive components in the RC and RL low-pass filters. In this section, we will discuss the operation of RC and RL high-pass filters. As you will see, most of the relationships for high-pass filters are nearly identical to those for low-pass filters, because the primary difference between a high-pass filter and its low-pass counterpart is component placement.

RC High-Pass Filters

An RC circuit acts as a high-pass filter when constructed as shown in Figure 19.22a. For comparison, an RC low-pass filter is shown in Figure 19.22b. As you can see, the capacitor and resistor positions are reversed between the two circuits. In the high-pass circuit, the capacitor is in the signal path and the resistor is the shunt component.

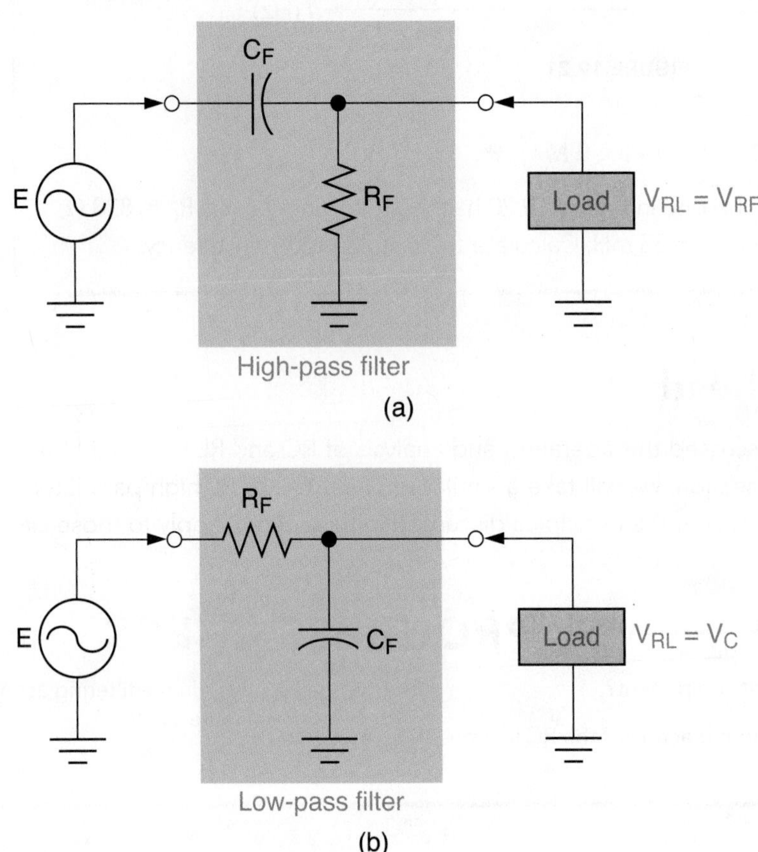

FIGURE 19.22 High-pass vs. low-pass RC filters.

The filtering action of the circuit in Figure 19.22a is a result of the capacitor's response to an increase in frequency. This response is illustrated in Figure 19.23. The reactance curve (which we used to describe low-pass filter operation) shows that capacitive reactance varies inversely with operating frequency. With this in mind, look at the equivalent circuit shown in Figure 19.23b. As you can see, the near-infinite reactance of the capacitor when the input frequency is 0 Hz is represented as a break in the signal path. Therefore, the source is isolated from the load, and $V_{RL} = 0$ V.

Assuming that the input frequency is near the high end of the reactance curve, X_C can be assumed to be approximately 0 Ω. With this in mind, look at the equivalent circuit shown in Figure 19.23c. In this case, the low-reactance capacitor is represented by a direct connection between the source and the load. As a result, the load voltage is approximately equal to the source voltage.

Between the extremes represented in Figure 19.23 lies a range of frequencies over which V_{RL} decreases from $V_{RL} \cong E$ to $V_{RL} = 0$ V.

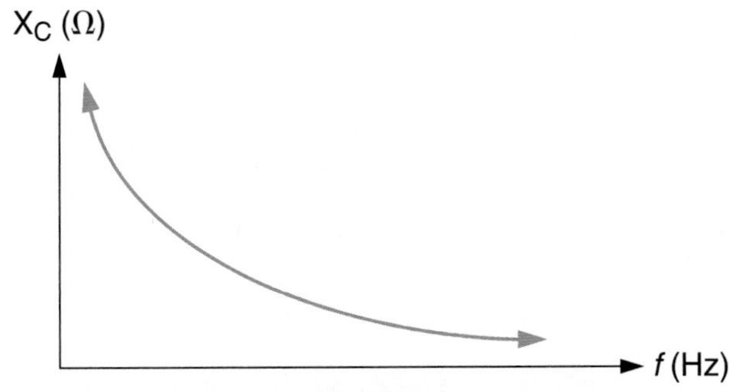

(a) Capacitive reactance vs. frequency

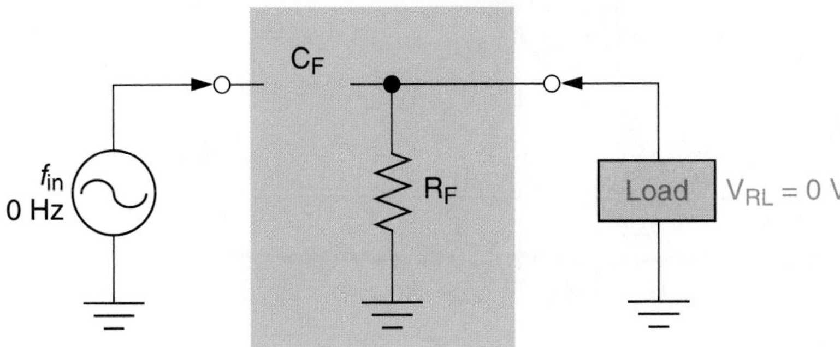

(b) Filter equivalent circuit when $f = 0$ Hz

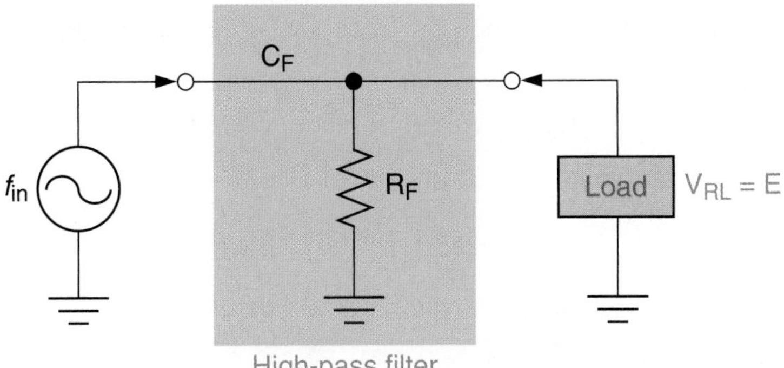

(c) Filter equivalent circuit when $f \to \infty$ Hz

FIGURE 19.23 High-pass RC filter operation.

Lower Cutoff Frequency (f_C)

The cutoff frequency for a RC high-pass filter is determined using the same relationship we established for the low-pass filter. By formula,

$$f_C = \frac{1}{2\pi RC}$$

This relationship was given as equation (19.7) in the last section. We can use this relationship because only the component positions within the filter have changed. In other words, the relationship among f_C, R, and C doesn't change simply because the component positions have changed. Example 19.8 demonstrates the complete approach to calculating the lower cutoff frequency for an RC high-pass filter.

EXAMPLE 19.8

Calculate the cutoff frequency for the circuit shown in Figure 19.24.

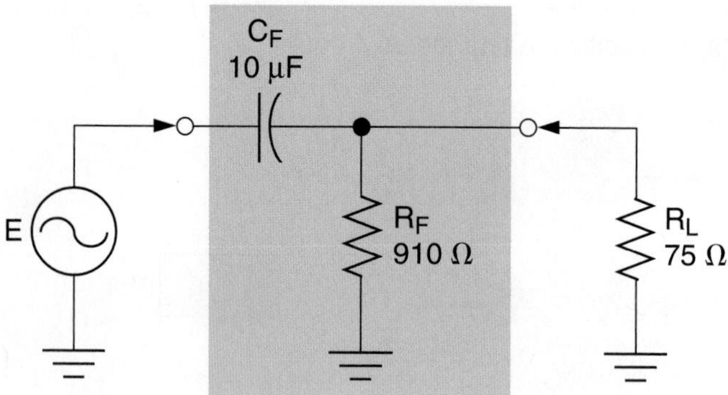

FIGURE 19.24

SOLUTION

The total resistance in the circuit equals the parallel combination of the filter resistor and the load resistance. The parallel equivalent of R_F and R_L is found as

$$R_{EQ} = R_F \parallel R_L$$

$$= \frac{1}{\frac{1}{R_F} + \frac{1}{R_L}}$$

$$= \frac{1}{\frac{1}{910\ \Omega} + \frac{1}{75\ \Omega}}$$

$$= 69.3\ \Omega$$

Using R_{EQ} as the value of R in equation (19.7), we get a cutoff frequency of

$$f_C = \frac{1}{2\pi RC}$$

$$= \frac{1}{2\pi(69.3\ \Omega)(10\ \mu F)}$$

$$= 230\ \text{Hz}$$

This result indicates that the power gain of the circuit is reduced to 50% of its maximum value when the operating frequency decreases to 230 Hz. The frequency response curve for the circuit is shown in Figure 19.25.

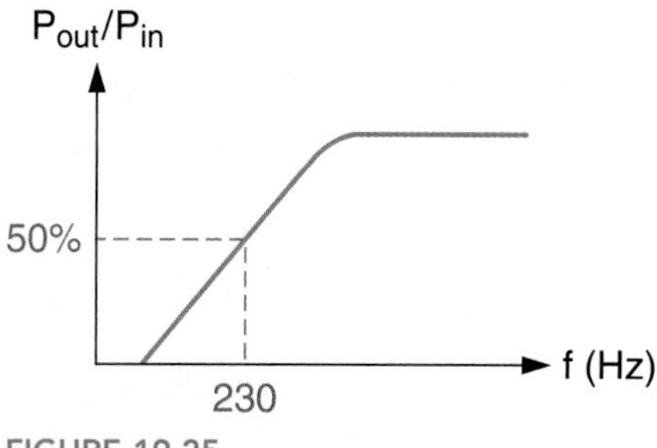

FIGURE 19.25

PRACTICE PROBLEM 19.8

A circuit like the one in Figure 19.24 has the following values: $C_F = 22\ \mu F$, $R_F = 750\ \Omega$, and $R_L = 30\ \Omega$. Calculate the value of the circuit cutoff frequency.

RL High-Pass Filters

An RL circuit acts as a high-pass filter when constructed as shown in Figure 19.26. In the circuit shown, the resistor is the series component and the inductor is the shunt component.

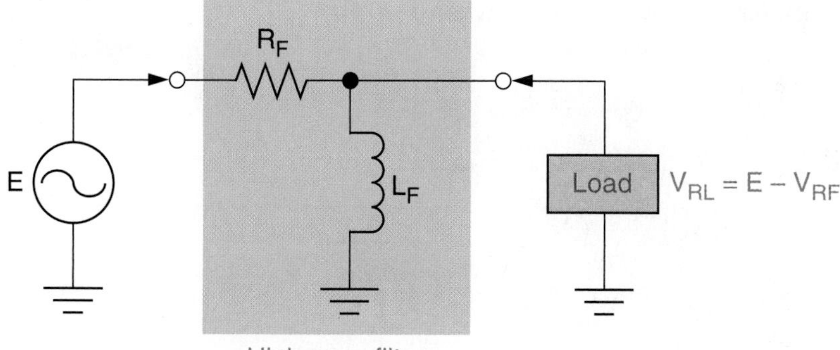

FIGURE 19.26 An RL high-pass filter.

The filtering action of the circuit in Figure 19.26 is a result of the inductor's response to a decrease in operating frequency. This response is illustrated in Figure 19.27. As the reactance curve indicates, X_L approaches infinity as frequency increases. If the input frequency decreases to 0 Hz, the inductive reactance decreases to $0\ \Omega$. In this case, the equivalent circuit in Figure 19.27b applies. Ignoring the small amount of winding resistance in the coil, the shunt inductor is represented as a shorted path to ground. In this case, $V_{RL} = 0$ V.

If the input frequency to the circuit increases, the reactance of the inductor increases until the component effectively acts as an open. If we assume that $X_L \cong \infty\ \Omega$, the equivalent circuit in Figure 19.27c applies. As you can see, the inductor is represented as a break in the shunt component path. In this case, $V_{RL} = E - V_{RF}$. Between the extremes represented in Figure 19.27 lies a range of frequencies over which V_{RL} decreases from $E - V_{RF}$ to 0 V.

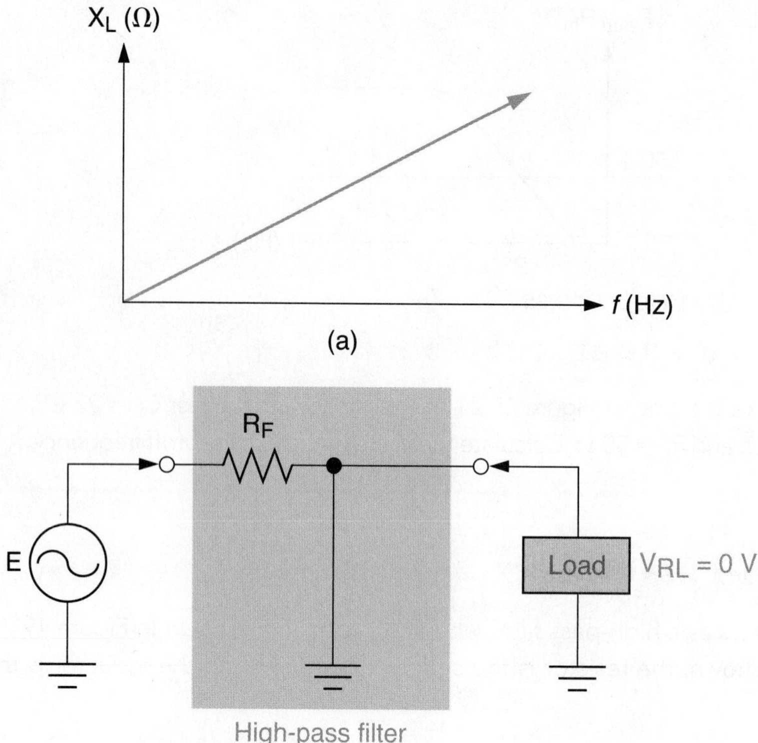

(b) Filter equivalent circuit when $f = 0$ Hz

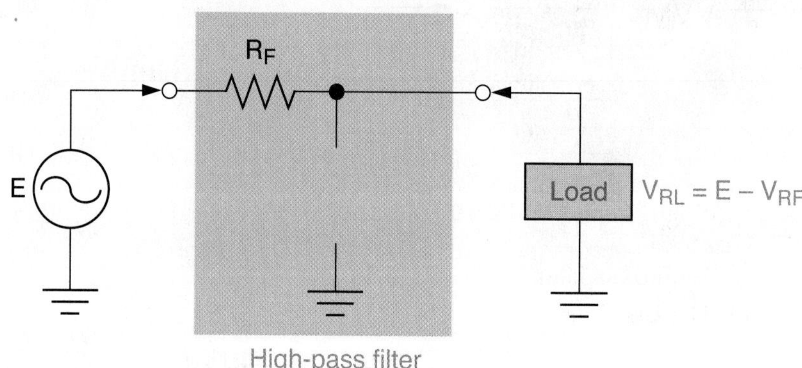

(c) Filter equivalent circuit when $f \rightarrow \infty$ Hz

FIGURE 19.27

Lower Cutoff Frequency (f_C)

The lower cutoff frequency for an RL high-pass filter is determined by the inductor and the parallel combination of R_F and R_L. By formula,

$$f_C = \frac{R_{EQ}}{2\pi L}$$

where $R_{EQ} = R_F \parallel R_L$. This relationship was introduced earlier in the chapter as equation (19.10).

The value of R_{EQ} in equation (19.10) is determined in the same manner we used to measure its value in an RL low-pass filter. Example 19.32 demonstrates the process used to calculate the cutoff frequency for an RL high-pass filter.

EXAMPLE 19.9

Calculate the cutoff frequency for the RL high-pass filter in Figure 19.28.

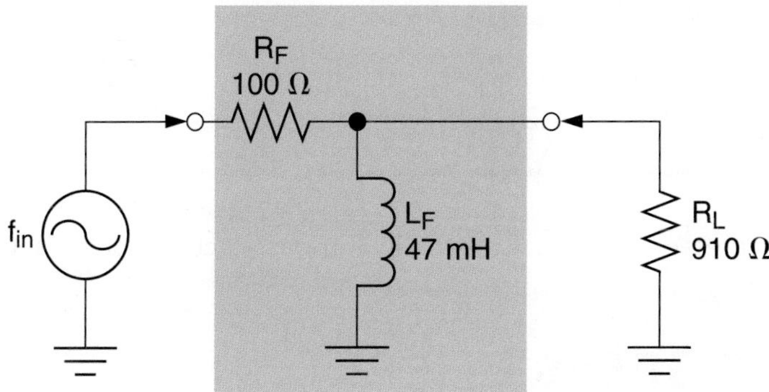

FIGURE 19.28

SOLUTION

First, the value of R_{EQ} is found as

$$R_{EQ} = R_F \parallel R_L$$

$$= \frac{1}{\frac{1}{R_F} + \frac{1}{R_L}}$$

$$= \frac{1}{\frac{1}{100\ \Omega} + \frac{1}{910\ \Omega}}$$

$$= 90.1\ \Omega$$

Now, the filter cutoff frequency is found as

$$f_C = \frac{R_{EQ}}{2\pi L}$$

$$= \frac{90.1\ \Omega}{2\pi(47\text{ mH})}$$

$$= 305\text{ Hz}$$

This result indicates that the power gain of the circuit is reduced to 50% of its maximum value when the operating frequency decreases to 305 Hz. The frequency response curve for the circuit is shown in Figure 19.29.

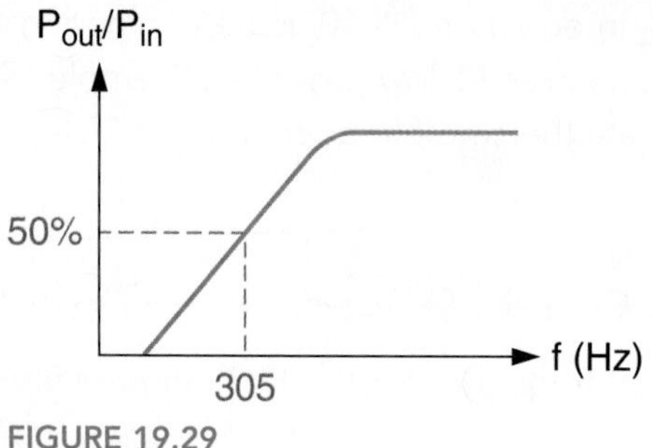

FIGURE 19.29

PRACTICE PROBLEM 19.9

A circuit like the one in Figure 19.28 has the following values: $R_F = 200\ \Omega$, $L_F = 1.5$ mH, and $R_L = 1.2$ kΩ. Calculate the filter cutoff frequency.

As you can see, the analysis of an RL high-pass circuit is identical to that of an RL low-pass filter.

PROGRESS CHECK

1. In terms of component placement, what is the difference between an RC high-pass filter and an RC low-pass filter?
2. Explain the filtering action of the RC high-pass filter in Figure 19.22.
3. In terms of component placement, what is the difference between an RL high-pass filter and an RL low-pass filter?
4. Explain the filtering action of the RL high-pass filter in Figure 19.26.

19.4 BANDPASS AND NOTCH FILTERS

Earlier in the chapter, you were told that bandpass and notch (or band-stop) filters are designed to pass or block a specified range of frequencies. As a review, the primary frequencies are identified on the frequency response curves in Figure 19.30. As you can see, each of these filters has two cutoff frequencies, designated f_{C1} and f_{C2}. The difference between the cutoff frequencies is referred to as the bandwidth (BW) of the filter and the geometric average of the cutoff frequencies is referred to as its *center frequency* (f_0).

There are many possible approaches to building bandpass and notch filters. Most common bandpass and notch filters are LC filters, like those shown in Figure 19.31. The circuits in Figure 19.31a are bandpass filters. The circuits in Figure 19.31b are notch filters. In each case, the filtering action is based on the resonant

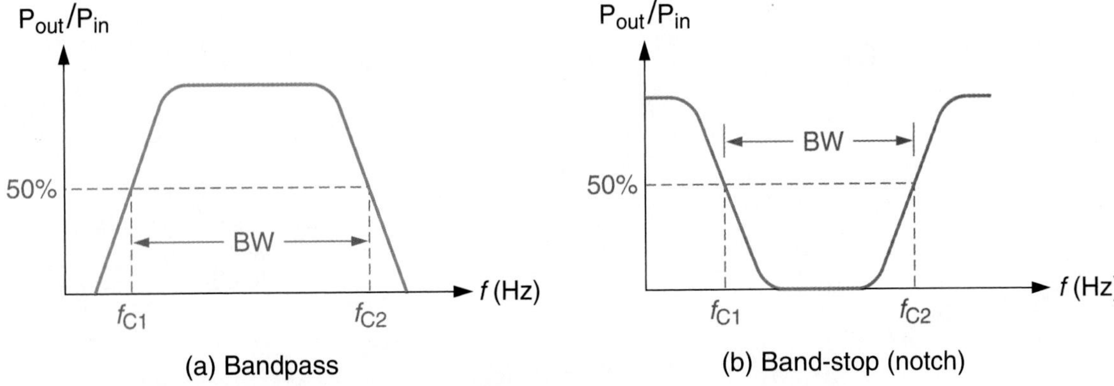

(a) Bandpass (b) Band-stop (notch)

FIGURE 19.30 Bandpass and band-stop filter frequency response.

Series Shunt

(a) Bandpass filters

Series Shunt

(b) Notch (band-stop) filters

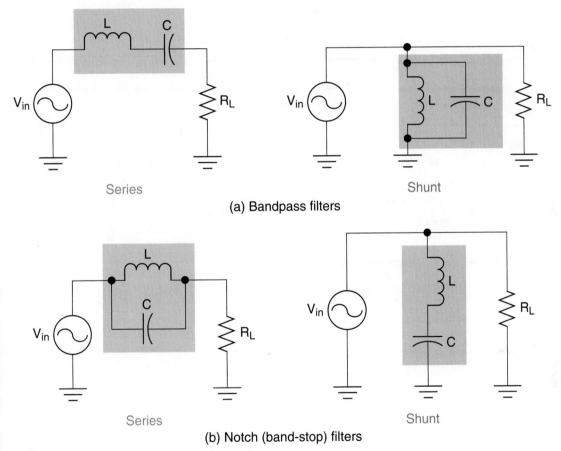

FIGURE 19.31 LC bandpass and band-stop filters.

characteristics of the LC circuits. The basic characteristics of series and parallel resonant circuits, which were introduced in Chapter 18, are listed in Figure 19.32.

Series LC Bandpass Filters

The operation of a series LC bandpass filter is easiest to understand when the filter is represented as an equivalent circuit like the one in Figure 19.33b. In this circuit,

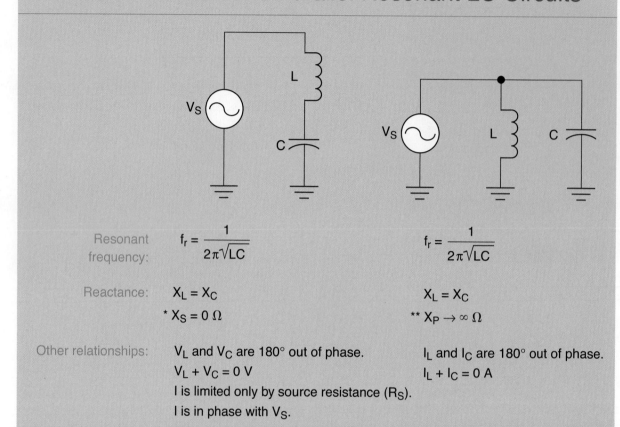

Series Resonant and Parallel Resonant LC Circuits

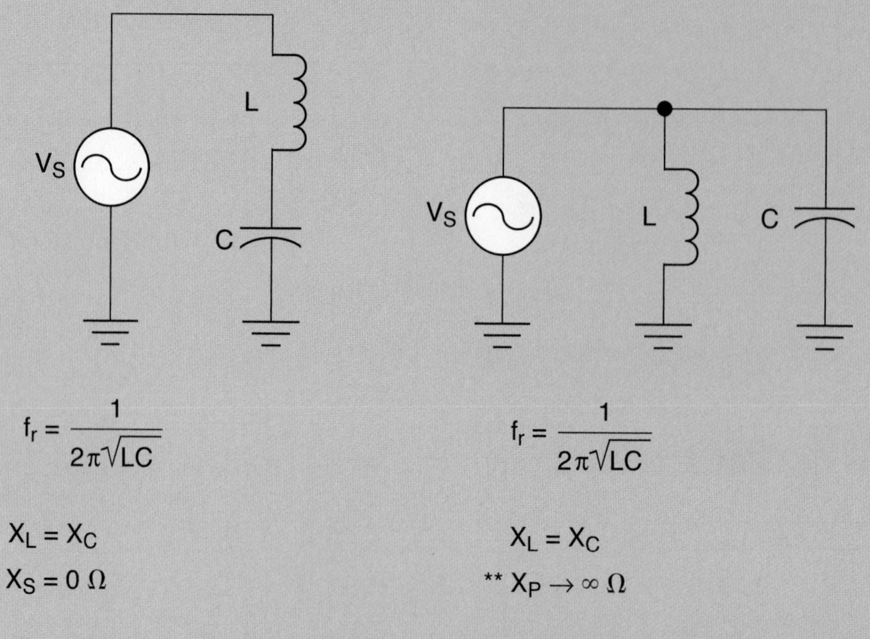

	Series	Parallel
Resonant frequency:	$f_r = \dfrac{1}{2\pi\sqrt{LC}}$	$f_r = \dfrac{1}{2\pi\sqrt{LC}}$
Reactance:	$X_L = X_C$ * $X_S = 0\ \Omega$	$X_L = X_C$ ** $X_P \rightarrow \infty\ \Omega$
Other relationships:	V_L and V_C are 180° out of phase. $V_L + V_C = 0$ V I is limited only by source resistance (R_S). I is in phase with V_S.	I_L and I_C are 180° out of phase. $I_L + I_C = 0$ A

* X_S represents the total series reactance.

**X_P represents the total parallel reactance.

FIGURE 19.32

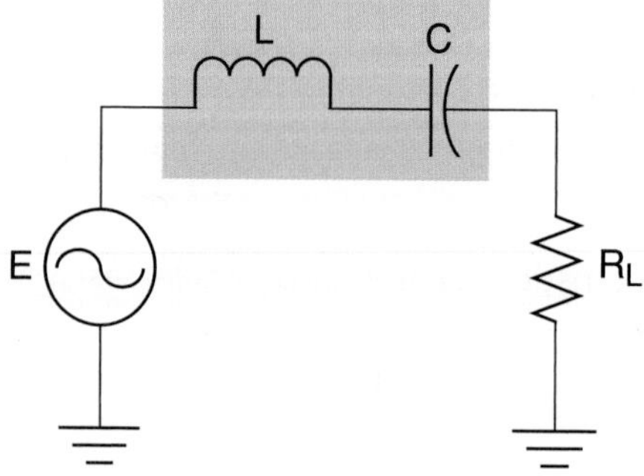

(a) A series LC bandpass filter

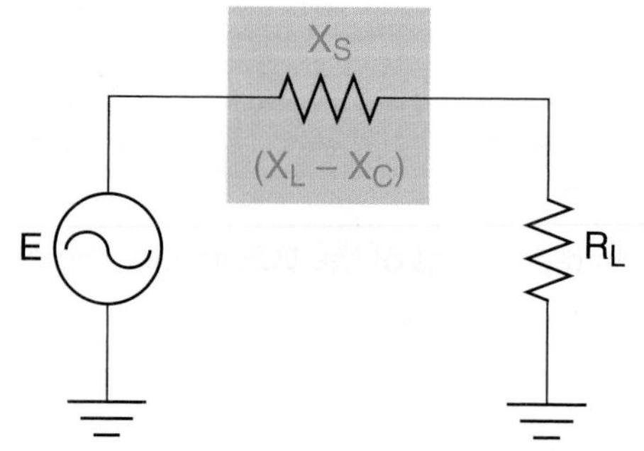

(b) The reactive equivalent of the filter

FIGURE 19.33 A series LC bandpass filter.

the net series reactance (X_S) of the filter is represented as a series component between the source and the load.

The operation of a series LC bandpass filter is based on the relationship between its input frequency and its resonant frequency, as follows:

- $f_{in} = f_r$. When the circuit is operating at resonance, the component reactances are equal and $X_S = X_L - X_C = 0\ \Omega$. This means that, ignoring R_W, the circuit current is determined by the source voltage and the load ($I_T = E/R_L$). The circuit is *resistive* in nature and the phase angle is 0°.
- $f_{in} < f_r$. When the input frequency drops below f_r, X_C increases and X_L decreases. As a result, the net reactance is *capacitive* in nature and the impedance phase angle is negative. As f_{in} continues to decrease, X_S increases in magnitude and the impedance phase angle becomes more negative. When f_{in} reaches 0 Hz, X_C and X_S approach infinity, and $I_T = 0$ A.
- $f_{in} > f_r$. When the input frequency rises above f_r, X_C decreases and X_L increases. As a result, the net reactance is *inductive* in nature and the impedance phase angle is positive. As f_{in} continues to increase, X_S increases in magnitude and the impedance phase angle becomes more positive. When f_{in} approaches infinity, X_L and X_S approach infinity, and $I_T = 0$ A.

The curve in Figure 19.34 illustrates the frequency response of a series LC bandpass filter.

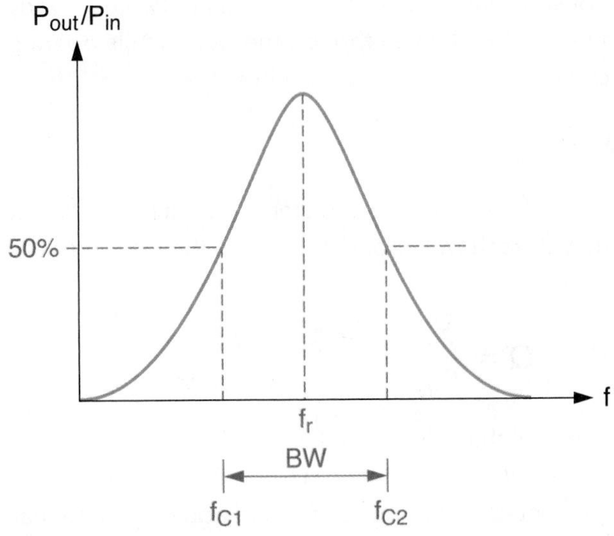

FIGURE 19.34 Frequency response curve for an LC bandpass filter.

Shunt LC Bandpass Filters

Once you understand the operation of the series LC bandpass filter, the operation of its shunt counterpart is relatively easy to visualize. A shunt LC bandpass filter is shown in Figure 19.35, along with its reactive equivalent circuit. Note that X_P in Figure 19.35b represents the parallel combination of X_L and X_C, as indicated by the equation.

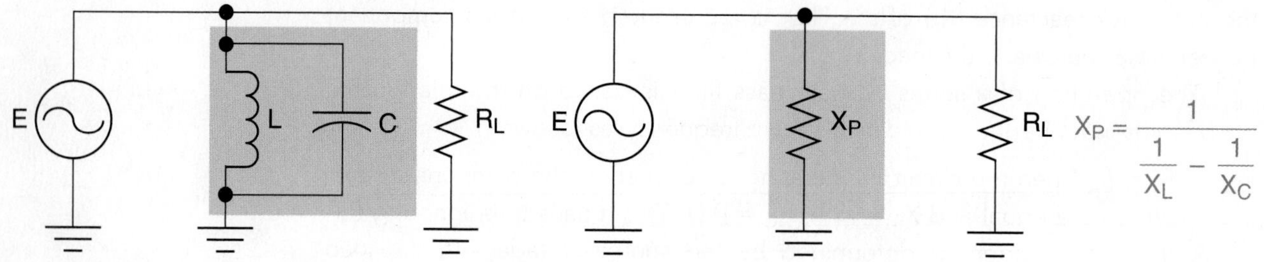

FIGURE 19.35 A shunt LC bandpass filter.

Here is how the parallel LC circuit responds to frequencies above, below, and at resonance:

- $f_{in} = f_r$. When the circuit is operating at resonance, $I_L = I_C$ and X_P approaches infinity. As a result, the filter is effectively removed from the circuit and $V_{RL} \cong E$. In this case, the circuit is *resistive* in nature and the phase angle is 0°.
- $f_{in} < f_r$. When the input frequency drops below f_r, X_L decreases. As f_{in} approaches 0 Hz, the reactance of the shunt inductor approaches 0 Ω, which shorts out the load. In this case, $V_{RL} \cong 0$ V.
- $f_{in} > f_r$. When the input frequency rises above f_r, X_C decreases. As f_{in} continues to increase, X_C approaches 0 Ω, which shorts out the load. Again, $V_{RL} \cong 0$ V.

Note that the frequency response curve for the circuit in Figure 19.35 is nearly identical to the one shown in Figure 19.34. Even though the two circuits operate differently, they produce the same overall frequency response.

Filter Quality (*Q*)

In Chapter 14, Q was introduced as a component characteristic. As a review, the Q of an inductor is a measure of how close the component comes to the *ideal* inductor, found as

$$Q = \frac{X_L}{R_W}$$

where R_W = the winding resistance of the coil.

The ideal inductor would have a winding resistance (R_W) of 0 Ω. As a result, the total opposition to current (Z) presented by the inductor would equal X_L. Note that the higher the Q of the component, the closer it comes to the ideal of $Z = X_L$.

It should be noted that the addition of a resistive load to a series or parallel LC filter reduces the Q of the circuit. We will address the effects of loading on filter Q later in this section.

Series Filter Frequency Analysis

In Section 19.1, you were shown how to calculate the bandwidth of a bandpass filter when the center frequency (f_0) and filter quality (Q) are known. The relationship between these filter characteristics was given as

$$BW = \frac{f_0}{Q}$$

This relationship was given as equation (19.5).

The center frequency of an LC bandpass filter equals the resonant frequency of the circuit. Therefore, the bandwidth of a series LC bandpass filter can be found as

(19.11)

$$BW = \frac{f_r}{Q_L}$$

where

f_r = the resonant frequency of the LC filter
Q_L = the *loaded-Q* of the filter

The loaded-Q (Q_L) of a filter is the quality (Q) of the circuit when a load is connected to its output terminals. The concept of Q loading is illustrated in Figure 19.36. As stated earlier, the Q of an inductor equals the ratio of inductive reactance to winding resistance. More precisely, it is the ratio of X_L to the total series resistance. As you can see, the total resistance in Figure 19.36 equals the sum of the winding resistance and the load resistance. Therefore, the loaded-Q of the circuit is found as

(19.12)

$$Q_L = \frac{X_L}{R_S + R_W + R_L}$$

The calculation of Q_L for the circuit in Figure 19.36 is shown in the figure. As you can see, the loaded-Q is significantly lower than the value of Q we would have obtained using only the inductor winding resistance.

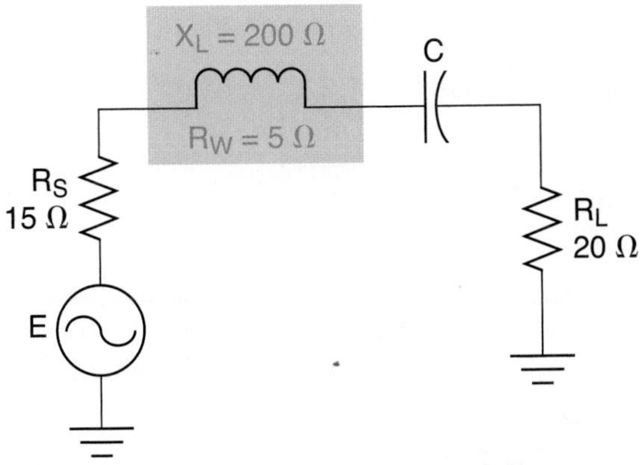

$X_L = 200\ \Omega$, $R_W = 5\ \Omega$, C, R_S 15 Ω, R_L 20 Ω, E

$$Q = \frac{X_L}{R_W} = \frac{200\ \Omega}{5\ \Omega} = 40$$

$$Q_L = \frac{X_L}{R_S + R_W + R_L} = \frac{200\ \Omega}{40\ \Omega} = 5$$

FIGURE 19.36 The values of Q and Q_L for an inductor.

To determine the bandwidth of a series LC bandpass filter, you must first calculate its resonant frequency. Then, you calculate the values of X_L and Q_L at the resonant frequency. Finally, the values of f_r and Q_L are used to determine the circuit bandwidth. Example 19.10 demonstrates this series of circuit calculations.

EXAMPLE 19.10

Determine the bandwidth of the series LC bandpass filter in Figure 19.37a.

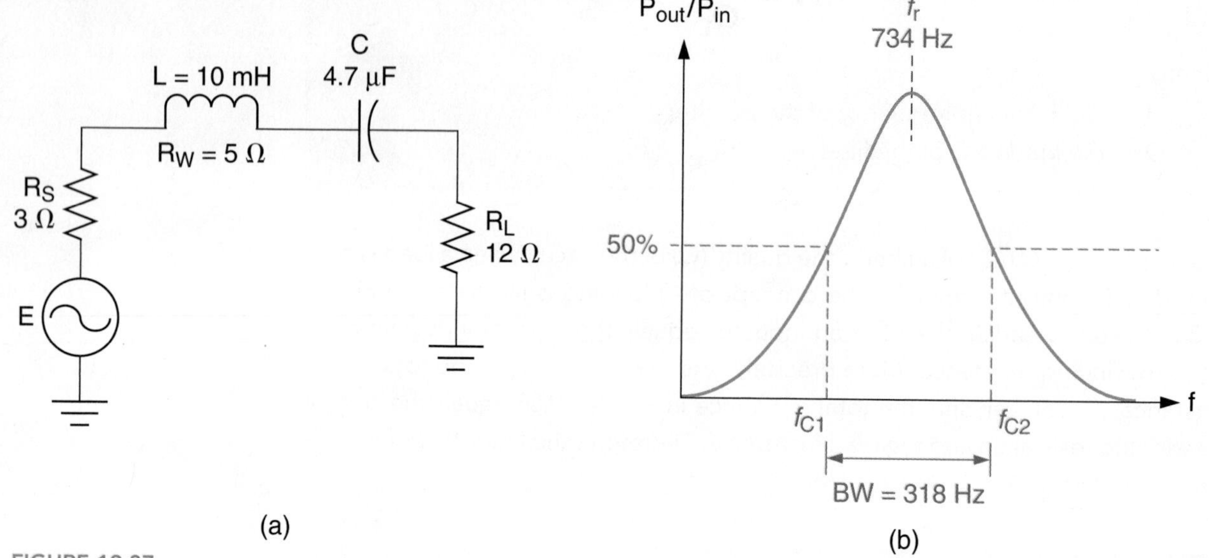

FIGURE 19.37

SOLUTION

First, the resonant frequency of the circuit is found as

$$f_r = \frac{1}{2\pi\sqrt{LC}}$$

$$= \frac{1}{2\pi\sqrt{(10 \text{ mH})(4.7 \text{ μF})}}$$

$$= 734 \text{ Hz}$$

When the circuit is operated at this frequency, the value of X_L is found as

$$X_L = 2\pi f L$$

$$= 2\pi(734 \text{ Hz})(10 \text{ mH})$$

$$= 46.1 \ \Omega$$

The circuit is shown to have values of $R_S = 3\ \Omega$, $R_W = 5\ \Omega$, and $R_L = 12\ \Omega$. Therefore, the loaded-Q of the component (at the resonant frequency) is found as

$$Q_L = \frac{X_L}{R_S + R_W + R_L} = \frac{46.1\ \Omega}{20\ \Omega} = 2.31$$

Finally, the bandwidth of the series bandpass filter can be found as

$$BW = \frac{f_r}{Q_L} = \frac{734\text{ Hz}}{2.31} = 318\text{ Hz}$$

The bandwidth and center frequency of the circuit are identified on the frequency response curve in Figure 19.37b.

PRACTICE PROBLEM 19.10

A filter like the one in Figure 19.37a has the following values: $L = 22\text{ mH}$, $R_W = 8\ \Omega$, $R_L = 330\ \Omega$, $R_S = 2\ \Omega$, and $C = 4.7\text{ nF}$. Calculate the bandwidth of the filter.

Once we know the bandwidth and center frequency of a series bandpass filter, we have the information needed to calculate the circuit cutoff frequencies. When $Q_L \geq 2$, the center (resonant) frequency (f_r) falls about halfway between the cutoff frequencies. As a result, the cutoff frequencies for a series bandpass filter can be found as

(19.13)

$$f_{C1} = f_r - \frac{BW}{2}$$

and

(19.14)

$$f_{C2} = f_r + \frac{BW}{2}$$

Example 19.11 applies these relationships to the circuit we analyzed in Example 19.10.

EXAMPLE 19.11

Determine the cutoff frequencies for the circuit in Figure 19.37a.

SOLUTION

In Example 19.10, the circuit was found to have the following values:

$$Q_L = 2.31 \qquad f_r = 734\text{ Hz} \qquad BW = 318\text{ Hz}$$

Since $Q_L > 2$, we can calculate the values of the cutoff frequencies as

$$f_{C1} = f_r - \frac{BW}{2}$$

$$= 734\text{ Hz} - \frac{318\text{ Hz}}{2}$$

$$= 575\text{ Hz}$$

and

$$f_{C2} = f_r + \frac{BW}{2}$$

$$= 734 \text{ Hz} + \frac{318 \text{ Hz}}{2}$$

$$= 893 \text{ Hz}$$

PRACTICE PROBLEM 19.11

Calculate the cutoff frequencies for the circuit described in Practice Problem 19.10.

When a filter has a value of $Q_L < 2$, the average frequency (f_{AVE}) for the circuit must be calculated using

(19.15)

$$f_{AVE} = f_r\sqrt{1 + \left(\frac{1}{2Q_L}\right)^2}$$

Once the value of f_{AVE} is determined, then the cutoff frequencies can be calculated using

(19.16)

$$f_{C1} = f_{AVE} - \frac{BW}{2}$$

and

(19.17)

$$f_{C2} = f_{AVE} + \frac{BW}{2}$$

These two equations are simply modified versions of the relationships we used in Example 19.11.

Shunt Filter Frequency Analysis

The analysis of a shunt LC bandpass filter is nearly identical to that of the series filter. The primary difference lies with the calculation of Q_L. For example, consider the circuit shown in Figure 19.38a. To determine the value of Q_L for this circuit, we

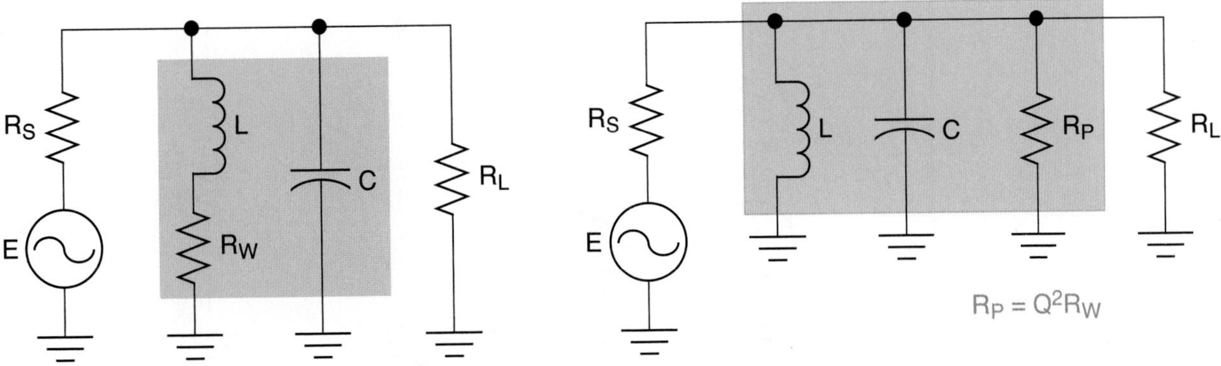

(a) A shunt LC bandpass filter (b) Filter parallel resistance (R_P)

FIGURE 19.38 A shunt LC bandpass filter and its equivalent circuit.

need to combine R_W with the other resistor values. In the configuration shown, calculating the total circuit resistance is extremely difficult. However, the winding resistance of the coil can be represented as an *equivalent parallel resistance* (R_P) as shown in Figure 19.38b. This parallel resistance, which represents the effective DC resistance of the filter, has a value that is found as

(19.18)

$$R_P = Q^2 R_W$$

where

Q = the *unloaded* Q of the filter

From the viewpoint of the filter, the three resistors in Figure 19.38b are in parallel. Using the parallel combination of the resistors, the *loaded-Q* of the circuit is found as

(19.19)

$$Q_L = \frac{R_S \parallel R_P \parallel R_L}{X_L}$$

Example 19.12 demonstrates the process used to calculate the loaded-Q of a shunt bandpass filter.

EXAMPLE 19.12

Determine the value of Q_L for the circuit shown in Figure 19.39.

SOLUTION

First, the resonant frequency of the circuit is found as

$$f_r = \frac{1}{2\pi\sqrt{LC}}$$

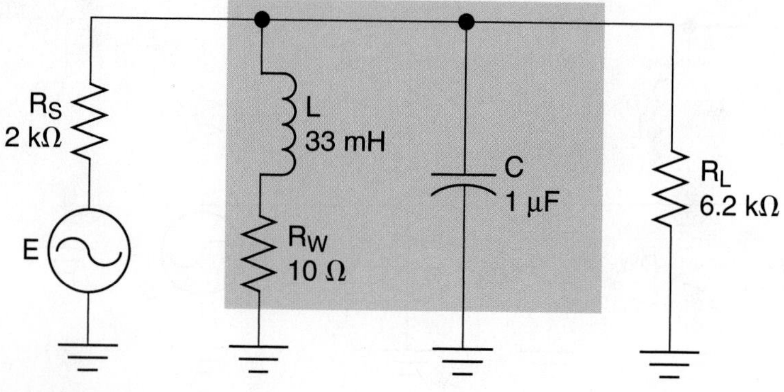

FIGURE 19.39

$$= \frac{1}{2\pi\sqrt{(33\text{ mH})(1\ \mu\text{F})}}$$
$$= 876\text{ Hz}$$

Now, the value of X_L at the resonant frequency is found as

$$X_L = 2\pi f L$$
$$= 2\pi(876\text{ Hz})(33\text{ mH})$$
$$= 182\ \Omega$$

The Q of the inductor can now be found as

$$Q = \frac{X_L}{R_W}$$
$$= \frac{182\ \Omega}{10\ \Omega}$$
$$= 18.2$$

Using the inductor Q, the value of R_P is found as

$$R_P = Q^2 R_W$$
$$= (18.2)^2(10\ \Omega)$$
$$= 3.31\text{ k}\Omega$$

Finally, the loaded Q of the filter is found as

$$Q_L = \frac{R_S \parallel R_P \parallel R_L}{X_L}$$
$$= \frac{2\text{ k}\Omega \parallel 3.31\text{ k}\Omega \parallel 6.2\text{ k}\Omega}{182\ \Omega}$$
$$= 5.7$$

PRACTICE PROBLEM 19.12

A circuit like the one in Figure 19.39 has the following values: $R_S = 3.3\text{ k}\Omega$, $L = 10\text{ mH}$, $R_W = 8\ \Omega$, $C = 2.2\ \mu\text{F}$, and $R_L = 5\text{ k}\Omega$. Calculate the value of Q_L for the filter.

Once the value of Q_L for a shunt bandpass filter has been determined, the rest of the frequency analysis proceeds just as it did for the series bandpass filter, as demonstrated in Example 19.13.

EXAMPLE 19.13

Calculate the bandwidth and cutoff frequencies for the filter in Figure 19.39.

SOLUTION

From Example 19.12, we know that the circuit has the following values:

$$f_r = 876 \text{ Hz}$$

and

$$Q_L = 5.7$$

Using these values, the bandwidth of the filter can be found as

$$\begin{aligned} BW &= \frac{f_r}{Q_L} \\ &= \frac{876 \text{ Hz}}{5.7} \\ &= 154 \text{ Hz} \end{aligned}$$

Since $Q_L > 2$, the filter cutoff frequencies can be found as

$$\begin{aligned} f_{C1} &= f_r - \frac{BW}{2} \\ &= 876 \text{ Hz} - \frac{154 \text{ Hz}}{2} \\ &= 799 \text{ Hz} \end{aligned}$$

and

$$\begin{aligned} f_{C2} &= f_r + \frac{BW}{2} \\ &= 876 \text{ Hz} + \frac{154 \text{ Hz}}{2} \\ &= 953 \text{ Hz} \end{aligned}$$

Using these values and those calculated in Example 19.12, the frequency response curve for the circuit is plotted as shown in Figure 19.40.

PRACTICE PROBLEM 19.13

Calculate the cutoff frequencies for the filter described in Practice Problem 19.12.

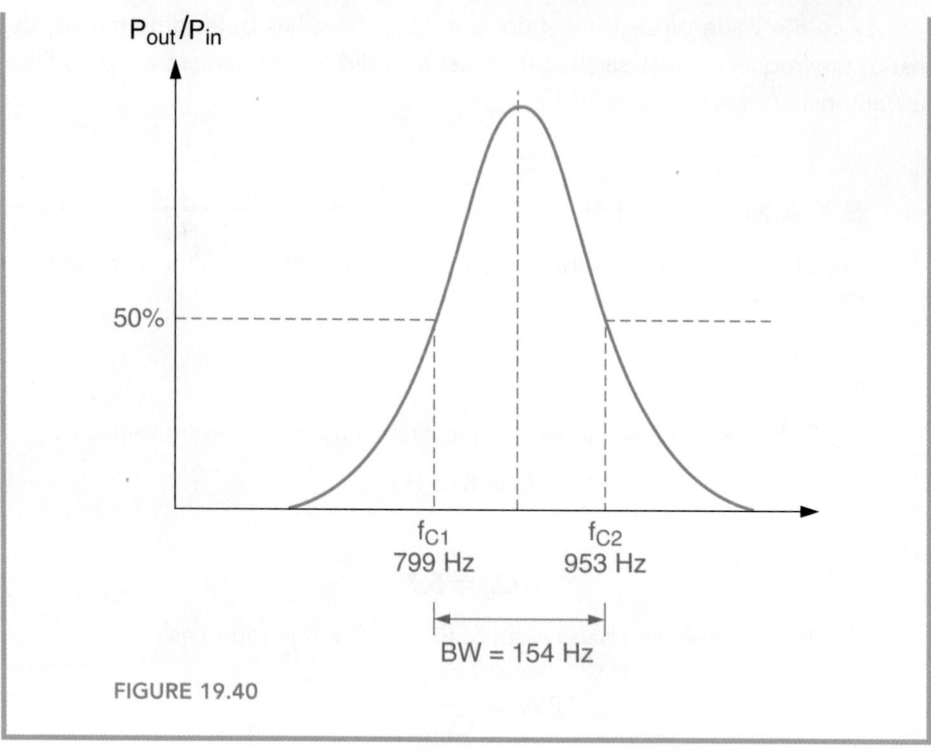

FIGURE 19.40

There is an interesting point to be made: While the *ideal* parallel LC circuit has infinite impedance at resonance, the impedance of the practical circuit is significantly lower. This impedance, which is resistive in nature, is represented by R_P.

Series LC Notch Filters

A series LC notch filter can be constructed by placing a parallel LC circuit in series between the signal source and the load. Such a circuit is shown in Figure 19.41, along with its frequency response curve.

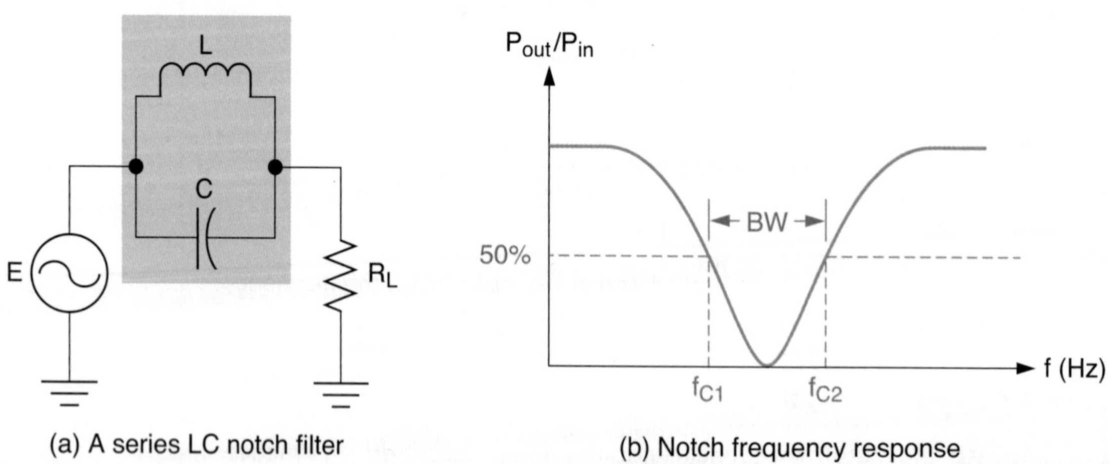

FIGURE 19.41 A series LC notch (band-stop) filter and its frequency response curve.

The operation of the series LC notch filter is easy to understand when you consider the circuit response to each of the following conditions:

$$f_{in} = 0\text{ Hz} \qquad f_{in} = f_r \qquad f_{in} = \infty\text{ Hz}$$

When $f_{in} = 0$ Hz, the reactance of the filter inductor is 0 Ω. In this case, the signal source is coupled directly to the load via the inductor, and $V_{RL} \cong E$. We say these values are approximately equal because some voltage is dropped across the winding resistance (R_W) of the coil. However, the value of R_W is typically much lower than the load resistance, so the approximation is valid in most cases.

When $f_{in} = f_r$, the impedance of the LC filter is infinite for all practical purposes. In this case, the LC circuit isolates the source from the load, and $V_{RL} = 0$ V. The increase in the impedance of the LC circuit as frequency increases from 0 Hz to the value of f_r accounts for the drop in load voltage. As the impedance of the LC filter increases, the filter drops more and more of the AC input signal. The resulting decrease in load voltage accounts for the slope of the frequency response curve as it goes from 0 Hz to f_r.

As f_{in} approaches ∞ Hz (a theoretical value), the reactance of the filter capacitor approaches 0 Ω. In this case, the signal source is coupled directly to the load via the capacitor. Again, $V_{RL} = E$.

The Shunt LC Notch Filter

A shunt LC notch filter can be constructed by placing a series LC circuit in parallel with a load. Such a circuit is shown in Figure 19.42. The frequency response curve of the circuit is identical, for all practical purposes, to the one shown in Figure 19.41b.

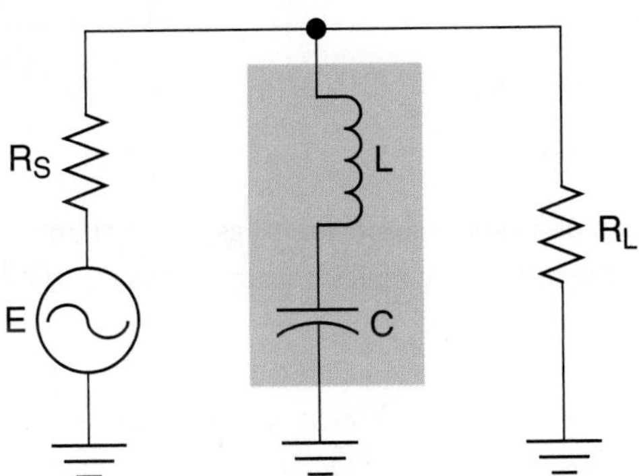

FIGURE 19.42 A shunt LC notch (band-stop) filter.

The operation of the shunt LC notch filter is easy to understand when you consider the circuit response to each of the following conditions:

$$f_{in} = 0\text{ Hz} \qquad f_{in} = f_r \qquad f_{in} = \infty\text{ Hz}$$

When $f_{in} = 0$ Hz, the reactance of the capacitor is infinite, so the LC circuit acts as an open. As a result, the circuit effectively consists of the signal source, R_S, and the load. In this case, $V_{RL} = E - V_{RS}$.

When $f_{in} = f_r$, the LC circuit essentially shorts out the load. In this case, the load voltage is approximately 0 V. The impedance of the LC circuit decreases from ∞ Ω to 0 Ω as the input frequency increases from 0 Hz to f_r. This decrease in impedance causes the voltage across the LC circuit to decrease. Since the LC circuit is in parallel with the load, load voltage also decreases. This accounts for the decrease shown in the frequency response curve (Figure 19.41b).

When f_{in} approaches ∞ Hz (a theoretical value), the reactance of the filter inductor is infinite. Therefore, the LC circuit acts as an open, just as it did when $f_{in} = 0$ Hz. The increase in filter impedance as frequency increases from f_r to ∞ Hz causes the voltage across the filter to increase. Since the load is in parallel with the LC filter, load voltage also increases. This accounts for the rise shown in the frequency response curve (Figure 19.41b).

PROGRESS CHECK

1. List the impedance, current, and voltage characteristics of a series resonant LC circuit.
2. List the impedance, current, and voltage characteristics of a parallel resonant LC circuit.
3. Describe the frequency response of the series LC bandpass filter in Figure 19.33a.
4. Describe the frequency response of the shunt LC bandpass filter in Figure 19.38a.
5. Explain the concept of *loaded-Q*.
6. Describe the operation of the series LC notch filter in Figure 19.41a.
7. Describe the operation of the shunt LC notch filter in Figure 19.42.

19.5 SUMMARY

Here is a summary of the major points made in this chapter:

Introduction to Frequency Response

- Signal loss caused by the frequency response of a circuit is referred to as *attenuation*.
  - Attenuation is normally described using the ratio of a circuit's output amplitude to its input amplitude.
  - In most cases, a power ratio is used.
- The frequency at which the power ratio of a circuit drops to 50% of its maximum value is referred to as the *cutoff frequency* (f_C).
- Filters are circuits that are designed for specific frequency response characteristics.
- There are four basic types of filters:
  - The low-pass filter is designed to pass all frequencies below its cutoff frequency.
  - The high-pass filter is designed to pass all frequencies above its cutoff frequency.
  - The bandpass filter is designed to pass the band of frequencies that lies between two cutoff frequencies.
  - The band-stop (or notch) filter is designed to block the band of frequencies that lies between two cutoff frequencies.
- The cutoff frequencies of a bandpass (or notch) filter are referred to as the *lower cutoff frequency* (f_{C1}) and the *upper cutoff frequency* (f_{C2}).

- Since bandpass and notch filters have two cutoff frequencies, two values are used to describe their operation that are rarely applied to the low-pass and high-pass filters: *bandwidth* and *center frequency*.
- The bandwidth of a filter is the range (or band) of frequencies between its cutoff frequencies.
- The bandwidth of a filter equals the difference between its cutoff frequencies.
- The center frequency f_0 of a filter is the geometric average of the cutoff frequencies.
- The ratio of f_{C2} to f_0 equals the ratio of f_0 to f_{C1}.
- Low-pass filters are normally described using their cutoff frequencies.
  - Example: A low-pass filter with a 100 Hz cutoff frequency.
- Bandpass and notch filters are normally described in terms of bandwidth and center frequency.
  - Example: A 20 kHz bandpass filter with a 150 kHz center frequency.
- The quality (Q) of a bandpass or notch filter equals the ratio of its center frequency to its bandwidth.
- The value of filter Q depends on circuit component values. The bandwidth of a filter depends on the values of center frequency and Q.
- The average frequency (f_{AVE}) of a filter lies halfway between its cutoff frequencies.
- Frequency response curves normally use logarithmic frequency scales.
  - The value of each increment on a logarithmic scale is a whole-number multiple of the previous increment.
  - An octave scale uses a frequency multiplier of 2 between increments.
  - A decade scale uses a frequency multiplier of 10 between increments.

RC and RL Low-Pass Filters

- When a capacitor (or other component) is connected from a signal path to ground, it is referred to as a *shunt* component.
- In an RC low-pass filter, the capacitor is the shunt component. (See Figure 19.13.)
  - The filtering action is provided by the capacitor, which shorts out more and more of the load voltage as the operating frequency increases.
- In an RL low-pass filter, the inductor is the series component. (See Figure 19.18.)
  - The inductor forms a voltage divider with the load.
  - The filtering action is provided by the inductor, which drops more and more of the input signal as the operating frequency increases.

RC and RL High-Pass Filters

- In an RC high-pass filter, the capacitor is the series component. (See Figure 19.22.)
  - The filtering action is provided by the capacitor, which allows more and more of the input signal to pass as the circuit frequency increases.
- In an RL high-pass filter, the inductor is the shunt component. (See Figure 19.26.)
  - The filtering action is provided by the inductor, which shorts out less and less of the load voltage as the circuit frequency increases.

Bandpass and Notch Filters

- The most common passive bandpass and notch filters are LC filters.
- There are two types of LC bandpass filters. (See Figure 19.31a.)
  - The series LC bandpass filter is positioned in series with the filter load.
  - The shunt LC bandpass filter is positioned in parallel with the filter load.

- The center frequency of an LC bandpass filter equals the resonant frequency of the LC circuit.
- The bandwidth of an LC bandpass filter depends on the loaded-Q of the circuit.
- There are two types of LC notch filters. (See Figure 19.31b.)
  - The series LC notch filter has a parallel LC circuit that is positioned in series with the filter load.
  - The shunt LC notch filter has a series LC circuit that is positioned in parallel with the filter load.
- The center frequency of an LC notch filter equals the resonant frequency of the LC circuit.
- The bandwidth of an LC notch filter depends on the loaded-Q of the circuit.

CHAPTER REVIEW

1. The frequency response characteristics of an RLC circuit are the result of ________.
   a) capacitive reactance increasing or decreasing
   b) inductive reactance increasing or decreasing
   c) both a and b
   d) the internal reactance of the source

2. The decrease in output from a circuit that is caused by a change in frequency is referred to as ________.
   a) the dropout voltage
   b) attenuation
   c) gain
   d) amplification

3. A frequency response curve usually represents ________.
   a) a decrease as frequency increases
   b) a decrease as frequency decreases
   c) the ratio of input amplitude to output amplitude
   d) the ratio of output amplitude to input amplitude

4. If a circuit or component has unity gain ________.
   a) it has a power output of 1 W
   b) it has a voltage output of 1 V
   c) its output amplitude equals its input amplitude
   d) it has a power gain of 1 W

5. Cutoff frequency (f_C) occurs ________.
   a) at the frequency where unity gain is reached
   b) at the frequency where power output drops to ½ P_{max}.
   c) at the frequency where voltage output drops to ½ V_{max}.
   d) all of the above

6. Cutoff frequency is determined by ________.
   a) the resistive and reactive components in the circuit
   b) the circuit's bandwidth
   c) the circuit's rejection factor
   d) the frequency where the reactances are equal

7. A filter that attenuates all frequencies *above* f_C is called a ________.
   a) low-pass filter
   b) high-pass filter
   c) notch filter
   d) reject filter

8. A filter that attenuates all frequencies between f_{C1} and f_{C2} is called ________.
   a) bandpass filter
   b) a brick-wall filter
   c) a high-Q filter
   d) a band-stop filter

9. The concept of bandwidth does not apply to ________ filters.
   a) notch filters
   b) band-reject
   c) bandpass
   d) low-pass

10. A bandpass filter has the following values: $f_{C1} = 80$ Hz and $f_{C2} = 450$ Hz. Given this information we can solve for the ________ of the circuit.
    a) gain
    b) Q
    c) attenuation
    d) bandwidth

11. The center frequency of a bandpass or notch filter is defined as ________.
    a) the midpoint between the cutoff frequencies
    b) the geometric average of the cutoff frequencies
    c) the difference between the cutoff frequencies
    d) the ratio of the cutoff frequencies

12. The quality of a filter can be defined as Q = ________.
    a) BW/f_0
    b) f_0/BW
    c) f_C/BW
    d) f_{ave}/BW

13. If the bandwidth of a notch filter increases we know that ________.
    a) the center frequency has shifted higher
    b) filter Q has increased
    c) the center frequency has shifted lower
    d) filter Q has decreased

14. The two most common log scales are ________.
    a) linear and semi-log
    b) octave and decibel
    c) octave and decade
    d) linear and decade

15. In an RL low-pass filter, the inductor is ________ with the load.
    a) in series
    b) in parallel
    c) either one depending on the desired gain
    d) either one depending on the desired Q

16. When the frequency applied to an RC low-pass filter increases, the output signal decreases because ________.
    a) of the RC time constant
    b) X_C decreases
    c) X_C increases
    d) the capacitor is in series with the load

17. A 200 Ω resistor is in series with the parallel combination of a 100 mH coil and a 1 kΩ load. This is a ________ filter.
    a) notch
    b) low-pass
    c) high-pass
    d) high-Q

18. The value of f_C for the circuit described in question 17 is ________.
    a) 265 Hz
    b) 318 Hz
    c) 126 Hz
    d) 418 Hz

19. If the value of a capacitor in a high-pass filter increases in value, the value of f_C ________.
    a) increases
    b) decreases
    c) remains the same
    d) f_C remains the same but Q increases

20. In an RL high-pass filter the inductor ________.
    a) is in parallel with the load
    b) is in series with the load
    c) determines the Q of the circuit
    d) determines the BW of the circuit

21. The filters that have lower cutoff frequencies are the ________.
    a) bandpass, notch, and low-pass filters
    b) bandpass, notch, and high-pass filters
    c) low-pass, high-pass, and notch filters
    d) low-pass, bandpass, and bandstop filters

22. The operation of a series LC bandpass filter is based upon ________.
    a) parallel resonance
    b) an inductor's ability to oppose a change in current
    c) series resonance
    d) either a or b

23. Within the passband of a shunt bandpass filter, the impedance of the LC circuit is ________.
    a) high
    b) low
    c) close to zero
    d) all of the above depending on the frequency

24. A series notch filter has a signal applied to it that is below the rejection band of the filter. This means that the reactance of the inductor is ________.
    a) greater than X_C
    b) very high
    c) less than X_C
    d) much greater than X_C

25. The loaded Q of a series LC filter is determined by ________.
    a) the coil's inductance, total resistance, and frequency
    b) X_L and the total resistance
    c) the coil's inductance, the total resistance, and f_r
    d) all of the above

26. Loaded Q is always ________.
    a) less than 1
    b) less than unloaded Q
    c) greater than 1
    d) greater than 2

27. If the resistance of the load increases, the ________ of a series LC filter will ________.
    a) Q, decrease
    b) BW, decrease
    c) Q, increase
    d) f_r, increase

28. A series LC bandpass filter has a loaded Q of 4.6 and a resonant frequency of 2.2 kHz. This means that f_{C1} = ________ and f_{C2} = ________.
    a) 2439 Hz, 1961 Hz
    b) 2.68 kHz, 1.72 kHz
    c) 2349 Hz, 1691 Hz
    d) 2.71 kHz, 1.86 kHz

29. If a resonant LC circuit has a Q that is less than 2, f_{ave} must be ________ f_r.
    a) less than
    b) greater than
    c) much less than
    d) at least 2 times greater than

30. A shunt LC filter has an unloaded Q of 46. The coil has a winding resistance of 6 Ω. The equivalent parallel resistance (R_P) for this coil is ________.
    a) 276 Ω
    b) 552 Ω
    c) 12.7 kΩ
    d) 25.4 kΩ

31. If the load resistance of shunt LC filter increases, Q ________ and BW ________.
    a) increases, increases
    b) decreases, decreases
    c) increases, decreases
    d) decreases, increases

32. A series resonant circuit is to be used as a notch filter. This means it must be ________.
    a) in parallel with the load
    b) a low-Q circuit
    c) a high-Q circuit
    d) in series with the load

PRACTICE PROBLEMS

1. Complete the table below.

	f_{C1}	f_{C2}	BW	f_0
a)	5 kHz	88 kHz	______	______
b)	640 Hz	1.8 kHz	______	______
c)	280 kHz	1.4 MHz	______	______
d)	60 Hz	300 Hz	______	______

2. Complete the table below.

f_{C1}	f_{C2}	BW	f_0
a) 2.2 kHz	63 kHz	______	______
b) 120 kHz	2 MHz	______	______
c) 400 Hz	800 Hz	______	______
d) 24 kHz	500 kHz	______	______

3. Complete the table below.

f_{C1}	f_{C2}	BW	f_0
a) 800 Hz	______	12 kHz	______
b) 5 kHz	______	______	10 kHz
c) ______	200 kHz	198 kHz	______

4. Complete the table below.

f_{C1}	f_{C2}	BW	f_0
a) 400 Hz	______	39.6 kHz	______
b) 50 kHz	______	______	200 kHz
c) ______	400 kHz	50 kHz	______

5. A filter has values of BW = 18 kHz and f_0 = 120 kHz. Determine the value of Q for the filter.
6. A filter has values of BW = 48 kHz and f_0 = 340 kHz. Determine the value of Q for the filter.
7. A filter has values of f_0 = 52 kHz and Q = 6.8. Determine the circuit bandwidth.
8. A filter has values of f_0 = 650 Hz and Q = 2.5. Determine the circuit bandwidth.
9. Determine the value of f_{AVE} for the filter described in Problem 7.
10. Determine the value of f_{AVE} for the filter described in Problem 8.
11. Calculate the cutoff frequency for the low-pass filter in Figure 19.43a.
12. Calculate the cutoff frequency for the low-pass filter in Figure 19.43b.

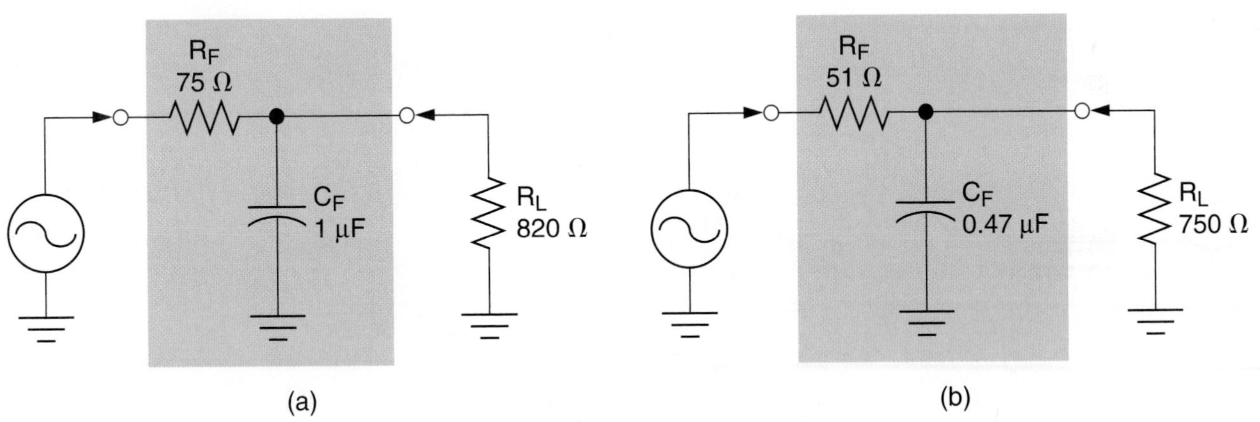

FIGURE 19.43

13. Calculate the cutoff frequency for the low-pass filter in Figure 19.44a.
14. Calculate the cutoff frequency for the low-pass filter in Figure 19.44b.

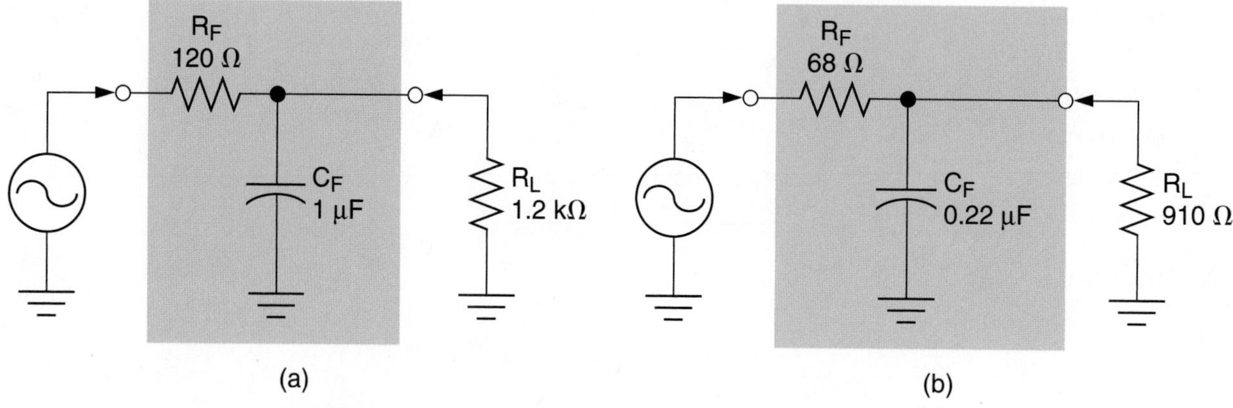

FIGURE 19.44

15. Calculate the cutoff frequency for the low-pass filter in Figure 19.45a.
16. Calculate the cutoff frequency for the low-pass filter in Figure 19.45b.

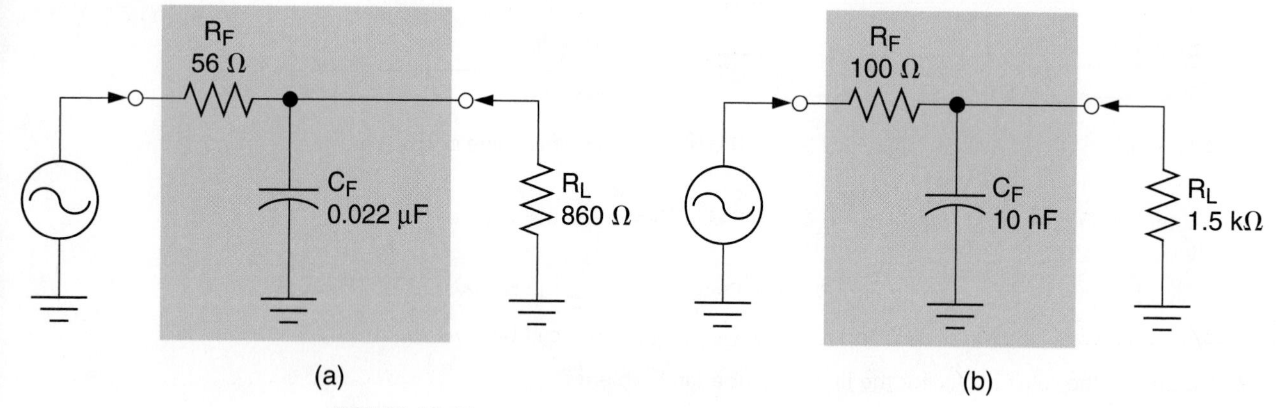

FIGURE 19.45

17. Calculate the cutoff frequency for the low-pass filter in Figure 19.46a.
18. Calculate the cutoff frequency for the low-pass filter in Figure 19.46b.

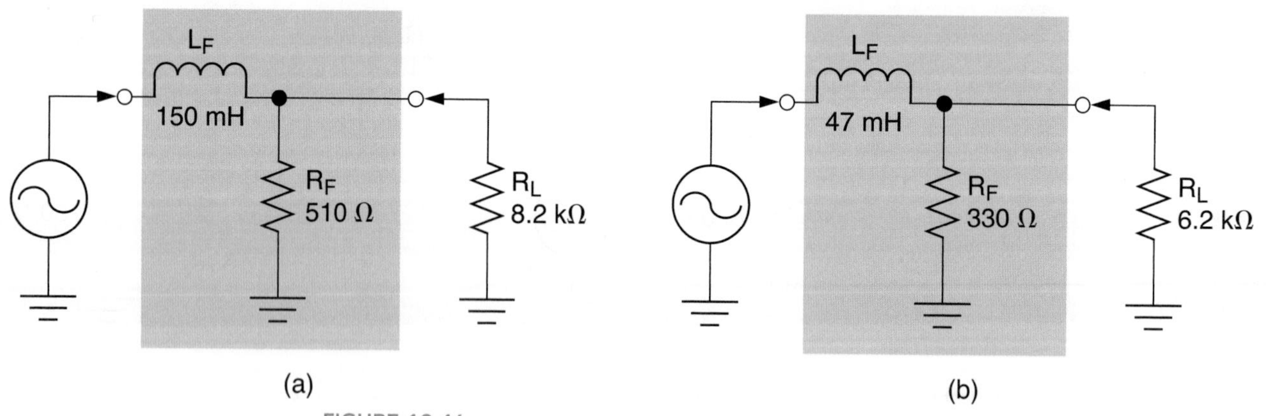

FIGURE 19.46

19. Calculate the cutoff frequency for the low-pass filter in Figure 19.47a.
20. Calculate the cutoff frequency for the low-pass filter in Figure 19.47b.

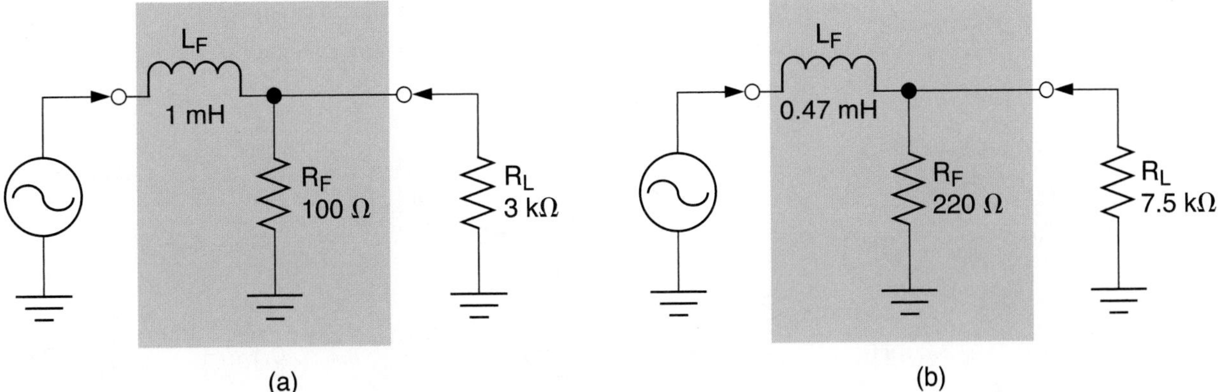

FIGURE 19.47

21. Calculate the cutoff frequency for the high-pass filter in Figure 19.48a.
22. Calculate the cutoff frequency for the high-pass filter in Figure 19.48b.

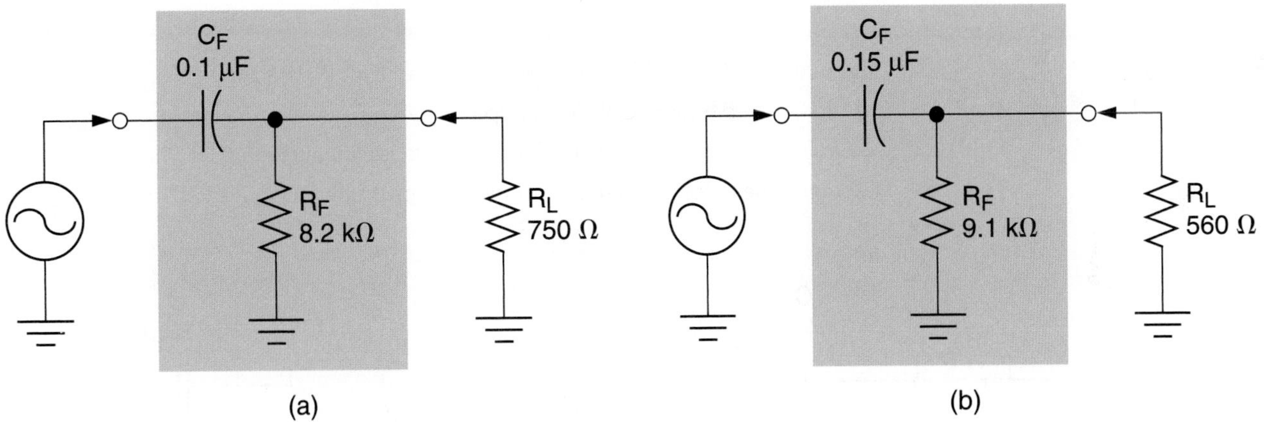

FIGURE 19.48

23. Calculate the cutoff frequency for the high-pass filter in Figure 19.49a.
24. Calculate the cutoff frequency for the high-pass filter in Figure 19.49b.

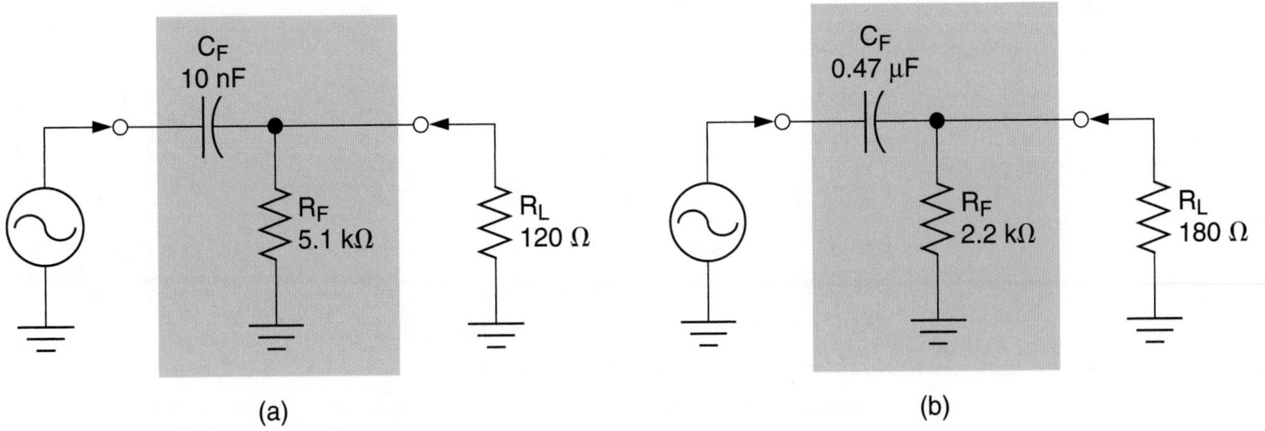

FIGURE 19.49

25. Calculate the cutoff frequency for the high-pass filter in Figure 19.50a.
26. Calculate the cutoff frequency for the high-pass filter in Figure 19.50b.

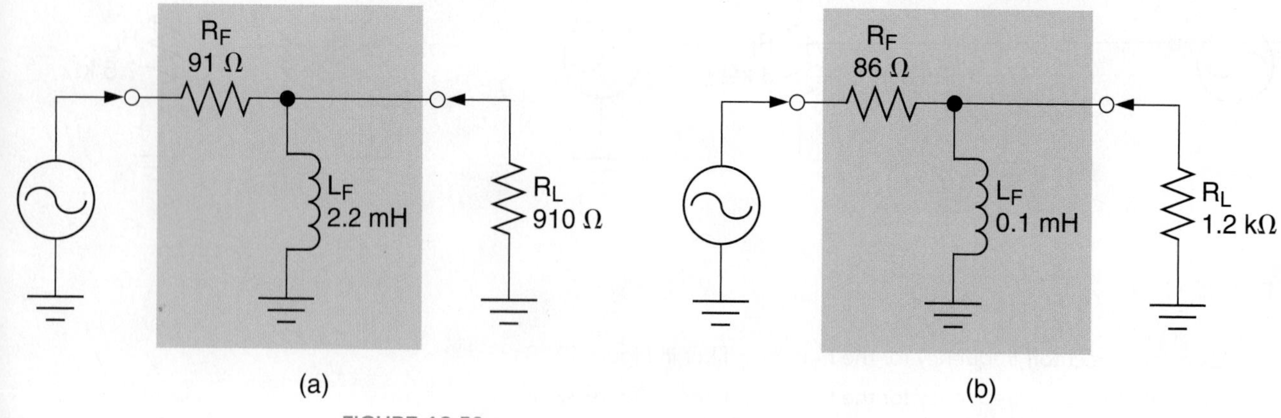

FIGURE 19.50

27. Determine the bandwidth of the bandpass filter in Figure 19.51a.
28. Determine the bandwidth of the bandpass filter in Figure 19.51b.
29. Determine the cutoff frequencies for the bandpass filter in Figure 19.51a.
30. Determine the cutoff frequencies for the bandpass filter in Figure 19.51b.

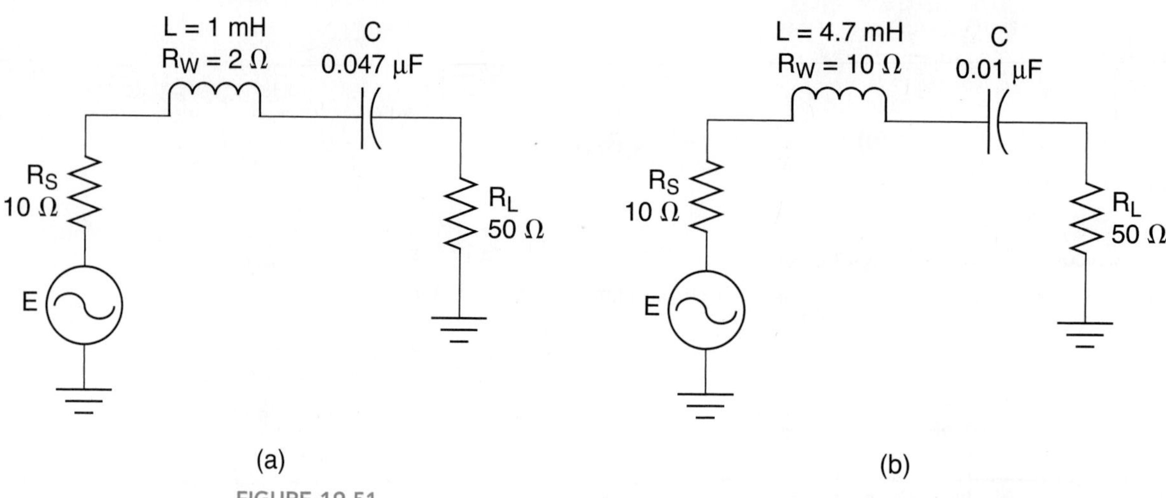

FIGURE 19.51

31. Determine the bandwidth and cutoff frequencies for the bandpass filter in Figure 19.52a.
32. Determine the bandwidth and cutoff frequencies for the bandpass filter in Figure 19.52b.
33. Determine the value of Q_L for the circuit in Figure 19.53a.
34. Determine the value of Q_L for the circuit in Figure 19.53b.
35. Calculate the bandwidth and cutoff frequencies for the bandpass filter in Figure 19.53a.
36. Calculate the bandwidth and cutoff frequencies for the bandpass filter in Figure 19.53b.

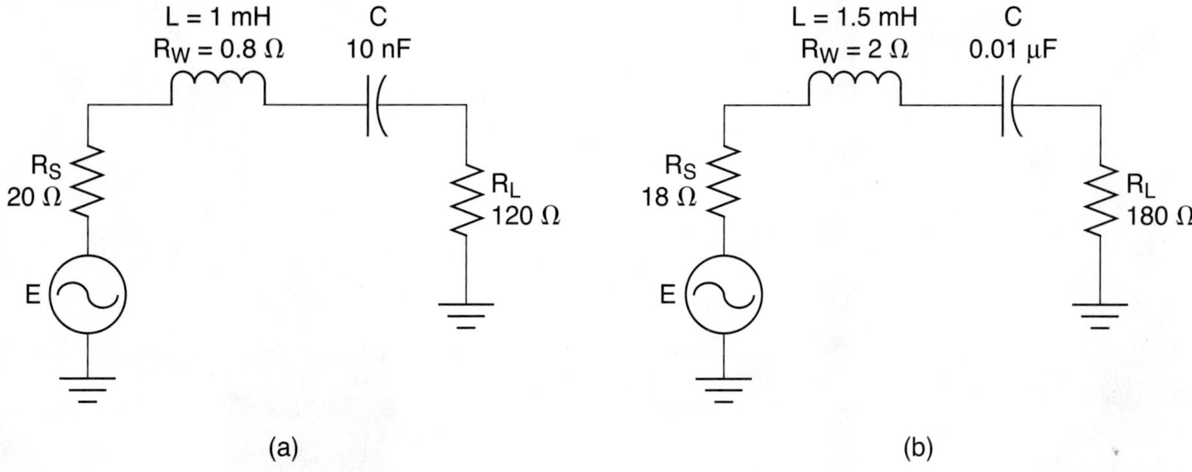

FIGURE 19.52

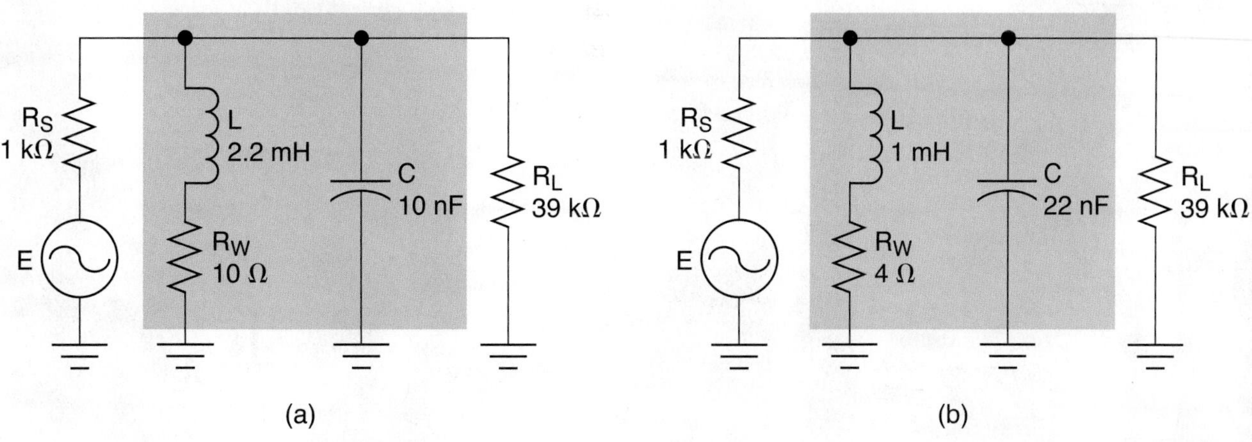

FIGURE 19.53

ANSWERS TO THE EXAMPLE PRACTICE PROBLEMS

19.1	818 Hz
19.2	490 Hz
19.3	5.33
19.4	50
19.5	f_0 = 150 Hz, f_{AVE} = 463 Hz
19.6	101 Hz
19.7	303 Hz
19.8	251 Hz
19.9	1.81 kHz
19.10	2.46 kHz
19.11	f_{C1} = 14.5 kHz, f_{C2} = 16.9 kHz
19.12	6.55
19.13	f_{C1} = 989 Hz, f_{C2} = 1.15 kHz

CHAPTER 20

THREE-PHASE POWER

PURPOSE

Throughout North America, commercial power is transmitted in the form of three-phase AC. Three-phase power is also used extensively in industrial applications, and three-phase AC motors have advantages over single-phase AC motors (as you will learn in upcoming chapters). For these reasons and more, it is important to understand the basics of three-phase power.

KEY TERMS

The following terms are introduced and defined in this chapter on the pages indicated:

Armature, 857
Balanced load, 861
Delta (Δ) circuit, 860
Delta-wye (Δ-Y) circuit, 877
Phase sequence, 859
Three-phase power, 856
Unbalanced load, 861
Wye (Y) circuit, 860
Wye-delta (Y-Δ) circuit, 873

OBJECTIVES

After completing this chapter, you should be able to:

1. Identify and describe three-phase AC waveforms.
2. Describe the concept of *phase sequence*.
3. List the advantages that three-phase power has over single-phase power.
4. Identify inductive *wye* and *delta* circuits.
5. Distinguish between *balanced* and *unbalanced* wye circuits.
6. Distinguish between *line* and *phase* voltages and currents.
7. Calculate the voltages and currents for a balanced wye circuit.
8. In terms of component connections, contrast delta and wye circuits.
9. Calculate the voltages and currents for a balanced delta circuit.
10. Calculate voltages and currents for balanced wye-delta and delta-wye circuits.
11. Calculate the power values for balanced three-phase circuits.
12. Determine the capacitor values needed for power factor correction in a three-phase circuit.

20.1 THREE-PHASE POWER

The Greek letter *phi* (ϕ) is often used to represent *phase*. Thus, the label 3ϕ translates as "three phase."

KEY CONCEPT

In any balanced three-phase circuit, power is equal to $\sqrt{3}$ times the product of the line voltage, line current, and power factor.

KEY CONCEPT

The term **THREE-PHASE POWER** is used to describe a means of generating and transmitting power using three sine waves that are 120° out of phase with each other.

A three-phase power source can be represented as shown in Figure 20.1a. The source has three "hot" conductors (labeled ϕA, ϕB, and ϕC) and a single neutral conductor (N). Each hot line carries a sine wave that is 120° out of phase with the other two sine waves. If we combine these sine waves on a single graph, we get the composite waveform shown in Figure 20.1b. The phase relationships among the three waves are highlighted in Figure 20.2.

THREE-PHASE POWER A means of generating and transmitting power using three sine waves that are 120° out of phase with each other.

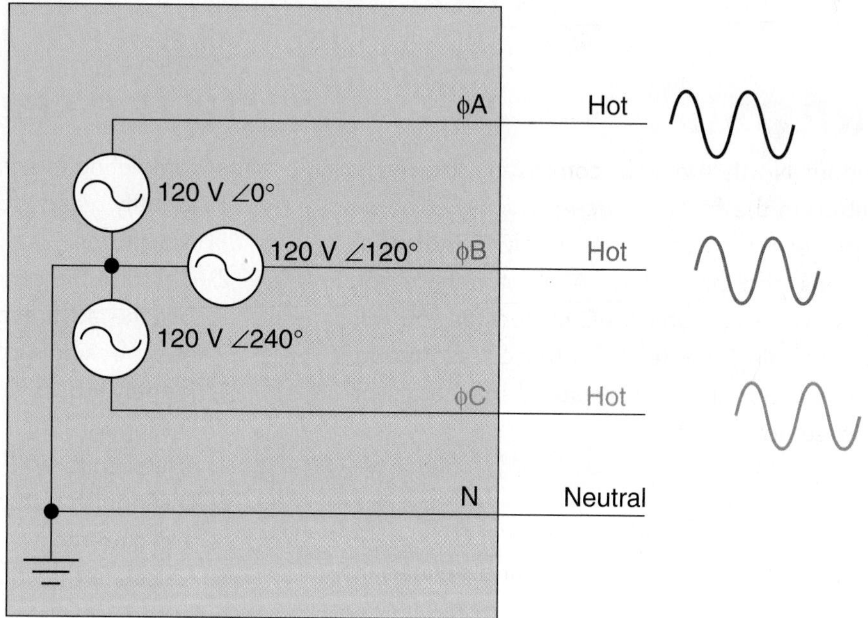

(a) Three-phase source block diagram

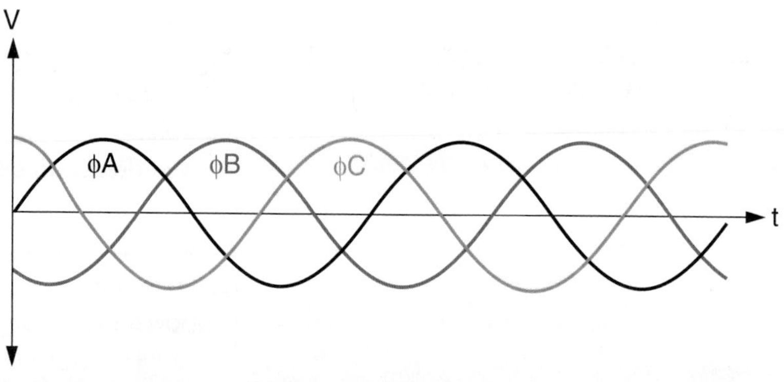

(b) Composite waveform

FIGURE 20.1 Three phase AC.

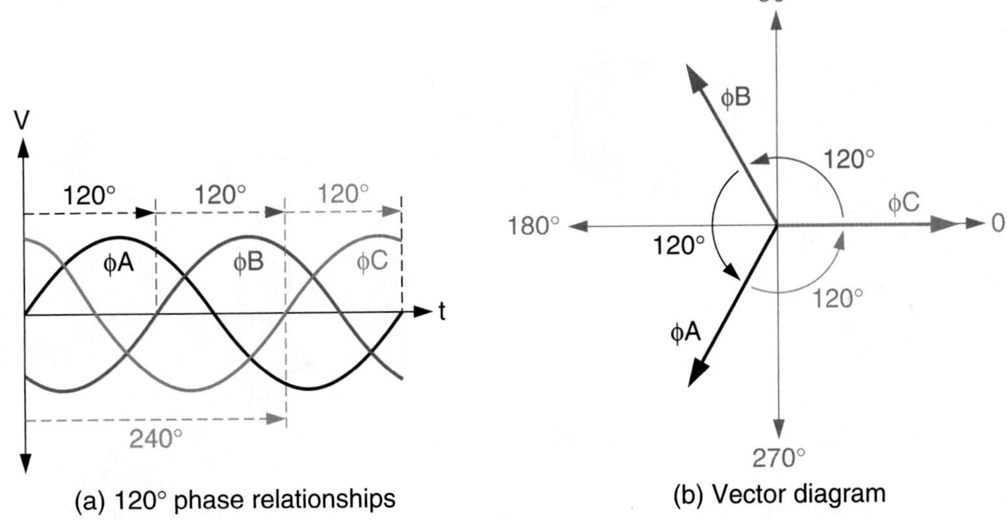

FIGURE 20.2 Phase relationships among 3ϕ AC waveforms.

The phase relationships in Figure 20.2a can be described as follows:

- At the *origin* of the graph (the point where the x-axis and *y*-axis meet), the ϕA waveform is at 0°.
- The ϕA waveform is at 120° when the ϕB waveform is at 0°. Therefore, the two waveforms are 120° out of phase.
- The ϕB waveform is at 120° when the ϕC waveform is at 0°. Therefore, the two waveforms are 120° out of phase.
- The ϕC waveform is at 120° when the ϕA waveform is at 0° (again). Therefore, the two waveforms are 120° out of phase.

As these points indicate, each waveform is 120° out of phase with the other two. Note that the green vertical line in Figure 20.2a indicates that ϕC = 0° when ϕB = 120° and ϕA = 240°. These values were used to generate the vector diagram in Figure 20.2b.

Generating Three-Phase Power

In Chapter 13, you were shown how a sine wave can be generated by rotating a conductor through a stationary magnetic field. As a review, the basic AC generator described in Chapter 13 is shown in Figure 20.3a. As the conductor rotates, it cuts through the magnetic field, generating a sine wave. The current-carrying conductor in any generator or motor (the rotor in this case) is referred to as its **ARMATURE**.

ARMATURE The current-carrying conductor in a generator or motor.

A sine wave can also be generated using a stationary armature and a rotating magnet, as shown in Figure 20.3b. Remember, all that is required to induce a voltage across a conductor is *relative motion* between the conductor and the magnetic field (as given in Faraday's first law). The armature can be the stationary element (or *stator*) and the magnet can be the rotating element (or *rotor*).

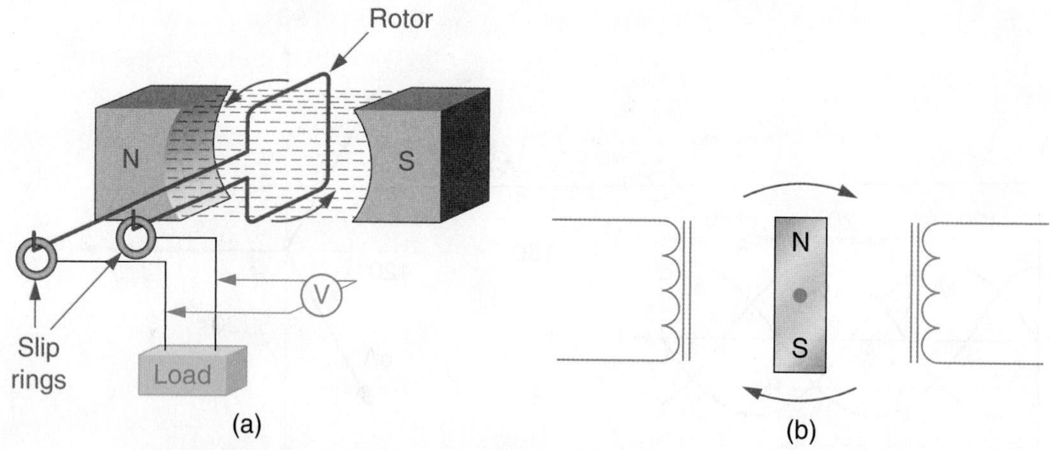

FIGURE 20.3 Sine wave generator configurations.

The three-phase generator in Figure 20.4 has a magnetic rotor and three stationary armatures that are positioned at 120° angles. As the magnet rotates on its axis, a voltage is generated across each coil. The 120° angle between the armatures produces the 120° phase angle between the three sine waves.

NOTE: The magnet in Figure 20.4 is the rotor. The three coils (conductors) and the structure that supports them form the stator.

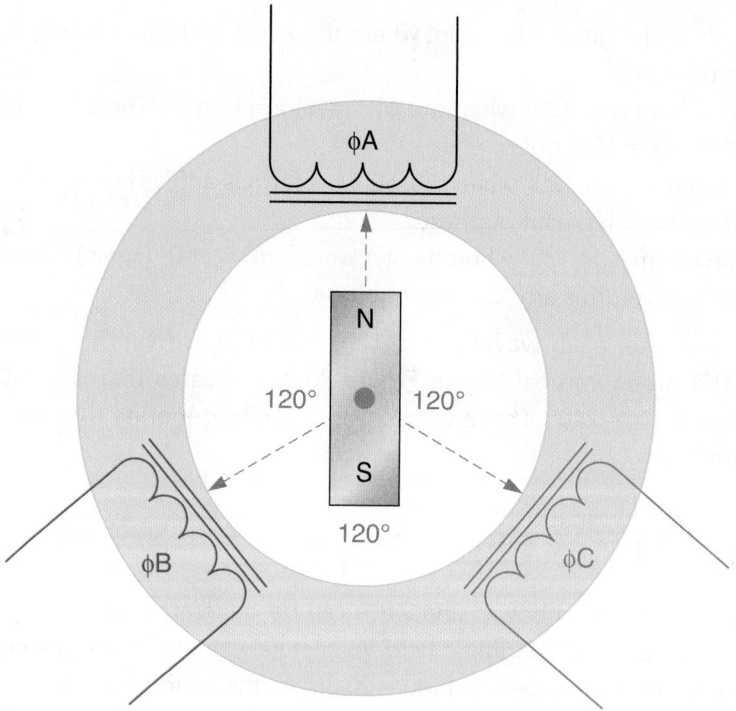

Rotation	*Phase Sequence*
Counterclockwise	A–B–C
Clockwise	C–B–A

FIGURE 20.4 Phase sequences for a 3ϕ power source.

The **PHASE SEQUENCE** of a three-phase power source is the order in which the three sine waves peak. For example, if the rotor in Figure 20.4 turns counter-clockwise (CCW), the phase sequence is A–B–C. This means that the ϕA sine wave peaks first, followed by the ϕB sine wave and the ϕC sine wave. If the rotor turns clockwise (CW), the phase sequence is C–B–A. Composite waveforms illustrating these phase sequences are shown in Figure 20.5.

PHASE SEQUENCE For a three-phase power source, the order in which the three output sine waves reach their peak values.

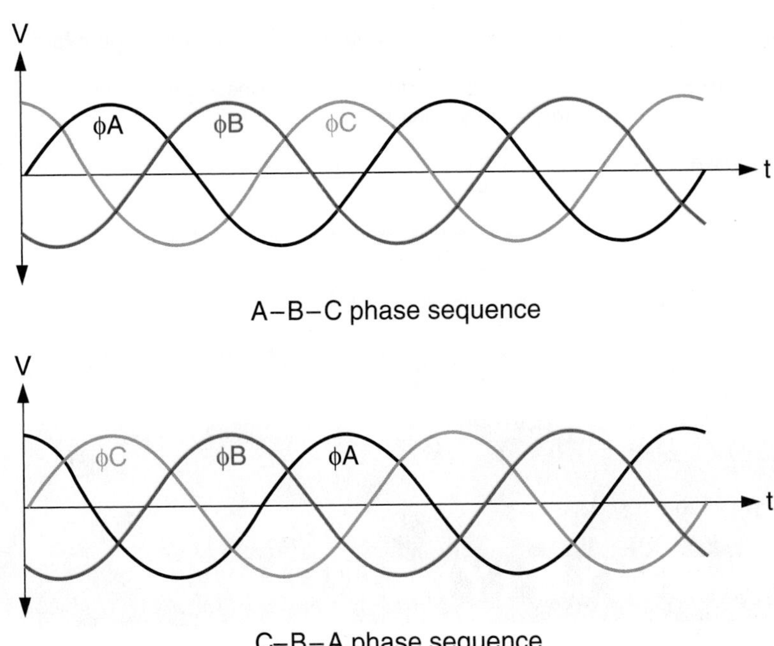

FIGURE 20.5 Phase sequences.

The phase sequence of a given three-phase generator is not important if the generator load is purely resistive. However, if the generator load is reactive (capacitive and/or inductive), the phase sequence of a three-phase generator becomes an important consideration.

The Advantages of Three-Phase Power

Three-phase power has several advantages that make it better suited for power transmission and industrial applications than single-phase power. Several of these advantages follow:

1. Three-phase motors are less complex and have higher horsepower ratings than comparable single-phase motors.
2. Three-phase transformers have greater output capabilities than comparable single-phase transformers.

A three phase motor or generator has approximately 1.5 times the horsepower rating of a comparable single phase machine.

3. All other factors being equal, single-phase circuits require larger diameter (heavier) conductors than three-phase circuits. A single-phase system uses approximately 1.15 times the copper of a comparable three-phase system, so three-phase systems are cheaper to build.

These advantages will be demonstrated in upcoming chapters.

PROGRESS CHECK

1. What is meant by the term *three-phase power*?
2. A three-phase generator has four output conductors. How many of the conductors are hot?
3. What is the phase relationship among the sine wave outputs of a three-phase generator?
4. What is the *phase sequence* of a three-phase generator?
5. List the advantages that three-phase circuits have over single-phase circuits.

20.2 INDUCTIVE WYE (Y) CIRCUITS

WYE (Y) CIRCUIT A circuit commonly found in three-phase systems with three components that are connected in the shape of the letter Y.

DELTA (Δ) CIRCUIT A circuit commonly found in three-phase systems with three components that are connected in the shape of the Greek letter Δ.

Two circuits that are used extensively in three-phase applications are the **WYE (Y)** and **DELTA (Δ) CIRCUITS**. As shown in Figure 20.6, the circuits resemble the letters Y and Δ. In this section, we examine the characteristics of wye inductive circuits. Delta inductive circuits are introduced in Section 20.3.

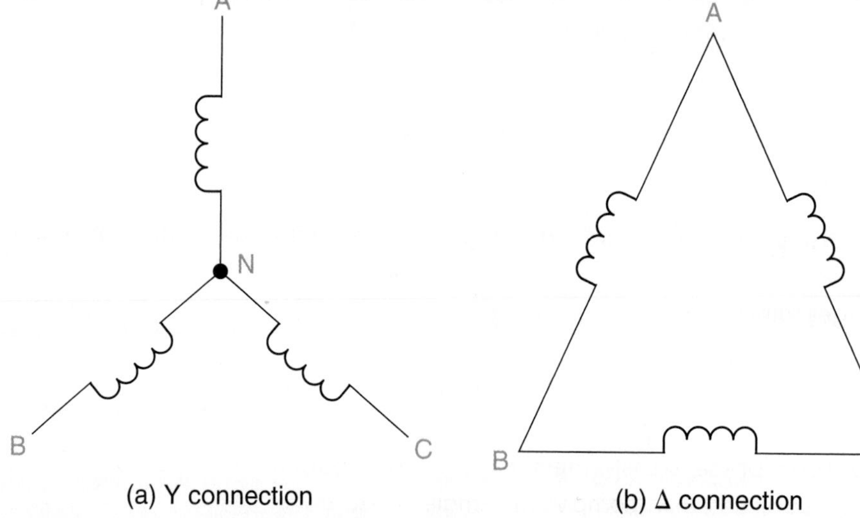

FIGURE 20.6 Wye (Y) and Delta (Δ) connections.

Before we begin our discussion of a wye system, we need to distinguish between *balanced* and *unbalanced* loads. A **BALANCED LOAD** is one in which the load currents are equal in value and 120° out of phase. An **UNBALANCED LOAD** is one in which the load currents are *not* equal and/or out of phase by some value other than 120°.

BALANCED LOAD One in which the load currents are equal in value and 120° out of phase.

UNBALANCED LOAD One in which the load currents are not equal and/or out of phase by some value other than 120°.

The analysis of a balanced wye system is relatively simple and straightforward. The analysis of an unbalanced wye system, on the other hand, can be difficult and complex. For this reason, all discussions in this chapter assume that the wye system is balanced. Fortunately, this is normally a valid assumption, as great care is taken during circuit design to produce balanced systems.

Wye (Y) Circuit Voltages

The voltages in a wye circuit are classified as *line voltages* or *phase voltages*. The phase voltages are identified in Figure 20.7a, the line voltages in Figure 20.7b. The subscripts in each case identify the terminals to which a voltmeter would be connected to measure that particular voltage. For example, E_{AN} is the voltage from the A terminal to the neutral (N) terminal. E_{AC}, on the other hand, is the voltage from the A terminal to the C terminal. The voltages in a wye circuit can also be represented using a vector diagram like the one shown in Figure 20.7c.

The wye connection is also referred to as a *tee* or *star connection*.

IN THE FIELD

According to Kirchhoff's voltage law (KVL), the voltage across the A and B terminals (E_{AB}) must equal the sum of E_{AN} and E_{BN}. By formula,

$$E_{AB} = E_{AN} + E_{BN} \tag{20.1}$$

By the same token,

$$E_{AC} = E_{AN} + E_{CN} \tag{20.2}$$

and

$$E_{BC} = E_{BN} + E_{CN} \tag{20.3}$$

In each of these equations, a line voltage is shown to equal the sum of two phase voltages. Because of the 120° phase relationship between the phase voltages, they must be added *geometrically* using vector addition or complex numbers. For example, look at the wye circuit in Figure 20.8a. In this circuit, E_{AN} and E_{BN} are 120° out of phase. Because of this phase relationship, they are added using vectors as shown in Figure 20.8b.

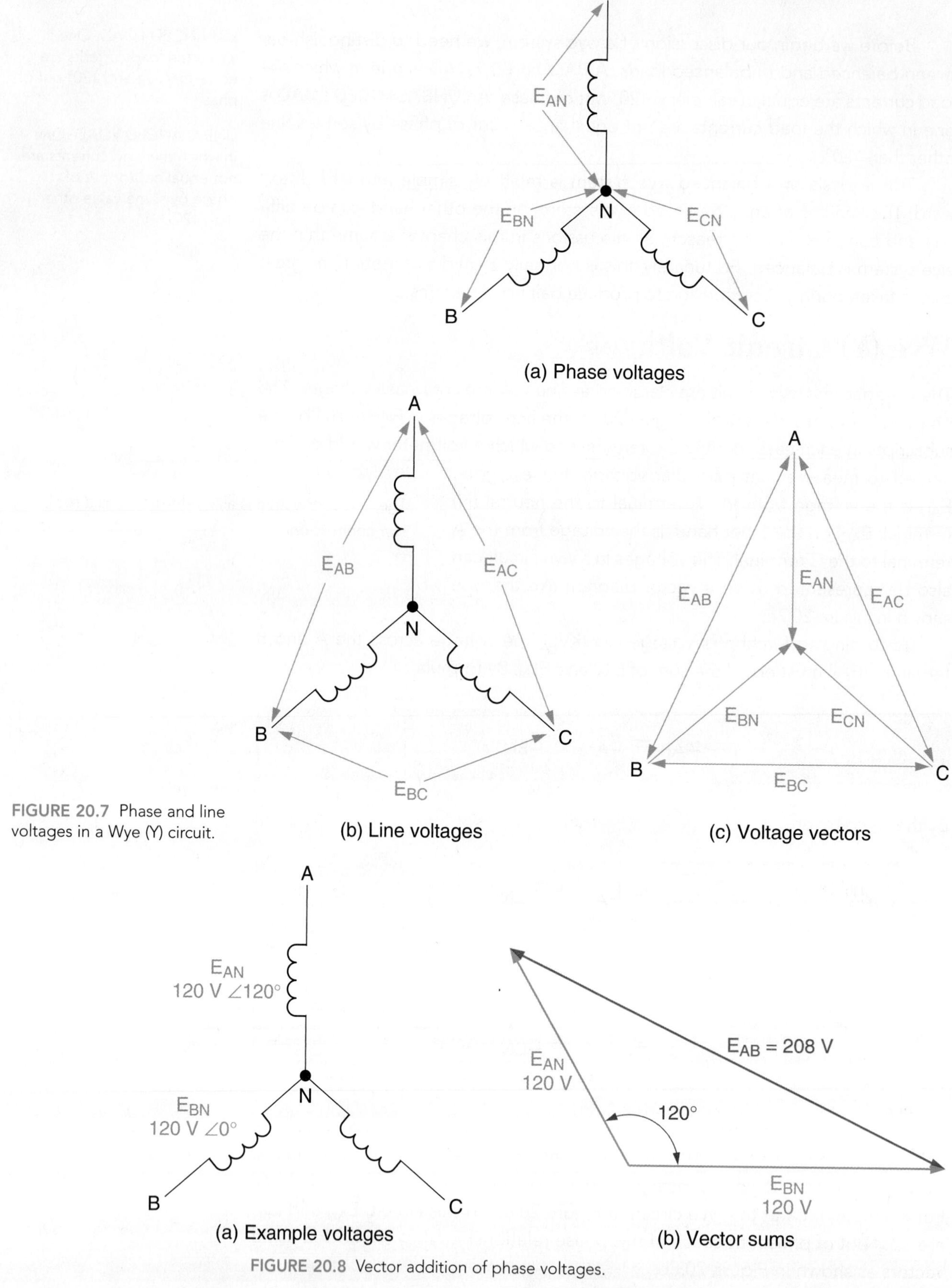

FIGURE 20.7 Phase and line voltages in a Wye (Y) circuit.

FIGURE 20.8 Vector addition of phase voltages.

Using vector addition to add the phase voltages in a balanced wye circuit is more complicated than need be. The relationship between any line voltage and its two phase voltages can be calculated using

(20.4)
$$E_{LINE} = \sqrt{3} \times E_{PHASE}$$

or

(20.5)
$$E_{LINE} = 1.732 \times E_{PHASE}$$

An application of these equations is demonstrated in the following example.

EXAMPLE 20.1

A balanced wye circuit has phase voltages of 120 V. What is the value of the line voltages?

SOLUTION

The line voltages for the circuit can be found as

$$\begin{aligned} E_{LINE} &= \sqrt{3} \times E_{PHASE} \\ &= 1.732 \times 120 \text{ V} \\ &= 208 \text{ V} \end{aligned}$$

Comparing this result to the vectors in Figure 20.8b, you can see that we obtained the same result you would get using vector addition or complex numbers. Note that $\sqrt{3}$ was shown to equal 1.732 in the problem, validating equation (20.5).

PRACTICE PROBLEM 20.1

A balanced wye circuit has phase voltages of 277 V. Determine the value of the line voltages.

A modified version of equation (20.4) can be used to calculate phase voltage when line voltage is known. This equation is

(20.6)
$$E_{PHASE} = \frac{E_{LINE}}{\sqrt{3}}$$

or

(20.7)

$$E_{PHASE} = \frac{E_{LINE}}{1.732}$$

Example 20.2 demonstrates the use of these equations.

EXAMPLE 20.2

A balance wye circuit has 4160-V line voltages. What is the value of the phase voltages?

SOLUTION

The value of the phase voltages is found as

$$E_{PHASE} = \frac{E_{LINE}}{\sqrt{3}}$$

$$= \frac{4160 \text{ V}}{1.732}$$

$$= 2.4 \text{ kV}$$

PRACTICE PROBLEM 20.2

A balanced wye circuit has 13,800-V line voltages. What is the value of the phase voltages?

It should be noted that, in practice, wye circuit phase voltages are internal to an electrical motor, generator, or transformer, and therefore cannot be measured. However, line voltages can be measured by connecting a true RMS voltmeter between any two lines. As a result, phase voltages are normally determined by measuring line voltage and using the measured value to calculate phase voltage.

Wye Circuit Current

Up to this point, we have been dealing with open wye circuits; that is, wye circuits that have no loads connected to the phase inductors. A loaded wye circuit is shown in Figure 20.9.

The loads in Figure 20.9 are labeled Z_{AN}, Z_{BN}, and Z_{CN}. In practice, these loads may be resistive, reactive (capacitive or inductive), or any combination of the three. The letters A, B, C, and N were included in the load circuit to indicate the connections to the source. For example, Z_{CN} is connected between the points

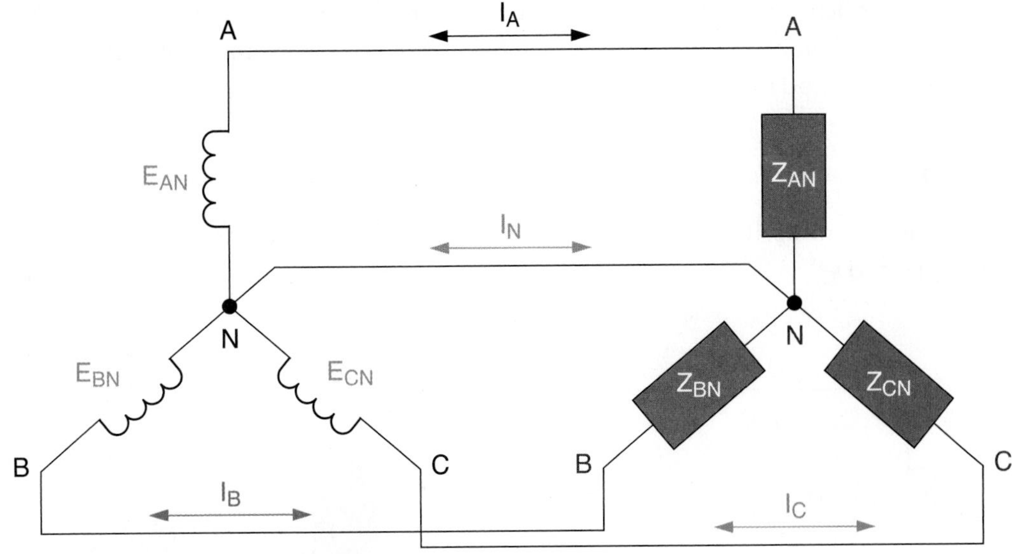

FIGURE 20.9 A loaded wye (Y) circuit.

labeled C and N. These letters indicate that the load is in parallel with phase inductor CN. As such, the voltage across Z_{CN} equals E_{CN}. By the same token,

- The voltage across Z_{AN} equals E_{AN}.
- The voltage across Z_{BN} equals E_{BN}.

The current arrows in Figure 20.9 indicate that the line currents pass through the loads. That is,

- I_A is the phase current through Z_{AN}.
- I_B is the phase current through Z_{BN}.
- I_C is the phase current through Z_{CN}

In other words, phase current equals line current in a wye-connected circuit. Based on the voltage and current relationships listed here, Ohm's law tells us that:

(20.8)

$$I_A = \frac{E_{AN}}{Z_{AN}}$$

(20.9)

$$I_B = \frac{E_{BN}}{Z_{BN}}$$

(20.10)

$$I_C = \frac{E_{CN}}{Z_{CN}}$$

When calculating the line currents in a wye circuit, it is important to consider the phase differences between the currents. Three-phase current calculations are demonstrated in Example 20.3.

EXAMPLE 20.3

The balanced wye circuit in Figure 20.10a has the following values: $E_{AN} = 120\ V$ and $Z_{AN} = 20\ \Omega$. Calculate the values of I_A, I_B, and I_C.

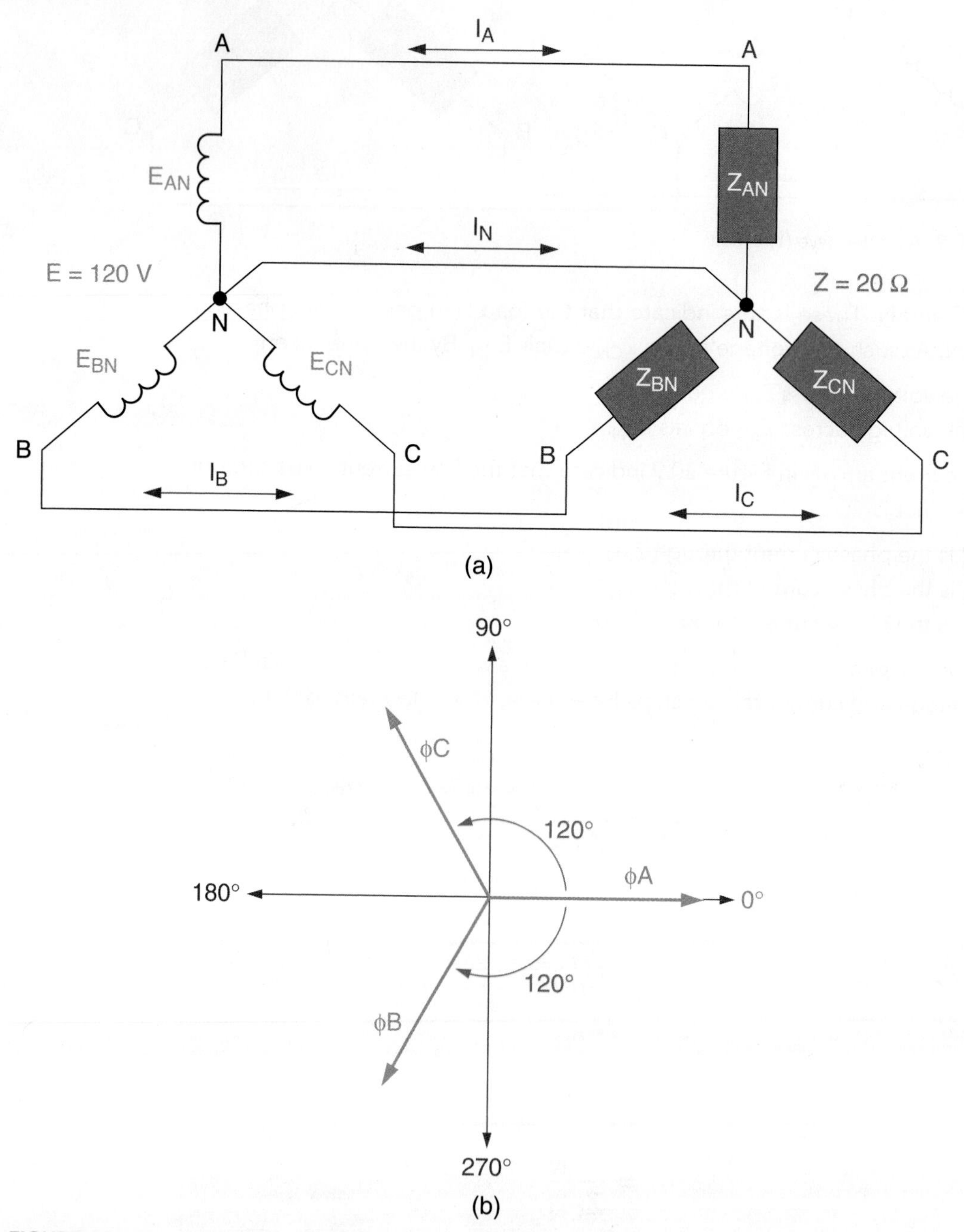

FIGURE 20.10

SOLUTION

The circuit phase relationships are shown in Figure 20.10b. Using the circuit values and phase angles, the phase current values are calculated as follows:

$$I_A = \frac{E_{AN}}{Z_{AN}} = \frac{120\text{ V }\angle 0°}{20\ \Omega} = 6\text{ A }\angle 0°$$

$$I_B = \frac{E_{BN}}{Z_{BN}} = \frac{120\text{ V }\angle -120°}{20\ \Omega} = 6\text{ A }\angle -120°$$

and

$$I_C = \frac{E_{CN}}{Z_{CN}} = \frac{120\text{ V }\angle 120°}{20\ \Omega} = 6\text{ A }\angle 120°$$

As you can see, the currents in a balanced wye circuit (like the voltages) are equal in value and 120° out of phase with each other.

PRACTICE PROBLEM 20.3

A circuit like the one in Figure 20.10a has the following values: $E_{CN} = 480$ V and $Z_{CN} = 160\ \Omega$. Calculate the values of the phase currents.

The currents calculated in Example 20.3 can be used to demonstrate one of the advantages that three-phase power systems have over others. According to Kirchhoff's current law (KCL), the current through the neutral (N) conductor must equal the sum of the other three currents. By formula,

(20.11)
$$I_N = I_A + I_B + I_C$$

Using our results from Example 20.3, we get

$$I_N = 6\text{ A }\angle 0° + 6\text{ A }\angle -120° + 6\text{ A }\angle 120°$$

Figure 20.11a shows the three current vectors representing the values in the I_N equation. When you position these vectors end-to-end (as shown in Figure 20.11b), you can see that the sum of the vectors is 0 A. Therefore,

$$I_N = 6\text{ A}\angle 0° + 6\text{ A}\angle -120° + 6\text{ A}\angle 120° = 0\text{ A}$$

This answer indicates that *there is no current (ideally) through the neutral conductor in a balanced wye circuit.* Even so, a neutral conductor is included in the circuit for grounding purposes.

Any current through the neutral conductor in a wye circuit is a result of minor variations in the loads, and normally has a very low value. As a result, even high-current wye circuits can use higher-gauge (lighter) neutral conductors than comparable single-phase circuits.

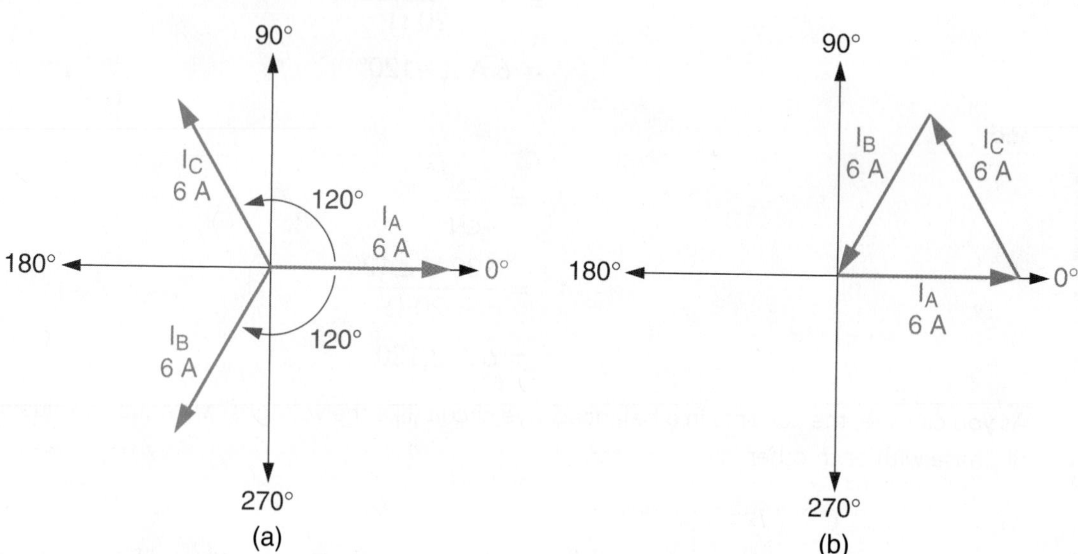

FIGURE 20.11 The vector sum of wye (Y) circuit phase currents.

One Final Point

Many relationships for balanced wye circuits have been established in this section. Among the most important are the following:

$$E_{LINE} = \sqrt{3} \times E_{PHASE}$$
$$I_{LINE} = I_{PHASE}$$
$$I_N = 0\text{ A}$$

In the next section, we will discuss the delta (Δ) inductive circuit. As you will see, a delta circuit has no neutral line, and E_{LINE} and I_{LINE} relationships that are the opposite of those listed for the wye circuit.

PROGRESS CHECK

1. Which two circuit configurations are used extensively in three-phase applications?
2. Explain the difference between a balanced and an unbalanced wye system.
3. Why can't you measure a phase voltage in a wye-connected circuit?
4. Where would you measure a line voltage in a wye-connected circuit?
5. What is the relationship between phase voltage and line voltage in a wye-connected circuit?
6. What is the relationship between phase current and line current in a wye-connected circuit?

20.3 INDUCTIVE DELTA (Δ) CIRCUITS

A *delta* (Δ) *connection* is formed by connecting the phase inductors in a three-phase generator as shown in Figure 20.12a. In this case, the phase inductors are connected end-to-end, rather than to a single common point (as in a wye-connected circuit). As a result, the phase voltages equal the line voltages. By formula,

$$E_{LINE} = E_{PHASE} \tag{20.12}$$

As with the wye circuit, the phase voltages (and therefore, the line voltages) in a delta circuit are 120° out of phase with each other. This phase relationship is illustrated by the vector diagram in Figure 20.12b.

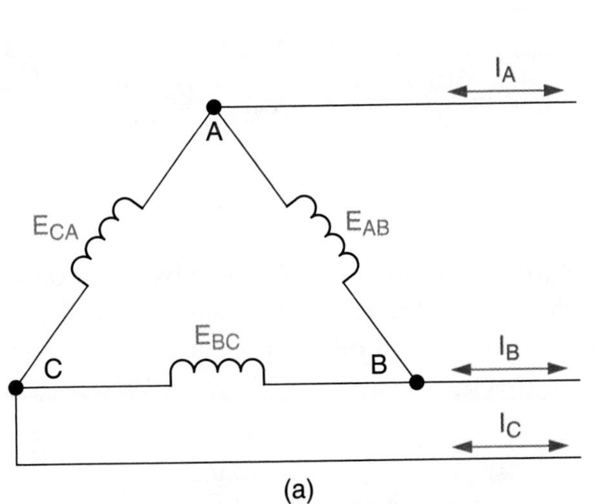

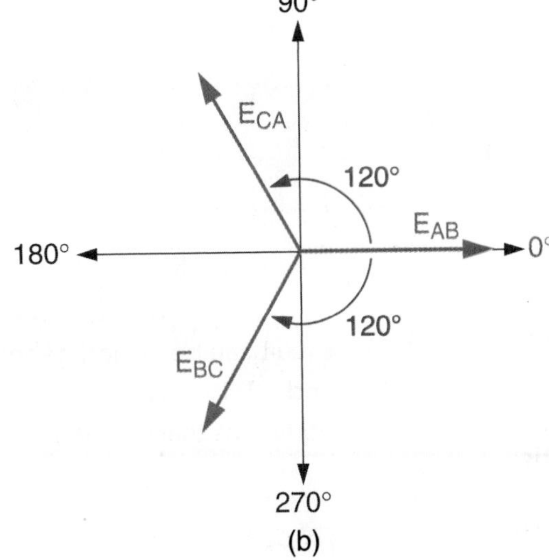

FIGURE 20.12 A delta (Δ) circuit and phase diagrams.

Circuit Currents

Figure 20.13 shows a delta source connected to delta-connected loads. The loads are labeled Z_{AB}, Z_{BC}, and Z_{CA}. In practice, these loads may be resistive, reactive (capacitive or inductive), or any combination of the three. The letters A, B, and C were included in the load circuit to identify the phase that is across each load. For example, Z_{CA} is connected between the C and A connection points in the source. As such, the voltage across Z_{CA} equals E_{CA}. By the same token,

- The voltage across Z_{AB} equals E_{AB}.
- The voltage across Z_{BC} equals E_{BC}.

The current arrows in Figure 20.13 identify the phase currents. That is,

- I_{AB} is the current through Z_{AB}.
- I_{BC} is the current through Z_{BC}.
- I_{CA} is the current through Z_{CA}.

Based on the voltage and current relationships listed here, Ohm's law tells us that:

(20.13)

$$I_{AB} = \frac{E_{AB}}{Z_{AB}}$$

(20.14)

$$I_{BC} = \frac{E_{BC}}{Z_{BC}}$$

(20.15)

$$I_{CA} = \frac{E_{CA}}{Z_{CA}}$$

Like the voltages in a balanced wye circuit, the currents in a balanced delta circuit can be added geometrically to obtain the values of the line currents (I_A, I_B, and I_C). However, we can save time and trouble by using either of two current equations that should look familiar. That is,

(20.16)

$$I_{LINE} = \sqrt{3} \times I_{PHASE}$$

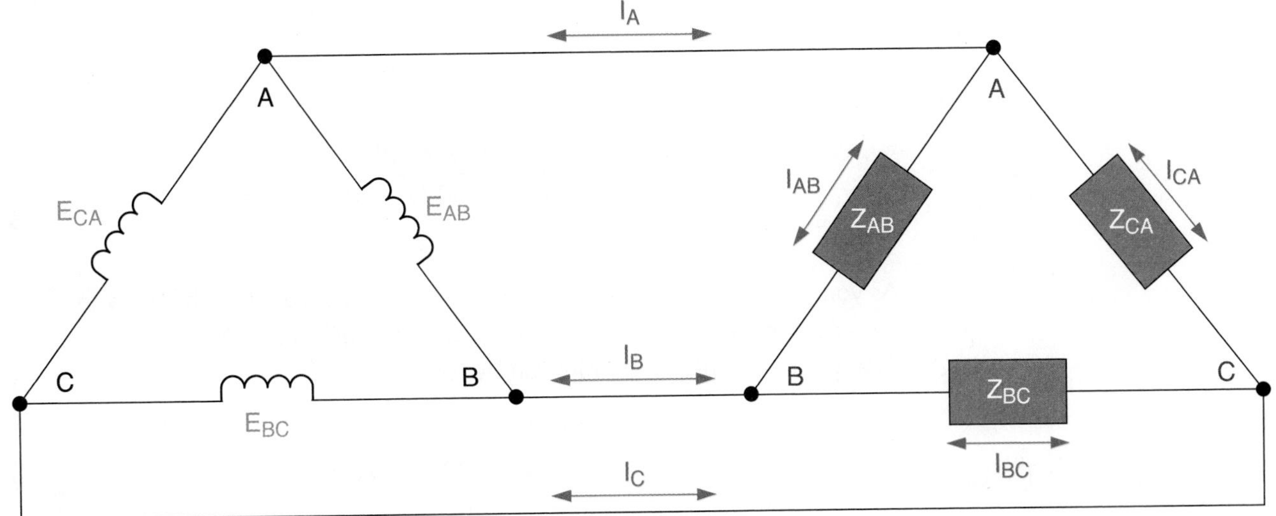

FIGURE 20.13 Currents through a loaded delta (Δ) circuit.

or

(20.17)
$$I_{LINE} = 1.732 \times I_{PHASE}$$

The current calculations for a three-phase delta circuit are demonstrated in Example 20.4.

EXAMPLE 20.4

A balanced delta circuit like the one in Figure 20.13 has the following values: $E_{AB} = 120$ V and $Z_{AB} = 30\ \Omega$. Calculate the values of I_{AB}, I_{BC}, and I_{CA}.

SOLUTION

The circuit phase relationships are shown in Figure 20.12b. Using the circuit values and phase angles, the current values are calculated as follows:

$$I_{AB} = \frac{E_{AB}}{Z_{AB}} = \frac{120\text{ V}\ \angle 0°}{30\ \Omega} = 4\text{ A}\ \angle 0°$$

$$I_{BC} = \frac{E_{BC}}{Z_{BC}} = \frac{120\text{ V}\ \angle -120°}{30\ \Omega} = 4\text{ A}\ \angle -120°$$

and

$$I_{CA} = \frac{E_{CA}}{Z_{CA}} = \frac{120\text{ V}\angle 120^\circ}{30\ \Omega} = 4\text{ A}\angle 120^\circ$$

As you can see, the currents in a balanced delta circuit (like the voltages) are equal in value and 120° out of phase with each other. The magnitude of any line current (I_A, I_B, or I_C) can be found as

$$I_{LINE} = \sqrt{3} \times I_{PHASE} = 1.732 \times 4\text{ A} = 6.93\text{ A}$$

PRACTICE PROBLEM 20.4

A circuit like the one in Figure 20.10a has the following values: $E_{CN} = 4.2$ kV and $Z_{CN} = 860\ \Omega$. Calculate the values of the phase currents.

As stated earlier, the line currents in a balanced delta circuit are 120° out of phase with each other. As such, the vectors for the line currents in Example 20.4 can be plotted as shown in Figure 20.14a. When you position these vectors end-to-end (as shown in Figure 20.14b), you can see that the sum of the vectors is 0 A. By formula,

(20.18) $$I_A + I_B + I_C = 0\text{ A}$$

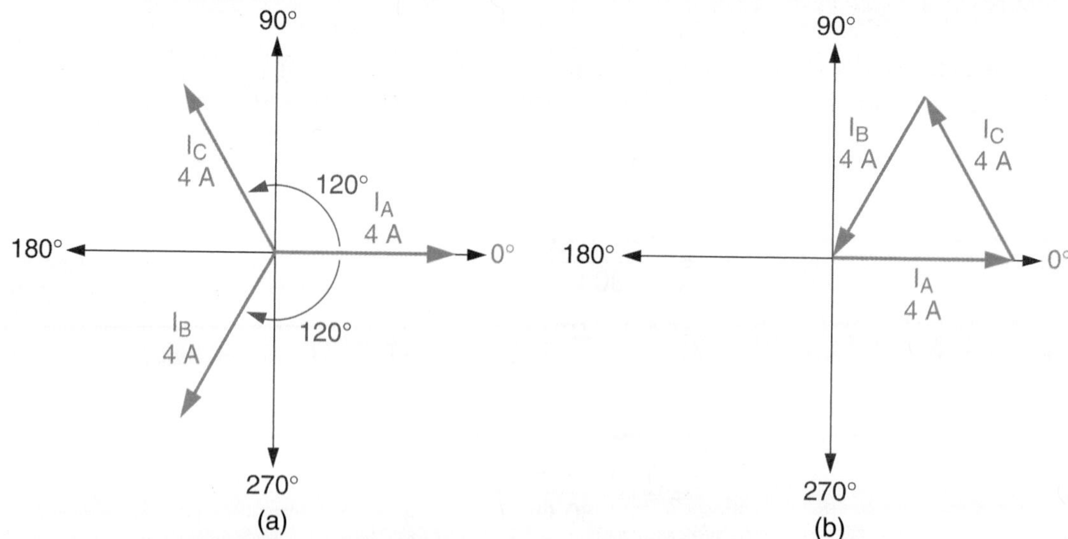

FIGURE 20.14 Current vectors for a balanced delta (Δ) circuit.

This relationship explains why a balanced delta circuit can work without a neutral conductor between the source and load. The currents between the source and load circuits add up to 0 A, so no return (neutral) is needed.

One Final Point

Earlier, you were told that the line voltage and current relationships for balanced wye and delta circuits are nearly opposites. For comparison, the primary voltage and current relationships for both circuits are listed below.

BALANCED WYE CIRCUIT	BALANCED DELTA CIRCUIT
$E_{LINE} = \sqrt{3} \times E_{PHASE}$	$E_{LINE} = E_{PHASE}$
$I_{LINE} = I_{PHASE}$	$I_{LINE} = \sqrt{3} \times I_{PHASE}$

As you can see, the current equation for each type of circuit resembles the voltage equation for the other. In the next section, we will combine wye and delta circuits.

PROGRESS CHECK

1. How is a delta connection configured? How does it differ from a wye connection?
2. What is the relationship between phase voltage and line voltage in a delta-connected circuit?
3. What is the relationship between phase current and line current in a delta-connected circuit?
4. What is the sum of the phase currents in a balanced delta-connected circuit?

20.4 WYE-DELTA AND DELTA-WYE CIRCUITS

Now that we have covered wye and delta circuits, we are going to take a moment to combine the two, starting with a **WYE-DELTA (Y-Δ) CIRCUIT**.

WYE-DELTA (Y-Δ) CIRCUIT A wye circuit with a delta load.

Balanced Wye-Delta Circuits

Figure 20.15 shows a wye circuit with a delta load. As you can see, there is no connection between the neutral (N) in the source and the load. This is because a delta load does not provide for a neutral connection, nor does it require one when the load is balanced.

The loads are connected across the phase inductors as follows:

- Z_{AB} is connected across ($E_{AN} + E_{BN}$).
- Z_{BC} is connected across ($E_{CN} + E_{BN}$).
- Z_{CA} is connected across ($E_{AN} + E_{CN}$).

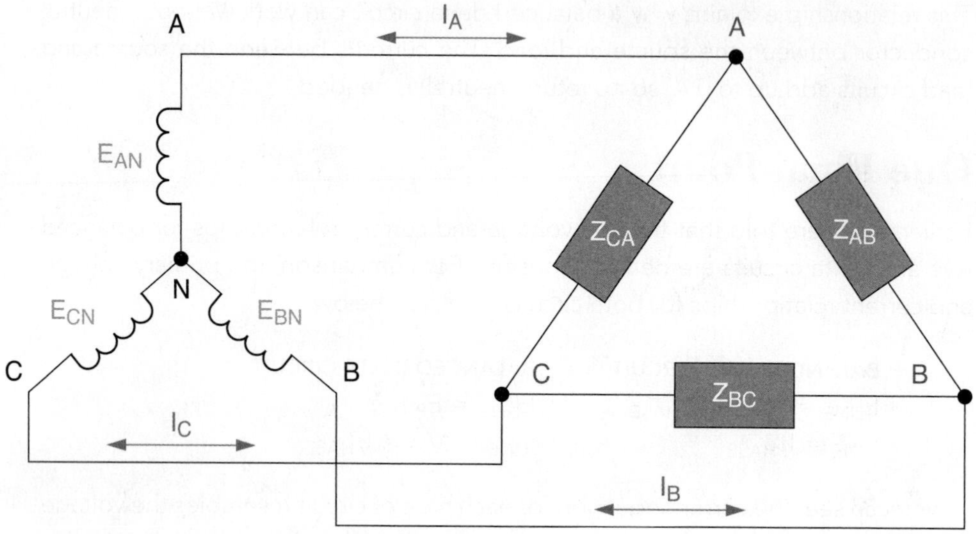

FIGURE 20.15 A wye-delta (Y-Δ) circuit.

Therefore, the load voltages can be found as

(20.19)
$$V_{AB} = E_{AN} + E_{BN} = E_{AB}$$

(20.20)
$$V_{BC} = E_{CN} + E_{BN} = E_{BC}$$

(20.21)
$$V_{CA} = E_{AN} + E_{CN} = E_{CA}$$

As always, E_{AB}, E_{BC}, and E_{CA} are the line voltages and are 120° out of phase with each other. Example 20.5 demonstrates a basic voltage and current analysis of the wye-delta circuit.

EXAMPLE 20.5

A circuit like the one in Figure 20.15 has the following values: $E_{LINE} = 480$ V and $Z = 18\ \Omega$ (each phase). Calculate the values of the load currents and voltages.

SOLUTION

The phase relationships among the line voltages are illustrated in Figure 20.16. Using these relationships, the load voltages are found as follows:

$$V_{AB} = E_{AB} = 480\text{ V} \angle 0°$$
$$V_{BC} = E_{BC} = 480\text{ V} \angle -120°$$

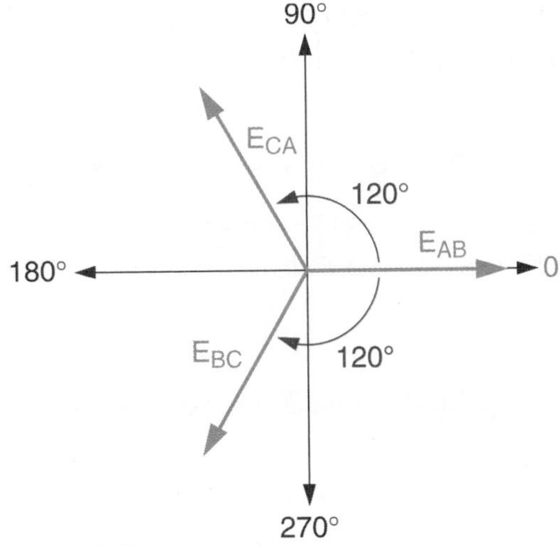

FIGURE 20.16

and

$$V_{CA} = E_{CA} = 480 \text{ V } \angle 120°$$

Now, the values of the load currents can be found as

$$I_{AB} = \frac{V_{AB}}{Z_{AB}}$$

$$= \frac{480 \text{ V } \angle 0°}{18 \ \Omega}$$

$$= 30 \text{ A } \angle 0°$$

$$I_{BC} = \frac{V_{BC}}{Z_{BC}}$$

$$= \frac{480 \text{ V } \angle -120°}{18 \ \Omega}$$

$$= 30 \text{ A } \angle -120°$$

and

$$I_{CA} = \frac{V_{CA}}{Z_{CA}}$$

$$= \frac{480 \text{ V } \angle 120°}{18 \ \Omega}$$

$$= 30 \text{ A } \angle 120°$$

PRACTICE PROBLEM 20.5

A circuit like the one in Figure 20.15 has the following values: $E_{LINE} = 208$ V and $Z = 52 \ \Omega$ (each load). Calculate the values of the load currents and voltages.

Up to this point, we have used only purely resistive loads in our examples. Let's take a look at what happens when the delta load contains some inductance.

EXAMPLE 20.6

A circuit like the one in Figure 20.15 has the following values: $E_{LINE} = 480\text{ V}$ and $Z = 24\ \Omega \angle 20°$ (each phase). Calculate the values of the load currents.

SOLUTION

Using the vector diagram in Figure 20.16, we calculate the values of load voltages as follows:

$$V_{AB} = E_{AB} = 480\text{ V} \angle 0°$$
$$V_{BC} = E_{BC} = 480\text{ V} \angle -120°$$

and

$$V_{CA} = E_{CA} = 480\text{ V} \angle 120°$$

Now, the values of the load phase currents are found as

$$I_{AB} = \frac{V_{AB}}{Z_{AB}} = \frac{480\text{ V} \angle 0°}{24\ \Omega \angle 20°} = 20\text{ A} \angle 0° - 20° = 20\text{ A} \angle -20°$$

$$I_{BC} = \frac{V_{BC}}{Z_{BC}} = \frac{480\text{ V} \angle -120°}{24\ \Omega \angle 20°} = 20\text{ A} \angle -120° - 20° = 20\text{ A} \angle -140°$$

and

$$I_{CA} = \frac{V_{CA}}{Z_{CA}} = \frac{480\text{ V} \angle 120°}{24\ \Omega \angle 20°} = 20\text{ A} \angle 120° - 20° = 20\text{ A} \angle 100°$$

The vectors representing the load voltages and currents are shown in Figure 20.17. The vector positions illustrate the 20° phase difference

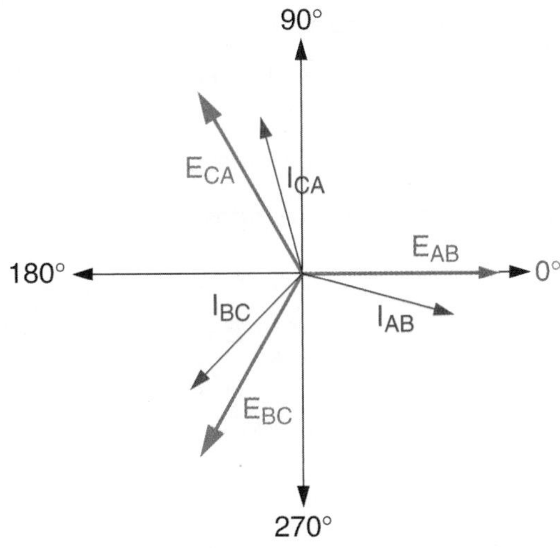

FIGURE 20.17

between the load voltages and currents, and the fact that the currents are *lagging* indicates that the loads contain some value of inductance.

PRACTICE PROBLEM 20.6

A circuit like the one in Figure 20.15 has the following values: $E_{LINE} = 208$ V and $Z = 24\ \Omega\ \angle 40°$ (each phase). Calculate the values of the phase currents.

Balanced Delta-Wye Circuits

Figure 20.18 shows a **DELTA-WYE (Δ-Y) CIRCUIT**. As was the case with the wye-delta circuit, there is no connection between the delta source and the neutral (N) in the wye-connected load.

DELTA-WYE (Δ-Y) CIRCUIT A delta circuit with a wye load.

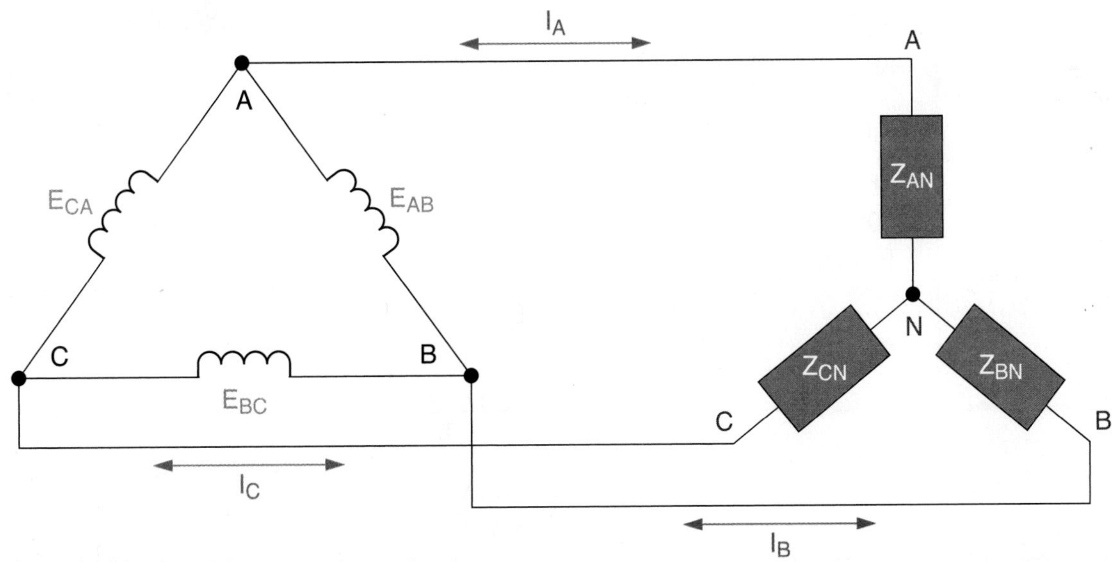

FIGURE 20.18 A delta-wye (Δ-Y) circuit.

The loads are connected to the source as follows:

- E_{AB} is the source for Z_{AB} (the combination of Z_{AN} and Z_{BN}).
- E_{BC} is the source for Z_{BC} (the combination of Z_{BN} and Z_{CN}).
- E_{CA} is the source for Z_{CA} (the combination of Z_{AN} and Z_{CN}).

Using equation (20.6), the load phase voltages can be found as

(20.22)

$$V_{BN} = \frac{E_{AB}}{\sqrt{3}}$$

(20.23)

$$V_{CN} = \frac{E_{BC}}{\sqrt{3}}$$

(20.24)

$$V_{AN} = \frac{E_{CA}}{\sqrt{3}}$$

As always, E_{AB}, E_{BC}, and E_{CA} are equal in value and 120° out of phase with each other. Example 20.7 demonstrates a basic voltage and current analysis of the wye-delta circuit.

EXAMPLE 20.7

A circuit like the one in Figure 20.18 has the following values: $E_{LINE} = 480$ V and $Z = 20\ \Omega$ (each load). Calculate the values of the load currents and voltages.

SOLUTION

The phase relationships among the line voltages are illustrated in Figure 20.16. Using these relationships, the load phase voltages are found as follows:

$$\begin{aligned} V_{BN} &= \frac{E_{AB}}{\sqrt{3}} \\ &= \frac{480\text{ V}\angle 0^\circ}{\sqrt{3}} \\ &= \frac{480\text{ V}}{1.732}\angle 0^\circ \\ &= 277\text{ V}\angle 0^\circ \end{aligned}$$

$$V_{CN} = \frac{E_{BC}}{\sqrt{3}}$$

$$= \frac{480 \text{ V} \angle -120°}{\sqrt{3}}$$

$$= \frac{480 \text{ V}}{1.732} \angle -120°$$

$$= 277 \text{ V} \angle -120°$$

and

$$V_{AN} = \frac{E_{CA}}{\sqrt{3}}$$

$$= \frac{480 \text{ V} \angle 120°}{\sqrt{3}}$$

$$= \frac{480 \text{ V}}{1.732} \angle 120°$$

$$= 277 \text{ V} \angle 120°$$

Now, the values of the load phase currents can be found as

$$I_B = \frac{V_{BN}}{Z_{BN}}$$

$$= \frac{277 \text{ V} \angle 0°}{20 \ \Omega}$$

$$= 13.9 \text{ A} \angle 0°$$

$$I_C = \frac{V_{CN}}{Z_{CN}}$$

$$= \frac{277 \text{ V} \angle -120°}{20 \ \Omega}$$

$$= 13.9 \text{ A} \angle -120°$$

and

$$I_A = \frac{V_{AN}}{Z_{AN}}$$

$$= \frac{277 \text{ V} \angle 120°}{20 \ \Omega}$$

$$= 13.9 \text{ A} \angle 120°$$

PRACTICE PROBLEM 20.7

A circuit like the one in Figure 20.18 has the following values: $E_{LINE} = 208$ V and $Z = 40 \ \Omega$ (each phase). Calculate the values of the load currents and voltages.

Let's take a look at what happens when the wye load contains some inductance.

EXAMPLE 20.8

A circuit like the one in Figure 20.18 has the following values: $E_{LINE} = 480$ V and $Z = 30\ \Omega \angle 30°$ (each phase). Calculate the values of the load phase currents.

SOLUTION

The phase relationships among the line voltages are illustrated in Figure 20.16. Using these relationships, the load phase voltages are found as follows:

$$V_{BN} = \frac{E_{AB}}{\sqrt{3}}$$
$$= \frac{480\text{ V} \angle 0°}{\sqrt{3}}$$
$$= \frac{480\text{ V}}{1.732} \angle 0°$$
$$= 277\text{ V} \angle 0°$$

$$V_{CN} = \frac{E_{BC}}{\sqrt{3}}$$
$$= \frac{480\text{ V} \angle -120°}{\sqrt{3}}$$
$$= \frac{480\text{ V}}{1.732} \angle -120°$$
$$= 277\text{ V} \angle -120°$$

and

$$V_{AN} = \frac{E_{CA}}{\sqrt{3}}$$
$$= \frac{480\text{ V} \angle 120°}{\sqrt{3}}$$
$$= \frac{480\text{ V}}{1.732} \angle 120°$$
$$= 277\text{ V} \angle 120°$$

Now, the values of the load phase currents can be found as

$$I_B = \frac{V_{BN}}{Z_{BN}}$$
$$= \frac{277\text{ V} \angle 0°}{30\ \Omega \angle 30°}$$

$$= 9.23\text{ A} \angle 0° - 30°$$
$$= 9.23\text{ A} \angle -30°$$

$$I_C = \frac{V_{CN}}{Z_{CN}}$$
$$= \frac{277\text{ V} \angle -120°}{30\ \Omega \angle 30°}$$
$$= 9.23\text{ A} \angle -120° - 30°$$
$$= 9.23\text{ A} \angle -150°$$

and

$$I_A = \frac{V_{AN}}{Z_{AN}}$$
$$= \frac{277\text{ V} \angle 120°}{30\ \Omega \angle 30°}$$
$$= 9.23\text{ A} \angle 120° - 30°$$
$$= 9.23\text{ A} \angle 90°$$

The vectors representing the load voltages and currents are shown in Figure 20.19. The vector positions illustrate the 30° phase difference between the load voltages and currents, and the fact that the currents are *lagging* indicates that the loads contain some value of inductance.

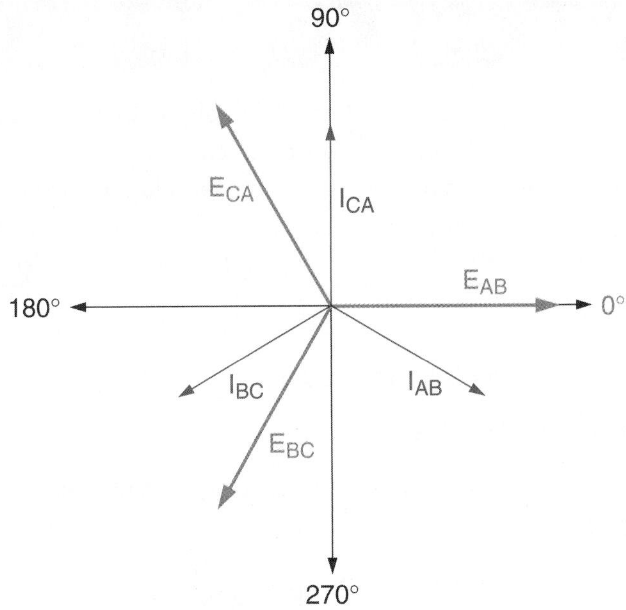

FIGURE 20.19

PRACTICE PROBLEM 20.8

A circuit like the one in Figure 20.18 has the following values: $E_{LINE} = 480$ V and $Z = 23\ \Omega \angle 36°$ (each phase). Calculate the values of the load currents.

One Final Note

In this section, you have been introduced to the voltage and current relationships in wye-delta and delta-wye circuits. In the next section, we will reinforce these relationships as we discuss power, power factor, and power factor correction in the various wye and delta circuits.

PROGRESS CHECK

1. Describe the line and phase voltage relationships in a wye-delta circuit.
2. Describe the line and phase current relationships in a wye-delta circuit.
3. Describe the line and phase voltage relationships in a delta-wye circuit.
4. Describe the line and phase current relationships in a delta-wye circuit?
5. When a load has a phase angle, what happens to the phase relationship between load current and voltage?

20.5 POWER AND POWER FACTOR

Once the voltage and current values for any wye or delta circuit (or combination of the two) are calculated, the power calculations fall right into place. In this section, we will examine the various power relationships in the circuit configurations we covered in previous sections. First, however, we need to establish some useful relationships among power, power factor, and the phase angle of an AC circuit.

Phase Angles, Power, and Power Factor

As you know, the power factor (PF) of a circuit is the ratio of true power to apparent power. By formula,

$$PF = \frac{P_R}{P_{APP}}$$

where P_R is resistive (true) power and P_{APP} is apparent power.

According to the power triangle in Figure 20.20, the cosine of the phase angle ($\cos \theta$) equals the ratio of resistive power (P_R) to apparent power (P_{APP}). Therefore, we can rewrite the equation above as follows:

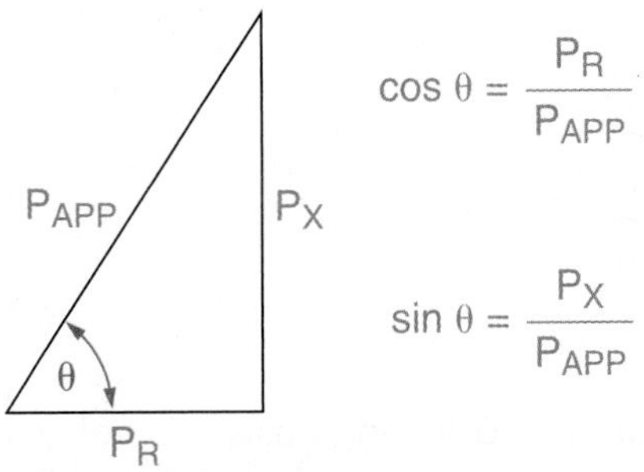

$$\cos \theta = \frac{P_R}{P_{APP}}$$

$$\sin \theta = \frac{P_X}{P_{APP}}$$

FIGURE 20.20 Power triangle.

(20.25)
$$PF = \frac{P_R}{P_{APP}} = \cos \theta$$

If we transpose the cos θ equation in Figure 20.20, we get the following useful relationship:

(20.26)
$$P_R = P_{APP} \cos \theta$$

If we transpose the sin θ equation shown in the figure, we get another useful relationship:

(20.27)
$$P_X = P_{APP} \sin \theta$$

Example 20.9 demonstrates the application of these relationships.

EXAMPLE 20.9

A three-phase circuit has the following values: $P_{APP} = 1200$ VA and $\theta = 30°$. Calculate the circuit resistive power, reactive power, and power factor.

SOLUTION

The circuit resistive power has a value of

$$\begin{aligned} P_R &= P_{APP} \cos \theta \\ &= 1200 \text{ VA} \times (\cos 30°) \\ &= 1200 \text{ VA} \times 0.866 \\ &= 1039 \text{ W} \end{aligned}$$

and the reactive power has a value of

$$P_X = P_{APP} \sin\theta$$
$$= 1200 \text{ VA} \times (\sin 30°)$$
$$= 1200 \text{ VA} \times 0.5$$
$$= 600 \text{ VAR}$$

Finally, the power factor (which we actually calculated already) has a value of

$$PF = \cos\theta$$
$$= \cos 30°$$
$$= 0.866$$

PRACTICE PROBLEM 20.9

A three-phase circuit has the following values: $P_{APP} = 1650$ VA and $\theta = 60°$. Calculate the circuit resistive power, reactive power, and power factor.

You may be wondering why we are taking the time to establish new power equations when we already have many we can use. As you will see, the equations introduced in this section make power calculations in three-phase circuits easier than they might be otherwise.

Balanced Wye Circuits

Figure 20.21 shows a balanced wye source and load, along with its voltage and current relationships. The equations shown were introduced in Section 20.2. As you will see, the power calculations can be made in a straightforward manner once the voltage and current values for a wye circuit are known.

In any three-phase load, apparent power equals the product of the voltage across each phase and the current through each phase. Because there are three phases, apparent power can be found for any wye- or delta-connected load using

(20.28)
$$P_{APP} = 3 \times E_P \times I_P$$

However, as noted earlier, phase voltage and current cannot be easily measured in most cases. For this reason line voltage and current values are used more often to calculate apparent power. For any wye-connected load, $E_P = E_L/\sqrt{3}$ and $I_P = I_L$. Therefore, we can rewrite equation (20.28) as

$$P_{APP} = 3\left(\frac{E_L}{\sqrt{3}} \times I_L\right)$$
$$= \sqrt{3} \times E_L \times I_L$$

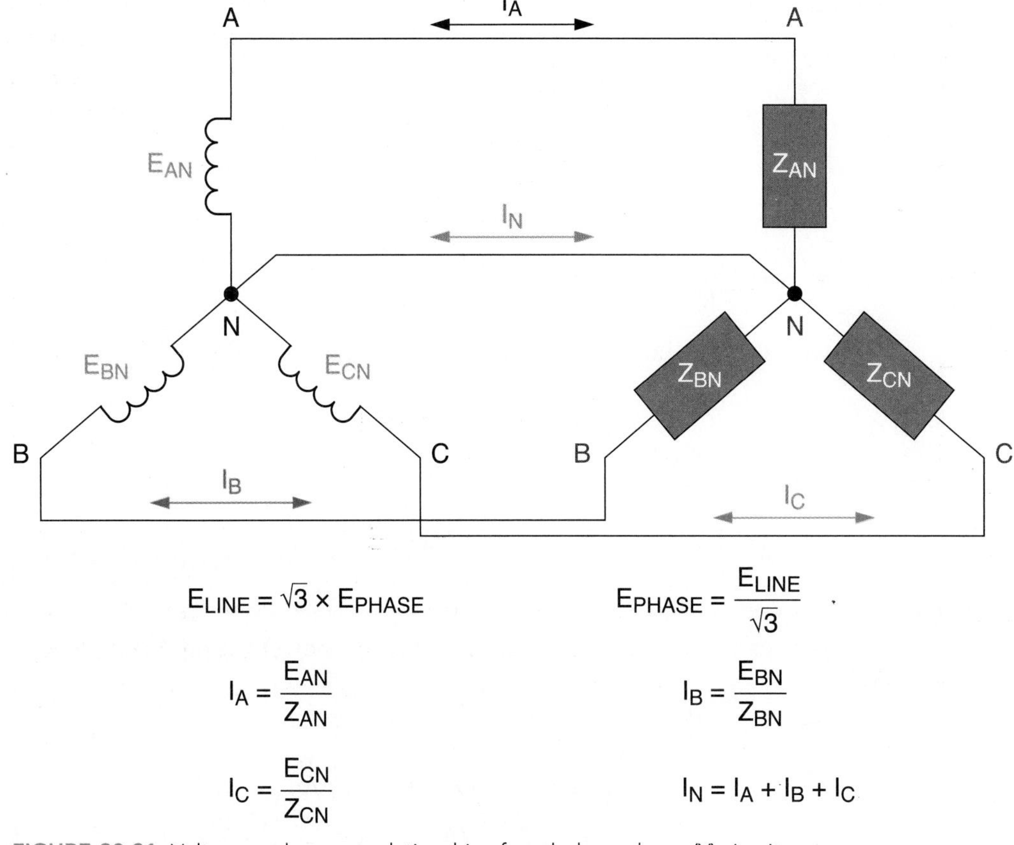

FIGURE 20.21 Voltage and current relationships for a balanced wye (Y) circuit.

For any delta-connected load, $E_P = E_L$ and $I_P = I_L/\sqrt{3}$. This means we can rewrite equation (20.28) as

$$P_{APP} = 3\left(E_L \times \frac{I_L}{\sqrt{3}}\right)$$
$$= \sqrt{3} \times E_L \times I_L$$

Thus, we can solve for apparent power in either a wye-connected or a delta-connected circuit using

$$P_{APP} = \sqrt{3} \times E_L \times I_L \qquad (20.29)$$

Example 20.10 demonstrates two approaches to calculating apparent power for a balanced wye circuit.

EXAMPLE 20.10

A balanced wye circuit has 480-V line voltages and a load value of $Z = 60\ \Omega$ for each phase. Calculate the apparent power for the circuit.

SOLUTION

First, we calculate the phase voltage as follows:

$$E_{PHASE} = \frac{E_{LINE}}{\sqrt{3}}$$
$$= \frac{480\text{ V}}{1.732}$$
$$= 277\text{ V}$$

Next, we calculate the value of the phase current.

$$I_{PHASE} = \frac{E_{PHASE}}{Z}$$
$$= \frac{277\text{ V}}{60\ \Omega}$$
$$= 4.62\text{ A}$$

Now, we can calculate the circuit apparent power using equation (20.28) as

$$P_{APP} = 3 \times E_P \times I_P$$
$$= 3 \times 277\text{ V} \times 4.62\text{ A}$$
$$= 3.84\text{ kVA}$$

Since $I_L = I_P$ for a wye-connected load, we can also calculate P_{APP} using equation (20.29) as

$$P_{APP} = \sqrt{3} \times E_L \times I_L$$
$$= 1.732 \times 480\text{ V} \times 4.62\text{ A}$$
$$= 3.84\text{ kVA}$$

PRACTICE PROBLEM 20.10

A balanced wye circuit has 208-V line voltages and a load value of $Z = 12\ \Omega$. Calculate the apparent power for the circuit using both phase and line values.

Before we continue, there is a point to be made regarding Example 20.10: Because the load in this circuit was purely resistive, the circuit power factor equals one (1), and $P_{APP} = P_R$. As such, our result would actually be measured in watts (W).

Power calculations get a bit more involved when dealing with a circuit containing some measurable amount of reactance, as demonstrated in Example 20.11.

EXAMPLE 20.11

The balanced wye-delta circuit in Figure 20.22 has a 120 V/60 Hz phase voltage. The delta-connected load has phase values of R = 20 Ω and L = 100 mH. Calculate the power values for the circuit.

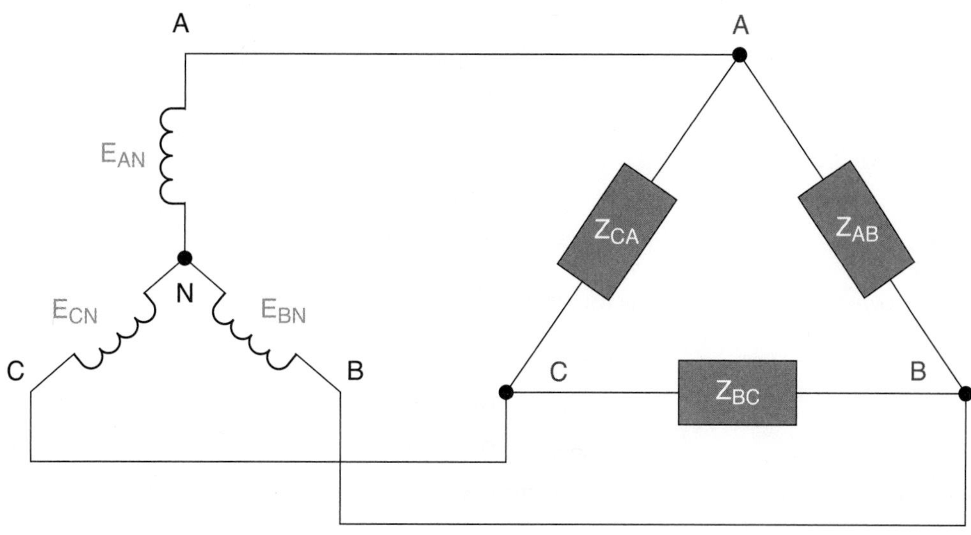

FIGURE 20.22

SOLUTION

We need to begin by calculating the phase impedance values, as follows:

$$X_L = 2\pi fL$$
$$= 2\pi \times 60 \text{ Hz} \times 100 \text{ mH}$$
$$= 37.7 \ \Omega$$
$$Z_P = \sqrt{X_L^2 + R^2}$$
$$= \sqrt{(37.7 \ \Omega)^2 + (20 \ \Omega)^2}$$
$$= 42.7 \ \Omega$$

and

$$\theta = \tan^{-1}\frac{X_L}{R}$$
$$= \tan^{-1}\frac{37.7 \ \Omega}{20 \ \Omega}$$
$$= \tan^{-1}(1.885)$$
$$= 62.1°$$

So, the delta-connected load has a value of $Z = 42.7 \ \Omega \angle 62.1°$. Because our source is wye-connected, the value of the line voltage (E_{LINE}) is found as

$$E_{LINE} = \sqrt{3} \times E_{PHASE}$$
$$= 1.732 \times 120 \text{ V}$$
$$= 208 \text{ V}$$

Because the load is delta-connected, its phase and line voltages are equal. Thus, the phase voltage for the load is 208 V, and the phase current can be found as

$$\begin{aligned} I_{PHASE} &= \frac{E_{PHASE}}{Z} \\ &= \frac{208\text{ V}}{42.7\ \Omega\ \angle 62.1°} \\ &= \frac{208\text{ V}}{42.7\ \Omega}\angle -62.1° \\ &= 4.87\text{ A}\ \angle -62.1° \end{aligned}$$

The apparent power for the load is now found as

$$\begin{aligned} P_{APP} &= 3 \times E_P \times I_P \\ &= 3 \times 208\text{ V} \times 4.87\text{ A} \\ &= 3 \times 1013\text{ VA} \\ &= 3039\text{ VA} \end{aligned}$$

The values of resistive and reactive power can now be found as

$$\begin{aligned} P_R &= P_{APP} \cos\theta \\ &= 3039\text{ VA} \times (\cos 62.1°) \\ &= 3039\text{ VA} \times 0.4679 \\ &= 1422\text{ W} \end{aligned}$$

and

$$\begin{aligned} P_X &= P_{APP} \sin\theta \\ &= 3039\text{ VA} \times (\sin 62.1°) \\ &= 3039\text{ VA} \times 0.8838 \\ &= 2686\text{ VAR} \end{aligned}$$

Finally, the power factor (which we have calculated already) has a value of

$$\begin{aligned} PF &= \cos\theta \\ &= \cos\ 62.1° \\ &= 0.4679 \end{aligned}$$

PRACTICE PROBLEM 20.11

A balanced wye-delta circuit like the one in Figure 20.22 has a 277 V/60 Hz phase voltage. The delta-connected load has phase values of $R = 30\ \Omega$ and $L = 330$ mH. Calculate the power values for the circuit.

Power Factor Correction

You may recall that circuits having a lower power factor require higher currents to produce the same amount of work. For this reason, utilities charge commercial customers a surcharge for low power factors, making power factor correction especially important when dealing with three-phase power circuits.

Most practical three-phase circuits have loads that are inductive to one extent or another (i.e., have lagging power factors). For this reason, most power factor correction circuits are capacitive. A capacitive power factor correction circuit is shown in Figure 20.23.

NOTE: The concept of power factor correction was introduced in Section 18.5.

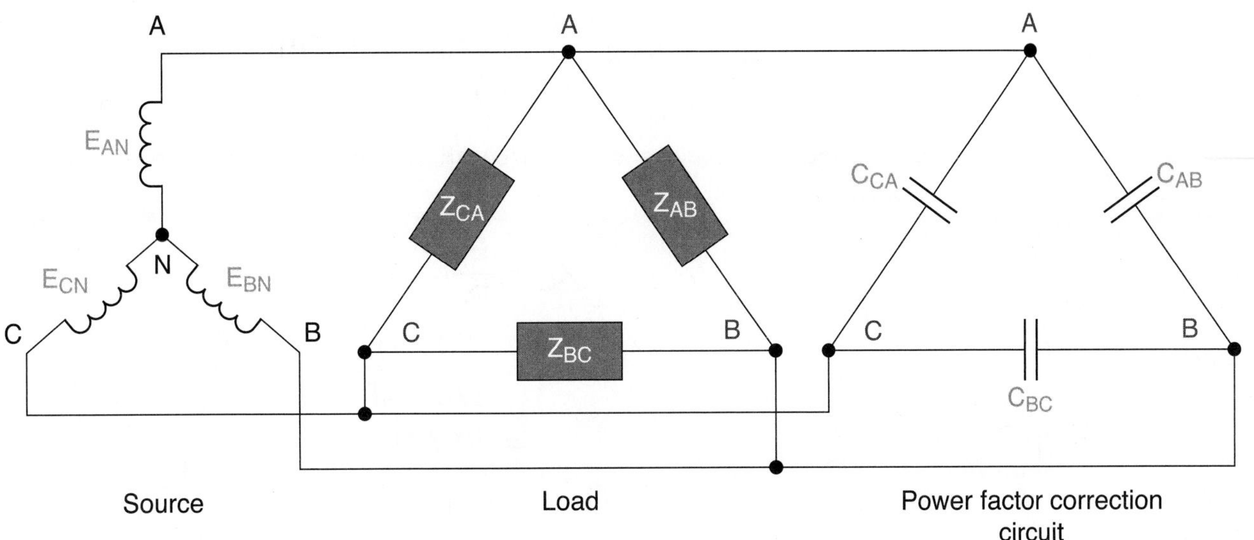

FIGURE 20.23 Power factor correction.

Using the values calculated in Example 20.11, we can go through the steps that are taken to select the values of the compensating capacitors. In the example, we determined that the balanced wye-delta circuit had the following values:

$$P_{APP} = 3039 \text{ VA} \qquad P_R = 1422 \text{ W}$$
$$P_X = 2686 \text{ VAR} \qquad PF = 0.4679$$

These values represent the combined effects of three matched loads (Z_{AB}, Z_{BC}, and Z_{CA}). Individually, the loads would have the following values (which are one third of the total values):

$$P_{APP} = 1013 \text{ VA} \qquad P_R = 474 \text{ W} \qquad P_X = 895 \text{ VAR}$$

Using the value of $P_R = 474$ W, the value of P_{APP} that will result in a 0.95 power factor can be found as

$$P_{APP} = \frac{P_R}{0.95}$$

$$= \frac{474\text{ W}}{0.95}$$
$$= 499\text{ VA}$$

To obtain a value of $P_{APP} = 499$ VA, reactive power (P_X) must be reduced to

$$P_X = \sqrt{P_{APP}^2 - P_R^2}$$
$$= \sqrt{(499\text{ VA})^2 - (474\text{ W})^2}$$
$$= 25\text{ VAR}$$

This result indicates that the value of P_X must be reduced by

$$\Delta P_X = 895\text{ VAR} - 25\text{ VAR}$$
$$= 870\text{ VAR}$$

Now, using $E_P = 208$ V and $P_X = 870$ VAR, the reactance of the compensating capacitor is calculated to be

$$X_C = \frac{E_P^2}{P_X}$$
$$= \frac{(208\text{ V})^2}{870\text{ VAR}}$$
$$= 49.7\ \Omega$$

Finally, the value of each compensating capacitor is found to be

$$C = \frac{1}{2\pi f X_C}$$
$$= \frac{1}{2\pi \times 60\text{ Hz} \times 49.7\ \Omega}$$
$$= 53.4\ \mu\text{F}$$

This result indicates that connecting a 53.4 μF capacitor across each load phase will increase the circuit power factor to 0.95.

PROGRESS CHECK

1. Given the values of P_{APP} and θ, explain how you would solve for P_R.
2. Given the values of P_{APP} and θ, explain how you would solve for P_X?
3. Explain how you would solve for P_{APP} using line voltage and current values.
4. Explain how you would solve for P_{APP} using phase voltage and current values.
5. Detail the steps you would take to calculate the value of compensating capacitors required to correct the power factor of a load to 0.95.

20.6 SUMMARY

Here is a summary of the major points made in this chapter:

Three-Phase Power

- Three-phase power consists of three sine waves 120° out of phase with each other.
  - If one phase is referenced at 0° then the other two phases would be at 120° and 240°. (See Figure 20.2.)
- One means of generating three-phase power is to rotate a magnetic rotor within three stationary armatures set 120° apart. (See Figure 20.4.)
  - The order in which the three sine waves peak is called the *phase sequence.*
  - The direction of rotor rotation determines the phase sequence.
- Three-phase power has several advantages over single-phase power:
  - Three-phase motors are less complex and have higher horsepower ratings than comparable single-phase motors.
  - Three-phase transformers have greater output capabilities than comparable single-phase transformers.
  - Single-phase systems require 1.15 times more copper than an equivalent three-phase system.

Inductive Wye (Y) Circuits

- Two circuits used extensively in three-phase circuits are the delta and wye circuits.
  - Delta circuits have each of the three conductors connected end-to-end forming a triangle shape like the Greek letter Δ.
  - Wye circuits have one end of each of the three conductors connected to a common point forming a shape similar to a capital Y.
  - Wye circuits are also called *tee* and *star* connections.
- A balanced wye circuit is one in which all load currents are equal and 120° out of phase with each other.
- An unbalanced wye circuit is one in which the load currents are not equal and/or 120° out of phase with each other.
- The voltages in a wye system are classified as phase voltages and line voltages.
  - The phase voltage is the voltage measured across any one arm in the wye to the common or neutral point. (See Figure 20.7a.)
  - The line voltage is the voltage measured between any two arms of the wye. (See Figure 20.7b.)
  - Any two phase or line voltages are still 120° out of phase with each other.
- In a balanced wye circuit, line voltage can be found using phase voltage as $E_{Line} = \sqrt{3} \times E_{Phase}$.
- In a balanced wye circuit, phase voltage can be found using line voltage as $E_{Phase} = \frac{E_{Line}}{\sqrt{3}}$.
  - $\sqrt{3}$ is approximately equal to 1.732.
- In a balanced wye circuit, phase and line currents are equal.
  - The vector sum of the three line or phase currents is 0 A in a balanced wye system, so neutral current equals 0 A.
  - Some wye circuits do not use a neutral conductor.

Inductive Delta (Δ) Circuits

- In a delta circuit the inductors are connected end-to-end rather than to a common point.
- In a balanced delta-connected circuit the phase and line voltages are equal, $E_{Line} = E_{Phase}$.
- In a balanced delta circuit, line current can be found using phase current as $I_{Line} = \sqrt{3} \times I_{Phase}$ or $1.732 \times I_{Phase}$.
- In a balanced delta circuit, phase current can be found using line current as $I_{Phase} = \frac{I_{Line}}{\sqrt{3}}$ or $\frac{I_{Line}}{1.732}$.
- In a balanced delta circuit, all three line or phase currents are 120° out of phase.
- In a balanced delta circuit, the vector sum of the three phase or line currents is 0 A.

Wye-Delta and Delta-Wye Circuits

- If the load in either a wye-delta circuit or a delta-wye circuit has a phase angle, both phase and line currents will shift by that same phase angle.
  - Even though the line and phase currents shift, they still maintain a relative phase relationship of 120°, line-to-line or phase-to-phase.

Power and Power Factor

- Power factor (PF) is the ratio of true power to apparent power. By formula $PF = \frac{P_R}{P_{APP}}$.
- Power factor is also equal to the cosine of the circuit's phase angle: $PF = \cos\theta$.
- Given the circuit phase angle and apparent power, real power can be found as $P_R = P_{APP} \cos\theta$.
- Given the circuit phase angle and apparent power, reactive power can be found as $P_X = P_{APP} \sin\theta$.
- The apparent power produced by any three-phase system can be calculated using voltage and current phase values using the equation $P_{APP} = 3 \times E_P \times I_P$.
- The apparent power produced by any three-phase system can be calculated using voltage and current line values using the equation $P_{APP} = \sqrt{3} \times E_L \times I_L$.
  - Line values are more commonly used to calculate P_{APP} since phase values are often difficult or impossible to measure.
- If a load is purely resistive, then $PF = 1$ and $P_{APP} = P_R$.
- Most practical three-phase loads have some amount of inductance.
  - When this is the case, the phase angle of the load must be taken into account when calculating all power values. (See Example 20.11.)
- A low power factor increases line current without any increase in work performed.
  - Utilities apply surcharges to businesses with low power factors.
- To correct the power factor of a three-phase load, capacitors may be connected across each phase of the load.
  - If the load is delta-connected, then the compensating capacitors must also be delta-connected.
  - If the load is wye-connected, then the compensating capacitors must be also.

CHAPTER REVIEW

1. The voltages in a three-phase circuit are always ________.
   a) 180° apart b) inductive
   c) 120° apart d) capacitive

2. Assume that one phase voltage in a three-phase circuit has a phase angle of 45°. If this phase is the reference, the phase angle of the other two phases will be ________ and ________.
   a) 120°, 240° b) 180°, 0°
   c) 0°, 120° d) 165°, 285°

3. The current-carrying conductor in an AC generator is referred to as a(n) ________.
   a) armature b) phasor
   c) stator d) rotor

4. The order in which each phase of a three-phase generator peaks is referred to as the ________.
   a) rotational order b) phase number
   c) phase sequence d) line sequence

5. Three-phase circuits ________ than single-phase circuits.
   a) require less copper
   b) have higher currents
   c) require less maintenance
   d) all of the above

6. In a balanced three-phase circuit ________.
   a) each phase voltage is separated by 120°
   b) the load on each phase is the same
   c) all line currents are equal
   d) all of the above

7. A wye-connected circuit has a line-to-line voltage of 480 V. The phase voltage of this circuit is ________.
   a) 277 V b) 208 V
   c) 831 V d) 480 V

8. A wye-connected circuit has a phase current of 68 A. The line current of this circuit is ________.
   a) 118 A b) 39 A
   c) 204 A d) 68 A

9. In a balanced wye-connected circuit, phase currents have relative phase angles of ________ while line currents have relative phase angles of ________.
   a) 120°, 120° b) 120°, 0°
   c) 0°, 120° d) 120°, 240°

10. In a balanced wye-connected circuit, neutral current is ________.
    a) equal to line current
    b) equal to phase current
    c) equal to $\sqrt{3} \times$ line current
    d) equal to zero

11. The neutral line of a delta-connected generator ________.
    a) requires a lower gage wire than a wye-connected generator
    b) does not exist
    c) carries more current than in a wye-connected generator
    d) is connected between ϕA and ϕB.

12. Phase voltage is ________ in a delta-connected circuit.
    a) equal to $\sqrt{3}$ times line voltage
    b) equal to line voltage
    c) equal to line voltage divided by $\sqrt{3}$
    d) 120° out of phase with line voltage

13. Line current is ________ in a delta-connected circuit.
    a) equal to $\sqrt{3}$ times phase current
    b) equal to phase current
    c) equal to phase current divided by $\sqrt{3}$
    d) 120° out of phase with phase current

14. A delta-connected generator and a wye-connected generator each have the same line current and line voltage. This means that the ________ has the highest phase voltage.
    a) wye-connected generator
    b) delta-connected generator
    c) neither of the above; they must have the same phase voltage
    d) either of the above; it depends on how the loads are connected

15. In a balanced wye-delta circuit the load phase current is ________ the line current.
    a) $\sqrt{3}$ times greater than
    b) equal to
    c) $\sqrt{3}$ times smaller than
    d) it depends on the phase voltage

16. In a balanced wye-delta circuit, the line voltage is ________ the source phase voltage.
    a) $\sqrt{3}$ times greater than
    b) equal to
    c) $\sqrt{3}$ times smaller than
    d) it depends on the phase voltage

17. A balanced wye-delta circuit has values of E_L = 480 V and I_L = 19 A. This means that the phase values for the load are E_P = ________ and I_P = ________.

    a) 277 V, 19 A b) 480 V, 19 A
    c) 480 V, 11 A d) 277 V, 11 A

18. The phase values for the source in question 17 are E_P = ________ and I_P = ________.

    a) 277 V, 19 A b) 480 V, 19 A
    c) 480 V, 11 A d) 277 V, 11 A

19. The source in a balanced delta-wye circuit has a value of E_P = 2.3 kV. The line current is measured at I_L = 62 A. This means that load phase voltage equals ________.

    a) 2.3 kV b) 3.98 kV
    c) 1.73 kV d) 1.33 kV

20. A balanced delta-wye circuit has source values of E_P = 13.8 kV and I_P = 108 A. The phase values for the load would be E_P = ________ and I_P = ________.

    a) 7.97 kV, 187 A b) 13.8 kV, 62.4 A
    c) 7.97 kV, 62.4 A d) 13.8 kV, 108 A

21. The more inductance in a load, ________.

    a) the lower the power factor
    b) the more positive the phase angle of the phase current
    c) the higher the power factor
    d) the lower the apparent power

22. A circuit has a current phase angle of 18° and apparent power of 2.3 kVA. The value of P_R for this circuit is ________.

    a) 2.19 kVA b) 711 W
    c) 2.19 kW d) 711 kVAR

23. A circuit has values of P_X = 430 VAR and P_R = 1.6 kW. The power factor for this circuit is ________.

    a) 1.66 kVA
    b) 0.966
    c) 1.04
    d) Cannot be determined with the information given

24. A delta source has values of E_L = 4.2 kV and I_P = 20 A. The value of P_{APP} = ________.

    a) 252 kVA b) 145 kVA
    c) 84 kVA d) 168 kVA

25. A balanced delta-wye circuit has a source phase current of 16.5 A and a load phase voltage of 120 V. The value of P_{APP} for this circuit is ________.

    a) 5.94 kVA b) 3.43 kVA
    c) 6.86 kVA d) 10.3 kVA

26. A delta-connected load produces 400 VARs lagging. The phase voltage is 277 V @ 60 Hz. You need to lower the VARs to 30 VARs lagging to meet PF requirements. The value of each capacitor needed is ________.

    a) 12.8 μF b) 4.26 μF
    c) 128 μF d) 9.3 μF

PRACTICE PROBLEMS

1. Complete the chart below. Assume in all cases that the circuit is perfectly balanced and the phase sequence is ABC.

ΦA	ΦB	ΦC
25°	____	____
____	190°	____
____	____	230°

2. Complete the chart below. Assume in all cases that the circuit is perfectly balanced and the phase sequence is ACB.

ΦA	ΦB	ΦC
220°	____	____
____	85°	____
____	____	30°

3. A balanced inductive wye circuit has values of E_P = 120 V and I_P = 12.5 A. Calculate the values of E_L and I_L.
4. A balanced inductive wye circuit has values of E_L = 450 V and I_P = 18 A. Calculate the values of E_P and I_L.
5. A balanced inductive wye circuit has values of E_P = 12 kV and I_L = 10.2 A. Calculate the values of E_L and I_P.
6. A balanced inductive wye circuit has values of E_L = 6.9 V and I_L = 9.6 A. Calculate the values of E_P and I_P.
7. A balanced inductive delta circuit has values of E_P = 277 V and I_P = 4.2 A. Calculate the values of E_L and I_L.
8. A balanced inductive delta circuit has values of E_L = 1.2 kV and I_P = 6.6 A. Calculate the values of E_P and I_L.
9. A balanced inductive delta circuit has values of E_P = 2.4 kV and I_L = 14.3 A. Calculate the values of E_L and I_P.
10. A balanced inductive delta circuit has values of E_L = 13.8 kV and I_L = 11.2 A. Calculate the values of E_P and I_P.
11. A circuit like the one in Figure 20.24 has values of E_L = 480 V and I_L = 12.2 A. Solve for the values of E_P and I_P.
12. A circuit like the one in Figure 20.24 has values of E_P = 4.2 kV and I_L = 4.4 A. Solve for the values of E_L and I_P.

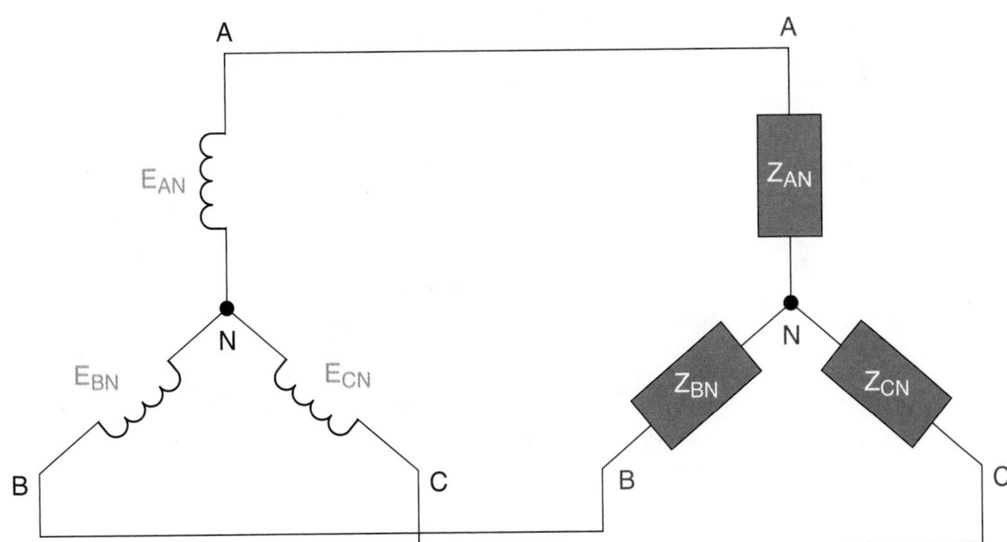

FIGURE 20.24

13. A circuit like the one in Figure 20.25 has values of $E_P = 1700$ V and $I_P = 3.6$ A. Solve for the values of E_L and I_L.
14. A circuit like the one in Figure 20.25 has values of $E_L = 128$ kV and $I_P = 13.5$ A. Solve for the values of E_P and I_L.

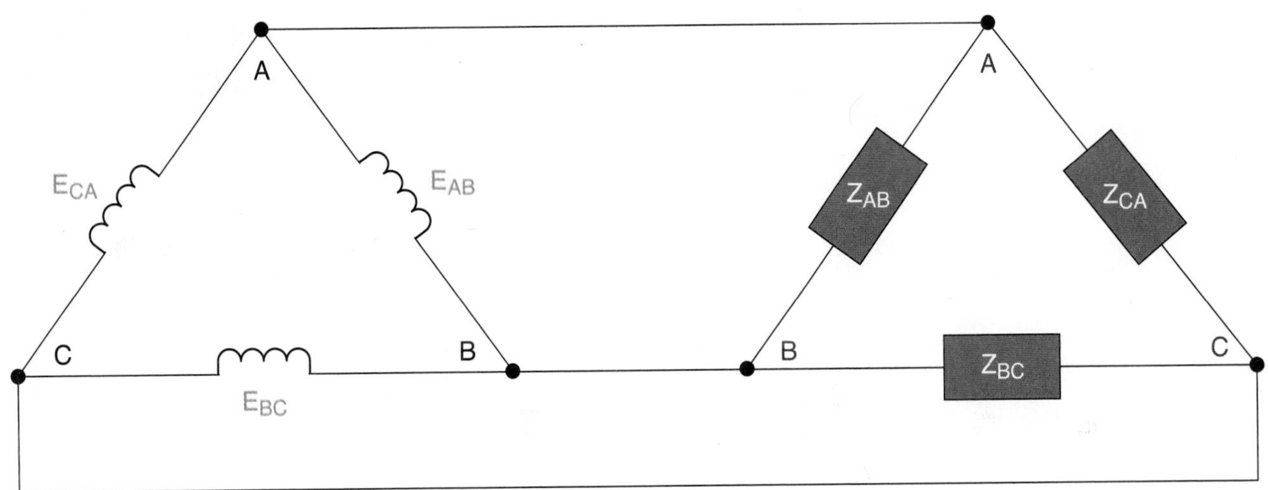

FIGURE 20.25

15. The source phase voltage for the circuit in Figure 20.26 is 2.2 kV. Each phase of the load has a resistance of 2.8 Ω. Solve for the load phase current and the circuit line current.
16. A circuit like the one in Figure 20.26 has $E_L = 480$ V and a resistance value of 8 Ω for each phase. Solve for the load phase current and the circuit line current.
17. A circuit like the one in Figure 20.26 has a source voltage of $E_P = 8$ kV and each phase of the load has an impedance of $Z = 8.2\ \Omega\angle 26°$. Solve for both the line current and the load phase current for this circuit. Include current phase angles.
18. Solve for the source phase current for the circuit described in question 16.
19. Solve for the source phase current for the circuit described in question 17.

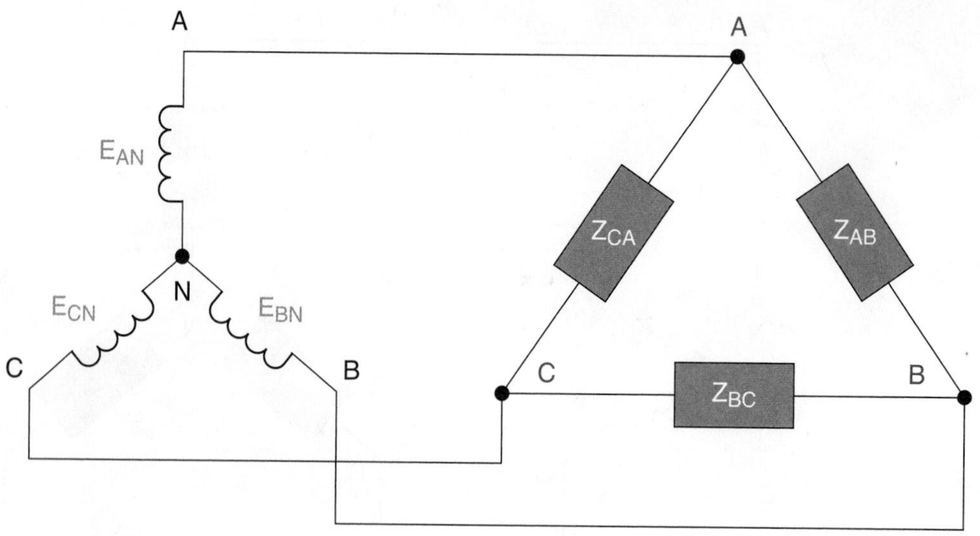

FIGURE 20.26

20. A circuit like the one in Figure 20.27 has values of E_L = 2.3 kV and I_L = 10.3 A. Solve for phase voltage and current values for both the source and the load.

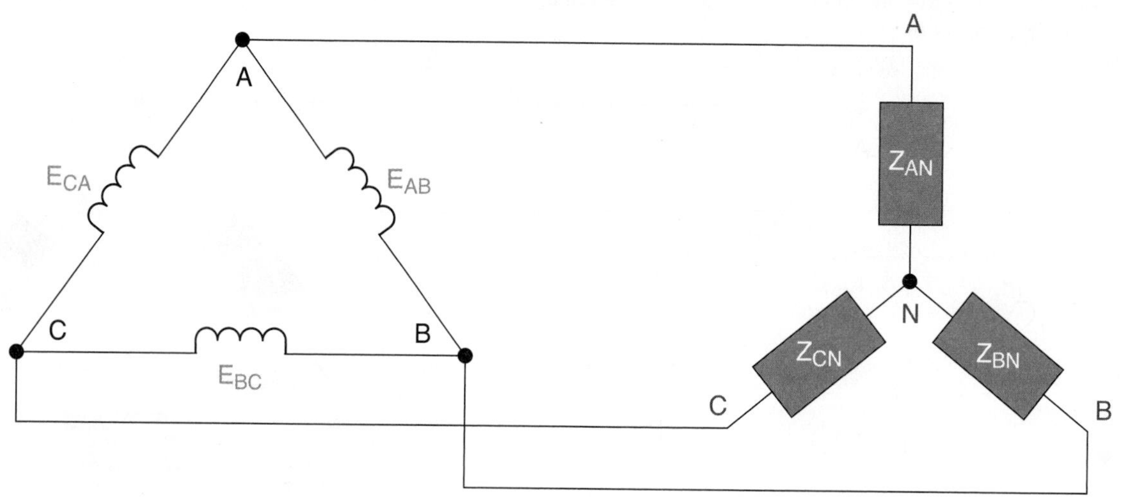

FIGURE 20.27

21. A circuit like the one in Figure 20.27 has source values of E_P = 1100 kV and I_P = 2.6 A. Solve for the line voltage and current values and the voltage and current phase values for the load.
22. A circuit like the one in Figure 20.27 has a source phase voltage of E_P = 240 V. The load has phase values of 18 Ω ∠16°. Solve for both line current and phase current values for the load. Include current phase angles.
23. A three-phase circuit produces an apparent power value of P_{APP} = 1380 VA. The current has a phase angle of θ = 22°. Calculate the circuit resistive power, reactive power, and power factor.
24. A three-phase circuit produces an apparent power value of P_{APP} = 2.1 kVA. The current has a phase angle of θ = 18°. Calculate the circuit resistive power, reactive power, and power factor.

25. A delta-wye circuit has the following values: $E_L = 560$ V and a load value of $Z = 8.2\ \Omega$ per phase. Solve for the apparent power supplied by this source.
26. Assume the phase angle of the load impedance for the circuit described in question 25 is $\theta = 26°$. Solve for P_R, P_X, and PF.
27. A wye-delta circuit has a source phase voltage of 17.6 kV and a load value of 26 Ω per phase. Solve for the apparent power supplied by this source.
28. Assume that the load for the circuit described in question 27 has a phase angle of $\theta = 31°$. Solve for P_R, P_X, and PF.
29. A delta-wye circuit has source values of $E_P = 120$ V and $I_L = 6.6$ A. Make the necessary calculations so that you can use both equations (20.28) and (20.29) to solve for P_{APP}.
30. A wye-delta circuit has a load phase current of 2.65 A per phase and a load impedance of 18 Ω per phase. Make the necessary calculations so that you can use both equations (20.28) and (20.29) to solve for P_{APP}.
31. The following measurements are made on a circuit driving a delta-connected motor. A voltmeter is used to measure the line voltage and finds that $E_L = 480$ V. Assume that this is a 60 Hz source. At full load a clamp ammeter measures a value of $I_L = 45$ A. A wattmeter inserted between the source and motor measures a value of 26.35 kW. Determine the values of P_{APP}, P_X, and the power factor for this circuit.
32. Using the values that you found in question 31, determine the value of each of the three capacitors needed to improve the power factor of this circuit to 97%.

ANSWERS TO THE EXAMPLE PRACTICE PROBLEMS

20.1	480 V
20.2	7967 V
20.3	$3\text{ A}\angle 0°$, $3\text{ A}\angle -120°$, and $3\text{ A}\angle 120°$
20.4	$4.88\text{ A}\angle 0°$, $4.88\text{ A}\angle -120°$, and $4.88\text{ A}\angle 120°$
20.5	$4\text{ A}\angle 0°$, $4\text{ A}\angle -120°$, and $4\text{ A}\angle 120°$
20.6	$8.67\text{ A}\angle -40°$, $8.67\text{ A}\angle -160°$, $8.67\text{ A}\angle 80°$
20.7	$V_{AN} = V_{BN} = V_{CN} = 120$ V, $3\text{ A}\angle 0°$, $3\text{ A}\angle -120°$, and $3\text{ A}\angle 120°$
20.8	$12\text{ A}\angle -36°$, $12\text{ A}\angle -156°$, $12\text{ A}\angle 84°$
20.9	$P_R = 825$ W, $P_X = 1429$ VARs, PF = 0.5
20.10	$P_{APP} = 3.6$ kVA
20.11	$E_L = 480$ V, $Z = 128\ \Omega\angle 76°$, $I_P = 3.75\text{ A}\angle -76°$, $P_{APP} = 1800$ VA, $P_R = 435$ W, $P_X = 1747$ VAR, PF = 0.242.

CHAPTER 21

INTRODUCTION TO TRANSFORMERS

PURPOSE

In Chapter 11, you were introduced to the concept of mutual inductance. At that time, you were told that mutual inductance is the process whereby the magnetic field produced by one coil induces a voltage across another coil. In this chapter, you are going to be introduced to the *transformer*, a component that operates on the principle of mutual inductance.

KEY TERMS

The following terms are introduced and defined in this chapter on the pages indicated:

OBJECTIVES

After completing this chapter, you should be able to:

1. State the purpose served by a transformer.
2. Describe the construction of a typical transformer.
3. List the three basic types of transformers.
4. Briefly describe transformer operation in terms of electromagnetic induction.
5. Describe the *turns ratio* of a transformer.

6. Perform the voltage calculations for a transformer using its turns ratio.
7. Identify a transformer using its turns ratio.
8. Discuss the relationship between transformer input power and output power.
9. List and describe the power losses that occur within a typical transformer.
10. Perform the current calculations for a transformer using its turns ratio.
11. Describe the relationship between the input and output voltages, currents, and power values for each type of transformer (step-up, step-down, and 1-to-1 transformers).
12. Describe the operation of a center-tapped transformer.
13. Discuss the construction and operation of a multiple-output transformer.
14. Describe the construction and operation of an autotransformer and compare buck and boost transformers.
15. List and describe the factors that affect transformer performance, and some of the methods used to minimize these factors.
16. List and describe the two main types of transformer core structures.
17. Explain the input/output phase relationship of a transformer and the use of polarity dots.
18. Discuss the importance of the transformer percent impedance (%Z) rating and transformer derating.
19. Describe the construction and application of current transformers.
20. List the *NEC*® transformer nameplate requirements for a transformer.

21.1 TRANSFORMER OPERATION: AN OVERVIEW

A **TRANSFORMER** is a component that uses electromagnetic induction to pass an AC signal from one circuit to another, while providing DC isolation between the two. The AC and DC input/output characteristics of a transformer are illustrated in Figure 21.1. As you can see:

1. An AC input to the transformer passes through to the output.
2. A DC input to the transformer is prevented from reaching the output.

TRANSFORMER A component that uses electromagnetic induction to pass an AC signal from one circuit to another, while providing DC isolation between the two.

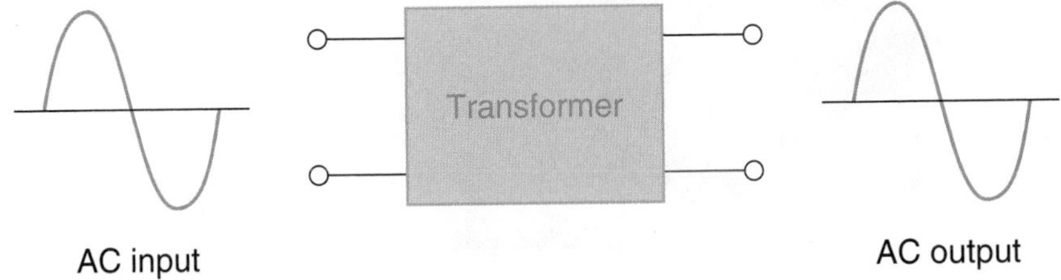

(a) The transformer couples a signal from its input to its output.

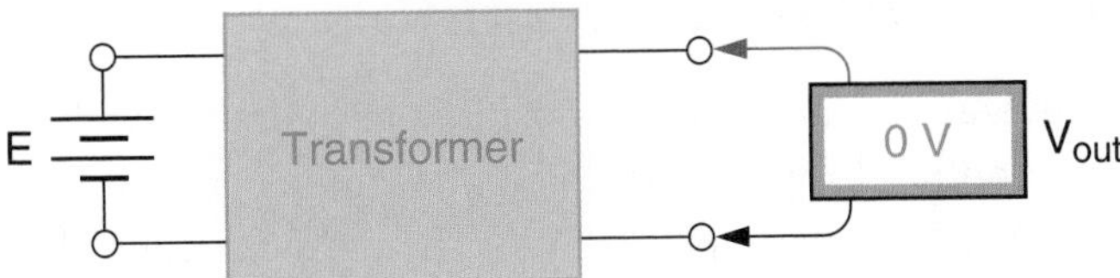

(b) The transformer isolates a DC input voltage (E) from its output.

FIGURE 21.1 Transformer AC coupling and DC isolation.

When a component (or circuit) allows any signal or potential to pass from one point to another, it is said to provide **COUPLING** between the points. When a component (or circuit) prevents a signal or potential from passing between points, it is said to provide **ISOLATION** between the two. Therefore, we can say that a transformer is a component that provides AC coupling and DC isolation between its input and output terminals. As you will see, this combination of characteristics makes the transformer extremely useful in many applications, such as electrical power transmission.

COUPLING When a component or circuit passes an AC signal from one point to another.

ISOLATION When a component or circuit prevents a signal from passing between points.

Transformer Construction

The simplest transformer is made up of two coils, called the **PRIMARY** and the **SECONDARY**. Note that the primary serves as the input to the transformer, and the secondary serves as its output.

Each of the transformers shown in Figure 21.2 contains the primary and secondary coils, and a **CORE** that is typically made of air or iron. Air-core transformers

PRIMARY The input coil of a transformer.

SECONDARY The output coil of a transformer.

CORE The space inside each coil in a transformer, usually containing air or iron.

are typically found in high-frequency, low-power electronic applications. Iron-core transformers are generally found in low-frequency, high-power electrical (and electronic) applications. As such, we will concentrate primarily on iron-core transformers in this chapter.

FIGURE 21.2 Some smaller transformers.

Transformer Schematic Symbols

Figure 21.3 shows several commonly used transformer schematic symbols. In each case, the coils are identified by the labels on the input and output *terminals* (leads). P_1 and P_2 are the primary terminals. S_1 and S_2 are the secondary terminals.

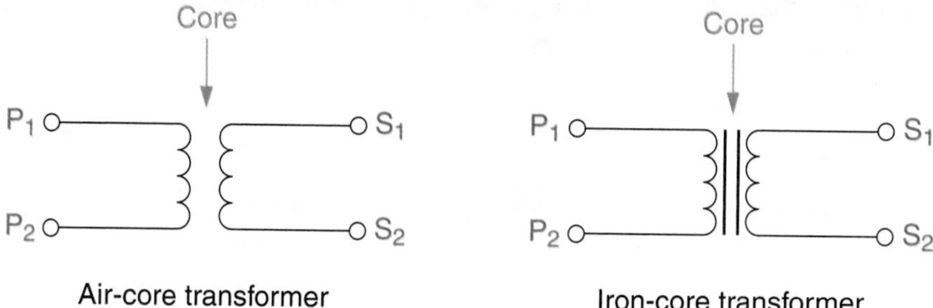

FIGURE 21.3 Transformer schematic symbols.

STEP-DOWN TRANSFORMER A transformer that has a lower output (secondary) voltage than input (primary) voltage.

STEP-UP TRANSFORMER A transformer that has a higher output (secondary) voltage than input (primary) voltage.

1-TO-1 TRANSFORMER A transformer that has equal output (secondary) and input (primary) voltages.

Voltage Classifications

A given transformer is commonly described in terms of the relationship between its input and output voltages. For example, Figure 21.4 shows three types of transformers, each with a unique relationship between the input and output voltages.

In Figure 21.4a, the secondary (output) voltage is lower than the primary (input) voltage. This type of transformer is called a **STEP-DOWN TRANSFORMER**. In contrast, the **STEP-UP TRANSFORMER** shown in Figure 21.4b provides a secondary voltage that is greater than its input voltage. The **1-TO-1 TRANSFORMER** shown

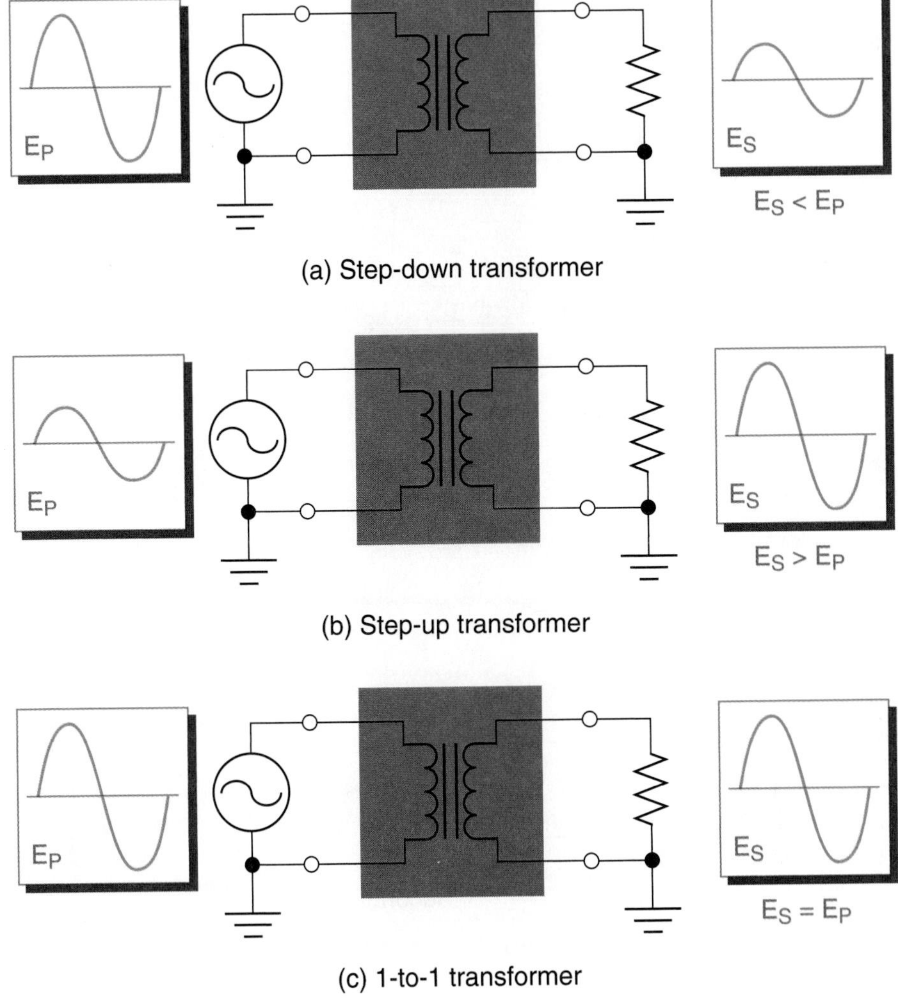

FIGURE 21.4 Types of transformers.

in Figure 21.4c has equal input and output voltages. (Later in this section, we will discuss the input/output power and current relationships for each type of transformer.)

Transformer Operation

Transformer operation is based on the principle of electromagnetic induction (which is sometimes referred to as **TRANSFORMER ACTION**). This principle is illustrated (as a review) in Figure 21.5a. When there is a changing current in the L_1 circuit, a changing magnetic field is generated. This changing magnetic field cuts through L_2, inducing a voltage across that coil. It is important to note that:

TRANSFORMER ACTION
A term used to describe the means by which a transformer couples an AC signal from its primary to its secondary by means of electromagnetic induction.

1. Any voltage waveform induced across L_2 normally has the same shape as the L_1 waveform. For example, if a sine wave is used to generate the changing current through L_1, the voltage induced across L_2 is also sinusoidal (provided the transformer input is not overdriven).

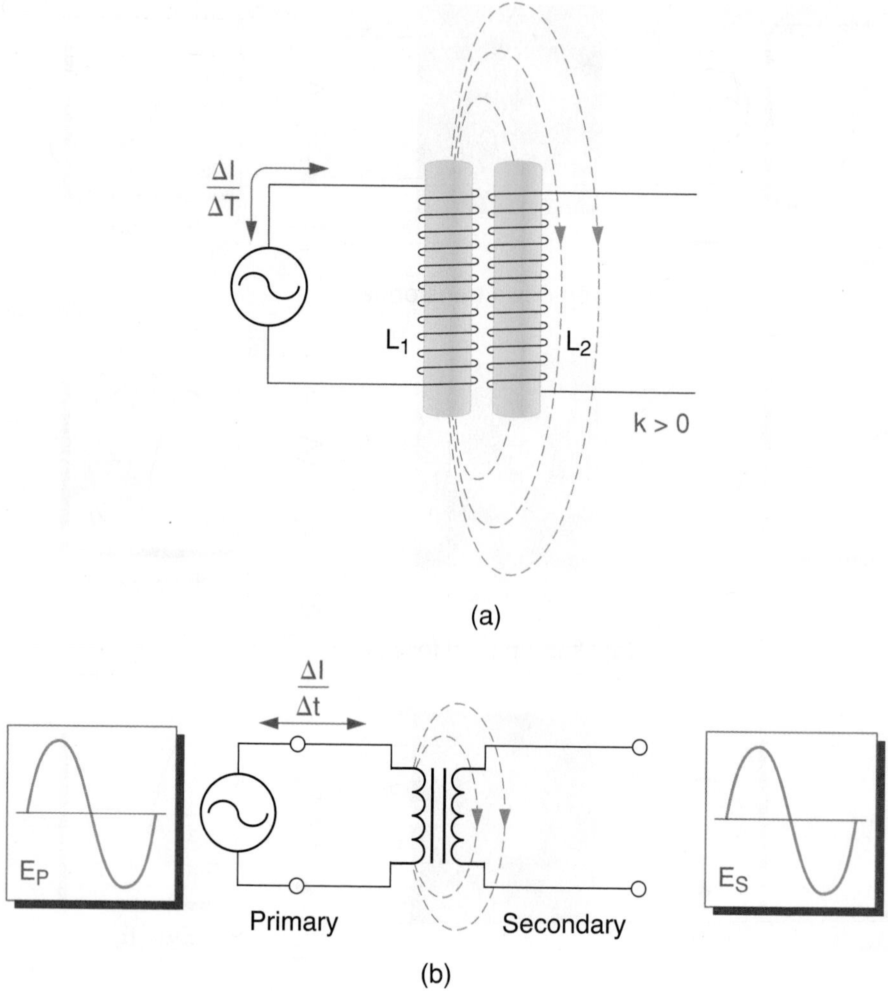

FIGURE 21.5 Transformer AC operation.

2. The inductors are not physically connected to each other. Therefore, any DC voltage applied to L_1 is not coupled to L_2.

As these statements indicate, the two-inductor circuit couples an AC signal from one inductor to the other, while providing DC isolation between the two.

Figure 21.5b shows how this principle of electromagnetic induction applies to a transformer. When an AC input is applied to the primary, a changing current is generated throughout the primary circuit. The resulting magnetic field cuts through the secondary coil, inducing a voltage across the secondary. This voltage is felt at the output terminals of the component. Assuming that the AC input to the transformer is a sine wave, the output is also a sine wave. As you can see, the AC input has been effectively coupled from the input of the transformer to its output.

While most transformers are used to change voltage and/or current levels, the 1-to-1 transformer is used to provide power to equipment (in hospital operating rooms, for example) so that all grounding can be properly isolated and detected.

If a DC voltage is applied to the primary of a transformer, a changing current is not generated in the primary coil. As a result, the primary coil does not generate a changing magnetic field, no voltage is induced across the secondary coil, and the output from the secondary is 0 V. This (again) is how the transformer provides DC isolation.

Turns Ratio

Each loop of wire in a given coil is referred to as a **TURN**. The **TURNS RATIO** of a transformer is the ratio of primary turns to secondary turns. For example, the transformer in Figure 21.6 is shown to have 320 turns in its primary coil and 80 turns in its secondary coil. Using these values, the turns ratio of the component is found as

TURN Each loop of wire in a coil.

TURNS RATIO The ratio of primary turns to secondary turns.

$$\frac{N_P}{N_S} = \frac{320 \text{ turns}}{80 \text{ turns}} = \frac{4}{1}$$

Note that the turns ratio of a given transformer is normally written in the form shown in Figure 21.6. Also note that the turns ratio of a transformer is a function of its construction. This means that, in most cases, it cannot be changed.

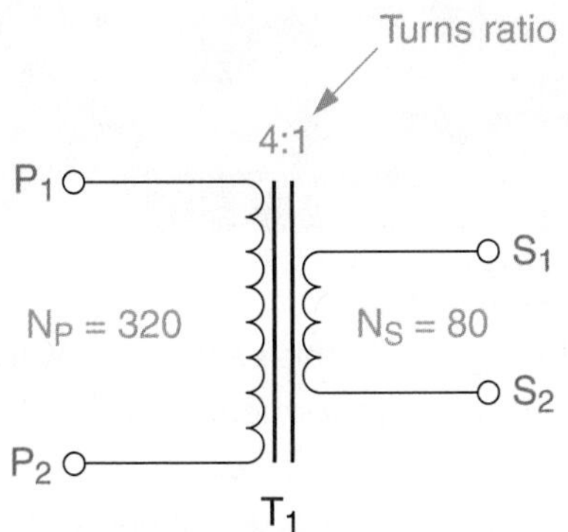

FIGURE 21.6 Transformer turns ratio.

The turns ratio of a transformer is important because it determines the ratio of primary voltage to secondary voltage. By formula,

(21.1)
$$\frac{N_P}{N_S} = \frac{E_P}{E_S}$$

For the transformer in Figure 21.6, this equation indicates that the voltage across the primary is four times the voltage across the secondary when the component has an AC input. As such, the transformer shown in the figure is a *step-down* transformer.

In the upcoming sections, you will learn how the turns ratio of a transformer is used in transformer voltage, current, and power calculations.

PROGRESS CHECK

1. What are the DC and AC characteristics of a transformer?
2. Define the terms *coupling* and *isolation*.
3. What is a transformer *primary*? What is a transformer *secondary*?
4. What is the most common core material for power transformers?
5. Describe characteristics of a step-up, a step-down, and a 1-to-1 transformer.
6. Define *turns ratio* and explain the relationship between turns ratio and transformer input and output voltage.

21.2 TRANSFORMER VOLTAGE, CURRENT, AND POWER

You have been shown that the relationship between transformer primary and secondary voltage is a function of the component's turns ratio. In this section, we will discuss the voltage, current, and power characteristics of transformers. As you will see, the turns ratio of a transformer also plays a role in primary and secondary current calculations. Power, on the other hand, is simply a function of current and voltage (as always).

Transformer Secondary Voltage (E_S)

Equation (21.1) can be used to define transformer secondary voltage in terms of the component's primary voltage and turns ratio, as follows:

(21.2)

$$E_S = E_P \times \frac{N_S}{N_P}$$

Example 21.1 demonstrates the use of this relationship.

EXAMPLE 21.1

Calculate the secondary voltage for the transformer shown in Figure 21.7.

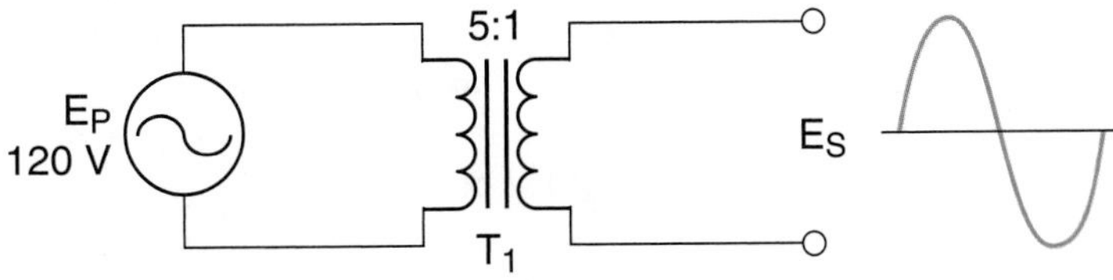

FIGURE 21.7

SOLUTION

The transformer shown has a 120 V input and a 5:1 turns ratio. Using these values, the secondary voltage is found as

$$E_S = E_P \times \frac{N_S}{N_P}$$

$$= 120\text{ V} \times \frac{1}{5}$$

$$= 24\text{ V}$$

PRACTICE PROBLEM 21.1

A transformer like the one in Figure 21.7 has a 480 V input and a 12:1 turns ratio. Calculate the secondary voltage for the transformer.

The transformer in Example 21.1 is a step-down transformer. The increase in voltage (from primary to secondary) produced by a *step-up* transformer is demonstrated in Example 21.2.

EXAMPLE 21.2

Determine the secondary voltage for the step-up transformer in Figure 21.8.

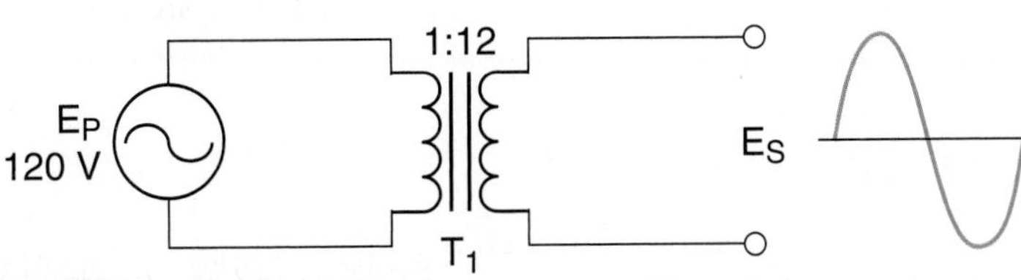

FIGURE 21.8

SOLUTION

The transformer has a 120 V input and a 1:12 turns ratio. Using these values, the secondary voltage is found as

$$E_S = E_P \times \frac{N_S}{N_P}$$

$$= 120\text{ V} \times \frac{12}{1}$$

$$= 1440\text{ V}$$

$$= 1.44\text{ kV}$$

PRACTICE PROBLEM 21.2

A transformer like the one in Figure 21.8 has a 480 V input and a 1:6 turns ratio. Calculate the secondary voltage for the transformer.

If we compare the components analyzed in Examples 21.1 and 21.2, we can establish a means of identifying a transformer by its turns ratio. For convenience, the values of interest from the examples are summarized as follows:

EXAMPLE	TYPE OF TRANSFORMER	N_P	N_S	N_P vs. N_S
21.1	Step-down	5	1	$N_P > N_S$
21.2	Step-up	1	12	$N_P < N_S$

As these results indicate, the type of transformer is determined by the relationship between the number of primary and secondary turns. When $N_P > N_S$, the component is a step-down transformer. When $N_P < N_S$, the component is a step-up transformer.

As you were told earlier, a 1-to-1 transformer has equal input and output voltages. This input/output voltage relationship can occur only when $N_P = N_S$.

The Phase Relationship Between E_P and E_S

The phase relationship between E_P and E_S depends on the wiring of the transformer. Some transformers are wired so that the primary and secondary voltages are in phase, while others are wired so that they are 180 degrees out of phase. The input/output phase relationship for a transformer is commonly identified using **POLARITY DOTS** like those shown in Figure 21.9. Note that the input and output voltages are assumed to be in phase when no polarity dots are shown on a transformer symbol.

POLARITY DOTS Dots drawn on the schematic symbol of a transformer that identify the input and output terminal voltages that are in phase.

Power Transfer

Under ideal conditions, all of the power applied to the primary (input) of a transformer is transferred to the secondary (output). By formula,

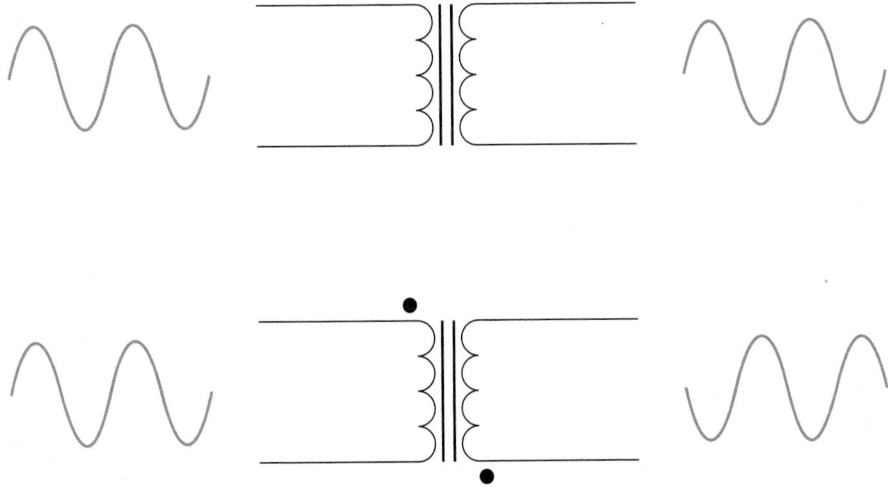

FIGURE 21.9 Transformer polarity dots.

(21.3)
$$P_P = P_S \quad [\text{ideal}]$$

or

(21.4)
$$E_P \times I_P = E_S \times I_S \quad [\text{ideal}]$$

As you know, power is a measure of energy used (or transferred) per second. With this in mind, look at the transformer in Figure 21.10. Under ideal conditions, the resistance of the primary coil is 0 Ω, so it does not dissipate any power. Therefore, all of the energy supplied to the transformer (by the source) is transferred to the magnetic flux. Assuming that $k = 1$ (which is also an ideal condition), all of the energy in the magnetic flux is transferred to the secondary coil. The coil in the secondary then converts this energy back into electrical energy, and delivers that energy to the load. Thus, all the transformer input power is transferred (via the magnetic flux) to the load.

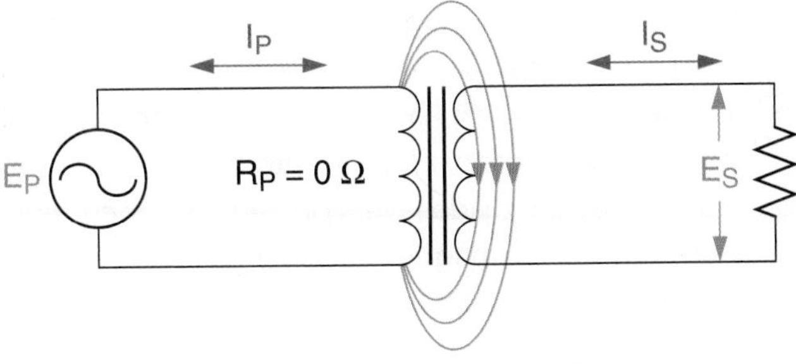

FIGURE 21.10 Transformer primary and secondary power.

Efficiency

In Chapter 4, we discussed the concept of *efficiency*. You may recall that efficiency is the ratio of output power to input power, given as a percent. The efficiency of a transformer is found as

(21.5)

$$\eta = \frac{P_S}{P_P} \times 100$$

where

η = the efficiency of the transformer (in percent)
P_S = the transformer secondary power
P_P = the transformer primary power

Since the ideal transformer has equal input and output power values, its efficiency is 100%.

The "Practical" Transformer

As you know, there are no ideal components. In practice, the value of secondary power is always slightly lower than the value of primary power. This is a result of power losses that occur in every practical transformer. At this point, we will take a brief look at these losses.

Copper Loss

All wire has some measurable amount of resistance. Even though this resistance is very low, it still causes some amount of power to be dissipated in both the primary and secondary coils. **COPPER LOSS**, also known as **I^2R LOSS**, is reduced by constructing transformers using the largest practical wire diameter. (Larger diameter wires have lower resistance, which reduces copper loss.)

COPPER LOSS Power that is dissipated when current passes through the primary and secondary coils of a transformer.

I^2R LOSS Another name for copper loss.

EDDY CURRENT Current that travels in a circular motion in the core of an iron-core transformer, generated by flux passing through the core.

LAMINATED CORE. An iron core that is broken up into thin layers with an oxide coating between each, preventing eddy currents from being generated.

Loss Due to Eddy Currents

This type of loss is unique to iron-core transformers like the one in Figure 21.11. When magnetic flux is generated in the primary, it passes through the iron core. Since iron is a conductor, the flux passing through the core generates a current within the core, called an **EDDY CURRENT**. As shown in Figure 21.11a, this current travels in a circular motion through the core. As eddy currents pass through an iron core, the resistance of the core causes some amount of power to be dissipated in the form of heat.

Eddy current loss can be reduced by using a **LAMINATED CORE**. This type of core (shown in Figure 21.11b) is broken up into thin layers, with an oxide coating between each. The insulation provided by the oxide layers prevents eddy currents from being generated (without interfering with magnetic coupling).

Hysteresis Loss

In Chapter 10, we defined *retentivity* as the ability of a material to retain its magnetic characteristics after a magnetizing force has been removed. Simply put, the term

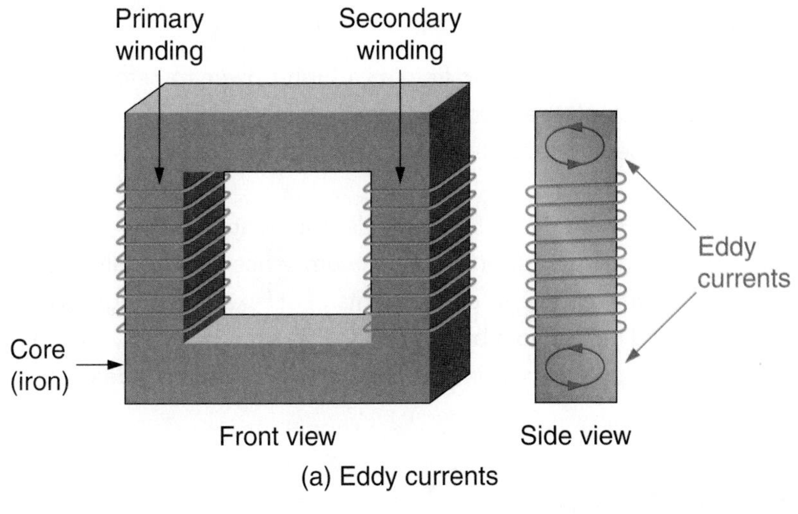

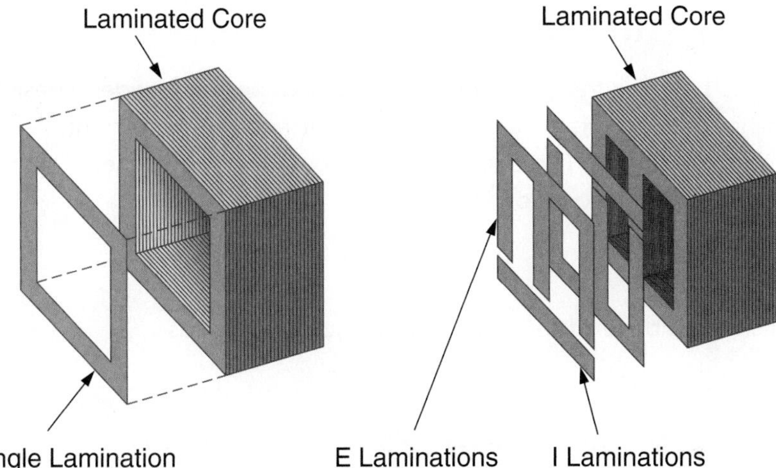

FIGURE 21.11 Transformer core eddy currents and laminations.
Source: Figure (b) is based on a figure in NEETS Manual 2, *Introduction to Alternating Current and Transformers*, 1984 edition.

HYSTERESIS LOSS The energy used to overcome the retentivity of a transformer (or inductor) core.

STRAY LOSS Power loss that results when flux generated in a transformer primary interacts with nearby conductors other than the transformer secondary.

HYSTERESIS LOSS refers to the energy expended to overcome the core's retentivity. Each time the transformer input reverses polarity, the polarity of the magnetic field must also reverse. To do this, it must overcome the retentivity of the core material. Hysteresis loss is reduced by using low-retentivity materials in the core, such as air or silicon steel (an alloy). It is also reduced by the use of a laminated core.

Many laminations are required to make up a core. There are usually 40 to 50 laminations per inch, with the core usually being one or two times the width of the core center leg. A very high grade of electrical steel is used, and the laminations must be very thin to reduce electrical losses in the core.

IN THE FIELD

Stray loss

No transformer has a perfect coefficient of coupling ($k = 1$), meaning that some of the magnetic flux generated by the primary does not cut into the secondary. The flux that does not reach the secondary may interact with nearby conductors, resulting in a power loss called **STRAY LOSS**.

Mechanical losses

The expanding and collapsing magnetic field in a high-power transformer can produce mechanical stresses on the component's wires, core, and support structure. In many cases, these stresses cause the component to vibrate, which wastes energy.

MECHANICAL LOSSES
Power loss that can occur when the expanding and collapsing magnetic field in a high-power transformer stresses the component's wires, core, and support structure.

Even though the **MECHANICAL LOSSES** just described are very real, each results in relatively little power loss under normal circumstances. Generally speaking, transformers are highly efficient components. For example, high-power transformers (100 MVA and higher) can be over 99% efficient. This is why the voltage and current relationships covered in this chapter are based on ideal transformer characteristics.

Transformer Primary and Secondary Current

The relationship between transformer input and output power provides a basis for the relationship between transformer input current and output current. As you know, the ideal transformer has equal input and output power values. This relationship was described in equation (21.4) as

$$E_P \times I_P = E_S \times I_S$$

This relationship can be rewritten to define secondary current as

(21.6)
$$I_S = I_P \frac{E_P}{E_S}$$

Equation (21.6) can be used to determine the value of I_S as demonstrated in Example 21.3.

EXAMPLE 21.3

Determine the value of secondary current (I_S) for the transformer in Figure 21.12.

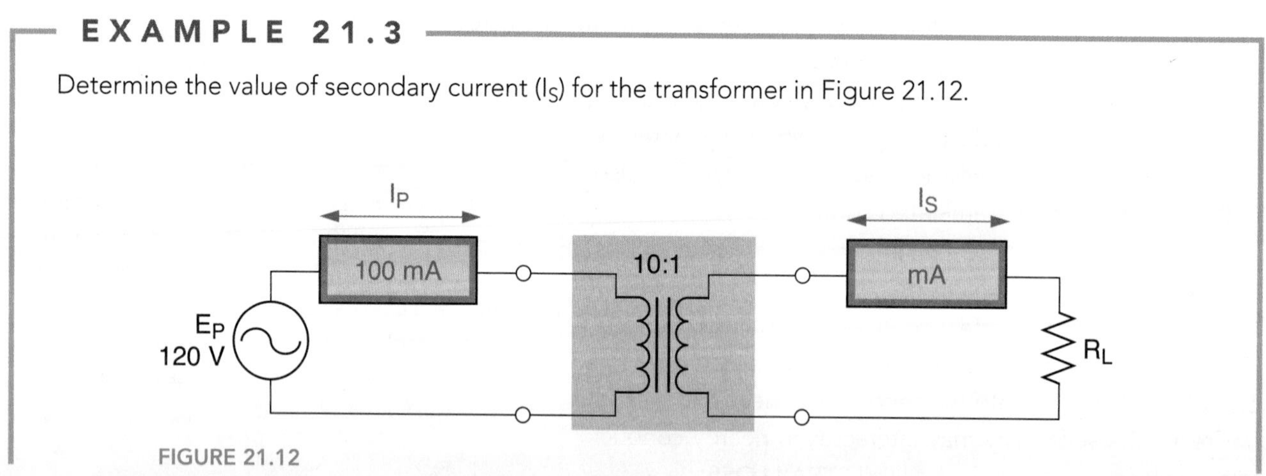

FIGURE 21.12

SOLUTION

First, the value of secondary voltage for the transformer is found as

$$E_S = E_P \times \frac{N_S}{N_P}$$

$$= (120\text{ V})\frac{1}{10}$$

$$= 12\text{ V}$$

Now, using E_S and the values shown in the figure, the value of I_S can be found as

$$I_S = I_P \times \frac{E_P}{E_S}$$

$$= 100\text{ mA} \times \frac{120\text{ V}}{12\text{ V}}$$

$$= 1\text{ A}$$

PRACTICE PROBLEM 21.3

A circuit like the one in Figure 21.12 has a 40-V input, a 2:1 turns ratio, and a value of $I_P = 1.5$ A. Calculate the value of the secondary current.

The voltage, current, and power values for the circuit analyzed in Example 21.3 are listed below. Note that the power values were found as $P = E \times I$.

CIRCUIT	VOLTAGE	CURRENT	POWER
Primary	120 V	100 mA	12 W
Secondary	12 V	1 A	12 W

Using the values listed, we can make several observations about the circuit:

1. The primary and secondary values of power equal each other.
2. Voltage *decreased* by a factor of 10 from primary to secondary.
3. Current *increased* by a factor of 10 from primary to secondary.

The first of these points simply reinforces the ideal power relationship described earlier in this section. The second two points, however, introduce a new and important concept: For any step-up or step-down transformer, the change in current from primary to secondary is inversely proportional to the change in voltage. That is,

1. A transformer that steps voltage *down* also steps current *up*.
2. A transformer that steps voltage *up* also steps current *down*.

Note that voltage and current change by the same factor for any transformer. For example, if a step-down transformer decreases voltage by a factor of *four*, it also increases current by a factor of *four*. (The same principle holds true for any step-up transformer.)

Earlier, you were told that voltage varies directly with the turns ratio of a transformer. That is, the change in voltage from input to output is determined by the turns ratio of the transformer. Since current and voltage vary inversely, current varies inversely with the turns ratio. By formula,

$$\frac{I_S}{I_P} = \frac{N_P}{N_S} \tag{21.7}$$

If we solve this relationship for secondary current (I_S), we get

$$I_S = I_P\frac{N_P}{N_S} \tag{21.8}$$

This equation is used in Example 21.4 to demonstrate the relationship between the values of voltage and current for a step-up transformer.

EXAMPLE 21.4

Calculate the output voltage and current values for the step-up transformer in Figure 21.13.

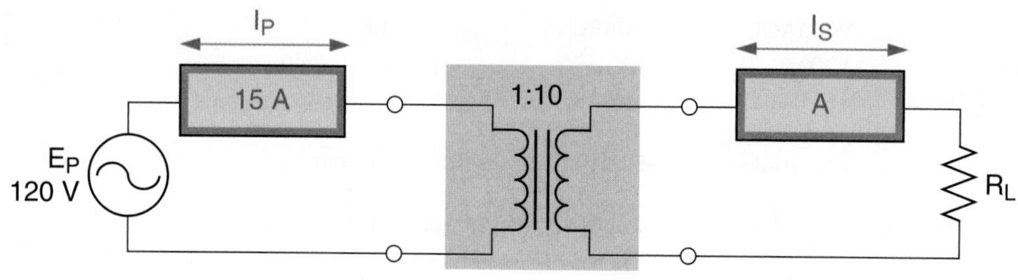

FIGURE 21.13

SOLUTION

The transformer secondary voltage is found as

$$\begin{aligned} E_S &= E_P \times \frac{N_S}{N_P} \\ &= 120\text{ V} \times \frac{10}{1} \\ &= 1200\text{ V} \\ &= 1.2\text{ kV} \end{aligned}$$

Using equation (21.8), the value of the secondary current can be found as

$$I_S = I_P\frac{N_P}{N_S}$$

$$= (15\text{ A})\frac{1}{10}$$

$$= 1.5\text{ A}$$

PRACTICE PROBLEM 21.4

A circuit like the one in Figure 21.13 has a 240-V input, a 1:5 turns ratio, and a value of $I_P = 5$ A. Calculate the value of the secondary current.

If we compare the results from Examples 21.3 and 21.4, we get a clearer picture of the relationships among the turns ratio, voltage values, and current values for step-up and step-down transformers. For convenience, the results from the two examples are listed in Table 21.1.

TABLE 21.1 • Results from Examples 21.3 and 21.4

TRANSFORMER	CIRCUIT	TURNS*	VOLTAGE	CURRENT	POWER
Step-down	Primary	10	120 V	100 mA	12 W
	Secondary	1	12 V	1 A	12 W
Step-up	Primary	1	120 V	15 A	1800 W
	Secondary	10	1.2 kV	1.5 A	1800 W

*These are the numbers given in the *turns ratio*. They do not equal the actual number of turns in the primary and secondary windings.

As shown in the table, the step-down transformer:

1. Has a turns ratio of 10:1.
2. Decreases voltage by a factor of 10.
3. Increases current by a factor of 10.

On the other hand, the step-up transformer:

1. Has a turns ratio of 1:10.
2. Increases voltage by a factor of 10.
3. Decreases current by a factor of 10.

These results agree with the statements made earlier regarding the effect of a transformer's turns ratio on the input and output voltage and current values.

In each of the last two examples, a value of primary current (I_P) was used to calculate secondary current (I_S). In practice, the value of I_S is calculated using E_S and load resistance (or power). Then, the transformer primary current is calculated using:

(21.9)

$$I_P = I_S\frac{N_S}{N_P}$$

The use of this relationship is demonstrated in the following example.

EXAMPLE 21.5

Calculate the value of primary current (I_P) for the circuit in Figure 21.14.

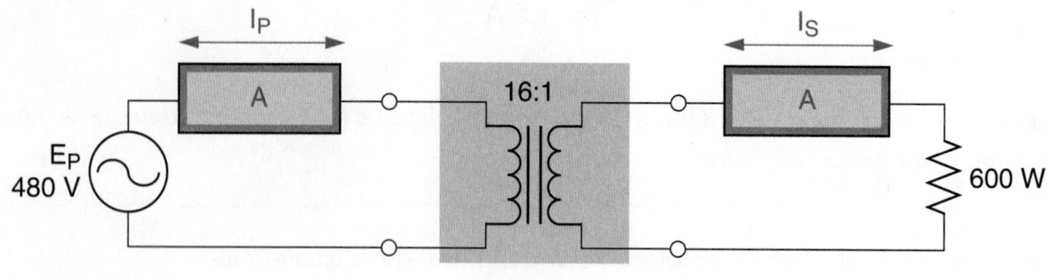

FIGURE 21.14

SOLUTION

First, the transformer secondary voltage is found as

$$E_S = E_P \times \frac{N_S}{N_P}$$

$$= 480\text{ V} \times \frac{1}{16}$$

$$= 30\text{ V}$$

Using Watt's law, the secondary current can now be found as

$$I_S = \frac{P_{LOAD}}{E_S}$$

$$= \frac{600\text{ W}}{30\text{ V}}$$

$$= 20\text{ A}$$

Finally, the current through the transformer primary can be found as

$$I_P = I_S \frac{N_S}{N_P}$$

$$= 20\text{ A} \times \frac{1}{16}$$

$$= 1.25\text{ A}$$

PRACTICE PROBLEM 21.5

A circuit like the one in Figure 21.14 has a 240 V line input, a 4:1 turns ratio, and a 1200 W load. Calculate the value of primary current for the circuit.

We have now established the basic voltage, current, and power relationships for transformers. For future reference, the input/output relationships established in this section are summarized in Figure 21.15. Note that the input and output values for the 1-to-1 transformer are equal in every category. As a result, this transformer is used strictly to provide DC isolation between two AC circuits without affecting the AC coupling from one to the other.

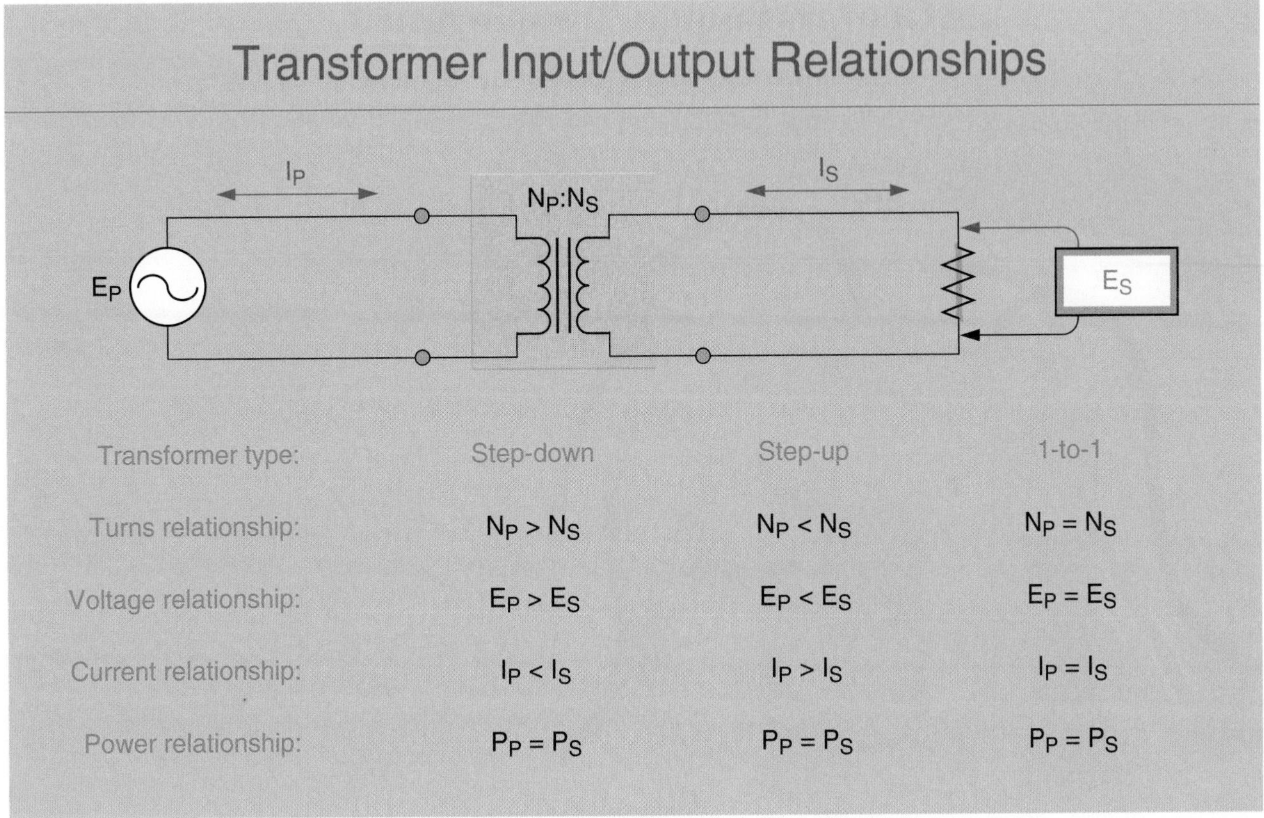

Transformer Input/Output Relationships
I_P
$N_P:N_S$
I_S
E_P
E_S
Transformer type: Step-down Step-up 1-to-1
Turns relationship: $N_P > N_S$ $N_P < N_S$ $N_P = N_S$
Voltage relationship: $E_P > E_S$ $E_P < E_S$ $E_P = E_S$
Current relationship: $I_P < I_S$ $I_P > I_S$ $I_P = I_S$
Power relationship: $P_P = P_S$ $P_P = P_S$ $P_P = P_S$

FIGURE 21.15

PROGRESS CHECK

1. Explain how you would calculate transformer secondary voltage given the primary voltage and turns ratio.
2. Explain how you can determine whether a transformer is a step-up or a step-down transformer based on the transformer's turns ratio.
3. Explain the use of polarity dots in a schematic.
4. What is the *efficiency* of an ideal transformer?
5. Define the following:
   a. copper loss
   b. eddy current loss
   c. hysteresis loss
   d. stray loss
   e. mechanical loss
6. Explain the relationship between primary and secondary voltage and current in both step-up and step-down transformers.

21.3 TRANSFORMER WINDING CONFIGURATIONS

Transformers can be wired in a variety of ways, each designed to accomplish some specific function. In this section, we will briefly examine some of the various transformer winding configurations.

CENTER-TAPPED TRANSFORMER A transformer that has an additional lead connected to the center of its secondary winding.

Center-Tapped Transformers

A **CENTER-TAPPED TRANSFORMER** is a transformer that has an additional lead connected to the center of its secondary winding. A center-tapped transformer (along with its schematic symbol) is shown in Figure 21.16.

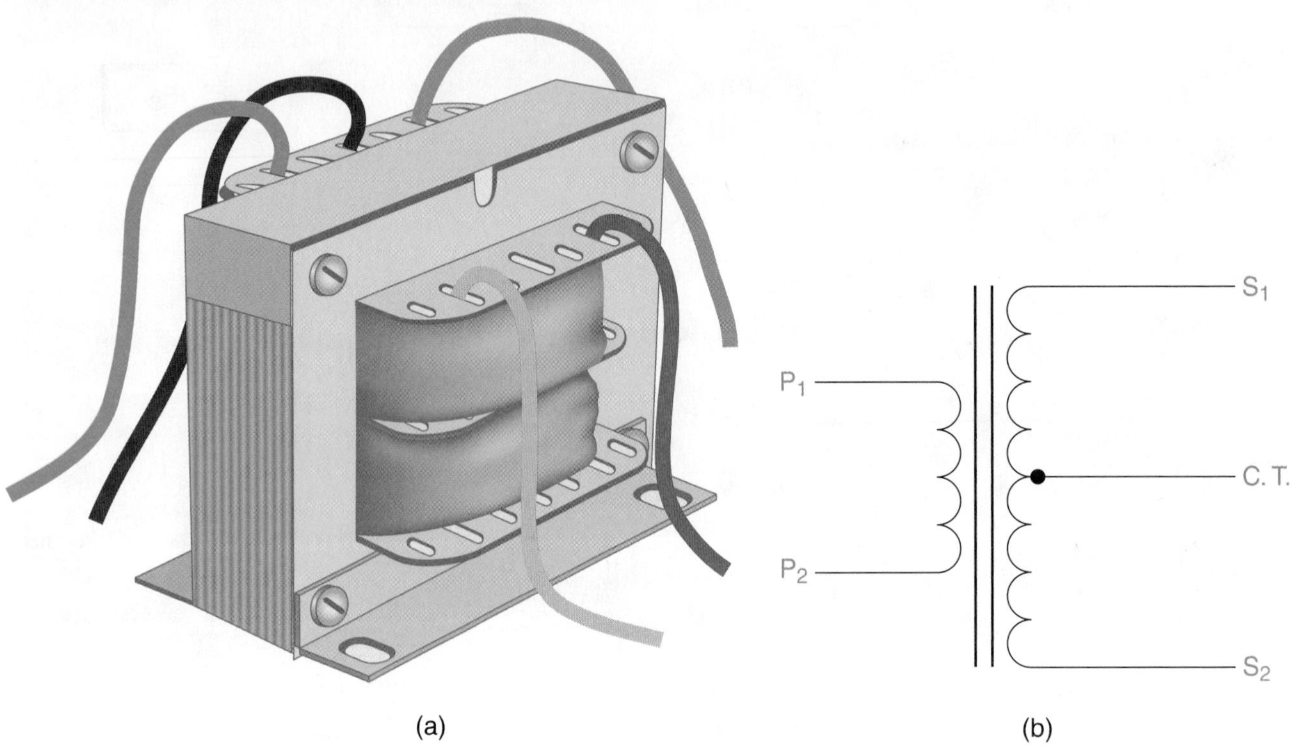

FIGURE 21.16 Center tapped transformer and symbol.

When a third lead is connected to the center of a transformer secondary, the voltage from either end-terminal of the secondary to the center tap is one-half the secondary voltage. This relationship is illustrated in Figure 21.17. The transformer shown is rated for a 240 V output. The voltage across either end-terminal (S_1 or S_2) and the center tap is one-half the secondary voltage, 120 V.

It should be noted that the end-terminal voltages, when measured with respect to the center tap, have opposite polarities. This principle is illustrated in Figure 21.18. If we assume that the secondary voltage has the polarity shown, the voltage at S_1 is positive with respect to the grounded center tap. At the same time, the voltage at S_2 is negative with respect to the grounded center tap. For

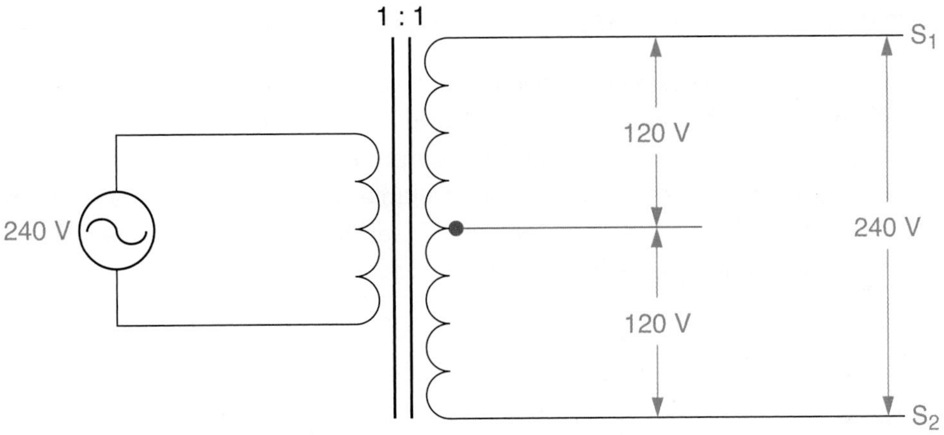

FIGURE 21.17

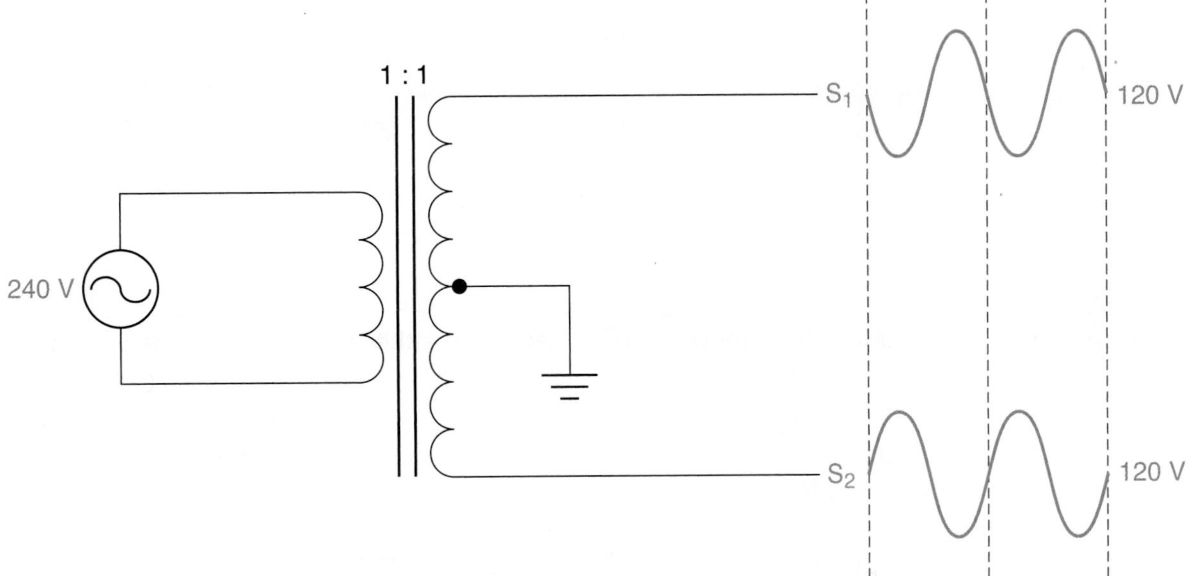

FIGURE 21.18 Center-tapped transformer operation.

this reason, the term **SPLIT PHASE** is often used to describe the output from the center-tapped transformer.

SPLIT PHASE A term that is often used to describe the output from the center-tapped transformer because the two output waveforms are 180° out of phase.

Balanced Loading

Most transformers with a secondary voltage of 120/240 V—like those used in residential installations—contain two 120 V secondary windings. Each 120 V winding is designed to handle one-half the nameplate kVA rating. It is important to make sure that the load between the two secondaries is *balanced.* If too much of the load is connected to one of the secondary windings, that winding may be overloaded, even though the total transformer rating has not been exceeded. This can lead to overheating and transformer failure.

Transformers with Multiple Outputs

MULTIPLE-OUTPUT TRANSFORMER A transformer that has two or more secondary coils.

When an electrical system needs two or more voltage levels to operate, they can usually be obtained from the AC line input using a **MULTIPLE-OUTPUT TRANSFORMER**. As the name implies, a multiple-output transformer is one that has two or more secondary coils. An example of a multiple-output transformer is shown in Figure 21.19. In this case, a single primary is used to couple the AC line input to *two* secondary windings, each having its own load.

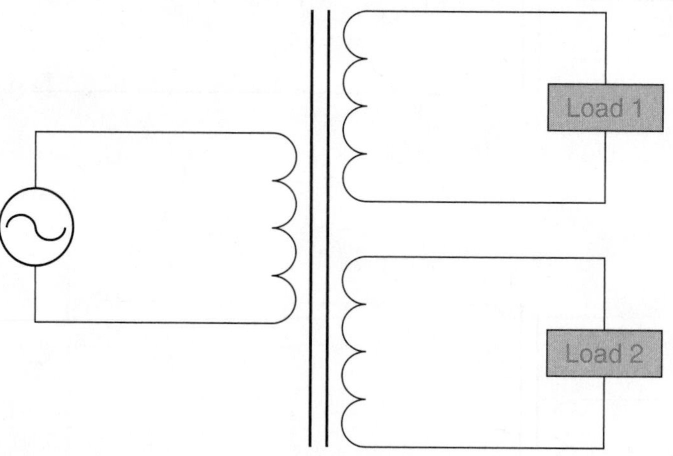

FIGURE 21.19 A multiple-output transformer.

Multiple outputs are produced using either separate windings (like those shown in Figure 21.19) or a single secondary winding with multiple taps. This wiring configuration is shown in Figure 21.20. As you can see, the circuit has been designed to provide three output voltages (25.2 V, 12.6 V, and 12.6 V) using two secondary taps.

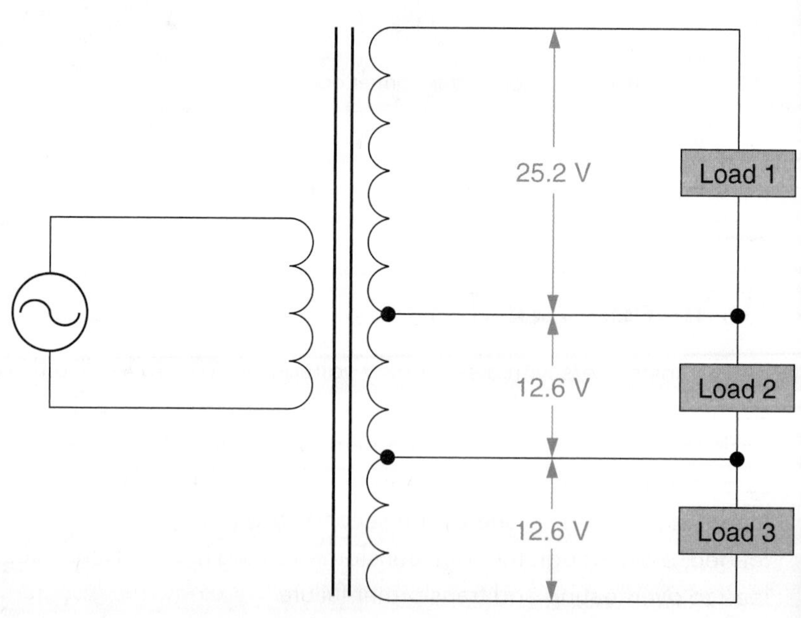

FIGURE 21.20

When working with multiple-secondary transformers, there are several things to keep in mind:

1. The transformer may have a single secondary winding with one or more taps (as in Figure 21.20), or it may contain two or more separate secondary windings (as in Figure 21.19).
2. Each secondary circuit is analyzed as if it were the only one. That is, each secondary circuit is analyzed independently of the other secondary circuits.
3. If a fault develops in one of the secondary circuits, it can affect the other secondary circuits, especially if the transformer has a single secondary winding with multiple taps. Even with multiple secondary windings, a short circuit in one of the secondary circuits can cause a primary fuse or breaker to open. In effect, the short causes power to be lost in all of the secondary circuits.

Transformers with Multiple Inputs

Transformers may also have more than one primary winding or several taps on the primary winding. One common use of multiple primary windings is the power supply transformers for devices designed for use in both North America and Europe (120 V/60 Hz and 240 V/50 Hz, respectively).

Transformer taps, on the other hand, are normally used to compensate for variations in the line voltage supplied to the primary. Taps are provided on the primary winding to correct for high or low line voltage conditions. With the correct tap, the transformer can still deliver its full rated secondary voltage. Standard tap increments are usually two and one-half or five percent of the rated primary voltage for both high- and low-voltage conditions.

Autotransformers

An **AUTOTRANSFORMER** is a transformer made up of a single coil that typically has three terminal connections. The schematic symbol for an autotransformer is shown in Figure 21.21. As you can see, the autotransformer is—for all practical purposes—a tapped inductor.

AUTOTRANSFORMER A transformer made up of a single coil that typically has three terminal connections.

The turns ratio of an autotransformer depends on:

1. The number of turns between terminals 1 and 3 ($N_{1,3}$).
2. The number of turns between terminals 2 and 3 ($N_{2,3}$).
3. The input/output designations of the terminal pairs.

The terminals listed above are identified in Figure 21.21. For standard autotransformers, the values of $N_{1,3}$ and $N_{2,3}$ are fixed. For a variable autotransformer, or **VARIAC**, the position of terminal 2 can be adjusted. This means that $N_{2,3}$ can be adjusted while $N_{1,3}$ remains fixed.

VARIAC A variable autotransformer.

The input/output designations of the terminal pairs depend on the circuit configuration. Because of its construction, the autotransformer can be wired in either a

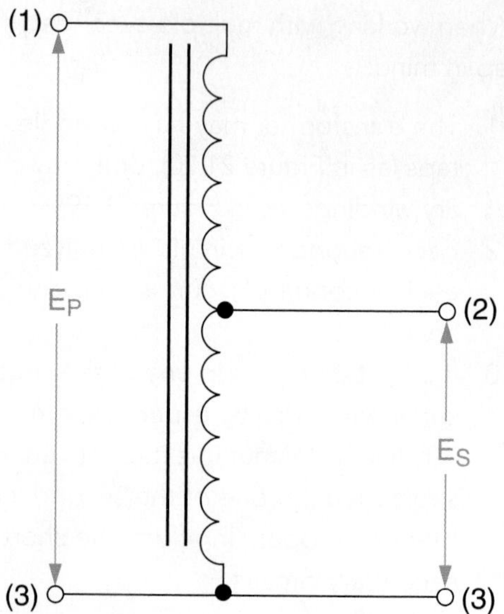

FIGURE 21.21 Autotransformer symbol.

step-up or a step-down configuration. These two configurations are illustrated in Figure 21.22. If we assume that terminal 2 center-taps the autotransformer, then

$$N_{1,3} = 2N_{2,3}$$

This equation indicates that:

a. The circuit in F 21.22a has a 2:1 turns ratio.
b. The circuit in Figure 21.22b has a 1:2 turns ratio.

An important point: Because the autotransformer uses the same coil for its primary and secondary circuits, it does not provide DC isolation between its source and load circuits. As a result, its use is limited to applications where DC isolation between the source and load is not required.

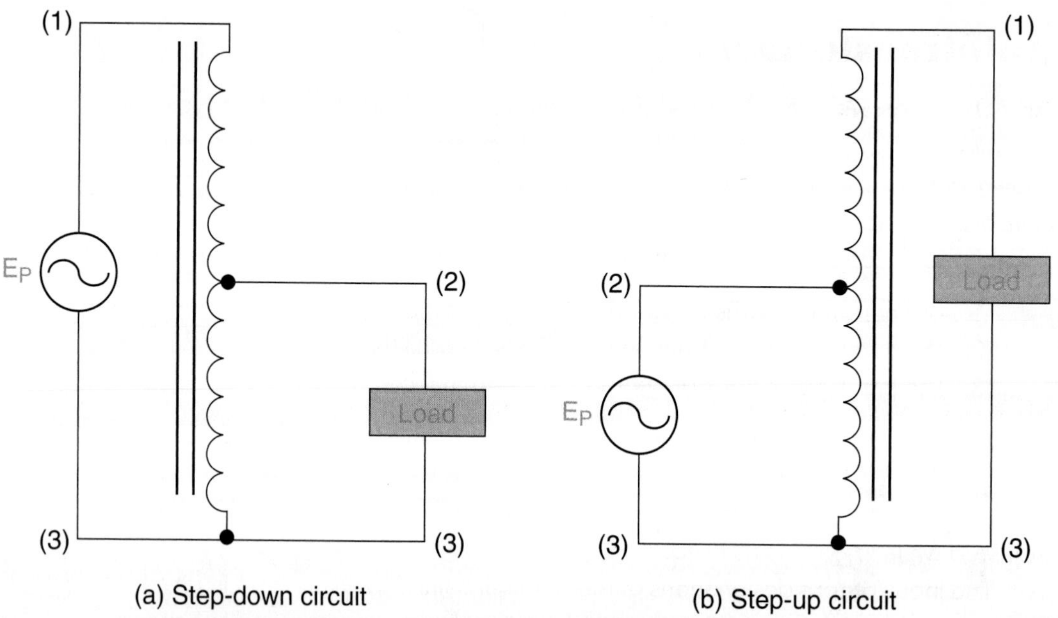

FIGURE 21.22 Autotransformer connections.

Buck and Boost Configurations

Figure 21.23a shows a variable autotransformer, or variac. As indicated, the output (E_S) can be varied between 0 V and 130 V when E_P = 120 V. If the component is wired so that E_S is slightly greater than E_P, it is referred to as a **BOOST TRANSFORMER**. For example, the boost transformer in Figure 21.23b is wired to provide a 130 V output when E_P = 120 V. If the component is wired so that E_S is slightly less than E_P, it is referred to as a **BUCK TRANSFORMER**. For example, the buck transformer in Figure 21.23c is wired to provide a 110 V output when E_P = 120 V.

BOOST TRANSFORMER One that increases (steps up) its input voltage by as much as 20%.

BUCK TRANSFORMER One that decreases (steps down) its input voltage by as much as 20%.

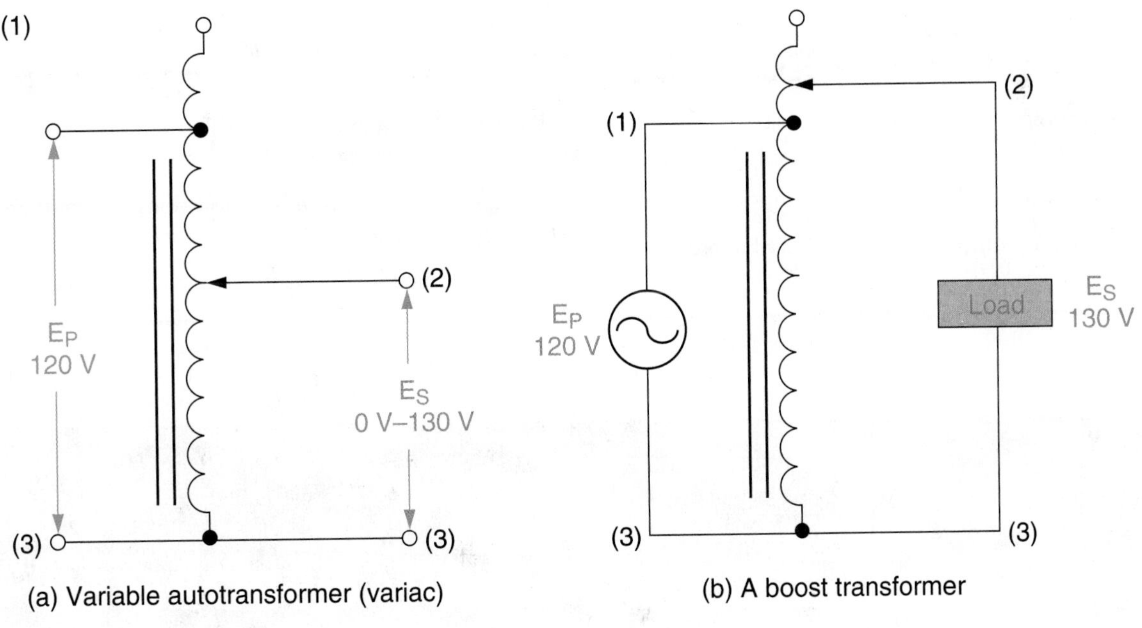

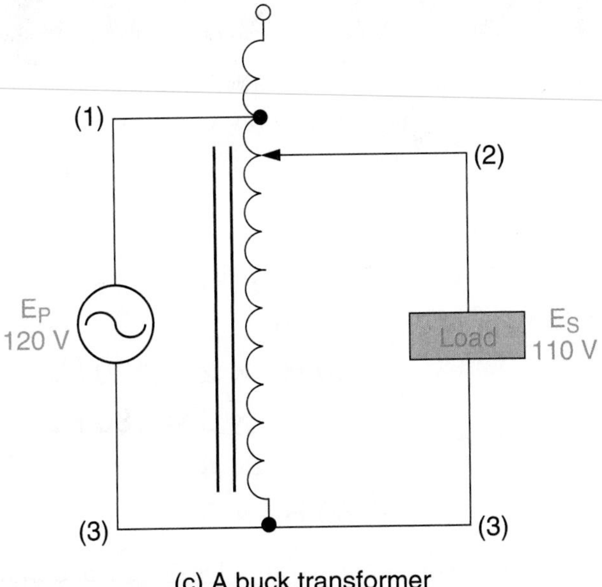

FIGURE 21.23 Autotransformer configurations.

As stated above, a buck or boost transformer provides an output voltage (E_S) that is slightly less than or greater than its primary voltage (E_P). What do we mean by *slightly*? There appears to be no industry standard, but most references agree that the output from a buck or boost transformer is within 20% of its primary voltage. Based on this value, we can make the following statements:

- A *buck* transformer is one that decreases (steps down) its input voltage by as much as 20%.
- A *boost* transformer is one that increases (steps up) its input voltage by as much as 20%.

Note that an autotransformer (like the one in Figure 21.23) can be adjusted to act as either a buck or a boost transformer.

PROGRESS CHECK

1. Define the term *split phase* as it applies to center-tapped transformers.
2. Define *balanced loading* and explain its importance.
3. What is a *multiple-output transformer* and what are its applications?
4. What is the reason that some transformers have taps on their primary windings?
5. What is an *autotransformer* and how does it differ from an isolation transformer?
6. Define and compare buck and boost transformers.

21.4 FACTORS AFFECTING TRANSFORMER PERFORMANCE

There are several factors that can adversely affect the operation of transformers, especially those with high kVA (or MVA) ratings. In this section, we will briefly examine several of these factors.

Harmonics

HARMONIC A whole-number multiple of a given frequency.

A **HARMONIC** is a whole-number multiple of a given frequency. For example, a 60 Hz sine wave has harmonics of

$$60\text{ Hz} \times 2 = 120\text{ Hz}$$
$$60\text{ Hz} \times 3 = 180\text{ Hz}$$
$$60\text{ Hz} \times 4 = 240\text{ Hz}$$
$$60\text{ Hz} \times 5 = 300\text{ Hz}$$

HARMONIC SERIES A group of frequencies that are whole-number multiples of the same frequency.

and so on. A group of related frequencies like the one shown here is referred to as a **HARMONIC SERIES**.

The reference frequency in a harmonic series (60 Hz in this case) is referred to as the **FUNDAMENTAL FREQUENCY**. The fundamental frequency is the lowest frequency in any harmonic series, and is most often the only desirable signal in the group. The other frequencies are generally identified as **N-ORDER HARMONICS**, as illustrated in Table 21.2.

FUNDAMENTAL FREQUENCY The reference (lowest) frequency in a harmonic series.

N-ORDER HARMONICS The frequencies in a harmonic series.

TABLE 21.2 • Frequency Identifiers

FUNDAMENTAL FREQUENCY	MULTIPLIER	HARMONIC FREQUENCY	IDENTIFIER
60 Hz	2	120 Hz	2nd-order harmonic
60 Hz	3	180 Hz	3rd-order harmonic
60 Hz	4	240 Hz	4th-order harmonic

In any harmonic series, amplitude decreases as frequency increases. This point is illustrated in Figure 21.24a. As you can see, the fundamental frequency has the highest amplitude. The amplitude of the 3rd-order harmonic is lower than

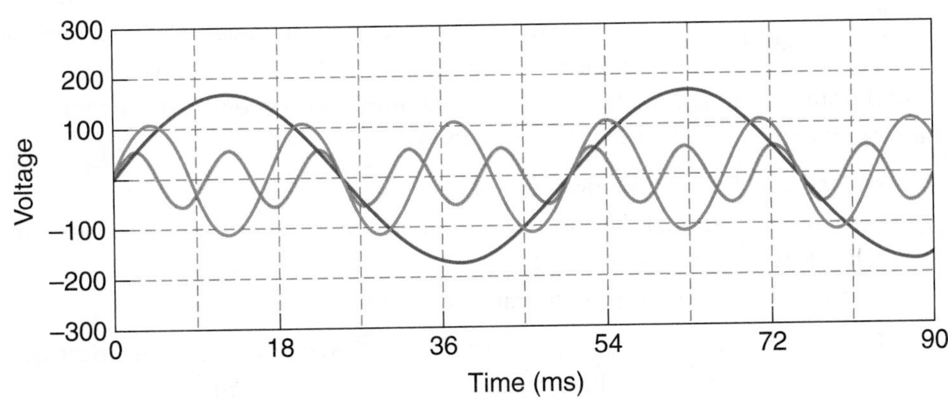

(a) A fundamental frequency with its 3rd and 5th harmonics

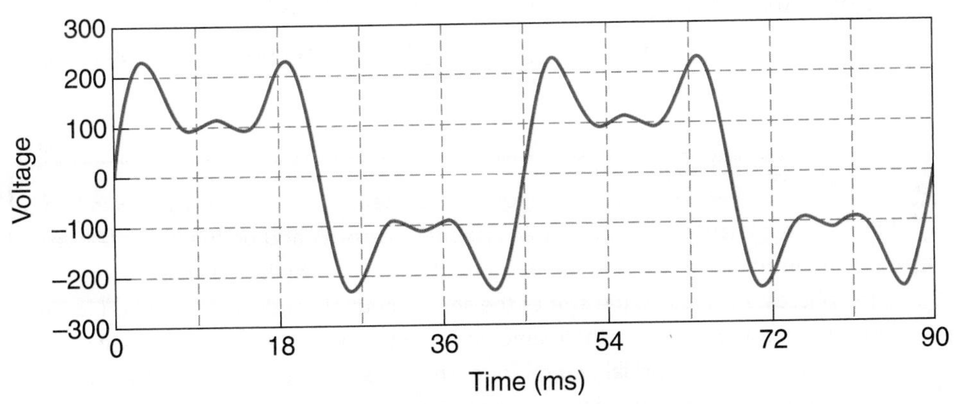

(b) Harmonic distortion (3rd and 5th harmonics)

FIGURE 21.24 Harmonics and harmonic distortion.

HARMONIC DISTORTION
An undesired change in the shape of a fundamental sine wave.

that of the fundamental frequency, and the amplitude of the 5th-order harmonic is even lower still. Even at low amplitudes, the harmonics can distort the fundamental frequency (desired signal) as shown in Figure 21.24b. This undesired change in the shape of the fundamental sine wave is called **HARMONIC DISTORTION**.

Harmonic currents in an AC power system combine with the fundamental to form distorted wave shapes. The amount of distortion is a function of the frequency and amplitude of the harmonic currents as well as the timing (phase relationships) of the respective waveforms. Algebraically adding the amplitudes of the fundamental and all harmonic currents present results in the distorted waveform.

KEY CONCEPT

Motors can be greatly impacted by *harmonic voltage distortion*. Decreased efficiency as well as overheating, vibration, and high-pitched noises in motors are all indicators of harmonic voltage distortion.

IN THE FIELD

Harmonics in electrical systems come from a number of nonlinear sources. In single-phase circuits, they can be produced by appliances such as air conditioners and dryers (when they first switch on), fax machines, copiers, medical diagnostic equipment, fluorescent light ballasts, and computer systems. (The last three of these can prove to be a challenge in hospitals and medical office buildings.) In three-phase industrial systems, harmonics are produced primarily by automation and environmental equipment. Left unchecked, harmonics can:

- Decrease the system power factor.
- Produce excessive heat in distribution transformers, shortening their lifespan.
- Produce voltage waveform distortion (as shown in Figure 21.24).
- Produce high currents in neutral conductors.

There are two general approaches to dealing with harmonics in electrical systems: The first is to deal with the symptoms rather than the harmonics themselves. Here are some of the steps that are commonly taken to minimize the negative effects of harmonics:

- Increase the size of neutral conductors to minimize the effects of harmonic-generated currents.

STAYING SAFE Neutral conductors can be negatively affected by *triplen* (3rd, 9th, 15th, etc.) harmonic currents because these currents add on the neutral—even though the nonlinear loads generating the harmonics may be balanced across all three phases. As a result, the size of the shared neutral conductor may need to be doubled to compensate for the harmonic currents and the resulting overheating of the conductor. An individual neutral for each phase conductor can also be used to compensate for the effects of these harmonics.

- Shorten the distance between the service panel and loads to reduce the voltage produced by harmonics.
- In three-phase systems:
  - Provide a separate return (neutral) for each phase.
  - Increase the size of the phase (hot) wires.

The second approach to minimizing the effects of harmonics in electrical systems is to deal with the harmonics themselves. One approach to reducing harmonics is through the use of **FERRORESONANT TRANSFORMERS**. A ferroresonant transformer is shown in Figure 21.25.

FERRORESONANT TRANSFORMER A transformer with two secondaries, one coupled to the load and the other used to "trap" harmonic frequencies.

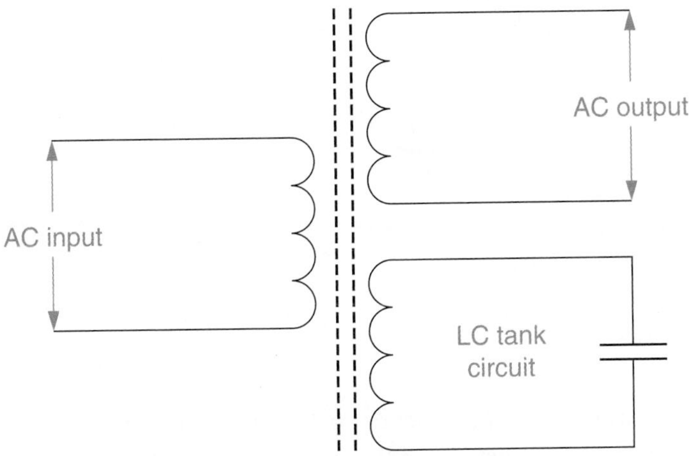

FIGURE 21.25 A ferroresonant transformer.

As shown in Figure 21.25, a ferroresonant transformer has two separate secondary coils. One couples the AC input (fundamental frequency) to the load, just like the secondary of any isolation transformer. The second coil is *tuned* using a capacitor. When connected as shown, the capacitor and its secondary winding form a parallel resonant circuit. This circuit "traps" the harmonic frequencies in the AC input, leaving a cleaner signal for the load.

Harmonics can also be reduced using a technique called **HARMONIC CANCELLATION**. A special type of transformer, called a *phase-shifting transformer*, is used to shift harmonics 180° and then combine them with harmonics from another source. Because the two harmonic waveforms are approximately 180° out of phase with each other, combining them results in a substantial reduction in their amplitude. In effect, they partially cancel each other out.

HARMONIC CANCELLATION A method of reducing the amplitude of harmonics by combining them with other harmonics that are 180° out of phase.

Transformer Hum

It is not uncommon to hear an audible hum coming from utility and other high-power distribution transformers. This humming sound is a result of a phenomenon called **MAGNETOSTRICTION**.

When the laminations in an iron-core transformer are exposed to expanding and contracting magnetic fields, they expand and contract slightly. As shown in

MAGNETOSTRICTION A phenomenon where the laminations in an iron-core transformer are exposed to expanding and contracting magnetic fields, causing them expand and contract slightly.

Figure 21.26, the expansions and contractions occur twice per AC cycle. When the core laminations expand and contract, the audible hum is produced. Because the expansion/contraction cycle occurs twice per input cycle, the hum frequency is 120 Hz.

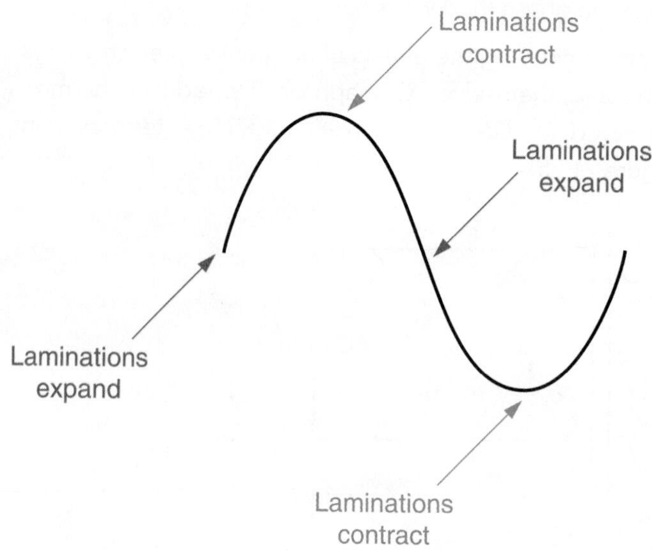

FIGURE 21.26 Lamination responses to a sine wave.

The volume of the transformer hum depends on the amplitude of the transformer AC input. The greater the amplitude of the transformer input, the greater the strength of the resulting magnetic field, and the louder the hum. Treating the hum produced by utility transformers usually amounts to minimizing its effects through mounting techniques and physical noise shields.

Core Saturation

There is a limit to the amount of flux that can be absorbed by any ferromagnetic transformer core. A high-amplitude AC input can cause the amount of flux produced by the transformer to exceed the limit of its core. When this happens, the output from the transformer is distorted.

From an engineering viewpoint, core saturation problems are avoided using transformer design techniques that make core saturation difficult to begin with. From an end-user viewpoint, core saturation problems are avoided by using transformers that can handle far more magnetic flux than any expected to be produced in the application.

Inrush Current

INRUSH CURRENT A high initial current in the primary of a transformer.

When power is first applied to a transformer, there is no induced voltage across the primary to limit the primary current. As a result, there is a current surge through the primary circuit called **INRUSH CURRENT**. Inrush current, which can easily be 10 (or more) times the normal primary current, and can be made worse by residual flux in the transformer core. Residual flux in the core can result in the core becoming saturated when power is first applied, increasing the amplitude of the inrush current and extending the length of time it takes to limit that current. The amplitude and maximum duration of any inrush current is an important consideration when selecting circuit overcurrent protection devices.

PROGRESS CHECK

1. Define the following terms:
   a. fundamental frequency
   b. harmonic and harmonic series
   c. harmonic distortion
2. List some of the common sources of harmonics.
3. What are the negative effects of harmonics on electrical systems?
4. List the two main approaches to controlling harmonics and give some examples of each.
5. What is a *ferroresonant transformer*?
6. What is *magnetostriction* and what does it cause?
7. Define the terms *core saturation* and *inrush current*.

21.5 RELATED TOPICS

In this section, we will complete the chapter by discussing some miscellaneous topics that relate to transformer operation and applications.

Iron Core Structures

Transformer cores are commonly made of steel and silicon. Adding silicon to the steel improves its magnetic properties and reduces core losses.

The two most common transformer core structures are the *core-type* and *shell-type*. These shell structures are shown in Figure 21.27. Both core structures

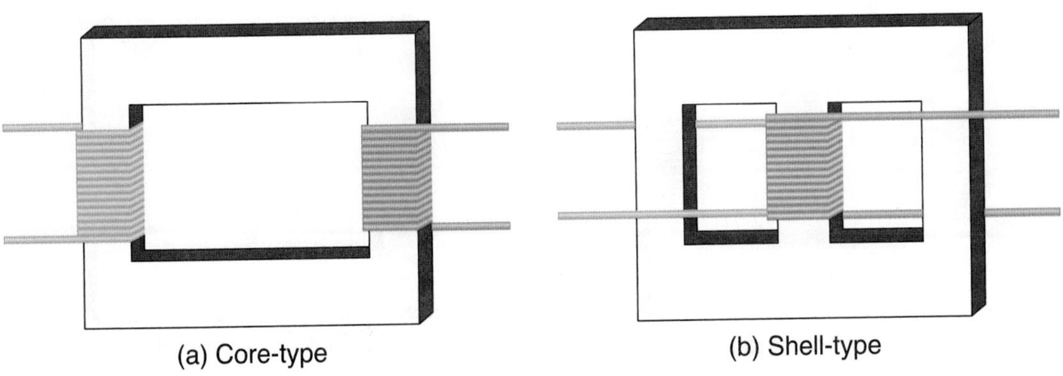

FIGURE 21.27 Iron core structures.

are normally laminated (as shown in Figure 21.11) to reduce eddy current losses. Low-voltage transformers tend to be shell-type, while high-voltage transformers are usually core-type.

The primary and secondary windings of a transformer are normally wrapped around opposite sides of a core-type structure as shown here. However, they *can* be wrapped around one end of the structure. When this is the case, the low-voltage coil is wrapped closest to the core, and the high-voltage coil is wrapped over the first. For example, a step-down transformer would be wound on the component so that the secondary is closest to the core structure and the primary covers the secondary.

There are several ways to wire a shell-type structure. In Figure 21.27b, the primary and secondary coils are wired one over the other, as described above. In Figure 21.28, the two coils are wired on the same leg of the structure, but next to each other (rather than one over the other).

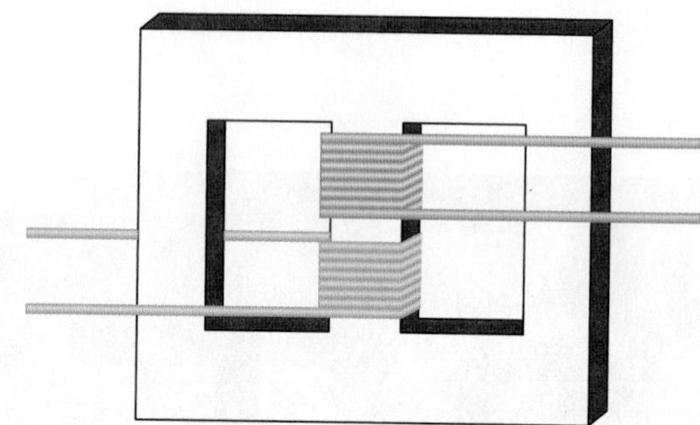

FIGURE 21.28 An alternate way to wire a shell-type transformer core.

Transformer Input/Output Phase Relationships

Earlier in the chapter, you were told that the input/output phase relationship of a transformer is indicated using polarity dots in its schematic symbol (as shown in Figure 21.9). What you weren't told at the time is that the phase relationship is determined by the relative directions of the primary and secondary windings, as shown in Figure 21.29.

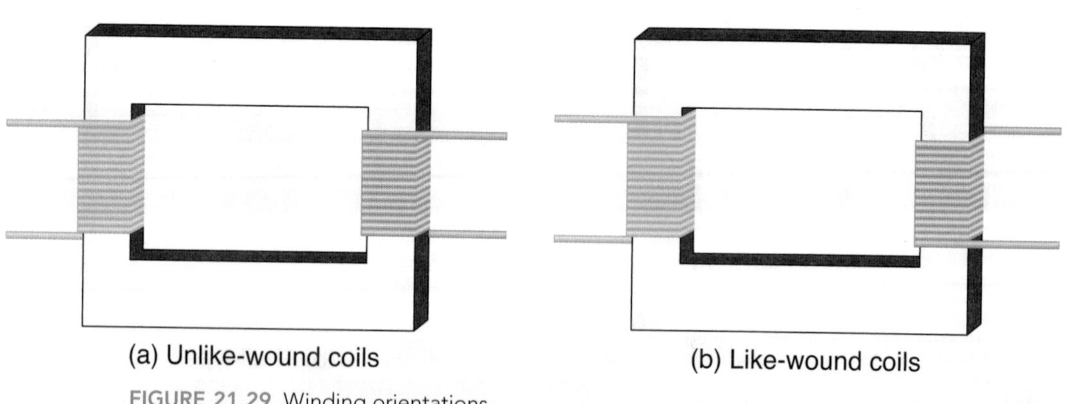

FIGURE 21.29 Winding orientations.

In Figure 21.29a, the coils are wound in opposite directions around the core. As a result, the primary and secondary currents are in opposite directions, and the induced voltages are 180° out of phase. In Figure 21.29b, the coils are wound in the same direction (CCW). As a result, the primary and secondary currents are in the same direction, and the induced voltages are in phase.

Transformer Percent Impedance (%Z)

There is a rating that is unique to transformers, called **PERCENT IMPEDANCE (%Z)**. This rating indicates the transformer primary voltage (as a percentage of its rated value) that generates maximum current through a shorted secondary. The percent impedance is measured as shown in Figure 21.30. With the secondary shorted, the source voltage is increased until the secondary current reaches its maximum value. The primary voltage is then measured, and the %Z is calculated using

PERCENT IMPEDANCE (%Z) The transformer primary voltage (as a percentage of its rated value) that generates maximum current through a shorted secondary.

(21.10)
$$\%Z = \frac{E_{P(measured)}}{E_{P(rated)}} \times 100$$

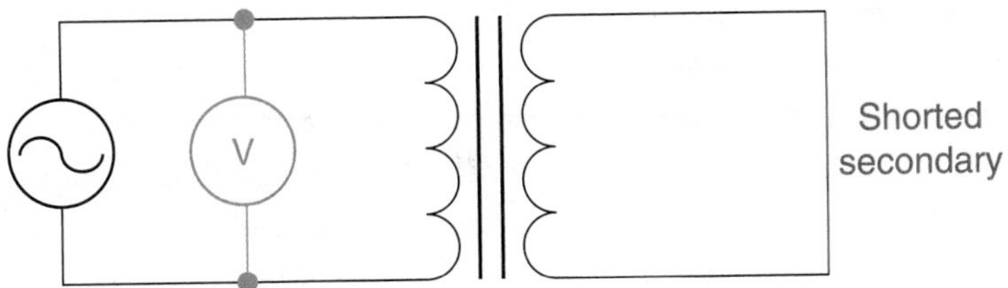

FIGURE 21.30 Measuring the percent impedance of a transformer.

The %Z of the transformer winding relates to its **VOLTAGE REGULATION**, that is, its ability to maintain a stable output voltage when the load demand changes. Transformer voltage regulation equals the difference between the *no-load voltage* (E_{NL}) and the *full-load voltage* (E_{FL}), usually expressed as a percentage. If a transformer delivers 200 volts at no load and the voltage drops to 190 volts at full load, its voltage regulation is

VOLTAGE REGULATION The ability of a transformer to maintain a stable output (secondary) voltage when the load demand changes.

$$\text{Regulation} = \frac{E_{NL} - E_{FL}}{E_{NL}} \times 100$$
$$= \frac{200\text{ V} - 190\text{ V}}{200\text{ V}} \times 100$$
$$= 5\%$$

The lower the %Z of the transformer, the better its voltage regulation characteristics, that is, the lower the change in output voltage when load demand increases.

Transformer impedance is also used to determine the required rating of a primary circuit breaker or fuse. For example, suppose a 5 kVA transformer with a 5% impedance rating has a 240 V line input. The short-circuit primary current is found as

$$
\begin{aligned}
I_{SC} &= \frac{VA}{E \times \%Z} \\
&= \frac{5\text{ kVA}}{240\text{ V} \times 5\%} \\
&= \frac{5\text{ kVA}}{240\text{ V} \times 0.05} \\
&= 417\text{A}
\end{aligned}
$$

Thus, the breaker or fuse in the primary must have a minimum interrupt rating of 417 A at 240 V. Note that the interrupt rating, not the blow rating, defines the maximum current that a fuse or breaker can interrupt. A fuse or breaker may have a blow rating in the tens of amperes, but an interrupt rating in the hundreds or even thousands of amperes.

Transformer Derating

As is the case with many devices, transformer ratings are valid only when operated under certain conditions. If the ambient temperature increases, the nameplate rating of the transformer may no longer be valid. The generally accepted standard is 40°C maximum (30°C average). Above this temperature the transformer must be derated. A representative transformer derating chart is shown below. Note that the chart below shows a derating factor of 0.8% per °C above 40°C.

STAYING SAFE Delta-wye transformers can be negatively impacted by the presence of triplen (3rd, 9th, 15th, etc.) harmonic currents because these currents pass through the secondary winding are trapped in the primary winding, possibly overheating the transformer. One method of coping with this situation is to derate the transformer; that is, reduce the nameplate kVA using calculations made from measurements of phase and secondary peak current values.

MAXIMUM AMBIENT TEMPERATURE	MAXIMUM PERCENTAGE OF LOADING
40°C (104°F)	100%
50°C (122°F)	92%
60°C (140°F)	84%

Current Transformers

A **CURRENT TRANSFORMER** is a component designed to produce a secondary current that is proportional to its primary current. A current transformer is shown in Figure 21.31.

CURRENT TRANSFORMER A component designed to produce a secondary current that is proportional to its primary current.

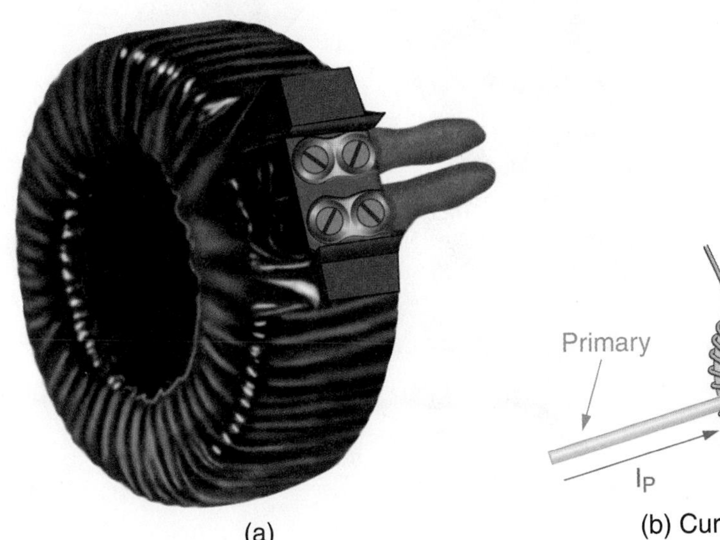

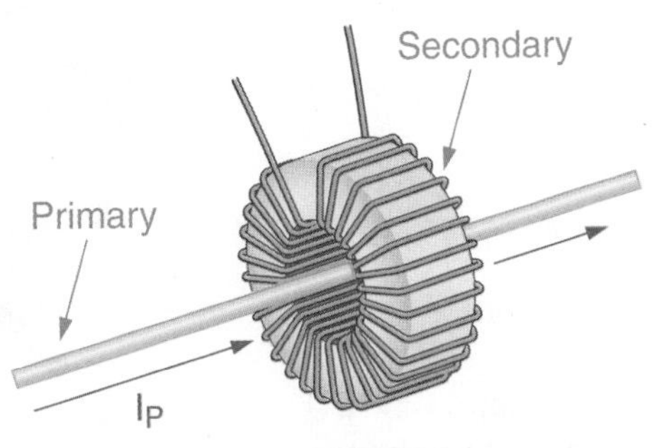

(a)

FIGURE 21.31 A current transformer.

As you can see, the current transformer resembles a toroid. However, the component shown in Figure 21.31a is only the transformer secondary. The transformer is complete when the secondary surrounds a current-carrying conductor as shown in Figure 21.31b. Here's how the component works:

- The conductor serves as the primary of the transformer.
- Current passing through the conductor (primary) generates a magnetic field.
- The magnetic field cuts through the secondary winding, generating a secondary current. The current generated is proportional to the current through the conductor.

The fact that the current transformer provides a secondary current that is proportional to the current through the primary makes the component useful as a current-sensing device. Note that current transformers are normally rated using the ratio of primary to secondary current. For example, a 500:1 current transformer generates 1 A of secondary current for every 500 A passing through the primary. Thus, current transformers make it possible to measure extremely high current values without having to interrupt the circuits under test.

A **SPLIT-CORE CURRENT TRANSFORMER** is one that can be split in two, placed around the conductor under test, and then closed. A split-core current transformer is shown in Figure 21.32.

SPLIT-CORE CURRENT TRANSFORMER One that can be split in two, placed around a conductor under test, and then closed.

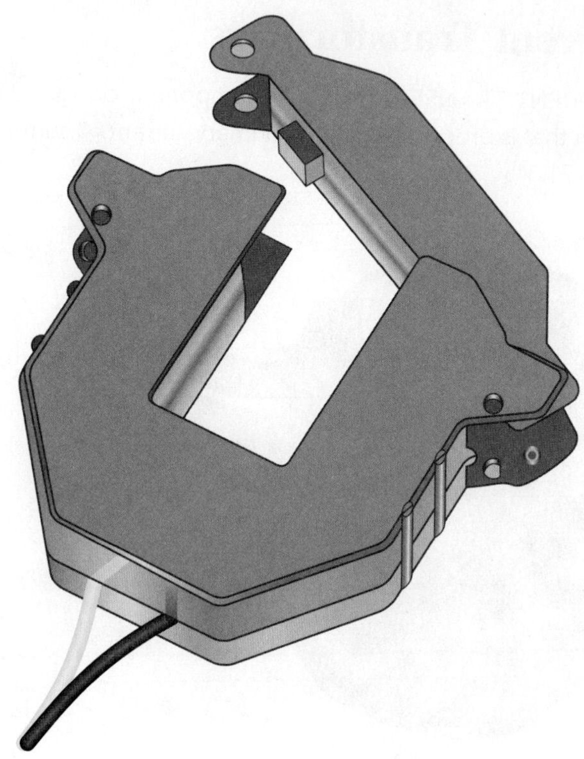

FIGURE 21.32 A split-core current transformer.

NEC® Transformer Nameplate Requirements

According to *Section 450.11* of the *2008 NEC*®, each transformer must have a nameplate indicating the following:

- Name of the transformer manufacturer
- Transformer KVA rating
- Frequency
- Primary and secondary voltages
- Transformer impedances (25 kVA and higher only)
- Required clearances for transformers with ventilation openings
- Amount and kind of insulating liquid (liquid-type transformers)
- Insulation temperature class (dry-type transformers)

PROGRESS CHECK

1. List the two main transformer core types.
2. Explain why primary and secondary windings are wound one on top of the other.
3. What determines which transformer winding is wound closest to the core?
4. What determines the input/output phase polarity of a transformer?
5. Explain how transformer impedance affects
   a. transformer voltage regulation
   b. the rating of the primary fuse or breaker
6. Define transformer derating and explain its importance.
7. List the information that must appear on a transformer nameplate.

21.6 SUMMARY

Here is a summary of the major points made in this chapter:

Transformer Operation: An Overview

- A transformer is a component that uses electromagnetic induction to couple an AC signal from one circuit to another.
  - Transformers provide coupling of AC signals.
  - Transformers provide isolation from DC voltages.
- The input coil of a transformer is called the *primary*, and the output is called the *secondary*.
- Transformer coils are wrapped around a core. For power transformers the core is usually an iron core.
- The three basic transformer types are step-up, step-down, and 1-to-1.
  - A step-up transformer produces a secondary voltage that is higher than the primary voltage.
  - A step-down transformer produces a secondary voltage that is lower than the primary voltage.
  - A 1-to-1 transformer produces a secondary voltage that is equal to the primary voltage.
- Transformer action is based on the principle of electromagnetic induction.
  - The changing current in the primary causes a changing magnetic field to be produced.
  - The changing magnetic field in the primary cuts the secondary coil, inducing a voltage. This couples the AC signal from the primary to the secondary.
- Each loop of wire in a coil is called a *turn*.
- The turns ratio is the ratio of primary turns to secondary turns.
  - The turns ratio determines the ratio of primary to secondary voltage defined by $\frac{N_P}{N_S} = \frac{E_P}{E_S}$.

Transformer Voltage, Current, and Power

- Transformer secondary voltage can be determined using the turns ratio and primary voltage as $E_S = E_P \times \frac{N_S}{N_P}$.
- The turns ratio can be used to determine if a transformer is a step-up or step-down device.
  - If $N_P > N_S$, the transformer is a step-down.
  - If $N_S > N_P$, the transformer is a step-up.
  - If $N_P = N_S$, the transformer is a 1-to-1, and input and output voltages are equal.
- Polarity dots are used in schematics to identify the relative phase between the primary and secondary voltages.
  - If no polarity dots are shown, the input and output are assumed to be in phase.
- For an ideal transformer, $P_P = P_S$ and efficiency is 100%.
- Transformer efficiency is the ratio of input power to output power found as $\eta = \frac{P_S}{P_P} \times 100$.
- A practical transformer's efficiency is less than 100% due to a variety of losses.
  - Copper loss is the I^2R loss due to the resistance of the wire in the primary and secondary coils.
  - Eddy current loss is the result of flux-generated currents in the core. Laminated cores are used to limit eddy current losses.

 - Hysteresis loss is a result of the energy expended to overcome the retentivity of the core material.
 - Stray loss is the result of a less-than-perfect coupling coefficient between the primary and secondary windings ($k < 1$).
 - Mechanical losses are caused by vibrations in the structure of the transformer and/or its housing.
- The relationship between transformer input power and output power provides a basis for the relationship between transformer input current and output current, defined as $I_S = I_P \frac{E_P}{E_S}$.
 - A transformer that steps up voltage steps down current.
 - A transformer that steps down voltage steps up current.
- Current varies inversely with the turns ratio, by formula, $\frac{I_S}{I_P} = \frac{N_P}{N_S}$.
 - This allows us to solve for secondary current as $I_S = I_P \frac{N_P}{N_S}$.

Transformer Winding Configurations

- A center-tapped transformer is one that has an additional lead connected to the center of its secondary winding.
 - This type of transformer is commonly used to provide 240/120 V split-phase residential service.
 - Measured with respect to the center tap, the two end-terminal voltages are 180° out of phase.
- Some transformers have multiple secondary windings rather than taps to provide more than one secondary voltage.
- Some transformers have multiple primary windings or taps.
 - Multiple primary windings are provided so that the transformer produces the rated secondary voltage for differing supply voltages. Devices designed to be used with 240 V European voltages and 120 V North American voltages are one example.
 - Taps on the primary winding are used to compensate for variations in line voltage. Tap increments are usually 2.5% or 5%, both above and below the rated line voltage.
- An autotransformer is comprised of a single coil with three terminals. It is basically a tapped inductor.
 - Autotransformers can be either step-up or step-down, depending on the turns ratio between the three terminals.
 - A variac is a variable autotransformer.
- A buck transformer is an autotransformer used to produce a secondary voltage slightly lower than the primary.
- A boost transformer is an autotransformer used to produce a secondary voltage slightly higher than the primary,

Factors Affecting Transformer Performance

- Harmonics are whole-number multiples of a given frequency. For example the second harmonic of 60 Hz is 120 Hz, the third is 180 Hz, and so on.
 - The group of harmonic frequencies is referred to as a *harmonic series*.

    - The reference frequency is referred to as the *fundamental frequency.*
    - The effect that harmonics have on the fundamental voltage waveform is referred to as *harmonic distortion.*
    - Harmonics decrease power factor, produce excessive heat, shorten the lifespan of equipment, and can cause high neutral conductor current.
- There are two general approaches to dealing with harmonics.
    - One approach is to deal with the symptoms by increasing neutral conductor size and/or shortening the distance between the service panel and the harmonic source.
    - A second approach is to minimize the harmonics themselves. One method is to use a ferrosresonant transformer. This transformer has an extra secondary coil with a capacitor that is tuned to act as a parallel resonant circuit at the harmonic frequency.
- Transformer hum is the result of the changing magnetic field causing the transformer core and/or housing to vibrate.
    - Since expansion/contraction occurs twice per cycle, hum has a frequency of 120 Hz.
    - Transformer hum is referred to as magnetostriction.
- There is a limit to how much magnetic flux a transformer core can absorb. If it becomes saturated, the output voltage from the transformer is distorted.
- Inrush current is the current that occurs when power is first applied to the transformer.
    - Inrush current can be 10 times (or higher) than the normal primary current.
    - Residual core flux can increase the amount of inrush current.

Related Topics

- Transformer cores are usually made of steel and silicon. The silicon helps to reduce core losses.
- The two most common core structures are the shell type and the core type.
    - In a shell-type structure the windings are on a single central post with the core surrounding the windings.
    - In a core-type structure the windings are on two outer posts, so the windings surround the core.
- Primary and secondary windings are wound one on top of the other to minimize flux leakage.
    - The low-voltage (fewest turns) coil is wound closest to the core and the high-voltage coil is wound over it.
- The input/output phase relationship of a transformer is determined by the relative directions of the primary and secondary windings.
    - If they are wound in the same direction, the input and output are in phase.
    - If they are wound in opposite directions, they are 180° out of phase.
- Transformer percent impedance (%Z) is important for two reasons:
    - It determines the voltage regulation of the transformer. Voltage regulation is the difference between full-load and no-load output voltage, usually expressed as a percentage.
    - Transformer percent impedance is also used to determine the sizing for the primary fuse or circuit breaker.
- Transformers must be derated if operated in environments with temperature above 40°C peak, or 30°C average. Derating means that the transformer must be operated at some kVA rating below its nameplate rating, determined by the ambient temperature.

- Current transformers produce a secondary current that is proportional to its primary current.
  - In a current transformer the primary is the current-carrying conductor and the secondary is a toroid-shaped coil.
  - Current transformers are usually used to measure very high currents and are rated based on the ratio of primary to secondary current.
  - A split-core current transformer is one that can be split in two, placed around the conductor, and then closed.
- *Section 450.11* of the *2008 NEC*® lists the specific information that must appear on a transformer nameplate (see page 936).

CHAPTER REVIEW

1. A transformer provides DC ________ and AC ________.
   a) coupling, isolation
   b) isolation, coupling
   c) gain, isolation
   d) isolation, gain

2. The output coil of a transformer is called the ________ and the input coil is called the ________.
   a) primary, secondary
   b) core, secondary
   c) gate, primary
   d) secondary, primary

3. Most power transformers are ________.
   a) step-up
   b) iron-core
   c) step-down
   d) phase-shifting

4. If the input and output voltages of a transformer are the same, it is a(n) ________ transformer.
   a) coupling
   b) unity
   c) 1-to-1
   d) auto

5. Transformer action is the result of ________.
   a) electromagnetic induction
   b) the turns ratio
   c) the core type
   d) all of the above

6. The turns ratio is the ratio of ________.
   a) primary to secondary turns
   b) secondary to primary turns
   c) output to input turns
   d) it depends on whether it is a step-up or step-down

7. A transformer has a line voltage of 208 V and a secondary voltage of 52 V. If the secondary has 400 turns, the primary has ________ turns.
   a) 100
   b) 1600
   c) 160
   d) 400

8. A transformer has a turns ratio of 1:16. If the primary voltage is 260 V, the secondary voltage is ________.
   a) 16.25 V
   b) 4.16 kV
   c) 162.5 V
   d) 416 V

9. A schematic often places a dot on the primary and secondary coils of a transformer to specify ________.
   a) a step-down transformer
   b) input and output terminals
   c) a step-up transformer
   d) relative phase polarity

10. In an ideal transformer, output power ________ input power.
    a) equals
    b) is directly proportional to
    c) is inversely proportional to
    d) is in phase with

11. The ratio of ________ defines transformer ________.
    a) input power to output power, effectiveness
    b) input voltage to output voltage, efficiency
    c) output power to input power, efficiency
    d) input power to output power, efficiency

12. I^2R loss is another name for ________.
    a) copper loss
    b) line loss
    c) inductor loss
    d) induction loss

13. Eddy current loss is minimized by ________.
    a) tuned primary windings
    b) center taps
    c) opposite phase secondaries
    d) laminated cores

14. Hysteresis loss is the result of overcoming ________.
    a) reverse polarity
    b) wire inductance
    c) core retentivity
    d) core inductance

15. A transformer has primary to secondary voltage values of 12 kV to 800 V. If the secondary current is 10.5 A, the primary current is ________.
    a) 700 mA
    b) 15.75 A
    c) 158 A
    d) 7 A

16. The voltages between the two secondary terminals of a transformer and ground are found to be 180° out of phase. Chances are this is a(n) ________ transformer.
    a) step-up
    b) center-tapped
    c) inverting
    d) step-down

17. Multiple-output transformers are used ________.
    a) in low-current applications
    b) when several line voltages may be available
    c) in high-current applications
    d) to provide several secondary voltages from one primary voltage

18. Taps on the primary winding of a transformer are used to ________.
    a) compensate for line voltage fluctuations
    b) provide multiple output voltages
    c) allow the connection of more than one input voltage
    d) all of the above

19. An autotransformer ________.
    a) is basically a single tapped inductor
    b) usually has three terminals
    c) can be either a step-up or a step-down
    d) all of the above

20. A variac is ________.
    a) a variable autotransformer
    b) a variable isolation transformer
    c) a variable current transformer
    d) all of the above, depending on how it is configured

21. Buck and boost transformers are used to compensate for ________ variations in line voltage.
    a) large
    b) small
    c) phase
    d) harmonic

22. The reference frequency in a harmonic series is called the ________.
    a) resonant frequency
    b) fundamental frequency
    c) center frequency
    d) unity harmonic

23. As the harmonic series rises ________.
    a) amplitude decreases
    b) amplitude increases
    c) frequency decreases
    d) the fundamental shifts

24. Ferroresonant transformers use ________ to help control harmonics.
    a) heavier neutrals
    b) phase shifting
    c) special core materials
    d) resonance

25. In North America you can assume that transformer hum has a frequency of ________ Hz.
    a) 60
    b) 120
    c) 30
    d) 180

26. Transformer inrush current occurs ________.
    a) as a result of core saturation
    b) when a load is connected
    c) when power is first applied
    d) when a load is disconnected

27. The two most common types of cores are ________.
    a) shell and core
    b) Y and T
    c) H and X
    d) iron and ferrite

28. The ________ winding is usually wrapped closest to the core.
    a) primary
    b) low-voltage
    c) secondary
    d) high-voltage

29. Primary and secondary phase polarity is a function of ________.
    a) which phase is closest to the core
    b) the direction in which each coil is wound
    c) harmonic content
    d) secondary taps

30. The primary of a current transformer ________.
    a) is out of phase with the secondary
    b) always has more turns
    c) is split
    d) is the current-carrying conductor

PRACTICE PROBLEMS

1. A transformer has a 15:1 turns ratio and a line voltage of 240 V. Solve for the secondary voltage.
2. A transformer has a 1:16 turns ratio and a line voltage of 120 V. Solve for the secondary voltage.
3. A transformer has a 2.5:1 turns ratio and a line voltage of 550 V. Solve for the secondary voltage.
4. A transformer has a 12:1 turns ratio and a secondary voltage of 12.6 V. Solve for the line voltage.
5. A transformer has a 1:3.6 turns ratio and a secondary voltage of 153 V. Solve for the line voltage.
6. Determine the secondary current (I_S) for the transformer shown in Figure 21.33a.
7. Determine the secondary current (I_S) for the transformer shown in Figure 21.33b.
8. Determine the secondary current (I_S) for the transformer shown in Figure 21.34a.
9. Determine the secondary current (I_S) for the transformer shown in Figure 21.34b.
10. A transformer has a line voltage of 208 V. It delivers 400 W to a 6 Ω resistive load. Solve for the secondary voltage.
11. A transformer has a line voltage of 115 V. It delivers 1.65 kW to a 5.5 Ω resistive load. Solve for the secondary voltage.

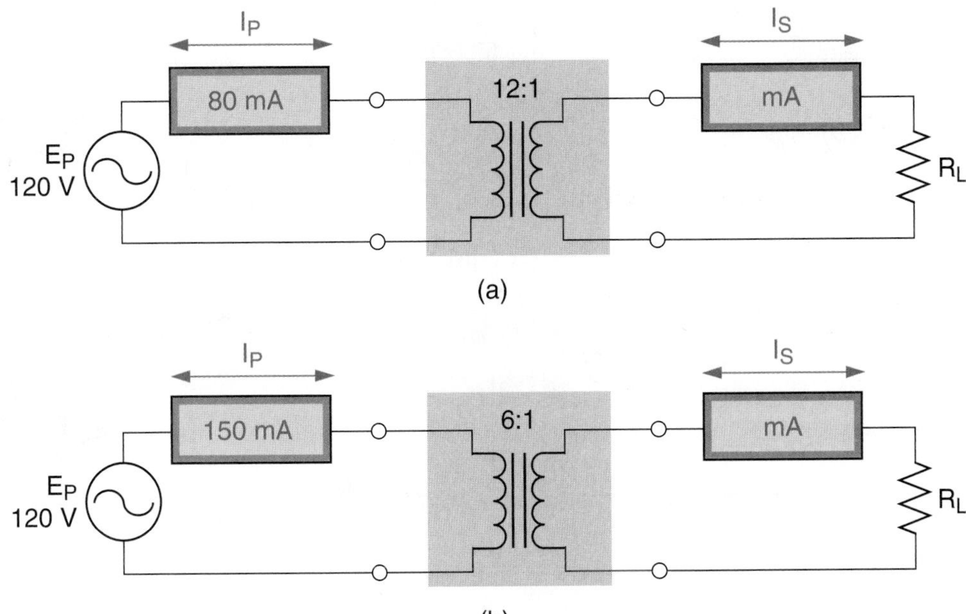

FIGURE 21.33

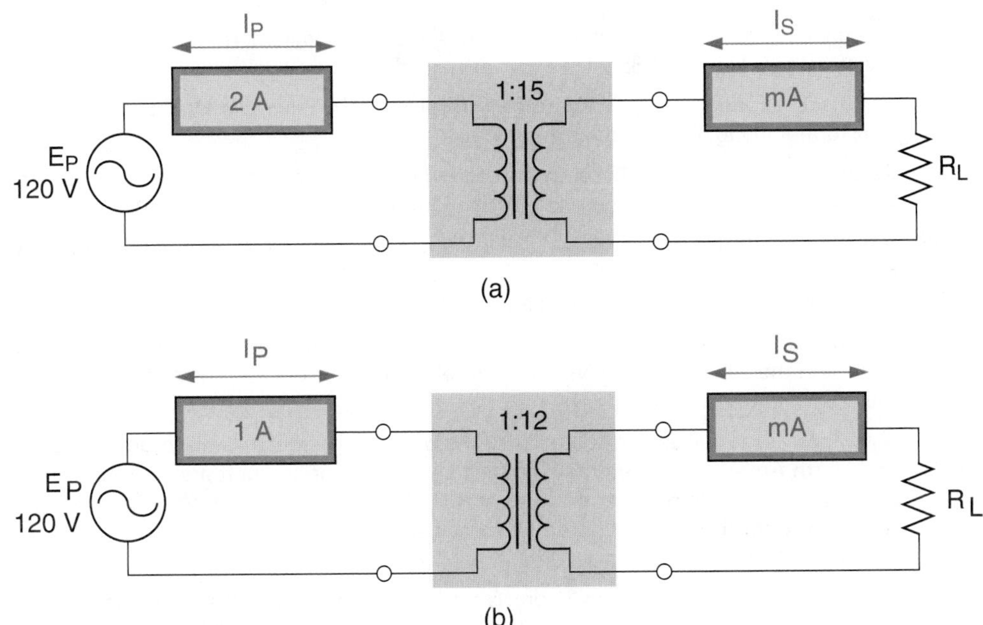

FIGURE 21.34

12. Solve for the turns ratio, primary current, and primary power for the circuit described in question 10.
13. Solve for the turns ratio, primary current, and primary power for the circuit described in question 11.
14. The transformer shown in Figure 21.35a has the line voltage, turns ratio, and resistive load value shown. Solve for the secondary voltage and current values.

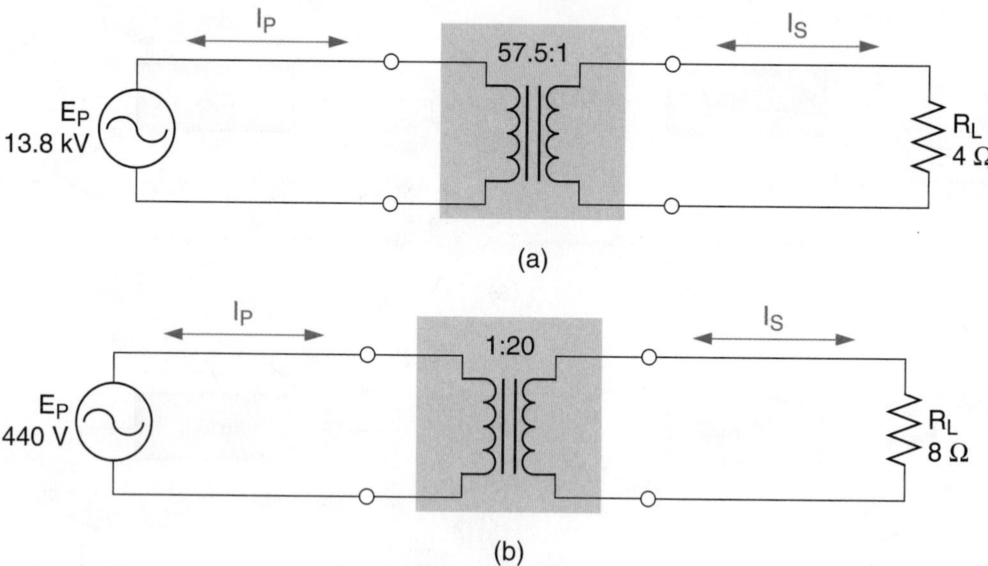

FIGURE 21.35

15. The transformer shown in Figure 21.35b has the line voltage, turns ratio, and resistive load value shown. Solve for the secondary voltage and current values.
16. Use the line voltage and primary current to solve for the minimum kVA rating for the transformer in Figure 21.35a.
17. Use the line voltage and primary current to solve for the minimum kVA rating for the transformer in Figure 21.35b.
18. Resolve problems 16 and 17 using the secondary voltage and current and compare to your results using line voltage and primary current.
19. A transformer has two secondary windings. It is to be installed on a 480 V line to supply both 120 V and 240 V loads. The primary winding has 800 turns. Solve for the number of turns in each secondary winding and the turns ratio for each.
20. The transformer described in question 19 must supply a maximum of 20 A each to the 120 V and the 240 V loads. Solve for the total primary current and the minimum kVA rating for the transformer.
21. A company that makes control units for the packaging industry sells in both Europe and North America. The internal voltages needed to run the unit are 48 V, 12 V, and 5 V. Assume that the number on turns on the North American (120 V) primary is 600 turns. Solve for the number of turns for each secondary and the European (240 V) primary.
22. The unit described in question 21 draws a maximum of 8.6 A on the 48 V load, 10.2 A on the 12 V load, and 5.2 A on the 5 V load. Solve for the maximum primary current for both European and North American installations.
23. A transformer delivers 240 V at no load and 222 V at full load. Solve for the impedance of this transformer expressed as a percent.
24. The transformer described in question 23 has a nameplate kVA rating of 8 kVA, and a line voltage of 480 V. Solve for the full-load current and short-circuit current.
25. Assume that the transformer described in question 24 has a derating factor of 0.82%/°C above 40°C. If it is to be used in an environment where maximum temperatures can reach 85°C, solve for the maximum full-load current that this transformer can safely supply.

ANSWERS TO THE EXAMPLE PRACTICE PROBLEMS

21.1 40 V
21.2 2.88 kV
21.3 3 A
21.4 1 A
21.5 80 A

CHAPTER 22

THREE-PHASE TRANSFORMERS

In the last two chapters, you were introduced to three-phase power and the basic operating principles of transformers. Now we will turn our attention to three-phase transformer circuits. Like single-phase transformers, three-phase transformers are most commonly used to change (step-up or step-down) values of voltage and/(or current. Many of the principles in this chapter will seem familiar, as they are simply new applications of principles that were covered earlier. Others are entirely new, most notably the use of groups (or *banks*) of single-phase transformers to operate as three-phase circuits.

KEY TERMS

The following terms are introduced and defined in this chapter on the pages indicated:

Capacity, 968
Distribution transformer, 965
Dry-type transformer, 965
Open-delta connection, 958
Pad-mounted transformer, 965
Pole-type transformer, 965
Submersible transformer, 967
Vault-type transformer, 967
Wet-type transformer, 967

OBJECTIVES

After completing this chapter, you should be able to:

1. Describe the construction of a shell-type three-phase transformer.
2. Identify a Y-Y connected transformer.
3. Explain how a three-phase transformer can be constructed using three single-phase transformers.
4. Identify a Δ-Δ connected transformer.
5. Identify Y-Δ connected and Δ-Y connected transformers.
6. List the advantages of configuring single-phase transformer banks as three-phase transformers.
7. Describe the effects of an open phase on the operation of Y-connected and Δ-connected transformers.
8. Calculate the values of apparent and resistive (true) power for a balanced three-phase transformer load.
9. Describe and contrast dry-type and wet-type transformers.
10. Identify the typical installation locations of vault-type and submersible transformers
11. Determine the configuration, line voltage, and phase voltage for a distribution transformer, given its rating.

22.1 THREE-PHASE TRANSFORMER CONFIGURATIONS

You were first introduced to delta (Δ) and wye (Y) connections in Chapter 20. In this section, we'll review the operating characteristics of basic delta, wye, delta-wye, and wye-delta circuits.

The construction of a three-phase transformer can be represented as shown in Figure 22.1. The shell-type core has three sets of primary and secondary windings. How these windings are connected together determines the configuration of the transformer (delta, wye, etc.).

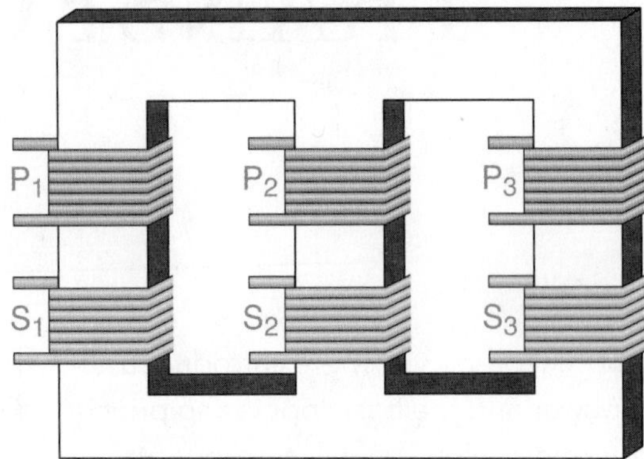

FIGURE 22.1 Three-phase transformer construction.

Y-Y Connected Transformer

The transformer in Figure 22.1 can be represented as shown in Figure 22.2. T_1, T_2, and T_3 represent the three primary/secondary coil pairs shown on the shell-type core. The lines labeled ϕA_1, ϕB_1, and ϕC_1 represent primary line conductors that

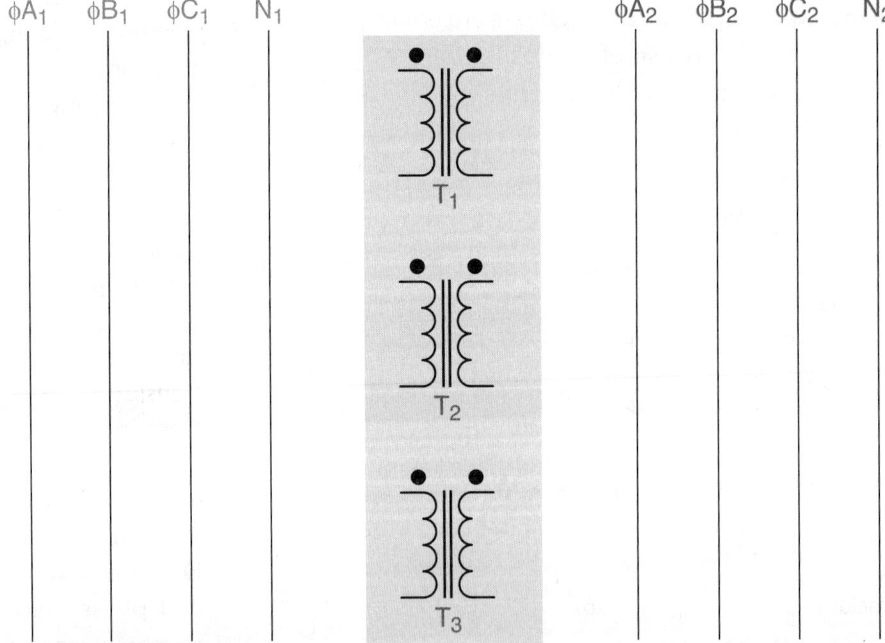

FIGURE 22.2 Elements of a transformer wiring diagram.

connect to the primary coils, and the line labeled N_1 represents a neutral conductor. Likewise, the lines labeled ϕA_2, ϕB_2, and ϕC_2 represent secondary line conductors and N_2 represents a neutral conductor.

When wired as shown in Figure 22.3, the shell-type transformer forms a Y-Y (wye primary–wye secondary) circuit. As such, the transformer primary and secondary current and voltage relationships are as follows:

$$E_L = \sqrt{3} \times E_P = 1.732 \times E_P \qquad I_L = I_P$$

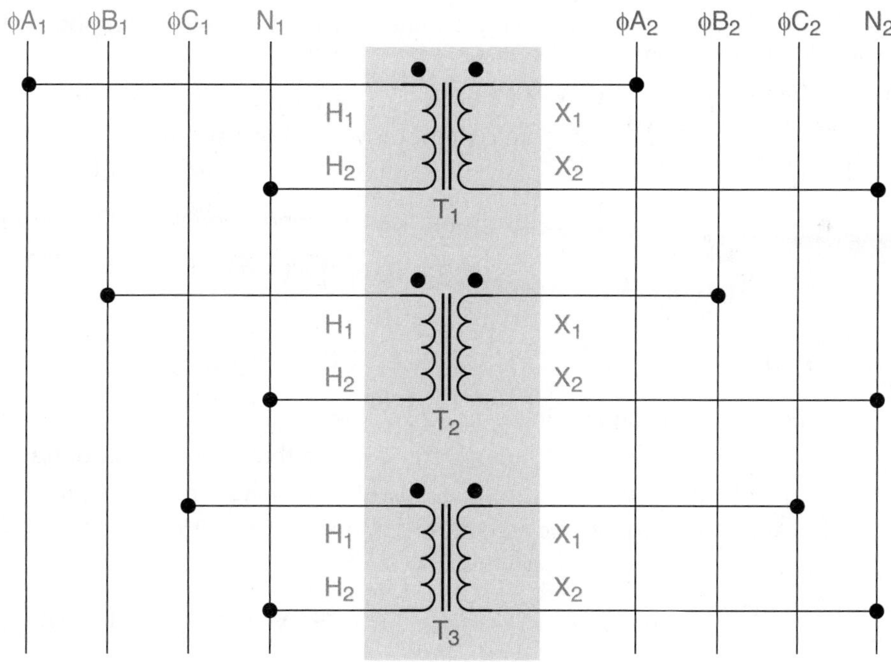

(a) Y–Y transformer wiring

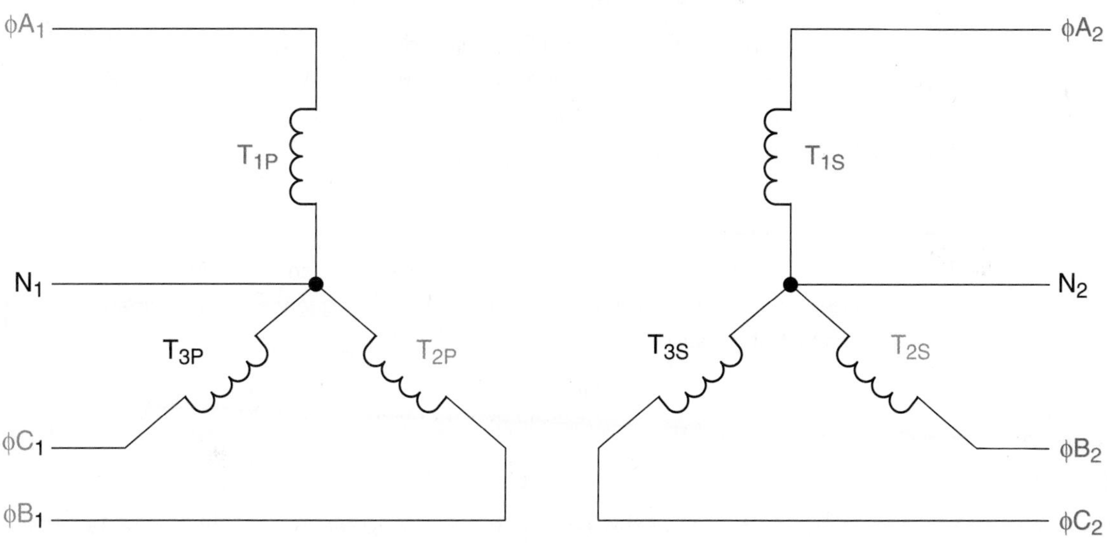

(b) Schematic diagram

FIGURE 22.3 Y-Y transformer wiring.

$$E_P = \frac{E_L}{\sqrt{3}} = \frac{E_L}{1.732} \qquad I_N = I_A + I_B + I_C = 0\text{ A}$$

where E_L and I_L are *line* values, and E_P and I_P are *phase* values. These relationships, which were introduced in Chapter 20, assume that the Y-Y circuit is balanced. (Before reading on, take a moment to trace out the circuit connections in Figure 22.3 to verify that the diagrams represent the same circuit.)

There are two points to be made:

- The wiring diagram in Figure 22.3a can be implemented using a bank (group) of three single-phase (1ϕ) transformers.
- Y-Y transformers are used in industrial applications, and are preferred over Δ-Δ transformers when it is critical to have a neutral connector in the secondary circuit.

The advantages of wiring a bank of 1ϕ transformers to form a 3ϕ transformer are explained later in this section.

STAYING SAFE To avoid excessive transformer heating, a wye-wye connected three-phase transformer must have both input and output neutrals grounded and a phase-to-neutral load imbalance that is less than 10%.

Δ-Δ Connected Transformer

When wired as shown in Figure 22.4, the shell-type transformer forms a Δ-Δ (delta primary–delta secondary) circuit. Note that there is no neutral line in the wiring diagram, which is consistent with the delta circuits in Chapter 20. The transformer primary and secondary current and voltage relationships are as follows:

$$E_L = E_P \qquad I_L = \sqrt{3} \times I_P = 1.732 \times I_P$$

$$I_P = \frac{I_L}{\sqrt{3}} = \frac{I_L}{1.732} \qquad I_A + I_B + I_C = 0\text{ A}$$

where E_L and I_L are *line* values, and E_P and I_P are *phase* values. These relationships, which were introduced in Chapter 20, assume that the Δ-Δ circuit is balanced. (Before reading on, take a moment to trace out the circuit connections in Figure 22.4 to verify that the diagrams represent the same circuit.)

The angular displacement (difference) between primary and secondary is zero in a delta-delta connected transformer. Also, the phase sequence of the primary is the same as that of the secondary.

IN THE FIELD

As was the case with the Y-Y circuit, the wiring diagram in Figure 22.4a can be implemented using a bank of single-phase transformers. Note that Δ-Δ transformers are most commonly found in industrial applications.

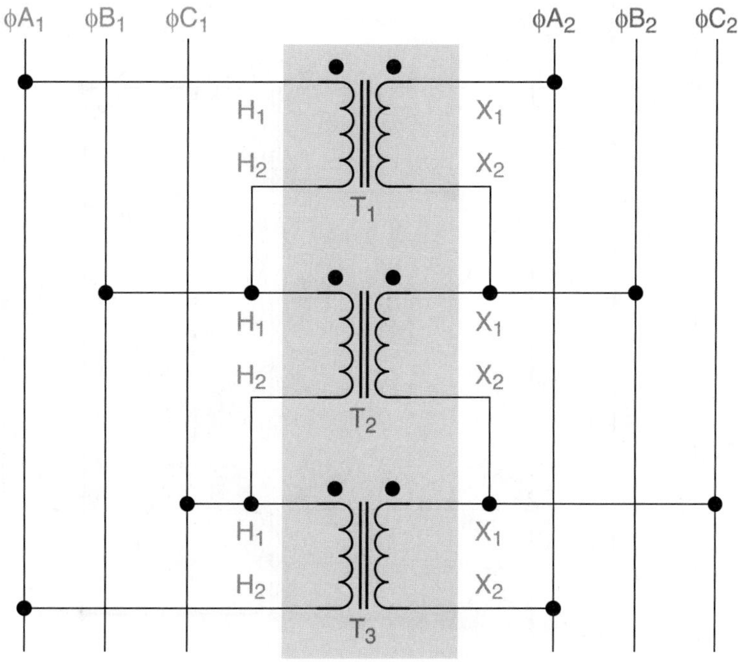

(a) Δ–Δ transformer wiring

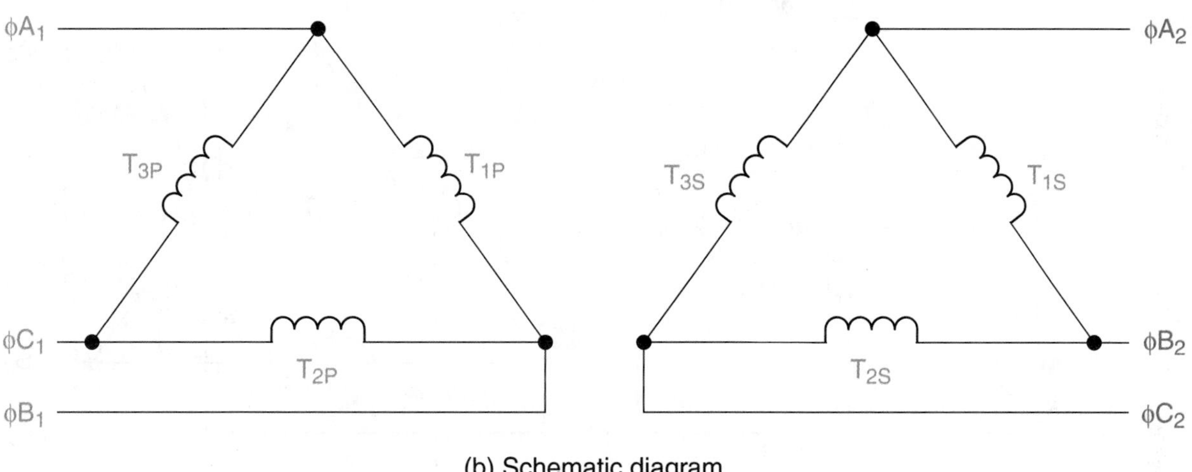

(b) Schematic diagram

FIGURE 22.4 Δ-Δ transformer diagrams.

Y-Δ Connected Transformer

When wired as shown in Figure 22.5, the shell-type transformer forms a Y-Δ (wye primary–delta secondary) circuit. Note that there is a neutral connection in the primary circuit, but none in the secondary circuit. This is consistent with the Y and Δ connections we've seen up to this point. (Before reading on, take a moment to trace out the circuit connections in Figure 22.5 to verify that the diagrams represent the same circuit.)

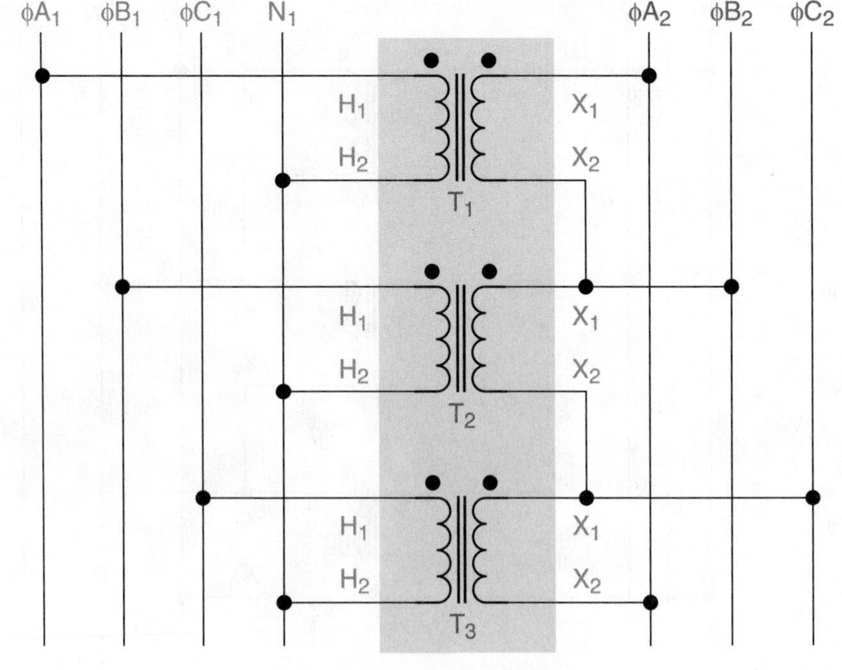

(a) Y–Δ transformer wiring

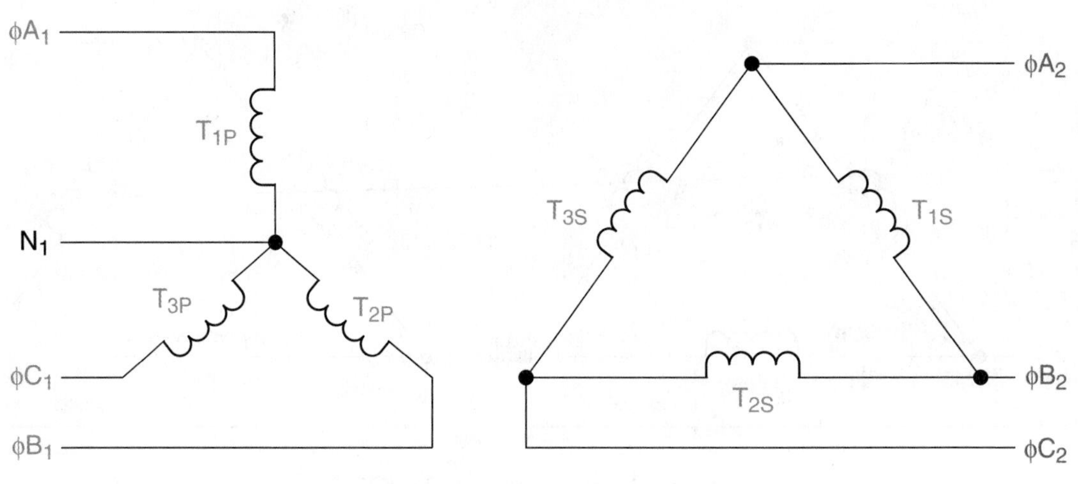

(b) Schematic diagram

FIGURE 22.5 Y-Δ transformer diagrams.

The angular displacement (difference) between primary and secondary is 30° in a delta-wye connected transformer. This is important if two such transformers are to be connected in parallel, since both must have the same angular phase displacement.

As was the case with the previous circuits, the wiring diagram in Figure 22.5a can be (and often is) implemented using a bank of single-phase (1ϕ) transformers.

Note that Y-Δ connected transformers are most commonly used in high-voltage transmission systems.

Δ-Y Connected Transformer

When wired as shown in Figure 22.6, the shell-type transformer forms a Δ-Y (delta primary–wye secondary) circuit. Note that there is a neutral connection in the secondary circuit, but none in the primary circuit. This is consistent with the Δ and Y connections we've seen up to this point. (Before reading on, take a moment to trace out the circuit connections in Figure 22.6 to verify that the diagrams represent the same circuit.)

> Triplen harmonic currents (those produced by 3rd, 9th, 15th, etc. harmonics), which are generated by common phase-to-neutral nonlinear loads such as PC's, photocopiers, and electronic lighting ballasts, pass through the secondary (wye) winding of delta-wye connected three-phase transformer, but are trapped by and circulate in the primary (delta) winding. This can cause overheating of the transformer if the population of such nonlinear loads is significant.
>
> IN THE FIELD

As was the case with the previous circuits, the circuit in Figure 22.6a can be implemented using single-phase (1ϕ) transformers. Note that Δ-Y connected transformers are most commonly found in commercial and industrial applications.

Why Use Single-Phase Transformer Banks?

As mentioned earlier, each transformer introduced in this section can be constructed using a bank (group) of single-phase transformers. Such a transformer bank is shown in Figure 22.7.

Why use three single-phase transformer banks in place of a single three-phase transformer? Two reasons: convenience and practicality. The most common failure in any three-phase system is a grounding fault where one phase fails (shorts) to ground. When a single three-phase transformer is being used, the failure of one phase necessitates replacement of the entire transformer. However, when a bank of single-phase transformers is used, the failure of any phase requires the replacement of that phase transformer only; and it is easier and cheaper to replace a single-phase transformer than a three-phase transformer. Also, a group of three single-phase transformers can be wired as any of the configurations that were introduced in this section. Three-phase transformers are manufactured in specific configurations, and therefore do not have this flexibility.

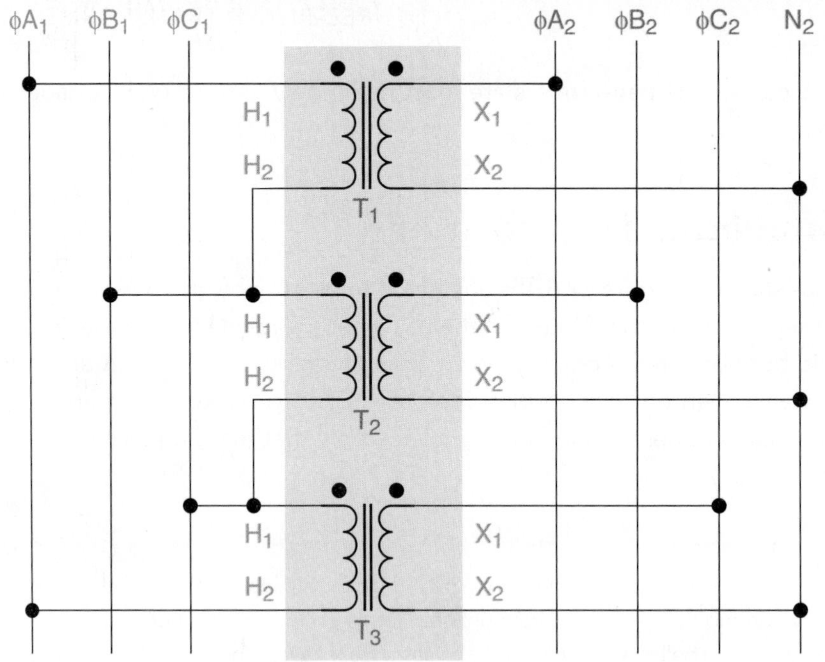

(a) Δ–Y transformer wiring

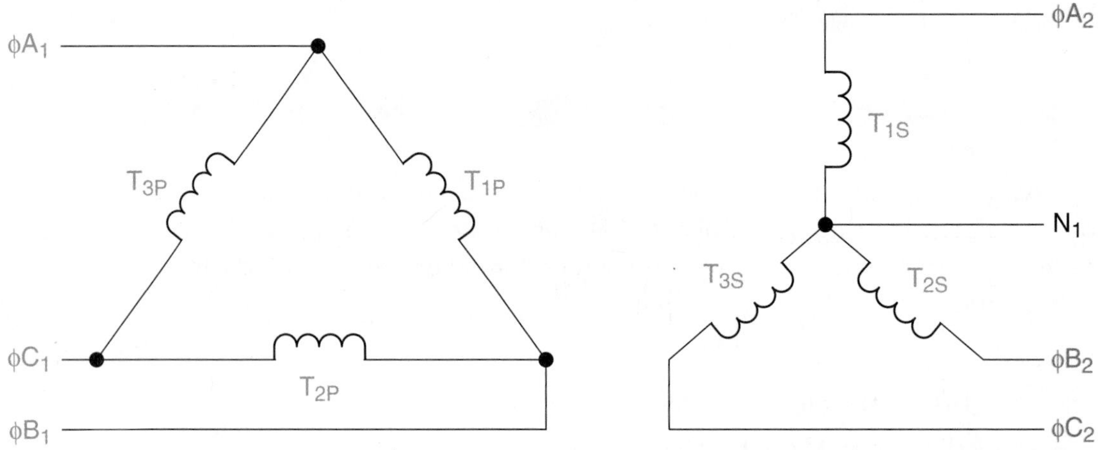

(b) Schematic diagram

FIGURE 22.6 Δ-Y transformer diagrams.

FIGURE 22.7 Three single-phase transformers wired as a three-phase transformer bank.
Shutterstock.com

Open Phases in Three-Phase Transformers

When one of the phase inductors in a wye-connected circuit opens, the entire circuit is effectively reduced to a single-phase circuit. This principle is illustrated in Figure 22.8a. When L_1 opens, ϕA is isolated from the circuit. When this occurs, there is no current through L_1, and only E_{BC} is unchanged. In effect, the three-phase circuit has been reduced to a single-phase circuit.

When one of the phase inductors in a delta-connected circuit opens, the circuit still operates as a three-phase circuit (at reduced capacity). This principle

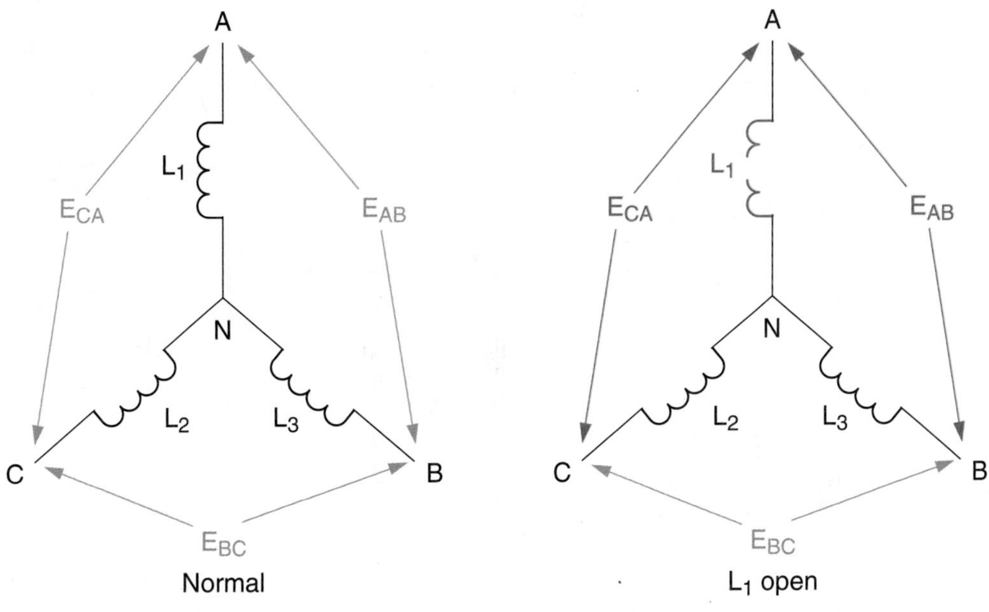

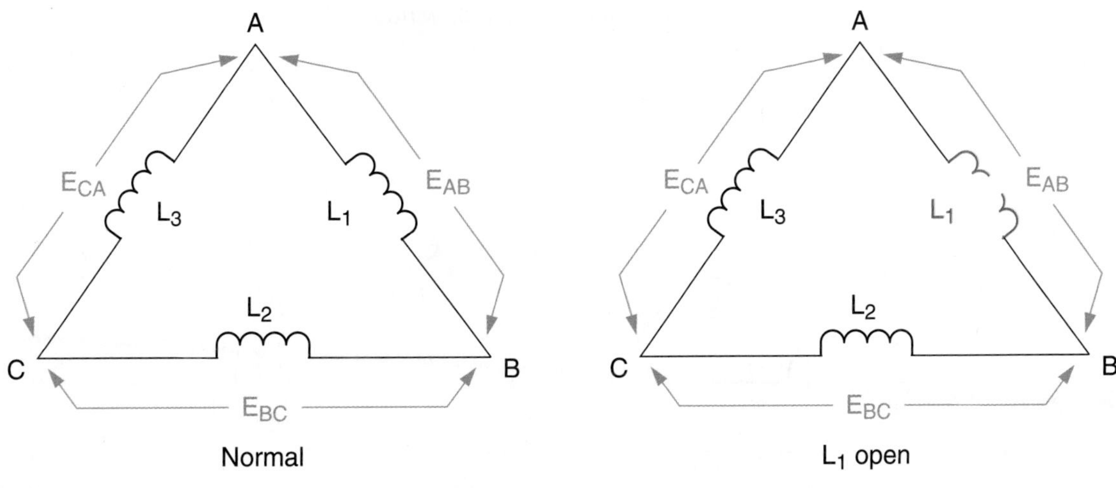

FIGURE 22.8 Y and Δ circuit voltages.

is illustrated in Figure 22.8b. When L_1 opens, none of the phase inputs is isolated from the circuit, so three-phase operation continues. However, there is no current through L_1, which *does* affect the overall operation of the delta circuit: The kVA rating of the transformer is reduced because the power handling capability of L_1 is reduced to 0 W. Even so, the circuit can continue three-phase operation at reduced capacity.

Open-Delta Connection

OPEN-DELTA CONNECTION
A delta circuit configuration that makes it possible to produce a three-phase transformer using only two single-phase transformers.

As stated earlier, a delta-connected transformer can operate at reduced capacity if one of its phases opens. This principle makes it possible to produce a three-phase circuit using only two single phase transformers. This **OPEN-DELTA CONNECTION**, which is rarely encountered anymore, is shown in Figure 22.9. Note that the kVA rating of an open-delta connection is limited to approximately 87% of the

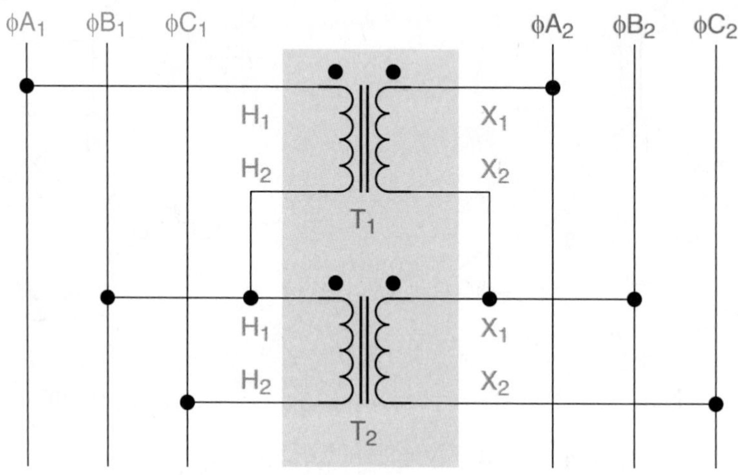

(a) Open delta transformer wiring

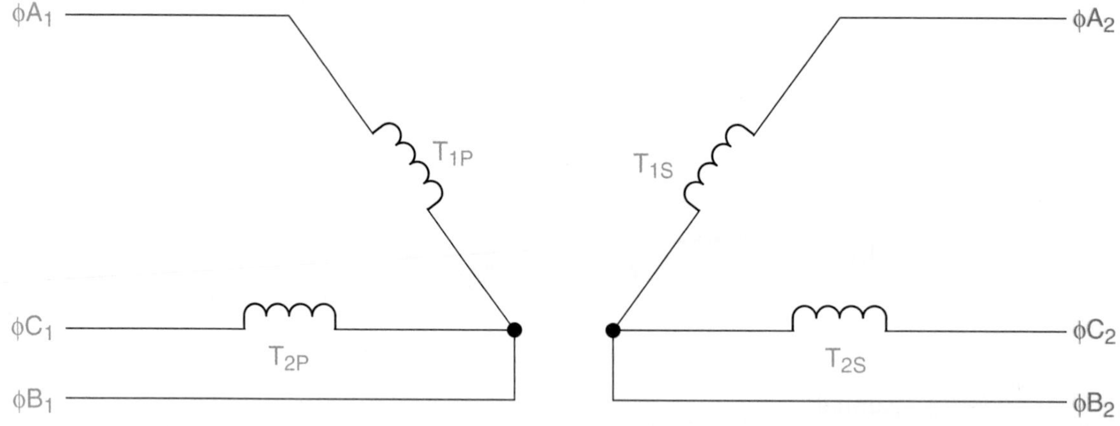

(b) Schematic diagram

FIGURE 22.9 Open-delta transformer diagrams.

sum of the nameplate kVA ratings of the two single-phase transformers. For example, if each transformer has a rating of 100 kVA, then the kVA rating of the open delta bank is 200 kVA × 87% = 174 kVA. This is due to the fact that only two transformers are carrying the load of three.

NOTE: 87% of the kVA capability of two single-phase transformers in an open-delta circuit is equal to 57.7% of the kVA capability of a comparable full delta transformer.

The two single-phase transformers connected open-delta to supply a three-phase load can handle only 57.7% of the load that can be handled if a third transformer is present.

IN THE FIELD

PROGRESS CHECK

1. Describe the basic construction of a three-phase transformer on a shell core.
2. Draw a Y-Y 3ϕ transformer and show how the primary phase coils connect to the line conductors and how the secondary phase coils connect to the load conductors.
3. What is the relationship between primary phase and line voltages in a Y-Y transformer? What are the phase and line current relationships?
4. Draw a Δ-Δ 3ϕ transformer and show how the primary phase coils connect to the line conductors and how the secondary phase coils connect to the load conductors?
5. What is the relationship between primary phase and line voltages in a Δ-Δ transformer? What are the phase and line current relationships?
6. Draw the line and phase connections for a Y-Δ and a Δ-Y transformer and explain the phase and line current and voltage relationships.
7. Explain why an open wye connection cannot be used in a 3ϕ circuit whereas an open-delta connection can.

22.2 TRANSFORMER POWER CALCULATIONS

In Chapter 20, you were introduced to three-phase circuit calculations. In this section, we will apply those calculations to three-phase transformer circuits. In the process, we will also take a look at some variations on the circuits covered in the last section.

Balanced Y-Connected Load

Figure 22.10 shows a Y-Y transformer with a resistive load. Example 22.1 demonstrates the calculation of load power for the circuit.

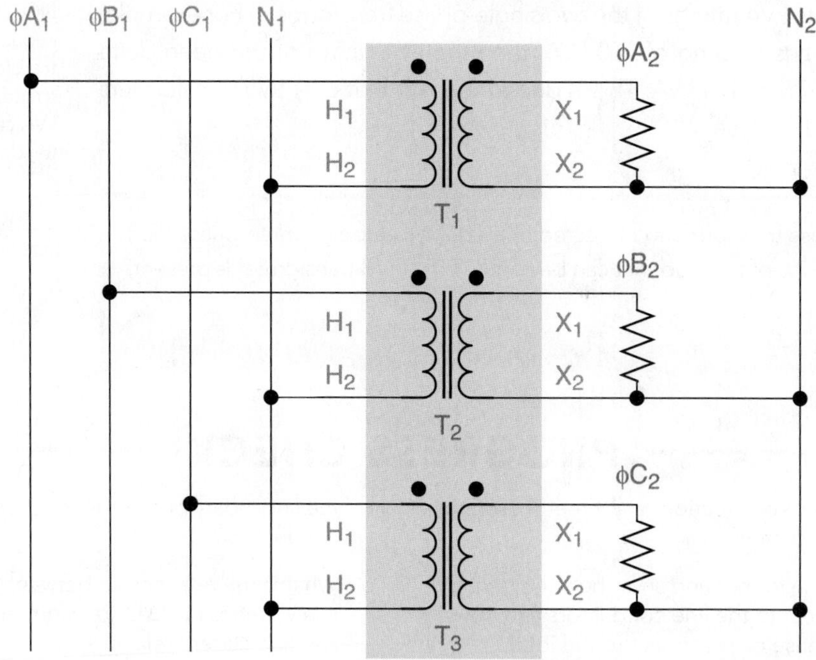

(a) Y–Y transformer wiring

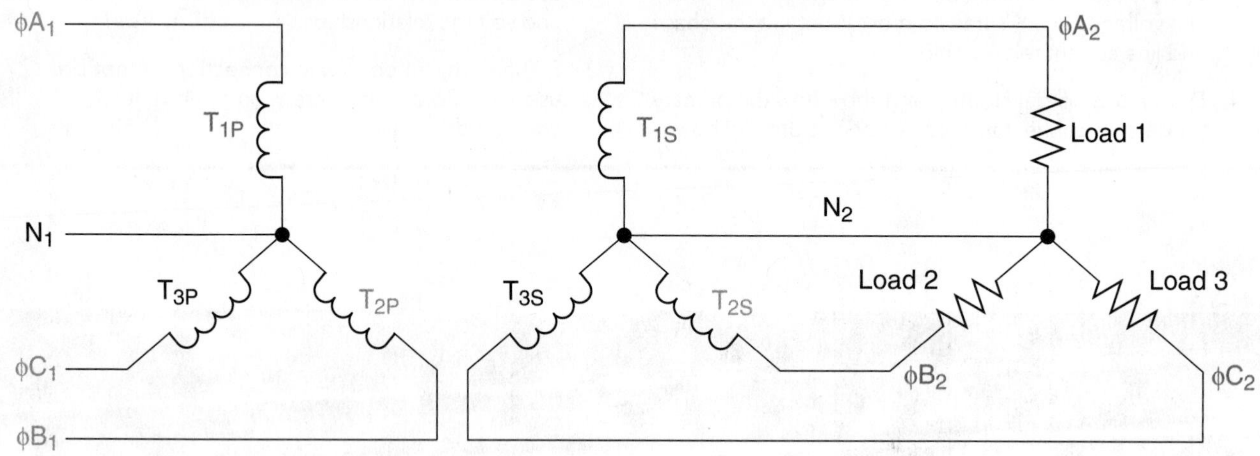

(b) Schematic diagram

FIGURE 22.10 A Y-Y transformer with a balanced load.

EXAMPLE 22.1

Assume that the secondary of the transformer in Figure 22.10 is rated at 208/120 V. The load is balanced with values of $R_L = 15\ \Omega$. Calculate the total power being supplied by the transformer.

SOLUTION

The transformer rating indicates that the transformer has values of $E_L = 208$ V and $E_P = 120$ V. According to Ohm's law, the current through each load has a value of

$$I_P = \frac{E_P}{R_L}$$

$$= \frac{120\ V}{15\ \Omega}$$

$$= 8\ A$$

The power drawn by each phase can now be found as

$$P_P = E_P \times I_P$$

$$= 120\ V \times 8\ A$$

$$= 960\ W$$

Finally, because the load circuit is purely resistive, the total power being drawn from the secondary is found as

$$P_T = P_{\phi A} + P_{\phi B} + P_{\phi C}$$

$$= 960\ W + 960\ W + 960\ W$$

$$= 2.88\ kW$$

PRACTICE PROBLEM 22.1

Assume that the secondary of the transformer in Figure 22.10 is rated at 480/277 V. The load is balanced with values of $R_L = 18\ \Omega$. Calculate the total power being drawn from the transformer.

In Chapter 20, you learned that resistive (true) power can be found as

$$P_R = P_{APP} \cos\theta$$

when apparent power (P_{APP}) and phase angle (θ) are known. In terms of phase voltages and currents, the equation can be rewritten as

(22.1)

$$P_R = 3E_P I_P \cos\theta$$

for a transformer with a Y-connected load. Example 22.2 demonstrates the use of this relationship.

EXAMPLE 22.2

The transformer secondary in Figure 22.11 is rated at 480/277 V. Calculate the circuit values of P_{APP} and P_R.

SOLUTION

The transformer rating indicates that that $E_L = 480$ V and $E_P = 277$ V. The value of phase current (I_P) can be found as

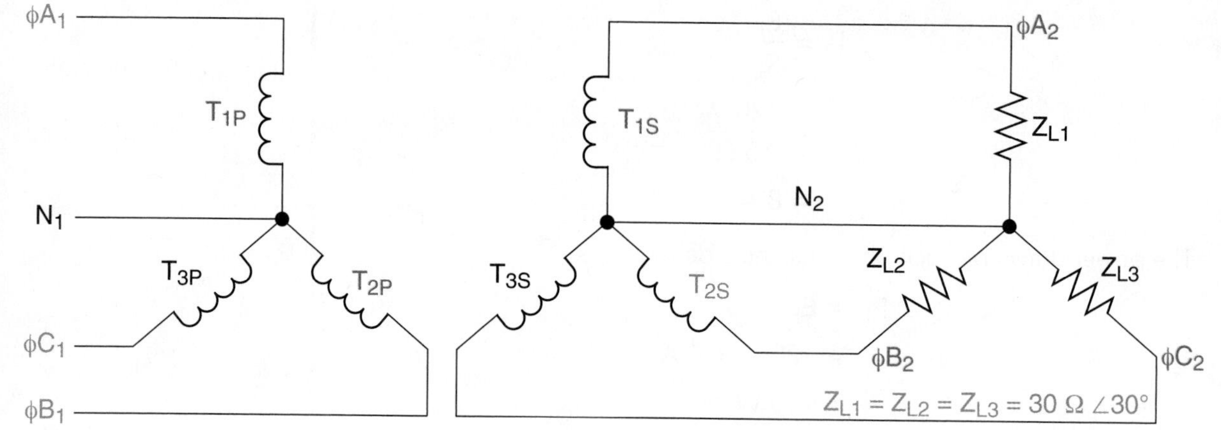

FIGURE 22.11

$$
\begin{aligned}
I_P &= \frac{E_P}{Z_L} \\
&= \frac{277\text{ V}}{30\ \Omega\ \angle 30^\circ} \\
&= 9.23\text{ A}\ \angle -30^\circ
\end{aligned}
$$

The total apparent power can now be found as

$$
\begin{aligned}
P_{APP} &= 3E_PI_P \\
&= 3 \times 277\text{ V} \times 9.23\text{ A} \\
&= 7.67\text{ kVA}
\end{aligned}
$$

and the actual load power dissipation can be found as

$$
\begin{aligned}
P_R &= 3E_PI_P\cos\theta \\
&= 7.67\text{ kVA} \times \cos(-30^\circ) \\
&= 7.67\text{ kVA} \times 0.866 \\
&= 6.64\text{ kW}
\end{aligned}
$$

PRACTICE PROBLEM 22.2

A circuit like the one in Figure 22.11 has a transformer secondary with a 208/120 V rating and a balanced load value of $Z_L = 15\ \Omega\ \angle\ 45^\circ$. Calculate the values of P_{APP} and P_R for the circuit.

Balanced Δ-Connected Load

When dealing with a transformer with a delta-connected load, the power calculations are essentially the same as those for a Y-connected load, as demonstrated in Example 22.3.

EXAMPLE 22.3

The transformer secondary in Figure 22.12 is rated at 240 V. Calculate the values of apparent and resistive (true) power for the circuit.

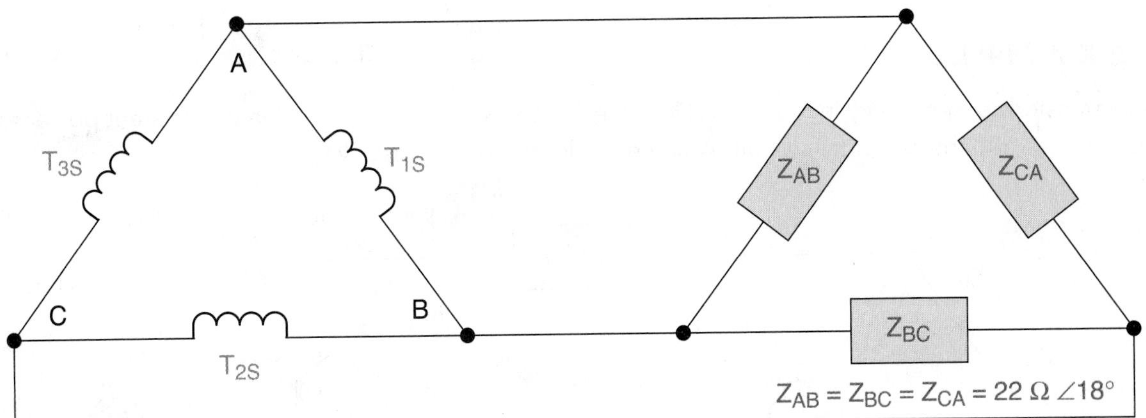

FIGURE 22.12

SOLUTION

The transformer rating indicates that the secondary has values of $E_P = E_L = 240$ V. The value of I_P can be found as

$$I_P = \frac{E_P}{Z_L}$$
$$= \frac{240\text{ V}}{10\ \Omega\ \angle 18^\circ}$$
$$= 24\text{ A}\ \angle -18^\circ$$

The total apparent power can now be found as

$$P_{APP} = 3E_PI_P$$
$$= 3 \times 240\text{ V} \times 24\text{ A}$$
$$= 17.3\text{ kVA}$$

and the actual load power dissipation can be found as

$$P_R = 3E_PI_P\cos\theta$$
$$= 17.3\text{ kVA} \times \cos(-18^\circ)$$
$$= 17.3\text{ kVA} \times 0.951$$
$$= 16.5\text{ kW}$$

PRACTICE PROBLEM 22.3

The transformer secondary in Figure 22.12 is rated at 240 V. Calculate the values of apparent and resistive (true) power for the circuit when $Z_L = 8\ \Omega\ \angle 12^\circ$.

The examples to this point have been based on voltage and impedance values. Example 22.4 demonstrates power calculations that are based on voltage and current values.

EXAMPLE 22.4

The transformer secondary in Figure 22.13 is rated at 240 V. Calculate the value of apparent power for each phase and the secondary circuit. Assume the load is purely resistive.

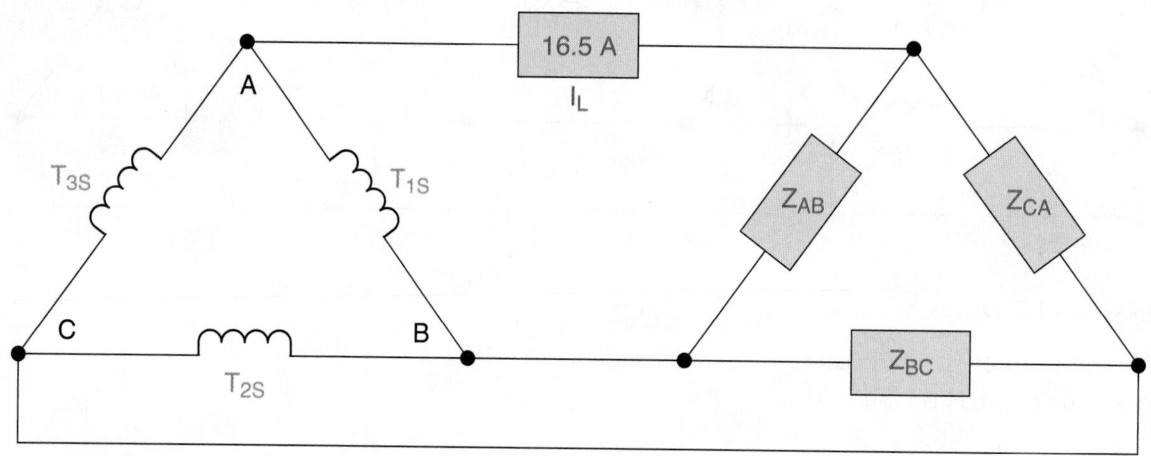

FIGURE 22.13

SOLUTION

The 240 V rating of the transformer secondary indicates that $E_P = E_L = 240$ V. With a line current of 8.5 A, the phase current (I_P) has a value of

$$I_P = \frac{I_L}{\sqrt{3}}$$
$$= \frac{16.5\text{ A}}{1.732}$$
$$= 9.53\text{ A}$$

The value of P_{APP} for each phase is

$$P_{APP} = E_p \times I_P$$
$$= 240\text{ V} \times 9.53\text{ A}$$
$$= 2.29\text{ kVA}$$

and the circuit value of P_{APP} is

$$P_{APP} = 3E_pI_P$$
$$= 3 \times 2.29\text{ kVA}$$
$$= 6.87\text{ kVA}$$

PRACTICE PROBLEM 22.4

The transformer secondary in Figure 22.13 is rated at 240 V. Calculate the value of apparent power for each phase and the secondary circuit if the measured line current (I_L) is 21 A. Assume the load is purely inductive.

As the examples in this section have shown, load power calculations for a Y-connected load are very similar to those for a Δ-connected circuit.

PROGRESS CHECK

1. Describe the procedure you would use to calculate P_{APP} for a Y-Y transformer connected to a Y-connected resistive load, given the line and phase voltage of the transformer secondary and the resistance of the load.
2. Given the phase angle of the load and the phase voltage and current values, explain how you would solve for P_{APP} and P_R.
3. If a load is delta-connected instead of wye-connected, how does this change your approach to solving for the load power values?
4. If you have values only for E_L and I_L, how would you would determine the power supplied by a given 3ϕ transformer to a load?

22.3 RELATED TOPICS

In this section, we will close out the chapter by examining some miscellaneous topics that relate to three-phase transformers.

Dry-Type versus Wet-Type Transformers

A **DISTRIBUTION TRANSFORMER** is used to step down the voltages transmitted by power utilities to values that are required by the customer. Two commonly seen distribution transformers—the **POLE-TYPE** and **PAD-MOUNTED** transformers—are shown in Figure 22.14.

Distribution transformers are classified as being either **DRY-TYPE** or **WET-TYPE**. These classifications indicate the means used to cool and/or insulate the component. Dry-type transformers are generally air-cooled and may have louvers to aid in cooling, as shown in Figure 22.15a.

Wet-type transformers are high-power transformers that generate too much heat for air cooling. A wet-type transformer is shown in Figure 22.15b.

DISTRIBUTION TRANSFORMER A transformer that is used to step down the voltages transmitted by power utilities to values that are required by the customers.

POLE-TYPE TRANSFORMER A distribution transformer that is mounted on a pole (such as a telephone pole).

PAD-MOUNTED TRANSFORMER A distribution transformer that is mounted on a rectangular concrete pad.

DRY-TYPE TRANSFORMER A transformer that is typically air-cooled and may have louvers to aid in cooling.

(a)

(b)

FIGURE 22.14 Pole and pad transformers.
Istock.com

(a)

(b)

FIGURE 22.15 Dry-type and wet-type transformers.
(a) Fotolia.com; (b) Shutterstock.com

Wet-type transformers are typically cooled using mineral oil or high-temperature hydrocarbons.

WET-TYPE TRANSFORMER A transformer that is typically cooled using mineral oil or high-temperature hydrocarbons.

> The true life expectancy of a transformer operating at varying temperatures cannot be accurately predicted. Fluctuating load conditions and changes in ambient temperatures make it difficult, if not impossible, to arrive at such definitive information.
>
> IN THE FIELD

Dry-type transformers are generally limited to outputs of 500 kVA or lower (though there are exceptions). In contrast, wet-type transformers can handle power outputs greater than 2 MVA. Wet-type transformers are also more efficient than dry-type, and are preferred for use outdoors. At the same time, wet-type transformers have their own drawbacks:

- In many cases, wet-type transformers contain fluids that are flammable.
- Wet-type transformers must be designed to ensure that fluid cannot leak from the transformer case. This increases the design and manufacturing costs of the components.

VAULT-TYPE TRANSFORMER A transformer designed for underground installation.

SUBMERSIBLE TRANSFORMER A transformer designed to be used in underwater power transmission systems.

Note that **VAULT-TYPE** and **SUBMERSIBLE** transformers are also manufactured. The vault-type is designed for underground installation, and the submersible is used in underwater power transmission systems.

Common Three-Phase Ratings

Three-phase transformers can be custom designed to meet specific input/output voltage requirements. At the same time, there are three-phase transformer configurations that are commonly available. Figure 22.16 shows the wiring for a 13.200Y/7620 to 208Y/120 volt transformer. This rating indicates that the transformer has:

- A Y-connected primary with $E_L = 13{,}200$ V and $E_P = 7620$ V.
- A Y-connected secondary with $E_L = 208$ V and $E_P = 120$ V.

These voltages can be measured as shown in Figure 22.17.

> The nameplate kVA rating of a transformer represents the kVA that will result in the rated average winding temperature rise when the unit is operated at 100 percent of rated kVA under normal service conditions.

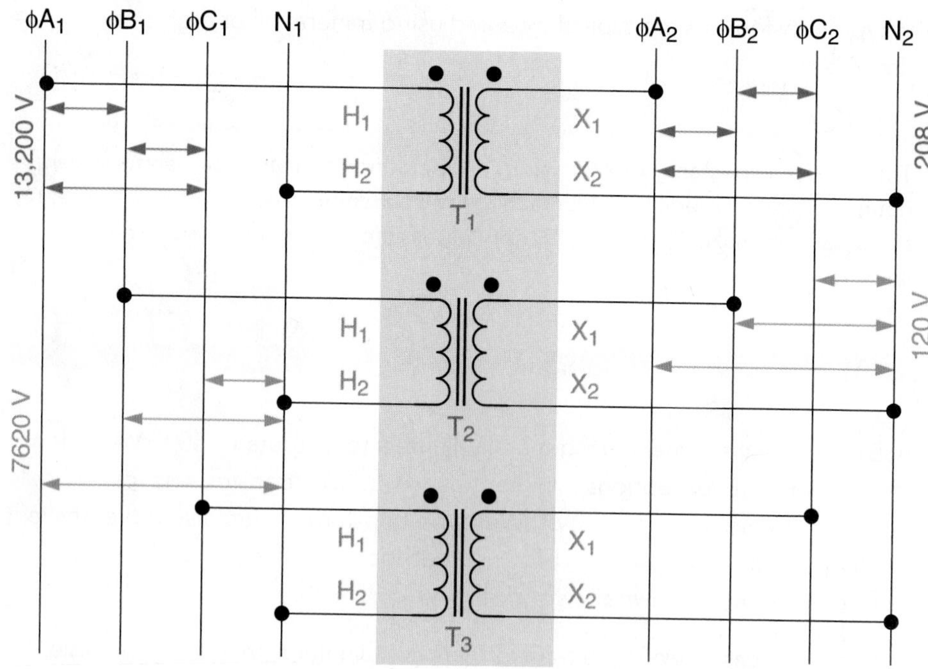

FIGURE 22.16 A 3-phase, 4-wire transformer.

The transformer represented in Figure 22.16 is a *3-phase, 4-wire* transformer. Some other common transformer ratings are listed below.

PRIMARY RATINGS	SECONDARY RATINGS
4160Y/2400	208Y/120
8320Y/4800	240
13,800Y/7970	480
22,860Y/13,200	480Y/277

Note that the 240 V and 480 V secondary ratings apply to transformers having a Δ-connected secondary. This is due to the fact that $E_L = E_P$ in any delta connection.

CAPACITY A transformer rating that indicates the amount of power the component can deliver to a load on a continual basis.

Distribution transformers also have standard **CAPACITY** ratings. The capacity of a transformer indicates the amount of power it can deliver to a load on a continual basis. Pole-mounted three-phase transformers have capacity ratings that range from 15 kVA to 500 kVA. Pad-mounted three-phase transformers have capacity ratings of 75 kVA to 2500 kVA (2.5 MVA).

PROGRESS CHECK

1. What is a distribution transformer? Name two common types of distribution transformers.
2. Compare dry-type and wet-type transformers. List the advantages and disadvantages of each.
3. What is a vault-type transformer? What is a submersible transformer?
4. List some of the common 3ϕ transformer ratings and explain what these ratings indicate.
5. Explain what is meant by transformer capacity.

22.4 SUMMARY

Here is a summary of the major points made in this chapter:

Three-Phase Transformer Configurations

- Y-Y connected transformers have Y-connected primaries and secondaries.
  - The transformer phases are connected to the line and load conductors as shown in Figure 22.3.
  - Primary and secondary voltage relationships are as follows: $E_L = \sqrt{3} \times E_P$ and $E_P = E_L/\sqrt{3}$.
  - Primary and secondary current relationships are simple since $I_L = I_P$ for both the primary and secondary.
- Δ-Δ connected transformers have Δ-connected primaries and secondaries.
  - The transformer phases are connected to the line and load conductors as shown in Figure 22.4.
  - Primary and secondary current relationships are as follows: $I_L = \sqrt{3} \times I_P$ and $I_P = I_L/\sqrt{3}$.
  - Primary and secondary voltage relationships are simple since $E_L = E_P$ for both the primary and secondary.
- Y-Δ connected transformers have Y-connected primaries and Δ-connected secondaries.
  - The transformer phases are connected to the line and load conductors as shown in Figure 22.5.
- Δ-Y connected transformers have Δ-connected primaries and Y-connected secondaries.
  - The transformer phases are connected to the line and load conductors as shown in Figure 22.6.
- Each of the four basic transformer configurations can be implemented by using three single-phase transformers (called a *bank*) rather than one three-phase unit.
- 1ϕ transformer banks are more common than 3ϕ transformers as they are more convenient and practical.
  - The most common fault is when one phase shorts to ground. When this happens with a 3ϕ transformer, the entire unit must be replaced. If a bank is used, only the faulted phase is replaced, which is easier and cheaper.
  - Three 1ϕ transformers can be wired in any configuration, but 3ϕ transformers are prewired. This makes 1ϕ transformer banks more flexible.
- If one phase of a Y-connected transformer opens, the transformer is reduced to 1ϕ operation (see Figure 22.8a).
- If one phase of a Δ-connected transformer opens, it can still operate as a 3ϕ transformer (see Figure 22.8b).
  - When a delta transformer is operated in this manner it is referred to as an open-delta connection.

Transformer Power Calculations

- The total power for a balanced Y-connected load is found as $P_{APP} = 3 \times E_P I_P$.
  - $E_P = E_L/\sqrt{3}$ and $I_P = I_L$.
  - True power is found as $P_R = P_{APP} \cos \theta$.

- The total power for a balanced Δ-connected load is also found as $P_{APP} = 3 \times E_P I_P$.
  - $E_p = E_L$ and $I_P = I_L/\sqrt{3}$.
  - True power is also found as $P_R = P_{APP} \cos \theta$.
- For both Y- and Δ-connected loads, total power can also be found using line values as $P_{APP} = \sqrt{3} \times E_L \times I_L$.

Related Topics

- Distribution transformers are used by utility companies to step-down voltages for their customers.
  - Two common types of distribution transformers are pole-mounted and pad-mounted.
- Distribution transformers are classified as wet-type and dry-type.
  - These classifications indicate how they are cooled.
- Dry-type transformers are air-cooled. Some use fans and/or louvers to aid in cooling. They are usually limited to kVA ratings below 500 kVA.
- Wet-type transformers are high-power transformers that use mineral oil or high-temperature hydrocarbons to aid in cooling.
  - They can have kVA ratings greater than 2 MVA.
  - In many cases these cooling fluids are flammable.
- Two other classifications of transformers are vault-type and submersible.
  - Vault-type transformers are designed for underground installations.
  - Submersibles are designed for underwater power transmission systems.
- Transformers are designed to meet specific input/output requirements.
- A 1,300Y/7620 to 208Y/120 transformer would have the following characteristics:
  - The primary is Y-connected to a 1300 V line with a phase voltage of 7620 V.
  - The secondary is also Y-connected with a line voltage of 208 V and a phase voltage of 120 V.
  - This is obviously a step-down transformer.
- Some other common ratings for transformers are:

PRIMARY RATINGS	SECONDARY RATINGS
4160Y/2400	208Y/120
8320Y/4800	240
13,800Y/7970	480
22,860Y/13,200	480Y/277

- Note that when there is a single value in either the primary or the secondary rating, the connection is a delta since $E_L = E_P$ for all delta connections.
- The capacity of a transformer indicates the amount of power the transformer can deliver, given in kVA or MVA.
  - Pole-mounted units have capacity ratings in the range of 15 kVA to 500 kVA.
  - Pad-mounted units have capacity ratings in the range of 75 kVA to 2500 kVA.

CHAPTER REVIEW

1. A balanced Y-Y transformer has primary line values of 115 kV and 25 A. The phase values for the primary are ________.

   a) 115 kV and 43.3 A

   b) 199 kV and 25 A

c) 66.4 kV and 25 A
d) 66.4 kV and 14.4 A

2. A balanced Y-Y transformer has secondary phase values of 13.8 kV and 9.5 A. The line values for the secondary are ________.
   a) 23.9 kVA and 9.5 A
   b) 13.8 kV and 16.5 A
   c) 13.8 V and 5.48 A
   d) 7.97 kV and 9.5 A

3. Three 1ϕ transformers connected together to form a 3ϕ transformer are commonly called a ________.
   a) star
   b) triplen
   c) triad
   d) bank

4. A Δ-Δ transformer has a primary line voltage of 13.8 kV and a secondary phase voltage of 2400 V. The turns ratio for this transformer is ________.
   a) 3.32:1
   b) 1:5.75
   c) 5.75:1
   d) 1:3.32

5. The transformer described in question 4 is connected to a Y-connected resistive load. Each phase of the load has a resistance of 5.4 Ω. The phase current in the load is ________.
   a 444 A
   b 257 A
   c 770 A
   d 643 A

6. If the load in question 5 was Δ-connected the secondary line current would be ________.
   a) 444 A
   b) 257 A
   c) 770 A
   d) 643 A

7. The neutral in a Δ-Δ transformer.
   a) is larger than in a Y-Y transformer
   b) carries the unbalanced current
   c) is larger than in a Y-Y transformer
   d) does not exist

8. Δ-Δ transformers are commonly found ________.
   a) in vaults
   b) on poles

c) in industrial applications

d) underwater

9. The primary line voltage is ________ the primary phase voltage in the Δ-Δ transformer.

a) equal to

b) out of phase by 180° with

c) always less than

d) always greater than

10. The secondary line voltage is ________ the secondary phase voltage in the Δ-Y transformer.

a) equal to

b) out of phase by 180° with

c) always less than

d) always greater than

11. The Y-Δ connected transformer is commonly used as a ________ transformer.

a) control

b) high-voltage distribution

c) pole-mounted

d) low-voltage distribution

12. The transformer configuration that *cannot* be wired as a bank of 1ϕ transformers is the ________.

a) Y-Y

b) Δ-Δ

c) Δ-Y

d) Y-Δ

e) none of the above

13. Transformer banks are preferred over 3ϕ transformers because they ________.

a) are more flexible and easier to service

b) are lighter

c) are initially cheaper

d) have higher power handling capacity

14. If one phase of a Y-Y transformer fails ________.

a) it must run at a reduced 3ϕ capacity

b) it functions as a 1ϕ transformer

c) it runs out of phase at a reduced capacity

d) line current increases by a factor of $\sqrt{3}$

15. The configuration in which two 1ϕ transformers are used as a 3ϕ transformer is called a(n) ________ transformer.

a) 2-to-3 delta

b) 3-to-2 delta

c) open-delta

d) closed delta

16. The two 1ϕ transformers in question 15 each have a kVA rating of 85 kVA. The maximum kVA rating of this two-transformer 3ϕ configuration is ________.

   a) 148 kVA

   b) 74 kVA

   c) 222 kVA

   d) 256 kVA

17. A Y-Y transformer is driving a Δ-connected resistive load rated at 8.2 Ω per phase. If the secondary line voltage of the transformer is 240 V, the total power dissipated by the load is ________.

   a) 50.7 W

   b) 7.02 kW

   c) 22.1 kW

   d) 63.2 kW

18. A delta-connected load has values of P_{APP} = 10.9 kVA and Z = 5.6 Ω ∠36° per phase. The total real power dissipated by the load is ________.

   a) 8.82 kW

   b) 26.5 kW

   c) 2.94 kW

   d) 13.3 kW

19. A Δ-Y transformer has secondary line values of E_L = 13.8 kV and I_L = 16.5 A. The total power supplied by this transformer is ________.

   a) 682 kVA

   b) 228 kVA

   c) 446 kVA

   d) 394 kVA

   e) cannot be solved; you need to know the Z of each load phase

20. A transformer used by a utility company to step-down voltages for its customer is called a ________ transformer.

   a) split-phase

   b) control

   c) distribution

   d) high-leg

21. Transformers are classified as ________ and ________ based on how they are cooled.

   a) dry-type, wet-type

   b) fin-cooled, fan-cooled

   c) oil-type, gas-type

   d) fan-cooled, louver-cooled

22. A transformer designed to be mounted underground is called a _______ transformer.
    a) BIL
    b) subunit
    c) vault-type
    d) buried

PRACTICE PROBLEMS

1. Complete the chart below for the missing values for the primary of each transformer.

Transformer Type	E_L	E_P	I_L	I_P
Y-Y	13.8 kV			8.2 A
Δ-Y		480 V	2.6 A	
Y-Δ		6.9 kV		11.5 A
Δ-Δ	9.6 kV		4.4 A	

2. Complete the chart below for the missing values for the secondary of each transformer.

Transformer Type	E_L	E_P	I_L	I_P
Y-Y		4400 V	6.6 A	
Δ-Y		480 V		3.2 A
Y-Δ	12.1 kV			15 A
Δ-Δ	2.2 kV		5.5 A	

3. A circuit like the one in Figure 22.17 has primary values of $E_L = 6.9$ kV and $I_L = 28$ A. Solve for E_P and I_P for the primary.

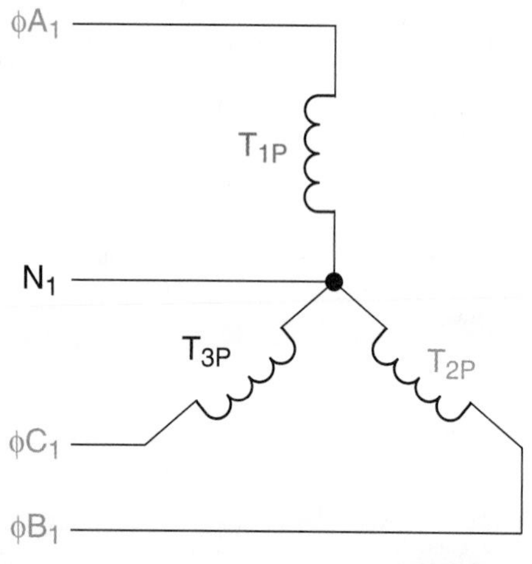

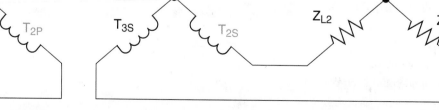

FIGURE 22.17

4. A circuit like the one in Figure 22.17 has primary values of $E_L = 2600$ V and $I_L = 36$ A. Solve for E_P and I_P for the primary.
5. A circuit like the one in Figure 22.17 has secondary values of $E_P = 660$ V and $I_P = 9$ A. Solve for E_L and I_L for the secondary.
6. A circuit like the one in Figure 22.17 has secondary values of $E_P = 550$ V and $I_P = 11.6$ A. Solve for E_L and I_L for the secondary.
7. For the circuit described in question 5, solve for the resistance value for each phase of the load.
8. For the circuit described in question 6, solve for the resistance value for each phase of the load.
9. A circuit like the one in Figure 22.18 has secondary values of $E_P = 2500$ V and $I_L = 45$ A. Solve for the values of E_L and I_P.

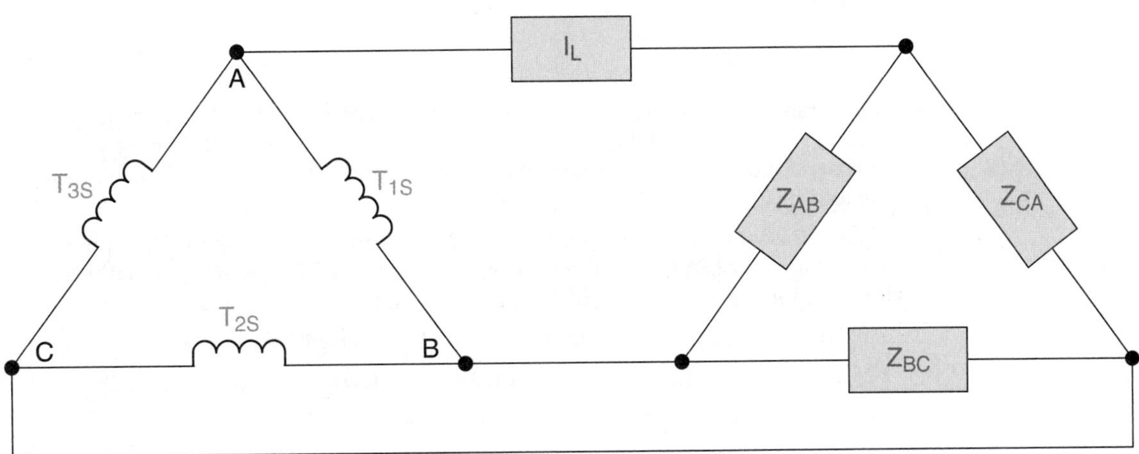

FIGURE 22.18

10. A circuit like the one in Figure 22.18 has secondary values of $E_L = 1600$ V and $I_P = 22$ A. Solve for the values of E_P and I_L.
11. How much power is being supplied to the transformer described in question 9?
12. How much power is the transformer described in question 10 supplying to the load?
13. If the secondary of the transformer described in question 9 was Y-connected and the given values remained the same, would it affect the amount of power supplied to the transformer? Explain your answer.
14. If the secondary of the transformer described in question 10 was Y-connected and the given values remained the same, would it affect the amount of power supplied to the load? Explain your answer.
15. The secondary of the transformer in Figure 22.19 has a value of $E_P = 260$ V. The load is purely resistive, and each phase has a resistance of 6.6 Ω. Solve for the power dissipated by each phase and the total load power.
16. A circuit like the one in Figure 22.19 has secondary values of $E_L = 505$ V and $I = 72.1$ A. Solve the resistance of each phase of the load.
17. A Y-Δ transformer has secondary line values of $E_L = 88$ V and $I_L = 2.65$ A. Solve for the total power dissipated by the load.
18. What is the configuration of the load for the circuit described in question 17? Explain your answer.

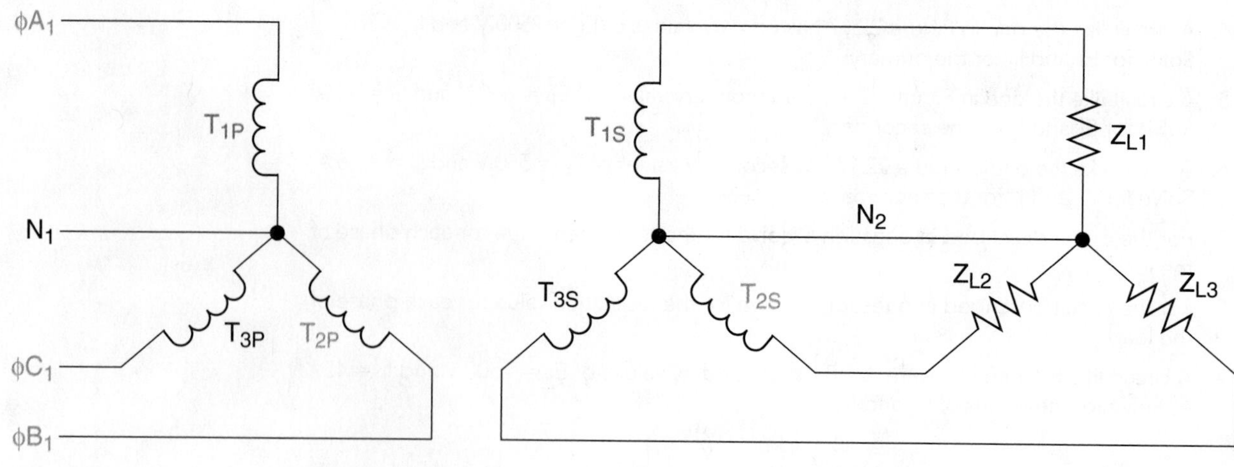

FIGURE 22.19

19. A Δ-Y transformer is driving a Y-connected load. The secondary line voltage is measured at 240 V and the load is dissipating 1200 W. Solve for the phase voltage of the secondary, the phase current of the load, and the resistance of each phase of the load.
20. Figure 22.20 shows the secondary of a Δ-Δ transformer and its load. The primary line voltage is 22.86 kV and the primary line current is 8.8 A. Assuming that the transformer has a turns ratio of 47.5, solve for the following values:
    a) The secondary line and phase values for both current and voltage.
    b) The value of resistance for each phase of the load.
    c) The total power supplied to the transformer.
    d) The total power supplied to the load.

(*Hint:* The turns ratio is the ratio of primary phase voltage to secondary phase voltage.)

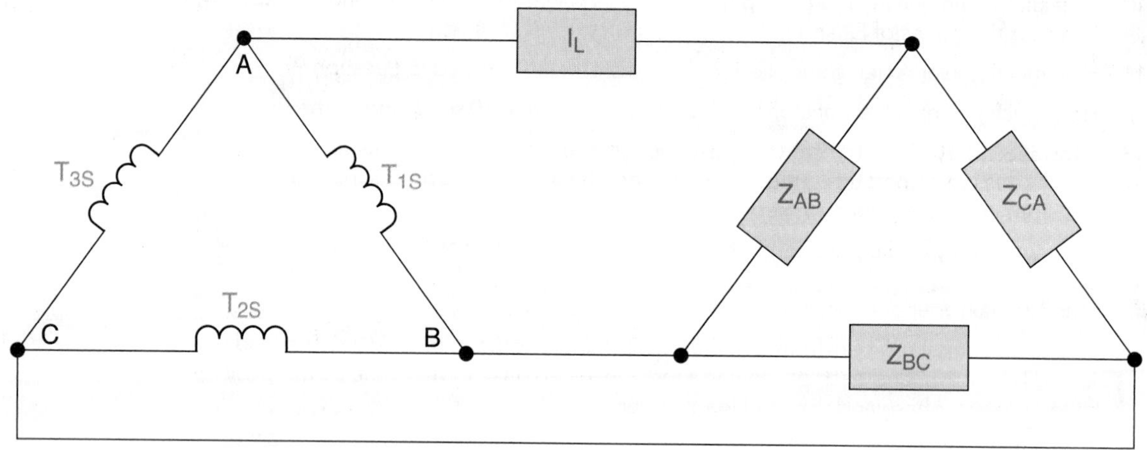

FIGURE 22.20

21. A circuit like the one in Figure 22.20 has load phase values of R = 7.7 Ω ∠28°. The secondary line voltage is 208 V. Solve for P_{APP} and P_R.

22. A circuit like the one in Figure 22.20 has load phase values of R = 4.2 Ω and L = 22 mH. The transformer secondary has a value of E_P = 128 V. Solve for the values of P_{APP}, P_R, and P_X. Assume that this is a North American installation with line frequency = 60 Hz.

ANSWERS TO THE EXAMPLE PRACTICE PROBLEMS

22.1 12.8 kW

22.2 2.88 kVA, 2.04 kW

22.3 7.2 kVA, 7.04 kVA

22.4 2.91 kVA, 8.73 kVA

CHAPTER 23

DC MACHINES: GENERATORS AND MOTORS

PURPOSE

DC generators produce single-polarity output voltages that are used by DC circuits and systems. DC generators are used primarily in industrial and automotive applications. DC motors are used primarily in applications where variable motor speeds are required.

In this chapter, you are introduced to the operating characteristics of DC generators and motors. Then, in Chapter 24, you will be introduced to their AC counterparts.

KEY TERMS

The following terms are introduced and defined in this chapter on the pages indicated:

OBJECTIVES

After completing this chapter, you should be able to:

1. Compare and contrast alternators and DC generators.
2. Explain how the commutator in a DC generator produces a single-polarity output.
3. Describe the relationship among dynamos, generators, and motors.

4. Describe the parts of a DC generator and the function of each.
5. Discuss *armature reaction* and the methods used to compensate for it.
6. Describe the construction of series DC generators.
7. List the factors that impact the output from a series DC generator and describe the effect of each.
8. Describe the construction of parallel DC generators.
9. Discuss parallel DC generator loading, voltage buildup, and output voltage control.
10. Describe the construction and operation of compound DC generators.
11. Describe and explain *compounding.*
12. List and describe the sources of DC generator power loss.
13. Explain the relationship between armature current and the direction of DC motor rotation.
14. Compare and contrast *torque* and *work* as each relates to DC motors.
15. Calculate the values of torque, full-load torque, and horsepower for a DC motor.
16. Describe series DC motor construction and operation.
17. Explain the relationship between series DC motor CEMF and startup current levels.
18. Describe shunt DC motor construction and operation.
19. Describe compound DC motor construction and operation.
20. Describe the operation of DC motor speed control circuits, direction control circuits, and kickback circuits.

23.1 GENERATORS AND ALTERNATORS: AN OVERVIEW

A **GENERATOR** is a machine that converts mechanical energy (motion) to electrical energy. In Chapter 13, you were introduced to the concept of using the motion of a conductor through a magnetic field to generate electrical energy. As a review, this method of producing an alternating current (AC) is illustrated in Figure 23.1. Note that an AC generator like the one represented in Figure 23.1 is called an **ALTERNATOR**.

GENERATOR A machine that converts mechanical energy (motion) to electrical energy.

ALTERNATOR An AC generator.

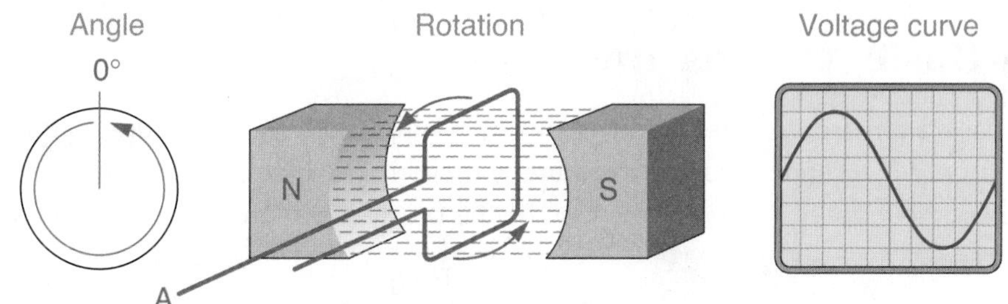

(a) One voltage cycle is produced by each rotation cycle of the rotor.

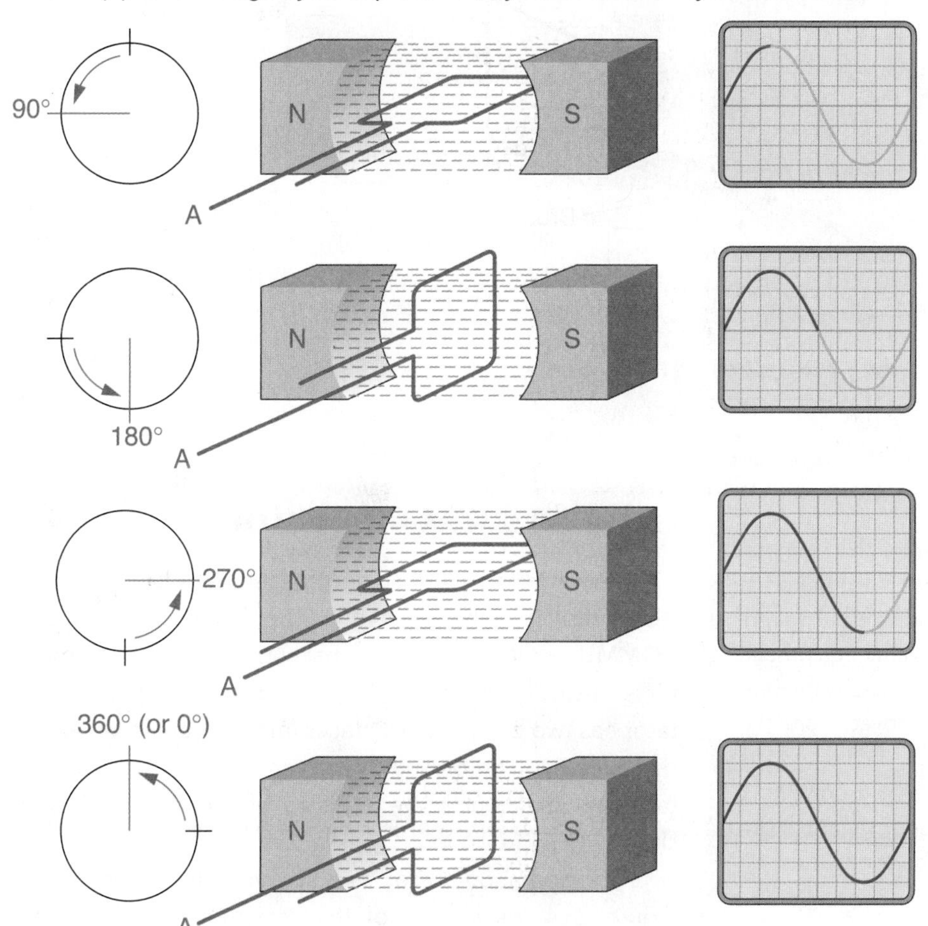

(b) Cycle development (in 90° increments).

FIGURE 23.1 Generating a sine wave.

A voltage is induced across the loop conductor (armature) in Figure 23.1 as it cuts through the stationary magnetic field. The magnitude of the voltage and its polarity (at any instant) depend on:

- The angle of rotation of the armature relative to the magnetic lines of force.
- The strength of the magnetic field (i.e., the flux density between the poles of the magnet).
- The speed at which the armature rotates.

The relationship between the armature angle and the induced voltage is highlighted in Figure 23.1b.

A Basic Alternator

A basic alternator (AC generator) is shown in Figure 23.2. The armature is attached to a pair of *slip rings*. As the armature and slip rings rotate, the voltage induced across the loop is coupled to the load by the *brushes* (stationary slip ring contacts).

COMMUTATOR A device that converts the output of an alternator to DC.

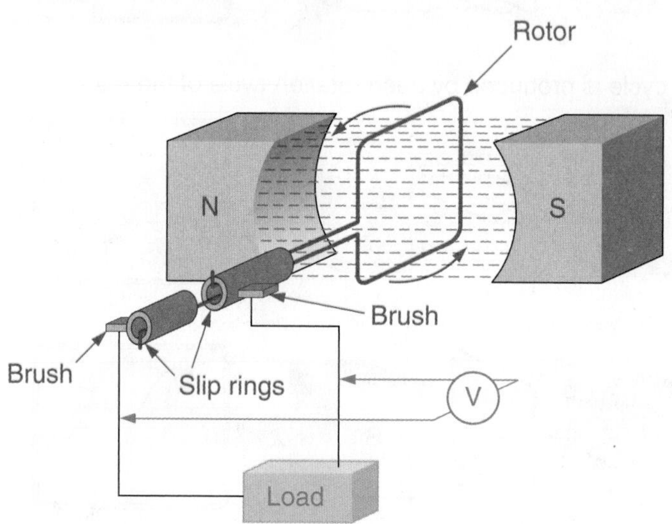

FIGURE 23.2 A basic alternator.

Surface wear is a major concern with commutators. The commutator on a small motor (rated at less than 1kVA) is not designed to be repaired. The commutator on large industrial equipment, on the other hand, can be resurfaced with abrasives, or the rotor can be removed from the frame, mounted in a large metal lathe, and the commutator resurfaced by cutting it down to a smaller diameter. The largest equipment may include a lathe turning attachment mounted directly over the commutator.

A Basic DC Generator

An alternator can be modified to generate DC instead of AC by replacing the slip rings with a similar device called a **COMMUTATOR**. The physical makeup of a simple commutator is illustrated in Figure 23.3a. As you can see, the commutator has two conductive surfaces (metal) that are separated by insulators. The brushes, which are stationary, make contact with the conductive surfaces of the commutator as shown in Figure 23.3b.

The operation of the basic DC generator is illustrated in Figure 23.4. As side A of the armature rotates upward through the magnetic field, it is connected to the load by the commutator and brush. This connection provides the output

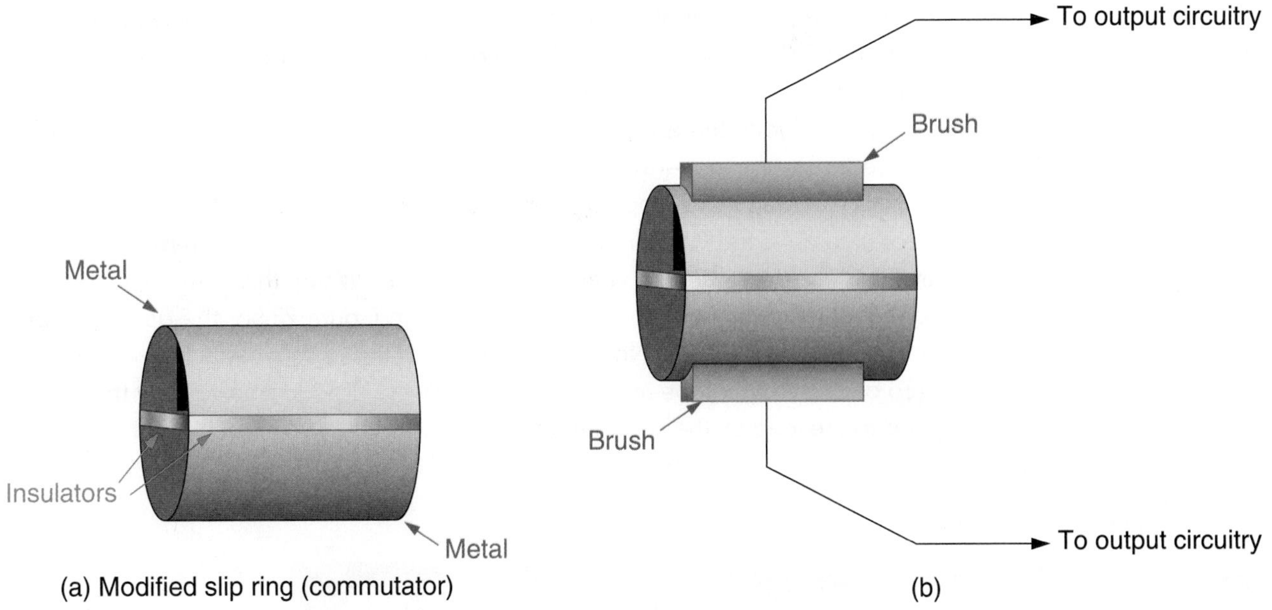

FIGURE 23.3 Commutator and brushes.

shown in Figure 23.4a. As side A reaches the 90° point in the armature rotation, its motion is parallel to the lines of magnetic flux. At this point, no voltage is generated and the generator output is 0 V. Note that the armature is said to be in the **NEUTRAL PLANE** when its motion is parallel to the lines of flux.

NEUTRAL PLANE The space above and below a stationary magnet where a spinning rotor is in parallel with the magnetic field.

As side B of the armature rotates upward through the magnetic field, it is connected to the load via the commutator and brush. This connection provides the output shown in Figure 23.4b. As side B reaches the 90° point in the armature

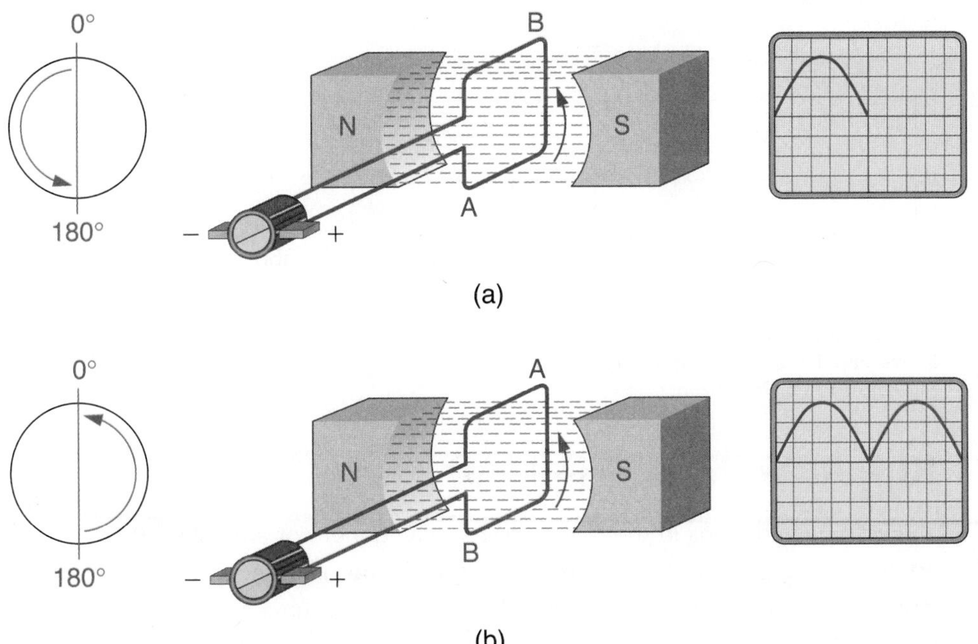

FIGURE 23.4 DC generator operation.

rotation (i.e., the neutral plane), no voltage is generated, and the output returns (again) to 0 V. As you can see, it is the commutator that produces the DC output from the generator.

If we modify the armature and commutator in a DC generator as shown in Figure 23.5a, the generator output improves significantly. As you can see, the armature contains two loop conductors at an angle of 90°. The two loops are connected to a 4-segment commutator, that is, a commutator that has four conductive surfaces that are separated by insulators. When this armature assembly rotates through the magnetic field as shown in Figure 23.5b, the generator produces the output shown. Note that the value of this output voltage never drops to zero because one of the armature loops is always producing a voltage that is coupled to the load by the commutator.

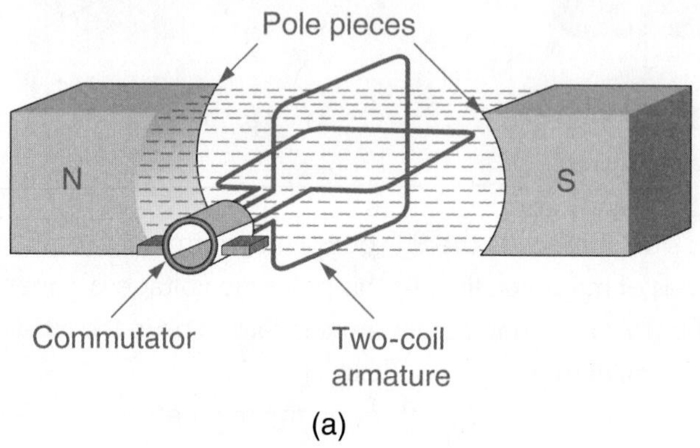

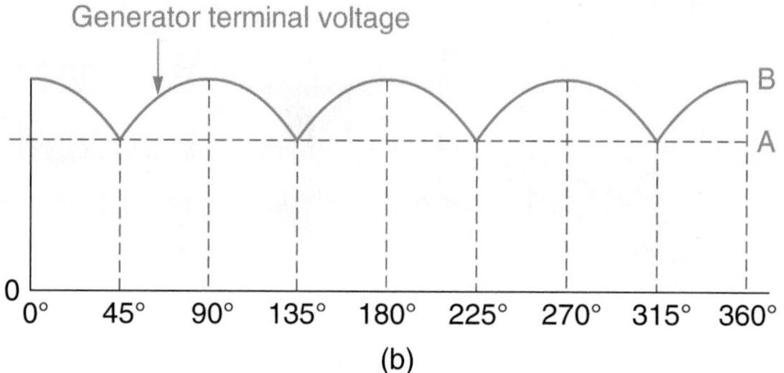

FIGURE 23.5 Two-conductor armature operation.
Source: Based on a figure in NEETS Manual 2, *Introduction to Alternating Current and Transformers*, 1984 edition

Ripple Voltage (V_{RIPPLE})

RIPPLE The voltage variations in the output from a DC generator.

The voltage variations in the output from a DC generator are called **RIPPLE**. The ripple voltage produced by a generator equals the difference between its maximum and minimum output voltages. For example, assume that the waveform in Figure 23.4 has a maximum value of +12 V. The ripple in this case has a value of

$$V_{RIPPLE} = 12\text{ V} - 0\text{ V} = 12\text{ V}$$

Now, assume that the waveform in Figure 23.5 has a maximum value of +12 V (labeled A) and a minimum value of +9 V (labeled B). In this case, the ripple has a value of

$$V_{RIPPLE} = 12\text{ V} - 9\text{ V} = 3\text{ V}$$

The ideal DC generator would produce an output with no variations at all. Therefore, the ideal value of V_{RIPPLE} is 0 V. Consequently, one of the goals of DC generator design and construction is to produce the least amount of ripple possible.

PROGRESS CHECK

1. What is a generator?
2. What is an alternator and how does it differ from a generator?
3. Explain the relationship between slip rings and brushes.
4. What is a commutator? How does it differ from the slip ring in an AC generator?
5. Explain the relationship between armature rotation and the neutral plane.
6. Define ripple voltage (V_{RIPPLE}) and explain its source.

23.2 DC GENERATOR ELEMENTS AND CONSTRUCTION

A **DYNAMO** is an electromechanical device that can convert mechanical energy to electrical energy, or electrical energy to mechanical energy. A *generator* is a dynamo that is configured to convert mechanical energy to electrical energy (as described in the last section). A **MOTOR** is a dynamo that is configured to convert electrical energy to mechanical energy.

Every dynamo has a stationary structure, or **STATOR**, and a rotating structure, or **ROTOR**. The rotor and stator of a basic dynamo are identified in Figure 23.6. The stator is made up of the frame, the pole cores, the interpoles, and windings. The rotor is made up of the armature conductors, the armature core, and the commutator.

DYNAMO An electromechanical device that can convert mechanical energy to electrical energy, or electrical energy to mechanical energy.

MOTOR A dynamo that is configured to convert electrical energy to mechanical energy.

STATOR The stationary structure of a dynamo.

ROTOR The rotating structure of a dynamo.

The Armature and Commutator

A typical generator armature, called a *drum-type armature*, is illustrated in Figure 23.7a. The armature is constructed on a shaft that allows it to spin freely. The *laminated iron core* (or *drum*) is etched with slots that hold the armature windings (conductors). Both ends of the windings are connected to the commutator. In other words, each armature winding begins and ends at the commutator. As highlighted in Figure 23.7b, the commutator has many segments that are separated

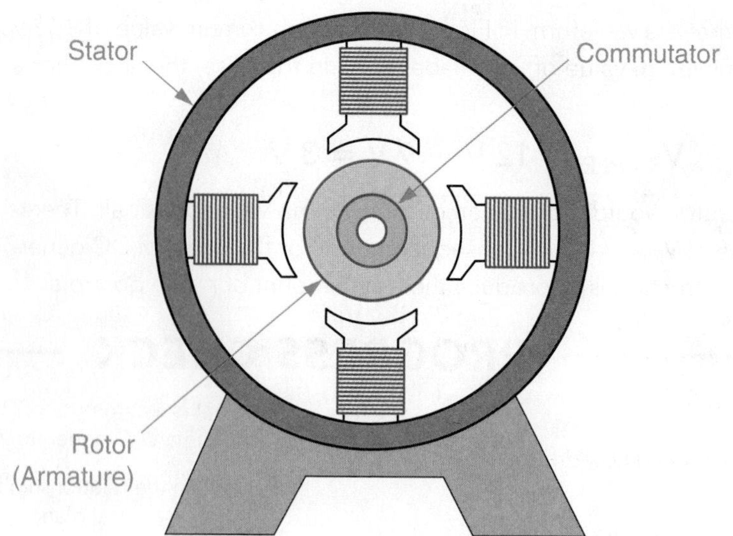

FIGURE 23.6 Dynamo construction.

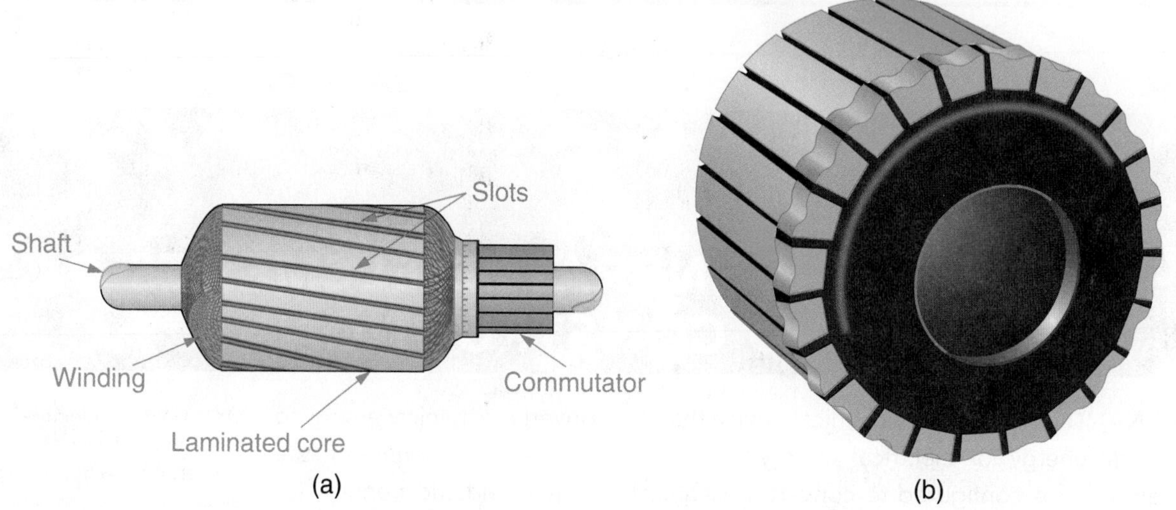

FIGURE 23.7 A drum-type armature and commutator.
Source: Figure (a) is based on a figure in NEETS Manual 5, *Introduction to Generators and Motors*, 1984 edition.

> The function of all DC generator winding configurations is to establish a conduction path between two commutator segments, the magnetic field (via the armature windings), and the load.
>
> IN THE FIELD

SIMPLEX WINDING A term used to describe an armature winding that consists of a single conductor.

MULTIPLEX WINDING A term used to describe an armature winding that consists of multiple conductors.

by insulating layers (rather than the two or four segments portrayed in the simplified generator shown earlier).

The generator armature may contain a single winding, or several independent windings connected in parallel. A single armature winding is referred to as a **SIMPLEX WINDING**. More than one winding is generally referred to as a **MULTIPLEX WINDING**.

Eddy Current Losses

You may recall that iron-core transformers are susceptible to power losses caused by *eddy currents*. Eddy currents are produced when an iron core is exposed to a magnetic field.

Like the iron core in a transformer, the armature drum is susceptible to power losses from eddy currents. Eddy currents (and the resulting losses) are minimized by manufacturing laminated armature drums, similar to the laminated iron cores used in transformers.

Generator Brushes

As you were told earlier, the commutator is in physical contact with two or more **BRUSHES** (stationary contacts) that couple the generated voltage from the armature to the load. A brush rigging is shown in Figure 23.8a.

BRUSHES The stationary contacts that couple the output from the rotor of a DC generator to the load.

(a)

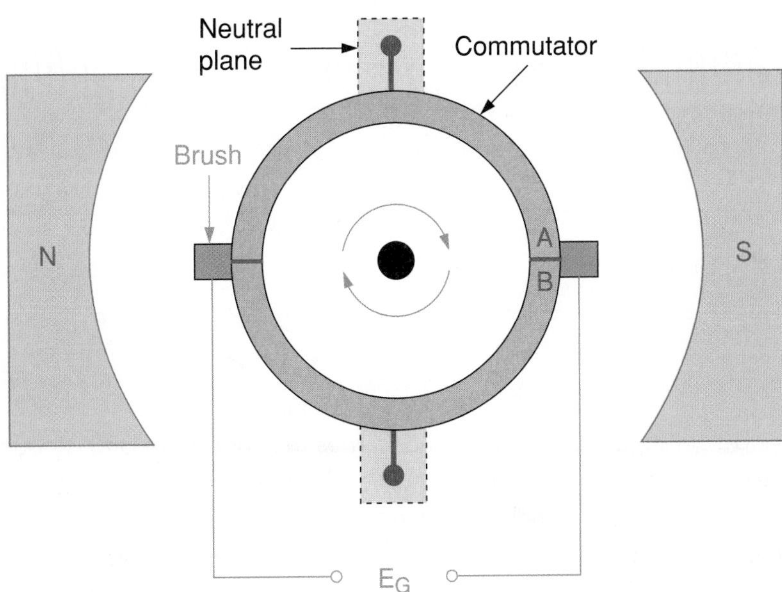

(b) End-view of a commutator and brushes

FIGURE 23.8 Generator commutator and brushes.
Source: Courtesy of Illinois Electric Works (www.illinoiselectric.com)

The switching function is shared by the commutator and the brushes, which normally contain carbon, carbon graphite, or a copper-filled carbon mixture. Low contact resistance, some controlled internal resistance, and good surface-wear qualities are essential for optimal brush operation.

IN THE FIELD

The end view in Figure 23.8b represents the commutator and brushes. As the commutator rotates, the brushes provide an electrical connection between two commutator segments. For example, in Figure 23.8b, each brush is in contact with both commutator segments. Because the brushes are conductors, each is shorting out the insulation between the two segments. However, because the armature loop conductors are positioned in the *neutral plane* (i.e., outside the magnetic field), there is no voltage across each segment pair at the instant they are shorted. In other words, there is little or no voltage across the segments at the instant they are physically connected by the brushes.

Field Poles

The magnetic field in a DC generator is typically produced by electromagnetic poles that are part of the stator.

In Chapter 10, you learned that iron exhibits magnetic properties when wrapped with a current-carrying coil as illustrated in Figure 23.9. The coil current produces lines of magnetic flux that are concentrated by the iron core. The flux produced by the iron-core coil varies directly with the number of turns (N) and the current through the coil (I). As such, the strength of the magnetic field can be controlled by varying the current through the coil. Note that the orientation of the magnetic field (N versus S) is determined by the direction of the coil current. With these points in mind, let's turn our attention to the electromagnetic poles used to produce the magnetic field in a DC generator.

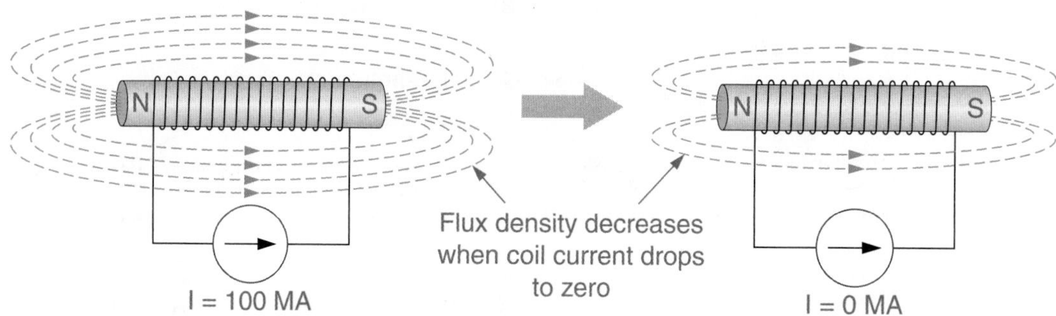

FIGURE 23.9 The relationship between current and its resulting magnetic field.

FIELD POLE A stator pole that supports a current-carrying winding used to produce a magnetic field.

The stator in Figure 23.10 contains four iron cores called **FIELD POLES**. Each field pole supports a winding that carries the current used to produce a magnetic field. The shape of the outer end of each pole is designed to produce a more stable magnetic field. They are curved to match the shape of the drum armature so the distance from the field pole to the armature windings is constant.

The field coils are wired in *series* (for high-voltage, low-current outputs), in *parallel* (for low-voltage, high-current outputs), or in *series-parallel* (for moderate

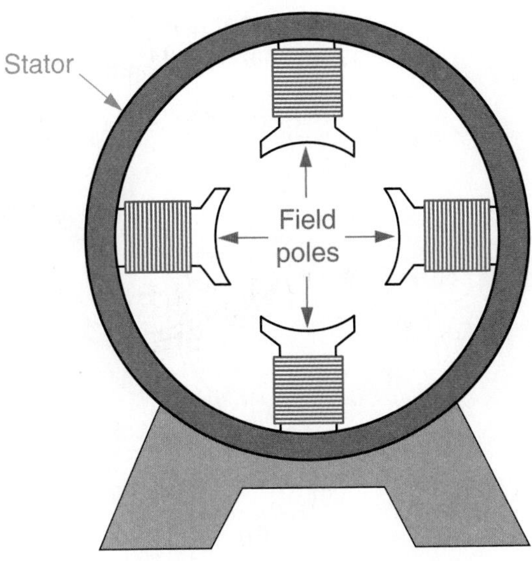

FIGURE 23.10 Field poles that make up the stator.

voltage and current outputs). Note that series-wound coils are described as **WAVE WOUND**, parallel-wound coils are described as **LAP WOUND**, and series-parallel coils are described as **FROG-LEG WOUND**. Of the three, series-parallel windings (frog-leg wound) are the most commonly used. We will examine field coil wiring in more detail later in this chapter.

WAVE WOUND A term used to describe a wiring scheme where the field pole windings are wired in series with each other.

LAP WOUND A term used to describe a wiring scheme where the field pole windings are wired in parallel with each other.

FROG-LEG WOUND A term used to describe a wiring scheme where the field pole windings are wired in series-parallel with each other.

> The armature in a DC generator must be wired so that each conductor originating at the commutator end of the device passes through one magnetic polarity (N or S), and returns to the commutator through the opposite magnetic polarity (i.e., on the opposite side of the armature core).
>
> IN THE FIELD

Compensating Windings

When current passes through the rotating armature windings in a generator, magnetic flux is generated around the wires. This magnetic flux can interfere with the stationary magnetic field generated by the field poles, as illustrated in Figure 23.11.

The small circles in Figure 23.11a represent two ends of an armature winding. The current through the coil produces magnetic flux around the rotating conductor. This flux interferes with the magnetic field produced by the field poles, as shown in Figure 23.11b. The interaction between the magnetic fields causes the neutral plane to shift in the direction of armature rotation. This phenomenon is known as **ARMATURE REACTION**. For the commutator to operate properly, one of two things must happen:

ARMATURE REACTION A term used to describe the interaction between the magnetic fields that causes the neutral plane to shift in the direction of armature rotation.

- The brushes must be shifted so that they are located in the repositioned neutral plane.
- The neutral plane must be shifted back to its original position so that the brushes are positioned properly.

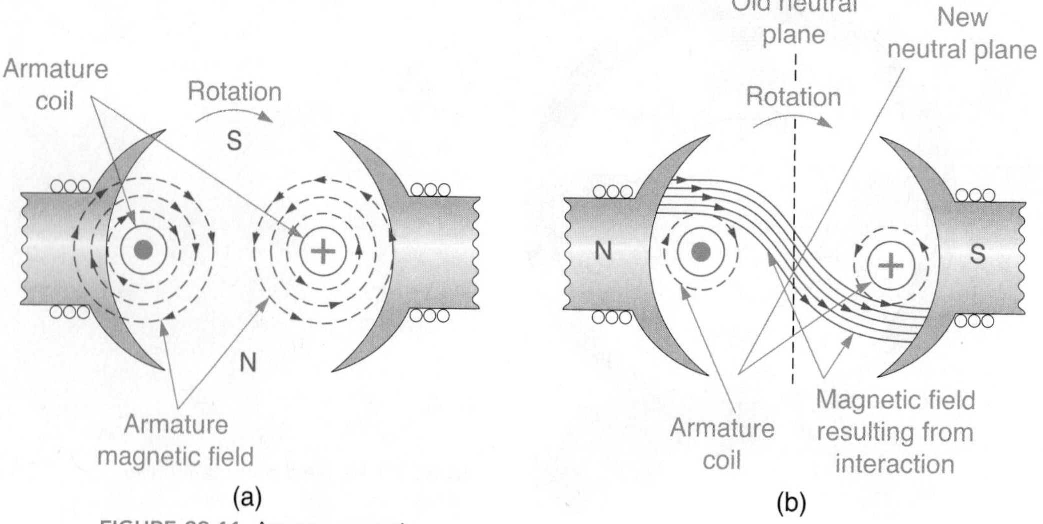

FIGURE 23.11 Armature reaction.
Source: Figure 23.11 is based on a figure in NEETS Manual 5, *Introduction to Generators and Motors*, 1984 edition.

COMMUTATING POLES Series-wound poles that produce magnetic flux that counters the flux produced by the armature current, restoring the neutral plane to its original orientation.

INTERPOLES Another name for the commutating poles.

COMPENSATING WINDINGS Short, series-wound coils wrapped at 90° angles to the field coils to help restore the neutral plane to its original orientation.

In smaller DC generators, the simplest solution is to position the brushes to match the altered neutral plane. Larger DC generators are designed so that the neutral plane is shifted to its "proper" position.

Shifting the neutral plane can be accomplished using coils that are series wound on **COMMUTATING POLES** (or **INTERPOLES**) as shown in Figure 23.12. The magnetic flux produced by the interpole windings counters the flux produced by the armature current, restoring the neutral plane to its original orientation. Neutral plane correction is also accomplished using **COMPENSATING WINDINGS**. These are short, series-wound coils that are wrapped at 90° angles to the field coils.

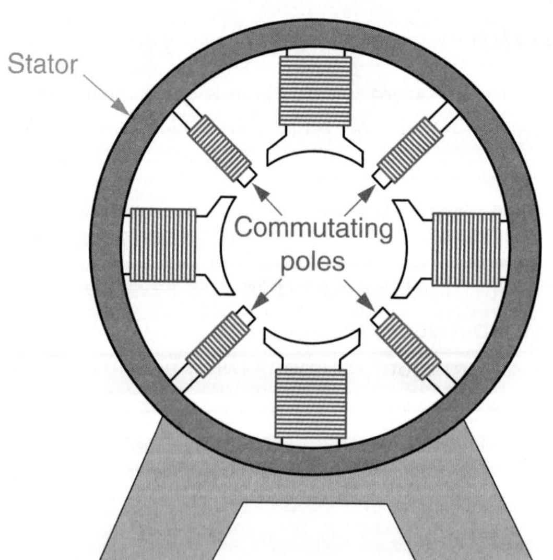

FIGURE 23.12 Stator commutating poles (interpoles).

Putting It All Together

A dynamo is a machine that can be manufactured to operate as a generator or a motor. A dynamo consists of a stator (stationary structure) and a rotor (rotating structure). The rotor contains the armature and commutator. The stator contains the field poles and commutating poles (or interpoles).

The field coils in a DC generator are electromagnets. The coils for these magnets are wired in series, parallel, or in series-parallel. Series (wave) wound generators are typically found in high-voltage, low-current applications. Parallel (lap) wound generators are typically found in high-current, low-voltage applications. Series-parallel (frog-leg) wound generators are the most common and are used in moderate current and/or voltage applications.

The main advantages of using electromagnetic poles are increased field strength and control of the strength of the fields. By varying the input voltage, the field strength is varied. By varying the field strength, the output voltage of the generator can be controlled.

As the rotor in a DC generator turns, a voltage is induced across the windings. The induced voltage generates currents that, in turn, generate magnetic flux. The flux generated by the armature windings interferes with the magnetic field produced by the field coils, causing the neutral plane to shift. This phenomenon, called *armature reaction*, can be countered using compensating windings that are series wound on commutating poles (or interpoles).

STAYING SAFE Additional maintenance must be performed on a DC machine due to the number of moving and wear parts that are in physical contact with each other. Very basic steps include insulation and circuit resistance testing. The insulation resistance value must be at least 100 MΩ, adjusted for temperature. (Resistance readings must also be corrected for temperature to the value that is often indicated on the motor nameplate.) Low resistance readings normally indicate a buildup of carbon dust and the need for blowing out the inside the motor.

PROGRESS CHECK

1. Define and compare a dynamo, a generator, and a motor.
2. Define and compare a rotor and a stator.
3. What are armatures and why are they laminated?
4. Define and compare simplex and multiplex windings.
5. What is a field pole?
6. Define and compare wave wound, lap wound, and frog-leg wound field coils.
7. What is armature reaction and how is it overcome?

23.3 SERIES GENERATORS

It was stated earlier that the field coils in a DC generator are wired in series, parallel, or series-parallel. In this section, we will examine series (wave) generators.

Series Generator Connections

SELF-EXCITED GENERATOR
One that provides its own field coil current via its armature.

The field poles of a series generator are wired as indicated in Figure 23.13. As the schematic indicates, the circuit is wired so that the armature current passes through the field coil. The generator in Figure 23.13 is classified as a **SELF-EXCITED GENERATOR** because it provides its own field coil current (via the armature).

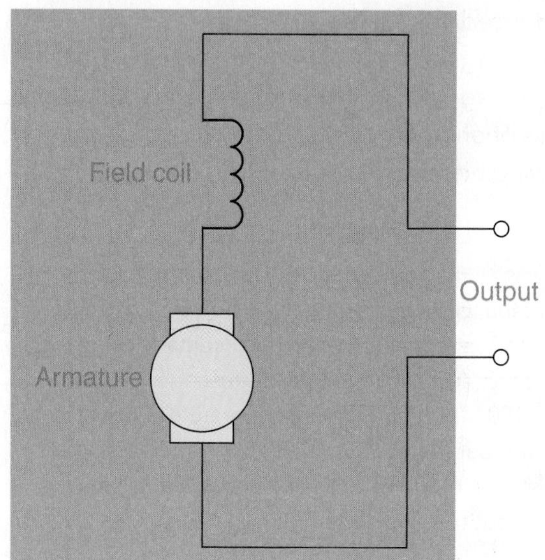

FIGURE 23.13 A series generator.

It should be noted that the coil symbol in the series generator schematic actually represents more than one field coil. For example, Figure 23.6 shows a DC generator that has four field cores, each supporting a field coil. Thus, the coil symbol in the schematic for that generator represents four field coils.

Generator Voltage

The output voltage from a DC generator depends on the number of armature conductors, the number of field poles, the flux per pole, and the rotational speed of the armature. By formula,

(23.1)

$$E_G = \frac{z}{a} \times \Phi \times P \times \frac{rpm}{60}$$

where

E_G = the generator voltage
z = the number of armature conductors
a = the number of parallel paths
Φ = the flux per pole, in webers (Wb)
P = the number of poles
rpm = the rotational speed, in revolutions per minute

Equation (23.1) is not used in any practical situations. However, it does establish several relationships. As the equation indicates, the output voltage from a DC generator varies directly with:

- The number of poles (field coils).
- The amount of flux per pole.
- The rotational speed of the armature.

If any of these values increases (or decreases), so does the output voltage from the generator.

The output voltage from a series generator can also be affected by the load current. For example, take a look at the loaded generator in Figure 23.14. The

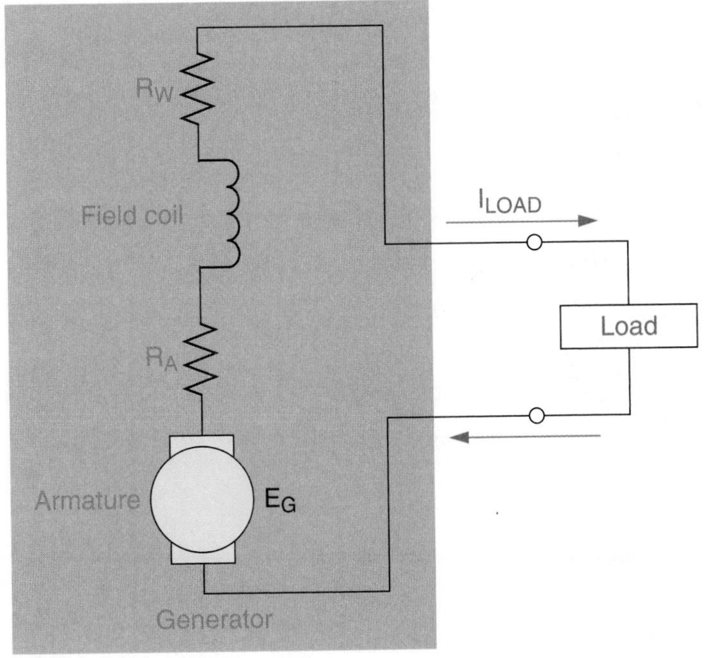

FIGURE 23.14 A loaded series generator.

schematic includes two resistors that represent the internal resistance of the armature (R_A) and the winding resistance of the field coil (R_W). Note that the armature resistance represents the combined effects of all the resistive elements in the armature, including the armature, commutator, and brush resistances.

According to Ohm's law, the current passing through R_A and R_W develops a voltage drop across each resistance. These combined voltage drops can be expressed as the product of the load current (I_L) and the total resistance, as follows:

$$I_L(R_A + R_W)$$

According to Kirchhoff's voltage law (KVL), the generator output voltage must equal the difference between the voltage generated by the armature (E_G) and the voltage drop across the generator's internal circuitry. By formula,

$$V_{OUT} = E_G - I_L(R_{INT} + R_W) \tag{23.2}$$

As this equation indicates, an increase in load demand causes a decrease in generator output voltage. By the same token, a decrease in load demand causes an increase in generator output voltage.

Open Load Condition

The series generator has another characteristic that distinguishes it from other DC generators: It must be connected to a load in order to generate an output voltage. The basis for this circuit requirement is illustrated in Figure 23.15.

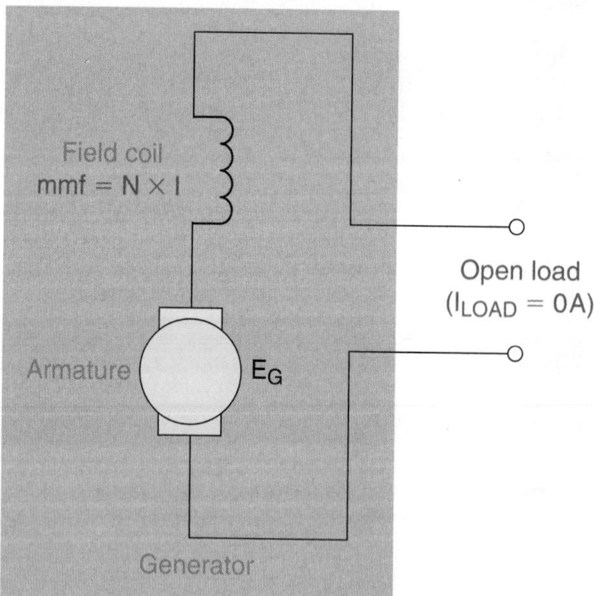

FIGURE 23.15 A series generator with an open load.

As you know, a complete path is required for current. When the load is open as shown in Figure 23.15, there is no path for current through the generator's internal circuit. As a result, the field coil does not generate the magnetomotive force (mmf) needed to produce a magnetic field, and the output voltage equals 0 V.

PROGRESS CHECK

1. How are the field coils wired in a DC series generator?
2. Explain why the series generator is classified as a self-excited generator.
3. List and explain all of the factors that determine the output voltage from a DC generator.
4. What effect does an open load have on the output voltage of a series generator? Explain why this is the case.

23.4 PARALLEL GENERATORS

It was stated earlier that the field coils in a DC generator are wired in series, parallel, or series-parallel. In this section, we will examine parallel (lap) generators.

Parallel Generator Connections

The field coil of a parallel generator is wired as shown in Figure 23.16. Like the coil in the series generator schematic, the coil symbol in Figure 23.16 represents more than one field coil.

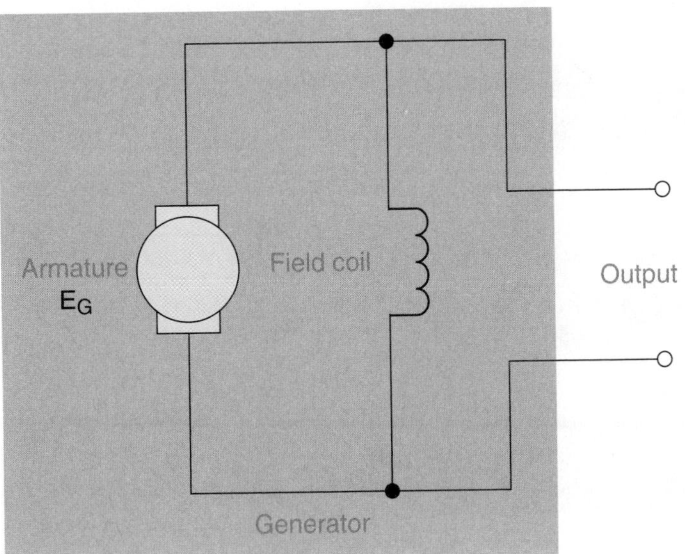

FIGURE 23.16 A parallel generator.

As the schematic shows, the circuit is wired so that the field coil is in parallel with the armature. As such,

- The voltage across the field coil equals the armature voltage.
- The current through the field coil is relatively independent of the armature current.

These statements need no explanation, as they are characteristic of any parallel circuit. At the same time, the relationship between load current and generator output voltage needs to be examined.

Figure 23.17 shows a load connected to the parallel generator output. If the load current demand increases, the generator output voltage decreases slightly. Here's why: The internal resistance of the armature (R_A) is included in Figure 23.17. When the load demand increases, so does the armature current (I_A). This current generates a voltage across R_A that equals $I_A \times R_A$. The generator output voltage (E_{OUT}) equals the difference between this voltage and the voltage generated by the armature (E_G), as follows:

(23.3)

$$E_{OUT} = E_G - I_A R_A$$

As this relationship indicates, the generator output voltage decreases when the load demand increases.

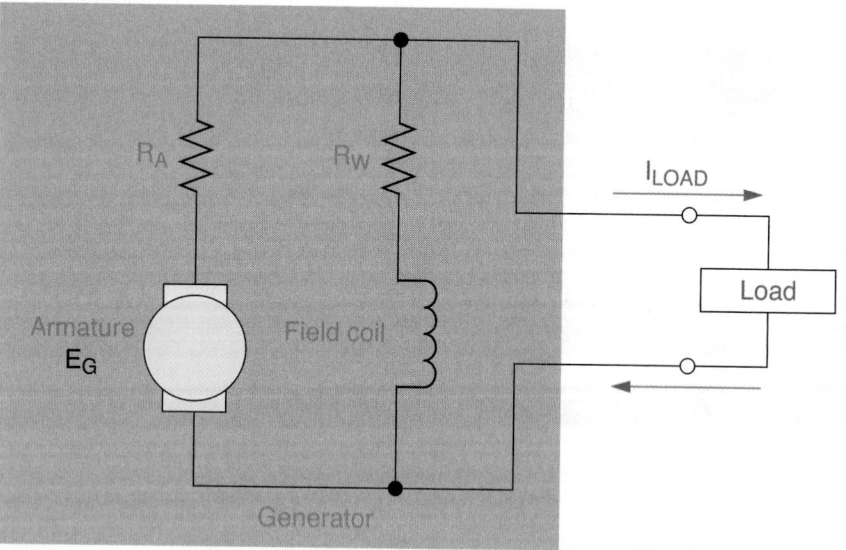

FIGURE 23.17 A loaded parallel generator.

Voltage Buildup

Like the series generator discussed earlier, the generator in Figure 23.17 is *self-excited*, meaning it provides its own field coil current. Here's how it works:

1. Current is generated in the generator armature as it spins.
2. The armature current passes through the field coils.
3. The current through the field coils generates a magnetic field.
4. The magnetic flux produced by the field coils permeates the armature, inducing a voltage across its windings.

As you can see, self-excitation is a continuing cycle with each step leading to the next. But how does the cycle begin?

When a DC generator begins to rotate (as the result of an external force), residual magnetism in the field cores induces a slight voltage across the armature windings. This voltage generates a low-level current that passes through the field coils. The current through the field coils increases the flux density, causing the voltage induced across the armature to increase. The increased armature voltage increases the current through the field coils, and so on. If the generator is operating properly, the cycle described here continues until the generator output voltage builds up to its rated value.

Output Voltage Control

The output voltage from a shunt generator can be controlled by placing a rheostat in series with the field coil, as shown in Figure 23.18. The control rheostat (R_{CON}) limits the current through the field coil and, therefore, the voltage induced across the armature. The higher the setting of R_{CON}, the lower the field coil current, and the lower the generator output voltage.

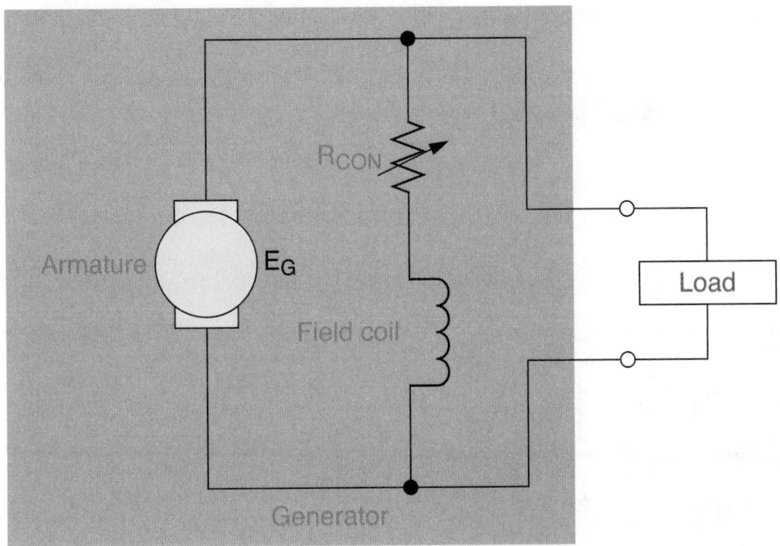

FIGURE 23.18 Parallel generator output voltage control.

PROGRESS CHECK

1. How are the field coils wired in a parallel DC generator?
2. Explain why the field coil current is relatively independent of armature current in a parallel DC generator.
3. Explain why generator output decreases when load demand increases.
4. Explain how a parallel DC generator produces its own field coil current.
5. How is the output voltage of a parallel DC generator usually controlled?

23.5 COMPOUND GENERATORS

The compound generator uses both series and parallel field coil windings to produce an output voltage that is more stable than that of either series or parallel generators. Because it provides the most stable output voltage, the series-parallel generator is used more often than the other two generators.

Compound Generator Winding

The field poles of a compound generator are wired as shown in Figure 23.19. As you can see, the generator contains both series-connected and parallel-connected (shunt) field coils. In practice:

- Each coil symbol in the schematic represents more than one field coil.
- Each generator field pole supports two coils; one series-connected coil and one parallel-connected coil.

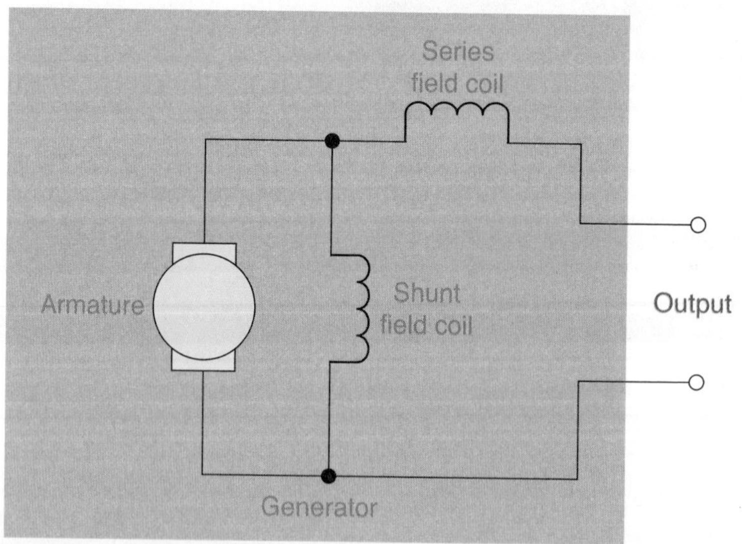

FIGURE 23.19 A compound generator.

Output Voltage Stability

When no load is connected to the output terminals of a compound generator, the armature current (I_A) passes through the shunt field coil as shown in Figure 23.20a. With no current through the series coil, the voltage at the output terminals of the generator equals the armature voltage (E_G).

When a load is connected to the generator output terminals, the armature current splits as shown in Figure 23.20b. In this case, the load current passes through the series field coil, increasing the magnetic flux produced by that coil. With the series and shunt field coils wrapped around the same field poles, the increased flux

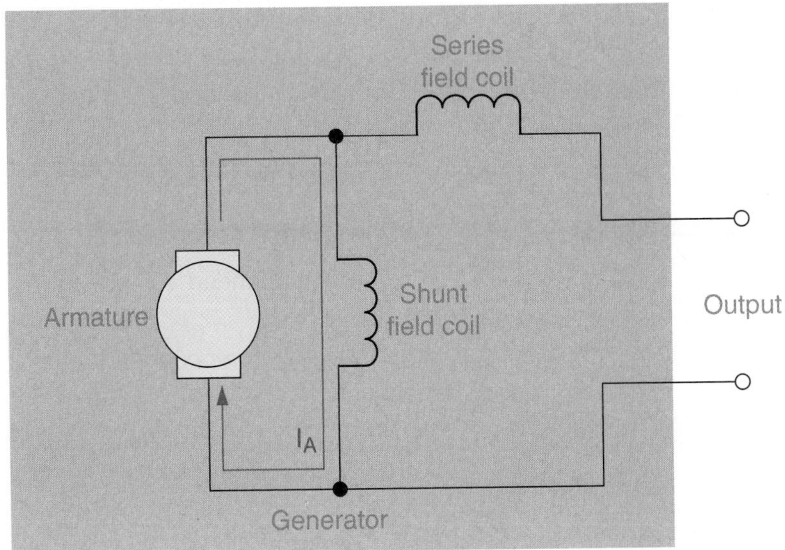

(a) Open load current

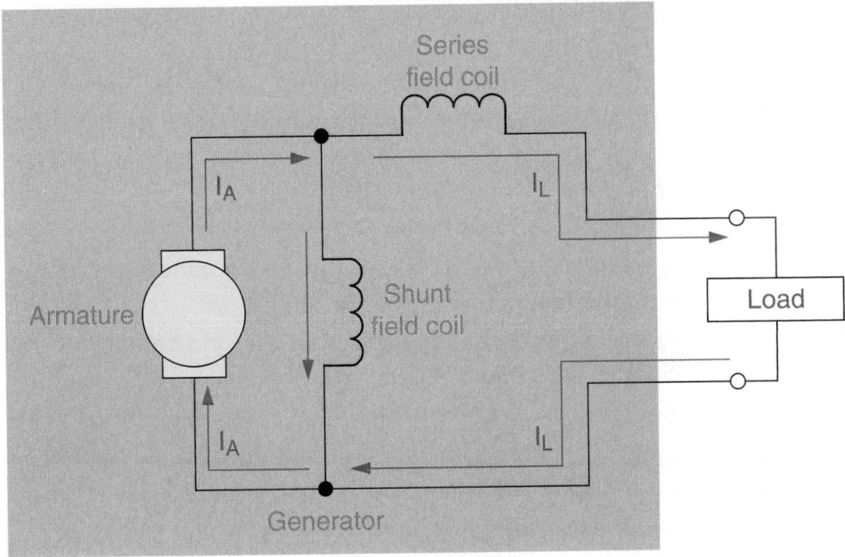

(b) Compound generator with load

FIGURE 23.20 Compound generator currents.

generated by the series coil adds to the flux generated by the shunt coil. As a result, the voltage induced across the armature increases. This offsets the voltage drop across the series field coil, keeping the generator output voltage stable.

Compounding

COMPOUNDING The process of compensating for the effects of predictable load demands on a generator.

COMPOUNDING is the process of compensating for the effects of predictable load demands on a generator. For example, the field coil in Figure 23.20 compensates for the decrease in generator output voltage that normally occurs when a load is connected to the circuit. The effect of this compounding is illustrated in Figure 23.21. The lower curve represents the change in output voltage that a parallel generator experiences when load demand increases. Note that the generator output is described as being **UNDER-COMPOUNDED** because the generator does not compensate for the effect of the load demand.

UNDER-COMPOUNDED A term used to describe the condition where a generator is not adequately compensating for changes in load demand.

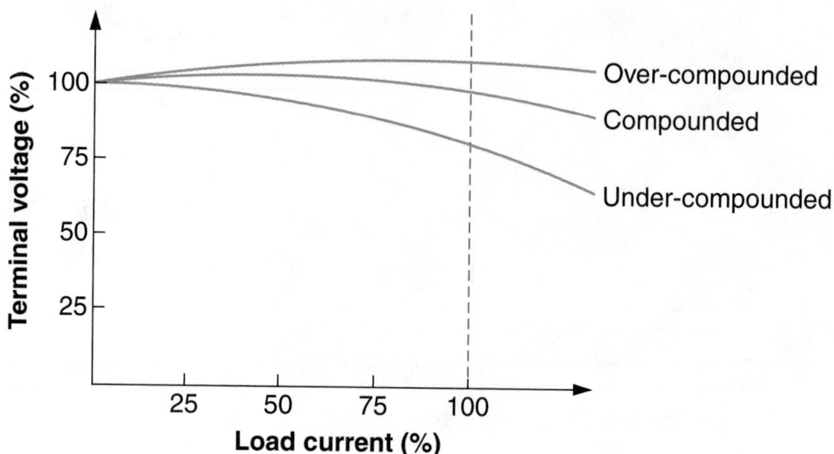

FIGURE 23.21 Generator compounding.

The middle curve in Figure 23.21 shows that the output from the compound generator is relatively stable against changes in load demand. In this case, the generator output is referred to as being **COMPOUNDED**.

COMPOUNDED A term used to describe the condition where a generator is adequately compensating for changes in load demand.

The upper curve in Figure 23.21 is labeled **OVER-COMPOUNDED**. Over-compounding is a design technique that provides added insurance against the effects of load demands on the generator output voltage.

OVER-COMPOUNDED A design technique that provides added insurance against the effects of load demands on the generator output voltage.

Over-compounding is accomplished by increasing the number of turns in the series field coil, as illustrated in Figure 23.22a. When current is drawn through the field coils, the magnetic flux produced by the added series coil turns actually causes the output from the generator to *increase* slightly. The **DIVERTER** in Figure 23.22b provides a means of adjusting the output from the generator to match the requirements of its load. The diverter provides a path for current to bypass the field coil. By adjusting the component (and therefore, the current through it), the generator can be set to maintain 100% of its output voltage over the range of load currents.

DIVERTER A circuit that provides a path for current to bypass the field coil, thereby helping a generator to match the requirements of its load.

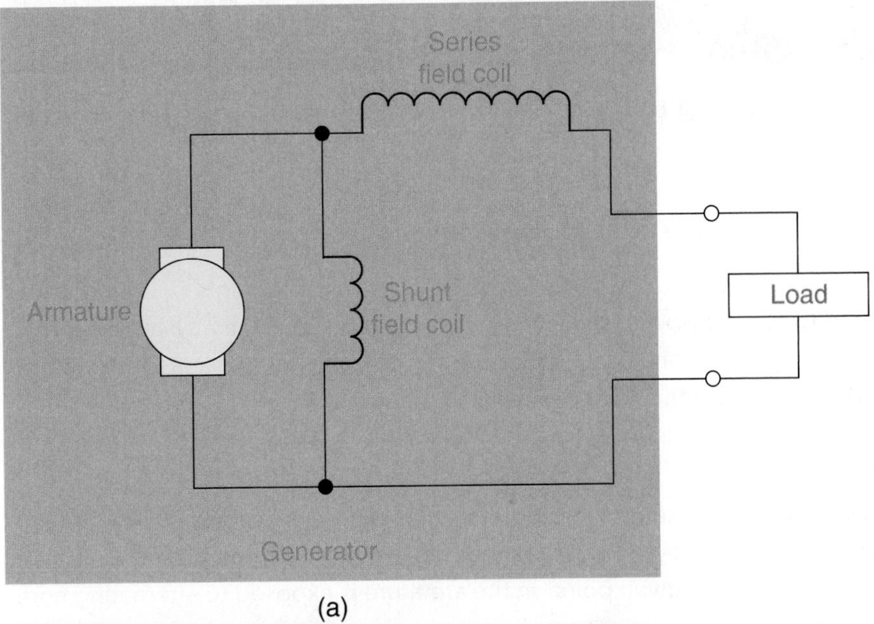

(a)

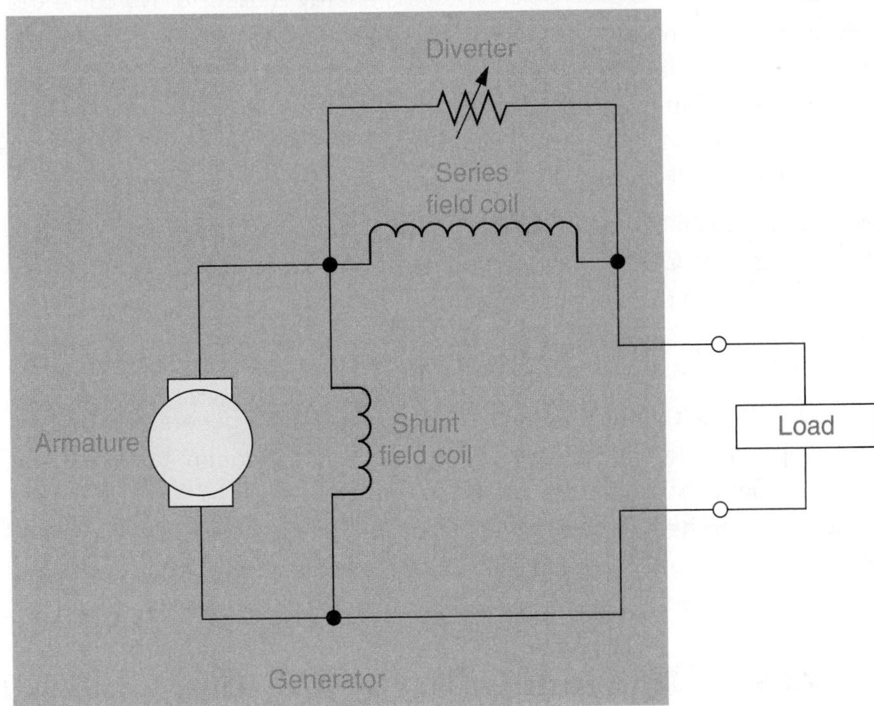

(b)

FIGURE 23.22 Over-compounding.

PROGRESS CHECK

1. Explain why the compound generator is used more often than the other two generator types.
2. Describe how the field coils of a compound generator are wound.
3. Explain why the output voltage of the compound generator is more stable than the other two designs.
4. Explain the concept of compounding. Describe over-compounding and under-compounding.
5. What is a diverter and what is it used for?

23.6 MORE ON DC GENERATORS

In this section, we will briefly discuss some miscellaneous topics relating to DC generator operation.

Generator Power Losses

Earlier in the chapter, you were told that the armature drum in a DC generator is laminated to reduce power losses from eddy currents. As with any component or circuit, there are additional factors that contribute to power loss in a DC generator. These factors are identified in the following paragraphs.

Hysteresis Loss

You may recall that hysteresis loss occurs when energy is used to change the orientation (polarity) of any residual magnetic flux in iron. As the armature in a DC generator rotates, any given point on the armature is exposed to alternating north (N) and south (S) poles. As each pole is passed, energy is used to overcome the magnetic orientation that was established by the previous pole (i.e., changing from N to S, or vice versa). Energy is also lost as the magnetic orientation changes in the field poles and interpoles.

Copper Loss

Like any electric circuit that has measurable values of current and resistance, power is lost in the form of heat. Contributors to copper loss (or I^2R loss) in a DC generator include the conductors in the armature, the brushes, and the magnetic pole windings.

Friction Loss

Energy is lost as a result of the friction that occurs when a DC generator is operating. For example, the friction between the brushes and the commutator causes a rise in temperature as the armature rotates.

The losses listed here all combine to cause the temperature of a DC generator to rise. The heat produced by smaller DC generators normally dissipates into the space around the generator without any additional cooling steps required.

Connecting DC Generators in Parallel

Two or more DC generators are often connected in parallel to provide the same load current capability as a larger generator. This is done for a number of reasons. First, a group of parallel-connected generators is more flexible. When current demand is low, the number of operating generators can be reduced. Conversely, the number of operating generators can be increased to meet higher current demands. Second, should one of the generators fail, it can be repaired and/or replaced without disrupting the other generators.

There are five requirements for connecting DC generators in parallel. These requirements are as follows:

- The generators should have nearly identical output voltage ratings and voltage versus load current characteristics.
- The generators must be connected positive-to-positive and negative-to-negative.
- A means of disconnecting each generator from the circuit must be provided.
- An **EQUALIZING CONNECTION** must be made between the generators to ensure that they respond equally to any change in load demand.
- The generators should have the same **PRIME MOVER** (external mechanical driver).

EQUALIZING CONNECTION A connection between the generators to ensure that they respond equally to any change in load demand.

PRIME MOVER The external mechanical driver for a generator.

The circuit in Figure 23.23 shows how the first four of these requirements can be met. As you can see, the two generators have the same voltage ratings. Assuming

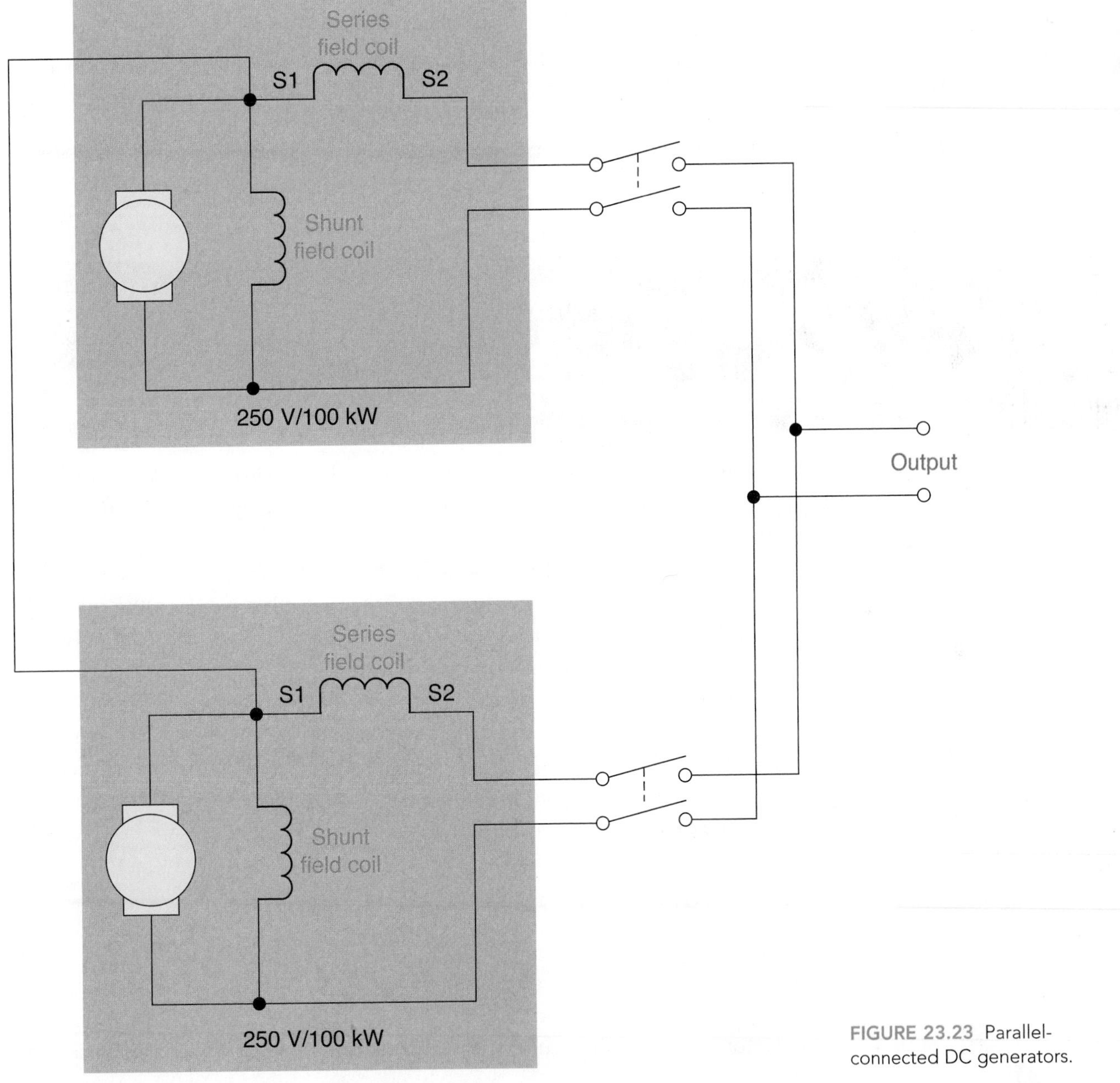

FIGURE 23.23 Parallel-connected DC generators.

the generators have near-identical voltage vs. load current curves, the two will provide nearly identical currents to a common load. The generators are connected positive-to-positive and negative-to negative, and each circuit also has a SPDT switch that can be used to disconnect the generator.

EQUALIZER Another name for an equalizing connection.

The connection between the S1 points in the two generators is the equalizing connection, or **EQUALIZER**. The equalizer alters the circuit so that the two series coils are in parallel with each other. With equal voltages across the two field coils, the currents through them must also be equal. As such, any change in load demand is split equally between the two generators.

PROGRESS CHECK

1. List and describe the main sources of power loss in DC generators.
2. Under what circumstances would DC generators be connected in parallel.
3. List and describe the requirements for connecting two or more DC generators in parallel.
4. Explain how an equalizing connection ensures that two (or more) generator outputs will respond equally to a change in load demand.

23.7 DC MOTORS: BASIC CONCEPTS

A DC motor is essentially a DC generator that is operated in reverse. Instead of converting mechanical energy to electrical energy (like a DC generator), a DC motor converts electrical energy to mechanical energy. This concept is illustrated in Figure 23.24. The armature in the DC generator is turned through a magnetic field by a prime mover (external mechanical source). The interaction between the magnetic field and the armature conductors generates electrical current. This current is coupled from the armature to the generator output via the commutator and brushes.

The DC motor, on the other hand, converts electrical energy from an external DC source to mechanical energy. The source current is coupled to the armature conductors via the commutator. The current through the armature conductors causes the armature to rotate through the magnetic field. The rotating armature provides the desired mechanical (rotational) energy.

Direction of Armature Rotation

The direction of armature rotation is determined by the direction of the current through its conductors. The relationship between the direction of current and conductor motion is illustrated in Figure 23.25. As the figure indicates, the force on a conductor is upward when current is in one direction, and downward when current

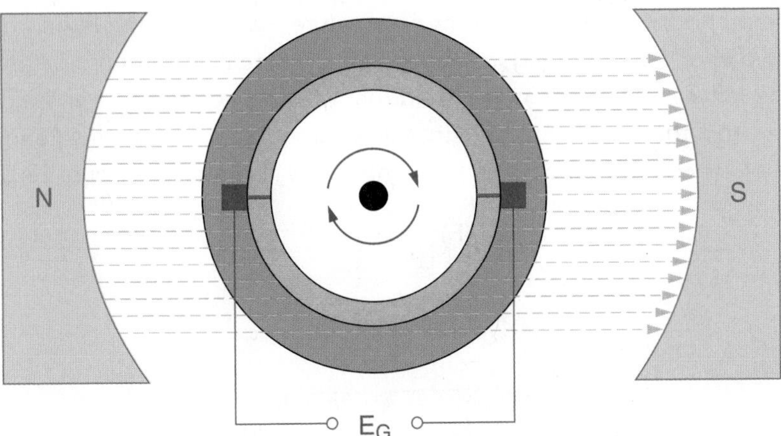

(a) A generator produces an output voltage (E_G) when turned by a primary mover.

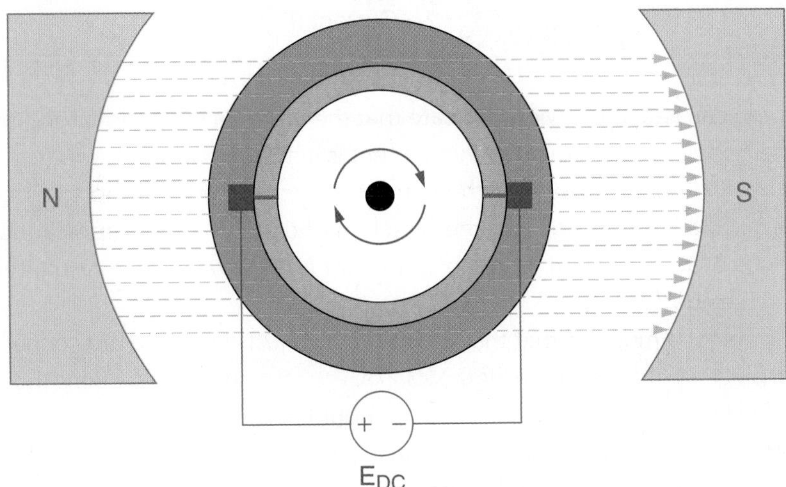

(b) A voltage source (E_{DC}) applied to the commutator causes a DC motor to turn.

FIGURE 23.24

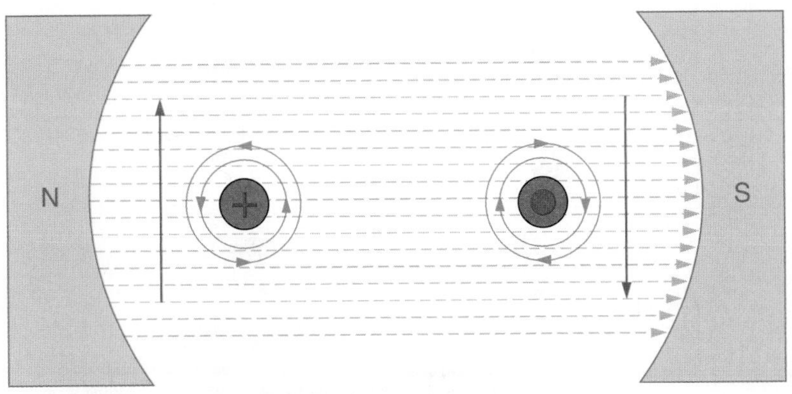

FIGURE 23.25 DC generator and motor functions.

is in the other direction. (The effects of armature reaction on the larger magnetic field have been left out of Figure 23.25 to simplify the illustration.)

The *left-hand rule* is a memory aid that will help you remember the relationship between the direction of electron flow through a magnetic field and the direction of conductor motion. The left-hand rule is illustrated in Figure 23.26.

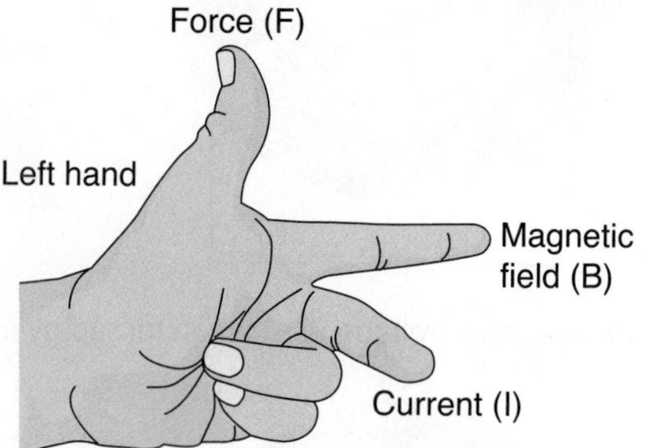

FIGURE 23.26

As you can see, the fingers indicate that the force on the conductor is upward when the middle finger indicates that current is in the direction shown. As a result, the conductor moves upward. If the current reverses, the left hand turns over so that the middle finger is again in the direction of current. When repositioned, the fingers indicate that the force on the conductor is downward. As a result, the conductor moves downward.

Assume for a moment that the two conductors just described are connected to a central pivot point as shown in Figure 23.27a. With the current directions shown, the left-hand conductor moves upward and the right-hand conductor moves downward until both reach the neutral plane as shown in Figure 23.27b. Now, assume that while the conductors are in the neutral plane, their current directions change as shown in Figure 23.27c. With the new current directions, the momentum of the conductors will

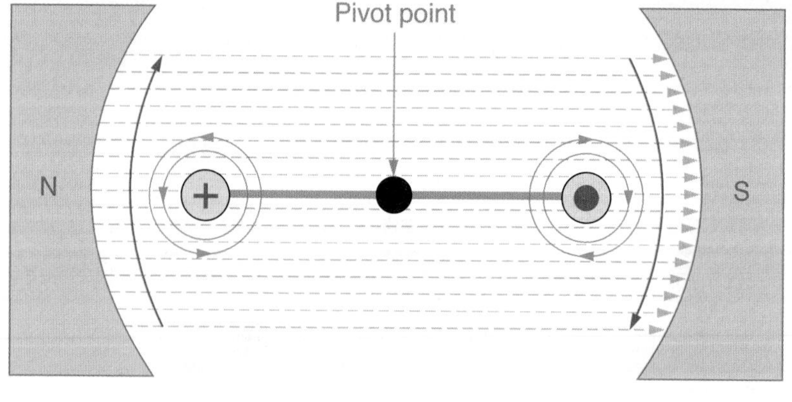

(a) Initial directions of current and motion

FIGURE 23.27

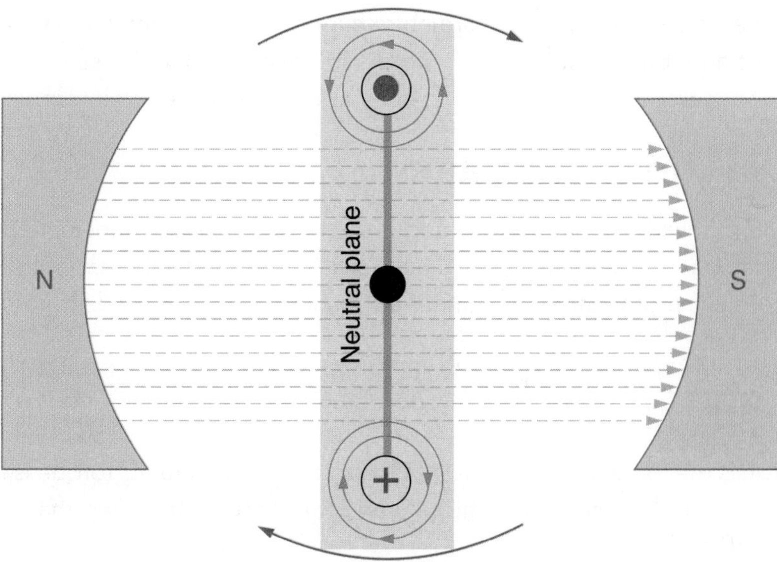

(b) Momentum carries the conductors through the neutral plane

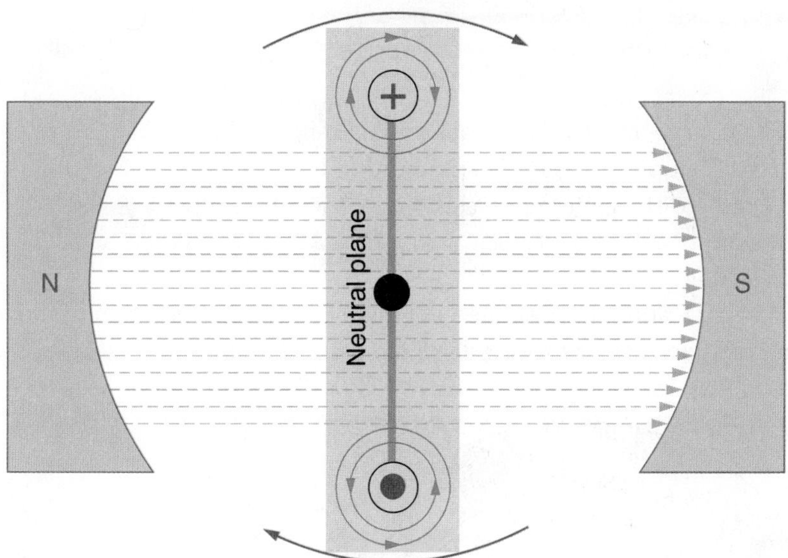

(c) Current directions are reversed while the conductors are in the neutral plane

FIGURE 23.27 *(Continued)*

keep them moving in a clockwise direction through the magnetic field (as shown in 23.27c). As you can see, the conductors can be made to spin about the pivot point by changing the direction of current while they are in the neutral plane. This is a fundamental principle of DC motor operation, and is referred to as **MOTOR ACTION**.

MOTOR ACTION Making a conductor (rotor) spin about a pivot point by changing the direction of current while it is in the neutral plane.

Armature Reaction

As is the case in DC generators, current through the armature conductors in a DC motor produces magnetic flux that:

- Alters the magnetic field produced by the field coils.
- Causes the neutral plane to shift.

Recall that this response to the armature current is called *armature reaction*. The effects of armature reaction in DC motors are minimized in the same manner as with DC generators: through the use of compensating coils and interpoles.

DC Motor Advantages

DC motors have two primary advantages over AC motors: First, the speed of a DC motor can be varied to suit specific applications by varying the armature voltage. Second, a DC motor can provide near-constant *torque* over a wide range of operating speeds.

Torque

TORQUE The force required to rotate an object about a point.

TORQUE is the force required to rotate an object about a point. Torque is found as the product of the force of rotation (F) and the distance between that force and the axis of rotation (r). By formula,

(23.4)
$$T = F \times r$$

where

T = torque, measured in pound-feet (lb-ft)
F = the force, measured in pounds (lb)
r = the distance from the axis of rotation, in feet (ft)

The values of F and r are identified in Figure 23.28. An application of the relationship in Equation (23.4) is provided in Example 23.1.

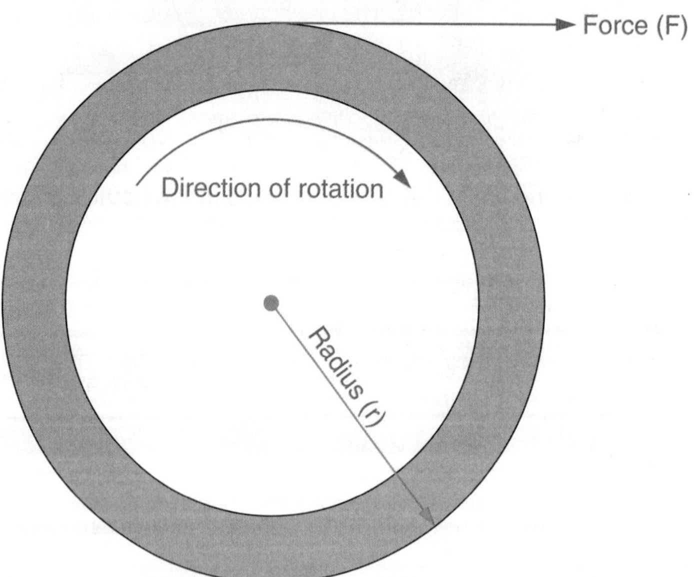

FIGURE 23.28 Rotational force and radius.

EXAMPLE 23.1

A motor like the one in Figure 23.28 is driving a winch with a 6-in diameter. Calculate the torque required for the winch to lift 180 pounds.

SOLUTION

The radius of a circle is half its diameter, so the winch has a value of

$$\begin{aligned} r &= \frac{6 \text{ inches}}{2} \\ &= 3 \text{ inches} \\ &= 0.25 \text{ feet} \end{aligned}$$

Now, the required torque is found as

$$\begin{aligned} T &= F \times r \\ &= 180 \text{ lb} \times 0.25 \text{ ft} \\ &= 45 \text{ lb-ft} \end{aligned}$$

PRACTICE PROBLEM 23.1

A motor like the one in Figure 23.28 is driving a winch with an 8-inch diameter. Calculate the torque required for the winch to lift 240 pounds.

Work

The **WORK** accomplished by a motor is the product of the force used to move an object and the distance the object is moved. By formula,

WORK For a motor, the product of the force used to move an object and the distance the object is moved.

(23.5)

$$W = F \times D$$

where

W = the work accomplished, in foot-pounds (ft-lb)
F = the force used, in pounds (lb)
D = the distance, in feet (ft)

The values of F and D used in equation (23.5) are illustrated in Figure 23.29. Example 23.2 demonstrates an application of the relationship.

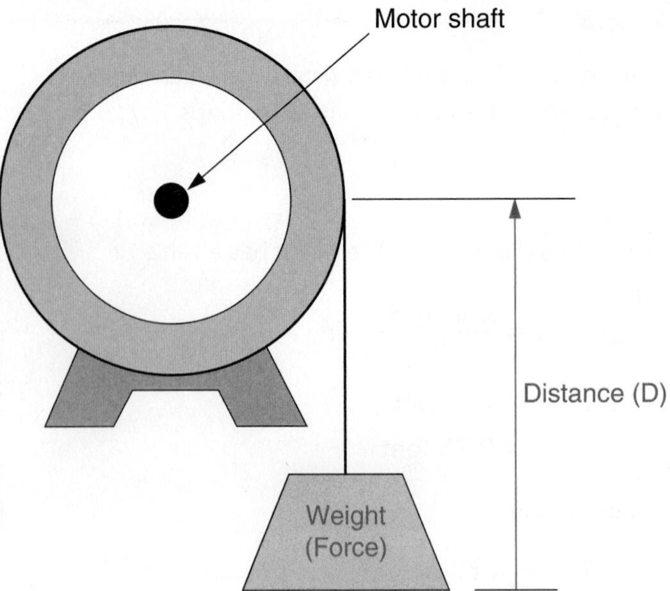

FIGURE 23.29 Work as a function of weight and distance.

EXAMPLE 23.2

Calculate the work accomplished by a motor-driven winch that lifts 620 pounds a distance of 30 inches.

SOLUTION

First, the distance the weight is being moved is converted to feet as follows:

$$D = \frac{30 \text{ in}}{12 \text{ in/ft}}$$
$$= 2.5 \text{ ft}$$

Now, the work performed by the motor is found as

$$W = F \times D$$
$$= 620 \text{ lb} \times 2.5 \text{ ft}$$
$$= 1550 \text{ ft-lb}$$

PRACTICE PROBLEM 23.2

Calculate the work accomplished by a motor-driven winch that lifts 750 pounds a distance of 20 inches.

Torque Versus Work

When you compare Examples 23.1 and 23.2, you'll see that the calculations for torque and work are nearly identical, as are their units of measure. For convenience, the equations and results from the two examples are provided here:

	Torque	Work
Equation:	T = F × r	W = F × D
Results:	45 lb-ft	1550 ft-lb

As you can see, the equations and units of measure are nearly identical. Because the two are so similar, it is important to take a moment to distinguish between them.

Units of Measure

Torque and work are both calculated using force (in pounds) and distance (in feet), so they must have the same unit of measure. To help distinguish one from the other, the abbreviations that make up the units of measure (ft and lb) are reversed. To provide an even clearer distinction between the two, some references rate torque in **FOOT-POUND FORCE (FT-LBF)**.

FOOT-POUND FORCE (FT-LBF) A unit of measure for torque.

Work Requires both Weight and Movement

In our work example, a 620-pound weight was being lifted 30 inches (2.5 ft). Had there been no weight (F = 0 lb), the amount of work would have been

$$\begin{aligned} W &= F \times D \\ &= 0 \text{ lb} \times 2.5 \text{ ft} \\ &= 0 \text{ ft-lb} \end{aligned}$$

Had the winch not lifted the weight (D = 0 ft), the amount of work would have been

$$\begin{aligned} W &= F \times D \\ &= 620 \text{ lb} \times 0 \text{ ft} \\ &= 0 \text{ ft-lb} \end{aligned}$$

As these calculations show, we have work only if a measureable weight is moved a measureable distance. Torque, on the other hand, does not have to accomplish anything.

In Figure 23.28, you can see that torque depends on the force that a motor can generate and the distance between the motor shaft and the outside of the winch. These values remain whether any work is actually accomplished. For example, the motor in Figure 23.30 produces 100 lb-ft of torque. However, 500 ft-lb of work is required to lift the weight. Because the weight provides a *counter torque* that is greater than the capability of the motor, the winch will not lift the weight and no work is accomplished. However, torque is still produced by the motor as it attempts to lift the weight.

Full-Load Torque

The torque produced by a motor when it is operating at its rated power and at full speed is referred to as **FULL-LOAD TORQUE**. The full-load torque for a motor can be found as

FULL-LOAD TORQUE The torque produced by a motor when it is operating at its rated power and at full speed.

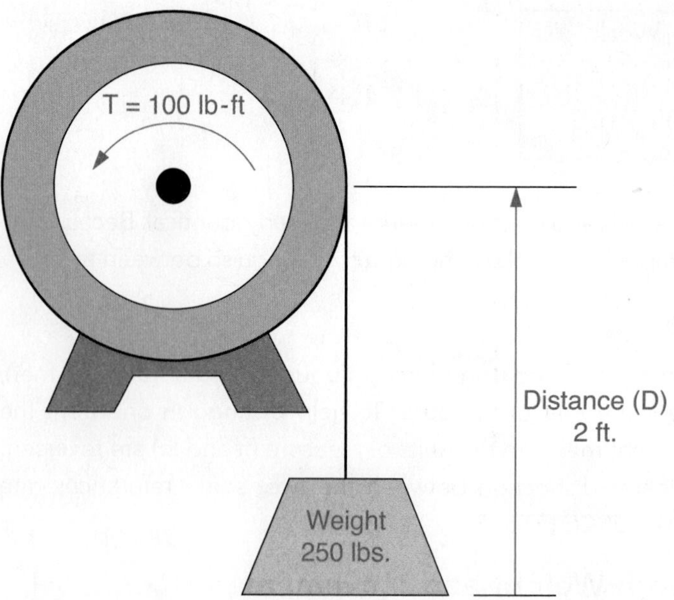

FIGURE 23.30

(23.6)

$$T = \frac{5252 \times HP}{rpm}$$

where

HP = engine horsepower
rpm = revolutions per minute
5252 = a constant

Example 23.3 demonstrates an application for this relationship.

EXAMPLE 23.3

Calculate the full-load torque for a 40 HP motor that is operating at 1200 rpm.

SOLUTION

Using the values given, the full-load torque for the motor is found as

$$T = \frac{5252 \times HP}{rpm}$$

$$= \frac{5252 \times 40}{1200}$$

$$= 175 \text{ lb-ft}$$

PRACTICE PROBLEM 23.3

Calculate the full-load torque for a 3 HP motor that is operating at 120 rpm.

It would seem that full-load torque would be the greatest value of torque for a motor, given that it is produced when the motor is operating at full power and full speed. However, this is not the case. The starting torque for a DC motor can be as much as five times its full-load torque.

AN ANALOGY: On a level surface, more torque is needed to accelerate a car to its cruising speed than is needed to maintain that speed.

Horsepower

Electrical power is the rate at which energy is used. Throughout this text, we have measured power in watts (W). You may recall that one watt of power is produced when one joule of energy is being used per second. By formula,

$$1 \text{ W} = 1 \text{ joule/second}$$

This relationship is presented here because it reinforces the concept that power is the rate at which energy is used.

Power, as it relates to motors, is the rate at which work is accomplished, measured in **FOOT-POUNDS PER MINUTE (FT-LB/MIN)**. By formula,

FOOT-POUNDS PER MINUTE (FT-LB/MIN) The unit of measure for motor *power* (the rate at which work is accomplished).

(23.7)

$$P = \frac{W}{T}$$

where

P = power, in ft-lb/min
W = work, in ft-lb
T = time, in min

If time is measured in seconds (rather than minutes), equation (23.7) yields a result in *foot-pounds per second (ft-lb/s)*.

Motor power is also measured in **HORSEPOWER (HP)**. Motors are typically rated in horsepower. Relating horsepower to the other units of measure for power,

HORSEPOWER (HP) A unit of measure for motor power, equal to 746 W (or 33,000 ft-lb/min)

(23.8)

$$1 \text{ HP} = 746 \text{ W}$$

(23.9)

$$1 \text{ HP} = 33{,}000 \text{ ft-lb/min}$$

(23.10)

$$1 \text{ HP} = 550 \text{ ft-lb/sec}$$

Based on these relationships, motor horsepower can be found as

(23.11)

$$HP = \frac{\text{Watts}}{746}$$

(23.12)

$$HP = \frac{\text{ft-lb/min}}{33{,}000}$$

and

(23.13)

$$HP = \frac{\text{ft-lb/s}}{550}$$

The following examples provide some applications for these relationships.

EXAMPLE 23.4

Calculate the horsepower of a 1200 W motor.

SOLUTION

Using the power rating of the motor, its power (in horsepower) is found as

$$HP = \frac{\text{Watts}}{746} = \frac{1200}{746} = 1.61$$

PRACTICE PROBLEM 23.4

Calculate the horsepower of a 22 kW motor.

EXAMPLE 23.5

Calculate the electrical power required by a 2.4 HP motor.

SOLUTION

One horsepower equals 746 watts. Therefore, the power required by a 2.4 HP motor can be found as

$$P = 2.4 \times 746 \text{ W} = 1790 \text{ W} = 1.79 \text{ kW}$$

PRACTICE PROBLEM 23.5

Calculate the power required by a 15 HP motor.

EXAMPLE 23.6

Calculate the ft-lb/min produced by a 12 HP motor.

SOLUTION

According to equation (23.9),

$$1\text{ HP} = 33{,}000\text{ ft-lb/min}$$

Therefore,

$$\begin{aligned}12\text{ HP} &= 12 \times 33{,}000\text{ ft-lb/min}\\ &= 396{,}000\text{ ft-lb/min}\end{aligned}$$

PRACTICE PROBLEM 23.6

Calculate the ft-lb/s produced by an 18 HP motor.

In this section, we discussed several topics that relate to all DC motors. At this point, we will turn our attention to some specific types of DC motors.

PROGRESS CHECK

1. Compare the operation of a DC generator and a DC motor.
2. Explain how the left-hand rule is used to determine the direction of motor rotation.
3. Define and contrast torque and work. How do their units of measure differ?
4. What is full-load torque? When does a motor produce the greatest amount of torque?
5. What is horsepower? List three units of measure that can be used to define the horsepower rating of a motor.

23.8 SERIES DC MOTORS

Because DC motors and generators are nearly identical, their schematics are very similar. The schematics for both machines are shown in Figure 23.31. The difference between the two schematics reflects the fact that the generator converts mechanical energy to electrical energy, whereas the motor converts electrical energy to mechanical energy.

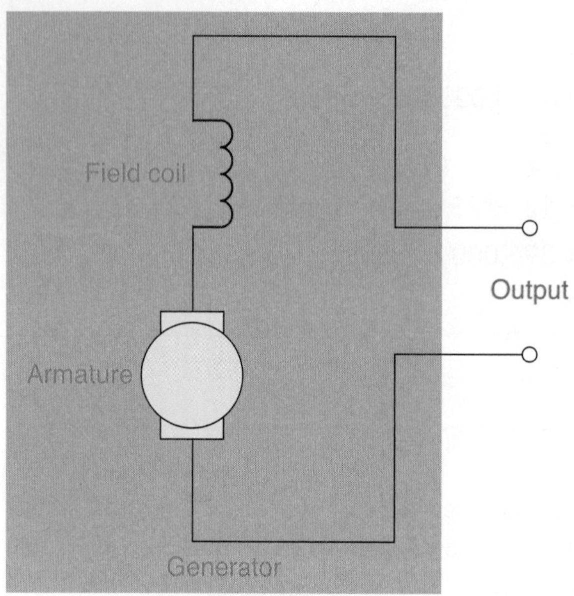

(a) Series DC generator

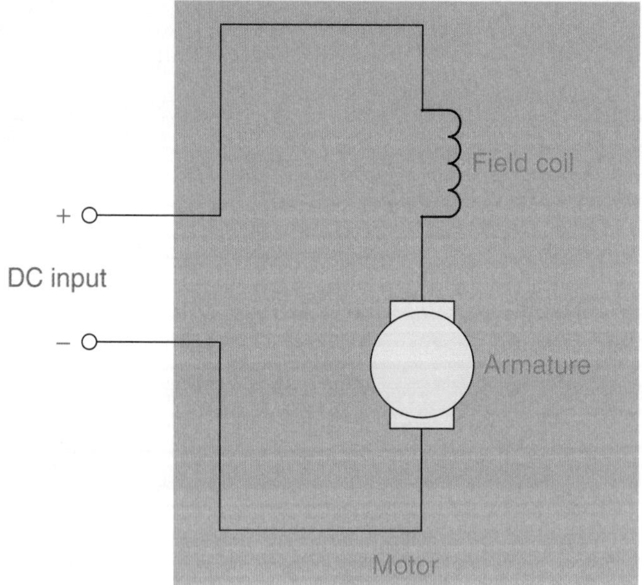

(b) Series DC motor

FIGURE 23.31 Series DC generator and motor configurations.

Counter EMF

When a voltage source is connected to the commutator of a DC motor (as shown in Figure 23.32), a current (I_A) is generated through the armature. As the armature rotates through the magnetic field, a counter EMF (CEMF) is induced across the armature (E_G) that opposes the DC input voltage (E_{DC}). As a result, I_A is limited to a value that can be found as

(23.14)

$$I_A = \frac{E_{DC} - E_G}{R_A}$$

where

I_A = the armature current
E_{DC} = the applied voltage
E_G = the CEMF generated by the armature rotating through the magnetic field
R_A = the resistance of the armature circuit (windings, brushes, and commutator)

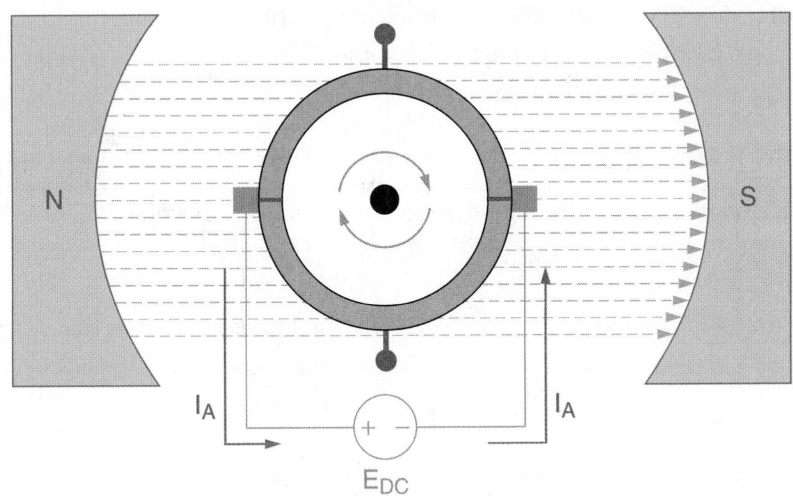

FIGURE 23.32 DC motor source voltage and current.

Example 23.7 demonstrates how the CEMF helps limit the armature current.

EXAMPLE 23.7

A circuit like the one in Figure 23.32 draws 12 A from a 120 V source. Assuming that the armature circuit resistance is 3.2 Ω, how much CEMF is being generated by the armature rotation?

SOLUTION

The I_A equation can be rewritten to provide a means of calculating the circuit CEMF, as follows:

$$E_G = E_{DC} - I_A R_A$$
$$= 120\text{ V} - (12\text{ A} \times 3.6\ \Omega)$$
$$= 120\text{ V} - 43.2\text{ V}$$
$$= 76.8\text{ V}$$

PRACTICE PROBLEM 23.7

A circuit like the one in Figure 23.32 draws 28 A from a 240 V source. Assuming that the armature circuit resistance is 4.8 Ω, how much CEMF is being generated by the armature rotation?

The result in Example 23.7 is significant because it indicates that the CEMF produced by the armature rotation helps limit the value of I_A when the motor is operating. However, this is not the case when the motor is first powered on.

STARTUP CURRENT The current through a motor when power is first applied.

When a DC motor is starting up, there is no CEMF across the armature. As a result, the motor **STARTUP CURRENT** is significantly higher than the full-load motor current. This point is demonstrated in Example 23.8.

EXAMPLE 23.8

Calculate the startup current through the motor described in Example 23.7.

SOLUTION

When the motor is starting, $E_G = 0$ V. Therefore, the armature current has a value of

$$I_A = \frac{E_{DC} - E_G}{R_A}$$
$$= \frac{120\text{ V} - 0\text{ V}}{3.6\ \Omega}$$
$$= 33.3\text{ A}$$

As this result indicates, the motor startup current (33.3 A) is almost three times its full-load current (12 A).

PRACTICE PROBLEM 23.8

Calculate the startup current for the circuit described in Practice Problem 23.7.

Motor startup current can be up to five times as great as full-load current. This is significant because of the relationship between motor current and torque.

Motor Current and Torque

The higher the armature current in a DC motor, the greater the torque generated by the motor. The relationship between armature current (I_A) and torque (T) is illustrated in Figure 23.33.

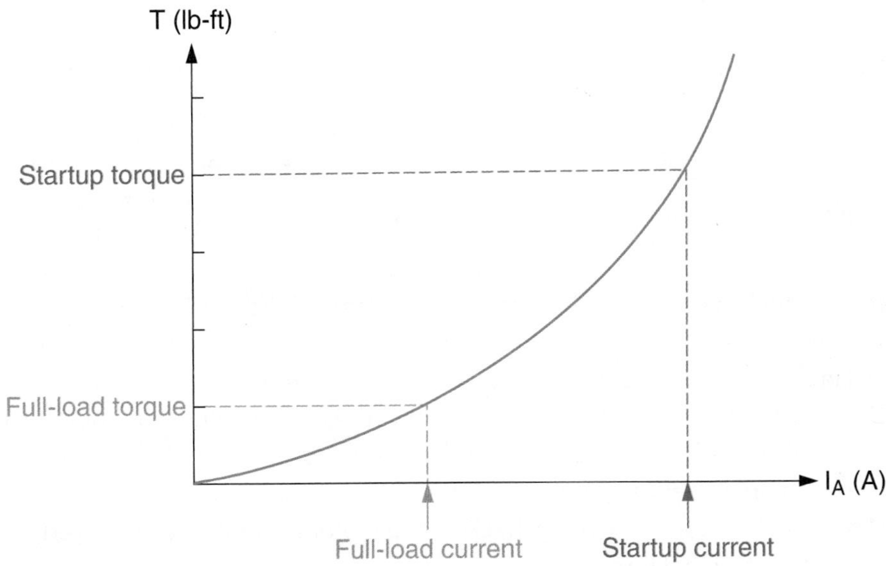

FIGURE 23.33 The relationship between series DC motor current and torque.

As the curve shows, the startup torque is much greater than the full-load torque produced by the motor. With this in mind, here's what happens when the motor is first powered on:

1. A high-value current (called *startup current*) is generated in the armature. At this point, there is no CEMF to limit the armature current, so its value is limited only by the armature conductors, brushes, and commutator and the resistance of the field coils.
2. The high startup current produces the high torque needed to increase the armature speed.
3. As the armature comes up to speed, the torque decreases. By the time the armature is rotating at full speed, the torque has decreased to its full-load value.

The relationship between motor torque and speed (in rpm) is illustrated further in Figure 23.34. As the graph shows, torque decreases as motor speed increases.

Motor Loading

The primary strength of the series DC motor is that it can provide high torque when heavily loaded and at startup. The primary drawback of the series DC motor

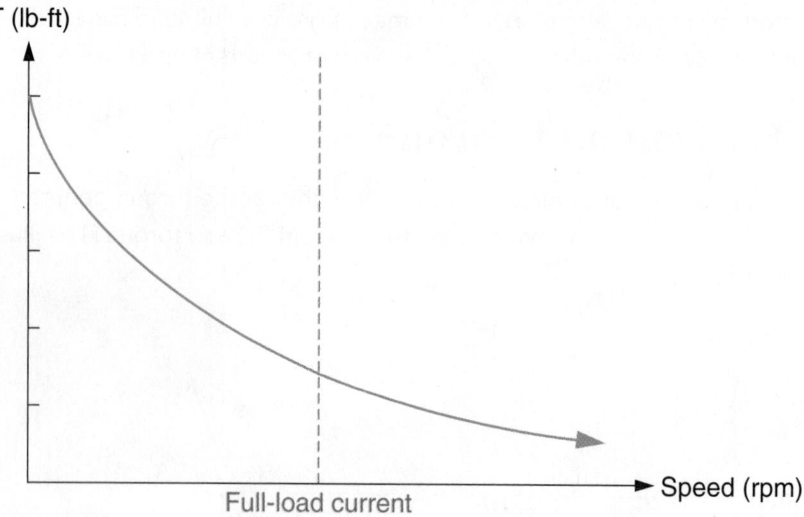

FIGURE 23.34 The relationship between motor torque and speed.

is that its output speed varies inversely with its load. If the load decreases, the motor speed increases, and vice versa. For this reason, series DC motors cannot be used in applications where constant motor speed is important.

Decreasing the load on a series DC motor causes the armature current to decrease. This reduces the field coil flux that limits the motor speed. As a result, the motor speeds up. If the load on a series motor is removed completely, the motor speed continually increases. Left unchecked, the motor is eventually destroyed by the effects of centrifugal force.

PROGRESS CHECK

1. Compare and contrast the schematics for a series DC motor and generator.
2. Explain how the armature produces counter EMF. Explain how this limits armature current.
3. What is motor startup current? Explain why it is higher than full-load motor current.
4. Explain the relationship between armature current and motor torque.
5. What is the relationship between motor speed and motor loading? Why should a series DC motor never be run without a load?

23.9 SHUNT DC MOTORS

A shunt DC motor is nearly identical to a shunt DC generator, as indicated by the schematics in Figure 23.35.

Unlike the series motor, shunt motor speed remains relatively constant despite variations in its load. The reason for this lies in the fact that the field coil is connected:

- In parallel with the armature.
- Across the DC voltage source.

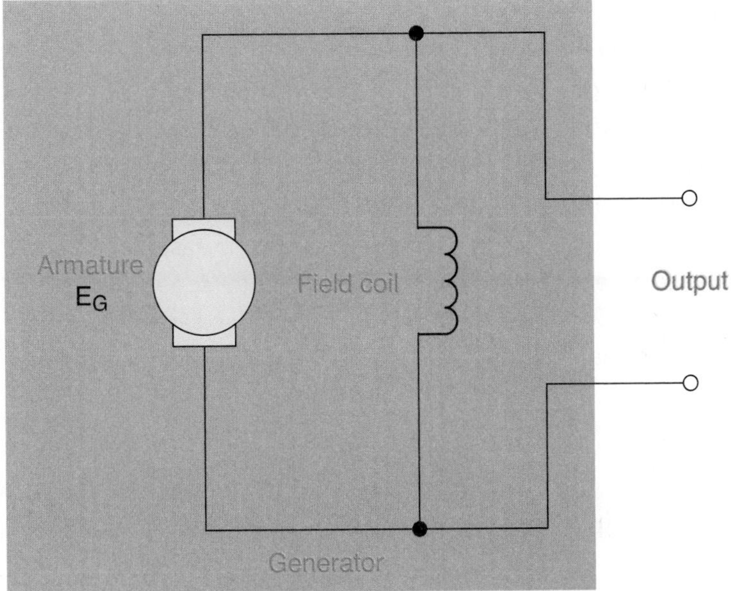

(a) Shunt generator

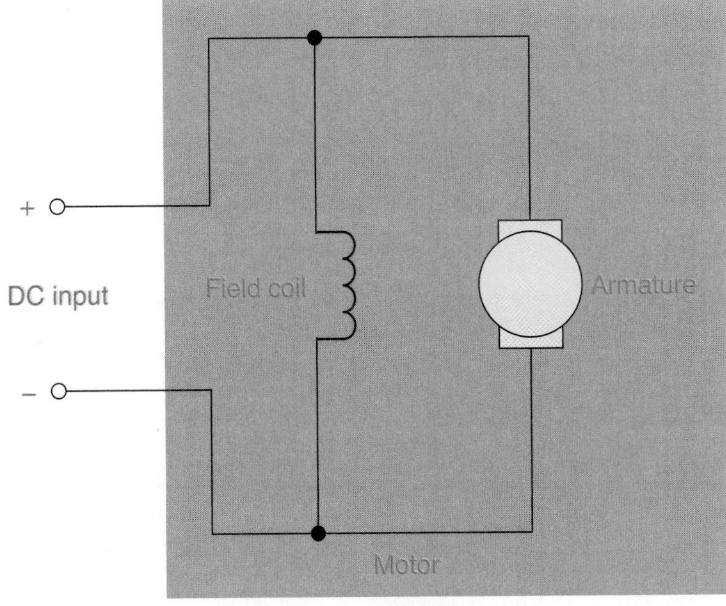

(b) Shunt motor

FIGURE 23.35 Shunt DC generator and motor configurations.

As such, any change in armature current (I_A) has little effect on the current through the field coil.

Motor Torque

The torque provided by a shunt DC motor varies directly with its load. That is:

- The torque increases proportionally to an increase in load.
- The torque decreases proportionally to a decrease in load.

These responses to any change in load hold the motor speed relatively constant over the motor's rated range of load values.

The operating principle described here is based on the relationship between armature current (I_A) and torque (T). A graph of this relationship is provided in Figure 23.36. As the graph shows, the torque produced by a shunt DC motor is directly proportional to the armature current. As such, the motor response to a change in load can be described as follows:

- When the motor load increases, the armature current increases. This causes a proportional increase in torque, which holds the motor speed constant.
- When the motor load decreases, the armature current decreases. This causes a proportional decrease in torque, which holds the motor speed constant.

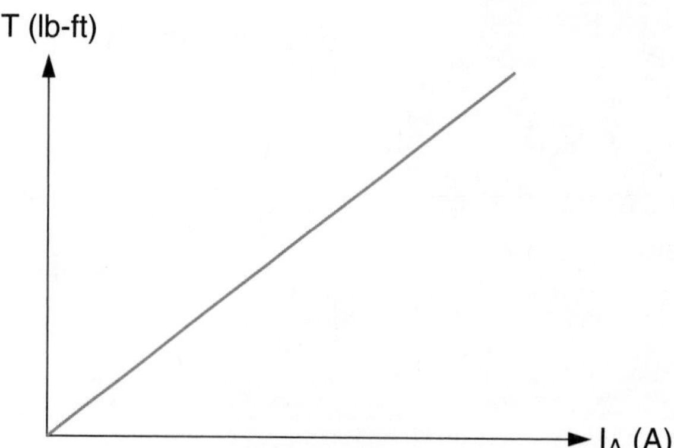

FIGURE 23.36 The relationship between shunt DC motor current and torque.

It should be noted that these are *ideal* relationships. In practice, motor speed can vary by as much as 10% over the rated range of loads. However, even this worst-case value is a significant improvement over the series DC motor.

PROGRESS CHECK

1. How is the field coil wired with respect to the armature and the external DC supply in a shunt DC motor?
2. How does the speed of a shunt motor react to a change in load?
3. How does the torque of a shunt motor react to a change in load?

23.10 COMPOUND DC MOTORS

A compound DC motor is nearly identical to a compound DC generator, as indicated by the schematics in Figure 23.37.

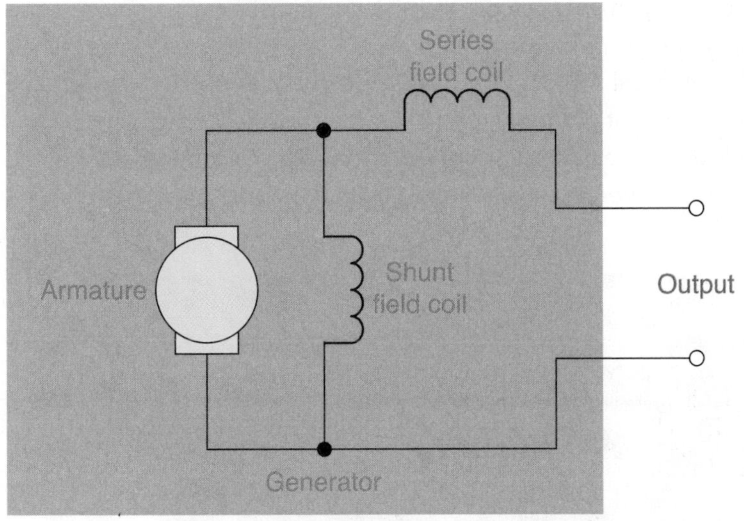

(a) Compound DC generator

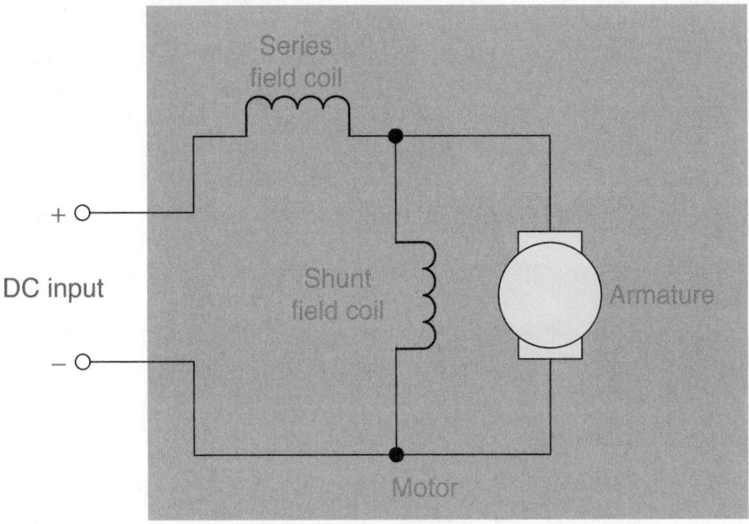

(b) Compound DC motor

FIGURE 23.37 Compound DC generator and motor configurations.

A compound DC motor has some of the characteristics of both series and shunt motors. Like a series motor, it can produce high torque when loaded and at startup. Like a shunt motor, it can maintain a relatively constant speed over its rated load range.

Cumulative Compound Motors

There are two common compound motor wiring configurations: cumulative and differential. The configuration of a compound motor is a function of the internal connections among its armature and field coils.

The shunt field coil in a *cumulative compound motor* is wired so that its flux aids the flux produced by the series coil and the armature. Two cumulative compound motor configurations are shown in Figure 23.38. (Note that the two terminals of each coil and the armature are labeled. These labels will be used to compare the wiring of these motors to that of other motors.) Of the two, the long-shunt motor is the more commonly used because the shunt current is relatively independent of the motor load, which gives it the best speed stability characteristics.

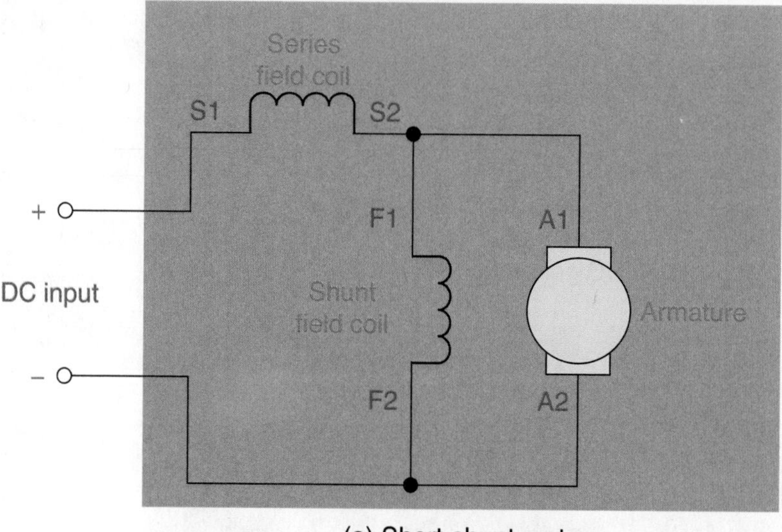

(a) Short-shunt motor

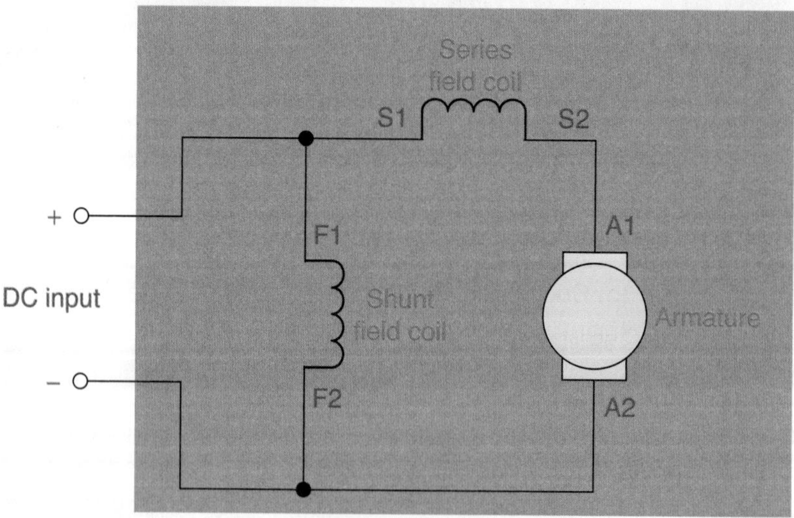

(b) Long-shunt motor

FIGURE 23.38 Cumulative compound motor configurations.

Differential Compound Motors

The shunt field coil in a *differential compound motor* is wired so that its flux opposes the flux produced by the series coil and the armature. The differential compound motor configurations are shown in Figure 23.39.

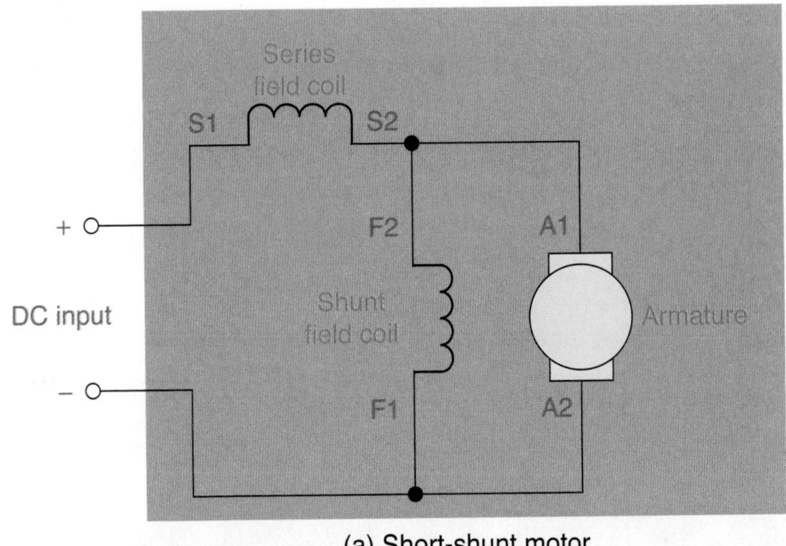

(a) Short-shunt motor

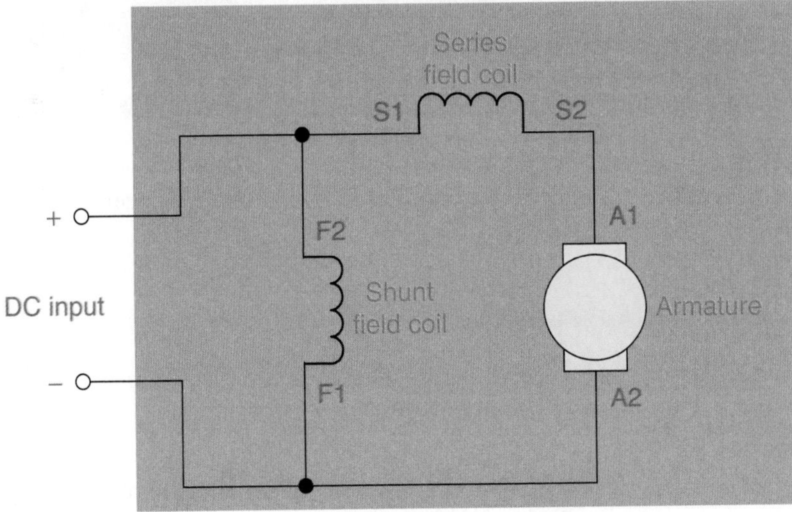

(b) Long-shunt motor

FIGURE 23.39 Differential compound motors configurations.

The circuits in Figure 23.39 may appear at first to be identical to the cumulative compound motors in Figure 23.38. However, if you compare the field coil labels in the two figures, you'll see that the shunt coil connections are reversed in the differential compound motor. Because of the reversed connection, the flux produced by the shunt coil opposes those produced by the field coil and armature. The benefit of this wiring scheme is that the motor speed is very stable over a wide range of load values. However, there are a couple of drawbacks:

- If the load increases to the point where motor speed begins to increase, the motor speed becomes unstable and could escalate out of control like a series DC motor.
- The torque characteristics are poorer than those of a cumulative compound motor.

For these reasons, cumulative compound motors are used far more often than differential compound motors.

PROGRESS CHECK

1. What are the characteristics of a compound DC motor?
2. What are the characteristics of a cumulative compound DC motor?
3. What are the characteristics of a differential compound DC motor?
4. Why are differential compound DC motors used less often than cumulative compound motors?

23.11 DC MOTOR CONTROL CIRCUITS

Now that you have been introduced to the most common types of DC motors, we are going to examine some of the circuits that are used to control their operation.

Speed Control Circuits

The speed of a DC motor can be controlled by varying either the armature current or the field current. Both of these methods of controlling DC motor speed are illustrated in Figure 23.40.

In Figure 23.40a, the motor speed is controlled using a rheostat that is in series with the armature. When the armature current passes through the rheostat, a voltage is developed across the component. Because the rheostat voltage (V_R) and the armature voltage (E_G) must add up to equal the DC input voltage, E_G decreases as V_R increases. By decreasing E_G, you reduce motor speed. Note that this control circuit can only reduce the motor speed below its rated value. The armature rheostat cannot be used to increase motor speed to a value that is greater than the motor's speed rating.

The shunt coil rheostat in Figure 23.40b can be adjusted to decrease or increase DC motor speed. When the rheostat setting is varied, the current

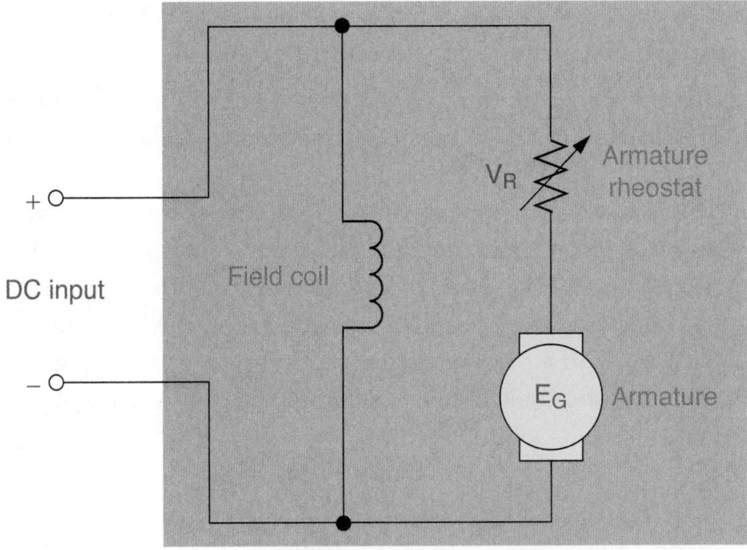

(a) Armature rheostat speed control

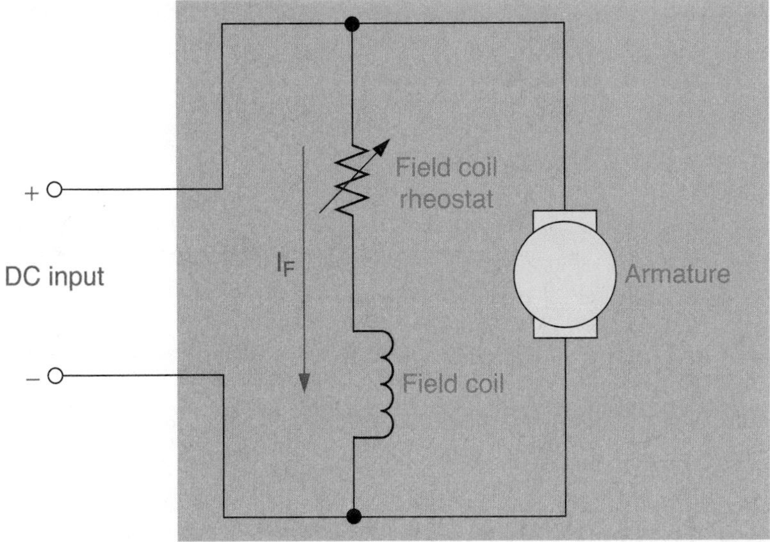

(b) Field coil rheostat speed control

FIGURE 23.40 Motor speed controls.

through the shunt coil varies inversely, which causes the motor speed to change. For example, here is what happens when the resistance of the rheostat *decreases*:

- The current through the shunt coil increases.
- The increase in current causes the flux produced by the shunt coil to increase.
- The increase in flux increases the CEMF across the armature.
- The increased CEMF causes the armature to slow down.

Likewise, when the resistance of the rheostat *increases*:

- The current through the shunt coil decreases.
- The decrease in current causes the flux produced by the shunt coil to decrease.

- The decrease in flux decreases the CEMF across the armature.
- The decreased CEMF causes the armature to speed up.

BASE SPEED The rated speed of a motor.

Unlike the armature speed control circuit, the speed control circuit in Figure 23.40b can be adjusted to allow the motor to exceed its rated speed (or **BASE SPEED**).

Though we have described shunt motors in this discussion, series motors may also require speed control circuitry to account for changes in motor speed that can result from a change in load. The speed of a series motor can be varied using a relatively low-value rheostat wired in parallel with the series coil, as shown in Figure 23.41. This resistor, called a *diverter*, causes a portion of the armature current to bypass the coil, reducing the CEMF produced by the coil. As a result, the motor speed increases.

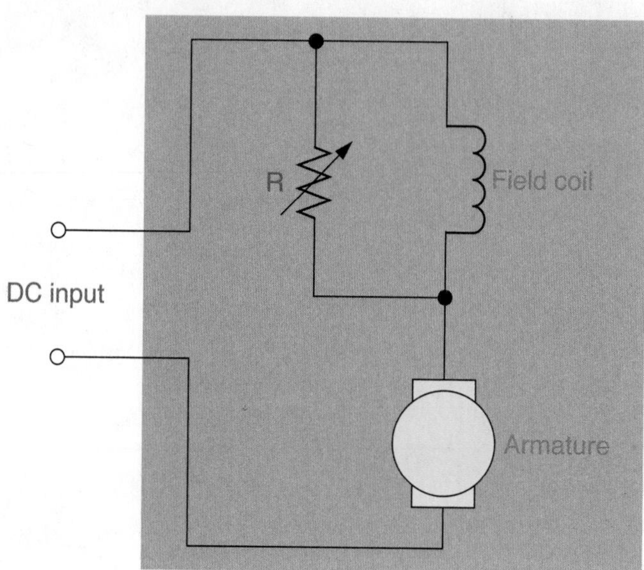

FIGURE 23.41 Series DC motor speed control.

Direction Control Circuits

The direction of armature rotation for a DC motor can be switched by a circuit like the one in Figure 23.42. The two switches in the schematic actually represent a single double-pull, double-throw (DPDT) switch. When this switch is in the up position:

- F1 is connected to A1.
- F2 is connected to A2.

and the motor operates normally, that is, the armature revolves clockwise (CW). When the switch is in the down position:

- F2 is connected to A1.
- F1 is connected to A2.

and the armature voltage polarity is reversed. As a result, the armature revolves counterclockwise (CCW). Note that the actual circuitry used to perform this function is far more complex than shown here. However, the function of the more complex

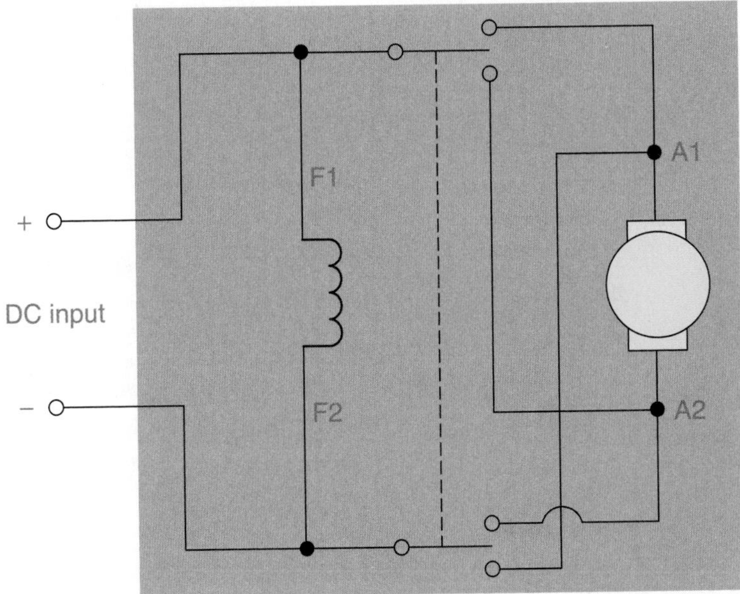

FIGURE 23.42 Direction control circuits for a shunt DC motor.

circuitry is the same: It reverses the polarity of the voltage applied to the armature, thereby reversing the direction of rotation.

A more practical direction control circuit is shown in Figure 23.43. The components labeled Q1 through Q4 are semiconductor components called *transistors*. Transistors are relatively complex components (compared to the components presented up to this point) that are normally covered in a course on solid-state electronic

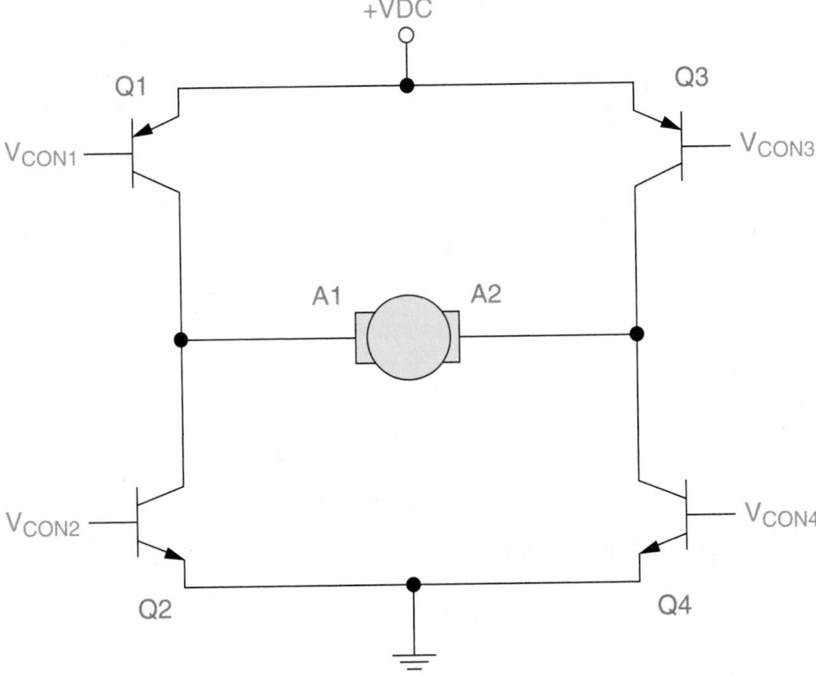

FIGURE 23.43 A transistor motor control circuit.

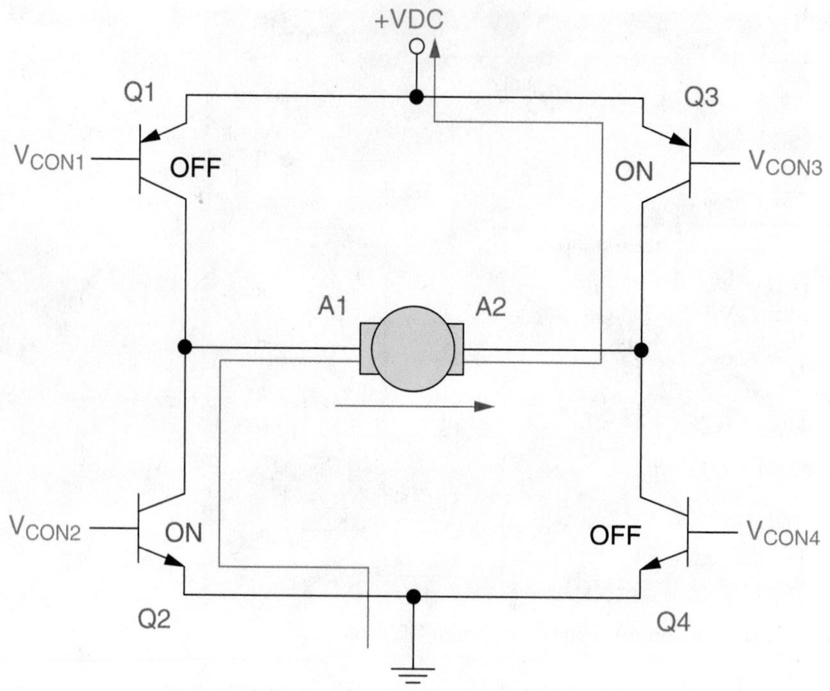

(b) Current path for CW armature rotation

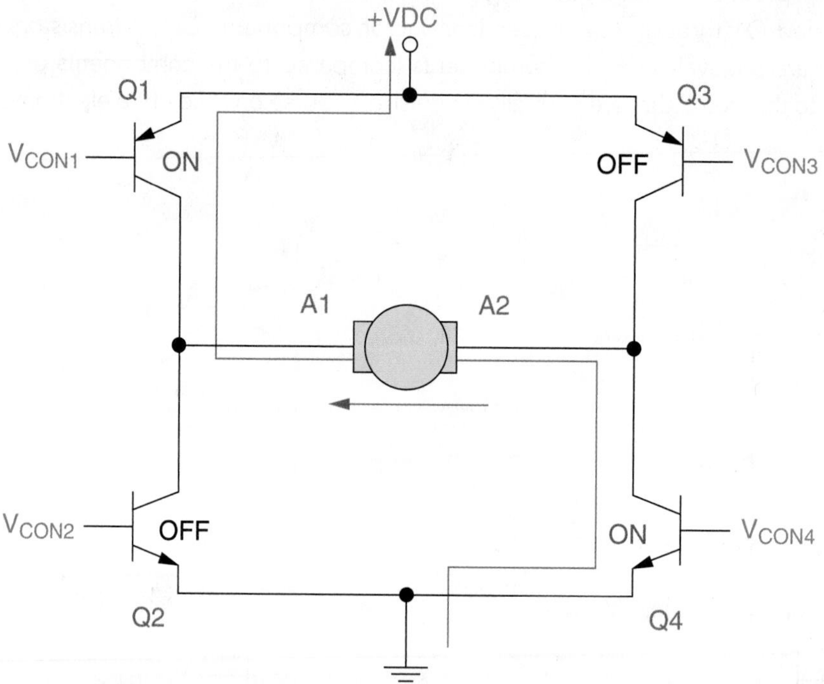

(c) Current path for CCW armature rotation

FIGURE 23.43 *(Continued)*

devices and circuits. As such, transistor theory of operation is outside the scope of this text. However, we can establish some operating characteristics that will demonstrate the function performed by the circuit in Figure 23.43.

The circuit in Figure 23.43 is called an **H-BRIDGE** because it resembles the letter H. Each transistor in the H-bridge functions as a closed switch when ON or an open switch when OFF. Whether a transistor is ON or OFF is determined by its control voltage (V_{CON}). The control voltages are determined by a computer or some other control circuit whose ultimate function is to control the operation of the motor.

H-BRIDGE A transistor circuit used to control the direction of motor rotation.

In Figure 23.43b, the control voltages are such that Q2 and Q3 are ON, while the other two transistors are OFF. When Q2 and Q3 are ON, they act as closed switches, providing the current path shown. Note that the current is passing through the motor in a direction that causes the armature to turn in the clockwise (CW) direction. In Figure 23.43c, the control voltages are such that Q1 and Q4 are ON, while the other two transistors are OFF. When Q1 and Q4 are ON, they act as closed switches, providing the current path shown. Note that the current is passing through the motor in a direction that causes the armature to turn in the counterclockwise (CCW) direction.

Kickback Circuits

As you know, the field coils and armature conductors in a DC motor both produce magnetic flux. When a DC motor is abruptly turned off, the energy contained in this flux is returned to the circuit.

In Chapter 11, you learned that the voltage induced across a coil can be found as

$$V_{IND} = L\frac{\Delta I}{\Delta t}$$

If the current through an inductor drops abruptly to 0 A, a high-value reverse voltage can be generated across the component. For example, let's say that we have 1 A of current through a 100 mH inductor. Then, in 500 ns, the current drops to 0 A. In this case,

$$\begin{aligned}\frac{\Delta I}{\Delta t} &= \frac{-1\text{ A}}{500\text{ ns}}\\ &= \frac{-1\text{ A}}{500 \times 10^{-9}\text{ s}}\\ &= -\,2\text{ MA/s}\end{aligned}$$

Note that the current has a negative value, indicating that the change represents a *decrease* in current. With this rate of change in current, the voltage across the 100 mH inductor has a value of

$$\begin{aligned}V_{IND} &= L\frac{\Delta I}{\Delta t}\\ &= (100\text{ mH})(-2\text{ MA/s})\\ &= -200\text{ kV}\end{aligned}$$

This calculation indicates that the sudden drop in current (1 A in 500 ns) causes a 200 kV reverse voltage to be developed across the coil.

The reverse voltage generated across the coils in a motor can wreak havoc in a motor control circuit. For example, refer back to Figure 23.43. When the circuit control voltages cause the motor to change direction, there is a brief instant when:

- All the transistors are off.
- The motor current is cut off.
- A reverse voltage is developed across the motor (the field coils and armature).

The transistors in the direction control circuit are not designed to withstand high voltages across their terminals when they are not conducting. Any (or all) of the transistors in the circuit could be damaged or destroyed as a result of the motor reverse voltage.

KICKBACK CIRCUIT A circuit that prevents the components in a motor control circuit from being damaged by any motor reverse voltage.

A **KICKBACK CIRCUIT** prevents the components in a motor control circuit from being damaged by any motor reverse voltage. A kickback circuit is shown in Figure 23.44a.

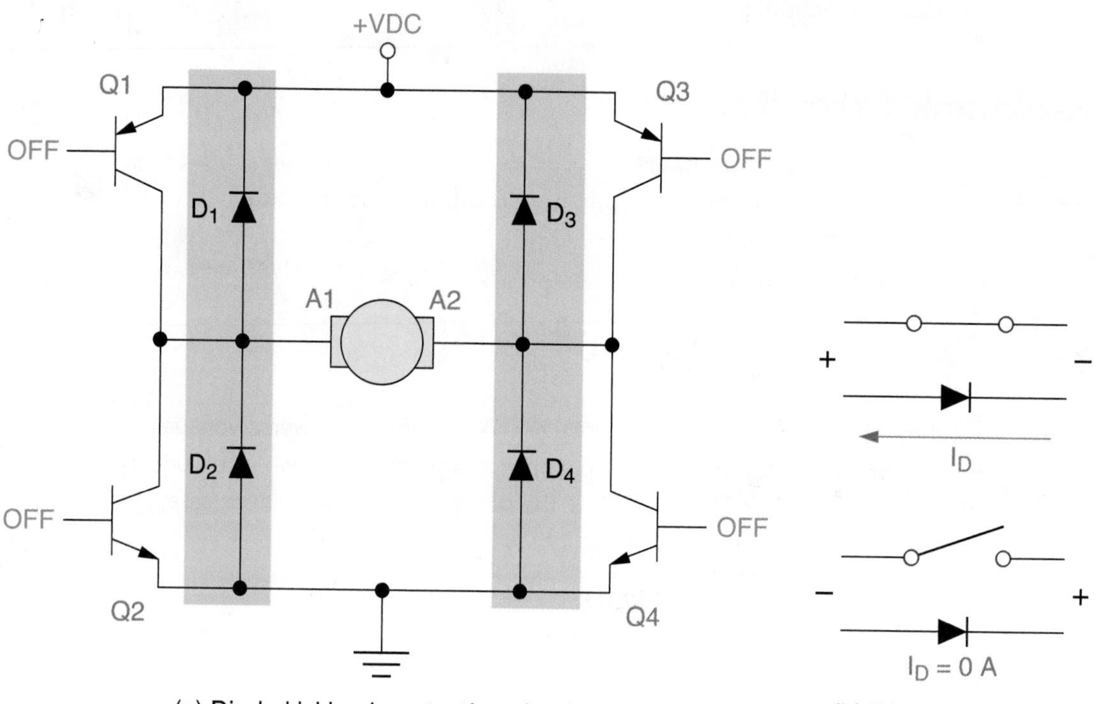

FIGURE 23.44 Diode kickback protection circuit.

The *diode* is another solid-state component whose theory of operation is outside the scope of this text. However, the circuit can be understood once we establish the fact that *the diode acts as a one-way conductor*. This characteristic of a basic diode is illustrated in Figure 23.44b. Note that the diode conducts when the diode symbol points toward the more negative voltage. When this is the case, diode current is against the arrow (as shown in Figure 23.44b). With this in mind, the operation of the kickback circuit is illustrated in Figure 23.45.

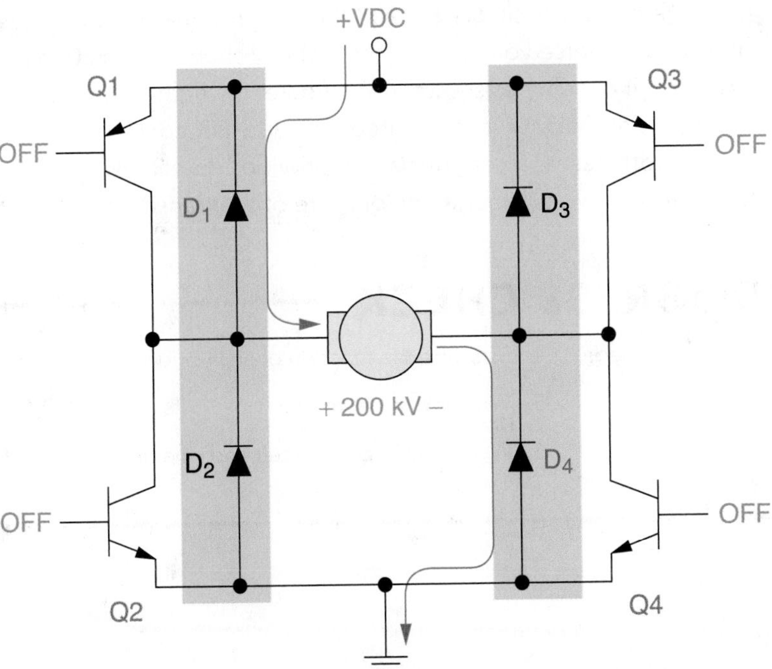

(a) Diode conduction when the motor reverse voltage has the polarity shown

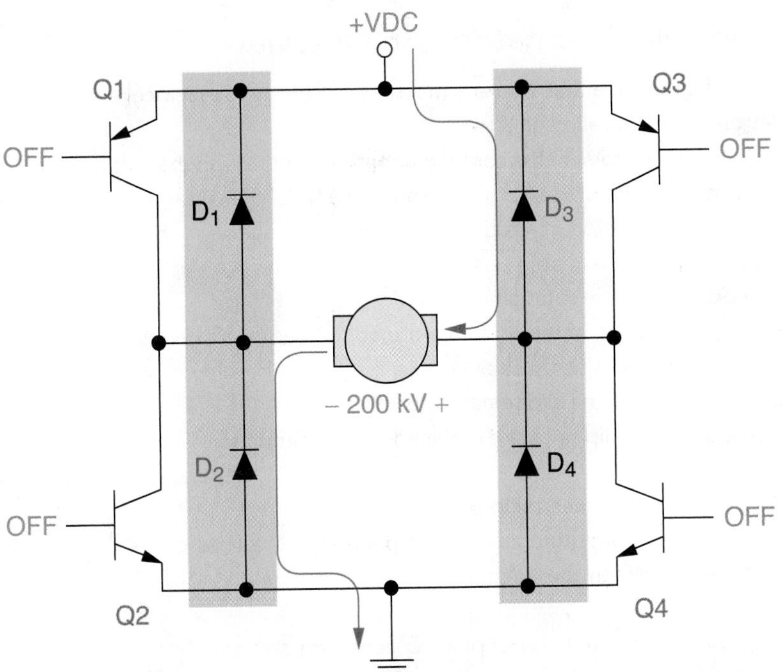

(b) Diode conduction when the motor reverse voltage has the polarity shown

FIGURE 23.45 Diode kickback protection circuit operation.

In Figure 23.45a, the transistors are all OFF. Assume the voltage across the motor is a high-value reverse voltage with the polarity shown. With 200 kV across the motor, D1 and D4 conduct as shown in the figure. If a high-value reverse voltage with the opposite polarity is generated, D2 and D3 conduct as shown in Figure 23.45b. In both cases, a current path is provided to eliminate the reverse voltage that might otherwise damage one or more of the transistors.

PROGRESS CHECK

1. List two currents that can be used to control motor speed.
2. Explain what happens when the resistance of a shunt coil rheostat is increased and decreased.
3. What is a diverter and what is its function?
4. Describe the function of an H-bridge.
5. What is a transistor and what is its function?
6. What is a kickback circuit and what is its function?

23.12 SUMMARY

Here is a summary of the major points made in this chapter:

Generators and Alternators: An Overview

- A generator is a machine that converts mechanical energy to electrical energy.
  - An AC generator is called an alternator.
- As the looped conductors on the rotating armature cut through the stationary magnetic field, a voltage is induced. The magnitude and polarity of the voltage are determined by:
  - The angle and direction of rotation of the armature relative to the magnetic field.
  - The flux density of the magnetic field.
  - The speed of armature rotation.
- In an alternator the coils of the armature are connected to slip rings, which in turn are connected to brushes. The brushes couple the induced voltage across the armature windings to the load via the slip rings.
- In a DC generator, the slip rings are replaced by a commutator.
  - The commutator has two or more conductive surfaces, depending on the number of armature loops, separated by insulators.
- When the coils of the armature are moving parallel to the lines of magnetic force zero voltage is induced. When the armature is in this position it is said to be in the neutral plane.
- The voltage variations in the output of a DC generator are referred to as *ripple*.
  - Ripple is the difference between its maximum and minimum output voltages

DC Generator Elements and Construction

- A dynamo converts mechanical energy to electrical energy, or electrical energy to mechanical energy.
  - A generator converts mechanical energy to electrical energy.
  - A motor converts electrical energy to mechanical energy.

- Every dynamo has a stationary structure called a *stator*, and a rotating structure called the *rotor*.
  - The stator is made up of the frame, the pole cores, the interpoles, and the windings.
  - The rotor is made up of the armature, its windings, and the commutator.
- A typical generator armature is a drum-type armature.
  - The armature is usually constructed using a laminated core with slots to hold the armature windings.
  - Both ends of the windings are connected to the commutator.
  - The commutator has as many segments as there are windings.
- A single armature winding is called a *simplex winding*.
- If there is more than one winding it is called a *multiplex winding*.
- Laminated armature cores are used to prevent eddy current losses.
- Brushes couple the generated voltage to the load by making contact with the commutator.
  - When the brushes couple two commutator segments across the insulator, the associated windings are positioned in or near the neutral plane so that little, if any, voltage is generated at that instant.
- Field poles are electromagnets that are part of the stator.
  - Coils of wire are wrapped around the poles. When current flows the poles are magnetized.
  - The strength and polarity of the magnetic field are determined by the number of turns, the amount of current through the coils, and the direction of the coil current.
- Field coils can be wound in three ways:
  - In series (called *wave wound*) for high-voltage, low-current outputs.
  - In parallel (called *lap wound*) for low-voltage, high-current outputs.
  - In series-parallel (called *frog-leg wound*) for moderate voltage and current outputs.
  - Frog-leg wound is the most common of the three.
- Armature reaction is the shifting of the neutral plane of the magnetic field due to magnetic flux generated by the armature windings.
  - The field shifts in the direction of armature rotation.
- Compensating windings are wound on commutating poles (or interpoles) to help control armature reaction and return the neutral plane to its original orientation.

Series Generators

- A series generator is wired so that the armature current passes through the field coil.
- A self-excited generator provides its own field coil current via the armature.
- Series generator output voltage varies directly with:
  - The magnetic flux per pole.
  - The number of poles.
  - The rotational speed of the armature.
- The output voltage of a series generator is also affected by field current.
  - Field current causes a voltage drop across the winding resistance of the field coil (R_W) as well as the internal resistance of the armature (R_A).
  - Just as the internal resistance of a DC voltage source affects its loaded output voltage, this voltage drop causes the generator output voltage to drop as field current increases.
- If the load on a series generator is open, then output voltage drops to zero as there is no path for field current to generate a magnetic field.

Parallel Generators

- In a parallel generator the voltage across the field coil equals the armature voltage since they are in parallel.
- The current through the field coil is relatively independent of armature current.
- Parallel generators are also self-excited generators since they provide their own field coil current. The process happens as follows:
  - Current is generated in the armature as it spins.
  - Armature current passes through the field coil.
  - The field coil current generates a magnetic field.
  - The field coil flux permeates the armature, inducing voltage across its windings.
- This process begins due to residual magnetism in the field cores, which induces a small voltage across the windings of the armature as it begins to rotate.
  - This generates a low-level current through the field coils.
  - This in turn increases the flux density across the armature.
  - The cycle continues until the generator reaches its rated output voltage.
- The output voltage of a shunt generator can be controlled by placing a rheostat in series with the field coil.
  - The rheostat limits the current through the field coil and therefore the voltage induced across the armature.

Compound Generators

- Compound generators use both series and parallel field coil windings.
  - This results in a more stable output voltage.
  - For this reason, the compound generator is the most common generator configuration.
- When no load is connected to a compound generator, armature current passes only through the parallel coil, and not the series coil.
  - This means that the output voltage equals the armature voltage.
- When a load is connected, armature current passes through both field coils.
  - Since both field coils are wrapped on the field poles, the increased flux overcomes the voltage drop across the series field coil, keeping the generator output voltage stable.
- Compounding is the process of compensating for the effects of load demands.
  - If there are too few turns on the series coil, the output voltage drops quickly as load demand increases. This is called *under-compounding.*
  - If there are too many turns on the series coil, the output voltage increases quickly as load demand increases. This is called *over-compounding.*
  - If the number of turns is correct, the output voltage remains relatively constant as load demand increases and the generator is said to be *compounded.*
- A diverter is a means of controlling output voltage as load demand changes.
  - The diverter provides a bypass of the field coil, which allows the operator to increase or decrease the magnetic flux and thus the induced voltage.

More on DC Generators

- The factors that contribute to generator power loss are as follows:
  - Hysteresis loss: The energy expended to change the orientation of the residual magnetic flux in the armature core.
  - Copper loss, also known as I^2R loss: The energy lost as a result of the heat produced by current through a resistance.

  - Friction loss: The energy lost due to heat generated by mechanical friction such as that between the brushes and commutator.
- DC generators can be connected in parallel to increase load current capacity.
- There are five requirements for parallel generators to operate properly:
  - They must have identical output voltage and voltage versus load current characteristics.
  - They must be connected positive-to-positive and negative-to-negative.
  - Some means of disconnecting each generator from the circuit must be provided.
  - Both generators must be driven by the same mechanical driver called the *prime mover.*

DC Motors: Basic Concepts

- A DC motor is basically a DC generator operated in reverse.
  - A motor converts electrical energy to mechanical energy, whereas a generator converts mechanical energy to electrical energy.
- The direction of armature rotation is determined by the direction of current through the armature's conductors.
  - The left-hand rule can be used to determine the direction of motor rotation.
- Compensating coils and interpoles are used to correct armature reaction in motors just as they are in generators.
- DC motors have two major advantages over AC motors:
  - Their speed can be varied.
  - They provide near-constant torque for a range of motor speeds.
- Torque is the force required to rotate an object about a point.
  - By formula, $T = F \times r$
- Work is the product of the force used to move an object times the distance moved.
  - By formula, $W = F \times D$.
- Torque is measured in lb-ft and work is measured in ft-lb.
- Work requires both weight and movement while torque does not.
  - A motor can produce torque even when it is not turning.
- Full-load torque is produced by a motor at rated power and full speed.
  - By formula, $T = \dfrac{5252 \times HP}{\text{rpm}}$
- The starting torque of a motor can be up to five times the full-load torque.
- Motor power is often measured in horsepower. By formula
  - 1 HP = 746 W
  - 1 HP = 33,000 ft-lb/min
  - 1 HP = 550 ft-lb/s
- Motor horsepower can be found as
  - $HP = \dfrac{\text{watts}}{746}$
  - $HP = \dfrac{\text{ft-lb/min}}{33{,}000}$
  - $HP = \dfrac{\text{ft-lb/s}}{550}$

Series DC Motors

- The schematics for DC motors and generators are almost identical (see Figure 23.31).
- Counter EMF (CEMF) is induced across the armature as it rotates through the magnetic field.
  - The CEMF opposes the DC input voltage, which limits armature current.

  - Armature current can be found as $I_A = \frac{E_{DC} - E_G}{R_A}$.
- Startup current can be significantly higher than full-load current. This is because there is no CEMF across the armature when the motor is initially powered up, and current is limited only by armature resistance.
- The higher the armature current, the greater the torque of the motor.
- The series DC motor has both strengths and weaknesses.
  - The primary strength is that series DC motors provide high torque both at startup and under heavy load.
  - The primary weakness is that series DC motors have poor speed control as motor speed varies with load.

Shunt DC Motors

- Shunt DC motors and generators are almost identical (see Figure 23.35).
- Unlike series DC motors, shunt DC motor speed remains relatively constant as load changes.
- Unlike series DC motors, shunt DC motor torque changes proportionally to changes in load.
  - As load current increases, armature current, and thus motor torque, increases, keeping the speed constant.
  - As load current decreases, armature current and torque decrease, keeping the speed constant.

Compound DC Motors

- A compound DC motor has some of the characteristics of both series and shunt motors.
  - It can produce high torque both at startup and under load.
  - Its speed control is much better than that of the series DC motor.
- The two common compound motor wiring configurations are cumulative and differential (see Figures 23.38 and 23.39).
  - In a cumulative compound DC motor, the flux generated by the shunt field coil aids the flux of the series coil and armature.
  - In a differential compound DC motor, the flux generated by the shunt field coil opposes the flux of the series coil and armature.
- Differential compound motors are not very common because,
  - As load increases, the motor may act like a series motor and run out of control.
  - Their torque characteristics are poor compared to those of the cumulative compound motor.

DC Motor Control Circuits

- DC motor speed can be controlled by varying either the armature or field current of a shunt DC motor (see Figure 23.40a).
- One method is to install a rheostat in series with the armature.
  - This method allows motor speed to be adjusted below the rated value, but not above.
- A rheostat in series with the shunt coil can be used to both increase and decrease motor speed (see Figure 23.40b).
  - The current through the shunt coil varies inversely with the resistance of the rheostat.
  - The shunt rheostat allows the motor to operate above its rated (or base) speed.

- Series motors may also require speed control (see Figure 23.41).
- A low-value variable resistor (called a *diverter*) is connected in parallel with the series coil.
    - This allows a portion of the armature current to bypass the coil, reducing CEMF and allowing motor speed to increase.
- The direction of motor rotation is controlled by the direction of current through the armature.
    - A simple control circuit is illustrated in Figure 23.42.
- A more practical control circuit uses solid-state components, called *transistors*, to change the direction of armature current (see Figure 23.43).
- An H-bridge uses four transistors to control armature current direction.
    - The transistors conduct in pairs. When one pair is on, the other pair is off.
- Kickback circuits are used to protect the components in a motor control circuit from being damaged by the high voltage produced when a motor changes direction or is turned off (see Figure 23.44).
    - Kickback circuits use solid-state components, called *diodes*, which act like one-way switches.
    - The diodes provide a current path that shorts out the kickback voltage and protects the control circuit components.

CHAPTER REVIEW

1. The voltage induced in the loop conductor in a generator depends on ________.
    a) the angle of armature rotation
    b) the magnetic field strength
    c) the speed of armature rotation
    d) all of the above

2. The armature of an AC generator is connected to ________, which in turn are in contact with stationary ________.
    a) slip rings, brushes
    b) brushes, commutators
    c) stators, rotors
    d) stators, brushes

3. A(n) ________ is found in a DC generator, but not in an AC generator.
    a) commutator
    b) armature
    c) stator
    d) rotor

4. When an armature coil is rotated to a point where it is moving parallel to the lines of magnetic force, it is said to be in ________.
    a) rotational order
    b) phase sequence
    c) the neutral plane
    d) linear mode

5. The difference between the minimum and maximum output voltages from a DC generator is called ________.
   a) ripple
   b) voltage regulation
   c) line regulation
   d) load regulation

6. The armature is a(n) ________ while the brushes are a(n) ________.
   a) conductor, insulator
   b) rotor, stator
   c) stator, rotor
   d) insulator, conductor

7. Armatures are laminated to ________.
   a) control hysteresis loss
   b) stabilize the neutral plane
   c) control eddy currents
   d) limit brush wear

8. When a brush comes in contact with more than one segment of the commutator the associated coils should be ________.
   a) moving parallel to the lines of magnetic force
   b) generating very little voltage
   c) in the neutral plane
   d) all of the above

9. Coils are wrapped around field poles to ________.
   a) control eddy current loss
   b) create electromagnets
   c) generate voltage in the pole
   d) control hysteresis loss

10. The most common field coil winding configuration is the ________ wound.
    a) wave b) frog-leg
    c) lap d) multiplex

11. Compensating windings are used to control ________.
    a) armature reaction
    b) interpoles
    c) commutating poles
    d) all of the above

12. The output voltage from a series DC generator varies directly with ________.
    a) the number of poles
    b) the magnetic flux per pole

c) the rotational speed of the armature
d) all of the above

13. The output voltage of a DC generator is affected by load demand due to ________.
    a) armature and winding resistance
    b) armature and field pole resistance
    c) hysteresis losses
    d) neutral plane losses

14. When a load opens, the output voltage for a series generator ________.
    a) increases slightly
    b) decreases slightly
    c) drops to near zero
    d) increases dramatically

15. The current in the field coil of a parallel DC generator is ________.
    a) relatively independent of armature current
    b) the same as armature current since they are in parallel
    c) much lower than armature current
    d) much higher than armature current

16. Like the series generator, the parallel generator is self-exciting since it ________.
    a) has its own field poles
    b) generates its own neutral plane
    c) provides its own field coil current
    d) provides its own armature flux

17. The output voltage of a parallel generator can be varied by ________ the field coil.
    a) installing a rheostat in series with
    b) changing the number of coils in
    c) using more wire in
    d) using less wire in

18. ________ generators have both series and parallel field coils.
    a) Wave-wound   b) Lap-wound
    c) Simplex   d) Compound

19. When no load is connected to a parallel DC generator, output voltage ________.
    a) equals the armature voltage
    b) drops to near zero
    c) increases slightly
    d) is equal to the parallel coil voltage

20. The process of compensating for predictable load demands is called ________.
    a) compounding   b) stabilization
    c) offsetting   d) multiplexing

21. In a generator, a diverter is used to ________.
    a) adjust for armature shift
    b) adjust for neutral plane shift
    c) adjust output voltage
    d) increase armature flux strength

22. Generators lose energy due to ________.
    a) hysteresis loss
    b) copper loss
    c) friction
    d) all of the above

23. Parallel-connected generators must always have the same ________.
    a) prime mover
    b) polarity
    c) voltage versus current rating
    d) all of the above

24. A DC motor converts ________ energy to ________ energy.
    a) DC, AC
    b) mechanical, electrical
    c) electrical, mechanical
    d) AC, DC

25. The left-hand rule is used in DC motors to determine ________.
    a) armature current
    b) armature rotation
    c) field current
    d) neutral plane shift

26. As in DC generators, interpoles are used in DC motors to compensate for ________.
    a) armature current
    b) armature rotation
    c) armature losses
    d) armature reaction

27. Torque is the product of ________ and is measured in ________.
    a) F × r, lb-ft
    b) F × D, ft-lb
    c) F × D, newtons
    d) F × r, joules/s

28. Work is the product of ________ and is measured in ________.
    a) F × r, lb-ft
    b) F × D, ft-lb
    c) F × D, newtons
    d) F × r, joules/s

29. Work requires ________ while torque does not.
    a) velocity
    b) movement
    c) force
    d) time

30. The torque produced by a motor at full speed and rated power is called ________.
    a) full-speed torque
    b) steady-state torque
    c) stable torque
    d) full-load torque

31. 1 HP is equal to ________.
    a) 33,000 ft-lb/min
    b) 746 W
    c) 550 ft-lb/s
    d) all of the above

32. The armature current develops ________ as the armature rotates through the magnetic field.
    a) neutral plane shift
    b) armature flux
    c) CEMF
    d) all of the above

33. A DC motor develops maximum torque at ________.
    a) full load
    b) startup
    c) no load
    d) shutdown

34. The counter EMF generated by the armature of a DC motor is the difference between the applied DC voltage and ________.
    a) the product of armature circuit current and resistance
    b) field voltage
    c) I^2R loss
    d) startup voltage

35. Startup current is ________.
    a) opposite to armature current
    b) in series with armature current
    c) less than full-load current
    d) up to five times higher than full-load current

36. If motor load decreases in a series DC motor ________.
    a) armature current decreases
    b) field coil flux decreases
    c) motor speed increases
    d) all of the above

37. One advantage of the shunt motor over the DC motor is ________.
    a) higher horsepower
    b) higher full-load torque
    c) more constant motor speed
    d) higher startup torque

38. The two wiring configurations of compound DC motors are ________.
    a) series and shunt
    b) cumulative and shunt
    c) differential and cumulative
    d) shunt and differential

39. The speed of a DC motor is controlled by varying ________.
    a) the armature flux
    b) either the field or armature current
    c) the field coil inductance
    d) the armature resistance

40. To control the speed of a shunt motor, a rheostat is installed in series with ________ or in parallel with ________.
    a) the field coil, the armature
    b) the DC supply, the field coil
    c) the armature, the DC supply
    d) the armature, the field coil

41. A diverter in a series motor controls speed by ________.
    a) partially bypassing the field coil
    b) partially bypassing the armature
    c) increasing armature flux
    d) increasing CEMF

42. The H-bridge uses four ________ to control the direction of current through the motor armature.
    a) diverters
    b) diodes
    c) transistors
    d) shunt relays

43. The H-bridge uses four ________ in a kickback circuit to protect the components of the motor control circuit from high induced voltages caused by reversing the direction of the motor.
    a) diverters
    b) diodes
    c) transistors
    d) shunt relays

PRACTICE PROBLEMS

1. A DC motor is driving a hoist with a diameter of 8.5 inches. Calculate the torque that the motor must produce to lift a 650-pound engine block.
2. A hoist with a 6.25 inch diameter must lift the same engine block described in question 1. Solve for the amount of torque the DC motor must produce.
3. A motor generates 82 ft-lb of torque while driving a winch that is lifting a 1250-pound crate. Solve for the diameter of the winch.
4. While lifting a sewer pipe, a motor driving a 9.2-inch winch generates 187 lb-ft of torque. Solve for the weight of the pipe.
5. The hoist in question 1 lifts the engine block 47 inches. Solve for the work accomplished by the motor.
6. The winch in question 4 must lift the sewer pipe 5.65 yards. Solve for the work accomplished by the motor.
7. A 75 HP DC motor is operated at 2100 rpm. Calculate the full-load torque generated by the motor.
8. A DC motor operating at 2300 rpm generates 23.835 lb-ft of torque. Solve for the HP rating of the motor.
9. Solve for the power (in ft-lb/s) used by the motor described in question 7.
10. Solve for the power (in ft-lb/min) used by the motor described in question 8.
11. Solve for the operating speed of a 1250 HP motor that generates 9118 lb-ft of torque.
12. A DC motor uses 2.984 kW of power when operated at 1750 rpm. Solve for the horsepower of this motor.
13. A DC motor has the following specifications: 125 HP @ 1750/2000 rpm. Solve for the number of watts required by this motor when run at full speed.
14. A DC motor has the following specifications: HP = 2 @ 1750 rpm. The supply voltage is 240 V. If the armature resistance is 4.4 Ω, solve for the CEMF generated by the armature.
15. A DC motor has the following specifications: HP = 3 @ 2300 rpm, amperage = 17.9 A. If the armature resistance is 2.63 Ω, solve for the CEMF generated by the armature.
16. Solve for the startup current for the motor described in question 14.
17. Solve for the startup current for the motor described in question 15.

ANSWERS TO THE EXAMPLE PRACTICE PROBLEMS

23.1	80 lb-ft
23.2	1250 ft-lb
23.3	131.3 lb-ft
23.4	29.5 HP
23.5	11.2 kW
23.6	9900 ft-lb/sec
23.7	105.6 V
23.8	50 A

VISITORS
GALLERY

CHAPTER 24

AC MACHINES: ALTERNATORS AND MOTORS

PURPOSE

Alternators are machines that generate alternating current (AC). Alternators can be as small as those found in automobiles or as large as the MVA generator pictured in Figure 24.1. Regardless of size and power output capability, alternators all operate according to the basic principles outlined in this chapter.

Structurally speaking, AC motors are very similar to alternators. In this chapter, we will examine the basics of alternators and AC motors.

FIGURE 24.1 A 33-ton MVA alternator.
Courtesy of Illinois Electric Works (www.illinoiselectric.com)

KEY TERMS

The following terms are introduced and defined in this chapter on the pages indicated:

OBJECTIVES

After completing this chapter, you should be able to:

1. State the function of an alternator.
2. Describe *rotating-armature* alternators.
3. Describe *rotating-field* alternators.
4. Compare and contrast *single-phase* alternators with *three-phase* alternators.
5. Describe the construction of a three-phase alternator.
6. Describe three-phase alternator winding configurations.
7. Determine the rotating speed of an alternator, given the number of poles it contains.
8. Calculate the percent of regulation for a given alternator.
9. State the function of an alternator voltage regulator.
10. Describe the operation of a basic AC motor.
11. Describe the operation of a *split-phase* AC motor.
12. Calculate the rotating speed of an AC motor.
13. Compare and contrast *synchronous* motors and *induction* motors.
14. Compare and contrast *wound* rotors and *squirrel-cage* rotors.
15. Describe rotor slip and calculate its value.
16. Describe the various types of split-phase induction motors.
17. Describe the construction and operation of a three-phase synchronous motor.

24.1 ALTERNATORS: AN OVERVIEW

An **ALTERNATOR** is a machine that converts mechanical energy (motion) to electrical energy. Like a DC generator, an alternator uses magnetic flux to induce a voltage across a conductor. You were first introduced in Chapter 13 to the concept of using the motion of a conductor through a magnetic field to generate electrical energy. As a review, this method of producing an alternating current (AC) is illustrated in Figure 24.2.

ALTERNATOR A machine that converts mechanical energy (motion) to electrical energy.

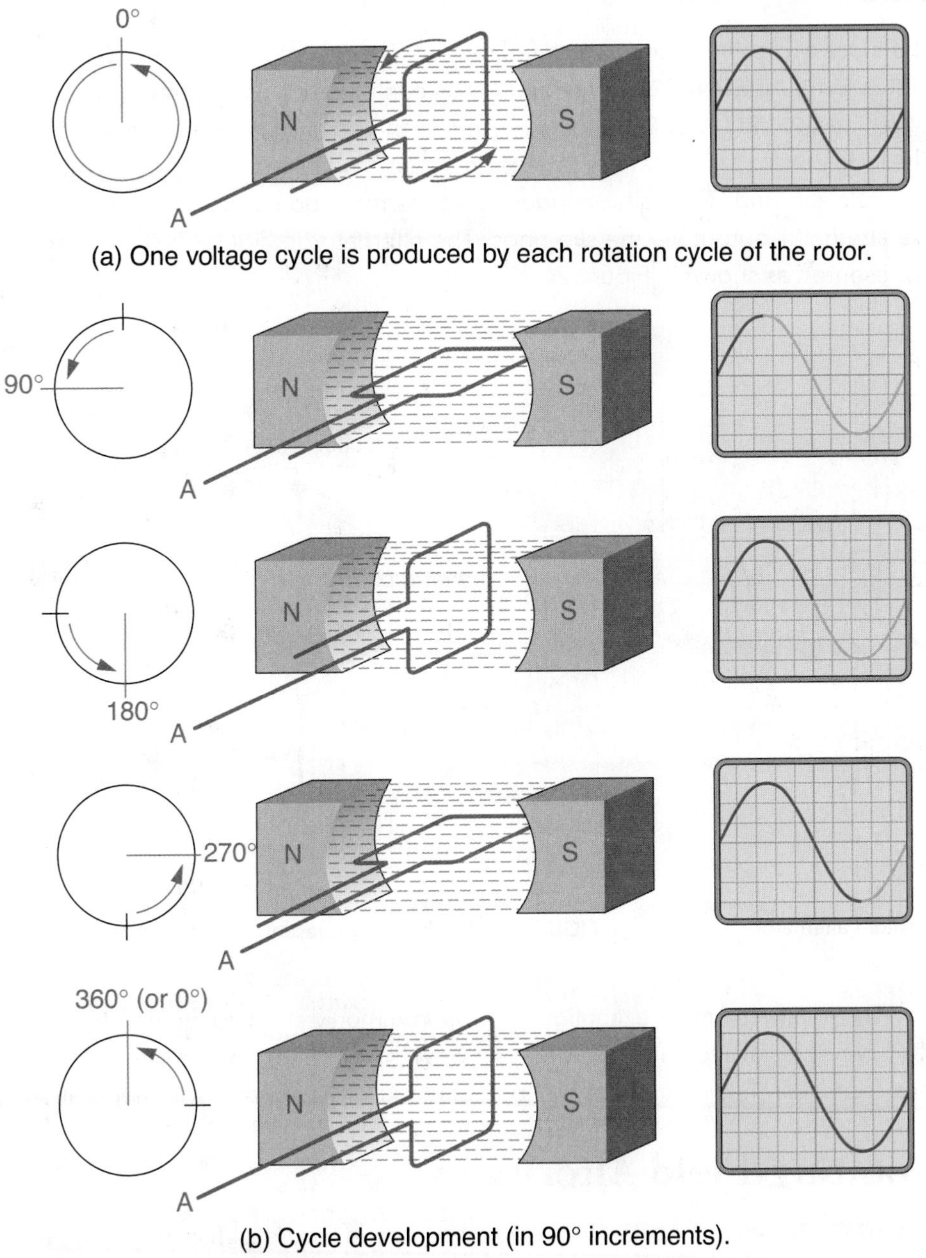

FIGURE 24.2 Generating a sine wave.

A voltage is induced across the loop conductor (armature) in Figure 24.2 as it cuts through the stationary magnetic field. The magnitude of the voltage and its polarity (at any instant) depend on:

- The angle and direction of rotation of the armature.
- The strength of the magnetic field (i.e., the flux density).
- The speed at which the armature rotates.

The relationship between the armature angle and the induced voltage is highlighted in Figure 24.2b.

Rotating-Armature Alternators

ROTATING-ARMATURE ALTERNATOR An alternator with rotating loop conductors and stationary magnetic fields.

DC FIELD SUPPLY A DC power supply that provides the current required by the stationary electromagnets in a rotating-armature alternator.

There are two basic alternator configurations. The first of these is the **ROTATING-ARMATURE ALTERNATOR**. This alternator configuration is illustrated in Figure 24.3. The **DC FIELD SUPPLY** provides the current required by the electromagnets. The loop conductor (armature) rotates on a central shaft and is connected on each end to a slip ring. The voltage induced across the loop conductor is coupled to the alternator output via the slip rings. The alternator in Figure 24.3 can also be represented as shown in Figure 24.4.

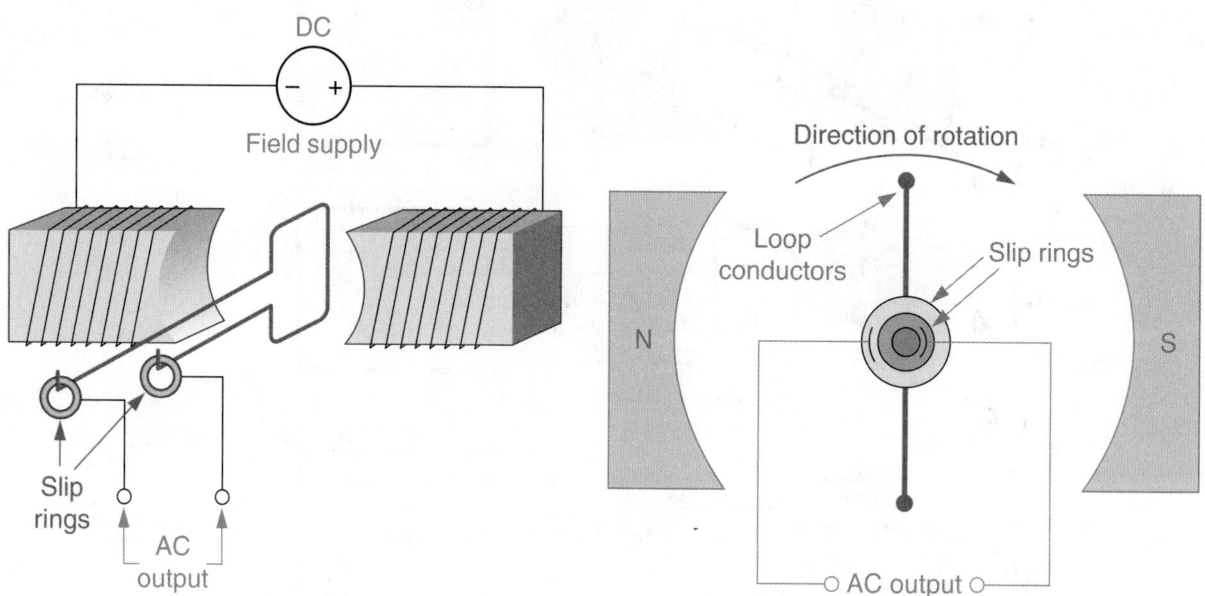

FIGURE 24.3 A rotating-armature alternator.

FIGURE 24.4 Another representation of a rotating-armature alternator.

The rotating-armature configuration is commonly used to demonstrate the basic principles of alternator operation. However, it is rarely used in practice because it can't provide the volt-ampere (VA) outputs required for most power applications.

Rotating-Field Alternators

ROTATING-FIELD ALTERNATOR An alternator with rotating magnetic fields and stationary conductors.

The **ROTATING-FIELD ALTERNATOR** represented in Figure 24.5 is the more practical and more commonly used of the two configurations. This type of alternator

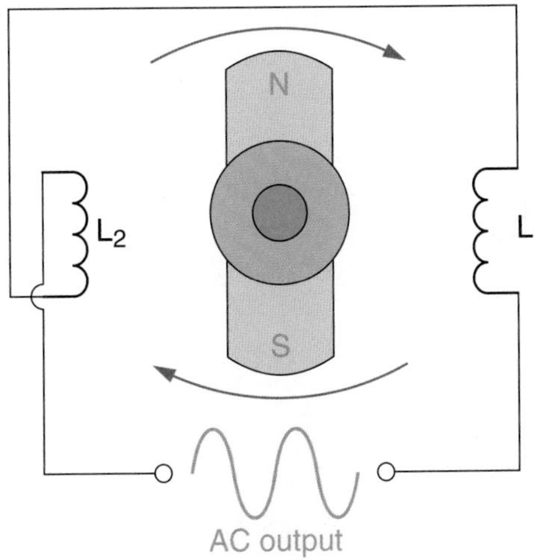

FIGURE 24.5 A rotating-field alternator.

uses one or more rotating magnets that induce a voltage across stationary coils (conductors). Note that the coil labeled L_2 is reverse wired so that the alternator provides the correct output polarity. This point is examined further later in this section.

Why are rotating fields (magnets) preferred over rotating armatures (conductors)? Rotating armatures require slip rings and brushes to couple the alternator output current to the load (as shown in Figure 24.3). The slip rings and brushes are difficult to insulate and can experience arcing in high-voltage applications. Stationary armatures, on the other hand, do not require slip rings and are easier to insulate (protect) against high-voltage arcing. Because rotating-field alternators are more commonly used, we will focus on them.

Single-Phase Alternators

A **SINGLE-PHASE ALTERNATOR** is one that produces a single sine wave output. The operation of a single-phase alternator is illustrated in Figure 24.6.

SINGLE-PHASE ALTERNATOR One that produces a single sine wave output.

The meter included in Figure 24.6 is a *galvanometer*. You may recall from Chapter 7 that a galvanometer is a current meter that indicates both the magnitude and direction of a current. Though a galvanometer would never be connected as shown here, we can use the operation of one to help you understand the operation of the single-phase alternator.

In Figure 24.6a, the armature flux is parallel to the windings of the stationary coils. Thus, the flux does not cut into the coils, and no voltage is induced across them. With no voltage across the coils, no current is generated in the output circuit and the galvanometer reading is 0 A.

In Figure 24.6b, the armature is perpendicular to the coils, and maximum voltage is induced across each. Note that the lower side of L_1 is positive and the upper side of L_2 is negative. These voltages generate a current through the galvanometer in the direction shown.

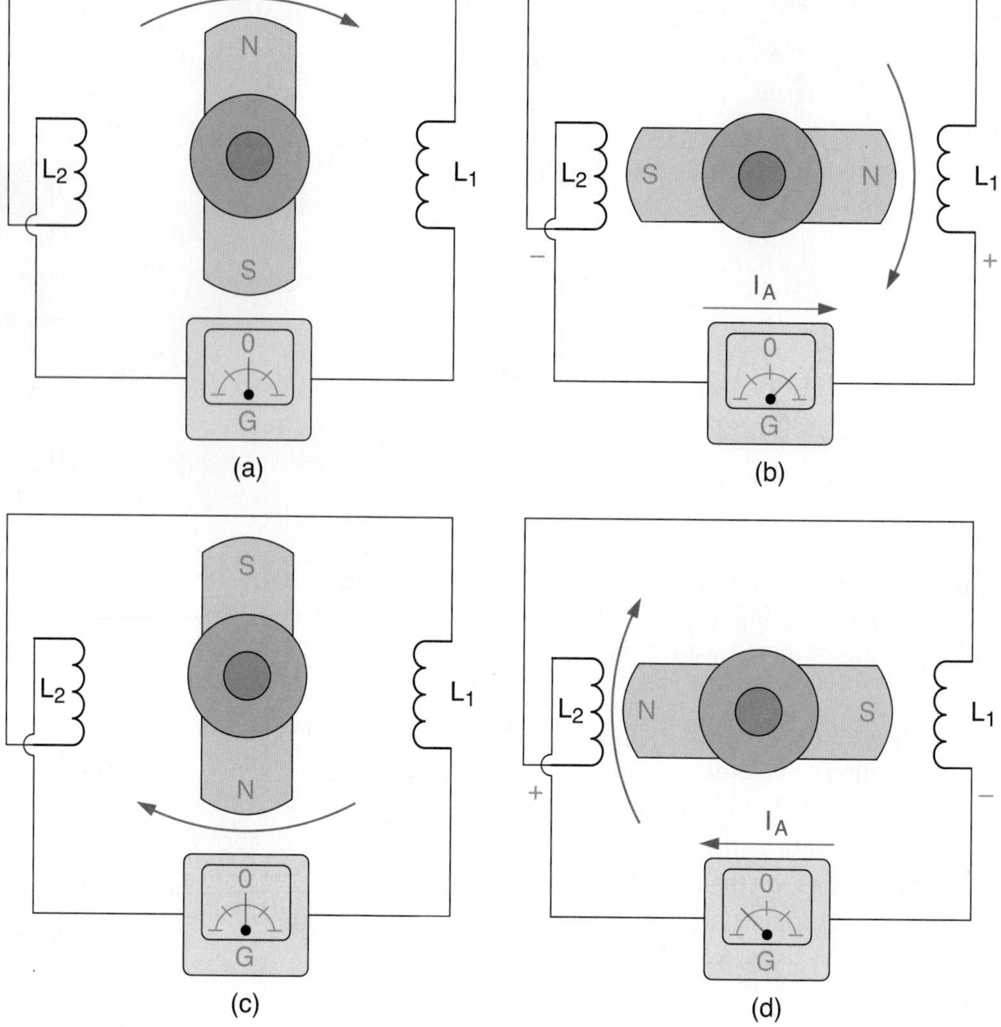

FIGURE 24.6 Single-phase alternator operation.

In Figure 24.6c, the armature flux (again) is parallel to the stationary coils, so no voltage is induced across them. With no induced voltage across the coils no current is generated in the output circuit and the galvanometer reading is 0 A.

In Figure 24.6d, the armature is once again perpendicular to the coils and maximum voltage is induced across each. Note that the lower side of L_1 is negative and the upper side of L_2 is positive. These voltages, which are the opposite of those in Figure 24.6b, generate a current through the galvanometer in the direction shown. Note that the galvanometer (load) current has reversed direction.

It is important to note that the circuit operation just described works only because of the "reverse wiring" of L_2. If L_2 was connected in the same fashion as L_1, they would form series opposing sources and would cancel each other out, therefore generating no load current. This is the reason for the reversed wiring of L_2 in the alternator.

Single-phase power is used only in relatively low-power applications (25 kW or less), such as 120 V backup electrical alternators.

Three-Phase Alternators

In Chapter 20, you were introduced to three-phase power. As part of that topic, you were introduced to a basic three-phase alternator. A basic three-phase alternator can be represented as shown in Figure 24.7a.

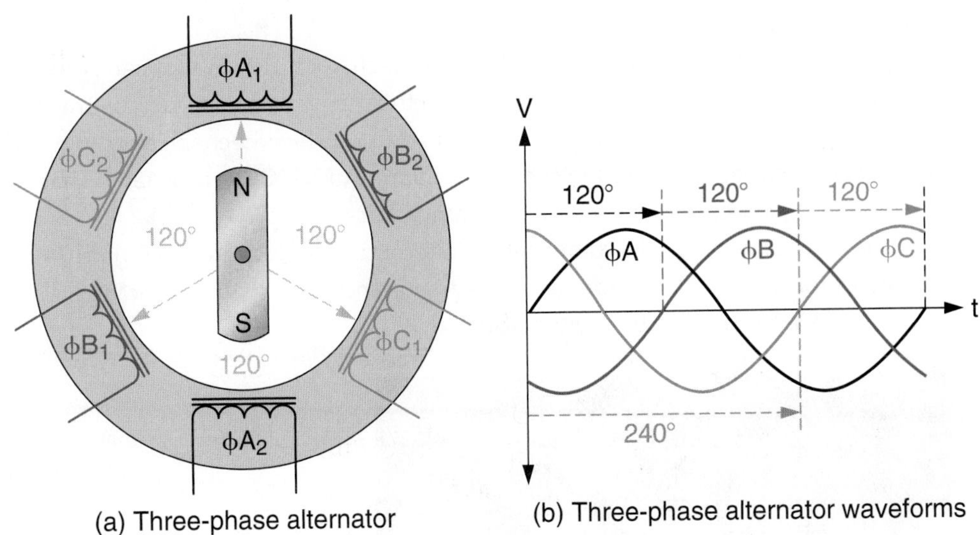

FIGURE 24.7 A three-phase alternator and its output waveforms.

Three-phase alternators are classified as *polyphase* alternators. A **POLYPHASE ALTERNATOR** is one that provides more than one sinusoidal output. As shown in Figure 24.7b, the three-phase alternator provides three sine wave outputs that are 120° out of phase.

POLYPHASE ALTERNATOR One that provides more than one sinusoidal output.

Like the single-phase alternator, each phase in the three-phase alternator has a pair of conducting coils spaced 180° apart. For example, the ϕA coils (labeled ϕA_1 and ϕA_2) are 180° out of phase. The six coils in the figure, spaced at equal intervals around the 360° stator, are separated by

$$\frac{360^\circ}{6} = 60^\circ$$

Note that the ϕA_1, ϕB_1, and ϕC_1 coils are 120° out of phase with each other. However, each coil is separated from the coil on either side of it by 60°.

The information here is intended only as an introduction to the concept of three-phase alternator operation. These alternators are examined further in the upcoming sections.

PROGRESS CHECK

1. What are the similarities and the differences between alternators and DC generators?
2. Name the two basic alternator configurations.
3. Which configuration is the most common and why?
4. What is a single-phase alternator? What is a polyphase alternator?

24.2 THREE-PHASE ALTERNATOR ELEMENTS AND CONSTRUCTION

Now that we have established some of the basic principles of alternator operation and construction, we are going to examine the elements that make up a three-phase alternator.

GENERATOR A dynamo that is configured to convert mechanical energy to electrical energy.

MOTOR A dynamo that is configured to convert electrical energy to mechanical energy.

You may recall that a *dynamo* is an electromechanical device that can convert mechanical energy to electrical energy, or electrical energy to mechanical energy. A **GENERATOR** is a dynamo that is configured to convert mechanical energy to electrical energy (as described in the last section). A **MOTOR** is a dynamo that is configured to convert electrical energy to mechanical energy.

The rotor and stator of a rotating-field alternator are illustrated in Figure 24.8. The stator (on the right) is made up of the frame and the armature windings. The rotor (on the left) contains the field magnets.

NOTE: The orange coating on the rotor and stator in Figure 24.8 is *insulating paint* that serves as a primer, sealer, adhesive, and protective finish, providing resistance to moisture, corrosion, oil, acid, heat, salt spray, and dust.

FIGURE 24.8 An alternator rotor and stator.
Source: Courtesy of Illinois Electric Works (www.illinoiselectric.com)

Rotor Construction

CYLINDRICAL ROTOR A rotor that is similar to that in a DC generator, but containing electromagnets (rather than armature conductors) that generate the alternator field.

SALIENT-POLE ROTOR A rotor containing electromagnets that protrude from the armature.

TURBINES High-speed generators that are typically driven by gas or steam.

TURBINE-DRIVEN ROTORS Cylindrical rotors that are used in turbines.

There are two types of alternator field rotors: the **CYLINDRICAL ROTOR** and the **SALIENT-POLE ROTOR**. Both are shown in Figure 24.9. If you compare the cylindrical rotor in Figure 24.9a to the DC generator rotor illustrated in Figure 23.7a, you can see that they are very similar to each other. Despite the similarity, the cylindrical rotor contains electromagnets (rather than armature conductors) that generate the alternator field. The electromagnets contained in the salient-pole rotor are more easily identified, as they protrude from the rotor surface as shown in Figure 24.9b.

Cylindrical rotors are used in **TURBINES** (high-speed generators that are typically driven by gas or steam). Because they are used in turbines, cylindrical rotors are often referred to as **TURBINE-DRIVEN ROTORS**. Cylindrical rotors are well suited to operate at speeds of 1200 rpm and higher because the field coils are wound within the body of the rotor. As such, they can withstand the centrifugal force produced at high speeds.

(a) (b)

FIGURE 24.9 Cylindrical and salient-pole rotors.
Source: Courtesy of Illinois Electric Works (www.illinoiselectric.com)

Salient-pole rotors are used in low-speed generators that operate below 1200 rpm, and are the more commonly used of the two rotor configurations. Salient-pole alternators are typically driven by water, wind, or internal-combustion engines.

A DC field supply provides the current required for the rotor to generate magnetic flux. The field supply is normally connected to the rotor via two slip rings.

Stator Construction

The stator of a rotating-field alternator serves as the support structure for its armature windings. As Figure 24.10a shows, the armature windings pass through slots in the stator assembly. These slots are shaped as shown in Figure 24.10b. Note that the stator assembly is subject to the same eddy current and copper (I^2R) power losses as the armature in a DC generator or a rotating-armature alternator.

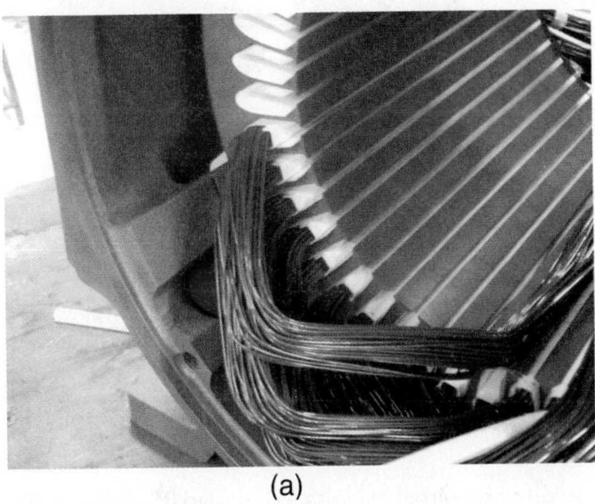

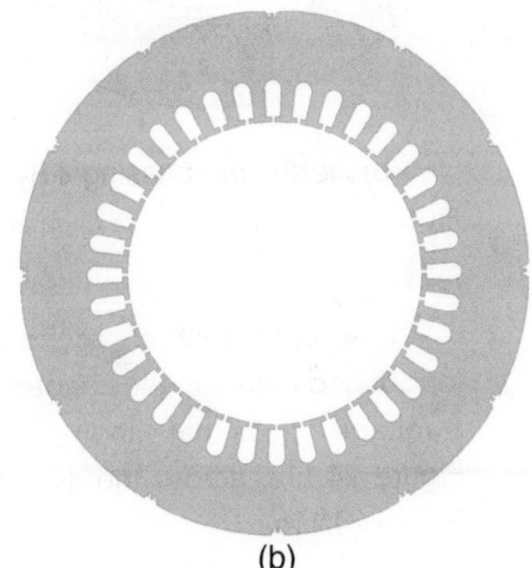

(a) (b)

FIGURE 24.10 An alternator stator.
Source: (a) Courtesy of Illinois Electric Works (www.illinoiselectric.com); (b) Courtesy of Siemens Industry, Inc. Used by permission.

Three-Phase Alternator Windings

As you know, a three-phase alternator provides sine wave outputs that are 120° out of phase. These phase relationships are the result of the armature configuration. This point is illustrated in Figure 24.11.

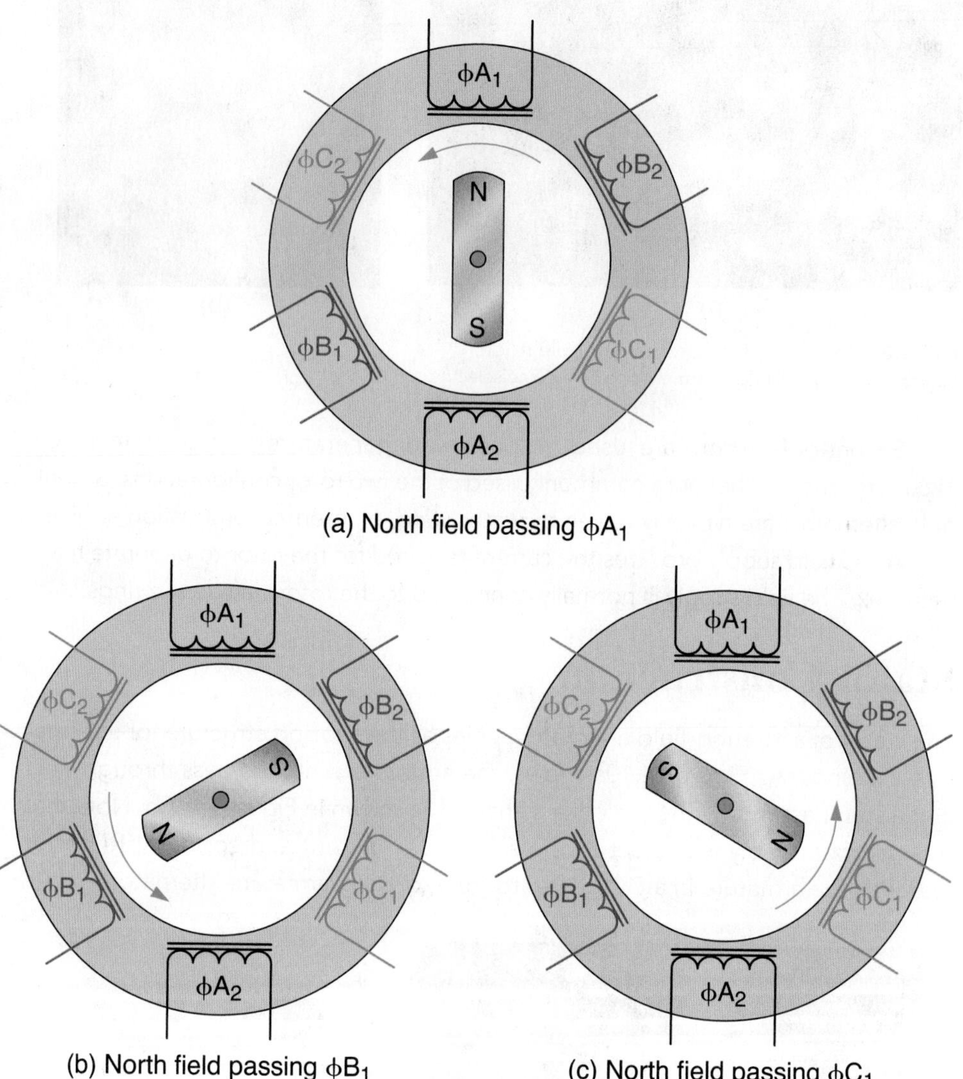

FIGURE 24.11 Three-phase alternator operation.

The representation of the three-phase alternator shown in Figure 24.11 was introduced earlier in the chapter. As the field passes each winding, the flux induces a voltage across that winding. Voltage is induced across the ϕA coils (windings) in Figure 24.11a, across the ϕB coils in Figure 24.11b, and across the ϕC coils in Figure 24.11c.

You may recall that three-phase circuits are wired in delta (Δ) or wye (Y) circuit configurations. As a review, delta (Δ) and wye (Y) circuits are summarized in Figure 24.12. Three-phase armature windings are wired in either of these configurations.

Summary Illustration

Inductive Wye (Y) and Delta (Δ) Circuits

Schematics:

A
N
B
C
Y connection

A
B
C
Δ connection

Basic relationships:
(Balanced circuits)

Y connection: $E_{LINE} = \sqrt{3} \times E_{PHASE}$, $I_{LINE} = I_{PHASE}$

Δ connection: $E_{LINE} = E_{PHASE}$, $I_{LINE} = \sqrt{3} \times I_{PHASE}$

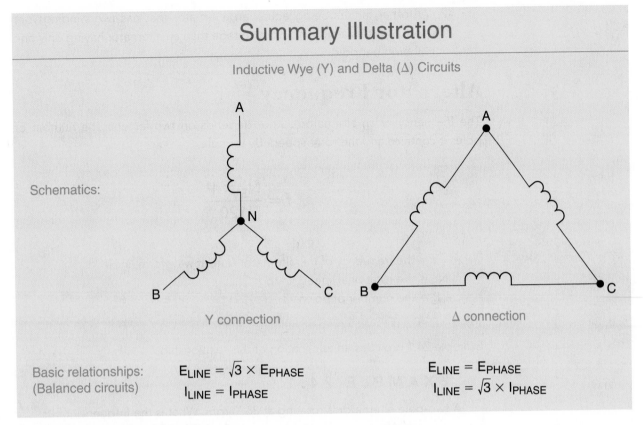

FIGURE 24.12 Wye (Y) and delta (Δ) circuits.

Alternators are generally wired so that there is either one winding per pole or two windings per pole. Using two windings per pole is the preferred approach for two reasons:

1. It makes more efficient use of the stator. Because the stator allows for two windings per pole, as shown in Figure 24.13, it makes sense to make use of that capability.

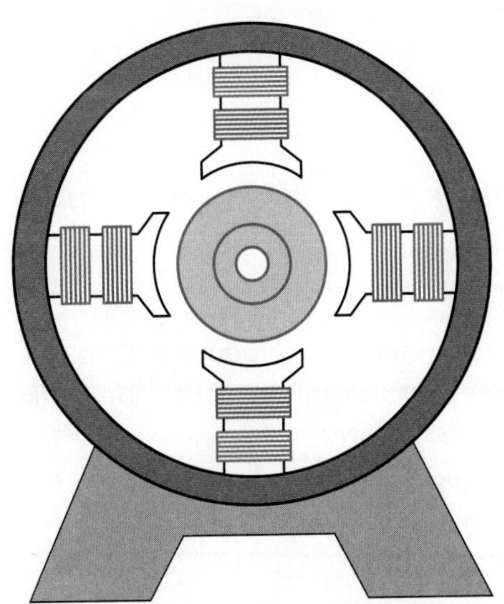

FIGURE 24.13 Stator pole wiring.

2. All other factors being equal, an alternator that has two windings per pole provides a higher output voltage than an alternator having only one winding per pole.

Alternator Frequency

The frequency of an alternator output depends on two factors: The number of poles it contains and the rotor speed. By formula,

(24.1)

$$f = \frac{N_S \times P}{120}$$

where

f = the frequency of the alternator output waveform, in Hz
N_S = the rotational speed, in rpm
P = the number of poles

Example 24.1 provides an application of this relationship.

EXAMPLE 24.1

A two-pole alternator is rotating at 3600 rpm. What is the frequency of its output waveform?

SOLUTION

Using the given values, the alternator output frequency is found as

$$f = \frac{N_S \times P}{120}$$

$$= \frac{3600 \times 2}{120}$$

$$= 60 \text{ Hz}$$

PRACTICE PROBLEM 24.1

A four-pole alternator is rotating at 1800 rpm. What is the frequency of its output waveform?

As equation (24.1) indicates, the rotation speed required to produce a 60 Hz sine wave depends (in part) on the number of alternator poles. The rotation speeds (in rpm) needed to produce a 60 Hz sine wave are as follows:

NUMBER OF POLES	ROTATIONAL SPEED (RPM)
2	3600
4	1800

NUMBER OF POLES	ROTATIONAL SPEED (RPM)
6	1200
8	900
10	720
12	600

Alternator Output Voltage

The output voltage generated by an alternator having two windings per pole can be calculated using

(24.2)

$$E = 4.44 \times \Phi \times f \times N$$

where

E = the effective (rms) alternator output voltage
Φ = the flux per pole, in webers (Wb)
f = the output frequency
N = the number of turns (per phase)

The frequency (f) and number of turns per phase (N) are fixed for a given alternator. However, the flux per pole (Φ), and thus the alternator output voltage, can be increased (or decreased) by adjusting the output from the alternator's DC field supply.

While frequency is controlled by the speed of rotation, the exact value of AC voltage produced by a synchronous machine is controlled by varying the current in its DC field windings. Output power is controlled by the torque applied to the alternator shaft by the prime mover.

IN THE FIELD

PROGRESS CHECK

1. List the two types of alternator field rotors and some of the common applications for each type.
2. What is a DC field supply and what is it used for?
3. How many windings can there be per pole in a three-phase alternator? Which is the most common approach?
4. What factors determine the frequency of an alternator's output voltage?
5. What factors determine the magnitude of an alternator's output voltage?
6. How can the output voltage of an alternator be adjusted?

24.3 THREE-PHASE ALTERNATOR CONNECTIONS AND LOADING

In this section, we conclude our examination of alternators by briefly discussing two important concepts: output voltage regulation and connecting alternators in parallel.

Voltage Regulation

You may recall that voltage regulation is a measure of a voltage source's ability to provide a constant output when the load demand varies, usually expressed as a percentage. By formula,

(24.3)

$$\text{Regulation}(\%) = \frac{E_{NL} - E_{FL}}{E_{NL}} \times 100$$

where

E_{NL} = the no-load alternator output voltage
E_{FL} = the full-load alternator output voltage

Example 24.2 demonstrates an application of this relationship.

EXAMPLE 24.2

The output from a 120-V three-phase alternator drops to 115 V under full-load conditions. Calculate the alternator's regulation.

SOLUTION

The alternator has values of $E_{NL} = 120$ V and $E_{FL} = 115$ V. Therefore, its regulation is

$$\text{Regulation}(\%) = \frac{E_{NL} - E_{FL}}{E_{NL}} \times 100$$

$$= \frac{120\text{ V} - 115\text{ V}}{120\text{ V}} \times 100$$

$$= 4.17\%$$

PRACTICE PROBLEM 24.2

The output from a 208-V three-phase alternator drops to 200 V under full-load conditions. Calculate the alternator's regulation.

The ideal alternator (if it existed) would generate the same output voltage regardless of the value of the load. In other words, its no-load and full-load output voltages would be equal, and its regulation would be 0%. For example, the regulation of an ideal alternator with $E_{NL} = E_{FL} = 120$ V would be found as

$$\text{Regulation}(\%) = \frac{E_{NL} - E_{FL}}{E_{NL}} \times 100$$

$$= \frac{120\text{ V} - 120\text{ V}}{120\text{ V}} \times 100$$

$$= 0\%$$

With 0% being the ideal value of voltage regulation, we can make the following statement: *The lower the voltage regulation of an alternator, the better.*

In practice, the terminal voltage of an alternator is affected by a number of factors, including armature winding reaction, which is the result of stator turns acting across the main field, and armature winding inductive reactance. As a result, its voltage regulation can be both variable and poor. For example, an alternator can display substantial voltage drop or significant voltage rise, depending on its load power factor. In practice, voltage regulation is controlled by external regulating circuits that vary the rotating field winding current so that a constant voltage is delivered to the load.

IN THE FIELD

The output voltage from an alternator is usually controlled by a voltage regulator circuit. Voltage regulator circuitry is rather complex and beyond the scope of this text, but its basic principles are straightforward. The output voltage of an alternator varies inversely with load demand. If load demand increases, output voltage decreases, and vice versa.

A voltage regulator circuit monitors the output voltage of the alternator. If the alternator output decreases from its rated value, the regulator circuit increases the output from the DC field supply. This causes an increase in the flux produced by each pole and thus the alternator's output voltage. If the output of the alternator increases, the DC field supply voltage decreases and the alternator output voltage returns to the desired value.

Connecting Alternators in Parallel

Like DC generators, alternators can be connected in parallel. Connecting a pair of alternators in parallel has two advantages:

1. The maximum possible output current (and therefore, power) is increased.
2. The load can operate (often at reduced capacity) while one alternator or the other is shut down for maintenance or repair.

When alternators are connected in parallel, several precautions must be taken. First, the alternators must provide the same output frequency and voltage. For example, two 120 V, 60 Hz alternators can be connected in parallel, but neither can be connected in parallel with a 480 V, 60 Hz alternator. Second, the motors must be synchronized to ensure that they reach their peak values at the same time.

When connecting either single- or three-phase alternators in parallel, a minor difference in the alternator output voltages is acceptable, since the parallel point voltages can be equalized using transformers. Also, the machines can rotate clockwise or counter clockwise and the phase sequence can be in any order within the respective machines. The primary requirement is that, by proper interconnection, the phase sequence must be identical at the actual paralleling point.

IN THE FIELD

The terms "phase sequence"and "in phase"*do not have the same meaning.* Phase sequence really means phase time sequence. For example, if one alternator has a *phase sequence* of A-B-C, the incoming alternator must also have the same phase sequence; otherwise their combined voltages will be different by as much as three quarters of the peak-to-peak sine wave voltage. These mismatched voltages will exist between two of the three phases, even though one phase is properly synchronized. The result is the generation of short-circuit current. On the other hand, the term "*in phase*" describes the action of two sine waves as they go through their maximum and minimum points at the same time and in the same direction. In other words, if two equal-frequency varying voltages are in phase, then each voltage is at the same absolute value at any instant.

KEY CONCEPT

PROGRESS CHECK

1. What is voltage regulation and how is it calculated?
2. What is the voltage regulation of an ideal alternator?
3. Explain how a voltage regulator circuit controls the output voltage of an alternator.
4. What are the advantages of operating alternators in parallel?
5. What are the precautions that must be taken when connecting alternators in parallel?

24.4 AC MOTORS: AN OVERVIEW

Like DC motors, AC motors are used to convert electrical energy to mechanical (rotational) energy. A basic single-phase AC motor can be represented as shown in Figure 24.14. The motor shown has a magnetic rotor. When the AC input is applied to the stator coils, they generate magnetic flux. The interaction between the stator field and the rotor field causes the rotor to make one 360° rotation per AC input cycle, as illustrated in Figure 24.15.

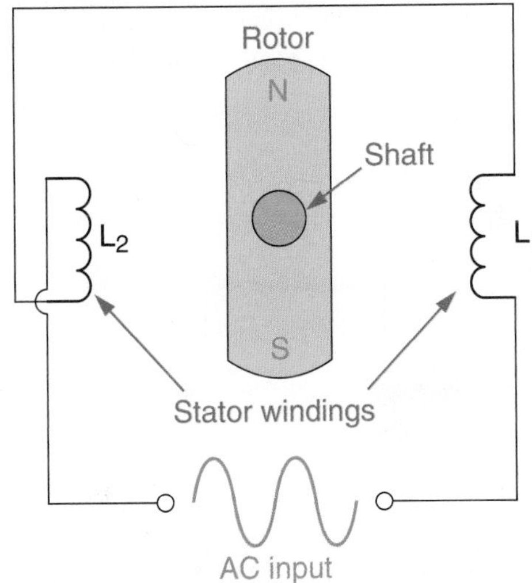

FIGURE 24.14 An AC motor.

In Figure 24.15a, the field current (I_F) through the stator coils produces the magnetic polarities shown. These polarities affect the rotor as follows:

- The N stator field repels the N end of the rotor and attracts the S end of the rotor.
- The S stator field repels the S end of the rotor and attracts the N end of the rotor.

As a result, the rotor moves in the direction indicated. When the AC input decreases to 0 V (as shown in Figure 24.15b), there is no current through the stator coils and therefore, no stator field. When the AC input is at its negative peak (as shown in Figure 24.15c), the stator field polarities are reversed. Again, the N stator field exerts a counterclockwise (CCW) force on the rotor, as does the S stator field, and the CCW rotation continues.

The operating cycle described above is illustrated further in Figure 24.16. When the AC input is at 0°, the stator coils do not produce a magnetic field. At the 90° point, the current generated by the AC input produces a magnetic field that has the polarity shown. At the 180° point, $I_F = 0$ A (again), and no magnetic field is produced. At the 270° point, the AC input is negative. As a result, the direction of the field current is reversed, as is the polarity of the magnetic field.

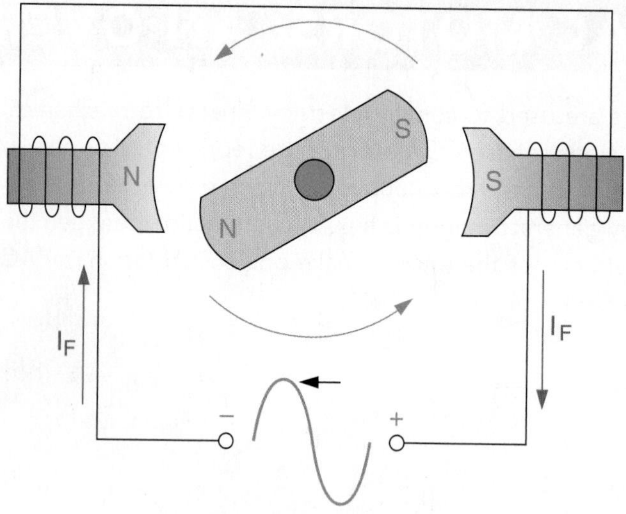

(a) AC input at its positive peak

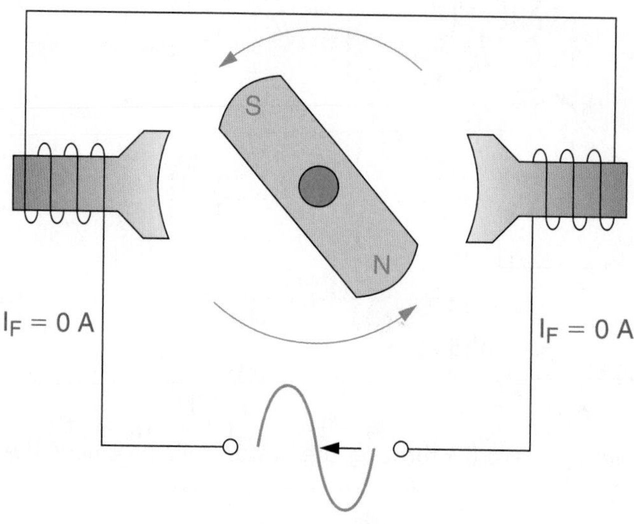

(b) AC input at 0 V

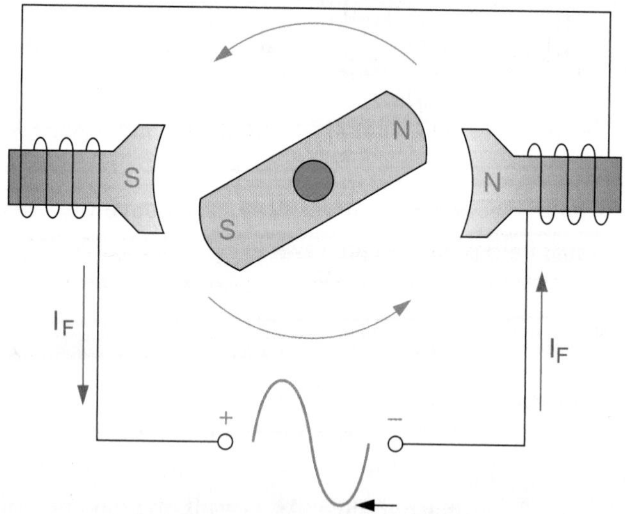

(c) AC input at its negative peak

FIGURE 24.15 AC motor operation.

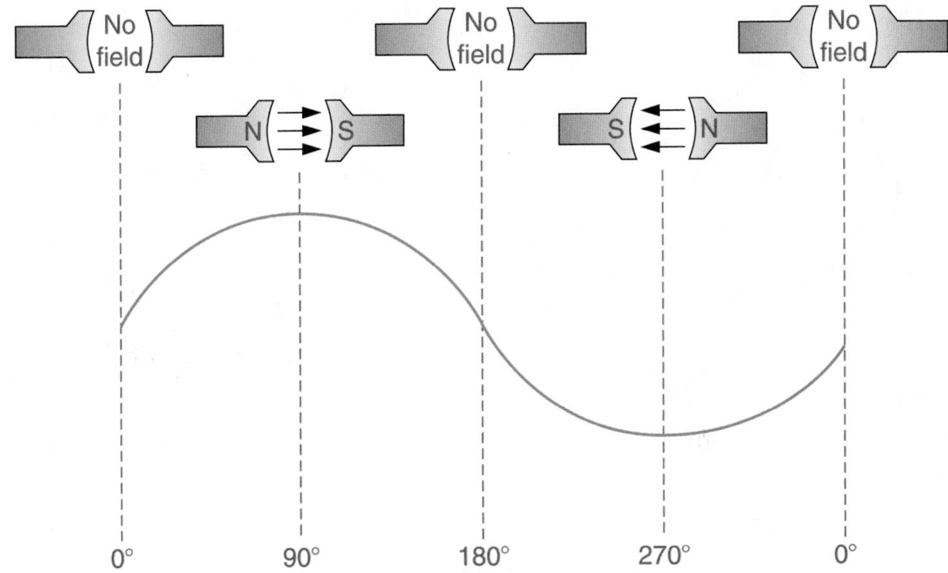

FIGURE 24.16 The relationship between current and the resulting magnetic field.

The motor represented in Figure 24.15 operates on the principle of *switching magnetic fields* back and forth between the stator field coils. In contrast, the split-phase motor represented in Figure 24.17a works on the principle of *rotating magnetic fields*.

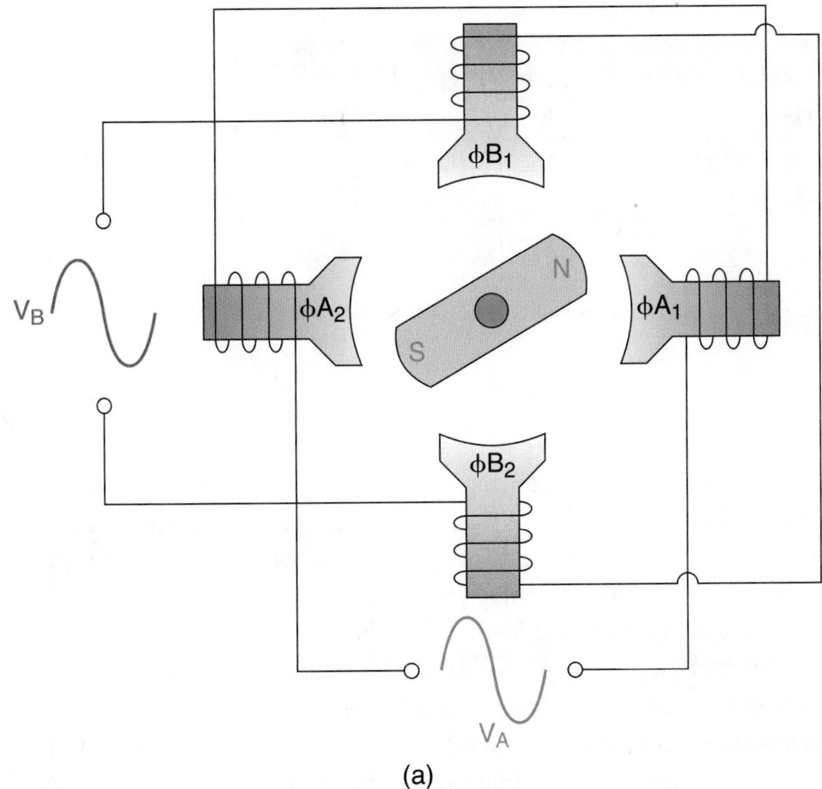

FIGURE 24.17 A split-phase motor.

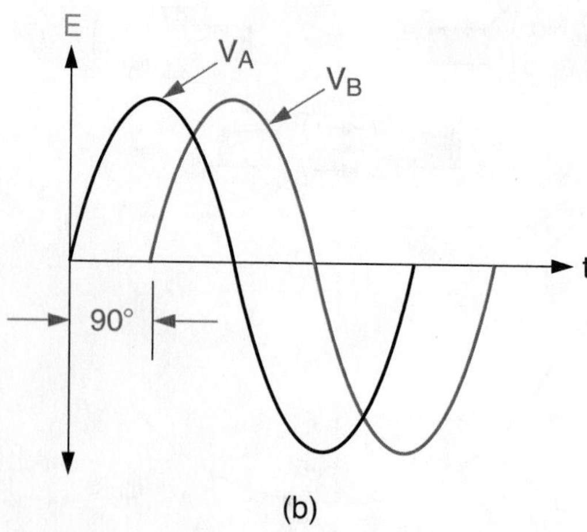

FIGURE 24.17 (*Continued*)

The split-phase motor has two sets of stator coils, labeled ΦA and ΦB, each with its own AC input. As shown in Figure 24.17b, the two AC inputs are 90° out of phase. The AC input to each set of stator coils produces a magnetic field as indicated in Figure 24.16. However, because the two AC inputs are 90° out of phase, they generate a magnetic field that effectively rotates through 360° as outlined in Table 24.1.

TABLE 24.1 • Split-Phase Motor Inputs and Phases

AC INPUT PHASES	ΦA_1	ΦA_2	ΦB_1	ΦB_2
V_A at 90°, V_B at 0°	S	N	No magnetic field	
V_A at 180°, V_B at 90°	No magnetic field		S	N
V_A at 270°, V_B at 180°	N	S	No magnetic field	
V_A at 0°, V_B at 270°	No magnetic field		N	S

The rotation of the stator fields can be demonstrated by tracing the progression of the north (N) magnetic pole through Table 24.1. As the table shows, the N pole moves from ΦA_2 to ΦB_2 to ΦA_1 to ΦB_1, indicating that the pole moves counterclockwise (CCW) from one stator coil to the next. As long as the input signals are present, the CCW rotation continues. Thus, the rotor in a split-phase motor is controlled using a rotating magnetic field.

There is an important point to be made: *The split-phase operation described here is actually a temporary condition that is often employed in single-phase motors using special circuitry.* Its purpose is to provide additional torque at startup to help bring the motor up to speed. Split-phase operation in single-phase motors is described in greater detail later in this chapter.

Motor Speed

Earlier in the chapter, you learned that the output frequency of an alternator can be calculated using

$$f = \frac{N_S \times P}{120}$$

where f is the frequency (in Hz), N_S is the rotational speed (in rpm), and P is the number of poles. This relationship can be transposed to provide us with the following equation for AC motor speed:

(24.4)
$$N_S = \frac{120f}{P}$$

Example 24.3 demonstrates an application of this relationship.

EXAMPLE 24.3

What is the rotational speed of a four-pole AC motor with a 60-Hz AC input?

SOLUTION

Using the given values of $f = 60$ Hz and $P = 4$, the motor speed is found as

$$N_S = \frac{120f}{P}$$
$$= \frac{120 \times 60\text{ Hz}}{4}$$
$$= 1800\text{ rpm}$$

PRACTICE PROBLEM 24.3

What is the rotational speed of a six-pole motor with a 120 Hz input frequency?

The speed of a motor (in rpm) varies inversely with the number of poles the motor contains. For example, the four-pole motor described in Example 24.3 rotates at 1800 rpm. At the same time, an eight-pole motor with a 60 Hz input rotates at a speed of

$$N_S = \frac{120f}{P}$$
$$= \frac{120 \times 60\text{ Hz}}{8}$$
$$= 900\text{ rpm}$$

Thus, a motor with twice the number of poles as our example motor rotates at half the speed.

Synchronous and Induction Motors

SYNCHRONOUS MOTOR
One that is synchronized to an alternator that acts as an energy and timing source.

There are two types of AC motors: *synchronous motors* and *induction motors*. A **SYNCHRONOUS MOTOR** is one that is synchronized to an alternator that acts as an energy and timing source. A synchronous AC motor is connected to an alternator as shown in Figure 24.18.

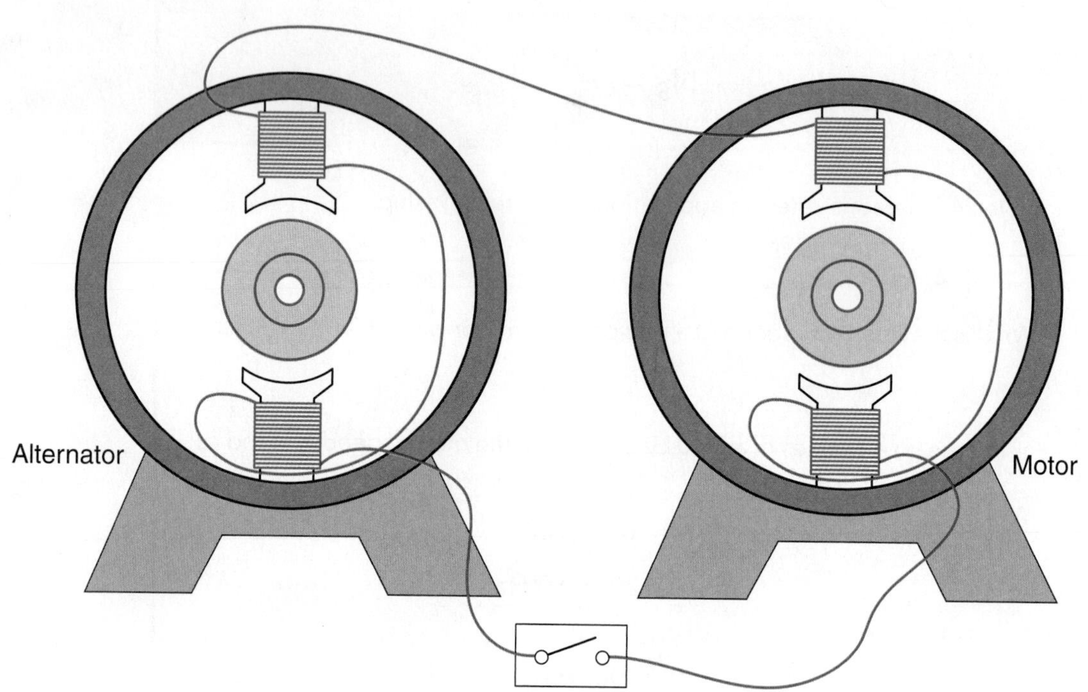

FIGURE 24.18 A synchronous motor and alternator.

Ideally, the motor in Figure 24.18 is perfectly in sync with the alternator rotor, meaning that the rotors of both machines are always at the same angle at the same time. When the alternator rotor is at 0°, so is the motor rotor; when the alternator is at 90°, so is the motor rotor, and so on. In practice, however, this is not the case. When a load is connected to the motor, the *torque* required to maintain its rotational speed causes the motor to *lag* the alternator slightly, as shown in Figure 24.19. As the load increases, so does the *torque angle* (α) between the motor and alternator rotors. If the load on the motor is sufficient, the torque angle becomes too great ($\alpha > 90°$) and synchronization is lost.

Even when the motor lags the alternator, the two continue to rotate at the same frequency as long as synchronization is maintained.

KEY CONCEPT

A synchronous motor most closely resembles an alternator with a salient-pole rotor. The rotor electromagnets (poles) are powered by a DC field supply. In

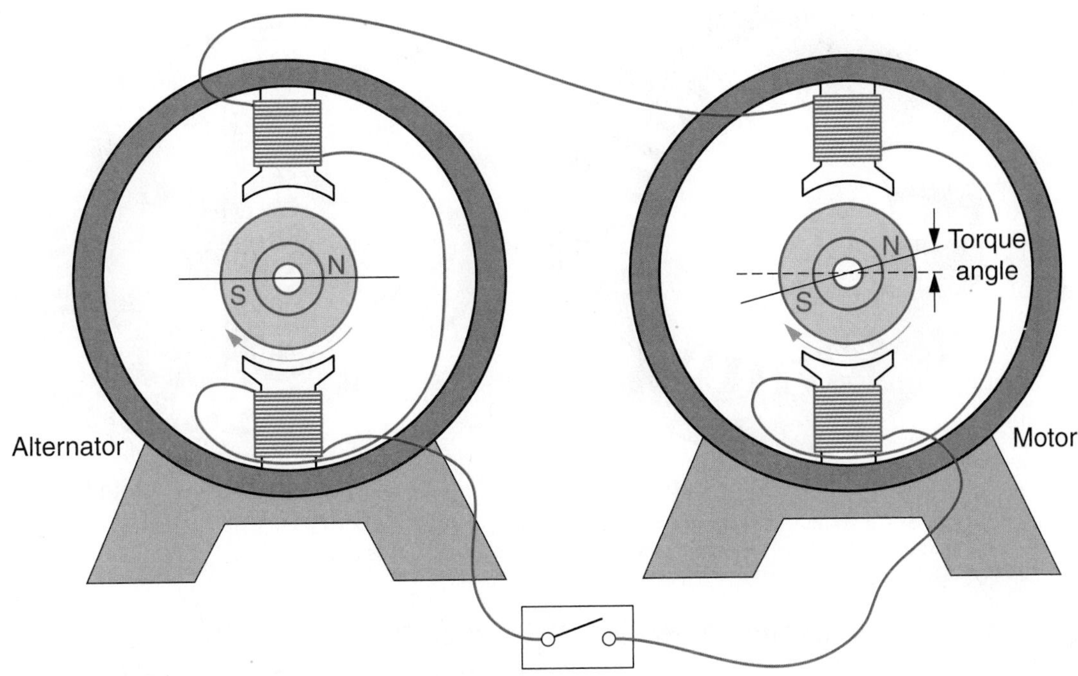

FIGURE 24.19 Synchronous motor torque angle.

contrast, an **INDUCTION MOTOR**, which is classified as an **ASYNCHRONOUS MOTOR**, does not have a DC field supply. Rather, rotor current is generated by electromagnetic induction, the same way power is transferred from the primary of a transformer to its secondary. In fact, an induction motor is sometimes described as a rotating transformer with the stator acting as the primary and the rotor acting as the secondary.

The source of rotor current is the primary distinction between synchronous and induction motors. In the upcoming sections, we will discuss the operating characteristics of synchronous and induction motors further.

INDUCTION MOTOR A motor with rotor current that is generated by electromagnetic induction rather than a DC field supply.

ASYNCHRONOUS MOTOR One that does not use an alternator as an energy and timing source.

Stator Windings

The stator of an induction motor is a cylindrical structure made of laminated steel segments like the one represented in Figure 24.20a. Using laminated segments reduces eddy current losses (as is the case in other motors and generators). The stator windings are wrapped as coils, covered with a protective layer of insulation, and mounted in the stator slots. Once the stator windings are mounted (Figure 24.20b), the stator assembly is coated with an epoxy resin that provides the coils with structural support and protection from harmful environmental conditions.

Rotor Construction

There are two types of rotors that are commonly found in AC motors: *wound* and *squirrel-cage*. The construction of a **WOUND ROTOR** is similar to that of the stator. That is, the rotor is made of laminated steel segments to reduce eddy current losses.

WOUND ROTOR A rotor made of laminated steel segments and wire conductors (windings) that are connected to slip rings.

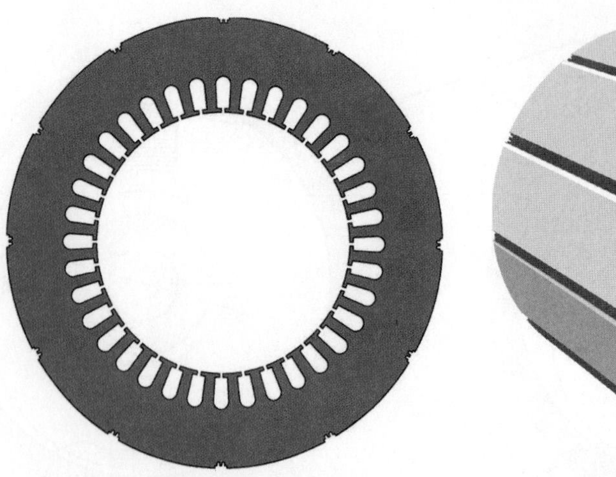

(a) Stator lamination (b) Stator windings completed

FIGURE 24.20 Stator lamination and windings.
Source: Courtesy of Siemens Industry, Inc. Used by permission.

A wound rotor is shown in Figure 24.21. The windings are connected to slip rings and brushes that can be used to couple the rotor to an external DC power source. Note that wound rotors are not used very often in practice because of their relatively high cost and maintenance requirements.

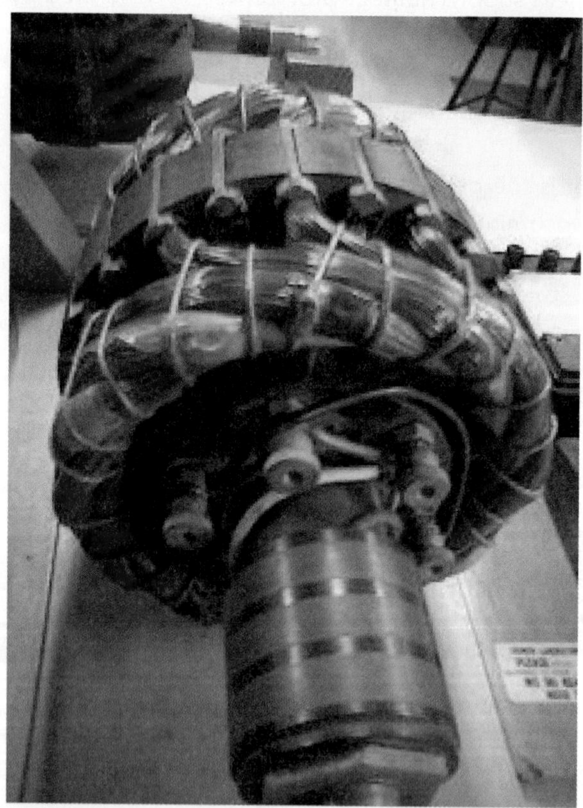

FIGURE 24.21 A wound rotor.
Source: Courtesy of Illinois Electric Works (www.illinoiselectric.com)

The **SQUIRREL-CAGE ROTOR** is the one most commonly used in AC motors. The rotor, which is shown in Figure 24.22a, contains conductor bars that extend through the length of the rotor in place of the wire conductors found in wound rotors. (A wound rotor is illustrated in Figure 24.22b for comparison.) The squirrel-cage rotor is easier to manufacture, requires less maintenance, and is more reliable than the wound motor.

SQUIRREL-CAGE ROTOR
One that contains conductor bars that extend through the length of the rotor in place of the wire conductors found in wound rotors.

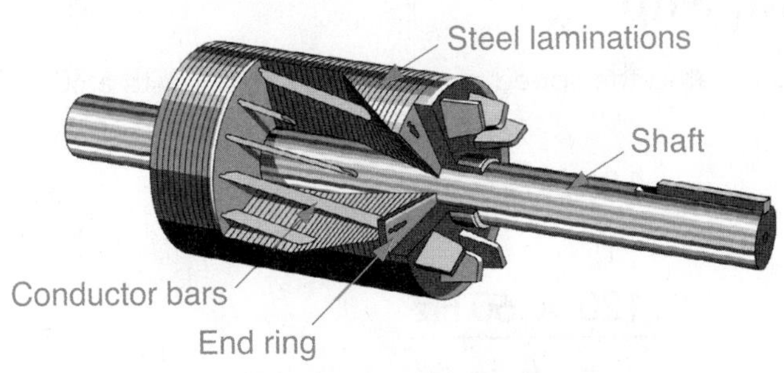

(a) Squirrel-cage rotor

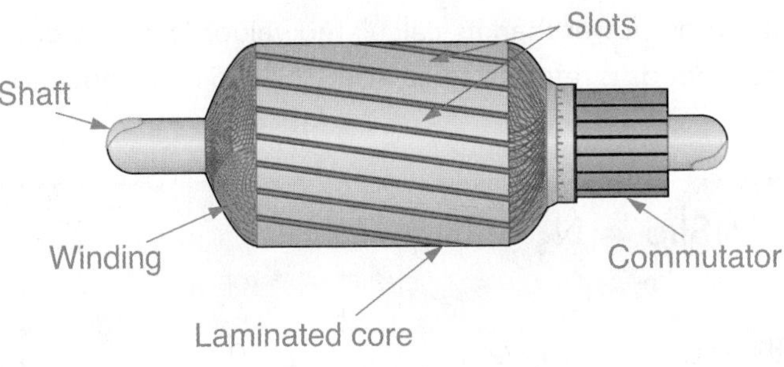

(b) Wound rotor

FIGURE 24.22
(a) Courtesy of Siemens Industry, Inc. Used by permission.

PROGRESS CHECK

1. Describe the interaction between an AC motor's stator field and the rotor field for one complete cycle of the AC input.
2. Describe and compare the operation of motors that use switching magnetic fields and rotating magnetic fields.
3. What is the phase difference between the two AC voltages in a split-phase motor?
4. What factors determine the rotational speed of a single-phase AC motor?
5. Describe and compare synchronous and induction single-phase AC motors.
6. What is the relationship between torque angle and motor load?
7. Describe the construction of wound rotors and squirrel-cage rotors. Which of the two is more common and why?

24.5 INDUCTION MOTORS

As stated earlier, the rotor current in an induction (or *asynchronous*) motor is generated by electromagnetic induction. In this section, we will take a closer look at the construction and operating characteristics of induction motors.

Motor Speed and Slip

Earlier in the chapter, we calculated the speed of a four-pole motor with a 60 Hz input as follows:

$$N_S = \frac{120f}{P} = \frac{120 \times 60\text{ Hz}}{4} = 1800\text{ rpm}$$

SLIP The difference between the speed of the rotating stator field and the rotor speed.

In reality, an induction motor runs *slower* than its calculated value. The difference between the stator and rotor speeds is referred to as **SLIP**. The slip of an induction motor is found as

(24.5)

$$\text{Slip} = N_S - N_R$$

where

N_S = the speed of the rotating stator field
N_R = the speed of the rotor

An application of this relationship is provided in the following example.

EXAMPLE 24.4

The rotor of a six-pole, 60 Hz induction motor spins at 1125 rpm. What is the slip of the motor?

SOLUTION

The speed of the rotating stator field is found as

$$N_S = \frac{120f}{P} = \frac{120 \times 60\text{ Hz}}{6} = 1200\text{ rpm}$$

and the slip is found as

$$\text{Slip} = N_S - N_R$$

$$= 1200 \text{ rpm} - 1125 \text{ rpm}$$
$$= 75 \text{ rpm}$$

This result indicates that the rotor is 75 rpm slower than the speed of the stator field.

PRACTICE PROBLEM 24.4

The rotor of a four-pole, 120 Hz induction motor spins at 3550 rpm. What is the slip of the motor?

The slip of a motor is often expressed as a percentage. For example, the slip calculated in Example 24.4, expressed as a percentage of the stator field speed, is found as

$$\text{Slip}(\%) = \frac{N_S - N_R}{N_S} \times 100$$
$$= \frac{1800 \text{ rpm} - 1725 \text{ rpm}}{1800 \text{ rpm}} \times 100$$
$$= 4.17\%$$

This result indicates that the rotor speed is 4.17% slower than the stator field speed.

Single-Phase Induction Motors

Single-phase induction motors are the most commonly used AC motors. You were introduced to the operation of this type of motor earlier in this chapter. As a review, single-phase motor operation is illustrated in Figure 24.23.

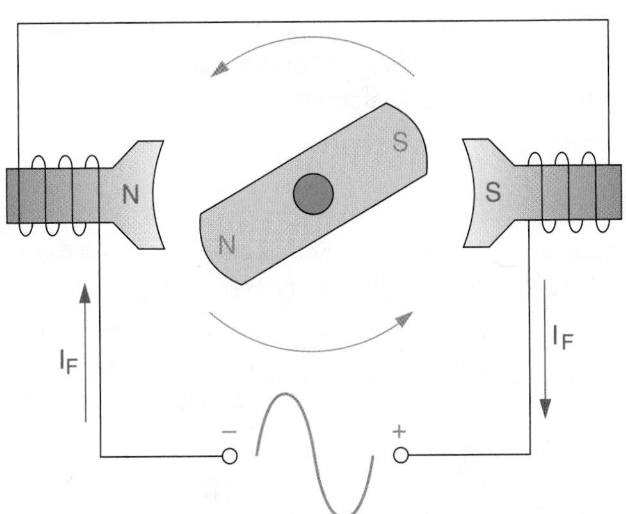

FIGURE 24.23 A single-phase induction motor.

When power is first applied to the motor represented in Figure 24.23, each stator coil attracts one of the motor poles, as follows:

- The south-seeking (S) stator pole attracts the north-seeking (N) rotor pole.
- The north-seeking (N) stator pole attracts the south-seeking (S) rotor pole.

As a result, the rotor begins to rotate about its axis.

Rotor Current

Earlier in the chapter, you learned that the rotor magnetic field is generated by current induced in the rotor. In order to generate this rotor current, there must be relative motion between the rotor conductors and the stator magnetic fields (in keeping with Lenz's law). If the rotor was turning in sync with the stator fields, there would be no relative motion between the stator fields and rotor, and there would be no rotor current. However, the difference between the speeds of the stator fields and the rotor (i.e., the slip) provides the relative motion required to generate the rotor current, as follows:

- The rotor is stationary at the moment when AC is applied to the stator fields.
- The stator fields quickly change polarity, providing relative motion between the stator fields and the rotor conductors.
- The relative motion between the stator fields and rotor conductors causes a voltage to be induced across the conductor, generating the rotor magnetic field.
- The rotor begins to turn, lagging behind the switching stator fields. The motor slip maintains the relative motion between the stator fields and the rotor, maintaining rotor motion.

Note that the motor slip is a result of the torque required to turn the rotor, and varies with the motor load. That is, motor slip increases when the load increases.

Split-Phase Induction Motors

SPLIT-PHASE INDUCTION MOTOR A single-phase induction motor with an added *starting component* that produces the torque required to start the motor and help bring the rotor up to speed.

CAPACITOR-START, INDUCTOR-RUN MOTOR A split-phase motor with a capacitive starting component.

A **SPLIT-PHASE INDUCTION MOTOR** is essentially a single-phase induction motor with an added *starting component* that produces the torque required to start the motor and help bring the rotor up to speed. The starting component is normally capacitive, resistive, or inductive.

Capacitor-Start, Inductor-Run

A split-phase motor with a capacitive starting component is commonly referred to as a **CAPACITOR-START, INDUCTOR-RUN MOTOR**. This type of motor can be represented as shown in Figure 24.24. The series combination of the *auxiliary winding* (or *starting winding*), the *starting capacitor* (C_S), and the switch is in parallel with the main stator winding. The auxiliary winding is physically at a 90° angle to the main stator winding.

The key to the operation of the capacitor start circuit is the relationship between the capacitor current and the stator coil current. This relationship is illustrated in Figure 24.25. As you can see, the capacitor current leads the AC input by 45°, and the

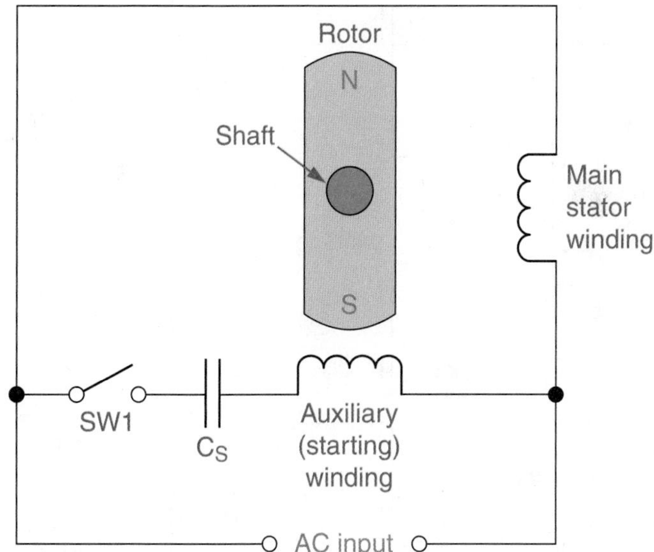

FIGURE 24.24 A capacitor-start, inductor-run motor.

stator coil current lags the AC input by 45°. As a result, the capacitor current leads the stator coil current by 90°. This provides the starting torque required by the motor (as described in the introduction to split-phase motors in Section 24.4).

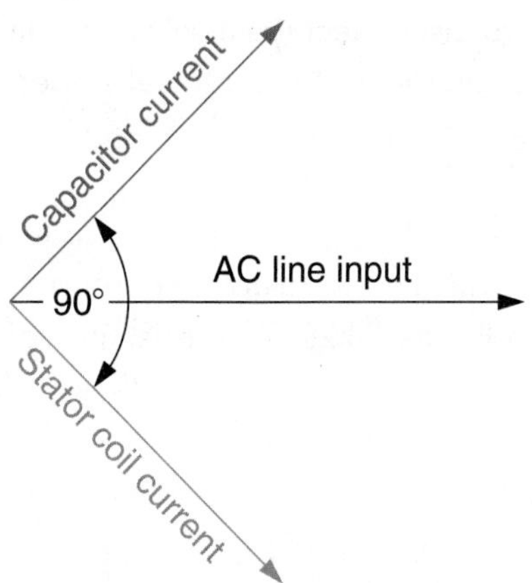

FIGURE 24.25 Capacitor and stator coil phase relationship.

Once the motor in Figure 24.24 approaches full-speed operation, a centrifugal switch (labeled SW1) opens automatically, effectively removing the auxiliary winding circuit from the motor. At this point, the machine operates like any other single-phase motor.

Capacitor-Start, Capacitor-Run

A **CAPACITOR-START, CAPACITOR-RUN MOTOR** contains an additional capacitor that is connected as shown in Figure 24.26. This circuit provides the same overall startup action as the capacitor-start, inductor-run motor in Figure 24.24. However, after the centrifugal switch opens, there is still a lower-value current through the

CAPACITOR-START, CAPACITOR-RUN MOTOR
A capacitor-start motor with an added capacitor that allows the motor to operate as a split-phase motor with a single-phase input.

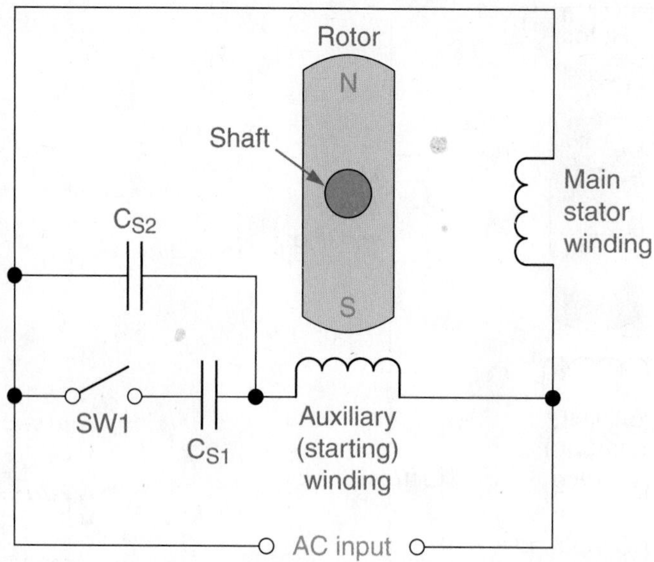

FIGURE 24.26 A capacitor-start, capacitor-run motor.

auxiliary coil and C_{S2}. This current is at a 90° angle to the higher-value current through the stator field coil. As such, the circuit continues to operate as a split-phase motor with a single-phase input.

A variation on the capacitor-start, capacitor-run circuit is shown in Figure 24.27. In this circuit, there is no centrifugal switch to disconnect the auxiliary circuit after the motor reaches its operating speed. As such, the motor continually operates as a split-phase motor with only a single-phase input.

INDUCTOR-START, INDUCTOR-RUN MOTOR One that has two windings (mounted at 90° angles) and a centrifugal switch.

Inductor-Start, Inductor-Run

The **INDUCTOR-START, INDUCTOR-RUN MOTOR** has two windings (mounted at 90° angles) and a centrifugal switch, as shown in Figure 24.28. The resistance of

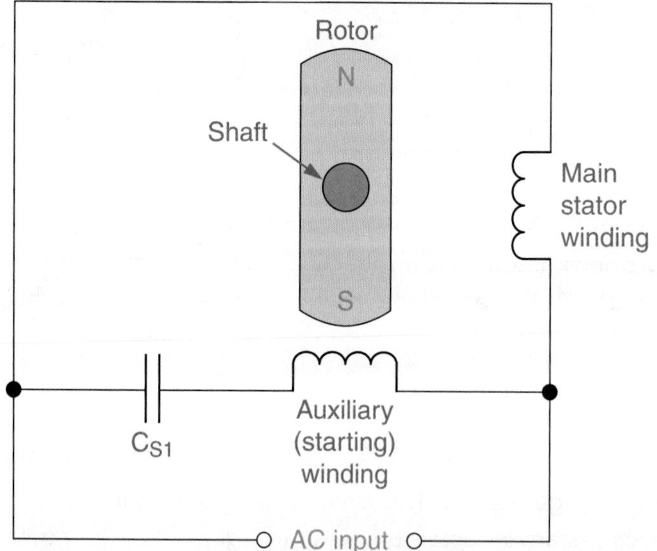

FIGURE 24.27 A modified capacitor-start, capacitor-run motor.

the auxiliary winding is significantly greater than that of the stator field winding. As a result, the auxiliary winding current *leads* the stator field winding current.

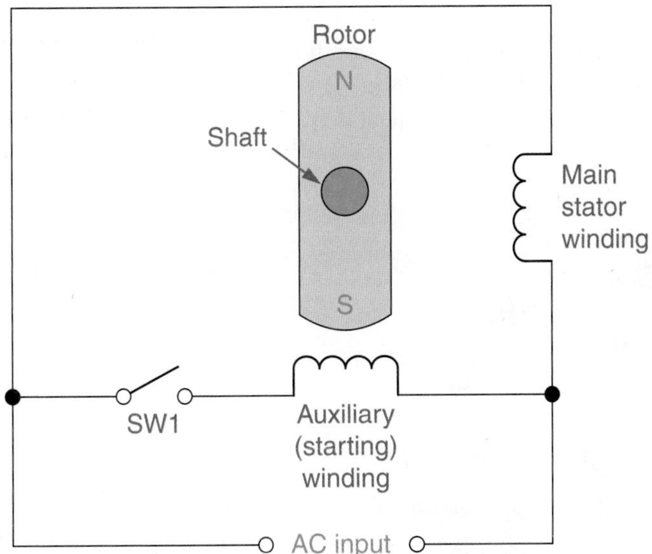

FIGURE 24.28 An inductor-start, inductor-run motor.

How does the difference in winding resistances produce a phase difference between the winding currents? First, remember that the fields are connected in parallel, so the winding voltages must be in phase with each other. Now, let's assume that the stator field winding is an *ideal* inductor. That being the case, the current through the stator field winding *lags* the component voltage by 90°. The auxiliary winding, on the other hand, is designed so that the winding resistance (R_W) is approximately equal to the winding reactance (X_L) at the motor operating frequency. That being the case, the phase angle between the auxiliary winding voltage and current is found as

$$\begin{aligned}\theta &= \tan^{-1}\left(\frac{-X_L}{R}\right) \\ &= \tan^{-1}(-1) \\ &= -45^\circ\end{aligned}$$

As this result indicates, the auxiliary winding current lags the component voltage by 45°. As such, it *leads* the stator field winding current by 45°. The differences in the winding fields that result from the phase difference between the two currents provide the extra torque needed when the motor starts. As the rotor comes up to speed, the centrifugal switch opens the auxiliary circuit and the motor runs as a single-phase motor.

As you can see, the primary difference between the various induction motors is the manner in which the rotor begins to revolve. In the next section, you will see that the synchronous motor does not have the same need for dedicated starting circuitry.

Three-Phase Induction Motors

A three-phase induction motor has three sets of stator field coils, as illustrated in Figure 24.29. Note that each pair of coils is connected internally; that is, coils ΦA_1 and ΦA_2 are connected in series, as are coils ΦB_1 and ΦB_2 and coils ΦC_1 and ΦC_2. These connections are illustrated in Figure 24.29b.

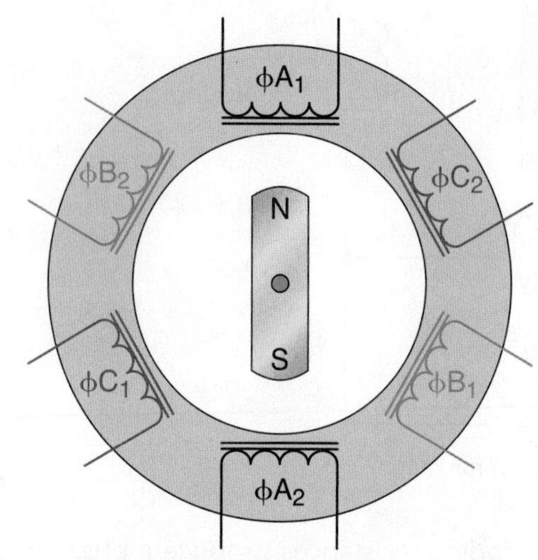

(a) Three-phase stator layout

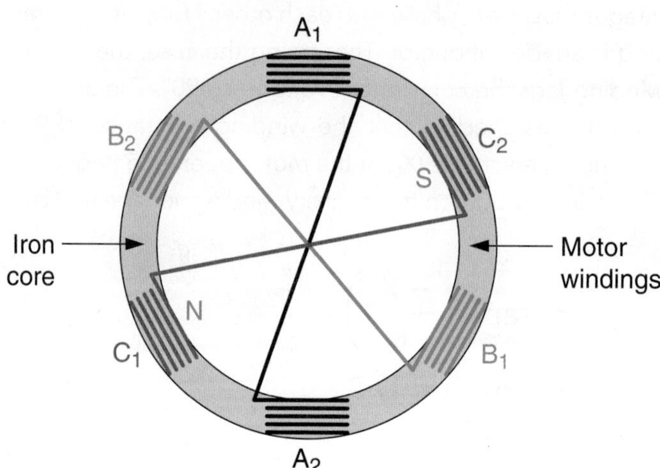

(b) Internal field coil connections

FIGURE 24.29 Three-phase field coil wiring.

The three-phase AC inputs to an induction motor are applied to the stator coil pairs. The currents generated by these AC inputs produce magnetic fields that rotate around the stator as shown in Figure 24.30a. Each stator field orientation occurs at the time indicated in Figure 24.30b. The rotor—represented by the arrow in the center of the stator—rotates with the stator magnetic fields as shown in the figure.

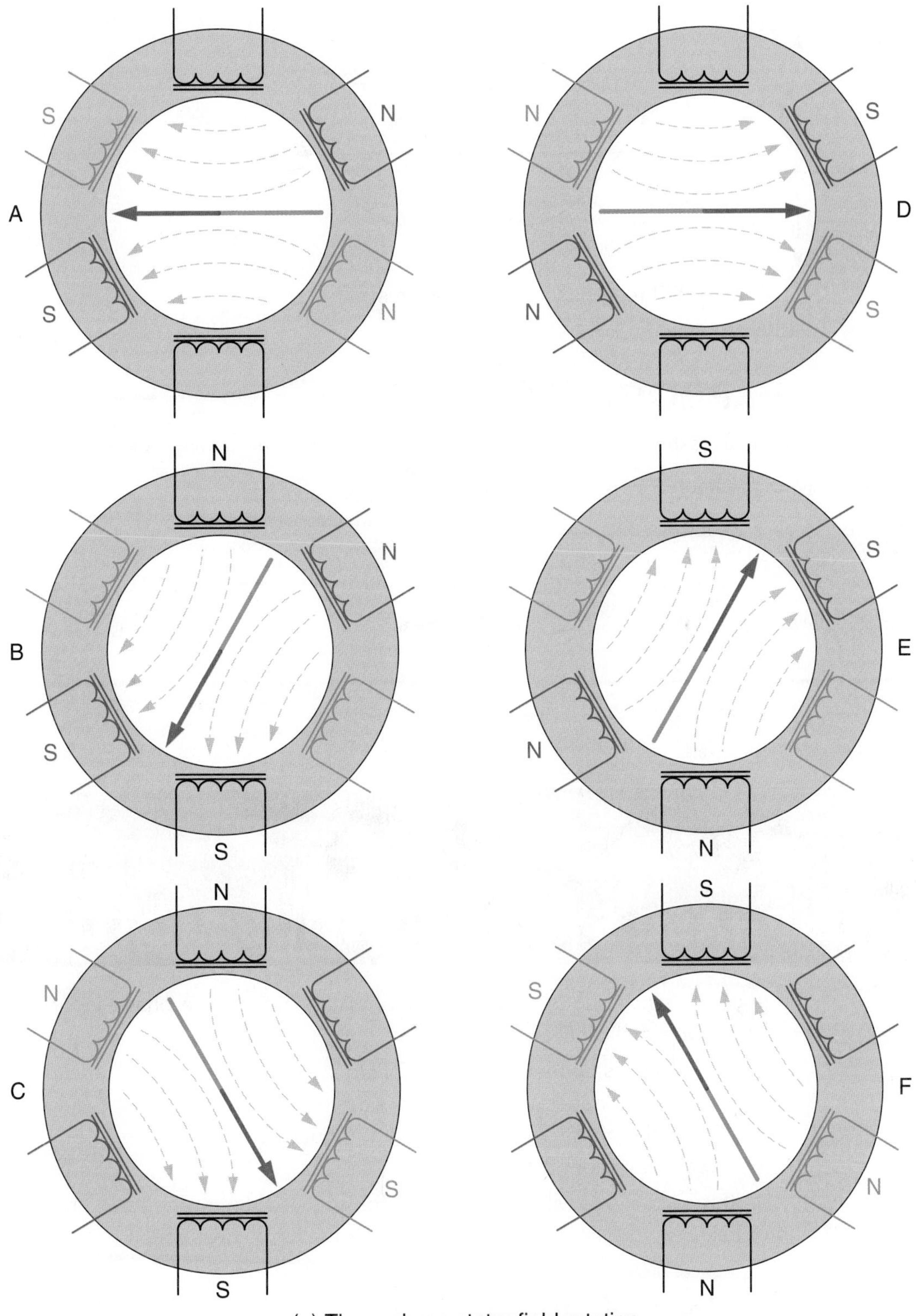

FIGURE 24.30 Three-phase motor operation and waveforms.

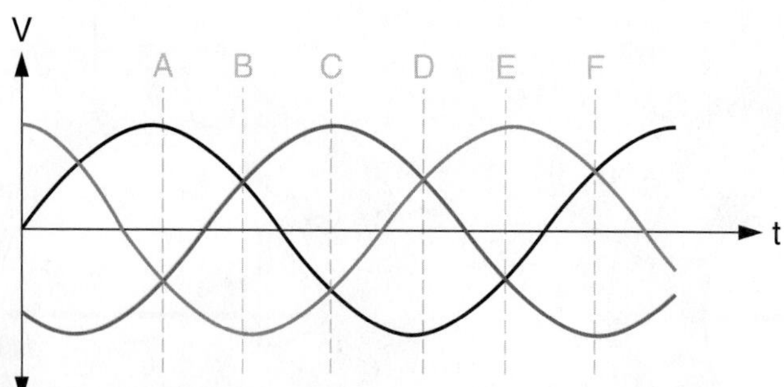

(b) Three-phase input waveforms

FIGURE 24.30 (*Continued*)

PROGRESS CHECK

1. What is motor slip and how is it usually expressed?
2. What is the most commonly used AC motor?
3. What is required in order to generate rotor current? How is motor slip related to this requirement?
4. What is the relationship between motor slip and motor load?
5. Describe and compare the following types of single-phase motors:
   a. Capacitor-start, inductor-run
   b. Capacitor-start, capacitor-run
   c. Inductor-start, inductor-run
6. What is a centrifugal switch and what is its use?
7. Describe the basic construction and operation of a three-phase induction motor.

24.6 THREE-PHASE SYNCHRONOUS MOTORS

Three-phase synchronous motors differ physically from three-phase induction motors in two significant ways:

1. The rotor is connected to a *DC field supply* that provides the current required to generate a rotor magnetic field.
2. The three-phase synchronous motor requires additional startup circuitry because it cannot start on its own.

In this section, we will take a closer look at these three-phase synchronous motor traits. As you will see, they both relate to the rotor construction.

Rotor Construction

The rotor of a three-phase synchronous motor closely resembles a salient-pole alternator rotor. As you may recall, the electromagnets of a salient-pole rotor extend outward from the rotor as illustrated in Figure 24.31.

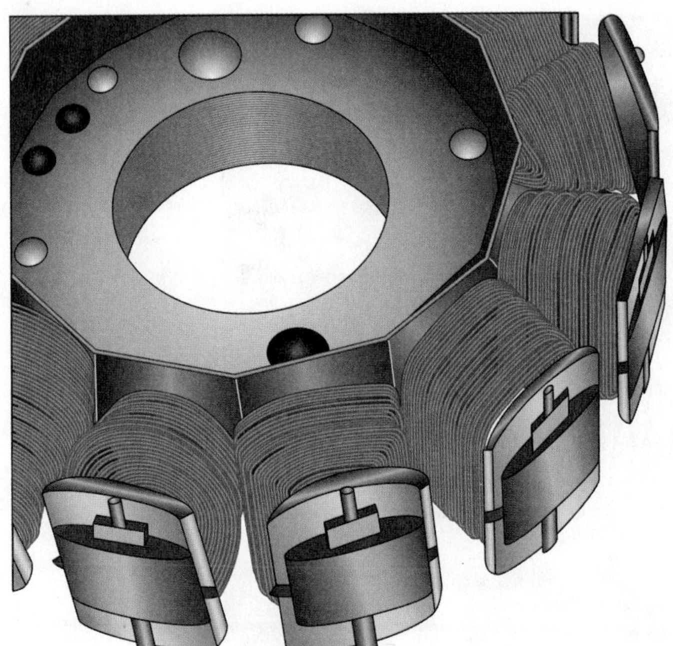

FIGURE 24.31 A salient-pole rotor.

If a rotor like the one in Figure 24.31 were installed in a three-phase synchronous motor, the motor would have difficulty starting. Here's the reason: When a three-phase input is applied to the stator coils in a synchronous motor, the stator magnetic field rotates as shown in Figure 24.30. Assuming that the stator input frequency is 60 Hz, the magnetic field makes 60 revolutions per second (rps). This corresponds to a rotation speed of

$$60 \times 60 \text{ rps} = 3600 \text{ rpm}$$

With the stator magnetic field rotating at 3600 rpm, the rotor cannot react quickly enough to the stator field to start.

The starting problem described here is overcome by modifying the rotor as illustrated in Figure 24.32. Squirrel-cage conductors are mounted around the outside of the salient-pole rotor. The squirrel-cage conductors (referred to as *amortisseur windings*) are able to respond much more quickly to the changes in the stator field, helping the motor to start. Here is how it works:

- Initially, the DC field supply is disconnected from the rotor field poles.
- When the three-phase input is applied to the stator, the squirrel-cage conductors cause the rotor to start rotating.
- As the rotor speed increases, a centrifugal switch closes—making the connection between the DC field supply and the rotor field poles.
- Once powered, the interaction between the stator and rotor magnetic fields causes the rotor to lock in sync with the rotating stator field.

Like the auxiliary winding in a split-phase motor, the squirrel-cage conductors become redundant once the motor comes up to speed. When the motor is up to speed and in sync, there is no cutting action between the squirrel-cage conductors

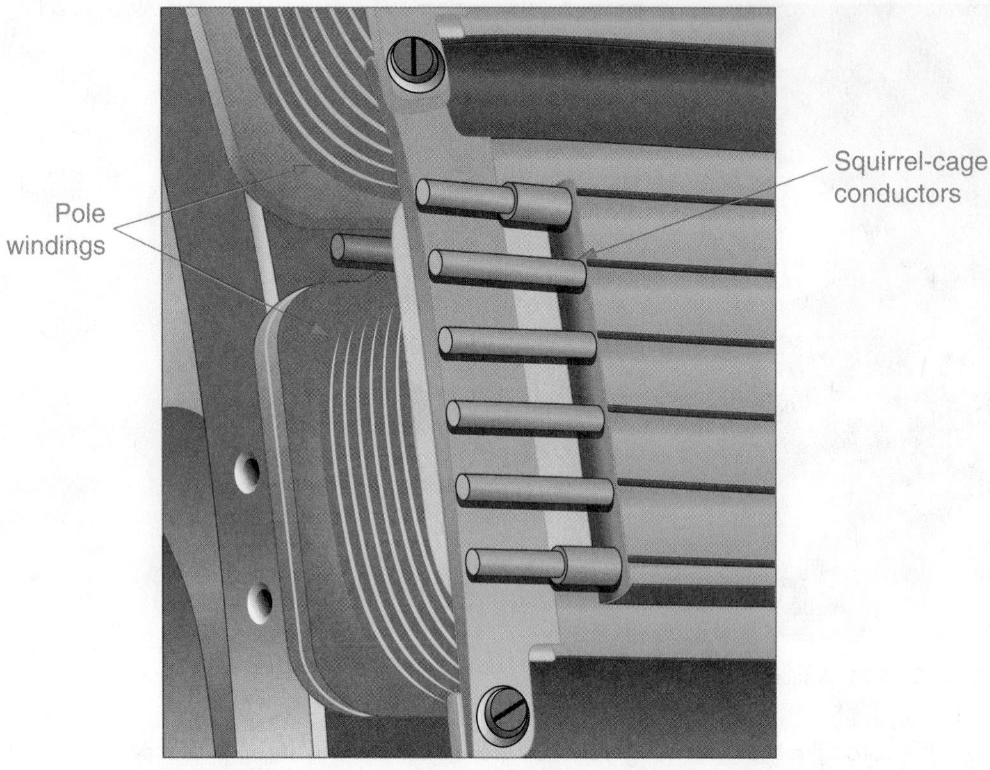

FIGURE 24.32 A salient-pole rotor with added squirrel-cage conductors.

and the stator field coils, so no voltage is induced across the squirrel-cage conductors. At this point, the DC-powered salient poles provide the magnetic fields required for the motor to operate at its rated capacity.

Synchronous Motor Speed

A three-phase synchronous motor is a fixed-speed machine, meaning that its no-load speed equals its full-load speed. The motor speed is determined by the motor input frequency and the number of poles in its rotor, as given by

$$N_S = \frac{120f}{P}$$

where N_S is the rotor speed (in rpm), f is the motor input frequency (in Hz), and P is the number of poles in the machine's rotor. A two-pole synchronous motor rotates at

$$\begin{aligned} N_S &= \frac{120f}{P} \\ &= \frac{120 \times 60\text{ Hz}}{2} \\ &= 3600\text{ rpm} \end{aligned}$$

which equals 60 revolutions per second, and a four-pole motor rotates at

$$N_S = \frac{120f}{P}$$

$$= \frac{120 \times 60\,\text{Hz}}{4}$$

$$= 1800 \text{ rpm}$$

which equals 30 revolutions per second. As these calculations indicate, a synchronous motor spins at the same speed as the stator magnetic field, or at some *submultiple* (fractional value) of that speed. In either case, the speed of the motor is constant as long as the motor is being operated within its established parameters (such as load, temperature, and so on).

PROGRESS CHECK

1. How do three-phase synchronous motors differ from three-phase induction motors?
2. Explain why a three-phase synchronous motor requires squirrel-cage conductors mounted outside the salient-pole rotor.
3. What factors determine the speed of a three-phase synchronous motor?

24.7 SUMMARY

Here is a summary of the major points that were made in this chapter:

Alternators: An Overview

- An alternator is a machine that converts mechanical energy (motion) into electrical energy.
- The magnitude and polarity of the output voltage from an alternator is determined by three factors:
  - The angle of rotation of the armature.
  - The strength of the magnetic field of the poles.
  - The speed of armature rotation.
- The rotating-armature alternator uses rotating conductors between fixed electromagnets (see Figure 24.3).
  - The DC field supply provides current for the electromagnets.
  - The output from the armature is coupled to the load via slip rings.
  - This type of alternator is rarely used as it cannot supply the high voltage required for most power applications.
- The rotating-field alternator is the more common and practical configuration (see Figure 24.5).
  - It is preferred over the rotating-armature type because it does not require brushes and slip rings, which are difficult to insulate and can arc under high-voltage conditions.

- A single-phase alternator produces a single sine wave output.
  - Figure 24.6 illustrates the operation of the single-phase alternator.
  - Single-phase alternators are usually limited to low-power applications, usually 25 kW or less.
- An alternator that produces more than one sine wave is called a *polyphase alternator.*
- A three-phase alternator is illustrated in Figure 24.7.
  - Each phase has two coils spaced 180° apart.
  - Each phase is spaced 120° from each of the other phases.
  - Each coil is spaced 60° apart.

Three-Phase Alternator Elements and Construction

- The two main types of alternator field rotors are the cylindrical rotor and the salient-pole rotor.
  - Cylindrical rotors are used in turbines, which are high-speed alternators typically driven by gas or steam.
  - Cylindrical rotors can operate at rotational speeds of 1200 rpm or higher since the field coils are wound within the body of the rotor.
- Salient-pole rotors are more commonly used than cylindrical rotors.
  - Salient-pole rotors are used in low-speed generators, usually driven by water, wind, or internal-combustion engines.
- The DC field supply provides the current required for the rotor to generate the necessary electromagnetic flux.
- The stator of the rotating-field alternator serves as the support structure for the armature windings (see Figure 24.10).
  - Stators are subject to the same eddy current and I^2R losses as the armature in a DC generator.
- Figure 24.11 illustrates the operation of a three-phase alternator.
  - Three-phase alternators can be wired as a delta or a wye.
- Alternators can have one or two windings per pole.
- Two windings per pole is preferred for two reasons:
  - They are more efficient.
  - All things being equal, they produce higher voltages than equivalent single-winding alternators.
- The frequency of an alternator's output voltage is determined by two factors:
  - The number of poles (P).
  - The rotational speed, in rpm (N_S).
- Alternator frequency is found as $f = \dfrac{N_S \times P}{120}$, where 120 is a constant.
- The effective (rms) output voltage for an alternator is determined by three factors:
  - The flux per pole in webers (Φ).
  - The output frequency in Hz (f).
  - The number of turns, per phase (N).
- Alternator output voltage is found as $E = 4.44 \times \Phi \times f \times N$
  - The flux per pole (Φ) can be increased or decreased by adjusting the DC field supply.

Three-Phase Alternator Connections and Loading

- The voltage regulation of an alternator is its ability to maintain a constant output voltage despite changes in load demand.

- The percent voltage regulation defines the percent variation of output voltage between full-load (E_{FL}) and no-load (E_{NL}) conditions.
  - By formula, Regulation (%) $= \frac{E_{NL} \times E_{FL}}{E_{NL}} \times 100$.
  - An ideal alternator would have 0% voltage regulation.
- Connecting alternators in parallel has two advantages:
  - Maximum output current (and power) are increased.
  - The system can be run at reduced capacity using just one alternator if the other fails or requires maintenance.
- Precautions must be taken when connecting alternators in parallel.
  - The output voltage and frequency must be the same.
  - The alternators must be synchronized, both in frequency and phase rotation.

AC Motors: An Overview

- Like DC motors, AC motors convert electrical energy into mechanical (rotational) energy (see Figure 24.15).
  - When AC is applied to the stator coils, they generate magnetic flux.
  - The interaction between the stator field and the rotor field causes the rotor to rotate 360° per AC input cycle.
- AC motors can operate on switching magnetic fields, or rotating magnetic fields.
- Motor speed can be found using the AC input frequency and the number of poles, by formula, $N_S = \frac{120f}{P}$.
- The two basic types of AC motors are the synchronous motor and the induction motor.
- A synchronous motor is synchronized to an alternator that acts as an energy and timing source.
  - Ideally, synchronous motors are perfectly in sync with the alternator, but in practice this is not the case.
  - When load is added to the motor it lags the alternator slightly, which is referred to as the torque angle (α).
  - A high enough load can cause α to increase beyond 90° and synchronization to be lost.
- Synchronous motors use an external DC field supply to power the electromagnets (poles).
- An induction motor is classified as an asynchronous motor.
  - Induction motors do not have an external field supply.
- The two types of rotors found in AC motors are the wound rotor and the squirrel-cage rotor.
  - The squirrel-cage rotor is the most common because it is cheaper, requires less maintenance, and is more reliable.

Induction Motors

- Rotor current, in an induction motor, is generated by electromagnetic induction, just like in a transformer.
- Actual motor speed is always slightly slower than calculated motor speed based on the input frequency.
- The difference between stator and rotor speeds is referred to as *slip*.
  - Slip is often expressed as a percentage.

- Single-phase induction motors are the most commonly used AC motors.
- As stated earlier, the rotor magnetic field in an induction motor is generated by induced current in the rotor.
  - In order to generate rotor current there must be relative motion between the rotor conductor and the stator pole magnetic fields.
  - Motor slip provides this relative motion.
  - Motor slip varies directly with motor load.
- A split-phase induction motor is basically a single-phase induction motor with an added starting component that provides the starting torque necessary to get the rotor moving.
  - The starting component can be capacitive, inductive, or resistive.
- The capacitor-start, inductor-run split-phase induction motor uses a capacitor as the starting component (see Figure 24.24).
  - The capacitor is in series with the auxiliary (or starting) winding and a centrifugal switch.
  - The auxiliary winding is positioned at a 90° angle from the main stator winding, and is in parallel with the capacitor/auxiliary coil/centrifugal switch circuit.
  - The capacitor/auxiliary coil circuit current leads the AC input by 45° while the main stator winding lags the AC input by 45°. This provides the necessary 90° phase difference covered in Section 24.4.
  - The centrifugal switch automatically opens when the motor reaches full operating speed. The motor then runs as a single-phase motor.
- The capacitor-start, capacitor-run motor uses an additional capacitor (see Figure 24.26).
  - The startup action is the same as for the capacitor-start, inductor-run motor.
  - When the motor reaches full speed and the centrifugal switch opens, there is still some current through the auxiliary coil and the second capacitor.
  - This current is at a 90° angle to the current through the stator field coil, so the motor continues to operate as a split-phase motor with a single-phase input.
  - One variation of this circuit does not use a centrifugal switch (see Figure 24.27).
- The inductor-start, inductor-run motor has two windings mounted at 90° angles and a centrifugal switch (see Figure 24.28).
  - The resistance of the auxiliary winding is much higher than that of the stator field winding.
  - As a result, the auxiliary winding current leads the stator field winding current.
  - This provides the necessary 90° phase angle between the two winding currents.
- The primary difference between the various induction motors is the manner in which the starting torque is produced.
- Three-phase induction motors have three sets of stator coils.
  - Each pair of stator coils is internally connected (see Figure 24.29).
- The currents generated by the three phases of the input voltage generate a rotating magnetic field. This causes the rotor to rotate with the stator magnetic field (see Figure 24.30).

Three-Phase Synchronous Motors

- Three-phase synchronous motors differ from three-phase induction motors in two ways.
  - The rotor is connected to a DC field supply.
  - It requires additional startup circuitry.
- The rotor of a three-phase synchronous motor resembles a salient-pole alternator rotor.
  - The squirrel-cage conductors can respond faster to the stator's rotating magnetic field, which starts the rotor.

  - As rotor speed increases, a centrifugal switch closes, which connects the DC field supply to the rotor field poles.
  - Once the motor is up to speed, the squirrel-cage conductors no longer function.
- Synchronous motor speed is a function of input frequency and the number of rotor poles. By formula, $N_S = \frac{120f}{P}$.

CHAPTER REVIEW QUESTIONS

1. An alternator is a machine that converts ________ energy to ________ energy.
   a) mechanical, potential
   b) mechanical, electrical
   c) electrical, mechanical
   d) electrical, kinetic

2. The magnitude and polarity of the output voltage of an alternator are determined by ________.
   a) the angle of armature rotation
   b) the flux density
   c) armature rotational speed
   d) all of the above

3. The ________ alternator is rarely used because it cannot produce the high voltages required for most applications.
   a) rotating-armature
   b) polyphase
   c) rotating-field
   d) single-phase

4. High-voltage arcing is caused by ________.
   a) field shift
   b) brushes and slip rings
   c) armature shift
   d) commutators

5. Stationary armatures do not require ________.
   a) rotors
   b) slip rings and brushes
   c) insulation
   d) load regulation

6. A galvanometer measures ________.
   a) gauss
   b) field strength
   c) current magnitude and direction
   d) field polarity

7. Alternators produce maximum voltage when the coils are ________.
   a) at 90°
   b) parallel to the magnetic line of flux
   c) at 120°
   d) perpendicular to the magnetic lines of flux

8. Alternators produce minimum voltage when the coils are ________.
   a) at 90°
   b) parallel to the magnetic line of flux
   c) at 120°
   d) perpendicular to the magnetic lines of flux

9. Alternators that generate more than one sine wave are called ________ alternators.
   a) polyphase
   b) dynamo
   c) compound
   d) multiphase

10. Cylindrical rotors are often referred to as ________ rotors.
    a) high-velocity
    b) turbine-driven
    c) lap-wound
    d) rotating field

11. Salient-pole rotors are ________.
    a) used at lower speeds than cylindrical rotors
    b) driven by water
    c) driven by wind
    d) all of the above

12. The current required for the electromagnets in the rotor is supplied by ________.
    a) residual flux
    b) the DC pole supply
    c) the DC field supply
    d) rotational flux

13. The output voltage of a DC generator is affected by load demand due to ________.
    a) armature and winding resistance
    b) armature and field pole resistance
    c) hysteresis losses
    d) neutral plane losses

14. The ________ of a rotating-field alternator serves as the support structure for the armature windings.
    a) salient pole
    b) stator
    c) cylinder
    d) rotor

15. Most alternators have ________ windings per pole.
    a) one
    b) two
    c) three
    d) four

16. If you increase the rotational speed of an alternator ________.
    a) the static field shifts
    b) the pole flux decreases
    c) the frequency of the output increases
    d) all of the above

17. A four-pole alternator is rotating at 1500 rpm. The output frequency of the alternator is ________.
    a) 50 Hz b) 120 Hz
    c) 60 Hz d) 80 Hz

18. At a rotational speed of 900 rpm you need an alternator with ________ poles to produce a 60 Hz output.
    a) 2 b) 4
    c) 6 d) 8

19. Alternator output voltage is directly proportional to ________.
    a) flux per pole
    b) frequency
    c) the number of turns
    d) all of the above

20. Voltage regulation (%) is equal to ________ minus ________ divided by ________ times 100.
    a) E_{NL}, E_{FL}, E_{NL} b) E_{FL}, E_{NL}, E_{NL}
    c) E_{NL}, E_{FL}, E_{FL} d) E_{FL}, E_{NL}, E_{FL}

21. An alternator has a voltage regulation rating of 3.33%. At full load it generates 232 V. This means that its no-load output voltage is ________.
    a) 242 V b) 213 V
    c) 256 V d) 240 V

22. Alternators connected in parallel must have the same ________.
    a) frequency b) voltage
    c) phase sequence d) all of the above

23. AC motors rotate as a result of the interaction between ________ and ________.
    a) stator current, field current
    b) the stator field, the rotor field
    c) the stator field, the DC supply
    d) the single-phase, the split-phase

24. A split-phase motor works on the principle of ________.
    a) rotating magnetic fields
    b) phase shift
    c) magnetic constriction
    d) switching magnetic fields

25. In a split-phase motor, the two AC inputs must be ________ out of phase.
    a) 45° b) 60°
    c) 90° d) 180°

26. Assuming the same input frequency, an AC motor with 6 poles will turn ________ percent ________ than one with 4 poles.
    a) 150, faster b) 67, slower
    c) 67, faster d) 150, slower

27. A synchronous motor is connected to a(n) ________, which provides timing.
    a) armature b) DC field supply
    c) sync switch d) alternator

28. In a synchronous motor, if ________ is too great, synchronization is lost.
    a) the torque angle b) motor lag
    c) the load d) all of the above

29. An induction motor does *not* have ________.
    a) a DC field supply
    b) a laminated stator
    c) an epoxy resin coating
    d) stator windings

30. Squirrel-cage rotors are preferred over wound rotors because they ________.
    a) are more reliable
    b) require less maintenance
    c) are easier to construct
    d) all of the above

31. Induction motors run ________ than the calculated value based on the number of poles and the input frequency.
    a) slower
    b) faster
    c) just slightly faster
    d) either one, depends on load

32. The most commonly used AC motors are ________ motors.
    a) three-phase induction
    b) single-phase induction
    c) wound-stator synchronous
    d) reversible

33. In order to generate rotor current in an induction motor, there must be ________ between the rotor conductors and the stator magnetic fields.
    a) the same frequency
    b) inverse polarity

c) a 90° phase difference

d) relative motion

34. A split-phase motor requires a ________ to generate initial torque.

    a) starting component

    b) centrifugal clutch

    c) DC field supply

    d) all of the above

35. The windings in an inductor-start, inductor-run motor are ________.

    a) equal b) set 90° apart

    c) inverted d) set 120° apart

36. The primary difference between different induction motors is ________.

    a) their maximum rotational speed

    b) how the rotor begins to rotate

    c) the stator housing

    d) the phase difference between the windings

37. Three-phase induction motors have ________ pairs of stator field coils.

    a) 3 b) 6

    c) 9 d) 12

38. Three-phase synchronous motors ________.

    a) require a DC field supply

    b) incorporate squirrel-cage conductors in the rotor

    c) use a centrifugal switch

    d) all of the above

39. The no-load speed of a three-phase synchronous motor is ________ its full-load speed.

    a) slower than

    b) faster than

    c) equal to

    d) depends on the number of poles

ANSWERS TO THE EXAMPLE PRACTICE PROBLEMS

24.1	60 Hz
24.2	3.85%
24.3	2400 rpm
24.4	50 rpm

CHAPTER 25

AC POWER TRANSMISSION AND DISTRIBUTION

PURPOSE

Electrical appliances and machines in homes and industry are of little use without a source of power. In this chapter, we will briefly examine the means by which utilities generate, transmit, and distribute electrical power.

KEY TERMS

The following terms are introduced and defined in this chapter on the pages indicated:

OBJECTIVES

After completing this chapter, you should be able to:

1. Briefly describe the overall process of power generation, transmission, and distribution.
2. Identify the elements that make up a power grid.
3. List the factors that are considered when locating a power generating plant.
4. List the three types of transmission lines.
5. Identify the elements that make up a power distribution system.
6. Identify the points where substations are typically found and the functions they perform.
7. Describe the construction and operation of hydroelectric plants.

8. Describe the Rankine cycle of a thermal power plant.
9. Describe the construction and operation of wind turbines.
10. Identify the common voltages found on transmission lines.
11. Describe how step-up transformers reduce the power loss experienced by transmission lines.
12. Compare and contrast the three types of substations.
13. Describe the elements typically found in a distribution grid.
14. State the two purposes served by local distribution substations.

25.1 POWER TRANSMISSION AND DISTRIBUTION: AN OVERVIEW

Generating and delivering electrical power to homes and industries can be divided into three processes. These processes are represented in Figure 25.1. **POWER GENERATION** is the process of converting one type of energy—most commonly *mechanical*—into electrical energy. **POWER TRANSMISSION** is the process of delivering generated power—usually over relatively long distances—to distribution circuits that are located in populated areas. As part of the transmission process, transformers are used to increase voltage levels to make long-range power transmission feasible. Finally, **POWER DISTRIBUTION** is the process of:

1. Reducing the transmitted power to customer-usable levels.
2. Delivering the electrical power to local homes and industries.

POWER GENERATION The process of converting one type of energy into electrical energy.

POWER TRANSMISSION The process of delivering generated power—usually over relatively long distances—to distribution circuits that are located in populated areas.

POWER DISTRIBUTION The process of reducing transmitted power to customer-usable levels and delivering that power to local homes and industries.

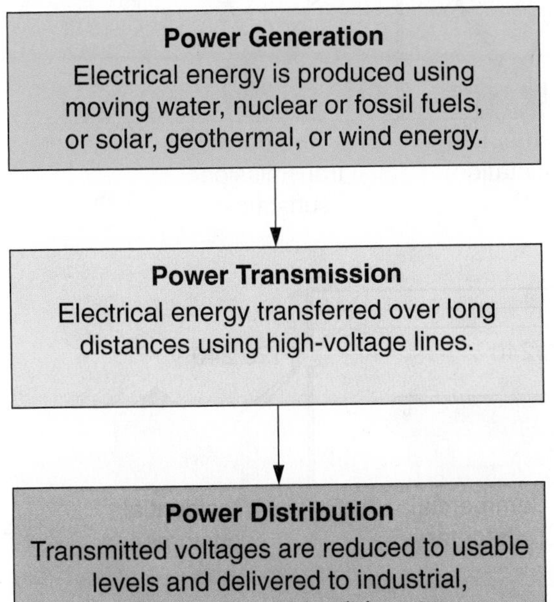

Power Generation
Electrical energy is produced using moving water, nuclear or fossil fuels, or solar, geothermal, or wind energy.
Power Transmission
Electrical energy transferred over long distances using high-voltage lines.
Power Distribution
Transmitted voltages are reduced to usable levels and delivered to industrial, commercial, and residential customers.

FIGURE 25.1 Generating, transmitting, and distributing electrical power.

Basic Transmission and Distribution System

An AC power transmission and distribution system includes a power generating plant, power transmission lines, power distribution systems, and substations. These elements are collectively referred to as a **GRID**. A generic power grid is illustrated in Figure 25.2.

GRID A term used to describe the power generating plant, power transmission lines, power distribution systems, and substations that make up an AC power distribution system.

The power generating plant is normally located outside densely populated areas. The factors considered when selecting the specific location of a given power generating plant include the specific type of plant (e.g., hydroelectric, fossil

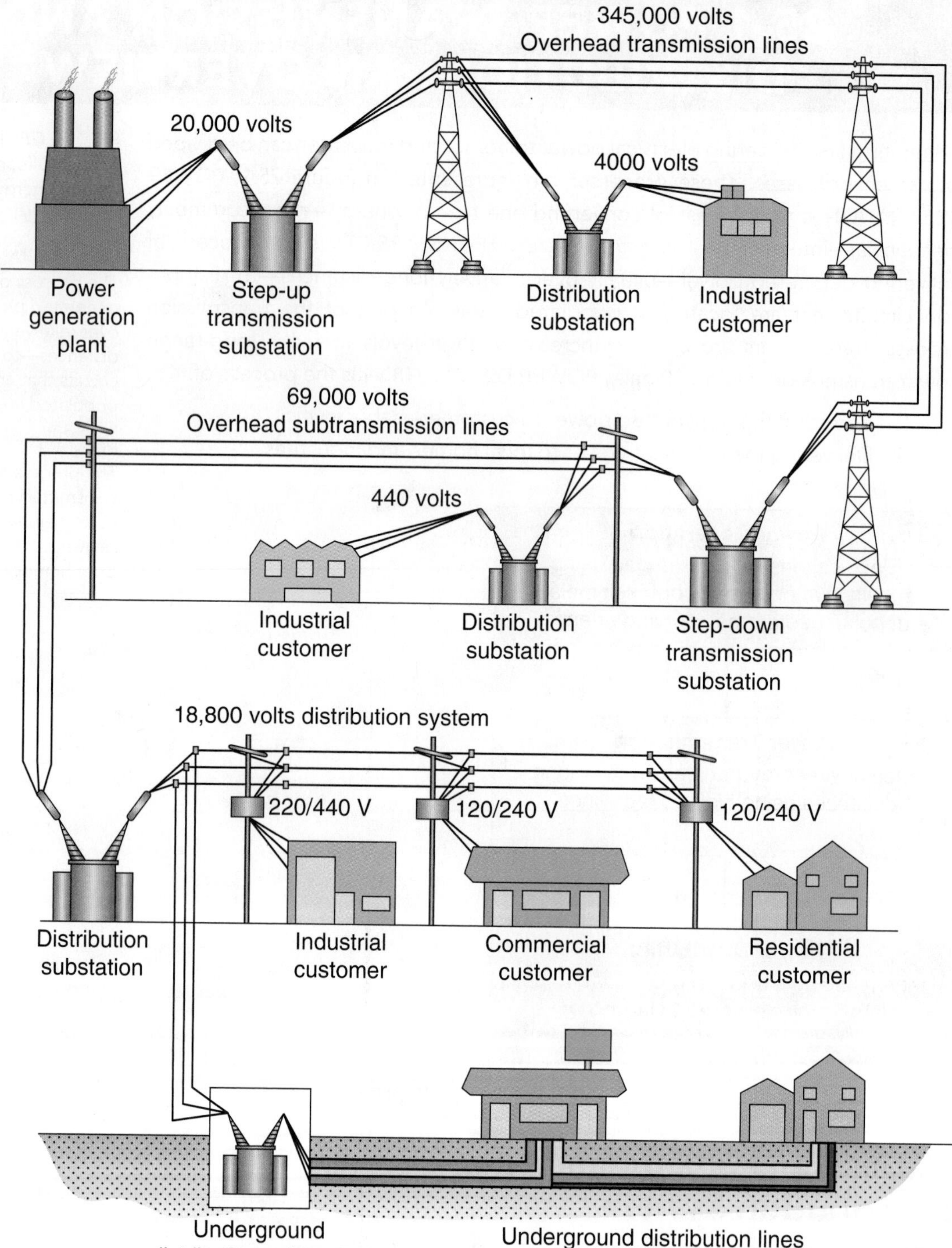

FIGURE 25.2 A power transmission and distribution system.
Source: Redrawn from the Occupational Safety and Health Administration (OSHA)

fuel, nuclear, etc.), construction and maintenance costs, the available labor pool, and other economic factors. Safety and environmental concerns also affect the decision-making process.

The **TRANSMISSION LINES** are the high-voltage lines used to transmit electrical energy over long distances. There are three types of transmission lines: *overhead* lines, *underground* lines, and *subtransmission* lines. The first two (overhead and underground) refer to the physical location of the transmission lines, while the third identifies a type of transmission line that carries lower voltages than the primary transmission lines. Note that subtransmission lines can be either overhead or underground lines.

TRANSMISSION LINES The high-voltage lines used to transmit electrical energy over long distances.

The **POWER DISTRIBUTION SYSTEM** includes the lines, poles, transformers, and switching and protection circuits that deliver electrical energy to customers at usable voltage levels. The distribution lines may be above ground or underground, and the transformers may be pole mounted or pad mounted. These distinctions are described further later in this chapter.

POWER DISTRIBUTION SYSTEM The lines, poles, transformers, and switching and protection circuits that deliver electrical energy to customers at usable voltage levels.

SUBSTATIONS can be found between the power generating plant and transmission lines, between the transmission and subtransmission lines, and between the transmission lines and distribution systems. Substations contain transformers that increase (step up) or decrease (step down) voltage levels, switching circuits, and overload protection circuits. The switching circuits allow the substation to:

SUBSTATIONS Utility stations that contain transformers that increase (step up) or decrease (step down) voltage levels, switching circuits, and overload protection circuits.

- Draw from multiple power sources.
- Direct the output electrical energy between multiple loads.

These capabilities are essential for the station to compensate for changes in load demand, and to reroute power when scheduled maintenance or grid faults make it necessary.

In the upcoming sections, we will take a closer look at each of the elements that make up a basic AC power transmission and distribution network.

PROGRESS CHECK

1. Define the following terms:
   a. power generation
   b. power transmission
   c. power distribution
2. Define a grid and list the elements that comprise the grid.
3. List and define the three types of transmission lines.
4. List the elements that comprise a power distribution system.
5. Describe a basic substation installation.

25.2 GENERATING ELECTRICAL POWER

We have examined the process of generating electrical power throughout the last five chapters. The common theme has been the concept of a *prime mover* (source of mechanical energy) driving a dynamo (alternator or generator) that converts mechanical energy to electrical energy. In this section, we take a look at the basic operating characteristics of common power generating plants.

Load factor and *capacity factor* are two mathematical ratios commonly used to measure utility power generation service. *Load factor* is the ratio of average load for a given period of time to the peak load for that same period. *Capacity factor* is the ratio of the average load for a given period of time to the output capacity of the power generation plant. Under ideal conditions, both the load and capacity factors would be *unity* (equal to 1).

IN THE FIELD

HYDROELECTRIC PLANT A power generating plant that converts the kinetic energy in flowing water to electrical energy.

TURBINE A rotor with blades that are driven by flowing water.

Hydroelectric Plants

A **HYDROELECTRIC PLANT** is one that converts the kinetic energy in flowing water to electrical energy. The basic structure of a hydroelectric plant is illustrated in Figure 25.3. The **TURBINE** is a rotor with blades that are driven by water flowing

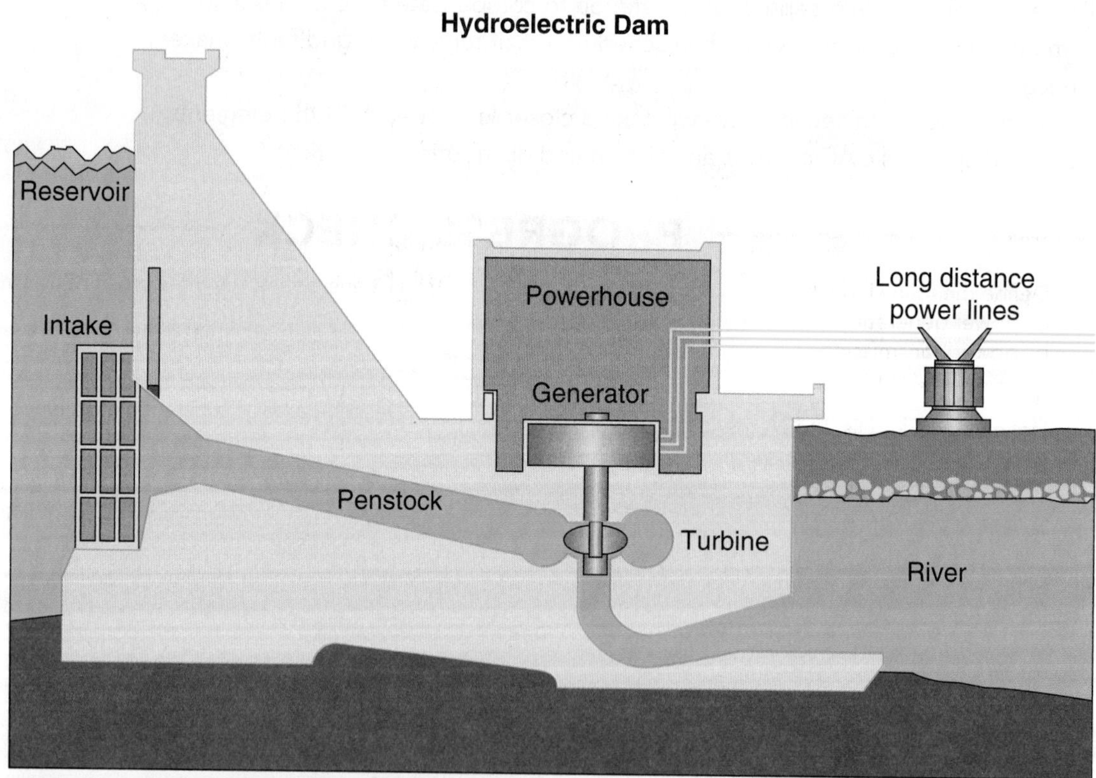

FIGURE 25.3 A hydroelectric plant.
Source: Redrawn from the U.S. Department of Energy

through the **PENSTOCK**, a conduit between the water reservoir and the power plant. The turbine drives the rotor of the alternator (generator). A step-up transformer in the power station increases the alternator output voltage to a level suitable for transmission over long distances and connects that output to the power transmission lines.

The dam represented in Figure 25.3 is referred to as an **IMPOUNDMENT DAM**. An impoundment dam completely blocks the natural flow of water from the reservoir to the river basin below the dam, the only path for water being through the penstock. In contrast, a **DIVERSION DAM**—which is simpler to design and build—does not block the natural flow of water through a waterway. Rather, it diverts a portion of the water through the penstock and turbine, and then returns the water to its natural path.

PENSTOCK A conduit between a water reservoir and the hydroelectric power plant.

IMPOUNDMENT DAM A dam that completely blocks the natural flow of water from a reservoir to the river basin below the dam, the only path for water being through the penstock.

DIVERSION DAM A dam that diverts a portion of natural flowing water through its penstock and turbine, and then returns the water to its natural path.

Thermal Power Plants

A **THERMAL POWER PLANT** uses heat to convert water to steam, and then uses the steam as the prime mover for its turbine. A thermal power plant operates as illustrated in Figure 25.4. The heat converts the water to steam. The steam is then applied to the turbine (which acts as the prime mover for the generator), a cooling structure condenses the steam back into water, and the cycle begins again. Note that the conversion of water to steam and back to water in a closed recirculating system is referred to as the **RANKINE CYCLE**.

Thermal power plants are classified by the type of fuel they use as a heat source. For example, a **FOSSIL-FUEL POWER PLANT** produces heat by burning coal, oil, or natural gas.

THERMAL POWER PLANT A power generating plant that uses heat to convert water to steam, and then uses the steam as the prime mover for its turbine.

RANKINE CYCLE The conversion of water to steam and back to water in a closed recirculating system.

FOSSIL-FUEL POWER PLANT A plant that produces heat for generating steam by burning coal, oil, or natural gas.

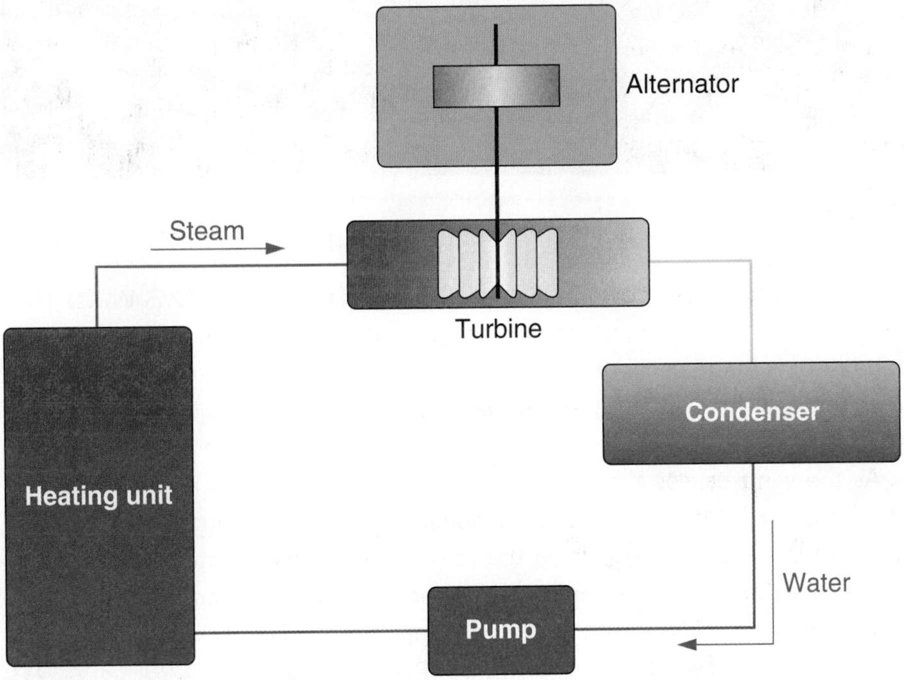

FIGURE 25.4 The Rankine cycle.

NUCLEAR POWER PLANT One that converts water to steam using heat produced by nuclear fission.

NUCLEAR FISSION The process of generating power by splitting the nucleus of an atom.

FUEL ROD A cylinder holding nuclear fuel that is used in nuclear power plants.

WIND POWER Energy in currents of air (wind) that can be converted to electrical energy using a wind turbine.

WIND FARM A group of wind turbines.

A **NUCLEAR POWER PLANT** uses heat produced through a process called **NUCLEAR FISSION**. The nuclear fuel is packaged in cylinders called **FUEL RODS**. The most commonly used fuels are uranium 235 (^{235}U) and plutonium 239 (^{239}Pu). As with any other type of thermal power plant, the heat produced by nuclear fission is used as the energy source in the Rankine cycle.

Wind Power

Nearly all the commercially produced electrical power in the United States currently comes from hydroelectric and thermal power plants. However, environmental issues—such as the production of greenhouse gases and the increasing cost of fossil fuels—are sparking greater interest in renewable sources of electrical energy, including wind power.

The term **WIND POWER** refers to the energy in currents of air (wind) that can be converted to electrical energy using a wind turbine. A group of wind turbines, referred to as a **WIND FARM**, is shown in Figure 25.5.

FIGURE 25.5
Courtesy of Siemens

HORIZONTAL-AXIS WIND TURBINE (HAWT) A wind turbine with a three-blade assembly that rotates on a horizontal shaft.

NACELLE The housing on a HAWT that contains the low-speed shaft and driving gears.

GEAR RATIO In a series of gears, the ratio of the speed of rotation of the first gear to that of the last gear.

YAW MOTOR AND YAW DRIVER The elements in a HAWT that turn the turbine toward the wind.

The wind turbines shown in Figure 25.5 are **HORIZONTAL-AXIS WIND TURBINES (HAWTS)**. Each turbine contains a generator as shown in Figure 25.6. The turbine rotor is made up of the three blades, the low-speed shaft, and the driving gear in the gear box. All but the rotor blades are contained in housing called the **NACELLE**.

As the wind causes the rotor to turn, the driving gear turns the generator rotor. The **GEAR RATIO** is such that the generator rotor spins much faster than the turbine rotor. Wires running down the tower transfer the electrical energy produced by the generator to an external pad-mounted transformer. Finally, the **YAW MOTOR** and **YAW DRIVER** turn the rotor so that it faces directly into the wind.

Another type of wind turbine, called a **VERTICAL-AXIS WIND TURBINE (VAWT)**, is shown in Figure 25.7. Though this type of wind turbine rotates on a vertical axis

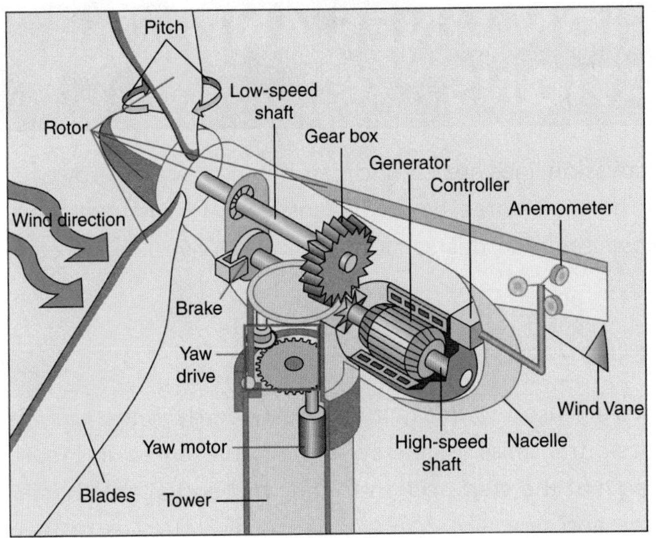

FIGURE 25.6 Horizontal-axis wind turbine (HAWT) construction.
Source: Redrawn from the Office of Energy, Efficiency, and Renewable Energy; U. S. Department of Energy.

FIGURE 25.7 A vertical-axis wind turbine (VAWT).
Source: U.S. Department of Energy.

rather than a horizontal one, the overall principle of using wind energy to generate electrical energy is the same.

VERTICAL-AXIS WIND TURBINE (VAWT) A type of wind turbine that rotates on a vertical axis rather than a horizontal one.

VAWTs have several advantages over HAWTs:

1. VAWTs are mounted so that the moving parts are closer to the ground. This makes it easier to perform maintenance and repair procedures.
2. VAWT rotation can be initiated at lower air speeds than HAWT rotation.
3. VAWTs are shorter than HAWTs, allowing them to be used in areas where height prohibits the use of HAWTs.

At the same time, VAWTs have several disadvantages:

1. VAWTs are less efficient (electrically) than HAWTs.
2. The moving parts of VAWTs are located at the lower end of the VAWT, which means that they are positioned under the weight of the structure. This makes them more difficult to design.
3. The blades of the VAWT are lower to the ground, where wind speed tends to be lower. Thus, a VAWT produces less electricity than a comparable HAWT.

PROGRESS CHECK

1. Define the following terms and explain their use in a hydroelectric plant:
   a. turbine
   b. penstock
2. What type of transformer connects the power station to the transmission lines?
3. What is the difference between an impoundment dam and a diversion dam?
4. What is the prime mover in a thermal power plant?
5. What are the yaw motor and yaw driver used for in a HAWT?
6. What is a VAWT?
7. What advantages does a VAWT have over a HAWT? Disadvantages?

25.3 TRANSMISSION LINES AND SUBSTATIONS

As mentioned earlier, transmission lines are used to transmit electrical energy over relatively long distances. There are three types of transmission lines: *overhead*, *underground*, and *subtransmission*. In this section, we will briefly examine each type of transmission line.

Overhead Transmission Lines

Overhead transmission lines are used to transmit electrical energy in the form of three-phase (3ϕ) AC. A basic transmission line system is represented in Figure 25.8. Transmission lines begin at the step-up transformer and end at the substation step-down transformer, and typically span a distance of 300 miles or less. Figure 25.9 will give you a good idea of the complex network of transmission lines that supplies power across the continental United States.

> For long distance overhead transmission lines, utilities typically use uninsulated aluminum conductors or aluminum conductor steel reinforced (ACSR) cables.
>
> IN THE FIELD

The most common transmission line voltages are 115 kV, 138 kV, 161 kV, 230 kV, 345 kV, 500 kV, and 765 kV. Note that voltages in the 345 kV to 765 kV range are sufficient to require extra transmission line precautions, such as increasing the distance between the lines to prevent **FLASHOVER** (electrical arcing from one line to another).

FLASHOVER Electrical arcing from one line to another.

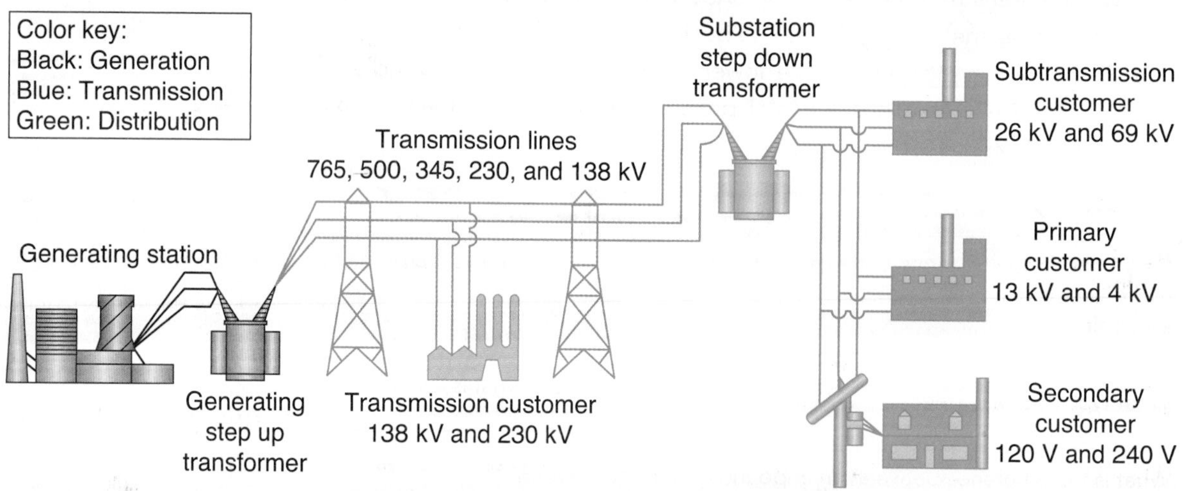

FIGURE 25.8 An overhead transmission and distribution system.
Source: Redrawn from the U. S. Department of Energy

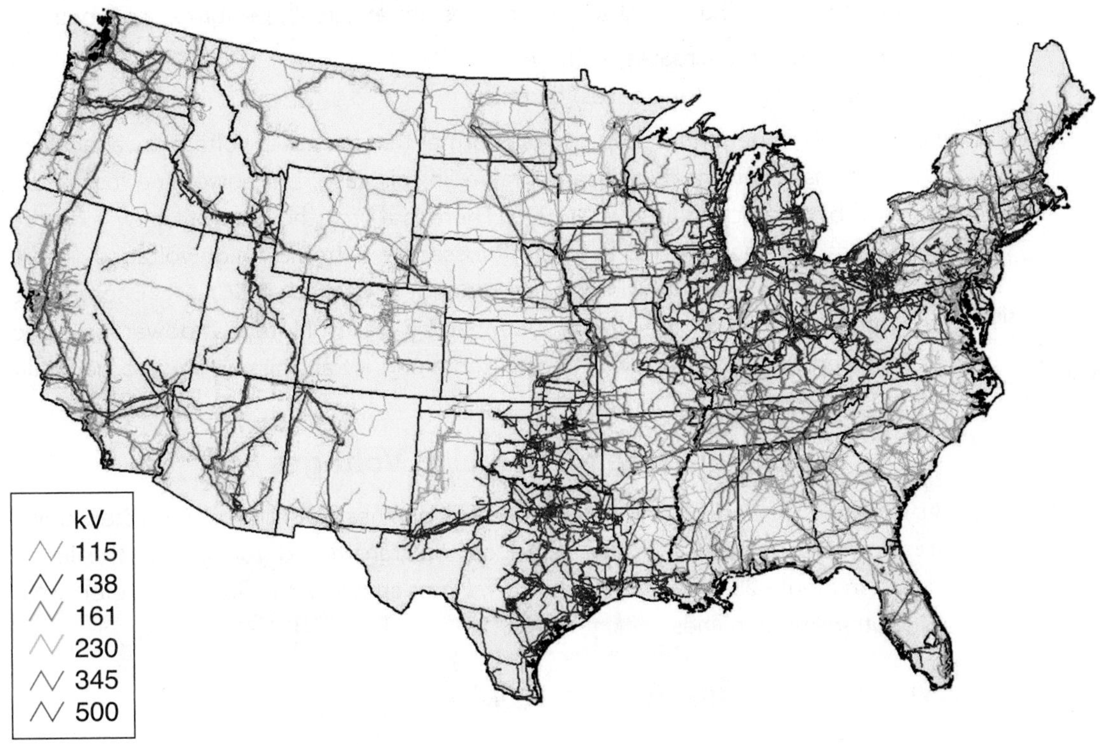

FIGURE 25.9 The U. S. power transmission grid.
Source: Redrawn from FEMA

Transmission line towers like those represented in Figure 25.8 are identified by their size and shape. Two high-voltage transmission line towers are shown in Figure 25.10.

FIGURE 25.10 Transmission line towers.
Istock.com

Reducing Power Line Losses

You may recall that the power dissipated by a resistance can be found as

$$P = I^2 \times R$$

As you learned in Chapter 20, a step-up transformer actually performs two functions:

- It steps up (increases) voltage.
- It steps down (decreases) current.

For example, a generator step-up transformer that increases voltage by a factor of 100 *decreases* current by the same value. In this case, the power line loss is decreased by a factor of $100^2 = 10{,}000$. There is a limit, however, to how high the voltage can be increased to reduce I^2R losses. At extremely high voltages—more than 2000 kV between conductor and ground—**CORONA DISCHARGE** losses are so high that they can offset the reduction in I^2R losses. Note that power losses up to 5 percent of the total power transmitted along a transmission line are generally considered acceptable (though not desirable).

CORONA DISCHARGE
Power loss along transmission lines that occurs when high voltages ionize the air surrounding the power lines.

Stepping Down Transmission Line Voltages

As shown in Figure 25.8, overhead transmission lines end at substation step-down transformers. These transformers reduce the transmitted voltages to the levels shown in Figure 25.8. The 26 kV and 69 kV lines shown in the figure are referred to as *subtransmission lines*.

Subtransmission Lines

SUBTRANSMISSION LINES carry voltages that are typically between 26 kV and 69 kV to *regional distribution substations*, that is, to substations that distribute power across a geographic region. Two subtransmission line towers are shown in Figure 25.11. Note that subtransmission lines may be tapped to provide power for local industries and substations.

SUBTRANSMISSION LINES
Lines that carry voltages—typically between 26 kV and 69 kV—to regional distribution substations.

Underground Transmission Lines

Underground transmission lines are used to transmit power through populated areas, under water, and in other places where overhead transmission lines cannot be used. Unlike overhead lines, underground transmission lines are insulated to protect them from water and other contaminants.

Underground transmission lines may be buried in trenches or conduit, or may be installed in underground transmission tunnels. Underground transmission tunnels are cooled to increase the current capacity of the transmission lines they contain. Due to the higher heat-related losses and higher costs of underground installations, overhead lines are far more common.

Underground distribution lines are found mostly in urban areas, as dictated by economics, congestion, and density of population. Although overhead lines are normally considered less expensive and easier to maintain, advancements in underground cable and construction technology have narrowed the cost gap to the point where such systems are competitive in many urban areas as well.

IN THE FIELD

(a)

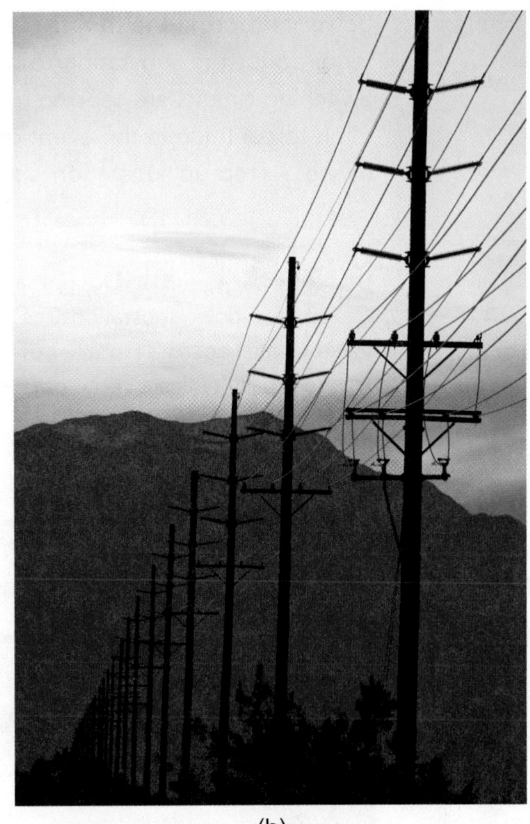

(b)

FIGURE 25.11 Subtransmission line towers.
(a) Shutterstock.com; (b) Istock.com

Substations

There are several types of electrical substations, each connecting two or more elements in the power transmission system. A **STEP-UP SUBSTATION** connects the power generating plant to the transmission lines. A **STEP-DOWN SUBSTATION** is connected between transmission lines and subtransmission or distribution lines. A **DISTRIBUTION SUBSTATION** connects subtransmission lines to the distribution lines that serve local utility customers.

STEP-UP SUBSTATION One that connects a power generating plant to the transmission lines.

STEP-DOWN SUBSTATION One that is connected between transmission lines and subtransmission or distribution lines.

DISTRIBUTION SUBSTATION One that connects subtransmission lines to local distribution lines.

Utility control and switching systems in step-up and step-down substations operate under demanding conditions, including high voltage and current levels, exposure to lightning, and 24-hour-a-day use. For reliable performance, a large margin of safety is built into the design of each element of these systems, along with extensive and documented testing.

IN THE FIELD

Electrical power generating plants typically generate voltages between 2.3 kV and 33 kV. *Step-up substations* contain high-voltage transformers that boost the generator output voltage to a value between 115 kV and 765 kV, and provide power outputs in the thousands of *kilovolt-amperes* (kVA) and higher. The power required for the substation to operate is drawn from the substation's input feed. Switching circuits are used to route power through the substation, allowing operation to continue in the event of a transformer failure or during required maintenance. A step-up substation is shown in Figure 25.12.

STAYING SAFE Do not attempt to enter a fenced-in distribution or step-up/step-down substation. The presence of open, energized parts (bussing, connectors, etc.) pose a severe risk of electrocution, even without direct contact, due to the possibility of flashover to objects at ground potential.

FIGURE 25.12 A step-up substation.
Istock.com

As mentioned earlier, *step-down substations* decrease transmission line voltages down to subtransmission line levels, typically between 26 kV and 69 kV. As such, these substations act as the interface between transmission lines and subtransmission lines. A step-down substation is shown in Figure 25.13.

FIGURE 25.13 A step-down substation.
Istock.com

Distribution substations decrease subtransmission voltages to customer-usable levels. As such, they are also step-down substations. Like other substations, distribution substations typically contain lightning arresters, overload protectors (circuit breakers), voltage regulators, input and output switching circuits, and control circuits. A distribution substation is shown in Figure 25.14.

Some distribution substations are entirely enclosed in buildings whereas others are built entirely in the open, with all equipment enclosed in one or more metal-clad units. The final design of the type of distribution substation depends on economic factors and future load growth as well as environmental, legal, and social issues.

IN THE FIELD

FIGURE 25.14 A distribution substation.

PROGRESS CHECK

1. List and describe the three types of transmission lines.
2. Explain how a step-up transformer at the generating site decreases power losses in transmission lines.
3. What are the two most common voltages carried by subtransmission lines?
4. List and define the various types of substations.

25.4 DISTRIBUTION GRIDS

DISTRIBUTION GRID The final element in the power transmission system; it begins at the distribution substation and ends at the customer's business or residence.

The **DISTRIBUTION GRID** is the final element in the power transmission system. It begins at the distribution substation and ends at the customer's place of business or residence.

The distribution grid may be overhead or underground. The overhead distribution grid includes poles, overhead wires, and transformers like those shown in Figure 25.15. An underground grid may include some poles and overhead wires; but at some point, the grid switches to underground conductors and pad transformers.

FIGURE 25.15 Overhead power distribution lines.
Istock.com

Distribution Grid Voltages

Distribution substations are typically fed from transmission lines (115 kV or higher) or subtransmission lines (26 kV to 69 kV). These voltages are stepped down to distribution levels (2.4 kV to 33 kV) that are transmitted over a network of *feeder lines*. The exact voltages found on the distribution grid depend on the needs of local industries and communities.

Distribution Grid Transformers

The feeder lines in the distribution grid often lead to transformers that serve two purposes:

1. They step down the distribution grid voltages to the levels required by the customers.
2. They provide *fault isolation* to prevent circuit overloads from affecting voltages at other points on the grid.

You may recall that a transformer provides electrical isolation between its primary and secondary circuits. When a fault occurs in the secondary (load side) of a transformer, the grid lines connected to the primary (source side) are protected from the fault.

The transformers used to perform the functions listed above are generally seen on poles (for overhead grids) or on pads (for underground grids). Both types of transformers are shown in Figure 25.16.

FIGURE 25.16 Pad and pole transformers.
Istock.com

Distribution grids may also contain smaller secondary substations that typically contain a transformer, one or more switches, and voltage regulator circuits. Such a substation, which is normally located near the area it services, is shown in Figure 25.17.

FIGURE 25.17 A distribution substation.

PROGRESS CHECK

1. What are the common voltage levels in a distribution grid?
2. What are feeder lines and what purpose do they serve?
3. Explain how the transformers used in the distribution grid provide fault isolation.

25.5 SUMMARY

Here is a summary of the major points that were made in this chapter:

Power Transmission and Distribution: An Overview

- The generation and delivery of electrical power can be divided into three processes:
  - AC power generation is the conversion of mechanical energy into electrical energy.
  - Power transmission is the delivery, over a distance, of the generated power to populated areas.
  - Power distribution is the process of stepping down the voltage to customer-usable levels and delivering power to local homes and industry.
- A power grid consists of the generating plant, the transmission lines, the power distribution system, and substations. (See Figure 25.2.)
- Power generating plants are usually located outside populated areas. Their location is determined by several factors:
  - The type of plant (hydroelectric, fossil fuel, nuclear, etc.).
  - Construction costs.
  - Labor pool and other economic factors.
  - Safety and environmental concerns.
- Transmission lines are the high-voltage lines that deliver power over long distances.
- There are three types of transmission lines:
  - Overhead lines are mounted on overhead towers.
  - Underground lines run underground as the name implies.
  - Subtransmission lines carry lower voltages than primary lines.
- The power distribution system includes the various power lines, transformers, and switching and protection circuitry.
- Substations are found between the generating plant and the transmission lines, and between the transmission lines and the distribution system.
  - Substations contain step-up or step-down transformers, switching gear, and overload protection circuits.

Generating Electrical Power

- All AC generation requires a primary mover, or source of mechanical energy.
- A hydroelectric plant converts the kinetic energy of flowing water to electrical energy. (See Figure 25.3.)
  - The turbine is a rotor with blades that is driven by the flowing water.
  - The penstock is the conduit that directs the flowing water through the turbine.
  - The step-up transformer between the power plant and the transmission lines increases the voltage to suitable transmission levels.

- There are two types of dams used in hydroelectric plants.
  - An impoundment dam totally blocks the natural flow of water, leaving the penstock as the only path.
  - A diversion dam, which is cheaper to build, diverts only a portion of the water through the penstock and turbine, and then returns it to its natural path.
- A thermal power plant heats water to convert it to steam and uses the steam to drive the turbine. (See Figure 25.4.)
  - The steam passes through a cooling structure and condenses back into water.
  - The water-to-steam-to-water cycle is called the Rankine cycle.
- Thermal power plants are classified by the fuel they use as a heat source.
- A fossil-fuel plant uses coal, oil, or natural gas.
- A nuclear plant uses nuclear fission.
  - The nuclear fuel is packaged in fuel rods.
  - The most common nuclear fuels are uranium 235 and plutonium 239.
- Increasing fuel costs and environmental concerns have resulted in a greater interest in wind power.
- Wind power uses air currents (wind) to generate electrical energy using a wind turbine.
  - A group of wind turbines is called a *wind farm.*
- The most common type of wind turbine is the horizontal-axis wind turbine (HAWT). (See Figure 25.6.)
- The vertical-axis wind turbine (VAWT) is used far less often than the HAWT. (See Figure 25.7.)

Transmission Lines and Substations

- As stated earlier, there are three types of transmission lines.
- High-voltage transmission lines are used to transmit 3ϕ power over distances that are typically 300 miles or less. (See Figure 25.8.)
  - Common transmission line voltages are 115 kV, 138 kV, 161 kV, 230 kV, 345 kV, 500 kV, and 765 kV.
  - Lines carrying voltage of 345 kV and higher require extra precautions, such as increasing line spacing to prevent flashover (arcing between lines).
- Step-up transformers are used to increase the voltage applied to the transmission lines to reduce power line losses.
  - Since $P = V \times I$, if the voltage is increased by a given factor, the current is reduced by the same factor for a given amount of power.
  - Since $P = I^2R$, if current is reduced by, say, 10 times, power losses are reduced by $10^2 = 100$ times.
  - Transmission line power losses in the 5 percent range are considered acceptable.
- Transmission lines end at the substation, where step-down transformers reduce the voltage to 26 kV to 69 kV. (See Figure 25.8.)
  - The lines carrying these lower voltages are referred to as subtransmission lines.
  - Subtransmission lines may be tapped to provide power to local substations or industry.
- Underground transmission lines are used where overhead lines are impractical, such as under water.
  - Underground lines may be buried in trenches or conduits, or may be in underground tunnels.
  - The underground tunnels may be cooled to increase current capacity.

- Substations connect two or more elements of the power transmission system. There are three main types of substations:
  - Step-up substations contain high-voltage transformers (rated in thousands of kVA or higher). Switching circuits are used to route power through the substation to allow for transformer failure or scheduled maintenance.
  - Step-down substations use transformers to decrease line voltages to subtransmission levels, typically between 26 kV and 69 kV. This is the interface between transmission lines and subtransmission lines.
  - Distribution substations decrease subtransmission voltages to customer-usable levels. Like other substations, distribution substations contain lightning arrestors, overload protectors (circuit breakers), voltage regulators, and switching and control circuits.

Distribution Grids

- The distribution grid is the final element in the power transmission system.
  - It begins at the distribution substation and ends at the customer.
  - Either overhead or underground lines are used, though overhead lines are more common.
- Distribution substations are typically fed from transmission or subtransmission lines, and these voltages are stepped down to levels in the 2.4 kV to 33 kV range.
  - These lower voltages are transmitted over a network of feeder lines.
  - The actual voltage is determined by local requirements.
- Feeder lines usually lead to transformers that serve two purposes:
  - They step down the distribution voltage to customer-required levels.
  - They provide fault isolation.
- Distribution grids may also contain smaller secondary substations that are located near the area they serve.

CHAPTER REVIEW

1. Power generation usually converts ________ energy to ________ energy.
   a) mechanical, potential
   b) mechanical, electrical
   c) electrical, mechanical
   d) electrical, kinetic

2. A(n) ________ is usually connected between the power plant and the transmission lines.
   a) step-up transformer
   b) step-down transformer
   c) inverter
   d) regulator

3. The power plant, transmission lines, distribution system, and substations comprise the power ________.
   a) shell
   b) elements
   c) field
   d) grid

4. The choice of the location for a power plant is usually determined by ________.
   a) environmental concerns
   b) economic concerns
   c) the type of power plant
   d) all of the above

5. Substations are found between ________.
   a) power generating plants and transmission lines
   b) transmission and subtransmission lines
   c) transmission lines and distribution systems
   d) all of the above

6. All types of AC generation require ________ in order to produce power.
   a) fuel
   b) kinetic energy
   c) a prime mover
   d) potential energy

7. The bladed rotor in a hydroelectric power plant is called a(n) ________.
   a) crankshaft
   b) turbine
   c) camshaft
   d) impeller

8. The water conduit between the reservoir and the plant's rotor is called the ________.
   a) sluice race
   b) penstock
   c) feed channel
   d) feeder race

9. A dam that completely blocks the natural flow of water from a reservoir is called a(n) ________ dam.
   a) diversion
   b) impediment
   c) impoundment
   d) intrusion

10. The steam-to-water-to-steam cycle in a thermal power plant is called the ________.
    a) Robarts cycle
    b) condensing cycle
    c) thermal cycle
    d) Rankine cycle

11. A nuclear power plant uses ________ to generate heat.
    a) nuclear fusion
    b) heavy water

c) nuclear fission

d) all of the above

12. HAWTs are ________.

a) the set of codes regulating wind farms

b) the most common types of wind turbines

c) four blade turbines

d) high-altitude wind turbines

13. Most wind turbines have ________ blades.

a) two

b) three

c) four

d) five

14. The ________ moves the wind turbine so that it is facing into the wind.

a) anemometer

b) nacelle

c) accelerometer

d) yaw motor/driver

15. The generator rotor in a wind turbine rotates ________ the turbine rotor.

a) faster than

b) slower than

c) at the same speed as

d) all of the above depending on the gear ratio

16. A value of ________ is not a common transmission line voltage.

a) 138 kV

b) 500 kV

c) 440 kV

d) 230 kV

17. Transmission line voltages between ________ require extra precautions.

a) 345 kV and 765 kV

b) 765 kV and 1 MV

c) 200 kV and 765 kV

d) 500 kV and 2 MV

18. Arcing between transmission lines is referred to as ________.

a) intermodulation

b) ground faulting

c) flashover

d) plasma ionization

19. A step-up transformer increases the voltage between the power plant and the transmission lines by a factor of 17.25:1. This means that I^2R losses would have been ________ times higher if the transformer was not used.
    a) 17.25 b) 34.5
    c) 298 d) 172.5

20. Subtransmission line voltages are typically between ________.
    a) 20 kV and 50 kV
    b) 26 kV and 69 kV
    c) 36 kV and 59 kV
    d) 34 kV and 65 kV

21. Unlike overhead transmission lines, underground lines are usually ________.
    a) insulated b) heated
    c) grounded d) waterproof

22. To provide for possible transformer failure, substations incorporate ________.
    a) arrestors
    b) switching circuits
    c) regulators
    d) all of the above

23. The final element in the power transmission system is the ________.
    a) step-up transformer
    b) distribution grid
    c) line stabilizer
    d) single-phase tap

24. Distribution grid voltages are usually in the range of ________.
    a) 2.4 kV to 33 kV
    b) 240 V to 3.3 kV
    c) 26 kV to 33 kV
    d) 26 kV to 69 kV

25. The lower voltage from distribution substations is transmitted over a network of ________.
    a) voltage regulators
    b) pad transformers
    c) feeder lines
    d) surge suppressors

26. Fault isolation in the distribution grid is provided by ________.
    a) step-up transformers
    b) surge suppressors
    c) ground-fault interrupters
    d) step-down transformers

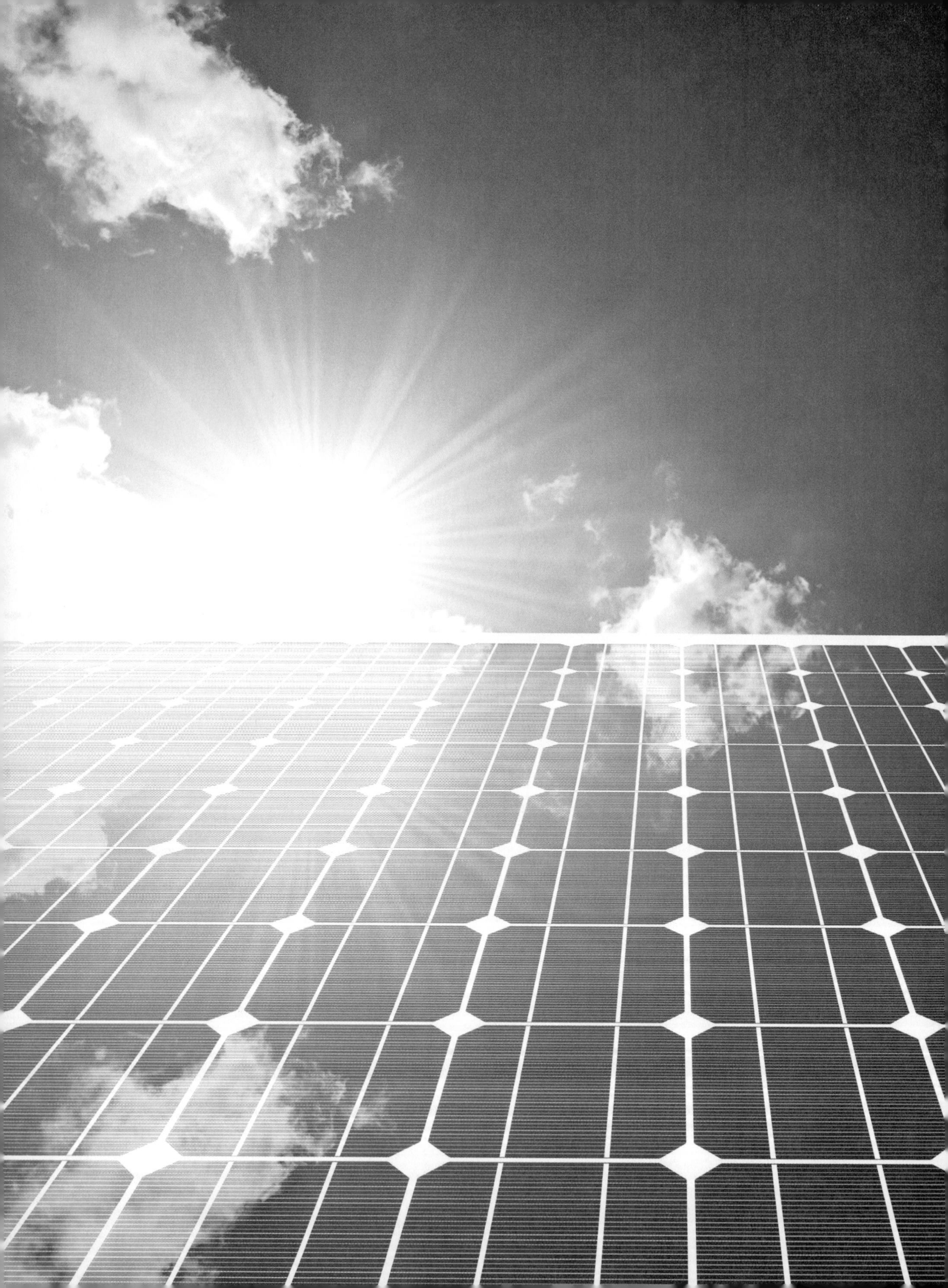

CHAPTER 26

GREEN POWER

PURPOSE

Concerns about global warming and other environmental issues are driving changes in many aspects of our lives, both at home and at work. In response to these concerns, **GREEN** (environmentally friendly) products and technologies are springing up everywhere.

Throughout your career, you will be exposed to an ever-growing variety of green technologies. This is due to the fact that electrical power generation is one of the country's major sources of greenhouse gas emissions.

In this chapter, we will look at the role that generating electrical power plays in the production of **GREENHOUSE GASES** (i.e., gases that contribute to global climate change). Then we will look at some alternative technologies that are in use today and others that are beginning to emerge. As you will learn, environmental concerns are changing the way we produce, distribute, and consume electrical power. This means that the role of the electrician will be changing as well.

KEY TERMS

The following terms are introduced and defined in this chapter on the pages indicated:

Fault-current limiter (FCL), 1136
Feed-in tariff (FIT), 1131
Fossil natural gas, 1121
Fuel Cell, 1131
Gas-to-liquid (GTL) technology, 1122
Gas turbine, 1127
Green, 1118
Greenhouse gases, 1118
Integrated gasification combined cycle (IGCC), 1120
Inverter, 1130
Light-emitting diode (LED), 1146
Liquefied natural gas (LNG), 1123
Microturbine, 1129
n-type material, 1130
Net metering, 1131
Organic light-emitting diodes (OLEDs), 1148
p-type material, 1130
Peak shaving, 1127
Peaker plants, 1138
Reciprocating internal combustion engine, 1127
Smart grid, 1133
Smart meters, 1138
Superconductors, 1134
Superconducting magnetic energy storage (SMES), 1136
Tidal power, 1125
Wave power, 1125

OBJECTIVES

After completing this chapter, you should be able to:

1. Describe the role that electricity generation plays in the production of greenhouse gases.
2. List and describe the different sources of power generation and their effect on the environment.
3. Describe coal gasification and explain how it reduces greenhouse gas emissions.
4. Define carbon capture and storage.
5. Explain the role of combined cycle power generation in natural gas-fired turbines.
6. Discuss some of the risks and benefits of nuclear power production.
7. List and compare other sources of large-scale power production, like wind, solar, hydro, and wave power.
8. Define distributed generation and explain its benefits.
9. List and compare the different types of distributed generation.
10. Define the Brayton cycle.
11. Discuss the basic operation of a photovoltaic cell.
12. Define building-integrated photovoltaics.
13. Discuss the government initiatives of net metering and feed-in tariffs.
14. Describe the basic operation of fuel cells.
15. Describe amorphous-core transformers.
16. List and compare the different types of superconductors.
17. Discuss a number of superconductor applications.
18. Discuss the Smart Grid and explain its benefits.
19. Define peaker plants and explain why they are needed.
20. Discuss the role that smart metering plays in the Smart Grid.
21. Discuss some of the technologies that are used in green buildings.
22. Describe the LEEDS® certification system.
23. Discuss the advantages of compact fluorescent lights (CFLs).
24. Compare LED lighting to CFLs.
25. Discuss the advantages and possible applications of organic light-emitting diodes.

GREEN A term used to describe environmentally friendly products and technologies.

GREENHOUSE GASES Gases that absorb and emit solar radiation in the infrared range, trapping heat in the earth's atmosphere.

26.1 LARGE-SCALE POWER GENERATION

Our economy and society both depend on abundant and easily accessible sources of electricity. Unfortunately, nothing comes without some kind of cost, and the environmental cost of power generation and distribution is significant. In this section, we take a look at the environmental impact of energy production and some steps that are being taken to limit that impact.

More Power

The need for increased electrical production is growing worldwide. The Energy Information Administration (EIA) projects a 1.7% yearly increase in electrical demand in the United States through 2030. Over the same period, the demand in China is expected to increase by nearly three times that rate—a staggering 4.8% annually.

Electrical power generation is one of the major sources of greenhouse gas emissions. The sources of CO_2e emissions in the United States are illustrated in Figure 26.1. As you can see, electric power generation is the single largest source of **CARBON DIOXIDE EQUIVALENT (CO_2e)** emissions. At just under 40% of the total U. S. CO_2e emissions, this represents approximately 10% of the world's CO_2e emissions.

> While the *electric vehicle* (EV) looks like a viable technology to help combat the greenhouse gas problem, both utilities and green energy policy experts have concerns. Utilities have expressed concern that EV charging could damage weak points in the distribution grid, and policy experts have expressed concern that this change in the nation's grid use could unintentionally detract from their green energy plans.
>
> IN THE FIELD

CARBON DIOXIDE EQUIVALENT (CO_2e) A unit of measure for greenhouse gas emissions that indicates their global warming impact referenced to equivalent amounts of carbon dioxide.

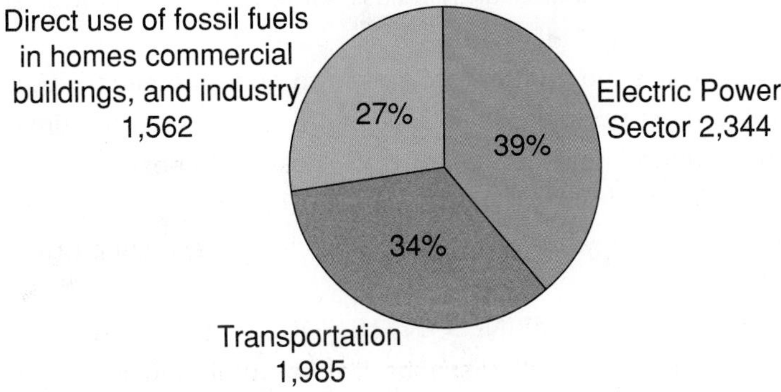

FIGURE 26.1 Sources of CO_2e.
Source: Redrawn from the Energy Information Administration.

Approximately 4.2 trillion kWh were generated in the United States in 2007. In the same year, we consumed about 3.93 trillion kilowatt-hours (kWh) of power—around 10.8 billion kWh per day. (The difference between the power generated and the power used is due to transmission line losses that occur between the utilities and the end users.)

> Utilities anticipate that future electric vehicle owners will use a 120 V charger that can draw up to 6.6 kW. Turning on one such charger is like adding up to three homes to a neighborhood, each with air conditioning, lights, and laundry running.
>
> IN THE FIELD

The fuel-source breakdown for the power generated in the United States is illustrated in Figure 26.2. As the chart shows, coal represents nearly half the fuel used to generate electricity, followed by natural gas and nuclear at approximately 20% each. As such, nearly 90% of the electrical power in the United States is generated using non-renewable energy sources. Only 8.3% of our energy currently comes from renewable energy sources.

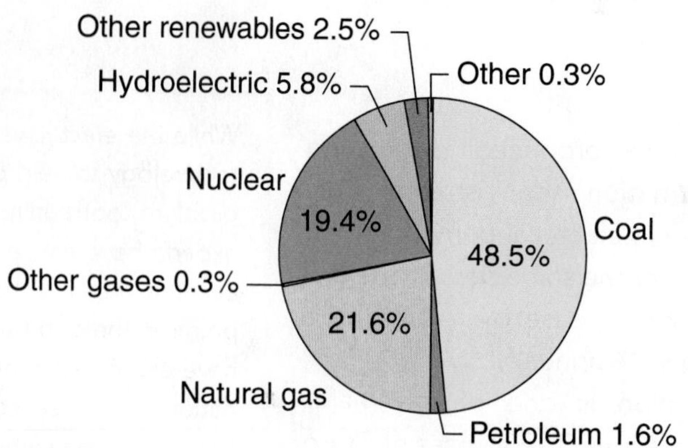

FIGURE 26.2 Fuels used to generate power in the U. S.
Source: Redrawn from the Energy Information Administration.

Coal Is King

CARBON PER UNIT ENERGY
The amount of CO_2 emitted per unit of energy produced.

Coal contains nearly 80% more **CARBON PER UNIT ENERGY** than natural gas. If we have to increase generating capacity by a projected 40% over the next 30 years, greenhouse gas emissions will rise significantly unless some major technological changes are made.

Despite the fact that coal is the "dirtiest" fuel used to generate electricity, it will most likely remain the most popular fuel in the United States for some time. This is primarily due to the abundance and relatively low cost of domestic coal resources. It is also very desirable, from a national security standpoint, to have an abundant domestic source of energy. The most promising clean coal technology is *coal gasification*.

INTEGRATED GASIFICATION COMBINED CYCLE (IGCC)
A process where coal is converted into a gas and purified before burning, resulting in reduced production of sulfur dioxide and other pollutants and improved efficiency.

INTEGRATED GASIFICATION COMBINED CYCLE (IGCC) is a relatively new technology that turns coal into a gas, called *syngas*, before burning. Gasification breaks down coal, or any other carbon-based fuel, into its chemical components. The coal is exposed to steam and controlled amounts of oxygen under high temperatures and pressures, which break apart the coal molecules. Figure 26.3 shows a block diagram of a basic gasification plant.

Gasification may be one of the most useful and flexible technologies for the future because it can be used to can burn a variety of fuels. It can also be used to produce hydrogen and other gases that can fuel hydrogen fuel cells or power-generating turbines. Gasification also produces chemicals such as sulfuric acid or elemental sulfur—both valuable byproducts—and keeps them out of the atmosphere.

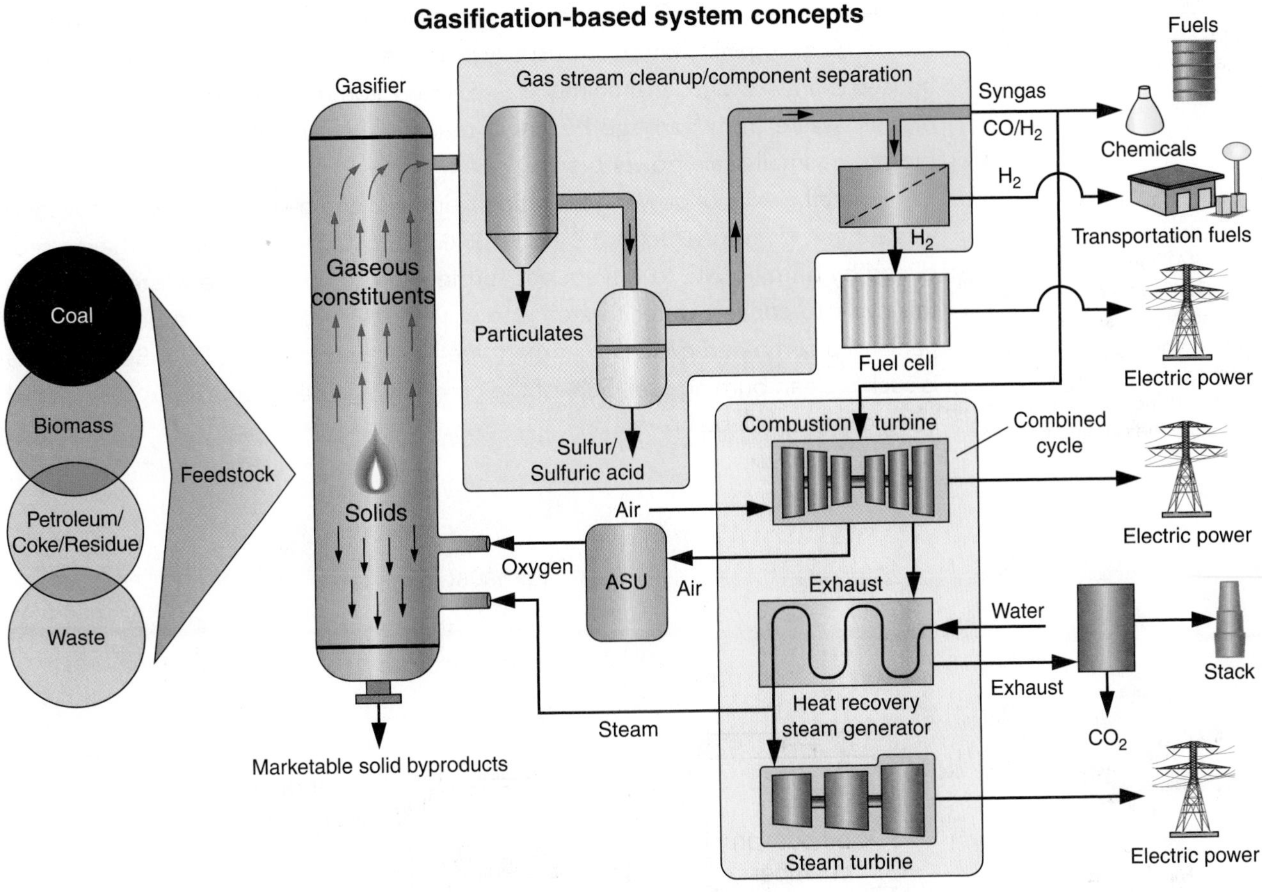

FIGURE 26.3 A representation of a gasification system.
Source: Redrawn from the U.S. Department of Energy.

It should be noted that gasification has its drawbacks. One aspect of the cycle that has yet to be completely implemented is carbon capture and storage. **CARBON CAPTURE AND STORAGE (CCS)**, as the name implies, involves removing the carbon left over from gasification before venting, and then storing that carbon in deep geological formations or the ocean. Carbon capture technology has been developed and tested, but carbon storage—also called **CARBON SEQUESTRATION**—is still in its infancy. CCS also requires a great deal of energy, which will increase the cost of power production while decreasing efficiency.

CARBON CAPTURE AND STORAGE (CCS) A means of collecting the carbon produced during gasification and then storing that carbon in deep geological formations or the ocean.

CARBON SEQUESTRATION Another name for carbon storage.

Natural Gas

Natural gas (NG) is the cleanest burning of the fossil fuels, but is not a renewable resource. Compared to coal, it produces half the carbon oxides, one-third the nitrogen oxides, and only one percent of the sulfur for the same amount of heat. The most common form of NG is called **FOSSIL NATURAL GAS**. It is found in oil wells, natural gas wells, and coal beds. NG is primarily methane, but there are many impurities in fossil NG, so it must be refined, which requires even more energy.

FOSSIL NATURAL GAS A common type of natural gas that consists primarily of methane, found in oil wells, natural gas wells, and coal beds.

BIOGAS A methane-rich gas that comes from the decay of organic matter.

Another form of NG is *biogas*. **BIOGAS** is a methane-rich gas that comes from the decay of organic matter. It is usually produced from agricultural waste, such as unused plant material and animal waste. Many communities are now separating organic waste from garbage before it reaches the landfill in order to produce biogas for small-scale power plants.

COMBINED CYCLE POWER GENERATION A process in which hot gases produced by burning NG drive a gas turbine, and then convert water to steam to drive a steam turbine using the waste heat from the burning process.

The efficiency of power generation using NG can be increased by recovering waste heat. **COMBINED CYCLE POWER GENERATION** uses the hot gases produced by burning NG to drive a gas turbine, and then uses the waste heat from this cycle to convert water to steam to drive a steam turbine. The basic components of a combined cycle power plant are illustrated in Figure 26.4. Because NG is such a clean burning fuel, 50% of the currently planned increases in U.S. generating capacity will be NG fired.

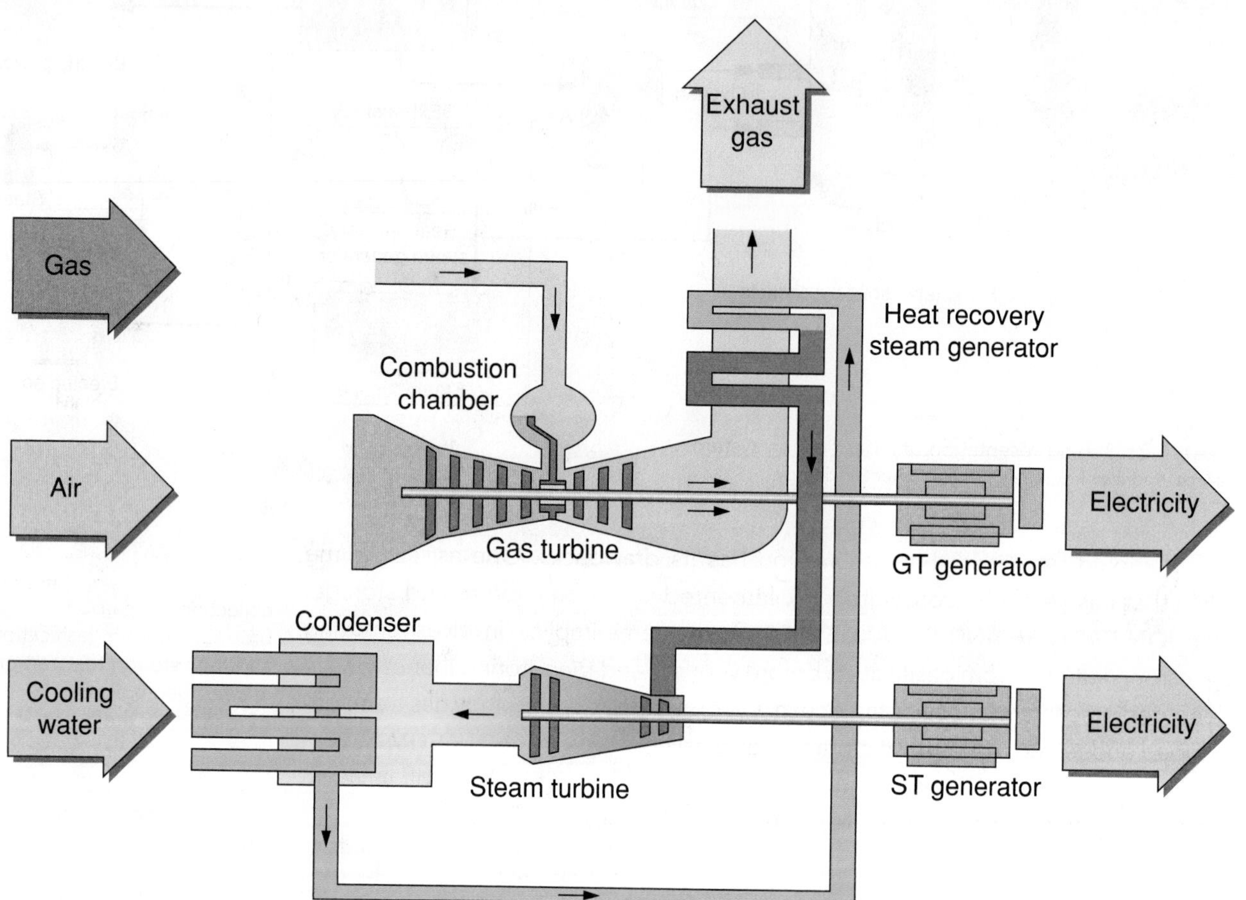

FIGURE 26.4 A combined cycle power plant.
Source: Redrawn with permission from Siemens.

GAS-TO-LIQUID (GTL) TECHNOLOGY Technology used to refine and cool NG to about −162°C (−260°F), allowing it be more easily transported.

One of the problems with NG is that it requires a great deal of infrastructure to deliver it to the end users (utility companies). Unlike coal and oil, which can be transported by truck or rail, dedicated pipelines must be used to deliver NG. Many producers are converting NG into a liquid. **GAS-TO-LIQUID (GTL) TECHNOLOGY** allows NG to be more easily transported. The GTL process involves refining raw gas and then cooling it to about −162°C (−260°F). In this form it is called **LIQUEFIED NATURAL GAS (LNG)**. The country of Qatar is scheduled to bring a 140,000 bbl/day (barrels per day) GTL plant online in 2010. An LNG cargo ship is shown in Figure 26.5.

LIQUEFIED NATURAL GAS (LNG) The form that NG takes as a result of the gas-to-liquid process.

FIGURE 26.5 A liquified natural gas (LNG) cargo ship.
Istock.com

Nuclear

There may be no more contentious energy issue than commercial nuclear power production. Proponents point to the fact that nuclear power produces no greenhouse gases and no particulates. Critics point to the environmental damage caused by past nuclear accidents like those at Chernobyl and Three Mile Island. Perhaps the biggest concern is how to safely store spent nuclear fuels that can remain deadly for literally hundreds of thousands of years. While the debate goes on, nuclear power produces 20% of the electricity in the United States, and that percentage may increase in the future.

The United States is the world's largest producer of nuclear generated electricity, accounting for 30% of global production. There are currently 104 nuclear power plants located in 31 states.

For nearly 30 years, very few nuclear power plants were commissioned. The last plant built in the U. S. was in 1996. But with increased concerns about greenhouse gases and global warming, the landscape is beginning to change. Since 2007, the U. S. government has received 17 applications for licenses to construct 26 new reactors.

There is no clear winner in the nuclear debate. Increasing the role of nuclear power production will lower greenhouse gas and particulate emissions, but increase the amount of radioactive waste that we have to safely store long beyond the foreseeable future. As of 2009, there are 270,000 tons of spent nuclear fuel in storage, almost all of it at the individual reactor sites.

Other Large-Scale Generating Technologies

Before we move on to small-scale generation, let's take a brief look at some other generating formats.

Oil

Petroleum-fired generating plants produce less than 2% of the electricity in the United States. Oil is not a clean burning fuel, it is expensive, and domestic reserves are low. For these reasons, oil may disappear as an electricity-generating fuel.

Hydro

Hydroelectric power represents just under 6% of the electricity produced in the United States. In contrast, Norway produces over 98% of its electricity using hydroelectric plants.

No new hydroelectric plants are currently under construction in the United States. The last proposed hydroelectric project—the Rampart Canyon dam in Alaska—was cancelled due to cost and concerns about loss of wildlife habitats. Hydro is a clean and highly efficient (over 90%) renewable source of energy; however, it does have a significant impact on the environment. The reservoir created by damming a river interferes with fish migration and impacts the upstream ecology. It is interesting to note that the cancelled Rampart Dam would have created a water reservoir the size of Lake Erie, which would have been the largest man-made body of water in the world.

Wind

You were introduced to wind power in Chapter 25. Wind generators impact the environment more subtly than most other sources of electricity. Significant clearing of forests may be necessary to build wind farms. Wind farms can also have a negative impact on bird populations. For example, 30 endangered golden eagles were killed or injured over a three-year period at the Altamont Pass wind farm in the Diablo Range in central California.

Generally speaking, the greatest opposition to wind farms comes from those who see them as unsightly and noisy structures that may lower property values. Despite these concerns, wind-generated electricity is increasing. According to the U.S. Department of Energy, wind power is the most rapidly growing source of electrical energy. From 2006 to 2007, wind-generated electricity production increased to 34.5 billion kWh from 26.6 billion kWh—an increase of nearly 30%.

Solar

CONCENTRATING SOLAR POWER (CSP) TECHNOLOGY Technology that uses mirrors or lenses to focus light, producing the heat required to convert water to steam to drive a steam turbine.

Large-scale solar power plants generate electricity in various ways. Until recently, they used **CONCENTRATING SOLAR POWER (CSP) TECHNOLOGY** exclusively. CSP technology uses mirrors or lenses to focus light to heat water and drive a steam turbine. At this time, the largest such solar power plant in the world is the 354 MW *Solar Energy Generating Systems* (SEGS) facility in California's Mojave desert, though larger ones are planned. A large-scale solar reflector installation is shown in Figure 26.6.

Large-scale photovoltaic (PV) arrays are becoming more popular as thin-film solar cell technology becomes cheaper and more efficient. A 550 MW photovoltaic facility in San Luis Obispo County, California should be fully operational by 2013.

FIGURE 26.6 A large-scale solar reflector installation.
Fotolia.com

Another type of solar generation combines CSP and PV. **CONCENTRATING PHOTOVOLTAIC (CPV) TECHNOLOGY**—like CSP installations—focuses the sun's energy onto photovoltaic cells, increasing their efficiency.

CONCENTRATING PHOTOVOLTAIC (CPV) TECHNOLOGY Technology that focuses the sun's energy onto photovoltaic cells, increasing their efficiency.

Solar power facilities do not have a significant impact on the environment when operating. The negative environmental impact of solar energy comes from the construction and disposal of the solar panels. Hazardous materials (such as cadmium and arsenic) are used in their production, so disposal of these materials will become an issue as older solar arrays are replaced. There is also some concern about the amount of land required to generate solar power—approximately one square kilometer (0.386 square mile) for every 20 to 60 MW generated.

Wave and Tidal

Although similar, these are two distinct technologies. **WAVE POWER** uses the constant rise and fall of surface waves or deep water waves and converts this mechanical energy into electrical energy. The world's first commercial wave farm is a 2.2 MW installation off the coast of Portugal that became operational in 2008. Wave power technology is still in its infancy. It has been estimated that deep water wave technology could someday produce several *terawatts* (TW) of power if we can learn to harvest it efficiently.

WAVE POWER Converting the mechanical energy produced by the constant rise and fall of surface waves or deep water waves into electrical energy.

TIDAL POWER installations have been around since the mid 1960s, but there are still only a handful of commercial plants. A tidal power plant uses a dam (called a **BARRAGE**) across a tidal **ESTUARY** (point where the mouth of a river meets the ocean). The barrage traps water in the estuary when the tide rises, and then channels the water through a turbine when the tide falls. In addition to high cost, there are significant environmental concerns with tidal power. Although they produce no pollutants, barrages block navigation, interfere with fish migration, and change the wet/dry cycle that supports the native plant life in the tidal basin.

TIDAL POWER Converting the mechanical energy produced by rising and falling tides into electrical energy.

BARRAGE The dam used in a tidal power plant.

ESTUARY The point where the mouth of a river meets the ocean.

The Ultimate Power Plant

There is a project under development by the DOE's Office of Fossil Energy to design and build a multifuel power plant that would be extremely efficient (greater than 75%), produce virtually no greenhouse gases or pollutants, and provide electricity at current rates. The basic modules of this project, called Vision 21, are slated to be developed by 2015. Here is how the DOE describes this facility:

> *"Vision 21 builds on a portfolio of technologies already being developed, including low-polluting combustion, gasification, high efficiency furnaces and heat exchangers, advanced gas turbines, fuel cells, and fuels synthesis, and adds other critical technologies and system integration techniques. When coupled with carbon dioxide capture and recycling or sequestration, Vision 21 systems would release no net carbon dioxide emissions and have no adverse environmental impacts."*

PROGRESS CHECK

1. What is the most popular fuel for electricity generation in the United States?
2. Describe coal gasification and explain the role carbon capture and storage (CCS) plays in this technology.
3. Which is the cleanest burning fossil fuel?
4. What is combined cycle power generation?
5. Explain why nuclear plant construction is increasing.
6. List and describe the three types of large-scale solar plants.
7. What is Vision 21?

26.2 DISTRIBUTED GENERATION

So far, we have concentrated on large-scale power plants. In this section, we look at small-scale power generation. As time goes on, you will encounter this approach to generating power, called *distributed generation*, more and more.

DISTRIBUTED GENERATION (DG) is the production of small-scale power near or at a customer's site. Distributed generation has many potential benefits, such as:

Studies conducted by Frost & Sullivan (a market study research firm) and reports from various investment companies indicate that distributed generation currently provides electricity to make 50 percent of our paper products, 25 percent of our chemicals, and 30 percent of our petroleum.

IN THE FIELD

- *Lower greenhouse gas emissions.* This benefit is the result of using more efficient and/or renewable energy sources.
- *Reduced power distribution infrastructure.* With power being generated nearby, less power distribution infrastructure is required.

DISTRIBUTED GENERATION (DG) The production of small-scale power at or near a customer's site.

- *Peak shaving.* **PEAK SHAVING** is the leveling out of power consumption over time. Peak shaving is discussed later in the chapter.
- *Fuel flexibility.* DG allows the customer to choose between several fuel sources based on cost and availability.

PEAK SHAVING The leveling out of power consumption over time.

Distributed generation can take many forms. It can be as simple as adding a couple of solar panels on a residence, or as complex as implementing a variety of power generating alternatives for an entire industrial complex. DG units can produce from a few kilowatts to hundreds of megawatts. Some DG installations include renewable and alternative energy sources, while others do not. Let's look at some of the more popular forms of distributed power generation.

Depending on the type of generating equipment, a distributed generation plant can be installed for $1200 to $3000 per kilowatt of generating capacity.

IN THE FIELD

Reciprocating Internal Combustion Engines

This technology has been around for years, and is still the most popular DG electricity source. There are many reasons for the popularity of **RECIPROCATING INTERNAL COMBUSTION ENGINES**, including low cost per kilowatt produced, low maintenance requirements, and high reliability. Diesel was once the fuel of choice, but natural gas (NG) took its place because it is cheaper and much cleaner. There are many dual-fuel generators that can run on either diesel or NG, and older diesel units can be retrofitted for dual-fuel operation at a reasonable cost. Lean-burn technologies and improved emission controls also make the reciprocating engine a reasonably green choice. The efficiency of these units varies depending on size, fuel, and whether *cogeneration* is employed. The larger the power plant, the higher the efficiency.

RECIPROCATING INTERNAL COMBUSTION ENGINE An engine known for its low cost per kilowatt produced, low maintenance requirements, and high reliability.

By taking advantage of process facility waste heat, internal combustion engines, microturbines and fuel cells can provide overall efficiencies approaching 80 percent.

IN THE FIELD

COGENERATION is the capture and use of the waste heat produced by a generator for space and/or water heating. When this technology is employed, reciprocating engines can approach 80% efficiency. Combined cycle power generation and the use of waste heat to drive a steam turbine to produce more electricity can also be employed with these units. Reciprocating engines can be used as a prime power source, as an onsite *peaker plant* (as described in Section 26.3), or as a *combined heat and power (CHP) source*. A lean-burn 1.25 MW NG generator is shown in Figure 26.7.

COGENERATION The capture and use of the waste heat produced by a generator for space and/or water heating.

Gas Turbines

GAS TURBINE operation is based on the same technology as a jet engine. The basic gas turbine consists of a *compressor*, a *combustion chamber*, and *turbine*, as shown in Figure 26.8.

GAS TURBINE A power generating machine, capable of burning any number of fuels, that in its simplest configuration consists of a compressor, a combustion chamber, and a turbine.

FIGURE 26.7 A lean-burn 1.25 MW NG generator.
Source: Pipeline and Gas Journal. June 2002, p. 34. Used by permission of Oildom Publishing and Foster Printing.

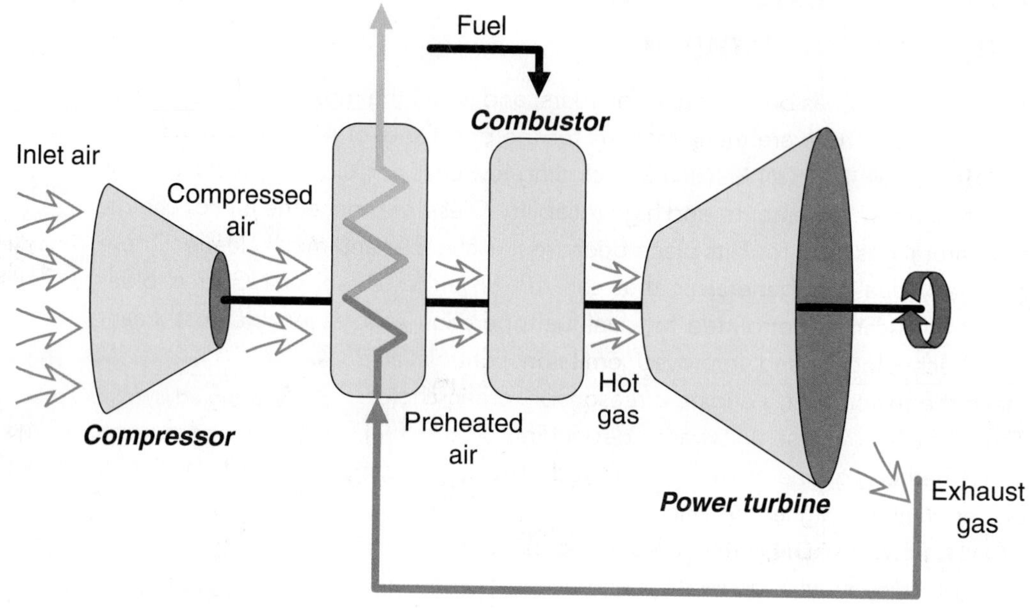

FIGURE 26.8 A basic gas turbine.

BIOMASS A term used in reference to the production of methane from the decay of vegetation, plant matter, or agricultural waste.

BRAYTON CYCLE The process of burning fuel with compressed air and then using the expanding gases to drive a turbine.

Gas turbines can burn a variety of fuels, from natural gas to methane produced using **BIOMASS** (vegetation, plant matter, or agricultural waste). Industrial gas turbines can generate from tens of kilowatts to hundreds of megawatts. They can reclaim heat and use it for combined cycle operation. They can also preheat the compressed air to increase combustion efficiency. *Intercoolers* added to the air intake can further increase combustion efficiency. This process of burning fuel with compressed air and then using the expanding gases to drive a turbine is called the **BRAYTON CYCLE**.

Gas turbines have advantages and disadvantages when compared to internal combustion engines. On the plus side, they are lighter, have a smaller footprint (require less square footage), have less down-time for maintenance, and have lower emissions. They can also burn a wider range of fuels than internal combustion engines, and new nitrogen-oxide emission-control technology promises to make them even greener. On the negative side, they take longer to power up and power down, and they are less efficient for outputs below 2 MW. Gas turbines are particularly well suited for large cogeneration applications.

A much smaller version of the industrial gas turbine is the *microturbine*. A **MICROTURBINE** works on the same principles as its larger cousin. It has output capabilities ranging from tens of kilowatts to a few hundred kilowatts. Microturbines are not quite as efficient as larger industrial units; but as demand increases, new technologies should bring costs down and efficiencies up.

MICROTURBINE A much smaller version of the industrial gas turbine that can provide outputs from tens of kilowatts to a few hundred kilowatts.

Microturbines are becoming popular in a variety of applications. For example, one major natural gas producer is diverting raw natural gas from the wellhead to power microturbines. The microturbines, in turn, supply the electrical power for all the onsite pumps and other equipment. Microturbines can burn even poor-quality fuels such as unrefined NG, low-methane biogas, and even vegetable oil. An industrial gas turbine and an array of 65 kW microturbines are shown in Figure 26.9.

(a)

(b)

FIGURE 26.9 An industrial gas turbine and an array of 65-kW microturbines.
Source: (a) Courtesy of Siemens.
Source: (b) Courtesy of Capstone Turbine Corporation.

Photovoltaic Cells

Photovoltaic (PV) cells convert solar energy (sunlight) into electrical energy. They produce no greenhouse gases or particulates, are quiet and reliable, and require very little maintenance. As a small-scale power generating alternative, PV cell arrays can be scaled to just about any size, depending on customer requirements. On the negative side, PV cells are still expensive and, of course, work only when the sun shines. However, costs are going down rapidly as new technologies are developed and mass production reduces costs. Currently, PV installations are increasing at an annual rate of close to 50% worldwide.

n-TYPE MATERIAL A semiconductor wafer treated with a five-valence-electron substance that causes it to have an excess of conduction band electrons.

p-TYPE MATERIAL A semiconductor wafer treated with a three-valence-electron substance that causes it to have an excess of valence-band holes.

A very basic description of PV cell operation follows: The base material for PV cells, is silicon—a semiconductor that has four valence electrons. Two wafers of silicon are treated with two different elements called *impurities*. One impurity changes silicon into an **N-TYPE MATERIAL** that has excess free electrons. The other converts silicon into a **P-TYPE MATERIAL** that has excess valence-band holes. When light hits the component, some photons release energy in the n-type material. Others pass through to the p-type material, generating free electrons that pass to the n-type material. When this occurs, an electron imbalance is created between the two materials. This imbalance causes electrons to flow through the lamp (load) and back to the p-type material. This flow of electrons (DC) continues as long as light hits the component.

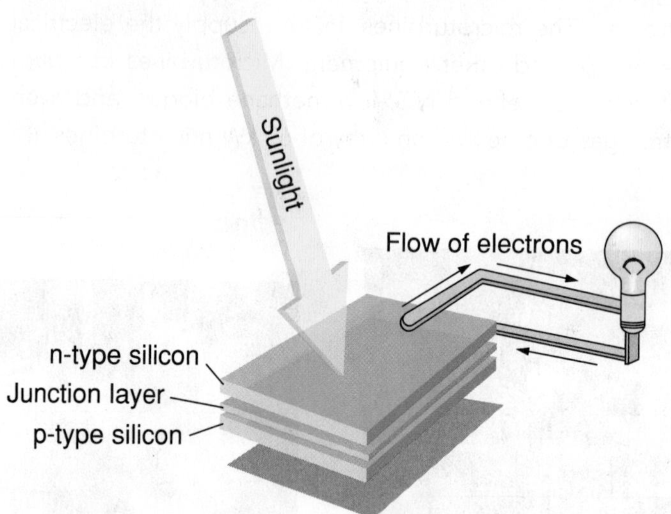

FIGURE 26.10 Photovoltaic cell operation.

BUILDING-INTEGRATED PV (BIPV) Integrating PV cells into building materials such as roofing shingles, driveways, and road paving.

Most current PV installations use crystalline silicon modules, but newer thin-film PV modules are becoming more common. Thin-film modules (which work on the same basic principles) use less material and are expected to be easier to mass produce. Most importantly, they offer the opportunity to integrate PV cells into building materials, such as roofing shingles or even driveways and road paving. This technology is referred to as **BUILDING-INTEGRATED PV (BIPV)**. In one experimental application, 45 miles of highway in Idaho had solar cells installed in the road surface. It is possible that in the future, a building could have all or part of its electricity needs met by PV cells in its roof and parking lot.

PV power generation is growing rapidly. It is a very green source of electricity, but has some inherent problems. Most obvious is that it works only when the sun is shining, so it has geographic, and possibly seasonal, limitations. Also, it produces direct current (DC), so an *inverter* is required if PV-generated power is to be integrated into a building's existing power distribution system.

INVERTER A circuit that converts DC to AC.

An **INVERTER** is a circuit that converts DC to AC. Inverters can cost as little as a hundred dollars or as much as several thousand dollars, depending on the inverter's current capacity and application. And like any device, they are not totally efficient. In some applications, efficiencies can approach 90%; but if excess PV power is sold back to the grid, losses can approach 50%. Interfacing with the power grid is discussed later in the chapter.

We'll close our discussion on PV power with a couple of government policy initiatives that are intended to help increase the use of PVs specifically, and DG power in general. **NET METERING** is a government policy that allows small-scale power suppliers to be paid for at least some of the power that they produce. Since 2005, federal law has required that all public utilities supply net metering upon request. A **FEED-IN TARIFF (FIT)** is a government policy that requires utilities to buy electricity from renewable energy sources at above-market rates. This policy has been adopted, in one form or another, by dozens of countries and in some U.S. states. Both of these policy initiatives are expected to help increase PV installations in the future.

NET METERING A government policy that allows small-scale power suppliers to be paid for at least some of the power that they produce.

FEED-IN TARIFF (FIT) A government policy that requires utilities to buy electricity from renewable energy sources at above-market rates.

Fuel Cells

FUEL CELLS hold a lot of promise, although they still have some problems. There are currently several thousand fuel cell installations worldwide. On the plus side, fuel cells are very quiet, they produce almost no greenhouse gases or other contaminants, and, if waste heat is captured and used, efficiencies can approach 80%. On the negative side fuel cells are expensive, since platinum is one of their critical components. Costs are expected to decrease over time as the technology evolves.

FUEL CELL A device that generates power through a chemical reaction involving hydrogen.

Fuel cells generate power through a chemical reaction involving hydrogen. There are several different types of fuel cells. They are named based on the electrolyte they use—phosphoric acid, alkaline, proton exchange membrane (PEM), solid oxide (SOFC), and molten carbonate (MCFC). PEM cells are the type favored by most hybrid-electric car manufacturers. They can respond very quickly to changes in load demand, but they are currently limited to outputs under 10 kW. The structure and operation of a PEM fuel cell are illustrated in Figure 26.11. SOFCs and MCFCs are slower to respond to changes in load, but they can produce tens of megawatts and more.

Balance of System

There are three subsystems in every DG installation: the *generator*, the *load*, and the rest of the equipment required to make the whole thing work. This third subsystem is usually referred to as the *balance of system* (BOS). Exactly what makes up the BOS is determined by the type of generation and whether it is connected to the grid. As you might expect, stand-alone systems have the simplest BOS requirements. The most complex BOS is found in installations that take power from the grid, and sell power back to the grid. Utilities usually have very strict protocols about voltage stability, frequency, and harmonic content. Some of the BOS requirements for a DG installation are:

- An inverter (if the installation is PV)
- A voltage regulator to maintain constant voltage
- Switching systems (if the system is grid connected)

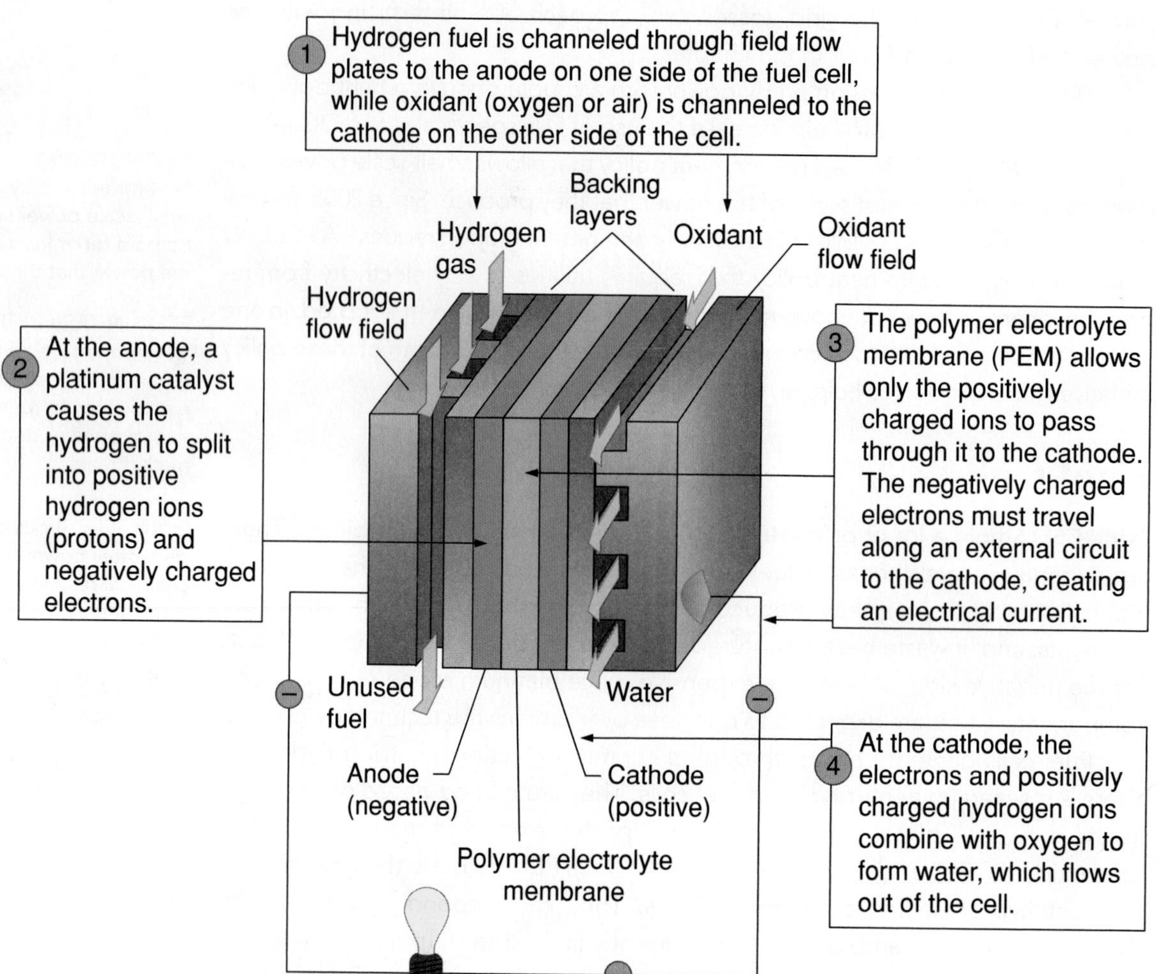

Proton exchange membrane fuel cell
1 Hydrogen fuel is channeled through field flow plates to the anode on one side of the fuel cell, while oxidant (oxygen or air) is channeled to the cathode on the other side of the cell.
Backing layers
Hydrogen gas
Oxidant
Oxidant flow field
Hydrogen flow field
2 At the anode, a platinum catalyst causes the hydrogen to split into positive hydrogen ions (protons) and negatively charged electrons.
3 The polymer electrolyte membrane (PEM) allows only the positively charged ions to pass through it to the cathode. The negatively charged electrons must travel along an external circuit to the cathode, creating an electrical current.
Unused fuel
Water
Anode (negative)
Cathode (positive)
Polymer electrolyte membrane
4 At the cathode, the electrons and positively charged hydrogen ions combine with oxygen to form water, which flows out of the cell.

FIGURE 26.11 PEM fuel cell structure and operation.
Source: Redrawn from FuelEconomy.gov.

- Power conditioning to keep the AC as pure as possible
- Power storage (batteries, etc.) if required
- Safety equipment, such as surge and overvoltage protectors and disconnects

This is only a very limited list. One supplier offers over 2000 BOS components.

PROGRESS CHECK

1. List and describe the benefits of distributed generation.
2. Define the following terms:
   a. Cogeneration
   b. Combined heat and power
3. Define and describe the Brayton cycle.
4. What fuels can a microturbine burn?
5. Explain the basic functioning of a photovoltaic cell.
6. Define building-integrated PV, and explain its advantages.
7. What fuel does a fuel cell utilize?
8. What is the balance of system?

26.3 A SMARTER, GREENER GRID

In Chapter 25 you were introduced to the electrical grid that distributes power to businesses, industry, and homes across the country. In this section we look at the grid from an environmental perspective.

The U.S. electric transmission grid currently consists of approximately 9,200 generating stations and over 300,000 miles of transmission lines. It is the largest interconnected machine on the planet—and it has some problems. For example:

- The average power plant in the United States was built using technologies from the 1960s (or earlier).
- The average age of the substation transformers in the United States is 40 years, which is beyond their life expectancy.
- Power outages currently cost consumers about $150 billion each year, and it is getting worse. There were 41% more outages in the second half of the 1990s than in the first half.
- Investment in the grid has dropped to about half what it was in the mid 1970s.

With the increasing demand for electrical power and growing environmental concerns, the DOE has proposed sweeping changes to the power grid. A full discussion of these changes is far beyond the scope of this text. Even so, we will examine two of these changes:

- Improvements to existing grid components, primarily transformers and conductors.
- The **SMART GRID**—an interactive grid in which the power producer, distributor, and end user are in constant real-time communication with each other.

SMART GRID An interactive grid in which the power producer, distributor, and end user are in constant real-time communication with each other.

New Technologies

Reducing any loss in the grid reduces the need to generate electricity, and thus lowers greenhouse gas emissions. Several new technologies are emerging. We will begin by discussing one that is already in wide use.

Amorphous-Core Transformers

The core material in most transformers is a type of silicon steel, known in the industry as **COLD-ROLLED GRAIN-ORIENTED (CRGO)** silicon steel. **AMORPHOUS-CORE TRANSFORMERS (ACTs)** use silicon steel cores that have been manufactured to reduce core-related losses. When molten silicon steel is cooled very quickly, it retains a chemical structure that resembles glass. This structure reduces hysteresis and eddy current losses. Rather than using stacked laminations, transformers with amorphous cores use ribbons of steel wound into a rectangular toroid-shaped core as shown in Figure 26.12.

COLD-ROLLED GRAIN-ORIENTED (CRGO) A type of silicon steel that is used in transformer cores.

AMORPHOUS-CORE TRANSFORMERS Transformers with cores of specially treated silicon steel wound into a rectangular toroid shape.

ACTs rated at 2500 kVA and higher have been on the market since the mid-1980s, and several hundred thousand are currently in use across the country. They are capable of lowering no-load core losses by up to 70%. For example, a typical

FIGURE 26.12 Amorphous cores.
Source: Courtesy of Hitachi Metals America—Metglas Inc.

250 kVA CRGO transformer has a no-load loss of about 100 W. An equivalent ACT has a no-load loss of about 25 W.

ACTs can cost 15% to 40% more than equivalent CEGOs, but utilities have found that savings in power generation costs more than offset this initial investment over the life of the transformer.

Superconductors

You may think this topic sounds like science fiction, but the U.S. DOE does not. The Office of Electricity Delivery and Energy Reliability (a department within the DOE) supports the use of *high-temperature superconducting* (HTS) power cables, motors, transformers, and generators to reduce energy losses in power line transmission and help lower related greenhouse gas emissions.

SUPERCONDUCTORS Man-made alloys that can pass electric current with virtually zero resistance when operated at extremely low temperatures.

SUPERCONDUCTORS are man-made alloys that can pass electric current with virtually zero resistance, but only if they are kept at very cold temperatures. Here is a brief history of the development of superconductors:

CRITICAL TEMPERATURE The maximum temperature at which superconductivity is maintained.

- **1970–1986:** During this period, the only superconductor materials available were *low-temperature superconductors (LTS)*. They had to be kept at temperatures below 20°K (−253°C). They are still used, but only in the laboratory, and only for superconducting magnets.
- **1986–2000:** In 1986 the first high-temperature superconductors were produced. They are referred to as *first generation high-temperature superconductors* (1G HTS). They have a **CRITICAL TEMPERATURE** (the maximum temperature at which superconductivity is maintained) of 115°K (−158°C). That may not seem like a "high" temperature, but it makes a huge difference in terms of practical applications. Low-temperature superconductors needed to be cooled with liquid helium, while 1G HTS could be cooled with liquid nitrogen at *one-fiftieth* the cost. This increased practicality

somewhat, but they were still very expensive (having silver in their makeup) and they could not be manufactured in long lengths.

- **1987–present:** Production of *second generation high-temperature superconductors* (2G HTS) began in 1987 and continues today. Here are the major improvements in 2G HTS:
  - They are more flexible.
  - They are much cheaper to produce (having 23 times less silver and 40 times less HTS material than 1G HTS).
  - They can be manufactured in usable lengths in an automated continuous process.

Currently there are several companies that offer 2G HTS wire commercially. Figure 26.13 illustrates one of the stages in 2G HTS wire production. The machine shown is the MOCD (metal organic chemical vapor deposition) system, which adds the all-important superconductor layer. Current manufacturing techniques allow for the production of 2 km long single-piece lengths.

- **2008–present:** Research is already underway on 3G HTS. The focus is on lowering costs and increasing critical current capacity. In 2008, a totally new class of 2G HTS was discovered, which is iron-based instead of copper-based. Scientists hope this new material will help them better understand the properties of superconductivity (no one really understands how it works), and it could lead to the production of the ultimate HTS—room temperature superconductors.

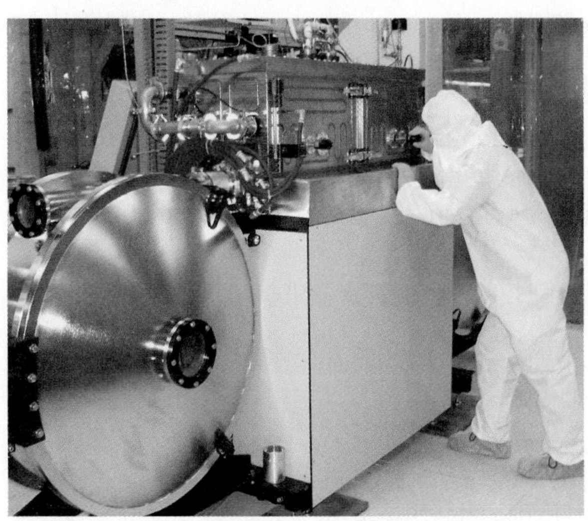

FIGURE 26.13 A MOCD (metal organic chemical vapor deposition) system.
Source: Courtesy of SuperPower Inc.

Benefits of High-Temperature Superconductors

The most obvious benefit of HTS is their near zero resistance. HTS transmission lines would have virtually no I^2R losses. The same would be true of HTS transformers and motors. Further, the size of transmission lines, transformers, and motors would decrease dramatically due to the increased current capacity of wires made of HTS. A 2G HTS conductor with a cross-sectional area of 1 cm can handle 300 kA of current (compared to 1 kA for standard copper wire). Also, wires made of HTS are installed in small underground trenches, so overhead wires and towers would be eliminated. This would dramatically reduce the footprint of transmission corridors and improve grid security.

CRITICAL CURRENT RATING The maximum current that a given HTS can pass before it loses superconductivity.

FAULT-CURRENT LIMITER (FCL) A circuit that is used to lower current in the event that a fault occurs.

There is another characteristic of 2G HTS material that is very useful. All HTS materials have a critical current rating. The **CRITICAL CURRENT RATING** is the maximum current that a given HTS can pass before it loses superconductivity. If this rating is exceeded, the material immediately becomes resistive. This makes 2G HTS perfect for use as a fault-current limiter. A **FAULT-CURRENT LIMITER (FCL)** is used to lower current in the event that a fault occurs. Due to its zero resistance characteristics, an HTS limiter is completely invisible to the system until there is an overcurrent event. When a fault current occurs, the HTS material very quickly increases in resistance, "quenching" the overcurrent. This provides protection to the grid components that are upstream from the fault.

Is This For Real?

By now you may be wondering what this has to do with working as an electrician. Here are some examples of current projects using 2G HTS wire:

- In June, 2007, the U.S. Department of Energy spent $9 million to help fund a prototype HTS installation for the Long Island Power Grid.
- In February of 2008, the DOE announced the completion of the first 2G HTS wire installation on the grid. It was a 350-meter run between two substations in Albany, New York.
- In 2007, the DOE spent $29 million on three FCL development projects.
- In July of 2009, a Massachusetts company announced it has been contracted to install a 2G HTS feed from a large-scale wind farm on the New England coast.
- In 2009, the DOE announced joint funding for four HTS FCL installations in California and Ohio.
- As of 2009 there are currently over 20 ongoing HTS installation projects worldwide.

As you can see, 2G HTS is already becoming part of the U.S. power grid. As such, it is quite possible that you will, at some point, find yourself working with 2G HTS.

SUPERCONDUCTING MAGNETIC ENERGY STORAGE (SMES) Storing electrical energy in the massive magnetic field of a closed superconducting inductor.

There is another very important use for HTS wire. As you know, electrical energy cannot be easily stored; it has to be used as it is generated. This is one of the major problems with inconsistent energy producers like solar and wind power. The Department of Energy supports a concept called **SUPERCONDUCTING MAGNETIC ENERGY STORAGE (SMES)** for storing electrical energy inductively. SMES is basically a very large air-core inductor buried underground. The energy is stored in the massive magnetic field of a closed superconducting inductor.

SMES is currently being used by some power utilities for power conditioning. There are several commercial 1 MW installations being used by plants that require extremely "clean" power, such as microchip manufacturers. In Wisconsin, several SMES units have been installed to improve grid stability.

There are also several large-scale experimental energy storage projects. One such SMES project is capable of supplying 10 MW of power for 2 hours. SMES is extremely efficient (over 95%) and extremely reliable because the critical parts are stationary. SMES may be one of the key components needed to make renewable energy sources, such as solar and wind, a more practical generating option.

Superconductors: The Bottom Line

During your professional careers, you will very likely be exposed to superconducting technology. The research in this field is progressing at an amazing rate, and practical HTS transmission lines, transformers, motors, and energy storage systems may soon become a commercial reality. You may be skeptical, but consider this: What do you think a mechanic would have said twenty years ago if you told him that he would have to know how to replace a GPS display in a Buick, or download software to update a car's engine management system?

The Smart Grid

The Smart Grid is seen by the U.S. DOE as one of the major technological changes needed to meet our future energy requirements while lowering the environmental impact of energy production. This is a huge undertaking, comparable to the development of the Internet or the construction of the interstate highway system. It is a process that will continue over time.

The Smart Grid will use digital communication technology to allow information to flow between the power generator, the distributor, and the consumer. There are many hurdles to overcome. Not only must new technology be developed, protocols set, and standards enforced, but the conflicting interests of a range of stakeholders must be reconciled as well. On top of that, all of these changes must occur while the grid runs at full capacity. We begin our discussion of Smart Grid development by looking at one of the biggest problems with current energy consumption.

Peaker Plants

Energy consumption is anything but constant. For example, a typical electricity demand profile for a 24-hour period is illustrated in Figure 26.14. Demand also changes with the seasons, and there are always unanticipated events. For instance, one day in Texas—a state with a lot of wind power capacity—the wind unexpectedly died out. The grid lost 1.3 GW of generating capacity in just 3 hours. We'll come back to this event later. The point is that power generation must meet consumer demand *all the time*.

With current technology, electrical energy cannot be stored on a large scale; it must be produced as needed. For this reason, there are many peaker plants across the country. **PEAKER PLANTS** generate electricity in response to peak load demands.

PEAKER PLANTS Power plants that generate electricity only in response to peak load demands.

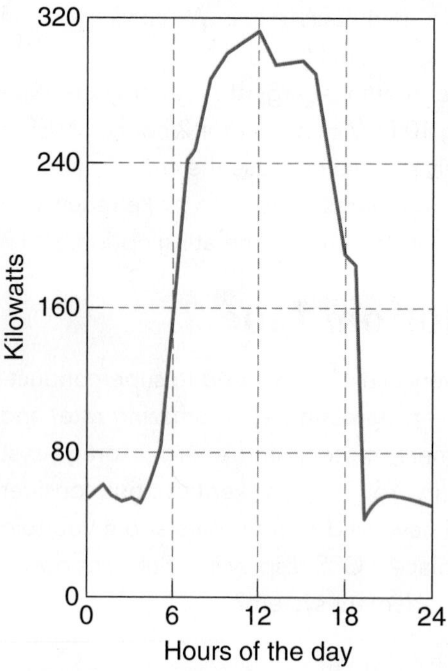

FIGURE 26.14 A typical electricity demand profile for a 24-hour period.
Source: Redrawn from the U.S. Department of Energy.

There are several problems with peaker plants. First, though not used all the time, they must still be maintained—along with their entire transmission infrastructure. (The DOE estimates that 10% of all the generating capacity, and 25% of the distribution infrastructure, is used only 5% of the time!) Peaker plants also have to get their fuel on the more expensive spot market, and many are older, less efficient, and higher polluting plants.

The Smart Grid will never completely eliminate peaker plants, but it is hoped that giving the grid real-time response capabilities and helping consumers make more intelligent power consumption choices can smooth out the demand profile, thereby reducing the need for peaker plant assets.

Smart Metering

SMART METERS Two-way communication devices that allow the electricity provider to monitor energy consumption in real time and price the electricity based on its real-time cost.

SMART METERS are two-way communication devices. They allow the electricity provider to monitor energy consumption in real time and price the electricity accordingly based on its real-time cost. They also allow consumers to track their own power consumption and follow the real-time cost of electricity. Smart meters provide a financial incentive to lower consumption during periods when the cost of electrical energy is at its peak.

By 2010, every home and business in Ontario, Canada will have a smart meter installed. Figure 26.15 shows one of several models of smart meters that are available.

FIGURE 26.15 A smart meter.
Source: Courtesy of Ontario's Independent Electricity System Operator.

Smart meters are only one component of the smart grid, but they are one of the first pieces of the puzzle that will be put in place. As one expert said, "*For a smarter grid, there is no silver bullet; there is instead silver buckshot.*" Many different technologies will need to be employed to make it work.

Electricity Today and Tomorrow

Consider how we buy electricity today. We get a bill once a month that is actually an estimate based on the amount of energy we consumed in the past. Most of us have no idea of what electrical energy actually costs. (One senior government official compared this to driving a car with no fuel gauge and buying gasoline without knowing the price.) If consumers are given more information, they have more choices. If *day-ahead pricing*, or even *hour-ahead pricing* is used, they will almost certainly change their consumption habits and lower peak demand.

Let's look a little further into the future. Wind and solar arrays will be seen on more commercial and residential buildings. Plug-in hybrid cars will also be more common. If individuals and businesses know the real-time price of electricity, they can choose when to sell their wind and solar electricity back to the grid for the highest price, and when to charge their plug-in cars at the lowest price.

There is a huge benefit for electricity providers as well. By helping energy consumers change their buying habits, that peak consumption graph in Figure 26.14 could be significantly leveled out. This means infrastructure construction can be pushed back, giving the utility time to introduce more efficient technologies. The DOE believes that the Smart Grid will have a major impact on capital costs for electricity providers. In one of their publications they make the following statement:

> *"For a smarter grid to benefit society, it must reduce utilities' capital and/or operating expenses today—or reduce costs in the future. It is estimated that Smart Grid enhancements will ease congestion and increase utilization (of full capacity), sending 50% to 300% more electricity through existing energy corridors."*

The Smart Grid and Distribution

Remember the incident in Texas that we discussed earlier, where the grid lost 1.3 GW of generating capacity in just 3 hours? Here is the rest of the story: As it turned out there was the beginning of a Smart Grid in this region. When the wind died out, the two-way communication technology between supplier and consumer described here was in place. An emergency response was initiated, and large industrial and commercial users were able to bring their distributed-energy generators on line and restore most of the lost capacity in just 10 minutes!

This is just one example of Smart Grid technology in action. In the future, it is envisioned that the Smart Grid will be able to control the distribution of electricity in real time. A document produced by the DOE presents their vision of how the Smart Grid will work by about 2020:

> *"An unusually destructive storm has isolated a community or region. Ten years ago, the wait for the appearance of a utility's "trouble trucks" would begin. The citizens would remain literally in the dark, their food spoiling, their security compromised and their families at risk.*
>
> *Instead, with full Smart Grid deployment, this future community is not waiting. Instead, it's able immediately to take advantage of distributed resources and standards that support a Smart Grid concept known as "islanding." Islanding is the ability of distributed generation to continue to generate power even when power from a utility is absent. Combining distributed resources of every description—rooftop PV (solar), fuel cells, electric vehicles—the community can generate sufficient electricity to keep the grocery store, the police department, traffic lights, the phone system and the community health center up and running. While it may take a week to restore the lines, the generation potential resident in the community means that citizens still have sufficient power to meet their essential needs."*

One Final Note

The Smart Grid is not wind and solar power, or cogeneration, or advanced energy storage, or green buildings. The Smart Grid is the interface that will let all of these new greener technologies work together seamlessly. It is much like a computer's operating system. We don't use the operating system directly, but it allows us to email, write documents, edit graphics, watch videos, or listen to music. It also supports any new applications that come on the market, making our computer a more powerful and flexible tool. The Smart Grid is a green technology enabling engine.

PROGRESS CHECK

1. List four current problems with the electrical grid.
2. What is an ACT and how will it lower grid losses?
3. Name and describe the differences between the three types of superconductors.
4. Define the following terms:
   a. 2G HTS
   b. Critical current rating
   c. Critical temperature rating
   d. Fault-current limiter
   e. SMES
5. What technology will the Smart Grid employ for communication between producers, distributors, and consumers?
6. What are peaker plants and why are they a concern?
7. What are smart meters and what are their benefits?

26.4 GREEN BUILDINGS

The *U.S. Green Building Council* (USGBC) is a nonprofit organization focused on the construction of green buildings and the retrofitting of older buildings to make them more energy efficient. The USGBC estimates that 65% of the electrical energy used in the United States is consumed by building operations. To this end they have developed a green rating system called *The Leadership in Energy and Environmental Design* (LEED) *Green Building Rating System*. LEEDS® is an internationally recognized certification system designed to help building owners and communities increase energy and water efficiency, reduce CO_2 emissions, and improve indoor environments.

As an electrician (or electrical technician) there are many new technologies coming on stream that you will likely be expected to install and maintain. One of the world's largest electrical supply companies carries more than a dozen green products or green technology applications to help reduce energy use in buildings. The following examples are illustrated in Figure 26.16:

- **Automatic Transfer Switches** These switches automatically switch between utility power and alternative energy sources like solar panels, wind turbines, and fuel cells.
- **Green Switchgear** Much of the construction of the switchgear manufactured today employs environmentally hazardous materials. For example, sulfur hexafluoride, a greenhouse gas 20,000 times more harmful than CO_2, is commonly used in the production of insulation. Companies are finding ways to manufacture switchgear that avoid the use of such materials.
- **Power Factor Correction** Sophisticated power factor correction systems and harmonic filters reduce energy consumption and increase equipment life.

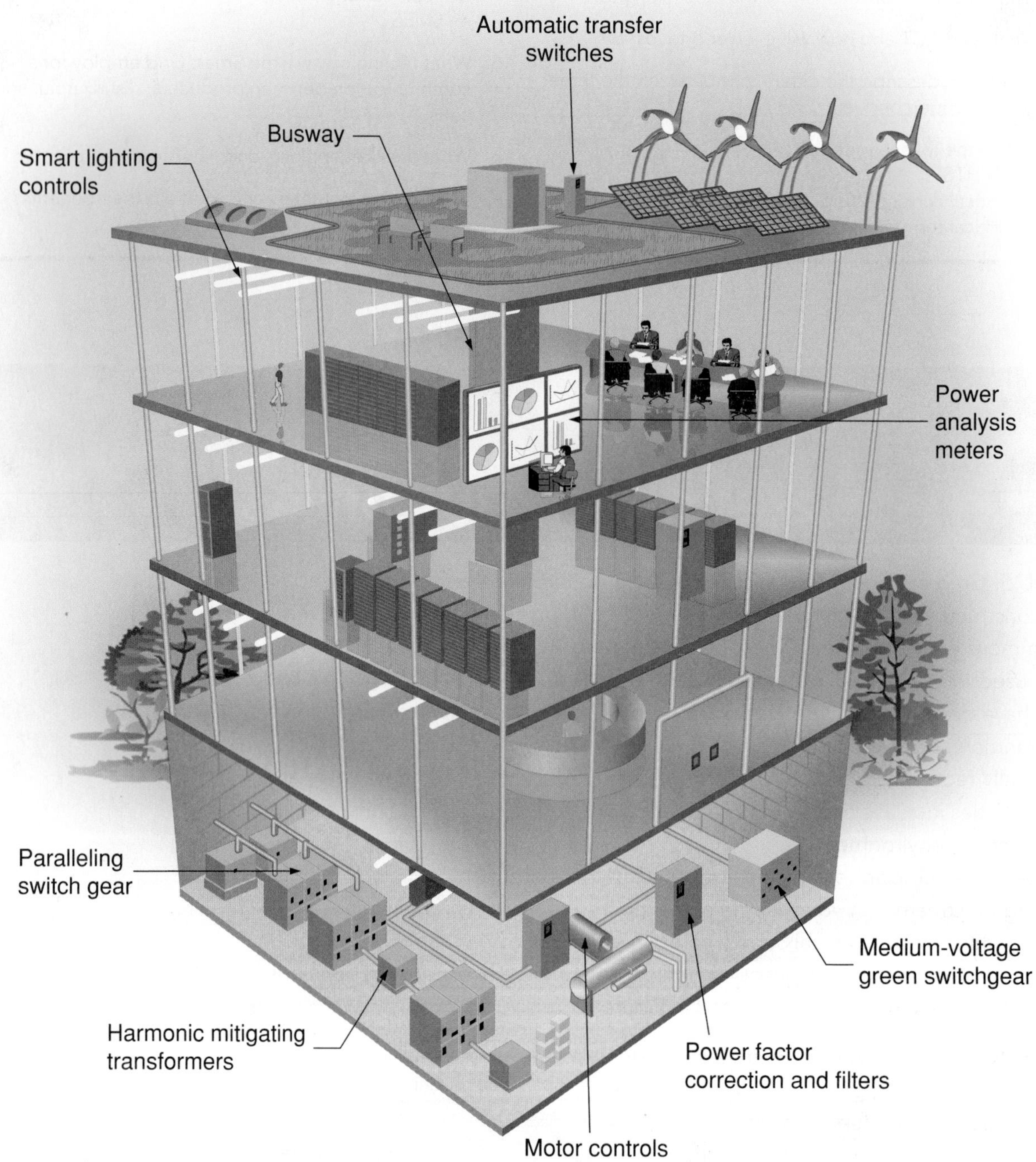

FIGURE 26.16 Green building technology.

- **Harmonic Mitigating Transformers (HMTs)** Most large buildings have a number of transformers in their power distribution network. Lower harmonic content means more efficient power consumption.
- **Variable Frequency Drive (VFD) Motor Controls and Soft-start Motor Controls** By matching motor speed to changing load demands (like for

elevators and HVAC) energy savings of 10 to 50% can be realized. Soft-start motor controls lower the power consumption of motors on start-up and increase motor life.

- **Intelligent Lighting Controls** Lights that turn off when a room is empty, and lower artificial light levels when natural light increases, can reduce lighting-related energy consumption by up to 30%. We will talk more about lighting in the next section.
- **Paralleling Switchgear** This type of switch gear manages the transition between on-site and utility power, and supports the use of alternative technologies like combined heat and power.
- **Busway** **BUSWAY** is defined by the National Electrical Manufacturers Association (NEMA) as "*a prefabricated electrical distribution system consisting of bus bars in a protective enclosure, including straight lengths, fittings, devices, and accessories.*" A Busway distribution network saves energy. Busway systems sized to carry the same current as cable and conduit have lower losses, use less copper and steel, and waste less material because they are built to exact lengths. One style of three-conductor Busway distribution system is shown in Figure 26.17.

BUSWAY A prefabricated distribution system with enclosed bus bars that include straight lengths, fittings, devices, and accessories.

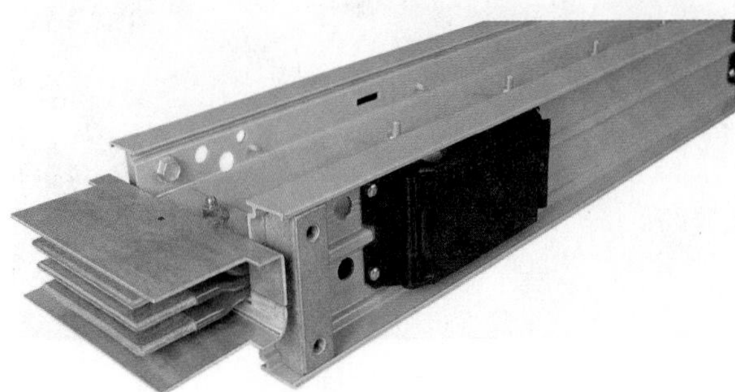

FIGURE 26.17
Source: Courtesy of GE Industrial.

- **Power Analysis Meters** Power analysis meters and their related software provide additional opportunities for lowering energy consumption. They can pinpoint power quality problems and energy inefficiencies in a building. Some can generate power consumption charts to help identify areas where energy savings can be realized.

This is only a partial list from only one company of the many green building technologies that are currently available.

PROGRESS CHECK

1. What is LEEDS®?
2. What is the function of an automatic transfer switch?
3. Explain how variable frequency drive motor controls save energy.
4. What is a Busway distribution system and what are its benefits?

26.5 GREEN LIGHTING

We are all familiar with air and water pollution—but light pollution? Yes, light pollution does exist. Astronomers are concerned about light pollution, as it restricts their ability to see the night sky. The DOE worries about light pollution because much of the energy that we turn into light is wasted. Figure 26.18 shows a NASA-generated composite view of the earth at night.

FIGURE 26.18 A NASA generated composite view of the earth at night.
Source: Courtesy of NASA.

Approximately one quarter of the world's energy consumption is used for lighting. In this section we talk about a variety of new lighting technologies that will reduce the amount of energy that we use to light our cities, homes, and commercial buildings.

Compact Fluorescent Lights

Compact fluorescent lights (CFLs) first hit the mainstream market in the early 1990s, but failed to catch on for several reasons, including relatively high cost and poor light quality (color temperature). Current models of CFLs seem to have overcome these early concerns. A number of CFL bulbs are shown in Figure 26.19. The bulbs in Figures (a) and (b) replace standard incandescent bulbs. Those in Figures (c) and (d) replace floodlights and halogen floods, respectively.

CFLs last seven to ten times longer than incandescent bulbs and use less energy for the same light output. One manufacturer of CFLs estimates that replacing just one incandescent bulb with a CFL in every home in the United States would save $600 million a year in energy costs nationally.

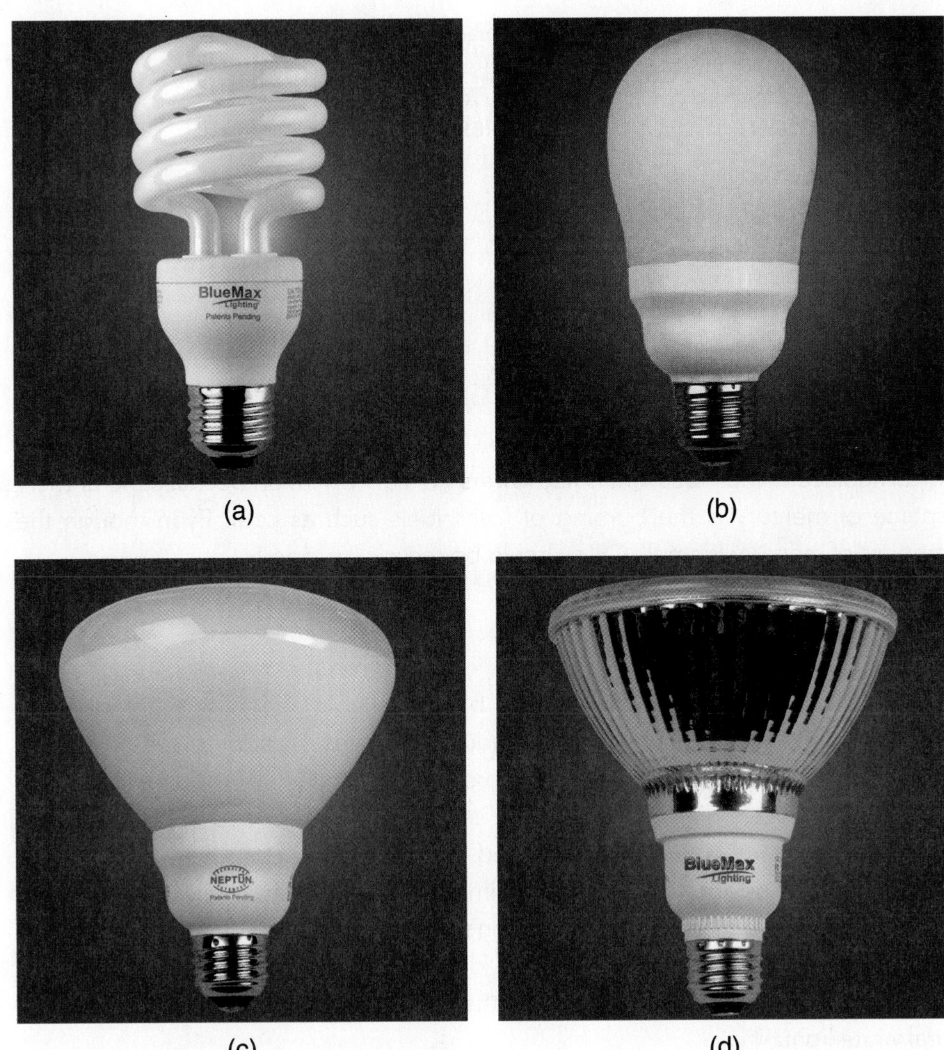

FIGURE 26.19 Common CFL bulbs.
Source: Courtesy of Full Spectrum Solutions.

Table 26.1 compares the power consumption of incandescent bulbs and CFLs for approximately the same light output. Incandescents produce about 14 lumens of light energy per watt of electrical power consumed. CFLs produce about 60 lumens per watt.

CFLs come in several forms, but all of them contain two main parts: one or more phosphor-filled tubes and a *ballast*. A ballast is a current-limiting circuit. Modern CFLs use electronic ballasts. The ballast can be integrated with the tube, like those in the fluorescent lamps commonly used in residential installations, or it can be separate, so that you only have to replace the tube. CFLs can be powered by DC or AC current. DC-powered CFLs are very useful in "off the grid" applications, such as small-scale solar, or even battery-powered installations.

One concern that has been raised with CFLs is that, like all fluorescent lamps, they contain mercury—a very hazardous element. CFLs contain only about 500 milligrams of mercury, about the size of the tip of a ballpoint pen. In fact, using

Table 26.1 • Power Consumption Comparison Between Incandescent Lamps and CFLs

INCANDESCENT	COMPACT FLUORESCENT
40 watts	9 watts
60 watts	13 watts
75 watts	20 watts
100 watts	26 watts
150 watts	39 watts

CFLs in place of incandescent bulbs lowers mercury pollution because the primary source of mercury is the burning of fossil fuels such as coal. Even though the amount of mercury is small, CFLs should be recycled properly.

LED Lighting

LIGHT-EMITTING DIODE (LED) A semiconductor component that emits light when the proper voltage is applied.

DOPING The process of adding impurity elements to pure semiconductors to determine their operating characteristics.

LIGHT-EMITTING DIODES (LEDS) have been available as practical components since the early 1960s. LEDs are semiconductor devices. By adding certain impurities to pure semiconductors (through a process called **DOPING**) and joining different types of doped material together, it is possible to produce a component that generates light when a voltage is applied.

When first developed, LEDs were limited to low-power applications (like the indicator light in a stereo), and could generate only certain frequencies (colors) of light. Then, higher-powered LEDs were developed and used in applications like traffic lights. It is only recently, however, that LEDs have been able to generate natural white light.

LED lighting is just beginning to become a practical light source. Several current styles of LED lights are shown in Figure 26.20. The lights in Figures (a) and (b) replace standard incandescent screw-in bulbs. The light in Figure (c) is a replacement for a standard fluorescent tube, and the one in (d) is an LED strip light.

LEDs use about 40% less power than comparable CFLs and last six to seven times longer. Unfortunately, they also cost much more than CFLs; however, costs are coming down as new manufacturing technologies are developed. LEDs are also very physically rugged as they have no filaments or gas-filled tubes. LEDs are used in a variety of applications, such as strip lighting, conventional residential space lighting, and a variety of outdoor and commercial lighting.

One characteristic of LEDs is that they are *point-source lights*. This means that LED lights are made up of multiple individual LED elements. LEDs are also very directional, which makes them suitable for street lighting.

Take another look at Figure 26.18. Note how much light is wasted as it radiates out into space. LED street lights help reduce this wasted illumination. Also, the light generated by LED street lights (unlike standard high-pressure

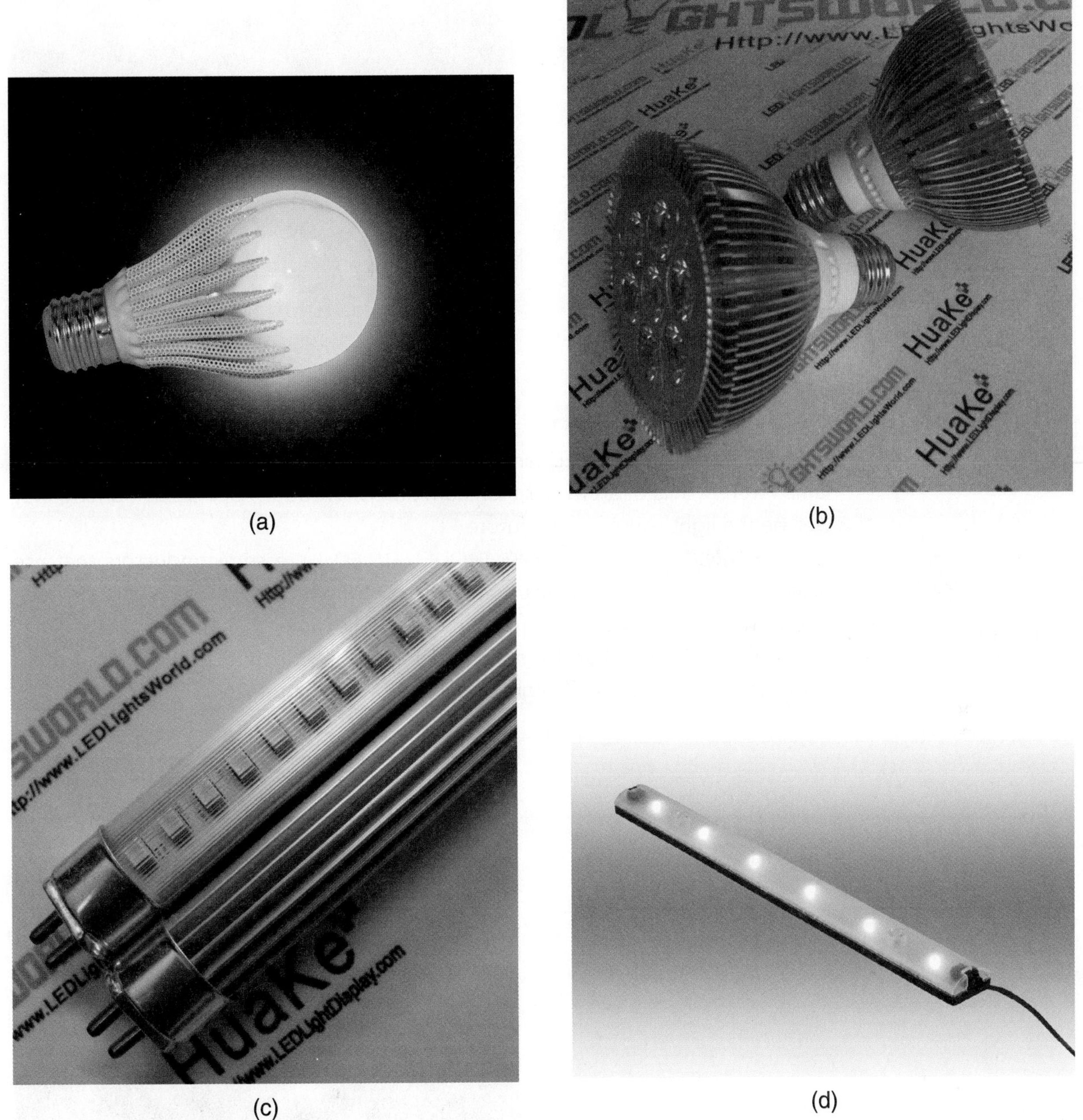

FIGURE 26.20 LED lights.
Source: (a, d) Courtesy of C. Crane Company Inc.
Source: (b, c) Courtesy of Huake Optoelectronics Co., Ltd.

sodium lights) can be dimmed or brightened based on the amount of ambient light. This can result in even more energy savings. For example, the city of Anchorage, Alaska is installing 16,000 LED street lights with an expected savings of $360,000 a year. Several other major cities, like New York and Los Angeles, have plans to install LED street lights. One type of LED street light is shown in Figure 26.21.

FIGURE 26.21 An LED street light.
Source: Courtesy of LED Roadway Lighting Ltd.

OLED Lighting

In April, 2008, the first OLED lamp was built in a limited quantity of 25 units. These lamps were works of art rather than commercial products, but OLED lighting may well be the light source of the future.

ORGANIC LIGHT-EMITTING DIODES (OLEDs) Light-producing semiconductors that contain organic materials.

ORGANIC LIGHT-EMITTING DIODES (OLEDs) are semiconductors that use organic materials in their construction. Like LED lighting, OLEDs are very efficient, but unlike LEDs, they are produced in thin sheets rather than as small bulbs. They can be very thin (less than 1 mm), are extremely flexible, and emit brilliant light of virtually any color. An OLED light panel is shown in Figure 26.22.

FIGURE 26.22 An organic LED (OLED) light panel.
Source: Courtesy of General Electric.

OLEDs may completely change the way we look at lighting a home or building. Because they are so thin and flexible, they can be attached directly to a ceiling or wall, or they could replace ceiling tiles. OLEDs can be manufactured using inkjet technology (yes, just like a computer printer) to apply the organic materials onto any suitable substrate. This means that, theoretically, the size of the panels is limitless.

Currently, OLEDs are mostly used in small applications, like camcorder and cell phone displays. Larger screens—like televisions—are on the market as well, but as with any new technology, they are expensive. Commercial lighting applications for OLEDs are still in their infancy, but at least six companies plan to begin mass-production of OLED lighting by 2012, and one as early as 2010. One market research company sees OLED sales at $1 billion by 2011.

One Final Point

The demand for electricity is growing at an accelerated rate every year, as is the concern about climate change. As technology changes, the job skills of electricians must change as well. Many of the products and technologies that we have discussed in this chapter are still in the development stage, but think about the astounding changes in communication technology that we have seen in the last decade. New Zealand is already certifying "green" master electricians. Let's end this chapter with a true story that was recently published in a Midwest newspaper.

Chris is a journeyman electrician in his mid-thirties. About five years ago he started taking courses on green technologies with the support of his local union and his employer. He was recently put in charge of supervising the electrical installations for a 35,000-square-foot green building. Among other green technologies, this building has smart lighting controls, energy storage systems, and a 550-panel solar grid that supplies 25% of the building's power. Chris saw the future coming and prepared for it, and this future is coming much faster than many of us realize.

PROGRESS CHECK

1. List two reasons why CFLs did not catch on when they first came out.
2. What is a ballast?
3. Why have LEDs not been used for lighting until very recently?
4. Why are LEDs more rugged than CFLs?
5. What are OLEDs and what are their potential benefits?

26.6 SUMMARY

Here is a summary of the major points that were made in this chapter:

Large-scale Power Generation

- The demand for electricity continues to increase.
- Electricity production is the single largest source of CO_2e emissions.
- A variety of fuels are used for electricity production.
    - Coal is the most popular at close to 40%.
    - Natural gas and nuclear are next at close to 20% each.
    - Only 8.3% comes from renewable energy sources.

- Integrated gasification combined cycle (IGCC) is a clean coal technology. (See Figure 26.3.)
  - Coal is turned into a gas called *syngas* by exposing it to high temperatures and pressure.
  - IGCC can burn a variety of fuels and produces useful chemicals such as sulfuric acid as byproducts.
- Carbon capture and storage (CCS) is an integral component in coal gasification.
  - Carbon capture is a proven technology, but carbon storage (also called sequestration) is still in the development stage.
- Natural gas (NG) is the cleanest burning fossil fuel, but is not renewable.
  - NG is found in oil wells, NG wells, and coal beds.
  - NG must be refined.
- Another form of NG is biogas.
  - Biogas comes from the decay of organic matter.
  - Communities are separating organic material from trash for use in small-scale power generation.
- Combined cycle power generation uses the hot gases from the burning of NG to drive a gas turbine, and the waste heat is used to drive a steam turbine.
- Much of the proposed increase in U.S. power generation will be NG fired.
- Nuclear power generation is a controversial subject.
  - Nuclear produces no greenhouse gases or particulates.
  - Spent nuclear fuel rods must be stored and can remain dangerous for hundreds of thousands of years.
- There are a number of other large-scale generating technologies.
  - Oil is not a clean-burning fuel, is expensive, and may disappear as a generating fuel.
  - Hydro is renewable and clean, but expensive, and it has a substantial ecological impact due to the reservoir flooding large amounts of land.
  - Wind power is clean, but inconsistent. It does have subtle environmental impacts on birds and bats, but the major concern is its effect on property values.
- Solar energy is also clean and renewable, but it produces energy only when the sun shines. It also requires a large amount of land.
  - Concentrating solar power (CSP) uses mirrors or lenses to concentrate the sun's energy and heat water to drive a steam turbine.
  - Photovoltaic (PV) arrays are becoming more popular as thin-film solar panels become more popular.
  - Concentrating photovoltaics is a combination of CSP and PV.
- The major environmental impact of PV will occur when older installations are decommissioned due to the use of dangerous materials like cadmium and arsenic in their construction.
- Wave and tidal power are two distinct technologies.
  - Wave power harnesses the rise and fall of either surface or deep water waves and converts that mechanical energy to electrical energy.
  - Tidal power dams a tidal estuary with a device called a *barrage*. The barrage traps water in the estuary when the tide rises, and channels the water through a turbine when the tide falls.
- The U.S. Department of Energy has a pilot project called Vision 21.
  - This project brings together a number of technologies to create a highly efficient power plant with virtually no environmental impact.
  - The modules of this project are to be developed by 2015.

Distributed Generation

- Distributed generation (DG) is the small-scale production of power at or near the customer's site. There are many benefits to DG.
  - Lower greenhouse gas emissions
  - Reduced infrastructure
  - Peak shaving
  - Fuel flexibility
- Reciprocating internal combustion engines are still the most common form of DG.
  - They are cost effective and reliable.
- The most common fuel is natural gas, but dual-fuel (diesel and NG) units also exist.
  - Older diesel units can be affordably converted to dual-fuel units.
- Cogeneration is the use of waste heat for space and/or water heating.
  - When used, this can boost efficiencies to close to 80%.
  - Combined heat and power (CHP) is another term for cogeneration.
- Gas turbines are based on the same technology as jet engines.
  - The basic components of a gas turbine are a compressor, a combustion chamber, and a turbine. (See Figure 26.8.)
  - Gas turbines can be used for cogeneration.
  - The process of burning fuel combined with compressed air to drive a turbine is called the *Brayton cycle*.
- Gas turbines are smaller and lighter than internal combustion generators, and have lower emissions. They also take longer to power up and down.
- A microturbine is a much smaller version of an industrial gas turbine. They come in sizes of up to a few hundred kilowatts.
  - Microturbines can run on a wide variety of fuels including raw NG, biogas, and even vegetable oil.
- Photovoltaic (PV) cells convert sunlight to electrical energy.
  - They produce no greenhouse gases and require little maintenance, but they are expensive and generate electricity only when the sun is shining.
- Costs are coming down rapidly, and PV installations are increasing at a rate of 50% per year.
- PV cells are silicon-based devices. N-type and p-type wafers of silicon are joined together to form the solar cell.
  - Photons striking the p-type wafer create free electrons that migrate to the n-type wafer. If an external current path is provided, the electrons will return to the p-type wafer, resulting in DC current. (See Figure 26.10.)
- Most current PV installations are crystalline silicon modules, but thin-film units are becoming more popular.
- Thin-film PV units can be used for building-integrated PV (BIPV).
  - BIPV is the integration of solar cells into common building materials like roofing shingles, or even paving.
- Solar panels produce DC current, so an inverter is required to integrate solar-generated power with a standard AC distribution system.
  - An inverter is a DC-to-AC converter circuit.
- There are two government policy initiatives that are designed to encourage the use of PV in particular and DG in general.
  - Net metering allows small-scale power suppliers to be paid for some of the power they produce.

  - A feed-in tariff (FiT) forces utilities to pay for energy from renewable sources at above-market rates.
- Fuel cells are very efficient and produce almost no pollutants, but they are still expensive as platinum is used in their construction.
  - Fuel cells use a chemical reaction rather than combustion to produce power. (See Figure 26.11.)
  - Fuel cells are named based on the type of electrolyte they use.
- The balance of system (BOS) is the third subsystem of a DG installation, with the generator and the load being the other two.
  - BOS components are things like inverters, voltage regulators, switching systems, power conditioning and storage, and safety equipment.

A Smarter, Greener Grid

- The U.S. grid consists of 9,200 generating stations and over 300,000 miles of transmission lines. It is the largest interconnected machine on the planet.
- The grid has a number of problems such as
  - The age of the power plants and components like substation transformers.
  - Increasing power outages that cost the economy hundreds of billions of dollars a year.
  - Decreasing investment in grid infrastructure.
- Amorphous core transformers (ACTs) use special core materials to lower eddy current and hysteresis loss.
  - ACTs can lower no-load losses by up to 70%, but they cost 15% to 40% more than standard cold-rolled grain-oriented (CRGO) core transformers.
  - This increased cost is recovered through energy savings over the life of the transformer.
- Superconductors are man-made alloys that can pass electric current with virtually no resistance if kept at low temperatures.
  - The first superconductors were developed in the 1970s and are called *low-temperature superconductors (LTS)*. They had to be cooled with liquid helium.
- First generation high-temperature superconductors (1G HTS) were first developed in 1986.
  - They could be cooled with liquid nitrogen, which is 50 times cheaper than helium.
  - 1G HTS were still not that practical as they were expensive and difficult to manufacture in usable lengths.
- Second generation superconductors (2G HTS) are cheaper, more flexible, and can be produced in long lengths using an automated process.
- HTS will allow smaller transmission lines as they can carry much more current than can conventional copper conductors for a given size.
  - Transformers and motors will also be much smaller.
  - Since HTS are buried in small trenches, transmission corridors will be much smaller and towers will not be needed.
- HTS can be used as a fault-current limiter (FCL).
  - All HTS have a critical current rating. If a current higher than that rating is forced through the HTS it becomes highly resistive, quenching a fault current and protecting equipment.
- Another use for HTS is in energy storage.
  - Superconducting magnetic energy storage (SMES) uses a superconducting inductor to store electrical energy in a magnetic field.

- The Smart Grid is a Department of Energy initiative to meet the future requirements that will be placed on the energy grid as demand increases.
  - It is as large a project as the Internet or the national highway system.
- The Smart Grid will use digital communication technology to allow energy producers, distributors, and consumers to communicate in real time.
- Peaker plants are needed to meet peak demand for electricity that is very uneven. (See Figure 26.14.)
  - 10% of generating capacity and 25% of distribution infrastructure is used only 5% of the time.
- Smart meters allow customers to see the real-time cost of electricity.
  - This should change consumption habits and even out peak demand.
  - Smart meters are only one part of the Smart Grid puzzle, but they are one of the first to be employed.
- It is hoped that the Smart Grid will eventually be able to control power production and distribution in real time in order to respond to any emergency.

Green Buildings

- The U.S. Green Building Council has developed an internationally recognized third-party rating system for green buildings called LEEDS®.
- It is estimated that 65% of electrical energy is used for building operations.
- There are a large number of green building technologies, such as:
  - Automatic transfer switches to switch between utility and alternative energy sources
  - Green switchgear that avoids the use of harmful materials
  - Advanced power factor correction and harmonic filters to increase efficiency
  - Variable speed motor drives and soft-start motor controls to lower power consumption and increase motor life
  - Intelligent lighting controls that respond to room usage and ambient light levels
  - Busway, an integrated electrical distribution systems that is more efficient than cable and conduit
  - Power analysis meters that can evaluate power usage and provide opportunities for savings

Green Lighting

- Compact fluorescent lights (CFLs) can replace incandescent lights in a number of applications.
  - They last seven to ten times longer and use one-third to one-quarter the energy of an equivalent incandescent.
  - CFLs produce 60 lumens per watt compared to 14 lumens for incandescents.
- The two main components in a CFL are the phosphor-filled tube and the ballast.
  - The ballast is a current-limiting circuit.
- CFLs can be powered by DC or AC current.
- CFLs do contain a small amount of mercury, but their use prevents much more mercury from entering the environment since the biggest source of mercury pollution is the burning of fossil fuels like coal.
- LEDs are semiconductor-based lights.
  - At first they were only very low-powered devices, like indicator lamps.
  - Then high-powered LEDs were developed, but they could not produce natural white light until recently.

- LEDs are even more energy efficient than CFLs, but they are currently more expensive. Costs are expected to decrease as the technology matures.
  - LEDs are point source lights, so one bulb is usually made up of many LEDs. (See Figure 26.20.)
  - LEDs are very directional, which makes them suitable for street lighting, since most of the light is directed to the street and not as much is radiated into space.
- Organic light-emitting diodes (OLEDs) are a newer technology that uses organic materials.
  - They are flat, very thin (less than 1 mm), flexible panels. (See Figure 26.22.)
  - OLEDs are currently used only in small applications like camcorder and phone displays, but may redefine how we light our world.

CHAPTER REVIEW

1. The EIA predicts that electricity demand in the United States will increase at an annual rate of ________ through 2030.
   a) 2.6%
   b) 1.7%
   c) 0.8%
   d) 4.8%

2. Electric power generation accounts for ________ of the greenhouse gas production in the United States.
   a) 35%
   b) 25%
   c) 10%
   d) 40%

3. The most common fuel used in the United States for electricity production is ________.
   a) oil
   b) coal
   c) natural gas
   d) nuclear

4. During gasification, coal is turned into ________.
   a) natural gas
   b) steam
   c) syngas
   d) a liquid

5. Carbon sequestration is another name for ________.
   a) carbon storage
   b) carbon pollution
   c) carbon capture
   d) greenhouse gas

6. Biogas is primarily ________.
   a) propane
   b) methane
   c) fossil NG
   d) diesel

7. The use of waste heat from a gas turbine to drive a steam turbine is called ________ generation.
   a) distributed
   b) combined cycle power
   c) Rankine cycle
   d) Brayton cycle

8. Nuclear power generation represents about ________ of the country's power generation.
   a) 10%
   b) 20%
   c) 15%
   d) 25%

9. Concentrating solar power uses mirrors or lenses to focus sunlight in order to ________.
   a) produce methane
   b) burn gas
   c) drive a gas turbine
   d) drive a steam turbine

10. A barrage is a type of ________ used in ________ power production.
    a) dam, tidal
    b) rotor, wind
    c) float, wave
    d) cell, solar

11. Distributed power is another name for ________.
    a) a wind farm
    b) small-scale power generation
    c) a solar array
    d) cogeneration

12. The most popular form of small-scale power generation is still ________.
    a) reciprocating internal combustion engines
    b) diesel generators
    c) solar panels
    d) wind turbines

13. Combined heat and power (CHP), the use of waste heat for space and/or water heating, is also known as ________.
    a) distributed power
    b) flexible fuel
    c) heat recycling
    d) cogeneration

14. The use of compressed air to burn a fuel, and then using the expanding gas to drive a turbine is called ________.
    a) the Rankine cycle
    b) turbocharging
    c) the Brayton cycle
    d) compressed combustion

15. One of the advantages of microturbines is that they can burn ________ as a fuel.
    a) unrefined natural gas
    b) vegetable oil
    c) low-methane content biogas
    d) all of the above

16. Photovoltaic installations are increasing at a rate of ________ a year.
    a) 25%
    b) 50%
    c) 75%
    d) 60%

17. The two wafers used in a PV cell are called n-type material and ________.
    a) the substrate
    b) o-type material
    c) the junction
    d) p-type material

18. The term BIPV refers to ________.
    a) photovoltaic cells in building materials
    b) bipolar photo cells
    c) the solar power regulating body
    d) solar panel accessories and mounting equipment

19. An inverter is a(n) ________.
    a) solar panel alignment system
    b) DC-to-AC converter
    c) current limiter
    d) AC-to-DC converter

20. The government policy that allows small-scale producers to be paid for some of the power they generate is called ________.
    a) net metering
    b) FiT
    c) smart metering
    d) all of the above

21. Fuel cells work on the basis of ________.
    a) burning hydrogen
    b) chemical reactions

c) electrolytic capacitance
d) phosphorous exchange

22. Balance of system refers to ________.
    a) voltage regulation
    b) DG equipment other than the load or generator
    c) frequency regulation
    d) power conditioning

23. The average age of a substation transformer in the United States is ________ years.
    a) 40
    b) 10
    c) 20
    d) 30

24. Amorphous-core transformers reduce ________.
    a) eddy current losses
    b) hysteresis losses
    c) no-load losses
    d) all of the above

25. High-temperature superconductors will be used in ________.
    a) transmission lines
    b) transformers
    c) motors
    d) all of the above

26. Low-temperature superconductors have to be cooled with liquid ________ while HTS can be cooled with liquid ________.
    a) hydrogen, helium
    b) hydrogen, oxygen
    c) helium, nitrogen
    d) oxygen, helium

27. The core of an ACT is ________.
    a) a shell core
    b) stacked laminates
    c) an H-type core
    d) a toroid core

28. One reason that HTS are so expensive is that they contain ________.
    a) platinum
    b) silver
    c) gold
    d) all of the above

29. 2G HTS are ________ than 1G HTS.
    a) cheaper
    b) more easily manufactured in longer lengths
    c) more flexible
    d) all of the above

30. If the critical current rating of a 2G HTS is exceeded it ________.
    a) is destroyed
    b) reverts to 1G HTS
    c) reverts to LTS
    d) becomes resistive

31. An FCL is a(n) ________.
    a) HTS current limiter
    b) energy storage device
    c) third-generation HTS
    d) room-temperature superconductor

32. An SMES is a(n) ________.
    a) HTS current limiter
    b) energy storage device
    c) third-generation HTS
    d) room-temperature superconductor

33. The Smart Grid uses ________ to allow the flow of information.
    a) the Internet
    b) cell phones
    c) digital communication technology
    d) all of the above

34. Power generating facilities that supply power during periods of high demand are called ________.
    a) peaker plants
    b) distributed generating plants
    c) under-supply plants
    d) standby plants

35. Smart meters ________.
    a) decrease power consumption
    b) are cell phones
    c) allow for real-time energy pricing
    d) all of the above

36. Smart meters should level peak demand through ________.
    a) power restriction
    b) changing consumption habits

c) power redistribution

d) all of the above

37. The U.S. Green Building Council has developed an independent third-party verified certification system called ________.

a) USGBC®

b) LEEDS®

c) STBC®

d) GVAC®

38. HMTs are used to ________.

a) reduce harmonics

b) correct power factor

c) shave peaking

d) all of the above

39. A 26 W CFL could be used to replace a ________ incandescent bulb.

a) 60 W

b) 100 W

c) 40 W

d) 150 W

40. LEDs last ________ than CFL and use ________ power.

a) longer, more

b) a shorter time, less

c) a shorter time, more

d) longer, less

41. LEDs ________.

a) are semiconductor devices

b) can be used indoors and outdoors

c) are very directional

d) all of the above

42. OLEDs are ________ in thickness.

a) less than 1 mm

b) less than 1 inch

c) less than 1 cm

d) just over 1 cm

43. Mass-production of OLEDs could start by ________.

a) 2012

b) 2015

c) 2010

d) 2020

CHAPTER 27

INTRODUCTION TO COPPER CABLING

PURPOSE

Telecommunications is the area of electricity/electronics that deals with transmitting audio (voice), video, and data over long distances. At one time, the term was used almost exclusively in reference to telephone, radio, and television systems. Now, however, it also refers to cable and satellite TV, the Internet, security systems, and cellular service networks. In this chapter, we will focus on the elements of telecommunications that involve copper cables and wires.

KEY TERMS

The following terms are introduced and defined in this chapter on the pages indicated:

66 termination block, 1176
110 termination block, 1180
BNC connector, 1184
Bridging clip, 1178
Cable, 1165
Coaxial cable (coax), 1182
Connecting block, 1181
Cross-talk, 1166
Decoding, 1163
Electromagnetic interference, 1166
Encoding, 1163
F connector, 1184
Foil twisted pair (FTP), 1168
Interference, 1166
Jack, 1171
M-block, 1177
Medium, 1163
N connector, 1184
Nonsplit block, 1177
Plenum, 1169
Plug, 1171
Punch-down tool, 1179
Receiver, 1163
Registered jack (RJ), 1172
Ring, 1168
Riser, 1169
Screened twisted pair (ScTP), 1168
Shielded twisted pair (STP), 1168
Skin effect, 1182
Split block, 1177
Telecommunications cabling, 1164
Telephony, 1164
Terminate, 1175
Termination block, 1175
Tip, 1168
Transmitter, 1163
Transmitting, 1163
Twist length, 1167
Twisted pair, 1165
Universal jack, 1173
Unshielded twisted pair (UTP), 1168
Wiring block, 1180

OBJECTIVES

After completing this chapter, you should be able to:

1. List the three elements that make up a telecommunications system and state the function of each.
2. Identify commercial telecommunications systems.
3. Identify the telecommunications applications for wire, coaxial cables, and fiber optic cables.

4. Describe twisted pairs and twisted-pair cables.
5. Discuss electromagnetic interference (EMI) and how its effects are reduced by twisted pairs.
6. Identify the *tip* and *ring* for a given twisted pair.
7. Contrast UTP, STP, and ScTP cables.
8. Identify the applications for each twisted-pair cable category.
9. Identify the wires in a four-pair or two-pair UTP cable using their color codes.
10. Discuss the registered jack (RJ) system for describing modular jacks.
11. Compare and contrast the T568A, T568B, and USOC modular jack standards.
12. Describe the construction of split and non-split 66 termination blocks.
13. Describe the procedure for connecting a wire to a 66 block using a punch-down tool.
14. Compare and contrast 66 and 110 termination blocks.
15. Describe the construction of a 110 termination block.
16. State the function of a *connecting block*.
17. Describe the physical construction of a coaxial cable.
18. Identify the RG ratings for a coax cable and the applications associated with each.
19. List and describe the common coax cable connectors.

27.1 INTRODUCTION TO TELECOMMUNICATIONS

Throughout history, the practice of transmitting information over distances has taken many forms. For example, semaphore flags were used to send messages from one naval vessel to another, telegraphs used Morse code to transmit information over long distances, and so on.

Despite their differences, all telecommunication systems contain the three elements shown in Figure 27.1. The **TRANSMITTER** performs two functions:

1. It converts the information to be sent into a series of electric signals. This conversion is referred to as **ENCODING**.
2. It transfers the encoded information to the *medium*. This transfer is referred to as **TRANSMITTING**.

The **MEDIUM** carries the encoded information from the source to its destination. The medium is most commonly air, copper wire or cable, or fiber optic cable. Finally, the **RECEIVER** accepts the transmitted information and restores it to its original form, usually audio, video, or data. The process of converting transmitted information back to its original form is referred to as **DECODING**.

TRANSMITTER The part of a telecommunications system that encodes and transmits information.

ENCODING The process of converting information into a series of electric signals so that it can be transmitted.

TRANSMITTING The process of transferring encoded information to a transmission medium.

MEDIUM The part of a telecommunications system that carries encoded information from the source to its destination, most commonly air, copper wire or cable, or fiber optic cable.

RECEIVER The part of a telecommunications system that accepts transmitted information and restores it to its original form, usually audio, video, or data.

DECODING The process of converting transmitted information back to its original form.

FIGURE 27.1 Block diagram of a basic telecommunications system.

The broadcast from a commercial radio station further illustrates the makeup of a telecommunications system. The radio station converts sound (music and/or voice) to a series of electromagnetic signals for transmitting through the air. These signals are transferred to the air by the transmitter antenna. Using its own antenna, your radio captures these signals and converts them back to sound. In this case, the radio station is the encoder/transmitter, air is the medium, and the radio is the receiver/decoder.

Commercial Telecommunications Systems

When we refer to *commercial telecommunications systems*, we are referring to public and commercial radio and television. These systems use one or more of the following transmission systems:

1. **Broadcast transmission.** Broadcast transmissions are accessible to anyone with a radio or digital television (DTV) receiver. The signals are transferred to and from the medium (air) using antennas, as indicated in the description of the commercial radio transmission.

NOTE: Cable and satellite systems are also used for Internet data communications.

2. **Cable transmission.** Coaxial (or fiber optic) cables are used as the medium in a cable radio and television system. The cable company receives signals for transmitting from a number of independent sources (stations), assigns the signals to the appropriate cable channels, and then retransmits the signals through the cable system. Legal reception of the cable transmission requires a subscription to the cable service.
3. **Satellite transmission.** Satellites are used—in one way or another—in all commercial telecommunications systems. The term *satellite transmission,* however, refers to a system in which the receiver requires a satellite dish and/or special decoding circuitry. For example, legal reception of a satellite TV transmission requires a satellite dish and a subscription with the satellite service provider.

Telephony

TELEPHONY The study of wireless and land telephone systems.

Commercial radio and television stations are not the only sources of transmitted information. Telephone systems allow for simultaneous two-way voice and data transmission using cable and/or air as the medium. Wireless telephone systems use air as the medium of transmission. Conventional telephone systems (sometimes referred to as *land lines*) use wire, coaxial cable, and/or fiber optic cable as the transmission medium. The study of wireless and land telephone systems is referred to as **TELEPHONY**.

Telecommunications Cabling

TELECOMMUNICATIONS CABLING The practice of preparing and installing the wires, cables, and connectors that are used in residential and commercial telecommunications circuits.

TELECOMMUNICATIONS CABLING is the practice of preparing and installing the wires, cables, and connectors that are used in residential and commercial telecommunications circuits. The common applications for these wires and cables are identified in Table 27.1.

TABLE 27.1 • Wire and Cable Applications

WIRE/CABLE TYPE	APPLICATION(S)
Wire	Telephony (voice, facsimile, and computer modem)
Copper (coaxial) cables	Telephony (voice, facsimile, and computer modem) Cable TV Satellite TV connections
Fiber optic cables	Telephony (voice, facsimile, and computer modem) Cable TV

The remainder of this chapter addresses wire and coaxial cables and connectors. Fiber optic cables are examined in Chapter 28.

PROGRESS CHECK

1. Define the following telecommunication terms:
   a. transmitter
   b. receiver
   c. encoding
   d. decoding
   e. medium
2. List and describe three types of telecommunication transmission systems.
3. Define the term *telephony*.
4. List the applications in telecommunications for the following:
   a. wire
   b. coaxial cables
   c. fiber optic cables

27.2 TWISTED-PAIR CABLES

A **CABLE** is an insulated electrical conductor—or group of insulated electrical conductors—that are used to transmit telecommunication signals from one point to another. In this chapter, we examine two types of copper cables: *twisted-pair* cables and *coaxial* cables. We will begin with a look at twisted-pair cables. Coaxial cables are introduced later in this chapter.

CABLE An insulated electrical conductor (or group of conductors) used to transmit telecommunication signals from one point to another.

Twisted Pairs

A **TWISTED PAIR** is simply two wires that are wrapped around each other (as shown in Figure 27.2) along their entire length. Two or four twisted pairs are typically grouped in a single cable. Note that:

- Two-pair cables, which are limited to relatively low frequencies, are used in telephone voice applications.
- Four-pair cables, which are capable of operating at much higher frequencies, are used in telephone voice and data transmission applications.

These points are discussed in more detail later in this section.

TWISTED PAIR Two wires that are wrapped around each other along their entire length.

FIGURE 27.2 Twisted pairs of wires.
Istock.com

Data transmissions are classified, in terms of their method of transmission, as simplex, half-duplex, or full-duplex. *Simplex* and *half-duplex* systems use only one pair of conductors to communicate. The simplex method transmits in one direction, while the half-duplex system can send signals in both directions, but not at the same time. The full-duplex system uses two pairs to communicate, with one pair always transmitting from Point A to Point B and the other pair transmitting from Point B to Point A.

KEY CONCEPT

The wires used in twisted pairs are solid or stranded, with diameters between 22 gauge and 26 gauge. Stranded wire is preferred because it is more flexible than solid wire. At the same time, solid wires are structurally stronger than stranded wires, and therefore better suited for use in certain physical environments.

Why Twist Wires into Pairs?

INTERFERENCE An undesired electrical signal that inhibits—or prevents the reception of—a transmitted signal.

The wires that make up a twisted pair carry low-amplitude signals from one point to another, typically in the low millivolt (mV) range. Signals in this range are particularly susceptible to the effects of *interference*. As you will learn, twisting pairs of wires around each other reduces (or eliminates) interference and its effects.

Balanced transmission helps to minimize electromagnetic emissions by sending equal but opposite signals down each wire. This cancels out the radiated signals and reduces electromagnetic interference.

KEY CONCEPT

INTERFERENCE is an undesired electrical signal that inhibits—or prevents the reception of—a transmitted signal. There are many possible causes of interference, each having its own source. We are primarily interested in *electromagnetic interference* (EMI) and *cross-talk*.

ELECTROMAGNETIC INTERFERENCE occurs when two wires are:

- Physically parallel to each other.
- In close proximity to each other.

ELECTROMAGNETIC INTERFERENCE (EMI) A type of interference that occurs when the electromagnetic field produced by a signal on one wire induces a similar signal in another wire.

CROSS-TALK EMI between wire *pairs* in a telecommunications cable.

When these conditions exist, the electromagnetic field produced by a signal on one wire induces a similar signal in the second wire. This phenomenon is similar to the operation of a *transformer*, which uses electromagnetic induction to couple a signal from its primary circuit to its secondary circuit.

When a cable contains two or more pairs of wires, the signal in one pair can be coupled to the other(s) as described above. EMI between wire *pairs* is known as **CROSS-TALK**.

Interference that is between two pairs and measured at the transmitter end of the cable is called Near End Crosstalk (NEXT). This phenomenon happens when there is some form of electromagnetic interference taking place relatively close to the point of measurement and can be caused by temporary twisted-pair cabling or worn spots in the cable insulation that allows the wiring of the two cables to come into close proximity.

IN THE FIELD

The effects of EMI can be reduced or eliminated by twisted pairs. When the wires are twisted as shown in Figure 27.2, the wires are no longer parallel, so EMI does not occur between them. To reduce the effects of cross-talk, the wire pairs in a given cable are twisted to different lengths as shown in Figure 27.3. Twisting wires to different lengths affects their values of inductance and capacitance. When

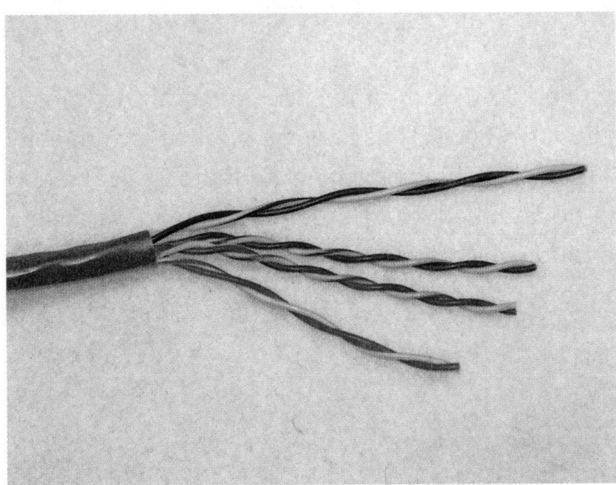

FIGURE 27.3 A twisted-pair cable with unequal twist lengths.

wire runs have different inductive and capacitive values, cross-talk is eliminated. Note that the distance from one twist to another in a twisted pair is called the **TWIST LENGTH** of the cable. As such, we can say that cross-talk is eliminated when wire pairs have unequal twist lengths (like those in Figure 27.3).

TWIST LENGTH The distance from one twist to the next in a twisted pair.

Another potential problem associated with twisted pair cabling is *pulse spreading*, which is the most common type of square-wave signal distortion. As a square wave travels down the cable, it distorts and begins to spread. If the signal spreads too much, it will be unintelligible to the receiver.

IN THE FIELD

Still another potential problem affecting twisted pair cabling performance is *attenuation*, which is a function of frequency and cable length. The higher the signal frequency, the greater the attenuation. Also, the longer the cable, the more energy a signal loses by the time it reaches the receiver.

KEY CONCEPT

Never replace a cable with one having a different characteristic impedance. For example, a 150-ohm cable should not be used as a cable replacement or addition in a network consisting of 100-ohm cable wiring because the signal will reflect off of the 100 ohm-to-150 ohm junction and become distorted.

IN THE FIELD

Tip and Ring

TIP and RING The two wires in a twisted pair.

The wires in a twisted pair are connected to the same circuit. One of the wires is called the **TIP** and the other is called the **RING**. These names originated with early telephone circuits that used 1/4″ plugs like the one shown in Figure 27.4. The end of the plug was referred to as the *tip* and the lower portion of the plug was called the *ring*. These two parts of the plug, which are electrically insulated from each other, were always connected to the two wires in a twisted pair.

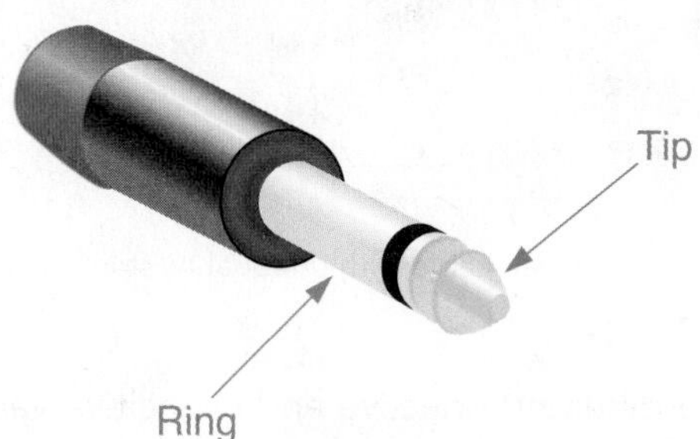

FIGURE 27.4 Plug tip and ring.

Shielded and Unshielded Cables

UNSHIELDED TWISTED PAIR (UTP) A term used to describe a cable that contains four twisted pairs.

SHIELDED TWISTED PAIR (STP) A term used to describe a cable that contains four twisted pairs with conductive foil surrounding each twisted pair, with an outer conductive foil wrapped around the group to provide added protection from EMI and cross-talk.

SCREENED TWISTED PAIR (SCTP) A term used to describe a cable that contains four twisted pairs with an outer conductive foil wrapped around the group to provide some protection from EMI.

FOIL TWISTED PAIR (FTP) Another name for a screened twisted pair.

There are two types of twisted-pair cables, as shown in Figure 27.5. The blue **UNSHIELDED TWISTED PAIR (UTP)** cable contains four twisted pairs. The wires in each pair are insulated and color coded (as described later in this section). UTPs are by far the most commonly used twisted-pair cable. The grey **SHIELDED TWISTED PAIR (STP)** cable has conductive foil surrounding each twisted pair, with an outer conductive foil wrapped around the group. This conductive foil provides additional shielding from external EMI and cross-talk, allowing the cable to be used in environments where UTPs cannot be used.

There is an additional type of cable. A **SCREENED TWISTED PAIR (SCTP)**, or **FOIL TWISTED PAIR (FTP)**, contains an outer conductive foil like the STP, but does not have conductive foil surrounding each twisted pair. As such, it provides some EMI shielding, but not as much as the shielded twisted pair.

FIGURE 27.5 STP and UTP cables.

Cable Categories

Twisted-pair cables are divided into categories, each having characteristic bandwidth limits and applications. These categories (or *cats*) and their characteristics are summarized in Table 27.2. Note that the categories listed in ***bold print*** are currently recognized by TIA/EIA.

TABLE 27.2 • Twisted-Pair Cable Categories

CATEGORY	CABLE TYPE(S)	BANDWIDTH DATA TRANSFER RATE		APPLICATION(S)
CAT1	UTP	< 1 MHz		Analog voice/Doorbell wiring
CAT2	UTP	4 Mb/s[1]		4 Mb/s token ring networks
CAT3	UPT, ScTP, STP	16 MHz	16 Mb/s	Voice & 10 Mb/s Ethernet
CAT4	UPT, ScTP, STP	20 MHz	20 Mb/s	16 Mb/s token ring networks
CAT5	UPT, ScTP, STP	100 MHz	1 Gb/s[2]	100 Mb/s token ring networks
CAT5e	UPT, ScTP, STP	100 MHz	1 Gb/s	100Base-T or Fast Ethernet
CAT6	UPT, ScTP, STP	250 MHz	10 Gb/s	Super-fast broadband
CAT6a	UPT, ScTP, STP	500 MHz	10 Gb/s	Suitable for 10GBase-T
CAT7	STP	600 MHz	10 Gb/s	CATV & 100 Gb/s Ethernet[3]

[1]Mb/s = megabits per second
[2]Gb/s = gigabits per second
[3]Not yet in practice

In the early 1990s, the Telecommunications Industry Association (TIA) and Electronics Industry Alliance (EIA)[1] joined forces to develop cable standards to ensure that cables would work with all commercial telecommunications equipment. Of the cables listed in Table 27.2, CAT5e is the most commonly used.

Plenum and Riser Cables

Twisted-pair cables are marked with a three-letter application code that indicates where the cable is used and what it is used for. The letters used in this code are identified in Table 27.3. Note that the term **PLENUM** is used to identify the air spaces above and below a room. The term **RISER** is used to identify the vertical spaces behind the walls of a room.

PLENUM The air spaces above and below a room.

RISER The vertical spaces behind the walls of a room.

Of the cables identified in Table 27.3, the plenum cable is the highest quality, followed by the riser cable and the general purpose cable. Note that a higher-quality cable can be substituted for a lower quality cable, but not the other way around. For example, a riser cable can be used in place of a general purpose cable, but not in place of a plenum cable.

[1]Prior to 1997, the Electronics Industry Alliance was known as the Electronics Industry Association.

TABLE 27.3 • Twisted-Pair Application Code

POSITION	LETTER	DESIGNATION
1st letter	C	Copper
2nd letter	C M	Communications Multipurpose
3rd letter	G P R	General Purpose Plenum Riser

Wire Color Coding

One of the TIA/EIA standards dictates the colors that are used to code the wires in a four-pair cable. This color code is presented in Table 27.4. The appearance of the various wires is illustrated in Figure 27.6.

TABLE 27.4 • Four-Pair Cable Color Code

PAIR	WIRE	POLARITY	WIRE COLOR	STRIPE COLOR
1	1	tip +	white	blue
	2	ring –	blue	white
2	3	tip +	white	orange
	4	ring –	orange	orange
3	5	tip +	white	green
	6	ring –	green	white
4	7	tip +	white	brown
	8	ring –	brown	white

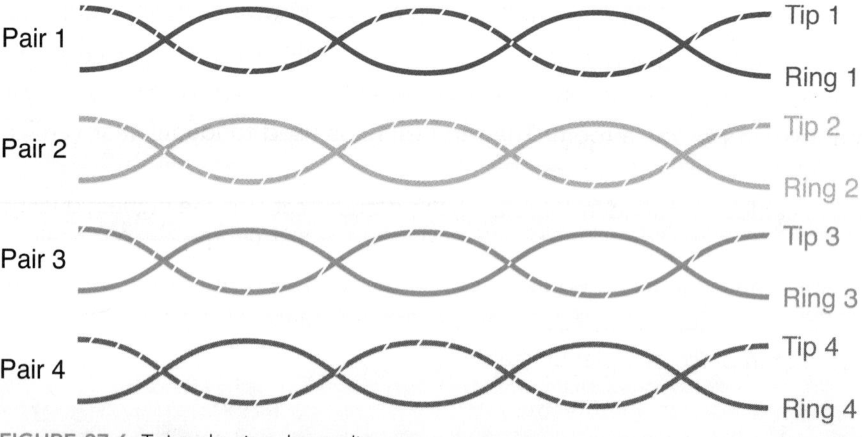

FIGURE 27.6 Twisted pair color coding.

CAT2 cables have their own color code. The CAT2 color code is presented in Table 27.5.

TABLE 27.5 • Two-Pair Cable Color Code

PAIR	WIRE	POLARITY	WIRE COLOR
1	1	tip +	green
	2	ring −	red
2	3	tip +	black
	4	ring −	yellow

PROGRESS CHECK

1. What are twisted pairs and how are they usually grouped?
2. What wire gauges are usually used in twisted pairs?
3. Why are wires twisted in pairs?
4. Define the following terms.
   a. EMI
   b. cross-talk
   c. twist length
   d. tip and ring
5. What is the difference between a shielded, an unshielded, and a screened twisted pair?
6. Define the terms *plenum* and *riser.*
7. Which category of twisted pairs has its own color code?

27.3 MODULAR JACKS AND PLUGS

Twisted-pair cables are commonly connected to telecommunications ports and equipment using *plugs* and *jacks*. A **PLUG** is a connector that is attached to the wires in a cable. A **JACK** is a docking port for a plug on an electronic device (such as a modem). A jack and its matching plug are shown (along with a twisted-pair cable) in Figure 27.7.

PLUG A connector that is attached to the wires in a cable.

JACK A docking port for a plug on an electronic device.

Modular Jacks

The jack shown in Figure 27.7 is called a *modular jack* (or *modular connector*). Modular jacks are often described using the number of wire positions (P) they contain and the number of conductors (C) that are connected to the jack. For example, the

FIGURE 27.7 A modular jack and matching plug.

6P4C jack illustrated in Figure 27.8 has six positions (6P) for wire connections. Only four conductors (4C) are actually connected to the jack. The other two jack positions (numbers 1 and 6) are not used.

NOTE: Figure 27.8 shows the RJ14 wiring standard for a 4-pair cable (upper) and a 2-pair cable (lower).

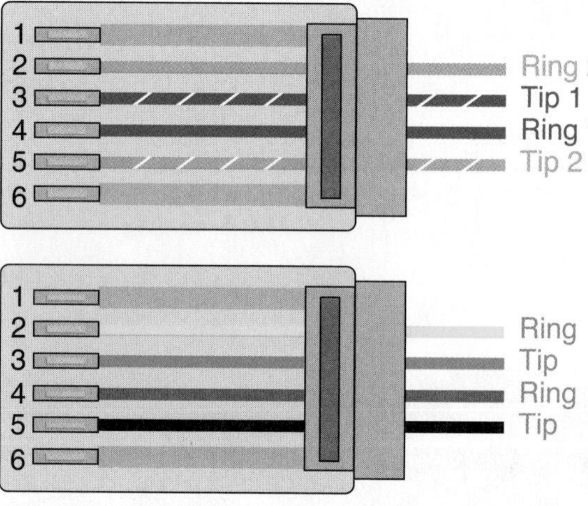

FIGURE 27.8 6P4C jacks.

REGISTERED JACK (RJ) A modular jack wired according to a standard set by the Federal Communications Commission (FCC).

Like other elements in telecommunications systems, the wiring of modular jacks has been standardized over the years to ensure that they are all wired in the same fashion. For example, when you encounter a jack like the one in Figure 27.8, you know how it is wired because a connection standard has been established. This standard, referred to as the **REGISTERED JACK (RJ)** standard, was established by the Federal Communications Commission (FCC). The jacks in Figure 27.8 are wired according to the RJ14 standard. The three wiring standards for six-position (6P) jacks are as follows:

- **RJ11** is a **6P2C** wiring standard.
- **RJ14** (Figure 27.8) is a **6P4C** wiring standard.
- **RJ25** is a **6P6C** wiring standard.

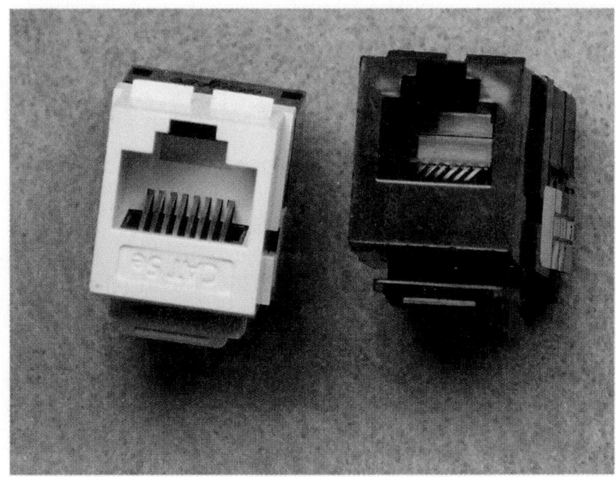

FIGURE 27.9 An 8P jack (left) and a 6P jack.

Figure 27.9 shows an eight position (8P) jack (on the left) and a six position (6P) jack. The eight position (8P) jack is described as a **UNIVERSAL JACK** because it can work with 4P, 6P, and 8P plugs. Note that, a plug designed for use with a lower-number jack will connect to a higher-number registered jack. However, the reverse is not true. The working combinations of plugs and jacks are listed in Table 27.6.

UNIVERSAL JACK A term used in reference to the 8P jack because it also works with 4P and 6P plugs.

TABLE 27.6 • Jack and Plug Combinations

JACK	COMPATIBLE PLUG(S)
4P	4P
6P	4P and 6P
8P[1]	4P, 6P, and 8P

[1]Referred to as the universal jack

4P jacks/plugs are commonly used to connect telephone handsets to their base sets. 6P jacks/plugs are used to connect telephones to telephone lines. 8P jacks/plugs are used to connect computers, laptops, and other computer peripherals to each other and telephone lines using four-pair cables.

Standard Pin Configurations

When we refer to *standard pin configurations*, we are referring to standards that dictate how the wires in a twisted-pair cable are connected to modular plugs and jacks. Three RJ45 (8P8C) wiring standards are shown in Figure 27.10.

T568A and T568B are the TIA/EIA standards for 8P8C plugs and jacks. The *T568A* configuration is used primarily in residential applications, and the *T568B* configuration is used primarily in commercial applications. The USOC (Universal Service Order Code) configuration shown in Figure 27.10 is limited to telephone (voice and low-speed data) applications.

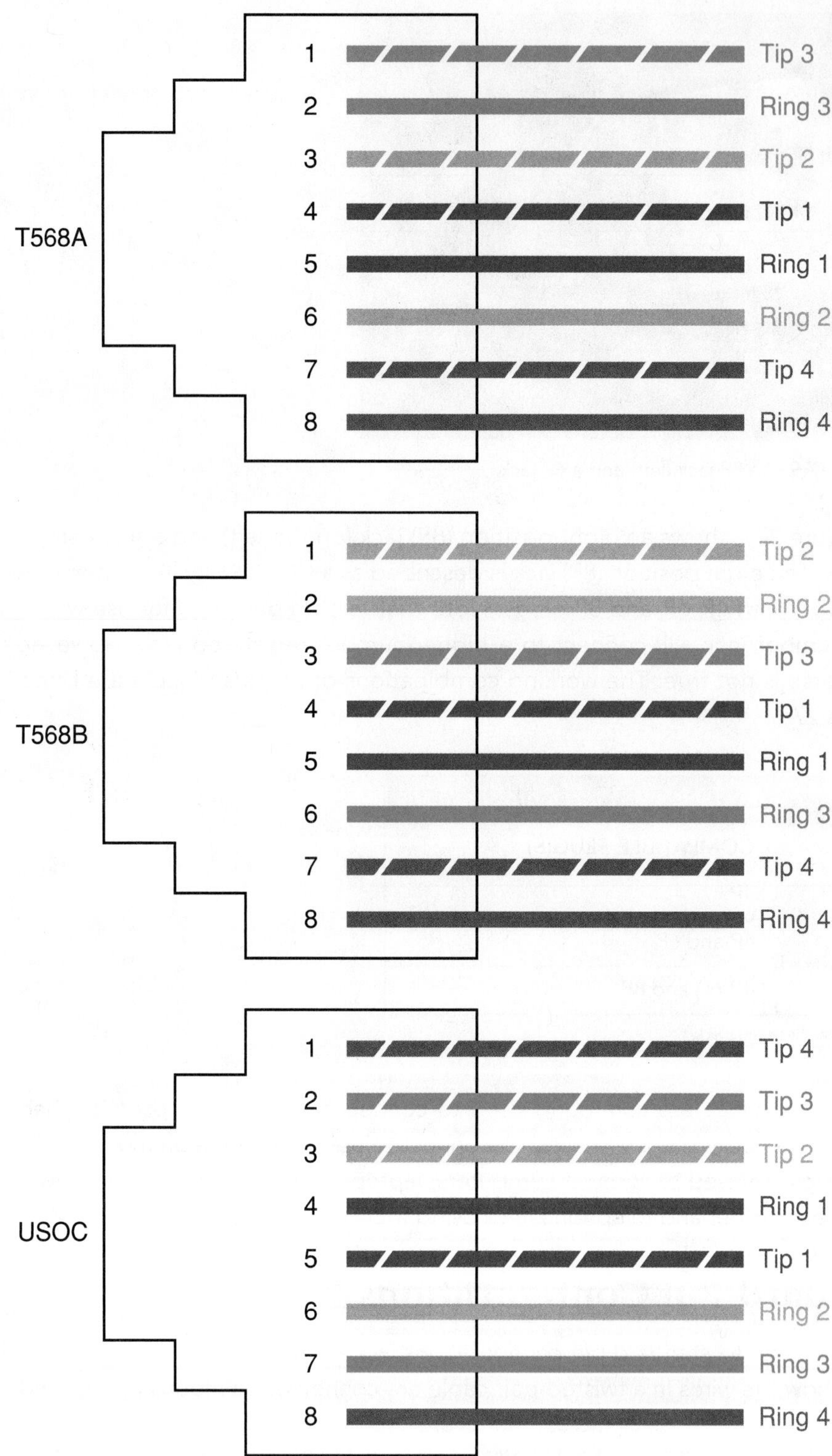

FIGURE 27.10 RJ45 wiring standards.

An Important Point

The RJ designations introduced in this section are sometimes mistakenly used to identify jacks and plugs, rather than the wiring configurations. For example, the designation RJ11 might be used to identify any six position (6P) jack or plug . . . even though 6P jacks and plugs can be wired in RJ11, RJ14, or RJ25 configurations. By the same token, the designation RJ45 is sometimes used to describe the following RJ wiring schemes:

- RJ49—an ISDN (Integrated Services Digital Network) basic rate interface. (This wiring scheme most commonly found in residential circuits.)
- RJ61—an 8P8C scheme used to carry four distinct telephone lines.

PROGRESS CHECK

1. Define and contrast plugs and jacks.
2. What criteria are used to describe modular jacks?
3. Which jack is referred to as a universal jack? Why?
4. List two standard pin configurations and state their most common uses.

27.4 TERMINATION BLOCKS

Cables always have some type of circuit connection at each end. The connector (or other connection device) on the end of a cable is said to **TERMINATE** the cable.

TERMINATE A term used to describe the connection(s) at the end of a cable.

In the previous section, you were introduced to the modular jacks and connectors that are used to terminate twisted pair cables. There are other devices that are used to terminate twisted-pair cables as well. In this section, we will look at two of these devices: the *66* and *110* termination blocks.

66 Termination Blocks

TERMINATION BLOCK A physical structure to which cable wires are individually connected.

66 TERMINATION BLOCK A block that is used to terminate telephone and data transmission lines.

A **TERMINATION BLOCK** is a physical structure to which cable wires are individually connected. The two most common termination blocks—the 66 and 110 blocks—are shown in Figure 27.11.

The **66 TERMINATION BLOCK** is used to terminate telephone and data transmission lines, usually in nonresidential facilities like businesses, schools, hospitals,

(a) 66 block

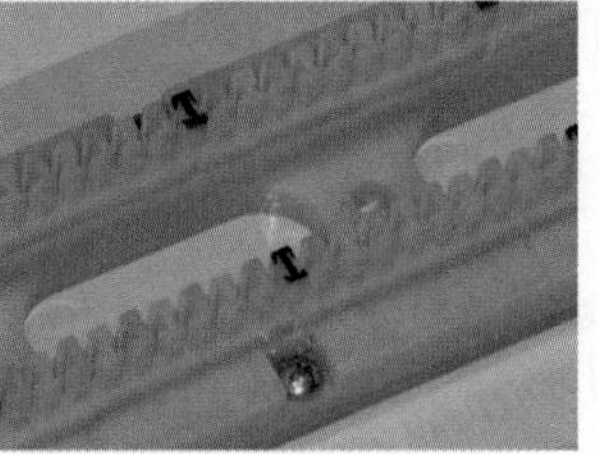

(b) 110 block

FIGURE 27.11 A 66 block and a 110 block.

and so on. These blocks are normally found in dedicated telecommunications cabinets and rooms like the one shown in Figure 27.12. As such, they are not normally visible to the general public.

FIGURE 27.12 A telecommunications cabinet.
Istock.com

At first glance, it might appear that the name *66* refers to the number of rows on the block, but this is not the case. There are actually 50 rows on the standard 66 block, as shown in Figure 27.13. The label *66* comes from the shape of the metal

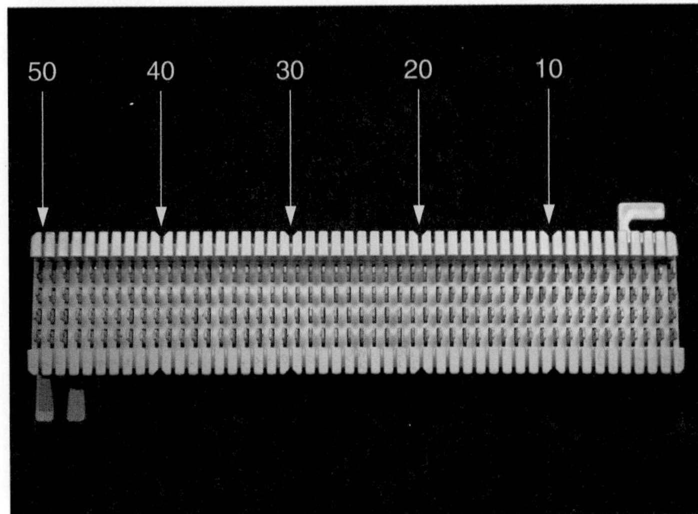

FIGURE 27.13 A 66 block.

M-BLOCK Another name for a 66 termination block.

clips on the block, which somewhat resemble the number 6. Note that the 66 block is also referred to as an **M-BLOCK**.

66 Block Construction

The 66 block has a plastic base that holds 50 rows of metal clips. There are four clips in each row, for a total of 200 clips. With one input wire connected to each row, the 66 block can terminate six four-pair UTP cables with two rows left over. Figure 27.14. shows the termination of a twisted pair cable on a 66 block. Note that the tip and ring of each twisted pair are always connected to adjacent rows.

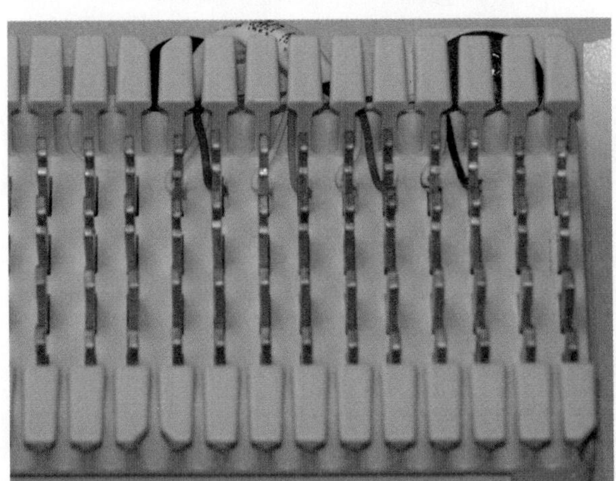

FIGURE 27.14 Twisted pair terminations on a 66 block.

NONSPLIT BLOCK A 66 block with rows containing four clips that are electrically connected to each other.

SPLIT BLOCK A 66 block with rows containing four clips that are physically and electrically split into two pairs.

The 66 block has two configurations: *nonsplit block* and *split block*. Each row in a **NONSPLIT BLOCK** contains four clips that are electrically connected to each other. In contrast, the two clips on the right side of a **SPLIT BLOCK** are physically and electrically isolated from the two clips on the left. Both blocks are shown in Figure 27.15. Note that the split block is more commonly used than the nonsplit block.

(a) Nonsplit block clip configuration

The center gap isolates the two pairs of clips

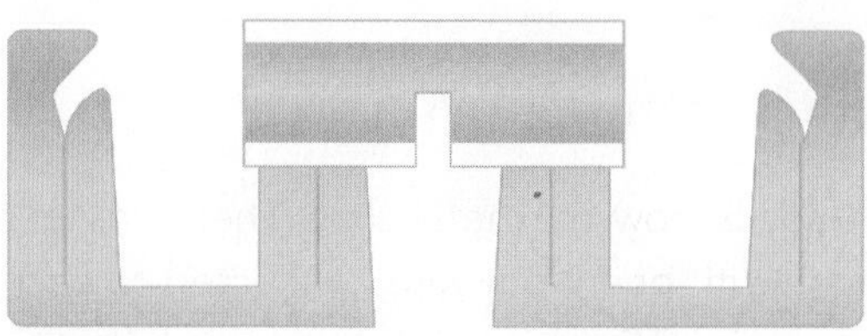
A *bridging clip* connects the two pairs of clips
(b) Split block clip configuration

FIGURE 27.15
Istock.com

BRIDGING CLIP A metal clip that is used to connect the left-hand clips to the right-hand clips on a split block.

A **BRIDGING CLIP** is used to connect the left-hand clips to the right-hand clips on a split block. Bridging clips can be seen on the block in Figure 27.16. When a bridging clip is present, the two wires that are fastened to the row (on the left and right ends) are electrically connected to each other.

FIGURE 27.16 A split block with bridging clips.

Wire Connections

Wires are connected to a 66 block using the **PUNCH-DOWN TOOL** shown in Figure 27.17a. Here is the procedure for connecting a four-pair UTP cable to a 66 block:

PUNCH-DOWN TOOL A tool used to connect wires to a termination block.

1. The cable is cut to expose approximately 2 inches of wire (Figure 27.17b).
2. The twisted pairs are unraveled (Figure 27.17c).

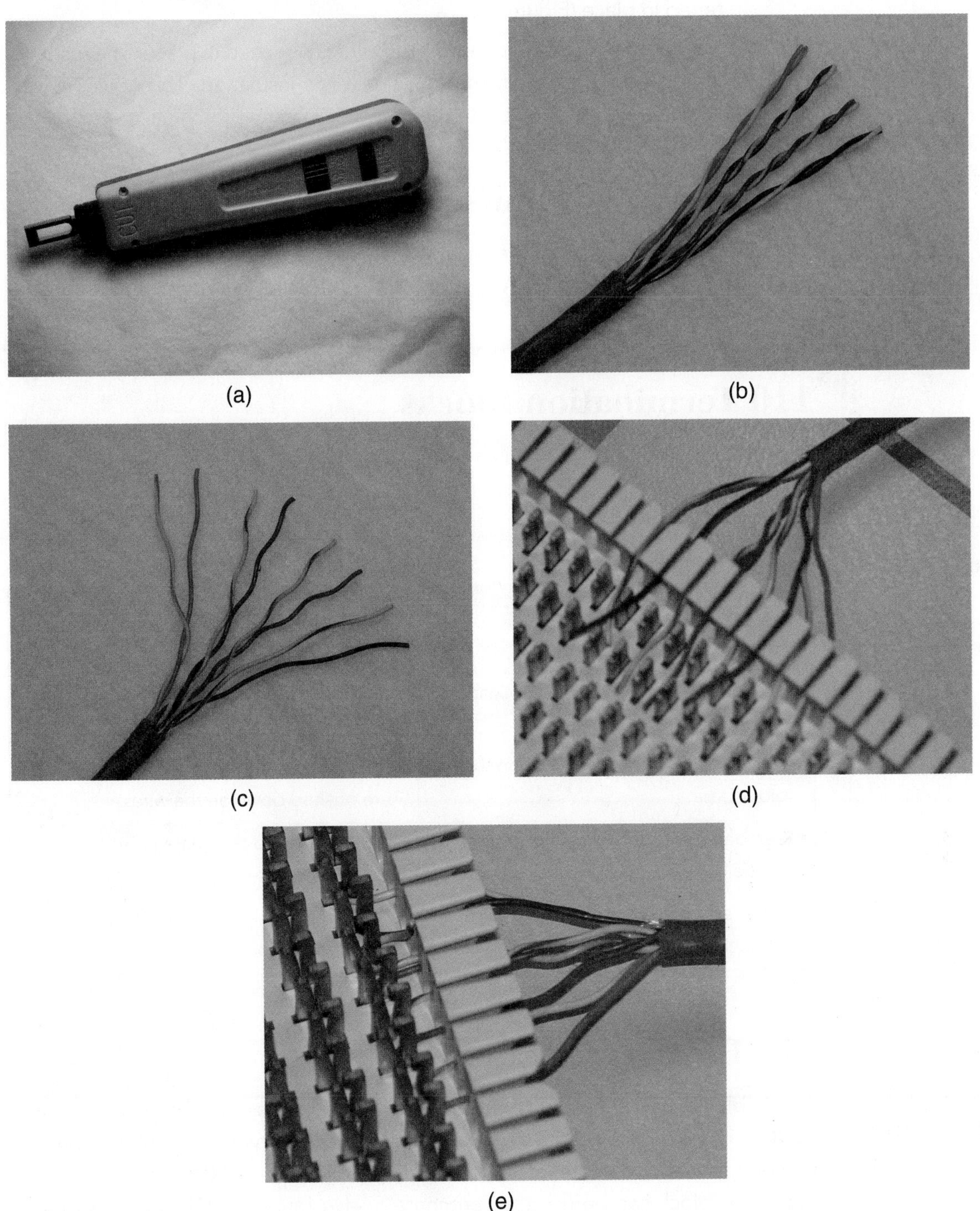

(a) (b) (c) (d) (e)

FIGURE 27.17

3. The wires are positioned in the side notches in the block. As stated earlier, the ring of each pair is positioned immediately following the tip of each pair (Figure 27.17d).
4. Each wire is positioned in the "V" at the top of its clip and the punch-down tool is placed over the clip. Pressure is applied on the punch down tool until it snaps down on the block. When this occurs, the wire is fastened in place (Figure 27.17e).

The electrical connection is made when the wire is punched in place. The inside of the clip is sharp enough to cut through the wire's insulation. This leaves an exposed electrical wire in contact with the clip.

Performance Limits

Because of its construction, the 66 block cannot be used for applications higher than CAT5. For CAT5e or CAT6 operation, a 110 termination block must be used.

110 Termination Blocks

110 TERMINATION BLOCK A termination block that is mostly plastic and is suited for higher-speed applications.

A **110 TERMINATION BLOCK** is a termination block that is mostly plastic and is suited for higher-speed applications. Table 27.7 provides a comparison of 66 and 110 blocks.

TABLE 27.7 • 110 Versus 66 Termination Blocks

66 BLOCKS	110 BLOCKS
• Limited to CAT3 and CAT5 applications.	• Suited for CAT5e and CAT6 applications.
• Connections between input and output wires are made by metal clips on the block base.	• Connections between input and output wires are made by *connecting blocks* that are pushed down on the wires.
• Can be used to terminate up to six four-pair (eight-wire) UTP cables.	• Can be used to terminate up to twelve four-pair (eight-wire) UTP cables.
• Wires are connected to the block using a punch-down tool.	• Wires are connected to the block using a connecting block.

110 Block Construction

WIRING BLOCK The base of a 110 termination block; it has two rows with 50 connection points each.

A 110 block makes use of two separate components. The base, or **WIRING BLOCK**, is shown in Figure 27.18. The wiring block has two rows with 50 connection points each, for a total of 100 connection points. With 100 connection points, the wiring block can be used to terminate twelve four-pair UTP cables with four positions left open, two on each side.

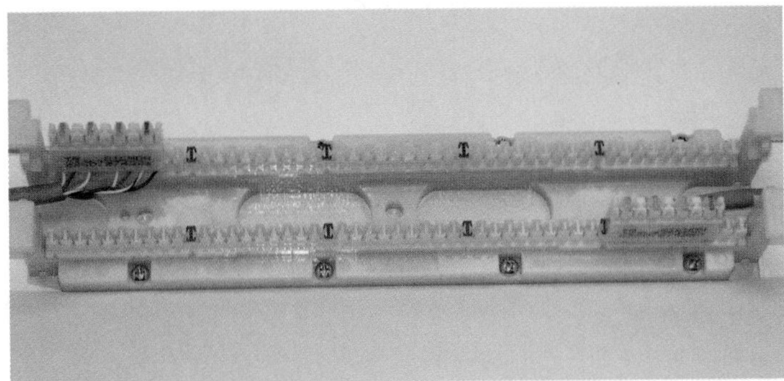

FIGURE 27.18 A 110 termination block.

The wiring block is structured to make the cable positions easy to identify. As shown in Figure 27.19:

- Even-numbered pins are taller than odd-numbered pins.
- A space is positioned between each group of ten pins.

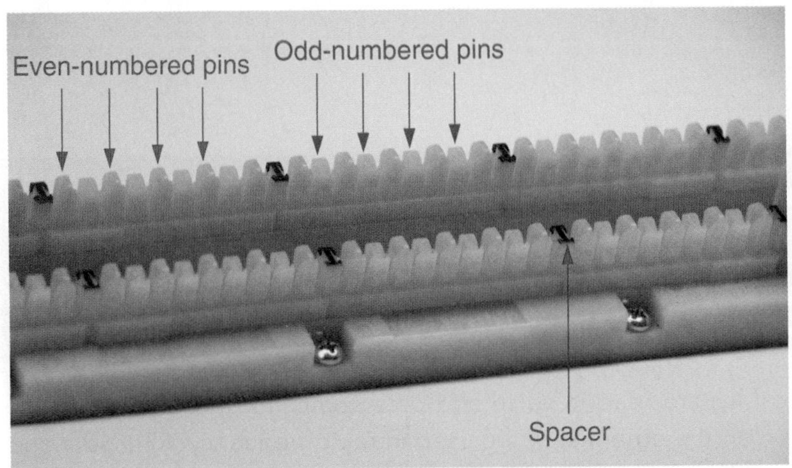

FIGURE 27.19 Cable positions and spacers on a 110 block.

The **CONNECTING BLOCK** performs two functions when pushed down onto the wiring block:

CONNECTING BLOCK A block that cuts through the cable wires and holds them in place on the wiring block.

1. It cuts through the insulation of the wires attached to the wiring block, making an electrical connection to those wires.
2. It holds the terminated wires in place on the wiring block.

A wiring block with an attached connecting block is shown in Figure 27.20. Here's how it works: The incoming cable is terminated at the plastic wiring block. Once the wires are connected to the wiring block, the connecting block is pushed down on the connections. Metal connectors inside the connecting block cut into the wires on the wiring block, making the needed electrical connections.

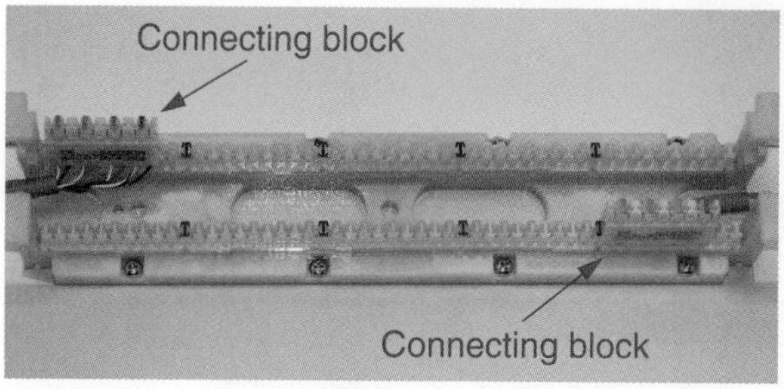

FIGURE 27.20 Connecting blocks on a 110 wiring block.

PROGRESS CHECK

1. What is a 66 termination block used for?
2. How many rows does a 66 termination block have? How many clips does it have?
3. What two configurations does the 66 termination block come in?
4. What is a bridging clip and what is it used for?
5. What type of tool is used to connect wires to a 66 block?
6. What are the performance limitations of the 66 block?
7. Describe a 110 block and compare it to a 66 block.

27.5 COAXIAL CABLES

Twisted-pair cables are used in voice and data communications applications. *Coaxial cables*, on the other hand, are used in radio-frequency (RF) and video applications. In this section, we will examine coaxial cable construction and applications.

Cable Construction

COAXIAL CABLE (COAX) A cable that contains two conductors separated by a *dielectric* (insulator).

SKIN EFFECT The tendency of current to move toward the outside surface (skin) of a conductor as frequency increases.

A **COAXIAL CABLE**, or ***coax***, contains two conductors separated by a *dielectric* (insulator). As shown in Figure 27.21, the inner conductor is surrounded by the dielectric. The inner or center conductor may be solid or stranded, with *seven-strand* being the most common type of stranded center conductor. Solid center conductors are used in permanent or *low-flex* applications. Stranded conductors are used in applications where flexibility is needed. The center conductor may be copper, copper-clad steel, or copper-clad aluminum. For high-frequency applications, silver-clad copper is used to offset *skin effect*. **SKIN EFFECT** is the tendency of current to move toward the outside surface (skin) of a conductor as frequency increases.

FIGURE 27.21 A coaxial cable.
Fotolia.com

The dielectric that separates the conductors is usually a material that has stable electrical properties over a wide range of frequencies. Polyethylene and polypropylene are two common dielectric materials. The outer conductor surrounds the dielectric and is constructed of braided copper and/or aluminum. In high-frequency applications, a second shield or braid may be added for better EMI shielding. Finally, the outer insulation surrounds the outer conductor. This insulation material, as well as providing environmental protection, may be used to improve the cable's fire-retardant properties.

Though not shown in Figure 27.21, the cable is normally terminated at both ends using one of the connectors that are introduced later in this section.

RG Ratings

Coax cables are assigned *RG ratings* that provide some indications of their construction and applications. The letters *RG* originally stood for *Radio Guide*—a long-outdated set of military specifications. Because coax cables no longer strictly adhere to these specifications, they are often referred to as "RG-XX type" cables, where XX is one or two numbers. The most common RG ratings are identified in Table 27.8. Of the cables listed, RG-58 and RG-60 are the most commonly encountered.

TABLE 27.8 • Coax Cable RG Ratings

RATING	IMPEDANCE	APPLICATIONS/NOTES
RG-6	75 Ω	Cable and satellite TV; cable modems; broadband Local Area Networks (LANs)
RG-8	50 Ω	Amateur radio
RG-11	75 Ω	Long cable runs
RG-58	50 Ω	Radio frequency communications and amateur radio
RG-59	75 Ω	Closed-circuit television (CCTV)
RG-60	50 Ω	High-definition cable TV and high-speed cable Internet

The RG rating of a coax cable does not relate to cable quality. For example, there are cheap RG-60 cables and there are high-quality RG-60 cables. Note also that the impedance referred to in Table 27.8 is the cable's *characteristic impedance*. Characteristic impedance is a complex topic that is outside the scope of this text, but here is a very brief explanation: As operating frequency increases, conductors begin to act like transmission lines. A *transmission line*, unlike a normal conductor, has constant impedance regardless of its physical length. Both a 1-meter length and a 100-meter length of RG-58 look like a 50 Ω load to a RF transmitter, as long as the cable is properly terminated and is undamaged.

BNC CONNECTOR One that is used primarily to connect coax cables to test equipment and in data processing applications.

N CONNECTOR One that is used primarily in land mobile, cellular, and pager infrastructure.

F CONNECTOR One that is used primarily in satellite, cable TV, and high-speed cable Internet connections.

Coax Cable Connectors

Three types of connectors are normally used to terminate coax cables, each well suited for one or more applications. The **BNC CONNECTOR** shown in Figure 27.22 is used primarily to connect coax cables to test equipment and in data processing applications. The **N CONNECTOR** is used primarily in land mobile, cellular, and pager infrastructure. The **F CONNECTOR** is used primarily in satellite and cable TV applications, as well as in high-speed cable Internet connections. The F and N connectors (not shown) are screw-on connectors, as opposed to the BNC connector, which is a twist-on connector.

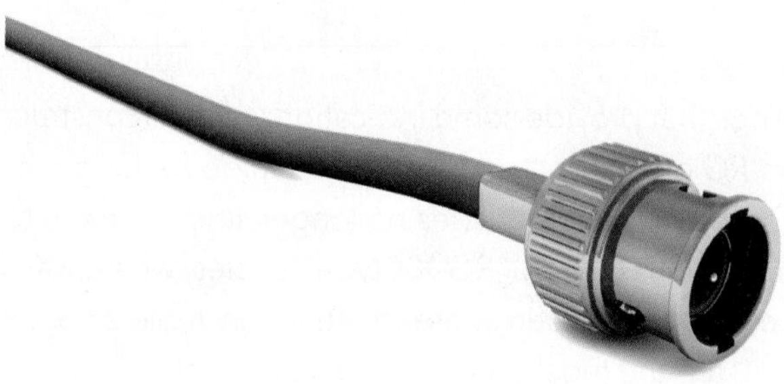

FIGURE 27.22
Fotolia.com

PROGRESS CHECK

1. Describe how a coaxial cable differs from a twisted pair. What are coaxial cables used for?
2. What is skin effect?
3. List the main RG-rated coaxial cables and their applications.
4. What is characteristic impedance? What are the circumstances under which a coaxial cable acts like a transmission line?
5. List the three main types of connectors used to terminate coax cables. What are each of these connectors used for?

27.6 SUMMARY

Here is a summary of the major points that were made in this chapter:

Introduction to Telecommunications

- All telecommunications systems contain three elements.
- A transmitter is used to convert the information into electrical signals. This process is called *encoding*.
- The encoded information is then transmitted through the medium.
- The medium carries the encoded information from the source to its destination.
  - The medium can be air, copper wire or cable, or fiber optic cable.
- The destination is called the *receiver.*
  - The receiver converts the electrical signals back to their original form. This process is called *decoding*.
- Commercial telecommunication systems refer to one or more of the following:
  - Broadcast transmission uses antennas to transmit and receive, and the air as the medium.
  - Cable transmission uses coax or fiber optic cable as the medium.
  - Although satellite transmission is used in one way or another by all commercial systems, the term *satellite transmission* describes a system in which a satellite dish is the receiver.
- Telephony is the study of wireless and land telephone systems.
- Three basic types of wire or cable are used in telecommunications.
  - Wire is used for telephony.
  - Coaxial cables are used for telephony, cable TV, and satellite TV connections.
  - Fiber optic cables are used for telephony and cable TV.

Twisted-Pair Cables

- A cable is an insulated electrical conductor, or group of conductors, that are used to transmit signals from one point to another.
- A twisted pair is simply two wires wrapped around each other.
  - Two or four twisted pairs are usually grouped in a single cable.
  - Two-pair cables are limited to low-frequency applications like telephone voice applications.
  - Four-pair cables are capable of operating at higher frequency. They can be used for telephone and data transmission.
- Twisted pairs can use solid or stranded wire.
  - The wire is usually between 22 gauge and 26 gauge.
- Twisting wires into pairs helps eliminate interference.
- Interference is the injection of unwanted electrical signals into the transmitted signal.
  - Electromagnetic interference (EMI) occurs when wires run parallel to each other in close proximity.
- When one pair of wires couples a signal into another pair of wires this is called cross-talk.
  - If the wires are wound as shown in Figure 27.2, they are no longer parallel.
  - To help reduce cross-talk, wire pairs in a given cable are twisted to different lengths. (See Figure 27.3.)
- The distance from one twist to another in a twisted pair is called the *twist length*.

- The two wires in a twisted pair are called the *tip* and the *ring*.
  - These names come from the use of ¼ inch jacks. (See Figure 27.4.)
- The two types of twisted pairs are the shielded twisted pair (STP) and the unshielded twisted pair (UTP).
  - The STP has conductive foil surrounding each pair and around the group, providing shielding from EMI and cross-talk.
  - A screened twisted pair (ScTP) has only one shield around the group, not around each pair.
- There are a number of categories (cats) of twisted pairs. (See Table 27.2.)
  - CAT5e is the most commonly used.
- Twisted-pair cables are marked with a three-letter application code. (See Table 27.3.)
- A plenum is the space above and below a room. A riser is the vertical space behind a wall.
  - Plenum cable is the highest quality cable.
  - You can always substitute a higher-quality cable for a lower, but not the other way around.
- The TIA/EIA dictates standardized color coding for twisted-pair cables.
  - Table 27.4 lists the color code for four-pair wire.
  - Table 27.5 lists the color code for two-pair wire.

Modular Jacks and Plugs

- A plug is a connector that attaches to the wire in a cable.
- A jack is the docking port for a plug. (See Figure 27.7.)
- Modular jacks are described based on the number of wire positions (P) and the number of conductors (c) that are connected to the jack.
  - For example, a 6P4C jack has six positions for wire connections, but only four wires are actually connected to the jack.
- The FCC has mandated a standard connection scheme for jacks. This standard is called the *registered jack* (RJ) standard.
  - Several types of jacks and their compatible plugs are listed in Table 27.7.
  - The RJ6 standard is used to connect telephone handsets to their bases.
  - The RJ11 standard is used to connect telephones to telephone lines.
  - The RJ45 standard is used to connect computers and their peripherals to each other and to telephone lines.
- Standard pin configurations refer to the standards that dictate how wires in a twisted pair are connected to modular plugs and jacks.
  - T568A and T568B are RJ45 standards.
- Some jacks and plugs are incorrectly referred to using RJ designations.

Termination Blocks

- The connector on the end of a cable is said to terminate the cable.
- A termination block is a physical structure to which cable wires are connected.
  - The two most common are the 66 and the 110 blocks.
- The 66 termination block is used to terminate telephone and data transmission lines, usually in nonresidential facilities.
  - They are usually found in dedicated cabinets. (See Figure 27.12.)
  - The 66 block actually has 50 rows of connectors, not 66.

    - The name *66 block* comes from the shape of the metal clips used, which look like the number 6.
    - The 66 block is also known as an *M-block*.
- The 66 block has 50 rows with 4 clips per row for a total of 200 clips.
    - The 66 block can terminate six four-pair UTP cables.
    - The tip and ring of each pair are connected to adjacent rows. (See Figure 27.14.)
- The 66 block has two configurations:
    - The nonsplit block contains four clips per row that are electrically connected to each other.
    - The two clips on the right in a split block are isolated from the two on the left.
    - The split block is the more common configuration.
- A bridging clip is used to connect the left-hand clips to the right-hand clips on a split block.
    - When a bridging clip is used, the two wires in that row are electrically connected.
- Wires are connected to a 66 block using a punch-down tool. (See Figure 27.17.)
    - The inside of the clip is sharp and cuts the wire's insulation so that the wire contacts the clip.
- 66 blocks cannot be used for applications higher than CAT5.
- For CAT5e or CAT6 a 110 termination block must be used.
- Table 27.8 provides a comparison of 66 and 110 blocks.
- 110 blocks use two separate components.
- The wiring block, or base, has two rows with 50 connection points each.
    - It can terminate twelve four-pair UTP cables. (See Figure 27.19.)
- The connecting block performs two functions.
    - It cuts through the insulation of the wires on the wiring block making the electrical connection.
    - It holds the wires in place.

Coaxial Cables

- A coaxial cable (coax) contains two conductors that are separated by a dielectric. (See Figure 27.21.)
    - The inner conductor may be solid or stranded.
    - For high-frequency applications, silver-clad copper is used due to skin effect.
- The dielectric material is chosen to have stable electrical properties over a wide range of frequencies.
    - Polypropylene and polyethylene are two common dielectrics.
- The outer conductor surrounds the dielectric. It is usually made of braided copper and/or aluminum.
    - For high-frequency applications, a second shield or braid may be added for better EMI protection.
- As well as providing environmental protection, the outer jacket may improve the cable's fire-retardant properties.
- RG ratings indicate the construction and application of specific cables. (See Table 27.8.)
    - The RG rating does not relate to quality.
    - RG-58 is the most commonly used cable.
- At high enough frequencies, coaxial cables can act as transmission lines.
- A transmission line has a characteristic impedance.
    - This means that an RF transmitter sees a constant cable impedance, regardless of length, so long as the cable is properly terminated and undamaged.

- Three common types of connectors are used to terminate coax cables.
  - The BNC connector is used in test equipment and data processing applications.
  - The N connector is used in land mobile, cellular, and pager infrastructure.
  - The F connector is used for satellite, cable TV, and high-speed cable Internet applications.

CHAPTER REVIEW

1. Converting information into electric signals is called ________ and converting it back is called ________.
   a) decoding, encoding
   b) modulating, demodulating
   c) encoding, decoding
   d) modulating, remodulating

2. The ________ is what carries the information that is being transmitted.
   a) format b) medium
   c) side band d) substrate

3. Normally, ________ twisted pairs are grouped together in a single cable.
   a) two or four b) six
   c) four or six d) four

4. One reason wires are twisted together is ________.
   a) to reduce capacitance
   b) to increase inductance
   c) to reduce interference
   d) to increase coupling

5. EMI between wire pairs is called ________.
   a) crosstalk
   b) radio frequency coupling
   c) reactive coupling
   d) damping factor

6. Twist length defines ________.
   a) the diameter of the twisted pair
   b) the cable length of a given twisted pair
   c) the length of wire for the rated impedance
   d) the distance from one twist to another

7. The terms *tip* and *ring* come from the use of ________ in early telephone circuits.
   a) mechanical switches
   b) ¼-inch jacks
   c) RCA jacks
   d) mechanical bell circuits

8. An STP has ________ than a UTP.
   a) better attenuation characteristics
   b) better EMI shielding
   c) poorer EMI shielding
   d) lower reactive losses

9. Screened twisted pair differs from foil twisted pair in ________.
   a) no way at all
   b) that it has wider bandwidth
   c) price
   d) that it has lower impedance

10. The twisted-pair cable with the widest bandwidth is ________.
    a) CAT5e b) CAT6a
    c) CAT7 d) CAT7b

11. The most commonly used twisted-pair cable is ________.
    a) CAT5e b) CAT6a
    c) CAT7 d) CAT7b

12. A plenum is a ________ while a riser is a ________.
    a) horizontal space, vertical space
    b) heat sink, support member
    c) conduit, bracket
    d) conduit, coupler

13. In the TIA/EIA four-pair standards, the tip ________ is the same as the ring ________.
    a) twist length, twist length
    b) impedance, impedance
    c) wire color, stripe color
    d) riser, plenum

14. In twisted-pair application codes, the highest-quality cable is ________.
    a) C b) P
    c) G d) R

15. It is acceptable to replace a ________ cable with a ________ cable.
    a) general purpose, riser
    b) riser, plenum
    c) copper, riser
    d) all of the above

16. A docking port for a plug is called a ________.
    a) termination b) jack
    c) crimp d) mount

17. Modular jacks are described by the number of ________ and ________.
    a) conductors, wire positions
    b) twisted pairs, bundles
    c) twisted pairs, wire positions
    d) wire positions, bundles

18. The ________ is considered to be a universal jack.
    a) 4P jack
    b) 6P jack
    c) 8P jack
    d) all of the above

19. An RJ11 ________.
    a) has six wire positions
    b) has four conductors connected to it
    c) is a 6P4C jack
    d) all of the above

20. The 66 termination block gets its name from ________.
    a) the number of wires connected to it
    b) the number of docking ports it has
    c) the number of bundles it can terminate
    d) the shape of the metal clips in the block

21. The number of rows on a 66 block is ________.
    a) 66
    b) 100
    c) 110
    d) 50

22. A 66 block is also known as a(n) ________.
    a) punch-down block
    b) data block
    c) M-block
    d) universal bus connector

23. In a 66 block the ________ of each twisted pair is (are) always connected to adjacent rows.
    a) tip and ring
    b) riser
    c) clip ring
    d) four clips

24. To connect the right-hand and left-hand clips on a split block you use a ________.
    a) jumper wire
    b) bridging clip
    c) split-block ring
    d) plenum jumper

25. Wires are connected to a 66 block using a ________.
    a) punch-down tool
    b) crimper
    c) ring splitter
    d) all of the above

26. 66 blocks can be used in applications up to and including ________.
    a) CAT5 b) CAT5e
    c) CAT7 d) CAT6a

27. A 110 block has ________ rows with ________ connection points.
    a) four, 50 b) two, 50
    c) two, 100 d) four, 100

28. In a 110 block, even-numbered pins are ________ odd-numbered pins.
    a) taller than
    b) on either side of
    c) underneath
    d) shorter than

29. Skin effect relates to the fact that current ________ as frequency ________.
    a) changes phase, increases
    b) group delay increases, decreases
    c) concentrates in the center of the conductor, increases
    d) moves towards the surface of a conductor, increases

30. A common dielectric material used in coax cable is ________.
    a) glass b) polystyrene
    c) polypropylene d) aluminum braid

31. The most common type of coax cable is RG-________.
    a) 8 b) 11
    c) 58 d) 60

32. A 10-meter length of RG-60 coax has a characteristic impedance of 50 Ω. This means that the characteristic impedance of 100 meters of RG-60 is ________.
    a) 50 Ω b) 100 Ω
    c) 500 Ω d) 5 Ω

33. Coax cables are usually terminated with ________ connectors.
    a) F b) N
    c) BNC d) all of the above

34. If the frequency is high enough, a cable can begin to act as a(n) ________.
    a) inductor
    b) transmission line
    c) capacitor
    d) resonant circuit

CHAPTER 28

INTRODUCTION TO FIBER OPTIC CABLING

PURPOSE

Copper-based telecommunications systems transmit information in the form of electric signals over solid (or stranded) wires. Fiber optic systems, on the other hand, transmit information in the form of light—most commonly infrared light. In this chapter, we take a brief look at optical fibers and fiber optic systems.

KEY TERMS

The following terms are introduced and defined in this chapter on the pages indicated:

Absorption, 1203
Buffer coating, 1200
Cladding, 1200
Coherent, 1209
Connector, 1205
Connector loss, 1205
Critical angle, 1201
Duplex connector, 1206
FC connector, 1207
Ferrule, 1206
Fiber optic cable, 1200
Fiber optics, 1195
Fusion splice, 1208
Index of refraction, 1199
LC connector, 1207
Light, 1195
Light-emitting diode (LED), 1208
Light intensity, 1198
Mechanical splice, 1208
Mode, 1200
Multimode fiber, 1200
Optical fiber, 1200
Photodetectors, 1197
Photodiode, 1210
Photoemitters, 1197
PIN diode, 1209
Refraction, 1199
Scattering, 1203
Simplex connector, 1205
Single-mode fiber, 1200
Snap coupling (SC) connector, 1206
Splice, 1205
ST connector, 1206
Wavelength (λ), 1196

OBJECTIVES

After completing this chapter, you should be able to:

1. Describe light.
2. Describe the relationship between frequency and wavelength.
3. Calculate the wavelength of any light frequency.
4. Describe the inverse-square law of light intensity.
5. Describe the physical construction of a basic fiber optic cable.
6. Contrast *single-mode* and *multimode* optical fibers.
7. List the three parts of a fiber optic telecommunications system.
8. Explain the concept of the *critical angle* of reflection.
9. List the advantages that fiber optic systems have over copper-based systems.
10. Discuss light attenuation and its causes.

11. Compare and contrast fiber *splices* and *connectors*.
12. Discuss connector loss and its causes.
13. Contrast *simplex* and *duplex* connectors.
14. List and describe the commonly used optical fiber connectors.
15. List and describe the types of optical fiber splices.
16. Describe light-emitting diodes (LEDs) and lasers.
17. Describe PIN diodes and photodiodes.

28.1 LIGHT CONCEPTS

FIBER OPTICS is the area of telecommunications that uses light to transmit information through glass or plastic cables. But before we begin discussing the specifics of fiber optics and fiber optic cabling, let's take a look at some of the properties of *light*.

LIGHT is electromagnetic energy that falls within a specific range of frequencies, as shown in Figure 28.1. The entire light spectrum falls within the range of 30 THz to 3 PHz and is further broken down as shown in Table 28.1. Note that infrared light is used in fiber optic telecommunications.

FIBER OPTICS The area of telecommunications that uses light to transmit information through glass or plastic cables.

LIGHT Electromagnetic energy that falls within a specific range of frequencies.

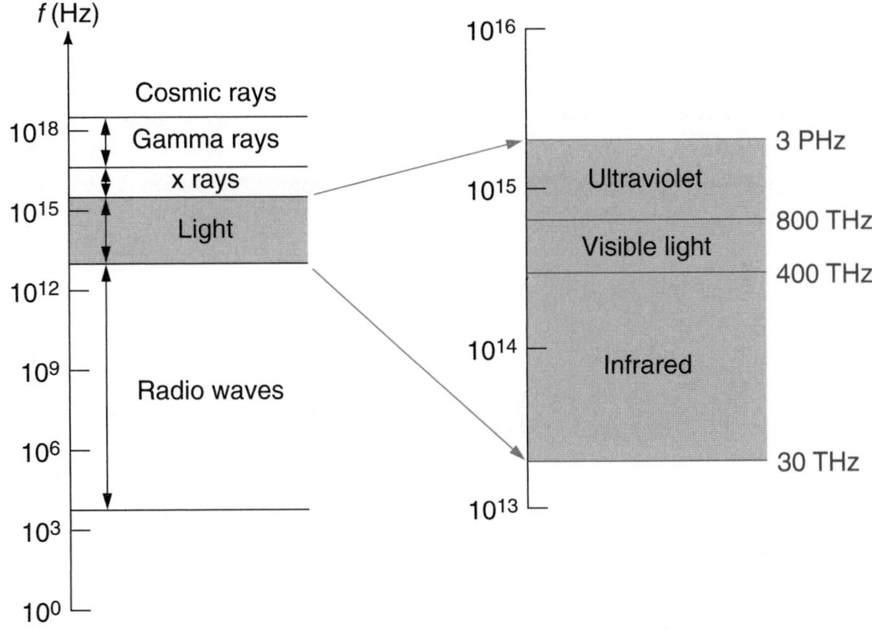

FIGURE 28.1 The light spectrum.

TABLE 28.1 • Light Frequency Ranges

TYPE OF LIGHT	APPROXIMATE FREQUENCY RANGE[1]
Infrared	30 THz to 400 THz
Visible	400 THz to 800 THz
Ultraviolet	800 THz to 3 PHz

[1]T stands for *tera* (10^{12}) and P stands for *peta* (10^{15})

The Speed of Light

In a vacuum, all electromagnetic waves travel at *the velocity of electromagnetic radiation* (c), which is more commonly referred to as *the speed of light*. Some commonly used speed-of-light measurements are:

- 186,000 mi/s (miles per second)
- 3×10^8 m/s (meters per second)

- 300 m/μs (meters per microsecond)
- 984 ft/μs (feet per microsecond)
- 3×10^{17} nm/s (nanometers/second)

We are interested in the speed of light because it is a constant used to calculate the *wavelength* of a transmitted signal. It should be noted that electromagnetic waves travel at lower speeds in any medium other than free space. For example, the velocity of light traveling through an optical fiber is approximately 30% lower than that of light traveling through free space. Even so, it is common practice to base wavelength calculations—and component ratings—on the free-space velocity of light.

Wavelength

WAVELENGTH The physical length of one cycle of a transmitted electromagnetic wave.

Two characteristics are commonly used to describe light. The first of these is **WAVELENGTH** (symbolized by λ, the Greek letter *lambda*). Wavelength is the physical length of one cycle of a transmitted electromagnetic wave. The wavelength of a transmitted signal depends on:

- The speed at which it travels (i.e., the speed of light).
- The frequency of the waveform.

The concept of wavelength (λ) is illustrated in Figure 28.2. The wavelength for a signal at a given frequency is found using

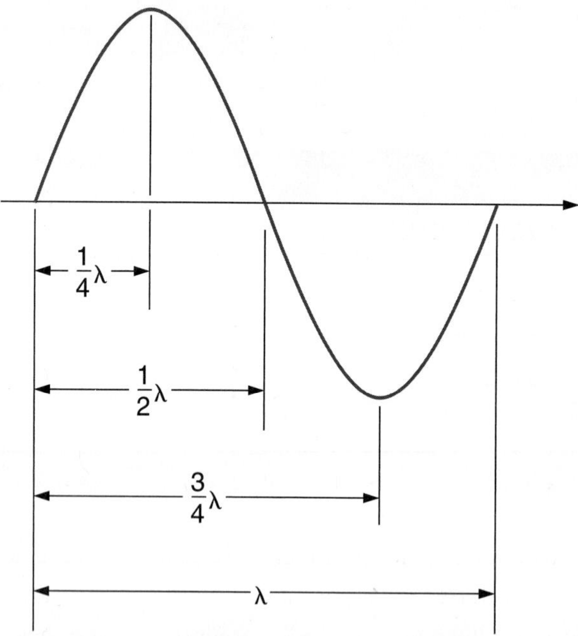

FIGURE 28.2 Wavelength.

(28.1)

$$\lambda = \frac{c}{f}$$

where

λ = the wavelength of the signal (in *nanometers*, nm)
c = the speed of light, given as 3×10^{17} nm/s
f = the frequency of the transmitted signal in Hz.

The following example demonstrates how to calculate the wavelength of a transmitted signal.

EXAMPLE 28.1

Determine the wavelength of a 150 THz infrared light signal.

SOLUTION

The wavelength of the signal is found as

$$\lambda = \frac{c}{f}$$

$$= \frac{3 \times 10^{17} \text{ nm/s}}{150 \times 10^{12} \text{ Hz}}$$

$$= 2000 \text{ nm}$$

Thus, the physical length of one cycle of a 150 THz signal is 2000 nm, or 2 μm.

PRACTICE PROBLEM 28.1

The infrared frequency spectrum falls between 30 THz and 400 THz. Determine the minimum and maximum wavelengths for this range of frequencies.

Wavelength is an important characteristic of light because **PHOTOEMITTERS** (devices that generate light) and **PHOTODETECTORS** (circuits that respond to light) are rated for specific wavelengths. You see, photoemitters are generally used to drive the inputs of photodetectors, as shown in Figure 28.3. For a specific emitter to be used with a specific detector, the two must be rated for the same approximate wavelength. Otherwise, the detector will not respond to the emitter output.

PHOTOEMITTERS Electronic devices that generate light.

PHOTODETECTORS Electronic devices that respond to light.

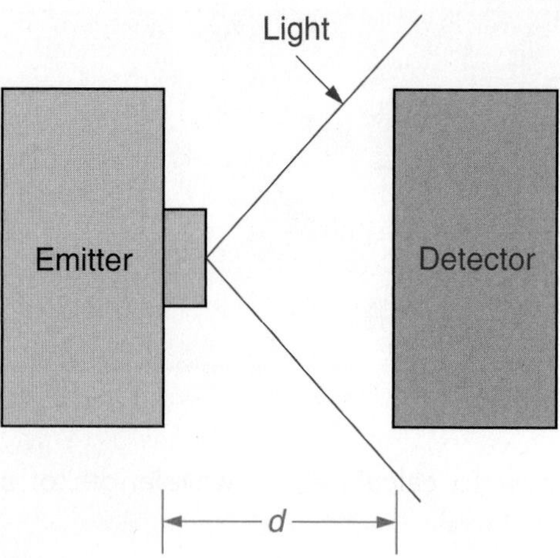

FIGURE 28.3 Light intensity.

Light Intensity

LIGHT INTENSITY The amount of light per unit area received by a given photodetector.

The second important light characteristic is *intensity*. **LIGHT INTENSITY** is the amount of light per unit area received by a given photodetector. For example, consider the emitter and detector in Figure 28.3. The light intensity is a measure of the amount of light received by the detector. Note that light intensity decreases as the distance (*d*) between the emitter and detector increases. Specifically, light intensity is inversely proportional to *the square* of the distance. For example, at 2 meters from its source, light intensity is

$$\frac{1}{2^2} = \frac{1}{4}$$

its original value. At 3 meters from its source, light intensity is

$$\frac{1}{3^2} = \frac{1}{9}$$

its original value. This relationship is described as the *inverse-square law*.

Reflection

When light strikes a reflective surface, it bounces off that surface as illustrated in Figure 28.4. When light is reflected as shown in the figure, there are two angles of

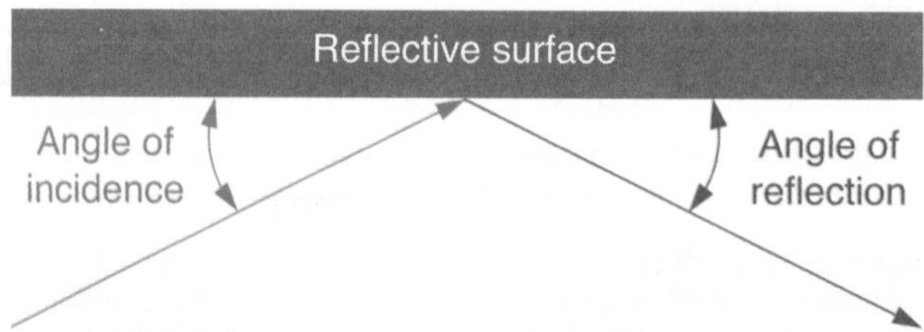

FIGURE 28.4 Angles of incidence and reflection.

interest: the *angle of incidence* (θ_I) and the *angle of reflection* (θ_R). These two angles are identified in the figure.

Normally, the angle of reflection equals the angle of incidence ($\theta_R = \theta_I$). For example, if light hits a reflective surface at a 12° angle, it bounces off the mirror at the same angle, 12°.

Refraction

REFRACTION is the bending that occurs when light passes between two mediums with different optical densities. A classic example of refraction is the way a pencil appears to bend when you stick it in a glass of water. The light bends as it passes from one medium (air) to the other (water), which makes the pencil appear to bend. This is a result of the velocity of light changing as it passes from one medium to another. The velocity of light in a given material is indicated by the material's **INDEX OF REFRACTION**. Free space has a refractive index of 1.0. All other materials have a higher index. For example, the refractive index of glass is approximately 1.5. Water has a refractive index of 1.33, and air is very close to 1.0.

REFRACTION The bending that occurs when light passes between two mediums with different optical densities.

INDEX OF REFRACTION The ratio of the speed of light in a vacuum (free space) to its velocity in a given medium.

As you will see, the concepts of reflection, refraction, and loss of intensity all play a role in the transmission of light through an optical fiber.

PROGRESS CHECK

1. What is the medium for fiber optic telecommunications?
2. What is the frequency range of light? Name three types of light.
3. Express the speed of light (c) using three common units of measure.
4. What is wavelength? What is its symbol? What equation is used to calculate wavelength?
5. What are photoemitters and photodetectors and how are they related?
6. Define light intensity.
7. Describe the *inverse-square law* of light intensity.
8. Define the angle of incidence (θ_I) and the angle of reflection (θ_R). How are they related?
9. Define and explain what causes refraction. What is the index of refraction?

28.2 INTRODUCTION TO FIBER OPTIC CABLES

OPTICAL FIBER A length of ultra-pure glass or plastic that provides a closed pathway for transmitting light between two points.

OPTICAL FIBER is a length of ultra-pure glass or plastic that provides a closed pathway for transmitting light between two points. Glass optical fibers are by far the most commonly used, so we will assume that the optical fibers in all our discussions are made of glass.

Because fiber cables are more delicate than copper wiring, they must be protected during construction, especially in situations where there will be a long lag time between installation work and final termination work.

IN THE FIELD

A **FIBER OPTIC CABLE** contains an optical fiber that is surrounded by a layer of glass called **CLADDING**, as shown in Figure 28.5. The cladding has reflective properties that help keep light in the core of the fiber. The optical fiber and cladding are surrounded by a **BUFFER COATING** that is similar to the insulation surrounding a wire: The buffer coating protects the optical fiber from water and mechanical stresses.

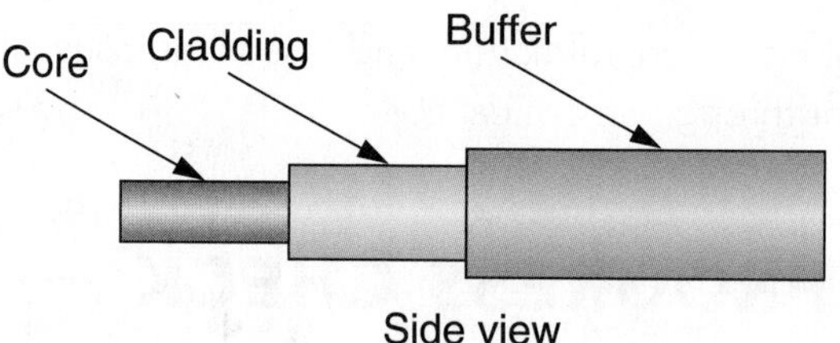

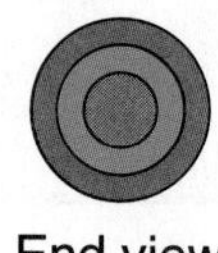

FIGURE 28.5 Parts of a fiber optic cable.

FIBER OPTIC CABLE A cable containing an optical fiber that is surrounded by a layer of glass called *cladding*.

CLADDING Glass that has reflective properties that help keep light in the core of an optical fiber.

BUFFER COATING A coating that protects an optical fiber from water and mechanical stresses.

SINGLE-MODE FIBER One with a diameter so narrow that it provides only one path, or *mode*, for light to pass through the cable.

MODE A path for light through an optical fiber.

MULTIMODE FIBER One with a diameter large enough to provide multiple paths (modes) for light to pass through the cable.

Types of Optical Fiber

There are essentially two types of optical fiber: *single-mode* and *multimode*. A **SINGLE-MODE FIBER** is one with a diameter so narrow that it provides only one path (or **MODE**) for light to pass through the cable. For example, one type of single-mode fiber has a core diameter of only 7.9 μm. A **MULTIMODE FIBER** is one with a diameter large enough to provide multiple paths (modes) for light to pass through the cable. A single-mode fiber, on the other hand, is so narrow that it can pass only one light signal. Multimode cables, even those having relatively small diameters, can support many modes. For example, one type of fiber with a core diameter of only 50 μm can support over 1000 modes.

Reflection in Optical Fibers

When light enters the core of an optical fiber at an angle, it strikes the cladding. Because of the cladding's high index of refraction, its surface acts as a reflective surface. As with any reflective surface, the angle of reflection (θ_R) normally equals the angle of incidence (θ_I). However, a point can be reached where the angle of incidence is so great that most of the light is no longer reflected. Instead, the light

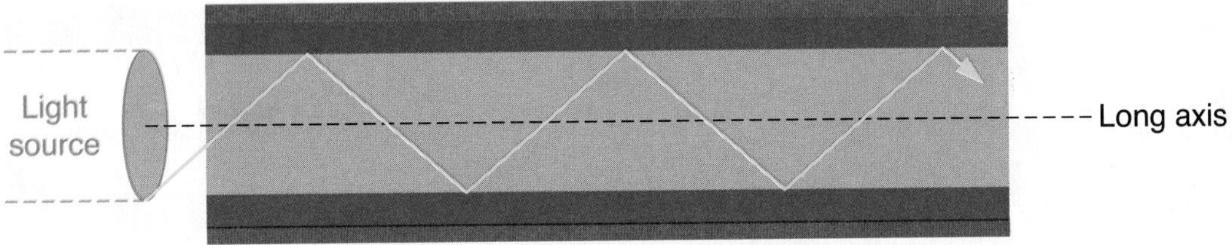

(a) Light reflecting through an optical fiber when the angle of incidence is lower than the critical angle

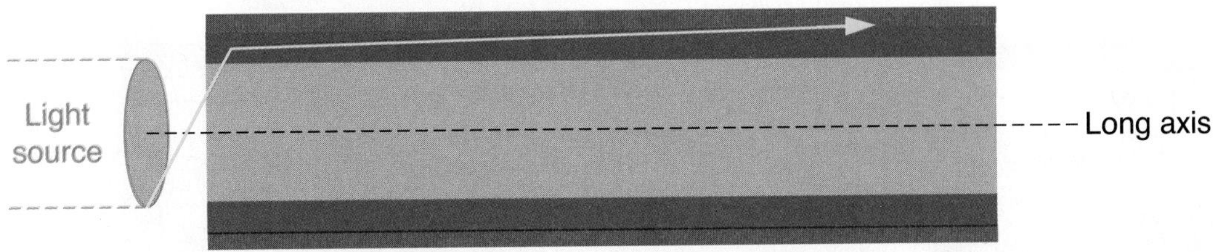

(b) Light passes into the cladding and is absorbed when the angle of incidence exceeds the critical angle

FIGURE 28.6 Fiber optic critical angle.

enters the cladding, where it is absorbed as shown in Figure 28.6. The angle of incidence at which this occurs is called the **CRITICAL ANGLE**.

CRITICAL ANGLE The angle of incidence where most of the light in an optical fiber enters the cladding rather than being reflected back into the core of the fiber.

A Basic Fiber Optic Telecommunications System

A fiber optic telecommunications system can be broken down into three primary components:

- A light source at the transmitting end
- The fiber optic cable
- A light detector at the receiving end

This system is represented in Figure 28.7. Here's how it works: In most cases, the transmitter converts (encodes) information to strings of ones and zeros. These ones and zeros are transmitted through the fiber optic cable as a series of light pulses. Finally, the receiver accepts the information and decodes it. This process is illustrated in Figure 28.8.

Fiber is becoming an economical alternative to copper for LAN wiring as a result of lower cost of fiber (vs. copper) and new fiber-optic connectors that offer savings in cost and space. Also, a low-cost standard for Fast Ethernet, called 100BASE-SX, has dramatically cut the cost of fiber-optic systems.

IN THE FIELD

Encoder
Light source
Long axis
Light detector
Decoder
Transmitter
Optical fiber
Receiver

FIGURE 28.7 A fiber optic telecommunication system

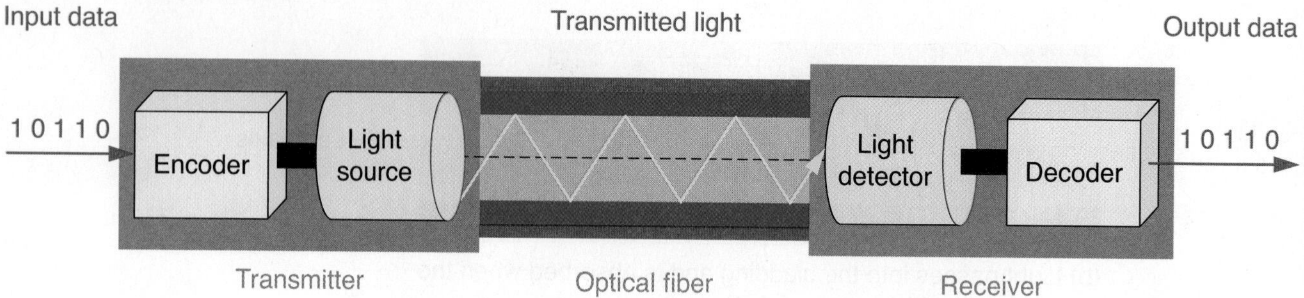

FIGURE 28.8 Data transfer through a fiber optic system.

New fiber systems are now sending 40 Gb/s (gigabits per second), and could theoretically go many times higher, while the greatest amount of data sent over copper is usually about 155 Mb/s (megabits per second).

IN THE FIELD

A fiber optic system has several advantages over the copper communications systems introduced in Chapter 27. Specifically, fiber optic cables and systems:

1. Are not susceptible to electromagnetic interference (EMI) or cross-talk like other telecommunications systems.
2. Can transmit much more information with far fewer errors than other telecommunications systems.
3. Are more secure than copper-based telecommunications systems.

Despite the advantages that a fiber optic system has over other telecommunications media, it is not without its own issues. We'll take a look at one of those issues now.

Due to the electrical properties of copper wiring, data signals undergo some corruption during transmission. Signal corruption within certain limits is acceptable, but if the electrical properties of the cable cause serious signal distortion, the cable must be repaired or replaced.

Attenuation

The *ideal* optical fiber would be structurally perfect, allowing light signals to pass without any negative effects. The ideal optical fiber, however, does not exist. Impurities left over from the manufacturing process cause a loss of light intensity (*attenuation*). The attenuation of a light signal as it passes through a fiber optic cable occurs when the light strikes impurity molecules in the fiber. When light hits an impurity molecule, one of two things happens:

1. The light reflects off the impurity molecule and heads in an unpredictable direction. This phenomenon, known as **SCATTERING**, is illustrated in Figure 28.9.
2. The molecule absorbs energy from the light and releases it in the form of heat. This phenomenon, known as **ABSORPTION**, is illustrated in Figure 28.10.

Of the two, scattering is the greater source of light attenuation in fiber optic cables.

SCATTERING The phenomenon in which light reflects off an impurity molecule in an optical fiber and heads in an unpredictable direction.

ABSORPTION The phenomenon in which an impurity molecule absorbs energy from the light and releases it in the form of heat.

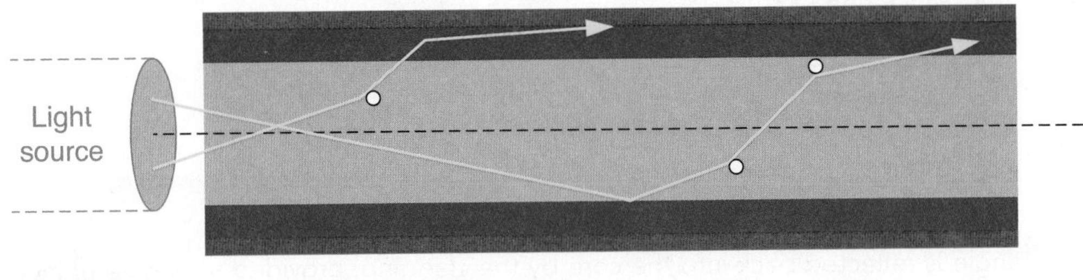

FIGURE 28.9

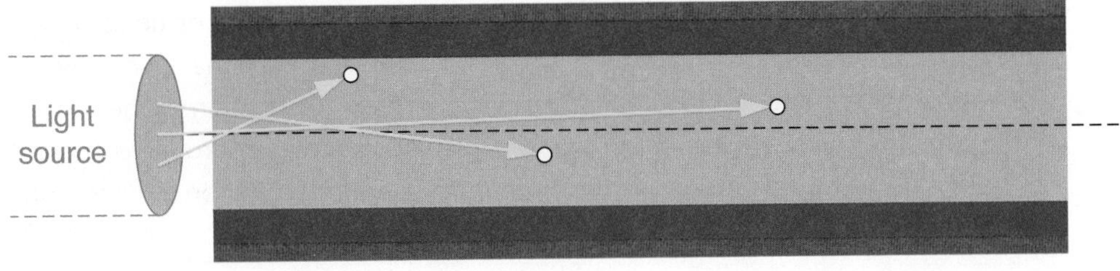

FIGURE 28.10

The effects of both scattering and absorption are more pronounced at some light wavelengths than at others. This point is illustrated by the graph shown in Figure 28.11.

Figure 28.11 shows that the attenuation caused by scattering decreases as wavelength increases frequency decreases, and that attenuation caused by absorption is significant over specific wavelength ranges that center on 1000 nm, 1400 nm, and 1600 nm. Little absorption occurs outside the ranges shown in the figure.

Putting It All Together

An optical fiber is a length of ultra-pure glass or plastic that provides a closed pathway for transmitting light between two points. Though sometimes made of plastic, optical fibers are far more often made of glass.

The center of a fiber optic cable is called the *core*. The glass that surrounds the core, called the *cladding*, has a reflective surface. Light entering the core at an

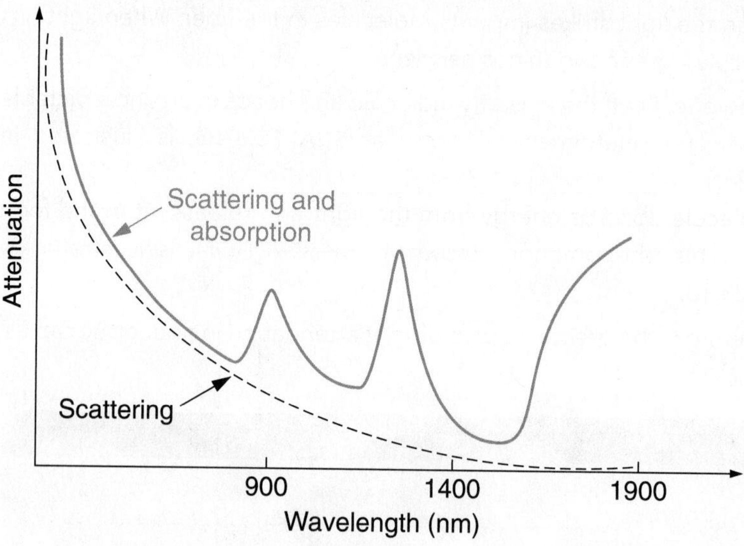

FIGURE 28.11 Attenuation in an optical fiber.

angle is reflected back into the core by the cladding, provided the angle of incidence is lower than the *critical angle* rating of the cable.

As light passes through an optical fiber, it loses some intensity (power) through attenuation. The primary causes of attenuation in a fiber optic cable are *scattering* and *absorption*. Scattering is the redirecting of light that occurs when light waves hit impurity molecules in the optical fiber. Absorption is the loss of light energy that occurs when impurity molecules absorb that energy and convert it to heat. Of the two, scattering is the primary cause of attenuation in fiber optic cables.

PROGRESS CHECK

1. Describe the basic construction of an optical fiber.
2. What is a mode? Describe the difference between single-mode and multimode fiber.
3. Define the term critical angle.
4. List and describe the three primary components in a fiber optic telecommunication system.
5. List three advantages that a fiber optic system has over a copper system.
6. Define the following terms:
   a. attenuation
   b. scattering
   c. absorption
7. Explain what effect frequency has on scattering and absorption.

28.3 FIBER OPTIC CONNECTORS AND SPLICES

Optical fibers are connected to each other, and to light emitter and detector circuits, using *splices* and *connectors*. A **SPLICE** is a mechanical connection that permanently joins two optical fibers. A **CONNECTOR**, on the other hand, connects one optical fiber to another—or to some device—in a way that allows the two to be disconnected.

When attaching a splice or a connector to an optical fiber, care must be taken to avoid introducing any *connector losses*. The term **CONNECTOR LOSS** describes any decrease in light intensity (measured in dB) that is introduced when a connector or splice is attached to an optical fiber. In this section, we identify some causes of connector loss. We will also look at some of the more common fiber optic connectors and splices.

SPLICE A mechanical connector that permanently joins two optical fibers.

CONNECTOR A mechanical connector that connects an optical fiber to another (or to some device) in a way that allows the two to be disconnected.

CONNECTOR LOSS Any decrease in light intensity that is introduced when a connector or splice is attached to an optical fiber.

Connector and Splice Losses

All connector losses can be avoided to one degree or another by an experienced technician. Connector losses are most often caused by contaminants (like dust and dirt), misalignment of the fiber(s) with the splice or connector, poor preparation and polishing of the fiber, or air gaps between two spliced fibers.

Contaminants are the primary cause of connector loss. The presence of contaminants can be reduced by:

- Working in clean areas away from heating and air conditioning vents.
- Cleaning connectors with lint-free pads.
- Keeping dust caps on anything that will be connected to an optical fiber.

Simplex and Duplex Connectors

Connectors are classified as *simplex* or *duplex*. A **SIMPLEX CONNECTOR** like the ones in Figure 28.12a are used to terminate a single fiber. In contrast, a

SIMPLEX CONNECTOR One that is used to terminate a single optical fiber.

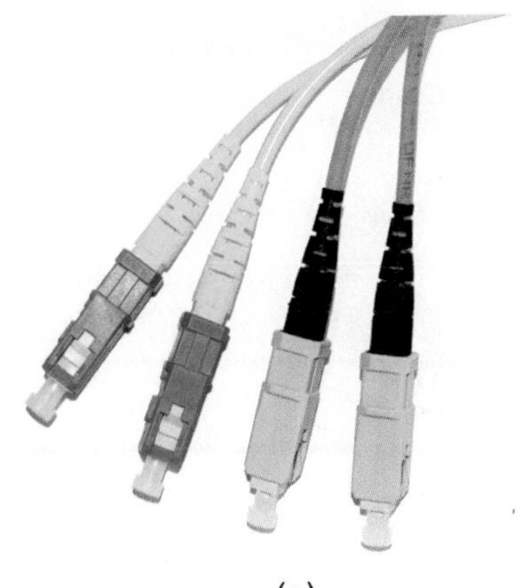

(a)

(b)

FIGURE 28.12 Simplex and duplex connectors.
(a) Fotolia.com; (b) Istock.com

DUPLEX CONNECTOR One that is used to terminate two optical fibers.

ST CONNECTOR A simplex, multimode, twist-on connector.

FERRULE A precision-drilled ceramic, metal, or plastic sleeve to which a bare optical fiber is permanently attached.

DUPLEX CONNECTOR like the ones in Figure 28.12b is used to terminate two fibers. Typically, one of the fibers is used to send (transmit) information while the other is used to receive information.

Common Connectors

The most common fiber connector is the *single tip* (ST) connector. The **ST CONNECTOR** is a simplex, multimode, twist-on connector like the one shown in Figure 28.13. The ST connector contains a 2.5 mm **FERRULE**—a precision-drilled ceramic, metal, or plastic sleeve to which a bare optical fiber is permanently attached using an adhesive. When attached to a ferrule, the end of the optical fiber is aligned with the end of the ferrule and polished smooth. ST connectors are widely used in optical LANs (Local Area Networks).

> It is important to leave extra fiber-optic cable at each access point. A minimum of nine feet (2.74 meters) in telecommunications closets and three feet (0.914 meter) at outlets is recommended.
>
> IN THE FIELD

As its name implies, the **SNAP COUPLING (SC) CONNECTOR** is a simplex or duplex connector that snaps into a docking jack. An SC connector is shown in Figure 28.14.

SNAP COUPLING CONNECTOR A simplex or duplex connector that snaps into a docking jack in a manner similar to an adaptor.

The SC connector is available in both single-mode and multimode configurations. Like the ST connector, the SC connector contains a 2.5 mm ferrule that protrudes slightly from the connector as shown in the figure. An *adaptor* (or

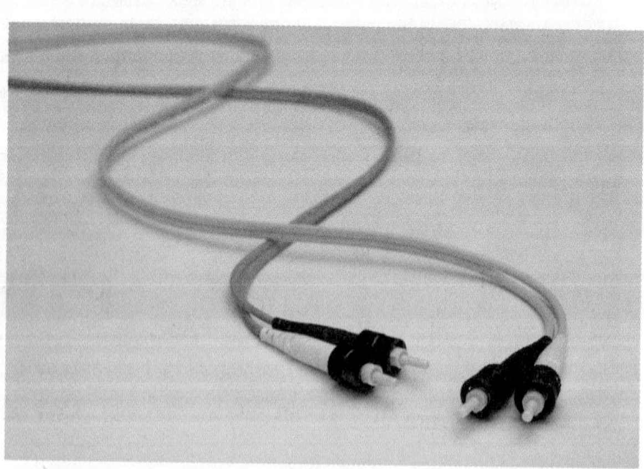

FIGURE 28.13 ST connectors.
Istock.com

FIGURE 28.14 An SC connector.
Istock.com

mating sleeve) like those shown in Figure 28.15 allows one SC connector to be mated to another.

An **FC CONNECTOR** is a simplex single-mode or multimode screw-on connector. An FC connector is shown in Figure 28.16. Like the ST and SC connectors, the FC connector contains a 2.5 mm ferrule. The FC connector is the oldest of the connector types introduced here.

FC CONNECTOR A simplex single-mode or multimode screw-on connector.

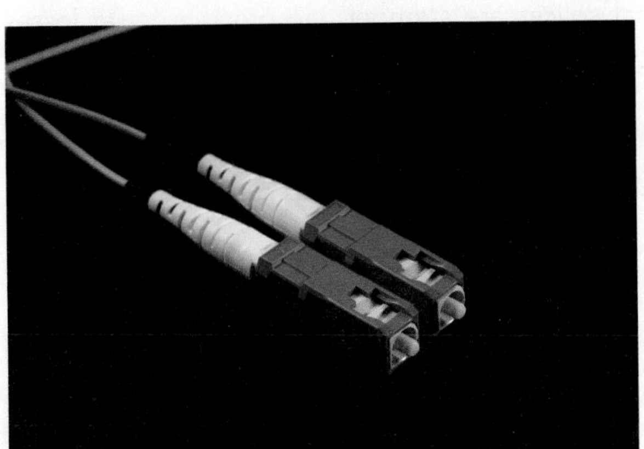

FIGURE 28.15 SC connector mating sleeves.
Fotolia.com

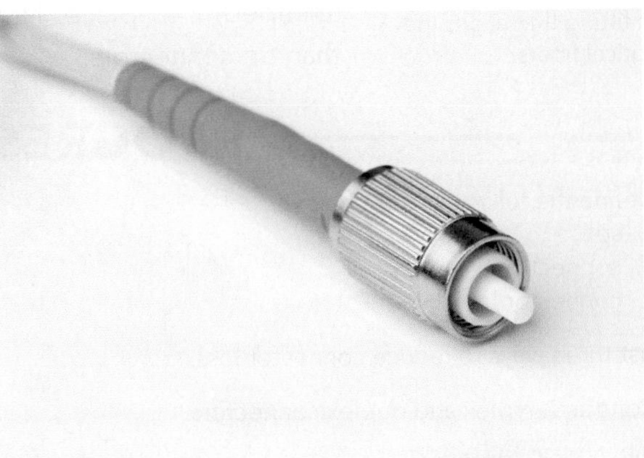

FIGURE 28.16 An FC connector.
Istock.com

The **LC CONNECTOR** shown in Figure 28.17 is a simplex or duplex connector that is available in both single-mode and multimode configurations. This snap-in connector resembles the plugs used with RJ-XX modular jacks. The LC connector contains a 1.25 mm ferrule.

LC CONNECTOR A simplex or duplex connector that is available in both single-mode and multimode configurations.

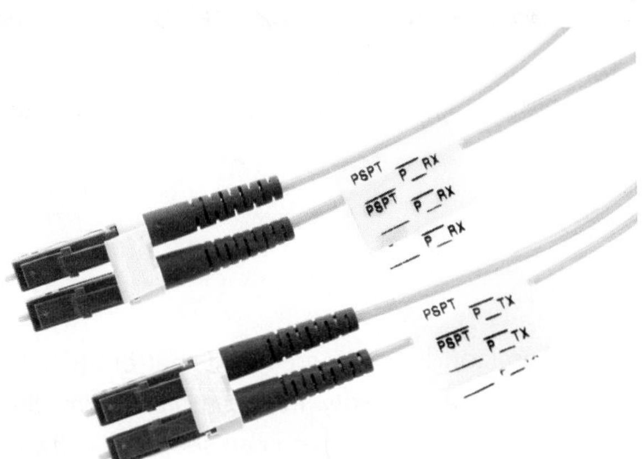

FIGURE 28.17 LC connectors.
Fotolia.com

There are many more optical fiber connectors than those presented here. The ones described in this section, however, are those most commonly found in *premises applications*; that is, in buildings and on campuses.

Splices

FUSION SPLICE A heat process that welds two optical fibers together.

MECHANICAL SPLICE A plastic fitting that aligns the two optical fibers.

As stated earlier, a splice is a permanent connection between two optical fibers. There are two types of splices: *fusion splices* and *mechanical splices*. A **FUSION SPLICE** is essentially a heat process that welds two optical fibers together. A **MECHANICAL SPLICE** is a plastic fitting that aligns the two optical fibers. Once the fibers are aligned, the mechanical splice is physically locked (crimped) into place. Note that splices are stronger and produce fewer losses than do connectors.

PROGRESS CHECK

1. Define the following terms:
   a. splice
   b. connector
   c. connector loss
2. List three ways to reduce connector loss.
3. Contrast simplex and duplex connectors.
4. What is a ferrule?
5. Define and compare the following connectors:
   a. ST connector
   b. SC connector
   c. FC connector
   d. LC connector
6. What is the difference between a fusion splice and a mechanical splice?

28.4 LIGHT SOURCES AND DETECTORS

Earlier in the chapter, you were told that optical fibers are used to connect light sources (emitters) and detectors. In this section, we take a very brief look at light emitters and light detectors.

Light Sources: Light-Emitting Diodes

LIGHT-EMITTING DIODE (LED) A semiconductor component that emits light when activated by a voltage having the proper polarity.

A **LIGHT-EMITTING DIODE (LED)** is a semiconductor component that emits light when activated by a voltage having the proper polarity. In its simplest form, the operation of an LED is as illustrated in Figure 28.18. As you can see, one voltage polarity does not cause the LED to light, but the opposite polarity does. The light generated by some LEDs is well suited for use with multimode optical fibers.

Light Sources: Lasers

COHERENT A term that is used to describe light that is highly directional and focused.

LASER is an acronym for *Light Amplification by Stimulated Emission of Radiation.* The light produced by a laser is described as being **COHERENT**. One aspect of

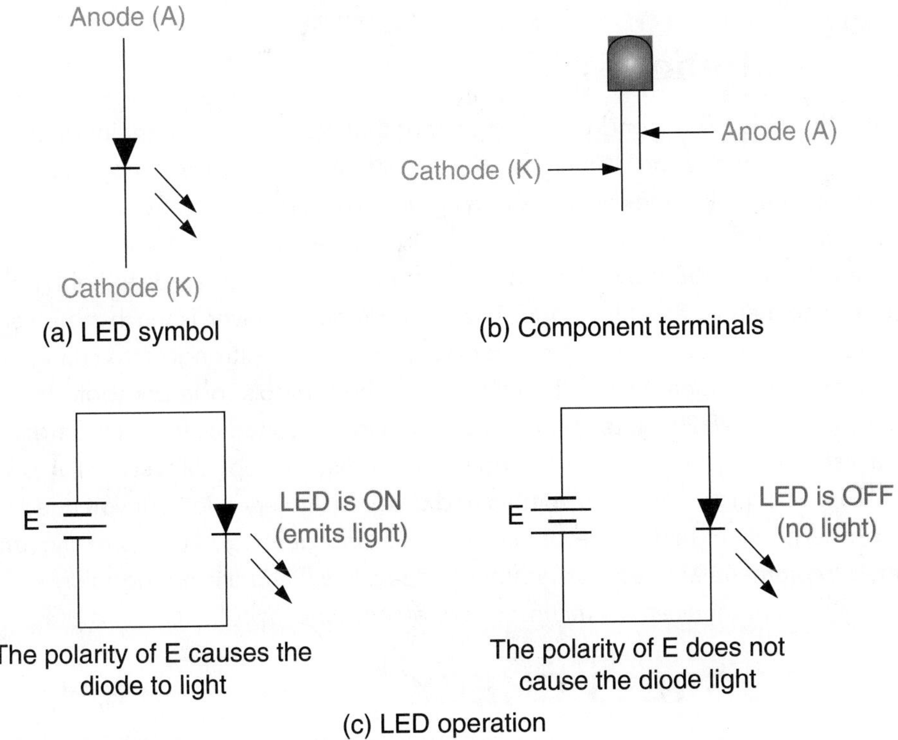

FIGURE 28.18 Light emitting diodes (LEDs).

coherent light that makes it very useful in fiber optic communication systems is that it is highly directional and focused. In fact, the light produced by a laser remains highly focused over significant distances.

Though once found almost exclusively in research facilities, lasers are now very common. Laser scanners are found in stores, CD players, and DVD players. Laser pointers are also widely available. The light generated by a small laser source can be seen passing through an optical fiber in Figure 28.19. Note that the light generated by some lasers is well suited for use with single-mode optical fibers.

FIGURE 28.19 Laser light showing through an optical connector.
Fotolia.com

PIN DIODE A semiconductor component that responds to a light input by increasing its conduction.

PHOTODIODE A semiconductor component that responds to light input in the same manner as a PIN diode.

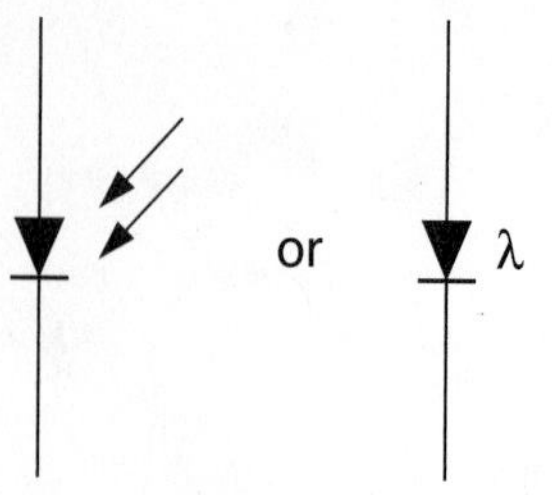

FIGURE 28.20 Photodiode symbols.

Light Detectors: PIN Diodes and Photodiodes

A **PIN DIODE** is a semiconductor component that responds to a light input by increasing its conduction. While an LED essentially works as a voltage-to-light converter, the PIN diode can serve as a light-to-voltage converter when properly wired into a circuit.

A **PHOTODIODE** is a semiconductor component that responds to light input in the same manner as a PIN diode. The photodiode, however, is much more sensitive to light inputs. That is, a photodiode will react more strongly to a change in a low-level light signal than will a PIN diode. The symbols for a photodiode are shown in Figure 28.20. Note that a photodiode is well suited as the light detector for a laser (which is normally employed in longer-distance optical systems).

This introduction to light emitters and detectors is very brief, but you need to keep one point in mind: *The optical emitters and detectors are normally contained in premanufactured transmitters and receivers, and are not normally serviced by telecommunications technicians or electricians.*

PROGRESS CHECK

1. Describe the basic operation of a light-emitting diode (LED).
2. What does the acronym LASER stand for?
3. Describe coherent light and explain why it is useful in fiber optic telecommunication.
4. What is a PIN diode? What is it used for?
5. What is a photodiode? How does it differ from a PIN diode?

28.5 SUMMARY

Here is a summary of the major points that were made in this chapter:

Light Concepts

- Fiber optics is the area of telecommunications that uses light to transmit information through glass or plastic cables.
- Light is electromagnetic energy that falls within the frequency range of 30 THz to 3 PHz. (See Figure 28.1.)
    - There are three types of light: infrared, visible, and ultraviolet.
    - Infrared light is used in fiber optic telecommunications.
- The speed of light (c) is a constant in a vacuum. It is expressed in several units of measure, such as 186,000 miles/second, 300 m/μs, and 3×10^{17} nm/s.
- Wavelength (λ) is the physical length of one cycle of an electromagnetic wave.
    - Wavelength is a function of the speed of light and the frequency of the waveform.
    - This is expressed mathematically as $\lambda = c/f$. (See Example 28.1.)
- Wavelength is an important characteristic since photoemitters and photodetectors are rated based on the wavelength they produce or respond to.

- Light intensity is the amount of light per unit area.
  - Light intensity decreases over distance.
- When light reflects off a surface, there are two angles of interest.
  - The angle of incidence (θ_I) is the angle at which the light strikes the surface.
  - The angle of reflection (θ_R) is the angle at which it reflects off the surface.
  - In most cases these angles are equal. (See Figure 28.4.)
- Refraction is the bending that occurs when light passes between two mediums with different optical densities.
  - A pencil in a glass of water is a common example of refraction.

Introduction to Fiber Optic Cables

- Optical fiber is ultra-pure glass or plastic that provides a closed pathway for transmitting light from one location to another.
  - Glass is by far the more common material.
- Optical fiber is surrounded by another layer of glass called *cladding*.
  - Cladding has reflective properties that keep the light inside the core of the fiber.
- The optical fiber and cladding are surrounded by a buffer coating, which is similar to the insulation around a copper wire. This coating protects the fiber and cladding from water and mechanical stress.
- There are two types of optical fiber.
  - Single-mode fiber is so narrow that it provides only one path (mode) for light signals.
  - Multimode fiber has a larger diameter and allows for multiple light paths.
- When light enters the fiber it reflects off the cladding, keeping it within the core.
- There is an angle of incidence, called the *critical angle*, at which light is not reflected and is absorbed by the cladding. (See Figure 28.6.)
- A basic fiber optic telecommunication system is comprised of three primary components.
  - The light source or transmitter, which encodes the signal into light pulses.
  - The fiber optic cable or medium.
  - The light detector, or receiver, which decodes the signal from light pulses into electrical signals.
- Fiber optic systems have advantages over copper systems.
  - They are not susceptible to electromagnetic interference (EMI) or cross-talk.
  - They can transmit more information with fewer errors.
  - They have better data security.
- Attenuation is the loss of light intensity, primarily caused by impurities in the fiber optic cable.
- Attenuation occurs when light strikes impurity molecules.
  - When light reflects off the molecules, it is known as scattering.
  - When light is absorbed by the molecules and turned into heat, it is known as absorption.
  - Scattering is the most common form of attenuation.
- Both scattering and absorption are more pronounced at specific wavelengths. (See Figure 28.11.)

Fiber Optic Connectors and Splices

- Optical cables are connected together, and to other components, using splices and connectors.
  - A splice is a mechanical connection that permanently joins two cables.
  - A connector allows connection and reconnection.

- Connector loss describes a loss of light intensity caused by either a connector or splice.
  - Connector loss describes any decrease in light intensity (measured in dB).
- This loss can result from contaminants. They can be reduced by:
  - Working in clean areas.
  - Cleaning connectors using lint-free cloths.
  - Keeping dust caps in place on all components.
- A simplex connector terminates a single fiber.
- A duplex connector terminates two fibers.
  - Typically one fiber is used to send, and the other to receive.
- The most common connector is the single tip (ST) connector.
  - It has a 2.5 mm ferrule (precision-drilled sleeve) to which an optical fiber is attached using adhesive.
  - ST connectors are widely used in LANs.
- A snap coupling (SC) connector snaps into a docking jack. (See Figure 28.14.)
  - SC connectors can be simplex or duplex, and single-mode or multimode.
  - Like the ST, the SC connector has a 2.5 mm ferrule.
- The FC connector is a single-mode or multimode screw-on connector. (See Figure 28.16.)
  - The FC connector also has a 2.5 mm ferrule.
  - It is the oldest connector design.
- A splice is a permanent connection between two optical fibers.
  - A fusion splice is a heat process that welds two fibers together.
  - A mechanical splice is a plastic fitting that aligns the fibers, and is then crimped.
  - Splices have fewer losses and are stronger than connectors.

Light Sources and Detectors

- A light-emitting diode (LED) is a semiconductor component that emits light when a voltage with the correct polarity is applied to it.
  - LEDs are suited for multimode optical fibers.
- LASER is an acronym for *Light **A**mplification by **S**timulated **E**missions of **R**adiation.*
- Lasers produce coherent light, which means that the light is highly focused and remains so over long distances.
  - Lasers are now commonly found in a variety of consumer products.
- A PIN diode is a semiconductor-based light detector.
  - Its conduction increases as light intensity increases.
  - PIN diodes act like light-to-voltage converters.
- A photodiode is another semiconductor-based light detector.
  - It is more sensitive than a PIN diode.
  - Photodiodes are used with lasers for longer-distance applications.

CHAPTER REVIEW

1. Light is electromagnetic energy within a frequency range of ________.
   a) 30 THz to 3 PHz
   b) 3 THz to 30 PHz
   c) 300 GHz to 3 THz
   d) 3 THz to 3 PHz

2. The three types of light are ________.
   a) ultrared, visible, and infraviolet
   b) ultraviolet, visible, and infrared
   c) LASER, LED, and PIN
   d) red, green, and blue

3. The speed of light in free space is ________.
   a) 300 m/μs
   b) 3×10^8 m/s
   c) 984 ft/s
   d) all of the above

4. The symbol for wavelength is ________.
   a) β
   b) α
   c) λ
   d) γ

5. A 985 THz signal has a wavelength of ________.
   a) 305 nm
   b) 3050 mm
   c) 3.05 mm
   d) 3050 nm

6. Photoemitters and detectors must be rated for the same ________ if they are to be used together.
   a) intensity
   b) modal length
   c) light velocity
   d) wavelength

7. The amount of light received by a light detector is called ________.
   a) reception rate
   b) detection rate
   c) light intensity
   d) capture ratio

8. The angle of reflection is usually equal to the angle of ________.
   a) transmission
   b) incidence
   c) reception
   d) attenuation

9. Refraction occurs when light ________.
   a) moves from one medium to another
   b) reflects
   c) critically reflects
   d) does not reflect at all

10. The core in a fiber optic cable is surrounded by a layer of glass called ________.
    a) refractory fiber
    b) cladding
    c) buffer glass
    d) incidence coating

11. The outer coating of an optical fiber is called a ________.
    a) buffer coating
    b) buffer cladding
    c) refractory coating
    d) critical coating

12. The number of paths for light that a fiber allows is called the number of ________.
    a) critical paths
    b) incident paths
    c) common modes
    d) modes

13. The angle at which light is no longer reflected, but is absorbed in the cladding, is called the ________ angle.
    a) lost
    b) absorption
    c) critical
    d) NR

14. An optical transmitter converts electrical signals to ________.
    a) analog
    b) light pulses
    c) hex
    d) reflections

15. Fiber optic systems are less susceptible to ________ than are copper cable systems.
    a) EMI
    b) data errors
    c) cross-talk
    d) all of the above

16. Losses in an optical fiber are the result of ________.
    a) reflections
    b) impurities
    c) refractions
    d) mode swings

17. The greatest source of attenuation in optical cables is ________.
    a) absorption
    b) scattering
    c) reflections
    d) mode interaction

18. Losses in optical cables are more pronounced ________.
    a) when the cable is hot
    b) at certain frequencies
    c) when the cable is cold
    d) in the infrared range

19. The main difference between a splice and a connector is that ________.
    a) the splice provides better data security
    b) the splice is permanent
    c) the connector is stronger
    d) all of the above

20. Connector loss is measured in ________.
    a) dB
    b) THz
    c) lumens
    d) candle power

21. The primary cause of conductor loss is ________.
    a) absorption
    b) reflective loss
    c) incidental loss
    d) contaminants

22. A connector designed for use with both receive and transmit fibers is called a ________.
    a) transmission coupler
    b) simplex connector
    c) duplex connector
    d) handshake connector

23. The precision-drilled sleeve in an ST connector is called a(n) ________.
    a) ST sleeve
    b) joining block
    c) ferrule
    d) critical sleeve

24. One type of screw-on connector is the ________ connector.
    a) FC
    b) SC
    c) LC
    d) SO

25. The connector that resembles an RJ-type modular jack plug is the ________ connector.
    a) FC
    b) SC
    c) LC
    d) SO

26. A mechanical splice uses a fitting that is ________ to hold the cables in place.
    a) punched down
    b) crimped
    c) fused
    d) adhered

27. Splices ________ compared to connectors.
    a) are stronger
    b) are weaker
    c) are cheaper
    d) introduce more loss when

28. An LED must have the correct ________ in order to emit light.
    a) receiver
    b) wavelength
    c) voltage polarity
    d) spectral balance

29. Lasers produce ________ light.
    a) coherent
    b) ultraviolet
    c) multimode
    d) single-mode

30. A PIN diode increases its ________ as light input increases.
    a) light output
    b) attenuation
    c) resistance
    d) conduction

31. A photodiode is more ________ than a PIN diode.
    a) neutral
    b) balanced
    c) sensitive
    d) resistive

ANSWER TO THE EXAMPLE PRACTICE PROBLEM

28.1 10 mm to 750 nm

APPENDIX A

Working with Units of Measure

One side of any equation must equal the other side. This is fairly self-evident when we are dealing only with numbers. But in electrical applications, we are often working with numbers that have one or more individual units of measure.

Units of measure give meaning to numbers. For example, let's say that a traffic report reads as follows:

$$\text{Average vehicle speed} = 70$$

You cannot accurately interpret this value without knowing the appropriate unit of measure. It could be miles per hour (mph), kilometers per hour (km/h), or some other unit of measure, and assuming the wrong unit of measure adversely affects the accuracy of the number (70 mph is about 1.6 times as fast as 70 km/h). As you can see, numbers are accurate and meaningful only when the correct unit of measure is identified.

Sometimes you will need to solve equations with several units of measure. For the result of any equation to be accurate and meaningful, both the number *and* its unit of measure must be correct. As such, it is important that you know how to determine the correct unit of measure for the result of any calculation. Before we get to some examples, let's review a few very basic rules of working with equations.

Rule #1: *You can multiply or divide any number by one (1) without changing the number.* This rule may seem self-evident, but when you use it in conjunction with Rule #2, it can be useful when simplifying equations.

Rule #2: *Any number divided by itself equals one (1).*

For example

$$\frac{6a}{4b} \times \frac{5c}{9a} = \frac{30c}{36b}$$

The variable "a" in the numerator (top) of the first fraction cancels out the variable "a" in the denominator (bottom) of the second fraction. This is because if you multiply this out you get

$$\frac{6a}{4b} \times \frac{5c}{9a} = \frac{30\cancel{a}c}{36\cancel{a}b}$$

Rule #2 tells us that dividing "a" by "a" gives us 1. Since a ÷ a = 1, the "a" variables cancel out. This principle applies to units of measure as well as to numbers or variables.

Rule #3: *If you perform the same mathematical operation to both sides of an equation, the equation is still valid.*

For example, if

$$X = Y$$

then

$$X \div 8 = Y \div 8$$

Rule #4: *If you are dividing any number by a fraction, you can "invert and multiply."*

For example,

$$y \div \frac{a}{b} = y \times \frac{b}{a}$$

As you can see, we *inverted* the fraction and *multiplied* the inverted fraction by y.

Now, let's take a look at how these rules can be used to effectively deal with units of measure.

EXAMPLE A.1

How long is 260 inches plus 12.5 feet?

SOLUTION

Because these two values have different units, they cannot be added together in their current form. We must either change the inches into feet, or the feet into inches. In this case we'll turn inches into feet.

We know that by definition,

$$1 \text{ ft} = 12 \text{ in}$$

We want to define inches in terms of feet, so we will apply Rule #3 and divide both sides of the equation by 12.

$$\frac{12 \text{ in}}{12} = \frac{1 \text{ ft}}{12}$$

Rule #2 tells us that this simplifies to 1 in = 1/12 ft = 0.083 ft. Stated another way, there are 0.083 feet per inch (0.083 ft/in). Now we can convert our inches into feet as

$$260 \cancel{\text{in}} \times \frac{0.0833 \text{ ft}}{1 \cancel{\text{in}}} = 21.67 \text{ ft}$$

and solve our original problem as

$$260 \text{ in} + 12.5 \text{ ft} = 21.67 \text{ ft} + 12.5 \text{ ft} = 34.17 \text{ ft}$$

Note how the inch units canceled out, just as if they were numbers or variables. This is a units-of-measure application of Rule #1 and Rule #2. Let's look at another example that applies to the study of electricity.

EXAMPLE A.2

6.55 A of current are measured in a circuit when 220 V are applied. How much power is dissipated by this circuit?

SOLUTION

We know from Watt's law that

$$\begin{aligned} P &= E \times I \\ &= 6.55\text{ A} \times 220\text{ V} \\ &= 1441\text{ W} \end{aligned}$$

But let's look a little closer at the units of measure. How do we get watts (W) by multiplying amperes (A) by volts (V)? The solution is to look at the definitions of the ampere, the volt, and the watt. The following definitions are provided in Chapter 2:

$$1\text{A} = 1\frac{\text{coulomb}}{\text{second}}$$

$$1\text{V} = 1\frac{\text{joule}}{\text{coulomb}}$$

and

$$1\text{W} = 1\frac{\text{joule}}{\text{second}}$$

If we use these relationships to rewrite our original equation, we get

$$P = \frac{6.55\,\cancel{\text{coulombs}}}{\text{second}} \times \frac{220\,\text{joules}}{\cancel{\text{coulomb}}} = \frac{1441\,\text{joules}}{\text{second}} = 1441\text{ W}$$

Note how the coulomb units cancel out.

Let's look at one more practical example. In Chapter 28 you will solve for the wavelength of an electromagnetic wave. Wavelength, symbolized by the Greek letter lambda (λ), is defined as the physical length of a transmitted electromagnetic wave. It is a function of the speed of light (symbolized by a lowercase "c") and the frequency of the waveform (f). Some people have trouble remembering exactly what the relationship is. Is it $\lambda = f/c$ or c/f? As demonstrated in the following example, you can always be certain when you pay close attention to the units of measure.

EXAMPLE A.3

The following information is given:
The velocity of light in a given fiber optic cable is found to be 2.03×10^{17} nanometers per second or 2.03×10^{17}nm/s. The frequency of the waveform is 162 THz or 162×10^{12} Hz. What is the wavelength of this electromagnetic wave?

SOLUTION

As we did in the previous example, let's define our units of measure a little more accurately:

- Velocity is measured in meters per second, or in this case, nanometers per second (nm/s).
- Frequency is measured in hertz (Hz), but the unit hertz really means cycles per second (cps).

$$1\text{Hz} = \frac{1\text{cycle}}{1\text{second}}$$

Wavelength is a measure of the distance traveled by one cycle, so its unit of measure in *meters per cycle*, or in this case *nanometers per cycle*. Using this information we can look at the two equations that may define wavelength. It is either

$$\lambda = \frac{\text{cycles}}{\text{second}} \div \frac{\text{meters}}{\text{second}} \quad \text{or} \quad \lambda = \frac{\text{meters}}{\text{second}} \div \frac{\text{cycles}}{\text{second}}$$

We'll use Rule #4 (the invert and multiply rule) to help figure out which equation is correct. If we apply Rule #4 to the first equation we get

$$\lambda = \frac{\text{cycles}}{\cancel{\text{second}}} \times \frac{\cancel{\text{second}}}{\text{meters}}$$

The seconds cancel out, leaving us with units of cycles per meter, which is not correct. If we apply Rule #4 to the second equation we get

$$\lambda = \frac{\text{meters}}{\cancel{\text{second}}} \times \frac{\cancel{\text{second}}}{\text{cycles}}$$

Again, the seconds cancel out, leaving us with meters per cycle, which is the correct unit for wavelength.

We can now solve the original equation as

$$\lambda = \frac{2.03 \times 10^{17}\text{nm}}{\cancel{\text{second}}} \times \frac{\cancel{\text{second}}}{162 \times 10^{12}\text{cycles}} = 1253\text{nm/cycle}$$

Summing Up

These have been just a few examples of working with units of measure. The bottom line is that units of measure are just as important as the numbers that they are related to. When a mathematical operation is done, the rules that apply to numbers apply to the units as well.

APPENDIX B

Working with Wire Tables

This appendix covers the following topics:

- Using the *NEC*®[1] wire tables to determine the current-carrying capacity (ampacity) of a conductor under specific conditions.
- Using the *NEC*® wire tables to determine the correct conductor to use under specific conditions.
- Calculating conductor resistance and voltage drops for long wire lengths.

Power and Heat

Choosing the correct conductor for a specific installation, or determining the current-carrying capacity (ampacity) of a given conductor, is usually related to the production of heat, and the ability of a conductor to dissipate that heat. There are other considerations (like whether the environment is wet or dry), but heat, and the ability to dissipate heat, is usually the primary reason for choosing one conductor over another.

All wire has some resistance. In any current-carrying conductor almost all of the absorbed power (I^2R) will be dissipated as heat. Back in Chapter 2 we discussed wire resistance and the factors that affect resistance. As you may recall, Equation (2.2) states that

$$R = \rho\frac{\ell}{A}$$

This equation tells us that wire resistance is affected by three primary factors:

- **The resistivity of the material (ρ)**: As we discussed back in Chapter 2, copper is the most commonly used conductor because it has the lowest resistivity of the common conductor materials. For example, a copper wire will absorb less power and generate less heat at a given current level than an equivalent aluminum wire because it has less resistance (see Table 2.3 on page 59).
- **The length of the wire (ℓ):** The longer a conductor, the higher its resistance (all other factors being equal).
- **The cross-sectional area (A)**: The American Wire Gauge Table is organized according to conductor cross-sectional area. A 6 AWG copper conductor has the same diameter as a 6 AWG aluminum conductor, even though they have different current-carrying capacities. The lower the gauge of a wire, the wider its diameter and the lower its resistance per unit length. This means that for a given current, a lower gauge conductor of the same material

[1]NFPA 70®, National Electrical Code and *NEC*® are registered trademarks of the National Fire Protection Association, Quincy, MA.

absorbs less power and produces less heat. This means that a wire with a greater cross-sectional area (lower wire gauge) can handle more current than a higher-gauge conductor with the same type of insulation.

It should be noted that the type of wire insulation is another important factor. Some types of wire insulation can withstand higher temperatures than others, and maintaining the integrity of a conductor's insulation is critical. We will discuss wire insulation in more detail later.

Factors That Affect Heat and Heat Dissipation

We have already mentioned the first two factors that affect the amount of heat that a conductor produces: the amount of current and the gauge of the wire (its resistance).

The higher the current through a given conductor, the greater the heat it produces. This should make sense. Because heat is a function of power, and power is found as I^2R, higher current increases the power absorbed by a conductor and the amount of heat it produces.

The smaller the cross-sectional area of a conductor (i.e., the higher the wire gauge), the greater the amount of heat it produces at a given current level. Again, this relates directly to the I^2R equation for power. A smaller diameter (higher gauge) wire has more resistance than a larger diameter (lower gauge) wire of the same material. For a given length of wire passing a given amount of current, more heat is produced by the higher gauge (higher resistance) wire.

Wire gauge and current are not the only factors that affect heat production and dissipation. There are several other factors that must be considered.

- **The number of conductors:** If several current-carrying conductors are run together, the heat produced by each conductor affects all the others. Each conductor heats up the other conductors in the same environment.
- **The ambient temperature:** The ambient temperature refers to the temperature around the conductor. This directly affects the ability of a conductor to dissipate the heat it produces. More heat can be dissipated by a conductor in a cooler environment than in a warmer environment.
- **The enclosure:** If a conductor is run in free air, it can dissipate heat more efficiently than if it is in a conduit or buried in a trench.
- **The insulation:** As mentioned earlier, different types of wire insulation can withstand different maximum temperatures before they begin to break down.

All of these factors come into play when using the *NEC*® wire tables to determine the ampacity of a given conductor and in choosing the proper conductor for a given installation.

Using the *NEC*® Wire Tables

NEC® *310* addresses conductors used in general wiring installations. *NEC*® *Tables 310.16* through *310.19* cover the selection of conductors based on size and insulation type under a specific set of conditions. These tables, along with a few others, will be the focus of our *NEC*® wire table discussions.

Determining Conductor Ampacity

Refer to *NEC*® *Table 310.16* in Table B.1. This table is divided vertically into two parts. The left side of the table covers copper conductors; the right side covers aluminum, or copper-clad aluminum. The far left column lists the wire sizes for copper, and the far right for aluminum or copper-clad aluminum. Note that wire size is specified in two different ways. Smaller diameter wires are specified using AWG ratings up to 4/0 AWG. Larger diameter conductors are specified in *thousands of circular mils* (kcmil). As you learned back in Chapter 2, a circular mil is the cross-sectional area of a wire having a diameter of one one-thousandth of an inch (0.001 in.). Refer to Figure 2.21 on page 58.

There are several installation conditions listed at the top of the table. These conditions are as follows:

- The maximum voltage of the installation cannot exceed 2000 V (2 kV).
- The temperature ratings of the conductors in this table are between 60°C and 90°C (140°F and 194°F). If you look at the six middle columns you will see that they are divided up into 60°C, 75°C, and 90°C columns depending on the type of insulation used. The various types of insulation are listed directly below the temperature designations. Higher temperature conductors are covered in *Tables 310.18* and *310.19* (Tables B-3 and B-4).
- There can be no more than three current-carrying conductors in a raceway, cable, or earth (directly buried) cable. If more than three current-carrying conductors are present, a correction factor must be employed.
- The ambient temperature (the temperature of the environment around the conductor) cannot exceed 30°C (86°F). If the ambient temperature is above this maximum value, a correction factor must be employed.

Now, look directly below the insulation types in each of the six middle columns. The three columns on the left list the ampacity of copper conductors, and the three columns on the right are for aluminum and copper-clad aluminum. For example, a Type TW 6 AWG copper conductor has an ampacity rating of 55 A, and the equivalent aluminum conductor has an ampacity rating of only 40 A. A copper conductor has less resistance, and therefore produces less heat at a given current level, than an equivalent aluminum conductor. Remember, these ratings are for ambient temperatures no greater than 30°C (86°F)—as stated at the top of the table. For higher ambient temperatures you must use a correction factor.

At the bottom of *Table 310.16* is a list of temperature correction factors. The column on the far left lists temperature ranges in degrees Celsius (°C) while the far-right column is in degrees Fahrenheit (°F). Let's look at a couple of examples that will help illustrate how to use *Table 310.16*.

TABLE B.1 • *NEC® Table 310.16* Allowable Ampacities of Insulated Conductors Rated 0 Through 2000 Volts, 60°C Through 90°C (140°F Through 194°F), Not More Than Three Current-Carrying Conductors in Raceway, Cable, or Earth (Directly Buried), Based on Ambient Temperature of 30°C (86°F)

	Temperature Rating of Conductor [See Table 310.13(A).]						
	60°C (140°F)	75°C(167°F)	90°C (194°F)	60°C (140°F)	75°C (167°F)	90°C (194°F)	
	Types TW, UF	Types RHW, THHW, THW, THWN, XHHW, USE, ZW	Types TBS, SA, SIS, FEP, FEPB, MI, RHH, RHW-2, THHN, THHW, THW-2, THWN-2, USE-2, XHH, XHHW, XHHW-2, ZW-2	Types TW, UF	Types RHW, THHW, THW, THWN, XHHW, USE	Types TBS, SA, SIS, THHN, THHW, THW-2, THWN-2, RHH, RHW-2, USE-2, XHH, XHHW, XHHW-2, ZW-2	
Size AWG or kcmil	COPPER			ALUMINUM OR COPPER-CLAD ALUMINUM			Size AWG or kcmil
18	—	—	14	—	—	—	—
16	—	—	18	—	—	—	—
14*	20	20	25	—	—	—	—
12*	25	25	30	20	20	25	12*
10*	30	35	40	25	30	35	10*
8	40	50	55	30	40	45	8
6	55	65	75	40	50	60	6
4	70	85	95	55	65	75	4
3	85	100	110	65	75	85	3
2	95	115	130	75	90	100	2
1	110	130	150	85	100	115	1
1/0	125	150	170	100	120	135	1/0
2/0	145	175	195	115	135	150	2/0
3/0	165	200	225	130	155	175	3/0
4/0	195	230	260	150	180	205	4/0
250	215	255	290	170	205	230	250
300	240	285	320	190	230	255	300
350	260	310	350	210	250	280	350
400	280	335	380	225	270	305	400
500	320	380	430	260	310	350	500
600	355	420	475	285	340	385	600
700	385	460	520	310	375	420	700
750	400	475	535	320	385	435	750
800	410	490	555	330	395	450	800
900	435	520	585	355	425	480	900
1000	455	545	615	375	445	500	1000
1250	495	590	665	405	485	545	1250
1500	520	625	705	435	520	585	1500
1750	545	650	735	455	545	615	1750
2000	560	665	750	470	560	630	2000
CORRECTION FACTORS							
Ambient Temp. (°C)	For ambient temperatures other than 30°C (86°F), multiply the allowable ampacities shown above by the appropriate factor shown below.						Ambient Temp. (°F)
21-25	1,08	1.05	1.04	1.08	1.05	1.04	70-77
26-30	1.00	1.00	1.00	1.00	1.00	1.00	78-86
31-35	0.91	0.94	0.96	0.91	0.94	0.96	87-95
36-40	0.82	0.88	0.91	0.82	0.88	0.91	96-104
41-45	0.71	0.82	0.87	0.71	0.82	0.87	105-113
46-50	0.58	0.75	0.82	0.58	0.75	0.82	114-122
51-55	0.41	0.67	0.76	0.41	0.67	0.76	123-131
56-60	—	0.58	0.71	—	0.58	0.71	132-140
61-70	—	0.33	0.58	—	0.33	0.58	141-158
71-80	—	—	0.41	—	—	0.41	159-176

*See 240.4(D).

TABLE B.2 • *NEC® Table 310.17* Allowable Ampacities of Single-Insulated Conductors Rated 0 Through 2000 Volts in Free Air, Based on Ambient Air Temperature of 30°C (86°F)

	Temperature Rating of Conductor [See Table 310.13(A).]						
	60°C (140°F)	75°C (167°F)	90°C (194°F)	60°C (140°F)	75°C (1167°F)	90°C (194°F)	
	Types TW, UF	Types RHW, THHW, THW, THWN, XHHW, USE, ZW	Types TBS, SA, SIS, FEP, FEPB, MI, RHH, RHW-2, THHN, THHW, THW-2, THWN-2, USE-2, XHH, XHHW, XHHW-2, ZW-2	Types TW, UF	Types RHW, THHW, THW, THWN, XHHW	Types TBS, SA, SIS, THHN, THHW, THW-2, THWN-2, RHH, RHW-2, USE-2, XHH, XHHW, XHHW-2, ZW-2	
Size AWG or kcmil	COPPER			ALUMINUM OR COPPER-CLAD ALUMINUM			Size AWG or kcmil
18	—	—	18	—	—	—	—
16	—	—	24	—	—	—	—
14*	25	30	35	—	—	—	—
12*	30	35	40	25	30	35	12*
10*	40	50	55	35	40	40	10*
8	60	70	80	45	55	60	8
6	80	95	105	60	75	80	6
4	105	125	140	80	100	110	4
3	120	145	165	95	115	130	3
2	140	170	190	110	135	150	2
1	165	195	220	130	155	175	1
1/0	195	230	260	150	180	205	1/0
2/0	225	265	300	175	210	235	2/0
3/0	260	310	350	200	240	275	3/0
4/0	300	360	405	235	280	315	4/0
250	340	405	455	265	315	355	250
300	375	445	505	290	350	395	300
350	420	505	570	330	395	445	350
400	455	545	615	355	425	480	400
500	515	620	700	405	485	545	500
600	575	690	780	455	540	615	600
700	630	755	855	500	595	675	700
750	655	785	885	515	620	700	750
800	680	815	920	535	645	725	800
900	730	870	985	580	700	785	900
1000	780	935	1055	625	750	845	1000
1250	890	1065	1200	710	855	960	1250
1500	980	1175	1325	795	950	1075	1500
1750	1070	1280	1445	875	1050	1185	1750
2000	1155	1385	1560	960	1150	1335	2000
CORRECTION FACTORS							
Ambient Temp. (°C)	For ambient temperatures other than 30°C (86°F), multiply the allowable ampacities shown above by the appropriate factor shown below.						Ambient Temp. (°F)
21–25	1.08	1.05	1.04	1.08	1.05	1.04	70–77
26–30	1.00	1.00	1.00	1.00	1.00	1.00	78–86
31–35	0.91	0.94	0.96	0.91	0.94	0.96	87–95
36–40	0.82	0.88	0.91	0.82	0.88	0.91	96–104
41–45	0.71	0.82	0.87	0.71	0.82	0.87	105–113
46–50	0.58	0.75	0.82	0.58	0.75	0.82	114–122
51–55	0.41	0.67	0.76	0.41	0.67	0.76	123–131
56–60	—	0.58	0.71	—	0.58	0.71	132–140
61–70	—	0.33	0.58	—	0.33	0.58	141–158
71–80	—	—	0.41	—	—	0.41	159–176

*See 240.4(D).

TABLE B.3 • *NEC® Table 310.18* Allowable Ampacities of Insulated Conductors Rated 0 Through 2000 Volts, 150°C Through 250°C (302°F Through 482°F). Not More Than Three Current-Carrying Conductors in Raceway or Cable, Based on Ambient Air Temperature of 40°C (104°F)

Size AWG or kcmil	Temperature Rating of Conductor [See Table 310.13(A).]				Size AWG or kcmil
	150°C (302°F)	200°C (392°F)	250°C (482°F)	150°C (302°F)	
	Type Z	Types FEP, FEPB, PFA, SA	Types PFAH, TFE	Type Z	
	COPPER		NICKEL OR NICKEL-COATED COPPER	ALUMINUM OR COPPER-CLAD ALUMINUM	
14	34	36	39	—	14
12	43	45	54	30	12
10	55	60	73	44	10
8	76	83	93	57	8
6	96	110	117	75	6
4	120	125	148	94	4
3	143	152	166	109	3
2	160	171	191	124	2
1	186	197	215	145	1
1/0	215	229	244	169	1/0
2/0	251	260	273	198	2/0
3/0	288	297	308	227	3/0
4/0	332	346	361	260	4/0
CORRECTION FACTORS					
Ambient Temp. (°C)	For ambient temperatures other than 40°C (104°F), multiply the allowable ampacities shown above by the appropriate factor shown below.				Ambient Temp. (°F)
41–50	0.95	0.97	0.98	0.95	105–122
51–60	0.90	0.94	0.95	0.90	123–140
61–70	0.85	0.90	0.93	0.85	141–158
71–80	0.80	0.87	0.90	0.80	159–176
81–90	0.74	0.83	0.87	0.74	177–194
91–100	0.67	0.79	0.85	0.67	195–212
101–120	0.52	0.71	0.79	0.52	213–248
121–140	0.30	0.61	0.72	0.30	249–284
141–160	—	0.50	0.65	—	285–320
161–180	—	0.35	0.58	—	321–356
181–200	—	—	0.49	—	357–392
201–225	—	—	0.35	—	393–437

TABLE B.4 • *NEC® Table 310.19* Allowable Ampacities of Single-Insulated Conductors, Rated 0 Through 2000 Volts, 150°C Through 250°C (302°F Through 482°F), in Free Air, Based on Ambient Air Temperature of 40°C (104°F)

	Temperature Rating of Conductor [See Table 310.13(A).]				
	150°C (302°F)	200°C (392°F)	250°C (482°F)	150°C (302°F)	
	Type Z	Types FEP, FEPB, PFA, SA	Types PFAH, TFE	Type Z	
Size AWG or kcmil	COPPER		NICKEL OR NICKEL-COATED COPPER	ALUMINUM OR COPPER-CLAD ALUMINUM	Size AWG or kcmil
14	46	54	59	—	14
12	60	68	78	47	12
10	80	90	107	63	10
8	106	124	142	83	8
6	155	165	205	112	6
4	190	220	278	148	4
3	214	252	327	170	3
2	255	293	381	198	2
1	293	344	440	228	1
1/0	339	399	532	263	1/0
2/0	390	467	591	305	2/0
3/0	451	546	708	351	3/0
4/0	529	629	830	411	4/0
CORRECTION FACTORS					
Ambient Temp. (°C)	For ambient temperatures other than 40°C (104°F), multiply the allowable ampacities shown above by the appropriate factor shown below.				Ambient Temp. (°F)
41–50	0.95	0.97	0.98	0.95	105–122
51–60	0.90	0.94	0.95	0.90	123–140
61–70	0.85	0.90	0.93	0.85	141–158
71–80	0.80	0.87	0.90	0.80	159–176
81–90	0.74	0.83	0.87	0.74	177–194
91–100	0.67	0.79	0.85	0.67	195–212
101–120	0.52	0.71	0.79	0.52	213–248
121–140	0.30	0.61	0.72	0.30	249–284
141–160	—	0.50	0.65	—	285–320
161–180	—	0.35	0.58	—	321–356
181–200	—	—	0.49	—	357–392
201–225	—	—	0.35	—	393–437

EXAMPLE B.1

Determine the ampacity of a Type THW, 4 AWG copper conductor operating in an environment with a maximum ambient temperature of 80°F.

SOLUTION

First, locate Type THW copper conductors in the second column (167°F) under the copper heading. Then locate 4 AWG in the far left-hand column. Following the 4 AWG row across to the 167°F copper column we see that this conductor has an ampacity rating of 85 A. Since the maximum ambient temperature is less than 86°F, no correction factor is required.

A couple of points need to be made before we do another example: There are a number of temperature values mentioned in Example B.1 that may cause some confusion. Let's deal with each one.

- The 80°F temperature given in the example is the maximum ambient (air) temperature that the conductor will be exposed to.
- The 86°F (30°C) temperature noted at the top of the table is the maximum ambient temperature allowed before a correction factor must be used.
- The 167°F (75°C) temperature rating of the column in which RHW insulation is listed is the maximum temperature that this type of insulation can withstand. It is the maximum temperature for this type of insulation due to current-related heat production. Stated another way, if any conductor in this column reaches temperatures above 167°F, then the insulation may begin to fail.

Let's work another example.

EXAMPLE B.2

Determine the ampacity of a Type RHW-2, 2/0 AWG aluminum conductor installed in an environment where the ambient temperature can reach 52°C.

SOLUTION

Type RHW-2 aluminum conductor is located in the right-hand 90°C (194°F) column. Locate 2/0 AWG in the far right-hand column and follow that row over to the aluminum 90°C column. This shows an uncorrected ampacity rating of 150 A.

Since the ambient temperature is above 30°C, we must refer to the correction factors at the bottom of *Table 310.16*. The far left-hand column lists ambient temperature ranges in °C. The fourth row from the bottom shows a range of 51°C to 55°C. Following this row across to the 90°C aluminum column we find a correction factor of 0.76. We can now solve for the ampacity of this conductor at 52°C as follows:

$$150\text{ A} \times 0.76 = 114\text{ A}$$

More Than Three Conductors

If there are more than three current-carrying conductors in a raceway or cable, a correction factor must be employed. Refer to *NEC® Table 310.15(B)(2)(a)* in Table B.5. This table lists the correction factors for multiple conductors in a raceway or cable. The use of this table is illustrated in Example B.3.

TABLE B.5 • *NEC® Table 310.15(B)(2)(a)* Adjustment Factors for More Than Three Current-Carrying Conductors in a Raceway or Cable

Number of Current-Carrying Conductors	Percent of Values in Tables 310.16 Through 310.19 as Adjusted for Ambient Temperature if Necessary
4–6	80
7–9	70
10–20	50
21–30	45
31–40	40
41 and above	35

FPN No. 1: See Annex B, Table B.310.11, for adjustment factors for more than three current-carrying conductors in a raceway or cable with load diversity.

EXAMPLE B.3

What is the ampacity of a Type TW 12 AWG conductor if 14 conductors are run in a conduit and the maximum ambient temperature is 130°F?

SOLUTION

First locate Type TW insulation in the right-hand aluminum (140°F) column and 12 AWG in the far right column of *Table 310.16*. Tracing the 12 AWG row, we find an uncorrected ampacity rating of 20 A. Since the temperature is above 86°F, a temperature correction factor must be used. Locate the proper temperature range in the far right column of the temperature correction factors. The range of 123–131°F yields a correction factor of 0.41 for Type TW insulation. This lowers the ampacity to

$$20\text{ A} \times 0.41 = 8.2\text{ A}$$

Now we must use *Table 310.15(B)(2)(a)* to find the correction factor for 14 conductors. For 10 to 20 conductors we must use a correction factor of 50% or 0.5. The fully corrected ampacity rating per conductor of 14 Type

TW, 12 AWG aluminum conductors in a conduit at an ambient temperature of 130°F is found as

$$8.2 \text{ A} \times 0.5 = 4.1 \text{ A}$$

It should be noted that we could have applied both correction factors at the same time to the uncorrected ampacity rating.

$$20 \text{ A} \times 0.41 \times 0.5 = 4.1 \text{ A}$$

Using NEC® *Table 310.17*

There is only one difference between *Table 310.16* and *Table 310.17* (Table B.2): *Table 310.16* deals with up to three conductors in a raceway, cable, or underground (directly buried) cable while *Table 310.17* deals with conductors in free air. The maximum voltage is still 2000 V and the maximum temperature allowed without correction is still 30°C. The same insulation materials are listed and the same temperature correction factors are used. Because the conductors are in free air, they can dissipate more heat and therefore have higher ampacity ratings. For example, look at the rating for Type XHH 1 AWG copper conductor. *Table 310.16* shows an uncorrected ampacity rating of 150 A. *Table 310.17* lists an ampacity rating for the same conductor in free air as 220 A. As we said earlier, it is all about heat and heat dissipation.

NEC® *Tables 310.18* and *310.19*

Refer to *NEC® Tables 310.18* and *310.19* (Tables B.3 and B.4). These two tables are basically the same as *Tables 310.16* and *310.17* respectively, but they apply to different insulating materials. The maximum voltage is still 2000 V, but the temperature range of the insulating materials listed is from 150°C (302°F) to 250°C (482°F). There are other types of wire listed as well, namely *nickel* and *nickel-coated copper*. The maximum ambient temperature is 40°C (104°F) before a correction factor must be used. Despite these differences, they are used in the same manner as *Tables 310.16* and *310.17*.

Wire Insulation

There are many different types of wire insulation. *NEC® Tables 310.13(A)* through *(D)* cover a variety of issues related to wire insulation. These include the trade name of the insulation, the letter designation, the thickness, and the actual insulating material. They also cover the maximum operating temperature and application provisions.

Insulation types and applications are identified using letter designations. For example:

- T stands for *thermoplastic.*
- H stands for *heat resistant,* and HH stands for *heat resistant to higher temperatures.*

- U stands for *underground applications.*
- W stands for use in *wet locations.*
- X stands for *thermoset,* a synthetic polymer.
- SE stands from *service entrance applications.*
- F stands for *feeder or branch-circuit cable.*
- N stands for a *nylon outer jacket.*

Any combination of these letters may be used to describe a particular insulation. For example, Type THW insulation is a thermoplastic that is heat resistant and can be used in both dry and wet locations. Type THHN is thermoplastic with a nylon outer jacket and a higher temperature rating, but is limited to dry or damp locations. Type USE is an underground service-entrance cable. Refer to *NEC® Article 100* (locations) for an explanation of dry, damp, and wet.

Selecting the Correct Conductor

So far we have discussed how to determine the ampacity of a given conductor under specific conditions. A far more common use of the *NEC®* tables is to select the correct conductor for a specific application. The process is similar, but some factors must be considered that were not a part of determining conductor ampacity. These factors include the following:

- **Continuous load:** *NEC® 2008 Article 100* defines a continuous load as "*A load where the maximum current is expected to continue for 3 hours or more.*" *NEC® 210.19 (A) (1)* states that a conductor that supplies a continuous load, or any combination of continuous and noncontinuous loads shall have an "*ampacity not less than the noncontinuous load plus 125% of the continuous load.*"
- **Termination Temperature Ratings:** *NEC® 110.14(C)* states "*The temperature rating associated with the ampacity of a conductor shall be selected and coordinated so as not to exceed the lowest temperature rating of any connected termination, conductor, or device.*" In other words, the temperature rating of a conductor cannot exceed the temperature rating of any device that it is connected to.

In many cases the temperature rating of a device may not be known. For this reason, for circuits rated 100 A or less, or marked for 14 AWG through 1 AWG conductors, conductors shall be chosen from the 60°C column. One exception is motors marked with the NEMA®[2] design letters B, C, or D. In this case conductors may be chosen from the 75°C column.

Almost any conductor is going to be terminated *inside* some component or device. This means that conductors must be chosen from *Table 310.16,* not *310.17,* which is for conductors in free air. For circuits rated above 100 A, or

[2]NEMA stands for National Electrical Manufacturers Association.

marked for conductors larger than 1 AWG, you may choose conductors from the 75°C column.

- **Ambient Temperature:** As was the case when determining ampacity, ambient temperature must be considered when choosing a conductor for a given application. The same temperature correction factors are used, but in a slightly different way.

Let's look at an example to help clarify the difference between solving for the ampacity of a conductor, and choosing the right conductor for a given application.

EXAMPLE B.4

Assume you must choose a conductor to supply four motors in a steel plant that will be required to run for more than three hours at a time (a continuous load). They are connected in parallel, and the full-load rating of each motor is 12.5 A. The motors do not show any NEMA code letters. The ambient temperature in the plant can reach 38°C. Type THW copper conductor is to be used. Solve for the correct conductor size.

SOLUTION

Since the motors will be connected in parallel, the uncorrected conductor current is found as

$$4 \text{ motors} \times 12.5 \text{ A/motor} = 50 \text{ A}$$

Since the load is continuous and below 100 A, we must adjust the ampacity by 125% as

$$50 \text{ A} \times 125\% = 62.5 \text{ A}$$

The ambient temperature is higher than 30°C, so we must employ a temperature correction factor from *Table 310.16*. For the range of 36°C to 40°C the correction factor for Type THW copper conductor (the 75°C column) is 0.88. The fully corrected conductor ampacity requirement is found as

$$\frac{62.5 \text{ A}}{0.88} = 71 \text{ A}$$

Note that we *divided* by the correction factor in this case, whereas we *multiplied* by the correction factor when determining conductor ampacity. Now we can find the actual conductor size. Type THW conductor is found in the 75°C column, but because the motors did not have the appropriate NEMA® code letters and we have no information about the termination temperature ratings of the motors, we must choose the conductor size from the 60°C column of *Table 310.16*. The smallest conductor with an ampacity rating above 71 A in this column is 3 AWG. It should be noted that a larger conductor may be used.

Termination and Lug Ratings

When making a termination inside any piece of equipment, such as a service panel or an enclosed circuit breaker, it is a mistake to choose a conductor based on the temperature rating of the termination lug rather than the temperature rating of the equipment itself. Manufacturers often use 90°C lugs in equipment that is rated for 60°C or 75°C.

Long Wire Lengths

Conductor length was never considered in our previous examples. The reason is that the *NEC*® wire tables assume that the length of the conductor is short enough so that its overall resistance does not significantly impact the circuit. This is not always the case. The most common concern when dealing with long wire lengths is *conductor voltage drop.*

Normally, the voltage drop across a conductor is negligible. However, the voltage drop across a very long length of wire may be sufficient to affect overall circuit operation, and some installations may require that the supply voltage not drop by more than a specified amount. When this is the case, you need to be able to determine the voltage drop across a given conductor. Once again we have to refer back to equation (2.2):

$$R = \rho\frac{\ell}{A}$$

This equation allows us to determine the resistance of a given conductor. A modified version of equation (2.2) allows us to determine the minimum cross-sectional area of a conductor of a given length for a given resistance, as demonstrated in Example B.5.

EXAMPLE B.5

A school is adding a new outdoor sports facility. The service panel for this facility is located in a shed 1600 feet from the main panel in the school. The supply voltage is 240 V, 1Φ and the peak current demand is 100 A. The design specifications state that the voltage at the new service panel can be no more than 5% below rated values. What is the smallest copper conductor that can be used in this installation?

SOLUTION

We begin by determining the total length of the conductor. Since this is a 1Φ supply, the actual length of the conductor is

$$2 \times 1600 \text{ ft} = 3200 \text{ ft}$$

We double the length because single phase requires both a hot and a neutral wire, and both are carrying an equal amount of current at any given time. Because both wires are in series with any load, the total voltage drop across both the hot and neutral conductors affects the load voltage. Next we solve for the amount of voltage drop that is allowed as

$$240 \text{ V} \times 5\% = 12 \text{ V}$$

Now we can solve for the resistance that will result in a voltage drop of 12 V at a current of 100 A. Using Ohm's law we get,

$$R = \frac{12 \text{ V}}{100 \text{ A}} = 0.12\ \Omega$$

Finally we can use a modified version of equation (2.2) to solve for the cross-sectional area of a 3200-ft copper conductor with 0.12 Ω of resistance. From Table 2.3 (page 59) we see that copper has a resistivity of 10.37 cmil-Ω/ft. Transposing equation (2.2) to solve for cross-sectional area we get

$$A = \rho\frac{\ell}{R} = (10.37 \text{ cmil-}\Omega/\text{ft})\left(\frac{3200 \text{ ft}}{0.12\ \Omega}\right) = 277 \text{ kcmil}$$

Referring to *NEC® Table 310.16*, we can see that the smallest copper conductor that has a cross-sectional area greater than 277 kcmil is 300 kcmil.

Three-Phase Conductor Voltage Drop

There is only one real difference in solving for the conductor-related voltage drop for a 3Φ system. Refer back to Example B.5. For the 1Φ system in this example, we doubled the length of the conductor as both the hot and neutral wires carry the same amount of current at the same time. This is not true for a 3Φ circuit.

At any given time, the three conductors in a 3Φ circuit are sharing the current load. At no time are all three conductors carrying the full load. For this reason, rather than multiply the distance by 2, we multiply by $\sqrt{3}$. This means that if the system described in Example B.5 was a 3Φ circuit, the conductor length would have been found as

$$\sqrt{3} \times 1600 \text{ ft} = 2771 \text{ ft}$$

and the required conductor cross-sectional area would have been

$$A = \rho\frac{\ell}{R} = 10.37 \text{ cmil-}\Omega/\text{ft}\left(\frac{2771 \text{ ft}}{0.12\ \Omega}\right) = 239 \text{ kcmil}$$

Referring to *Table 310.16* we see that the smallest conductor we could use is 250 kcmil.

Conductors in Parallel

Refer back to Example B.5. We found that the smallest conductor that could be used in that application was 300 kcmil. This large a conductor can be difficult to work with, so it may be advantageous to use two or more smaller conductors connected in parallel rather than one large conductor. *NEC® Article 310.4* covers conductors in parallel.

Conductors sized 1/0 AWG and larger may be connected in parallel (electrically joined at each end) provided that each conductor

- Is the same length.
- Has the same cross-sectional area.
- Has the same insulation type.
- Is terminated in the same manner.

Further, if the conductors are run in separate cables or raceways, each cable or raceway must have the same number of conductors and the same electrical characteristics.

Conductors smaller than 1/0 AWG may be connected in parallel to supply control power provided that

- They are contained in the same raceway or cable.
- The ampacity of each conductor is sufficient to handle the entire load current shared by the parallel-connected conductors.
- The overcurrent protection is such that the ampacity of no individual cable is exceeded if one or more of the parallel-connected conductors should accidentally become disconnected.

Another situation where parallel conductors are permitted is covered in *NEC® Article 310.4, Exception No. 3*. This exception states that under engineering supervision, grounded conductors sized 1 AWG or 2 AWG may be connected in parallel for existing installations. This is to help alleviate overheating due to high-content triplen harmonic currents in the neutral conductors of 3Φ circuits.

Parallel Conductors in Metal Raceways

In Chapter 21, you learned that transformer cores are usually composed of thin laminated layers that are electrically isolated from each other. This is to prevent losses due to the eddy currents that are induced in the iron cores of the transformer. A similar type of problem can arise when a conductor is run in a ferrous metal raceway or enclosure.

If a single conductor is run in a ferrous metal enclosure, the expanding and collapsing magnetic field can induce eddy currents in the metal enclosure. This can result in significant heating of the enclosure, which would have a negative impact on the ampacity of the conductor and could pose a potential fire hazard. For this reason, *NEC® Article 300.20* specifies that conductors for each phase of the circuit must be grouped together in the enclosure. For 1Φ systems, both the hot and neutral lines must be in any ferrous metal enclosure. For 3Φ systems, all three phases must be in the enclosure. The phase difference between the currents in each of the conductors tends to cancel out the induced eddy currents, limiting the induction heating of the enclosure.

APPENDIX C

Glossary

1-to-1 transformer A transformer that has equal output (secondary) and input (primary) voltages.
110 termination block A termination block that is mostly plastic and is suited for higher-speed applications.
66 termination block A block that is used to terminate telephone and data transmission lines.

Absorption The phenomenon in which an impurity molecule absorbs energy from the light and releases it in the form of heat.
AC adaptor A device that converts an AC line voltage to the DC operating voltage required to power an electronic system.
Alternating current (AC) Current that periodically changes direction.
Alternations The positive and negative halves of a waveform.
Alternator An AC generator; a machine that converts mechanical energy (motion) to electrical energy.
American Wire Gauge (AWG) A system that uses numbers to identify industry-standard wire sizes.
Ammeter A meter used to measure current.
Amorphous-core transformers Transformers with cores of silicon steel wound into a rectangular toroid shape.
Ampacity The maximum allowable current that can be safely carried by a given conductor, measured in amperes.
Ampere (A) The unit of measure for current; equal to one coulomb per second.
Ampere-hour (Ah) The product of current (in amperes) and time (in hours); the unit of measure of battery capacity.
Ampere-turn (A · t) For a given coil, the product of coil current and the number of turns.
Analog Multimeter A meter with a pointer that moves over a stationary scale to indicate the result of a measurement.
Apparent power (P_{APP}) The product of $E \times I_T$ in an AC circuit; the combination of resistive and reactive power, measured in volt-amperes (VA).
Arc flash A sudden, explosive arc of electrical energy.
Armature The current-carrying conductor in any generator or motor.
Armature reaction A term used to describe the interaction between the magnetic fields that causes the neutral plane to shift in the direction of armature rotation.
Asynchronous motor One that does not use an alternator as an energy and timing source.
Atom The smallest particle of matter that retains the physical characteristics of an element.

Atomic number A unique number for every element on the periodic table, equal to the number of protons in the nucleus of the atom.

Attenuation Any signal loss caused by the frequency response of a circuit.

Autotransformer A transformer made up of a single coil that typically has three terminal connections.

Average AC power The value that falls midway between 0 W and the peak power value (P_{Pk}) for a sine wave.

Average breakdown voltage The voltage that will cause an insulator to break down and conduct.

Averaging meter An AC meter that measures the half-cycle average of a waveform and then converts that value to an RMS readout.

Avionics The area of electronics that deals with aircraft communications and navigation systems, aircraft and weather radar systems, and other on-board instruments and data systems.

Balanced load One in which the load currents are equal in value and 120° out of phase.

Band-stop filter A circuit designed to block the band of frequencies that lies between its cutoff frequencies, labeled f_{C1} and f_{C2}.

Bandpass filter A circuit designed to pass the band of frequencies that lies between two cutoff frequencies, labeled f_{C1} and f_{C2}.

Bandwidth The range (or *band*) of frequencies between the cutoff frequencies of a filter.

Barrage The dam used in a tidal power plant.

Base speed The rated speed of a motor.

Battery A device that converts chemical, thermal, or light energy into electrical energy.

Biogas A methane-rich gas that comes from the decay of organic matter.

Biomass A term used in reference to vegetation, plant matter, or agricultural waste.

Biomedical electronics The area of electronics that deals with systems designed for use in diagnosing, monitoring, and treating medical problems.

BNC connector One that is used primarily to connect coax cables to test equipment and in data processing applications.

Bohr model The simplest model of the atom.

Bonding The permanent joining of conductive non–current-carrying materials in a structure.

Boost transformer One that increases (steps up) its input voltage by as much as 20%.

Branch Each current path in a parallel circuit.

Branch circuits The circuits that distribute power throughout a residence or other building.

Brayton cycle The process of burning fuel with compressed air and then using the expanding gases to drive a turbine.

Bridging clip A metal clip that is used to connect the left-hand clips to the right-hand clips on a split block.

Brushes The stationary contacts that couple the output from the rotor of a DC generator to the load.

Buck transformer One that decreases (steps down) its input voltage by as much as 20%.

Buffer coating A coating that protects an optical fiber from water and mechanical stresses.
Building-integrated PV (BIPV) Integrating PV cells into building materials such as roofing shingles, driveways, and road paving.
Bulk power factor correction Another term for centralized correction.
Busway A prefabricated distribution system with enclosed bus bars that include straight lengths, fittings, devices, and accessories.

Cable An insulated electrical conductor (or group of conductors) that are used to transmit telecommunication signals from one point to another.
Capacitance The ability of a component to store energy in the form of an electrostatic charge; the electrical property that opposes any change in voltage.
Capacitive reactance (X_C) The opposition that a capacitor presents to an alternating current, measured in ohms (Ω).
Capacitor A component that stores energy in an electric field; also called a **condenser**. A component that is designed to provide a specific value of capacitance.
Capacitor-start, capacitor-run motor A capacitor-start motor with an added capacitor that allows the motor to operate as a split-phase motor with a single-phase input.
Capacitor-start, inductor-run motor A split-phase motor with a capacitive starting component.
Capacity A battery rating that indicates how long the battery will last at a given output current, measured in ampere-hours (Ah). The ability of a capacitor to store a specific amount of charge (per volt applied). A transformer rating that indicates the amount of power the component can deliver to a load on a continual basis.
Carbon capture and storage (CCS) A means of collecting the carbon produced during gasification and then storing that carbon in deep geological formations or the ocean.
Carbon-composition resistor A component that uses carbon to provide a desired value of resistance.
Carbon dioxide equivalent (CO_2e) A unit of measure for greenhouse gas emissions that indicates their global warming potential.
Carbon per unit energy The amount of CO_2 emitted per unit of energy produced.
Carbon sequestration Another name for *carbon capture and storage* (CCS).
Cartridge fuse A fuse that contains a thin conductor that melts at relatively low temperatures.
Catastrophic failure The complete (and sometimes violent) failure of a component or circuit.
Cell A single unit in a battery that is designed to produce electrical energy through thermal (heat), chemical, or optical (light) means; most batteries contain multiple cells.
Center frequency (f_0) The frequency that equals the geometric average of the cutoff frequencies.
Center-tapped transformer A transformer that has an additional lead connected to the center of its secondary winding.
Centralized power factor correction A power factor correction technique in which compensating capacitor banks are located at the line input to a plant.

Charge An electrical force that causes two particles to be attracted to, or repelled from, each other.

Chassis ground A 0 V reference for the components in an electronic system that does not provide a connection between the system ground and the earth.

Choke Another name for an inductor.

Circuit A group of components that performs a specific function.

Circuit breaker A component that can be reset and used again after breaking a current path.

Circular-mil (cmil) A measure of area found by squaring the diameter (in mils) of the conductor.

Circular-mil-foot (cmil-ft) A measure of volume with a diameter of one mil and a length of one foot.

Cladding Glass that has reflective properties that help keep light in the core of an optical fiber.

Clevis The mechanical connectors that fasten utility conductors to the side of a building.

Coaxial cable (or coax) A cable that contains two conductors separated by a *dielectric* (insulator).

Coefficient of coupling (*k*) The degree of coupling that takes place between two or more coils.

Coercive force The magnetizing force required to return the residual flux density (B) in a material to zero.

Cogeneration The capture and use of the waste heat produced by a generator for space and/or water heating.

Coherent A term that is used to describe light that is highly directional and focused.

Coil A wire wrapped into a series of loops for the purpose of concentrating magnetic flux; another name for an *inductor.*

Cold-rolled grain-oriented (CRGO) steel A type of silicon steel that is used in transformer cores.

Combination circuit Another name for a series-parallel circuit.

Combined cycle power generation A process in which hot gases produced by burning NG drive a gas turbine, and then convert water to steam to drive a steam turbine using the waste heat from the burning process.

Commercial electricians Electricians who install, maintain, and repair the circuits found in industrial and commercial buildings.

Common A point that serves as the voltage reference in the circuit; a power supply terminal that is connected to both DC outputs.

Communications systems Electronic systems that transmit and/or receive information.

Commutating poles Series-wound poles that produce magnetic flux that counters the flux produced by the armature current, restoring the neutral plane to its original orientation.

Commutator A device that converts the output of an alternator to DC.

Compensating capacitor A capacitor connected in parallel with a motor for power factor correction.

Compensating windings Short, series-wound coils wrapped at 90° angles to the field coils to help restore the neutral plane to its original orientation.

Compounded A term used to describe the condition where a generator is adequately compensating for changes in load demand.

Compounding The process of compensating for the effects of predictable load demands on a generator.

Computer systems Electronic systems that store and process information.

Concentrating photovoltaic (CPV) technology Technology that focuses the sun's energy onto photovoltaic cells, increasing their efficiency.

Concentrating solar power (CSP) Technology that uses mirrors or lenses to focus light, producing the heat required to convert water to steam to drive a steam turbine.

Condenser Another name for a **capacitor**.

Conductivity The conductance of a specific volume of an element or compound.

Conductor Any element or material that has extremely low resistance.

Connecting block A block that cuts through the cable wires and holds them in place on the wiring block.

Connector A mechanical connector that connects an optical fiber to another (or to some device) in a way that allows the two to be disconnected.

Connector loss Any decrease in light intensity that is introduced when a connector or splice is attached to an optical fiber.

Conventional current A theory that defines current as *the flow of charge from positive to negative.*

Copper loss Power that is dissipated when current passes through the primary and secondary coils of a transformer.

Core The space inside each coil in a transformer, usually containing air or iron.

Corona discharge Power loss along transmission lines that occurs when high voltages ionize the air surrounding the power lines.

Coulomb (C) The total charge on 6.25×10^{18} electrons.

Counter emf The induced voltage across an inductor that opposes any change in coil current.

Coupled A term used to describe two components (or circuits) that are connected so that energy is transferred from one to the other.

Coupling When a component (or circuit) passes an AC signal from one point to another.

Creep The loosening of connecting materials due to heating.

Critical angle The angle of incidence where most of the light in an optical fiber enters the cladding rather than being reflected.

Critical current rating The maximum current that a given HTS can pass before it loses superconductivity.

Critical temperature The maximum temperature at which superconductivity is maintained.

Cross-talk EMI between wire *pairs* in a telecommunications cable.

Current The directed flow of charge through a conductor.

Current divider Another name for a parallel circuit, which divides the source current among its branches.

Current source A source that provides an output current value that remains relatively constant over a wide range of load resistance values.

Current transformer A component designed to produce a secondary current that is proportional to its primary current.

Cutoff frequency (f_C) The frequency at which the power gain of a circuit drops to 50% of its maximum value.

Cycle The complete transition through one positive and one negative alternation.

Cylindrical rotor A rotor that is similar to that in a DC generator, but containing electromagnets (rather than armature conductors) that generate the alternator field.

DC field supply A DC power supply that provides the current required by the stationary electromagnets in a rotating-armature alternator.

DC offset When a sine wave (or any other waveform) is centered around a voltage other than 0 V.

DC power supply A device with variable DC outputs that can be adjusted to any value within its design limits.

Decade A frequency multiplier of 10.

Decade box A device containing several series-connected potentiometers that allow the resistance of the device to be set to any value within its limits, usually between 1 Ω and 9.999999 MΩ.

Decade scale A scale where the value of each increment is 10 times the value of the previous increment.

Decay curve The portion of the RL curve that represents the decrease in current that occurs after power is disconnected from the circuit; the portion of the RC curve that represents the decrease in voltage that occurs after power is disconnected from the circuit.

Decoding The process of converting transmitted information back to its original form.

Delta (Δ) circuit A circuit commonly found in three-phase systems with three components that are connected in the shape of the Greek letter Δ.

Delta-wye (Δ-Y) circuit A delta circuit with a wye load.

Dielectric The insulating layer in a capacitor.

Difference of potential The difference between the voltages at any two points in a circuit.

Digital meter A meter that uses a digital readout to display the result of a measurement.

Digital multimeter (DMM) A digital meter that measures three of the most basic electrical properties: voltage, current, and resistance.

Direct current (DC) Current that is in one direction only.

Distributed generation (DG) The production of small-scale power at or near the customer's site.

Distribution grid The final element in the power transmission system; it begins at the distribution substation and ends at the customer's business or residence.

Distribution substation One that connects subtransmission lines to local distribution lines.

Distribution transformer A transformer that is used to step down the voltages transmitted by power utilities to values that are required by the customers.

Diversion dam A dam that diverts a portion of natural flowing water through its penstock and turbine, and then returns the water to its natural path.

Diverter A circuit that provides a path for current to bypass the field coil, thereby helping a generator to match the requirements of its load.

Domain theory The theory that atoms with like magnetic fields join to form *magnetic domains*.
Doping The process of adding impurity elements to pure semiconductors to determine their operating characteristics.
Drip loops Another term for the tails on a service entrance (see *Tails*).
Dry cell Another name for a primary (voltaic) cell; a cell that cannot be recharged.
Dry-type transformer A transformer that is typically air-cooled and may have louvers to aid in cooling.
Duplex connector One that is used to terminate two optical fibers.
Duplex winding A term used to describe an armature winding that consists of multiple conductors.
Duty cycle The ratio of pulse width to waveform period, expressed as a percent.
Dynamic A value that changes when specific circuit conditions change.
Dynamic value A value that changes when specific circuit conditions change.
Dynamo An electromechanical device that can convert mechanical energy to electrical energy, or electrical energy to mechanical energy.

Earth ground A physical (and electrical) connection to the earth.
Eddy current Current that travels in a circular motion in the core of an iron-core transformer, generated by flux passing through the core.
Edison base A fuse base that is similar to the base of a light bulb; a fuse base that fits into a standard fuse socket.
Effective values RMS values, which produce the same heating effect as their equivalent DC values.
Efficiency The ratio of a circuit's output power to its input power, given as a percent.
Electric meter A meter that measures the amount of energy that is provided to a residence.
Electrical force (E) The difference of potential across the terminals of a battery or other voltage source.
Electrician Someone who has been trained to install, operate, maintain, and repair electrical circuits and systems.
Electrodes The terminals of a chemical cell.
Electrolyte A chemical that interacts with the electrodes and serves as a conductor between them.
Electrolytic capacitor A capacitor that contains a conducting liquid that serves as one of its plates.
Electromagnetic interference (EMI) A type of interference that occurs when the electromagnetic field produced by a signal in one wire induces a similar signal in another wire.
Electromotive force (EMF) Another name for electrical force; the difference of potential across the terminals of a battery or other voltage source.
Electron flow A theory that defines current as *the flow of electrons from negative to positive*.
Electronics The field that deals with the design, manufacturing, installation, maintenance, and repair of electronic circuits and systems.

Electronics engineer Someone who designs electronic components, circuits, and systems.
Electronics technician Someone who installs, maintains, and repairs electronic systems and circuits using electronic test equipment.
Electrons Particles that orbit the nucleus of an atom.
Electrostatic charge The attraction between charges that are stationary.
Element Any substance that cannot be broken down into two or more simpler substances.
Encoding The process of converting information into a series of electric signals so that it can be transmitted.
Engineering notation A shorthand method of representing large and small numbers.
Equalizer Another name for an *equalizing connection.*
Equalizing connection A connection between the generators to ensure that they respond equally to any change in load demand.
Equivalent circuit A circuit derived by combining groups of parallel and/or series components in one circuit to obtain an equivalent, but simpler, circuit.
Estuary The point where the mouth of a river meets the ocean.
Externally operated (EXO) switch A main disconnect switch that is positioned between the service entrance and the electrical panel.

F connector One that is used primarily in satellite, cable TV, and high-speed cable Internet connections.
Farad The amount of capacitance that stores one coulomb of charge for each 1-V difference of potential across a capacitor's plates.
Fault-current limiter (FCL) A circuit that is used to lower current in the event that a fault occurs.
FC connector A simplex single-mode or multimode screw-on connector.
Feed-in tariff (FiT) A government policy that requires utilities to buy electricity from renewable energy sources at above-market rates.
Feeder circuits The lines that bring power to the electrical panel.
Ferrite core A magnetic core that consists primarily of iron oxide.
Ferromagnetic High-permeability materials, considered to be magnetic.
Ferroresonant transformers A transformer with two secondaries, one coupled to the load and the other used to "trap" harmonic frequencies.
Ferrule A precision-drilled ceramic, metal, or plastic sleeve to which a bare optical fiber is permanently attached.
Fiber optic cable A cable containing an optical fiber that is surrounded by a layer of glass called *cladding.*
Fiber optics The area of telecommunications that uses light to transmit information through glass or plastic cables.
Field pole A stator pole that supports a current-carrying winding used to produce a magnetic field.
Field strength Another name for flux density.
Filter A circuit designed to pass a specific range of frequencies from input to output while blocking others.
Fixed resistor One that has a specific ohmic value that cannot be changed by the user.

Flashover Electrical arcing from one line to another.

Flux density A measure of *flux per unit area*, and therefore, an indicator of magnetic strength.

Foil twisted pair (FTP) Another name for a screened twisted pair.

Foot-pound force (ft-lbf) A unit of measure for torque.

Foot-pounds per minute (ft-lb/min) The unit of measure for motor *power* (the rate at which work is accomplished).

Fossil-fuel power plant A plant that produces heat for generating steam by burning coal, oil, or natural gas.

Fossil natural gas A common type of natural gas that consists primarily of methane, found in oil wells, natural gas wells, and coal beds.

Free electron One that is not bound to an atom and therefore is free to drift from one atom to another.

Freezing When an electrical shock causes muscles to contract and a person cannot let go of a live circuit.

Frequency The rate at which waveform cycles repeat themselves, in cycles per second.

Frequency counter A piece of equipment that indicates the number of times certain events occur every second.

Frequency response The reaction a circuit has to a change in frequency.

Frequency response curve A graph that shows the effect that frequency has on circuit gain.

Frog-leg wound A term used to describe a wiring scheme in which the field pole windings are wired in series-parallel with each other.

Fuel cell A device that generates power through a chemical reaction involving hydrogen.

Fuel rod A cylinder holding nuclear fuel that is used in nuclear power plants.

Full load One that draws maximum current from the source.

Full-load torque The torque produced by a motor when it is operating at its rated power and at full speed.

Function generator A piece of equipment that generates a variety of waveforms.

Function select A control that sets a multimeter to measure volts (V), amperes (A), or ohms (Ω).

Fundamental frequency The reference (lowest) frequency in a harmonic series.

Fuse A component that automatically breaks an electrical connection if the current increases beyond a certain value.

Fusion splice A heat process that welds two optical fibers together.

Gain The ratio of a circuit's output amplitude to its input amplitude.

Galvanometer A current meter that indicates both the magnitude and the direction of a low-value current.

Gang-mounted potentiometers Two potentiometers that share a common shaft that controls both.

Gas-to-liquid (GTL) technology Technology used to refine and cool NG to about −162°C (−260°F), allowing it be more easily transported.

Gas turbine A power generating machine consisting of a compressor, a combustion chamber, and a turbine that burns any number of fuels.

Gear ratio In a series of gears, the ratio of the speed of rotation of the first gear to that of the last gear.

Generator A machine (dynamo) that converts mechanical energy (motion) to electrical energy.

Glowing connection A circuit connection that literally glows as a result of current passing through a high-resistance connection, usually the end result of creep and/or oxidation.

Green A term used to describe environmentally friendly products and technologies.

Greenhouse gases Gases shown to contribute to global climate change.

Grid A term used to describe the power generating plant, power transmission lines, power distribution systems, and substations that make up an AC power distribution system.

Ground (or common) A point that serves as the reference for all voltages in the circuit.

Ground clamp A clamp used to connect a ground wire to a water pipe, establishing a connection between earth ground and the neutral bar in the service panel.

Ground fault circuit interrupter (GFCI) A special type of receptacle designed to switch off in the event that there is a measurable difference between the current levels on the hot and neutral lines.

Ground rod A metal rod, driven into the ground outside a residence, that provides an electrical ground connection for the neutral bar in the service panel.

H-bridge A transistor circuit used to control the direction of motor rotation.

Half-cycle Another way of referring to either alternation of a waveform.

Half-cycle average The average of all the instantaneous voltage or current values in either of a waveform's alternations.

Harmonic A whole-number multiple of a given frequency.

Harmonic cancellation A method of reducing the amplitude of harmonics by combining them with other harmonics that are 180° out of phase.

Harmonic distortion An undesired change in the shape of a fundamental sine wave.

Harmonic series A group of frequencies that are whole-number multiples of the same frequency.

Henry (H) The value of inductance that generates a 1-V counter emf when inductor current changes by one ampere per second (A/s).

Hertz (Hz) The unit of measure for frequency, equal to 1 cycle per second.

High-pass filter A circuit designed to pass all frequencies above its cutoff frequency.

High-resistance connection A condition that results from a poor electrical connection.

Horizontal-axis wind turbine (HAWT) A wind turbine with a three-blade assembly that rotates on a horizontal shaft.

Horsepower (HP) A unit of measure for motor power, equal to 746 W (or 33,000 ft-lb/min).

Hydroelectric plant A power generating plant that converts the kinetic energy in flowing water to electrical energy.

Hysteresis The time lag between the removal of a magnetizing force and the resulting drop in flux density.

Hysteresis loss The energy used to overcome the retentivity of a transformer (or inductor) core.

I^2R loss Another name for *copper loss.*

Ideal current source A current source that maintains a constant output current regardless of its load resistance.

Imaginary power Another term for *reactive power.*

Impedance (Z) The total opposition to current in an AC circuit, consisting of resistance and/or reactance.

Impoundment dam A dam that completely blocks the natural flow of water from a reservoir to the river basin below the dam, the only path for water being through the penstock.

Impurity Any element that is added to a previously pure element to alter its electrical characteristics.

Index of refraction The ratio of the speed of light in a vacuum (free space) to its velocity in a given medium.

Inductance The ability of a component to induce a voltage across itself or a nearby circuit by generating a changing magnetic field; the electrical property that opposes any change in current.

Induction Producing an artificial magnet by aligning magnetic domains using an external force.

Induction motor A motor with rotor current that is generated by electromagnetic induction rather than a DC field supply.

Inductive reactance (X_L) The opposition that an inductor presents to an alternating current, measured in ohms (Ω).

Inductor A component that stores energy in a magnetic field; also called a **coil** or a **choke**. A component designed to provide a specific measure of inductance.

Inductor-start, inductor-run motor One that has two windings (mounted at 90° angles) and a centrifugal switch.

Industrial electricians Electricians who install, maintain, and repair the circuits found in industrial and commercial buildings.

Industrial electronics The area of electronics that deals with systems designed for use in industrial (manufacturing) environments.

Inrush current A high initial current in the primary of a transformer.

Instantaneous value The voltage or current magnitude at a specific point on a waveform.

Insulator Any element or material that has extremely high resistance.

Integrated gasification combined cycle (IGCC) A process where coal is converted into a gas and purified before burning, resulting in reduced production of sulfur dioxide and other pollutants and improved efficiency.

Integrated resistors Micro-miniature components that are made using semiconductors other than carbon.

Intelligent power factor controller (IPFC) Another name for a KVAR-sensitive controller.

Interference An undesired electrical signal that inhibits—or prevents the reception of—a transmitted signal.

Interpoles Another name for the commutating poles.

Inverter A circuit that converts DC to AC.

Ionize The process of becoming electrically charged.

Isolation When a component (or circuit) prevents electrical energy from passing between two points.

Jack A docking port for a plug on an electronic device.

Kickback circuit A circuit that prevents the components in a motor control circuit from being damaged by any motor reverse voltage.
Kilowatt-hour (kWh) The amount of energy used by a 1000 W device that is run for one hour.
Kirchhoff's Current Law (KCL) A law stating that the current leaving any point in a circuit must equal the current entering the point.
Kirchhoff's Voltage Law (KVL) A law stating that the sum of the voltages in a series circuit must equal the source voltage.
KVAR-sensitive power factor controller A power factor correction circuit that measures changes in the power factor as motors are started or stopped, and switches capacitors in or out as needed to keep the power factor at a preset level.

Lagging VAR Another term for inductive VAR.
Laminated core An iron core that is broken up into thin layers with an oxide coating between each, preventing eddy currents from being generated.
Lanyard A short length of rope that is used to secure a safety harness to some structure in order to keep a worker from being injured in a fall.
Lap wound A term used to describe a wiring scheme in which the field pole windings are wired in parallel with each other.
LC connector A simplex or duplex connector that is available in both single-mode and multimode configurations.
Leading VAR Another term for capacitive VAR.
Leakage current The current that "leaks" through a capacitor's dielectric, typically in the low μA range.
Left-hand rule A memory aid that helps you remember the relationship between the direction of electron flow and the direction of the resulting magnetic field.
Lenz's law A law stating that an induced voltage always opposes its source.
Light Electromagnetic energy that falls within a specific range of frequencies.
Light-emitting diode (LED) A semiconductor component that emits light when activated by a voltage having the proper polarity.
Light intensity The amount of light per unit area received by a given photodetector.
Linear taper Potentiometer resistance that changes at a linear (constant) rate as the shaft is turned.
Liquefied natural gas (LNG) The form that NG takes as a result of the gas-to-liquid process.
Load The current demand on the output of a circuit.
Load analysis Analyzing the effect that a change in load has on the output from a voltage or current source.
Load-specific fault In a circuit that has two or more loads, a fault that affects only one load.
Lockout/tagout A common method of making sure that a circuit is deenergized and that it stays that way until work on the circuit is complete.

Logarithmic scale A scale with a frequency spread between increments that increases at a geometric rate.
Low-pass filter A circuit designed to pass all frequencies below its cutoff frequency.

M-block Another name for a 66 termination block.
Machine action Making a conductor (rotor) spin about a pivot point by changing the direction of current while it is in the neutral plane.
Magnetic field The area of space surrounding a magnet that contains magnetic flux.
Magnetic flux The lines of force produced by a magnet.
Magnetic induction Using a magnetic field to align magnetic domains, thereby producing an artificial magnet.
Magnetic north The earth's magnetic north pole.
Magnetic shielding Insulating an instrument from the influence of magnetic flux.
Magnetic south The earth's magnetic south pole.
Magnetization curve A curve that plots flux density (B) as a function of magnetic field strength (H).
Magnetomotive force (mmf) The force that generates magnetic flux in a material that has reluctance; the magnetic counterpart to electromotive force (emf).
Magnetostriction A phenomenon in which the laminations in an iron-core transformer are exposed to expanding and contracting magnetic fields, causing them expand and contract slightly.
Magnitude The value of a voltage or current.
Main breaker A double-pole, double-throw (DPDT) switch that serves as the main disconnect for an electrical panel.
Main disconnect A switch that is used to disconnect power from a residence in case of an emergency.
Mast In a mast-type service drop, the conduit and weather head that extend upward from the roof.
Mast knob On a mast-type service drop, the structure on the mast to which the service lines are attached.
Matter Anything that has mass and occupies space.
Maxwell (Mx) A unit of measure of magnetic flux, equal to one line of flux.
Mechanical losses Power loss that can occur when the expanding and collapsing magnetic field in a high-power transformer stresses the component's wires, core, and support structure.
Mechanical splice A plastic fitting that aligns the two optical fibers.
Medium The part of a telecommunications system that carries encoded information from the source to its destination, most commonly air, copper wire or cable, or fiber optic cable.
Meter movement A structure that moves a meter pointer in response to an electric current.
Microturbine A much smaller version of the industrial gas turbine that can provide outputs from tens of kilowatts to a few hundred kilowatts.
Microweber (μWb) A practical unit of measure of magnetic flux, equal to 100 Mx (lines of flux).
Mil A unit of length, equal to 0.001 inch.

Milliampere (mA) One one-thousandth of an ampere.

Mode A path for light through an optical fiber.

Momentary switch A switch that makes or breaks a connection only as long as the button is pushed.

Motor A dynamo that is configured to convert electrical energy to mechanical energy.

Motor action Making a conductor (rotor) spin about a pivot point by changing the direction of current while it is in the neutral plane.

Multimeter An instrument used to measure current, voltage, and resistance.

Multimode fiber One with a diameter large enough to provide multiple paths (modes) for light to pass through the cable.

Multiple-output transformer A transformer that has two or more secondary coils.

Multiplex winding A term used to describe an armature winding that consists of multiple conductors.

Multiside board A PC board designed so that components are mounted on both sides.

Multisource circuit A circuit with multiple voltage and/or current sources.

Multiturn potentiometers A potentiometer whose control shaft can be rotated more than 360°.

Mutual inductance When the expanding and collapsing flux produced by an inductor induces a voltage across another inductor that is in close proximity.

N connector One that is used primarily in land mobile, cellular, and pager infrastructure.

n-order harmonics The frequencies in a harmonic series.

n-type material A semiconductor wafer treated with the five-valence-electron substance that causes it to have an excess of conduction band electrons.

Nacelle The housing on a HAWT that contains the low-speed shaft and driving gears.

Negative charge One of two electrical charges; the other is **positive charge**.

Negative temperature coefficient A value that indicates how much the resistance of a material decreases as temperature increases.

Negative voltage A term indicating that a voltage is more negative than the circuit reference point.

Net metering A government policy that allows small-scale power suppliers to be paid for at least some of the power that they produce.

Network theorems Analysis techniques that are used to solve specific types of circuit problems.

Neutral bar The common connection point for all the neutral lines in a residence.

Neutral plane The space above and below a stationary magnet where a spinning rotor is in parallel with the magnetic field.

Neutron One of two particles found in the nucleus of an atom; the other is the **proton**.

No-load output voltage The voltage measured at the load terminals of a circuit with the load removed.

Node A point connecting three or more current paths.

Nonlinear taper Potentiometer resistance that changes at a nonlinear (changing) rate as the shaft is turned.

Nonmetallic (NM) cable A group of insulated conductors housed in a protective plastic jacket; also called *Romex*.

Nonsplit block A 66 block with rows containing four clips that are electrically connected to each other.
Normally-closed (NC) switch A push-button switch that normally *makes* a connection between its terminals.
Normally-open (NO) switch A push-button switch that normally *breaks* the connection between its terminals.
North-seeking pole The pole where lines of force leave a magnet.
Norton current (I_N) The current source in a Norton equivalent circuit.
Norton equivalent circuit For any circuit, an equivalent circuit that contains a current source (I_N) in parallel with a source resistance (R_N).
Norton resistance (R_N) The source resistance in a Norton equivalent circuit.
Norton's theorem A theorem stating that any resistive circuit or network, no matter how complex, can be represented as a current source in parallel with a source resistance.
Notch filter Another name for a *band-stop filter.*
Nuclear fission The process of generating power by splitting the nucleus of an atom.
Nuclear power plant One that converts water to steam using heat produced by nuclear fission.
Nucleus The central core of an atom.

Octave A frequency multiplier of two (2).
Octave scale A scale where the value of each increment is two times the value of the previous increment.
Ohm The unit of measure for resistance.
Ohm-centimeter (Ω-cm) The unit of measure for the resistivity of one cubic centimeter of a material.
Ohmmeter A meter used to measure resistance.
Ohm's law A law stating that current is directly proportional to voltage and inversely proportional to resistance.
Open circuit A complete break in the current path through all or part of a circuit.
Open-delta connection A delta circuit configuration that makes it possible to produce a three-phase transformer using only two single-phase transformers.
Optical fiber A length of ultra-pure glass or plastic that provides a closed pathway for transmitting light between two points.
Organic light-emitting diodes (OLEDs) Semiconductors that contain organic materials.
Oscilloscope A piece of test equipment that provides a visual display for a variety of voltage and time measurements.
Outlet A term that is often used in reference to a receptacle even though it more properly refers to the receptacle, its mounting box, and wiring.
Output voltage The voltage at the load terminals of a circuit.
Outside lineman Someone who installs and repairs power transmission and distribution lines and circuits.
Over-compounded A design technique that provides added insurance against the effects of load demands on the generator output voltage.
Overload A load that exceeds the design limits of a circuit or the current rating of its overcurrent protective element.

p-type material A semiconductor wafer treated with a three-valence-electron substance that causes it to have an excess of valence-band holes.

Pad-mounted transformer A distribution transformer that is mounted on a rectangular concrete pad.

Parallax error Any change in readings that results from looking at a scale from an angle.

Parallel circuit A circuit that provides more than one current path between any two points.

Parallel-equivalent circuit An equivalent circuit that is made up entirely of components connected in parallel.

Peak shaving The leveling out of power consumption over time.

Peak-to-peak value The difference between the positive and negative peak values of a waveform.

Peak value The maximum value reached on either alternation of a waveform.

Peaker plants Distribution generators that generate electricity in response to peak load demands.

Penstock A conduit between a water reservoir and the hydroelectric power plant.

Percent impedance (%Z) The transformer primary voltage (as a percentage of its rated value) that generates maximum current through a shorted secondary.

Period The time required to complete one cycle of an AC waveform.

Periodic table of the elements A table that identifies all the known elements.

Periodic waveform A waveform that repeats itself continuously.

Permeability (μ) A measure of the ease with which lines of magnetic force are established within a material.

Permittivity A measure of the ease with which lines of electrical force are established within a given material.

Personal protective equipment Clothing and equipment that are used to protect a worker from injury on the job site.

Phase The position of a point on a waveform relative to the start of the waveform, usually given in degrees.

Phase angle The phase difference between two waveforms.

Phase sequence For a three-phase power source, the order in which the three output sine waves reach their peak values.

Photodetectors Electronic devices that that respond to light.

Photodiode A semiconductor component that responds to light input in the same manner as a PIN diode.

Photoemitters Electronic devices that generate light.

PIN diode A semiconductor component that responds to a light input by increasing its conduction.

Plates The conductive surfaces of a capacitor.

Plenum The air spaces above and below a room.

Plug A connector that is attached to the wires in a cable.

Polarity The electrical orientation of a voltage, identified as *negative* or *positive*.

Polarity dots Dots drawn on the schematic symbol of a transformer that identify the input and output terminal voltages that are in phase.

Pole The moving contact(s) in a switch.

Pole-type transformer A distribution transformer that is mounted on a pole (such as a telephone pole).
Poles The points where magnetic lines of force leave (and return to) a magnet.
Polyphase alternator One that provides more than one sinusoidal output.
Positive charge One of two electrical charges; the other is **negative charge**.
Positive ion An atom with a positive net charge.
Positive temperature coefficient A value that indicates how much the resistance of a material increases as temperature increases.
Positive voltage A term indicating that a voltage is more positive than the circuit reference point.
Potentiometer A three-terminal resistor whose value can be adjusted by the user.
Power The rate at which a component, circuit, or system uses energy.
Power buses Two metal structures that provide the 120 V connections and mounting structures for the branch circuit breakers.
Power distribution The process of reducing transmitted power to customer-usable levels and delivering that power to local homes and industries.
Power distribution system The lines, poles, transformers, and switching and protection circuits that deliver electrical energy to customers at usable voltage levels.
Power factor The ratio of resistive power to apparent power.
Power factor correction Using a circuit to improve the power factor of a circuit.
Power gain The ratio of a circuit's output power to its input power.
Power generation The process of converting one type of energy into electrical energy.
Primary The input coil of a transformer.
Primary cell A cell that cannot be recharged; also called a *voltaic cell* or *dry cell*.
Prime mover The external mechanical driver for a generator.
Printed circuit (PC) board A board that provides mechanical (physical) support for components and conductors that connect those components.
Proton One of two particles found in the nucleus of an atom; the other is the **neutron**.
Prototype The first working model of a circuit or system.
Pulse width (PW) The positive alternation of a rectangular waveform.
Punch-down tool A tool used to connect wires to a termination block.

Quality (Q) For a bandpass or notch filter, the ratio of its center frequency to its bandwidth.

Rankine cycle The conversion of water to steam and back to water in a closed recirculating system.
Reactance The opposition to alternating current provided by an inductor or a capacitor.
Reactive power (P_X) A value that indicates the rate at which energy is transferred to and from an inductor's magnetic field, measured in volt-amperes-reactive (VAR); also called *imaginary power*.
Receiver The part of a telecommunications system that accepts transmitted information and restores it to its original form, usually audio, video, or data.

Receptacle An electrical connection point for such items as lamps, home electronics, and appliances; that is, for various loads on the system.
Rechargeable cell Another name for a secondary (wet) cell; a cell that can be recharged.
Reciprocating internal combustion engine An engine known for its low cost per kilowatt produced, low maintenance requirements, and high reliability.
Rectifier A circuit that converts AC to pulsating DC.
Refraction The bending that occurs when light passes between two mediums with different optical densities.
Registered jack (RJ) A modular jack wired according to a standard set by the Federal Communications Commission (FCC).
Rejection base A fuse base that requires the use of an adaptor to fit into a standard fuse socket.
Relative permeability (μ_r) The ratio of the material's permeability to that of free space.
Relative permittivity The ratio of a material's permittivity to that of a vacuum.
Relay A switch that is opened or closed by an input signal.
Reluctance The opposition that a material presents to magnetic lines of force.
Residential electrician Someone who installs, maintains, and repairs all the electrical circuits found in houses, apartments, and condominiums.
Residual flux density The flux density that remains in a material after a magnetizing force is removed.
Resistance The opposition to current provided by a material, component, or circuit.
Resistive power (P_R) The power actually dissipated by any resistance (such as the winding resistance of an inductor), measured in watts (W); also referred to as **true power**.
Resistivity The resistance of a specific volume of an element or compound.
Resistor A component that is used to restrict (limit) current.
Resolution A measure of the rate at which potentiometer resistance changes, ohms per degrees (Ω/°).
Resonance For an LC circuit, the operating state where $X_L = X_C$.
Resonant frequency (f_r) For an LC circuit, the frequency at which $X_L = X_C$.
Retentivity The ability of an artificial magnet to retain its magnetic characteristics.
Return Another name for a neutral wire.
Rheostat A two-terminal variable resistor that is used to control current.
Ring and **Tip** The two wires in a twisted pair.
Ring magnet A magnet that does not have any identifiable poles or air gaps, and does not generate an external magnetic field.
Ripple The voltage variations in the output from a DC generator.
Rise curve The portion of the RL curve that represents the increase in current that occurs after power is applied to the circuit; the portion of the RC curve that represents the increase in voltage that occurs after power is applied to the circuit.
Riser The vertical spaces behind the walls of a room.
Root-mean-square (RMS) value
Rotary switch A switch with one or more poles and multiple throws.
Rotating-armature alternator An alternator with rotating loop conductors and stationary magnetic fields.
Rotating-field alternator An alternator with rotating magnetic fields and stationary conductors.

Rotor The rotating plates in an interleaved capacitor; the rotating structure of a dynamo.
Rowland's law The magnetic counterpart of Ohm's law, it defines the relationship between mmf, flux, and reluctance.
Salient-pole rotor A rotor containing electromagnets that protrude from the armature.
Scattering The phenomenon where light reflects off an impurity molecule in an optical fiber and heads in an unpredictable direction.
Schematic diagrams Drawings that show you how the components in a circuit are interconnected.
Screened twisted pair (ScTP) A term used to describe a cable that contains four twisted pairs with an outer conductive foil wrapped around the group to provide added protection from EMI.
Secondary The output coil of a transformer.
Secondary cell A cell that can be recharged; also called a *rechargeable cell* or *wet cell.*
Self-excited generator One that provides its own field coil current via its armature.
Self-inductance The process whereby a coil induces a voltage across itself by generating a changing magnetic field.
Semiconductor Any element or material with resistance that falls about midway between that of a conductor and that of an insulator.
Series circuit A circuit that contains only one current path.
Series-equivalent circuit An equivalent circuit that is made up entirely of components connected in series.
Series-parallel circuit One that contains both series and parallel connections.
Service drop An overhead power connection from the utility lines to the service entrance.
Service entrance The point where power from the utility company enters a residence.
Service lateral An underground service entrance.
Service meter A meter that measures the energy (in kWh) provided by the power company.
Service panel The box that houses circuit breakers and/or fuses and distributes power to the various circuits in a residence.
Shells Electron orbital paths that circle the nucleus of an atom.
Shielded twisted pairs A term used to describe a cable that contains four twisted pairs with conductive foil surrounding each twisted pair, with an outer conductive foil wrapped around the group to provide added protection from EMI and cross-talk.
Short circuit An unintentional extremely low resistance that allows an abnormally high current through all or part of a circuit.
Shorted-load output current (I_{SL}) The maximum output from a current source, measured with its output terminals shorted.
Shorting tool A conductor with a double-insulated handle that is used to discharge a capacitor.
Simplex connector One that is used to terminate a single optical fiber.
Simplex winding A term used to describe an armature winding that consists of a single conductor.
Sine wave Another name for a sinusoidal waveform.
Single-mode fiber One with a diameter so narrow that it provides only one path, or *mode*, for light to pass through the cable.

Single-phase alternator An alternator that produces a single sine wave output.
Sinusoidal waveform A waveform whose magnitude varies with the sine of its phase.
Skin effect The tendency of current to move toward the outside surface (skin) of a conductor as frequency increases.
Slip The difference between the speed of the rotating stator field and the rotor speed.
Smart grid An interactive grid in which the power producer, distributor, and end user are in constant real-time communication with each other.
Smart meters Two-way communication devices that allow the electricity provider to monitor energy consumption in real time and price the electricity based on its real-time cost.
Snap coupling connector A simplex or duplex connector that snaps into a docking jack in a manner similar to an adaptor.
Solid wire A wire that contains a single, solid conductor.
Solid-state relay A semiconductor relay that is used to control an AC signal.
South-seeking pole The pole where lines of force return to a magnet.
Space width (SW) The negative alternation of a rectangular waveform.
Splice A mechanical connector that permanently joins two optical fibers.
Split block A 66 block with rows containing four clips that are physically and electrically split into two pairs.
Split-core current transformer One that can be split in two, placed around a conductor under test, and then closed.
Split phase A term that is often used to describe the output from the center-tapped transformer because the two output waveforms are 180° out of phase.
Split-phase induction motor A single-phase induction motor with an added starting component that produces the torque required to start the motor and help bring the rotor up to speed.
Square wave A special-case rectangular waveform that has equal pulse width and space width values.
Squirrel-cage rotor One that contains conductor bars that extend through the length of the rotor in place of the wire conductors found in wound rotors.
ST connector A simplex, multimode, twist-on connector.
Startup current The current through a motor when power is first applied.
Static A value that does not change in response to circuit conditions.
Static compensation Another term for *static power factor correction.*
Static correction Power factor correction that is implemented for each individual motor.
Static power factor correction Power factor correction that is implemented for each individual motor.
Static value A value that does not change in response to circuit conditions.
Stator The stationary plates in an interleaved capacitor; the stationary structure of a dynamo.
Step-down transformer A transformer that has a lower output (secondary) voltage than input (primary) voltage.
Step-up transformer A transformer that has a higher output (secondary) voltage than input (primary) voltage.

Stranded wire A wire that contains any number of thin wires wrapped to form a larger wire.
Stray loss Power loss that results when flux generated in a transformer primary interacts with nearby conductors other than the transformer secondary.
Step-down substation One that is connected between transmission lines and subtransmission or distribution lines.
Step-up substation One that connects a power generating plant to the transmission lines.
Submersible transformer A transformer that is designed to be used in underwater power transmission systems.
Substations Utility stations that contain transformers that increase (step up) or decrease (step down) voltage levels, switching circuits, and overload protection circuits.
Subtransmission lines Lines that carry voltages—typically between 26 kV and 69 kV—to regional distribution substations.
Superconducting magnetic energy storage (SMES) Storing electrical energy in the massive magnetic field of a closed superconducting inductor.
Superconductors Man-made alloys that can pass electric current with virtually zero resistance when operated at extremely low temperatures.
Superposition theorem A theorem stating that the response of a circuit to more than one source can be determined by analyzing the circuit's response to each source and combining the results.
Surge A sudden and momentary current overload that is commonly caused by the startup of an electrical machine, such as a motor or compressor.
Switch A component that allows you to manually make or break the connection between two or more points in a circuit.
Symmetrical waveform A waveform that has identical positive and negative alternations.
Synchronous motor One that is synchronized to an alternator that acts as an energy and timing source.

Tails The wire loops at the service entrance that prevent mechanical stresses on the power lines and prevent any water (due to rain) from traveling along the lines into the service drop conduit. Also known as **drip loops**.
Taper A measure of how potentiometer resistance changes as the control shaft is rotated between its extremes.
Telecommunications The area of communications that deals with sending and receiving voice, data, and/or video information.
Telecommunications cabling The practice of preparing and installing the wires, cables, and connectors that are used in residential and commercial telecommunications circuits.
Telecommunications technician Someone who installs, maintains, troubleshoots, and repairs residential and commercial data, audio, video, and security circuits.
Telephony The study of wireless and land telephone systems.
Terminals Wire connection points.
Terminate A term used to describe the connection(s) at the end of a cable.

Termination block A physical structure to which cable wires are individually connected.

Tesla (T) A unit of measure of flux density, equal to one weber per square meter (Wb/m^2).

Thermal power plant A power generating plant that uses heat to convert water to steam, and then uses the steam as the prime mover for its turbine.

Thermistor A component whose resistance varies with temperature.

Thevenin equivalent circuit For any circuit, an equivalent circuit that contains a voltage source (E_{TH}) in series with a source resistance (R_{TH}).

Thevenin resistance (R_{TH}) The source resistance in a Thevenin equivalent circuit.

Thevenin's theorem A theorem stating that any resistive circuit or network, no matter how complex, can be represented as a voltage source in series with a source resistance.

Thevenin voltage (E_{TH}) The voltage source in a Thevenin equivalent circuit.

Three-phase power A means of generating and transmitting power using three sine waves that are 120° out of phase with each other.

Throw The stationary contact(s) in a switch.

Tidal power Converting the mechanical energy produced by rising tides into electrical energy.

Time base The oscilloscope control that determines the time represented by the interval between any two adjacent major divisions along the *x*-axis.

Time constant (τ) One of five equal time periods in the rise and decay curves for an RL circuit.

Tip and Ring The two wires in a twisted pair.

Tolerance The possible range of values for a component, given as a percentage of its nominal (rated) value.

Toroid A doughnut-shaped magnetic core.

Torque The force required to rotate an object about a point.

Traces The conductors that connect the components on a PC board.

Transducer A device that converts energy from one form to another.

Transformer A component that contains one or more inductors that are wrapped around a single physical structure that is used to raise or lower a voltage level; a component that uses electromagnetic induction to pass an AC signal from one circuit to another, while providing DC isolation between the two.

Transformer action A term used to describe the means by which a transformer couples an AC signal from its primary to its secondary by means of electromagnetic induction.

Transition A sudden change in current (or voltage) from one level to another.

Transmission lines The high-voltage lines used to transmit electrical energy over long distances.

Transmitter The part of a telecommunications system that encodes and transmits information.

Transmitting The process of transferring encoded information to a transmission medium.

Troubleshooting The process of locating faults in a circuit or system.

True power Another term for *resistive power.*

True-RMS meter An AC meter that uses relatively complex circuitry to directly measure the RMS value of any true AC waveform.

Turbine A high-speed generator that is typically driven by gas or steam.
Turbine-driven rotors Cylindrical rotors that are used in turbines.
Turn Each loop of wire in a coil.
Turns ratio The ratio of primary turns to secondary turns.
Twist length The distance from one twist to the next in a twisted pair.
Twisted pair Two wires that are wrapped around each other along their entire length.
Type S fuse A fuse with a *rejection base* (one that requires the use of an adaptor to fit into a standard fuse socket).
Type T fuse A fuse with an *Edison base* (a base similar to the base of a light bulb).

Unbalanced load One in which the load currents are not equal and/or out of phase by some value other than 120°.
Under-compounded A term used to describe the condition in which a generator is not adequately compensating for changes in load demand.
Unity coupling An ideal condition in which all of the flux produced by one coil passes through another.
Unity gain A gain of one (1).
Universal jack A term used in reference to the RJ-45 because it also works with RJ-6 and RJ-11 plugs.
Unshielded twisted pairs A term used to describe a cable that contains four twisted pairs that are not shielded from EMI and cross-talk.

Variable resistor One that can be adjusted to any ohmic value within a specified range.
Variac A variable autotransformer.
Valence shell The outermost orbital path for a given atom.
Vault-type transformer A transformer that is designed for underground installation.
Vertical-axis wind turbine (VAWT) A type of wind turbine that rotates on a vertical axis rather than a horizontal one.
Vertical sensitivity The value represented by each major vertical division on an oscilloscope, expressed in *volts/division.*
Volt (V) The unit of measure for voltage; the amount of electrical force that uses one joule of energy to move one coulomb of charge.
Volt-amperes (VA) The unit of measure for apparent power.
Volt-amperes reactive (VAR) The unit of measure for reactive power.
Voltage The force that generates the flow of electrons (current) through a circuit.
Voltage divider A term that is often used to describe a series circuit because the source voltage is divided among the components in the circuit.
Voltage-divider equation An equation that allows us to calculate the voltage drop across a resistor without first calculating the value of the circuit current.
Voltage drop A term that is commonly used to describe a change from one potential to a *lower* potential.
Voltage gain The ratio of a circuit's output voltage to its input voltage.
Voltage regulation The ability of a transformer to maintain a stable output (secondary) voltage when the load demand changes.
Voltaic cell Another name for a primary (dry) cell; a cell that cannot be recharged.

Voltmeter A meter used to measure voltage.

Voltmeter loading The reduction in component voltage that can result from connecting a voltmeter across a relatively high-resistance (or open) component.

Volt-ohm-milliammeter (VOM) A meter that uses an analog scale (rather than a digital scale) to provide a readout.

Watt The unit of measure for power, equal to one joule per second.

Wave power Converting the mechanical energy produced by the constant rise and fall of surface waves or deep water waves into electrical energy.

Wave wound A term used to describe a wiring scheme in which the field pole windings are wired in series with each other.

Wavelength The physical length of one cycle of a transmitted electromagnetic wave.

Weber (Wb) A unit of measure of magnetic flux, equal to 10^8 Mx (lines of flux).

Wet cell Another name for a rechargeable (secondary) cell; a cell that can be recharged.

Wet-type transformer A transformer that is typically cooled using mineral oil or high-temperature hydrocarbons.

Wheatstone bridge A circuit containing four resistors and a meter "bridge" that provides extremely accurate resistive measurements.

Wind farm A group of wind turbines.

Wind power Energy in currents of air (wind) that can be converted to electrical energy using a wind turbine.

Winding resistance (R_W) The resistance of a coil (inductor).

Wiper arm The potentiometer terminal that is connected to a sliding contact.

Wire-wound resistor A resistor that uses the resistivity of a length of wire as the source of its resistance.

Wiring block The base of a 110 termination block; it has two rows with 50 connection points each.

Work For a motor, the product of the force used to move an object and the distance the object is moved.

Wound rotor A rotor made of laminated steel segments and wire conductors (windings) that are connected to slip rings.

Wye (Y) circuit A circuit commonly found in three-phase systems with three components that are connected in the shape of the letter Y.

Wye-delta (Y–Δ) circuit A wye circuit with a delta load.

Yaw motor and yaw driver The elements in a HAWT that turn the turbine toward the wind.

APPENDIX D

Answers to Odd-Numbered Practice Problems

Chapter 1

PRACTICE PROBLEM ANSWERS

1. a) 38,400 m = 38.4 km b) 234,000 W = 234 kW
   c) 44,320,000 Hz = 44.32 MHz d) 175,000 V = 175 kV
   e) 4,870,000,000 Hz = 4870 MHz = 4.87 GHz
3. a) 0.22 H = 220 mH b) 0.00047 F = 470 μF c) 0.00566 m = 5.66 mm
   d) 0.000045 A = 450 μA e) 0.288 A = 288 mA

Chapter 2

PRACTICE PROBLEM ANSWERS

1. 50 mA
3. 4 μA
5. 69.1 mΩ
7. 646 μΩ

Chapter 3

PRACTICE PROBLEM ANSWERS

1. R_{max} = 112.2 kΩ R_{min} = 107.8 kΩ
3. R_{max} = 378 kΩ R_{min} = 342 kΩ
5. a) R = 160 Ω b) R = 2.7 MΩ c) R = 390 kΩ d) R = 51 Ω
   e) R = 4.7 Ω f) R = 56 kΩ
7. a) orange, orange, orange b) brown, green, black c) white, brown, yellow
   d) red, red, red e) brown, red, yellow f) brown, black, blue
   g) brown, gray, gold h) red, yellow, silver
9. a) R_{max} = 252 kΩ, R_{min} = 228 kΩ b) R_{max} = 3.67 Ω, R_{min} = 3.53 Ω
   c) R_{max} = 90.2 Ω, R_{min} = 73.8 Ω d) R_{max} = 0.6936 Ω, R_{min} = 0.6664 Ω

Chapter 4

PRACTICE PROBLEM ANSWERS

1. a) I = 81.8 mA b) I = 4.44 A c) I = 255 mA
   d) I = 485 mA e) I = 22.1 A

3. a) $E = 82$ V b) $E = 6.5$ V c) $E = 35.1$ V d) $E = 26.4$ V e) $E = 2.88$ V
5. a) $R = 150\ \Omega$ b) $R = 470\ \Omega$ c) $R = 220\ \Omega$
   d) $R = 27.3\ \Omega$ e) $R = 100\ k\Omega$
7. a) $P = 2400$ W b) $P = 18.8$ W c) $P = 55$ W
   d) $P = 132$ W e) $P = 10.9$ mW
9. a) $P = 820$ W b) $P = 423$ mW c) $P = 456$ W
   d) $P = 634$ mW e) $P = 2.3$ W
11. a) $P = 14.7$ W b) $P = 356$ W c) $P = 30.6$ W
   d) $P = 7.76$ W e) $P = 3.31$ kW
13. $P_{min} = 480$ W
15. $P_{min} = 303$ W
17. a) $\eta = 24\%$ b) $\eta = 7.42\%$ c) $\eta = 7.88\%$
   d) $\eta = 34\%$ e) $\eta = 4\%$
19. $P_L = 222$ W
21. $E_{kWh} = 8.7$ kWh
23. a) $E = 400$ V and $R = 40\ k\Omega$ b) $I = 500\ \mu A$ and $R = 64\ k\Omega$
   c) $I = 8.37$ mA and $E = 27.6$ V d) $R = 3.0\ k\Omega$ and $P = 675$ mW
   e) $E = 28.8$ V and $P = 691$ mW

Chapter 5

PRACTICE PROBLEM ANSWERS

1. a) 2.65 kΩ b) 275 c) 1241 Ω d) 106.4 kΩ
3. a) $R_3 = 470\ \Omega$ b) $R_1 = 27\ k\Omega$ c) $R_2 = 5.1\ k\Omega$ d) $R_T = 148\ \Omega$
5. $R_3 = 7\ k\Omega$
7. $R_T = 55\ \Omega$, $I_T = 2.18$ A
9. $R_T = 100\ \Omega$, $I_T = 1.2$ A
11. $R_T = 538\ \Omega$, $R_2 = 68\ \Omega$
13. $R_T = 20.4\ k\Omega$, $R_2 = 2.6\ k\Omega$
15. $E = 18$ V
17. $E = 5.5$ V
19. $I_T = 100$ mA, $P_T = 600$ mW, $P_1 = 270$ mW, $P_2 = 330$ mW, $P_T = 600$ mW
21. $I_T = 15$ mA, $P_T = 225$ mW, $P_1 = 115$ mW, $P_2 = 87.8$ mW, $P_3 = 22.5$ mW, $P_T = 225$ mW
23. $V_{R2} = 3.3$ V, $I_T = 100$ mA
25. $V_{R2} = 8.9$ V, $I_T = 22.8$ mA
27. $V_A = 5.45$ V
29. $V_{R1} = 6.82$ V $V_{R2} = 8.18$ V
31. $V_{R1} = 2.2$ V $V_{R2} = 7.8$ V $V_{R3} = 20$ V

33. $V_A = 4.36$ V

35. $V_A = 12.6$ V

37. $R_T = 770$ kΩ, $I_T = 312$ μA, $P_T = 74.9$ mW, $V_{R1} = 193$ V, $P_{R1} = 60.2$ mW,
$V_{R2} = 46.8$ V, $P_{R1} = 14.6$ mW

39. $R_T = 2.82$ kΩ, $I_T = 85.1$ mA, $P_T = 20.4$ W, $V_{R1} = 10.2$ V, $P_{R1} = 868$ mW
$V_{R2} = 17$ V, $P_{R2} = 1.45$ W, $V_{R3} = 128$ V, $P_{R3} = 10.9$ W
$V_{R4} = 85.1$ V, $P_{R4} = 7.24$ W

41. $V_M = 59.81$ V.

43. $V_M = 118.7$ V.

45. I_T should be 455 μA. The wrong value resistors are used, the supply voltage is too high, or there is a partial short.

Chapter 6

PRACTICE PROBLEM ANSWERS

1. a) $I_T = 4.6$ A, b) $I_T = 6.9$ A, c) $I_T = 15.55$ A, d) 1.806 A

3. $I_{R1} = 600$ mA, $I_{R2} = 400$ mA

5. $I_{R1} = 667$ mA, $I_{R2} = 1$ A, $I_{R1} = 1.2$ A, $I_T = 2.867$ A

7. $R_T = 120$ Ω

9. $R_T = 41.9$ Ω

11. $R_T = 6.43$ Ω

13. $R_T = 21.6$ Ω

15. $I_{R1} = 400$ mA, $I_{R2} = 100$ mA, $R_2 = 1.2$ kΩ

17. For $R_L = 1.2$ kΩ, $V_L = 216$ V, For $R_L = 3.3$ kΩ, $V_L = 594$ V.

19. $R_T = 10$ Ω, $I_{R1} = 400$ mA, $I_{R2} = 200$ mA.

21. $R_T = 40.5$ Ω, $I_{R1} = 298$ mA, $I_{R2} = 203$ mA.

23. $R_T = 36.1$ Ω, $I_T = 3.32$ A, $I_{R1} = 856$ mA, $P_{R1} = 103$ W, $I_{R2} = 999$ mA,
$P_{R2} = 120$ W, $I_{R3} = 1.46$ A, $P_{R3} = 175$ W, $P_T = 398$ W.

25. $R_T = 6.32$ Ω, $I_T = 3.8$ A, $I_{R1} = 2$ A, $P_{R1} = 48$ W, $I_{R2} = 667$ mA, $P_{R2} = 16$ W,
$I_{R3} = 889$ mA, $P_{R3} = 21.3$ W, $I_{R4} = 240$ mA, $P_{R4} = 5.76$ W, $P_T = 91.2$ W

27. a) $E = 3$ V, $I_{R2} = 100$ mA, $I_T = 430$ mA, $I_{R3} = 30$ mA, $R_3 = 100$ Ω.
b) $R_3 = 150$ Ω, $V_{R3} = 6$ V, $R_T = 50$ Ω, $I_S = 120$ mA

Chapter 7

PRACTICE PROBLEM ANSWERS

1. $R_1 + R_2 = 30$ Ω, $I_{R1} = I_{R2} = 800$ mA, $V_{R1} = 9.6$ V, $V_{R2} = 14.4$ V, $P_{R1} = 7.68$ W,
$P_{R2} = 11.5$ W, $R_3 + R_4 = 62$ Ω, $I_{R3} = I_{R4} = 387$ mA, $V_{R3} = 18.2$ V, $V_{R4} = 5.81$ V,
$P_{R3} = 7.04$ W, $P_{R4} = 2.25$ W, $R_T = 20.2$ Ω, $I_T = 1.19$ A, $P_T = 28.6$ W.

3. $R_1 + R_2 = 400$ Ω, $I_{R1} = I_{R2} = 300$ mA, $V_{R1} = 30$ V, $V_{R2} = 90$ V, $P_{R1} = 9$ W,
$P_{R2} = 27$ W, $R_3 + R_4 + R_5 = 1$ kΩ, $I_{R3} = I_{R4} = I_{R5} = 120$ mA, $V_{R3} = 24$ V,

$V_{R4} = 56.4$ V, $V_{R5} = 39.6$ V, $P_{R3} = 2.88$ W, $P_{R4} = 6.77$ W, $P_{R5} = 4.75$ W, $R_T = 286\ \Omega$, $I_T = 420$ mA, $P_T = 50.4$ W.

5. $R_1 \parallel R_2 = 6.88\ \Omega$, $R_3 \parallel R_4 = 24.5\ \Omega$, $R_T = 31.4\ \Omega$, $V_{R1} = V_{R2} = 10.5$ V, $V_{R3} = V_{R4} = 37.5$ V, $I_{R1} = 477$ mA, $I_{R2} = 1.05$ A, $I_{R3} = 735$ mA, $I_{R4} = 798$ mA, $P_{R1} = 5.01$ W, $P_{R2} = 11$ W, $P_{R3} = 27.6$ W, $P_{R4} = 29.9$ W, $I_T = 1.53$ A, $P_T = 73.4$ W.
7. $R_1 \parallel R_2 = 60\ \Omega$, $R_3 \parallel R_4 = 120\ \Omega$, $R_5 \parallel R_6 = 120\ \Omega$, $R_T = 300\ \Omega$, $V_{R1} = V_{R2} = 9.6$ V, $V_{R3} = V_{R4} = 19.2$ V, $V_{R4} = V_{R5} = 19.2$ V, $I_{R1} = 96$ mA, $I_{R2} = 64$ mA, $P_{R1} = 922$ mW, $P_{R2} = 614$ mW, $I_{R3} = 53.3$ mA, $I_{R4} = 107$ mA, $P_{R3} = 1.02$ W, $P_{R4} = 2.05$ W, $I_{R5} = 64$ mA, $I_{R6} = 96$ mA, $P_{R5} = 1.23$ W, $P_{R6} = 1.84$ W, $I_T = 160$ mA, $P_T = 7.69$ W.
9. The equivalent of the parallel network = $23.2\ \Omega$.

   The series equivalent is $23.2\ \Omega$ in series with $10\ \Omega$ and $15\ \Omega$.
11. The equivalent of the second branch = $48\ \Omega$.

    The parallel equivalent is $150\ \Omega$ in parallel with $48\ \Omega$.
13. R_L is in parallel with the source, so $V_{RL} = E = 120$ V.
15. $R_2 \parallel R_L = 24\ \Omega$. $V_{RL} = 33.9$ V, $I_{RL} = 283$ mA.
17. $R_2 \parallel R_L = 32.7\ \Omega$. $V_{RL} = 87.8$ V, $P_{RL} = 21.4$ W.
19. $R_2 \parallel R_L = 16.7\ \Omega$. $V_L = 31.6$ V, $V_{NL} = 34.3$ V.
21. $V_{R2} = 35.64$ V, $R_{EQ} = 94.7$ kΩ. $V_{R2} = 35.32$ V.
23. $R_X = 10$ kΩ.
25. $R_2 \parallel R_3 = 12\ \Omega$. $R_1 + R_2 = 22\ \Omega$. $I_{R1} = 2.18$ A. $V_{R1} = 21.8$ V. $V_{R2} = V_{R3} = 26.2$ V, $I_{R2} = 873$ mA, $I_{R3} = 1.31$ mA, $R_4 + R_L = 33\ \Omega$. $I_{R4} = I_{RL} = 1.45$ A. $V_{R4} = 31.9$ V, $V_{RL} = 15.95$ V.
    a) If R_2 opens then $I_{R1} = I_{R3} = 1.6$ A. $V_{R1} = 16$ V and $V_{R3} = 32$ V.
    b) If R_4 is shorted then there is no change in the first branch. $V_{RL} = E = 48$ V, $I_{R4} = I_{RL} = 4.36$ A and $V_{R4} = 0$ V.
27. a) The most likely cause is R_3 open.
    b) Most likely the fuse is open. This could result from a short in any branch.

Chapter 9

PRACTICE PROBLEM ANSWERS

1. For the Thevenin circuit $I_{RL} = 3.69$ A

   For the Norton circuit $R_T = 23.1\ \Omega$ and $I_{RL} = 3.70$ A
3. $I_N = 4$ A, $R_N = R_S = 30\ \Omega$.
5. $I_N = 480$ mA, $R_N = R_S = 200\ \Omega$.

   For $R_L = 100\ \Omega$, $R_T = 66.7\ \Omega$. $I_{RL} = 320$ mA,

   For $R_L = 330\ \Omega$, $R_T = 125\ \Omega$. $I_{RL} = 182$ mA
7. $V_{TH} = 30$ V, $R_{TH} = R_S = 7.5\ \Omega$.
9. $V_{TH} = E = 24$ V, $R_{TH} = = 33.2\ \Omega$.

11. $V_{TH} = 14.6$ V, $R_{TH} = 8.78\ \Omega$.
13. $V_{TH} = 10$ V, $R_{TH} = 5\ \Omega$, $P_{Lmax} = 5$ W.
15. $V_{TH} = 62.9$ V, $R_{TH} = 62.7\ \Omega$, $P_{Lmax} = 15.8$ W.
17. With Load 1 removed, $R_B = 146\ \Omega$, $V_B = 47.9$ V
    $V_{TH} = 27.4$ V, $R_{TH} = 114\ \Omega$
19. $V_{TH} = 2$ V, $R_{TH} = 87\ \Omega$, $P_{Lmax} = 11.5$ mW.
21. $I_N = 4.22$ A, $R_N = 4.52\ \Omega$.
23. $I_N = 722$ mA, $R_N = 33.2\ \Omega$.
25. For source A, $R_2 \parallel R_3 = 17.65\ \Omega$, $V_{R2} = V_{R3} = 0.9$ V
    $V_{R1} = 5.1$ V, positive on the left of R_1 and R_2 and positive on the top for R_3.
    For source B, $R_1 \parallel R_3 = 60\ \Omega$, so $V_{R1} = V_{R3} = 9$ V, $V_{R2} = 3$ V. Positive on the right of both R_1 and R_2 and positive on the top for R_3.
    $V_{R1} = 3.9$ V with the positive sign on the right.
    $V_{R2} = 2.1$ V with the positive sign on the right.
    $V_{R3} = 9.9$ V with the positive sign on the top.
27. For source A, $R_2 \parallel R_3 = 12\ \Omega$, $V_{R2} = V_{R3} = 3.56$ V, $V_{R1} = 4.44$ V.
    Positive on the left of both R_1 and R_2 and positive on the top for R_3.
    For source B, $R_1 \parallel R_3 = 10\ \Omega$, so $V_{R1} = V_{R3} = 3.33$ V, $V_{R2} = 6.67$ V.
    Positive on the right of both R_1 and R_2 and positive on the top for R_3.
    $V_{R1} = 1.11$ V with the positive sign on the left.
    $V_{R2} = 3.11$ V with the positive sign on the right.
    $V_{R3} = 6.89$ V with the positive sign on the top.
29. For source A, $R_2 \parallel (R_3 + R_4) = 65.2\ \Omega$,
    $V_{R2} = 4.74$ V positive to the left.
    $V_{R3} = 3.70$ V and $V_{R4} = 1.04$ V both positive on top
    $V_{R1} = 7.26$ V positive on the left.
    For source B, $R_1 \parallel (R_3 + R_4) = 83.3\ \Omega$,
    $V_{R1} = 6.31$ V positive on the right
    $V_{R3} = 4.92$ V and $V_{R4} = 1.39$ V, both positive on top
    $V_{R2} = 5.69$ V positive on the right.
    $V_{R1} = 0.95$ V positive sign on the left.
    $V_{R2} = 0.95$ V with the positive sign on the right.
    $V_{R3} = 8.62$ V with the positive sign on the top
    $V_{R4} = 2.43$ V with the positive sign on the top.
31. Here is one approach, though there are several. Start by treating everything to the right as a load and Thevenize the circuit to the left,
    $V_{TH} = 80$ V and $R_{TH} = 66.7\ \Omega$
    Assuming $L_1 = 150\ \Omega$, $R_3 + (R_4 \parallel R_{L2}) = 105\ \Omega$

$V_{L1} = 48.9$ V, $V_{R4} = V_{L2} = 21$ V and $V_{R3} = 27.9$ V

Assuming $L_1 = 300\ \Omega$ the equivalent resistance $= 161.5\ \Omega$

$V_{L1} = 56.6$ V, $V_{R4} = V_{L2} = 24.3$ V and $V_{R3} = 32.3$ V.

33. The combination of I_B, R_B, and R_2 is converted to a Thevenin equivalent, $E_{B(TH)} = -8$ V, $R_{B(TH)} = 300\ \Omega$.

The no-load output voltages are found using superposition

$V_{NL(A)} = 3.57$ V, $V_{NL(B)} = -2.29$ V

$E_{TH} = 1.28$ V, $R_{TH} = 85.7\ \Omega$, $P_{L(max)} = 4.78$ mW.

Chapter 10

PRACTICE PROBLEM ANSWERS

1. B = =1.8 T
3. B = =2.0 T
5. $\mu_r = 600$
7. $\mu_r = 79.6 \times 10^3$
9. B = 268 μT
11. B = 53.3 mT

Chapter 11

PRACTICE PROBLEM ANSWERS

1. $V_{IND} = 6.6$ mV.
3. $V_{IND} = 2.5$ V.
5. $A = 3.14 \times 10^{-6}\ m^2$, L = 2.27 μH.
7. $A = 3.14 \times 10^{-6}\ m^2$, L = 452 μH.
9. Assuming $k = 0$, $L_T = 103.77$ mH.
11. Assuming $k = 0$, $L_T = 87.3$ μH.
13. $\tau = 133$ ns.
15. $\tau = 16.7$ μs.

Chapter 12

PRACTICE PROBLEM ANSWERS

1. C = 8000 μF.
3. Q = 396 μC.
5. $$C_T = \cfrac{1}{\cfrac{1}{10\ \mu F} + \cfrac{1}{220\ \mu F} + \cfrac{1}{330\ \mu F}} = 9.3\ \mu F.$$
7. $C_T = 329$ pF.

9. $C_T = 55\ \mu F$.
11. $\tau = 1.8$ ms.
13. $\tau = 15.5$ ms.
15. $\tau = = 330$ ms, $5\tau = 1.65$ s to completely charge.

Chapter 13

PRACTICE PROBLEM ANSWERS

1. T = 200 ms
3. T = 80 μs
5. For waveform a) $f = 12.5$ Hz., for waveform b) $f = 2$ kHz.
7. T = = 30 μs and $f = 33.3$ kHz.
9. a) T = 2 ms, b) T = 400 μs, c) T = 1 μs, d) T = 13.3 μs
11. $V_{AVE} = 7.64$ V
13. $V_{PP} = 36$ V and $V_{AVE} = 11.5$ V
15. $V_{L(RMS)} = 67.9$ V and $I_{L(RMS)} = 144$ mA
17. $V_{L(RMS)} = 110$ V and $I_{L(RMS)} = 500$ mA
19. $P_L = 48$ W
21. $V_{PK} = 150$ mV and $V_{PP} = 300$ mV
23. $V_{PK} = 20$ V and $V_{PP} = 40$ V
25. a) v = 13 V, b) v = 7.5 V, c) v = −13 V, d) v = −2.6 V
27. The instantaneous value of A = 10.6 V.
29. Duty cycle (%) = 18.8%
31. Duty cycle (%) = 8.33%
33. $V_{AVE} = 0$ V.

Chapter 14

PRACTICE PROBLEM ANSWERS

1. $X_L = 754\ \Omega$.
3. $I_{RMS} = 4.8$ A, $V_{RMS} = 36$ V, $X_L = 5\ \Omega$.
5. E = 4.12 V, and $\theta = 14°$.
7. $X_L = 124\ \Omega$, $Z_T = 760\ \Omega$, and $\theta = 9.39°$.
9. $X_L = 37.7\ \Omega$, $\theta = 2.88°$, $Z_T = 751\ \Omega$ and $I_T = 63.9$ mA.
   $V_L = 2.41$ V, $V_R = 47.9$ V.
11. $X_{L1} = 82.9\ \Omega$, $X_{L2} = 37.7\ \Omega$, $X_{LT} = 121\ \Omega$, and $\theta = 13.3°$
    $Z_T = 524\ \Omega$, $I_T = 45.8$ mA, $V_{L1} = 3.80$ V, $V_{L2} = 1.73$ V, $V_R = 23.4$ V
13. $Z_T = 20.4\ \Omega$, $I_T = 3.04$ A, $P_X = 37$ VAR, $P_R = 185$ W.
15. $P_{APP} = 189$ VA, $\theta = 11.3°$.

17. $Z_T = 104\ \Omega$, $I_T = 1.15$ A, $V_L = 34.5$ V, $V_R = 115$ V, $P_X = 39.7$ VAR, $P_R = 132$ W, $P_{APP} = 138$ VA, and $\theta = 16.7°$, $P_F = 0.957$.

Chapter 15

PRACTICE PROBLEM ANSWERS

1. $I_T = 6.35$ A, $\theta = -28.2°$.
3. $I_T = 6.71$ A, $Z = 17.9\ \Omega$.
5. $Z_T = 111\ \Omega$, $\theta = 21.8°$.
7. $X_L = 177\ \Omega$, $I_L = 678$ mA, $I_R = 160$ mA, $I_T = 697$ mA, $\theta = -76.7°$, $Z = 172\ \Omega$, $P_X = 81.4$ VAR, $P_R = 19.2$ W, $P_{APP} = 83.6$ VA.
9. $X_{L1} = 82.9\ \Omega$, $I_{L1} = 1.45$ A, $X_{L2} = 56.5\ \Omega$, $I_{L2} = 2.12$ A, $I_{LT} = 3.57$ A, $I_R = 400$ mA, $I_T = 3.59$ A, $\theta = -83.6°$, $Z_T = 33.4\ \Omega$, $P_{L1} = 174$ VAR, $P_{L2} = 254$ VAR, $P_{XT} = 428$ VAR, $P_R = 48$ W, $P_{APP} = 431$ VA.
   $P_{APP} = \sqrt{(381\ \text{VAR})^2 + (84\ \text{W})^2} = 390$ VA.

Chapter 16

PRACTICE PROBLEM ANSWERS

1. $X_C = 80\ \text{k}\Omega$.
3. $X_C = 60.3\ \Omega$.
5. $X_C = 121\ \Omega$, $I = 992$ mA.
7. $I_{RMS} = 6.67$ A, $V_{RMS} = 108$ V, $X_C = 4.67\ \Omega$.
9. $X_{CT} = 580\ \Omega$.
11. $E = 12.6$ V, and $\theta = -71.6°$.
13. $X_C = 88.4\ \Omega$, $Z_T = 174\ \Omega$, $\theta = -30.5°$.
15. $X_C = 265\ \Omega$, $Z_T = 795\ \Omega$, $\theta = -19.5°$, $I_T = 151$ mA, $V_C = 40$ V, $V_R = 113$ V.
17. $X_{C1} = 282\ \Omega$, $X_{C2} = 402\ \Omega$, $X_{CT} = 684\ \Omega$, $Z_T = 923\ \Omega$, $\theta = -56.1°$, $I_T = 260$ mA, $V_{C1} = 73.3$ V, $V_{C2} = 105$ V, $V_R = 161$ V.
19. $X_C = 177\ \Omega$, $Z_T = 587\ \Omega$, $I_T = 204$ mA, $P_X = 7.37$ VAR, $P_R = 23.3$ W.
21. $P_{APP} = 24.4$ VA, $\theta = -17.6°$.
23. $Z_T = 192\ \Omega$, $I_T = 625$ mA, $V_C = 75$ V, $V_R = 93.8$ V, $P_X = 46.9$ mVAR, $P_R = 58.6$ mVAR, $P_{APP} = 75$ VA, $\theta = -38.7°$.

Chapter 17

PRACTICE PROBLEM ANSWERS

1. $I_T = 6.35$ A, $\theta = 28.2°$.
3. $Z_T = 179\ \Omega$.
5. $Z_T = 71\ \Omega$, $\theta = 53.7°$.

7. $X_C = 804\ \Omega$, $I_C = 149$ mA, $I_R = 364$ mA, $I_T = 393$ mA, $\theta = 20.8°$, $Z_T = 305\ \Omega$, $P_X = 17.8$ VAR, $P_R = 43.7$ W, $P_{APP} = 47.2$ VA, PF $= 0.926$.
9. $X_C = 1.77$ kΩ, $I_C = 67.8$ mA, $I_R = 235$ mA, $I_T = 245$ mA, $\theta = 16.1°$, $Z_T = 490\ \Omega$, $P_X = 8.14$ VAR, $P_R = 28.2$ W, $P_{APP} = 29.4$ VA, PF $= 0.959$.

Chapter 18

PRACTICE PROBLEM ANSWERS

1. $E = 7$ V, $\theta = +90°$.
3. $X_S = 120\ \Omega$, $\theta = +90°$.
5. $X_S = 340\ \Omega$, $\theta = +90°$.
7. $V_L = 384$ V, $V_C = 264$ V, $E = 120$ V, $\theta = +90°$.
9. $X_L = 17.7\ \Omega$, $X_C = 26.5\ \Omega$, $V_L = 66.4$ V, $V_C = 99.4$ V, $E = -33$ V, $\theta = -90°$.
11. $I_T = -4.7$ A, $\theta = -90°$.
13. $X_P = -450\ \Omega$.
15. $f_r = 1.59$ kHz.
17. $X_L = 999$ mΩ, $X_C = 1.00\ \Omega$.
19. $X_S = 0\ \Omega$, $I_T = 667$ mA, $V_L = 120$ V, $V_C = 120$ V, and $V_{LC} = 0$ V.
21. $X_S = 85\ \Omega$, $Z = 86.9\ \Omega$, $\theta = 78°$.
23. $V_{LC} = 7$ V, $E = 13.9$ V, $\theta = 30.3°$.
25. $X_L = 113\ \Omega$, $X_C = 26.5\ \Omega$, $X_S = 86.5\ \Omega$, $Z = 114\ \Omega$, $\theta = 49.1°$, $I_T = 1.05$ A, $V_L = 119$ V, $V_C = 27.8$ V, $V_R = 78.8$ V, $P_L = 125$ VAR, $P_C = 29.2$ VAR , $P_R = 82.7$ W, $P_{LC} = 95.5$ VAR, $P_{APP} = 126$ VA, PF $= 0.656$.
27. $I_{LC} = -34$ A, $I_T = 38.5$ A, $\theta = -62.1°$.
29. $X_L = 113\ \Omega$, $X_C = 56\ \Omega$, $I_L = 1.06$ A, $I_C = 2.14$ A, $I_{LC} = 1.08$ A, $I_R = 800$ mA, $I_T = 1.34$ A, $\theta = 53.5°$, $Z = 89.6\ \Omega$, $P_L = 127$ VAR, $P_C = 256$ VAR, $P_X = -129$ VAR, $P_R = = 96$ W, $P_{APP} = 161$ VA, PF $= 0.6$.

Chapter 19

PRACTICE PROBLEM ANSWERS

1. a) BW $= 83$ kHz, $f_0 = 21$ kHz. b) BW $= = 1.16$ kHz, $f_0 = 1.07$ kHz.
   c) BW $= 1.12$ MHz, $f_0 = 626$ kHz. d) BW $= 240$ Hz, $f_0 = 134$ Hz.
3. a) $f_{C2} = 12.8$ kHz, $f_0 = 3.2$ kHz. b) $f_{C2} = 20$ kHz, BW $= 15$ kHz
   c) $f_{C1} = 2$ kHz, $f_0 = 20$ kHz.
5. 6.67.
7. BW $= 7.65$ kHz.
9. Since $Q > 2$ $f_{AVE} = f_0 = 52$ kHz.
11. $R_{EQ} = 68.7\ \Omega$, $f_C = 2.32$ kHz.
13. $R_{EQ} = 109\ \Omega$, $f_C = 14.6$ kHz.

15. $R_{EQ} = 52.6\ \Omega$, $f_C = 138$ kHz.
17. $R_{EQ} = 480\ \Omega$, $f_C = 509$ Hz.
19. $R_{EQ} = 96.8\ \Omega$, $f_C = 15.5$ kHz.
21. $R_{EQ} = 687\ \Omega$, $f_C = 2.32$ kHz.
23. $R_{EQ} = 117\ \Omega$, $f_C = 136$ kHz.
25. $R_{EQ} = 82.7\ \Omega$, $f_C = 5.98$ kHz.
27. $f_r = 23.2$ kHz, $X_L = 146\ \Omega$, $Q_L = 2.35$, BW = 9.87 kHz.
29. $f_{C1} = 18.3$ kHz, $f_{C2} = 28.1$ kHz.
31. $f_r = 50.3$ kHz, $X_L = 316\ \Omega$, $Q_L = 2.24$, BW = 22.5 kHz, $f_{C1} = 39.1$ kHz, $f_{C2} = 61.6$ kHz.
33. $f_r = 33.9$ kHz, $X_L = 469\ \Omega$, Q = 46.9, $R_P = 22$ kΩ, $Q_L = 1.99$
35. From Question 33, $f_r = 33.9$ kHz and $Q_L = 1.99$, BW = 17 kHz,
Since $Q_L < 2$, $f_{AVE} = 34.95$ kHz, $f_{C1} = 26.45$ kHz, $f_{C2} = 43.45$ kHz.

Chapter 20

PRACTICE PROBLEM ANSWERS

1.

ΦA	ΦB	ΦC
25°	265°	145°
310°	**190°**	70°
110°	350°	**230°**

3. $E_L = 208$ V and $I_L = I_P = 12.5$ A.
5. $E_L = 20.8$ kV and $I_L = I_P = 10.2$ A.
7. $I_L = 7.27$ A and $E_L = E_P = 277$ V.
9. $I_P = 8.26$ A and $E_L = E_P = 2.4$ kV.
11. $E_P = 277$ V, and $I_P = I_L = 12.2$ A.
13. $I_L = 6.24$ A, and $E_L = E_P = 1700$ V.
15. For the source, $E_L = 3.81$ kV. For the load, $E_P = E_L$, $I_P = 1.36$ kA.
17. For the source, $E_L = 13.9$ kV. For the load, $E_P = E_L$, $I_P = 1.70\ \text{kA}\angle -26°$, and $I_L = I_P = 1.70\ \text{kA}\angle -26°$.
19. For the load, $I_L = 2.94\ \text{kA}.\angle -26°$, For the source, $I_P = I_L = 2.94\ \text{kA}.\angle -26°$.
21. For the source, $E_L = E_P = 1100$ kV, and $I_L = 4.5$ A.
23. $P_R = 1280$ W, $P_X = 517$ VAR, PF = 0.927.
25. For the load, $E_P = 323$ V, $I_P = 39.4$ A, $P_{APP} = 38{,}179$ VA.
27. For the source, $E_L = 30.5$ kV, For the load, $E_P = E_L = 30.5$ kV, $I_P = 1.17$ kA, and $P_{APP} = 107$ MVA
29. For the source, $E_L = E_P = 120$ V, $P_{APP} = 1372$ VA.
For the load, $I_P = I_L = 6.6$ A, and $E_P = 69.3$ V, $P_{APP} = 1372$ VA.
31. $P_{APP} = 37.4$ kVA, $P_X = 26.5$ kVAR, PF = 0.705, or 70.5%.

Chapter 21

PRACTICE PROBLEM ANSWERS

1. $E_S = 16$ V.
3. $E_S = 220$ V.
5. $E_P = 42.5$ V.
7. $I_S = 900$ mA.
9. $I_S = 83.3$ mA.
11. $E_S = 95.3$ V.
13. From question 11, turns ratio = 1.21:1, $I_S = 17.3$ A, $I_P = 14.3$ A, and $P_P = 1.64$ kW.
15. $E_S = 8.8$ kV, $I_S = 1.1$ kA.
17. $I_P = 1.1 \text{ kA} \times 20 = 22$ kA., minimum kVA rating = 9680 kVA.
19. 200 turns on the 120 V secondary, 400 turns on the 240 V secondary
21. 48 V secondary = 120 turns, 12 V secondary = 30 turns,
    5 V secondary = 12.5 turns.
    Primary turns on the European primary = 300 turns.
23. %Z = 7.5%
25. The derated value of $I_{FL} = 10.5$ A.

Chapter 22

PRACTICE PROBLEM ANSWERS

1. Solving for the primary of each transformer,
   For the Y − Y, $E_P = 7.97$ kV, $I_L = I_P = 8.2$ A.
   For the Δ − Y, $E_L = E_P = 480$ V, $I_P = 1.5$ A.
   For the Y − Δ, $E_L = 11.95$ kV. $I_L = I_P = 11.5$ A.
   For the $\Delta - \Delta$, $E_P = E_L = 9.6$ kV, $I_P = 2.54$ A.
3. $E_P = 3.98$ kV, $I_P = I_L = 28$ A.
5. $E_L = 1143$ V, $I_P = I_L = 9$ A.
7. $R = 73.3\ \Omega$.
9. $E_P = E_L = 2500$ V and $I_P = \dfrac{I_L}{\sqrt{3}} = 26$ A.
11. Assuming no significant transformer losses, $P_{Primary} = P_{Secondary} = 195$ kVA
13. If the secondary was Y connected then $I_L = I_P = 45$A and $E_P = 2500$ V,
    $P_T = 3 \times E_P \times I_P = 338$ kVA, so with a Y-connected secondary, the power supplied to the transformer is 143 kVA higher.
15. $E_{P(secondary)} = E_{P(load)} = 260$ V, $I_{P(load)} = 39.4$ A,
    Power dissipated by each phase = 10, 244 W,
    Total load power = 30.7 kW.

17. Total power = 404 W.
19. E_P = 139 V, $E_{P(Load)}$ = 139 V. P_{load} = 400 W, $I_{P(Load)}$ = 2.88 A, and R_{Load} = 48.3 Ω.
21. $E_P = E_L$ = 208 V, $I_{P(Load)}$ = 27 A∠−28°, P_{APP} = 16.8 kVA∠−28°, P_R = 14.8 kW.

Chapter 23

PRACTICE PROBLEM ANSWERS

1. r = 4.25 = 0.354 ft. T = 230 lb-ft.
3. r = 0.0656 ft, D = 0.1312 ft = 1.57 in.
5. D = 3.92 ft, W = 2548 ft-lb
7. T = 186 lb-ft.
9. P = 41,250 W.
11. rpm = 720 rpm.
13. 1 HP = 746 W so 125 HP = 125 HP × 746 W/HP = 93.25 kW.
15. 3 HP = 3 HP × 746 W/HP = 2238 W, so $E_{DC} = \frac{P}{I}$ = 125 V,
    $E_G = E_{DC} - I_A R_A$ = 77.9 V.
17. At startup E_G = 0 V, so $I_A = \frac{E_{DC}}{R_A}$ = 47.5 A.

INDEX